Lauren Richards
106 Carr St.
Chapel Hill, N.C.

967-5460

D0880968

BK04759734

INTERMEDIATE CALCULUS

MULTIVARIABLE FUNCTIONS AND DIFFERENTIAL EQUATIONS WITH APPLICATIONS

JAMES F. HURLEY
University of Connecticut

SAUNDERS COLLEGE Philadelphia

Saunders College
West Washington Square
Philadelphia, PA 19105

Library of Congress Cataloging in Publication Data

Hurley, James Frederick, 1941-
Intermediate Calculus

Includes bibliographical references and index.
1. Calculus. I. Title.
QA303.H86 515 78-65378
ISBN 0-03-056783-1

Text art: John Hackmaster, Dmitri Karetnikov, Jay Freedman, Linda Savalli. Thanks to the Mathematics Department of Drexel University for use of the IBM 5100 computer and plotter.

Cover illustration: Computer art (Precision Visuals)

Intermediate Calculus: Multivariable Functions and Differential Equations with Applications ISBN 0-03-056783-1

© 1980 by Saunders College/Holt, Rinehart and Winston. Copyright under the International Copyright Union. All rights reserved. This book is protected by copyright. No part of it may be reproduced, stored in a retrieval system, or transmitted in any form or by any means, electronic, mechanical, photocopying, recording, or otherwise, without written permission from the publisher. Made in the United States of America. Library of Congress card number 78-65378.

0123 000 9 8 7 6 5 4 3 2

To Cecile, JoJo, and Gigi

PREFACE

This book is designed to introduce the calculus of functions of several variables and differential equations to students who have completed a one-year course in elementary calculus. Care has been taken to motivate each new topic prior to its study. This is usually done by relating it to notions the student has met previously or by showing that it is an appropriate tool for solving some naturally arising problem or for extending some basic idea.

Multivariable calculus is developed by searching for analogues of familiar results and techniques of elementary calculus. To make the analogies as sharp as possible, vector notation such as $\boldsymbol{y} = \boldsymbol{f}(\boldsymbol{x})$ is used. This not only helps to bring out the essence of the mathematical ideas, but also formulates them in the notation most commonly used in the science and engineering courses that many of the students take concurrently. Considerable attention is given to geometric ideas in the early chapters. It is my belief that students who are thoroughly familiar with curves and surfaces in three dimensions are well equipped to set up the multiple, line, and surface integrals that occur later.

In discussing both multivariable calculus and differential equations, ideas and techniques of linear algebra have been used wherever they contribute to a clearer understanding of the material being presented. No previous experience with linear algebra is needed. Rather, an attempt has been made to integrate linear algebra into the presentation in much the same way that elementary calculus texts integrate analytic geometry. This approach can serve to introduce the main computational techniques to students who do not take a formal course in linear algebra. For students who have taken or are taking such a course, it can serve to illustrate the wide applicability of the subject.

Recommendations of the Committee on the Undergraduate Program in Mathematics (CUPM) of the Mathematical Association of America and of the National Research Council Committee on Applied Mathematics Training, in particular those urging realistic applications from a variety of fields in lower-division courses, have played a major role in structuring the book. A large number of examples and exercises involve applications of the mathematical ideas to other disciplines. The chapters on differential equations particularly emphasize mathematical modeling; they present applications not only to the physical sciences and engineering but also to fields such as psychology, ecology, demography, economics, criminology, and political science. No background in any of the applied fields is needed or assumed.

Numerous historical notes appear, both to give credit to those who have built the subject and to make the student aware of mathematics as a continually developing area of knowledge.

Differentiation is approached by way of the twentieth century notion of the total derivative. This is introduced heuristically as what ought to result from putting together in an appropriate way all the partial derivatives of a real valued function of several real variables. After a close look at the nature of the derivative in elementary calculus, the total derivative of f is defined from the point of view of local linear approximation of f. Early experience with this important concept illuminates the nature of differentiation and points the way to linear algebra, modern advanced calculus, and differential equations for students continuing in mathematics.

Double and triple integrals over rectangles and rectangular parallelepipeds are discussed first because of their simplicity and direct analogy to definite integrals over closed intervals. Fubini's Theorem is the central computational tool used in evaluating multiple integrals, and sections on polar, cylindrical, and spherical coordinate integrals lead ultimately to change of variables in multiple integrals. The treatment of vector integration includes the Theorems of Green, Gauss, and Stokes, as well as several classical applications to engineering and physics.

Chapter 8 presents first order and linear second order differential equations, with a review of complex numbers. The constant coefficient case is emphasized, but there is a section devoted to power series and numerical solutions. Chapter 9 considers n-th order linear equations and Chapter 10 treats first order systems. The Laplace transform is introduced as a means to solve initial value problems in both chapters. This material can be studied before the calculus chapters are begun. An independent chapter introduces infinite series, which are referred to in several places in the rest of the text.

I am grateful to a great many individuals and institutions who have contributed to the evolution and improvement of this work. Cecile N. Hurley not only supplied and assisted with numerous applied examples, but also patiently and faithfully typed the the entire manuscript from handwritten copy whose legibility often left much to be desired. Robert J. Weber showed me a number of interesting applications and referred me to others during many stimulating conversations and lectures on mathematical modeling. Professors Beverly Henderson of Cornell University, Robert Martin of the University of Pennsylvania, and Kenneth Bogart of Dartmouth University made encouraging and helpful reviews of an early version of the book. Professors Bogart, John Scheick of Ohio State University, and John Schiller of Temple University carefully read the semi-final manuscript, and their extraordinarily thorough reviews and suggestions helped eliminate many errors and significantly improve the exposition. Professor Philip Gillett of the University of Wisconsin Marathon Center checked the final manuscript and is responsible for the correction of other errors and for several additional improvements. (Remaining defects are, of course, solely my responsibility.) David Mohrman and Anthony Chiodo assisted in compiling and checking answers to the exercises. Joseph Hurley helped with final manuscript preparations.

I am also indebted to my department head, John V. Ryff, who permitted me to class-test two earlier versions of the manuscript. My classes of 1976–77 and 1977–78 contributed countless valuable suggestions while cheerfully coping with a textbook-in-the-making. To these classes, and to earlier ones at the University of Connecticut, the University of the Philippines, Ateneo de Manila University, De La Salle University, and the University of California campuses at Riverside and Los Angeles,

I am indebted for enthusiastic reception of my ideas, for the many suggestions to put them into written form, and for the encouragement to see this project through to completion.

Over the past fifteen years, I have taught sophomore mathematics from texts authored by many mathematicians, and have consulted numerous others in the course of preparing lectures. All of these have helped me to shape my view of the material contained here, and their contribution is gratefully acknowledged. I have also benefitted during this period from grants under the Fulbright-Hays program and the National Science Foundation's Scientists and Engineers in Economic Development program, which afforded me the opportunity to teach students with a wide range of backgrounds. The influence of Louis Leithold, my own intermediate calculus instructor, will be evident to anyone familiar with Professor Leithold's exemplary texts. Finally, it is a pleasure to thank the editorial and production staffs of W.B. Saunders Company, whose professionalism and skill have been so helpful. In particular, Jay Freedman's contributions in both developmental and copy editing have significantly raised the quality of the book. The enthusiastic faith of Saunders' Mathematics Editor, Bill Karjane, in this project over the past three years has truly been the major contributing force to its publication.

JAMES F. HURLEY

FOREWORD TO THE INSTRUCTOR

This text is written with flexibility in mind, and it can be used for a variety of courses enrolling students who have completed a year of calculus. Chapter 1 of the accompanying *Instructor's Manual* explores various possible courses that can be given from the text, and includes sample outlines. At the University of Connecticut, Chapters 1 through 5 (omitting Section 1.7) have been taught in the first semester and Chapters 6 through 10 (plus Section 1.7) in the second semester. Chapter 11 can replace parts of Chapters 1 and 2 for students with previous knowledge of vectors.

Each section normally says more that you can, or will want to, cover in the classroom. The *Instructor's Manual* contains suggestions about specific items that can be left for student reading. Sections to which more than one class period can be devoted include 1.8, 4.3, 4.5, 7.2, 7.6 to 7.8, 8.2, 8.8, 9.3, 10.2, and 10.4.

The flow chart on the following page presents section and chapter dependency in detail, and the *Instructor's Manual* discusses several alternatives to following the natural order of chapters. Note particularly that the text can be used for two separate sophomore courses: a 4- or 5-credit third semester (or fourth quarter) calculus course, and a 3- or 4-credit course in linear algebra and differential equations. The former would cover as much of Chapters 1 to 7 as time allows. To save time, Chapters 4 and 6 can be done together in the order 4.1 to 4.3, 6.1 to 6.5, 4.4, 4.5–6.6, 4.6–6.7, 4.7 to 4.10. Sections 6.8 and 6.9 can be omitted or integrated into Sections 5.5, 5.7, and 5.8. The differential equations course can be constructed from Sections 1.6 to 1.8, 4.2, 6.1 to 6.4, and Chapters 8, 9, and 10. Note also that Chapter 8 can be covered independently of linear algebra and at virtually any point of a year course taught from the book. Chapter 9 requires Sections 1.6 to 1.8 and 4.2 in addition to Chapter 8. After Section 10.1, Sections 6.1 to 6.4 are needed for full appreciation of Chapter 10.

Section 5.4 on polar coordinates and most of Chapter 11 on infinite sequences and series are written without proofs. The reason for this is that many elementary calculus classes will have discussed this material. For those students coming from such classes, these portions can serve as ready reference and review. You can, however, teach this material to students without previous background in it by filling in some elementary proofs.

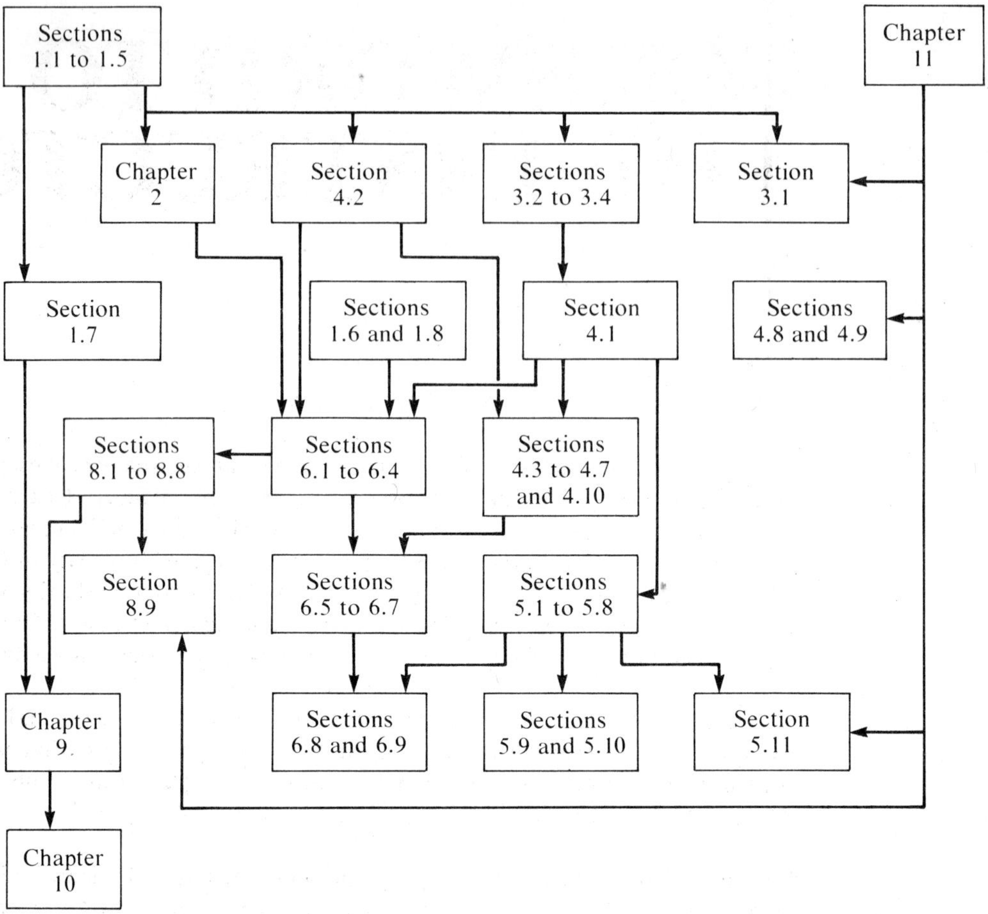
Sections 1.1 to 1.5
Chapter 11
Chapter 2
Section 4.2
Sections 3.2 to 3.4
Section 3.1
Section 1.7
Sections 1.6 and 1.8
Section 4.1
Sections 4.8 and 4.9
Sections 8.1 to 8.8
Sections 6.1 to 6.4
Sections 4.3 to 4.7 and 4.10
Section 8.9
Sections 6.5 to 6.7
Sections 5.1 to 5.8
Chapter 9
Sections 6.8 and 6.9
Sections 5.9 and 5.10
Section 5.11
Chapter 10

FOREWORD TO THE STUDENT

This text is written for you to read and learn from. Since there is no such thing as a typical student, material has been included to meet the needs of a variety of students who enroll in courses following elementary calculus.

If you are a prospective mathematics major, then you will want to study not only the computational techniques and applications, but also the proofs and discussions of a more theoretical nature. On the other hand, if you are an engineering or physical, social, life, or management science major, then you will probably want to concentrate primarily on the computational techniques and illustrated uses of them in applied settings.

In any case, you will find the sections organized as follows. An introductory portion sets the scene, usually by reminding you of an important idea from your previous mathematical study which is related to the material to be presented. This is intended to help you to get your bearings and to understand *why* what follows is being done. Next come the definitions needed to give precise statements of the results of the section. These are important because they specify under exactly what circumstances the techniques can be used. If you take note of this, you can avoid the kind of frustration that often comes to students who try to apply machinery in contexts where it is inapplicable! Proofs are given in enough detail for you to follow step-by-step should you wish to. Your instructor will probably indicate to you the degree to which he thinks you should study these proofs. They are not required in order to do most exercises or follow the development of subsequent sections. In the References you can find sources containing omitted proofs that require more machinery or sublety than is appropriate to a book at this level. If you are ambitious, you can consult these references to obtain a more complete account of the subject. After the proofs, illustrations are given of how the material presented in the section is used in computational and/or applied problems. Study these carefully, because most of the exercises are based on ideas presented in these examples.

The exercises themselves are *essential* to mastering the material. Don't kid yourself—anybody who has passed an elementary calculus course can *follow* the solutions to the examples. But that is NOT the same as learning to *do* them! The latter ability can come only as a result of working the exercises. In many sections, you will find exercises that ask you to apply the techniques of the section to other mathematical and nonmathematical contexts. It is important to become adept at this, since much of the use given the material of this book lies in such applications.

Some exercises, generally toward the end of each list, are not just straightforward repetitions of the examples. Such exercises are designed to enable you to dig beneath the surface, to think about the material more deeply, and hence to improve your grasp of it. Hints are provided in many cases to help you get started. Many of these exercises are not proofs, and if you really want to *thoroughly* master intermediate calculus, you should look for such exercises and try them whether or not your instructor assigns them.

Answers to most odd-numbered computational problems appear at the end of the book. Every attempt has been made to ensure their accuracy, but they were all done and checked by human beings, many by students like you. If your answer differs from the one given, treat that as you would a discrepancy in answers with a classmate. Check your work, but don't assume in advance that the book's answers are infallible.

Definitions, lemmas, theorems, propositions (*i.e.*, little theorems), collaries, and examples are numbered sequentially within each section with a double number. Thus, Example 4.3 is the third numbered item in Section 4. A reference to Theorem 7.4 is to the fourth numbered item in the seventh section of the chapter being read. The presence of a third number in a reference tells which chapter (other than the one being read) contains it. Thus, Theorem 4.5.3 refers to the third numbered item in Section 5 of Chapter 4. Terms being defined are printed in **boldface** for emphasis. Proofs end with QED, which is the abbreviation for the Latin phrase *quod erat demonstrandum,* meaning *which was to be proved.* Until the nineteenth century, most mathematics was written in Latin, and this particular phrase is the Latin translation of the Greek words with which Euclid ended his proofs more than 2000 years ago, *ὅπερ ἔδει δεῖξαι.*

Finally, a word about how to read a mathematics book. In order to *really* understand, and learn, what you are reading, it is necessary to work through the development actively. That is, you should read with a pencil in your hand and a pad of paper at your side. Whenever the text claims that something follows from something else, the proper thing to do is to verify this for yourself by performing whatever algebraic calculation may be needed to derive the expression in question. **DON'T** make the two common mistakes of (a) simply taking the author's word for everything without trying to understand why it is true or (b) giving up in despair the first time you don't instantaneously see in your head how a particular expression results from another. The book has been written so that any steps not explicitly shown can be filled in by performing straightforward algebraic calculations. Doing such calculations will help you understand the material better and build confidence.

CONTENTS

1 VECTOR GEOMETRY AND ALGEBRA

0 INTRODUCTION

So far in your study of calculus, the main objects of discussion have been *real valued functions f of one real variable x*. For such functions we will use the notation

$$f: \boldsymbol{R} \to \boldsymbol{R} \quad \text{given by} \quad y = f(x)$$

where x is a real number in the domain D of f. (We generally will not write $f: D \to \boldsymbol{R}$ unless there is some need to emphasize the nature of the exact domain of f.) You have learned about graphs, limits, continuity, and integrals of these functions. In this text you will learn about the same concepts for *vector valued functions of several real variables*. These are functions $\boldsymbol{F}$ defined on domains D contained in some *n-dimensional Cartesian* (or *Euclidean*) *space* $\boldsymbol{R}^n$ and which take values in some space $\boldsymbol{R}^m$ (where m may or may not be the same as n). Our notation for such functions will be

$$\boldsymbol{F}: \boldsymbol{R}^n \to \boldsymbol{R}^m,$$

where $\boldsymbol{R}^1$ will denote the familiar set $\boldsymbol{R}$ of real numbers. (Again this means that the domain of $\boldsymbol{F}$ is some set $D \subseteq \boldsymbol{R}^n$ which may or may not be all of $\boldsymbol{R}^n$ and which we will usually not specify further.)

As you might imagine, a prerequisite to the study of functions defined on $\boldsymbol{R}^n$ is a working familiarity with such spaces. The goal of Chapter 1, then, is to develop that familiarity.

The first section introduces $\boldsymbol{R}^n$ to you via the real line $\boldsymbol{R}^1$, the Cartesian plane $\boldsymbol{R}^2$ (which should be old friends of yours by now), and Euclidean three-space $\boldsymbol{R}^3$, which in many ways is just $\boldsymbol{R}^2$ with a new perpendicular axis drawn to it.

In Sections 2 and 5, useful tools are developed for studying the *geometry* of $\boldsymbol{R}^n$, which is the main business of Sections 3, 4, and 5. The last three sections illustrate how certain geometric questions can be answered with purely algebraic tools. At the same time the questions will motivate an entirely new area of study, *linear algebra*. One of the themes of this book will be the repeated return to linear algebra as a source of methods to solve the problems of geometry and calculus which arise as we proceed.

1 THE SPACE R^n

The formal algebraic study of vectors goes back to the great nineteenth century Irish mathematician and physicist William Rowan Hamilton (1805–1865) and the German philosopher and mathematician Hermann G. Grassmann (1809–1877). But much earlier Sir Isaac Newton (1642–1727), one of the co-inventors of modern calculus, in his laws of motion dealt informally with geometric vectors as objects which possessed both length and direction. And in studying velocity, the ancient Greeks seem to have used vectors and even found resultant vectors by means of the parallelogram law. If you have studied physics, then you have no doubt dealt with vectors as geometric quantities with length and direction, and perhaps also as algebraic quantities. It is the latter point of view which goes back to Hamilton and Grassmann and which represents a profound contribution in the modern development of mathematics, science, and engineering. The description of vectors algebraically is our goal in this section.

Let us start with the geometric idea. A vector is a quantity which has both *length* and *direction*. Suppose we look at $\boldsymbol{R}=\boldsymbol{R}^1$ and ask how we might represent such a quantity. In $\boldsymbol{R}^1$ there are only two directions—positive (toward the right) and negative (toward the left)—and the assignment of coordinates to $\boldsymbol{R}^1$ affords a natural measurement of length. Thus, a vector three units long directed toward the right is quite simply described by (+3), while a vector of the same length directed to the left can be described by (−3). Notice that the length of the vector is simply its absolute value:

$$|(+3)|=3, \qquad |(-3)|=3.$$

In general, a vector $\vec{v}$ represented by (x) has length $|(x)|=\sqrt{x^2}$. This simple observation is the basis for our subsequent notation for the length of a vector.

Something else you may notice is that this description of vectors by real numbers carries with it no information about where a vector begins (or ends). Thus in Figure 1.1, (+2) describes both $\vec{v}_1$ and $\vec{v}_2$. While this may seem to be a

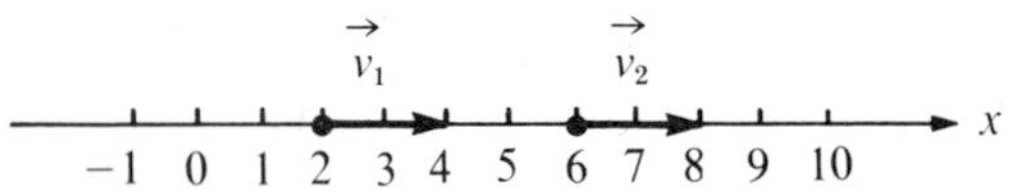

FIGURE 1.1

flaw in our representation scheme, it really is not. For in physics (and elsewhere), geometric vectors with the same length and same direction are identical, regardless of whether their initial points coincide. For this reason, the term "free vector" is often used to emphasize that one is free to draw a vector from any initial point one likes.

A representation of the vector drawn from a to b is $(b-a)$, since $b-a$ gives the directed distance from a to b (positive if $b>a$, negative if $b<a$, 0 if $b=a$). See Figure 1.2.

FIGURE 1.2

The resultant of two vectors in $\boldsymbol{R}^1$ is easily computed by algebraic addition. The equation

$$(+2)+(-3)=(-1)$$

is readily seen to say that the resultant of a vector of length two pointing to the right and a vector of length three pointing to the left is a vector of length one pointing to the left. Moreover, comparison of the space taken up by the equation and the statement in words suggests that the algebraic description we have developed is a valuable simplification.

Finally, notice that we can multiply vectors by real numbers in a natural way. The vector $a\vec{v}$ will be $\vec{0}$ if $a=0$, and otherwise will point in the same direction as $\vec{v}$ if $a>0$ or in the opposite direction from $\vec{v}$ if $a<0$. Its length will be $|a|$ times the length of $\vec{v}$. See Figure 1.3.

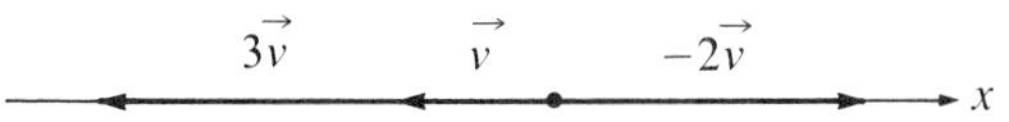

FIGURE 1.3

What about vectors in the plane $\boldsymbol{R}^2$? If we regard vectors as objects with length and direction, and identify those which have the same length and direction (no matter where their initial point happens to be), then how can we represent them algebraically? Since in $\boldsymbol{R}^1$ single real numbers gave us an adequate algebraic representation, we expect that in $\boldsymbol{R}^2$ we can use *pairs* of real numbers to represent vectors. Let's see how.

Since we are concerned only with length and direction, we can translate any given vector $\vec{v}$ parallel to itself until its beginning point is at the origin (Figure 1.4), and then use the coordinates (a, b) of its endpoint to represent $\vec{v}$. We know how to measure distances in $\boldsymbol{R}^2$ using the distance formula, so we see that the length of $\vec{v}$ is given by

$$|\vec{v}|=|(a, b)|=\sqrt{(a-0)^2+(b-0)^2}=\sqrt{a^2+b^2}.$$

Note the similarity between this and the formula for length of a vector in $\boldsymbol{R}^1$.

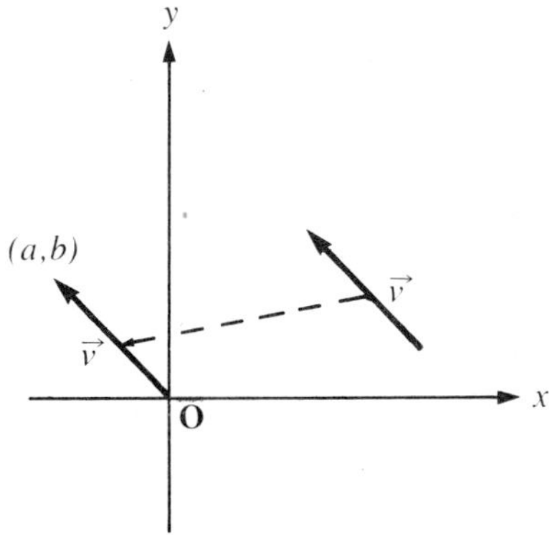

FIGURE 1.4

The analogy with one-dimensional vectors suggests that we should add vectors algebraically by adding corresponding coordinates of the representing ordered pairs, i.e.,

(*) $$(a, b)+(c, d)=(a+c, b+d).$$

Similarly, the natural way to carry out multiplication by a real number r is to use the formula

(**) $$r(a, b) = (ra, rb).$$

1.1 **DEFINITION.** If $\vec{v} = (a, b)$ and $\vec{w} = (c, d)$ are vectors in $\boldsymbol{R}^2$ and $r \in \boldsymbol{R}$, then $\vec{v} + \vec{w}$ and $r\vec{v}$ are defined by (*) and (**). The **vector difference** $\vec{v} - \vec{w}$ is defined to be $\vec{v} + (-\vec{w})$, where $-\vec{w} = (-1)\vec{w}$.

Is this definition satisfactory? Well, we arrived at it reasonably, using analogy with the situation in $\boldsymbol{R}^1$. Does it, though, give us the geometric properties we expect of vectors in the plane? It is easy to see that (**) is compatible with our geometric expectations. First,

$$|r(a, b)| = |(ra, rb)| = \sqrt{r^2a^2 + r^2b^2} = |r|\sqrt{a^2 + b^2}$$
$$= |r|\,|(a, b)|.$$

Thus $r\vec{v}$ has length $|r|$ times $|\vec{v}|$, as we expect. Also (Figure 1.5), it is easy to see that *$r\vec{v}$ points in the same direction as $\vec{v}$* if $r > 0$ or *in the opposite direction if* $r < 0$.

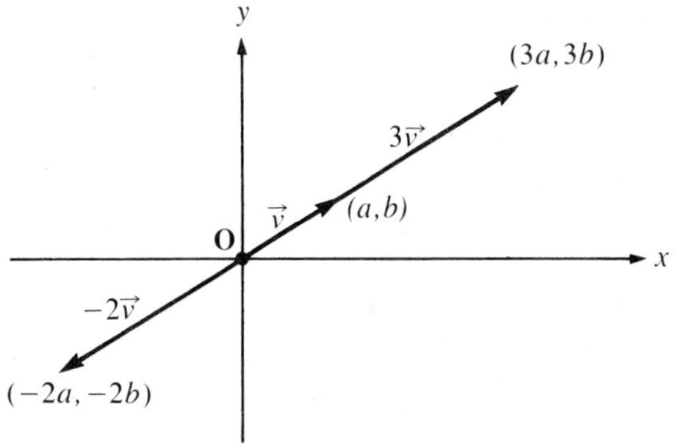

FIGURE 1.5

For (*) to be satisfactory geometrically, addition performed according to that formula should obey the *parallelogram law* for addition of geometric vectors. As you probably recall, this law states that if $\vec{v}$ and $\vec{w}$ are not collinear vectors, then $\vec{v} + \vec{w}$ is the diagonal of the parallelogram formed by $\vec{v}$ and $\vec{w}$ when $\vec{v}$ and $\vec{w}$ are drawn from the same initial point. See Figure 1.6. Let's verify that this holds if we add using (*). We need to show that *OADB* in Figure 1.6 is a parallelogram.

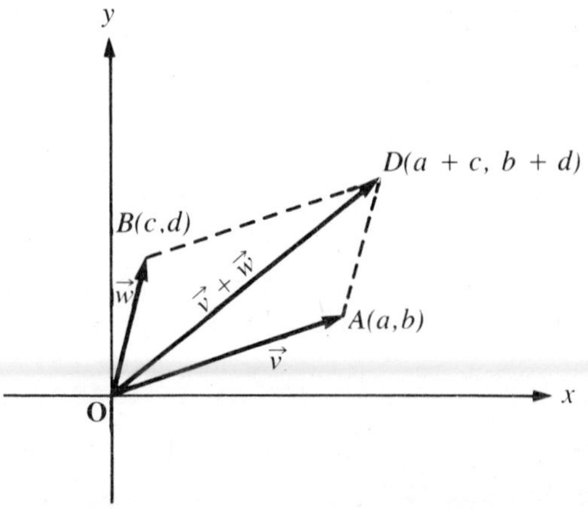

FIGURE 1.6

First, if $c=0$, then both AD and OB are horizontal, hence parallel. If $c\neq 0$, then OB has slope

$$\frac{d}{c}=\frac{d-0}{c-0}=\frac{(b+d)-b}{(a+c)-a}=\text{slope } AD.$$

Hence, AD and OB are parallel. Next, if $a=0$, then OA and BD are both vertical. Otherwise, both have slope

$$\frac{b}{a}=\frac{(b+d)-d}{(a+c)-c},$$

so are parallel. Thus $OADB$ *is* a parallelogram, as desired.

It can actually be further shown that if we want the parallelogram law for vector addition to hold, then we have *no choice* but to define algebraic addition in $\boldsymbol{R}^2$ by (*). (You are asked to carry this out in Exercise 16 below.)

The parallelogram law enables us to easily obtain the algebraic representation of the *position vector* $\overrightarrow{AB}$ drawn from a point $A(a_1, a_2)$ to $B(b_1, b_2)$ (Figure 1.7). For we see that vector $\overrightarrow{OB}$ is the sum of $\overrightarrow{OA}$ and $\overrightarrow{AB}$. Since $\overrightarrow{OA}=(a_1, a_2)$

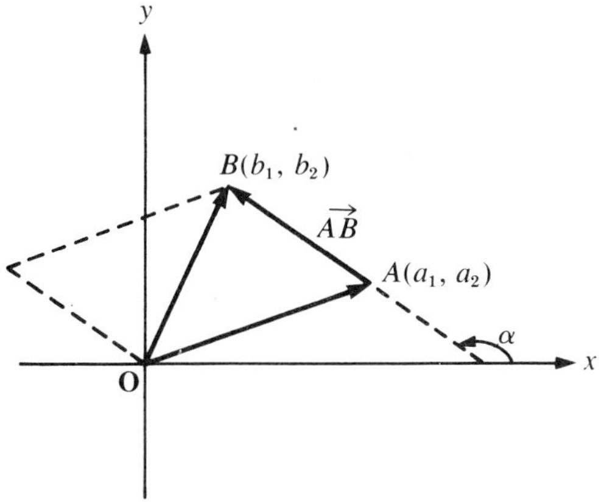

FIGURE 1.7

and $\overrightarrow{OB}=(b_1, b_2)$, we have then

$$(b_1, b_2)=(a_1, a_2)+\overrightarrow{AB}.$$

Thus, subtracting (a_1, a_2) from each side, we get

$$\begin{aligned}-(a_1, a_2)+(b_1, b_2)&=-(a_1, a_2)+(a_1, a_2)+\overrightarrow{AB},\\ (b_1-a_1, b_2-a_2)&=(0, 0)+\overrightarrow{AB}\\ &=\overrightarrow{AB}.\end{aligned}$$

The position vector $\overrightarrow{AB}$ is obtained by subtracting the respective coordinates of A from the corresponding coordinates of B.

Observe that in Figure 1.7 the quotient $\dfrac{b_2-a_2}{b_1-a_1}$ gives the slope m of the line through AB, and $m=\tan\alpha$ where α is the inclination of that line, i.e., the counterclockwise angle from the positive x-axis to the line.

This representation thus seems to contain enough information about the vector from A to B to specify both its length $(=\sqrt{(b_1-a_1)^2+(b_2-a_2)^2})$ and its direction (unless $A=B$, in which case $\overrightarrow{AB}=(0,0)$ is called the *zero vector,* the one and only vector which has no direction; note also that the zero vector has length 0).

1.2 EXAMPLE. Find the position vector from (a) $A(-2,1)$ to $B(1,-\frac{3}{2})$; (b) $C(1,-\frac{2}{3})$ to $D(1,\frac{1}{3})$; (c) $E(2,5)$ to $F(5,\frac{5}{2})$. Sketch, and compute the length of each vector.

Solution.

$$\overrightarrow{AB}=(1-(-2),-\tfrac{3}{2}-1)=(3,-\tfrac{5}{2})$$
$$\overrightarrow{CD}=(1-1,\tfrac{1}{3}-(-\tfrac{2}{3}))=(0,1)$$
$$\overrightarrow{EF}=(5-2,\tfrac{5}{2}-5)=(3,-\tfrac{5}{2})$$
$$|\overrightarrow{AB}|=\sqrt{3^2+(-5/2)^2}=\sqrt{9+25/4}=\sqrt{61/4}=\sqrt{61}/2$$
$$|\overrightarrow{CD}|=\sqrt{0^2+1^2}=1$$
$$|\overrightarrow{EF}|=\sqrt{3^2+(-5/2)^2}=\sqrt{61}/2$$

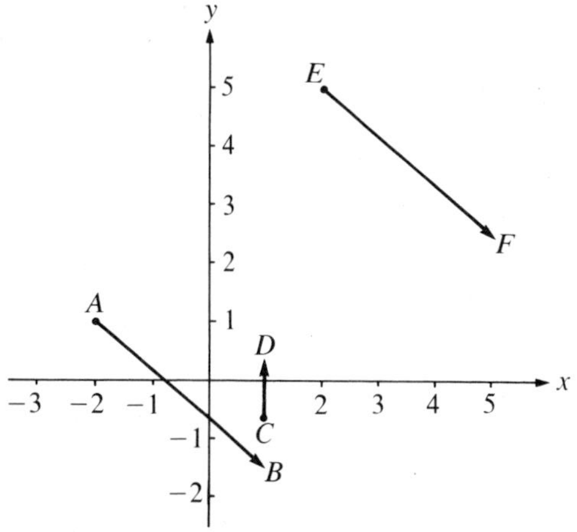

FIGURE 1.8

Example 1.2 illustrates that our representation scheme for vectors in the plane still identifies vectors with the same length and direction (e.g., $\overrightarrow{AB}$ and $\overrightarrow{EF}$).

By now a common thread is discernible running through the algebraic description of geometric vectors in $\boldsymbol{R}^1$ and $\boldsymbol{R}^2$. In both spaces we represent vectors by sequences: one-element sequences in $\boldsymbol{R}^1$, two-element sequences (i.e., ordered pairs) in $\boldsymbol{R}^2$. We add vectors by adding corresponding entries of the sequences. We multiply a vector by a real number r by multiplying each entry of the sequence by r. We compute the length of a vector by taking the square root of the sum of the squares of its entries. These observations suggest that to represent vectors in $\boldsymbol{R}^3$, three-dimensional Cartesian space, we should use ordered triples of real numbers.

We impose coordinates on three-space as follows. Three mutually perpendicular lines (called the x-, y-, and z-axes or x_1-, x_2-, and x_3-axes, etc. and labeled as in Figure 1.9) are drawn from a common point $\boldsymbol{O}$ selected as the origin. Since two lines determine a plane, we have determined three **coordinate planes,** called the *xy-plane*, *yz-plane*, and *xz-plane*. The coordinates of a point P in space are then (x_0, y_0, z_0), where x_0 is the directed distance from the yz-plane to the point, y_0 is the directed distance from the xz-plane to the point, and z_0 is the

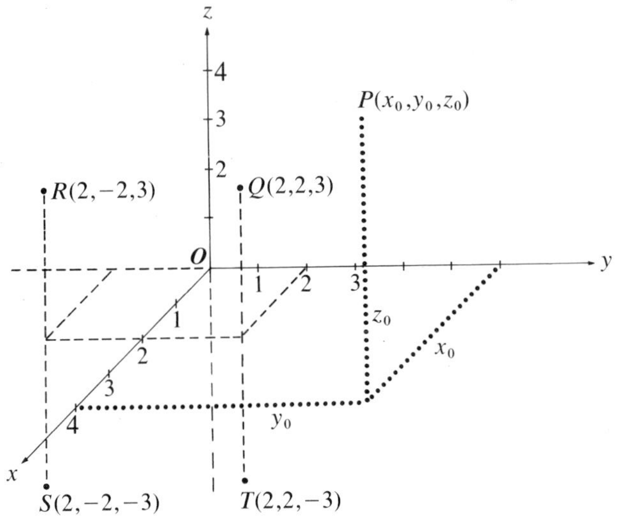

FIGURE 1.9

directed distance from the xy-plane to the point. Here directed distances are positive when measured upward, to the right, or to the front (i.e., in the direction of the positive x-axis, positive y-axis, or positive z-axis). As a practical matter, to locate P, we measure x_0 and y_0 as usual in the xy-plane, and then go up or down to P according to whether z_0 is positive or negative. See Figure 1.9.

Just as the coordinate axes in $\boldsymbol{R}^2$ divide the plane into four quadrants, so the three coordinate axes divide $\boldsymbol{R}^3$ into eight **octants.** Customarily the only octant assigned a number is the *first* octant, all of whose points have all three coordinates positive. In drawing three-dimensional pictures, clutter is often minimized by showing only the first octant portion completely, and either omitting the rest or drawing it with dotted lines.

Notice that just as in $\boldsymbol{R}^2$, the *orientation* of the axes in $\boldsymbol{R}^3$ has to be prescribed. The orientations shown in Figures 1.10 and 1.11 give different coordinatizations of $\boldsymbol{R}^2$ and $\boldsymbol{R}^3$.

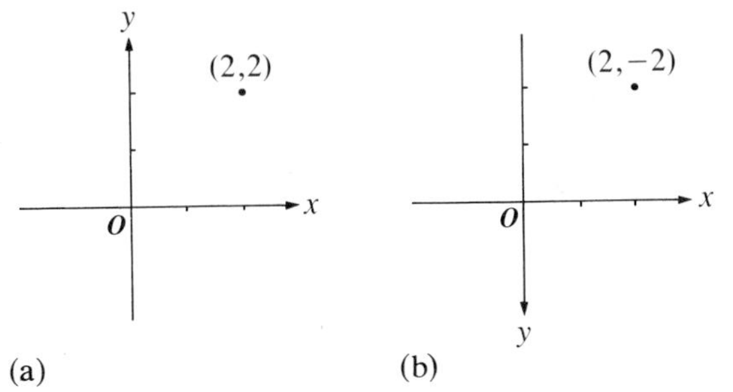

FIGURE 1.10

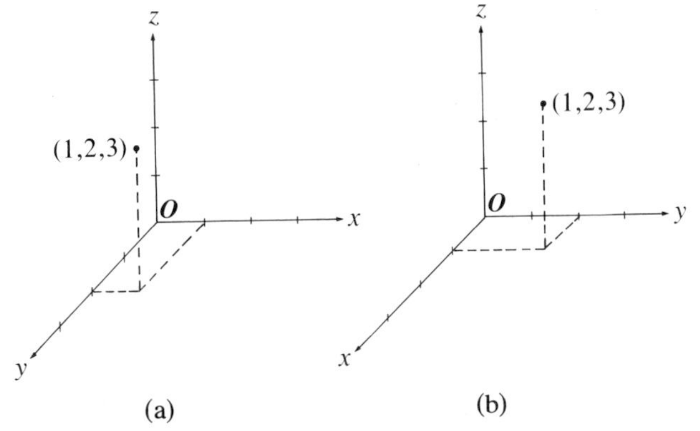

FIGURE 1.11

The orientation used throughout this book (and most others, also) is the *right-handed* one of Figure 1.9 and the right half of Figure 1.11. The terminology results from the following model. If you orient your right index finger in the direction of the positive x-axis and your right middle finger in the direction of the positive y-axis, then your right thumb will point upward in the direction of the positive z-axis. (Try it, remembering that the positive x-axis is to be visualized as being directed outward from the book.) See Figure 1.12.

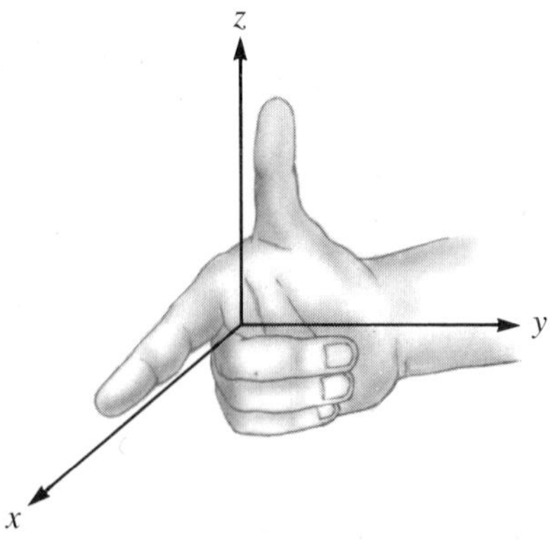

FIGURE 1.12

Now we can introduce vectors in $\boldsymbol{R}^3$ by analogy to the procedure used in $\boldsymbol{R}^2$. We thus again translate a given vector $\vec{v}$ in $\boldsymbol{R}^3$ parallel to itself until its initial point is at the origin. We then use the coordinates (a, b, c) of the endpoint P of the translated $\vec{v}$ as its algebraic representation. See Figure 1.13. Addition and scalar multiplication are defined as in Definition 1.1.

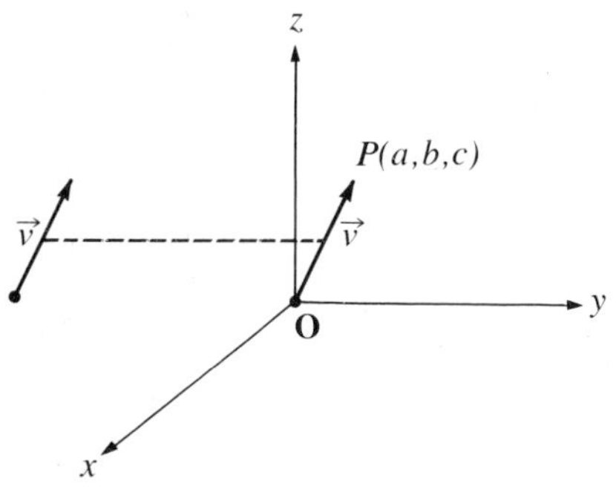

FIGURE 1.13

1.3 DEFINITION. If $\vec{v} = (a, b, c)$ and $\vec{w} = (d, e, f)$ are vectors in $\boldsymbol{R}^3$, then

(1) $$\vec{v} + \vec{w} = (a + d, b + e, c + f)$$

and

(2) $$r\vec{v} = (ra, rb, rc)$$

for any $r \in \boldsymbol{R}$. The **vector difference** $\vec{v} - \vec{w}$ means $\vec{v} + (-\vec{w})$, where $-\vec{w} = (-1)\vec{w}$.

It can again be shown (Exercise 17 below) that (1) is unavoidable if we want the parallelogram law to hold for vector addition. The parallelogram law, in turn, once more gives us the algebraic representation of the *position vector* $\overrightarrow{PQ}$ joining any point $P \in \boldsymbol{R}^3$ to another point $Q \in \boldsymbol{R}^3$.

1.4 PROPOSITION. If $\vec{v}=\overrightarrow{OP}$ where $P=(x_1, y_1, z_1)$, and $\vec{w}=\overrightarrow{OQ}$ where $Q=(x_2, y_2, z_2)$, then the position vector from P to Q is

(3)
$$\overrightarrow{PQ}=\overrightarrow{OQ}-\overrightarrow{OP}=(x_2-x_1, y_2-y_1, z_2-z_1)=\vec{w}-\vec{v}.$$

Proof. Referring to Figure 1.14, we see that $\overrightarrow{PQ}$ is the diagonal of the parallelogram formed by $\vec{w}$ and $-\vec{v}$. Since $\vec{w}=(x_2, y_2, z_2)$ and $-\vec{v}=(-x_1, -y_1, -z_1)$, we have

$$\begin{aligned}\overrightarrow{PQ}&=\vec{w}+(-\vec{v})=(x_2-x_1, y_2-y_1, z_2-z_1)\\&=\vec{w}-\vec{v}.\end{aligned}$$

QED

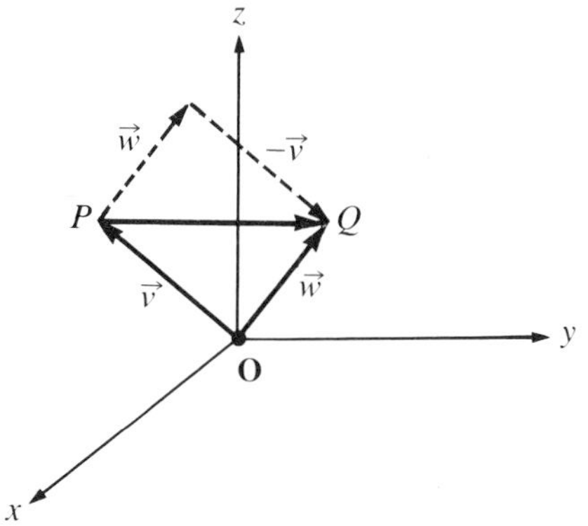

FIGURE 1.14

As before, then, the position vector $\overrightarrow{PQ}$ is obtained by subtracting the respective coordinates of P from those of Q. Notice then that

$$\overrightarrow{QP}=\vec{v}-\vec{w}=-\overrightarrow{PQ}.$$

To complete our analogy with $\boldsymbol{R}^2$, we need a distance formula for $\boldsymbol{R}^3$.

1.5 THEOREM (DISTANCE FORMULA). If $P(x_1, y_1, z_1)$ and $Q(x_2, y_2, z_2)$ are two points in $\boldsymbol{R}^3$, then the distance between them is

$$|\overrightarrow{PQ}|=\sqrt{(x_2-x_1)^2+(y_2-y_1)^2+(z_2-z_1)^2}.$$

Proof. Construct the rectangular parallelepiped with faces parallel to the coordinate planes and vertices P and Q (see Figure 1.15). The vertex directly above Q is labeled B and has coordinates (x_2, y_2, z_1). We have that

$$|\overrightarrow{BQ}|=|z_2-z_1|=\sqrt{(z_2-z_1)^2}$$

and

$$|\overrightarrow{PB}|=\sqrt{(x_2-x_1)^2+(y_2-y_1)^2}$$

by the distance formulas for $\boldsymbol{R}^1$ and $\boldsymbol{R}^2$. ($\overrightarrow{BQ}$ can be viewed as a vector in the z-axis, a copy of $\boldsymbol{R}^1$; and $\overrightarrow{PB}$ can be viewed as a vector in the plane through P, A, and B, which is a copy of the xy-plane.) Triangle PBQ is a right triangle with hypotenuse $|\overrightarrow{PQ}|$. Hence

$$\begin{aligned}|\overrightarrow{PQ}|&=\sqrt{|\overrightarrow{PB}|^2+|\overrightarrow{BQ}|^2}\\|\overrightarrow{PQ}|&=\sqrt{(x_2-x_1)^2+(y_2-y_1)^2+(z_2-z_1)^2}.\end{aligned}$$

QED

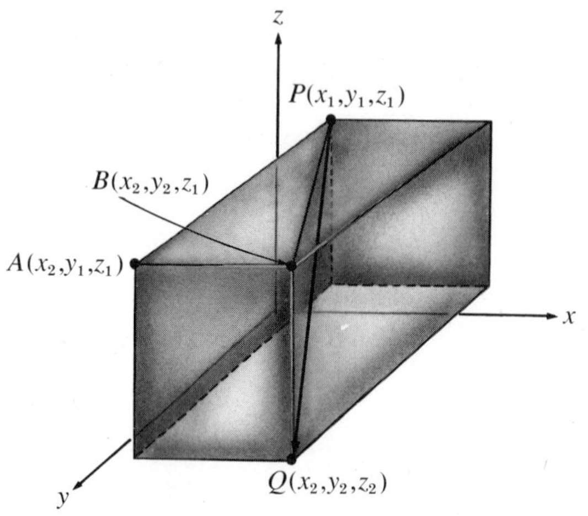

FIGURE 1.15

1.6 COROLLARY. The position vector $\overrightarrow{OP}$ drawn from the origin $\boldsymbol{O}$ to any point $P(x, y, z)$ has length

$$|\overrightarrow{OP}| = \sqrt{x^2+y^2+z^2}.$$

It is now easy to see that, just as in $\boldsymbol{R}^1$ and $\boldsymbol{R}^2$, if $r \in \boldsymbol{R}$, then $r\vec{v}$ is a vector of length $|r|\,|\vec{v}|$ which points in the same direction as $\vec{v}$ if $r>0$ or in the opposite direction if $r<0$. (See Exercise 18 below.)

1.7 EXAMPLE. (a) Find the length of the vector from $P(1, -\frac{1}{2}, 3)$ to $Q(-2, \frac{3}{2}, 5)$. (b) Show that

$$M\left(\frac{a_1+b_1}{2}, \frac{a_2+b_2}{2}, \frac{a_3+b_3}{2}\right)$$

is the midpoint of the segment from $A(a_1, a_2, a_3)$ to $B(b_1, b_2, b_3)$.

Solution. (a) $|\overrightarrow{PQ}| = \sqrt{(1+2)^2+(-\frac{1}{2}-\frac{3}{2})^2+(3-5)^2}$

$$= \sqrt{3^2+(-2)^2+(-2)^2}$$

$$= \sqrt{9+4+4} = \sqrt{17}.$$

(b)
$$|\overrightarrow{AM}| = \sqrt{\left(\frac{a_1+b_1}{2}-a_1\right)^2+\left(\frac{a_2+b_2}{2}-a_2\right)^2+\left(\frac{a_3+b_3}{2}-a_3\right)^2}$$

$$= \sqrt{\left(\frac{b_1-a_1}{2}\right)^2+\left(\frac{b_2-a_2}{2}\right)^2+\left(\frac{b_3-a_3}{2}\right)^2}.$$

$$|\overrightarrow{MB}| = \sqrt{\left(\frac{a_1+b_1}{2}-b_1\right)^2+\left(\frac{a_2+b_2}{2}-b_2\right)^2+\left(\frac{a_3+b_3}{2}-b_3\right)^2}$$

$$= \sqrt{\left(\frac{a_1-b_1}{2}\right)^2+\left(\frac{a_2-b_2}{2}\right)^2+\left(\frac{a_3-b_3}{2}\right)^2}.$$

Note that

$$|\overrightarrow{AM}| = |\overrightarrow{MB}| = \tfrac{1}{2}\sqrt{(b_1-a_1)^2+(b_2-a_2)^2+(b_3-a_3)^2}$$
$$= \tfrac{1}{2}\,|\overrightarrow{AB}|.$$

Thus M is the midpoint of the segment from A to B.

1.8 EXAMPLE. Find an equation for the sphere of radius 5 centered at $C(3, -1, 2)$.

Solution. If $P(x, y, z)$ is any point on the sphere, then $|\overrightarrow{PC}| = 5$. Hence we have

$$\sqrt{(x-3)^2+(y+1)^2+(z-2)^2} = 5.$$

Squaring both sides, we get

$$(x-3)^2+(y+1)^2+(z-2)^2 = 25.$$

It is very easy to see that Example 1.8 generalizes immediately.

1.9 PROPOSITION. The sphere with center $C(a, b, c)$ and radius r has the equation

$$(x-a)^2+(y-b)^2+(z-c)^2 = r^2.$$

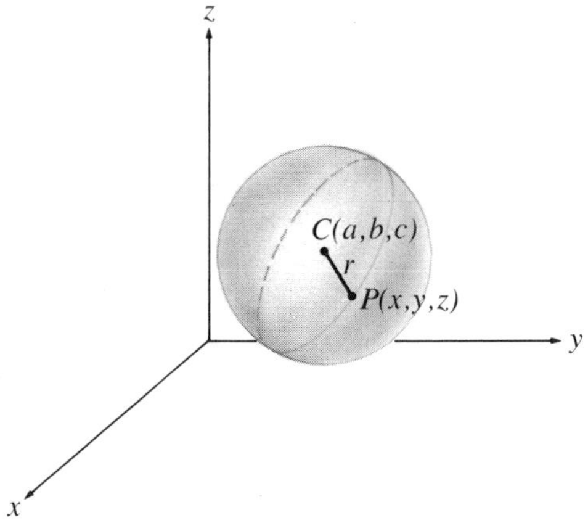

FIGURE 1.16

1.10 EXAMPLE. Find the center and radius of the sphere whose equation is $x^2+y^2+z^2-3x+6y=1$.

Solution. Grouping terms of the same variable together and completing the square, we have

$$x^2-3x+y^2+6y+z^2=1,$$
$$x^2-3x+\tfrac{9}{4}+y^2+6y+9+z^2=\tfrac{9}{4}+9+1,$$
$$(x-\tfrac{3}{2})^2+(y+3)^2+z^2=\tfrac{9}{4}+\tfrac{36}{4}+\tfrac{4}{4}=\tfrac{49}{4}.$$

So the center is $(\frac{3}{2}, -3, 0)$ and the radius is $\frac{7}{2}$.

The following properties of vector addition and multiplication by real numbers (which are called *scalars*, since the real numbers populate the three axes, or scales, we use to measure vectors) are all direct consequences of Definition 1.3.

1.11 THEOREM. For all vectors $\vec{u}$, $\vec{v}$, $\vec{w}$ in $\boldsymbol{R}^3$ and all scalars s and $t \in \boldsymbol{R}$,
(1) $\vec{v}+\vec{w}=\vec{w}+\vec{v}$. (Addition is **commutative**.)
(2) $\vec{u}+(\vec{v}+\vec{w})=(\vec{u}+\vec{v})+\vec{w}$. (Addition is **associative**.)
(3) $\vec{v}+\vec{0}=\vec{v}$, where $\vec{0}=(0,0,0)\in\boldsymbol{R}^3$ is the **additive identity**.
(4) $\vec{v}+(-\vec{v})=\vec{0}$, where $-\vec{v}=(-a,-b,-c)$ is the **additive inverse** of $\vec{v}=(a,b,c)$.
(5) $s(\vec{v}+\vec{w})=s\vec{v}+s\vec{w}$. (A kind of **distributive law**.)

(6) $(s+t)\vec{v} = s\vec{v} + t\vec{v}$. (Another kind of distributive law.)
(7) $s(t\vec{v}) = (st)\vec{v}$. (A kind of associative law for scalar multiplication.)
(8) $1\vec{v} = \vec{v}$. (Thus, 1 acts as an **identity** for scalar multiplication.)
(9) $0\vec{v} = \vec{0}$.

Partial Proof. Since many of these properties are evident, we verify only (2) and (5). Suppose

$$\vec{v} = (a, b, c), \qquad \vec{w} = (p, q, r), \qquad \vec{u} = (x, y, z).$$

Then

$$\begin{aligned} \vec{u} + (\vec{v} + \vec{w}) &= (x, y, z) + (a+p, b+q, c+r) \\ &= (x+(a+p), y+(b+q), z+(c+r)) \end{aligned}$$

and

$$\begin{aligned} (\vec{u} + \vec{v}) + \vec{w} &= (x+a, y+b, z+c) + (p, q, r) \\ &= ((x+a)+p, (y+b)+q, (z+c)+r). \end{aligned}$$

Each coordinate of $\vec{u} + (\vec{v} + \vec{w})$ equals the corresponding coordinate of $(\vec{u} + \vec{v}) + \vec{w}$ by the associative law of addition of real numbers. Hence (2) follows. As for (5),

$$\begin{aligned} s(\vec{v} + \vec{w}) &= s(a+p, b+q, c+r) \\ &= (s(a+p), s(b+q), s(c+r)) \\ &= (sa+sp, sb+sq, sc+sr) \\ &= (sa, sb, sc) + (sp, sq, sr) \\ &= s(a, b, c) + s(p, q, r) \\ &= s\vec{v} + s\vec{w} \end{aligned}$$

where we used the distributive law for real numbers in the third line. QED

1.12 **EXAMPLE.** If $\vec{v} = (-4, 2, 0)$ and $\vec{w} = (6, -3, 4)$, then find $\vec{v} + \vec{w}$, $\vec{v} - \vec{w}$, and $2\vec{v} + \vec{w}$. Plot.

Solution.

$$\vec{v} + \vec{w} = (-4+6, 2+(-3), 0+4) = (2, -1, 4).$$

$$\vec{v} - \vec{w} = \vec{v} + (-1)\vec{w} = (-4, 2, 0) + (-6, 3, -4) = (-10, 5, -4).$$

$$2\vec{v} + \vec{w} = (-8, 4, 0) + (6, -3, 4) = (-2, 1, 4).$$

In Figure 1.17, $\vec{v} - \vec{w}$ was drawn both with its initial point at the tip of $\vec{w}$ and with its initial point at $\boldsymbol{O}$ (cf. Proposition 1.4). Given any vector, we are, of course, free to draw it from any initial point we please. It is often convenient to use $\boldsymbol{O}$ as initial point, but in Figure 1.17 the tip of $\vec{w}$ was just as natural a place to begin $\vec{v} - \vec{w}$.

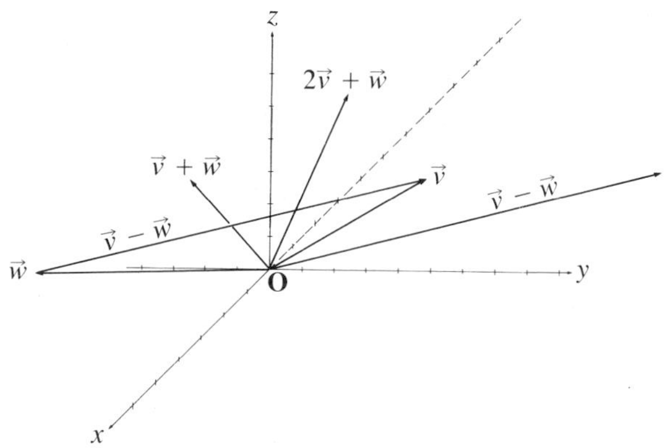

FIGURE 1.17

Finally, let us give a general description of what we have done so far.

1.13 DEFINITION. $\boldsymbol{R}^n = \{(x_1, x_2, \ldots, x_n) \mid x_i \in \boldsymbol{R}\}$, the set of real n-tuples, is called **n-dimensional Cartesian** (or **Euclidean**) **space**. If $\vec{x} = (x_1, x_2, \ldots, x_n)$ and $\vec{y} = (y_1, y_2, \ldots, y_n)$, then addition of vectors, multiplication of vectors by real numbers, and length of vectors are defined by

$$\vec{x}+\vec{y} = (x_1+y_1, x_2+y_2, \ldots, x_n+y_n),$$
$$a\vec{x} = (ax_1, ax_2, \ldots, ax_n),$$
$$\vec{x}-\vec{y} = \vec{x}+(-1)\vec{y},$$
$$|\vec{x}| = \sqrt{x_1^2+x_2^2+\ldots+x_n^2}.$$

If $|\vec{x}| = 1$, then $\vec{x}$ is called a **unit vector.**

Notice that for $n = 1$, 2, or 3, Definition 1.13 includes the separate algebraic descriptions we have given for $\boldsymbol{R}^1$, $\boldsymbol{R}^2$, and $\boldsymbol{R}^3$. It is also easy to show that Theorem 1.11 is true for $\boldsymbol{R}^n$ (Exercise 12 below). This suggests the advantages inherent in working with $\boldsymbol{R}^n$. As long as n is not specified, discussions about $\boldsymbol{R}^n$ apply to $\boldsymbol{R}^1$, $\boldsymbol{R}^2$, $\boldsymbol{R}^3$, in fact to *any* Cartesian space of any dimension. In much of the rest of this book we will develop machinery for use primarily in $\boldsymbol{R}^2$ or $\boldsymbol{R}^3$. To do this with just one discussion (rather than separate discussions for each space, $\boldsymbol{R}^2$ and $\boldsymbol{R}^3$), we will work in $\boldsymbol{R}^n$ as much as possible, without specifying n. As a bonus, much of our theory will apply to *all* Cartesian spaces, which play a prominent role in much of higher mathematics and applied mathematics. But we will seldom have use here for this generality. To visualize the contexts of our discussions, it is perfectly proper to draw pictures in $\boldsymbol{R}^2$ or $\boldsymbol{R}^3$ representing the situation under discussion in $\boldsymbol{R}^n$ when $n = 2$ or 3. However, you will maximize your appreciation of the discussions if you don't just rely on the illustrations in the book, but draw your own pictures to illustrate how you visualize the situation in question.

The list of properties in Theorem 1.11 is used to give meaning to the concept of vector in settings even more general than $\boldsymbol{R}^n$.

1.14 DEFINITION. A collection V of objects (called **vectors**) is called a **real vector space** if there is a sum $\vec{v}+\vec{w} \in V$ defined for members of V and also a product $r\vec{v} \in V$ of a real number r and a vector $\vec{v}$ such that properties (1) through (9) of Theorem 1.11 hold for V.

1.15 EXAMPLE. The set V of all continuous real valued functions f defined on an interval $[a, b]$ is a real vector space.

Proof. There are well known operations of addition and multiplication by scalars for real valued functions. Namely, $f+g$ is defined by $(f+g)(x)=f(x)+g(x)$, and rf is defined for $r\in \boldsymbol{R}$ by $(rf)(x)=rf(x)$. Moreover, basic theorems of elementary calculus assure that $f+g$ and rf are continuous on $[a, b]$ if f and g are continuous. Therefore, the sum of two elements of V is in V, and also the product rf is in V, for any real number r, as required. Properties (1), (2), (5), (6), (7), (8), and (9) follow directly from the definitions of $f+g$ and rf. As for (3) and (4), $\mathbf{0}$ is the function defined by $\mathbf{0}(x)=0$ for all x, and $-f$ is given by $(-f)(x)=-f(x)$ for all x in $[a, b]$. Thus properties (1) through (9) of Theorem 1.11 hold, so V is a real vector space. QED

In this section we have used the notation $\vec{v}$ for a vector. This notation is appropriate for geometric vectors, and is convenient to use when writing on paper or blackboards. For algebraic vectors (n-tuples) in print, however, boldface is the more conventional notation. Since we have shown how to represent geometric vectors as algebraic vectors, we will henceforth use the notation $\boldsymbol{v}$ instead of $\vec{v}$. For the position vector $\overrightarrow{OP}=(x, y, z)$ we will use the generic symbol $\boldsymbol{x}$. The position vector from A to B will still be denoted as $\overrightarrow{AB}$ in most contexts, but we will frequently simplify this notation by letting $\boldsymbol{u}$ (or $\boldsymbol{v}$ or $\boldsymbol{w}$, etc.)$=\overrightarrow{AB}$.

While all vectors in this section have been written algebraically as *row vectors* $(x_1, x_2, \ldots, x_n)$, there would seem to be no compelling reason (aside from making the printing simpler) for preferring $(x_1, x_2, \ldots, x_n)$ to the column

$$\begin{pmatrix} x_1 \\ x_2 \\ \cdot \\ \cdot \\ \cdot \\ x_n \end{pmatrix}$$

as a representation for a vector in $\boldsymbol{R}^n$. Later we shall not hesitate to write vectors in column form as well as in row form and will use whichever notation is more appropriate to a given context.

Finally, we introduce a useful collection of vectors in $\boldsymbol{R}^n$ which will appear frequently throughout this book.

1.16 **DEFINITION.** The **standard basis** for $\boldsymbol{R}^n$ is the ordered set $\{\boldsymbol{e}_1, \boldsymbol{e}_2, \ldots, \boldsymbol{e}_n\}$ where $\boldsymbol{e}_1=(1, 0, 0, \ldots, 0)$, $\boldsymbol{e}_2=(0, 1, 0, \ldots, 0), \ldots, \boldsymbol{e}_n=(0, 0, 0, \ldots, 1)$. If $n=2$ (or 3), one writes $\boldsymbol{i}$ in place of $\boldsymbol{e}_1$ and $\boldsymbol{j}$ in place of $\boldsymbol{e}_2$ (and $\boldsymbol{k}$ in place of $\boldsymbol{e}_3$). See Figure 1.18.

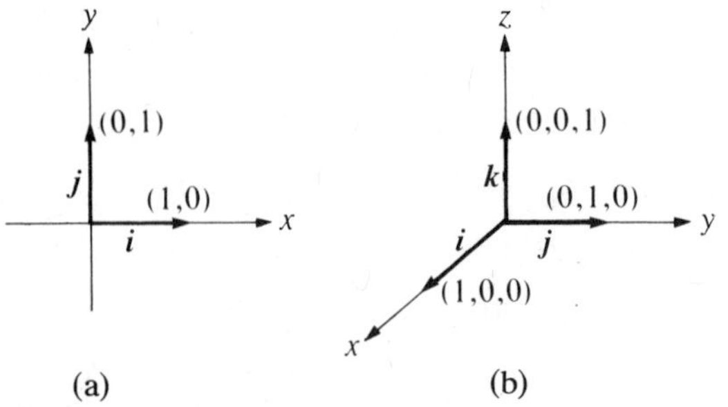

FIGURE 1.18

These vectors permit an alternate representation for vectors $\boldsymbol{v}$ in $\boldsymbol{R}^n$. For example, we can write the vector $(3, -2, 2)$ as $3\boldsymbol{i}-2\boldsymbol{j}+2\boldsymbol{k}$ in $\boldsymbol{R}^3$, since $(3, -2, 2)=3(1, 0, 0)-2(0, 1, 0)+2(0, 0, 1)$. For more on this, see Exercise 19 below.

EXERCISES 1.1

1 Plot accurately:
(a) $(3, 7, -2)$
(b) $(-2, 6, -4)$
(c) $(2, -6, 4)$
(d) $(2, 6, -4)$
(e) $(-2, -6, 4)$
(f) $(-2, -6, -4)$

2 Draw the rectangular parallelepiped with faces parallel to the coordinate planes and with the given points as vertices.
(a) $(2, -3, 4)$ and $(-1, 2, 1)$
(b) $(3, 1, -2)$ and $(5, -1, 3)$

3 Find the length and midpoint of the segments joining the given points.
(a) $(1, -2, 3)$ and $(-2, 4, 5)$
(b) $(0, \frac{3}{2}, \sqrt{2})$ and $(2, \frac{7}{2}, -\sqrt{2})$

4 Determine whether the given points are collinear.
(a) $(-\frac{1}{2}, -\frac{3}{2}, 5)$, $(\frac{5}{2}, -\frac{7}{2}, 0)$, and $(-\frac{7}{2}, \frac{13}{2}, 5)$
(b) $(2, -2, -3)$, $(-1, 1, 1)$, and $(5, -5, -7)$

5 If $\boldsymbol{v} = (-\frac{1}{2}, 1)$ and $\boldsymbol{w} = (\frac{5}{2}, -3)$, then compute (a) $\boldsymbol{v}+\boldsymbol{w}$, (b) $\boldsymbol{v}-\boldsymbol{w}$, (c) $\boldsymbol{w}-\boldsymbol{v}$, (d) $4\boldsymbol{v}-2\boldsymbol{w}$, (e) $-2\boldsymbol{v}+4\boldsymbol{w}$, (f) $|\boldsymbol{v}|$, (g) $|\boldsymbol{w}|$. (h) Find the unit vectors in the direction of $\boldsymbol{v}$ and $\boldsymbol{w}$.

6 If $\boldsymbol{v} = (1, 2, -2)$ and $\boldsymbol{w} = (3, -1, 1)$, then compute (a) $\boldsymbol{v}+\boldsymbol{w}$, (b) $\boldsymbol{v}-\boldsymbol{w}$, (c) $\boldsymbol{w}-\boldsymbol{v}$, (d) $-2\boldsymbol{v}+3\boldsymbol{w}$, (e) $\frac{1}{3}\boldsymbol{v}-\frac{4}{3}\boldsymbol{w}$, (f) $|\boldsymbol{v}|$, (g) $|\boldsymbol{w}|$. (h) Find the unit vectors in the direction of $\boldsymbol{v}$ and $\boldsymbol{w}$.

7 Repeat Exercise 6 if $\boldsymbol{v} = (1, -2, 0, 3)$ and $\boldsymbol{w} = (-1, 1, 2, 1)$ in $\boldsymbol{R}^4$.

8 Find the position vector $\overrightarrow{AB}$ and compute its length if
(a) $A = (1, 2, -3, 1)$, $B = (-1, 2, 1, 3)$
(b) $A = (0, 2, 1, -1)$, $B = (3, -2, 4, 1)$

9 Find the position vector $\overrightarrow{AB}$, compute its length, and plot if
(a) $A = (1, -1, 3)$ and $B = (-2, -1, 1)$
(b) $A = (0, 2, -1)$ and $B = (1, -1, 1)$

10 (a) Prove (1) of Theorem 1.11.
(b) Prove (4) of Theorem 1.11.

11 (a) Prove (6) of Theorem 1.11.
(b) Prove (7), (8), and (9) of Theorem 1.11.

12 For the analogue of Theorem 1.11 in $\boldsymbol{R}^n$, write out the proof of parts (2) and (5). State parts (3) and (4).

13 Find the equation of the sphere
(a) with center $(1, -2, -2)$, radius 2
(b) with center $(2, 0, 1)$, radius 3
(c) with center $(1, -1, 4)$, radius $\frac{1}{2}$

14 Find the center and radius of the sphere whose equation is

$$x^2+y^2+z^2-3x+2y-4z = \frac{7}{4}$$

15 Show that any vector $\boldsymbol{v}$ in $\boldsymbol{R}^3$ can be written as $\boldsymbol{v} = |\boldsymbol{v}|\,\boldsymbol{u}$, where $\boldsymbol{u}$ is a unit vector.

16 Suppose we represent $\boldsymbol{v}$ in $\boldsymbol{R}^2$ by (a, b) and $\boldsymbol{w}$ in $\boldsymbol{R}^2$ by (c, d). If addition is defined by the parallelogram law, then show that $\boldsymbol{v}+\boldsymbol{w}$ *must* be the vector $(a+c, b+d)$. [*Hint*: Let $\boldsymbol{v}+\boldsymbol{w} = (x, y)$, and solve for x and y by projecting the sides of the parallelogram formed by $\boldsymbol{v}$ and $\boldsymbol{w}$ onto the coordinate axes.]

17 Suppose we represent $\boldsymbol{v}$ in $\boldsymbol{R}^3$ by (a_1, a_2, a_3) and $\boldsymbol{w}$ in $\boldsymbol{R}^3$ by (b_1, b_2, b_3). If addition is defined by the parallelogram law, then show that $\boldsymbol{v}+\boldsymbol{w}$ *must* be the vector $(a_1+b_1, a_2+b_2, a_3+b_3)$. [See the hint for Exercise 16.]

18 If $\boldsymbol{v} \in \boldsymbol{R}^3$ and $r \in \boldsymbol{R}$, then show that $|r\boldsymbol{v}| = |r|\,|\boldsymbol{v}|$ and $r\boldsymbol{v}$ points in the same direction as $\boldsymbol{v}$ if $r > 0$, or in the opposite direction if $r < 0$.

19 Show that any vector in $\boldsymbol{R}^n$ is uniquely expressible in the form $\sum_{i=1}^{n} r_i \boldsymbol{e}_i$, for suitable $r_i \in \boldsymbol{R}$.

20 In applications, vectors are often written as columns. Suppose that a manufacturing company produces four products, which are shipped to four regional distributors located in the Northeast, Southeast, Northwest, and Southwest. Each distributor submits his order to the parent company on a form with Product A listed first, Product B second, Product C third, and Product D fourth. We can represent such an order by a vector

$$\boldsymbol{x} = \begin{pmatrix} x_1 \\ x_2 \\ x_3 \\ x_4 \end{pmatrix}$$

for the Northeast distributor, and vectors

$$\boldsymbol{y} = \begin{pmatrix} y_1 \\ y_2 \\ y_3 \\ y_4 \end{pmatrix}, \quad \boldsymbol{z} = \begin{pmatrix} z_1 \\ z_2 \\ z_3 \\ z_4 \end{pmatrix}, \quad \text{and} \quad \boldsymbol{w} = \begin{pmatrix} w_1 \\ w_2 \\ w_3 \\ w_4 \end{pmatrix}$$

for the Southeast, Northwest, and Southwest distributors, respectively.

(a) What do the vectors $\boldsymbol{x}+\boldsymbol{y}$, $\boldsymbol{x}+\boldsymbol{z}$, $\boldsymbol{y}+\boldsymbol{w}$, and $\boldsymbol{z}+\boldsymbol{w}$ represent?

(b) If next month the distributors' orders change according to the following table, then compute for each the vector giving his new order.

Distributor	Change in order
NE	+2%
SE	−2.2%
NW	+1.1%
SW	−0.2%

(c) Does the set of all possible orders for A, B, C, and D form a real vector space? Why or why not?

21 If a force $\boldsymbol{F}$ acts in space, then at any point $\boldsymbol{x}$ we have

$$\boldsymbol{F}(\boldsymbol{x}) = (F_1(\boldsymbol{x}), F_2(\boldsymbol{x}), F_3(\boldsymbol{x})).$$

(a) Express $\boldsymbol{F}$ as the resultant of three forces acting parallel to the three coordinate axes.

(b) If a particle of mass m has acceleration $\boldsymbol{a}(\boldsymbol{x})$ at $\boldsymbol{x}$, then Newtonian physics tells us that $\boldsymbol{F}(\boldsymbol{x}) = m\boldsymbol{a}(\boldsymbol{x})$. Show that this *single* vector equation is equivalent to *three* scalar equations, each having the same general form. (This suggests some of the simplification that the use of vectors produces in physics and engineering.)

22 Show that any vector $\boldsymbol{v}$ in $\boldsymbol{R}^n$ can be written as $\boldsymbol{v} = |\boldsymbol{v}|\,\boldsymbol{u}$, where $\boldsymbol{u}$ is a unit vector. [*Hint:* See Exercise 15.]

23 Show that the set of all polynomials with real coefficients is a real vector space, if we add and multiply by real scalars in the usual ways.

24 Show that the set of all differentiable real valued functions on an interval $[a, b]$ is a real vector space.

25 Show that the set of all integrable real valued functions on an interval $[a, b]$ is a real vector space.

26 Show that the set of all functions $f:[0, 1]\to[0, 1]$ is *not* a real vector space under the usual operations of addition and multiplication by a real scalar. [*Hint*: Show that $f+g$ need not be such a function even if f and g are.]

27 Show that the set $\mathscr{F}(\boldsymbol{R}^n)$ of all functions $f:\boldsymbol{R}^n \to \boldsymbol{R}$ with a common domain D is a real vector space. **(This is needed in Sections 5.1, 8.3, and 9.4.)**

2 DOT PRODUCT AND ANGLES

In the preceding section we introduced the algebraic representation of geometric vectors by n-tuples of real numbers, and we gave a simple formula for computing the length of a vector. We still lack an algebraic tool for dealing with the other essential ingredient of a geometric vector, its direction. The goal of this section is the development of such an algebraic tool for measuring the direction of a (nonzero) vector. Let's see if we can discover such a tool in $\boldsymbol{R}^2$ and then apply it to the general case of vectors in $\boldsymbol{R}^n$.

Given a vector $\boldsymbol{x}=(x_1, x_2)\neq(0, 0)$ in $\boldsymbol{R}^2$, then its direction is described completely by the angle α between $\boldsymbol{x}$ and $\boldsymbol{i}=(1, 0)$, the unit vector in the direction of the positive x-axis. Here it is sufficient to restrict α to be a first or second quadrant angle. See Figure 2.1.

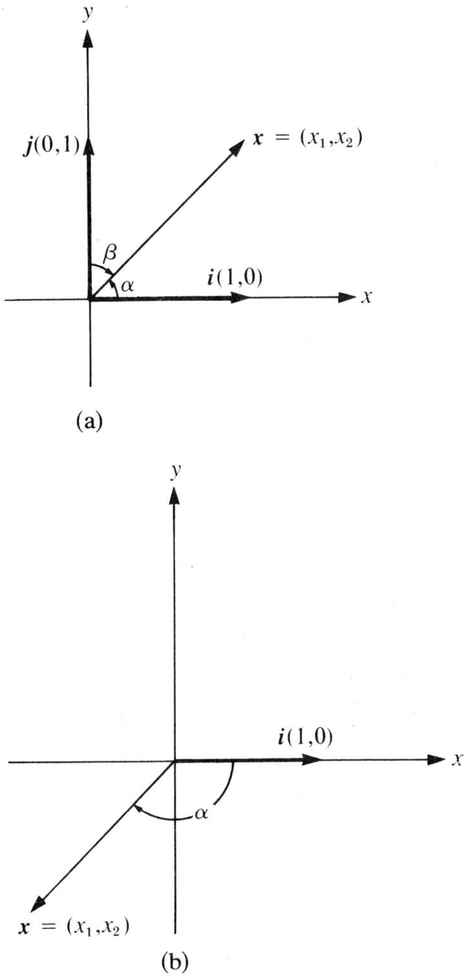

FIGURE 2.1

Note also that we could just as well describe the direction of $\boldsymbol{x}$ in terms of the angle β between $\boldsymbol{x}$ and $\boldsymbol{j}$. A simple formula we could use to find α is

$$\cos\alpha = \frac{x_1}{\sqrt{x_1^2+x_2^2}} = \frac{x_1\cdot 1 + x_2\cdot 0}{|\boldsymbol{x}|\,|\boldsymbol{i}|}.$$

Similarly, there is a simple formula for β,

$$\cos\beta = \frac{x_2}{\sqrt{x_1^2+x_2^2}} = \frac{x_1\cdot 0 + x_2\cdot 1}{|\boldsymbol{x}|\,|\boldsymbol{j}|}.$$

In each case, the cosine of the angle between $\boldsymbol{x}$ and the unit vector is given by a quotient. The denominator is the product of the lengths of $\boldsymbol{x}$ and the unit vector. The numerator is the sum of the products of the corresponding coordinates of $\boldsymbol{x}$ and the unit vector. This numerator is an important geometric quantity, as we see from the following more general problem.

2.1 **EXAMPLE.** If $\boldsymbol{x}=(x_1, x_2)$ and $\boldsymbol{y}=(y_1, y_2)$ are nonzero, then find the angle θ between $\boldsymbol{x}$ and $\boldsymbol{y}$. See Figure 2.2.

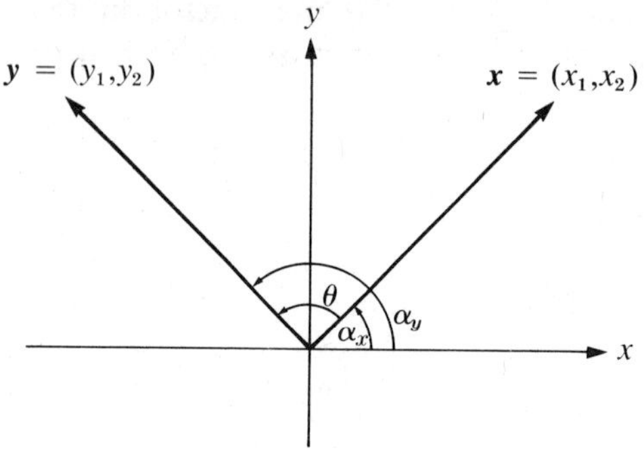

FIGURE 2.2

Solution. We have $\theta=\alpha_y-\alpha_x$, so

$$\cos\theta = \cos(\alpha_y-\alpha_x) = (\cos\alpha_y)(\cos\alpha_x)+(\sin\alpha_y)(\sin\alpha_x)$$

(from trigonometry)

$$= \frac{y_1}{\sqrt{y_1^2+y_2^2}}\cdot\frac{x_1}{\sqrt{x_1^2+x_2^2}} + \frac{y_2}{\sqrt{y_1^2+y_2^2}}\cdot\frac{x_2}{\sqrt{x_1^2+x_2^2}}$$

$$= \frac{x_1y_1+x_2y_2}{|\boldsymbol{x}|\,|\boldsymbol{y}|}.$$

[Since $\cos\theta=\cos(-\theta)$, it doesn't matter whether θ is measured from $\boldsymbol{x}$ to $\boldsymbol{y}$ or from $\boldsymbol{y}$ to $\boldsymbol{x}$.] We notice that the numerator of the expression for $\cos\theta$ again is the sum of the products of the corresponding coordinates of $\boldsymbol{x}$ and $\boldsymbol{y}$. This quantity is of such great importance that we single it out, give it a name, and develop some of its properties.

2.2 **DEFINITION.** If $\boldsymbol{x}=(x_1, x_2, \ldots, x_n)$ and $\boldsymbol{y}=(y_1, y_2, \ldots, y_n)$, then the **dot product** (or **inner product**) of $\boldsymbol{x}$ and $\boldsymbol{y}$ is the scalar

$$\boldsymbol{x}\cdot\boldsymbol{y} = x_1y_1+x_2y_2+\ldots+x_ny_n.$$

Thus our result in Example 2.1 takes the form $\cos\theta = \dfrac{\boldsymbol{x}\cdot\boldsymbol{y}}{|\boldsymbol{x}|\,|\boldsymbol{y}|}$, and our formulas for the angles α and β between $\boldsymbol{x}$ and the coordinate axes become

$$\cos\alpha = \frac{\boldsymbol{x}\cdot\boldsymbol{i}}{|\boldsymbol{x}|\,|\boldsymbol{i}|}, \qquad \cos\beta = \frac{\boldsymbol{x}\cdot\boldsymbol{j}}{|\boldsymbol{x}|\,|\boldsymbol{j}|}.$$

2.3 EXAMPLE. (a) Find the angle between $\boldsymbol{x}=(1, 2)$ and $\boldsymbol{y}=(3, 1)$. (b) Find $\boldsymbol{x}\cdot\boldsymbol{y}$ if $\boldsymbol{x}=(1, -1, 0)$ and $\boldsymbol{y}=(-1, 1, \sqrt{2})$.

Solution. (a) $\cos\theta = \dfrac{\boldsymbol{x}\cdot\boldsymbol{y}}{|\boldsymbol{x}|\,|\boldsymbol{y}|} = \dfrac{1\cdot 3+2\cdot 1}{\sqrt{5}\sqrt{10}}$

$$= \frac{5}{5\sqrt{2}} = \frac{1}{\sqrt{2}}.$$

So $\theta = \dfrac{\pi}{4}$ radians (45°).

(b) $\boldsymbol{x}\cdot\boldsymbol{y} = 1(-1)+(-1)1+0\sqrt{2} = -2.$

Before applying the dot product in $\boldsymbol{R}^n$, we need to know some more about it. Each of the properties in the following theorem is important enough to have its own name.

2.4 THEOREM. For all vectors $\boldsymbol{x}$, $\boldsymbol{y}$, $\boldsymbol{z}$ and all real numbers a,

(1) $\boldsymbol{x}\cdot\boldsymbol{y}=\boldsymbol{y}\cdot\boldsymbol{x}$. (The dot product is *symmetric*.)

(2) $\boldsymbol{x}\cdot(\boldsymbol{y}+\boldsymbol{z})=\boldsymbol{x}\cdot\boldsymbol{y}+\boldsymbol{x}\cdot\boldsymbol{z}$ and $(a\boldsymbol{x})\cdot\boldsymbol{y}=\boldsymbol{x}\cdot(a\boldsymbol{y})=a(\boldsymbol{x}\cdot\boldsymbol{y})$.
(The dot product is a *bilinear form*.)

(3) $\boldsymbol{0}\cdot\boldsymbol{x}=0$ for all $\boldsymbol{x}$. If $\boldsymbol{v}\cdot\boldsymbol{x}=0$ for all $\boldsymbol{x}$, then $\boldsymbol{v}=\boldsymbol{0}$.
(The dot product is *nondegenerate*.)

(4) $\boldsymbol{x}\cdot\boldsymbol{x}\geq 0$ for all $\boldsymbol{x}$, and $\boldsymbol{x}\cdot\boldsymbol{x}=0$ only when $\boldsymbol{x}=\boldsymbol{0}$.
(The dot product is *positive definite*.)

Proof. Each of these properties is a simple consequence of the definition and properties of the real numbers. We leave verification of (1) and (2) for you (Exercise 12b below). As for (3), $\boldsymbol{0}\cdot\boldsymbol{x}=0$ is clear. If $\boldsymbol{v}\cdot\boldsymbol{x}=0$ for all $\boldsymbol{x}$, where $\boldsymbol{v}=(v_1, v_2, \ldots, v_n)$, then let's compute $\boldsymbol{v}\cdot\boldsymbol{e}_i$ for each i [cf. Definition 1.16]. We have $\boldsymbol{e}_i=(0, 0, \ldots, 0, 1, 0, \ldots, 0)$ where the 1 occurs in the i-th coordinate. Thus, $\boldsymbol{v}\cdot\boldsymbol{e}_i=v_i=0$. So for each i, the i-th coordinate of $\boldsymbol{v}$ is 0. Hence $\boldsymbol{v}=\boldsymbol{0}$.

(4) If $\boldsymbol{x}=(x_1, x_2, \ldots, x_n)$, then

$$\boldsymbol{x}\cdot\boldsymbol{x}=x_1^2+x_2^2+\ldots+x_n^2\geq 0$$

since it is a sum of squares. Moreover, the only time a sum of squares x_i^2 can be 0 is when each summand $x_i^2=0$. Thus $\boldsymbol{x}\cdot\boldsymbol{x}=0$ only when each $x_i=0$, i.e., only when $\boldsymbol{x}=\boldsymbol{0}$. QED

2.5 COROLLARY. $|\boldsymbol{x}|=\sqrt{\boldsymbol{x}\cdot\boldsymbol{x}}$.

Proof. This is a corollary of Definitions 2.2 and 1.13:

$$\sqrt{\boldsymbol{x}\cdot\boldsymbol{x}}=\sqrt{x_1^2+x_2^2+\ldots+x_n^2}=|\boldsymbol{x}|.$$ QED

We come now to an important result which carries the names of two famous Europeans of the past century, Augustin-Louis Cauchy (1789–1857) of France, one of the leaders in the development of modern calculus, and Hermann A. Schwarz (1843–1921) of Poland, who made important contributions to advanced calculus and higher mathematics. [This result also appears to have been discovered independently by the Russian mathematician Victor Bunyakovsky (1804–1889).]

2.6 THEOREM (CAUCHY–SCHWARZ INEQUALITY). For all $\boldsymbol{x}$ and $\mathbf{y}$ in $\boldsymbol{R}^n$,

$$|\boldsymbol{x} \cdot \mathbf{y}| \leq |\boldsymbol{x}|\,|\mathbf{y}|.$$

(Note that $|\boldsymbol{x} \cdot \mathbf{y}|$ is the *absolute value* of the real number $\boldsymbol{x} \cdot \mathbf{y}$, while $|\boldsymbol{x}|$ and $|\mathbf{y}|$ are the *lengths* of the vectors $\boldsymbol{x}$ and $\mathbf{y}$.)

Proof. This inequality clearly holds (as an equality) if either $\boldsymbol{x}=\mathbf{0}$ or $\mathbf{y}=\mathbf{0}$. So we can confine our attention to the case where $\boldsymbol{x}\neq\mathbf{0}$ and $\mathbf{y}\neq\mathbf{0}$. Now we consider the unit vectors $\boldsymbol{u}=\frac{1}{|\boldsymbol{x}|}\boldsymbol{x}$ and $\boldsymbol{v}=\frac{1}{|\mathbf{y}|}\mathbf{y}$. We will first prove that the inequality holds for $\boldsymbol{u}$ and $\boldsymbol{v}$, and then show that it must also hold for $\boldsymbol{x}$ and $\mathbf{y}$ as a consequence. We have

$$\begin{aligned} 0 \leq (\boldsymbol{u}-\boldsymbol{v}) \cdot (\boldsymbol{u}-\boldsymbol{v}) &= (\boldsymbol{u}-\boldsymbol{v}) \cdot \boldsymbol{u} + (\boldsymbol{u}-\boldsymbol{v}) \cdot (-\boldsymbol{v}) \\ &= \boldsymbol{u} \cdot \boldsymbol{u} - \boldsymbol{v} \cdot \boldsymbol{u} - \boldsymbol{u} \cdot \boldsymbol{v} + \boldsymbol{v} \cdot \boldsymbol{v} \quad \text{by Theorem 2.4(1) and (2)} \\ &= 2 - 2(\boldsymbol{u} \cdot \boldsymbol{v}) \quad \text{by Theorem 2.4(1)} \end{aligned}$$

Thus $2(\boldsymbol{u} \cdot \boldsymbol{v}) \leq 2$, so $(\boldsymbol{u} \cdot \boldsymbol{v}) \leq 1 = |\boldsymbol{u}|\,|\boldsymbol{v}|$. Hence

$$\frac{1}{|\boldsymbol{x}|}\boldsymbol{x} \cdot \frac{1}{|\mathbf{y}|}\mathbf{y} \leq 1,$$

so by Theorem 2.4(2)

$$\frac{1}{|\boldsymbol{x}|\,|\mathbf{y}|}\boldsymbol{x} \cdot \mathbf{y} \leq 1.$$

Thus

(1) $$\boldsymbol{x} \cdot \mathbf{y} \leq |\boldsymbol{x}|\,|\mathbf{y}|.$$

This holds true for any nonzero vectors $\boldsymbol{x}$ and $\mathbf{y}$, so it still holds if we replace $\boldsymbol{x}$ by $-\boldsymbol{x}$:

$$(-\boldsymbol{x}) \cdot \mathbf{y} \leq |-\boldsymbol{x}|\,|\mathbf{y}|.$$

Thus

$$-(\boldsymbol{x} \cdot \mathbf{y}) \leq |\boldsymbol{x}|\,|\mathbf{y}|$$

by Theorem 2.4(2), and the fact that $|\boldsymbol{x}|=|-\boldsymbol{x}|$. Hence

(2) $$\boldsymbol{x} \cdot \mathbf{y} \geq -|\boldsymbol{x}|\,|\mathbf{y}|.$$

Combining (1) and (2), we have

$$-|\boldsymbol{x}|\,|\mathbf{y}| \leq \boldsymbol{x} \cdot \mathbf{y} \leq |\boldsymbol{x}|\,|\mathbf{y}|,$$

which is equivalent to $|\boldsymbol{x} \cdot \mathbf{y}| \leq |\boldsymbol{x}|\,|\mathbf{y}|$. QED

It is worthwhile to write the Cauchy–Schwarz inequality in coordinate form:

$$\left|\sum_{i=1}^{n} x_i y_i\right| \le \sqrt{\left(\sum_{i=1}^{n} x_i^2\right)\left(\sum_{i=1}^{n} y_i^2\right)},$$

or, equivalently,

$$\left(\sum_{i=1}^{n} x_i y_i\right)^2 \le \left(\sum_{i=1}^{n} x_i^2\right)\left(\sum_{i=1}^{n} y_i^2\right).$$

Observe that in $\boldsymbol{R}^2$ or $\boldsymbol{R}^3$, we could have proved the Cauchy–Schwarz inequality from the formula

$$\cos\theta = \frac{\boldsymbol{x}\cdot\boldsymbol{y}}{|\boldsymbol{x}|\,|\boldsymbol{y}|}$$

in Example 2.1 (for the case $\boldsymbol{x} \neq \boldsymbol{0}$ and $\boldsymbol{y} \neq \boldsymbol{0}$). For we know that $|\cos\theta| \le 1$, so

$$\frac{\boldsymbol{x}\cdot\boldsymbol{y}}{|\boldsymbol{x}|\,|\boldsymbol{y}|} \le 1, \quad \text{hence} \quad |\boldsymbol{x}\cdot\boldsymbol{y}| \le |\boldsymbol{x}|\,|\boldsymbol{y}|.$$

The more general proof we have given has the advantage of applying to $\boldsymbol{R}^n$ for $n > 3$, where we have no geometric idea of angle as yet. In fact, let's use the Cauchy–Schwarz inequality to *define* what angle means in $\boldsymbol{R}^n$. First, the angle between any vector $\boldsymbol{x}$ and $\boldsymbol{0}$ is defined to be 0. For the angle between two nonzero vectors $\boldsymbol{x}$ and $\boldsymbol{y}$, we simply extend the formula we know in $\boldsymbol{R}^2$.

2.7 DEFINITION. If $\boldsymbol{x} \neq \boldsymbol{0}$ and $\boldsymbol{y} \neq \boldsymbol{0}$, then **the angle $\boldsymbol{\theta}$ between $\boldsymbol{x}$ and $\boldsymbol{y}$** is defined by

$$\boxed{\cos\theta = \frac{\boldsymbol{x}\cdot\boldsymbol{y}}{|\boldsymbol{x}|\,|\boldsymbol{y}|}.}$$

Thus,

$$\theta = \operatorname{Cos}^{-1}\frac{\boldsymbol{x}\cdot\boldsymbol{y}}{|\boldsymbol{x}|\,|\boldsymbol{y}|}$$

is a *first or second quadrant angle.*

Since, by the Cauchy–Schwarz inequality,

$$-|\boldsymbol{x}|\,|\boldsymbol{y}| \le \boldsymbol{x}\cdot\boldsymbol{y} \le |\boldsymbol{x}|\,|\boldsymbol{y}|,$$

we have

$$-1 \le \frac{\boldsymbol{x}\cdot\boldsymbol{y}}{|\boldsymbol{x}|\,|\boldsymbol{y}|} \le 1.$$

Thus, there is a unique angle θ between 0 and π such that

$$\cos\theta = \frac{\boldsymbol{x}\cdot\boldsymbol{y}}{|\boldsymbol{x}|\,|\boldsymbol{y}|}.$$

So this definition is sensible, and has the added virtue of being the exact analogue for $\boldsymbol{R}^n$ of the formula we found in $\boldsymbol{R}^2$ from our knowledge of angles in the plane. It can similarly be shown to agree with our notion of angle in $\boldsymbol{R}^3$.

2.8 EXAMPLE. (a) Find the angle between $\boldsymbol{x}=(1,-1,0)$ and $\boldsymbol{y}=(-1,1,\sqrt{2})$. (b) Show that the vectors $\boldsymbol{z}=(1,2,3)$ and $\boldsymbol{w}=(-1,-1,1)$ are perpendicular.

Solution. (a) In Example 2.3 we already computed $\boldsymbol{x}\cdot\boldsymbol{y}=-2$. So

$$\cos\theta=\frac{\boldsymbol{x}\cdot\boldsymbol{y}}{|\boldsymbol{x}|\,|\boldsymbol{y}|}=\frac{-2}{\sqrt{1^2+(-1)^2+0^2}\,\sqrt{(-1)^2+1^2+2}}$$
$$=\frac{-2}{\sqrt{2}\cdot 2}=-\frac{1}{\sqrt{2}}.$$

Thus, $\theta=\dfrac{3\pi}{4}$ (135°).

(b) $\boldsymbol{z}\cdot\boldsymbol{w}=-1-2+3=0$. Hence $\cos\theta=0$, so $\theta=\dfrac{\pi}{2}$. Thus the vectors are perpendicular. See Figure 2.3.

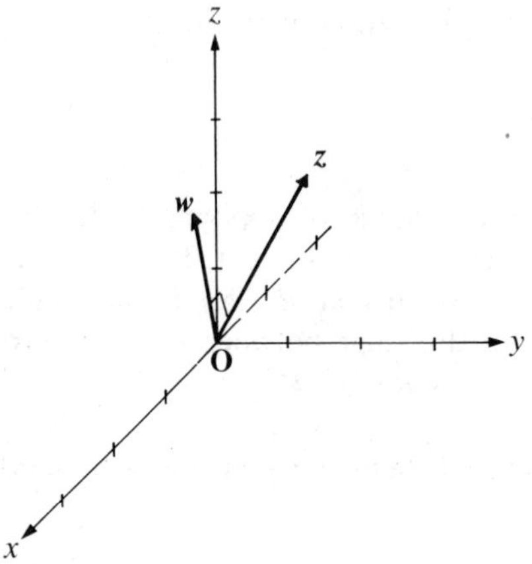

FIGURE 2.3

Example 2.8(b) generalizes immediately.

2.9 COROLLARY. Two nonzero vectors $\boldsymbol{x}$ and $\boldsymbol{y}$ are perpendicular if and only if $\boldsymbol{x}\cdot\boldsymbol{y}=0$.

Proof. $\boldsymbol{x}\perp\boldsymbol{y}$ holds if and only if

$$\cos\theta=\frac{\boldsymbol{x}\cdot\boldsymbol{y}}{|\boldsymbol{x}|\,|\boldsymbol{y}|}=0,$$

which in turn holds if and only if $\boldsymbol{x}\cdot\boldsymbol{y}=0$. QED

The term **orthogonal** is widely used as a synonym for perpendicular. We shall use both terms freely.

It is clear geometrically (at least in $\boldsymbol{R}^2$) that the length of $(\boldsymbol{x}+\boldsymbol{y})$ is no greater than $|\boldsymbol{x}|+|\boldsymbol{y}|$, since $\boldsymbol{x}$, $\boldsymbol{y}$, and $\boldsymbol{x}+\boldsymbol{y}$ in general form a triangle. (See Figure 2.4.)

It is easy to prove this algebraically for $\boldsymbol{R}^n$.

2.10 THEOREM (TRIANGLE INEQUALITY). For all vectors $\boldsymbol{x}$ and $\boldsymbol{y}$ in $\boldsymbol{R}^n$,

$$|\boldsymbol{x}+\boldsymbol{y}|\le|\boldsymbol{x}|+|\boldsymbol{y}|.$$

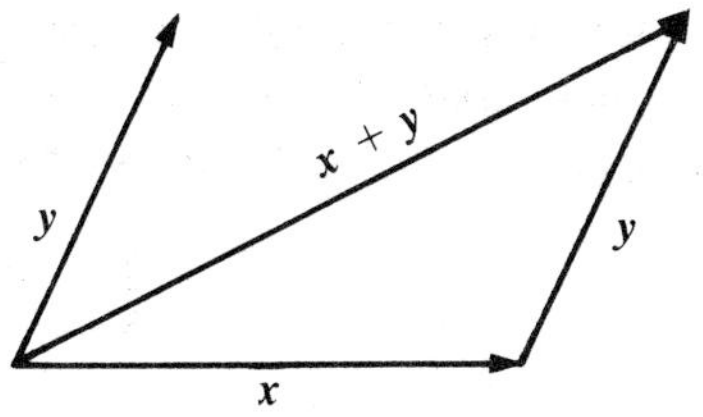

FIGURE 2.4

Proof.

$$|x+y|^2 = (x+y)\cdot(x+y) = x\cdot x + y\cdot x + x\cdot y + y\cdot y$$
by Theorem 2.4(1) and (2)
$$= x\cdot x + 2x\cdot y + y\cdot y$$
by Theorem 2.4(1)
$$\leq |x|\,|x| + 2\,|x|\,|y| + |y|\,|y|$$
by Cauchy–Schwarz.

Thus, $|x+y|^2 \leq (|x|+|y|)^2$. Now extract the positive square roots: $|x+y| \leq |x|+|y|$. QED

The triangle inequality looks like the one you learned for absolute value in elementary calculus, and is used in much the same way. Note in Theorem 2.10 that when $x \perp y$, equality holds. This permits us to state the Pythagorean Theorem in the following vector form.

2.11 PYTHAGOREAN THEOREM. If x and y are perpendicular, then $|x+y|^2 = |x|^2 + |y|^2$.

We have reached a point where we can settle the algebraic meaning of the direction of a vector in a way whose reasonableness can now be explained.

2.12 DEFINITION. If $x = (x_1, x_2, \ldots, x_n) \neq 0 \in R^n$, then the **direction of x** is

$$u = \frac{1}{|x|}x.$$

Thus the direction of a vector x is the unit vector u which points toward the tip of x when x and u are drawn with the origin as initial point (Figure 2.5).

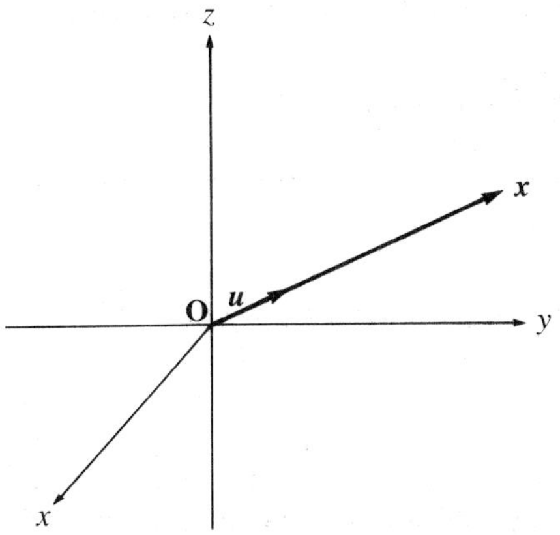

FIGURE 2.5

Why is this reasonable? Because $\boldsymbol{u}$ carries in its coordinates enough information to recover the angles $\boldsymbol{x}$ makes with every coordinate axis. To see this, let $\alpha_1, \alpha_2, \alpha_3, \ldots, \alpha_n$ be the angles formed by $\boldsymbol{x}=(x_1, x_2, x_3, \ldots, x_n)$ and the standard basis vectors $\boldsymbol{e}_1, \boldsymbol{e}_2, \ldots, \boldsymbol{e}_n$. Then we have

$$\cos\alpha_1 = \frac{\boldsymbol{x}\cdot\boldsymbol{e}_1}{|\boldsymbol{x}|\,|\boldsymbol{e}_1|} = \frac{x_1}{\sqrt{x_1^2+\ldots+x_n^2}} = u_1,$$

$$\cos\alpha_2 = \frac{\boldsymbol{x}\cdot\boldsymbol{e}_2}{|\boldsymbol{x}|\,|\boldsymbol{e}_2|} = \frac{x_2}{\sqrt{x_1^2+\ldots+x_n^2}} = u_2,$$

$$\vdots$$

$$\cos\alpha_n = \frac{\boldsymbol{x}\cdot\boldsymbol{e}_n}{|\boldsymbol{x}|\,|\boldsymbol{e}_n|} = \frac{x_n}{\sqrt{x_1^2+\ldots+x_n^2}} = u_n.$$

So $\boldsymbol{u}=(\cos\alpha_1, \cos\alpha_2, \cos\alpha_3, \ldots, \cos\alpha_n)$ is a vector whose coordinates are the cosines of the **direction angles** formed by $\boldsymbol{x}$ and the coordinate axes. Hence the terminology "direction of $\boldsymbol{x}$ is $\boldsymbol{u}$" is appropriate and reasonable.

Now for any $a \neq 0$, $a\boldsymbol{v}$ has the same direction as $\boldsymbol{v}$ (if $a>0$) or the opposite direction (if $a<0$), since

$$\frac{1}{|a\boldsymbol{v}|}(a\boldsymbol{v}) = \frac{1}{|a|\,|\boldsymbol{v}|}a\boldsymbol{v} = \frac{\pm 1}{|\boldsymbol{v}|}\boldsymbol{v}.$$

Thus we make the following definition.

2.13 DEFINITION. The vectors $\boldsymbol{x}$ and $\boldsymbol{y}$ are said to be **parallel** if either they have the same direction or their directions differ by a factor of -1.

It is easy to see that $\boldsymbol{x}$ is parallel to $\boldsymbol{y}$ if and only if $\boldsymbol{x}=a\boldsymbol{y}$ for some $a\neq 0$.

2.14 EXAMPLE. Find the direction and direction angles of the vector drawn from $P(1,-1,2)$ to $Q(-1,0,4)$.

Solution. Here $\boldsymbol{x}=\overrightarrow{PQ}=(-2,1,2)$. Thus

$$|\boldsymbol{x}| = \sqrt{(-2)^2+1^2+2^2} = \sqrt{9} = 3,$$

so the direction of $\boldsymbol{x}$ is $\boldsymbol{u}=\frac{1}{3}(-2,1,2)=(-\frac{2}{3},\frac{1}{3},\frac{2}{3})$. The direction angles are $\alpha = \mathrm{Cos}^{-1}(-\frac{2}{3})$, $\beta=\mathrm{Cos}^{-1}\frac{1}{3}$, $\gamma=\mathrm{Cos}^{-1}\frac{2}{3}$. From a table of cosines, $\alpha\approx 2.30$ radians (about 132°), $\beta\approx 1.23$ radians (about 70°), and $\gamma\approx 0.84$ radians (about 48°).

NOTE: The symbol $\approx$ is read "**is approximately equal to**."

As a further application of the dot product, consider the problem of resolving a force vector $\boldsymbol{F}$ into components.

2.15 EXAMPLE. In Figure 2.6, $\boldsymbol{F}$ acts on a particle at $\boldsymbol{x}$ which moves on the unit circle $x^2+y^2=1$ in the direction shown. Find the components of $\boldsymbol{F}$ in the directions tangent to and normal to the circle.

Solution. Let $\boldsymbol{T}$ and $\boldsymbol{N}$ be unit vectors tangent to and normal to the circle at $\boldsymbol{x}$. Let us label the tangential component of $\boldsymbol{F}$ as $\boldsymbol{F}_{\boldsymbol{T}}$ and the normal component of $\boldsymbol{F}$

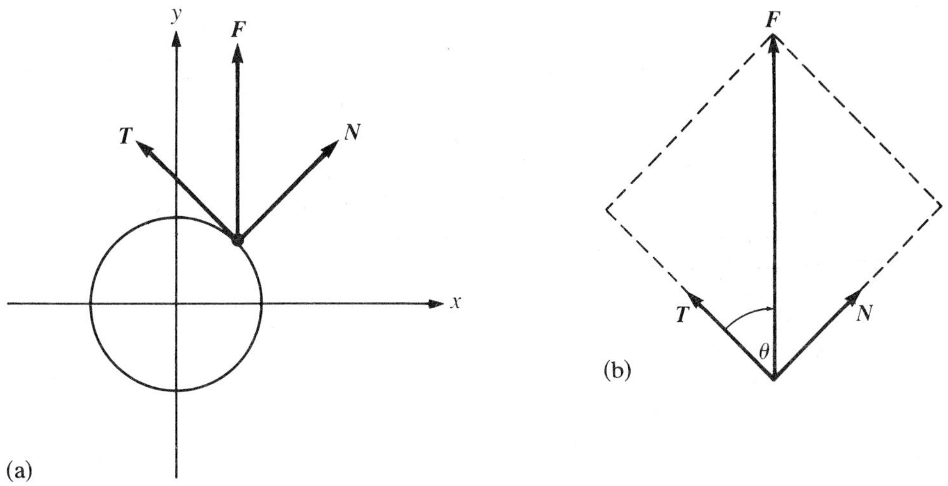

FIGURE 2.6

as $\boldsymbol{F}_{\boldsymbol{N}}$, where $\cos\theta = \dfrac{|\boldsymbol{F}_{\boldsymbol{T}}|}{|\boldsymbol{F}|}$ and $\sin\theta = \dfrac{|\boldsymbol{F}_{\boldsymbol{N}}|}{|\boldsymbol{F}|}$. Thus

$$|\boldsymbol{F}_{\boldsymbol{T}}| = |\boldsymbol{F}|\cos\theta = |\boldsymbol{F}|\frac{\boldsymbol{F}\cdot\boldsymbol{T}}{|\boldsymbol{F}|\,|\boldsymbol{T}|} = \boldsymbol{F}\cdot\boldsymbol{T}$$

$$|\boldsymbol{F}_{\boldsymbol{N}}| = |\boldsymbol{F}|\sin\theta = |\boldsymbol{F}|\cos\left(\frac{\pi}{2}-\theta\right) = |\boldsymbol{F}|\frac{\boldsymbol{F}\cdot\boldsymbol{N}}{|\boldsymbol{F}|\,|\boldsymbol{N}|} = \boldsymbol{F}\cdot\boldsymbol{N}.$$

We have $\boldsymbol{F} = (\boldsymbol{F}\cdot\boldsymbol{T})\boldsymbol{T} + (\boldsymbol{F}\cdot\boldsymbol{N})\boldsymbol{N}$.

Example 2.15 suggests the following definition.

2.16 DEFINITION. If $\boldsymbol{x}$ is a vector, then the **coordinate of $\boldsymbol{x}$ in the direction of $\boldsymbol{v}$** is

$$\boxed{x_{\boldsymbol{v}} = \boldsymbol{x}\cdot\frac{\boldsymbol{v}}{|\boldsymbol{v}|}.}$$

The **component of $\boldsymbol{x}$ in the direction of $\boldsymbol{v}$** is $x_{\boldsymbol{v}}\boldsymbol{u}$ where $\boldsymbol{u} = \dfrac{\boldsymbol{v}}{|\boldsymbol{v}|}$ is the unit vector in the direction of $\boldsymbol{v}$.

2.17 EXAMPLE. Find the vector obtained when $\boldsymbol{x} = (-2, 4, 10)$ is projected onto $\boldsymbol{v} = (12, 0, 9)$.

Solution. Refer to Figure 2.7.

The vector sought is the component of $\boldsymbol{x}$ in the direction of $\boldsymbol{v}$. Here the direction of $\boldsymbol{v}$ is

$$\boldsymbol{u} = \frac{1}{|\boldsymbol{v}|}\boldsymbol{v} = \frac{1}{\sqrt{144+0+81}}(12, 0, 9)$$
$$= \tfrac{1}{15}(12, 0, 9) = (\tfrac{4}{5}, 0, \tfrac{3}{5}).$$

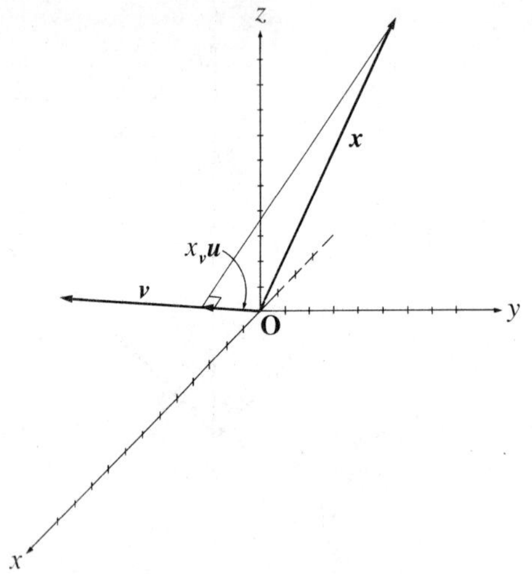

FIGURE 2.7

Then the projection of $\boldsymbol{x}$ onto $\boldsymbol{v}$ is

$$\begin{aligned} x_v\boldsymbol{u} &= (\boldsymbol{x}\cdot\boldsymbol{u})\boldsymbol{u} \\ &= (-\tfrac{8}{5}+0+\tfrac{30}{5})(\tfrac{4}{5}, 0, \tfrac{3}{5}) \\ &= \tfrac{22}{5}(\tfrac{4}{5}, 0, \tfrac{3}{5}) = (\tfrac{88}{25}, 0, \tfrac{66}{25}). \end{aligned}$$

EXERCISES 1.2

In Exercises 1 to 4, find the angle between each pair of vectors.

1 (a) $(3, -1)$ and $(2, 1)$ (b) $\left(\sqrt{3}, \frac{\sqrt{3}}{2}\right)$ and $(1, 2)$

2 (a) $(-1, 3)$ and $(2, -1)$ (b) $(\frac{1}{7}, -3)$ and $(28, \frac{4}{3})$

3 (a) $(1, -1, 0)$ and $(-1, 2, 1)$ (b) $(-1, 1, 0)$ and $(0, 1, -1)$

4 (a) $(\sqrt{2}, 1, 1)$ and $(0, 1, 1)$ (b) $(-\frac{3}{2}, 0, 1)$ and $(2, -8, 3)$

5 Under what circumstances does equality hold in the Cauchy–Schwarz inequality? Explain your answer thoroughly.

6 Under what circumstances can equality hold in the triangle inequality? Explain your answer.

In Exercises 7 and 8, find the direction and direction angles of the given vectors.

7 (a) $(1, -1, 0)$ (b) $(\sqrt{3}, 0, 1)$

8 (a) $(-4, 3, 5)$ (b) $(1, -2, 2)$

9 (a) If $\cos\alpha_1, \cos\alpha_2, \ldots, \cos\alpha_n$ are the direction cosines of $\boldsymbol{v}$, then show that $\cos^2\alpha_1 + \cos^2\alpha_2 + \cos^2\alpha_3 + \ldots + \cos^2\alpha_n = 1$.

(b) Specialize to $\boldsymbol{R}^2$ and show how part (a) gives the trigonometric identity $\sin^2\theta + \cos^2\theta = 1$.

10 Find a unit vector orthogonal to $(1, 1)$. Is there only one such unit vector? If not, how many are there?

11 Find a unit vector orthogonal to $(-1, 3, 2)$. How many such vectors are there?

12 (a) Show that $\boldsymbol{e}_i \cdot \boldsymbol{e}_j = 0$ where $i \neq j$, and $\boldsymbol{e}_i \cdot \boldsymbol{e}_i = 1$.
(b) Prove (1) and (2) of Theorem 2.4.

13 Find x so that $(2, x, -3)$ is orthogonal to $(-1, 3, -2)$. Is there any value of x for which $(2, x, -3)$ is parallel to $(-1, 3, -2)$?

14 Find y so that $(-2, 1, y)$ is orthogonal to $(\frac{6}{5}, -\frac{3}{5}, 1)$. Is there a value of y for which $(-2, 1, y)$ is parallel to $(\frac{6}{5}, -\frac{3}{5}, 1)$?

15 Find the component of $(2, -3, 4)$ in the direction of $(1, 1, \sqrt{2})$.

16 Find the component of $(3, 1, -2)$ in the direction of $(-3, 6, 2)$.

17 In physics, the *work* done by a constant force $\boldsymbol{F}$ in moving a particle along a straight line segment is defined as the product of the distance moved times the magnitude of the force exerted in the direction of the motion. A constant force $\boldsymbol{F} = (1, 4)$ acts on a particle as it moves from the origin to the point $(3, 3)$. Find the work done. See Figure 2.8.

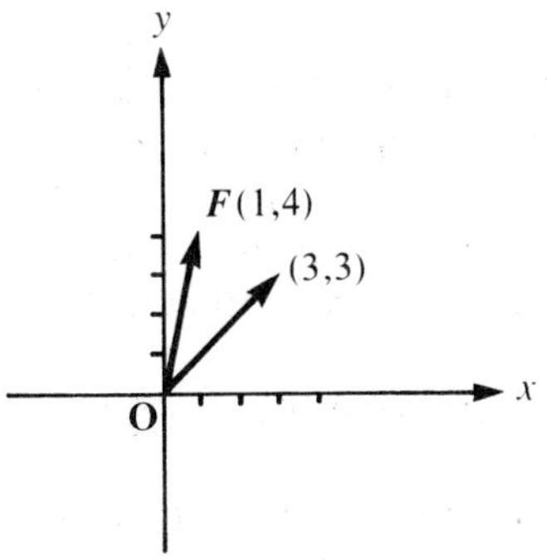

FIGURE 2.8

18 Prove the *Law of Cosines* (a generalization of the Pythagorean Theorem):

$$|\mathbf{x} - \mathbf{y}|^2 = |\mathbf{x}|^2 + |\mathbf{y}|^2 - 2\,|\mathbf{x}|\,|\mathbf{y}| \cos \theta$$

where θ is the angle between $\mathbf{x}$ and $\mathbf{y}$. Refer to Figure 2.9.

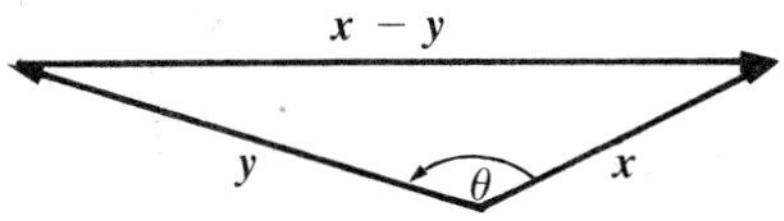

FIGURE 2.9

19 Use vectors to prove the *parallelogram identity:* the sum of the squares of the lengths of the diagonals of a parallelogram is the same as the sum of the squares of the lengths of the sides. (*Hint:* Consider the parallelogram formed by $\mathbf{x}$ and $\mathbf{y}$.)

20 Prove the *polarization identity:*

$$|\mathbf{x} + \mathbf{y}|^2 - |\mathbf{x} - \mathbf{y}|^2 = 4\mathbf{x} \cdot \mathbf{y}.$$

21 (a) If $\mathbf{x}$ and $\mathbf{y}$ have the same length, then show that $\mathbf{x} + \mathbf{y}$ and $\mathbf{x} - \mathbf{y}$ are orthogonal.
(b) Use this to prove that an angle inscribed in a semicircle is a right angle. See Figure 2.10.

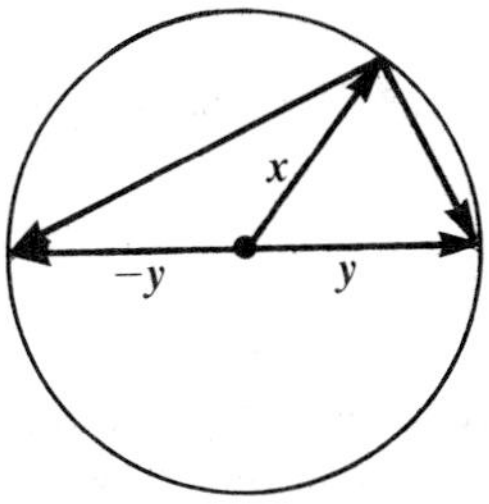

FIGURE 2.10

22 Use vectors to show that the line segment connecting the midpoints of two sides of a triangle is parallel to the third side and has length half the length of the third side.

23 If $\boldsymbol{x}$ is any vector and $\boldsymbol{u}$ is a unit vector (any direction), then show that $\boldsymbol{x}-(\boldsymbol{x}\cdot\boldsymbol{u})\boldsymbol{u}$ is perpendicular to $\boldsymbol{u}$. Use this and the component of $\boldsymbol{x}$ in the direction of $\boldsymbol{u}$ to resolve $\boldsymbol{x}$ into two vectors, one of which is perpendicular to $\boldsymbol{u}$, and the other of which is parallel to $\boldsymbol{u}$.

3 LINES IN $\boldsymbol{R}^n$

In the elementary analytic geometry of $\boldsymbol{R}^2$, the curves with simplest equations in the Cartesian coordinate system are straight lines, whose equations are polynomial equations $ax+by=c$ of degree 1. (Hence the use of the term "linear" for polynomial equations of first degree.) In $\boldsymbol{R}^3$ we still have a clear picture of what a line is, and in this section our goal is to develop an algebraic description of lines not only in $\boldsymbol{R}^3$ but more generally in $\boldsymbol{R}^n$.

Let us begin with another look at lines in $\boldsymbol{R}^2$, this time from the vector point of view.

Suppose that ℓ is a given line in $\boldsymbol{R}^2$ and is determined by two points $P_0(x_0, y_0)$ and $P_1(x_1, y_1)$. See Figure 3.1. Let $P(x, y)$ be an arbitrary point on ℓ. If we let

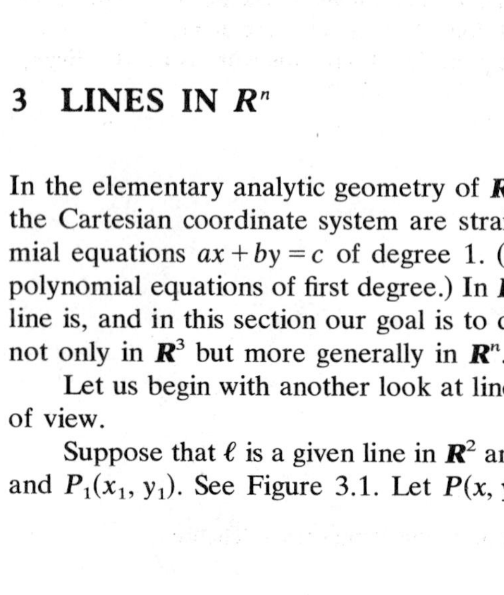

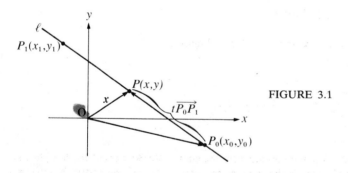

FIGURE 3.1

$\boldsymbol{x}=\overrightarrow{OP}$, then how can we give a formula for $\boldsymbol{x}$ in terms of known vectors? Well, first we see that $\overrightarrow{P_0P_1}=(x_1-x_0, y_1-y_0)$ is a vector in the direction of the line. Next, we see that we can get to P if we travel first to P_0 and then add a suitable multiple of $\overrightarrow{P_0P_1}$ to $\overrightarrow{OP_0}$, for we will then be moving along ℓ from P_0. In Figure 3.1 it appears that

(1) $$\boldsymbol{x}=\overrightarrow{OP}=\overrightarrow{OP_0}+t\overrightarrow{P_0P_1}$$

where t is approximately $\frac{1}{2}$. Let $\boldsymbol{v}=\overrightarrow{P_0P_1}$ and $\boldsymbol{x}_0=\overrightarrow{OP_0}$. In terms of these known vectors, (1) becomes

(2) $$\boldsymbol{x}=\boldsymbol{x}_0+t\boldsymbol{v}$$

Our discussion shows that for any P on ℓ, $\boldsymbol{x}=\overrightarrow{OP}$ is given by (2) for some choice of $t\in\boldsymbol{R}$. Conversely, if P is the endpoint of any vector $\boldsymbol{x}$ given by (2), then

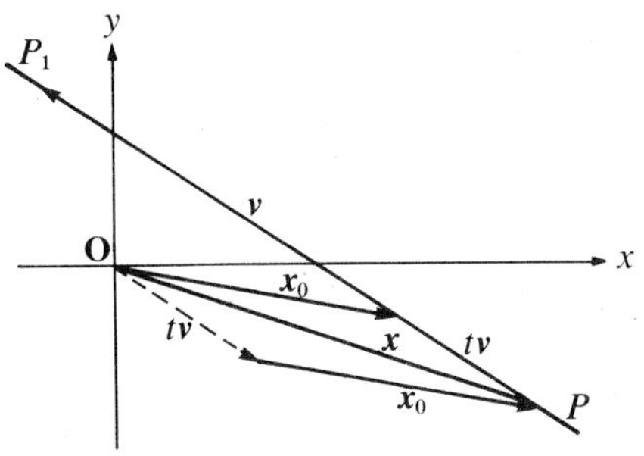

FIGURE 3.2

the parallelogram law shows that P is on ℓ. See Figure 3.2. Thus, (2) holds if and only if the endpoint P of $\mathbf{x}$ lies on the line ℓ. Therefore it is natural to call (2) a vector equation of ℓ.

3.1 DEFINITION. **A vector equation of the line ℓ through two points P_0 and P_1** is

(2)
$$\boxed{\mathbf{x} = t\mathbf{v} + \mathbf{x}_0}$$

where $\mathbf{v} = \overrightarrow{P_0P_1}$, $\mathbf{x}_0 = \overrightarrow{OP_0}$, and t is a real variable called a **parameter**.

Take note of the fact that in (2), t is the *only* variable. Both $\mathbf{x}_0$ and $\mathbf{v}$ are *fixed* vectors. You may like to think of (2) as the vector analogue of the scalar equation

$$y = mx + b$$

in which m and b are fixed in $\mathbf{R}^1$ and x is a real variable.

If we put (2) in coordinate form, then we have

(3)
$$\begin{aligned}(x, y) &= (x_0, y_0) + t(x_1 - x_0, y_1 - y_0),\\ &= (x_0 + t(x_1 - x_0), y_0 + t(y_1 - y_0)).\end{aligned}$$

From (3) we can obtain scalar equations of ℓ.

3.2 DEFINITION. If $P_0(x_0, y_0)$ and $P_1(x_1, y_1)$ are two points on ℓ, then **parametric (scalar) equations** of ℓ are

(4)
$$\boxed{\begin{cases} x = x_0 + t(x_1 - x_0) \\ y = y_0 + t(y_1 - y_0) \end{cases}}$$

You will notice that in neither Definition 3.1 nor Definition 3.2 have we suggested that a line has *unique* vector or parametric scalar equations. There is good reason for this—the equations are *not* uniquely determined by ℓ, as the following example illustrates.

3.3 EXAMPLE. Find vector and parametric equations of the line ℓ through $(-2, 5)$ and $(1, 1)$ using (a) $P_0 = (1, 1)$, (b) $P_0 = (-2, 5)$. Reconcile the two answers.

Solution. See Figure 3.3. A vector in ℓ is $\mathbf{v} = (-2-1, 5-1) = (-3, 4)$.

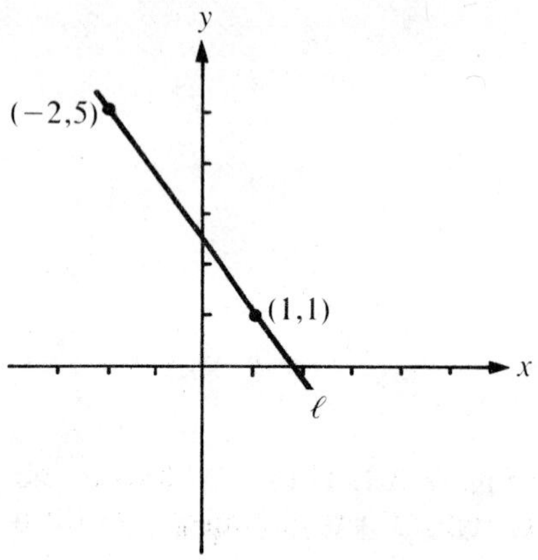

FIGURE 3.3

(a) $\boldsymbol{x} = t(-3, 4) + (1, 1)$ is a vector equation, since $\overrightarrow{OP_0} = (1, 1)$. The parametric equations are obtained from

$$(x, y) = (-3t, 4t) + (1, 1) = (-3t + 1, 4t + 1),$$

(*) $$x = -3t + 1, \qquad y = 4t + 1.$$

(b) $\boldsymbol{x} = s(-3, 4) + (-2, 5)$ is a vector equation.

$$(x, y) = (-3s, 4s) + (-2, 5) = (-3s - 2, 4s + 5).$$

The parametric equations are

(**) $$x = -3s - 2, \qquad y = 4s + 5.$$

The Equations (*) and (**) appear different. But what has really changed is the parameter. For if we eliminate t and s, we get in each case the familiar scalar equation of ℓ:

(*) $x = -3t + 1$ gives

$$3t = -x + 1, \qquad t = -\tfrac{1}{3}x + \tfrac{1}{3}.$$

Then $y = 4t + 1$ gives $y = 4(-\tfrac{1}{3}x + \tfrac{1}{3}) + 1$ or, equivalently,

$$y = -\tfrac{4}{3}x + \tfrac{4}{3} + 1 = -\tfrac{4}{3}x + \tfrac{7}{3}.$$

(**) $x = -3s - 2$ gives

$$3s = -x - 2, \qquad s = -\tfrac{1}{3}x - \tfrac{2}{3}.$$

Then $y = 4s + 5$ gives $y = 4(-\tfrac{1}{3}x - \tfrac{2}{3}) + 5$ or, equivalently,

$$y = -\tfrac{4}{3}x - \tfrac{8}{3} + 5 = -\tfrac{4}{3}x + \tfrac{7}{3}.$$

We can express s in terms of t if we equate the two expressions for x (or y):

$$x = -3t + 1 = -3s - 2$$

so

$$-3s = -3t + 3,$$

$$s = t - 1.$$

This example shows why the equation of a line in $\boldsymbol{R}^2$ is often given in the form $y = mx + b$, for such an equation *is* unique. The parametric equations and the vector equation are *not unique*, but differ when either the parameter t or the point P_0 is changed. A procedure like that in Example 5.2 is needed to see if two different-looking vector equations give the same line. This should warn you when checking your answers that more than a cursory glance may be needed!

Now what in our discussion can be carried over to lines in $\boldsymbol{R}^n$? Everything! In fact, if you take a close look at Definition 3.1 you will see no explicit mention of the number of coordinates of the points. We will thus use Definition 3.1 as the definition of a vector equation of the line through two points P_0 and P_1 in $\boldsymbol{R}^n$. For purposes of illustration, let us use Definition 3.1 in $\boldsymbol{R}^3$ and see what we come up with.

Suppose we have two points $P_0(x_0, y_0, z_0)$ and $P_1(x_1, y_1, z_1)$ in $\boldsymbol{R}^3$ and want to find the equation of the line through these two points. Just as in $\boldsymbol{R}^2$, $\boldsymbol{v} = \overrightarrow{P_0P_1} = (x_1 - x_0, y_1 - y_0, z_1 - z_0)$ is a vector in this line. Then Definition 3.1 gives us the vector equation

$$\boldsymbol{x} = t\boldsymbol{v} + \boldsymbol{x}_0 = t(x_1 - x_0, y_1 - y_0, z_1 - z_0) + (x_0, y_0, z_0).$$

So the parametric equations are

$$(5) \qquad \begin{cases} x = t(x_1 - x_0) + x_0 \\ y = t(y_1 - y_0) + y_0. \\ z = t(z_1 - z_0) + z_0 \end{cases}$$

In $\boldsymbol{R}^2$ we saw in Example 3.3 how elimination of the parameter t enables us to obtain a *single* scalar coordinate equation of a line ℓ from its two parametric scalar equations. In $\boldsymbol{R}^3$, elimination of the parameter t will give us 2 $(= 3 - 1)$ scalar equations. If $x_1 - x_0$, $y_1 - y_0$, and $z_1 - z_0$ are all nonzero, then we can eliminate t in the parametric equations (5) as follows:

$$x - x_0 = t(x_1 - x_0), \qquad y - y_0 = t(y_1 - y_0), \qquad z - z_0 = t(z_1 - z_0).$$

Thus

$$\frac{x - x_0}{x_1 - x_0} = \frac{y - y_0}{y_1 - y_0} = \frac{z - z_0}{z_1 - z_0} = t.$$

3.4 DEFINITION. The **symmetric scalar equations** of a line in $\boldsymbol{R}^3$ through two points $P_0(x_0, y_0, z_0)$ and $P_1(x_1, y_1, z_1)$ are

$$\boxed{\frac{x - x_0}{x_1 - x_0} = \frac{y - y_0}{y_1 - y_0} = \frac{z - z_0}{z_1 - z_0}}$$

if $x_1 - x_0$, $y_1 - y_0$, $z_1 - z_0$ are all nonzero.

If one or two of the numbers $x_1 - x_0$, $y_1 - y_0$, $z_1 - z_0$, are zero (not all three can be zero or else $P_0 = P_1$), then no standard convention exists on what form the symmetric scalar equations should take. If, say, $x_1 - x_0$ and $y_1 - y_0$ are zero, then the parametric equations $x = x_0$, $y = y_0$, $z = t(z_1 - z_0) + z_0$ are probably the only scalar equations you need to concern yourself with. Notice that in $\boldsymbol{R}^n$, if the parameter t is eliminated, then we are left with $n - 1$ *scalar equations* which assert the equality of

$$\frac{x-x_0}{x_1-x_0}, \qquad \frac{y-y_0}{y_1-y_0}, \qquad \frac{z-z_0}{z_1-z_0}, \quad \text{etc.}$$

In particular in $\boldsymbol{R}^2$, *and only there*, we get exactly one scalar equation

$$\frac{y-y_0}{y_1-y_0}=\frac{x-x_0}{x_1-x_0}$$

which is easily seen to be equivalent to the *point-slope* equation

$$y-y_0=\frac{y_1-y_0}{x_1-x_0}(x-x_0)=m(x-x_0).$$

Take another look now at Example 3.3 in light of these ideas.

3.5 EXAMPLE. Find vector and scalar equations of the line through the two points $P(1, 2, 3)$ and $Q(-1, 1, 4)$. See Figure 3.4.

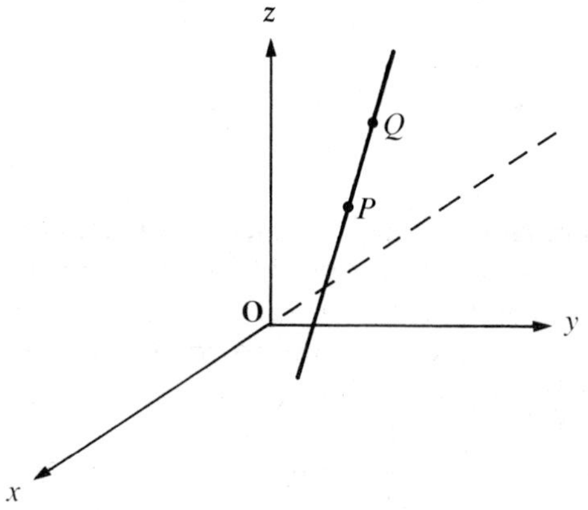

FIGURE 3.4

Solution. Here $\boldsymbol{v}=\overrightarrow{PQ}=(-1-1, 1-2, 4-3)=(-2, -1, 1)$. So a vector equation for the line is

$$\boldsymbol{x}=(1, 2, 3)+t(-2, -1, 1).$$

To get the scalar equations, we simply combine the terms on the right side of the vector equation.

$$\begin{aligned}\boldsymbol{x}=(x, y, z)&=(1, 2, 3)+(-2t, -t, t)\\ &=(1-2t, 2-t, 3+t).\end{aligned}$$

Thus the parametric scalar equations are

$$x=1-2t, \qquad y=2-t, \qquad z=3+t.$$

The symmetric scalar equations are obtained from these.

$$x-1=-2t, \qquad y-2=-t, \qquad z-3=t,$$

$$\frac{x-1}{-2}=t, \qquad \frac{y-2}{-1}=t, \qquad \frac{z-3}{1}=t.$$

So

$$\frac{x-1}{-2}=\frac{y-2}{-1}=\frac{z-3}{1}.$$

3.6 DEFINITION. Two lines ℓ and ℓ' with equations $\boldsymbol{x}=s\boldsymbol{v}+\boldsymbol{x}_0$ and $\boldsymbol{x}=t\boldsymbol{w}+\boldsymbol{x}_1$ are **parallel** if and only if $\boldsymbol{v}$ and $\boldsymbol{w}$ are parallel vectors. The lines are **perpendicular** if and only if they intersect and $\boldsymbol{v}$ and $\boldsymbol{w}$ are perpendicular vectors (cf. Definition 2.13 and Corollary 2.9).

Note that even though the vectors $(1, 1, 0)$ and $(1, -1, 3)$ are perpendicular, still the lines $\ell_1 : \boldsymbol{x}=\boldsymbol{0}+s(1, 1, 0)$ and $\ell_2 : \boldsymbol{x}=(0, 0, 3)+t(1, -1, 3)$ are not perpendicular, since they don't intersect. We can see this by noting that at a point of intersection we would have, on equating corresponding coordinates,

$$s=t, \qquad s=-t, \qquad \text{and} \qquad 0=3+3t.$$

Subtraction of the first two gives $t=0$, but the third says $t=-1$. So these equations are inconsistent. Hence there can be no point of intersection. See Figure 3.5.

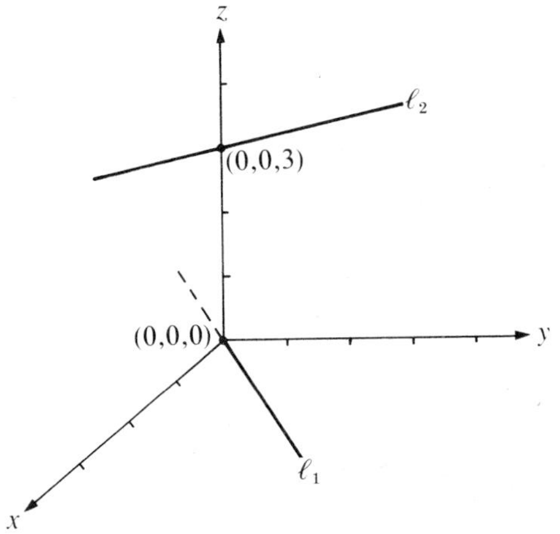

FIGURE 3.5

The following theorem is a simple consequence of Definition 3.6, and it will be put to great use in succeeding sections.

3.7 THEOREM. Two lines ℓ and ℓ' with equations $\boldsymbol{x}=t\boldsymbol{v}+\boldsymbol{x}_0$ and $\boldsymbol{x}=s\boldsymbol{w}+\boldsymbol{x}_1$ are parallel if and only if $\boldsymbol{v}=a\boldsymbol{w}$ for some real number a. They are perpendicular if and only if they intersect and $\boldsymbol{v}\cdot\boldsymbol{w}=0$.

Proof. The vectors are parallel (as remarked after Definition 2.13) if and only if $\boldsymbol{v}=a\boldsymbol{w}$. Similarly, the condition that $\boldsymbol{v}$ be perpendicular to $\boldsymbol{w}$ is that $\boldsymbol{v}\cdot\boldsymbol{w}=0$.

QED

We close this section with an often useful representation of the line segment joining two points P and Q.

3.8 PROPOSITION. If $P(x_1, x_2, \ldots, x_n)$ and $Q(y_1, y_2, \ldots, y_n)$ are two points in $\boldsymbol{R}^n$, then the line segment joining P to Q consists of all endpoints of the vectors $(1-t)\boldsymbol{x}_0+t\boldsymbol{y}_0$, for $0\le t\le 1$, where $\boldsymbol{x}_0=\overrightarrow{OP}$ and $\boldsymbol{y}_0=\overrightarrow{OQ}$.

Proof. The line through PQ has equation

$$\mathbf{x}=\mathbf{x}_0+t(\mathbf{y}_0-\mathbf{x}_0)=\mathbf{x}_0+t\mathbf{y}_0-t\mathbf{x}_0=(1-t)\mathbf{x}_0+t\mathbf{y}_0.$$

When $t=0$, we get $\mathbf{x}_0$; when $t=1$, we get $\mathbf{y}_0$. For $0<t<1$, we get $\mathbf{x}_0+t(\mathbf{y}_0-\mathbf{x}_0)$, a vector whose endpoint is on the segment from P to Q, since we are adding to $\mathbf{x}_0$ a fractional part of the vector $\mathbf{y}_0-\mathbf{x}_0$ which joins P to Q. See Figure 3.6. QED

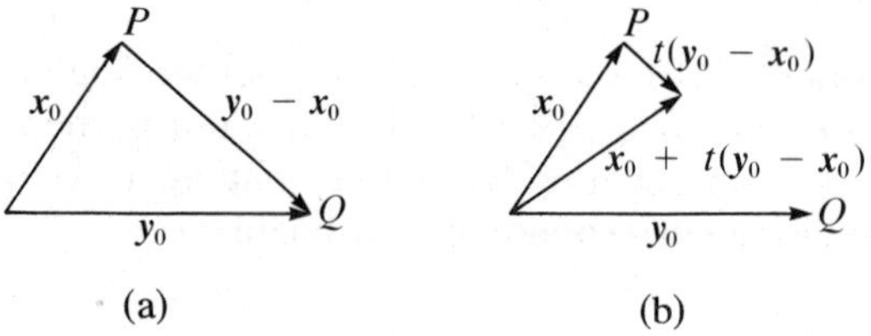

FIGURE 3.6

EXERCISES 1.3

In Exercises 1 to 4, find a vector equation for the given line.

1 (a) The line through (2, 1) and (−1, 5). (b) The y-axis in $\mathbf{R}^2$.
(c) The line through (−7, −2) and (4, 4).

2 (a) The line through (1, −2) in the direction $\left(\frac{1}{\sqrt{5}}, -\frac{2}{\sqrt{5}}\right)$.

(b) The line through (0, 0) in the direction $\left(\frac{1}{\sqrt{2}}, \frac{1}{\sqrt{2}}\right)$.

3 (a) The line through (1, 2, 3) and (−1, −2, −3).
(b) The line through (0, 1, −2) and (3, 1, −2). (c) The y-axis in $\mathbf{R}^3$.
(d) The line through (1, −1, 4) and (7, −1, −1).

4 (a) The line through (0, 0, 0) in the direction $\left(\frac{1}{\sqrt{5}}, \frac{2}{\sqrt{5}}, 0\right)$.

(b) The line through (1, 2, −1) in the direction $\left(\frac{1}{\sqrt{7}}, \frac{2}{\sqrt{7}}, \frac{\sqrt{2}}{\sqrt{7}}\right)$.

In Exercises 5 to 8, find a vector equation, a symmetric scalar equation, and parametric scalar equations for the given line.

5 The line through (1, 2, 3) and (−1, −3, 5).

6 The line through (1, 4, −2) and (3, 4, 5).

7 The line through (−1, 5, 4) in the direction of $\mathbf{v}=(1, -1, -2)$.

8 The line through (2, 5, 4) in the direction of $\mathbf{v}=(-1, 1, 3)$.

9 Find parametric equations for the line through (1, 1, 2) which is parallel to the line

$$\frac{x-1}{3}=\frac{y}{-1}=\frac{2z-3}{4}.$$

In Exercises 10 and 11, determine whether the two given lines intersect. If they do, then find the point of intersection. If they do not, explain why not.

10 The line $\mathbf{x}=(2, -3, 1)+t(1, 2, -3)$ and the line $x=1-2s$, $y=2+3s$, $z=4+6s$.

11 The line $x=1+3t$, $y=-2-t$, $z=2+2t$ and the line $\frac{x+1}{-1}=\frac{y-3}{2}=\frac{z-5}{-1}$.

12 Find parametric equations for the line through $(1, 1, 3)$ which is parallel to the line $x=2-t,\ y=-t,\ z=3+3t$.

13 Are the lines $x=-2+t,\ y=2t,\ z=-3-t$ and $x=5-\frac{1}{2}s,\ y=-1-s,\ z=7+\frac{1}{2}s$ parallel?

14 Show that the lines $\boldsymbol{x}=(0, -2, 0)+t(1, -2, -3)$ and $\boldsymbol{x}=(0, 1, 0)+s(-1, -2, 1)$ are not perpendicular even though their directions are.

15 William Tell was imprisoned for refusing to salute the Duke of Austria's hat. To escape death, he promised to shoot an apple from his son's head with a bow and arrow. Suppose you find yourself in a predicament like young Tell's. Knowing that an arrow shot from a nearby bow travels approximately a straight line, and that the top of your head is exactly 6 feet above the floor when you stand perfectly straight, you ask to be given the coordinates of the tip of Tell's arrow (a) as it leaves the bow and (b) as it reaches a point a quarter of the way toward you.

If your feet are at $(0, 0, 0)$, and if you are told that $A=(100, 50, 5)$ and that $B=(75, 37.5, 5.2)$, then should you slouch down to make sure the arrow gets the apple and not your head?

4 A FIRST LOOK AT PLANES IN $\boldsymbol{R}^3$

While a complete treatment of planes must await the development of another vector operation, we can easily describe the "normal form" of the equation of a plane in $\boldsymbol{R}^3$. We can think of a plane in $\boldsymbol{R}^3$ as a two dimensional surface, a copy of $\boldsymbol{R}^2$ with the coordinate axes erased if you like. Let such a plane be given. Let $\boldsymbol{n}$ be any vector perpendicular to the plane (called a **normal vector**), by which we mean that when $\boldsymbol{n}$ is drawn with its initial point at $P_0(x_0, y_0, z_0)$ in the plane, then $\boldsymbol{n}$ is perpendicular to any vector $\overrightarrow{P_0P}$ for P a point in the plane. See Figure 4.1.

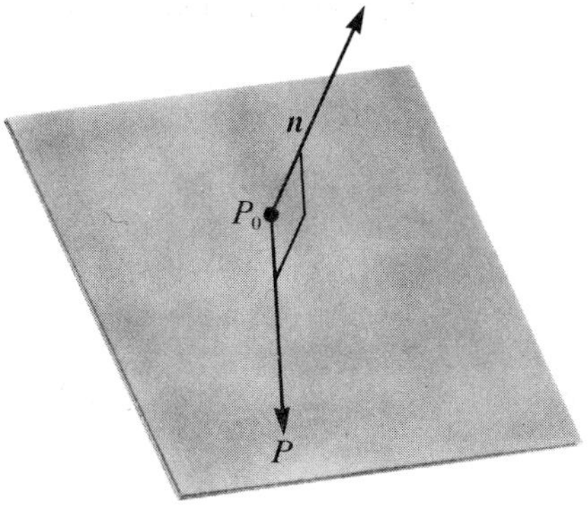

FIGURE 4.1

To get an equation for the plane, let $P(x, y, z)$ be any point in the plane. If P_0 is the point (x_0, y_0, z_0), then $\overrightarrow{P_0P}=(x-x_0, y-y_0, z-z_0)$. Thus $\boldsymbol{n}\cdot\overrightarrow{P_0P}=0$.

4.1 DEFINITION. The **normal form equation** of the plane through $P(x_0, y_0, z_0)$ with normal vector $\boldsymbol{n}$ is

$$\boldsymbol{n}\cdot(\boldsymbol{x}-\boldsymbol{x}_0)=0,$$

where $\boldsymbol{x}=(x, y, z)$ and $\boldsymbol{x}_0=(x_0, y_0, z_0)$.

If $\boldsymbol{n}=(a, b, c)$, then we can get a nice scalar equation for the plane as follows.

$$(a, b, c)\cdot(x-x_0, y-y_0, z-z_0)=0$$
$$a(x-x_0)+b(y-y_0)+c(z-z_0)=0$$
$$ax+by+cz=d, \quad \text{where} \quad d=ax_0+by_0+cz_0.$$

Thus, in $\boldsymbol{R}^3$ *the equation of a plane is a linear equation in the variables x, y, and z.* Conversely, given any equation $ax+by+cz=d$, we can write it in the form

$$\boldsymbol{n}\cdot\boldsymbol{x}=(a, b, c)\cdot(x, y, z)=d.$$

If (x_0, y_0, z_0) is any point satisfying this equation, then $\boldsymbol{n}\cdot\boldsymbol{x}_0=ax_0+by_0+cz_0=d$. Hence we have

$$\begin{aligned}(a, b, c)\cdot(x, y, z)&=ax_0+by_0+cz_0\\&=(a, b, c)\cdot(x_0, y_0, z_0).\end{aligned}$$

Then by Theorem 2.4(2) we have

$$(a, b, c)\cdot[(x, y, z)-(x_0, y_0, z_0)]=0$$

Therefore,

$$\boldsymbol{n}\cdot(\boldsymbol{x}-\boldsymbol{x}_0)=0.$$

> *Any first degree equation $ax+by+cz=d$ is the equation of a plane with normal vector (a, b, c).*

4.2 **EXAMPLE.** Find vector and scalar equations for the plane through $(-2, 1, 3)$ with normal vector $(3, 1, 5)$. Draw a picture.

Solution. The vector form is given by Definition 4.1: $[\boldsymbol{x}-(-2, 1, 3)]\cdot(3, 1, 5)=0$. The scalar equation is easily obtained from this.

$$\begin{aligned}[(x, y, z)-(-2, 1, 3)]\cdot(3, 1, 5)&=0\\(x+2, y-1, z-3)\cdot(3, 1, 5)&=0\\3(x+2)+y-1+5(z-3)&=0\\3x+6+y-1+5z-15&=0\\3x+y+5z&=10.\end{aligned}$$

To draw this plane it is helpful to show the **traces** of the plane in the coordinate planes. These are the lines in which the plane intersects the xy-, yz-, and xz-planes. To determine these lines, we find the points where the plane intersects each coordinate axis. When $y=0=z$, we have $x=\frac{10}{3}$. So $(\frac{10}{3}, 0, 0)$ is on the plane. Similarly, we find that $(0, 10, 0)$ and $(0, 0, 2)$ are on the plane. See Figure 4.2, where the first octant portion of the plane is shown.

4.3 **EXAMPLE.** Put in normal form: $2x-y-z=5$.

Solution. We must find a point (x_0, y_0, z_0) which satisfies the given equation. We could proceed by trial and error, but it is more efficient to assign y_0 and z_0

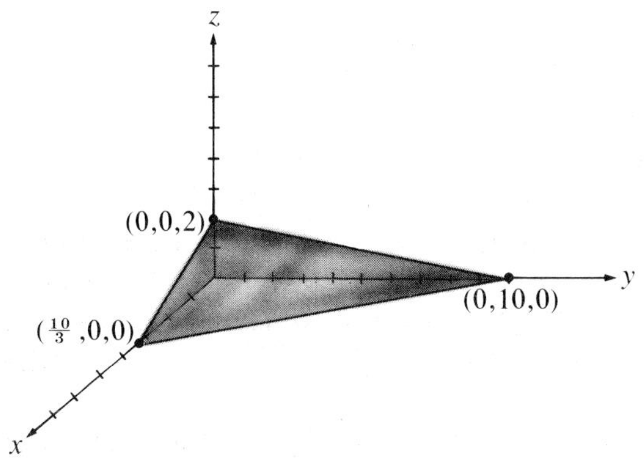

FIGURE 4.2

arbitrary values, and determine x_0 from the equation. So let $y_0 = 1$, $z_0 = 0$. Then

$$2x_0 - 1 - 0 = 5,$$
$$2x_0 = 6$$
$$x_0 = 3$$

So a point on the plane is (3, 1, 0). Thus we can rewrite the given equation in the form

$$(2, -1, -1) \cdot (x, y, z) = (2, -1, -1) \cdot (3, 1, 0)$$
$$(2, -1, -1) \cdot [(x, y, z) - (3, 1, 0)] = 0$$
$$(2, -1, -1) \cdot [\boldsymbol{x} - (3, 1, 0)] = 0.$$

Hence a normal vector to this plane is $\boldsymbol{n} = (2, -1, -1)$ (as is any nonzero multiple of $\boldsymbol{n}$).

We can easily decide whether two planes are parallel or perpendicular.

4.4 **THEOREM.** Let $\boldsymbol{n}_1 \cdot (\boldsymbol{x} - \boldsymbol{x}_0) = 0$ and $\boldsymbol{n}_2 \cdot (\boldsymbol{x} - \boldsymbol{x}_1) = 0$ be the equations of two planes. Then:

(1) the planes are parallel if and only if for some $a \neq 0$

$$\boldsymbol{n}_1 = a\boldsymbol{n}_2 \quad \text{but} \quad \boldsymbol{n}_1 \cdot \boldsymbol{x}_0 \neq a\boldsymbol{n}_2 \cdot \boldsymbol{x}_1;$$

(2) the planes are perpendicular if and only if $\boldsymbol{n}_1 \cdot \boldsymbol{n}_2 = 0$.

Proof. (1) The planes are parallel if and only if they do not coincide and any two respective normal vectors are parallel. Thus, they are parallel if and only if both $\boldsymbol{n}_1 = a\boldsymbol{n}_2$ for some nonzero $a \in \boldsymbol{R}$ and also the equations $\boldsymbol{n}_1 \cdot \boldsymbol{x} = \boldsymbol{n}_1 \cdot \boldsymbol{x}_0$ and $\boldsymbol{n}_2 \cdot \boldsymbol{x} = \boldsymbol{n}_2 \cdot \boldsymbol{x}_1$ are not *identical.* The second condition means that $\boldsymbol{n}_1 \cdot \boldsymbol{x} = \boldsymbol{n}_1 \cdot \boldsymbol{x}_0$ and $a\boldsymbol{n}_2 \cdot \boldsymbol{x} = a\boldsymbol{n}_2 \cdot \boldsymbol{x}_1$ *differ,* i.e., $\boldsymbol{n}_1 \cdot \boldsymbol{x}_0 \neq a\boldsymbol{n}_2 \cdot \boldsymbol{x}_1$.

(2) The planes are perpendicular if and only if $\boldsymbol{n}_1$ and $\boldsymbol{n}_2$ are orthogonal, i.e., $\boldsymbol{n}_1 \cdot \boldsymbol{n}_2 = 0$. QED

4.5 **EXAMPLE.** Show that the planes with equations $2x - 3y + 5z = 7$ and $-3x + \frac{9}{2}y - \frac{15}{2}z = 5$ do not intersect.

Solution. Put the equations in normal form:

$$(2, -3, 5) \cdot (x, y, z) = 7 = \boldsymbol{n}_1 \cdot \boldsymbol{x}_0$$
$$(-3, \tfrac{9}{2}, -\tfrac{15}{2}) \cdot (x, y, z) = 5 = \boldsymbol{n}_2 \cdot \boldsymbol{x}_1.$$

Then we have $\boldsymbol{n}_1 = (2, -3, 5)$ and $\boldsymbol{n}_2 = (-3, \frac{9}{2}, -\frac{15}{2})$. Since $\boldsymbol{n}_1 = -\frac{2}{3}\boldsymbol{n}_2$, $\boldsymbol{n}_1$ and $\boldsymbol{n}_2$ are parallel. Thus by Theorem 4.4, the given planes are also parallel or coincident. But since $7 \neq -\frac{2}{3} \cdot 5$, we see that the planes are not coincident. Hence they are parallel, so they have no points of intersection. QED

Two planes which *do* intersect but don't coincide will intersect in a line. The following example illustrates how we can find equations for the line of intersection from the equations for the planes.

4.6 **EXAMPLE.** Find vector and scalar equations for the line of intersection of the two planes

(1) $$3x - 2y + z = 1$$

(2) $$-2x + y + 3z = 2$$

Solution. The most efficient technique is to eliminate one of the variables. (It is not important *which* one is eliminated.) Here we can easily eliminate y by multiplying the second equation by 2 and adding to the first. (Another equally easy technique is to solve (2) for y in terms of x and z, and then substitute that expression into (1).) We get $-x + 7z = 5$, so $x = 7z - 5$. Then we can express y in terms of z from (2) also:

$$\begin{aligned} y &= 2 + 2x - 3z \\ &= 2 + 14z - 10 - 3z, \\ y &= 11z - 8. \end{aligned}$$

We thus have a set of parametric scalar equations for the line if we put $z = t$. For then

(3) $$x = 7t - 5$$

(4) $$y = 11t - 8$$

(5) $$z = t$$

In vector form

$$\begin{pmatrix} x \\ y \\ z \end{pmatrix} = t\begin{pmatrix} 7 \\ 11 \\ 1 \end{pmatrix} + \begin{pmatrix} -5 \\ -8 \\ 0 \end{pmatrix},$$

i.e., $\boldsymbol{x} = t\boldsymbol{v} + \boldsymbol{x}_0$ where $\boldsymbol{v} = (7, 11, 1)$ and $\boldsymbol{x}_0 = (-5, -8, 0)$.

Later on (in Section 6) we will learn how this elimination technique can be formulated generally to find solutions to systems of linear equations.

EXERCISES 1.4

1 Find the scalar equation of the plane through $(1, -2, 5)$ perpendicular to $\boldsymbol{n} = (3, -4, 1)$. Sketch.

2 Give the equations of the xy-, yz-, and xz-planes. Sketch.

3 What is the equation of the plane 5 units below the xy-plane? 3 units to the right of the xz-plane? 1 unit behind the yz-plane? Sketch.

4 The line through the origin perpendicular to a plane intersects it at the point $(2, -1, 1)$. Find the scalar equation of the plane. Sketch.

5 A plane through $(2, -2, 5)$ is perpendicular to the line through $(1, 1, 1)$ and $(-2, 3, 1)$. Find a scalar equation of the plane.

6 The line through $(2, -1, -3)$ perpendicular to a certain plane intersects it at the point $(3, -3, 2)$. Find a scalar equation for the plane.

7 Give the equation of the plane through $(1, -2, -1)$ which
(a) is perpendicular to the x-axis.
(b) is parallel to the xz-plane.

8 Find the equation of the plane through $(1, 2, 3)$ which is parallel to the plane whose equation is $-x+3y-4z=2$.

9 Which of the following planes are coincident, parallel, perpendicular?
(a) $-x+2y-2z=3$ (b) $\frac{1}{2}x-y+z=-\frac{3}{2}$
(c) $4x-2y-4z=7$ (d) $3x-6y+6z=11$

10 Which of the following planes are coincident, parallel, perpendicular?
(a) $\frac{2}{3}x-y-5z=4$ (b) $-4x+6y+30z=21$
(c) $3x-3y+z=7$ (d) $2x-3y-15z=12$

11 Find vector and scalar equations for the line of intersection of the planes

$$x-2y+2z=4$$
$$2x-y+3z=5.$$

12 Find vector and scalar equations for the line of intersection of the two planes

$$x-2y+3z=5$$
$$8x+7y+z=2.$$

13 Find parametric equations for the traces of the plane $3x-5y+2z=15$ in the three coordinate planes. Draw a picture.

14 Find parametric equations for the traces of the plane $x+4y+5z=10$ in the three coordinate planes. Draw a picture.

15 Is the line $\boldsymbol{x}=(2, 1, 1)+t(-1, 3, 2)$ normal to the plane $x-3y-2z=11$? Find the point of intersection.

16 Show that the line $\boldsymbol{x}=(3, 1, -2)+t(1, -1, 3)$ does not intersect the plane $2x-y-z=5$.

17 Prove that the line $\boldsymbol{x}=\boldsymbol{x}_0+t\boldsymbol{v}$ is parallel to the plane $\boldsymbol{n}\cdot(\boldsymbol{x}-\boldsymbol{x}_1)=0$ if and only if $\boldsymbol{n}\cdot\boldsymbol{v}=0$.

18 Let $\boldsymbol{y}_0=\overrightarrow{OQ}\in \boldsymbol{R}^3$. Let π be the plane $\boldsymbol{n}\cdot(\boldsymbol{x}-\boldsymbol{x}_0)=0$, and let $\boldsymbol{N}=\boldsymbol{n}/|\boldsymbol{n}|$. Derive the formula

$$(*)\quad d=|\boldsymbol{N}\cdot\boldsymbol{x}_0-\boldsymbol{N}\cdot\boldsymbol{y}_0|$$

for the perpendicular distance from Q to π. See Figure 4.3. [*Hint:* Let P be the point of intersection of π and the perpendicular drawn from Q. Show that the line through

P and Q has equation $\mathbf{x}=\mathbf{y}_0+t\mathbf{N}$. Since P is on π, $\overrightarrow{OP}=\mathbf{y}_0+t_0\mathbf{N}$, so $d=|\overrightarrow{PQ}|=|t_0|$. Show from the equation of π that $|t_0|$ is given by the right side of (*).]

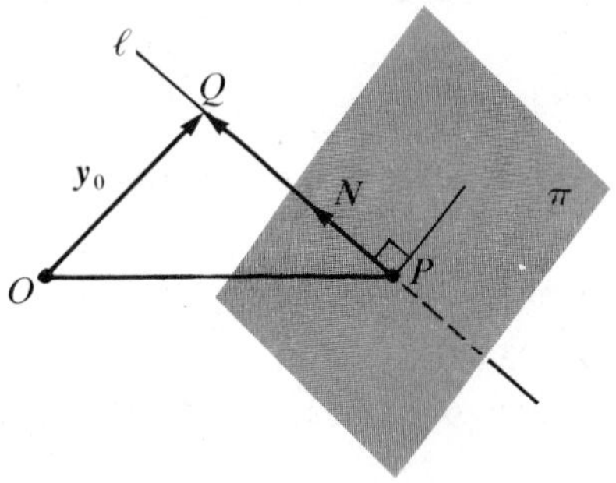

FIGURE 4.3

19 Find the distance from the point $(2, 1, -1)$ to the plane $5x-2y+3z=2$. Use (*) in Exercise 18.

20 Find the distance from the point $(2, 3, 1)$ to the plane $2x-y+2z=9$. Use (*) in Exercise 18.

5 A SECOND LOOK AT PLANES. THE CROSS PRODUCT

In the last section we studied planes as sets of points orthogonal to normal vectors $\boldsymbol{n}$. But just as a line is determined by two points, so a plane is determined by two (nonskew) lines (or three noncollinear points). We ought to be able to find the equation of a plane described in terms of two lines (or vectors) in it or in terms of three noncollinear points on it. So let us see what we can do to find the equation of a plane given (a) two lines ℓ_1 and ℓ_2 in the plane, or (b) three points P, Q, R on the plane.

First suppose we are given $\ell_1:\mathbf{x}=\mathbf{x}_1+t\boldsymbol{v}$ and $\ell_2:\mathbf{x}=\mathbf{x}_2+s\mathbf{w}$, where $\boldsymbol{v}$ and $\mathbf{w}$ are vectors giving the directions of ℓ_1 and ℓ_2 respectively. In order to determine the normal form equation of the plane, we need to find a vector $\boldsymbol{n}$ perpendicular to the plane. For then the plane will have equation $\boldsymbol{n}\cdot(\mathbf{x}-\mathbf{x}_0)=0$, where the endpoint of $\mathbf{x}_0$ is a point in the plane. See Figure 5.1. Then $\boldsymbol{n}$ must be perpendicular to *both* $\boldsymbol{v}$ and $\mathbf{w}$, i.e.,

(*) $$\boldsymbol{n}\cdot\boldsymbol{v}=0, \qquad \boldsymbol{n}\cdot\mathbf{w}=0.$$

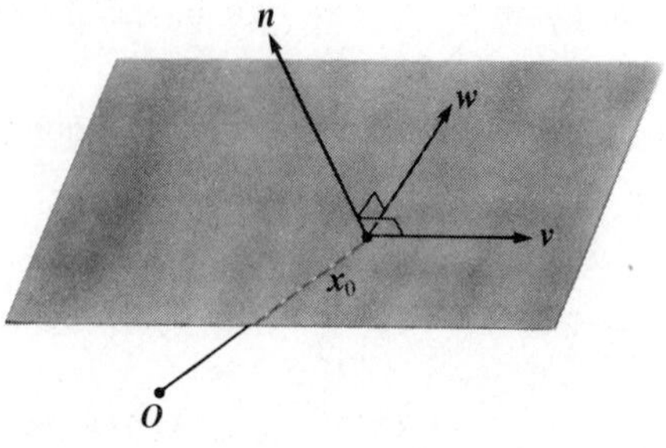

FIGURE 5.1

So if we can find a vector which is perpendicular to *both* $\boldsymbol{v}$ and $\mathbf{w}$, then we will have a normal vector to the plane. Before proceeding, we remark that *case (b) above brings us to this same problem*. For if P, Q, and R are given points in the plane, then let $\boldsymbol{v}=\overrightarrow{PQ}$ and $\mathbf{w}=\overrightarrow{PR}$. These are two vectors in the plane, so a normal vector $\boldsymbol{n}$ to the plane must satisfy (*). See Figure 5.2.

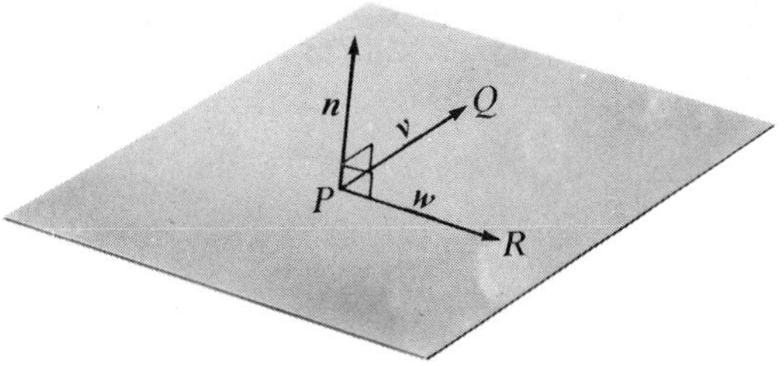

FIGURE 5.2

Now let's see if we can find a vector $\boldsymbol{n}=(x, y, z)$ which satisfies (*). Suppose that $\boldsymbol{v}=(v_1, v_2, v_3)$ and $\boldsymbol{w}=(w_1, w_2, w_3)$. Then the vector equations (*) translate into the scalar equations

(1) $$v_1 x + v_2 y + v_3 z = 0$$

(2) $$w_1 x + w_2 y + w_3 z = 0.$$

Geometrically, (1) and (2) represent two planes perpendicular to $\boldsymbol{v}$ and $\boldsymbol{w}$. We expect these two planes to intersect in a line which will then be perpendicular to *both* $\boldsymbol{v}$ and $\boldsymbol{w}$. Let us use the technique of elimination from Example 4.6 to find an equation of this line, or at least some vector lying in the line. To eliminate y, we multiply (1) by w_2 and (2) by v_2, and subtract:

$$\begin{aligned} w_2 \cdot (1)&: v_1 w_2 x + v_2 w_2 y + v_3 w_2 z = 0 \\ v_2 \cdot (2)&: v_2 w_1 x + v_2 w_2 y + v_2 w_3 z = 0 \\ \hline w_2 \cdot (1) - v_2 \cdot (2)&: (v_1 w_2 - v_2 w_1)x + (v_3 w_2 - v_2 w_3)z = 0 \end{aligned}$$

Thus,

(3) $$(v_1 w_2 - v_2 w_1)x = (v_2 w_3 - v_3 w_2)z.$$

Similarly, we eliminate x if we multiply (1) by w_1 and (2) by v_1, and subtract:

$$\begin{aligned} w_1 \cdot (1)&: v_1 w_1 x + v_2 w_1 y + v_3 w_1 z = 0 \\ v_1 \cdot (2)&: v_1 w_1 x + v_1 w_2 y + v_1 w_3 z = 0 \\ \hline w_1 \cdot (1) - v_1 \cdot (2)&: (v_2 w_1 - v_1 w_2)y + (v_3 w_1 - v_1 w_3)z = 0. \end{aligned}$$

Hence,

(4) $$(v_3 w_1 - v_1 w_3)z = (v_1 w_2 - v_2 w_1)y$$

Now, suppose we set $z = v_1 w_2 - v_2 w_1$. Then we see that $x = v_2 w_3 - v_3 w_2$ will satisfy (3). From (4) we similarly see that $y = v_3 w_1 - v_1 w_3$ will be a solution. Hence a vector orthogonal to both $\boldsymbol{v}=(v_1, v_2, v_3)$ and $\boldsymbol{w}=(w_1, w_2, w_3)$ is $\boldsymbol{n} = (v_2 w_3 - v_3 w_2, v_3 w_1 - v_1 w_3, v_1 w_2 - v_2 w_1)$. This normal vector is of great importance, so it is given a special name.

5.1 DEFINITION. If $\boldsymbol{v}=(v_1, v_2, v_3)$ and $\boldsymbol{w}=(w_1, w_2, w_3)$ are two vectors in $\boldsymbol{R}^3$, then their **cross product** $\boldsymbol{v} \times \boldsymbol{w}$ is defined by

$$\begin{aligned} \boldsymbol{v} \times \boldsymbol{w} &= (v_2 w_3 - v_3 w_2, v_3 w_1 - v_1 w_3, v_1 w_2 - v_2 w_1) \\ &= \det\begin{pmatrix} v_2 & v_3 \\ w_2 & w_3 \end{pmatrix}\boldsymbol{i} + \det\begin{pmatrix} v_3 & v_1 \\ w_3 & w_1 \end{pmatrix}\boldsymbol{j} + \det\begin{pmatrix} v_1 & v_2 \\ w_1 & w_2 \end{pmatrix}\boldsymbol{k}. \end{aligned}$$

Recall that the 2-by-2 *determinant* function is given by

$$\det\begin{pmatrix} a & b \\ c & d \end{pmatrix} = ad - bc.$$

In computing $\boldsymbol{v} \times \boldsymbol{w}$ it is helpful to write the coordinates of $\boldsymbol{v}$ above those of $\boldsymbol{w}$. One then computes each coordinate in turn. For the first coordinate,

$$\begin{array}{l} \boldsymbol{v} = (v_1, v_2, v_3) \\ \qquad\quad \times \\ \boldsymbol{w} = (w_1, w_2, w_3) \end{array}$$

where a downward pointing arrow multiplies the two entries it connects and prefixes a plus sign, while an upward pointing arrow multiplies the two entries it connects and prefixes a minus sign. Thus, for the second and third coordinates, we have

$$\begin{array}{l} (v_1, v_2, v_3) \\ \qquad \times \\ (w_1, w_2, w_3) \end{array} \quad \text{and} \quad \begin{array}{l} (v_1, v_2, v_3) \\ \quad \times \\ (w_1, w_2, w_3). \end{array}$$

5.2 **EXAMPLE.** Find $\boldsymbol{v} \times \boldsymbol{w}$ if $\boldsymbol{v} = (2, -1, 3)$ and $\boldsymbol{w} = (1, 5, -3)$.

Solution.

$$\boldsymbol{v} \times \boldsymbol{w} = (3-15, 3+6, 10+1) = (-12, 9, 11).$$

5.3 **EXAMPLE.** Find the equation of the plane determined by the two lines $\boldsymbol{x} = (2, 0, 3) + t(2, -1, 3)$ and $\boldsymbol{x} = (-4, 3, -6) + s(1, 5, -3)$.

Solution. The vector $\boldsymbol{v} = (2, -1, 3)$ is in the direction of the first line, and $\boldsymbol{w} = (1, 5, -3)$ is in the direction of the second line. So a vector normal to the plane is $\boldsymbol{v} \times \boldsymbol{w}$, which we just computed in the preceding example: $\boldsymbol{v} \times \boldsymbol{w} = (-12, 9, 11)$. Taking $t = 0$, we see that $(2, 0, 3)$ is a point on the plane. So an equation is

$$(-12, 9, 11) \cdot [(x, y, z) - (2, 0, 3)] = 0,$$

which becomes

$$-12x + 24 + 9y + 11z - 33 = 0,$$
$$-12x + 9y + 11z = 9.$$

In this example, we have tacitly assumed that the given lines really *do* determine a plane. But how can we be sure that they were not *skew* lines? The simplest way is usually to *check* that the coordinates of any point on either line actually satisfy the equation of the plane we obtained. Here we can do that easily. Since any point (x, y, z) on the first line has coordinates $x = 2 + 2t$, $y = -t$, $z = 3 + 3t$ for some t, we have

$$-12x + 9y + 11z = -24 - 24t - 9t + 33 + 33t = 9.$$

So (x, y, z) is indeed on our plane. Similarly, a point (x, y, z) on the second line has coordinates $x = -4 + s$, $y = 3 + 5s$, $z = -6 - 3s$ for some s. Then

$$-12x + 9y + 11z = 48 - 12s + 27 + 45s - 66 - 33s = 9.$$

See Exercise 20 for what happens in case the lines are skew, and fail to determine a plane.

5.4 **EXAMPLE.** Find the equation of the plane through the three points $A(1, -2, 3)$, $B(-2, 1, 1)$, and $C(1, 3, -2)$.

Solution. The vectors $\boldsymbol{v} = \overrightarrow{AB}$ and $\boldsymbol{w} = \overrightarrow{AC}$ lie in the plane, so $\boldsymbol{v} \times \boldsymbol{w}$ is normal to the plane. We have

$$\boldsymbol{v} = (-3, 3, -2) \quad \text{and} \quad \boldsymbol{w} = (0, 5, -5),$$

so

$$\begin{aligned} \boldsymbol{v} \times \boldsymbol{w} &= (-15 + 10, 0 - 15, -15 - 0) \\ &= (-5, -15, -15) = -5(1, 3, 3). \end{aligned}$$

Thus for a normal vector we can take $\boldsymbol{n} = (1, 3, 3)$. Then an equation for the plane is $\boldsymbol{n} \cdot (\boldsymbol{x} - \overrightarrow{OA}) = 0$, i.e.,

$$\begin{aligned} (1, 3, 3) \cdot [(x, y, z) - (1, -2, 3)] &= 0, \\ (1, 3, 3) \cdot (x - 1, y + 2, z - 3) &= 0 \\ x - 1 + 3y + 6 + 3z - 9 &= 0 \\ x + 3y + 3z &= 4. \end{aligned}$$

You should verify for yourself that the coordinates of all the points A, B, and C satisfy this equation.

We turn now to some algebraic properties of the cross product. In the following result, note particularly that the cross product *fails* to be commutative or associative. It is doubtless the first kind of multiplication you have met which lacks these familiar properties. (See Exercise 21 below for more on the associative law. Exercise 16 below gives a kind of substitute for it, and part (2) of the next Theorem gives a substitute for the commutative law.)

5.5 **THEOREM.** For all vectors $\boldsymbol{u}$, $\boldsymbol{v}$, and $\boldsymbol{w}$ in $\boldsymbol{R}^3$,

(1) $(\boldsymbol{v} \times \boldsymbol{w}) \cdot \boldsymbol{v} = (\boldsymbol{v} \times \boldsymbol{w}) \cdot \boldsymbol{w} = 0$
(2) $\boldsymbol{v} \times \boldsymbol{w} = -\boldsymbol{w} \times \boldsymbol{v}$
(3) $(\boldsymbol{u} + \boldsymbol{v}) \times \boldsymbol{w} = \boldsymbol{u} \times \boldsymbol{w} + \boldsymbol{v} \times \boldsymbol{w}$
(4) $\boldsymbol{u} \cdot (\boldsymbol{v} \times \boldsymbol{w}) = \boldsymbol{v} \cdot (\boldsymbol{w} \times \boldsymbol{u}) = \boldsymbol{w} \cdot (\boldsymbol{u} \times \boldsymbol{v})$
(5) $\boldsymbol{u} \times (\boldsymbol{v} \times \boldsymbol{w}) = (\boldsymbol{u} \cdot \boldsymbol{w})\boldsymbol{v} - (\boldsymbol{u} \cdot \boldsymbol{v})\boldsymbol{w}$.

Partial Proof. Statement (1) is just a restatement of the fact that $\boldsymbol{v} \times \boldsymbol{w}$ is perpendicular to both $\boldsymbol{v}$ and $\boldsymbol{w}$. Let $\boldsymbol{u} = (u_1, u_2, u_3)$, $\boldsymbol{v} = (v_1, v_2, v_3)$, and $\boldsymbol{w} = (w_1, w_2, w_3)$. Then (2) follows easily: by Definition 5.1,

$$\boldsymbol{v} \times \boldsymbol{w} = (v_2 w_3 - v_3 w_2, v_3 w_1 - v_1 w_3, v_1 w_2 - v_2 w_1).$$

Also,

$$\begin{aligned}\boldsymbol{w}\times\boldsymbol{v}&=(w_2v_3-w_3v_2,\ w_3v_1-w_1v_3,\ w_1v_2-w_2v_1)\\&=(-(v_2w_3-v_3w_2),\ -(v_3w_1-v_1w_3),\ -(v_1w_2-v_2w_1))\\&=-\boldsymbol{v}\times\boldsymbol{w}.\end{aligned}$$

The verification of (3) is left as an exercise for you (Exercise 10 below). We verify the first equality in (4) and leave the second for you (Exercise 11 below). First,

$$\begin{aligned}\boldsymbol{u}\cdot(\boldsymbol{v}\times\boldsymbol{w})&=(u_1,u_2,u_3)\cdot(v_2w_3-v_3w_2,\ v_3w_1-v_1w_3,\ v_1w_2-v_2w_1)\\&=u_1v_2w_3-u_1v_3w_2+u_2v_3w_1-u_2v_1w_3+u_3v_1w_2-u_3v_2w_1.\end{aligned}$$

Next,

$$\begin{aligned}\boldsymbol{v}\cdot(\boldsymbol{w}\times\boldsymbol{u})&=(v_1,v_2,v_3)\cdot(w_2u_3-w_3u_2,\ w_3u_1-w_1u_3,\ w_1u_2-w_2u_1)\\&=u_3v_1w_2-u_2v_1w_3+u_1v_2w_3-u_3v_2w_1+u_2v_3w_1-u_1v_3w_2\\&=\boldsymbol{u}\cdot(\boldsymbol{v}\times\boldsymbol{w}).\end{aligned}$$

Finally, we consider (5). We have $\boldsymbol{u}=(u_1,u_2,u_3)$ and

$$\boldsymbol{v}\times\boldsymbol{w}=(v_2w_3-v_3w_2,\ v_3w_1-v_1w_3,\ v_1w_2-v_2w_1),$$

so the first coordinate of $\boldsymbol{u}\times(\boldsymbol{v}\times\boldsymbol{w})$ is

$$u_2(v_1w_2-v_2w_1)-u_3(v_3w_1-v_1w_3)=u_2v_1w_2-u_2v_2w_1+u_3v_1w_3-u_3v_3w_1.$$

The first coordinate of $(\boldsymbol{u}\cdot\boldsymbol{w})\boldsymbol{v}$ is

$$(u_1w_1+u_2w_2+u_3w_3)v_1=u_1v_1w_1+u_2v_1w_2+u_3v_1w_3.$$

The first coordinate of $(\boldsymbol{u}\cdot\boldsymbol{v})\boldsymbol{w}$ is

$$(u_1v_1+u_2v_2+u_3v_3)w_1=u_1v_1w_1+u_2v_2w_1+u_3v_3w_1.$$

Subtracting these last two expressions, we see that the right-hand side of (5) has as first coordinate

$$u_2v_1w_2-u_2v_2w_1+u_3v_1w_3-u_3v_3w_1,$$

which is the first coordinate of $\boldsymbol{u}\times(\boldsymbol{v}\times\boldsymbol{w})$ calculated above. Similar computations verify that the second and third coordinates of $\boldsymbol{u}\times(\boldsymbol{v}\times\boldsymbol{w})$ are respectively equal to the second and third coordinates of $(\boldsymbol{u}\cdot\boldsymbol{w})\boldsymbol{v}-(\boldsymbol{u}\cdot\boldsymbol{v})\boldsymbol{w}$. We leave these for you (Exercise 12 below).

In Section 8 we will make heavy use of properties (2) and (4). Note that in (4) we see that we may permute $\boldsymbol{u}$, $\boldsymbol{v}$, and $\boldsymbol{w}$ *cyclically* without changing the value of $\boldsymbol{u}\cdot(\boldsymbol{v}\times\boldsymbol{w})$, but we cannot permute them arbitrarily. See Figure 5.3. By way of an

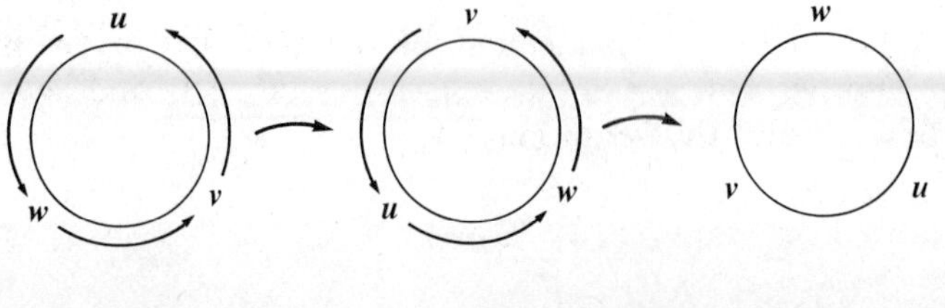

FIGURE 5.3
Cyclic permutations of $\boldsymbol{u}$, $\boldsymbol{v}$, and $\boldsymbol{w}$.

example we have the following, as a direct consequence of Theorems 5.5(2) and 2.4(2).

5.6 COROLLARY. $\boldsymbol{u} \cdot (\boldsymbol{w} \times \boldsymbol{v}) = -\boldsymbol{u} \cdot (\boldsymbol{v} \times \boldsymbol{w})$.

The expression $\boldsymbol{u} \cdot (\boldsymbol{v} \times \boldsymbol{w})$ is called the **triple scalar product** of $\boldsymbol{u}$, $\boldsymbol{v}$, and $\boldsymbol{w}$ in that order.

We turn now to an important geometric property of the cross product, which can be thought of as paralleling the fact that $\boldsymbol{x} \cdot \boldsymbol{y} = |\boldsymbol{x}|\,|\boldsymbol{y}| \cos\theta$ (cf. Definition 2.7).

5.7 LEMMA. $|\boldsymbol{v} \times \boldsymbol{w}|^2 = |\boldsymbol{v}|^2\,|\boldsymbol{w}|^2 - (\boldsymbol{v} \cdot \boldsymbol{w})^2$.

Proof.

$$\begin{aligned}|\boldsymbol{v} \times \boldsymbol{w}|^2 &= (v_2w_3 - v_3w_2)^2 + (v_3w_1 - v_1w_3)^2 + (v_1w_2 - v_2w_1)^2.\\ &= (v_2w_3)^2 - 2v_2v_3w_2w_3 + (v_3w_2)^2 + (v_3w_1)^2 - 2v_1v_3w_1w_3\\ &\quad + (v_1w_3)^2 + (v_1w_2)^2 - 2v_1v_2w_1w_2 + (v_2w_1)^2.\end{aligned}$$

$$\begin{aligned}|\boldsymbol{v}|^2\,|\boldsymbol{w}|^2 - (\boldsymbol{v} \cdot \boldsymbol{w})^2 &= (v_1^2 + v_2^2 + v_3^2)(w_1^2 + w_2^2 + w_3^2) - (v_1w_1 + v_2w_2 + v_3w_3)^2\\ &= (v_1w_1)^2 + (v_1w_2)^2 + (v_1w_3)^2 + (v_2w_1)^2 + (v_2w_2)^2\\ &\quad + (v_2w_3)^2 + (v_3w_1)^2 + (v_3w_2)^2 + (v_3w_3)^2 - [(v_1w_1)^2\\ &\quad + (v_2w_2)^2 + (v_3w_3)^2 + 2v_1v_2w_1w_2 + 2v_1v_3w_1w_3 + 2v_2v_3w_2w_3]\\ &= (v_1w_2)^2 + (v_1w_3)^2 + (v_2w_1)^2 + (v_2w_3)^2 + (v_3w_1)^2\\ &\quad + (v_3w_2)^2 - 2v_1v_2w_1w_2 - 2v_1v_3w_1w_3 - 2v_2v_3w_2w_3\\ &= |\boldsymbol{v} \times \boldsymbol{w}|^2.\end{aligned}$$

QED

5.8 THEOREM. $|\boldsymbol{v} \times \boldsymbol{w}| = |\boldsymbol{v}|\,|\boldsymbol{w}| \sin\theta$, where θ is the angle between $\boldsymbol{v}$ and $\boldsymbol{w}$.

Proof.

$$\begin{aligned}|\boldsymbol{v} \times \boldsymbol{w}| &= \sqrt{|\boldsymbol{v}|^2\,|\boldsymbol{w}|^2 - (\boldsymbol{v} \cdot \boldsymbol{w})^2} && \text{from Lemma 5.7}\\ &= \sqrt{|\boldsymbol{v}|^2\,|\boldsymbol{w}|^2 - |\boldsymbol{v}|^2\,|\boldsymbol{w}|^2 \cos^2\theta} && \text{from Definition 2.7}\\ &= |\boldsymbol{v}|\,|\boldsymbol{w}| \sqrt{1 - \cos^2\theta}\\ &= |\boldsymbol{v}|\,|\boldsymbol{w}| \sqrt{\sin^2\theta}\\ &= |\boldsymbol{v}|\,|\boldsymbol{w}| \sin\theta\end{aligned}$$

since θ is a first or second quadrant angle (Definition 2.7), so that $\sin\theta$ is nonnegative. QED

5.9 COROLLARY. The area of the parallelogram formed by $\boldsymbol{v}$ and $\boldsymbol{w}$ is $|\boldsymbol{v} \times \boldsymbol{w}|$.

Proof. The area of a parallelogram is its base times its altitude. Referring to Figure 5.4, we see that if we take the base to be $|\boldsymbol{v}|$, then the altitude is $|\boldsymbol{w}| \sin\theta$. So the area is $|\boldsymbol{v}|\,|\boldsymbol{w}| \sin\theta = |\boldsymbol{v} \times \boldsymbol{w}|$. QED

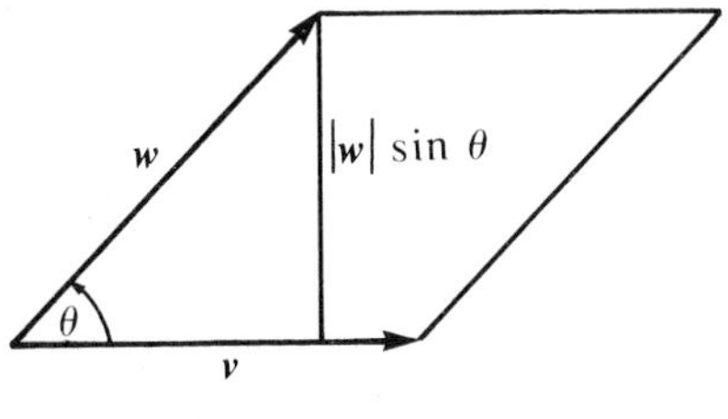

FIGURE 5.4

5.10 EXAMPLE. Find the area of the parallelogram in the plane determined by the vectors (2, 5) and (−1, 1). Refer to Figure 5.5.

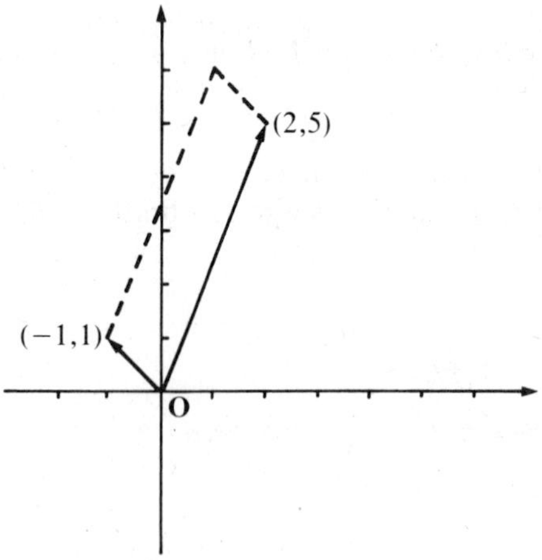

FIGURE 5.5

Solution. We can regard these vectors as vectors in $\boldsymbol{R}^3$, namely, $\boldsymbol{v}=(2, 5, 0)$ and $\boldsymbol{w}=(-1, 1, 0)$. Then the area of the parallelogram they determine is

$$|\boldsymbol{v}\times \boldsymbol{w}| = |(0, 0, 2+5)| = 7.$$

5.11 COROLLARY. Two nonzero vectors $\boldsymbol{v}$ and $\boldsymbol{w}$ are parallel if and only if $\boldsymbol{v}\times\boldsymbol{w}=\boldsymbol{0}$.

Proof. We know that $\boldsymbol{v}$ and $\boldsymbol{w}$ are parallel if and only if the angle between them is 0 or π (Definition 2.13). But $|\boldsymbol{v}\times \boldsymbol{w}| = |\boldsymbol{v}|\,|\boldsymbol{w}| \sin\theta = 0$ if and only if $\sin\theta = 0$, where θ is a first or second quadrant angle. Thus $|\boldsymbol{v}\times\boldsymbol{w}| = 0$ if and only if $\theta = 0$ or π, i.e., if and only if $\boldsymbol{v}$ and $\boldsymbol{w}$ are parallel. QED

We close this section by mentioning one simple physical application of the cross product. In fact, one of the origins of the cross product lies in the problem of giving a quantitative definition of the *torque* at a point P_0 resulting from a force $\boldsymbol{F}$ acting at a point P. From a qualitative point of view, we think of the torque as a measure of the tendency for the mass at P to rotate about P_0. See Figure 5.6.

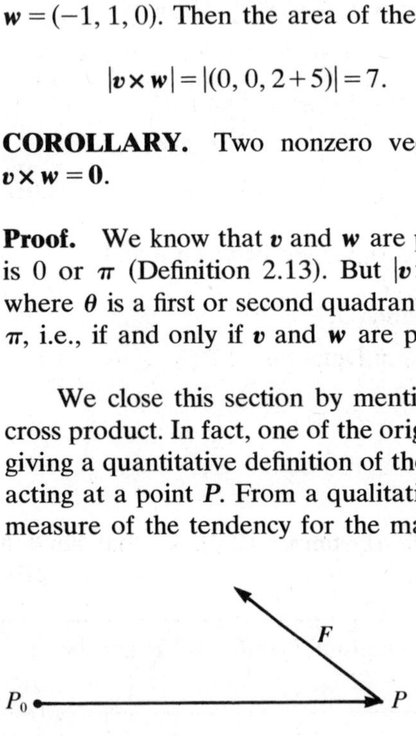

FIGURE 5.6

For the case of a force acting perpendicular to $\overrightarrow{P_0P}$, we have an ancient definition for the magnitude of this torque. Archimedes discovered that this magnitude is simply $|\boldsymbol{x}|\,|\boldsymbol{F}|$, where $\boldsymbol{x} = \overrightarrow{P_0P}$. In the case of a general force $\boldsymbol{F}$, we can resolve $\boldsymbol{F}$ into components $\boldsymbol{F}_1$ and $\boldsymbol{F}_2$ as shown in Figure 5.7. Then the component

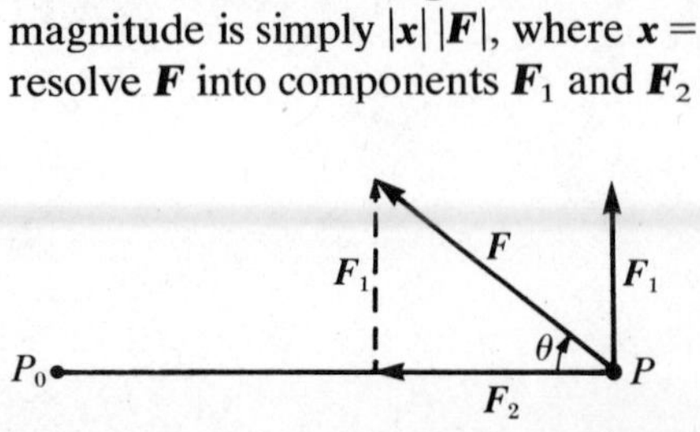

FIGURE 5.7

$\boldsymbol{F}_1$ is the one which produces torque, and the magnitude of this is easily computed if we draw $\boldsymbol{F}_1$ with its initial point at the tip of $\boldsymbol{F}_2$. For then we have $|\boldsymbol{F}_1| = |\boldsymbol{F}| \sin\theta$, so the magnitude of the torque due to $\boldsymbol{F}$ is $|\boldsymbol{x}|\,|\boldsymbol{F}| \sin\theta = |\boldsymbol{x} \times \boldsymbol{F}|$. It is then natural to give the following definition.

5.12 **DEFINITION.** The **torque** $\boldsymbol{L}$ at P_0 produced by a force $\boldsymbol{F}$ acting on a mass at point P is $\boldsymbol{L} = \overrightarrow{P_0P} \times \boldsymbol{F}$.

Note that according to this definition the torque is a *vector* perpendicular to both $\overrightarrow{P_0P}$ and $\boldsymbol{F}$, i.e., a vector in the axis of rotation which the force tends to produce. The direction of this vector, it so happens, is the direction of advance of a screw with a right-handed thread acted on by $\boldsymbol{F}$. See Figure 5.8. (This accounts in part for the term "right-handed" as applied to the coordinate system we have

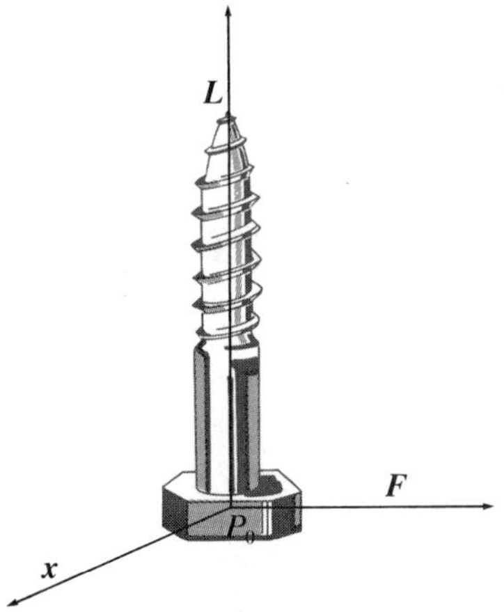

FIGURE 5.8

adopted.) Notice that $\boldsymbol{k} = \boldsymbol{i} \times \boldsymbol{j}$ (Exercise 14 below), so the positive z-axis is in the direction of advance of a screw with right-handed thread whose head is at the origin and which is turned in the direction from the positive x-axis to the positive y-axis.

To summarize the geometry of the cross product $\boldsymbol{v} \times \boldsymbol{w}$: its magnitude is the area of the parallelogram formed by $\boldsymbol{v}$ and $\boldsymbol{w}$; its direction is that determined by the right hand rule; it is perpendicular to both $\boldsymbol{v}$ and $\boldsymbol{w}$ and hence to the plane of $\boldsymbol{v}$ and $\boldsymbol{w}$.

EXERCISES 1.5

1 Find the equation of the plane through the three points $P(1, -1, 3)$, $Q(0, 1, 3)$, and $R(-2, 1, -1)$. Check by verifying that the coordinates of P, Q, and R satisfy your equation.

2 Find the equation of the plane through the two lines $\boldsymbol{x} = (2, 0, -1) + s(3, -3, -1)$ and $\boldsymbol{x} = (3, -1, 1) + t(1, -1, 2)$. Check by verifying that the coordinates of all points on the lines satisfy your equation.

3 Find vector and scalar equations for the plane through the points $P(3, 0, 1)$, $Q(2, -1, 3)$, and $R(-1, 4, 1)$.

4 A constant force $\boldsymbol{F} = (3, -1, 2)$ acts on a particle located at $(2, 1, 1)$. Find the resulting torque at (a) the origin, (b) $(-1, 1, 0)$.

5 Find the area of the parallelogram in space formed by the vectors $\boldsymbol{u} = (1, 5, 1)$ and $\boldsymbol{v} = (-2, 1, 3)$.

6 Find the area of the parallelogram in the plane formed by the origin and the points (2, 3), (−1, 2), and (1, 5).

7 Show that the area of the triangle whose vertices are the endpoints of the vectors $\boldsymbol{u}$, $\boldsymbol{v}$, and $\boldsymbol{w}$ is

$$A = \tfrac{1}{2}|(\boldsymbol{v}-\boldsymbol{u})\times(\boldsymbol{w}-\boldsymbol{u})|.$$

8 Find the area of the triangle with vertices (−2, 1, 5), (4, 0, 6) and (3, −3, 2). (See Exercise 7.)

9 Find the area of the triangle with vertices (0, 0, 0), (2, 3, −2) and (−1, 1, 4). (See Exercise 7.)

10 Prove (3) of Theorem 5.5.

11 Prove that $\boldsymbol{u}\cdot(\boldsymbol{v}\times\boldsymbol{w}) = (\boldsymbol{u}\times\boldsymbol{v})\cdot\boldsymbol{w}$, the second equality in Theorem 5.5(4).

12 Verify the equality of the respective second and third coordinates in Theorem 5.5(5).

13 Show that $\boldsymbol{u}\cdot(\boldsymbol{v}\times\boldsymbol{w}) = -\boldsymbol{v}\cdot(\boldsymbol{u}\times\boldsymbol{w}) = -\boldsymbol{w}\cdot(\boldsymbol{v}\times\boldsymbol{u})$.

14 Show that $\boldsymbol{i}\times\boldsymbol{j}=\boldsymbol{k}$, $\boldsymbol{j}\times\boldsymbol{k}=\boldsymbol{i}$, and $\boldsymbol{k}\times\boldsymbol{i}=\boldsymbol{j}$.

15 Show that

$$(\boldsymbol{a}\times\boldsymbol{b})\cdot(\boldsymbol{c}\times\boldsymbol{d}) = \det\begin{pmatrix}\boldsymbol{a}\cdot\boldsymbol{c} & \boldsymbol{b}\cdot\boldsymbol{c}\\ \boldsymbol{a}\cdot\boldsymbol{d} & \boldsymbol{b}\cdot\boldsymbol{d}\end{pmatrix}.$$

16 Prove the *Jacobi identity:* For all $\boldsymbol{u}$, $\boldsymbol{v}$, and $\boldsymbol{w}$ in $\boldsymbol{R}^3$,

$$\boldsymbol{u}\times(\boldsymbol{v}\times\boldsymbol{w})+\boldsymbol{v}\times(\boldsymbol{w}\times\boldsymbol{u})+\boldsymbol{w}\times(\boldsymbol{u}\times\boldsymbol{v})=\boldsymbol{0}.$$

17 A collection L of vectors is called a *Lie algebra* if Theorem 1.11 holds for L and if there is a multiplication of vectors defined on L which satisfies, for all $\boldsymbol{u}$, $\boldsymbol{v}$, and $\boldsymbol{w}$ in L and real numbers a and b,
(a) $\boldsymbol{v}\boldsymbol{v}=0$
(b) $(\boldsymbol{u}+\boldsymbol{v})\boldsymbol{w}=\boldsymbol{u}\boldsymbol{w}+\boldsymbol{v}\boldsymbol{w}$ and $\boldsymbol{w}(\boldsymbol{u}+\boldsymbol{v})=\boldsymbol{w}\boldsymbol{u}+\boldsymbol{w}\boldsymbol{v}$
(c) $(a\boldsymbol{v})(b\boldsymbol{w})=ab(\boldsymbol{v}\boldsymbol{w})$
(d) $\boldsymbol{u}(\boldsymbol{v}\boldsymbol{w})+\boldsymbol{v}(\boldsymbol{w}\boldsymbol{u})+\boldsymbol{w}(\boldsymbol{u}\boldsymbol{v})=\boldsymbol{0}$
Show that $\boldsymbol{R}^3$ with the vector cross product is a Lie algebra. (Lie algebras have found application in higher mathematics, nuclear particle theory, quantum mechanics, materials science, and control theory.)

18 If $\boldsymbol{x}=\boldsymbol{x}_0+s\boldsymbol{u}$ and $\boldsymbol{x}=\boldsymbol{x}_1+t\boldsymbol{v}$ are two skew (i.e., nonintersecting, nonparallel lines) in $\boldsymbol{R}^3$, then show that
(a) $\boldsymbol{N}=\dfrac{\boldsymbol{u}\times\boldsymbol{v}}{|\boldsymbol{u}\times\boldsymbol{v}|}$ is a unit vector perpendicular to both lines.
(b) the distance between the lines is $|\boldsymbol{N}\cdot(\boldsymbol{x}_0-\boldsymbol{x}_1)|$. See Figure 5.9.

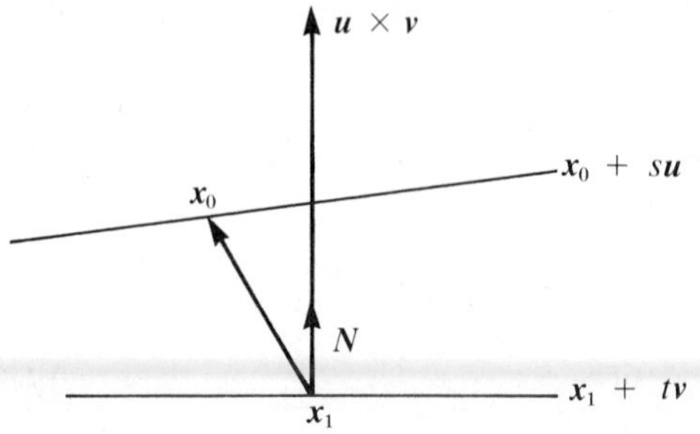

FIGURE 5.9

19 Use Exercise 18 to find the distance between the skew lines $\boldsymbol{x}=s(1,1,1)$ and $\boldsymbol{x}=(-1,0,2)+t(1,1,2)$.

20 For the lines $\boldsymbol{x} = \boldsymbol{0} + t(1, 1, 0)$ and $\boldsymbol{x} = (0, 0, 3) + s(1, -1, 3)$, show that the approach of Example 5.3 would produce the equation $3x - 3y - 2z = 0$. Then show that the points of the second line fail to lie on this plane. (This allows you to conclude that the lines are skew.)

21 Use Theorem 5.5(5) to produce an example showing that the cross product fails to be associative.

6 SYSTEMS OF LINEAR EQUATIONS

In the preceding sections we have seen that the equation of a plane has the form $ax + by + cz = d$, i.e., $\boldsymbol{n} \cdot \boldsymbol{x} = d$, where $\boldsymbol{n} = (a, b, c)$ and $\boldsymbol{x} = (x, y, z)$. By analogy with the case of two variables, a polynomial equation of degree 1 in x, y, and z is called a *linear equation.* The algebra growing out of the study of linear equations is called *linear algebra,* and we begin to meet linear algebra as soon as we try to find algebraic equations for the intersections of planes. We have thus already had a foretaste of this section in Example 4.6 and in the introduction to the cross product in Section 5.

Given two planes $\boldsymbol{n}_1 \cdot \boldsymbol{x} = d_1$ and $\boldsymbol{n}_2 \cdot \boldsymbol{x} = d_2$, we would expect, on geometric grounds, that they would either be coincident or parallel, or intersect in a line. Similarly, we would expect that three given planes $\boldsymbol{n}_1 \cdot \boldsymbol{x} = d_1$, $\boldsymbol{n}_2 \cdot \boldsymbol{x} = d_2$, and $\boldsymbol{n}_3 \cdot \boldsymbol{x} = d_3$ either would not intersect (because two or three of the planes happen to be parallel), would coincide, or would intersect in a line or point. In view of what we saw in Example 4.6, we expect that we ought to be able to go through some kind of algebraic calculation to determine just *which* of the general possibilities actually holds for any given example. We present such a method of calculation in this section, called **Gaussian elimination in matrix form.** This method takes its name from the great German mathematician Karl Friedrich Gauss (1777–1855), who invented this approach as a tool for his work in astronomical calculations. Even with the advent of computers and special numerical algorithms for solving systems of linear equations, Gauss's method remains the most efficient direct method of solution.

For definiteness, we introduce the method for the case of *three* linear equations in three unknowns, although *the method applies equally well to general systems of m linear equations in n unknowns.* Given the system

$$(*) \qquad \begin{cases} a_1x + b_1y + c_1z = d_1 \\ a_2x + b_2y + c_2z = d_2 \\ a_3x + b_3y + c_3z = d_3, \end{cases}$$

the aim of Gaussian elimination is to reduce this system to a system with the same set of solutions and in a simpler form, such as

$$(**) \qquad \begin{cases} \bar{a}_1x \qquad\qquad\qquad = \bar{d}_1 \\ \qquad\quad \bar{b}_2y \qquad\quad = \bar{d}_2 \\ \qquad\qquad\qquad \bar{c}_3z = \bar{d}_3. \end{cases}$$

The motive is clear: a system like (**) can be solved immediately. The question is then, how can we reduce (*) to (**)?

To answer this, we use the following facts.

6.1 THEOREM. The set of solutions to (*) is not changed if
(1) any equation is multiplied through by a nonzero real number;

(2) any two equations are interchanged;
(3) any equation is replaced by the result of adding to it any nonzero multiple of another equation in the system.

The processes (1), (2), and (3) are called **elementary operations.** We will say that two systems with the same solution set are **equivalent.**

Proof. It is easy to see that (1) and (2) will not affect the set of solutions, since they result in systems virtually identical to the original system. To see that (3) does not change the set solutions, consider the equations

(i) $$a_i x + b_i y + c_i z = d_i$$

and

(j) $$a_j x + b_j y + c_j z = d_j$$

If $k \neq 0$, then we claim that a system involving (i) and (j) will have the same set of solutions as a system involving (j) and

$$(\text{i}'):\ (a_i + ka_j)x + (b_i + kb_j)y + (c_i + kc_j)z = d_i + kd_j.$$

For if (x, y, z) is a solution for the system involving (i) and (j), then (x, y, z) is a solution for the system involving (j) and (i′), since (x, y, z) automatically satisfies (j), and we also have

$$(a_i + ka_j)x + (b_i + kb_j)y + (c_i + kc_j)z$$
$$= a_i x + b_i y + c_i z + k(a_j x + b_j y + c_j z) = d_i + kd_j.$$

So (i′) is satisfied. Conversely, if (x, y, z) is a solution of the system involving (i′) and (j), then (x, y, z) automatically satisfies (j), and we also have

$$a_i x + b_i y + c_i z = (a_i + ka_j)x + (b_i + kb_j)y + (c_i + kc_j)z - k(a_j x + b_i y + c_j z)$$
$$= d_i + kd_j - kd_j = d_i.$$

So (i) is satisfied. Thus operation (3) does not affect the solution of a system of linear equations. QED

You may have noticed that it is a bit cumbersome to write the entire system out fully when applying elementary operations to it. To streamline the notation, we write (*) in following matrix form, due to the French mathematician Camille Jordan (1838–1922):

(***) $$\begin{pmatrix} a_1 & b_1 & c_1 & d_1 \\ a_2 & b_2 & c_2 & d_2 \\ a_3 & b_3 & c_3 & d_3 \end{pmatrix}$$

An ***m*-by-*n* matrix** is just an array of numbers having m rows and n columns. (Here $m = 3$ and $n = 4$.) Note that we can regard the rows as vectors in $\boldsymbol{R}^4$ and the columns as vectors in $\boldsymbol{R}^3$. Here we have put the first equation into the first row, the second equation into the second row, and the third equation into the third row. We have written only the *essential* information from each equation: the coefficients. We put the coefficients of x into the first column, the coefficients of y into the second column, the coefficients of z into the third column, and the constants into the last column. We further agree not to operate on the columns in any way, so that when we have completed our reduction of (***), we can easily write down the corresponding system of linear equations (**).

To carry out the reduction process, we use the three **elementary row operations** on matrices which correspond to the elementary operations (1), (2), and (3) on a system of equations in Theorem 6.1. These are:

(1′) The i-th row is multiplied by a nonzero real number c (abbreviation: cR_i).
(2′) Rows i and j are interchanged (abbreviation: R_{ij}).
(3′) Some row vector R_i is replaced by $R_i + cR_j$, the result of adding to R_i a nonzero multiple of the row vector R_j.

The strategy employed in using these operations is the following. First, get ±1 as the entry in row 1, column 1 by using these operations. Then use type (3′) operations to reduce to a matrix with 0 in each entry of column 1 below the first row. Next, get ±1 as the entry in row 2, column 2 or, failing that, as far left as possible. Then use type (3′) operations to obtain zeros above and below that ±1. Next, get ±1 in row 3 as far to the left as possible, and continue as long as you can. When you can no longer proceed without undoing some zeros previously obtained, the resulting matrix will be in what is called *reduced row echelon form.* The corresponding system of equations will be such that its solution set can be read off from the matrix.

Before illustrating the procedure with examples, we mention that any system of linear equations has either a *unique solution, no solution,* or *infinitely many solutions.* The reason for this involves more linear algebra than we can go into here, but can be seen easily in the case of system (*). If we have three equations in x, y, and z, then geometrically they represent three planes. Those planes intersect either not at all or in a unique point, a line, or a plane (see Figure 6.1). Hence the system has either no solution, one solution, or infinitely many solution points. The following three examples illustrate how to determine algebraically precisely which alternative holds true in a particular circumstance.

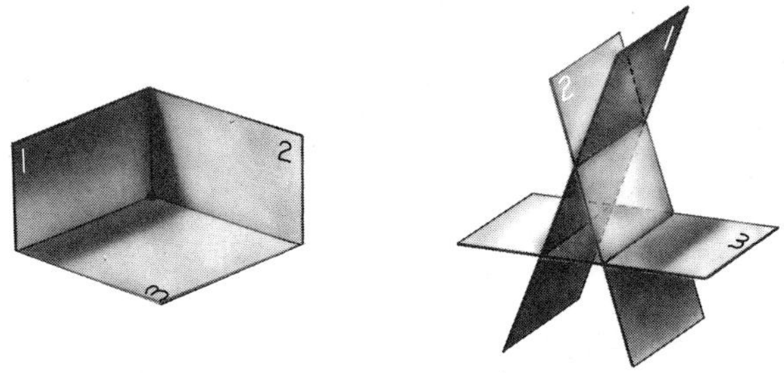

(a) Unique solution; unique point of intersection

(b) No solution; no intersection

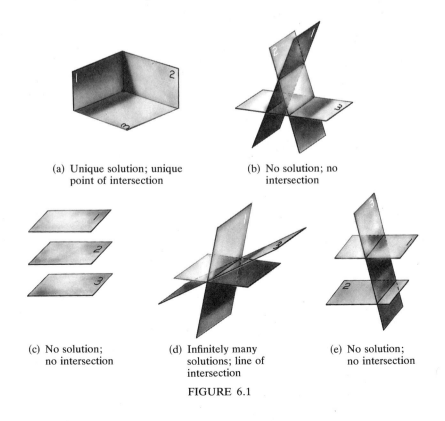

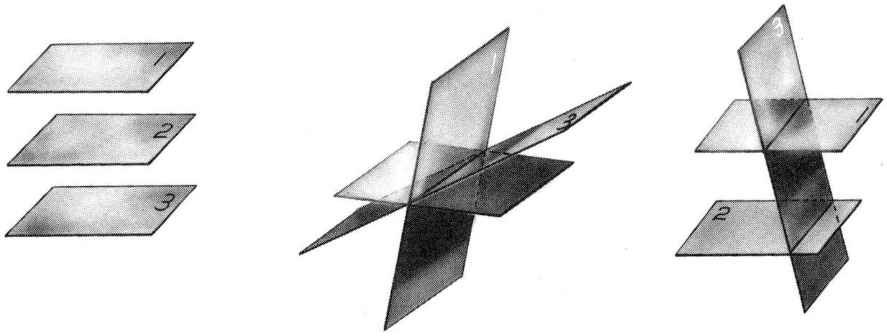

(c) No solution; no intersection

(d) Infinitely many solutions; line of intersection

(e) No solution; no intersection

FIGURE 6.1

6.2 EXAMPLE. Solve the system

$$\begin{cases} 2x-2y+3z=\ \ 9 \\ -2x+3y+5z=-2. \\ -3x+2y+5z=-3 \end{cases}$$

Solution. We carry out the strategy outlined above, *taking care to show what is done at each step. You should be sure to do this in your work, so that it can be followed and checked.* We can obtain a -1 in the $(1, 1)$ position of the matrix by adding R_1 to R_3 and then interchanging the first and third rows. (We could also multiply R_1 by $\frac{1}{2}$, but this would introduce the fractions $\frac{3}{2}$ and $\frac{9}{2}$ into the first row. Your calculations will be simplest if you *avoid* fractions as much as possible.)

$$\begin{pmatrix} 2 & -2 & 3 & 9 \\ -2 & 3 & 5 & -2 \\ -3 & 2 & 5 & -3 \end{pmatrix} \xrightarrow{\text{Add } R_1 \text{ to } R_3} \begin{pmatrix} 2 & -2 & 3 & 9 \\ -2 & 3 & 5 & -2 \\ -1 & 0 & 8 & 6 \end{pmatrix} \xrightarrow{R_{13}}$$

$$\begin{pmatrix} -1 & 0 & 8 & 6 \\ -2 & 3 & 5 & -2 \\ 2 & -2 & 3 & 9 \end{pmatrix}.$$

Now we can obtain 0's below the $(1, 1)$ entry by adding $-2R_1$ to R_2 and $2R_1$ to R_3.

$$\begin{pmatrix} -1 & 0 & 8 & 6 \\ -2 & 3 & 5 & -2 \\ 2 & -2 & 3 & 9 \end{pmatrix} \xrightarrow[\text{Add } -2R_1 \text{ to } R_2]{\text{Add } 2R_1 \text{ to } R_3}$$

$$\begin{pmatrix} -1 & 0 & 8 & 6 \\ 0 & 3 & -11 & -14 \\ 0 & -2 & 19 & 21 \end{pmatrix} \xrightarrow{\text{Add } R_3 \text{ to } R_2} \begin{pmatrix} -1 & 0 & 8 & 6 \\ 0 & 1 & 8 & 7 \\ 0 & -2 & 19 & 21 \end{pmatrix}.$$

Having obtained a 1 in the $(2, 2)$ entry by adding R_3 to R_2, we can obtain a 0 in the $(3, 2)$ entry by adding $2R_2$ to R_3.

$$\begin{pmatrix} -1 & 0 & 8 & 6 \\ 0 & 1 & 8 & 7 \\ 0 & -2 & 19 & 21 \end{pmatrix} \xrightarrow{\text{Add } 2R_2 \text{ to } R_3} \begin{pmatrix} -1 & 0 & 8 & 6 \\ 0 & 1 & 8 & 7 \\ 0 & 0 & 35 & 35 \end{pmatrix} \xrightarrow{\frac{1}{35}R_3}$$

$$\begin{pmatrix} -1 & 0 & 8 & 6 \\ 0 & 1 & 8 & 7 \\ 0 & 0 & 1 & 1 \end{pmatrix}.$$

Now with a 1 in the $(3, 3)$ position, it is easy to obtain 0's above it by adding $-8R_3$ to R_1, and then to R_2.

$$\begin{pmatrix} -1 & 0 & 8 & 6 \\ 0 & 1 & 8 & 7 \\ 0 & 0 & 1 & 1 \end{pmatrix} \xrightarrow[\text{Add } -8R_3 \text{ to } R_2]{\text{Add } -8R_3 \text{ to } R_1} \begin{pmatrix} -1 & 0 & 0 & -2 \\ 0 & 1 & 0 & -1 \\ 0 & 0 & 1 & 1 \end{pmatrix}.$$

Thus a system with the same solution as the original system is $-x=-2$, $y=-1$,

$z=1$. Hence $\boldsymbol{x}_0=(2,-1,1)$ is the solution to the given system. Geometrically, the three planes $(2,-2,3)\cdot\boldsymbol{x}=9$, $(-2,3,5)\cdot\boldsymbol{x}=-2$, and $(-3,2,5)\cdot\boldsymbol{x}=-3$ intersect in the point $(2,-1,1)$.

We next consider an example which has no solution.

6.3 EXAMPLE. Find the intersection of the three planes $(3,2,-1)\cdot\boldsymbol{x}=1$, $(-1,-\frac{2}{3},\frac{1}{3})\cdot\boldsymbol{x}=2$, and $(\frac{3}{2},1,-\frac{1}{2})\cdot\boldsymbol{x}=6$.

Solution. We have

$$\begin{pmatrix} 3 & 2 & -1 & 1 \\ -1 & -\frac{2}{3} & \frac{1}{3} & 2 \\ \frac{3}{2} & 1 & -\frac{1}{2} & 6 \end{pmatrix} \xrightarrow{R_{12}} \begin{pmatrix} -1 & -\frac{2}{3} & \frac{1}{3} & 2 \\ 3 & 2 & -1 & 1 \\ \frac{3}{2} & 1 & -\frac{1}{2} & 6 \end{pmatrix} \xrightarrow[\text{Add } \frac{3}{2}R_1 \text{ to } R_3]{\text{Add } 3R_1 \text{ to } R_2} \begin{pmatrix} -1 & -\frac{2}{3} & -\frac{1}{3} & 2 \\ 0 & 0 & 0 & 7 \\ 0 & 0 & 0 & 9 \end{pmatrix}.$$

Thus the original system has the same set of solutions as the system $-x-\frac{2}{3}y-\frac{1}{3}z=2$, $0x+0y+0z=7$, and $0x+0y+0z=9$. This latter system of equations has only the empty set as its set of solutions. Hence the original system has no solution. (All three planes are parallel, as you should verify.)

Now we consider an example in which the three planes intersect in a line.

6.4 EXAMPLE. Solve the system

$$\text{(1)}\qquad \begin{cases} x-y+\ \ z=2 \\ x+y+2z=3 \\ 2x+\qquad 3z=5 \end{cases}$$

Solution. Our reduction strategy gives

$$\begin{pmatrix} 1 & -1 & 1 & 2 \\ 1 & 1 & 2 & 3 \\ 2 & 0 & 3 & 5 \end{pmatrix} \xrightarrow[\text{Add } -2R_1 \text{ to } R_3]{\text{Add } -R_1 \text{ to } R_2} \begin{pmatrix} 1 & -1 & 1 & 2 \\ 0 & 2 & 1 & 1 \\ 0 & 2 & 1 & 1 \end{pmatrix} \xrightarrow{\text{Add } -R_2 \text{ to } R_3}$$

$$\begin{pmatrix} 1 & -1 & 1 & 2 \\ 0 & 2 & 1 & 1 \\ 0 & 0 & 0 & 0 \end{pmatrix} \xrightarrow{\frac{1}{2}R_2} \begin{pmatrix} 1 & -1 & 1 & 2 \\ 0 & 1 & \frac{1}{2} & \frac{1}{2} \\ 0 & 0 & 0 & 0 \end{pmatrix} \xrightarrow{\text{Add } R_2 \text{ to } R_1} \begin{pmatrix} 1 & 0 & \frac{3}{2} & \frac{5}{2} \\ 0 & 1 & \frac{1}{2} & \frac{1}{2} \\ 0 & 0 & 0 & 0 \end{pmatrix}.$$

So the original system is equivalent to the system

$$\begin{cases} x \qquad +\frac{3}{2}z=\frac{5}{2} \\ \qquad y+\frac{1}{2}z=\frac{1}{2}. \\ \qquad\qquad 0=0 \end{cases}$$

This last system is easy to solve. The first two equations say that

$$\begin{cases} x = \frac{5}{2} - \frac{3}{2}z, \\ y = \frac{1}{2} - \frac{1}{2}z. \end{cases}$$

None of the equations restrict z at all, so z is *arbitrary*. If we set $z = t$, then we obtain

$$\begin{cases} x = \frac{5}{2} - \frac{3}{2}t \\ y = \frac{1}{2} - \frac{1}{2}t. \\ z = \quad t \end{cases}$$

In vector form, $\mathbf{x} = (x, y, z) = (\frac{5}{2}, \frac{1}{2}, 0) + t(-\frac{3}{2}, -\frac{1}{2}, 1)$, which we recognize as the equation of the line ℓ through $(\frac{5}{2}, \frac{1}{2}, 0)$ in the direction of $(-\frac{3}{2}, -\frac{1}{2}, 1)$.

In Example 6.4, we actually have more information in the solution than may at first be apparent. Geometrically, it is clear that if we translate the three given planes parallel to themselves until they pass through the origin, then the translated planes intersect in a line through the origin parallel to the line ℓ. (See Figure 6.2.)

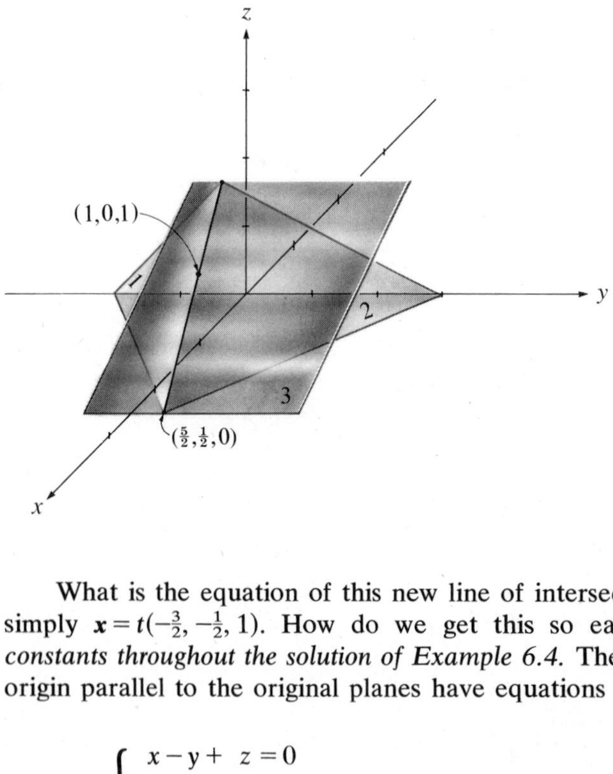

FIGURE 6.2

What is the equation of this new line of intersection? Its vector equation is simply $\mathbf{x} = t(-\frac{3}{2}, -\frac{1}{2}, 1)$. How do we get this so easily? We simply *delete the constants throughout the solution of Example 6.4.* The planes passing through the origin parallel to the original planes have equations

$$(2) \qquad \begin{cases} x - y + z = 0 \\ x + y + 2z = 0 \\ 2x \quad + 3z = 0 \end{cases}$$

since these planes all pass through (0, 0, 0) and have the same respective normal vectors as those in Example 6.4. So if we solved the system (2) by our reduction scheme, then everything would be as in Example 6.4, except that the last column would be changed to a column of zeros. This column would not be altered in the reduction process. Thus we would obtain the solution

$$\boldsymbol{x} = (0, 0, 0) + t(-\tfrac{3}{2}, -\tfrac{1}{2}, 1) = t(-\tfrac{3}{2}, -\tfrac{1}{2}, 1).$$

Geometrically, this is a line through the origin. *Algebraically*, it is a *real vector space*. This follows easily from Definition 1.14; observe that the sum of two vectors of this form and all real multiples of such vectors are again of this form:

$$t_1(-\tfrac{3}{2}, -\tfrac{1}{2}, 1) + t_2(-\tfrac{3}{2}, -\tfrac{1}{2}, 1) = (t_1 + t_2)(-\tfrac{3}{2}, -\tfrac{1}{2}, 1)$$
$$r[t(-\tfrac{3}{2}, -\tfrac{1}{2}, 1)] = rt(-\tfrac{3}{2}, -\tfrac{1}{2}, 1).$$

Since $\mathbf{0} = 0(-\tfrac{3}{2}, -\tfrac{1}{2}, 1)$ and $-t(-\tfrac{3}{2}, -\tfrac{1}{2}, 1)$ are also of this form and since properties (1), (2), and (5) through (9) of Theorem 1.11 can be shown to hold, $\{t(-\tfrac{3}{2}, -\tfrac{1}{2}, 1) \mid t \in \boldsymbol{R}\}$ is a real vector space by Definition 1.14.

The foregoing is not accidental, but rather illustrates general behavior which we will describe carefully.

6.5 DEFINITION. A **homogeneous system** of linear equations is a system such that every equation has the form

$$a_1x_1 + a_2x_2 + \ldots + a_nx_n = 0.$$

Notice that a homogeneous system always has at least the one solution $x_1 = x_2 = \ldots = x_n = 0$, called the *trivial solution.*

The next theorem is proved by using exactly the same reasoning set forth above for the homogeneous system corresponding to Example 6.4. (Exercise 19 below asks you to write out the proof.)

6.6 THEOREM. The set K of solutions to any homogeneous system of linear equations forms a real vector space.

Recall in Example 6.4 that the general solution had the form $\boldsymbol{x} = \boldsymbol{x}_p + \boldsymbol{k}$, where $\boldsymbol{x}_p = (\tfrac{5}{2}, \tfrac{1}{2}, 0)$ was one particular solution and $\boldsymbol{k} = t(-\tfrac{3}{2}, -\tfrac{1}{2}, 1)$ was any vector in K, the real vector space of solutions to the corresponding homogeneous system. We can now show that the general solution *always* has that form.

6.7 THEOREM. Let

$$\text{(3)} \quad \begin{cases} a_{11}x_1 + a_{12}x_2 + \ldots + a_{1n}x_n = d_1 \\ a_{21}x_1 + a_{22}x_2 + \ldots + a_{2n}x_n = d_2 \\ \vdots \qquad \vdots \qquad \vdots \qquad \vdots \quad \vdots \\ a_{m1}x_1 + a_{m2}x_2 + \ldots + a_{mn}x_n = d_m \end{cases}$$

be a system of m linear equations in n unknowns. Let $\boldsymbol{x}_p = (x_{p1}, x_{p2}, \ldots, x_{pn})$ be a (fixed) particular solution. Let K be the set of all solutions to the associated homogeneous system which results when each d_i is replaced by 0, for $i = 1, 2, \ldots, m$. Then the set S of all solutions to (3) is given by

$$\boxed{S = \{\boldsymbol{x} = \boldsymbol{x}_p + \boldsymbol{k} \mid \boldsymbol{k} \in K\}.}$$

Proof. We again give the proof in the case of $n=3$ variables, but the same reasoning applies in general. Here (3) takes the form

$$\begin{cases} a_1x+b_1y+c_1z=d_1 \\ a_2x+b_2y+c_2z=d_2. \\ a_3x+b_3y+c_3z=d_3 \end{cases} \tag{4}$$

Let $\boldsymbol{x}_p=(x_p, y_p, z_p)$. We must first show that if $\boldsymbol{k}=(k_1, k_2, k_3)\in K$, then $\boldsymbol{x}_p+\boldsymbol{k}$ actually is a solution to (4). That is easy, since for any equation (i), $1\le i\le m$, we have

$$\begin{aligned} &a_i(x_p+k_1)+b_i(y_p+k_2)+c_i(z_p+k_3) \\ &\quad=a_ix_p+a_ik_1+b_iy_p+b_ik_2+c_iz_p+c_ik_3 \\ &\quad=a_ix_p+b_iy_p+c_iz_p+a_ik_1+b_ik_2+c_ik_3 \\ &\quad=d_i+0=d_i. \end{aligned}$$

So $\boldsymbol{x}_p+\boldsymbol{k}$ is a solution to each equation (i). Conversely, we must show that *every* solution to (4) has the form $\boldsymbol{x}_p+\boldsymbol{k}$, where $\boldsymbol{k}$ is some member of K. For this let $\boldsymbol{x}_1=(x_1, y_1, z_1)$ be *any* solution to (4). Let $\boldsymbol{k}=\boldsymbol{x}_1-\boldsymbol{x}_p=(x_1-x_p, y_1-y_p, z_1-z_p)$. Then we certainly have $\boldsymbol{x}_1=\boldsymbol{x}_p+\boldsymbol{k}$. Thus our proof will be complete if we can show that $\boldsymbol{k}$ belongs to K. To do this, we check that $\boldsymbol{k}$ satisfies each equation (i) of the homogeneous system, $1\le i\le m$:

$$\begin{aligned} &a_i(x_1-x_p)+b_i(y_1-y_p)+c_i(z_1-z_p) \\ &\quad=a_ix_1-a_ix_p+b_iy_1-b_iy_p+c_iz_1-c_iz_p \\ &\quad=a_ix_1+b_iy_1+c_iz_1-(a_ix_p+b_iy_p+c_iz_p) \\ &\quad=d_i-d_i=0 \end{aligned}$$

(since $\boldsymbol{x}_1$ and $\boldsymbol{x}_p$ are both solutions to the given system). Thus $\boldsymbol{k}$ *is* a solution to the homogeneous system. So we have shown that any solution $\boldsymbol{x}_1$ of the given system has the required form $\boldsymbol{x}_p+\boldsymbol{k}$, where $\boldsymbol{k}$ is a solution to the homogeneous system. QED

In Example 6.4 we have $\boldsymbol{x}_p=(\frac{5}{2}, \frac{1}{2}, 0)$, and $\boldsymbol{k}$ is any multiple of $(-\frac{3}{2}, -\frac{1}{2}, 1)$. Notice that $\boldsymbol{x}_p$ is not unique: it can be *any* solution to the given system. In Example 6.4 one could use $(1, 0, 1)$ for $\boldsymbol{x}_p$ instead of $(\frac{5}{2}, \frac{1}{2}, 0)$. We used the latter since our technique brought us to this particular solution. A good guesser might have gotten $(1, 0, 1)$ much faster than we got $(\frac{5}{2}, \frac{1}{2}, 0)$, but our technique has the advantage of general applicability. Note also that $(-\frac{3}{2}, -\frac{1}{2}, 1)$ is not unique either. *Any* nonzero multiple of it (such as $(3, 1, -2)$) would serve just as well. But again, $(-\frac{3}{2}, -\frac{1}{2}, 1)$ is the vector our technique produced, so we might as well make use of it. The next two examples illustrate Theorem 6.7.

6.8 EXAMPLE. For the system

$$\begin{cases} 2x-3y-z=1 \\ x+y+2z=-1 \end{cases}$$

(a) Find a particular solution.
(b) Find the complete solution to the associated homogeneous system.

(c) Find the complete solution to the given system, i.e., the line of intersection of the two given planes.

Solution. We can do all three parts at once.

$$\begin{pmatrix} 2 & -3 & -1 & 1 \\ 1 & 1 & 2 & -1 \end{pmatrix} \xrightarrow{R_{12}} \begin{pmatrix} 1 & 1 & 2 & -1 \\ 2 & -3 & -1 & 1 \end{pmatrix} \xrightarrow{\text{Add } -2R_1 \text{ to } R_2}$$

$$\begin{pmatrix} 1 & 1 & 2 & -1 \\ 0 & -5 & -5 & 3 \end{pmatrix} \xrightarrow{-\frac{1}{5}R_2} \begin{pmatrix} 1 & 1 & 2 & -1 \\ 0 & 1 & 1 & -\frac{3}{5} \end{pmatrix} \xrightarrow{\text{Add } -R_2 \text{ to } R_1}$$

$$\begin{pmatrix} 1 & 0 & 1 & -\frac{2}{5} \\ 0 & 1 & 1 & -\frac{3}{5} \end{pmatrix}.$$

So the original system is equivalent to the system

$$\begin{cases} x + z = -\frac{2}{5} \\ y + z = -\frac{3}{5}, \end{cases} \qquad \text{i.e.,} \begin{cases} x = -\frac{2}{5} - t \\ y = -\frac{3}{5} - t \\ z = \phantom{-\frac{3}{5} -{}} t \end{cases}.$$

Thus the answer to (c) is

$$\boldsymbol{x} = (-\tfrac{2}{5}, -\tfrac{3}{5}, 0) + t(-1, -1, 1), \quad \text{a line.}$$

An answer to (a) is $(-\frac{2}{5}, -\frac{3}{5}, 0)$, and an answer to (b) is $\{t(-1, -1, 1) \mid t \in \boldsymbol{R}\}$.

6.9 EXAMPLE. Solve the following system of linear equations in $\boldsymbol{R}^5$.

$$\begin{cases} x_1 + 2x_2 - 3x_3 + 6x_4 - 11x_5 = 0 \\ -2x_1 - 3x_2 + 4x_3 - 10x_4 + 15x_5 = 0 \\ -x_1 - 3x_4 - x_5 = 1 \\ x_1 + 3x_2 - 3x_3 + 6x_4 - 14x_5 = 2 \\ -x_1 - x_2 + x_3 - 4x_4 + 4x_5 = 0 \end{cases}$$

Solution. We use our reduction strategy.

$$\begin{pmatrix} 1 & 2 & -3 & 6 & -11 & 0 \\ -2 & -3 & 4 & -10 & 15 & 0 \\ -1 & 0 & 0 & -3 & -1 & 1 \\ 1 & 3 & -3 & 6 & -14 & 2 \\ -1 & -1 & 1 & -4 & 4 & 0 \end{pmatrix} \xrightarrow[\substack{\text{Add } -R_1 \text{ to } R_4 \\ \text{Add } R_1 \text{ to } R_5}]{\substack{\text{Add } 2R_1 \text{ to } R_2 \\ \text{Add } R_1 \text{ to } R_3}}$$

$$\begin{pmatrix} 1 & 2 & -3 & 6 & -11 & 0 \\ 0 & 1 & -2 & 2 & -7 & 0 \\ 0 & 2 & -3 & 3 & -12 & 1 \\ 0 & 1 & 0 & 0 & -3 & 2 \\ 0 & 1 & -2 & 2 & -7 & 0 \end{pmatrix} \xrightarrow{R_{24}}$$

$$\begin{pmatrix} 1 & 2 & -3 & 6 & -11 & 0 \\ 0 & 1 & 0 & 0 & -3 & 2 \\ 0 & 2 & -3 & 3 & -12 & 1 \\ 0 & 1 & -2 & 2 & -7 & 0 \\ 0 & 1 & -2 & 2 & -7 & 0 \end{pmatrix} \xrightarrow[\text{Add } -R_4 \text{ to } R_5]{\substack{\text{Add } -2R_2 \text{ to } R_1 \\ \text{Add } -2R_2 \text{ to } R_3}}$$

$$\begin{pmatrix} 1 & 0 & -3 & 6 & -5 & -4 \\ 0 & 1 & 0 & 0 & -3 & 2 \\ 0 & 0 & -3 & 3 & -6 & -3 \\ 0 & 1 & -2 & 2 & -7 & 0 \\ 0 & 0 & 0 & 0 & 0 & 0 \end{pmatrix} \xrightarrow[\text{Add } -R_2 \text{ to } R_4]{-\frac{1}{3}R_3}$$

$$\begin{pmatrix} 1 & 0 & -3 & 6 & -5 & -4 \\ 0 & 1 & 0 & 0 & -3 & 2 \\ 0 & 0 & 1 & -1 & 2 & 1 \\ 0 & 0 & -2 & 2 & -4 & -2 \\ 0 & 0 & 0 & 0 & 0 & 0 \end{pmatrix} \xrightarrow[\text{Add } 2R_3 \text{ to } R_4]{\text{Add } 3R_3 \text{ to } R_1}$$

$$\begin{pmatrix} 1 & 0 & 0 & 3 & 1 & -1 \\ 0 & 1 & 0 & 0 & -3 & 2 \\ 0 & 0 & 1 & -1 & 2 & 1 \\ 0 & 0 & 0 & 0 & 0 & 0 \\ 0 & 0 & 0 & 0 & 0 & 0 \end{pmatrix}.$$

So we have as an equivalent reduced system

$$\begin{cases} x_1 \qquad\qquad + 3x_4 + x_5 = -1 \\ \qquad x_2 \qquad\qquad - 3x_5 = 2 \\ \qquad\qquad x_3 - x_4 + 2x_5 = 1. \end{cases}$$

This is equivalent to

$$\begin{cases} x_1 = -1 - 3x_4 - x_5 \\ x_2 = 2 \qquad + 3x_5 \\ x_3 = 1 + x_4 - 2x_5 \\ x_4 = \qquad x_4 \\ x_5 = \qquad\qquad x_5. \end{cases}$$

Since there is no restriction on x_4 and x_5, we may set $x_4 = s$ and $x_5 = t$, where s and t are real parameters. If we write the above in vector form, then we have

$$\begin{pmatrix} x_1 \\ x_2 \\ x_3 \\ x_4 \\ x_5 \end{pmatrix} = \begin{pmatrix} -1 \\ 2 \\ 1 \\ 0 \\ 0 \end{pmatrix} + s\begin{pmatrix} -3 \\ 0 \\ 1 \\ 1 \\ 0 \end{pmatrix} + t\begin{pmatrix} -1 \\ 3 \\ -2 \\ 0 \\ 1 \end{pmatrix}.$$

We can imagine this as the parametric equation of a plane through the point $(-1, 2, 1, 0, 0)$ in $\boldsymbol{R}^5$ determined by the vectors $(-3, 0, 1, 1, 0)$ and $(-1, 3, -2, 0, 1)$. The homogeneous problem has solution space

$$K = \{s(-3, 0, 1, 1, 0) + t(-1, 3, -2, 0, 1) \mid s, t \in \mathbf{R}\}$$

and a particular solution is $(-1, 2, 1, 0, 0)$.

While we may have a hard time visualizing the geometry of Example 6.9, problems of this very sort often arise in applications. In many circumstances in physics, engineering, business, and economics, the number of variables cannot be held to two or three. In fact, problems involving hundreds or even thousands of linear equations and variables can arise. Such systems must be handled by computers. We indicate some of the applications of systems of linear equations in Exercises 12 to 18 below. For simplicity, we limit ourselves to three or four variables.

EXERCISES 1.6

Find for each system (a) the complete solution to the associated homogeneous system, (b) a particular solution if any exists, and (c) the complete solution. (If the system has no solution, then just remark that the solution set in (c) is empty.)

1 $\begin{cases} x - y + 2z = -2 \\ 3x - 2y + 4z = -5 \\ 2y - 3z = 2 \end{cases}$

2 $\begin{cases} x + z = 1 \\ y - z = -1 \\ 2x + 2y + z = 2 \end{cases}$

3 $\begin{cases} x + 2y + z = -1 \\ y + 2z = 0 \\ x + 3y + 3z = -4 \end{cases}$

4 $\begin{cases} x + y + 2z = 2 \\ x - y + z = 4 \\ 3x + y + 5z = 6 \end{cases}$

5 $\begin{cases} x - y + 2z = 7 \\ 2x + 3y - z = -11 \end{cases}$

6 $\begin{cases} x + y + 3z = 8 \\ y + 3z = 6 \\ x + 2y + 6z = 14 \end{cases}$

7 $\begin{cases} x - 2y - 3z = 2 \\ x - 4y - 13z = 14 \\ -3x + 5y + 4z = 0 \end{cases}$

8 $\begin{cases} 2x - 3y + z = -1 \\ -3x + \frac{9}{2}y - \frac{3}{2}z = \frac{3}{2} \\ -6x + 9y - 3z = 3 \end{cases}$

9 $\begin{cases} x + y - 3z + w = 1 \\ 2x - 4y + 2w = 2 \\ 3x - 4y - 2z = 0 \\ x - 2z + 3w = 3 \end{cases}$

10 $\begin{cases} x_1 + 2x_2 - 2x_3 + 3x_4 - 4x_5 = -3 \\ 2x_1 + 4x_2 - 5x_3 + 6x_4 - 5x_5 = -1 \\ -x_1 - 2x_2 - 3x_4 + 11x_5 = 15 \end{cases}$

11 $\begin{cases} x_1 + 2x_2 - x_3 - 5x_4 + 2x_5 = -3 \\ x_2 + x_3 - 2x_4 - 4x_5 = 1 \\ 2x_1 - 3x_2 + 2x_3 + 4x_4 - x_5 = 9 \end{cases}$

12 A tool plant makes three kinds of tools, each of which requires certain amounts of iron, aluminum, and nickel, as specified in the table.

	T_1	T_2	T_3
Iron	2	4	3
Aluminum	2	1	2
Nickel	2	1	3

An industrial spy working for a competing plant reports that the plant buys 72,000 units of iron, 48,000 units of aluminum, and 60,000 units of nickel per week. Set up and solve a system of equations which will enable the competing plant to calculate the weekly production of T_1, T_2, and T_3. (Assume that all the raw materials are used each week.)

13 A hot dog manufacturing plant makes its product from pork, beef, and a cereal filler. The amounts of protein and fat, in grams per ounce, are given in the table.

	Protein	Fat
Pork	5	3
Beef	6	2
Filler	3	2

The company wants its hot dogs to provide 65 grams of protein and 36 grams of fat per pound. Find the number of ounces of pork, beef, and cereal that should go into each pound of hot dogs to achieve this.

14 In Exercise 13, suppose that only pork and beef are used to make *all meat* hot dogs. Is it possible to achieve exactly 65 grams of protein and 36 grams of fat per pound of hot dogs?

15 In electrical theory, *Kirchhoff's Laws* state that in a closed circuit (a) the algebraic sum of all currents i meeting at any branch point is 0, and (b) the sum of the voltage drops R_i in any loop is the algebraic sum of the voltages E in that branch. Here current i is in amps, R is in ohms, and E is in volts. Consider the circuit in Figure 6.3. Apply Kirchhoff's law (a) to branchpoints A and B and (b) to each branch to get a system of equations in i_1, i_2, and i_3. Solve this system to find i_1, i_2 and i_3.

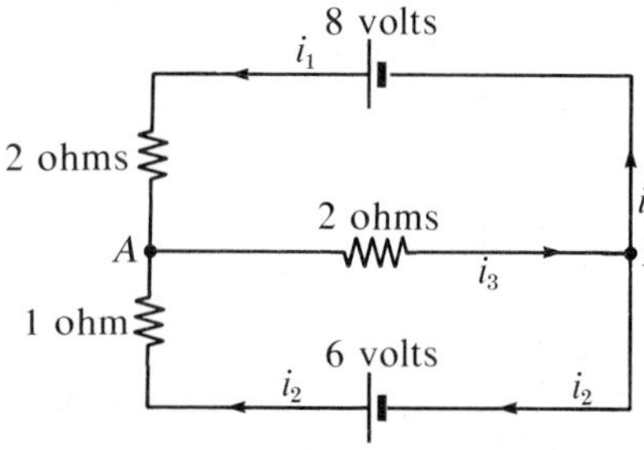

FIGURE 6.3

16 In traffic flows, it is desirable to avoid congestion at intersections. A condition analogous to Kirchhoff's Laws for electricity is that *the number of cars entering an intersection every hour must equal the number of cars leaving that intersection per hour.* In Figure 6.4 is shown a grid of one-way streets in the center of a city. At rush-hour,

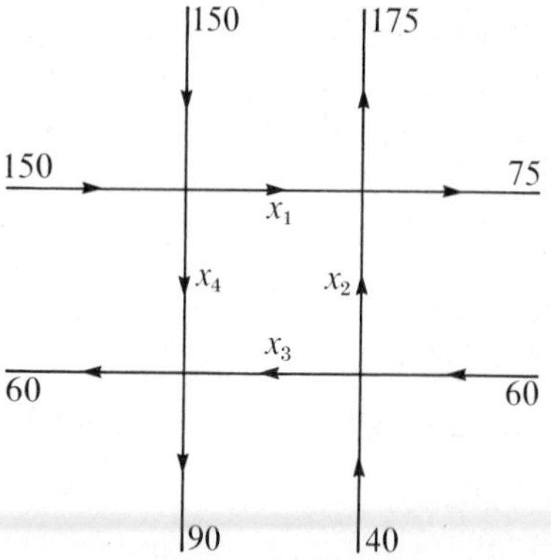

FIGURE 6.4

traffic counters establish the loads shown. What is the possible range of values for x_4 in order for x_1, x_2, and x_3 to all be positive? (If any x_i becomes negative, then disaster has struck, because drivers are proceeding, in their desperation, down one-way streets the wrong way!)

17 The *Leontief economic model* (named for Wassily W. Leontief (1906–) of Harvard University, the 1973 Nobel laureate in Economics) can be described as follows for an economy with three interdependent industries. Each industry's production (output) is used by itself and possibly the other two industries as well. In addition, the consuming public buys each industry's output. Let a_{ij} be the number of dollars worth of production from industry i needed to produce one dollar's worth of the product of industry j. Let x_i be the output of industry i, measured also in dollars. Then $a_{ij}x_j$ is the total dollar value of the production from industry i used by industry j.

(a) What total dollar value of production from each industry is needed by all industries in the economy?

(b) If the consuming public buys b_i dollars worth of industry i's output, then what is the total output in dollars required from industry i to meet consumer and industrial demand?

(c) By equating each x_i to the expression in (b), generate a system of linear equations which describe this economy, and write it in standard form.

18 Refer to Exercise 17. Suppose that the coefficients a_{ij} are given by

$$\begin{pmatrix} 0.2 & 0.4 & 0.2 \\ 0 & 0.2 & 0.4 \\ 0 & 0 & 0.4 \end{pmatrix}$$

and the consuming public demand is given by

$$\begin{pmatrix} 6 \\ 4 \\ 12 \end{pmatrix}$$

in thousands of dollars. Find the output levels required to meet industrial and consumer demand.

19 Prove Theorem 6.6.

7 LINEAR INDEPENDENCE AND DIMENSION[1]

We have an intuitive geometric idea of dimension. A line is one-dimensional, the plane two-dimensional, and space three-dimensional. Can some kind of algebraic definition of dimension be given? It turns out that one *can* be given for real vector spaces, and its determination involves systems of linear equations. This notion is also of wide application outside mathematics. (See Exercises 12 to 18 below.)

7.1 **DEFINITION.** A nonempty set $S=\{\boldsymbol{v}_1, \boldsymbol{v}_2, \ldots, \boldsymbol{v}_n\}$ of vectors in a real vector space V is **linearly dependent** if one of the vectors $\boldsymbol{v}_i$ can be expressed as a **linear combination** of the other vectors, i.e., for some $i \in \{1, 2, \ldots, n\}$,

$$\boldsymbol{v}_i = a_1\boldsymbol{v}_1 + \ldots + a_{i-1}\boldsymbol{v}_{i-1} + a_{i+1}\boldsymbol{v}_{i+1} + \ldots + a_n\boldsymbol{v}_n.$$

Otherwise, the set of vectors is **linearly independent.**

For instance, if $\boldsymbol{v}_1=(1, 2, 3)$, $\boldsymbol{v}_2=(-1, 1, 0)$, and $\boldsymbol{v}_3=(-3, 1, -2)$, then

1. This section is optional at this stage. It may be deferred until you start to read Chapter 9.

$\{\boldsymbol{v}_1, \boldsymbol{v}_2, \boldsymbol{v}_3\}$ is linearly dependent. This is so because $\boldsymbol{v}_1 = \frac{7}{2}\boldsymbol{v}_2 - \frac{3}{2}\boldsymbol{v}_3$, as you should verify on scratch paper.

In this section we will develop a computational procedure to test sets of vectors for linear dependence. It will also enable you to find coefficients (like $\frac{7}{2}$ and $-\frac{3}{2}$ above) which give one vector (like $\boldsymbol{v}_1$ above) in a linearly dependent set as a linear combination of the others. The next result is fundamental to that computational technique.

7.2 **THEOREM.** Linear dependence of a set $S = \{\boldsymbol{v}_1, \boldsymbol{v}_2, \ldots, \boldsymbol{v}_n\}$ of distinct vectors is equivalent to the existence of a solution other than the trivial solution $(0, 0, \ldots, 0)$ to the homogeneous linear equation

(1) $$x_1\boldsymbol{v}_1 + x_2\boldsymbol{v}_2 + \ldots + x_n\boldsymbol{v}_n = \boldsymbol{0}.$$

Proof. First, if we can express $\boldsymbol{v}_i = \sum_{j \neq i} a_j \boldsymbol{v}_j$ for some i, then a nontrivial solution to (1) is given by

$$\boldsymbol{x}_0 = (a_1, \ldots, a_{i-1}, -1, a_{i+1}, \ldots, a_n).$$

Conversely, suppose that there is a nontrivial solution $\boldsymbol{x}_1 = (b_1, b_2, \ldots, b_n)$ to (1). Then at least one of the real numbers $b_i \neq 0$. Then we have

(2) $$b_i\boldsymbol{v}_i = -(b_1\boldsymbol{v}_1 + \ldots + b_{i-1}\boldsymbol{v}_{i-1} + b_{i+1}\boldsymbol{v}_{i+1} + \ldots + b_n\boldsymbol{v}_n).$$

Since $b_i \neq 0$, we can multiply both sides of (2) by $1/b_i$, and use Theorem 1.11 to get

$$\boldsymbol{v}_i = -\left(\frac{b_1}{b_i}\boldsymbol{v}_1 + \ldots + \frac{b_{i-1}}{b_i}\boldsymbol{v}_{i-1} + \frac{b_{i+1}}{b_i}\boldsymbol{v}_{i+1} + \ldots + \frac{b_n}{b_i}\boldsymbol{v}_n\right).$$

Thus we have $\boldsymbol{v}_i$ expressed as a linear combination of the other vectors in S, so S is linearly dependent. Thus S is linearly dependent if and only if (1) has a nontrivial solution.
QED

7.3 **COROLLARY.** A set S of vectors is linearly independent if and only if the sole solution to (1) is the trivial solution $\boldsymbol{x}_0 = (0, 0, \ldots, 0)$.

7.4 **EXAMPLE.** Show that $\{\boldsymbol{i}, \boldsymbol{j}, \boldsymbol{k}\}$ is linearly independent.

Proof. Consider (1) in the form $x\boldsymbol{i} + y\boldsymbol{j} + z\boldsymbol{k} = \boldsymbol{0}$. This is simply $(x, y, z) = (0, 0, 0)$, which clearly has only the trivial solution. Hence, by Corollary 7.3, $\{\boldsymbol{i}, \boldsymbol{j}, \boldsymbol{k}\}$ is linearly independent.
QED

7.5 **DEFINITION.** A real vector space V is said to **have dimension n** (or **be n-dimensional**) if there is some linearly independent set of n vectors in V, but no linearly independent set of more than n vectors.

7.6 **EXAMPLE.** Show that $\boldsymbol{R}^3$ has dimension 3.

Proof. We saw in Example 7.4 that $\boldsymbol{R}^3$ has at least one linearly independent set of three vectors, namely $\{\boldsymbol{i}, \boldsymbol{j}, \boldsymbol{k}\}$. It remains to show that any set of more than three vectors must be linearly dependent. For this let $S = \{\boldsymbol{v}_1, \boldsymbol{v}_2, \boldsymbol{v}_3, \boldsymbol{v}_4, \ldots\}$ be any set of more than three vectors. Here let $\boldsymbol{v}_1 = (a_1, a_2, a_3)$, $\boldsymbol{v}_2 = (b_1, b_2, b_3)$, $\boldsymbol{v}_3 = (c_1, c_2, c_3)$, $\boldsymbol{v}_4 = (d_1, d_2, d_3)$, etc. Then to test S for linear dependence, we must consider (1) in the form

(3) $$x_1\boldsymbol{v}_1 + x_2\boldsymbol{v}_2 + x_3\boldsymbol{v}_3 + x_4\boldsymbol{v}_4 + \ldots = \boldsymbol{0}, \text{ i.e.,}$$
$$x_1(a_1, a_2, a_3) + x_2(b_1, b_2, b_3) + x_3(c_1, c_2, c_3) + x_4(d_1, d_2, d_3) + \ldots = (0, 0, 0).$$

If any of the vectors $\boldsymbol{v}_i$ is $\boldsymbol{0}$, then we get a nontrivial solution to (3) by taking $\boldsymbol{x}_0 = (0, 0, \ldots, 0, 1, 0, \ldots, 0)$, where the 1 is in position i. So now suppose that no $\boldsymbol{v}_i = \boldsymbol{0}$. The matrix form of the system (3) is

$$\begin{pmatrix} a_1 & b_1 & c_1 & d_1 & \ldots & 0 \\ a_2 & b_2 & c_2 & d_2 & \ldots & 0 \\ a_3 & b_3 & c_3 & d_3 & \ldots & 0 \end{pmatrix}$$

Now we can see that (3) has a nontrivial solution. For we can reduce its matrix to one of the forms

$$\begin{pmatrix} 1 & 0 & 0 & \overline{d_1} & \ldots & 0 \\ 0 & 1 & 0 & \overline{d_2} & \ldots & 0 \\ 0 & 0 & 1 & \overline{d_3} & \ldots & 0 \end{pmatrix}, \quad \begin{pmatrix} 1 & 0 & \overline{c_1} & \overline{d_1} & \ldots & 0 \\ 0 & 1 & \overline{c_2} & \overline{d_2} & \ldots & 0 \\ 0 & 0 & 0 & 0 & \ldots & 0 \end{pmatrix},$$

$$\begin{pmatrix} 1 & \overline{b_1} & \overline{c_1} & \overline{d_1} & \ldots & 0 \\ 0 & 0 & 0 & 0 & \ldots & 0 \\ 0 & 0 & 0 & 0 & \ldots & 0 \end{pmatrix}, \quad \begin{pmatrix} 1 & \overline{b_1} & 0 & \overline{d_1} & \ldots & 0 \\ 0 & 0 & 1 & \overline{d_2} & \ldots & 0 \\ 0 & 0 & 0 & 0 & \ldots & 0 \end{pmatrix},$$

$$\begin{pmatrix} 1 & \overline{b_1} & \overline{c_1} & 0 & \overline{e_1} & \ldots & 0 \\ 0 & 0 & 0 & 1 & 0 & \ldots & 0 \\ 0 & 0 & 0 & 0 & 0 & \ldots & 0 \end{pmatrix}, \quad \text{etc.}$$

In each of the possible cases, at least one variable is arbitrary, so we have infinitely many solutions, in particular nontrivial solutions. Thus S is linearly dependent. Hence, dim $\boldsymbol{R}^3 = 3$. QED

A useful general observation to be made from the proof is that, in order to test a given set of vectors for linear dependence, we can put the coordinates of the vectors into the *columns* of a matrix and reduce it by Gaussian elimination. This saves the work of first writing down explicitly the homogeneous system

$$x_1\boldsymbol{v}_1 + x_2\boldsymbol{v}_2 + \ldots + x_n\boldsymbol{v}_n = \boldsymbol{0}$$

to get the matrix.

Using exactly the same reasoning employed in Example 7.6, we can show that $\boldsymbol{R}^n$ has dimension n. As part of this, the following basic fact about homogeneous systems is established, just as was done for $n = 3$ in Example 7.6. We omit the proof because its ideas are already presented in Example 7.6.

7.7 THEOREM. Any homogeneous system of m linear equations in n unknowns has infinitely many solutions if $m < n$.

7.8 EXAMPLE. Determine whether $\{\boldsymbol{v}_1, \boldsymbol{v}_2, \boldsymbol{v}_3\}$ is a linearly independent set if $\boldsymbol{v}_1 = (1, 2, 3)$, $\boldsymbol{v}_2 = (-1, 1, 0)$, and $\boldsymbol{v}_3 = (-3, 1, -2)$.

Solution. We must investigate the homogeneous linear system

$$x\boldsymbol{v}_1 + y\boldsymbol{v}_2 + z\boldsymbol{v}_3 = \mathbf{0}.$$

Its matrix is

$$\begin{pmatrix} 1 & -1 & -3 & 0 \\ 2 & 1 & 1 & 0 \\ 3 & 0 & -2 & 0 \end{pmatrix}.$$

We reduce it as follows:

$$\begin{pmatrix} 1 & -1 & -3 & 0 \\ 2 & 1 & 1 & 0 \\ 3 & 0 & -2 & 0 \end{pmatrix} \xrightarrow[\text{Add } -3R_1 \text{ to } R_3]{\text{Add } -2R_1 \text{ to } R_2} \begin{pmatrix} 1 & -1 & -3 & 0 \\ 0 & 3 & 7 & 0 \\ 0 & 3 & 7 & 0 \end{pmatrix} \xrightarrow{\text{Add } -R_2 \text{ to } R_3}$$

$$\begin{pmatrix} 1 & -1 & -3 & 0 \\ 0 & 3 & 7 & 0 \\ 0 & 0 & 0 & 0 \end{pmatrix}.$$

We could stop now, because we know there will be infinitely many solutions, so S must be linearly dependent. But if we are curious to discover a relation of linear dependence among $\boldsymbol{v}_1$, $\boldsymbol{v}_2$, and $\boldsymbol{v}_3$, then our reduction method will produce it.

$$\begin{pmatrix} 1 & -1 & -3 & 0 \\ 0 & 3 & 7 & 0 \\ 0 & 0 & 0 & 0 \end{pmatrix} \xrightarrow{\frac{1}{3}R_2} \begin{pmatrix} 1 & -1 & -3 & 0 \\ 0 & 1 & \frac{7}{3} & 0 \\ 0 & 0 & 0 & 0 \end{pmatrix} \xrightarrow{\text{Add } R_2 \text{ to } R_1}$$

$$\begin{pmatrix} 1 & 0 & -\frac{2}{3} & 0 \\ 0 & 1 & \frac{7}{3} & 0 \\ 0 & 0 & 0 & 0 \end{pmatrix}.$$

So we have

$$\begin{cases} x - \frac{2}{3}z = 0 \\ y + \frac{7}{3}z = 0 \end{cases} \quad \text{i.e.,} \begin{cases} x = \frac{2}{3}t \\ y = -\frac{7}{3}t \\ z = t \end{cases}$$

If we take $t = 3$, we have $(2, -7, 3)$ as a solution, i.e.,

$$2\boldsymbol{v}_1 - 7\boldsymbol{v}_2 + 3\boldsymbol{v}_3 = 0.$$

So S is linearly dependent.

7.9 DEFINITION. A **basis** for a real vector space of dimension n is any linearly independent set of n vectors in V.

Thus the standard basis for $\boldsymbol{R}^n$ is a basis for $\boldsymbol{R}^n$ according to this definition. In particular, $\{\boldsymbol{i}, \boldsymbol{j}, \boldsymbol{k}\}$ is a basis for $\boldsymbol{R}^3$. The notion of a basis will be of great importance in our study of linear differential equations later.

Our technique of Gaussian elimination always produces a basis for the space K of solutions to the associated homogeneous system. For instance, in Example

6.9, $\boldsymbol{v}=(-3, 0, 1, 1, 0)$ and $\boldsymbol{w}=(-1, 3, -2, 0, 1)$ are linearly independent in $\boldsymbol{R}^5$, since if $x\boldsymbol{v}+y\boldsymbol{w}=\boldsymbol{0}$, then consideration of the fourth and fifth coordinates shows that $x=0$ and $y=0$.

7.10 EXAMPLE. Show that $\{\boldsymbol{v}_1, \boldsymbol{v}_2, \boldsymbol{v}_3\}$ is a basis for $\boldsymbol{R}^3$ if $\boldsymbol{v}_1=(1, 1, 0)$, $\boldsymbol{v}_2=(0, 1, 1)$, and $\boldsymbol{v}_3=(1, 0, 1)$.

Proof. We have

$$\begin{pmatrix} 1 & 0 & 1 & 0 \\ 1 & 1 & 0 & 0 \\ 0 & 1 & 1 & 0 \end{pmatrix} \xrightarrow{\text{Add } -R_1 \text{ to } R_2} \begin{pmatrix} 1 & 0 & 1 & 0 \\ 0 & 1 & -1 & 0 \\ 0 & 1 & 1 & 0 \end{pmatrix} \xrightarrow{\text{Add } -R_2 \text{ to } R_3}$$

$$\begin{pmatrix} 1 & 0 & 1 & 0 \\ 0 & 1 & -1 & 0 \\ 0 & 0 & 2 & 0 \end{pmatrix} \xrightarrow{\frac{1}{2}R_3} \begin{pmatrix} 1 & 0 & 1 & 0 \\ 0 & 1 & -1 & 0 \\ 0 & 0 & 1 & 0 \end{pmatrix} \xrightarrow[\text{Add } -R_3 \text{ to } R_1]{\text{Add } R_3 \text{ to } R_2}$$

$$\begin{pmatrix} 1 & 0 & 0 & 0 \\ 0 & 1 & 0 & 0 \\ 0 & 0 & 1 & 0 \end{pmatrix}.$$

So the homogeneous system has only the trivial solution $x=0$, $y=0$, $z=0$. Hence the given set is linearly independent. Since it has three elements, it is a basis for $\boldsymbol{R}^3$. QED

EXERCISES 1.7

In Exercises 1 to 8, solve equation (1) to determine whether the given set of vectors is linearly independent. If it is linearly dependent, then give a relation of linear dependence for the set as in Example 7.8.

1 $\{(1, 3), (2, 4)\}$. Is this a basis for $\boldsymbol{R}^2$?

2 $\{(1, \frac{1}{2}), (-2, -1)\}$. Is this a basis for $\boldsymbol{R}^2$?

3 $\{(1, -1, 3), (2, 1, 4)\}$. Is this a basis for $\boldsymbol{R}^3$?

4 $\{(4, 4, -6), (-10, -10, 15)\}$. Is this a basis for $\boldsymbol{R}^3$?

5 $\{(1, -2, 3), (4, -5, 6), (7, 8, -9)\}$. Is this a basis for $\boldsymbol{R}^3$?

6 $\{(2, 3, 1), (-1, 1, 0), (5, 10, 3)\}$. Is this a basis for $\boldsymbol{R}^3$?

7 $\{(1, -1, 1, -1), (2, 3, -4, 1), (0, -5, 6, -3)\}$. Is this a basis for $\boldsymbol{R}^4$?

8 $\{(1, 1, 1, 0), (1, 0, 1, 1), (0, 0, 1, 1)\}$. Is this a basis for $\boldsymbol{R}^4$?

9 Suppose that $S=\{\boldsymbol{v}_1, \boldsymbol{v}_2, \ldots, \boldsymbol{v}_n\}$ is a linearly independent set in V and suppose that $\boldsymbol{w}$ in V is expressed in two ways as a linear combination of the vectors in S:

$$\boldsymbol{w}=a_1\boldsymbol{v}_1+a_2\boldsymbol{v}_2+\ldots+a_n\boldsymbol{v}_n \quad \text{and} \quad \boldsymbol{w}=b_1\boldsymbol{v}_1+b_2\boldsymbol{v}_2+\ldots+b_n\boldsymbol{v}_n.$$

Then show that $a_i=b_i$ for each $i=1, 2, \ldots, n$.

10 If B is a basis for the real vector space V, then show that every vector $\boldsymbol{v}$ in V is a *unique* linear combination of the vectors in B. The scalars occurring in this linear combination are called the *coordinates* of $\boldsymbol{v}$ relative to the basis B. [*Hint:* Consider the cases $\boldsymbol{v}\in B$ and $\boldsymbol{v}\notin B$ separately. For the second case, what can you say about $B\cup\{\boldsymbol{v}\}$? For the uniqueness part, use Exercise 9.]

11 If $S=\{\boldsymbol{v}_1, \boldsymbol{v}_2, \ldots, \boldsymbol{v}_n\}$ is linearly independent and *spans* V in the sense that every vector in V is a linear combination of the vectors in S, then show that S is a basis for V, so V has dimension n.

12 It has been determined experimentally that any hue of paint except white can be produced by properly mixing the primary colors blue, yellow, and red. Let us denote a mixed color C by (x, y, z), where x is the number of gallons of blue paint, y is the number of gallons of yellow paint, and z is the number of gallons of red paint mixed to produce $x+y+z$ gallons of mixture of color C.

(a) What are the representations for pure blue, yellow and red paint in this scheme?

(b) Give a meaning to $(x, y, z)+(x', y', z')$ and $a(x, y, z)$. Interpret in terms of paint. Use Exercise 10 to show that any given amount of paint of any color C has a *unique* recipe with ingredients pure blue, yellow, and red.

13 Refer to Exercise 12. Given that purple is $(1, 0, 1)$, orange is $(0, 1, 1)$, and green is $(1, 1, 0)$.

(a) Show that $B=\{(1, 0, 1), (0, 1, 1), (1, 1, 0)\}$ is a basis for $\boldsymbol{R}^3$.

(b) Find an expression for $(0, 0, 1)$ as a linear combination of the vectors in B.

(c) Is it possible to obtain the primary color red by mixing purple, orange, and green in the proper amounts?

(d) Do the paint colors constitute a vector subspace of $\boldsymbol{R}^3$?

14 A chemical reaction written $2a+b \rightleftharpoons c+d$ means that 2 molecules of substance a react with one molecule of substance b to form one molecule of substance c and one molecule of substance d. We can write this as a homogeneous linear equation, $2a+b-c-d=0$. What is the chemical significance of the one-dimensional vector space of multiples of $(2, 1, -1, -1)$? (The elements of the space are called *stoichiometric coefficients.*)

15 Refer to Exercise 14. To produce sulfuric acid (H_2SO_4) by the contact process, iron pyrites, a mineral containing iron sulfide (Fe_2S_3) and sulfur (S), is burned in air to produce sulfur dioxide (SO_2) and sulfur trioxide (SO_3). (The latter forms sulfuric acid when added to water.) In the combustion process some of the reactions involved are:

$$S+O_2 \rightarrow SO_2$$
$$2S+3O_2 \rightarrow 2SO_3$$
$$2Fe_2S_3+9O_2 \rightarrow 2Fe_2O_3+6SO_2$$
$$Fe_2S_3+6O_2 \rightarrow Fe_2O_3+3SO_3$$

(a) Let a, b, c, d, e, and f stand for the respective substances S, O_2, SO_2, SO_3, Fe_2S_3, and Fe_2O_3 (ferric oxide). Write linear equations in these six variables for the above four reactions.

(b) Write down the 6-dimensional vectors of stoichiometric coefficients $\boldsymbol{v}_1$, $\boldsymbol{v}_2$, $\boldsymbol{v}_3$, $\boldsymbol{v}_4$.

(c) Show that $\{\boldsymbol{v}_1, \boldsymbol{v}_2, \boldsymbol{v}_3\}$ is linearly independent but $\{\boldsymbol{v}_1, \boldsymbol{v}_2, \boldsymbol{v}_3, \boldsymbol{v}_4\}$ is linearly dependent. Thus only three *independent* reactions occur: the last one, for example, is a consequence of the first three.

16 Refer to Exercise 15. Show that $\{\boldsymbol{v}_1, \boldsymbol{v}_2, \boldsymbol{v}_4\}$ is also linearly independent. Hence, if a computer were being programmed to describe the process, either the third or the fourth reaction could be omitted.

17 Refer to Exercise 14. When nitrogen reacts with oxygen, different oxides of nitrogen (which are among the components of *smog*) are formed, depending on the circumstances of the reaction. Among the possible reactions taking place are

$2N_2+O_2 \rightarrow 2N_2O$ (nitrous oxide—"laughing gas")

$N_2+O_2 \rightarrow 2NO$ (nitric oxide)

$N_2+2O_2 \rightarrow 2NO_2$ (nitrogen dioxide—unstable)

$N_2+2O_2 \rightarrow N_2O_4$ (nitrogen tetroxide)

$NO_2+NO_2 \rightarrow N_2O_4$

(a) Let a, b, c, d, e, and f stand for the respective substances N_2, O_2, N_2O, NO, NO_2, and N_2O_4. Write linear equations in these six variables for the above five reactions.
(b) Write the 6-dimensional vectors of stoichiometric coefficients $\boldsymbol{v}_1$, $\boldsymbol{v}_2$, $\boldsymbol{v}_3$, $\boldsymbol{v}_4$, $\boldsymbol{v}_5$. Show that $\{\boldsymbol{v}_1, \boldsymbol{v}_2, \boldsymbol{v}_3, \boldsymbol{v}_4\}$ is linearly independent but that $\{\boldsymbol{v}_1, \boldsymbol{v}_2, \boldsymbol{v}_3, \boldsymbol{v}_4, \boldsymbol{v}_5\}$ is linearly dependent.
(c) Show that in fact $\{\boldsymbol{v}_3, \boldsymbol{v}_4, \boldsymbol{v}_5\}$ is linearly dependent, so that the last reaction is a consequence of the preceding two.

18 Refer to Exercise 17. Show that $\{\boldsymbol{v}_1, \boldsymbol{v}_2, \boldsymbol{v}_3, \boldsymbol{v}_5\}$ is linearly independent also. Hence a complete list of independent reactions *must* include the first two reactions.

8 DETERMINANTS

We have seen in Section 5 (Corollary 5.9) that if $\boldsymbol{v} = (v_1, v_2)$ and $\boldsymbol{w} = (w_1, w_2)$ are two vectors in the plane, then the area of the parallelogram formed by them is $|\boldsymbol{v} \times \boldsymbol{w}|$, where we identify $\boldsymbol{v} = (v_1, v_2, 0)$ and $\boldsymbol{w} = (w_1, w_2, 0)$ with vectors in $\boldsymbol{R}^3$. We have in this case

$$\boldsymbol{v} \times \boldsymbol{w} = (0, 0, v_1 w_2 - v_2 w_1),$$

so the area of the parallelogram formed is

$$A = |\boldsymbol{v} \times \boldsymbol{w}| = \left|\det\begin{pmatrix} v_1 & v_2 \\ w_1 & w_2 \end{pmatrix}\right|. \tag{1}$$

The determinant function is defined not just for 2 by 2 matrices like

$$\begin{pmatrix} v_1 & v_2 \\ w_1 & w_2 \end{pmatrix},$$

but also for 3 by 3, 4 by 4, in fact for n by n matrices. In this section we are concerned primarily with the 3 by 3 determinant, which arises geometrically as a natural analogue of (1).

We can easily pose the three-dimensional analogue to the above problem of determining the *area of the parallelogram formed by two vectors.* Given three vectors $\boldsymbol{u}$, $\boldsymbol{v}$, and $\boldsymbol{w}$ in $\boldsymbol{R}^3$, we can ask for the *volume of the parallelepiped they form.* Of course, they might *fail* to form a parallelepiped (if two of the vectors are collinear, for instance). But in most cases (see Figure 8.1) we expect that they will

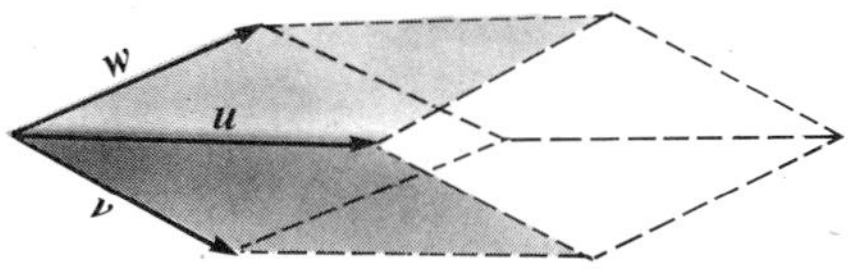

FIGURE 8.1

form one. We recall from Euclidean solid geometry that the volume of a parallelepiped is defined to be the area of a base times the altitude drawn to that base. If we consider the base formed by $\boldsymbol{u}$ and $\boldsymbol{v}$, then the corresponding altitude is the absolute value of the coordinate of $\boldsymbol{w}$ in the direction of $\boldsymbol{u} \times \boldsymbol{v}$ (Figure 8.2).

Let θ be the angle between $\boldsymbol{u} \times \boldsymbol{v}$ and $\boldsymbol{w}$. Then this altitude is $|\boldsymbol{w}|\,|\cos\theta|$. The area of the parallelogram formed by $\boldsymbol{u}$ and $\boldsymbol{v}$ is $|\boldsymbol{u} \times \boldsymbol{v}|$. So we have for the volume of the parallelepiped $V = |\boldsymbol{u} \times \boldsymbol{v}|\,|\boldsymbol{w}|\,|\cos\theta| = |(\boldsymbol{u} \times \boldsymbol{v}) \cdot \boldsymbol{w}|$. This is the absolute value

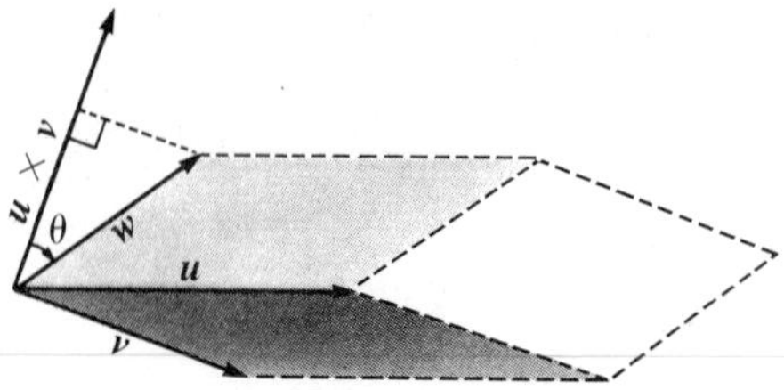

FIGURE 8.2

of the *triple scalar product* of $\boldsymbol{u}$, $\boldsymbol{v}$, and $\boldsymbol{w}$ (cf. Theorem 5.5(4)). We thus have proved the following.

8.1 **THEOREM.** If $\boldsymbol{u}, \boldsymbol{v}, \boldsymbol{w} \in \boldsymbol{R}^3$, then the volume of the parallelepiped formed by them is the absolute value of their triple scalar product.

Since we take the absolute value, we see that we can not only compute the triple scalar product

$$(\boldsymbol{u}\times\boldsymbol{v})\cdot\boldsymbol{w}=\boldsymbol{u}\cdot(\boldsymbol{v}\times\boldsymbol{w})=\boldsymbol{v}\cdot(\boldsymbol{w}\times\boldsymbol{u})$$

but could just as well compute its negative (Corollary 5.6)

$$(\boldsymbol{v}\times\boldsymbol{u})\cdot\boldsymbol{w}=\boldsymbol{u}\cdot(\boldsymbol{w}\times\boldsymbol{v})=\boldsymbol{v}\cdot(\boldsymbol{u}\times\boldsymbol{w})$$

instead. This corresponds to the observation that we could just as well consider the base of the parallelepiped to be the parallelogram formed by $\boldsymbol{u}$ and $\boldsymbol{w}$ (or $\boldsymbol{w}$ and $\boldsymbol{u}$) or the parallelogram formed by $\boldsymbol{v}$ and $\boldsymbol{w}$ (or $\boldsymbol{w}$ and $\boldsymbol{v}$). We also observe that even if $\boldsymbol{u}$, $\boldsymbol{v}$, and $\boldsymbol{w}$ fail to form a parallelepiped, we can *still* apply Theorem 8.1 sensibly. For if two of the vectors are collinear, then either $\boldsymbol{u}\times\boldsymbol{v}$ or $\boldsymbol{v}\times\boldsymbol{w}$ or $\boldsymbol{w}\times\boldsymbol{u}$ will be $\boldsymbol{0}$ by Corollary 5.1. So Theorem 8.1 tells us that $\boldsymbol{u}$, $\boldsymbol{v}$, and $\boldsymbol{w}$ fail to form a parallelepiped by giving us 0 as its volume.

8.2 **EXAMPLE.** Determine whether $\boldsymbol{u}=(1,3,-1)$, $\boldsymbol{v}=(-2,0,3)$, and $\boldsymbol{w}=(2,1,-2)$ form a parallelepiped in space. If they do, what is its volume?

Solution. Here $\boldsymbol{u}\times\boldsymbol{v}=(9-0,2-3,0+6)=(9,-1,6)$. So

$$(\boldsymbol{u}\times\boldsymbol{v})\cdot\boldsymbol{w}=(9,-1,6)\cdot(2,1,-2)=18-1-12=5.$$

Therefore, the given vectors *do* form a parallelepiped and its volume is 5.

You have probably studied two-by-two and three-by-three determinants before. We have already seen in (1) how the two-by-two determinant arises as the third coordinate of a cross product of two plane vectors $(v_1, w_1, 0)$ and $(v_2, w_2, 0)$. You may be surprised to learn that the three-by-three determinant

$$\det\begin{pmatrix}\boldsymbol{u}\\ \boldsymbol{v}\\ \boldsymbol{w}\end{pmatrix}=\det\begin{pmatrix}u_1 & u_2 & u_3\\ v_1 & v_2 & v_3\\ w_1 & w_2 & w_3\end{pmatrix}$$

is precisely the triple scalar product $(\boldsymbol{u}\times\boldsymbol{v})\cdot\boldsymbol{w}=\boldsymbol{u}\cdot(\boldsymbol{v}\times\boldsymbol{w})$. For in the proof of Theorem 5.5(4), we obtained the formula

(2) $$\boldsymbol{u}\cdot(\boldsymbol{v}\times\boldsymbol{w})=u_1v_2w_3-u_1v_3w_2+u_2v_3w_1-u_2v_1w_3+u_3v_1w_2-u_3v_2w_1$$

(3) $$= u_1\left(\det\begin{pmatrix} v_2 & v_3 \\ w_2 & w_3 \end{pmatrix}\right) - u_2\left(\det\begin{pmatrix} v_1 & v_3 \\ w_1 & w_3 \end{pmatrix}\right) + u_3\left(\det\begin{pmatrix} v_1 & v_2 \\ w_1 & w_2 \end{pmatrix}\right)$$

which is the defining equation for

$$\det\begin{pmatrix} \boldsymbol{u} \\ \boldsymbol{v} \\ \boldsymbol{w} \end{pmatrix}.$$

We will use the triple scalar product as the definition of the three-by-three determinant.

8.3 DEFINITION. Given a three-by-three matrix

$$A = \begin{pmatrix} a_1 & a_2 & a_3 \\ b_1 & b_2 & b_3 \\ c_1 & c_2 & c_3 \end{pmatrix},$$

then its **determinant** is defined by

$$\det A = \boldsymbol{a} \cdot (\boldsymbol{b} \times \boldsymbol{c})$$

where $\boldsymbol{a} = (a_1, a_2, a_3)$, $\boldsymbol{b} = (b_1, b_2, b_3)$, $\boldsymbol{c} = (c_1, c_2, c_3)$.

We remark that this definition can be used to cover the two-by-two determinant function as well.

8.4 COROLLARY

$$\det\begin{pmatrix} a_1 & a_2 \\ b_1 & b_2 \end{pmatrix} = \det\begin{pmatrix} a_1 & a_2 & 0 \\ b_1 & b_2 & 0 \\ 0 & 0 & 1 \end{pmatrix}.$$

Proof. Let us evaluate the right side. We have

$$\det\begin{pmatrix} a_1 & a_2 & 0 \\ b_1 & b_2 & 0 \\ 0 & 0 & 1 \end{pmatrix} = (a_1, a_2, 0) \cdot [(b_1, b_2, 0) \times (0, 0, 1)]$$

$$= (a_1, a_2, 0) \cdot (b_2, -b_1, 0)$$

$$= a_1 b_2 - a_2 b_1$$

$$= \det\begin{pmatrix} a_1 & a_2 \\ b_1 & b_2 \end{pmatrix}.$$ QED

We can now use 8.3 and 8.4 to derive many of the important properties of two-by-two and three-by-three determinants. You will notice that this is facilitated by our knowledge of the cross product from Section 5. In the following, A can be assumed to be a three-by-three matrix in view of Corollary 8.4.

8.5 THEOREM. (a) $\det A = \det A^t$, where A^t is the **transpose** matrix obtained by interchanging the rows of A with the columns of A:

$$A^t = \begin{pmatrix} a_1 & b_1 & c_1 \\ a_2 & b_2 & c_2 \\ a_3 & b_3 & c_3 \end{pmatrix} \quad \text{if} \quad A = \begin{pmatrix} a_1 & a_2 & a_3 \\ b_1 & b_2 & b_3 \\ c_1 & c_2 & c_3 \end{pmatrix}.$$

(b) If B is obtained from A by multiplying any row or column by a real number r, then $\det B = r \det A$.

(c) If A has $\mathbf{0}$ as a row or column, or if two rows or columns are equal or proportional, then $\det A = 0$.

Proof. (a) Use formula (2) to calculate

$$\begin{aligned} \det A^t &= a_1 \det\begin{pmatrix} b_2 & c_2 \\ b_3 & c_3 \end{pmatrix} - b_1 \det\begin{pmatrix} a_2 & c_2 \\ a_3 & c_3 \end{pmatrix} + c_1 \det\begin{pmatrix} a_2 & b_2 \\ a_3 & b_3 \end{pmatrix} \\ &= a_1(b_2c_3 - b_3c_2) - b_1(a_2c_3 - a_3c_2) + c_1(a_2b_3 - a_3b_2) \\ &= a_1b_2c_3 - a_1b_3c_2 - a_2b_1c_3 + a_3b_1c_2 + a_2b_3c_1 - a_3b_2c_1. \end{aligned}$$

From (2) we also have

$$\begin{aligned} \det A &= a_1b_2c_3 - a_1b_3c_2 + a_2b_3c_1 - a_2b_1c_3 + a_3b_1c_2 - a_3b_2c_1 \\ &= \det A^t. \end{aligned}$$

In proving (b) and (c), it is enough to consider only the rows, since the rows of A are the columns of A^t, and $\det A = \det A^t$.

(b) First let us assume that row $\boldsymbol{a}$ of A is multiplied by r to obtain B. Then

$$\begin{aligned} \det B &= (r\boldsymbol{a}) \cdot (\mathbf{b} \times \mathbf{c}) \\ &= r[\boldsymbol{a} \cdot (\mathbf{b} \times \mathbf{c})] \\ &= r \det A \qquad \text{by Definition 8.3.} \end{aligned}$$

If either row $\boldsymbol{b}$ or row $\mathbf{c}$ is the one multiplied by r, then we use Theorem 5.5(4) to place that row to the left of the dot in the calculation above; the same theorem then guarantees that the triple scalar product obtained is equal to $\det B$.

(c) If two rows are proportional, say $\boldsymbol{a} = k\mathbf{c}$, then

$$\begin{aligned} \det A &= \boldsymbol{a} \cdot (\mathbf{b} \times \mathbf{c}) = k\mathbf{c} \cdot (\mathbf{b} \times \mathbf{c}) \\ &= k[\mathbf{b} \cdot (\mathbf{c} \times \mathbf{c})] \qquad \text{by Theorem 5.5(4)} \\ &= k(\mathbf{b} \cdot \mathbf{0}) \qquad \text{by Corollary 5.11} \\ &= 0. \end{aligned}$$

The case $k = 1$ covers the equality of the two rows. And $k = 0$ covers the case of $\mathbf{0}$ as a row. QED

8.6 THEOREM. (a) If B is obtained from A by interchanging two rows or columns, then $\det B = -\det A$.

(b)

$$\det\begin{pmatrix} \mathbf{a} + \mathbf{a}' \\ \mathbf{b} \\ \mathbf{c} \end{pmatrix} = \det\begin{pmatrix} \mathbf{a} \\ \mathbf{b} \\ \mathbf{c} \end{pmatrix} + \det\begin{pmatrix} \mathbf{a}' \\ \mathbf{b} \\ \mathbf{c} \end{pmatrix},$$

with similar equations holding if **b** is replaced by $\boldsymbol{b}+\boldsymbol{b}'$, or **c** is replaced by $\boldsymbol{c}+\boldsymbol{c}'$.

(c) If B is obtained from A by replacing a row of A by the result of adding to it a multiple of some other row of A, then $\det B = \det A$.

(d)

$$\det\begin{pmatrix} \boldsymbol{i} \\ \boldsymbol{j} \\ \boldsymbol{k} \end{pmatrix} = 1.$$

(e)

$$\det\begin{pmatrix} a_1 & a_2 & a_3 \\ 0 & b_2 & b_3 \\ 0 & 0 & c_3 \end{pmatrix} = a_1 b_2 c_3$$

Proof. (a) Suppose that rows (1) and (3) are interchanged. Then

$$\begin{aligned}
\det\begin{pmatrix} \boldsymbol{c} \\ \boldsymbol{b} \\ \boldsymbol{a} \end{pmatrix} &= \boldsymbol{c}\cdot(\boldsymbol{b}\times\boldsymbol{a}) = (\boldsymbol{b}\times\boldsymbol{a})\cdot\boldsymbol{c} \\
&= -(\boldsymbol{a}\times\boldsymbol{b})\cdot\boldsymbol{c} \quad \text{by Theorem 5.5(2)} \\
&= -\boldsymbol{a}\cdot(\boldsymbol{b}\times\boldsymbol{c}) \quad \text{by Corollary 5.6} \\
&= -\det\begin{pmatrix} \boldsymbol{a} \\ \boldsymbol{b} \\ \boldsymbol{c} \end{pmatrix}.
\end{aligned}$$

(b)

$$\begin{aligned}
\det\begin{pmatrix} \boldsymbol{a}+\boldsymbol{a}' \\ \boldsymbol{b} \\ \boldsymbol{c} \end{pmatrix} &= (\boldsymbol{a}+\boldsymbol{a}')\cdot(\boldsymbol{b}\times\boldsymbol{c}) \\
&= \boldsymbol{a}\cdot(\boldsymbol{b}\times\boldsymbol{c}) + \boldsymbol{a}'\cdot(\boldsymbol{b}\times\boldsymbol{c}) \\
&= \det\begin{pmatrix} \boldsymbol{a} \\ \boldsymbol{b} \\ \boldsymbol{c} \end{pmatrix} + \det\begin{pmatrix} \boldsymbol{a}' \\ \boldsymbol{b} \\ \boldsymbol{c} \end{pmatrix}.
\end{aligned}$$

(c) We consider the case in which the first row $\boldsymbol{a}$ is replaced by $\boldsymbol{a}+k\boldsymbol{c}$. We have

$$\begin{aligned}
\det\begin{pmatrix} \boldsymbol{a}+k\boldsymbol{c} \\ \boldsymbol{b} \\ \boldsymbol{c} \end{pmatrix} &= \det\begin{pmatrix} \boldsymbol{a} \\ \boldsymbol{b} \\ \boldsymbol{c} \end{pmatrix} + \det\begin{pmatrix} k\boldsymbol{c} \\ \boldsymbol{b} \\ \boldsymbol{c} \end{pmatrix} \quad \text{by (b)} \\
&= \det A + 0 \quad \text{by 8.5 (c)} \\
&= \det A.
\end{aligned}$$

(d)

$$\begin{aligned}
\det\begin{pmatrix} \boldsymbol{i} \\ \boldsymbol{j} \\ \boldsymbol{k} \end{pmatrix} &= \boldsymbol{i}\cdot(\boldsymbol{j}\times\boldsymbol{k}) \\
&= \boldsymbol{i}\cdot\boldsymbol{i} \\
&= 1
\end{aligned}$$

(e)

$$\det\begin{pmatrix} a_1 & a_2 & a_3 \\ 0 & b_2 & b_3 \\ 0 & 0 & c_3 \end{pmatrix} = a_1 b_2 c_3 + 0 + 0 - 0 - 0 - 0$$

$$= a_1 b_2 c_3.$$ QED

Theorem 8.6(a) to (c) lets us use elementary row (or column) operations to evaluate 3-by-3 determinants. We simply reduce to a "triangular" form and use Theorem 8.6(e).

8.7 EXAMPLE. Use elementary row operations to compute

$$\det\begin{pmatrix} 1 & -2 & 1 \\ -2 & 5 & 0 \\ 3 & -1 & 1 \end{pmatrix}.$$

Solution. We can add $2R_1$ to R_2 and $-3R_1$ to R_3 without changing the value of the determinant.

$$\det\begin{pmatrix} 1 & -2 & 1 \\ -2 & 5 & 0 \\ 3 & -1 & 1 \end{pmatrix} = \det\begin{pmatrix} 1 & -2 & 1 \\ 0 & 1 & 2 \\ 0 & 5 & -2 \end{pmatrix}.$$

Similarly, we can add $2R_2$ to R_1 and $-5R_2$ to R_3 without changing the value of the determinant.

$$\det\begin{pmatrix} 1 & -2 & 1 \\ 0 & 1 & 2 \\ 0 & 5 & -2 \end{pmatrix} = \det\begin{pmatrix} 1 & 0 & 5 \\ 0 & 1 & 2 \\ 0 & 0 & -12 \end{pmatrix}$$

by Theorem 8.5(c). Now we can use Theorem 8.6(e) to conclude that

$$\det\begin{pmatrix} 1 & -2 & 1 \\ -2 & 5 & 0 \\ 3 & -1 & 1 \end{pmatrix} = -12.$$

Given a sequence of three non-collinear vectors $(\boldsymbol{u}, \boldsymbol{v}, \boldsymbol{w})$, one sometimes defines their **orientation** to be **right-handed** if

$$\det\begin{pmatrix} \boldsymbol{u} \\ \boldsymbol{v} \\ \boldsymbol{w} \end{pmatrix} > 0.$$

The orientation is **left-handed** if

$$\det\begin{pmatrix} \boldsymbol{u} \\ \boldsymbol{v} \\ \boldsymbol{w} \end{pmatrix} < 0.$$

Theorem 8.6(d) then says that $(\boldsymbol{i}, \boldsymbol{j}, \boldsymbol{k})$ is a right-handed system. This is in agreement with our use of the term "right-handed" for our coordinate system in $\boldsymbol{R}^3$.

While elementary row operations can be used to evaluate 3-by-3 determinants, Example 8.7 fails to show much gain in so using them. Indeed, direct use of (2) or (3) is often quicker and easier. The main value of elementary row operations lies in their ability to reduce evaluation of n-by-n determinants to the computation of 3-by-3 determinants. First, how are determinants of 4-by-4, 5-by-5, . . . , n-by-n matrices defined? There are a number of equivalent ways to approach this, but the simplest is a "boot-strap" operation which extends (3). More precisely, given a 4-by-4 matrix

$$A = \begin{pmatrix} a_{11} & a_{12} & a_{13} & a_{14} \\ a_{21} & a_{22} & a_{23} & a_{24} \\ a_{31} & a_{32} & a_{33} & a_{34} \\ a_{41} & a_{42} & a_{43} & a_{44} \end{pmatrix}$$

we define

(4) $$\det A = a_{11}M_{11} - a_{12}M_{12} + a_{13}M_{13} - a_{14}M_{14}$$

where M_{1j} is the **(1, *j*)-minor of A.** By definition, M_{1j} is the determinant of the 3-by-3 matrix that results when the row and column containing a_{1j} (that is, row 1 and column j) are deleted from A. Although we shall not do so here, it can be shown that $\det A$ can be expanded by minors of *any row or column*, not just row 1. That is, for any i and j, we have

(5) $$\begin{aligned} \det A &= (-1)^{i+1}a_{i1}M_{i1} + (-1)^{i+2}a_{i2}M_{i2} + (-1)^{i+3}a_{i3}M_{i3} + (-1)^{i+4}a_{i4}M_{i4} \\ &= (-1)^{1+j}a_{1j}M_{1j} + (-1)^{2+j}a_{2j}M_{2j} + (-1)^{3+j}a_{3j}M_{3j} + (-1)^{4+j}a_{4j}M_{4j} \end{aligned}$$

It can also be shown that Theorems 8.5 and 8.6 hold for 4-by-4 determinants. Then, to evaluate a 4-by-4 determinant, $\det A$ say, work can be saved by using elementary row operations to obtain a column with just *one* nonzero entry. For then $\det A$ can be evaluated by computing just *one* 3-by-3 determinant instead of the *four* 3-by-3 determinants which occur in (4) and (5).

8.8 EXAMPLE. Compute $\det \begin{pmatrix} 2 & -2 & 0 & 0 \\ -1 & 2 & -1 & 0 \\ 0 & -1 & 2 & -1 \\ 0 & 0 & -1 & 2 \end{pmatrix}$.

Solution. Adding twice R_2 to R_1, we have by Theorem 8.6(c)

$$\det \begin{pmatrix} 2 & -2 & 0 & 0 \\ -1 & 2 & -1 & 0 \\ 0 & -1 & 2 & -1 \\ 0 & 0 & -1 & 2 \end{pmatrix} = \det \begin{pmatrix} 0 & 2 & -2 & 0 \\ -1 & 2 & -1 & 0 \\ 0 & -1 & 2 & -1 \\ 0 & 0 & -1 & 2 \end{pmatrix}$$

$$= 0 - (-1)\det \begin{pmatrix} 2 & -2 & 0 \\ -1 & 2 & -1 \\ 0 & -1 & 2 \end{pmatrix} + 0 - 0$$

expanding by minors of the first column

$$= (+1)(8 + 0 + 0 - 0 - 2 - 4) = 2$$

where we used (2) to compute the 3-by-3 determinant.

Now, 5-by-5 determinants can be defined in terms of 4-by-4 determinants by using the analogue of (4). In general, if A is an n-by-n matrix, then we define

(6)
$$\det A = a_{11}M_{11} - a_{12}M_{12} + a_{13}M_{13} - \ldots + (-1)^{1+j}a_{1j}M_{1j} + \ldots + (-1)^{1+n}a_{1n}M_{1n}.$$

It can again be proved that Theorems 8.5 and 8.6 hold and that $\det A$ can be evaluated by using minors of any row or column, that is,

(7)
$$\begin{aligned}\det A &= (-1)^{i+1}a_{i1}M_{i1} + (-1)^{i+2}a_{i2}M_{i2} + \ldots + (-1)^{i+j}a_{ij}M_{ij} \\ &\quad + \ldots + (-1)^{i+n}a_{in}M_{in} \\ &= (-1)^{1+j}a_{1j}M_{1j} + \ldots + (-1)^{i+j}a_{ij}M_{ij} + \ldots + (-1)^{n+j}a_{nj}M_{nj}.\end{aligned}$$

To evaluate such higher order determinants, we use elementary row operations repeatedly as in Example 8.8. This ultimately will reduce the evaluation of $\det A$ to the computation of a *single* 3-by-3 determinant, since we can first reduce to a single $(n-1)$ by $(n-1)$ determinant, which can then be reduced to a single $(n-2)$ by $(n-2)$ determinant, etc.

In your previous study of determinants, you may have met *Cramer's Rule*, named for the eighteenth century Swiss mathematician Gabriel Cramer (1704–1752). Given a system of n linear equations in n unknowns, Cramer's Rule gives formulas for the coordinates $x_1, x_2, \ldots, x_n$ of a unique solution if there is a unique solution. This rule is thus applicable **only in case there is a unique solution.** Even though it is of no use when infinitely many solutions or no solutions exist, it is useful enough in theory and practice for us to state it formally. Again we don't prove it, but the proof for $n = 2$ or 3 is not hard. (See Exercises 23(a) and 23(b) below.)

8.9 CRAMER'S RULE. If the system

$$\begin{cases} a_{11}x_1 + a_{12}x_2 + \ldots + a_{1j}x_j + \ldots + a_{1n}x_n = d_1 \\ a_{21}x_1 + a_{22}x_2 + \ldots + a_{2j}x_j + \ldots + a_{2n}x_n = d_2 \\ \quad\vdots \qquad\quad \vdots \qquad\quad \vdots \qquad\qquad\quad \vdots \qquad\qquad \vdots \\ a_{n1}x_1 + a_{n2}x_2 + \ldots + a_{nj}x_j + \ldots + a_{nn}x_n = d_n \end{cases}$$

has $\det A \neq 0$ (where A is the matrix of coefficients), then there is a unique solution

$$\boldsymbol{x}_0 = (x_1, x_2, \ldots, x_j, \ldots, x_n).$$

Here each unknown x_j is given by

$$x_j = \frac{\det\begin{pmatrix} a_{11} & a_{12} & \ldots & d_1 & \ldots & a_{1n} \\ a_{21} & a_{22} & \ldots & d_2 & \ldots & a_{2n} \\ \vdots & \vdots & & \vdots & & \vdots \\ a_{n1} & a_{n2} & \ldots & d_n & \ldots & a_{nn} \end{pmatrix}}{\det\begin{pmatrix} a_{11} & a_{12} & \ldots & a_{1j} & \ldots & a_{1n} \\ a_{21} & a_{22} & \ldots & a_{2j} & \ldots & a_{2n} \\ \vdots & \vdots & & \vdots & & \vdots \\ a_{n1} & a_{n2} & \ldots & a_{nj} & \ldots & a_{nn} \end{pmatrix}}$$

for $j=1,2,\ldots n$. Thus each unknown x_j is the quotient of two determinants: the denominator is the determinant of the coefficients, and the numerator is identical except that column j is replaced by the column of constants d_i.

8.10 EXAMPLE. Use Cramer's Rule to solve

$$\begin{cases} x+2y+3z=1 \\ 2x+\ \ y+3z=2 \\ 3x+\ \ y+\ \ z=1. \end{cases}$$

Solution. We have

$$\det\begin{pmatrix} 1 & 2 & 3 \\ 2 & 1 & 3 \\ 3 & 1 & 1 \end{pmatrix} = 1+18+6-9-3-4=9.$$

Thus

$$x=\tfrac{1}{9}\det\begin{pmatrix} 1 & 2 & 3 \\ 2 & 1 & 3 \\ 1 & 1 & 1 \end{pmatrix} = \tfrac{1}{9}(1+6+6-3-3-4)=\tfrac{3}{9}=\tfrac{1}{3}.$$

$$y=\tfrac{1}{9}\det\begin{pmatrix} 1 & 1 & 3 \\ 2 & 2 & 3 \\ 3 & 1 & 1 \end{pmatrix} = \tfrac{1}{9}(2+9+6-18-3-2)=-\tfrac{6}{9}=-\tfrac{2}{3}.$$

$$z=\tfrac{1}{9}\det\begin{pmatrix} 1 & 2 & 1 \\ 2 & 1 & 2 \\ 3 & 1 & 1 \end{pmatrix} = \tfrac{1}{9}(1+12+2-3-2-4)=\tfrac{6}{9}=\tfrac{2}{3}.$$

So the unique solution is $\boldsymbol{x}_0=(\tfrac{1}{3},-\tfrac{2}{3},\tfrac{2}{3})$.

While Cramer's Rule works well when $n=2$ or 3, it is usually easier to use Gaussian elimination to solve systems of more than three equations. The latter method has the added advantage of applying not only when a unique solution fails to exist, but also to systems in which the numbers of unknowns and equations can be different.

EXERCISES 1.8

In Exercises 1 to 4, determine whether the given vectors form a parallelepiped in R^3. If they do, find its volume.

1 $(4,1,-1)$, $(-1,2,1)$, $(0,3,2)$.

2 $(-3,1,1)$, $(1,-2,-2)$, $(5,-1,1)$.

3 $(1,3,-1)$, $(2,-1,2)$, $(-1,11,-7)$.

4 $(2,0,1)$, $(-1,3,2)$, $(3,3,4)$.

In Exercises 5 to 12, evaluate the determinant of the given matrix.

5 $\begin{pmatrix} 1 & 5 \\ -1 & 1 \end{pmatrix}$

6 $\begin{pmatrix} 10 & 3 \\ -1 & 1 \end{pmatrix}$

7 $\begin{pmatrix} 2 & -1 & 0 \\ -1 & 2 & -1 \\ 0 & -1 & 2 \end{pmatrix}$ **8** $\begin{pmatrix} 3 & -2 & 1 \\ 1 & 1 & 2 \\ 2 & 2 & -3 \end{pmatrix}$

9 $\begin{pmatrix} 2 & -1 & 0 & 0 \\ -1 & 2 & -1 & 0 \\ 0 & -1 & 2 & -1 \\ 0 & 0 & -1 & 2 \end{pmatrix}$ **10** $\begin{pmatrix} 2 & -1 & 0 & 0 \\ -1 & 2 & -2 & 0 \\ 0 & -1 & 2 & -1 \\ 0 & 0 & -1 & 2 \end{pmatrix}$

11 $\begin{pmatrix} 2 & -1 & 0 & 0 & 0 \\ -1 & 2 & -1 & 0 & 0 \\ 0 & -1 & 2 & -1 & 0 \\ 0 & 0 & -1 & 2 & -1 \\ 0 & 0 & 0 & -1 & 2 \end{pmatrix}$ **12** $\begin{pmatrix} 2 & -2 & 0 & 0 & 0 \\ -1 & 2 & -1 & 0 & 0 \\ 0 & -1 & 2 & -1 & 0 \\ 0 & 0 & -1 & 2 & -1 \\ 0 & 0 & 0 & -1 & 2 \end{pmatrix}$

13 Give the proof of Theorem 8.5(b) in case the second row ($\boldsymbol{b}$) or third row ($\boldsymbol{c}$) is multiplied by r.

14 Give the proof of Theorem 8.5(c) in case the second and third rows are proportional.

15 Prove Theorem 8.6(a) in case rows (1) and (2) or (2) and (3) are interchanged.

16 Prove Theorem 8.6(b) in case $\boldsymbol{b}$ is replaced by $\boldsymbol{b}+\boldsymbol{b}'$.

17 Prove Theorem 8.6(c) in case the second row, $\boldsymbol{b}$, is replaced by $\boldsymbol{b}+k\boldsymbol{a}$.

18 Prove Theorem 8.6(c) in case the third row, $\boldsymbol{c}$, is replaced by $\boldsymbol{c}+k\boldsymbol{b}$.

19 Is $(-\boldsymbol{i}, -\boldsymbol{j}, -\boldsymbol{k})$ a right-handed or left-handed system?

20 Write out Cramer's Rule if $n=2$. Use it, and then Gaussian elimination, to solve

$$\begin{cases} 2x+3y=-2 \\ 3x-5y=3. \end{cases}$$

Which method is faster?

21 Write out Cramer's Rule if $n=3$. Use it, and then Gaussian elimination, to solve

$$\begin{cases} 2x-3y+z=4 \\ x+y-z=1 \\ x-2y+2z=7 \end{cases}$$

Which method is faster?

22 Write out Cramer's Rule if $n=4$. Use it, and then Gaussian elimination to solve

$$\begin{cases} 2x-y+3z-w=-4 \\ x-y-3z+w=3 \\ -x+2y+2z-w=0 \\ 3x-y-z+2w=4 \end{cases}.$$

Which method is faster?

23 (a) Prove Cramer's Rule if $n=2$.
(b) Prove Cramer's Rule if $n=3$.
Hint: If the system is

$$\begin{cases} \boldsymbol{a}_1\cdot\boldsymbol{x}=b_1 \\ \boldsymbol{a}_2\cdot\boldsymbol{x}=b_2, \\ \boldsymbol{a}_3\cdot\boldsymbol{x}=b_3 \end{cases}$$

then

$$x_1 \det A = \det\begin{pmatrix} a_{11}x_1 & a_{12} & a_{13} \\ a_{21}x_1 & a_{22} & a_{23} \\ a_{31}x_1 & a_{32} & a_{33} \end{pmatrix} = \det\begin{pmatrix} \boldsymbol{a}_1 \cdot \boldsymbol{x} & a_{12} & a_{13} \\ \boldsymbol{a}_2 \cdot \boldsymbol{x} & a_{22} & a_{23} \\ \boldsymbol{a}_3 \cdot \boldsymbol{x} & a_{32} & a_{33} \end{pmatrix}.$$

24 Use the second line of the defining formula (Equation (2) preceding Definition 8.3) for a three-by-three determinant to show that the cross product of $\boldsymbol{v} = (v_1, v_2, v_3)$ and $\boldsymbol{w} = (w_1, w_2, w_3)$ can be computed as the formal determinant

$$\det\begin{pmatrix} \boldsymbol{i} & \boldsymbol{j} & \boldsymbol{k} \\ v_1 & v_2 & v_3 \\ w_1 & w_2 & w_3 \end{pmatrix}.$$

REVIEW EXERCISES 1.9

1 If $\boldsymbol{v} = (2, -1, 2)$ and $\boldsymbol{w} = (-3, -1, 1)$, then compute
(a) $-2\boldsymbol{v} + 3\boldsymbol{w}$, (b) $\frac{1}{3}\boldsymbol{v} - \frac{4}{5}\boldsymbol{w}$, (c) $|\boldsymbol{v}|$.
(d) Find the unit vectors in the direction of $\boldsymbol{v}$ and $\boldsymbol{w}$.

2 Find the center and radius of the sphere whose equation is

$$x^2 + y^2 + z^2 + 7y - 5z = \tfrac{7}{4}.$$

3 Show that the set of all functions $f: \boldsymbol{R} \to \boldsymbol{R}$ with domain $\boldsymbol{R}$ is a real vector space.

4 Find the angles between
(a) $(\frac{1}{2}, -\frac{3}{2})$ and $(-\frac{9}{2}, -\frac{3}{2})$ (b) $(1, -1, 3)$ and $(-4, 2, 2)$

5 Find the direction and direction angles of $(2, 2, 2\sqrt{2})$.

6 (a) Find a unit vector orthogonal to $(3, 2, 6)$. How many such vectors are there?
(b) Repeat part (a) for $(3, 1)$.

7 Find the component of $(-2, 3, 1)$ in the direction of $(2, -3, 6)$.

8 Find vector, parametric scalar, and symmetric scalar equations of the line the line
(a) through $(-1, 1, 3)$ and $(1, -1, -3)$. (b) through $(2, -1, 3)$ in the direction $(\frac{1}{3}, \frac{2}{3}, \frac{2}{3})$.

9 Find parametric equations for the line through $(2, 0, -1)$ which is parallel to the line

$$\frac{x}{-2} = \frac{y-3}{-2} = \frac{2z+1}{3}.$$

10 Do the lines

$$\boldsymbol{x} = (1, -1, 0) + t(-1, 3, 1)$$

and

$$x = s,\ y = 6 + s,\ z = -4 - 6s$$

intersect? If so, find their point of intersection. If not, explain why not.

11 Are the lines

$$\frac{x+2}{-3} = \frac{y}{2} = \frac{z-1}{-1}$$

and

$$x = 1 + t,\ y = -2 - 2t,\ z = \tfrac{7}{2} - 7t$$

perpendicular?

12 The line through the origin perpendicular to a plane intersects the plane at the point $(1, -3, -1)$. Find the scalar equation of the plane. Sketch.

13 The line through $(3, 0, -2)$ perpendicular to a certain plane intersects it at the point $(3, -3, 2)$. Find a scalar equation for the plane.

14 A plane through $(-1, 5, -2)$ is perpendicular to the line through $(-1, 1, 4)$ and $(2, -1, 3)$. Find a scalar equation of the plane.

15 Find the equation of the plane through $(-1, 4, 1)$ which is parallel to the plane whose equation is $5x - 2y + z = 3$.

16 Find vector and scalar equations for the line of intersection of the planes

$$\begin{cases} -x + 3y - 2z = -4 \\ 2x - 5y + z = 3. \end{cases}$$

17 Find the distance from the point $(-1, 2, 3)$ to the plane $2x - y + 3z = 5$. (See problems 18 to 20 in Exercises 1.4.)

18 Is the line $\boldsymbol{x} = (-1, 3, 1) + t(1, 0, 4)$ normal to the plane $12x - 7y + 8z = 19$? Find the point of intersection.

19 Find the equation of the plane through $P(2, -1, 2)$, $Q(-1, 3, 2)$, $R(5, 0, -3)$. Check by verifying that the coordinates of P, Q, and R satisfy your equation.

20 Find the equation of the plane through the two lines $\boldsymbol{x} = (1, -3, 2) + s(1, -5, -1)$ and $\boldsymbol{x} = (0, 2, 3) + t(1, 1, 3)$. Check by verifying that the coordinates of all points on the lines satisfy your equation.

21 Find parametric vector and scalar equations for the plane through the points $P(-1, -1, 2)$, $Q(0, 3, -1)$, and $R(2, -3, 5)$. Find also the normal form equation of this plane.

22 A constant force $\boldsymbol{F} = (1, -1, 5)$ acts on a particle located at $(-1, 3, -1)$. Find the resulting torque (a) at the origin and (b) at $(2, 3, 0)$.

23 Find the area of the parallelogram in the plane with vertices $(0, 0)$, $(-3, 1)$, $(5, 2)$ and $(2, 3)$.

In Exercises 24 to 26, solve the system completely, find the solution to the associated homogeneous problem, and give a particular solution.

24 $\begin{cases} 3x - y + 2z = 3 \\ 2x + 2y + z = 2 \\ x - 3y + z = 4. \end{cases}$

25 $\begin{cases} x - 2y - 3z = 2 \\ x - 4y - 13z = 14 \\ -3x + 5y + 4z = 0. \end{cases}$

26 $\begin{cases} x + 2y - 3z + w = 10 \\ x - 2y + 3z + 2w = 4 \\ x + y + z + w = 0 \\ -x + 2y - 3z - 2w = -4. \end{cases}$

Test the sets of vectors in Exercises 27 and 28 for linear independence. If they are linearly dependent, then give a relation of linear dependence.

27 (a) $\{(1, 3, 4), (4, 0, 1), (3, 1, 2)\}$. Is this a basis for $\boldsymbol{R}^3$?
(b) $\{(2, 1, 3), (3, 3, 2), (1, 0, 1)\}$. Is this a basis for $\boldsymbol{R}^3$?

28 (a) $\{(-3, 2, 1, 4), (4, 1, 0, 2), (-10, 3, 2, 6)\}$. Is this a basis for $\boldsymbol{R}^4$?
(b) $\{(-1, 0, 1, 1), (0, 0, -1, 1), (1, 0, -1, 0)\}$. Is this a basis for $\boldsymbol{R}^4$?

29 Evaluate

$$\det\begin{pmatrix} 2 & -1 & 2 \\ 3 & 2 & -4 \\ 4 & 1 & 0 \end{pmatrix}.$$

30 Evaluate

$$\det\begin{pmatrix} 2 & -1 & 0 & 0 \\ -1 & 2 & -1 & 0 \\ 0 & -1 & 2 & -1 \\ 0 & 0 & -1 & 2 \end{pmatrix}.$$

31 Do the vectors $(-2, 0, 1)$, $(0, 3, 1)$, and $(1, 5, -3)$ form a parallelepiped in $\boldsymbol{R}^3$?

32 Use Cramer's Rule to solve the system.

$$\begin{cases} x - 2y + z = 1 \\ 3x + y - 2z = 2. \\ x - y - z = -2 \end{cases}$$

2 VECTOR FUNCTIONS

0 INTRODUCTION

Having acquired a working knowledge of $\boldsymbol{R}^n$, we are now ready to start studying functions whose domains or ranges (or both) are sets in $\boldsymbol{R}^n$. In this chapter we consider functions $\boldsymbol{f}:\boldsymbol{R}^1 \to \boldsymbol{R}^m$. If t is a real variable in the domain D of f (recall that our notation means that $D \subseteq \boldsymbol{R}$ and $\boldsymbol{f}(t) \in \boldsymbol{R}^m$ for all $t \in D$), then we write

$$\boldsymbol{f}(t) = (f_1(t), f_2(t), \ldots, f_m(t))$$

where each *coordinate function* $f_i(t)$ is a real valued function of the real variable t. Since $\boldsymbol{f}$ is obtained by putting these real valued functions together, we find that we can develop the calculus of $\boldsymbol{f}$ by simply doing the usual elementary calculus that we know with the coordinate functions $f_1, f_2, \ldots, f_m$, and then putting all those results together in a vector. Later on, this same trick will be employed to reduce the calculus of functions $\boldsymbol{F}:\boldsymbol{R}^n \to \boldsymbol{R}^m$ to the calculus of the coordinate functions $f_i:\boldsymbol{R}^n \to \boldsymbol{R}$, which we consider in the next three chapters.

In this chapter we will find that we have a natural analogue of a *curve* in the plane. We will find *tangent vectors* and *lines* to the curve by differentiation, just as we found tangent lines by differentiation in first year calculus. We will also develop a formula for *arc length* that corresponds to the formula in elementary calculus. Finally, we will see how our point of view has a natural application to the study of *moving particles* in $\boldsymbol{R}^3$, and hence to *Newtonian mechanics.* [Sir Isaac Newton (1642–1727), the co-founder of modern calculus, was also one of the giants of physics.]

1 LIMITS, CONTINUITY, DIFFERENTIATION

The objects of study in this chapter are vector valued functions of a single real variable.

1.1 DEFINITION. A **vector function** (or **vector valued function**) on $\boldsymbol{R}$ is a function $\boldsymbol{f}:\boldsymbol{R}\to\boldsymbol{R}^m$ for some fixed m (often $m=2$ or 3). We write

$$\boldsymbol{f}(t)=(x_1(t),\ldots,x_m(t))=\boldsymbol{x}(t)\in\boldsymbol{R}^m.$$

The real valued functions f_i given by

$$f_i(t)=x_i(t),\qquad i=1,2,\ldots,m,$$

are called the **coordinate functions** of $\boldsymbol{f}$.

Vector functions occur frequently in applications, such as the two cited in the next example.

1.2 EXAMPLE. (a) The path described by a moving particle in $\boldsymbol{R}^3$ is given by a vector function $\boldsymbol{p}$. The position at time t is given by

$$\boldsymbol{p}(t)=(x(t),y(t),z(t))=\boldsymbol{x}(t).$$

To study such motion, Newton assumed that $\boldsymbol{p}$ is continuous and differentiable. Before getting to a discussion of moving particles, then, we have to define continuity and differentiability of vector functions.

(b) In studying population growth, demographers in each country try to develop a mathematical model to describe the population at time t. As an example, the English economist, Thomas R. Malthus (1766–1834) constructed the *exponential growth* model described by $\dfrac{dP}{dt}=kP$, where k is the constant called the *continuous growth rate*. Then (see Section 8.2), $P(t)=P_0e^{kt}$. A *world* population growth model would just be a function $\boldsymbol{f}:\boldsymbol{R}\to\boldsymbol{R}^m$, where m is the number of countries in the world (about 150), and where $f_i(t)$ is the formula for the population in country i at time t. The total world population at time t, which is the sum of the $f_i(t)$ over all i, might be less interesting than the nature of certain $f_i(t)$ corresponding to countries where the growth rate is high. (It turns out that $f_i(t)$ is usually more complicated than the simple function P_0e^{kt} of Malthus. We will return to this in our study of differential equations. See Example 8.2.3.)

In both parts of the preceding example, it seems natural to study the vector function $\boldsymbol{f}$ by studying its coordinate functions $f_i(t)$. It is these coordinate functions that we will use to develop the calculus of a vector function $\boldsymbol{f}$.

1.3 DEFINITION. If $\boldsymbol{f}:\boldsymbol{R}\to\boldsymbol{R}^m$ is a vector function, then $\lim_{t\to t_0}\boldsymbol{f}(t)=\boldsymbol{v}\in\boldsymbol{R}^m$ means $\lim_{t\to t_0} f_i(t)=v_i$ for each $i=1,2,\ldots,m$. Here $\boldsymbol{v}=(v_1,v_2,\ldots,v_m)$.

1.4 EXAMPLE. If $\boldsymbol{f}(t)=(\cos t,\sin t,t)$, then find $\lim_{t\to\pi/2}\boldsymbol{f}(t)$.

Solution.

$$\lim_{t\to\pi/2} f(t) = \left(\lim_{t\to\pi/2} \cos t, \lim_{t\to\pi/2} \sin t, \lim_{t\to\pi/2} t\right)$$
$$= (0, 1, \pi/2).$$

It is now clear how continuity and differentiability for a vector function f should be defined.

1.5 DEFINITION. A vector function $f: \boldsymbol{R} \to \boldsymbol{R}^m$ is **continuous** at the point $t = t_0$ if and only if each of its coordinate functions $f_i(t)$ is continuous at the point $t = t_0$. The function f is **differentiable** at $t = t_0$ if and only if each of its coordinate functions is differentiable at t_0. In this case we write

$$\frac{df}{dt}(t_0) = \dot{x}(t_0) = \left(\frac{df_1}{dt}(t_0), \frac{df_2}{dt}(t_0), \ldots, \frac{df_m}{dt}(t_0)\right).$$

The notation $\dot{x}(t)$ is used in place of $x'(t)$ according to the custom, which goes back to Newton, of denoting differentiation with respect to *time* by a dot instead of a prime. Since one of our principal applications will be to Newtonian mechanics, we follow this custom.

1.6 EXAMPLE. Is f in Example 1.4 continuous at $t_0 = \pi/2$? Is it differentiable there? If so, find $\dot{f}(t_0)$.

Solution. Each coordinate function of f is not only continuous, but also differentiable at every point $t \in \boldsymbol{R}$. Hence, f is continuous and differentiable at $t_0 = \pi/2$. We have

$$\dot{f}(t) = (-\sin t, \cos t, 1),$$

so

$$\dot{f}(\pi/2) = (-1, 0, 1).$$

We remark that in view of Definition 1.3, we could write the condition for f to be continuous at $t = t_0$ as

(1) $$\lim_{t\to t_0} f(t) = f(t_0),$$

since our definition requires for each i that

$$\lim_{t\to t_0} f_i(t) = f_i(t_0).$$

Equation (1) looks *exactly* like the definition of continuity given in elementary calculus, as well it should. We emphasize again that we extend the concept to vector functions by putting together the notion of continuity for the coordinate functions, just as we create vector functions by putting together the real valued coordinate functions. If we wanted to take the time, we could prove that the sum of continuous functions is continuous, and other results akin to theorems of first year calculus. Some of these are in the exercises, but in this section we will content ourselves with the following collection of properties of differentiation.

1.7 THEOREM. Let f and g be differentiable vector functions, and h a differentiable scalar function. Put $\boldsymbol{x}(t)=f(t)$, $\boldsymbol{y}(t)=g(t)$. Then

(1) $\dfrac{d}{dt}[\boldsymbol{x}(t)+\boldsymbol{y}(t)]=\dot{\boldsymbol{x}}(t)+\dot{\boldsymbol{y}}(t)$.

(2) $\dfrac{d}{dt}[\boldsymbol{x}(t)\cdot\boldsymbol{y}(t)]=\dot{\boldsymbol{x}}(t)\cdot\boldsymbol{y}(t)+\boldsymbol{x}(t)\cdot\dot{\boldsymbol{y}}(t)$.

(3) $\dfrac{d}{dt}[|\boldsymbol{x}(t)|]=\dfrac{1}{|\boldsymbol{x}(t)|}[\boldsymbol{x}(t)\cdot\dot{\boldsymbol{x}}(t)]$ if $\boldsymbol{x}(t)\neq\boldsymbol{0}$.

(4) $\dfrac{d}{dt}[h(t)\boldsymbol{x}(t)]=\dot{h}(t)\boldsymbol{x}(t)+h(t)\dot{\boldsymbol{x}}(t)$.

(5) If $m=3$, then

$$\frac{d}{dt}[\boldsymbol{x}(t)\times\boldsymbol{y}(t)]=\dot{\boldsymbol{x}}(t)\times\boldsymbol{y}(t)+\boldsymbol{x}(t)\times\dot{\boldsymbol{y}}(t).$$

(6) *Chain rule:* $\dfrac{d}{ds}\boldsymbol{x}(t(s))=$ [redacted] where $t=t(s)$ is a scalar function.

Partial Proof. We leave (1), (2), and (4) for Exercises 15, 16, and 17 below.

(3) If $\boldsymbol{x}(t)=(f_1(t), f_2(t), \ldots, f_m(t))$, then

$$|\boldsymbol{x}(t)|=[(f_1(t))^2+(f_2(t))^2+\ldots+(f_m(t))^2]^{1/2}.$$

This is just a real valued function of the real variable t, so we can differentiate it using elementary calculus.

$$\frac{d}{dt}|\boldsymbol{x}(t)|=\tfrac{1}{2}[(f_1(t))^2+(f_2(t))^2+\ldots+(f_m(t))^2]^{-1/2}\left[2f_1(t)\frac{df_1}{dt}(t)\right.$$
$$\left.+2f_2(t)\frac{df_2}{dt}(t)+\ldots+2f_m(t)\frac{df_m}{dt}(t)\right].$$
$$=\frac{1}{|\boldsymbol{x}(t)|}[\boldsymbol{x}(t)\cdot\dot{\boldsymbol{x}}(t)],$$

since $\frac{1}{2}$ cancels all the 2's.

(5) Here we can write $\boldsymbol{x}(t)=(f_1(t), f_2(t), f_3(t))$ and $\boldsymbol{y}(t)=(g_1(t), g_2(t), g_3(t))$. Then

$$\boldsymbol{x}(t)\times\boldsymbol{y}(t)=(f_2(t)g_3(t)-f_3(t)g_2(t), f_3(t)g_1(t)-f_1(t)g_3(t), f_1(t)g_2(t)-f_2(t)g_1(t)).$$

To differentiate this, we differentiate each term and use the rule for the derivative of the product of real valued functions. We get

$$\begin{aligned}
\frac{d}{dt}(\boldsymbol{x}(t)\times\boldsymbol{y}(t))&=(f_2'(t)g_3(t)-f_3'(t)g_2(t)+f_2(t)g_3'(t)-f_3(t)g_2'(t),\\
&\quad f_3'(t)g_1(t)-f_1'(t)g_3(t)+f_3(t)g_1'(t)-f_1(t)g_3'(t),\\
&\quad f_1'(t)g_2(t)-f_2'(t)g_1(t)+f_1(t)g_2'(t)-f_2(t)g_1'(t))\\
&=(f_2'(t)g_3(t)-f_3'(t)g_2(t), f_3'(t)g_1(t)-f_1'(t)g_3(t),\\
&\quad f_1'(t)g_2(t)-f_2'(t)g_1(t))\\
&\quad+(f_2(t)g_3'(t)-f_3(t)g_2'(t), f_3(t)g_1'(t)-f_1(t)g_3'(t),\\
&\quad f_1(t)g_2'(t)-f_2(t)g_1'(t))\\
&=(\dot{\boldsymbol{x}}(t)\times\boldsymbol{y}(t))+(\boldsymbol{x}(t)\times\dot{\boldsymbol{y}}(t)).
\end{aligned}$$

(6) We have

$$\frac{d}{ds}[\boldsymbol{x}(t(s))] = \left(\frac{df_1}{ds}(t(s)), \frac{df_2}{ds}(t(s)), \ldots, \frac{df_m}{ds}(t(s))\right)$$
$$= \left(\frac{df_1}{dt}\frac{dt}{ds}, \frac{df_2}{dt}\frac{dt}{ds}, \ldots, \frac{df_m}{dt}\frac{dt}{ds}\right)$$
$$= \left(\frac{df_1}{dt}, \frac{df_2}{dt}, \ldots, \frac{df_m}{dt}\right)\frac{dt}{ds}$$
$$= \frac{d\boldsymbol{x}(t)}{dt}\frac{dt}{ds}.$$

QED

Differentiation also has an interpretation involving tangent lines, just as it did in elementary calculus. First we need the notion of a curve in $\boldsymbol{R}^m$.

1.8 **DEFINITION.** A **curve in $\boldsymbol{R}^m$** is the range of some continuous vector function $f: \boldsymbol{R} \to \boldsymbol{R}^m$.

If we write

(*) $$f(t) = (f_1(t), f_2(t), \ldots, f_m(t)) = (x_1, x_2, \ldots, x_m) = \boldsymbol{x}(t),$$

then we imagine the curve being traced out as t varies. See Figure 1.1, which suggests a natural *orientation*: we proceed along the curve in the direction of

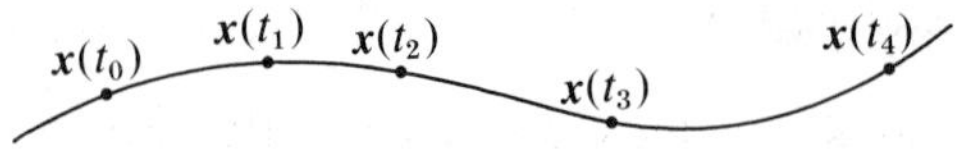

FIGURE 1.1

increasing t. That is, $\boldsymbol{x}(t_1)$ precedes $\boldsymbol{x}(t_2)$ if $t_1 < t_2$. This orientation will be particularly convenient when discussing the motion of a particle moving along the curve.

The real variable t in Definition 1.8 is called a **parameter,** and the equations $x_i = f_i(t)$ are called **parametric equations** for the curve. The process of describing a curve by giving a vector function of the form (*) is called **parametrization** of the curve. In Chapter 7 we will need to be able to parametrize common curves. Here we want to learn to recognize common curves given in parametric form. This usually involves *eliminating the parameter t,* as the following example illustrates.

1.9 **EXAMPLE.** Identify the plane curve defined by

$$\boldsymbol{x}(t) = \left(\frac{t^2}{4p}, t\right), \qquad p > 0.$$

Solution. We can calculate as many points as we like on the curve by assigning values to t. We show some of these points in tabular form. If we plot these points,

t	0	1	-1	2	-2	$2\sqrt{p}$	$-2\sqrt{p}$	4	-4
x	0	$\frac{1}{4p}$	$\frac{1}{4p}$	$\frac{1}{p}$	$\frac{1}{p}$	1	1	$\frac{4}{p}$	$\frac{4}{p}$
y	0	1	-1	2	-2	$2\sqrt{p}$	$-2\sqrt{p}$	4	-4

we get a picture like Figure 1.2(a), where we have taken $p = \frac{1}{2}$. The curve looks parabolic. Can we confirm this? We can, if we eliminate t from the equations.

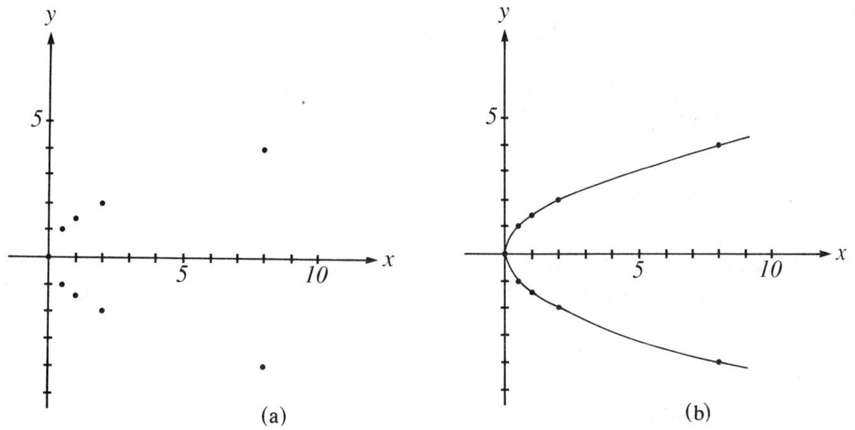

FIGURE 1.2

Since $y=t$ and $x=\dfrac{t^2}{4p}$, we have $x=\dfrac{y^2}{4p}$, i.e., $y^2=4px$. We recognize this as the equation of a parabola with vertex at (0, 0). Thus we can fill in Figure 1.2(a) to get Figure 1.2(b).

You probably recall from elementary calculus that the derivative was used to find both velocities of particles moving along straight lines and also tangent lines to curves. The derivative of a vector function can be put to the same kind of use, as follows.

Suppose that $\boldsymbol{f}:\boldsymbol{R}\to\boldsymbol{R}^m$ is differentiable at t_0. Then it is continuous (Exercise 19 below), so for small values of $h\in\boldsymbol{R}$, the endpoints of $\boldsymbol{x}(t_0)$ and $\boldsymbol{x}(t_0+h)$ are close. See Figure 1.3(a). The vector $\boldsymbol{x}(t_0+h)-\boldsymbol{x}(t_0)$ is a *secant* vector from the endpoint of $\boldsymbol{x}(t_0)$ to the endpoint of $\boldsymbol{x}(t_0+h)$. Then for $h>0$,

$$\boldsymbol{s}_h=\frac{1}{h}[\boldsymbol{x}(t_0+h)-\boldsymbol{x}(t_0)]$$

points in the same direction, which approximates the direction of motion of a particle moving along the curve. See Figure 1.3(b). Moreover, the magnitude of $\boldsymbol{s}_h$

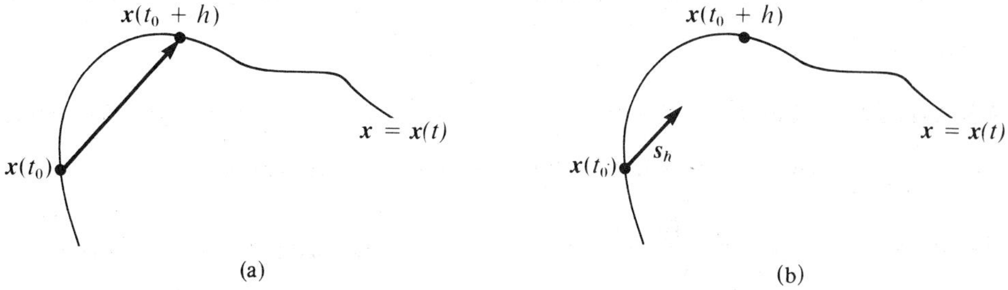

FIGURE 1.3

approximates the speed of motion along the curve, because it measures the average displacement from $\boldsymbol{x}(t_0)$ per unit time. Now, what happens as $h\to 0$? Geometrically, $\boldsymbol{s}_h$ seems to approach a tangent vector, and its magnitude seems to approach what we would reasonably think of as the instantaneous speed of the particle at $t=t_0$. Since $\boldsymbol{f}$ is differentiable, the i-th coordinate function approaches

$\frac{dx_i}{dt}(t_0)$, for $i = 1, 2, \ldots, m$. Thus s_h approaches

$$\dot{x}(t_0) = \left(\frac{dx_1}{dt}(t_0), \ldots, \frac{dx_m}{dt}(t_0)\right).$$

This leads to the following definition.

1.10 DEFINITION. Let a particle move on the curve $x = x(t)$ in $\mathbf{R}^m$. If $\dot{x}(t_0)$ exists, then the **velocity vector** along the curve at $x(t_0)$ (or at time $t = t_0$) is

$$\boxed{v(t_0) = \dot{x}(t_0) = \frac{dx}{dt}(t_0).}$$

The **speed** of motion at $x(t_0)$ is the magnitude $|v(t_0)|$ of the velocity at t_0. If $\dot{x}(t_0) \neq \mathbf{0}$, then the **unit tangent vector** at $x(t_0)$ is

$$\boxed{T(t_0) = \frac{1}{|\dot{x}(t_0)|}\dot{x}(t_0).}$$

In Definition 1.10, we emphasize that the velocity vector $v(t_0) = \dot{x}(t_0)$ is *itself* a tangent vector to the curve at each point t_0 where $x(t)$ is differentiable. In fact, our development gives precisely the notion of tangent you saw in elementary calculus in case $m = 2$. A parametric representation of the graph of $y = g(x)$ is obtained by letting $x = t$ and $y = g(t)$, that is,

$$x(t) = (t, g(t)).$$

A tangent vector at (x_0, y_0) is then

$$v(t_0) = (1, \dot{g}(t_0)) = \left(1, \frac{dg}{dx}(x_0)\right)$$

since $t = x$. Because the tangent line ℓ is determined by the direction of $v(t_0)$, we see that ℓ has slope $\frac{dg}{dx}(x_0)$, which indeed agrees with the elementary calculus definition of tangent line!

1.11 EXAMPLE. Find the velocity and speed of a particle moving on the curve $x(t) = \left(\frac{t^2}{4p}, t\right)$ when $t = p$. What is the unit tangent at that point?

Solution. Here $\dot{x}(t) = \left(\frac{t}{2p}, 1\right)$. When $t = p$, we have $\dot{x}(p) = (\frac{1}{2}, 1)$. This is the velocity. The speed is

$$|\dot{x}(p)| = \sqrt{\tfrac{1}{4} + 1} = \tfrac{1}{2}\sqrt{5}.$$

The unit tangent vector is

$$\frac{1}{|\dot{x}(p)|}\dot{x}(p) = \frac{2}{\sqrt{5}}(\tfrac{1}{2}, 1) = \left(\frac{1}{\sqrt{5}}, \frac{2}{\sqrt{5}}\right).$$

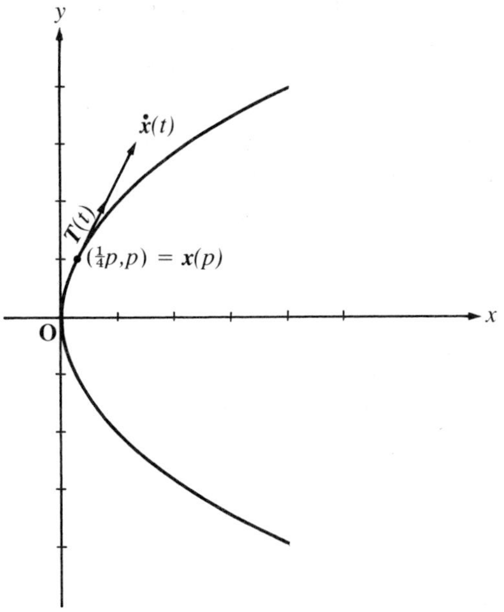

FIGURE 1.4

We show these in Figure 1.4, where we have taken $p = 1$.

We now present an example of an application of Theorem 1.7. We will use (3) of that theorem to establish the *reflection property* for parabolas. This property is at the heart of several important technological applications. For example, the headlamp reflectors on most American-made automobiles are in the shape of a paraboloid, the surface that results when a parabola, such as that in Examples 1.9 and 1.11, is revolved about its principal axis. The reason for this design is that if a tiny bulb is placed at the focus, then every light beam the bulb emits is reflected off the parabolic surface parallel to the principal axis. To an oncoming driver, the entire surface appears illuminated. See Figure 1.5. The same principle, in reverse, is utilized in making reflecting telescopes and radar scanners. For all practical purposes, for example, the light rays from a star strike the parabolic reflecting

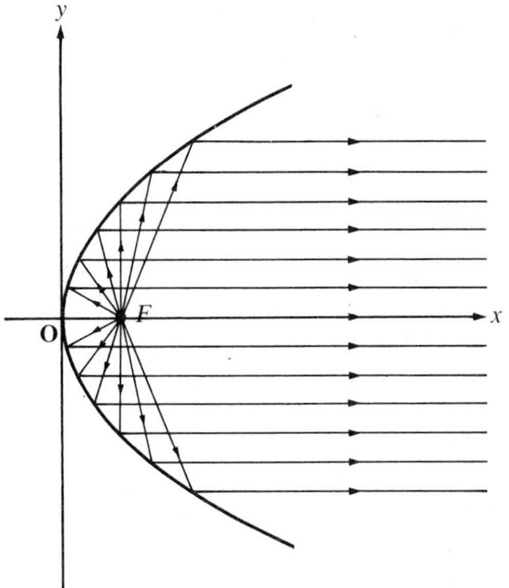

FIGURE 1.5

surface of a telescope parallel to the principal axis. They are then all reflected to the focus for viewing. The mathematical basis for these reflecting properties is contained in the following example.

1.12 EXAMPLE. Show that the angles α and β in Figure 1.6 are equal, where F is the focus of the parabola, P is any point on the parabola, and ℓ is the tangent line at P. (Note that α is the angle of incidence and β is the angle of reflection of a light beam emanating from the focus F.)

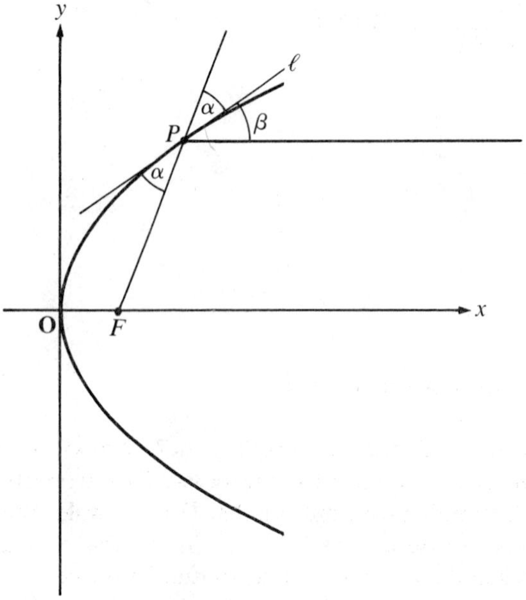

FIGURE 1.6

Proof. Set up coordinate axes so that the parabola is the graph of $y^2=4px$, i.e., $\boldsymbol{x}(t)=\left(\frac{t^2}{4p}, t\right)$ (see Example 1.9). In Figure 1.7, we see that α is the angle between $\dot{\boldsymbol{x}}(t)$ and the vector $\boldsymbol{v}=\overrightarrow{FP}$ (which we have drawn with the initial point P). Also, β is the angle between $\dot{\boldsymbol{x}}(t)$ and $\boldsymbol{w}=\overrightarrow{QP}$ (which we have also drawn from P). The point Q is on the directrix $x=-p$ at the same height as P, so it has coordinates $(-p, t)$. Thus $\overrightarrow{OQ}=(-p, t)$, so $\frac{d}{dt}(\overrightarrow{OQ})=(0, 1)=\boldsymbol{j}$. From the defining condition of a parabola, we have

$$|\overrightarrow{QP}|=|\overrightarrow{FP}|,$$

i.e.,

$$|\boldsymbol{x}(t)-\overrightarrow{OQ}|=|\boldsymbol{x}(t)-\overrightarrow{OF}|$$

Differentiating this with respect to t using Theorem 1.7(3) and (6), we get

(2) $$\frac{1}{|\boldsymbol{x}(t)-\overrightarrow{OQ}|}(\boldsymbol{x}(t)-\overrightarrow{OQ})\cdot(\dot{\boldsymbol{x}}(t)-\boldsymbol{j})=\frac{1}{|\boldsymbol{x}(t)-\overrightarrow{OF}|}(\boldsymbol{x}(t)-\overrightarrow{OF})\cdot(\dot{\boldsymbol{x}}(t)-\boldsymbol{0}).$$

Since $\boldsymbol{x}(t)-\overrightarrow{OQ}$ is parallel to the x-axis, it is perpendicular to $\boldsymbol{j}$, so that

$$\begin{aligned}(\boldsymbol{x}(t)-\overrightarrow{OQ})\cdot(\dot{\boldsymbol{x}}(t)-\boldsymbol{j})&=(\boldsymbol{x}(t)-\overrightarrow{OQ})\cdot\dot{\boldsymbol{x}}(t)-(\boldsymbol{x}(t)-\overrightarrow{OQ})\cdot\boldsymbol{j}\\&=(\boldsymbol{x}(t)-\overrightarrow{OQ})\cdot\dot{\boldsymbol{x}}(t)-0\\&=(\boldsymbol{x}(t)-\overrightarrow{OQ})\cdot\dot{\boldsymbol{x}}(t).\end{aligned}$$

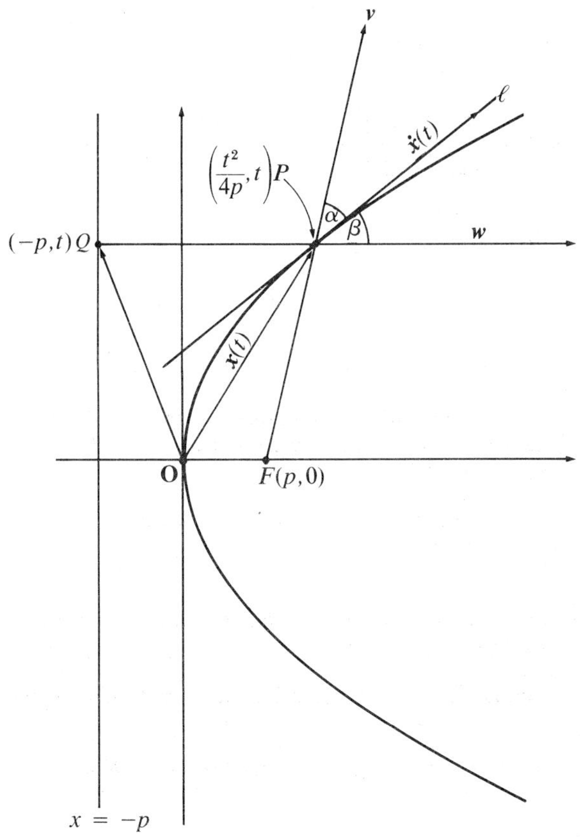

FIGURE 1.7

Thus (2) becomes

$$\frac{\boldsymbol{x}(t)-\overrightarrow{OQ}}{|\boldsymbol{x}(t)-\overrightarrow{OQ}|}\cdot\dot{\boldsymbol{x}}(t)=\frac{\boldsymbol{x}(t)-\overrightarrow{OF}}{|\boldsymbol{x}(t)-\overrightarrow{OF}|}\cdot\dot{\boldsymbol{x}}(t).$$

Dividing both sides by $|\dot{\boldsymbol{x}}(t)|$, we obtain

$$\frac{(\boldsymbol{x}(t)-\overrightarrow{OQ})\cdot\dot{\boldsymbol{x}}(t)}{|\boldsymbol{x}(t)-OQ|\,|\dot{\boldsymbol{x}}(t)|}=\frac{(\boldsymbol{x}(t)-\overrightarrow{OF})\cdot\dot{\boldsymbol{x}}(t)}{|\boldsymbol{x}(t)-\overrightarrow{OF}|\,|\dot{\boldsymbol{x}}(t)|}$$

i.e.,

$$\cos\alpha=\cos\beta.$$

Since α and β are between 0 and π, we can conclude that $\alpha=\beta$. QED

To complete the extension of the main ideas of elementary calculus to vector functions, we now introduce the notion of the integral of a vector function f over a closed interval of $\boldsymbol{R}$.

1.13 DEFINITION. If $f:\boldsymbol{R}\to\boldsymbol{R}^m$ is a vector function such that each coordinate function $f_i(t)$ is integrable over $[a, b]$, then

$$\int_a^b f(t)\,dt=\left(\int_a^b f_1(t)\,dt, \int_a^b f_2(t)\,dt, \ldots, \int_a^b f_m(t)\,dt\right).$$

1.14 EXAMPLE. If $f(t)=\left(\frac{t^2}{4p}, t\right)$, then

$$\int_0^1 f(t)\,dt=\left(\int_0^1 \frac{t^2}{4p}\,dt, \int_0^1 t\,dt\right)=\left(\frac{t^3}{12p}\Big]_0^1, \frac{t^2}{2}\Big]_0^1\right)$$
$$=\left(\frac{1}{12p}, \frac{1}{2}\right).$$

EXERCISES 2.1

1 (a) Show that $f(t)=(\cos t, \sin t)$ is continuous and differentiable for all t.
(b) What is the curve $x(t)=(\cos t, \sin t)$? (c) Find a tangent vector at $t=0$.
(d) Show that $\dot{x}(t)$ is orthogonal to $x(t)$ for all t.

2 (a) Show that $f(t)=(2\cos t, 3\sin t)$ is continuous and differentiable for all t.
(b) Identify the curve $x(t)=(2\cos t, 3\sin t)$.
(c) Find a tangent vector at $t=\pi/4$.

3. If $x(t)=(t-1, t^2+1, t^3-1)$, then find the equation of the tangent line to the curve $x=x(t)$ when $t=2$. What is the speed at $t=2$?

4 If $x(t)=(\cos t, \sin t, t)$, then find the equation of the tangent line at $t=\pi/2$. What is the speed at $t=\pi/2$?

5 Find the equation of the tangent line to the curve $x(t)=(2t^2, t, t^3)$ at $t=1$. What is the speed at $t=1$?

6 Find the equation of the tangent line to the curve $x(t)=(t\sin t, 3t, t\cos t)$ when $t=\pi$. What is the speed at that point?

7 Find the equation of the tangent line to the curve $x=x(t)=(t, \sin 4t, \cos 4t)$ at the point $t=\pi/8$.

8 Find the equation of the tangent line to the curve $x=x(t)=(2e^t, 2e^{-t}, 5\cos t)$ at the point $t=0$.

9 If

$$f(t)=\left(\frac{\sin t}{t}, \frac{1-\cos t}{t}, t^2\right),$$

then find $\lim_{t\to 0} f(t)$. Is f continuous at $t=0$?

10 If

$$f(t)=\left(\frac{t^2}{\sin t}, \frac{t}{1-\cos t}, t^2\right),$$

then find $\lim_{t\to 0} f(t)$. Is f continuous at $t=0$?

11 If $x(t)=(\cos t, \sin t, t)$ and $y(t)=(2, t^2, -4t)$, then find

(a) $\frac{d}{dt}[x(t)\times y(t)]$ (b) $\frac{d}{dt}[x(t)\cdot y(t)]$

12 If $x(t)=(e^t, \ln t, t)$ and $y(t)=(\cos t, \sin t, t)$, then find

(a) $\frac{d}{dt}[x(t)\times y(t)]$ (b) $\frac{d}{dt}[x(t)\cdot y(t)]$

13 Establish the *reflecting property for ellipses:* If F_1 and F_2 are the foci of an ellipse and P is any point on the ellipse, then the line segments F_1P and F_2P make equal angles with the tangent line at P. (*Hint:* The defining equation of the ellipse is

$$|\boldsymbol{x}(t)-\overrightarrow{OF_1}|+|\boldsymbol{x}(t)-\overrightarrow{OF_2}|=2a.)$$

14 Establish the *reflecting property for hyperbolas*: If F_1 and F_2 are the foci of a hyperbola and P is any point on the hyperbola, then the tangent vector $\dot{\boldsymbol{x}}(t)$ at P makes equal angles with the vectors $\overrightarrow{F_1P}$ and $\overrightarrow{F_2P}$. (*Hint:* The defining equation of the hyperbola is

$$|\boldsymbol{x}(t)-\overrightarrow{OF_1}|-|\boldsymbol{x}(t)-\overrightarrow{OF_2}|=-2a$$

if $\boldsymbol{x}(t)$ is on the branch nearer F_1.)

15 Prove Theorem 1.7(1). **16** Prove Theorem 1.7(2). **17** Prove Theorem 1.7(4).

18 If $\boldsymbol{f}$ and $\boldsymbol{g}$ are continuous at t_0, then prove that any linear combination $a\boldsymbol{f}+b\boldsymbol{g}$ is continuous at t_0, where $a, b \in \boldsymbol{R}$ are constants.

19 Prove that if $\boldsymbol{f}:\boldsymbol{R}\to\boldsymbol{R}^m$ is differentiable at t_0, then it is continuous there.

20 If $\boldsymbol{f}(t)=(\cos t, \sin t)$, then find

$$\int_0^{\pi} \boldsymbol{f}(t)\,dt.$$

21 If $\boldsymbol{f}(t)=(\cos t, \sin t, t)$, then find

$$\int_{\pi/2}^{2\pi} \boldsymbol{f}(t)\,dt.$$

22 If $\boldsymbol{f}$ and $\boldsymbol{g}$ are integrable over $[a, b]$ and $c, k \in \boldsymbol{R}$, then show that

$$\int_a^b [c\boldsymbol{f}(t)+k\boldsymbol{g}(t)]\,dt = c\int_a^b \boldsymbol{f}(t)\,dt + k\int_a^b \boldsymbol{g}(t)\,dt.$$

23 If a particle moves on a curve so that $|\boldsymbol{x}(t)|$ is constant, then show that $\boldsymbol{x}(t)$ is perpendicular to $\boldsymbol{v}(t)$ for all t. (*Hint:* Express $|\boldsymbol{x}(t)|$ in terms of the dot product and differentiate.)

24 If $\boldsymbol{x}=\boldsymbol{x}(t)$ is a curve whose tangent vector is always perpendicular to $\boldsymbol{x}(t)$, then show that $|\boldsymbol{x}(t)|$ is constant. (*Hint:* Differentiate $|\boldsymbol{x}(t)|^2$.)

25 If $\boldsymbol{x}=\boldsymbol{x}(t)$ is the path of a particle with variable mass $m(t)$, then the *linear momentum* of the particle is the vector $\boldsymbol{p}(t)=m(t)\dot{\boldsymbol{x}}(t)$. Show that if $\dfrac{d}{dt}(m(t)\dot{\boldsymbol{x}}(t))=\boldsymbol{0}$ for all t, then $\boldsymbol{p}(t)$ is constant. (This is one form of the *Law of Conservation of Linear Momentum.*)

26 If $\boldsymbol{x}=\boldsymbol{x}(t)$ is the path of a particle with variable mass $m(t)$, then the torque about $\boldsymbol{0}$ is defined to be

$$\boldsymbol{L}(t)=\boldsymbol{x}(t)\times\frac{d}{dt}(m(t)\dot{\boldsymbol{x}}(t)).$$

The *angular momentum* about $\boldsymbol{0}$ is defined to be $\boldsymbol{M}(t)=\boldsymbol{x}(t)\times m(t)\dot{\boldsymbol{x}}(t)$. Show that $\dfrac{d\boldsymbol{M}(t)}{dt}=\boldsymbol{L}(t)$.

27 Refer to Exercise 26. Show that if the torque is identically zero, then the angular momentum is constant. (This is one form of the *Law of Conservation of Angular Momentum.*)

2 VELOCITY, ACCELERATION, ARC LENGTH

If a particle moves in $\boldsymbol{R}^m$ along the path $\boldsymbol{x}=\boldsymbol{x}(t)$, then we have already seen (Definition 1.10) that its velocity at time t_0 is defined by

$$\boldsymbol{v}(t_0)=\dot{\boldsymbol{x}}(t_0)=\lim_{t\to t_0}\frac{\boldsymbol{x}(t)-\boldsymbol{x}(t_0)}{t-t_0}.$$

The velocity vector $\boldsymbol{v}(t_0)$ is a tangent vector to the curve $\boldsymbol{x}=\boldsymbol{x}(t)$ at $t=t_0$, as we also have seen. Thus, at any instant of time, the velocity vector points in the direction of motion. We have also defined the speed at $t=t_0$ to be the magnitude of the velocity,

$$|\boldsymbol{v}(t_0)|=\sqrt{\dot{x}_1(t)^2+\dot{x}_2(t)^2+\ldots+\dot{x}_n(t)^2}=\sqrt{\boldsymbol{v}(t_0)\cdot\boldsymbol{v}(t_0)}.$$

To complete the Newtonian model for the motion, we need two additional definitions.

2.1 DEFINITION. If a particle of mass m moves on the curve $\boldsymbol{x}=\boldsymbol{x}(t)$, then its **acceleration** at any time t_0 is

$$\boldsymbol{a}(t_0)=\lim_{t\to t_0}\frac{\boldsymbol{v}(t)-\boldsymbol{v}(t_0)}{t-t_0}=\dot{\boldsymbol{v}}(t_0)=\ddot{\boldsymbol{x}}(t_0).$$

(In this notation, the double dots indicate the second derivative with respect to t.) The **force** acting on the particle at time t_0 is defined to be

$$\boldsymbol{F}(t_0)=m\boldsymbol{a}(t_0).$$

(This latter definition is sometimes referred to as *Newton's second law of motion.*)

2.2 EXAMPLE. Describe the path in $\boldsymbol{R}^3$ of a particle of mass $m\neq 0$ on which there is an equilibrium of forces, in the sense that $\boldsymbol{F}(t)=\boldsymbol{0}$.

Solution. Here, since $m\neq 0$, we have $\boldsymbol{a}(t)=\boldsymbol{0}=(0, 0, 0)$. Thus,

$$\left(\frac{d^2x}{dt^2}, \frac{d^2y}{dt^2}, \frac{d^2z}{dt^2}\right)=(0, 0, 0).$$

Hence $\boldsymbol{v}(t)=(c_1, c_2, c_3)$ is constant. We thus have

$$\left(\frac{dx}{dt}, \frac{dy}{dt}, \frac{dz}{dt}\right)=(c_1, c_2, c_3).$$

Integration of this gives $x=c_1t+k_1$, $y=c_2t+k_2$, $z=c_3t+k_3$. In vector form, then,

$$\boldsymbol{x}=t(c_1, c_2, c_3)+(k_1, k_2, k_3)=t\boldsymbol{c}+\boldsymbol{k}.$$

This is the vector equation of a straight line. Hence the particle moves on a straight line with constant velocity.

2.3 EXAMPLE. A particle moves with constant speed. Show that its acceleration vector, when not zero, is always perpendicular to its velocity vector.

Proof. We have $\sqrt{\boldsymbol{v}(t)\cdot\boldsymbol{v}(t)}=c$. Thus $\boldsymbol{v}(t)\cdot\boldsymbol{v}(t)=c^2$. If we differentiate each side of the latter equation, and use Theorem 1.7(2), then we get $2\boldsymbol{v}(t)\cdot\dot{\boldsymbol{v}}(t)=0$. That is, $2\boldsymbol{v}(t)\cdot\boldsymbol{a}(t)=0$. Hence $\boldsymbol{v}(t)\cdot\boldsymbol{a}(t)=0$. So either $\boldsymbol{a}(t)=0$ or else $\boldsymbol{a}(t)\perp\boldsymbol{v}(t)$. QED

2.4 **EXAMPLE.** A particle moves on the ellipse

$$\frac{x^2}{a^2}+\frac{y^2}{b^2}=1.$$

Find the velocity, speed, and acceleration.

Solution. We can give a vector equation of the ellipse if we let t be the central angle shown in Figure 2.1. For then $x=a\cos t$ and $y=b\sin t$, since

$$\cos^2 t+\sin^2 t=\frac{x^2}{a^2}+\frac{y^2}{b^2}=1.$$

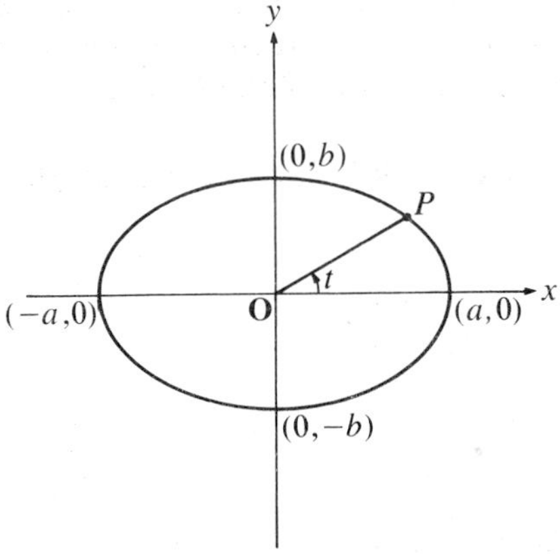

FIGURE 2.1

Thus $\boldsymbol{x}(t)=(a\cos t, b\sin t)$. Then by differentiation $\boldsymbol{v}(t)=(-a\sin t, b\cos t)$, and $\boldsymbol{a}(t)=(-a\cos t, -b\sin t)$. The speed is $|\boldsymbol{v}(t)|=\sqrt{a^2\sin^2 t+b^2\cos^2 t}$.

Now we turn to the problem of calculating the length of arc along a curve $\boldsymbol{x}=\boldsymbol{x}(t)$ between $t=a$ and $t=b$. To begin with, there is no problem if the speed is a constant, r. For then the formula $d=r(b-a)$ should apply, since with constant speed r, the total length covered in a time interval of length $b-a$ should be the product of the speed r and the time elapsed.

In the general case, we assume that the path is *smooth*. By this we mean that $\boldsymbol{x}=\boldsymbol{x}(t)$ has a continuous derivative $\dot{\boldsymbol{x}}(t)$ on $[a, b]$. Following an idea from elementary calculus, we partition the interval into a large number of subintervals $[t_{i-1}, t_i]$, each of very short length. Then, since $\boldsymbol{x}(t)$ is smooth, the speed $|\dot{\boldsymbol{x}}(t)|$ is continuous on each short interval $[t_{i-1}, t_i]$. The speed therefore will not vary much over each interval. Hence we won't err by much if we say $|\dot{\boldsymbol{x}}(t)|$ is approximately $|\dot{\boldsymbol{x}}(u_i)|$ for any point $u_i\in[t_{i-1}, t_i]$. Then the arc length over $[t_{i-1}, t_i]$ will be approximated by

$$|\dot{\boldsymbol{x}}(u_i)|\,(t_i-t_{i-1})=|\dot{\boldsymbol{x}}(u_i)|\,\Delta t_i.$$

We can then reason that the arc length over the *entire* interval $[a, b]$ is the sum of the arc lengths over the subintervals $[t_{i-1}, t_i]$. See Figure 2.2. Hence the arc length over the interval $[a, b]$ should be approximated by $\Sigma_i\,|\dot{\boldsymbol{x}}(u_i)|\,\Delta t_i$. The latter sum is just a *Riemann sum* for the speed function over the interval $[a, b]$. As we make

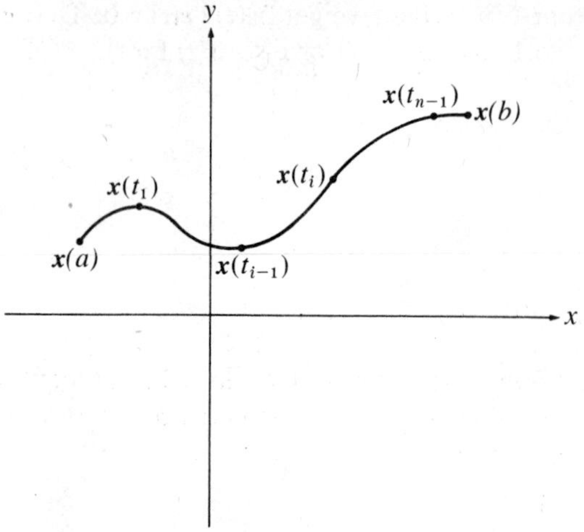

FIGURE 2.2

subintervals of smaller and smaller length, we expect to get a closer and closer approximation to the actual arc length over $[a, b]$. Hence it seems reasonable to *define* the arc length over $[a, b]$ to be the definite integral of the speed.

2.5 DEFINITION. If $\boldsymbol{x}=\boldsymbol{x}(t)$ defines a smooth curve in $\boldsymbol{R}^m$, then the **arc length** of the curve between $t=a$ and $t=b$ is

$$L=\int_a^b |\dot{\boldsymbol{x}}(t)|\,dt.$$

The **arc length function** is the real valued function $s:[a, b]\rightarrow \boldsymbol{R}$ defined by

$$s(t)=\int_a^t |\dot{\boldsymbol{x}}(u)|\,du$$

Definition 2.5 appears to suggest that the length of arc of a curve might depend on *how* the curve is parametrized. Fortunately, this is not the case, and we can easily show this as follows. Suppose that we have

$$\boldsymbol{x}=\boldsymbol{f}(t), \qquad t\in[a, b]$$

and also

$$\boldsymbol{x}=\boldsymbol{g}(u), \qquad u\in[c, d]$$

where $t=h(u)$ is such that $a=h(c)$ and $b=h(d)$, and h is continuous. (This assumption guarantees that all integrals below are well defined.) Then we have

$$\boldsymbol{x}=\boldsymbol{f}(t)=\boldsymbol{f}(h(u)).$$

Thus from Theorem 1.7(6),

$$\frac{d\boldsymbol{x}}{du}=\frac{d\boldsymbol{f}}{dt}\frac{dt}{du}.$$

The technique of change of variables from elementary calculus gives us

$$\int_a^b |\dot{\boldsymbol{x}}(u)|\,du = \int_a^b \left|\frac{d\boldsymbol{x}}{du}\right| du = \int_{u=a}^{u=b} \left|\frac{df}{dt}\frac{dt}{du}\right| du$$
$$= \int_c^d |\dot{\boldsymbol{x}}(t)|\,dt.$$

We therefore get the same arc length no matter which parametrization of the curve is used in Definition 2.5. (We will return to this point in Theorem 7.1.8.)

Since we assume that $\boldsymbol{x}(t)$ is smooth, $\dot{\boldsymbol{x}}(t)$ is continuous, and hence so is its length, $|\dot{\boldsymbol{x}}(t)|$. So every smooth curve has an arc length over a finite interval. Notice that if $\boldsymbol{x}(t) = (x_1(t), x_2(t), \ldots, x_m(t))$, then we have

$$L = \int_a^b \sqrt{\dot{x}_1(t)^2 + \dot{x}_2(t)^2 + \ldots + \dot{x}_m(t)^2}\,dt.$$

2.6 EXAMPLE. Find the length of arc of one quarter of the circle $x^2 + y^2 = a^2$. See Figure 2.3.

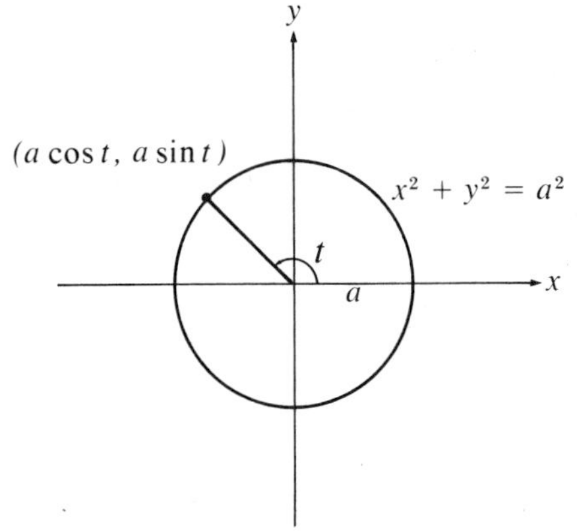

FIGURE 2.3

Solution. We write the curve in vector form as $\boldsymbol{x} = \boldsymbol{x}(t) = (a \cos t, a \sin t), 0 \le t \le 2\pi$. Then $\dot{\boldsymbol{x}}(t) = (-a \sin t, a \cos t)$. Hence the speed is

$$|\dot{\boldsymbol{x}}(t)| = \sqrt{a^2 \sin^2 t + a^2 \cos^2 t} = a.$$

So the required arc length is

$$\int_0^{\pi/2} a\,dt = \frac{\pi}{2}a.$$

Notice that the entire circle has circumference (arc length) $\int_0^{2\pi} a\,dt = 2\pi a$, in agreement with the formula you have known since elementary school.

2.7 EXAMPLE. Find the length of arc along the parabola $y^2 = 4px$ between $x = 0$ and $x = p$. See Figure 2.4.

Solution. Here we have, as in Example 1.9, $\boldsymbol{x}(t) = \left(\frac{t^2}{4p}, t\right)$ as a vector equation. For $x = 0$ we have $t = 0$. For $x = p$, we need $\frac{t^2}{4p} = p$, that is, $t^2 = 4p^2$, so $t = 2p$ will

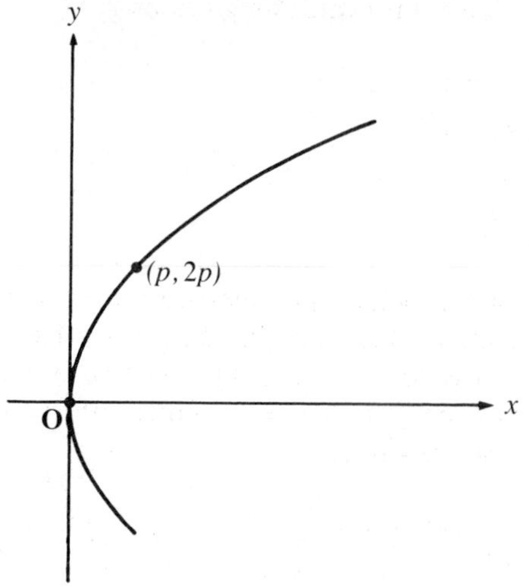

FIGURE 2.4

do. We have $\dot{\boldsymbol{x}}(t)=\left(\frac{t}{2p}, 1\right)$, so

$$|\dot{\boldsymbol{x}}(t)|=\sqrt{\frac{t^2}{4p^2}+1}=\frac{\sqrt{t^2+4p^2}}{2p}.$$

Thus the arc length desired is

$$\int_0^{2p}\frac{1}{2p}\sqrt{t^2+4p^2}\,dt=\frac{1}{2p}\int_0^{2p}\sqrt{t^2+4p^2}\,dt.$$

To evaluate this integral, we use Formula #23 in the Table of Integrals. We obtain

$$\begin{aligned}\frac{1}{2p}\int_0^{2p}\sqrt{t^2+4p^2}\,dt&=\frac{1}{2p}\left[\frac{t}{2}\sqrt{t^2+4p^2}+2p^2\ln|t+\sqrt{t^2+4p^2}|\right]_0^{2p}\\&=\frac{1}{2p}[p\sqrt{8p^2}+2p^2\ln|2p+\sqrt{8p^2}|-0\\&\quad-2p^2\ln|0+\sqrt{4p^2}|]\\&=\frac{1}{2p}\left[2\sqrt{2}p^2+2p^2\ln\frac{2p+2\sqrt{2}p}{2p}\right]\\&=\sqrt{2}p+p\ln(1+\sqrt{2}).\end{aligned}$$

How does Definition 2.5 compare with the elementary calculus formula for the length of arc of a smooth curve given as the graph of an equation $y=f(x)$ between $x=a$ and $x=b$? In this case, a natural parametrization gives the vector equation $\boldsymbol{x}(t)=(t, f(t))$. Then $\dot{\boldsymbol{x}}(t)=(1, f'(t))$, so the speed is $|\dot{\boldsymbol{x}}(t)|=\sqrt{1+f'(t)^2}$. Thus Definition 2.5 gives as arc length

$$L=\int_a^b\sqrt{1+f'(t)^2}\,dt=\int_a^b\sqrt{1+f'(x)^2}\,dx,$$

which you may recall is *exactly* the formula given in elementary calculus for the length of arc of the graph $y=f(x)$ between $x=a$ and $x=b$. Hence our Definition 2.5 is a natural generalization of the former notion of arc length.

Since we have defined a real valued arc length function s in Definition 2.5, by $s(t) = \int_a^t |\dot{\boldsymbol{x}}(u)|\, du$, we have from elementary calculus that s is a differentiable function on $[a, b]$ and

$$\frac{ds}{dt} = |\dot{\boldsymbol{x}}(t)|.$$

This is a completely natural result which simply says that the rate of change of arc length with respect to time is the speed of travel along the curve $\boldsymbol{x} = \boldsymbol{x}(t)$. A notion that is sometimes useful can be described in terms of the arc length function.

2.8 **DEFINITION.** The curve $\boldsymbol{x} = \boldsymbol{x}(t)$ is said to be **parametrized by arc length** in case the speed $|\dot{\boldsymbol{x}}(t)|$ is identically equal to one.

Let's see why the terminology "parametrized by arc length" is reasonable. If $|\dot{\boldsymbol{x}}(t)|$ is constantly one, then the arc length function

$$s(t) = \int_a^t 1\, du = t - a \qquad \text{on } [a, b].$$

So as t varies from a to b, the arc length varies with it, from 0 at $t = a$, to $b - a$ at $t = b$. In particular, if we start our clocks when we leave a, then $s(t) = t$ since $a = 0$. Then the distance we move along the curve is the time elapsed in moving.

2.9 **EXAMPLE.** Show that $\boldsymbol{x}$ is parametrized by arc length if

$$\boldsymbol{x}(t) = (\tfrac{1}{3}t, 2\sqrt{2}\cos\tfrac{1}{3}t, 2\sqrt{2}\sin\tfrac{1}{3}t), \quad 0 \le t \le 2\pi.$$

Proof. We find that

$$\dot{\boldsymbol{x}}(t) = \left(\frac{1}{3}, -\frac{2\sqrt{2}}{3}\sin\frac{1}{3}t, \frac{2\sqrt{2}}{3}\cos\frac{1}{3}t\right),$$

so

$$|\dot{\boldsymbol{x}}(t)| = \sqrt{\tfrac{1}{9} + \tfrac{8}{9}} = 1.$$

Hence the curve is parametrized by arc length. QED

2.10 **EXAMPLE.** For the curve $\boldsymbol{x} = \boldsymbol{x}(t) = (\cos t, \sin t, t)$, $0 \le t \le 2\pi$, find a parametrization by arc length.

Solution. Let us first find the arc length function $s = s(t)$. We have $\dot{\boldsymbol{x}}(t) = (-\sin t, \cos t, 1)$ so the speed is $|\dot{\boldsymbol{x}}(t)| = \sqrt{\sin^2 t + \cos^2 t + 1} = \sqrt{2}$. Hence $s(t) = \int_0^t |\dot{\boldsymbol{x}}(u)|\, du = \int_0^t \sqrt{2}\, dt = \sqrt{2}t$. Thus $t = s/\sqrt{2}$ gives t as a function of s. So if we parametrize the curve in terms of the new parameter $t(s) = s/\sqrt{2}$, then we should obtain a parametrization by arc length. We have

$$\boldsymbol{x}(t(s)) = \left(\cos\frac{s}{\sqrt{2}}, \sin\frac{s}{\sqrt{2}}, \frac{s}{\sqrt{2}}\right) = \boldsymbol{y}(s).$$

The speed is then

$$\begin{aligned}\left|\frac{d\mathbf{y}}{ds}\right| &= \left|\left(-\frac{1}{\sqrt{2}}\sin\frac{s}{\sqrt{2}}, \frac{1}{\sqrt{2}}\cos\frac{s}{\sqrt{2}}, \frac{1}{\sqrt{2}}\right)\right| \\ &= \sqrt{\tfrac{1}{2}\sin^2\frac{s}{\sqrt{2}} + \tfrac{1}{2}\cos^2\frac{s}{\sqrt{2}} + \frac{1}{2}} \\ &= \sqrt{\tfrac{1}{2}+\tfrac{1}{2}} = 1.\end{aligned}$$

Hence the parametrization $\mathbf{x} = \mathbf{y}(s) = \mathbf{x}(t(s))$ is a parametrization by arc length. The curve is shown in Figure 2.5.

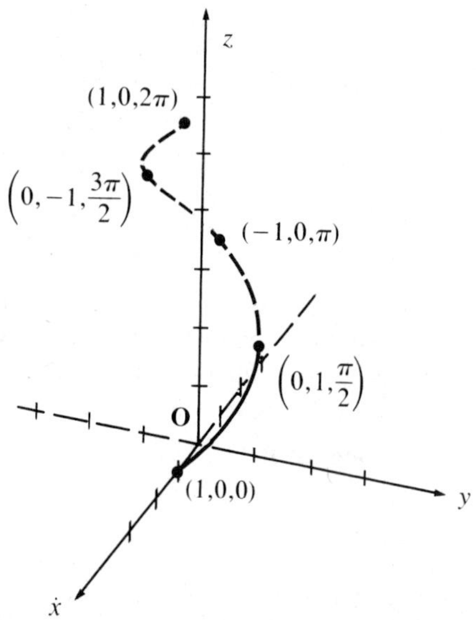

FIGURE 2.5

The curve in Example 2.10 is called a **circular helix.** One of the greatest achievements of modern biochemistry was the characterization of the molecular structure of deoxyribonucleic acid (DNA, the genetic material of living cells). This was achieved in 1953 at Cambridge University by the American biologist James D. Watson (1928–) working with the English biochemist Francis H. C. Crick (1916–), and their work was recognized by the Nobel Foundation in 1962 when they shared the Nobel prize for physiology and medicine. They first proposed and then verified experimentally that the DNA molecule consists of two parallel almost-circular helices of sugar and phosphate linked by certain organic bases called purines and pyrimidines. The two helices can be approximately parametrized as

$$\mathbf{x}(t) = (\cos t, \sin t, t)$$

and

$$\begin{aligned}\mathbf{y}(t) &= (\cos(t-\pi), \sin(t-\pi), t-\pi) \\ &= (\cos t, -\sin t, t-\pi)\end{aligned}$$

where t varies over an interval $[0, 2k\pi]$ for some k, a positive integer which varies from molecule to molecule. As t increases, the helices ascend and wind around the z-axis in a counterclockwise sense, as viewed from above. See Figure 2.6,

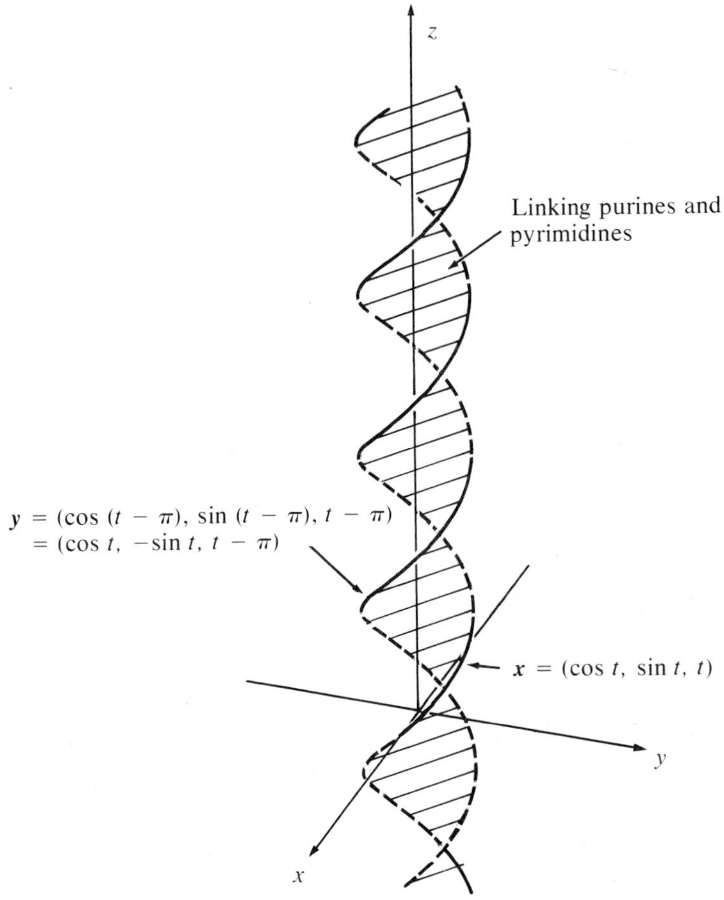

FIGURE 2.6
The DNA molecule.

where some of the linking bases are shown. As Example 2.10 suggests, the calculus of the circular helix is relatively simple since the speed of traversal is constant.

EXERCISES 2.2

In Exercises 1 to 4, a particle moves on the given path in $\boldsymbol{R}^3$. Find the velocity, acceleration, and speed. Sketch the path between $t=0$ and $t=2\pi$.

1 The helix $\boldsymbol{x}=\boldsymbol{x}(t)=(\cos\frac{1}{2}t, \sin\frac{1}{2}t, t)$.

2 The elliptical spiral $\boldsymbol{x}=\boldsymbol{x}(t)=(\cos t, 2\sin t, 3t)$.

3 The expanding spiral $\boldsymbol{x}=\boldsymbol{x}(t)=(t\cos t, t\sin t, t)$.

4 A particle moves counterclockwise on the circle $x^2+y^2=9$. Parametrize the path and then calculate the velocity, acceleration, and speed.

5 Describe the path in $\boldsymbol{R}^3$ of a particle subjected to a constant force $\boldsymbol{F}=(c_1, c_2, c_3)$.

In Exercises 6 and 7, find the length of arc along the given curve over the given interval.

6 $\boldsymbol{x}(t)=(t, \sin 4t, \cos 4t)$ between 0 and π. **7** $\boldsymbol{x}(t)=(t^2, t^3)$ between 0 and 2.

8 Parametrize the curve $\boldsymbol{x}(t)=(2t, t, 1-t)$, $t\in[0, 1]$, by arc length.

9 Parametrize the curve in Exercise 6 by arc length.

10 Show that the curve

$$\boldsymbol{x}(t) = \left(\tfrac{1}{6}\cos 3t, \tfrac{1}{6}\sin 3t, \frac{1}{\sqrt{2}}t\right)$$

is parametrized by arc length.

11 A curve in the plane given in polar coordinates as the graph of $r = f(\theta)$, $a \le \theta \le b$, can be parametrized by

$$\begin{aligned}\boldsymbol{x} = \boldsymbol{x}(\theta) &= (r\cos\theta, r\sin\theta) = (f(\theta)\cos\theta, f(\theta)\sin\theta)\\ &= f(\theta)(\cos\theta, \sin\theta).\end{aligned}$$

(a) Find the velocity (as a function of θ). (b) Find the speed.
(c) Show that the arc length is given by

$$L = \int_a^b \sqrt{r^2 + \left(\frac{dr}{d\theta}\right)^2}\, d\theta.$$

(d) What happens in (a), (b), and (c) if we regard $\theta = \theta(t)$ as a function of t?

12 Find the length of arc of the logarithmic spiral $r = e^\theta$ between 0 and π. (Use Exercise 11c.)

13 A shell of mass m is fired from a cannon whose angle of elevation with the ground is α (see Figure 2.7). The initial velocity is $\boldsymbol{v}_0 = (v_0\cos\alpha, v_0\sin\alpha)$, where v_0 is the muzzle speed. Assume that the only force acting on the shell is $\boldsymbol{F} = (0, -mg)$, where g is the acceleration due to gravity.

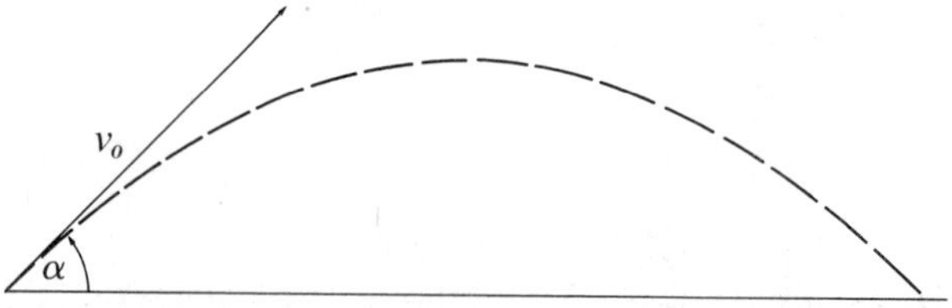

FIGURE 2.7

(a) Describe the path $\boldsymbol{x} = \boldsymbol{x}(t)$ of the shell. Identify the curve. (*Hint:* Use $\boldsymbol{F} = m\boldsymbol{a} = m\ddot{\boldsymbol{x}}(t)$.)
(b) Find the linear distance from the cannon to the point of impact of the shell, assuming level ground.

14 Assume that the cannon in Exercise 13 is located at the base of a hillside which has an angle of elevation of 30° (Figure 2.8). Show that the shell hits the hillside at point P whose x-coordinate is

$$\frac{2v_0^2}{g}\left(\sin\alpha\cos\alpha - \frac{1}{\sqrt{3}}\cos^2\alpha\right).$$

$\left(\textit{Hint:}\text{ The shell hits the hillside at the point where } \frac{y}{x} = \tan 30° = \frac{1}{\sqrt{3}}.\right)$

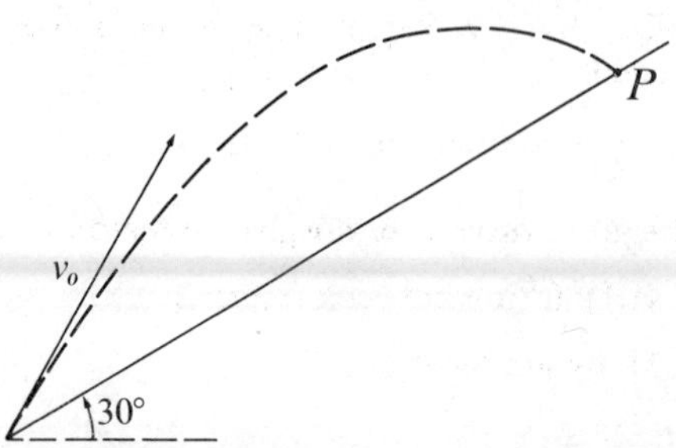

FIGURE 2.8

15 If a wire in space lies along the curve $\boldsymbol{x}=(x(t), y(t), z(t))$ between $t=a$ and $t=b$, and the mass per unit length at $\boldsymbol{x}$ is $\delta(\boldsymbol{x}(t))$, then the total mass of the wire is defined to be $M=\int_a^b \delta(\boldsymbol{x}(t))\,|\dot{\boldsymbol{x}}(t)|\,dt$. Find the mass of $\boldsymbol{x}(t)=(\cos t, t, \sin t)$, $0\le t\le 2\pi$, if $\delta(\boldsymbol{x}(t))=y(t)^2$.

16 Use the definition given in Exercise 15 to find the mass of a wire lying along the graph of $\boldsymbol{x}(t)=(t, t^2, t^2)$ between $t=0$ and $t=1$ if the density at $(x(t), y(t), z(t))$ is $6t$.

17 When a body rotates about the z-axis through the origin, its angular velocity is defined as $\boldsymbol{\omega}=\dot{\theta}\boldsymbol{k}$, where $\dot{\theta}$ is the number of radians per unit time in the rotation. See Figure 2.9. The linear velocity is defined to be $\boldsymbol{v}=\boldsymbol{\omega}\times\boldsymbol{x}$, where $\boldsymbol{x}(t)$ is the position of the body at time t. Show that the linear speed is the product of the angular speed, the displacement $|\boldsymbol{x}(t)|$ from the origin, and $\cos\phi$, where ϕ is the angle formed by $\boldsymbol{x}(t)$ and the xy-plane.

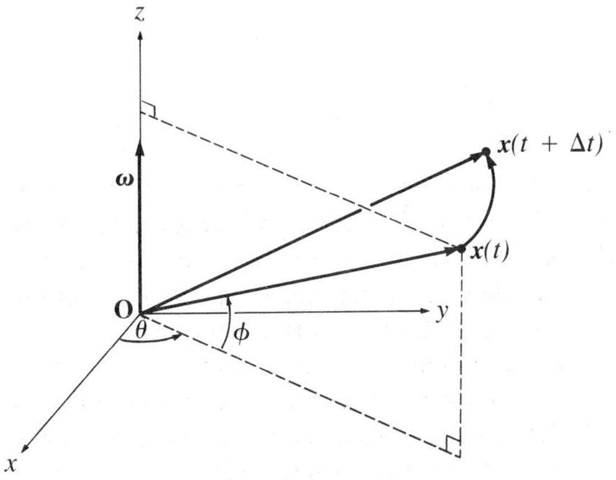

FIGURE 2.9

18 The earth rotates about its axis at a speed of 2π radians per day. Assume that the earth is a ball of radius 4000 miles, and put the origin at its center. Find the linear speed in miles per hour of a point on the surface of the earth at (a) the equator, (b) Manila, Philippines (located at 15° North $=\pi/12$ radians north of the equator), (c) Bordeaux, France (located at 45° North), (d) the North Pole. (See Exercise 17.)

3 CURVATURE, UNIT TANGENT, AND NORMAL

In this section we will capitalize on the arc length function introduced in Definition 2.5 to obtain some further properties of curves which turn out to be important in studying the motion of a particle in $\boldsymbol{R}^m$.

Recall (Definition 1.10) that we have defined the *unit tangent vector* to a curve $\boldsymbol{x}=\boldsymbol{x}(t)$ as

(1) $$\boldsymbol{T}(t)=\frac{1}{|\dot{\boldsymbol{x}}(t)|}\dot{\boldsymbol{x}}(t), \quad \text{if} \quad \dot{\boldsymbol{x}}(t)\neq \boldsymbol{0}.$$

This definition seems (as did Definition 2.5) to depend on the parametrization of the curve. However, the unit tangent vector to a curve *ought* to be something intrinsically determined by the curve itself rather than by the nature of the particular parametrization. Can we then give a more intrinsic description of $\boldsymbol{T}$? We can, if we use the arc length variable $s=s(t)$; as we saw following Definition 2.5, $s(t)$ is independent of the parametrization. Recall that

$$s(t)=\int_a^t |\dot{\boldsymbol{x}}(u)|\,du,$$

and so, if $\dot{\boldsymbol{x}}(t) \neq \mathbf{0}$,

$$\frac{ds}{dt} = |\dot{\boldsymbol{x}}(t)| > 0.$$

We can then use the Chain Rule (Theorem 1.7(6)) to obtain from (1)

$$\boldsymbol{T}(t) = \frac{\dot{\boldsymbol{x}}(t)}{\dfrac{ds}{dt}} = \frac{\dfrac{d\boldsymbol{x}}{ds}\dfrac{ds}{dt}}{\dfrac{ds}{dt}},$$

i.e.,

(2) $$\boldsymbol{T}(t) = \frac{d\boldsymbol{x}}{ds}.$$

This description of the unit tangent vector is more intrinsic to the curve than is (1), since it says that $\boldsymbol{T}$ is the rate of change of the curve with respect to its arc length. We will soon find formula (2) useful in measuring the curvature of a curve.

Intuitively, we have a notion of the degree of curvature that a given curve manifests. This is reflected in speed limit signs that are often seen on highways, warning of how fast a car can safely travel when negotiating a curve. You have surely acquired the knowledge that the *sharper* the curve (i.e., the greater its curvature), the *lower* the speed limit. This seems to reflect an increasing tendency for the car to leave the road as the curvature becomes greater. How can we assign a mathematical measure to this intuitive idea? First we have to try to see what mathematically distinguishes a curve in the road from a straight patch of road. For this, refer to Figure 3.1, where we represent a particularly formidable curve. Let us consider its vector equation $\boldsymbol{x} = \boldsymbol{x}(t)$ for various t.

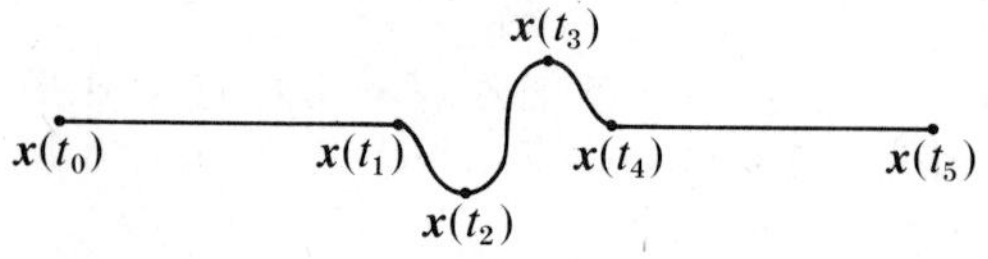

FIGURE 3.1

First we notice that between t_0 and t_1 as well as between t_4 and t_5, $\boldsymbol{x}(t)$ is approximately a straight line segment. So in each of these intervals, $\boldsymbol{x}(t) = t\boldsymbol{a} + \boldsymbol{b}$ for appropriate $\boldsymbol{a}$ and $\boldsymbol{b}$. Thus $\dot{\boldsymbol{x}}(t) = \boldsymbol{a}$ is constant, and so $\boldsymbol{T}(t)$ is constant with length one. Between t_1 and t_2, however, $\boldsymbol{x}(t)$ is given by a non-linear expression. Hence $\dot{\boldsymbol{x}}(t)$ is no longer constant, and neither is $\boldsymbol{T}(t)$. While the length of $\boldsymbol{T}(t)$ is always one, it is constantly turning, in fact quite sharply at the points $\boldsymbol{x}(t_2)$ and $\boldsymbol{x}(t_3)$. We seem, then, to have a good measure of curvature: how much the unit tangent vector turns per unit length of arc as we move along the curve. This latter turning rate is something we can measure mathematically.

3.1 **DEFINITION.** The **curvature** K of a curve $\boldsymbol{x} = \boldsymbol{x}(t)$ at a point $\boldsymbol{x}(t_0)$ is the magnitude of the rate of change of the direction of the curve with respect to arc length. That is,

$$K = \left|\frac{d\boldsymbol{T}(t)}{ds}\right|_{t=t_0}$$

where $\boldsymbol{T}(t)$ is the unit tangent vector.

Definition 3.1 has the conceptual advantage of being phrased in terms of the arc length of the curve, which we know is independent of parametrization. However, since we have defined the curvature in terms of a derivative with respect to s rather than with respect to t, we are faced with the problem of *computing* K for a given curve $\boldsymbol{x} = \boldsymbol{x}(t)$. The following result gives us a formula for K in terms of the parameter t.

3.2 **THEOREM.** At any point $\boldsymbol{x}(t)$ where $\dot{\boldsymbol{x}}(t) \neq \boldsymbol{0}$,

$$K = \frac{|\dot{\boldsymbol{T}}(t)|}{|\dot{\boldsymbol{x}}(t)|}.$$

Proof. The Chain Rule (Theorem 1.7(6)) tells us that

$$\frac{d\boldsymbol{T}}{ds} = \frac{d\boldsymbol{T}}{dt}\frac{dt}{ds}.$$

Since $\dfrac{ds}{dt} = |\dot{\boldsymbol{x}}(t)| > 0$, we know that s is an increasing function of t. Then from elementary calculus we know that the inverse function $t = t(s)$ is differentiable with

$$\frac{dt}{ds} = \frac{1}{\dfrac{ds}{dt}}.$$

We therefore have

(*) $$\frac{d\boldsymbol{T}}{ds} = \frac{d\boldsymbol{T}}{dt}\frac{1}{\dfrac{ds}{dt}} = \frac{\dot{\boldsymbol{T}}}{|\dot{\boldsymbol{x}}(t)|}.$$

Hence,

$$K = \left|\frac{d\boldsymbol{T}}{ds}\right| = \frac{|\dot{\boldsymbol{T}}|}{|\dot{\boldsymbol{x}}(t)|}.$$

QED

3.3 **EXAMPLE.** Show that the curvature of a line in $\boldsymbol{R}^m$ is 0.

Proof. Represent the line as $\boldsymbol{x}(t) = t\boldsymbol{a} + \boldsymbol{b}$. See Figure 3.2. Then $\dot{\boldsymbol{x}}(t) = \boldsymbol{a}$. Thus $\boldsymbol{T} = \dfrac{\boldsymbol{a}}{|\boldsymbol{a}|}$. Hence $\dot{\boldsymbol{T}} = \boldsymbol{0}$, since $\boldsymbol{a}$ is constant. Therefore,

$$K = \frac{|\dot{\boldsymbol{T}}|}{|\dot{\boldsymbol{x}}(t)|} = \frac{0}{|\boldsymbol{a}|} = 0,$$

in conformity with our intuitive notion of curvature derived from highway driving.

QED

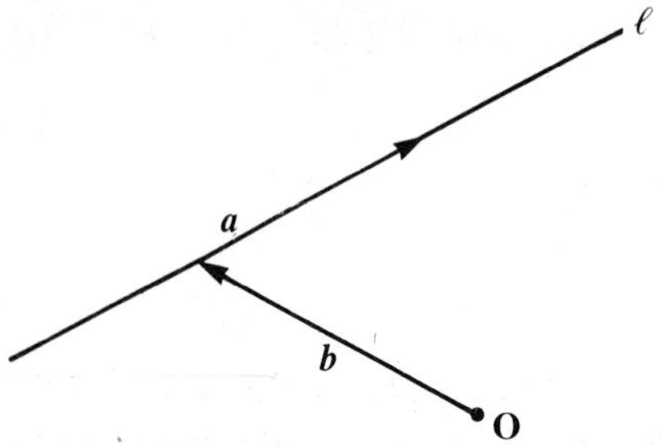

FIGURE 3.2

3.4 EXAMPLE. Find the curvature of a circle of radius a.

Solution. We might as well view the circle as being in $\boldsymbol{R}^2$, where we can

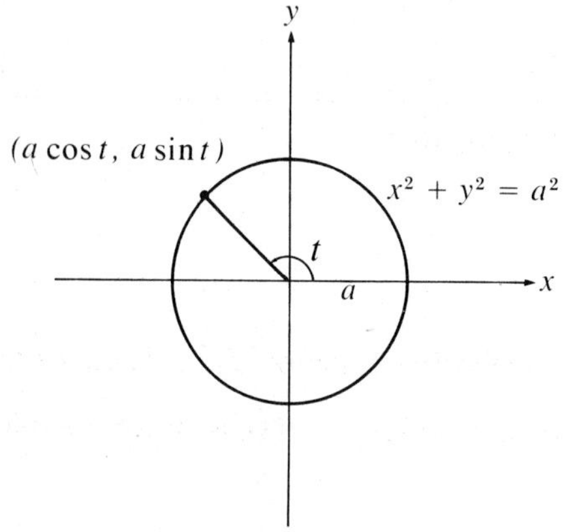

FIGURE 3.3

represent it as $\boldsymbol{x}(t)=(a \cos t, a \sin t)$. (See Figure 3.3.) Then $\dot{\boldsymbol{x}}(t)=(-a \sin t, a \cos t)$. So

$$|\dot{\boldsymbol{x}}(t)|=\sqrt{a^2 \sin^2 t+a^2 \cos^2 t}=a,$$

and

$$\boldsymbol{T}=\frac{\dot{\boldsymbol{x}}(t)}{|\dot{\boldsymbol{x}}(t)|}=\frac{1}{a}\dot{\boldsymbol{x}}(t)=(-\sin t, \cos t).$$

Then $\dot{\boldsymbol{T}}=(-\cos t, -\sin t)$. Hence by Theorem 3.2,

$$K=\frac{|\dot{\boldsymbol{T}}|}{|\dot{\boldsymbol{x}}(t)|}=\frac{\sqrt{\cos^2 t+\sin^2 t}}{a}=\frac{1}{a}.$$

We see from Example 3.4 that the curvature of a circle is simply the reciprocal of its radius. This is consistent with our intuitive picture of circles with small radii curving more sharply than circles with large radii.

Now, the unit tangent vector $\boldsymbol{T}$ has constant length one. Thus $\boldsymbol{T}\cdot\boldsymbol{T}=1$. If we differentiate with respect to t and use Theorem 1.7(2), we get $\dot{\boldsymbol{T}}\cdot\boldsymbol{T}+\boldsymbol{T}\cdot\dot{\boldsymbol{T}}=0$, i.e., $2\boldsymbol{T}\cdot\dot{\boldsymbol{T}}=0$; hence $\boldsymbol{T}\cdot\dot{\boldsymbol{T}}=0$. Thus, $\dot{\boldsymbol{T}}$ either is $\boldsymbol{0}$ or is a *normal* vector to the curve $\boldsymbol{x}=\boldsymbol{x}(t)$.

3.5 **DEFINITION.** If $\dot{\boldsymbol{T}} \neq \boldsymbol{0}$, then the **unit normal vector** to the curve $\boldsymbol{x} = \boldsymbol{x}(t)$ is

$$\boxed{\boldsymbol{N} = \frac{1}{|\dot{\boldsymbol{T}}|}\dot{\boldsymbol{T}}.}$$

You might expect that $|\dot{\boldsymbol{T}}| = 1$, since $|\boldsymbol{T}| = 1$. This happened in Example 3.4, where $\dot{\boldsymbol{T}} = (-\cos t, -\sin t)$, but it does not happen in general, as we see in the following example.

3.6 **EXAMPLE.** Find K and the unit tangent and normal vectors to the ellipse $\boldsymbol{x}(t) = (2\cos t, 3\sin t)$, at $t = \pi/2$.

Solution. Here $\dot{\boldsymbol{x}}(t) = (-2\sin t, 3\cos t)$. Then

$$\begin{aligned}|\dot{\boldsymbol{x}}(t)| &= \sqrt{4\sin^2 t + 9\cos^2 t} = \sqrt{4(1-\cos^2 t) + 9\cos^2 t}\\ &= \sqrt{4 + 5\cos^2 t}.\end{aligned}$$

Thus

$$\boldsymbol{T} = \left(\frac{-2\sin t}{\sqrt{4+5\cos^2 t}}, \frac{3\cos t}{\sqrt{4+5\cos^2 t}}\right).$$

Then $|\dot{\boldsymbol{x}}(\pi/2)| = \sqrt{4} = 2$. In general,

$$\begin{aligned}\dot{\boldsymbol{T}} &= \left(\frac{\sqrt{4+5\cos^2 t}(-2\cos t) + (2\sin t)\frac{1}{2}(4+5\cos^2 t)^{-1/2}(-10\cos t\sin t)}{4+5\cos^2 t},\right.\\ &\qquad \left.\frac{\sqrt{4+5\cos^2 t}(-3\sin t) - (3\cos t)\frac{1}{2}(4+5\cos^2 t)^{-1/2}(-10\cos t\sin t)}{4+5\cos^2 t}\right)\\ &= \left(\frac{(4+5\cos^2 t)(-2\cos t) - 10\cos t\sin^2 t}{(4+5\cos^2 t)^{3/2}},\right.\\ &\qquad \left.\frac{(4+5\cos^3 t)(-3\sin t) + 15\cos^2 t\sin t}{(4+5\cos^2 t)^{3/2}}\right)\end{aligned}$$

When $t = \pi/2$ we get

$$\dot{\boldsymbol{T}}\left(\frac{\pi}{2}\right) = \left(0, \frac{-12}{8}\right) = \left(0, -\frac{3}{2}\right).$$

Then

$$\boldsymbol{N} = \frac{(0, -\frac{3}{2})}{|(0, -\frac{3}{2})|} = (0, -1).$$

When $t = \dfrac{\pi}{2}$, we have $\boldsymbol{T} = \left(-\dfrac{2}{2}, 0\right) = (-1, 0)$. See Figure 3.4. Since $\left|\dot{\boldsymbol{T}}\left(\dfrac{\pi}{2}\right)\right| = \dfrac{3}{2}$, we finally find $K = \dfrac{\left|\dot{\boldsymbol{T}}\left(\frac{\pi}{2}\right)\right|}{\left|\dot{\boldsymbol{x}}\left(\frac{\pi}{2}\right)\right|} = \dfrac{\frac{3}{2}}{2} = \dfrac{3}{4}$.

Once more, the unit normal appears to be more intrinsic to the geometry of a

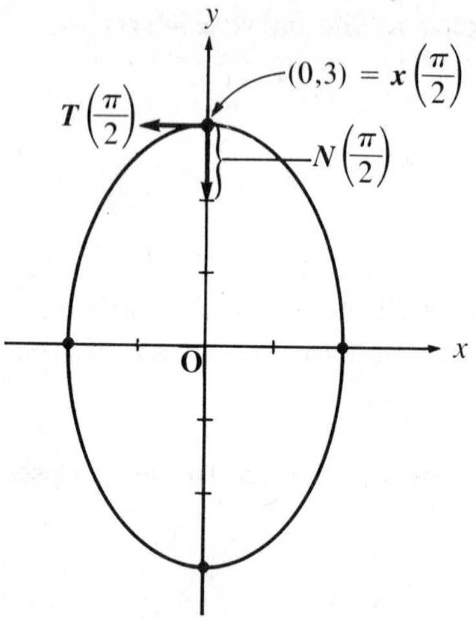

FIGURE 3.4

given curve than Definition 3.5 would make it appear. Our next result gives a formula for $\boldsymbol{N}$ that does not involve the parameter t but instead involves the arc length variable s, which we have seen is independent of the parametrization.

3.7 **THEOREM.** If $K \neq 0$, then

$$\boldsymbol{N} = \frac{1}{K}\frac{d\boldsymbol{T}}{ds}.$$

Proof. From (*) in the proof of Theorem 3.2, we have

$$\frac{d\boldsymbol{T}}{ds} = \frac{\dot{\boldsymbol{T}}}{|\dot{\boldsymbol{x}}(t)|}.$$

Using this and the formula for K derived in Theorem 3.2, we obtain

$$\frac{1}{K}\frac{d\boldsymbol{T}}{ds} = \frac{|\dot{\boldsymbol{x}}(t)|}{|\dot{\boldsymbol{T}}|}\frac{\dot{\boldsymbol{T}}}{|\dot{\boldsymbol{x}}(t)|} = \frac{\dot{\boldsymbol{T}}}{|\dot{\boldsymbol{T}}|} = \boldsymbol{N}.$$

QED

Theorem 3.7 puts us in a position to give a mathematical explanation of the tendency of an automobile to leave the road on a sharp curve.

From (1) we have $\dot{\boldsymbol{x}}(t) = |\dot{\boldsymbol{x}}(t)|\,\boldsymbol{T} = \frac{ds}{dt}\boldsymbol{T}$. Let's differentiate with respect to t and use Theorem 1.7(4, 6). We get

$$\begin{aligned}\boldsymbol{a}(t) = \ddot{\boldsymbol{x}}(t) &= \frac{d^2s}{dt^2}\boldsymbol{T} + \frac{ds}{dt}\frac{d\boldsymbol{T}}{dt}\\ &= \frac{d^2s}{dt^2}\boldsymbol{T} + \frac{ds}{dt}\frac{d\boldsymbol{T}}{ds}\frac{ds}{dt}\\ &= \frac{d^2s}{dt^2}\boldsymbol{T} + \left(\frac{ds}{dt}\right)^2\frac{d\boldsymbol{T}}{ds}\end{aligned}$$

(3)
$$\boldsymbol{a}(t) = \frac{d^2s}{dt^2}\boldsymbol{T} + \left(\frac{ds}{dt}\right)^2 K\boldsymbol{N}$$

by Theorem 3.7.

We thus have resolved the acceleration vector $\boldsymbol{a}(t)$ into two components, $\boldsymbol{a_T}$ tangent to the curve, and $\boldsymbol{a_N}$ normal to the curve (Figure 3.5). How should you

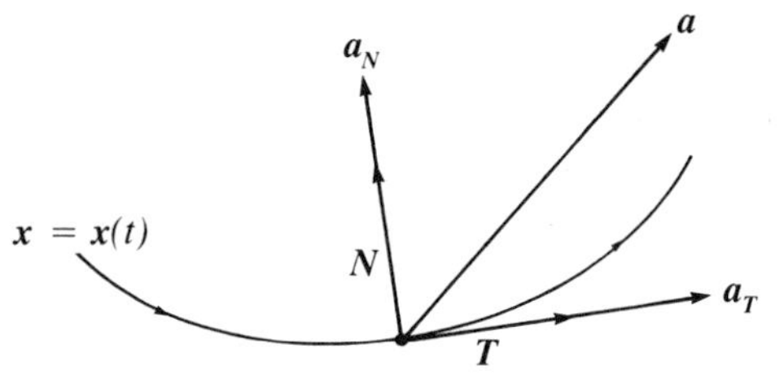

FIGURE 3.5

now control your car so as to help it stay on the road through the curve? Since the direction of the curve at each point is given by $\boldsymbol{T}$, you want $\boldsymbol{a_T}$ to be as large as possible relative to $\boldsymbol{a_N}$. If K is large (i.e., the curve is sharp), then since the magnitude of $\boldsymbol{a_N}$ is $\left(\frac{ds}{dt}\right)^2 K$, the only obvious way to make $\boldsymbol{a_N}$ small is to make $\frac{ds}{dt}$ *small relative to* K. This is the reason that speed limit signs are posted warning drivers to *reduce speed before entering a curve.* In fact, if you wait until *entering* the curve to brake, then in the curve you will be decreasing your speed. That is,

$$\frac{d^2s}{dt^2} = \frac{d}{dt}\left(\frac{ds}{dt}\right) < 0.$$

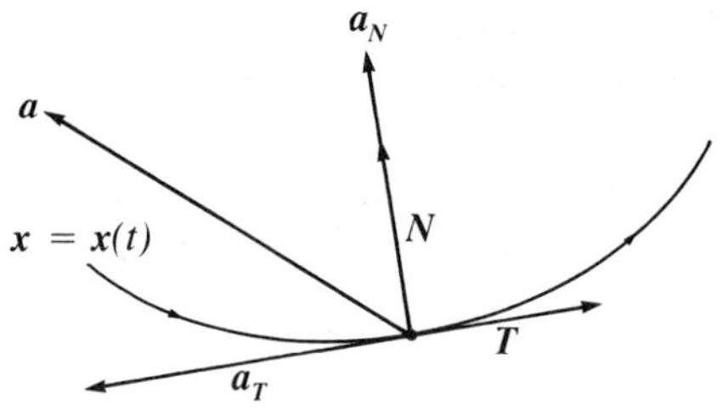

FIGURE 3.6

As you can see from Figure 3.6, $\boldsymbol{a_T}$ will no longer be pointing in the direction of motion, but in the opposite direction. Then $\boldsymbol{a}$ points in the direction of the resultant force $\boldsymbol{F} = m\boldsymbol{a}$ acting on the car. That force *hinders* rather than helps the car in negotiating the curve, since it is far removed from the direction of motion. On the other hand, if you slow below the speed limit *before* entering the curve and then increase your speed slightly (i.e., engage the accelerator moderately) in the curve, then $\frac{d^2s}{dt^2} > 0$ in the curve. Hence the tangential component $\boldsymbol{a_T}$ is as in Figure 3.5; i.e., it is a positive factor in keeping your car on the road.

While we have seen a practical implication of equation (3), its importance in dynamics is far greater still. For this reason the two summands in (3) are given special names.

3.8 DEFINITION. The **tangential** and **normal components of acceleration** of a particle moving along the path $\boldsymbol{x} = \boldsymbol{x}(t)$ are

$$\boldsymbol{a_T} = \frac{d^2s}{dt^2}\boldsymbol{T} \quad \text{and} \quad \boldsymbol{a_N} = \left(\frac{ds}{dt}\right)^2 K\boldsymbol{N}.$$

We remark that alternatively we can compute $\boldsymbol{a_N}$ from the formula

(4) $$\boldsymbol{a_N} = |\boldsymbol{v}(t)|\,|\dot{\boldsymbol{T}}|\,\boldsymbol{N} = \frac{ds}{dt}|\dot{\boldsymbol{T}}|\,\boldsymbol{N}$$

since $|\dot{\boldsymbol{T}}|\,\boldsymbol{N} = \dot{\boldsymbol{T}} = \frac{d\boldsymbol{T}}{ds}\frac{ds}{dt} = \frac{ds}{dt}K\boldsymbol{N}$ by Theorem 3.7. So

$$\frac{ds}{dt}|\dot{\boldsymbol{T}}|\,\boldsymbol{N} = \left(\frac{ds}{dt}\right)^2 K\boldsymbol{N} = \boldsymbol{a_N}.$$

While formulas (3) and (4) are very important in contexts like that of a car rounding a curve in the road, the fact that they do not involve t can make them a chore to use for a curve given parametrically as $\boldsymbol{x} = \boldsymbol{x}(t)$. When confronted with such a parametrized curve, it is often easier to compute $\boldsymbol{a_T}$ directly from $\boldsymbol{x}$ and $\boldsymbol{v}$, as the component of $\boldsymbol{a}$ in the direction of $\boldsymbol{v}$ (Definition 1.2.16). We have, in fact,

(5) $$\boldsymbol{a_T} = (\boldsymbol{a} \cdot \boldsymbol{T})\boldsymbol{T} = \frac{\boldsymbol{a} \cdot \boldsymbol{v}}{|\boldsymbol{v}|}\frac{\boldsymbol{v}}{|\boldsymbol{v}|}$$

Since $\boldsymbol{a} = \boldsymbol{a_T} + \boldsymbol{a_N}$, we also have

(6) $$\boldsymbol{a_N} = \boldsymbol{a} - \boldsymbol{a_T}.$$

A real advantage inherent in (6) is that it permits us to compute $\boldsymbol{N}$ without recourse to either Definition 3.5 or Theorem 3.7. We can thereby save considerable labor in differentiation, as illustrated by the next example.

3.9 EXAMPLE. A particle moves on a circle of radius a with angular speed $\omega(t) = \dot{\theta}(t)$, where $\theta(t)$ is the central angle shown in Figure 3.7. Find the tangential and normal components of the acceleration. Also find $\boldsymbol{N}$.

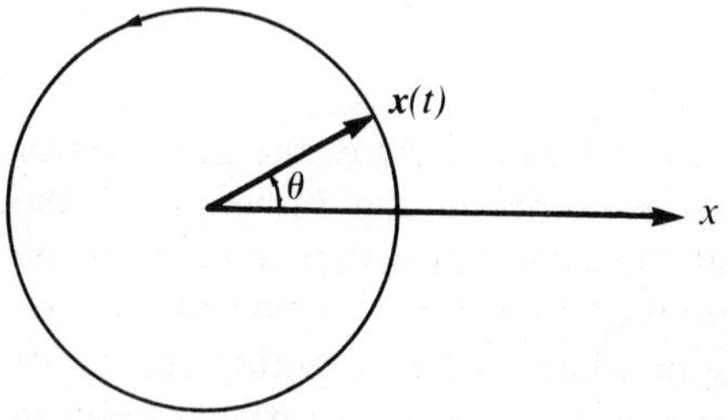

FIGURE 3.7

Solution. We have $\boldsymbol{x}(t) = (a \cos \theta(t), a \sin \theta(t))$. So

$$\begin{aligned} \boldsymbol{v}(t) &= a(-\omega \sin \theta(t), \omega \cos \theta(t)) \quad \text{by the Chain Rule,} \\ &= a\omega(-\sin \theta(t), \cos \theta(t)). \end{aligned}$$

Also

$$\begin{aligned} \boldsymbol{a}(t) &= a\dot{\omega}(-\sin \theta(t), \cos \theta(t)) + a\omega(-\omega \cos \theta(t), -\omega \sin \theta(t)) \\ &= a(-\dot{\omega} \sin \theta(t) - \omega^2 \cos \theta(t), \dot{\omega} \cos \theta(t) - \omega^2 \sin \theta(t)). \end{aligned}$$

Further,

$$\boldsymbol{T} = \frac{\dot{\boldsymbol{x}}(t)}{|\dot{\boldsymbol{x}}(t)|} = \frac{\boldsymbol{v}(t)}{|\boldsymbol{v}(t)|} = (-\sin \theta(t), \cos \theta(t)).$$

Thus, from (5),

$$\begin{aligned}\boldsymbol{a_T} &= a(\dot{\omega}\sin^2\theta(t)+\omega^2\sin\theta(t)\cos\theta(t)+\dot{\omega}\cos^2\theta(t)\\&\quad-\omega^2\sin\theta(t)\cos\theta(t))\boldsymbol{T}\\&= a\dot{\omega}\boldsymbol{T}\end{aligned}$$

Since $K=1/a$ is known from Example 3.4, we can easily compute $\boldsymbol{a_N}$ from (3). We get

(7) $$\boldsymbol{a_N}=\left(\frac{ds}{dt}\right)^2 K\boldsymbol{N}=a^2\omega^2\frac{1}{a}\boldsymbol{N}=a\omega^2\boldsymbol{N}.$$

To obtain an explicit formula for $\boldsymbol{a_N}$ in terms of $\theta(t)$, we can use (6) as follows.

$$\begin{aligned}\boldsymbol{a_N} &= \boldsymbol{a}-\boldsymbol{a_T}\\&= a(-\dot{\omega}\sin\theta-\omega^2\cos\theta, \dot{\omega}\cos\theta-\omega^2\sin\theta)\\&\quad-a(-\dot{\omega}\sin\theta, \dot{\omega}\cos\theta)\\&= -a\omega^2(\cos\theta, \sin\theta).\end{aligned}$$

Using (7), we can see immediately that $\boldsymbol{N}=-(\cos\theta, \sin\theta)$ without having to compute $\dot{\boldsymbol{T}}$.

In $\boldsymbol{R}^3$ there is a formula for K that can be derived from (3), which is stated purely in terms of the velocity and acceleration along a parametrized curve $\boldsymbol{x}=\boldsymbol{x}(t)$.

3.10 THEOREM. If $\boldsymbol{x}=\boldsymbol{x}(t)$ is a curve in $\boldsymbol{R}^3$, then at any point where the speed is not zero,

$$K=\frac{|\boldsymbol{v}(t)\times\boldsymbol{a}(t)|}{|\boldsymbol{v}(t)|^3}.$$

Proof. We have

$$\boldsymbol{v}=\frac{ds}{dt}\boldsymbol{T}\quad\text{and}\quad\boldsymbol{a}=\frac{d^2s}{dt^2}\boldsymbol{T}+\left(\frac{ds}{dt}\right)^2K\boldsymbol{N}$$

Hence

$$\begin{aligned}\boldsymbol{v}\times\boldsymbol{a} &= \frac{ds}{dt}\frac{d^2s}{dt^2}\boldsymbol{T}\times\boldsymbol{T}+\left(\frac{ds}{dt}\right)^3K(\boldsymbol{T}\times\boldsymbol{N})\\&= \left(\frac{ds}{dt}\right)^3K(\boldsymbol{T}\times\boldsymbol{N}).\end{aligned}$$

Then by Theorem 1.5.8,

$$|\boldsymbol{v}\times\boldsymbol{a}|=\left(\frac{ds}{dt}\right)^3K\,|\boldsymbol{T}|\,|\boldsymbol{N}|\sin\frac{\pi}{2}=\left(\frac{ds}{dt}\right)^3K\quad\text{since}\quad\boldsymbol{T}\perp\boldsymbol{N}.$$

Then

$$\frac{|\boldsymbol{v}\times\boldsymbol{a}|}{|\boldsymbol{v}(t)|^3}=K.$$

QED

3.11 EXAMPLE. Find the curvature of $\boldsymbol{x}(t)=te^t\boldsymbol{i}-t^2e^{2t}\boldsymbol{j}+t\boldsymbol{k}$ at $t=0$.

Solution. We have

$$\begin{aligned}\boldsymbol{v}(t)&=\dot{\boldsymbol{x}}(t)\\&=(e^t+te^t,\,-2te^{2t}-2t^2e^{2t},\,1)\\\boldsymbol{a}(t)&=\ddot{\boldsymbol{x}}(t)=(2e^t+te^t,\,-2e^{2t}-8te^{2t}-4t^2e^{2t},\,0)\end{aligned}$$

So

$$\boldsymbol{v}(0)=(1,0,1),\ \boldsymbol{a}(0)=(2,-2,0),$$

and

$$|\boldsymbol{v}(0)|=\sqrt{2}.$$

Thus,

$$\boldsymbol{v}(0)\times\boldsymbol{a}(0)=(2,2,-2),$$

and

$$|\boldsymbol{v}(0)\times\boldsymbol{a}(0)|=2\sqrt{3}.$$

So

$$K=\frac{2\sqrt{3}}{(\sqrt{2})^3}=\sqrt{\tfrac{3}{2}}.$$

EXERCISES 2.3

In Exercises 1 and 2, find K, $\boldsymbol{T}$, and $\boldsymbol{N}$.

1 $\boldsymbol{x}(t)=(\cos t,\sin t,t)$. **2** $\boldsymbol{x}(t)=(e^t\cos t,\,e^t\sin t,\,0)$.

In Exercises 3 and 4 find K and $\boldsymbol{T}$ at the given point.

3 $\boldsymbol{x}(t)=(t,t^2,t^3)$ at $t=1$. **4** $\boldsymbol{x}(t)=(t^2-1,\,1+t,\,t^3)$ at $t=1$.

5 If $\boldsymbol{x}(t)=(3\sin t,\,3\cos t,\,4t)$, then find $\boldsymbol{T}$, $\boldsymbol{N}$, and K.

6 If $\boldsymbol{x}(t)=(e^t\cos t,\,e^t\sin t,\,e^t)$, then find $\boldsymbol{T}$, $\boldsymbol{N}$, and K.

7 For the curve of Exercise 1, find the tangential and normal components of acceleration.

8 For the curve of Exercise 2, find the tangential and normal components of acceleration.

9 For the parabola $\boldsymbol{x}(t)=(t^2,t)$, find $\boldsymbol{v}(t)$, $\boldsymbol{a}(t)$, and the tangential and normal components of $\boldsymbol{a}(0)$ and $\boldsymbol{a}(1)$.

10 For the curve of Exercise 5, find $\boldsymbol{v}(t)$, $\boldsymbol{a}(t)$, and the tangential and normal components of $\boldsymbol{a}(\pi/2)$.

11 For the curve $\boldsymbol{x}(t)=(a\cos t,\,a\sin t,\,bt)$, show that the tangential component of acceleration is $\boldsymbol{0}$ and the normal component is $|\boldsymbol{a}|\boldsymbol{N}$.

12 If $\boldsymbol{x}=\boldsymbol{x}(t)$, then show that

$$K=\frac{\sqrt{|\dot{\boldsymbol{x}}|^2\,|\ddot{\boldsymbol{x}}|^2-(\dot{\boldsymbol{x}}\cdot\ddot{\boldsymbol{x}})^2}}{|\dot{\boldsymbol{x}}|^3},\quad \text{if } \dot{\boldsymbol{x}}\neq\boldsymbol{0}.$$

13 If $\boldsymbol{x}=(x(t), y(t))$ is a curve in $\boldsymbol{R}^2$, then show that

$$K=\frac{|\dot{x}\ddot{y}-\dot{y}\ddot{x}|}{(\dot{x}^2+\dot{y}^2)^{3/2}};$$

here assume that $\dot{\boldsymbol{x}}(t)\neq\boldsymbol{0}$.

14 If a plane curve is the graph of $y=f(x)$, then show that

$$K=\frac{|f''(x)|}{[1+f'(x)^2]^{3/2}}$$

(*Hint:* Parametrize the curve in the form $\boldsymbol{x}(t)=(t, f(t))$.)

15 Use the formula in Exercise 14 to find the curvature of $y=\sin x$ at $x=\pi/2$.

16 What is the curvature of a plane curve $y=f(x)$ at a point of inflection where f'' exists?

17 Prove that if a curve in $\boldsymbol{R}^m$ has curvature identically 0, then the curve is a straight line. (This is the converse of Example 3.3.)

REVIEW EXERCISES 2.4

1 If $\boldsymbol{x}(t)=(\cos 2t, \sin 2t, t)$, then find the equation of the tangent line at $t=\pi/4$. What is the speed there?

2 If

$$\boldsymbol{f}(t)=\left(\frac{t^2-4}{t-2}, 4t, 5\right),$$

then is $\boldsymbol{f}$ continuous at $t=2$? Can you define $\boldsymbol{f}(2)$ so that $\boldsymbol{f}$ becomes continuous at $t=2$?

3 If $\boldsymbol{f}(t)=(\cos 2t, \sin 2t, t)$, then find

$$\int_0^{\pi/4} \boldsymbol{f}(t)\,dt.$$

4 A particle moves on the parabola $y^2=4x$. Find the velocity, acceleration, and speed at any time t.

5 Describe the path in $\boldsymbol{R}^2$ of a particle subjected to a force $\boldsymbol{F}(t)=(-\cos t, -\sin t)$.

6 Find the length of arc along $\boldsymbol{x}(t)=(2t, t^2, t)$ between $t=0$ and $t=1$.

7 Show that the curve $\boldsymbol{x}(t)=(\frac{1}{3}t, 4-\frac{2}{3}t, \frac{2}{3}t)$ is parametrized by arc length.

8 If $\boldsymbol{x}(t)=(1, e^t\cos t, e^t\sin t)$, then find K, $\boldsymbol{T}$, and $\boldsymbol{N}$.

9 Repeat Exercise 8 for $\boldsymbol{x}(t)=(\sin 2t, t, \cos 2t)$.

10 If $\boldsymbol{x}(t)=(t, \sin 4t, \cos 4t)$, $0\leq t\leq\pi$, then find:
(a) The velocity, speed, and acceleration at any time t.
(b) The length of arc of the curve.
(c) A parametrization of the curve by arc length.
(d) $\boldsymbol{T}$, $\boldsymbol{N}$, and K at $t=\pi/2$.
(e) The tangential and normal components of acceleration.

11 A particle moves on the curve

$$\boldsymbol{x} = \boldsymbol{x}(t) = (t \sin t, t \cos t, t^2 - 1), \quad t \in [0, 2\pi].$$

Find:

(a) The velocity, speed and acceleration in general and at $t = \pi$.
(b) The arc length.
(c) The tangential and normal components of acceleration at $t = \pi$.

3 REAL VALUED FUNCTIONS

0 INTRODUCTION

Having considered functions $f: \boldsymbol{R} \to \boldsymbol{R}^n$ in the last chapter, we are ready now to turn to functions $f: \boldsymbol{R}^n \to \boldsymbol{R}$.

First, when $n = 2$, there is a natural *graph* of a function $f: \boldsymbol{R}^2 \to \boldsymbol{R}$. This is a subset of $\boldsymbol{R}^3$ analogous to the graph of a function $f: \boldsymbol{R} \to \boldsymbol{R}$ in elementary analytic geometry. But this time instead of being a one-dimensional set (a curve), the graph generally comprises a two-dimensional set, a *surface* in $\boldsymbol{R}^3$. See Figure 0.1. In this chapter we introduce the appropriate methods for graphing such surfaces, as well as slightly more general surfaces called *quadric surfaces*, which are three-dimensional analogues of conic sections.

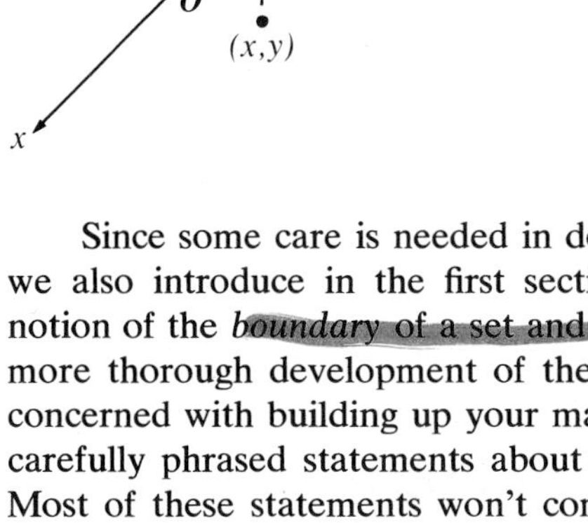

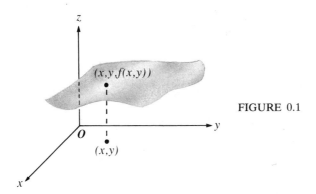

FIGURE 0.1

Since some care is needed in doing calculus for real valued functions on $\boldsymbol{R}^n$, we also introduce in the first section some *topological* ideas built around the notion of the *boundary* of a set and the *interior* of the set. In advanced calculus a more thorough development of these ideas is made, but here we are primarily concerned with building up your mathematical vocabulary enough to understand carefully phrased statements about the calculus of real valued functions on $\boldsymbol{R}^n$. Most of these statements won't come until the next two chapters, although the second section here discusses *limits* and *continuity* of such functions.

1 OPEN AND CLOSED SETS

As mentioned in the introduction, we need to exercise some care in developing the calculus of functions $f: \mathbf{R}^n \to \mathbf{R}$. It turns out to be of importance, for instance, to know what kind of set the domain $D \subseteq \mathbf{R}^n$ of such a function is. In this section we introduce those attributes of D which have the greatest impact on the calculus of f. The detailed study of these attributes is given the name *topology*, and advanced calculus requires quite a bit more topology than you will see here. For the purposes of this text, a good goal to work toward is the ability to recognize open and closed sets on sight.

These concepts hinge on the notion of *closeness*, which we can easily measure in $\mathbf{R}^n$. After all, given two points $\mathbf{x}$ and $\mathbf{y}$ in $\mathbf{R}^n$, we can compute the distance between them by calculating the length of $\mathbf{y}-\mathbf{x}$. This affords us a reasonable concept of *convergence* of a sequence of points in $\mathbf{R}^n$.

1.1 DEFINITION. A sequence $(\mathbf{x}_k)$ of points $\mathbf{x}_k = (x_{k1}, x_{k2}, \ldots, x_{kn})$ in $\mathbf{R}^n$ **converges to** (or **approaches**) the **limit** $\mathbf{x}_0 = (x_{01}, x_{02}, \ldots, x_{0n}) \in \mathbf{R}^n$ if $\lim_{k\to\infty} |\mathbf{x}_k - \mathbf{x}_0| = 0$.

In other words, the points $\mathbf{x}_k$ approach $\mathbf{x}_0$ if the distance between $\mathbf{x}_k$ and $\mathbf{x}_0$ approaches 0. In this case we write

$$\lim_{k\to\infty} \mathbf{x}_k = \mathbf{x}_0,$$

or $\mathbf{x}_k \to \mathbf{x}_0$ as $k \to \infty$.

Since each $|\mathbf{x}_k - \mathbf{x}_0|$ is a real number, convergence in $\mathbf{R}^n$ reduces to the notion of convergence of a sequence of real numbers, which you have already studied in elementary calculus (or Chapter 11, Section 1). We can make this reduction even more striking. *Convergence in $\mathbf{R}^n$ is equivalent to convergence in each coordinate position.*

1.2 THEOREM. Let $\mathbf{x}_k = (x_{k1}, x_{k2}, \ldots, x_{kn}) \in \mathbf{R}^n$. Then $\mathbf{x}_k \to \mathbf{x}_0$ if and only if for each coordinate $i = 1, 2, \ldots, n$

$$\lim_{k\to\infty} x_{ki} = x_{0i},$$

where x_{0i} is the i-th coordinate of $\mathbf{x}_0 = (x_{01}, x_{02}, \ldots, x_{0n})$.

Proof. Suppose first that each $\lim_{k\to\infty} x_{ki} = x_{0i}$, $i = 1, 2, \ldots, n$. Since

$$|\mathbf{x}_k - \mathbf{x}_0| = \sqrt{\sum_{i=1}^{n} [x_{ki} - x_{0i}]^2},$$

as $k \to \infty$ we have

$$|\mathbf{x}_k - \mathbf{x}_0| \to \sqrt{\sum_{i=1}^{n} 0^2} = 0.$$

Thus $\mathbf{x}_k \to \mathbf{x}_0$. Conversely, suppose that $\mathbf{x}_k \to \mathbf{x}_0$. Then

$$|\mathbf{x}_k - \mathbf{x}_0| = \sqrt{\sum_{i=1}^{n} [x_{ki} - x_{0i}]^2}$$

approaches 0. Then each $x_{ki}-x_{0i}$ must approach zero, since for each i,

$$0 \le |x_{ki}-x_{0i}| = \sqrt{(x_{ki}-x_{0i})^2} \le \sqrt{\sum_{i=1}^{n}(x_{ki}-x_{0i})^2} = |\boldsymbol{x}_k - \boldsymbol{x}_0|.$$

Hence $x_{ki} \to x_{0i}$. QED

This theorem affords a simple test for convergence in $\boldsymbol{R}^n$, as the next two examples illustrate.

1.3 EXAMPLE. Determine whether the sequence $(\boldsymbol{x}_k)$ converges if $\boldsymbol{x}_k = \left(\dfrac{2^k}{2^k-1}, 3-\dfrac{1}{k}\right)$.

Solution. How can we obtain a candidate for $\boldsymbol{x}_0$? In light of Theorem 1.2, $\boldsymbol{x}_k = (x_k, y_k)$ converges if and only if x_k and y_k both converge. Here $\lim\limits_{k\to\infty} x_k = \lim\limits_{k\to\infty} \dfrac{2^k}{2^k-1} = 1$ and $\lim\limits_{k\to\infty} y_k = \lim\limits_{k\to\infty}\left(3-\dfrac{1}{k}\right) = 3$. Hence by Theorem 1.2, the given sequence converges to the limit (1, 3).

1.4 EXAMPLE. Determine whether the sequence

$$\boldsymbol{x}_k = \left(\sin\frac{k\pi}{2}, \frac{k-1}{k^2+1}, \frac{\ln k}{k}\right)$$

converges in $\boldsymbol{R}^3$.

Solution. By Theorem 1.2, we have convergence if and only if $\lim\limits_{k\to\infty} x_k$, $\lim\limits_{k\to\infty} y_k$, and $\lim\limits_{k\to\infty} z_k$ all exist. But note that $\sin\dfrac{k\pi}{2}$ oscillates between the two values 1 and -1. Hence $\lim\limits_{k\to\infty} x_k$ fails to exist. So even though $\lim\limits_{k\to\infty} y_k = 0 = \lim\limits_{k\to\infty} z_k$, $\boldsymbol{x}_k$ fails to converge. We say it diverges. (By the way, recall that to find $\lim\limits_{k\to\infty} z_k = \lim\limits_{k\to\infty}\dfrac{\ln k}{k}$ we can use L'Hôpital's Rule:

$$\lim_{k\to\infty}\frac{\ln k}{k} = \lim_{x\to+\infty}\frac{\ln x}{x} = \lim_{x\to+\infty}\frac{\frac{1}{x}}{1} = 0.)$$

Now we want to describe the important concepts of open sets and closed sets. Loosely speaking, a set is *open* if none of its points are *boundary points*. To make this statement precise, we have to define the notion of boundary point of a set. The technical definition says a point P is a boundary point of a set S if P is the limit of some convergent sequence $(\boldsymbol{x}_k)$ of points *in* S and also the limit of some convergent sequence $(\boldsymbol{y}_k)$ of points *not* in S. This sounds somewhat involved, but intuitively we can recognize the boundary points of many familiar sets with no difficulty. The regions

$$\{(x, y) \mid x^2+y^2<1\}, \quad \{(x, y) \mid x^2+y^2\le 1\},$$
$$\{(x, y) \mid x^2+y^2\ge 1\} \quad \text{and} \quad \{(x, y) \mid x^2+y^2>1\}$$

all have the unit circle

$$\{(x, y) \mid x^2+y^2=1\}$$

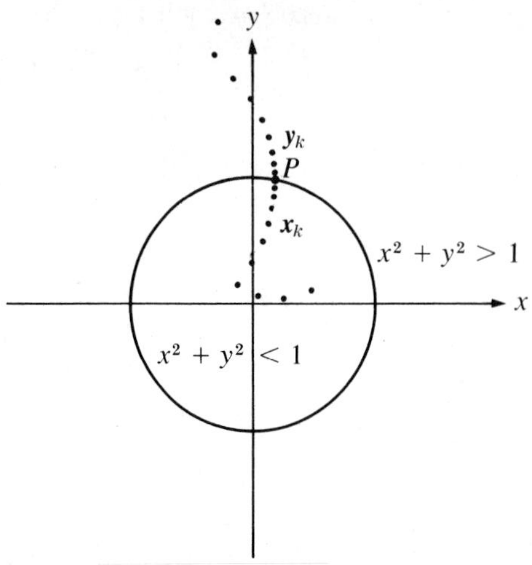

FIGURE 1.1

as boundary, for example. See Figure 1.1, where we show illustrations of the convergent sequences $(\boldsymbol{x}_k)$ and $(\boldsymbol{y}_k)$. The first and fourth of these sets do not include any of their boundary points and so are open. Can we think of some way to express this mathematically? Notice that every point (x, y) in the first or fourth set seems to lie at some *positive* distance δ from the nearest boundary point. Thus the disk

$$B(0, \delta) = \{(x, y) \mid x^2 + y^2 < \delta^2\}$$

is wholly contained in the set, as is any disk of radius smaller than δ. See Figure 1.2. It is this property which we single out for the formal definition of an open set

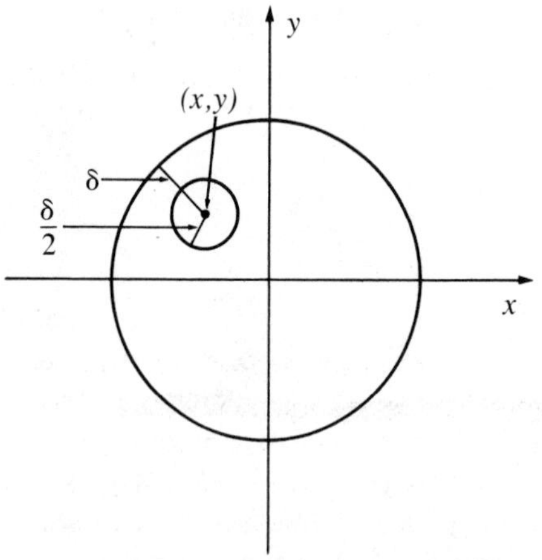

FIGURE 1.2

in $\boldsymbol{R}^n$. First, let us establish some notation. The **open ball** about P of radius δ is

$$B(P, \delta) = \{Q \in \boldsymbol{R}^n \mid d(P, Q) < \delta\},$$

where $d(P, Q) = |\overrightarrow{PQ}|$ is the distance between P and Q.

1.5 DEFINITION. A point $P(x_1, x_2, \ldots, x_n) \in \boldsymbol{R}^n$ is an **interior point** of a set S if there is some $\delta > 0$ such that $B(P, \delta) \subseteq S$. The set of all interior points of S is denoted Int S. S is an **open set** if $S = \text{Int } S$, i.e., if *every* point of S is an interior point.

Our previous discussion shows that $S_1 = \{(x, y) \mid x^2 + y^2 < 1\}$ and $S_2 = \{(x, y) \mid x^2 + y^2 > 1\}$ are both open sets.

1.6 EXAMPLE. Decide whether the following sets are open.
(a) $\{(x, y) \mid x > 5\}$, (b) $\{(x, y) \mid x > 5, y \geq 1\}$, (c) $\{(x, y) \mid x^2 + y^2 \geq 1\}$.

Solution. (a) This set is open since none of its points lie on the boundary line $x = 5$. So we can always surround a point (x, y) of the set by a disk that is completely in the set. See Figure 1.3.

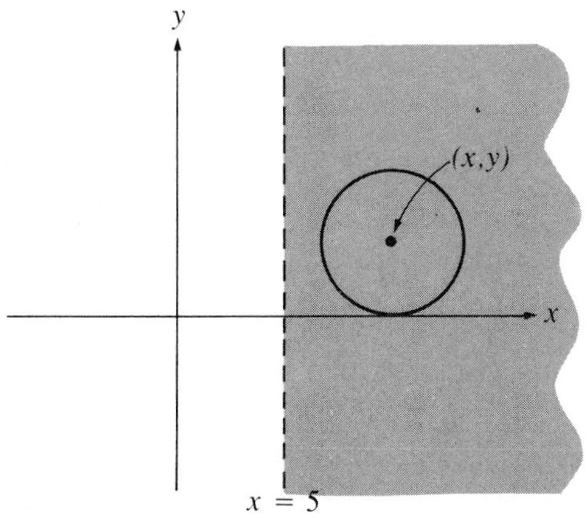

FIGURE 1.3

(b) This set is *not* open. If we select a point $(x, 1)$ in the set, then no ball about the point will be completely contained in the set, since the ball will include points (x, y) with $y < 1$. See Figure 1.4.

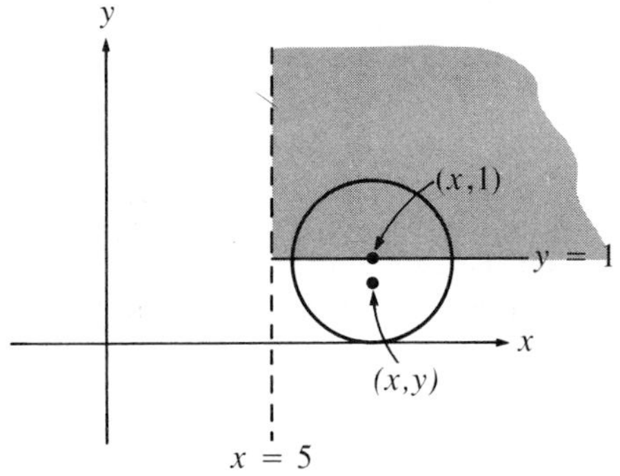

FIGURE 1.4

(c) This set is *not* open. If we select any point (x_0, y_0) such that $x_0^2 + y_0^2 = 1$, then *any* ball about the point will include points (x, y) such that $x^2 + y^2 < 1$. See Figure 1.5.

The set in Example 1.6(c) seems to miss being open in the worst way: *all* of its boundary points are in the set. A set with this property is called *closed*. We are again faced with the task of giving a mathematical formulation of this. Let us try to approach this backwards. If a set S *isn't* closed, then we want it to omit at least one boundary point. If this boundary point is $\mathbf{x}_0$, then since $\mathbf{x}_0$ is on the boundary, we can find a sequence of points $(\mathbf{x}_k)$ in S which approaches $\mathbf{x}_0$. See Figure 1.6. So a

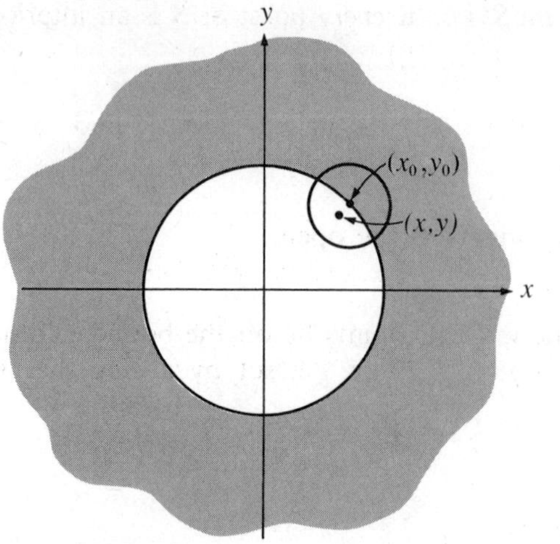

FIGURE 1.5

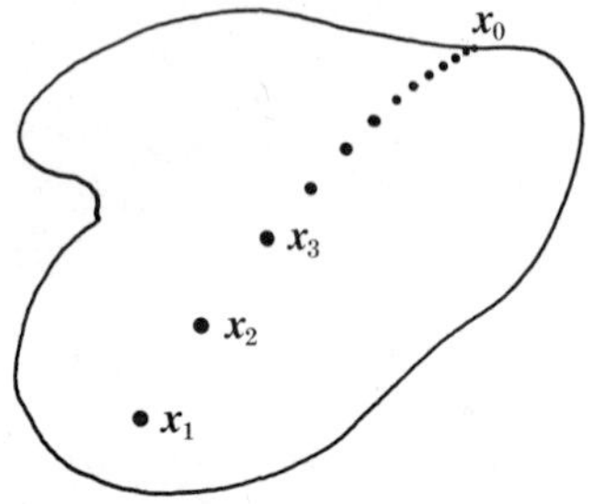

FIGURE 1.6

non-closed set has the property that some point $\mathbf{x}_0 \notin S$ is the limit of a sequence of points $\mathbf{x}_k \in S$. We define a set to be closed then if *all limits* $\mathbf{x}_0$ *of convergent sequences of points in S must be in S.*

1.7 DEFINITION. A set $S \subseteq \mathbf{R}^n$ is **closed** if, whenever a sequence $(\mathbf{x}_k)$ of points of S converges to a limit $\mathbf{x}_0$, then $\mathbf{x}_0$ is a point of S.

1.8 EXAMPLE. Decide whether the following sets are closed.

(a) $\left\{(x, y) \,\middle|\, \dfrac{x^2}{a^2} - \dfrac{y^2}{b^2} \le 1\right\}$, (b) $\{(x, y) \mid x > 5, y \ge 1\}$.

Solution. (a) This is the region on and between the two branches of a hyperbola. (See Figure 1.7.) The boundary is the hyperbola itself, which is included in the set.

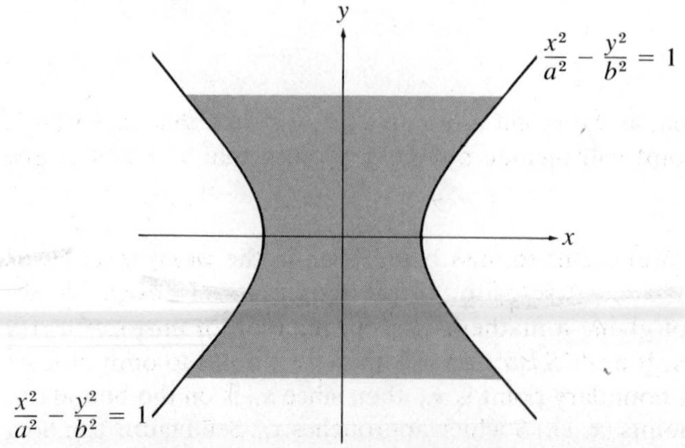

FIGURE 1.7

Intuitively, this set seems to contain all its boundary points, so that it is closed. We can even *prove* that it is closed from Definition 1.7. If (x_k, y_k) is a sequence of points in the set, then

$$\frac{x_k^2}{a^2}-\frac{y_k^2}{b^2}\leq 1.$$

Hence

$$\lim_{k\to\infty}\left(\frac{x_k^2}{a^2}-\frac{y_k^2}{b^2}\right)\leq 1$$

whenever this limit exists. Thus, if (x_0, y_0) is the limit of a sequence of points (x_k, y_k) in the set, then

$$\frac{x_0^2}{a^2}-\frac{y_0^2}{b^2}\leq 1,$$

so (x_0, y_0) is in the set.

(b) The set is shown in Figure 1.4 and fails to contain some of its boundary points, namely the points $(5, y)$ where $y\geq 1$. In fact, $(5, y)$ is the limit of the sequence of points $\left(5+\frac{1}{n}, y\right)$, all of which are in the set; but $(5, y)$ is *not* in the set. So the set is not closed by Definition 1.7.

Notice then that some sets (like the one in Examples 1.6(b) and 1.8(b)) are *neither open nor closed*! Moreover, some sets are *both open and closed*! Two examples of this are the empty set $\varnothing$, and $\boldsymbol{R}^n$ itself. Note that $\boldsymbol{R}^n$ certainly has no boundary points outside $\boldsymbol{R}^n$, so it is closed. But every point P certainly has a ball $B(P, 1)$ about it which is completely contained in $\boldsymbol{R}^n$, so $\boldsymbol{R}^n$ is also open! The empty set is always a bit pathological. It is open because you cannot find any point $P\in\varnothing$ which *doesn't* have a ball $B(P, \delta)$ around it completely contained in $\varnothing$ (because you cannot find any points $P\in\varnothing$ of *any* kind!). It is also closed, since you cannot find a sequence of points $\boldsymbol{x}_k\in\varnothing$ converging to a point outside $\varnothing$ (because you cannot find any sequence of points $\boldsymbol{x}_k\in\varnothing$ of *any* kind!).

Is there any reasonably appealing connection between open and closed sets? There is, and we close this section with it.

1.9 **THEOREM.** A set $S\subseteq\boldsymbol{R}^n$ is open if and only if its complement $\boldsymbol{R}^n-S=\{\boldsymbol{x}\in\boldsymbol{R}^n \mid \boldsymbol{x}\notin S\}$ is closed.

Proof. Suppose first that S is open. We want to prove that $\boldsymbol{R}^n-S$ is closed. We do this by showing the absurdity of the alternative, that $\boldsymbol{R}^n-S$ is not closed. For if $\boldsymbol{R}^n-S$ isn't closed, then we can find a sequence of points $\boldsymbol{x}_k\in\boldsymbol{R}^n-S$ whose limit $\boldsymbol{x}_0$ is not in $\boldsymbol{R}^n-S$. But if $\boldsymbol{x}_0\notin\boldsymbol{R}^n-S$, then $\boldsymbol{x}_0\in S$. Then there is a ball $B(\boldsymbol{x}_0, \delta)\subseteq S$ since S is open. But then each $\boldsymbol{x}_k$ must be *at least* δ units from the limit $\boldsymbol{x}_0$. (See Figure 1.8.) But this is absurd: we chose the $\boldsymbol{x}_k$ to approach $\boldsymbol{x}_0$ as a limit: Hence, if S is open, then $\boldsymbol{R}^n-S$ must be closed.

Conversely, suppose that $\boldsymbol{R}^n-S$ is closed. To show that S is open, we have to show that every $\boldsymbol{x}_0\in S$ has an open ball $B(\boldsymbol{x}_0, \delta)\subseteq S$. We show this by showing that the alternative is absurd. For suppose that some $\boldsymbol{x}_0\in S$ has no such ball $B(\boldsymbol{x}_0, \delta)\subseteq S$. Then if we successively take $\delta=1, \frac{1}{2}, \frac{1}{3}, \ldots, \frac{1}{n}, \ldots$ it must be that

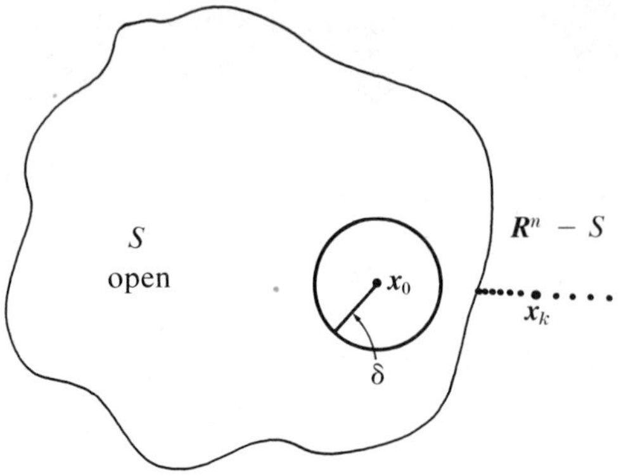

FIGURE 1.8

every $B\left(x_0, \frac{1}{i}\right) \nsubseteq S$. That means that each $B\left(x_0, \frac{1}{i}\right)$ contains at least *some* points of $\boldsymbol{R}^n - S$. In each $B\left(x_0, \frac{1}{i}\right)$, then, we could choose an $x_i \in \boldsymbol{R}^n - S$. If we did, then we would have a sequence (x_i) of points in $\boldsymbol{R}^n - S$ which converged to x_0 $\left(\text{since } \lim_{i\to\infty} \frac{1}{i} = 0\right)$. But since $\boldsymbol{R}^n - S$ is closed, x_0 would *have* to be in $\boldsymbol{R}^n - S$. But this is absurd: we *chose* $x_0 \in S$! Hence, if $\boldsymbol{R}^n - S$ is closed, then S is open. QED

If we employ this theorem, then whenever we show that a particular set S is open, we have also shown that its complement $\boldsymbol{R}^n - S$ is closed. By the same token, since the complement of the complement of S is S, any time we show that a set S is closed, then we have shown that its complement $\boldsymbol{R}^n - S$ is open, as follows: $\boldsymbol{R}^n - (\boldsymbol{R}^n - S) = S$ is closed; so by Theorem 1.9, $\boldsymbol{R}^n - S$ is open.

1.10 EXAMPLE. Show that

$$S = \left\{(x, y) \,\middle|\, \frac{x^2}{a^2} - \frac{y^2}{b^2} > 1\right\}$$

is open.

Solution. Note that

$$\boldsymbol{R}^2 - S = \left\{(x, y) \,\middle|\, \frac{x^2}{a^2} - \frac{y^2}{b^2} \leq 1\right\}$$

which we showed to be closed in Example 1.8. Thus by Theorem 1.9, the complement of $\boldsymbol{R}^2 - S$, i.e., S, is open. QED

EXERCISES 3.1

In Exercises 1 to 24, classify each set as open, closed, both, or neither. Explain your answers without giving formal proofs.

1 $[a, b] \subseteq \boldsymbol{R}$.

2 $(a, b] \subseteq \boldsymbol{R}$.

3 $(a, b) \subseteq \boldsymbol{R}$.

4 $[0, +\infty) \subseteq \boldsymbol{R}$.

5 $(-\infty, 0) \subseteq \boldsymbol{R}$.

6 The parabola $y^2 = 4x$ in $\boldsymbol{R}^2$.

7 The sphere $x^2 + y^2 + z^2 = 9$ in $\boldsymbol{R}^3$.

8 $\{(x, y) \mid x^2 > 1\}$ in $\boldsymbol{R}^2$.

9 $\{(x, y) \mid x^2 \le 1, y = 3\}$ in $\boldsymbol{R}^2$.

10 $\{(x, y) \mid x^2 + y^2 < 4\}$ in $\boldsymbol{R}^2$.

11 $\{(x, y, z) \mid x^2 + y^2 < 4\}$ in $\boldsymbol{R}^3$.

12 $\{(x, y, z) \mid x^2 + y^2 < 4, z = 0\}$ in $\boldsymbol{R}^3$

13 $\left\{(x, y) \;\middle|\; \frac{x^2}{4} + \frac{y^2}{9} > 1\right\}$ in $\boldsymbol{R}^2$.

14 $\left\{(x, y, z) \;\middle|\; \frac{x^2}{4} + \frac{y^2}{9} > 1, z = 0\right\}$ in $\boldsymbol{R}^3$.

15 $\{(x, y, z, w) \mid 3x - y + 2z + w \le 3\}$ in $\boldsymbol{R}^4$.

16 $\{(x, y, z, w) \mid x^2 + y^2 + z^2 + w^2 > 9\}$ in $\boldsymbol{R}^4$.

17 $\{(x, y) \mid 2x + 3y \le 6\}$ in $\boldsymbol{R}^2$.

18 $\{(x, y) \mid y < x\}$ in $\boldsymbol{R}^2$.

19 $\{(x, y) \mid x^2 + y^2 \ge 0\}$ in $\boldsymbol{R}^2$.

20 $\{(x, y) \mid x^2 + y^2 + 1 \le 0\}$ in $\boldsymbol{R}^2$.

21 $\{(x, y) \mid x^2 + y^2 + 1 < 0\}$ in $\boldsymbol{R}^2$.

22 $\{(x, y) \mid x^2 + y^2 > 0\}$ in $\boldsymbol{R}^2$.

23 $\{(2, 1, -3)\}$ in $\boldsymbol{R}^3$.

24 $\{(x, y, z) \mid x \text{ is an integer}\}$ in $\boldsymbol{R}^3$.

25 If S and T are open sets, then show that $S \cup T$ and $S \cap T$ are open.

26 If S and T are closed sets, then show that $S \cup T$ and $S \cap T$ are closed.

27 (a) Prove that $\left\{(x, y) \;\middle|\; \frac{x^2}{4} + \frac{y^2}{9} = 1\right\}$ is closed.
(b) What set have you therefore shown is open?

28 (a) Prove that $\{(x, y, z) \mid 3x + 2y - z \ge 4\}$ is a closed set.
(b) What set have you therefore shown is open?

In Exercises 29 to 38, determine whether the given sequence $(\boldsymbol{x}_k)$ converges. If it does, then give the limit.

29 $\boldsymbol{x}_k = \left(\frac{1}{1-k}, \sin(2k+1)\pi\right)$

30 $\boldsymbol{x}_k = \left(\frac{1}{1-k}, \cos 2k\pi\right)$

31 $\boldsymbol{x}_k = \left(\operatorname{Tan}^{-1} \frac{k^3}{2k^2 - k}, 3 - \frac{1}{1+k}\right)$

32 $\boldsymbol{x}_k = \left(\operatorname{Tan}^{-1} \frac{2k^2}{2k^2 - 3k + 1}, e^{1/k}\right)$

33 $\boldsymbol{x}_k = \left(\frac{k^2 - 5k + 1}{k^2 + 3k + 3}, \frac{\sin k}{k}\right)$

34 $\boldsymbol{x}_k = \left(k \sin \frac{1}{k}, \frac{\ln k^2}{k}\right)$

35 $\boldsymbol{x}_k = \left(\tan \frac{\pi k^2}{2k^2 - 3}, \frac{1}{k}, \frac{k+1}{k-1}\right)$

36 $\boldsymbol{x}_k = \left(\ln \frac{1}{k}, e^{1/k}, \frac{1}{k}\right)$

37 $\boldsymbol{x}_k = \left(\sin(-1)^{2k+1} \frac{\pi}{2}, \frac{\cos 3k}{k}, e^{-k}\right)$

38 $\boldsymbol{x}_k = \left(\cos(-1)^k \frac{\pi}{2}, \tan \frac{k}{k+1} \pi, e^{-k+6}\right)$

39 In economics, a *commodity* is a good or service that can be purchased. In an economy with n commodities, a *commodity bundle* is a vector $\boldsymbol{x} \in \boldsymbol{R}^n$ such that each $x_j \ge 0$. Show that the set C of all commodity bundles is a closed subset of $\boldsymbol{R}^n$.

40 Refer to Exercise 39. A household of consumers establishes a *preference relation* and an *indifference relation* on the set C. (For example, in a simple economy with three commodities such as food, clothing, and housing, (2, 1, 1) might be preferred to (3, 1, 0), while the household might be indifferent between (2, 1, 1) and (1, 2, 1).) For $\boldsymbol{x} \in C$, $P_{\boldsymbol{x}}$ represents the set of $\boldsymbol{y} \in C$ that are as desired as or preferred to $\boldsymbol{x}$, while $NP_{\boldsymbol{x}}$ represents the set of $\boldsymbol{y}$ such that $\boldsymbol{x}$ is as desired as or preferred to $\boldsymbol{y}$. A basic axiom of the economic theory of the household is that $P_{\boldsymbol{x}}$ and $NP_{\boldsymbol{x}}$ are closed sets. Assuming this axiom, show that the indifference set $I_{\boldsymbol{x}}$ of all $\boldsymbol{y} \in C$ such that the household is indifferent between $\boldsymbol{x}$ and $\boldsymbol{y}$ must be a closed set also.

2 LIMITS AND CONTINUITY

We now turn to a formulation of the notions of limit and continuity for a function $f: \boldsymbol{R}^n \to \boldsymbol{R}$. Such a function assigns to each vector $\boldsymbol{x}$ in its domain $D \subseteq \boldsymbol{R}^n$ one and only one real number $f(\boldsymbol{x}) \in \boldsymbol{R}$. You have probably had experience with such functions before and have certainly had experience already with such functions in this book. For instance, the *length* function $|\ \ |$ for vectors assigns to each vector $\boldsymbol{x}=(x_1,\ldots,x_n)\in\boldsymbol{R}^n$ a nonnegative real number $|\boldsymbol{x}|=\sqrt{x_1^2+\cdots+x_n^2}$. We used this function in the last section to describe convergence, for example. Other examples abound. The "*ideal gas*" *law* says that the volume of a gas is a function of two variables P (pressure) and T (temperature): $V=nR\dfrac{T}{P}$ (where n is the number of moles of the gas, and R is the gas constant).

The daily weather map of the National Weather Service is a compilation of the values of the *barometric pressure function* (and other meteorological functions such as temperature) at various longitudes ℓ and latitudes m. Meteorologists use the reported values $B(\ell, m)$ of the barometric pressure function as a key ingredient in weather prediction.

In economics, an important quantity prominently reported every month is the *cost of living index*. The prices p_i of certain key commodities (goods and services) are monitored each month. A fixed weight w_i is assigned to each commodity in an effort to reflect its relative importance in a person's life. For example, housing costs are weighted more heavily than new car prices. The cost of living index is then the weighted average of the $p_i: C=f(p_1, p_2,\ldots,p_n)=w_1p_1+\cdots+w_np_n$. (The weights are chosen so that C averages 100 for some period such as 1967–69, for purposes of ready comparison of today's cost of living index with that of several years ago.)

We want to assign an appropriate meaning to a statement like $\lim\limits_{\boldsymbol{x}\to\boldsymbol{x}_0} f(\boldsymbol{x})=L$ for a function $f:\boldsymbol{R}^n\to\boldsymbol{R}$. Here it is natural to imitate the definition given in elementary calculus for functions $f:\boldsymbol{R}\to\boldsymbol{R}$. Thus we should be able to make $f(\boldsymbol{x})$ as close to L as we please by taking $\boldsymbol{x}$ close enough to $\boldsymbol{x}_0$.

2.1 DEFINITION. Suppose that $f:\boldsymbol{R}^n\to\boldsymbol{R}$ has domain D and $D\supseteq U-\{\boldsymbol{x}_0\}$ for some open set U containing $\boldsymbol{x}_0$. Then $\lim\limits_{\boldsymbol{x}\to\boldsymbol{x}_0} f(\boldsymbol{x})=L$ means that for every $\varepsilon>0$, there is a $\delta>0$ such that $|f(\boldsymbol{x})-L|<\varepsilon$ whenever $0<|\boldsymbol{x}-\boldsymbol{x}_0|<\delta$. We can rephrase this as: For every $\varepsilon>0$, there is a $\delta>0$ such that $|f(\boldsymbol{x})-L|<\varepsilon$ whenever $\boldsymbol{x}$ $(\neq\boldsymbol{x}_0)$ is in $B(\boldsymbol{x}_0,\delta)$.

Why have we required that f be defined throughout an *open* set U containing $\boldsymbol{x}_0$ except possibly at $\boldsymbol{x}_0$? This is to insure that there is some open ball $B(\boldsymbol{x}_0,\delta)$ about $\boldsymbol{x}_0$ on which f is defined, so that the condition $|f(\boldsymbol{x})-L|<\varepsilon$ is meaningful for all $\boldsymbol{x}$ near $\boldsymbol{x}_0$. One sometimes calls $B(\boldsymbol{x}_0,\delta)-\{\boldsymbol{x}_0\}$ the *deleted open ball* about $\boldsymbol{x}_0$ of radius δ.

2.2 EXAMPLE. Does f defined by

$$f(x, y)=\frac{x^4-y^4}{x^2+y^2}$$

have a limit at $(0, 0)$?

Solution. We have

$$f(x, y)=\frac{(x^2+y^2)(x^2-y^2)}{x^2+y^2}=x^2-y^2$$

as long as $(x, y)\neq(0, 0)$. So f is defined on any (deleted) open set about $(0, 0)$.

Moreover,

$$\lim_{(x,y)\to(0,0)} f(x, y) = \lim_{(x,y)\to(0,0)} (x^2 - y^2) = 0,$$

so the limit exists and equals 0. (This example illustrates why we required f to be defined *near* rather than *at* $\boldsymbol{x}_0$ in Definition 2.1.)

We won't emphasize the technical aspects of Definition 2.1, but will rely on our intuition about limits based on our experience from elementary calculus. We will also rely on the following theorem, whose proof we largely omit. It is proved (using Definition 2.1) in the same way that the corresponding result for real valued functions of a real variable is proved. (See Exercises 13 and 14 below.)

2.3 THEOREM. Let f and g be defined on a deleted open set about $\boldsymbol{x}_0$. If $\lim_{\boldsymbol{x}\to\boldsymbol{x}_0} f(\boldsymbol{x}) = L$ and $\lim_{\boldsymbol{x}\to\boldsymbol{x}_0} g(\boldsymbol{x}) = M$, then

(1) $\lim_{\boldsymbol{x}\to\boldsymbol{x}_0} (f \pm g)(\boldsymbol{x}) = L \pm M$,

(2) $\lim_{\boldsymbol{x}\to\boldsymbol{x}_0} kf(\boldsymbol{x}) = kL$ for any $k \in \boldsymbol{R}$,

(3) $\lim_{\boldsymbol{x}\to\boldsymbol{x}_0} f(\boldsymbol{x})g(\boldsymbol{x}) = LM$,

(4) $\lim_{\boldsymbol{x}\to\boldsymbol{x}_0} \dfrac{f(\boldsymbol{x})}{g(\boldsymbol{x})} = \dfrac{L}{M}$ if $M \neq 0$.

Partial Proof. (2) If $k = 0$, then clearly $\lim_{\boldsymbol{x}\to\boldsymbol{x}_0} kf(\boldsymbol{x}) = kL = 0$. If $k \neq 0$, then given $\varepsilon > 0$, choose $\delta > 0$ so that $|f(\boldsymbol{x}) - L| < \dfrac{\varepsilon}{|k|}$ when $\boldsymbol{x} \neq \boldsymbol{x}_0$ is in $B(\boldsymbol{x}_0, \delta)$. Then for $\boldsymbol{x} \neq \boldsymbol{x}_0$ in $B(\boldsymbol{x}_0, \delta)$,

$$|kf(\boldsymbol{x}) - kL| = |k|\,|f(\boldsymbol{x}) - L| < |k|\frac{\varepsilon}{|k|} = \varepsilon.$$

So by Definition 2.1, $\lim_{\boldsymbol{x}\to\boldsymbol{x}_0} kf(\boldsymbol{x}) = kL$. QED

The next definition gives the natural analogue of the notion of a continuous function $f: \boldsymbol{R} \to \boldsymbol{R}$.

2.4 DEFINITION. A function $f: \boldsymbol{R}^n \to \boldsymbol{R}$ is **continuous** at $\boldsymbol{x}_0$ if $\lim_{\boldsymbol{x}\to\boldsymbol{x}_0} f(\boldsymbol{x}) = f(\boldsymbol{x}_0)$.

Thus, a continuous function is one which *preserves limits:* if $\boldsymbol{x} \to \boldsymbol{x}_0$, then $f(\boldsymbol{x}) \to f(\boldsymbol{x}_0)$. So limits of continuous functions may be correctly computed by substitution. In the same way as for real valued functions of one real variable, the next theorem follows immediately from Theorem 2.3 and the definition of continuity. Notice that a function which is continuous at $\boldsymbol{x}_0$ must, in light of Definitions 2.1 and 2.4, be defined on an open set containing $\boldsymbol{x}_0$. Thus $\boldsymbol{x}_0$ is in the domain of f.

2.5 THEOREM. If f and g are continuous at $\boldsymbol{x}_0$, then so are

(1) $f \pm g$, (2) kf for any $k \in \boldsymbol{R}$,

(3) fg, and (4) $\dfrac{f}{g}$ provided $g(\boldsymbol{x}_0) \neq 0$.

This theorem provides us with lots of continuous functions. First, note that in $\boldsymbol{R}^2$ the projection functions P_x and P_y are everywhere continuous, where

$P_x(x, y) = x$ and $P_y(x, y) = y$. Then by Theorem 2.5(1, 2, 3), any polynomial function in x and y is everywhere continuous. Then by Theorem 2.5(4) any rational function in x and y is continuous wherever the denominator is nonzero. This latter restriction provides us with a wealth of problems. Given polynomials $f(x, y)$ and $g(x, y)$, we can form the "schizophrenic" rational function

$$r(x, y) = \begin{cases} \dfrac{f(x, y)}{g(x, y)} & \text{when} \quad g(x, y) \neq 0 \\ 0 & \text{whenever} \quad g(x, y) = 0, \end{cases}$$

and we can ask where this schizophrenic function is continuous.

2.6 **EXAMPLE.** Decide whether the function f is continuous at $(0, 0)$, where

$$f(x, y) = \begin{cases} \dfrac{x^2 - y^2}{x^2 + 2y^2} & \text{if} \quad (x, y) \neq (0, 0) \\ 0 & \text{at} \quad (0, 0) \end{cases}$$

Solution. This schizophrenic function is certainly continuous at all points (x, y) such that $x^2 + 2y^2 \neq 0$, i.e., everywhere except possibly at $(0, 0)$, by Theorem 2.5(4). Now, to settle the question at $(0, 0)$, we must first guess an answer and then try to verify our guess. A sometimes useful rule of thumb is:

> *Guess continuity if we can divide out of the numerator and denominator all factors that are* 0 *at the point of schizophrenia* (as in Example 2.2) or *if the degree of every term of the numerator exceeds the degree of the denominator,* and *guess discontinuity otherwise.*

So here we would guess discontinuity. Now, how could we verify this? If our guess is correct, then we can usually find a path of approach to $(0, 0)$ such that $f(x, y)$ tends to a limit *other* than $f(0, 0)$ along this path of approach. Here let's consider $(x, y) \to (0, 0)$ along the line $y = x$. Then $(x, y) = (t, t)$, so along this line

$$\lim_{(x,y)\to(0,0)} f(x, y) = \lim_{t\to 0} \frac{t^2 - t^2}{t^2 + 2t^2} = 0 = f(0, 0).$$

Well, our first try failed because we *did* approach $f(0, 0)$ along it. Let's not give up yet, though. Suppose we try to approach $(0, 0)$ along the line $y = 2x$, i.e., $(x, y) = (t, 2t)$. Then along this line

$$\lim_{(x,y)\to(0,0)} f(x, y) = \lim_{t\to 0} \frac{t^2 - 4t^2}{t^2 + 8t^2} = \lim_{t\to 0} \frac{-3t^2}{9t^2} = \lim_{t\to 0} -\tfrac{1}{3}$$
$$= -\tfrac{1}{3} \neq 0 = f(0, 0).$$

Hence $\lim_{(x,y)\to(0,0)} f(x, y)$ fails to exist, so the function is discontinuous. (Note that if $\lim_{\mathbf{x}\to\mathbf{0}} f(\mathbf{x})$ did exist, then we would *always* obtain its value no matter *how* we approached $(0, 0)$.

2.7 EXAMPLE. Decide whether the function f is continuous at $(0, 0)$, if

$$f(x, y) = \begin{cases} \dfrac{x^3y - xy^3}{x^2 + y^2} & \text{if } (x, y) \neq (0, 0) \\ 0 & \text{if } (x, y) = (0, 0). \end{cases}$$

Solution. Here each term of the numerator has higher degree than the denominator, so we expect that f *is* continuous at $(0, 0)$. To verify this, we must show that $f(x, y) \to 0$ as $(x, y) \to (0, 0)$. The best way to accomplish that is to find an expression whose absolute value is *larger* than $|f(x, y)|$ and which clearly approaches zero as $(x, y) \to (0, 0)$. We can obtain that sort of larger expression by first observing that $|x| \leq |\boldsymbol{x}|$ and $|y| \leq |\boldsymbol{x}|$. Then

$$|f(x, y)| = \frac{|xy(x^2 - y^2)|}{x^2 + y^2} = \frac{|x|\,|y|\,|x + y|\,|x - y|}{x^2 + y^2}$$

$$\leq \frac{|x|\,|y|\,(|x| + |y|)(|x| + |y|)}{x^2 + y^2} \quad \text{by the Triangle Inequality (Theorem 1.2.10)}$$

$$\leq \frac{|\boldsymbol{x}|\,|\boldsymbol{x}|\,(|\boldsymbol{x}| + |\boldsymbol{x}|)(|\boldsymbol{x}| + |\boldsymbol{x}|)}{|\boldsymbol{x}|^2} = 4\,|\boldsymbol{x}|^2 = 4(x^2 + y^2).$$

Now as $(x, y) = \boldsymbol{x} \to (0, 0)$, we have $4(x^2 + y^2) \to 0$, and so $|f(x, y)| \to 0$ also. We conclude that $\lim_{\boldsymbol{x} \to \boldsymbol{0}} f(\boldsymbol{x}) = 0 = f(\boldsymbol{0})$, and hence f is continuous at $(0, 0)$.

The illustration on the cover of this book offers a preview of the next section, on graphing, by showing a computer-generated graph of the function f of the preceding example.

You will notice that it is usually more challenging to verify a guess that a function *is* continuous at $(0, 0)$ than it is to verify that a function is *not* continuous there. In the latter case, it is usually fairly simple to find a path of approach along which $f(x, y)$ does not approach $f(0, 0)$. But to prove continuity we have to come up with a bit of clever reasoning which *proves* that $f(x, y) \to f(0, 0)$ *no matter how* we approach $(0, 0)$.

Assuming that we can establish that a number of elementary functions are continuous, then by forming composite functions from these and ordinary continuous real valued functions in $\boldsymbol{R}$, we ought to produce continuous functions. The next result formally guarantees this expectation.

2.8 THEOREM. Suppose that $f: \boldsymbol{R} \to \boldsymbol{R}$ is continuous on an open interval containing t_0, and suppose that $g: \boldsymbol{R}^n \to \boldsymbol{R}$ is continuous on an open set containing $\boldsymbol{x}_0$, where $g(\boldsymbol{x}_0) = t_0$. Then the composite function $f \circ g: \boldsymbol{R}^n \to \boldsymbol{R}$ is continuous at $\boldsymbol{x}_0$.

Proof. We need to show that

$$\lim_{\boldsymbol{x} \to \boldsymbol{x}_0} f \circ g(\boldsymbol{x}) = f \circ g(\boldsymbol{x}_0) = f(t_0).$$

For this, we can reason as follows. As $\boldsymbol{x} \to \boldsymbol{x}_0$, $t = g(\boldsymbol{x}) \to t_0 = g(\boldsymbol{x}_0)$ since g is continuous. But as $t \to t_0$, $f(t) \to f(t_0)$ since f is continuous, i.e., $f(g(\boldsymbol{x})) \to f(g(\boldsymbol{x}_0))$. If we want to be formal, let $\varepsilon > 0$ be given. Then find $\delta_1 > 0$ such that $|f(t) - f(t_0)| < \varepsilon$ whenever $|t - t_0| < \delta_1$. And find $\delta_2 > 0$ such that $|g(\boldsymbol{x}) - g(\boldsymbol{x}_0)| < \delta_1$

whenever $|\mathbf{x}-\mathbf{x}_0|<\delta_2$. Then for $|\mathbf{x}-\mathbf{x}_0|<\delta_2$, we have $|f(g(\mathbf{x}))-f(g(\mathbf{x}_0))|<\varepsilon$. So

$$\lim_{\mathbf{x}\to\mathbf{x}_0} f\circ g(\mathbf{x}) = f\circ g(\mathbf{x}_0). \qquad \text{QED}$$

As a consequence of the foregoing theorem, we can be sure, for instance, that the function f defined by $f(x, y)=\sin e^{xy}$ is continuous on $\mathbf{R}^2$, since the polynomial function xy is continuous on $\mathbf{R}^2$ and the exponential and sine functions are continuous on $\mathbf{R}$.

In our study of extreme values in the next chapter, in order to have faith in the existence of the extreme values we seek, we will need one more result about continuous functions. This result is quite a bit deeper than the others of this section, so we don't attempt to prove it. Before we can even state it, we need a definition.

2.9 DEFINITION. A set $S\subseteq\mathbf{R}^n$ is **bounded** if there is some number $\delta>0$ such that $S\subseteq B(\mathbf{0}, \delta)$.

Hence in $\mathbf{R}^3$, the set

$$\{(x, y, z)\mid 0\le x\le 3,\ y^2+z^2\le 1000\}$$

is bounded, but the xy-plane is not bounded.

2.10 THEOREM. If C is a closed and bounded set in $\mathbf{R}^n$ and if $f:\mathbf{R}^n\to\mathbf{R}$ has a domain containing C, then f has an absolute maximum and an absolute minimum on C. That is, there are points $\mathbf{x}_0$ and $\mathbf{x}_1$ in C such that $f(\mathbf{x}_0)\le f(\mathbf{x})\le f(\mathbf{x}_1)$ for all $\mathbf{x}\in C$.

From the theorem, we can be sure, for example, that the function

$$f(x, y, z)=\frac{\sin xy+e^{yz}}{3x^2+5y^2+z^2}$$

has an absolute maximum and an absolute minimum on the sphere $x^2+y^2+z^2=1$ (which is a closed and bounded set in $\mathbf{R}^3$).

The theorem gives absolutely no hint of where to find $\mathbf{x}_0$ and $\mathbf{x}_1$, however. The development of efficient techniques for locating $\mathbf{x}_0$ and $\mathbf{x}_1$ must await a study of differentiation of functions of several variables, as you would suspect from your experience in elementary calculus.

EXERCISES 3.2

In Exercises 1 to 4, evaluate the limit if it exists. If it does not, try to explain why.

1 $\displaystyle\lim_{(x,y)\to(2,0)} \ln\frac{x^2-y^2}{x-y}$

2 $\displaystyle\lim_{(x,y)\to(2,1)} \frac{x^3-y^3}{x^2-y^2}$

3 $\displaystyle\lim_{(x,y)\to(2,0)} \frac{\ln xy}{x^2+y^2}$

4 $\displaystyle\lim_{(x,y)\to(2,1)} \frac{x^2+y^2}{xy^2-2}$

In Exercises 5 to 12, find all points at which the given schizophrenic function is continuous.

5 $f(x, y)=\begin{cases} \dfrac{xy}{x^2+y^2} & \text{if } (x, y)\neq(0, 0) \\ 0 & \text{if } (x, y)=(0, 0) \end{cases}$

6 $g(x, y)=\begin{cases} \dfrac{x^2y}{x^2+y^2} & \text{if } (x, y)\neq(0, 0) \\ 0 & \text{if } (x, y)=(0, 0) \end{cases}$

7 $h(x, y)=\begin{cases} \dfrac{x^2y+xy^2}{x^2+y^2} & \text{if } (x, y)\neq(0, 0) \\ 0 & \text{if } (x, y)=(0, 0) \end{cases}$

8 $f(x, y)=\begin{cases} \dfrac{\sin xy}{xy} & \text{if } xy\neq 0 \\ 1 & \text{if } xy=0 \end{cases}$

9 $g(x, y)=\begin{cases} \dfrac{x^2y+3x^2}{x^2+y^2} & \text{if } (x, y)\neq(0, 0) \\ 0 & \text{if } (x, y)=(0, 0) \end{cases}$

10 $h(x, y)=\begin{cases} \dfrac{xy^2-y^2}{x^2+y^2} & \text{if } (x, y)\neq(0, 0) \\ 0 & \text{if } (x, y)=(0, 0) \end{cases}$

11 $f(x, y)=\begin{cases} \dfrac{x^2}{x^2+y^2} & \text{if } (x, y)\neq(0, 0) \\ \frac{1}{2} & \text{if } (x, y)=(0, 0) \end{cases}$

12 $g(x, y)=\begin{cases} \dfrac{x^2-y^2}{x^2+y^2} & \text{if } (x, y)\neq(0, 0) \\ 0 & \text{if } (x, y)=(0, 0) \end{cases}$

13 Prove Theorem 2.3(1) for the case of a plus sign.

14 Prove Theorem 2.3(1) for the case of a minus sign.

15 Prove that the length function on $\boldsymbol{R}^n$ is everywhere continuous.

16 Discuss the continuity of V in the ideal gas equation $PV=nRT$ with reference to the physical possibility of $P=0$.

17 Is the cost of living index a continuous function of $p_1, p_2, \ldots, p_n$?

18 Prove that the function f is continuous on $\boldsymbol{R}^3$ if $f(x, y, z)=\dfrac{x \sin yz - e^x \cos yz}{1+x^2+y^2+z^2}$.

19 If $f:\boldsymbol{R}^2\to\boldsymbol{R}$ is continuous at (x_0, y_0), then show that f_{x_0} defined by $f_{x_0}(y)=f(x_0, y)$ is a continuous function at $y=y_0$. Also show that f_{y_0} defined by $f_{y_0}(x)=f(x, y_0)$ is continuous at $x=x_0$.

20 The continuity of f_{x_0} at $y=y_0$ and f_{y_0} at $x=x_0$ is *not* enough to conclude that f is continuous at (x_0, y_0). Show this by considering the function in Exercise 5 again, with $(x_0, y_0)=(0, 0)$.

21 In the economic theory of the household (cf. Problems 39 and 40 of Exercises 3.1), we can show that there is a continuous *utility* function $u:C\to\boldsymbol{R}$ (C is the commodity space) such that $u(\boldsymbol{x})\geq u(\boldsymbol{y})$ holds if and only if the commodity bundle $\boldsymbol{x}$ is as desired as or preferred to $\boldsymbol{y}$. The utility function is a measure of the happiness derived from purchasing a commodity bundle. Show that if D is the set of all commodity bundles $\boldsymbol{x}$ such that $x_j\leq 1000$ for all j, then the utility function has a maximum and a minimum on D.

22 Refer to Exercise 21. Can you conclude that u has a maximum and a minimum on C? Why or why not?

3 GRAPHING

Very early in your study of functions $f:\boldsymbol{R}\to\boldsymbol{R}$ you learned how to draw graphs of such functions. As you know, the graph of a function $f:\boldsymbol{R}\to\boldsymbol{R}$ is the set of all points $P(x, y)$ in the plane $\boldsymbol{R}^2$ such that $y=f(x)$. See Figure 3.1. In a similar vein, you learned to draw the graph of a functional equation $g(x, y)=0$. The graph again consisted of all points $P(x, y)$ in $\boldsymbol{R}^2$ for which $g(x, y)=0$. For example, if $g(x, y)=x^2+4y^2-16$, the graph is an ellipse. See Figure 3.2.

Much of elementary differential calculus was devoted to developing techniques to help you draw graphs accurately. Since our professed goal is to develop

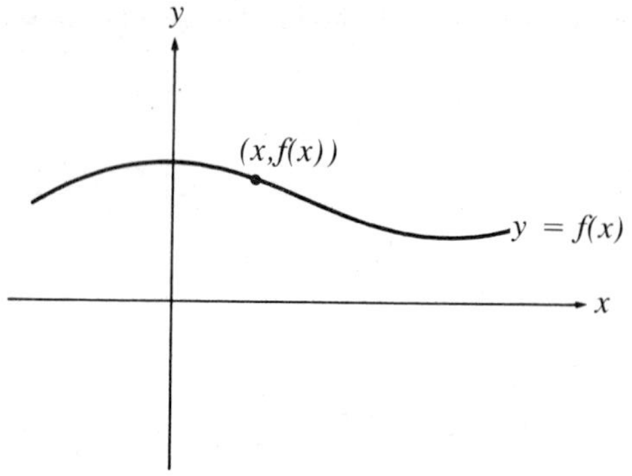

FIGURE 3.1

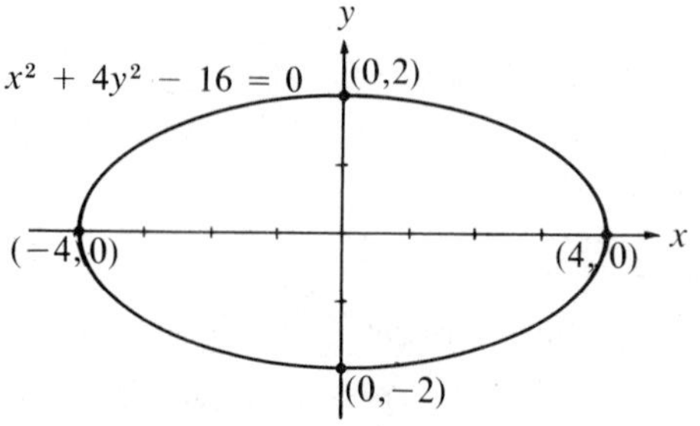

FIGURE 3.2

the calculus of functions $f: \boldsymbol{R}^n \to \boldsymbol{R}$, it is natural to consider the graphs of functions $f: \boldsymbol{R}^n \to \boldsymbol{R}$ and of functional equations $g(x_1, x_2, \ldots, x_n, y) = 0$. In both cases a notion of graph suggests itself.

3.1 DEFINITION. The **graph of a function** $f: \boldsymbol{R}^n \to \boldsymbol{R}$ is the subset of $\boldsymbol{R}^{n+1}$ consisting of all points $(x_1, x_2, \ldots, x_n, y)$ such that $y = f(x_1, x_2, \ldots, x_n)$. The **graph of a functional equation** $g(x_1, x_2, \ldots, x_n, y) = 0$ consists of all points $(x_1, x_2, \ldots, x_n, y) \in \boldsymbol{R}^{n+1}$ such that $g(x_1, x_2, \ldots, x_n, y) = 0$.

Notice that this definition includes the graphs discussed above—they are just the ones that arise in case $n = 1$. Notice also that we are confronted with an immediate problem in trying to *draw* a graph. We have pictorial representations only for $\boldsymbol{R}^2$ and $\boldsymbol{R}^3$ among all the Cartesian spaces in which graphs may lie. If we want to draw graphs on paper or blackboard, then, we can draw graphs only for functions $f: \boldsymbol{R}^2 \to \boldsymbol{R}$ and functional equations $g(x, y, z) = 0$ in addition to those we already are experienced with. In this section we briefly discuss some of the important kinds of graphs in $\boldsymbol{R}^3$, and in the next section we will study analogues of the conic sections of $\boldsymbol{R}^2$.

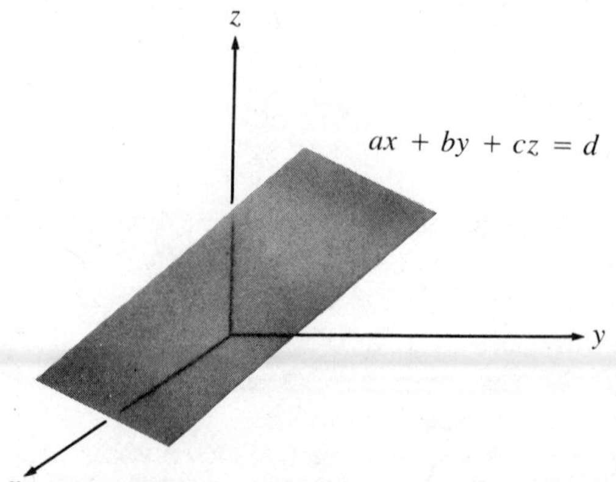

FIGURE 3.3

To start with, we already are familiar with some graphs in $\boldsymbol{R}^3$. For example, a functional equation $ax+by+cz-d=0$ has a plane as its graph. See Figure 3.3. And a polynomial function $f:\boldsymbol{R}^2 \to \boldsymbol{R}$ of degree one (say $f(x, y)=ax+by+d$) likewise has a plane as its graph. Why? Because for such a function, a point (x, y, z) on the graph satisfies $z=f(x, y)=ax+by+d$. See Figure 3.4.

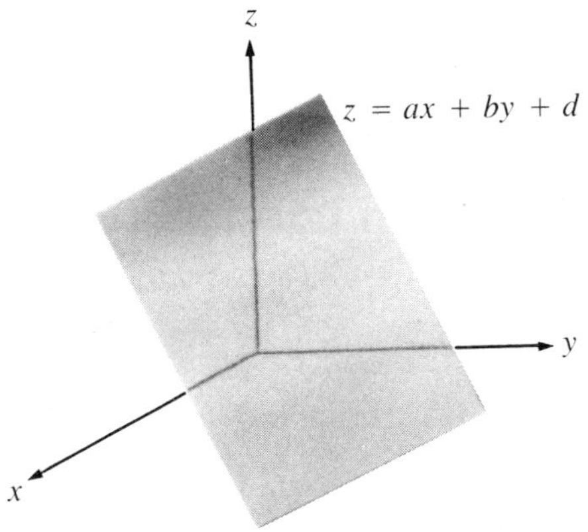

FIGURE 3.4

We also already know the graph of one quadratic polynomial equation: $x^2+y^2+z^2=a^2$ has for its graph a sphere of radius a and center $\mathbf{0}$. See Figure 3.5. Let us now consider some other important surfaces which arise as graphs.

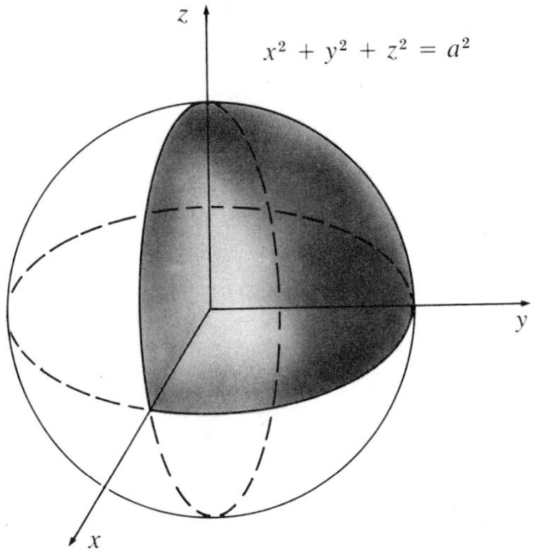

FIGURE 3.5

3.2 DEFINITION. Let ℓ be a line in $\boldsymbol{R}^3$ which intersects a plane curve γ but is not in the plane of γ. A **cylinder** or **cylindrical surface** is the set of all points $P \in \boldsymbol{R}^3$ such that P lies on a line ℓ' parallel to ℓ which also intersects γ. (See Figure 3.6.)

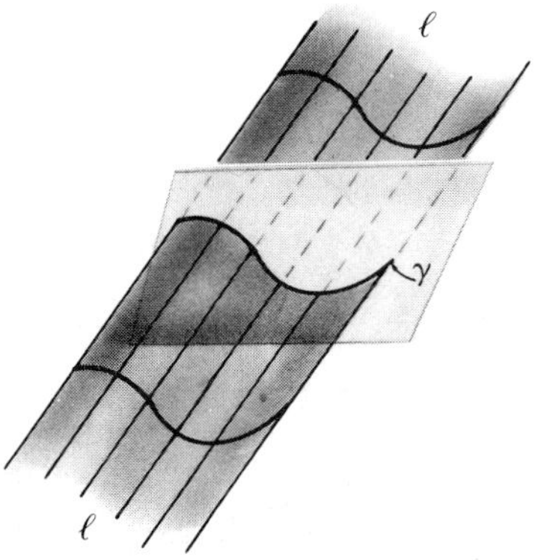

FIGURE 3.6

The general definition we have given means that a cylinder is unbounded in $\boldsymbol{R}^3$. The kinds of cylinders with which you probably have the most experience are **right circular cylinders**. For these, γ is a circle and ℓ is perpendicular to the plane

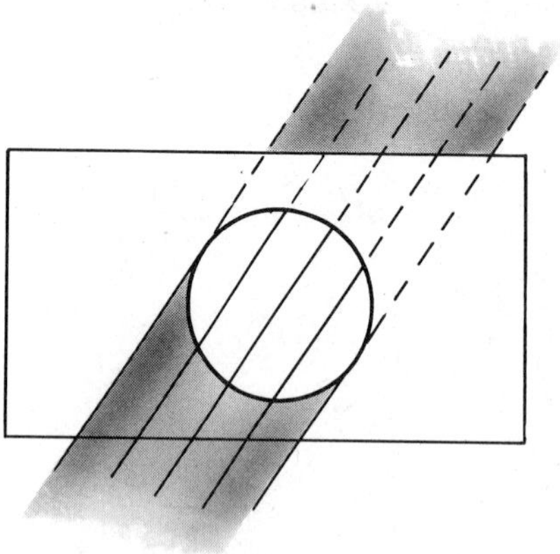

FIGURE 3.7

of γ. See Figure 3.7. In real-life situations you have probably dealt with *truncated* right circular cylinders (e.g., soup cans; see Figure 3.8). The general definition does not require γ to be a circle (or even a closed curve) or ℓ to be perpendicular to γ. In practice, however, we will deal only with *right* cylinders for which ℓ *is* perpendicular to the plane of γ.

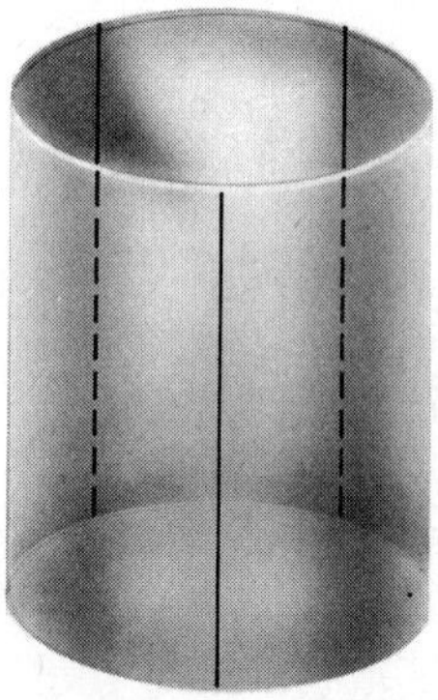

FIGURE 3.8

In general, the graphs in $\boldsymbol{R}^3$ of functional equations $f(x, y)=0$ or $g(y, z)=0$ or $h(x, z)=0$ are right cylinders. You can usually get a good picture of how they look by drawing the curves of intersection with the coordinate plane corresponding to the two variables present.

3.3 **EXAMPLE.** Draw the graph of $\dfrac{x^2}{4}-\dfrac{y^2}{9}=1$ in $\boldsymbol{R}^3$.

Solution. The intersection of this surface with the xy-plane is the hyperbola $\dfrac{x^2}{4}-\dfrac{y^2}{9}=1$ with vertices $(\pm 2, 0)$ and asymptotes $y=\pm\frac{3}{2}x$ (Figure 3.9). The cylinder is thus called a *hyperbolic cylinder.* We have pictured it in Figure 3.10.

An important kind of surface, which you probably saw in elementary calculus, is the next type we consider.

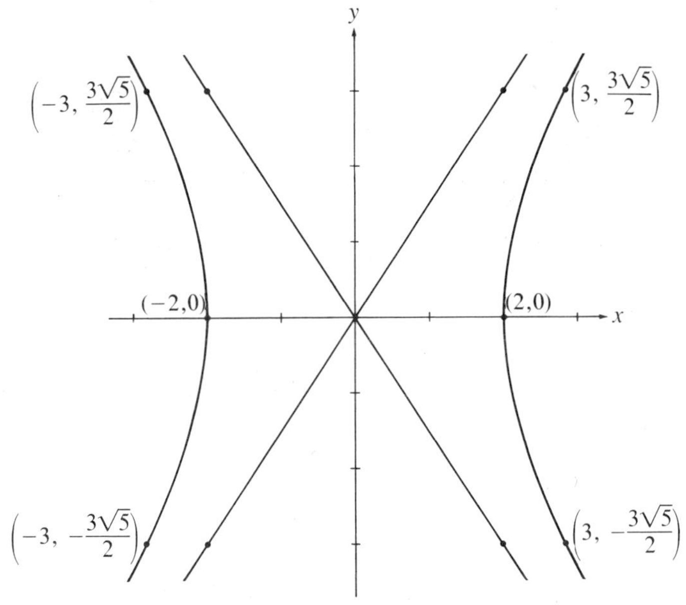

FIGURE 3.9

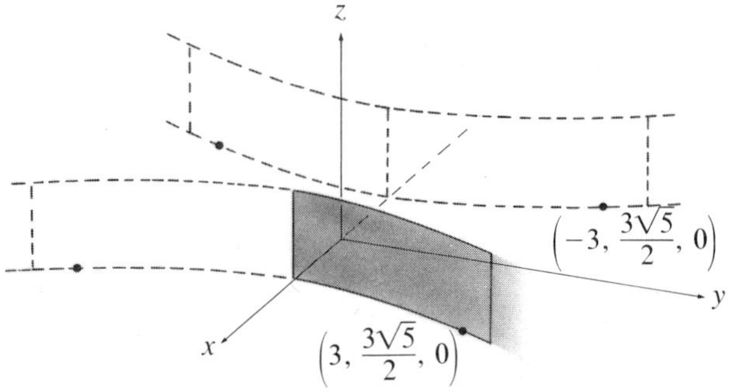

FIGURE 3.10

3.4 DEFINITION. A **surface of revolution** is the surface which results when a plane curve γ is revolved about a line ℓ in the plane of the curve. (The line is called the *axis of revolution.*)

If the curve γ is a circle and the line ℓ passes through a diameter, then the resulting surface is a sphere. If the curve γ is a parabola, hyperbola, or ellipse, and the line ℓ is the principal axis or a minor axis, then the resulting surface is called a *paraboloid, hyperboloid,* or *ellipsoid of revolution* respectively. See Figure 3.11.

Suppose we revolve the plane curve which is the graph of $y=f(x)$ about the axis $y=0$ in the xy-plane. What is the equation of the resulting surface? See Figures 3.12 and 3.13, where we picture the rotation as taking place in $\boldsymbol{R}^3$.

Let $P(x, y, z)$ be a point on the surface. Then $Q(x, y_1, 0)$ is on the curve in the xy-plane, as shown in Figure 3.13. Let P_0 be the point $(x, 0, 0)$ on the x-axis. Since we revolve the curve about the x-axis, the radii $|\overrightarrow{P_0P}|=|\overrightarrow{P_0Q}|$. Hence $\sqrt{y^2+z^2}=|y_1|$. Squaring this we have $y^2+z^2=y_1^2$. But $y_1=f(x)$ since (x, y_1) is on the curve that we revolve. So we get $y^2+z^2=f(x)^2$. A similar derivation could have been given if we revolved any curve in *any* coordinate plane about one of the plane's coordinate axes.

We summarize in a general theorem about revolving curves in any coordinate plane about a coordinate axis in that plane. Suppose that the curve is the graph of

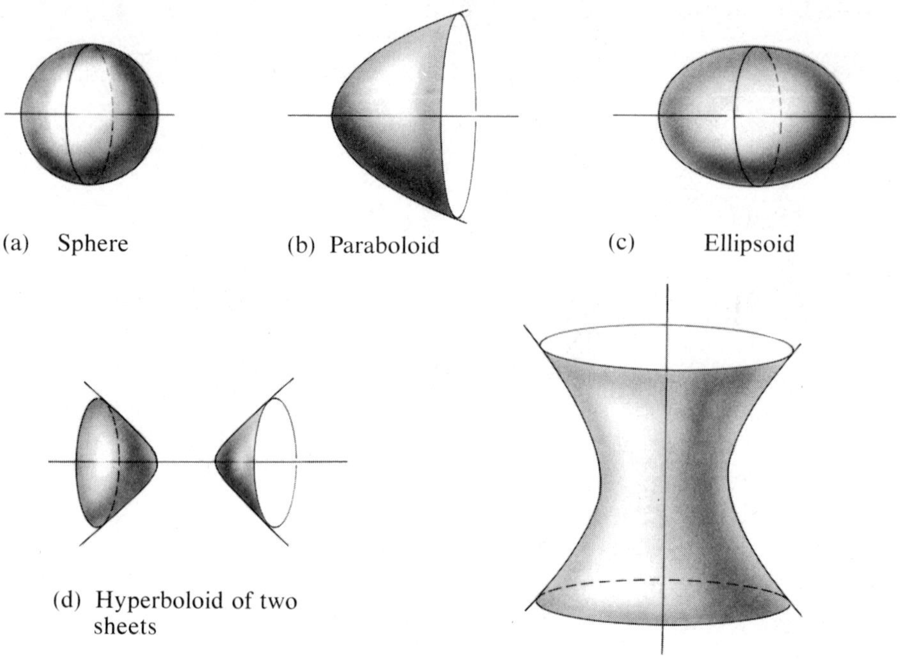

(a) Sphere (b) Paraboloid (c) Ellipsoid

(d) Hyperboloid of two sheets

(e) Hyperboloid of one sheet

FIGURE 3.11

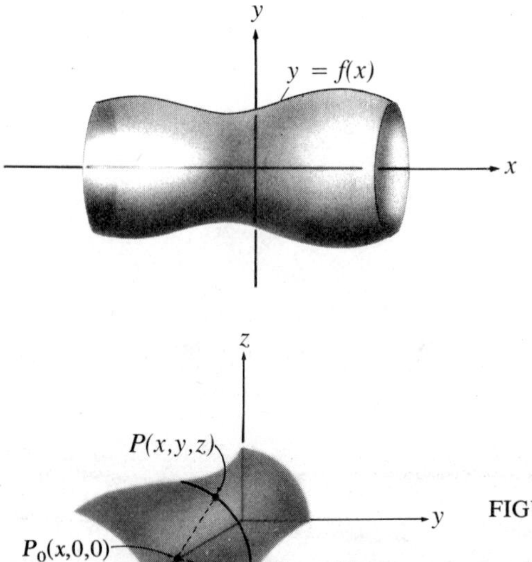

FIGURE 3.12

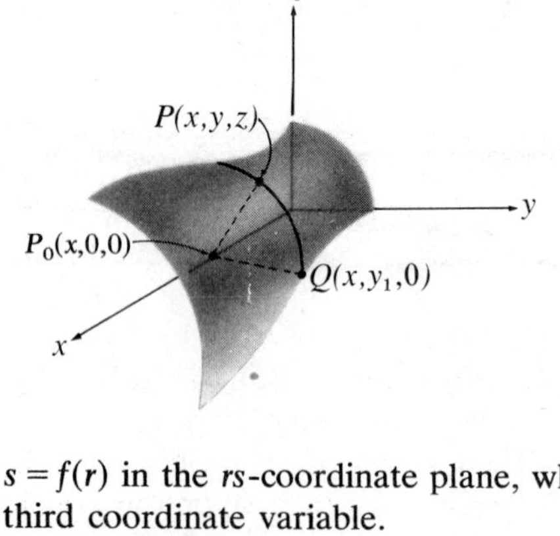

FIGURE 3.13

$s = f(r)$ in the rs-coordinate plane, where r and s are x, y, or z. Let t denote the third coordinate variable.

3.5 **THEOREM.** If the graph of $s = f(r)$ in the rs-plane is revolved about the r-axis, then the surface of revolution generated has the equation

$$s^2 + t^2 = f(r)^2.$$

3.6 EXAMPLE. Find the equation of the surface of revolution formed by revolving the given curve about the given coordinate axis.

(a) $\dfrac{x^2}{4}+y^2=1$ about the x-axis,

(b) $\dfrac{x^2}{4}+y^2=1$ about the y-axis,

(c) $x^2=8z$ about the z-axis.

Solution. We apply Theorem 3.5.

(a) We can solve for $y=f(x): y=\pm\sqrt{1-\dfrac{x^2}{4}}$. Clearly, revolving the top half, $y=\sqrt{1-\dfrac{x^2}{4}}$, about the x-axis generates the entire surface. Here, in the notation of Theorem 3.5, we can take $r=x$, $s=y$, and $t=z$. So the equation is $y^2+z^2=(\sqrt{1-(x^2/4)})^2=1-(x^2/4)$. We can rewrite this as $x^2+4y^2+4z^2=4$. The surface is an *ellipsoid* of revolution. See Figure 3.14.

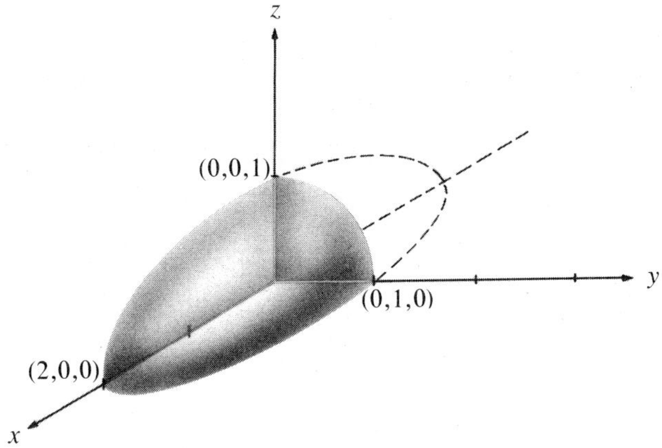

FIGURE 3.14

(b) Here we solve for $x=\pm\sqrt{4-4y^2}$. Again it is enough to revolve the positive x half of the curve about the y-axis. In the theorem we have $r=y$, $s=x$, and $t=z$. So the equation is $x^2+z^2=(\sqrt{4-4y^2})^2$, which simplifies to $x^2+4y^2+z^2=4$. The surface is an ellipsoid of revolution. See Figure 3.15.

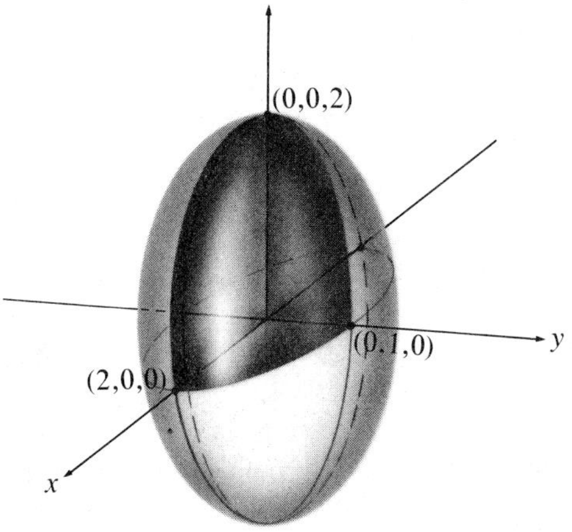

FIGURE 3.15

(c) Here $x=\pm2\sqrt{2z}$. It is sufficient to revolve $x=2\sqrt{2z}$. In the theorem we have $r=z$, $s=x$, and $t=y$. So the equation is $x^2+y^2=8z$. The surface is a paraboloid of revolution. See Figure 3.16.

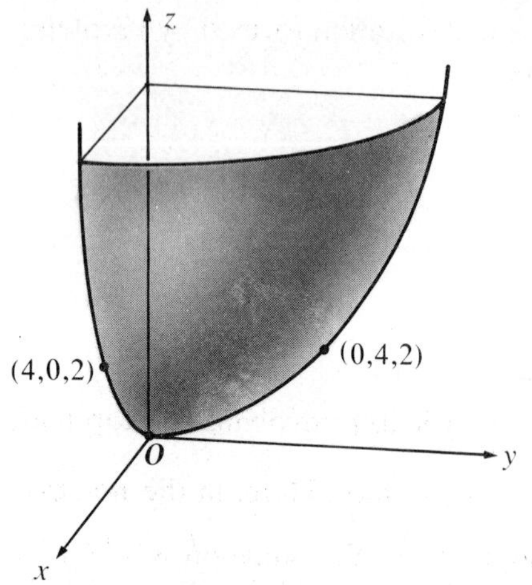

FIGURE 3.16

3.7 EXAMPLE. Identify the surface of revolution whose equation is $x^2-9y^2+z^2=36$.

Solution. We have $x^2+z^2=36+9y^2$. So in the theorem, x^2+z^2 corresponds to s^2+t^2, and $36+9y^2$ corresponds to $(f(r))^2$. Thus $f(r)=\sqrt{36+9y^2}$. We have then the curve $x=\pm\sqrt{36+9y^2}$ revolved about the y-axis. The curve is $x^2-9y^2=36$, a hyperbola. This surface is a hyperboloid of revolution. It is shown in Figure 3.17.

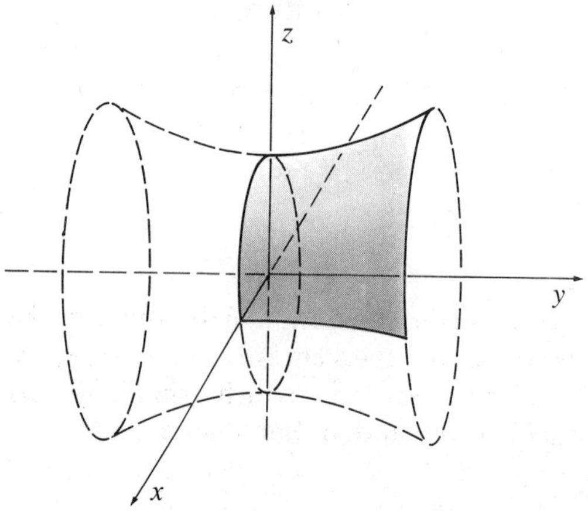

FIGURE 3.17

Besides the kind of graphing we have discussed so far, there is another important graphical representation of functions $f: \mathbf{R}^2 \to \mathbf{R}$. The method is used in making weather maps and also in making contour maps of a geographical region.

The procedure is the following. In the *two*-dimensional real plane, the curves $f(x, y)=C$ are drawn for various values of the constant C. The idea is that by looking at these **level curves** (called *contour lines* in geography, *isothermal* or *isobaric* lines in meteorology, etc.) you can visualize the behavior of the function being mapped. Refer to Figure 3.18, where we show a contour map of a region with two mountain peaks at A and B and a pass between them at C. The elevations shown are in feet.

On a weather map, isobaric lines are used to locate centers of high pressure or low pressure. In Figure 3.19 we show a weather map with low pressure centers near New York and Seattle, and a high pressure center over the Rocky Mountains.

3.8 EXAMPLE. Plot level curves of the sphere $x^2+y^2+z^2=9$ and of the paraboloid $z=x^2+y^2$. What conclusions can you draw?

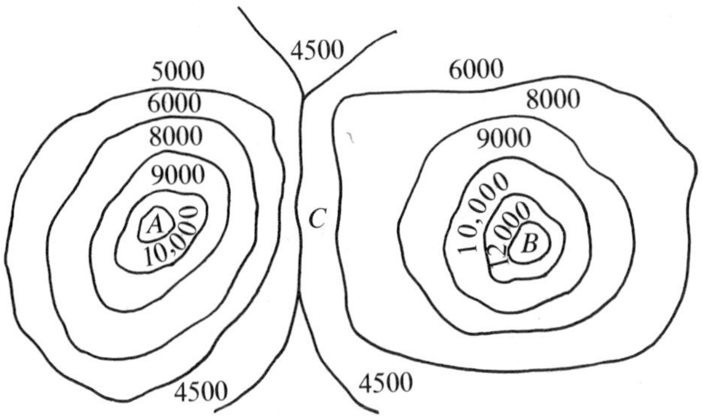

FIGURE 3.18

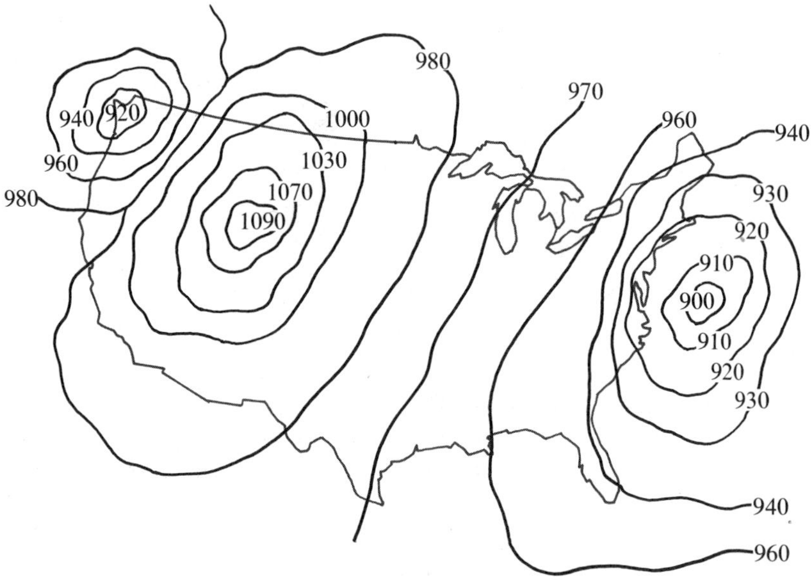

FIGURE 3.19

Solution. In Figure 3.20(a) we show the level curves $z=C$ for the sphere. In Figure 3.20(b) we show level curves $z=C$ for the paraboloid.

In the first case, the level curves are of the form $x^2+y^2=9-C^2$. Here, C can vary between -3 and $+3$. In the case of the paraboloid, the level curves are of the form $x^2+y^2=C$, and so C can be any positive number. In both cases, the level curves are circles. How, then, can we distinguish between the surfaces on the basis of their level curves?

Notice that the level curves for the sphere *contract* from the circle $x^2+y^2=9$ $(C=0)$ to the point $(0, 0)$ $(C=3)$, while those of the paraboloid *expand* outward. So even though in both cases the level curves are circles, we can see that the surfaces they derive from are quite different. The sphere has no points outside the circle of radius 3. As the elevation increases toward 3, there are smaller and smaller circular configurations at that elevation. See Figure 3.21(a). The paraboloid, however, has points for *all* values of $z>0$, and as the elevation increases, there is always a circular configuration of points at that elevation. The radii of the circular configurations grow as the elevation increases. We get a picture of a surface with a circular cross section parallel to the xy-plane which becomes larger as z grows. See Figure 3.21(b).

While level curves afford an *alternative* graphical representation of functions $f:\mathbf{R}^2\to\mathbf{R}$, we can use the idea behind them to produce our *only* kind of graphical

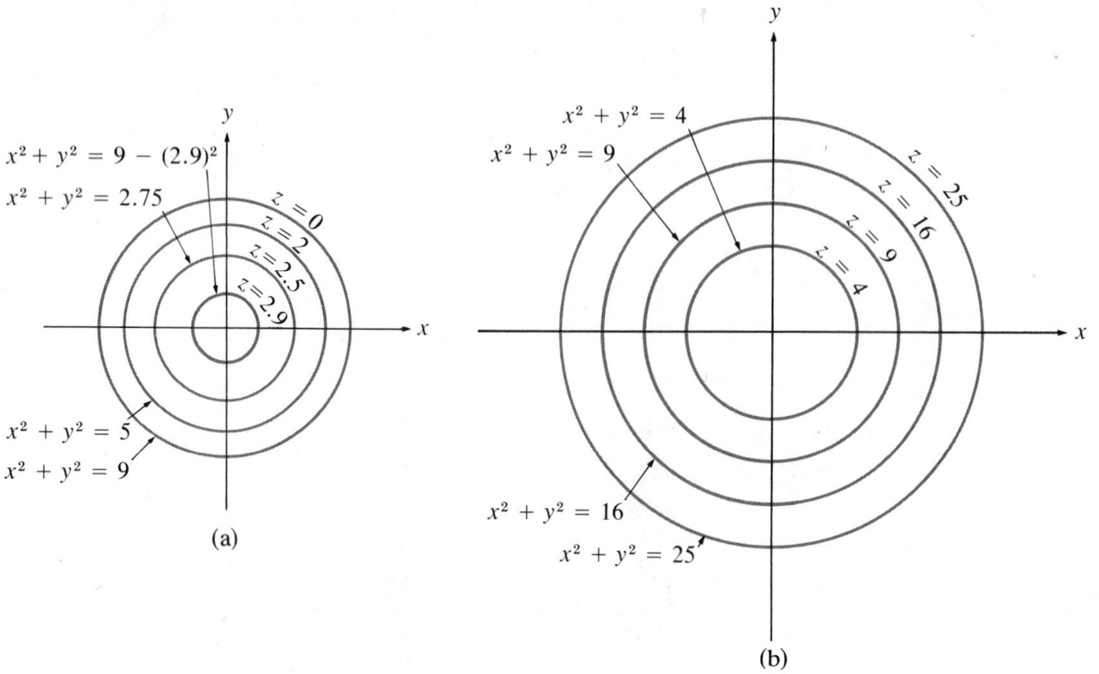

FIGURE 3.20

representation for functions $f: \mathbf{R}^3 \to \mathbf{R}$. The graph of such a function is a subset of $\mathbf{R}^4$ for which no pictorial representation on a two dimensional surface like a page or a blackboard is available. But we *can* draw the **level surfaces** $f(x, y, z) = C$ for such a function.

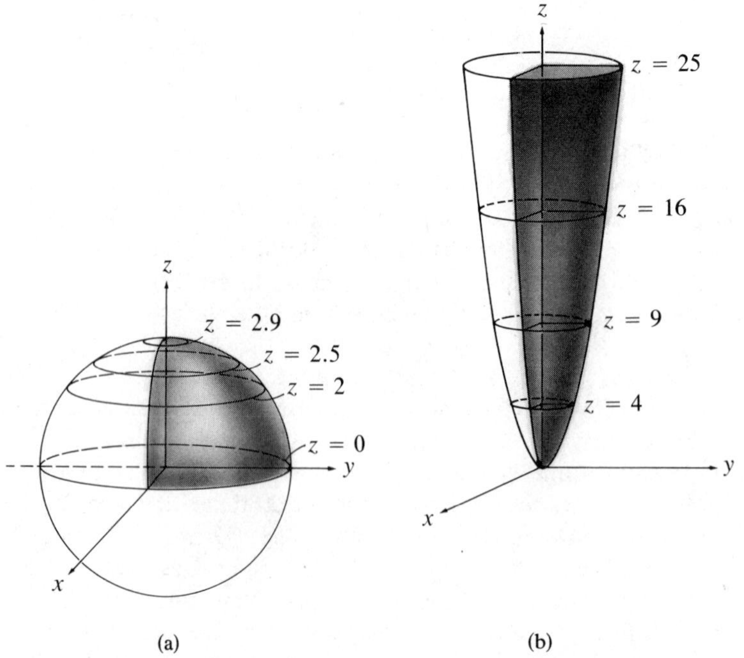

FIGURE 3.21

3.9 EXAMPLE. Describe the level surfaces of the function $f(x, y, z) = x^2 + y^2 + z^2$.

Solution. If C is a positive constant, then the level surface $f(x, y, z) = C$ is the

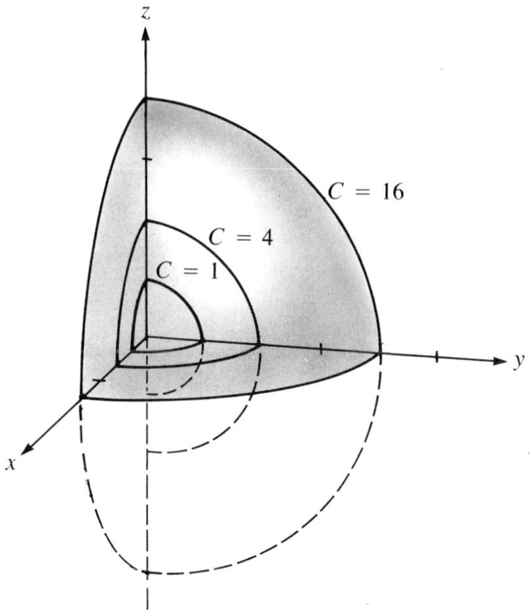

FIGURE 3.22

sphere $x^2+y^2+z^2=C$ of radius $\sqrt{C}$ centered at the origin. We get a sequence of expanding spheres (shown in Figure 3.22 for $C=1$, 4, and 16) as C increases. So we can think of the graph of $w=x^2+y^2+z^2$ as a hypersurface in $\boldsymbol{R}^4$ analogous to the paraboloid in $\boldsymbol{R}^3$ that is the graph of $z=x^2+y^2$. (Take another look at Example 3.8 to bring the analogy into sharper focus.)

EXERCISES 3.3

In Exercises 1 to 8, draw the graphs of the given functions $f:\boldsymbol{R}^2\to\boldsymbol{R}$. Name the type of surface.

1 $z=1-x^2$

2 $z=y^2+4$

3 $x^2+y^2=4$

4 $3x^2+4y^2=12$

5 $4x^2-9y^2=36$

6 $x^2+y^2=4z$

7 $4x^2+9y^2+4z^2=36$

8 $x^2+y^2-4z^2=4$

In Exercises 9 to 20, find the equation of the surface of revolution.

9 The graph of $y=x^2$ in the xy-plane revolved about the x-axis.

10 The graph of $y=x^2$ in the xy-plane revolved about the y-axis.

11 The graph of $y^2=4z$ in the yz-plane revolved about the y-axis.

12 The graph of $y^2=4z$ in the yz-plane revolved about the z-axis.

13 The graph of $\frac{x^2}{4}-\frac{y^2}{9}=1$ in the xy-plane revolved about the x-axis.

14 The graph of $\frac{x^2}{4}-\frac{y^2}{9}=1$ in the xy-plane revolved about the y-axis.

15 The graph of $\frac{y^2}{16}-\frac{z^2}{9}=1$ in the yz-plane revolved about the y-axis.

16 The graph of $\frac{x^2}{4}-\frac{z^2}{16}=1$ in the xz-plane revolved about the z-axis.

17 The graph of $x^2+y^2=4$ in the xy-plane revolved about the y-axis.

18 The graph of $y=3x$ in the xy-plane revolved about the y-axis.

19 The graph of $y=3x$ in the xy-plane revolved about the x-axis.

20 The graph of $y=\sin x$ in the xy-plane revolved about the x-axis.

21 Is the graph of $9x^2-4y^2+z^2=36$ a surface of revolution? Of what curve revolved about what axis?

22 Is the graph of $x^2+4y^2+16z^2=64$ a surface of revolution? Of what curve revolved about what axis?

23 Is the graph of $4x^2-3y^2+4z^2=9$ a surface of revolution? Of what curve revolved about what axis?

24 Is the graph of $9x^2+4y^2+9z^2=36$ a surface of revolution? Of what curve revolved about what axis?

In Exercises 25 to 28, plot the level curves of the given surface for the values suggested.

25 $z=y^2-x^2$ for $z=1, 4, 9, 16$.

26 $z=y^2+4x$ for $z=1, 4, 9, 16$

27 $z=x^2+9y^2$ for $z=1, 9, 81$

28 $z=4x^2+y^2$ for $z=1, 4, 16$

29 Suppose that the potential at a point in the plane (other than the origin) is given by

$$P(x, y)=\frac{1}{\sqrt{x^2+y^2}}.$$

Draw the *equipotential curves* for $P=\frac{1}{2}$, 1, 2, and 4.

30 Suppose that the potential at a point in the plane (other than (1, 2)) is given by

$$P(x, y)=\frac{1}{\sqrt{(x-1)^2+(y-2)^2}}.$$

Draw the equipotential curves for $P=\frac{1}{2}$, 1, 2, and 4.

In Exercises 31 to 34, describe the level surfaces of the function and draw those for the indicated values of w.

31 $x^2+y^2+z^2+w^2=9$ $w=0, 2, 3$

32 $x^2+y^2+z^2-w^2=9$ $w=0, 1, 3$

33 $x^2+\frac{y^2}{9}+\frac{z^2}{9}+w^2=4$ $w=0, 1, 2$

34 $x^2-y+z^2-w=9$ $w=0, 1, 3$

4 QUADRIC SURFACES

We saw in Section 1.4 that the graph of a first degree linear equation $ax+by+cz=d$ turned out to be a plane in $\mathbf{R}^3$. Recalling that the graph in $\mathbf{R}^2$ of a second degree polynomial in two variables is a (possibly degenerate) conic section, we might naturally inquire about the possible graphs in $\mathbf{R}^3$ of second degree polynomial equations in x, y, and z. In this section we describe a family of surfaces (and degenerate special cases) which arise as such graphs. These surfaces possess a number of symmetry properties which give the graphs a pleasing appearance. The surfaces all have equations formed by adding or subtracting terms of the form

$$\frac{x^2}{a^2}, \frac{y^2}{b^2}, \frac{z^2}{c^2}, \frac{x}{d}, \frac{y}{e}, \text{ and } \frac{z}{f},$$

and equating to a constant.

Before cataloging these surfaces, we mention simple criteria for the graph of an equation $f(x, y, z)=0$ to be symmetric with respect to a coordinate plane. First, *if the equation is unchanged when z is replaced by $-z$*, then $(x, y, -z)$ is a point of the graph whenever (x, y, z) is. *Hence the graph is symmetric relative to the xy-plane.* Similarly, *if the equation is unchanged when x is replaced by $-x$, then the graph is symmetric relative to the yz-plane.* And, *if the equation is unchanged when y is replaced by $-y$, then the graph is symmetric relative to the xz-plane.* In drawing the graphs of surfaces which possess symmetry relative to coordinate planes, it is a good idea to take advantage of symmetry as much as possible to reduce the number of points you have to plot.

Now we can catalog the most common types of quadric surfaces. By reference to this catalog you should be able to name and draw rough sketches of most quadric surfaces you are apt to encounter. We number the catalog entries for ease of reference.

4.1 RIGHT CYLINDERS. We have already seen (following Definition 3.2) that the graph of a quadratic polynomial function involving only two of the variables x, y, z is a cylinder perpendicular to the coordinate plane of the two variables which occur. See Example 3.3 and Problems 3 to 5 of Exercises 3.3 for illustrations.

4.2 RIGHT CONES. The graph of the equation

(i) $$\frac{x^2}{a^2}+\frac{y^2}{b^2}-\frac{z^2}{c^2}=0$$

is a *right elliptical cone.* Its axis is the z-axis. The cross sections made by planes $z=\pm k$ parallel to the xy-plane (i.e., the level curves) are ellipses

$$\frac{x^2}{a^2}+\frac{y^2}{b^2}=\frac{k^2}{c^2}$$

centered on the z-axis. See Figure 4.1.

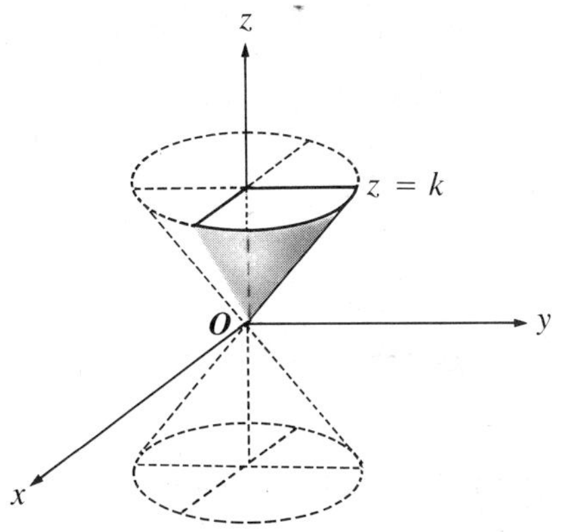

FIGURE 4.1

Notice that the graph is symmetric in all three coordinate planes. If $a=b$ we have circular cross sections. (In this case, then, we have a *right circular cone.*)

Variations:

(ii) $$\frac{x^2}{a^2}-\frac{y^2}{b^2}+\frac{z^2}{c^2}=0$$

is a right elliptical cone whose axis is the y-axis. (Rotate Figure 4.1 90° clockwise.)

(iii) $$-\frac{x^2}{a^2}+\frac{y^2}{b^2}+\frac{z^2}{c^2}=0$$

is a right elliptical cone whose axis is the x-axis. (Rotate the top of Figure 4.1 90° toward the front.)

(iv) $$\frac{x^2}{a^2}+\frac{y^2}{b^2}+\frac{z^2}{c^2}=0$$

is a degenerate cone consisting of the point (0, 0, 0).

4.3 ELLIPTICAL PARABOLOIDS. The graph of

(i) $$\frac{x^2}{a^2}+\frac{y^2}{b^2}=\frac{\pm z}{c^2}$$

is an *elliptical paraboloid* with the z-axis as the axis. The cross sections made by planes $z=\mp k$ are again ellipses which expand as k increases. Note that when $x=0$, we have $z=\pm\frac{c^2}{b^2}y^2$, and when $y=0$ we have $z=\pm\frac{c^2}{a^2}x^2$. These parabolas both open upward if the plus sign is present, or downward if the minus sign is present, so the same is true of the surface. See Figure 4.2, where we have taken

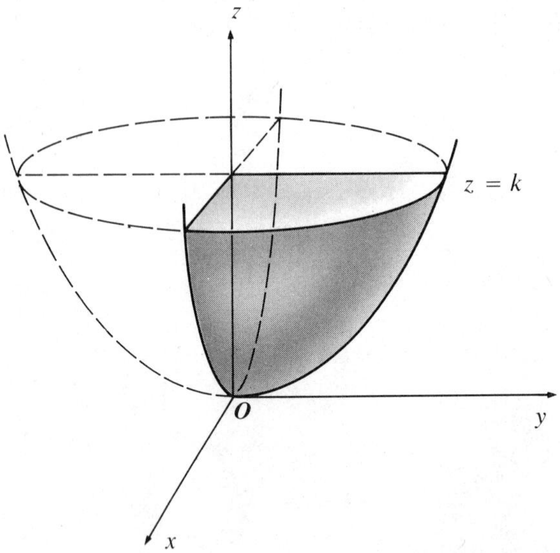

FIGURE 4.2

the plus sign for $\frac{z}{c^2}$. Note that we have symmetry relative to the yz- and xz-planes. If $a=b$ we have a paraboloid of revolution.

Variations:

(ii) $$\frac{x^2}{a^2}+\frac{z^2}{c^2}=\pm\frac{y}{b^2}$$

is an elliptical paraboloid with the y-axis as its axis. It opens to the right if the positive sign is associated with y/b^2. (Rotate Figure 4.2 90° clockwise.) It opens to the left if the negative sign is associated with y/b^2. (Rotate Figure 4.2 90° counterclockwise.)

(iii) $$\frac{y^2}{b^2}+\frac{z^2}{c^2}=\pm\frac{x}{a^2}$$

is an elliptical paraboloid with the x-axis as its axis. It opens to the front if the positive sign is associated with x/a^2. (Rotate the top of Figure 4.2 90° to the front.) It opens to the rear if the negative sign is associated with x/a^2. (Rotate the top of Figure 4.2 90° to the rear.)

(iv) $$\frac{x^2}{a^2}+\frac{y^2}{b^2}+d=\frac{\pm z}{c^2}$$

has the same shape as the surface described by Equation (i), but its vertex is now at $(0, 0, \pm c^2 d)$ instead of $(0, 0, 0)$.

4.4 HYPERBOLIC PARABOLOIDS. The graph of

(i) $$-\frac{x^2}{a^2}+\frac{y^2}{b^2}=\frac{z}{c^2}$$

is a *hyperbolic paraboloid* (or *saddle*). The horizontal cross sections $z=k>0$ are hyperbolas

$$\frac{y^2}{b^2}-\frac{x^2}{a^2}=\frac{k}{c^2}$$

whose vertices lie above the y-axis. Planes $z=k<0$ intersect the surface in hyperbolas

$$\frac{x^2}{a^2}-\frac{y^2}{b^2}=-\frac{k}{c^2}$$

whose vertices lie below the x-axis. The cross section made by the yz-plane $(x=0)$ is the parabola $z=\frac{c^2}{b^2}y^2$, which opens upward. The cross section made by the xz-plane $(y=0)$ is the parabola $z=-\frac{c^2}{a^2}x^2$, which opens downward. We have symmetry relative to the xz- and yz-planes, but not relative to the xy-plane. See Figure 4.3.

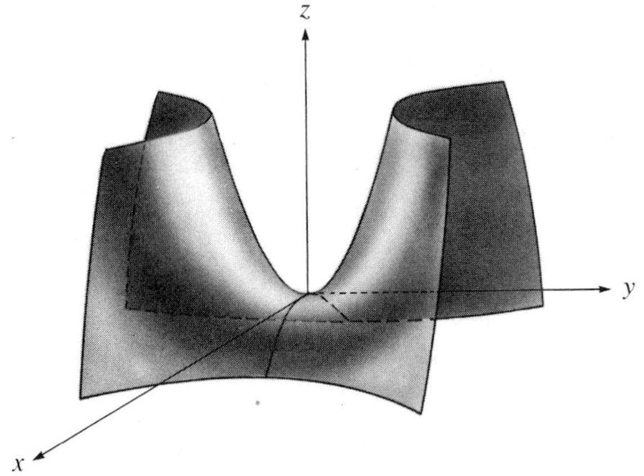

FIGURE 4.3

Variations:

(ii)
$$\frac{x^2}{a^2}-\frac{y^2}{b^2}=\frac{z}{c^2}$$

is a hyperbolic paraboloid rotated 90° about the z-axis from the one shown in Figure 4.3. (Alternatively, you can turn Figure 4.3 upside down, since z has been replaced by $-z$ in (i).)

(iii)
$$\pm\frac{y^2}{b^2}\mp\frac{z^2}{c^2}=\frac{x}{a^2}$$

is a hyperbolic paraboloid rotated 90° about the y-axis from the one shown in Figure 4.3.

(iv)
$$\pm\frac{x^2}{a^2}\mp\frac{z^2}{c^2}=\frac{y}{b^2}$$

is a hyperbolic paraboloid rotated 90° about the z-axis from Case (iii).

4.5 ELLIPSOIDS. The graph of

$$\frac{x^2}{a^2}+\frac{y^2}{b^2}+\frac{z^2}{c^2}=1$$

is an *ellipsoid*. The cross sections with planes $z=k$ are of the form

$$\frac{x^2}{a^2}+\frac{y^2}{b^2}=1-\frac{k^2}{c^2},$$

which are ellipses as long as $\frac{k^2}{c^2}<1$. When $k=\pm c$ we get a single point. When $k^2>c^2$, we get no intersection. So the graph lies between $z=-c$ and $z=+c$. Notice that the cross sections made by planes $x=k$ and $y=k$ are of the same sort. If any two of a^2, b^2, and c^2 are equal, then we get an ellipsoid of revolution (which looks something like an American football). If $a^2=b^2=c^2$, we get a sphere of radius a. In all cases the graph is symmetric relative to all three coordinate planes. See Figure 4.4.

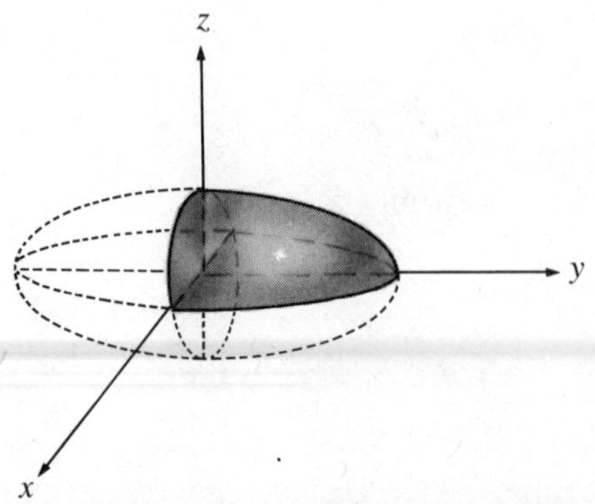

FIGURE 4.4

The 1975 Nobel Prize for physics was awarded to Aage Bohr (1922– , the son of Niels Bohr [1885–1962], the 1922 Nobel laureate in physics for his pioneering work in nuclear physics) and Benjamin Mottelson (1926–) of the Niels Bohr Institute in Denmark and James Rainwater (1917–) of Columbia University for work done in the 1940's and 1950's that determined that the nucleus of the atom has the shape of an ellipsoid.

4.6 HYPERBOLOID OF ONE SHEET. The graph of

(i) $$\frac{x^2}{a^2}+\frac{y^2}{b^2}-\frac{z^2}{c^2}=1$$

is a hyperboloid of one sheet with the z-axis as its axis. The cross sections made by the two planes $z=k$ have equations

$$\frac{x^2}{a^2}+\frac{y^2}{b^2}=1+\frac{k^2}{c^2}, \qquad z=k$$

and so are ellipses for all values of k. The cross section with the yz-plane is the hyperbola

$$\frac{y^2}{b^2}-\frac{z^2}{c^2}=1, \qquad x=0.$$

The cross section with the xz-plane is the hyperbola

$$\frac{x^2}{a^2}-\frac{z^2}{c^2}=1, \qquad y=0.$$

If $a^2=b^2$, then we have a hyperboloid of revolution about the z-axis. If $a^2=c^2$, then we have a hyperboloid of revolution about the y-axis. If $b^2=c^2$, then we have a hyperboloid of revolution about the x-axis. (See Theorem 3.5.) For the picture, see Figure 4.5. Notice that the graph is symmetric relative to all three coordinate planes.

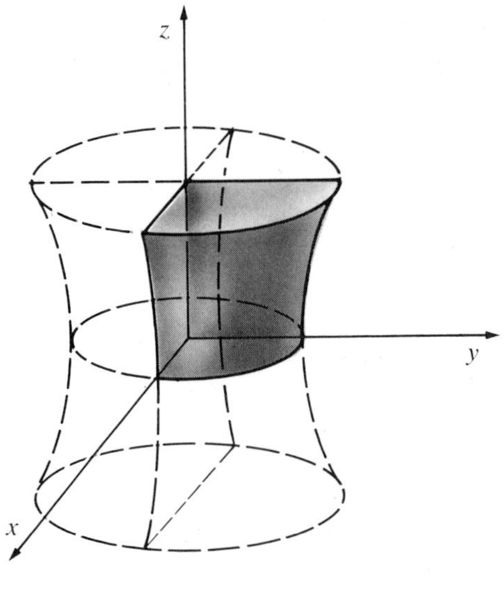

FIGURE 4.5

Variations:

(ii) $$\frac{x^2}{a^2}-\frac{y^2}{b^2}+\frac{z^2}{c^2}=1$$

is a hyperboloid of one sheet having the y-axis as axis. (Rotate Figure 4.5 90° clockwise.)

(iii) $$-\frac{x^2}{a^2}+\frac{y^2}{b^2}+\frac{z^2}{c^2}=1$$

is a hyperboloid of one sheet having the x-axis as axis. (Rotate Figure 4.5 90°, top toward the front.)

4.7 HYPERBOLOID OF TWO SHEETS. The graph of

(i) $$-\frac{x^2}{a^2}+\frac{y^2}{b^2}-\frac{z^2}{c^2}=1$$

is a hyperboloid of two sheets with the y-axis as axis. This is the only coordinate axis on which the graph has any points. The cross sections made by planes $z=k$ are hyperbolas

$$\frac{y^2}{b^2}-\frac{x^2}{a^2}=1+\frac{k^2}{c^2}, \qquad z=k.$$

Similarly, the cross sections made by planes $x=k$ are hyperbolas

$$\frac{y^2}{b^2}-\frac{z^2}{c^2}=1+\frac{k^2}{a^2}, \qquad x=k.$$

In both cases the hyperbolas have vertices on the y-axis. Now consider cross sections made by the planes $y=k$. We get

$$\frac{x^2}{a^2}+\frac{z^2}{c^2}=\frac{k^2}{b^2}-1, \qquad y=k.$$

So we get *no* cross section if $k^2<b^2$, a single point if $k^2=b^2$, and an ellipse if $k^2>b^2$. If $a^2=c^2$, we have a hyperboloid of revolution about the y-axis. Notice that the graph is symmetric relative to all three coordinate planes. See Figure 4.6 for the picture.

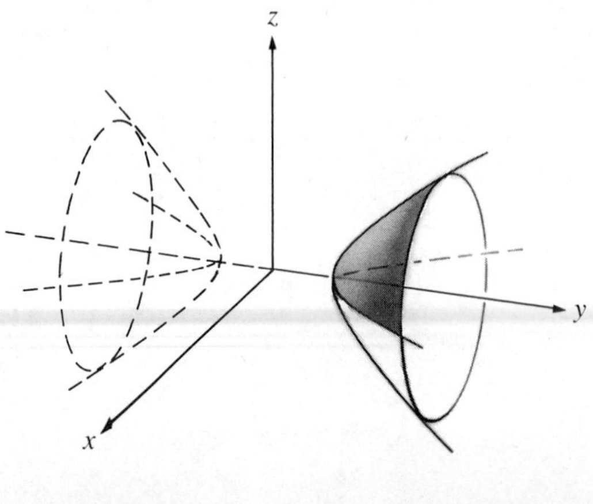

FIGURE 4.6

Variations:

(ii) $$-\frac{x^2}{a^2}-\frac{y^2}{b^2}+\frac{z^2}{c^2}=1$$

is a hyperboloid of two sheets with the z-axis as axis. (Rotate Figure 4.6 90° counterclockwise.)

(iii) $$\frac{x^2}{a^2}-\frac{y^2}{b^2}-\frac{z^2}{c^2}=1$$

is a hyperboloid of two sheets with the x-axis as axis. (Rotate Figure 4.6 90° about the z-axis.)

We close this section by illustrating how the information listed above can be used in practice.

4.8 EXAMPLE. Name and draw a rough sketch of the quadric surface $4x^2-9y^2+z^2=36$.

Solution. First, divide through by 36 to put the equation in standard form:

$$\frac{x^2}{9}-\frac{y^2}{4}+\frac{z^2}{36}=1.$$

This is an equation of type 4.6(ii), so it is a hyperboloid of one sheet having as axis the y-axis. This in itself is enough to draw a very rough sketch. To make the sketch a bit less rough, we can find the intersection points of the surface with the coordinate axes. There is no intersection with the y-axis, of course, as we can see in two ways: (1) The y-axis is the axis of the hyperboloid; (2) when $x=z=0$, we get $y^2=-4$, so no such y exists. When x and y are 0, $z=\pm 6$. When y and z are zero, $x=\pm 3$. A rough sketch is given in Figure 4.7.

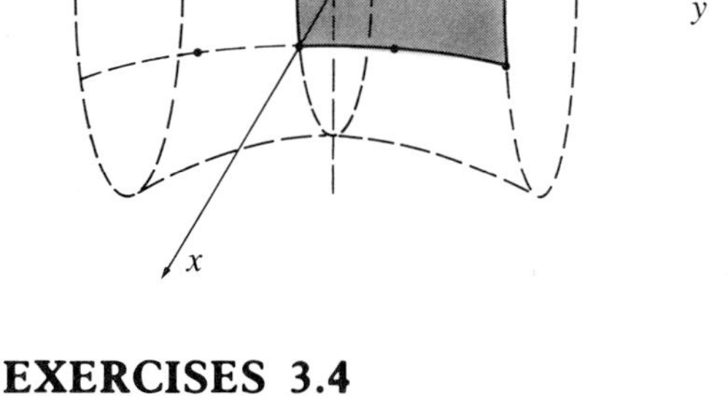

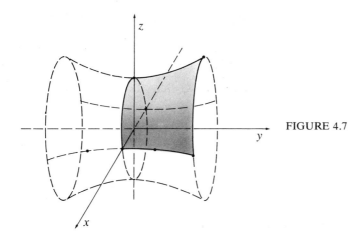

FIGURE 4.7

EXERCISES 3.4

Identify and draw a sketch of the graph of each equation.

1 $y^2+3z^2=9$

2 $y^2-x^2=4$

3 $y^2+4z=4$

4 $x^2-2y^2+4z^2=0$

5 $z^2=x^2+y^2$

6 $4x^2+y^2-z=0$

7 $2x^2+4y+4z^2=0$

8 $z=x^2+y^2$

9 $x^2-2y^2-4z=0$

10 $y^2=z+x^2$

11 $y^2-2x^2-z=0$

12 $5x^2+15y^2+10z^2=30$

13 $3x^2+y^2+3z^2=9$

14 $x^2+y^2-2z^2=4$

15 $2x^2-y^2+4z^2=4$

16 $y^2=1+x^2+z^2$

17 $2x^2-y^2-4z^2=4$

18 $z^2=2+2x^2+y^2$

REVIEW EXERCISES 3.5

1 Classify as open, closed, both, or neither:

(a) The line $y=5$ in $\boldsymbol{R}^2$.

(b) The plane $x+2y-3z=7$ in $\boldsymbol{R}^3$.

(c) $\{(x, y) \mid x>1, y^2\leq 9\}$ in $\boldsymbol{R}^2$.

(d) $\left\{(x, y, z) \,\middle|\, \frac{x^2}{4}+\frac{y^2}{16}+\frac{z^2}{9}\leq 1\right\}$ in $\boldsymbol{R}^3$.

(e) $\{(x, y, z) \mid x>1, y^2+2z^2<9\}$ in $\boldsymbol{R}^3$.

(f) $\{(x, y, z) \mid x^2+y^2+z^2+1\leq 0\}$ in $\boldsymbol{R}^3$.

(g) $\{(x, y, z, w) \mid 2x+y\geq 1, 3z-w>5\}$ in $\boldsymbol{R}^4$.

2 Determine whether the given sequence converges. If it does, then find the limit.

(a) $\boldsymbol{x}_k=\left(\frac{2k^2-k+5}{9k^2+2k+1}, \frac{\sin k+\cos k}{k}\right)$

(b) $\boldsymbol{x}_k=\left(\tan\frac{\pi k}{2k+5}, \frac{2-k}{k}, e^{-1/k}\right)$

(c) $\boldsymbol{x}_k=\left(\operatorname{Tan}^{-1}\frac{k^2+1}{k+1}, e^{-k}, \frac{k+1}{k^2+1}\right)$

(d) $\boldsymbol{x}_k=\left(\sin(-1)^k\frac{\pi}{2}, \frac{\cos k}{k}, \frac{1}{k^2+5}\right)$

In Exercises 3 to 5, find all points where the given schizophrenic function is continuous.

3 $f(x, y)=\begin{cases} \dfrac{x^2y}{x^3+y^3} & \text{if } x+y\neq 0 \\ 0 & \text{if } x+y=0. \end{cases}$

4 $g(x, y)=\begin{cases} \dfrac{x^2y^2}{x^2+y^2} & \text{if } (x, y)\neq(0, 0) \\ 0 & \text{if } (x, y)=(0, 0). \end{cases}$

5 $f(x, y)=\begin{cases} (x^2+y^2)\sin\dfrac{1}{x} & \text{if } x\neq 0 \\ 0 & \text{if } x=0. \end{cases}$

In Exercises 6 to 9, name and draw the graphs of the given functions $f:\boldsymbol{R}^2\to\boldsymbol{R}$ or functional equations $g(x, y, z)=0$.

6 $z=x+y^2$

7 $9z^2-16y^2=144$

8 $x^2+4y^2+z^2=4$

9 $4x^2-9y^2+4z^2=36$

In Exercises 10 and 11, find the equation of the given surface of revolution.

10 The graph of $y=x^2+4$ in the xy-plane is revolved about (a) the x-axis, (b) the y-axis.

11 The graph of $\frac{y^2}{4}-\frac{z^2}{9}=1$ in the yz-plane is revolved about (a) the y-axis, (b) the z-axis.

12 Plot the level curves of $z=x^2+4y$ for $z=1, 4, 9, 16$.

13 Plot the level surfaces of $\frac{x^2}{4}+y^2+\frac{z^2}{4}+w^2=4$ for $w=0, 1, 2$.

In Exercises 14 to 22, identify and draw a sketch of each equation.

14 $2z^2-y^2=4$

15 $x^2-4y=8$

16 $2y^2-x^2+3z^2=0$

17 $y^2=x^2-z^2$

18 $3x-6y^2-9z^2=0$

19 $y^2=4x^2+2z$

20 $2x^2+4y^2+2z^2=8$

21 $x^2-3y^2+6z^2=6$

22 $3y^2-5x^2-6z^2=30$

4 DIFFERENTIATION OF SCALAR FUNCTIONS

0 INTRODUCTION

Now that we have introduced functions $f: \boldsymbol{R}^n \to \boldsymbol{R}$ and studied their graphs in some detail, we are ready to continue the development of the calculus of these functions that we started in Section 3.2. The differential calculus of these functions is the business of this chapter, and their integral calculus is the subject of the next chapter.

Differentiation of functions $f: \boldsymbol{R}^n \to \boldsymbol{R}$ is carried out in two stages. The first *freezes* $n-1$ of the variables and allows only one of them to vary. This brings us back to the situation of elementary calculus, because we have essentially reduced f to a real valued function of a *single* real variable. This first stage is developed in Section 1, where the *partial derivatives* of f are defined. The second stage involves giving a meaning to the *total derivative* of f. For this we need to go back to the notion of the derivative in elementary calculus and see what the essence of the idea of differentiation is. Then we can transfer that to the case of $f: \boldsymbol{R}^n \to \boldsymbol{R}$. The essence turns out to involve the notion of *linear function*, and in Section 2 we study that notion; then we put it to work in Section 3 to develop the idea of the total derivative and differentiability.

The rest of the chapter is concerned with developing differentiation techniques and applications analogous to those you learned in elementary calculus. For instance, in elementary calculus you learned how to find equations of tangent lines to curves. In Sections 1 and 4 you will learn how to find equations of *tangent planes to surfaces*. In elementary calculus you learned the mean value theorem for derivatives, the chain rule, implicit differentiation, Taylor polynomials, higher derivatives, and the use of the first and second derivatives to find maxima and minima. As the titles of Sections 5 (*Chain Rule*), 6 (*Implicit Functions*), 7 (*Higher Partial Derivatives*), 8 (*Taylor Polynomials*), and 9 (*Extreme Values*) suggest, these topics will be presented here for functions of several variables. The final section presents the technique of *Lagrange multipliers*, a powerful tool for finding the extreme values of a function that is subject to some constraint. These kinds of problems arise very frequently in applications where physical, economic, or other factors constrain the functions under study.

The pervasive theme of this chapter is the use of the simplest functions (linear functions) to approximate arbitrary differentiable functions. This idea you probably met in elementary calculus in the notion of the differential. While it may not have seemed particularly central to elementary calculus, it turns out that this idea is paramount in calculus of several variables. You will see it recur prominently in Chapters 6 and 7.

1 PARTIAL DERIVATIVES

You probably first saw differentiation presented for functions $f:\mathbf{R}\to\mathbf{R}$ in connection with the problem of defining a tangent line to a curve at a point or the problem of defining the instantaneous velocity of a particle moving on a straight line. We turn now to the question of defining the *tangent plane* to a surface that is the graph of $f:\mathbf{R}^2\to\mathbf{R}$ at a point $\mathbf{x}_0=(x_0, y_0, z_0)$ where $z_0=f(x_0, y_0)$. Consider such a surface, as in Figure 1.1 where we have drawn a representation of what a

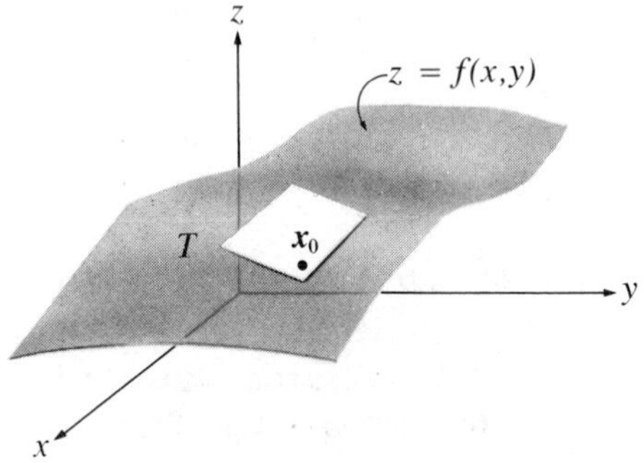

FIGURE 1.1

tangent plane T at $\mathbf{x}_0$ might look like. We are going to define the tangent plane to be the plane determined by the tangent *vectors* to the two curves $\mathbf{x}_1(t)=(x_0, t, f(x_0, t))$ and $\mathbf{x}_2(t)=(t, y_0, f(t, y_0))$. These curves are the curves of intersection of the surface with the planes $x=x_0$ and $y=y_0$. See Figure 1.2. The tangent vectors to these curves at $\mathbf{x}_0$ are easily computed as in Section 2.2.

$$
(1)\qquad \begin{cases} \mathbf{t}_1=\dot{\mathbf{x}}_1(t)\Big|_{t=y_0}=\left(0, 1, \left.\dfrac{df(x_0, t)}{dt}\right|_{t=y_0}\right), \\ \mathbf{t}_2=\dot{\mathbf{x}}_2(t)\Big|_{t=x_0}=\left(1, 0, \left.\dfrac{df(t, y_0)}{dt}\right|_{t=x_0}\right). \end{cases}
$$

The third coordinates of these tangent vectors are of great importance.

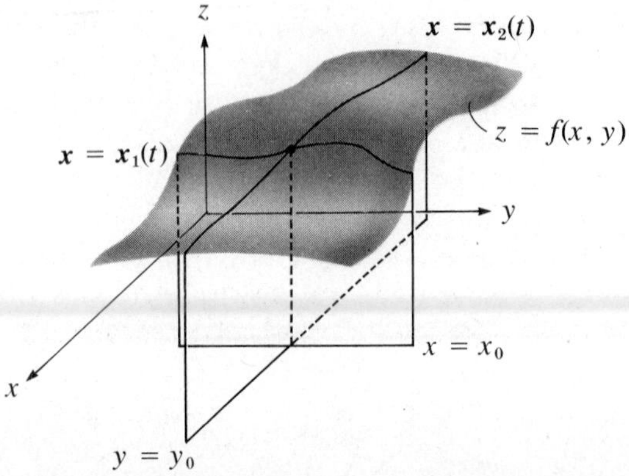

FIGURE 1.2

1.1 **DEFINITION.** The **partial derivatives** of $f:\boldsymbol{R}^2 \to \boldsymbol{R}$ at a point $\boldsymbol{x}_0 = (x_0, y_0, f(x_0, y_0))$ are

$$\frac{\partial f}{\partial x}(x_0, y_0) = \frac{df(x, y_0)}{dx}\bigg|_{x=x_0} = \lim_{h\to 0} \frac{f(x_0+h, y_0) - f(x_0, y_0)}{h}$$

and

$$\frac{\partial f}{\partial y}(x_0, y_0) = \frac{df(x_0, y)}{dy}\bigg|_{y=y_0} = \lim_{h\to 0} \frac{f(x_0, y_0+h) - f(x_0, y_0)}{h},$$

provided these limits exist.

The symbols $\frac{\partial f}{\partial x}(x_0, y_0)$ and $\frac{\partial f}{\partial y}(x_0, y_0)$ are read "the partial (derivative) of f with respect to x (respectively, y) at (x_0, y_0)."

Other notations in use for $\frac{\partial f}{\partial x}(x_0, y_0)$ are

$$f_x(x_0, y_0), \quad f_1(x_0, y_0), \quad \frac{\partial z}{\partial x}(x_0, y_0), \quad z_x(x_0, y_0), \quad D_x f(x_0, y_0), \quad \text{and} \quad D_1 f(x_0, y_0).$$

And $\frac{\partial f}{\partial y}(x_0, y_0)$ is also denoted

$$f_y(x_0, y_0), \quad f_2(x_0, y_0), \quad \frac{\partial z}{\partial y}(x_0, y_0), \quad z_y(x_0, y_0), \quad D_y f(x_0, y_0) \quad \text{and} \quad D_2 f(x_0, y_0).$$

Here $z = f(x, y)$. We shall use most of these, but not $f_1(x_0, y_0)$ and $f_2(x_0, y_0)$, which will be used for another purpose in Chapter 6.

Notice that to compute a partial derivative of f with respect to a variable, the other variable is *frozen* so that f becomes a function just of *one* real variable. The partial derivative is then just calculated as the *ordinary* derivative of a real valued function of a single real variable.

1.2 **EXAMPLE.** If $f(x, y) = x^2 + y^2$, then find $\frac{\partial f}{\partial x}$ and $\frac{\partial f}{\partial y}$ at (1, 2, 5).

Solution.

$$\frac{\partial f}{\partial x}(1, 2) = \frac{d}{dx}(x^2+4)\bigg|_{x=1} = 2x\bigg|_{x=1} = 2.$$

$$\frac{\partial f}{\partial y}(1, 2) = \frac{d}{dy}(1+y^2)\bigg|_{y=2} = 2y\bigg|_{y=2} = 4.$$

We could have computed f_x and f_y at *any* point (x, y, z) by simply regarding y (respectively, x) as a constant and then differentiating with respect to x (respectively, y). Then we would have obtained formulas

$$f_x = \frac{\partial}{\partial x}(x^2+y^2) = 2x$$

$$f_y = \frac{\partial}{\partial y}(x^2+y^2) = 2y.$$

Then to evaluate $f_x(1,2)$ and $f_y(1,2)$ we could have substituted 1 for x and 2 for y.

Now let's return to our discussion of the tangent plane to the graph of $z=f(x, y)$ at $\boldsymbol{x}_0$. We can write formulas (1) in the form

(2)
$$\begin{cases} \boldsymbol{t}_1=\dot{\boldsymbol{x}}_1(t)|_{t=y_0}=\left(0, 1, \dfrac{\partial f}{\partial y}(x_0, y_0)\right)=(0, 1, f_y(x_0, y_0)), \\ \boldsymbol{t}_2=\dot{\boldsymbol{x}}_2(t)|_{t=x_0}=\left(1, 0, \dfrac{\partial f}{\partial x}(x_0, y_0)\right)=(1, 0, f_x(x_0, y_0)). \end{cases}$$

We now give our promised definition of the tangent plane at $\boldsymbol{x}_0$.

1.3 DEFINITION. Let $z=f(x, y)$ where $\dfrac{\partial f}{\partial x}$ and $\dfrac{\partial f}{\partial y}$ exist on an open set U containing (x_0, y_0) and are continuous at (x_0, y_0). Then the **tangent plane** to the surface that is the graph of f is the plane determined by the tangent vectors $\boldsymbol{t}_1$ and $\boldsymbol{t}_2$ given by (2).

The reason for the continuity assumptions in Definition 1.3 will be seen in Section 3 (see Theorem 3.10). Assuming that they hold, it is easy to find the equation of the tangent plane. A normal vector is

$$\boldsymbol{n}=\boldsymbol{t}_1\times\boldsymbol{t}_2=(f_x(x_0, y_0), f_y(x_0, y_0), -1).$$

So a vector equation of the tangent plane is

(3)
$$\boldsymbol{n}\cdot(\boldsymbol{x}-\boldsymbol{x}_0)=0,$$

i.e.,

$$(f_x(x_0, y_0), f_y(x_0, y_0), -1)\cdot(x-x_0, y-y_0, z-z_0)=0,$$
$$f_x(x_0, y_0)(x-x_0)+f_y(x_0, y_0)(y-y_0)-(z-z_0)=0.$$

Thus we have established the following theorem.

1.4 THEOREM. If the surface $z=f(x, y)$ has a tangent plane at $\boldsymbol{x}_0=(x_0, y_0, z_0)$, then the normal form vector equation of the tangent plane is

$$(f_x(x_0, y_0), f_y(x_0, y_0), -1)\cdot(\boldsymbol{x}-\boldsymbol{x}_0)=0$$

and its scalar equation is

(4)
$$z-z_0=\frac{\partial f}{\partial x}(x_0, y_0)(x-x_0)+\frac{\partial f}{\partial y}(x_0, y_0)(y-y_0).$$

The similarity between Equation (4) and the equation of the *tangent line* to the curve $y=f(x)$ from elementary calculus is striking. For the tangent line has equation

(5)
$$y-y_0=\frac{df}{dx}(x_0)(x-x_0).$$

We can make (4) look even more like (5) if we write (4) in dot product form as

(6) $$z - z_0 = \left(\frac{\partial f}{\partial x}(x_0, y_0), \frac{\partial f}{\partial y}(x_0, y_0)\right) \cdot (x - x_0, y - y_0).$$

Indeed, putting the "partial derivatives" together, if the terminology is well chosen, ought to give us the "total derivative" so that (6) would then be an exact analogue of (5). For we would get the equation of the tangent plane by equating $z - z_0$ ("dependent variable") to the result of taking the (dot) product of the total derivative and the vector "independent variable" $\boldsymbol{x} - \boldsymbol{x}_0 = (x - x_0, y - y_0)$. We will see in Section 3 how to make these ideas precise.

1.5 EXAMPLE. Find the scalar equation of the tangent plane to the paraboloid $z = x^2 + y^2$ at the point $(1, 2, 5)$.

Solution. In Example 1.2 we already calculated $f_x(1, 2) = 2$ and $f_y(1, 2) = 4$. So we can substitute directly into Equation (4): $z - 5 = 2(x - 1) + 4(y - 2)$. If we like, we can rewrite this as $2x + 4y - z = 5$.

This chapter is advertised to be about differentiation of functions $f: \boldsymbol{R}^n \to \boldsymbol{R}$, but so far n has only been 2! The reason is that the case $n = 2$ illustrates perfectly the notion of partial derivative. We have *two* partial derivatives, one for each independent variable. In the case of $f: \boldsymbol{R}^n \to \boldsymbol{R}$, we get n partial derivatives, obtained in each case by freezing all the independent variables except one, and then differentiating the resulting function of that one variable.

1.6 DEFINITION. If $f: \boldsymbol{R}^n \to \boldsymbol{R}$, then the n **partial derivatives** of f at $\boldsymbol{x}_0 = (x_{01}, x_{02}, x_{03}, \ldots, x_{0n})$ are defined by

$$\begin{aligned}\frac{\partial f}{\partial x_i}(\boldsymbol{x}_0) &= \left.\frac{df(x_{01}, \ldots, x_{0,i-1}, x_i, x_{0,i+1}, \ldots, x_{0n})}{dx_i}\right|_{x_i = x_{0i}} \\ &= \lim_{h \to 0}\left\{\frac{f(x_{01}, \ldots, x_{0,i-1}, x_{0i} + h, x_{0,i+1}, \ldots, x_{0n})}{h}\right. \\ &\qquad \left. - \frac{f(x_{01}, \ldots, x_{0,i-1}, x_{0i}, x_{0,i+1}, \ldots, x_{0n})}{h}\right\}\end{aligned}$$

for $i = 1, 2, \ldots, n$.

We shall seldom have to treat cases where $n > 3$. In case $n = 3$, we write $\boldsymbol{x}_0 = (x_0, y_0, z_0)$ and the three partial derivatives take the form

$$\frac{\partial f}{\partial x}(x_0, y_0, z_0) = \left.\frac{df(x, y_0, z_0)}{dx}\right|_{x = x_0}$$

$$\frac{\partial f}{\partial y}(x_0, y_0, z_0) = \left.\frac{df(x_0, y, z_0)}{dy}\right|_{y = y_0}$$

$$\frac{\partial f}{\partial z}(x_0, y_0, z_0) = \left.\frac{df(x_0, y_0, z)}{dz}\right|_{z = z_0}$$

1.7 EXAMPLE. Find the partial derivatives of $f: \boldsymbol{R}^3 \to \boldsymbol{R}$ defined by $f(x, y, z) = e^{xy} \sin^2 yz + y^2 \ln xyz$.

Solution.

$$\frac{\partial f}{\partial x} = ye^{xy} \sin^2 yz + y^2 \frac{1}{xyz} yz$$

$$= ye^{xy} \sin^2 yz + \frac{y^2}{x}.$$

$$\frac{\partial f}{\partial y} = (xe^{xy})\sin^2 yz + e^{xy}(2z \sin yz \cos yz) + 2y \ln xyz + y^2 \frac{1}{xyz} xz$$

$$= xe^{xy} \sin^2 yz + 2ze^{xy} \sin yz \cos yz + 2y \ln xyz + y.$$

$$\frac{\partial f}{\partial z} = 2ye^{xy} \sin yz \cos yz + y^2 \frac{1}{xyz} xy$$

$$= 2ye^{xy} \sin yz \cos yz + \frac{y^2}{z}.$$

These ideas have considerable application in economics. Many economic quantities are functions of several variables. For instance, production of an industry is a function of the number x of workers employed in the industry and also the number of dollars y of capital invested in the industry (and perhaps of other variables, too). One attempt to describe this relationship mathematically used the modeling function $p(x, y) = ax^{1-e}y^e$ where a and e are constants and e is between 0 and 1. As another example, the cost of producing one unit of a given commodity can be modeled by a function $c(x, y) = a + bx + ey$, where a is a constant (which can be interpreted as "plant overhead"—utility costs, upkeep, taxes on the property, etc.), x is the number of dollars in labor costs needed to produce one unit of the commodity, and y is the number of dollars in the cost of raw materials needed to produce one unit. Here b and e are constants. In economics, an important tool in studying a variable like production or cost is the *marginal production* or *cost* relative to each independent variable.

1.8 **DEFINITION.** If z is a function of x and y, then the **marginal value of** z **relative to** x is $\frac{\partial z}{\partial x}$ and the **marginal value of** z **relative to** y is $\frac{\partial z}{\partial y}$.

1.9 **EXAMPLE.** Compute the marginal costs for the model $c(x, y) = a + bx + ey$ mentioned before Definition 1.8. Interpret the results.

Solution. $\frac{\partial c(x, y)}{\partial x} = b$ and $\frac{\partial c(x, y)}{\partial y} = e$. These are the marginal costs relative to labor and relative to raw materials, respectively. What the constants give is the rate of increase of the cost with respect to labor and raw materials, respectively. Roughly speaking, if labor costs rise one dollar per unit, the cost of the product will increase b dollars per unit. If the raw materials cost goes up one dollar per unit, then the cost of the product will increase e dollars per unit. While we have considered an extremely simple model in which only two variables contribute to cost, you should be able to see that there is nothing to prevent calculation of marginal cost for cost functions of more variables by taking partial derivatives. Economists can, using these ideas, calculate such quantities as the effect on the price of a new automobile resulting from a collective bargaining agreement increasing labor salaries, or from an increase in the price of steel used to make cars, or from an increase in the price of petroleum used in transporting steel to the automobile factory.

EXERCISES 4.1

In Exercises 1 to 6, find all the first partial derivatives at a general point, and evaluate them at the given point.

1 $f(x, y) = \sqrt{x^2+y^2}$; $(1, 2)$

2 $f(x, y) = \sqrt{4-x^2-y^2}$; $(1, 1)$

3 $f(x, y) = 6e^y \cos x - 5e^{xy} \ln x^2y^2$; $\left(\frac{\pi}{2}, 1\right)$

4 $f(x, y) = (\sin xy)\sqrt{x^2+y^2}$; $\left(1, \frac{\pi}{2}\right)$

5 $f(x, y, z) = \sqrt{x^2+y^2+z^2}$; $(1, 1, 1)$

6 $f(x, y, z) = x^2y \sin xyz - e^{z^2y} \cos xyz$; $(1, 1, \pi)$

In Exercises 7 to 10, find the scalar equation of the tangent plane to the surface at the given point.

7 $z = 3x^2+5y^2$ at $(1, 1, 8)$

8 $z = \sqrt{9-x^2-y^2}$ at $(2, 1, 2)$

9 $z = -\frac{x^2}{4}+\frac{y^2}{9}$ at $(2, 3, 0)$

10 $z = \sqrt{x^2+y^2}$ at $(4, 3, 5)$

In Exercises 11 to 14, find a unit normal vector to the surface at the given point.

11 The surface of Exercise 7.

12 The surface of Exercise 8.

13 The surface of Exercise 9.

14 The surface of Exercise 10.

15 Find the distance from the origin to the tangent plane to the surface $z = x^2+y$ at $(1, 1, 2)$. (*Hint:* Use Exercise 18, Exercises 1.4.)

16 Find the distance from the origin to the tangent plane to the surface $z = x^2+y^2$ at $(1, 1, 2)$. (See Exercise 15.)

17 For the ideal gas law $P = nR\frac{T}{V}$, where n is the number of moles, R is a proportionality constant, T is the temperature in degrees Kelvin, V is the volume in cubic centimeters, and P is the pressure in atmospheres, find $\frac{\partial P}{\partial T}$ and $\frac{\partial P}{\partial V}$. Interpret in terms of the effect on the pressure of a unit increase or decrease in V and T (cf. Example 1.9).

18 Find the marginal production relative to labor cost and capital investment in the model $P(x, y) = ax^{1-e}y^e$ mentioned after Example 1.7. Interpret (cf. Example 1.9).

19 Suppose that a beer company tries to model the demand for its product by $d(p, x, y) = a - 10p + 5x + y$, where a is a constant (representing those unshakeably loyal to the firm's beer), p is the number of cents in the price of a bottle of beer, x is the number of cents in the average price per bottle of competing beer, and y is the number of cents per bottle spent in aging the beer. Discuss the marginal demand for the beer, and the sensitivity of demand to changes in p, x, and y.

20 Suppose that in Exercise 19 the beer company becomes dissatisfied with the simple linear model of demand and changes to a nonlinear model $d(p, x, y) = a^{-10}px^5y$ where p, x, and y are as in Exercise 19. Discuss the marginal demand and the sensitivity of the demand to change in p, x, and y.

21 In a certain market situation, the demands for two goods are given by $d_1 = f(x, y)$ and $d_2 = g(x, y)$, where x is the price of the first commodity and y is the price of the second. Assume $\frac{\partial d_1}{\partial x} < 0$ and $\frac{\partial d_2}{\partial y} < 0$. If the goods compete, then what do you expect of the signs of $\frac{\partial d_1}{\partial y}$ and $\frac{\partial d_2}{\partial x}$? Explain.

22 Refer to Exercise 21. Suppose that the two goods complement each other (such as new housing construction and furniture). Again assuming $\frac{\partial d_1}{\partial x}<0$ and $\frac{\partial d_2}{\partial y}<0$, what do you expect of the signs of $\frac{\partial d_1}{\partial y}$ and $\frac{\partial d_2}{\partial x}$? Explain.

23 In the economic theory of consumption, the preferences of a household are measured by a *utility function* $u:\mathbf{R}^n \to \mathbf{R}$. The domain of u is the set $D=\{\mathbf{x}=(x_1, \ldots, x_n) \mid x_i \geq 0$ for $i=1, 2, \ldots, n\}$ where each x_i represents the quantity of some commodity i (good or service) that can be purchased by the household. A vector $\mathbf{x}$ in D is called a *commodity bundle*, and u is supposed to assign some numerical measure to the attractiveness of commodity bundles to the household—the higher the utility, the more attractive the bundle is. The "Axiom of Greed" assumes that $\frac{\partial u}{\partial x_i}>0$ for each i. Show that under this assumption, u is an increasing function in each individual variable x_i.

24 Refer to Exercise 23. If $\mathbf{x}=(x_1, \ldots, x_n)$ and $\mathbf{y}=(y_1, \ldots, y_n)$ satisfy $x_i \geq y_i$ for each i and if $\mathbf{x} \neq \mathbf{y}$, then show that $u(\mathbf{x})>u(\mathbf{y})$. Thus, the bigger the commodity bundle, the greater its appeal.

25 Discuss the accuracy of the economic model in Exercises 23 and 24 as a reflection of real life. Is it reasonable to assume that "bigger is always better"? Are there *limits* to greed? Should an axiom be added that states that u is a *bounded* function?

26 Let $p(x_1, x_2, \ldots, x_n)$ be the production of a certain plant as a function of the number x_1 of workers and other variables. The average **productivity** of a worker is defined to be $\frac{p(\mathbf{x})}{x_1}$. Show that as x_1 increases, the productivity also increases if the marginal production $\frac{\partial p}{\partial x_1}$ exceeds the average productivity $\frac{p(\mathbf{x})}{x_1}$. In such a situation, the plant hires new workers.

27 Refer to Exercise 26. Show that if the marginal productivity $\frac{\partial p}{\partial x_1}<\frac{p(\mathbf{x})}{x_1}$, then productivity increases as x_1 decreases. In such a situation, a plant lays off workers.

28 In determining dosage levels of cortisone prescribed for cortisone-deficient children, endocrinologists use the *body surface area* as a guide. For a child of weight x kilograms and height y centimeters, the approximation

$$A=0.007x^{0.425}y^{0.725}$$

is used for body surface area in square meters. (See, for example, J. I. Routh, *Mathematical Preparation for the Health Sciences*, 2nd edition. W. B. Saunders Company, Philadelphia, 1976, p. 79.) Compute $\frac{\partial A}{\partial x}$ and $\frac{\partial A}{\partial y}$ when $x=30$ and $y=125$. What is the physical significance of these partial derivatives? (If you don't have an electronic calculator with an x^y key, use the approximations $(30)^{-0.575} \approx 0.141$, $(125)^{-0.275} \approx 0.265$, $(125)^{0.725} \approx 33.133$, and $(30)^{0.425} \approx 4.244$.)

2 LINEAR FUNCTIONS

We have introduced the notion of partial derivative and even remarked that the very terminology itself suggests that there should be such a thing as the total derivative. In this section and the next we turn to the problem of how such a concept should be defined. It seems reasonable that a good place to start looking for a clue is the derivative of a function $f:\mathbf{R} \to \mathbf{R}$. So let's take a close look at that notion of differentiability in elementary calculus.

Why do we consider the derivative at all? From one point of view, to measure the rate of change of a function at a point. But as we remarked in Section 1, your first introduction to the derivative was probably from the point of view of describing the tangent line to a curve $y = f(x)$ at a point $x = x_0$. Why was this a problem worth considering? Because if the function f is differentiable at $x = x_0$, then we can *closely approximate f near x_0 by the tangent line to the graph of f at $x = x_0$*. And after all, the simplest kinds of functional equations are those for lines. So there is a definite conceptual simplification to be had by replacing an arbitrary differentiable function by a function of the form $y = mx + b$ near the point x_0. Let's make these ideas more explicit.

If f is differentiable at $x_0 \in \mathbf{R}$, then

$$\lim_{h\to 0} \frac{f(x_0+h)-f(x_0)}{h} = f'(x_0).$$

Put $x_0 + h = x$. Then as $h \to 0$, $x \to x_0$ and conversely. Hence we can write the defining condition in the form

$$\lim_{x\to x_0} \frac{f(x)-f(x_0)}{x-x_0} = f'(x_0).$$

So in a small interval around x_0, we can approximate $\dfrac{f(x)-f(x_0)}{x-x_0}$ by $f'(x_0)$. We then have

$$\frac{f(x)-f(x_0)}{x-x_0} \approx f'(x_0),$$

$$f(x)-f(x_0) \approx f'(x_0)(x-x_0)$$

$$f(x) \approx f(x_0) + f'(x_0)(x-x_0).$$

The right side of this last approximation is just the formula for the y-coordinate of a point (x, y) on the tangent line ℓ to the graph of f at x_0. See Figure 2.1,

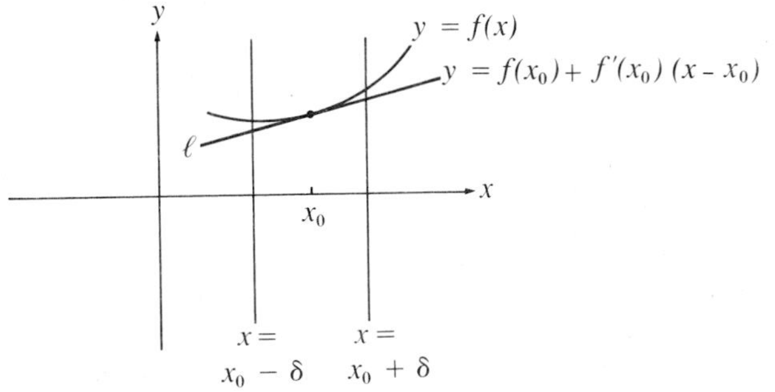

FIGURE 2.1

where we show graphically how in a neighborhood $(x_0 - \delta, x_0 + \delta)$ the tangent line value $f(x_0) + f'(x_0)(x - x_0)$ is close to the function value $f(x)$. The term $f'(x_0) \cdot (x - x_0)$ is often used to define the *differential* df_{x_0} of f at x_0. This is regarded as a function of the increment $x - x_0$ whose value at $x - x_0$ is $f'(x_0)(x - x_0)$. That is,

$$df_{x_0}(x-x_0) = f'(x_0)(x-x_0),$$

and one thinks of df_{x_0} as a function of "dx" (which is defined to be $x - x_0$).

Now let's focus attention on the nature of this function df_{x_0}. It is a particularly simple kind of function: to each value of the independent variable $x - x_0$ it assigns the real number $f'(x_0)(x - x_0)$ gotten by multiplying the value of the independent variable by the constant $f'(x_0)$. Functions of this sort, it turns out, are of fundamental importance not only in calculus but also in linear algebra and abstract algebra.

2.1 **DEFINITION.** A function $\ell: \boldsymbol{R} \to \boldsymbol{R}$ is called a **linear function** if

$$\ell(x+y) = \ell(x) + \ell(y)$$

and

$$\ell(ax) = a\ell(x)$$

for all $x, y \in \boldsymbol{R}$, and any $a \in \boldsymbol{R}$.

These are precisely the functions that *preserve* the vector space operations on $\boldsymbol{R}^1$. For a linear function ℓ preserves the addition structure in the sense of giving the same answer whether we add x and y and then map under ℓ, or we map to $\ell(x)$ and $\ell(y)$, and then add. (See Figure 2.2.) In the same way ℓ preserves the scalar product: we can apply ℓ before or after multiplying by the scalar a, and we get the same answer in each case. In Figure 2.2, either path we take around the diagram brings us to the same answer.

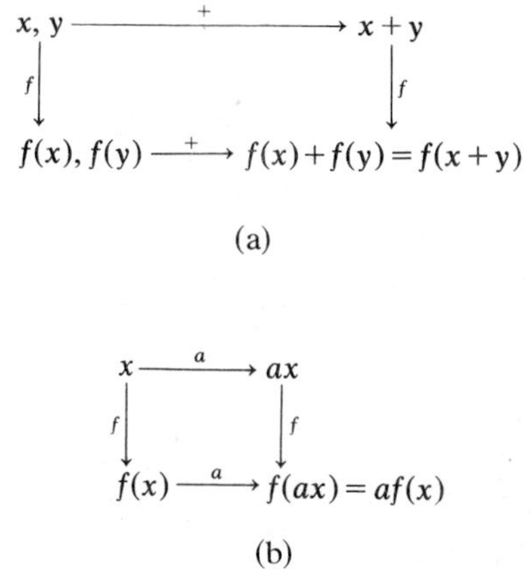

FIGURE 2.2

This preservation phenomenon seems to satisfy a basic human desire. If you think back to your early experience in elementary algebra, you may recall wanting all kinds of nonlinear functions to be linear. Did you ever wish (or even believe), for example, that $\sqrt{a+b} = \sqrt{a} + \sqrt{b}$? If you did, then you were wishing for the square root function to be linear!

What kinds of functions $f: \boldsymbol{R} \to \boldsymbol{R}$ *are* linear? This is very easy to answer.

2.2 **THEOREM.** A function $f: \boldsymbol{R} \to \boldsymbol{R}$ is linear if and only if there is a constant m such that $f(x) = mx$ for all $x \in \boldsymbol{R}$. The linear functions are thus the functions whose graphs are lines through the origin. (Hence the name "linear.")

Proof. First, any function $f: \boldsymbol{R} \to \boldsymbol{R}$ of this form is linear. If $f(x) = mx$, then

$$\begin{aligned} f(x+y) &= m(x+y) = mx + my \\ &= f(x) + f(y), \end{aligned}$$

and

$$f(ax) = m(ax) = (ma)x = (am)x$$
$$= a(mx) = af(x).$$

Conversely, suppose that $f: \boldsymbol{R} \to \boldsymbol{R}$ is linear. Let $m = f(1)$. Then for any $x \in \boldsymbol{R}$ we certainly have $x = x1$. But then since f is linear,

$$f(x) = xf(1) = xm = mx. \qquad \text{QED}$$

Take particular note of the fact that the geometric objects corresponding to the term *linear* are not just lines, but *lines through the origin.* Although the graph of the function $g(x) = mx + b$ is a line, it is not a line through the origin, unless $b = 0$. Hence g is *not* a linear function in the sense of Definition 2.1 unless $b = 0$. As you might expect, functions which are as close to being linear as g is when $b \neq 0$ are of interest and importance. We will see more of these functions, which are called *affine.* (See Exercise 17 below.)

Now we see that the differential function df_{x_0} is linear, since it just multiplies each value of its independent variable $x - x_0$ $(= dx)$ by $m = f'(x_0)$ when it is applied.

Linear functions $f: \boldsymbol{R} \to \boldsymbol{R}$ are certainly simple. Let's next consider linear functions $f: \boldsymbol{R}^n \to \boldsymbol{R}$ of several real variables.

2.3 **DEFINITION.** A function $f: \boldsymbol{R}^n \to \boldsymbol{R}$ is called a **linear function** if f preserves the vector space operations in $\boldsymbol{R}^n$, i.e., if

$$f(\boldsymbol{x} + \boldsymbol{y}) = f(\boldsymbol{x}) + f(\boldsymbol{y})$$

and

$$f(a\boldsymbol{x}) = af(\boldsymbol{x})$$

for all $\boldsymbol{x}$, $\boldsymbol{y} \in \boldsymbol{R}^n$, and all $a \in \boldsymbol{R}$.

For example, $f(x, y) = 3x + 2y$ defines a linear function $f: \boldsymbol{R}^2 \to \boldsymbol{R}$, as we can easily verify.

$$\begin{aligned} f(\boldsymbol{x}_1 + \boldsymbol{x}_2) &= f((x_1, y_1) + (x_2, y_2)) \\ &= f(x_1 + x_2, y_1 + y_2) \\ &= 3(x_1 + x_2) + 2(y_1 + y_2) \\ &= 3x_1 + 3x_2 + 2y_1 + 2y_2 \\ &= 3x_1 + 2y_1 + 3x_2 + 2y_2 \\ &= f(\boldsymbol{x}_1) + f(\boldsymbol{x}_2) \end{aligned}$$

and

$$\begin{aligned} f(a\boldsymbol{x}) &= f(ax, ay) \\ &= 3(ax) + 2(ay) \\ &= a(3x) + a(2y) \\ &= a(3x + 2y) = af(\boldsymbol{x}). \end{aligned}$$

Notice that this $f(x, y)$ has a formula of the form $m_1x+m_2y=(m_1, m_2)\cdot(x, y)$. Just as every linear function $f:\boldsymbol{R}\to\boldsymbol{R}$ turned out to be the *product* of the independent variable x with a fixed constant m, so it turns out that every linear function $f:\boldsymbol{R}^n\to\boldsymbol{R}$ is the *dot product* of the vector $\boldsymbol{x}=(x_1, x_2, \ldots, x_n)$ of independent variables with a fixed vector $\boldsymbol{m}=(m_1, m_2, \ldots, m_n)$.

2.4 **THEOREM.** A function $f:\boldsymbol{R}^n\to\boldsymbol{R}$ is linear if and only if there is a constant vector $\boldsymbol{m}=(m_1, m_2, \ldots, m_n)\in\boldsymbol{R}^n$ such that

$$f(\boldsymbol{x})=\boldsymbol{m}\cdot\boldsymbol{x}=m_1x_1+m_2x_2+\ldots+m_nx_n$$

for all $\boldsymbol{x}=(x_1, x_2, \ldots, x_n)\in\boldsymbol{R}^n$.

Proof. First, any such function $f:\boldsymbol{R}^n\to\boldsymbol{R}$ is linear. For any $\boldsymbol{x}, \boldsymbol{y}\in\boldsymbol{R}^n$, we have

$$f(\boldsymbol{x}+\boldsymbol{y})=\boldsymbol{m}\cdot(\boldsymbol{x}+\boldsymbol{y})=\boldsymbol{m}\cdot\boldsymbol{x}+\boldsymbol{m}\cdot\boldsymbol{y}=f(\boldsymbol{x})+f(\boldsymbol{y}),$$

using Theorem 1.2.4(2). Also for any $a\in\boldsymbol{R}$ and $\boldsymbol{x}\in\boldsymbol{R}^n$,

$$f(a\boldsymbol{x})=\boldsymbol{m}\cdot(a\boldsymbol{x})=a(\boldsymbol{m}\cdot\boldsymbol{x})=af(\boldsymbol{x})$$

by Theorem 1.2.4(2). So any f given by a formula $f(\boldsymbol{x})=\boldsymbol{m}\cdot\boldsymbol{x}$ is linear.

For the converse, we assume that $f:\boldsymbol{R}^n\to\boldsymbol{R}$ is linear and show that $f(\boldsymbol{x})=\boldsymbol{m}\cdot\boldsymbol{x}$ for all $\boldsymbol{x}\in\boldsymbol{R}^n$. We have to somehow produce the right constant vector $\boldsymbol{m}$. A natural candidate suggests itself from the proof of Theorem 2.2 and the example we considered after Definition 2.3. What vectors play a role analogous to $1\in\boldsymbol{R}$? They are the standard basis vectors $\boldsymbol{e}_1, \boldsymbol{e}_2, \ldots, \boldsymbol{e}_n$ with 0 in every entry except for a 1 as the i-th coordinate. We can imitate the proof of Theorem 2.2, then, as follows. For any $\boldsymbol{x}\in\boldsymbol{R}^n$,

$$\boldsymbol{x}=(x_1, x_2, \ldots, x_n)=x_1\boldsymbol{e}_1+x_2\boldsymbol{e}_2+\ldots+x_n\boldsymbol{e}_n.$$

Since f is linear,

$$\begin{aligned}f(\boldsymbol{x})&=f(x_1\boldsymbol{e}_1)+f(x_2\boldsymbol{e}_2+\ldots+x_n\boldsymbol{e}_n)\\&=f(x_1\boldsymbol{e}_1)+f(x_2\boldsymbol{e}_2)+\ldots+f(x_n\boldsymbol{e}_n)\\&=x_1f(\boldsymbol{e}_1)+x_2f(\boldsymbol{e}_2)+\ldots+x_nf(\boldsymbol{e}_n)\\&=(x_1, x_2, \ldots, x_n)\cdot(f(\boldsymbol{e}_1), f(\boldsymbol{e}_2), \ldots, f(\boldsymbol{e}_n))\\&=\boldsymbol{x}\cdot\boldsymbol{m}\\&=\boldsymbol{m}\cdot\boldsymbol{x}\end{aligned}$$

where $\boldsymbol{m}=(f(\boldsymbol{e}_1), f(\boldsymbol{e}_2), \ldots, f(\boldsymbol{e}_n))$ is a constant vector. QED

Observe that Theorem 2.4 looks identical to Theorem 2.2. In fact, Theorem 2.2 is a *special case* of Theorem 2.4, for $n=1$. For in the vector space $\boldsymbol{R}$, the dot product is just the *ordinary* product of real numbers.

We have completely characterized linear functions $f:\boldsymbol{R}^n\to\boldsymbol{R}$ in Theorem 2.4, and we will find it helpful to work with linear functions in the representation of Theorem 2.4.

2.5 **EXAMPLE.** Decide whether $f:\boldsymbol{R}^3\to\boldsymbol{R}$ is linear if $f(x, y, z)=-\pi x+\frac{3}{2}y-17z$. If it is linear, then give its representation in the form of Theorem 2.4.

Solution. We have

$$f(\boldsymbol{x}) = (-\pi, \tfrac{3}{2}, -17) \cdot (x, y, z)$$
$$= \boldsymbol{m} \cdot \boldsymbol{x}$$

where $\boldsymbol{m} = (-\pi, \frac{3}{2}, -17)$. So f is linear and we already have it represented in the form of Theorem 2.4.

2.6 EXAMPLE. Decide whether $f: \boldsymbol{R}^3 \to \boldsymbol{R}$ given by

$$f(x, y, z) = 2xy - 3y + z$$

is linear. If so, then give its representation in the form of Theorem 2.4.

Solution. We have

$$f(x, y, z) = (2y, -3, 1) \cdot (x, y, z)$$

where $(2y, -3, 1)$ is *not* a constant vector. So the function is not linear, by Theorem 2.4. We can also show that it is not linear directly from the definition by showing that it fails to preserve one of the operations of addition or scalar multiplication (since, in order to be linear, it has to preserve *both* of these operations). Consider

$$f(1, 2, 3) = 2(1)(2) - 3(2) + 3$$
$$= 4 - 6 + 3 = 1$$
$$f(5(1, 2, 3)) = f(5, 10, 15) = 2(5)(10) - 3(10) + 15$$
$$= 100 - 30 + 15 = 85$$
$$\neq 5f(1, 2, 3)(= 5),$$

so f fails to preserve multiplication by scalars.

We turn now to the calculus of linear functions. It is easy to show that these functions are continuous.

2.6 THEOREM. If $f: \boldsymbol{R}^n \to \boldsymbol{R}$ is a linear function, then f is continuous at all points of $\boldsymbol{R}^n$.

Proof. If $f: \boldsymbol{R}^n \to \boldsymbol{R}$ is linear, then we have $f(\boldsymbol{x}) = \boldsymbol{m} \cdot \boldsymbol{x}$ for all $\boldsymbol{x} \in \boldsymbol{R}^n$, where $\boldsymbol{m} = (f(\boldsymbol{e}_1), f(\boldsymbol{e}_2), \ldots, f(\boldsymbol{e}_n))$. We need to show that $\lim_{x \to x_0} f(\boldsymbol{x}) = f(\boldsymbol{x}_0)$. By Definition 3.2.1, this means that we have to show that $\lim_{\boldsymbol{x} \to \boldsymbol{x}_0} |f(\boldsymbol{x}) - f(\boldsymbol{x}_0)| = 0$. We have

$$|f(\boldsymbol{x}) - f(\boldsymbol{x}_0)| = |\boldsymbol{m} \cdot \boldsymbol{x} - \boldsymbol{m} \cdot \boldsymbol{x}_0|$$
$$= |\boldsymbol{m} \cdot (\boldsymbol{x} - \boldsymbol{x}_0)|$$
$$\leq |\boldsymbol{m}|\, |\boldsymbol{x} - \boldsymbol{x}_0|$$

by the Cauchy–Schwarz Inequality (Theorem 1.2.6). Thus

$$\lim_{\boldsymbol{x} \to \boldsymbol{x}_0} |f(\boldsymbol{x}) - f(\boldsymbol{x}_0)| \leq \lim_{\boldsymbol{x} \to \boldsymbol{x}_0} |\boldsymbol{m}|\, |\boldsymbol{x} - \boldsymbol{x}_0|$$
$$= |\boldsymbol{m}| \lim_{\boldsymbol{x} \to \boldsymbol{x}_0} |\boldsymbol{x} - \boldsymbol{x}_0| = |\boldsymbol{m}|\, 0 = 0.$$

Hence

$$\lim_{x \to x_0} |f(x) - f(x_0)| = 0. \qquad \text{QED}$$

EXERCISES 4.2

In Exercises 1 to 16, decide whether the function defined by the given formula is linear. Explain your reasoning. If it is linear, represent it in the form given in Theorem 2.2 or Theorem 2.4.

1 $f(x) = 3 \sin x$

2 $f(x) = e^x$

3 $f(x) = 3x$

4 $f(x) = 5 + 2x$

5 $f(x) = \sqrt{e}x$

6 $f(x, y) = 5x + 2y$

7 $f(x, y) = 4x$

8 $f(x, y) = 2x - 5y + 7$

9 $f(x, y) = 3 \cos x + 2y$

10 $f(x, y, z) = 3x - 4y + 2z$

11 $f(x, y, z) = 2x + 3y - z + 1$

12 $f(x, y, z) = xy + 2z$

13 $f(x, y, z) = ex + 2y - z$

14 $f(x, y, z) = 3x - y + xz$

15 $f(x, y, z, w) = 3x - 2y + 4z - 5w$

16 $f(x, y, z, w) = 4x - 3y + 4z - 5w + 1$

17 A function $f: \mathbf{R}^n \to \mathbf{R}$ is called *affine* if $f(\mathbf{x}) = \mathbf{m} \cdot \mathbf{x} + b$ for some nonzero constant $b \in \mathbf{R}$, where $\mathbf{m} \in \mathbf{R}^n$ is a constant vector. Which of the functions in Exercises 1 to 16 are affine?

18 If $f: \mathbf{R}^n \to \mathbf{R}$ is linear, then what can you say about the n partial derivatives $\dfrac{\partial f}{\partial x_i}$?

19 In *linear interpolation*, one approximates the value of $f(x)$ between the known values of $f(x_0)$ and $f(x_1)$ by using the proportion

$$\frac{x - x_0}{x_1 - x_0} \approx \frac{f(x) - f(x_0)}{f(x_1) - f(x_0)}.$$

(This process is usually employed when dealing with tables of values of trigonometric, exponential, and logarithmic functions, and you may have used it.) Show that the process is actually an approximation of f by an affine function in the interval $[x_0, x]$.

20 Since it is a basic human desire that functions be linear, people sometimes *assume* functions that they are working with are linear, without bothering to check. A group of students in a chemistry laboratory had to heat some water first to 40°C, then to 80°C. The class ran out of centigrade thermometers, so the students were given Fahrenheit thermometers and told to convert. They correctly calculated 40°C to be 104°F. They reasoned that they should heat their water to 104°F first, and then, since $80 = 2 \cdot 40$, they should heat their water to 208°F. Was this all right? Explain. (This is actually a true story.)

3 DIFFERENTIABLE FUNCTIONS

Having examined linear functions $f: \mathbf{R}^n \to \mathbf{R}$, we are now equipped to discuss the notion of a *differentiable* function $f: \mathbf{R}^n \to \mathbf{R}$. The idea of partial differentiation goes back to the origins of modern calculus three centuries ago, but the notions of differentiability and the total derivative belong to the twentieth century. The total derivative was introduced by the English mathematician William H. Young

(1863–1942) in 1909. He applied the theory of vectors of Hamilton and Grassmann, and the theory of matrices and linear algebra developed in the last half of the nineteenth century by the English mathematicians Sir Arthur Cayley (1821–1895) and James J. Sylvester (1814–1897). The present section may thus mark your first experience with modern mathematics.

Let's begin by returning to the idea we were considering at the start of Section 2. We want to make more precise the connection between linear functions $\ell:\boldsymbol{R}\to\boldsymbol{R}$ and differentiable functions $f:\boldsymbol{R}\to\boldsymbol{R}$.

To start with, recall that if f is differentiable at x_0 then we can approximate f near x_0 by its tangent line ℓ at $(x_0, f(x_0))$. Hence,

(*) $$f(x)\approx f(x_0)+f'(x_0)(x-x_0).$$

If we write

$$e(x)=f(x)-f(x_0)-f'(x_0)(x-x_0),$$

then $e(x)$ measures the directed distance from the tangent line ℓ to the graph of f at the value x. See Figure 3.1. (Note that if $e(x)>0$, then the graph of f lies above

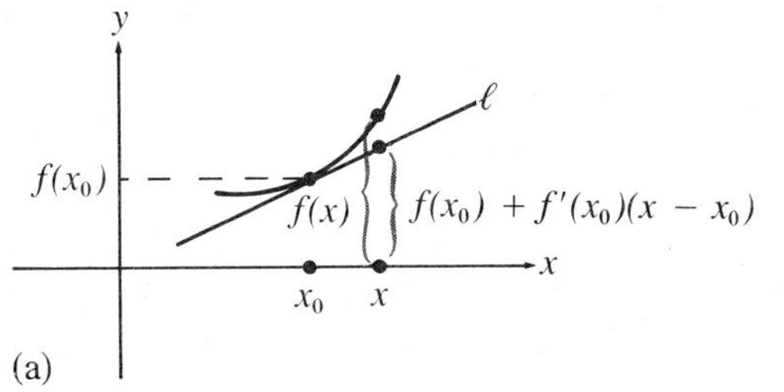

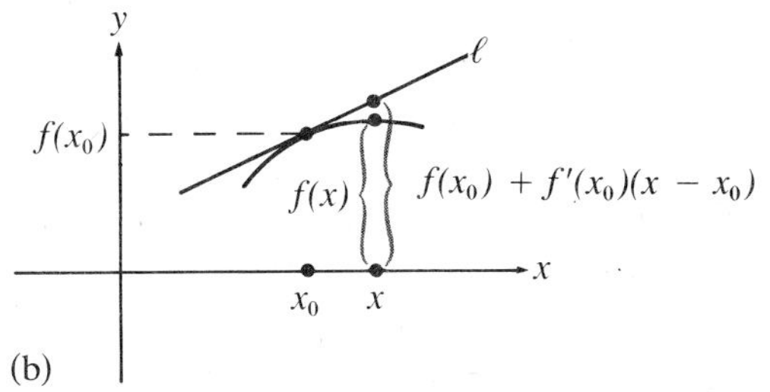

FIGURE 3.1

ℓ; if $e(x)<0$, then the graph of f lies below ℓ.) Thus $e(x)$ measures the *error* made in using the approximation (*) for f near $x=x_0$.

We can then measure how good the approximation (*) is by measuring $e(x)$. It turns out that (*) is a *good* approximation in the sense that as $x\to x_0$, the tangent line value $f(x_0)+f'(x_0)(x-x_0)$ approaches $f(x_0)$ *more rapidly* than x approaches x_0. By this we mean that

$$\frac{e(x)}{x-x_0}\to 0 \quad \text{as} \quad x\to x_0.$$

That is, even after division by $x-x_0$ the error $e(x)$ made in using (*) is still so small that it tends to 0 as x approaches x_0. To see this, we calculate

$$\begin{aligned}\lim_{x\to x_0}\frac{e(x)}{x-x_0} &= \lim_{x\to x_0}\left[\frac{f(x)-f(x_0)}{x-x_0}-f'(x_0)\right]\\ &= f'(x_0)-f'(x_0)=0.\end{aligned}$$

If we introduce the linear function $\ell: \boldsymbol{R} \rightarrow \boldsymbol{R}$ by $\ell(x) = f'(x_0)x$, then we can write our approximation in the form

$$f(x) = f(x_0) + \ell(x - x_0) + e(x),$$

where

$$\lim_{x \to x_0} \frac{e(x)}{x - x_0} = 0.$$

We have then established half of the following fundamental theorem.

3.1 THEOREM. A function $f: \boldsymbol{R} \rightarrow \boldsymbol{R}$ is differentiable at x_0 if and only if there is a linear function $\ell: \boldsymbol{R} \rightarrow \boldsymbol{R}$ such that

$$f(x) = f(x_0) + \ell(x - x_0) + e(x),$$

where

$$\lim_{x \to x_0} \frac{e(x)}{x - x_0} = 0.$$

In case f is differentiable at x_0, then ℓ is given by

$$\ell(x) = f'(x_0)x.$$

Proof. We have just shown that if f is differentiable at x_0, then we have the required ℓ and e. Suppose conversely that a linear function $\ell: \boldsymbol{R} \rightarrow \boldsymbol{R}$ exists such that $f(x) = f(x_0) + \ell(x - x_0) + e(x)$ where

$$\lim_{x \to x_0} \frac{e(x)}{x - x_0} = 0.$$

Then we must show that f is differentiable at x_0. We have

$$f(x) - f(x_0) = \ell(x - x_0) + e(x).$$

Moreover, since ℓ is a linear function, $\ell(x - x_0) = m(x - x_0)$ for some constant $m \in \boldsymbol{R}$. Then

$$\begin{aligned} \lim_{x \to x_0} \frac{f(x) - f(x_0)}{x - x_0} &= \lim_{x \to x_0} \frac{m(x - x_0) + e(x)}{x - x_0} \\ &= \lim_{x \to x_0} m + \lim_{x \to x_0} \frac{e(x)}{x - x_0} \\ &= m + 0 = m. \end{aligned}$$

Thus f is differentiable at x_0, and $f'(x_0) = m$. Therefore, $\ell(x) = f'(x_0)x$. QED

At first glance, Theorem 3.1 may seem a rather roundabout way to describe differentiability of a function $f: \boldsymbol{R} \rightarrow \boldsymbol{R}$. But it merely says, in a precise way, that the existence of the derivative at x_0 is *equivalent* to the ability to approximate f near x_0 by its tangent line in such a way that the error made approaches 0 faster than x approaches x_0.

3.2 **EXAMPLE.** Use Theorem 3.1 to show that $f:\mathbf{R}\to\mathbf{R}$ given by $f(x)=\frac{1}{x}$ is differentiable at $x=2$.

Solution. Here $x_0=2$, and we know from elementary calculus that $f'(2)$ should be $-\frac{1}{4}$ since $f'(x)=-\frac{1}{x^2}$ as long as $x\neq 0$. To use Theorem 3.1, we set $m=-\frac{1}{4}$ so that

$$\ell(x-x_0)=-\tfrac{1}{4}(x-2).$$

Then we have

$$\frac{1}{x}=\tfrac{1}{2}-\tfrac{1}{4}(x-2)+e(x).$$

To show that f is differentiable, we need to show that $\frac{e(x)}{x-2}$ approaches 0 as $x\to 2$. Since

$$e(x)=\frac{1}{x}-\tfrac{1}{2}+\tfrac{1}{4}(x-2)=\frac{4-2x+x(x-2)}{4x}$$

$$=\frac{(x-2)^2}{4x},$$

we obtain

$$\lim_{x\to 2}\frac{e(x)}{x-2}=\lim_{x\to 2}\frac{x-2}{4x}=0.$$

Therefore, Theorem 3.1 tells us that f *is* differentiable at $x=2$ and that $f'(2)=-\frac{1}{4}$, as expected. (Note: Theorem 3.1 does not tell us how to *find* $f'(2)$. This is why in elementary calculus $f'(x_0)$ is *not* defined via the property in Theorem 3.1: it is not useful in calculating derivatives or generating formulas.)

How can we translate the notion of a differentiable function $f:\mathbf{R}\to\mathbf{R}$ to a function $f:\mathbf{R}^n\to\mathbf{R}$? By using the property of Theorem 3.1, which we have seen is *equivalent* to the elementary calculus notion of differentiability. (Exercise 28 below shows that the familiar definition of differentiability given in elementary calculus fails to extend to functions $f:\mathbf{R}^n\to\mathbf{R}$.)

3.3 **DEFINITION.** Suppose that $f:\mathbf{R}^n\to\mathbf{R}$ is defined on an open set containing $\mathbf{x}_0\in\mathbf{R}^n$. Then f is **differentiable** at $\mathbf{x}_0$ if there is a linear function $\ell:\mathbf{R}^n\to\mathbf{R}$ such that

(1) $$f(\mathbf{x})=f(\mathbf{x}_0)+\ell(\mathbf{x}-\mathbf{x}_0)+e(\mathbf{x}),$$

where

(2) $$\lim_{\mathbf{x}\to\mathbf{x}_0}\frac{e(\mathbf{x})}{|\mathbf{x}-\mathbf{x}_0|}=0.$$

The condition (2) is *exactly* the condition of Theorem 3.1 with one exception. Since we can't divide the real valued function $e(\mathbf{x})$ by a vector, we have expressed the condition that $e(\mathbf{x})\to 0$ more rapidly than $\mathbf{x}\to\mathbf{x}_0$ by requiring that the ratio of

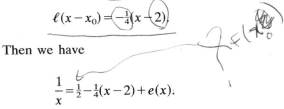

$e(\boldsymbol{x})$ to the *length* of $\boldsymbol{x}-\boldsymbol{x}_0$ approach 0 as $\boldsymbol{x}\to\boldsymbol{x}_0$. (We have required f to be defined on an open set containing $\boldsymbol{x}_0$ so that Equation (1) is meaningful for all $\boldsymbol{x}$ close to $\boldsymbol{x}_0$.)

In Theorem 3.1 we gave a simple formula for $\ell(x-x_0)$ in case $f:\boldsymbol{R}\to\boldsymbol{R}$ is differentiable:

(3) $$\ell(x-x_0)=f'(x_0)(x-x_0).$$

There should be an analogously simple formula for the present case. To see what it should be, let's look at the case $n=2$ again. Here we have, from Section 1, a notion of tangent plane analogous to the tangent line in the $n=1$ case. On page 151 we saw that we could express the equation of the tangent plane to the graph $z=f(x, y)$ at (x_0, y_0, z_0) in the form

(4) $$z=f(\boldsymbol{x}_0)+\left(\frac{\partial f}{\partial x}(\boldsymbol{x}_0), \frac{\partial f}{\partial y}(\boldsymbol{x}_0)\right)\cdot(\boldsymbol{x}-\boldsymbol{x}_0).$$

This looks so analogous to the equation

(5) $$y=f(x_0)+f'(x_0)(x-x_0)$$

of the tangent line to the graph of $y=f(x)$ at (x_0, y_0) that we were led to expect that the *total derivative* of $f:\boldsymbol{R}^2\to\boldsymbol{R}$ at $\boldsymbol{x}_0=(x_0, y_0)$ should be obtained by *putting the partial derivatives of f together* to produce a vector. For if we denote that total derivative by $\boldsymbol{f}'(\boldsymbol{x}_0)$, then (4) would become

(6) $$z=f(\boldsymbol{x}_0)+\boldsymbol{f}'(\boldsymbol{x}_0)\cdot(\boldsymbol{x}-\boldsymbol{x}_0),$$

which corresponds exactly to (5). If such perfect analogy seems too compelling to be wrong, then your instincts are good. Now let's see how we can convince ourselves that the vector of partial derivatives really *is* what we should call the derivative of $f:\boldsymbol{R}^2\to\boldsymbol{R}$.

If f is differentiable, then we know from Theorem 2.4 that there is *some* $\boldsymbol{m}=(m_1, m_2)$ in $\boldsymbol{R}^2$ such that

(7) $$\ell(\boldsymbol{x}-\boldsymbol{x}_0)=\boldsymbol{m}\cdot(\boldsymbol{x}-\boldsymbol{x}_0).$$

In view of Theorem 3.1 and Equation (3), the vector $\boldsymbol{m}$ is what should be called the (total) derivative of $f:\boldsymbol{R}^2\to\boldsymbol{R}$ at $\boldsymbol{x}_0$. We need to show then that $\boldsymbol{m}$ in (7) is actually

$$\left(\frac{\partial f}{\partial x}(\boldsymbol{x}_0), \frac{\partial f}{\partial y}(\boldsymbol{y}_0)\right).$$

That is, we need to show that (2) in Definition 3.3 is satisfied if we use ℓ given by (7) in Definition 3.3. Our next result establishes precisely this fact.

3.4 THEOREM. If $f:\boldsymbol{R}^n\to\boldsymbol{R}$ is differentiable at $\boldsymbol{x}_0$, then the linear function ℓ in Definition 3.3 is given by $\ell(\boldsymbol{x}-\boldsymbol{x}_0)=\boldsymbol{m}\cdot(\boldsymbol{x}-\boldsymbol{x}_0)$ where $\boldsymbol{m}=(m_1, m_2, \ldots, m_n)$ with

$$m_i=\frac{\partial f}{\partial x_i}(\boldsymbol{x}_0).$$

Proof. Since ℓ is linear, we know that (7) holds for *some* fixed vector $\boldsymbol{m} = (m_1, m_2, \ldots, m_n)$ by Theorem 2.4. The only thing we need to prove, then, is that $m_i = \dfrac{\partial f}{\partial x_i}(\boldsymbol{x}_0)$. For this we use Equation (1) of Definition 3.3 with $\boldsymbol{x} = (x_{01}, \ldots, x_{0,i-1}, x_i, x_{0,i+1}, \ldots, x_{0n})$. Notice that every coordinate of $\boldsymbol{x}$ is frozen to its value at $\boldsymbol{x}_0 = (x_{01}, \ldots, x_{0i}, \ldots, x_{0n})$ except the i-th coordinate. Thus $\boldsymbol{x} \to \boldsymbol{x}_0$ is equivalent to $x_i \to x_{0i}$. From Equation (1) we have

$$\begin{aligned} f(\boldsymbol{x}) &= f(\boldsymbol{x}_0) + \ell(\boldsymbol{x} - \boldsymbol{x}_0) + e(\boldsymbol{x}) \\ &= f(\boldsymbol{x}_0) + \boldsymbol{m} \cdot (\boldsymbol{x} - \boldsymbol{x}_0) + e(\boldsymbol{x}) \\ &= f(\boldsymbol{x}_0) + m_i(x_i - x_{0i}) + e(\boldsymbol{x}). \end{aligned}$$

Hence,

(8) $$e(\boldsymbol{x}) = f(\boldsymbol{x}) - f(\boldsymbol{x}_0) - m_i(x_i - x_{0i}).$$

Since $|\boldsymbol{x} - \boldsymbol{x}_0| = |x_i - x_{0i}|$, Definition 3.3 tells us that

$$0 = \lim_{\boldsymbol{x} \to \boldsymbol{x}_0} \frac{e(\boldsymbol{x})}{|\boldsymbol{x} - \boldsymbol{x}_0|} = \lim_{x_i \to x_{0i}} \frac{e(\boldsymbol{x})}{|x_i - x_{0i}|}.$$

Thus,

$$\lim_{x_i \to x_{0i}} \frac{e(\boldsymbol{x})}{x_i - x_{0i}} = \lim_{x_i \to x_{0i}} \frac{e(\boldsymbol{x})}{\pm|x_i - x_{0i}|} = \pm 0 = 0.$$

From (8), this gives

$$\begin{aligned} 0 = \lim_{x_i \to x_{0i}} \frac{e(\boldsymbol{x})}{x_i - x_{0i}} &= \lim_{x_i \to x_{0i}} \frac{f(\boldsymbol{x}) - f(\boldsymbol{x}_0) - m_i(x_i - x_{0i})}{x_i - x_{0i}} \\ &= \lim_{x_i \to x_{0i}} \left[\frac{f(\boldsymbol{x}) - f(\boldsymbol{x}_0)}{x_i - x_{0i}} - m_i\right] \\ &= \frac{\partial f}{\partial x_i}(\boldsymbol{x}_0) - m_i. \end{aligned}$$

Hence,

$$m_i = \frac{\partial f}{\partial x_i}(\boldsymbol{x}_0).$$ QED

3.5 **DEFINITION.** If $f: \boldsymbol{R}^n \to \boldsymbol{R}$ is differentiable at $\boldsymbol{x}_0 \in \boldsymbol{R}^n$, then the **total derivative** $f'(\boldsymbol{x}_0)$ of f at $\boldsymbol{x}_0$ is the vector of partial derivatives:

$$f'(\boldsymbol{x}_0) = \left(\frac{\partial f}{\partial x_1}(\boldsymbol{x}_0), \frac{\partial f}{\partial x_2}(\boldsymbol{x}_0), \ldots, \frac{\partial f}{\partial x_n}(\boldsymbol{x}_0)\right).$$

3.6 **EXAMPLE.** Show that $f: \boldsymbol{R}^2 \to \boldsymbol{R}$ given by $f(x, y) = x^2 + y^2$ is differentiable at $\boldsymbol{x}_0 = (2, 3)$.

Solution. We have only one candidate for the total derivative by Theorem 3.4, namely

$$\left(\frac{\partial f}{\partial x}(\boldsymbol{x}_0), \frac{\partial f}{\partial y}(\boldsymbol{x}_0)\right) = (2x, 2y)|_{(2,3)} = (4, 6).$$

We have to show that this works in Definition 3.3. So put

$$f(\boldsymbol{x}) = f(\boldsymbol{x}_0) + (4, 6) \cdot (\boldsymbol{x} - \boldsymbol{x}_0) + e(\boldsymbol{x}).$$

This equation defines $e(\boldsymbol{x})$, and we must show that

$$\lim_{\boldsymbol{x} \to \boldsymbol{x}_0} \frac{e(\boldsymbol{x})}{|\boldsymbol{x} - \boldsymbol{x}_0|} = 0.$$

Here

$$f(\boldsymbol{x}_0) = f(2, 3) = 2^2 + 3^2 = 13$$

and

$$\boldsymbol{x} - \boldsymbol{x}_0 = (x, y) - (2, 3) = (x - 2, y - 3).$$

So we have

$$x^2 + y^2 = 13 + (4, 6) \cdot (x - 2, y - 3) + e(\boldsymbol{x}).$$

Hence

$$\begin{aligned} e(\boldsymbol{x}) &= x^2 + y^2 - 13 - 4(x - 2) - 6(y - 3) \\ &= x^2 + y^2 - 13 - 4x + 8 - 6y + 18 \\ &= x^2 - 4x + y^2 - 6y + 13 \\ &= x^2 - 4x + 4 + y^2 - 6y + 9 \\ &= (x - 2)^2 + (y - 3)^2. \end{aligned}$$

Now

$$|\boldsymbol{x} - \boldsymbol{x}_0| = |(x - 2, y - 3)| = \sqrt{(x - 2)^2 + (y - 3)^2}.$$

So

$$\begin{aligned} \lim_{\boldsymbol{x} \to \boldsymbol{x}_0} \frac{e(\boldsymbol{x})}{|\boldsymbol{x} - \boldsymbol{x}_0|} &= \lim_{\boldsymbol{x} \to \boldsymbol{x}_0} \frac{(x - 2)^2 + (y - 3)^2}{\sqrt{(x - 2)^2 + (y - 3)^2}} \\ &= \lim_{\boldsymbol{x} \to \boldsymbol{x}_0} \sqrt{(x - 2)^2 + (y - 3)^2} = 0. \end{aligned}$$

Hence by Definition 3.3, f is differentiable at $(2, 3)$. So by Definition 3.5, $f'(2, 3) = (4, 6)$. QED

We can very easily prove that differentiable functions $f: \boldsymbol{R}^n \to \boldsymbol{R}$ are continuous, as was the case in elementary calculus.

3.7 THEOREM. If $f: \boldsymbol{R}^n \to \boldsymbol{R}$ is differentiable at $\boldsymbol{x}_0$, then it is continuous at $\boldsymbol{x}_0$.

Proof. We need to show that

$$\lim_{\boldsymbol{x}\to\boldsymbol{x}_0} f(\boldsymbol{x}) = f(\boldsymbol{x}_0)$$

or, equivalently, that

$$\lim_{\boldsymbol{x}\to\boldsymbol{x}_0} [f(\boldsymbol{x}) - f(\boldsymbol{x}_0)] = 0.$$

Since f is differentiable, we have from Equation (1) in Definition 3.3

$$f(\boldsymbol{x}) - f(\boldsymbol{x}_0) = \ell(\boldsymbol{x} - \boldsymbol{x}_0) + e(\boldsymbol{x})$$

for some linear function ℓ. Now from Definition 3.3,

$$\lim_{\boldsymbol{x}\to\boldsymbol{x}_0} \frac{e(\boldsymbol{x})}{|\boldsymbol{x} - \boldsymbol{x}_0|} = 0.$$

Hence,

$$\lim_{\boldsymbol{x}\to\boldsymbol{x}_0} e(\boldsymbol{x}) = \lim_{\boldsymbol{x}\to\boldsymbol{x}_0} |\boldsymbol{x} - \boldsymbol{x}_0| \frac{e(\boldsymbol{x})}{|\boldsymbol{x} - \boldsymbol{x}_0|} = 0 \cdot 0 = 0.$$

Also, from Theorem 2.6, ℓ is continuous at $\boldsymbol{x}_0$. Thus,

$$\begin{aligned}\lim_{\boldsymbol{x}\to\boldsymbol{x}_0} [f(\boldsymbol{x}) - f(\boldsymbol{x}_0)] &= \lim_{\boldsymbol{x}\to\boldsymbol{x}_0} \ell(\boldsymbol{x} - \boldsymbol{x}_0) + \lim_{\boldsymbol{x}\to\boldsymbol{x}_0} e(\boldsymbol{x}) \\ &= \ell(\boldsymbol{x}_0 - \boldsymbol{x}_0) + 0 \\ &= \boldsymbol{m} \cdot \boldsymbol{0} \\ &= 0\end{aligned}$$

for the constant vector $\boldsymbol{m} = f'(\boldsymbol{x}_0)$ corresponding to ℓ (Theorem 2.4). QED

You may naturally be theorizing now that all we have to do to compute the total derivative of any function f at any point $\boldsymbol{x}_0$ is to simply compute the vector of partial derivatives at $\boldsymbol{x}_0$. This is *almost* right (Theorem 3.10), but not quite. The problem is that while the vector of partial derivatives is indeed the *only candidate* for $f'(\boldsymbol{x}_0)$, it still may not *be* $f'(\boldsymbol{x}_0)$. The (perhaps disturbing) problem is that all the partial derivatives of f may exist at $\boldsymbol{x}_0$ *without* f being differentiable at $\boldsymbol{x}_0$, so $f'(\boldsymbol{x}_0)$ may not exist! This is somewhat akin to continuity in each variable separately not being enough to give continuity for a function of several variables. (Take a second look at Exercises 17 and 18 in Exercises 3.2.)

3.8 EXAMPLE. Determine whether the schizophrenic function f defined by

$$f(x, y) = \begin{cases} \dfrac{xy}{x^2 + y^2} & \text{if} \quad (x, y) \neq (0, 0) \\ 0 & \text{if} \quad (x, y) = (0, 0) \end{cases}$$

is differentiable at $(0, 0)$.

Solution. There *is* a candidate for $f'(0, 0)$. If we freeze y at 0, we see $\frac{\partial f}{\partial x}(0, 0) = 0$; and if we freeze x at 0, we see $\frac{\partial f}{\partial y}(0, 0) = 0$. But our candidate $(0, 0) = \left(\frac{\partial f}{\partial x}(0, 0), \frac{\partial f}{\partial y}(0, 0)\right)$ is *not* the total derivative of f at $(0, 0)$ because f *isn't* differentiable at $(0, 0)$. For if f were differentiable at $(0, 0)$, then by Theorem 3.7 it would have to be continuous at $(0, 0)$, which it is *not*. For example, if $(x, y) \to (0, 0)$ along the line $x = y$, then $f(x, y) \to \frac{1}{2}$, not $f(0, 0)$ (cf. Example 3.2.6).

We can also see that Definition 3.3 fails for the candidate $(0, 0)$. We have

$$f(x, y) = f(0, 0) + (0, 0) \cdot (x - 0, y - 0) + e(\boldsymbol{x}) = e(\boldsymbol{x}).$$

So for f to be differentiable at 0,

$$\lim_{\boldsymbol{x} \to \boldsymbol{x}_0} \frac{f(\boldsymbol{x})}{|\boldsymbol{x} - \boldsymbol{0}|} = \lim_{(x,y)\to(0,0)} \frac{f(x, y)}{\sqrt{x^2 + y^2}} = \lim_{(x,y)\to(0,0)} \frac{xy}{(x^2 + y^2)^{3/2}}$$

must be 0. But this limit fails to be 0 (it even fails to exist!) as we can see by letting $(x, y) \to (0, 0)$ again along the line $x = y$.

As a result of this kind of behavior, it is appropriate to create notation specifically for the vector of partial derivatives, which may exist and yet *not* be $f'(\boldsymbol{x}_0)$.

3.9 **DEFINITION.** If $f: \boldsymbol{R}^n \to \boldsymbol{R}$ has partial derivatives with respect to each of its variables at $\boldsymbol{x}_0$, then the **gradient of** f at $\boldsymbol{x}_0$ is

$$(\textbf{grad } f)(\boldsymbol{x}_0) = \nabla f(\boldsymbol{x}_0) = \left(\frac{\partial f}{\partial x_1}(\boldsymbol{x}_0), \frac{\partial f}{\partial x_2}(\boldsymbol{x}_0), \ldots, \frac{\partial f}{\partial x_n}(\boldsymbol{x}_0)\right).$$

(The symbol ∇ is read "del.")

Example 3.8 shows that ∇f can exist at a point $\boldsymbol{x}_0$ where f is not even continuous, let alone differentiable. This suggests why, in defining the tangent plane to $z = f(x, y)$ (Definition 1.3), we required more than the mere existence of $\frac{\partial f}{\partial x}$ and $\frac{\partial f}{\partial y}$ at $\boldsymbol{x}_0$. From our experience in elementary calculus, we expect that only *differentiable* functions ought to have graphs possessing tangent planes. The question then is, how can we tell whether a given function $f: \boldsymbol{R}^2 \to \boldsymbol{R}$ is differentiable at $\boldsymbol{x}_0$? That is, when does the easily computed gradient $\nabla f(\boldsymbol{x}_0)$, the sole candidate for $f'(\boldsymbol{x}_0)$, actually *give* the total derivative of f at $\boldsymbol{x}_0$? Our next result is comforting in that it says that $\nabla f(\boldsymbol{x}_0)$ is the total derivative for *most* functions we meet in practice, and explains the continuity conditions we imposed in Definition 3.1. We give the proof just for the case $n = 2$, but all the essential ideas extend directly to the general case.

3.10 **THEOREM.** Suppose that $f: \boldsymbol{R}^n \to \boldsymbol{R}$ is defined on an open set U containing $\boldsymbol{x}_0$. If all the partial derivatives $\frac{\partial f}{\partial x_i}$, $i = 1, 2, \ldots, n$, exist on U and are continuous at $\boldsymbol{x}_0$, then f is differentiable at $\boldsymbol{x}_0$, so $\nabla f(\boldsymbol{x}_0)$ is the total derivative $f'(\boldsymbol{x}_0)$.

Proof for $n = 2$. Let $\boldsymbol{x} = (x, y) \in U$ and $\boldsymbol{x}_0 = (x_0, y_0)$. We have

(9) $$f(\boldsymbol{x}) - f(\boldsymbol{x}_0) = [f(x, y) - f(x, y_0)] + [f(x, y_0) - f(x_0, y_0)]$$

where we have merely subtracted and then added $f(x, y_0)$. Now look at each term in brackets. It is the difference between two real valued functions of a *single* real variable: y in the first bracket and x in the second. We can then apply the Mean Value Theorem of elementary calculus to each of these functions of a single real variable, since each can be partially differentiated on U. We get

$$f(x, y) - f(x, y_0) = \frac{\partial f}{\partial y}(x, y_1)(y - y_0), \text{ where } y_1 \text{ is between } y_0 \text{ and } y;$$

$$f(x, y_0) - f(x_0, y_0) = \frac{\partial f}{\partial x}(x_1, y_0)(x - x_0), \text{ where } x_1 \text{ is between } x_0 \text{ and } x.$$

Substituting these into (9), we obtain

(10) $$f(\boldsymbol{x}) - f(\boldsymbol{x}_0) = \frac{\partial f}{\partial y}(x, y_1)(y - y_0) + \frac{\partial f}{\partial x}(x_1, y_0)(x - x_0)$$

Now put $f(\boldsymbol{x}) = f(\boldsymbol{x}_0) + \nabla f(\boldsymbol{x}_0) \cdot (\boldsymbol{x} - \boldsymbol{x}_0) + e(\boldsymbol{x})$. Then

$$\begin{aligned} e(\boldsymbol{x}) &= f(\boldsymbol{x}) - f(\boldsymbol{x}_0) - \nabla f(\boldsymbol{x}_0) \cdot (\boldsymbol{x} - \boldsymbol{x}_0) \\ &= \frac{\partial f}{\partial y}(x, y_1)(y - y_0) + \frac{\partial f}{\partial x}(x_1, y_0)(x - x_0) \\ &\quad - \frac{\partial f}{\partial x}(x_0, y_0)(x - x_0) - \frac{\partial f}{\partial y}(x_0, y_0)(y - y_0) \\ &\qquad \text{from (10) and Definition 3.9} \\ &= (x - x_0)\left[\frac{\partial f}{\partial x}(x_1, y_0) - \frac{\partial f}{\partial x}(x_0, y_0)\right] + (y - y_0)\left[\frac{\partial f}{\partial y}(x, y_1) - \frac{\partial f}{\partial y}(x_0, y_0)\right] \end{aligned}$$

Then

$$|e(\boldsymbol{x})| \le |x - x_0| \left|\frac{\partial f}{\partial x}(x_1, y_0) - \frac{\partial f}{\partial x}(x_0, y_0)\right| + |y - y_0| \left|\frac{\partial f}{\partial y}(x, y_1) - \frac{\partial f}{\partial y}(x_0, y_0)\right|$$

by the Triangle Inequality (Theorem 1.2.10). Therefore,

$$|e(\boldsymbol{x})| \le |\boldsymbol{x} - \boldsymbol{x}_0| \left[\left|\frac{\partial f}{\partial x}(x_1, y_0) - \frac{\partial f}{\partial x}(x_0, y_0)\right| + \left|\frac{\partial f}{\partial y}(x, y_1) - \frac{\partial f}{\partial y}(x_0, y_0)\right|\right],$$

since $|x - x_0| \le |\boldsymbol{x} - \boldsymbol{x}_0|$ and $|y - y_0| \le |\boldsymbol{x} - \boldsymbol{x}_0|$. Hence,

$$\frac{|e(\boldsymbol{x})|}{|\boldsymbol{x} - \boldsymbol{x}_0|} \le \left|\frac{\partial f}{\partial x}(x_1, y_0) - \frac{\partial f}{\partial x}(x_0, y_0)\right| + \left|\frac{\partial f}{\partial y}(x, y_1) - \frac{\partial f}{\partial y}(x_0, y_0)\right|.$$

As $\boldsymbol{x} \to \boldsymbol{x}_0$, each absolute value on the right approaches 0 since $\dfrac{\partial f}{\partial x}$ and $\dfrac{\partial f}{\partial y}$ are continuous at $\boldsymbol{x}_0$. Thus

$$\lim_{\boldsymbol{x}\to\boldsymbol{x}_0}\frac{|e(\boldsymbol{x})|}{|\boldsymbol{x}-\boldsymbol{x}_0|}=0;\quad\text{hence}\quad\lim_{\boldsymbol{x}\to\boldsymbol{x}_0}\frac{e(\boldsymbol{x})}{|\boldsymbol{x}-\boldsymbol{x}_0|}=0.$$

Thus, f is differentiable at $\boldsymbol{x}_0$ by Definition 3.3. QED

Now, in order to decide if a given function f is differentiable at $\boldsymbol{x}_0$, we can simply calculate $\nabla f(\boldsymbol{x})$ and look at its coordinates. If they are all continuous at $\boldsymbol{x}_0$, as will usually be the case, then f is differentiable at $\boldsymbol{x}_0$ and $f'(\boldsymbol{x}_0)=\nabla f(\boldsymbol{x}_0)$. If not all the partials are continuous at $\boldsymbol{x}_0$, then we need to take a close look at Definition 3.3 to determine whether it holds for f at $\boldsymbol{x}_0$, as we did in Example 3.8. Note in that example that

$$\frac{\partial f}{\partial x}=\frac{(x^2+y^2)y-xy(2x)}{(x^2+y^2)^2}=\frac{y^3-x^2y}{(x^2+y^2)^2}\quad\text{and}$$

$$\frac{\partial f}{\partial y}=\frac{x^3-xy^2}{(x^2+y^2)^2}.$$

So both $\dfrac{\partial f}{\partial x}$ and $\dfrac{\partial f}{\partial y}$ fail to be continuous at $(0, 0)$, since neither one has limit 0 (in fact, neither one has a limit at all!) at $(x_0, y_0)=(0, 0)$.

When f *is* differentiable at $\boldsymbol{x}_0$, we can use this fact to approximate f by the *tangent function*

$$f(\boldsymbol{x}_0)+f'(\boldsymbol{x}_0)\cdot(\boldsymbol{x}-\boldsymbol{x}_0)$$

just as we used it in elementary calculus. Although we can actually *draw* the graph of the tangent function only in the case of $f:\boldsymbol{R}^2\to\boldsymbol{R}$, when we have the tangent plane to the graph of f at $(x_0, y_0, f(x_0, y_0))$, we can still think of the expression $f(\boldsymbol{x}_0)+f'(\boldsymbol{x}_0)\cdot(\boldsymbol{x}-\boldsymbol{x}_0)$ as representing a tangent "plane" to the "surface" in $\boldsymbol{R}^{n+1}$ which is the graph of f. By analogy with the case of $f:\boldsymbol{R}\to\boldsymbol{R}$, the expression $f'(\boldsymbol{x}_0)\cdot(\boldsymbol{x}-\boldsymbol{x}_0)$ is sometimes called the (total) *differential* of f near x_0, and denoted $df_{\boldsymbol{x}_0}(\boldsymbol{x}-\boldsymbol{x}_0)$.

While Equation (2) of Definition 3.3 leads us to expect the tangent approximation to approximate f pretty closely near $\boldsymbol{x}_0$, we have not said anything about just *how* accurately it approximates f. That is, we have no upper bound on $|e(\boldsymbol{x})|$ for $\boldsymbol{x}$ near $\boldsymbol{x}_0$. For careful work with approximations, such estimates on the size of $e(\boldsymbol{x})$ are essential. But they turn out to involve higher order derivatives, so that we cannot pursue this until after Section 7. (See Theorem 8.9.) In the meantime, we will simply assume that (2) guarantees that the tangent approximation to $f(\boldsymbol{x})$ is quite good as long as we keep $\boldsymbol{x}$ near $\boldsymbol{x}_0$. (One says that $e(\boldsymbol{x})$ has a *smaller order of magnitude* than $|\boldsymbol{x}-\boldsymbol{x}_0|$ because (2) holds.) Our next example shows the details involved in using the tangent approximation.

3.11 EXAMPLE. *Boyle's Law* for an ideal gas of constant weight maintained at a constant temperature is $PV=c$, where P is the pressure (in pounds per square inch, say) and V is the volume (in cubic feet). In a certain gas chamber, the pressure is measured as 100.00 pounds per square inch. This is accurate to within 2%. The chamber is manufactured to be 4.00 cubic feet in interior volume, subject to an error of 0.5%. Determine c and use the tangent approximation to estimate its degree of accuracy.

Solution. We have $c=(100.00)(4.00)=400.00$. Now, the maximum possible error in the measurement $P_0=100.00$ is 2%. Since 2% of 100 is 2, P lies in the interval [98, 102]. The maximum possible error in the measurement $V_0=4.00$ is 0.5%, so V lies in the interval $[4-(0.005)4, 4+(0.005)4]=[4-0.02, 4+0.02]=[3.98, 4.02]$. Since $c=PV$, we have $\frac{\partial c}{\partial P}=V$ and $\frac{\partial c}{\partial V}=P$. So at $(P_0, V_0)=(100, 4)$, we have $\left(\frac{\partial c}{\partial P}, \frac{\partial c}{\partial V}\right)=(4, 100)$. The tangent approximation for c is

$$\begin{aligned} A(P, V) &= c(P_0, V_0)+\left(\frac{\partial c}{\partial P}(P_0, V_0), \frac{\partial c}{\partial V}(P_0, V_0)\right)\cdot(P-P_0, V-V_0) \\ &= 400+(4, 100)\cdot(P-100, V-4) \\ &= 400+4(P-100)+100(V-4). \end{aligned}$$

The maximum error in c is then approximated by

$$\begin{aligned} |A(P, V)-400| &= |4(P-100)+100(V-4)| \\ &\le 4\,|P-100|+100\,|V-4| \quad \text{by the Triangle Inequality} \\ &\le 4(2)+100(0.02) \\ &\le 8+2=10. \end{aligned}$$

Therefore, an upper bound for the error in c is approximately 10. So we estimate that c lies in the interval [390, 410]. The maximum *per cent error* in c is about 10/400, i.e., 2.5%. (Note that this error is *greater* than the error in either P or V.)

EXERCISES 4.3

In Exercises 1 to 8, find the total derivatives of the given functions at the given points.

1 $f(x, y)=x^2+3y$ at (2, 1)

2 $f(x, y)=\cos xy$ at $\left(1, \frac{\pi}{3}\right)$

3 $f(x, y)=e^{x^2+y^2}$ at (0, 1)

4 $f(x, y)=\ln(x^2+y^2)$ at (1, 2)

5 $f(x, y)=\sqrt{x^2+y^2}$ at (1, 1)

6 $f(x, y, z)=3x^2+y^2-4z^2$ at (1, 0, 1)

7 $f(x, y, z)=e^{x^2+y^2}\sin z$ at $\left(1, 1, \frac{\pi}{3}\right)$

8 $f(x, y, z)=\ln(x^2+xy)+x^2y^2z$ at (2, 1, 1)

9 Use Definition 3.3 to show that $f(x, y)=x^2+y^2$ is differentiable at (2, 1).

10 Use Definition 3.4 to show that $f(x, y)=3x+2y$ is differentiable at (2, 1).

11 Use the tangent approximation to find an approximate value for $\sqrt{(3.01)^2+(3.98)^2}$. (*Hint:* Let $f(x, y)=\sqrt{x^2+y^2}$.)

12 Use the tangent approximation to find an approximate value for $\sqrt{(1.1)^3+(1.98)^3}$.

13 Use the tangent approximation to find an approximate value for $(1.99)(8.03)^{4/3}$.

14 Use the tangent approximation to find an approximate value for $\frac{1}{2}(3+\frac{1}{12})^2\sin\left(\frac{\pi}{6}+0.04\right)$.

15 The dimensions of a rectangular room are measured as 9.00, 12.00, and 8.00 feet respectively, with possible errors of ±0.01, ±0.01, and ±0.04 feet respectively. Calculate the length of the diagonal across the room and estimate the maximum error possible in this. (The *diagonal* means the line drawn from the floor in one corner to the ceiling in the opposite corner.)

16 The dimensions of a box are measured as 12.00, 4.00, and 3.00 inches respectively, with possible errors of ±0.03, ±0.01, and ±0.01 inches respectively. Calculate the length of the diagonal of the box and find the maximum possible error in this.

17 Use the tangent approximation to approximate $\sqrt{(2.01)^2+(1.94)^2+(0.98)^2}$.

18 Use the tangent approximation to approximate $\sqrt{(1.99)^2+(2.97)^3+(1.01)^2}$.

19 An aluminum beer can in the shape of a right circular cylinder has inner radius 2 inches and interior height 6 inches. The aluminum sides, top, and bottom are 0.02 inch thick. Use the tangent approximation to compute the approximate amount of aluminum in each can. (*Hint:* $V=\pi x^2 y$, where x is the radius and y is the height.)

20 A cylindrical oil storage tank has inner radius of 20 feet and inner height of 30 feet. The sides, floor, and top are a metallic alloy 2 inches thick. Find the approximate amount of alloy needed to make one such oil tank.

21 Show that the schizophrenic function

$$f(x, y)=\begin{cases}\dfrac{3xy}{x^2+y^2} & \text{for } (x, y)\neq(0, 0)\\ 0 & \text{for } (x, y)=(0, 0)\end{cases}$$

is not differentiable at (0, 0).

22 Show that the schizophrenic function

$$f(x, y)=\begin{cases}\dfrac{x^2+y^2}{xy^2-2} & \text{for } xy^2\neq 2\\ 0 & \text{for } xy^2=2\end{cases}$$

is not differentiable at (2, 1).

23 Show that the schizophrenic function

$$f(x, y)=\begin{cases}\dfrac{x^2y+xy^2}{x^2+y^2} & \text{for } (x, y)\neq(0, 0)\\ 0 & \text{for } (x, y)=(0, 0)\end{cases}$$

is not differentiable at (0, 0) even though it is continuous there.

24 Prove Theorem 3.10 for $n=3$.

25 Show that every linear function is differentiable. Find its total derivative.

26 Prove that every affine function is differentiable. What is the total derivative?

27 At which points $\boldsymbol{x}_0$ is the length function $f(\boldsymbol{x})=|\boldsymbol{x}|$ differentiable?

28 Let $f:\boldsymbol{R}^n\to\boldsymbol{R}$. By considering Definition 1.6, show that we cannot simply define $f'(\boldsymbol{x}_0)$ to be

$$\lim_{|\boldsymbol{h}|\to 0}\frac{f(\boldsymbol{x}_0+\boldsymbol{h})-f(\boldsymbol{x}_0)}{|\boldsymbol{h}|}$$

in imitation of the usual elementary calculus definition of the derivative.

4 DIRECTIONAL DERIVATIVES

In Section 1 we defined the partial derivatives of $f:\boldsymbol{R}^n\to\boldsymbol{R}$ with respect to each separate variable. They measure the rate of change of f in the direction of each standard basis vector $\boldsymbol{e}_i$. For example, recall that if $f:\boldsymbol{R}^2 \to \boldsymbol{R}$, then $\dfrac{\partial f}{\partial x}(\boldsymbol{x}_0)$ is the rate of change of f in the x-direction, i.e., in the direction of $\boldsymbol{e}_1=(1, 0)$. Similarly, $\dfrac{\partial f}{\partial y}(\boldsymbol{x}_0)$ is the rate of change of f in the direction of $\boldsymbol{e}_2=(0, 1)$. These are simply the slopes of the curves $\boldsymbol{x}_2=(t, y_0, f(t, y_0))$ and $\boldsymbol{x}_1=(x_0, t, f(x_0, t))$ at $\boldsymbol{x}_0=(x_0, y_0)$. (Look back at page 148.)

A natural question to raise is whether we can measure the rate of change of f in any *arbitrary* direction $\boldsymbol{u}=(u_1, u_2)$ in $\boldsymbol{R}^2$. Here $\boldsymbol{u}$ is a unit vector. This rate of change would be the slope of the curve

$$\boldsymbol{x}(t)=(x_0+tu_1, y_0+tu_2, f(x_0+tu_1, y_0+tu_2))$$

of intersection of the graph of $z=f(x, y)$ and the plane shown in Figure 4.1. That plane is perpendicular to the xy-plane and passes through the vector $\boldsymbol{u}$ drawn from the point (x_0, y_0).

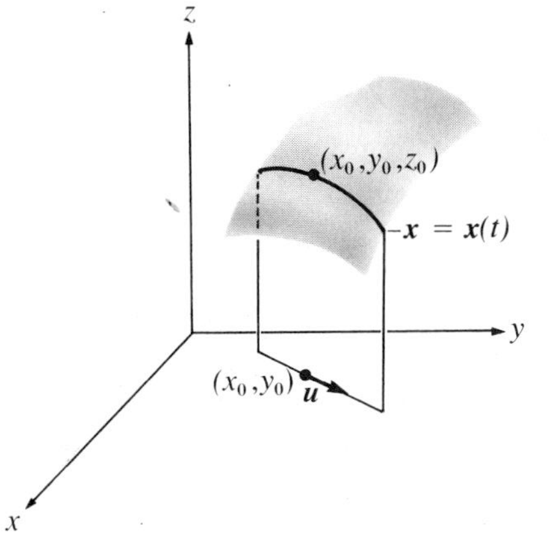

FIGURE 4.1

To answer this question we will use the idea of Definition 1.1 in a slightly different way. Recall that in Definition 1.1 we turned $f:\boldsymbol{R}^n \to \boldsymbol{R}$ into a function of just one real variable, so that we could define the partial derivative of f with respect to this variable as the ordinary elementary calculus derivative. We did this by freezing the values of $n-1$ of its variables and allowing the one remaining variable to vary. We can interpret this in terms of the unit vectors $\boldsymbol{e}_i$ as follows. Given $\boldsymbol{x}_0 \in \boldsymbol{R}^n$, we considered $f(\boldsymbol{x}_0+t\boldsymbol{e}_i)$ for t a real variable. As t varies, the i-th coordinate variable $x_i=x_{0i}+t$ is the only variable which actually changes. In these terms, we can rewrite the partial derivative $\dfrac{\partial f}{\partial x_i}(\boldsymbol{x}_0)$ as

$$\begin{aligned}\frac{\partial f}{\partial x_i}(\boldsymbol{x}_0) &= \lim_{x_i \to x_{0i}} \frac{f(x_{0i}, \ldots, x_{0,i-1}, x_i, x_{0,i+1}, \ldots, x_{0n})-f(\boldsymbol{x}_0)}{x_i-x_{0i}} \\ &= \lim_{t\to 0} \frac{f(\boldsymbol{x}_0+t\boldsymbol{e}_i)-f(\boldsymbol{x}_0)}{t}.\end{aligned}$$

This gives us the form in which to define the partial derivative of f in the direction of an arbitrary unit vector $\boldsymbol{u}$.

4.1 **DEFINITION.** Let $f:\boldsymbol{R}^n \to \boldsymbol{R}$ and $\boldsymbol{x}_0 \in \boldsymbol{R}^n$. Let $\boldsymbol{u} \in \boldsymbol{R}^n$ be a unit vector. Then the **partial derivative of f in the direction $\boldsymbol{u}$ at $\boldsymbol{x}_0$** or the **directional derivative of f in**

the direction u at x_0 is

(1) $$\frac{\partial f}{\partial \boldsymbol{u}}(\boldsymbol{x}_0)=\lim_{t\to 0}\frac{f(\boldsymbol{x}_0+t\boldsymbol{u})-f(\boldsymbol{x}_0)}{t}$$

provided this limit exists.

We have seen that this coincides with the partial derivative relative to x_i if $\boldsymbol{u}$ is the basis vector $\boldsymbol{e}_i$. So we have still another notation for the partial derivative $\frac{\partial f}{\partial x_i}(\boldsymbol{x}_0)$ if we would want to use it: $\frac{\partial f}{\partial \boldsymbol{e}_i}(\boldsymbol{x}_0)$. In other texts you will find various notations employed for the partial derivative of f in the direction of $\boldsymbol{u}$ at $\boldsymbol{x}_0$, among the most popular of which is $D_{\boldsymbol{u}}f(\boldsymbol{x}_0)$.

Whereas we could easily compute the partial derivatives of f relative to each variable x_i , we are faced with an immediate problem of coming up with a means of computing $\frac{\partial f}{\partial \boldsymbol{u}}(\boldsymbol{x}_0)$ without recourse to (1), which our instincts tell us is more work than should be needed to find a derivative. The next result provides us with a simple formula for $\frac{\partial f}{\partial \boldsymbol{u}}(\boldsymbol{x}_0)$ when f is differentiable at $\boldsymbol{x}_0$.

4.2 **THEOREM.** If f is differentiable at $\boldsymbol{x}_0$, then

$$\frac{\partial f}{\partial \boldsymbol{u}}(\boldsymbol{x}_0)=f'(\boldsymbol{x}_0)\cdot\boldsymbol{u}=\nabla f(\boldsymbol{x}_0)\cdot\boldsymbol{u}.$$

That is, the partial derivative of f in the direction of $\boldsymbol{u}$ at $\boldsymbol{x}_0$ is the coordinate of the total derivative of f at $\boldsymbol{x}_0$ in the direction of $\boldsymbol{u}$. (See Definition 1.2.16.)

Proof. We want to show that

$$\lim_{t\to 0}\frac{f(\boldsymbol{x}_0+t\boldsymbol{u})-f(\boldsymbol{x}_0)}{t}=f'(\boldsymbol{x}_0)\cdot\boldsymbol{u},$$

which is equivalent to showing that

$$\lim_{t\to 0}\left[\frac{f(\boldsymbol{x}_0+t\boldsymbol{u})-f(\boldsymbol{x}_0)}{t}-f'(\boldsymbol{x}_0)\cdot\boldsymbol{u}\right]=0.$$

This limit is

$$\lim_{t\to 0}\frac{f(\boldsymbol{x}_0+t\boldsymbol{u})-f(\boldsymbol{x}_0)-tf'(\boldsymbol{x}_0)\cdot\boldsymbol{u}}{t}=\lim_{t\to 0}\frac{f(\boldsymbol{x}_0+t\boldsymbol{u})-f(\boldsymbol{x}_0)-f'(\boldsymbol{x}_0)\cdot t\boldsymbol{u}}{t}.$$

To put this into the form of Definition 3.3, let

$$\boldsymbol{x}=\boldsymbol{x}_0+t\boldsymbol{u}.$$

Then $t\boldsymbol{u}=\boldsymbol{x}-\boldsymbol{x}_0$. Equating lengths and using $|\boldsymbol{u}|=1$, we obtain

$$|t|=|\boldsymbol{x}-\boldsymbol{x}_0|,\quad \text{i.e.,}\quad t=\pm|\boldsymbol{x}-\boldsymbol{x}_0|.$$

Thus $t \to 0$ is equivalent to $\boldsymbol{x} \to \boldsymbol{x}_0$. Hence

$$\lim_{t\to 0} \frac{f(\boldsymbol{x}_0+t\boldsymbol{u})-f(\boldsymbol{x}_0)-f'(\boldsymbol{x}_0)\cdot t\boldsymbol{u}}{t} = \lim_{\boldsymbol{x}\to\boldsymbol{x}_0} \frac{f(\boldsymbol{x})-f(\boldsymbol{x}_0)-f'(\boldsymbol{x}_0)\cdot(\boldsymbol{x}-\boldsymbol{x}_0)}{\pm|\boldsymbol{x}-\boldsymbol{x}_0|}$$

$$= \lim_{\boldsymbol{x}\to\boldsymbol{x}_0} \frac{\pm e(\boldsymbol{x})}{|\boldsymbol{x}-\boldsymbol{x}_0|} = 0$$

by Definition 3.3 since f is differentiable at $\boldsymbol{x}_0$. QED

Figure 4.2 illustrates the fact that $\dfrac{\partial f}{\partial \boldsymbol{u}}(\boldsymbol{x}_0)\boldsymbol{u}$ is the component of the total derivative $f'(\boldsymbol{x}_0) = \nabla f(\boldsymbol{x}_0)$ in the direction $\boldsymbol{u}$ for a differentiable function f. Thus the term "partial derivative of f in the direction $\boldsymbol{u}$" is quite an appropriate one for $\dfrac{\partial f}{\partial \boldsymbol{u}}(x_0)$.

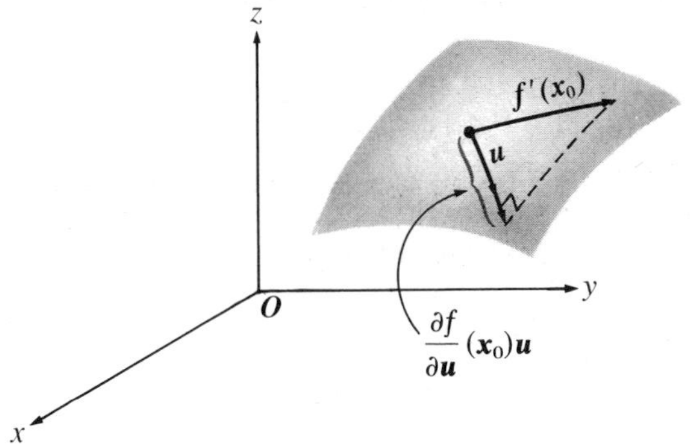

FIGURE 4.2

4.3 EXAMPLE. Find the partial derivative of f in the direction of $\boldsymbol{v} = (2, 1)$ at $\boldsymbol{x}_0 = (-1, 2)$ if $f(x, y) = x^2 + y^2 - x \cos \pi y - y \sin \pi x$.

Solution. First we compute the total derivative.

$$f'(x, y) = \nabla f(x, y)$$
$$= (2x - \cos \pi y - \pi y \cos \pi x,\ 2y + \pi x \sin \pi y - \sin \pi x).$$

At $\boldsymbol{x}_0$,

$$f'(\boldsymbol{x}_0) = (-2-1+2\pi, 4-0-0)$$
$$= (-3+2\pi, 4).$$

Since $\boldsymbol{v}$ is *not* a unit vector, we must find its direction: $\boldsymbol{u} = \dfrac{1}{\sqrt{5}}(2, 1) = \left(\dfrac{2}{\sqrt{5}}, \dfrac{1}{\sqrt{5}}\right)$. Then by Theorem 4.2,

$$\frac{\partial f}{\partial \boldsymbol{u}}(\boldsymbol{x}_0) = (-3+2\pi, 4)\cdot\left(\frac{2}{\sqrt{5}}, \frac{1}{\sqrt{5}}\right)$$

$$= \frac{1}{\sqrt{5}}(-6+4\pi+4)$$

$$= \frac{1}{\sqrt{5}}(4\pi - 2).$$

Theorem 4.2 says that for differentiable f

$$\frac{\partial f}{\partial \boldsymbol{u}}(\boldsymbol{x}_0) = \boldsymbol{f}'(\boldsymbol{x}_0) \cdot \boldsymbol{u} = \boldsymbol{\nabla} f(\boldsymbol{x}_0) \cdot \boldsymbol{u}.$$

We can use this formula to determine the maximum and minimum values of the directional derivative at $\boldsymbol{x}_0$ when $\boldsymbol{\nabla} f(\boldsymbol{x}_0)$ is not the zero vector. Definition 1.2.7 gives

(2) $$\begin{aligned}\frac{\partial f}{\partial \boldsymbol{u}}(\boldsymbol{x}_0) &= \boldsymbol{f}'(\boldsymbol{x}_0) \cdot \boldsymbol{u} \\ &= |\boldsymbol{f}'(\boldsymbol{x}_0)|\, |\boldsymbol{u}| \cos \theta \\ &= |\boldsymbol{f}'(\boldsymbol{x}_0)| \cos \theta\end{aligned}$$

where θ is the angle between $\boldsymbol{f}'(\boldsymbol{x}_0)$ and $\boldsymbol{u}$. The maximum value of $\cos \theta$ is 1, when $\theta = 0$. In this case, $\boldsymbol{u}$ is the direction of $\boldsymbol{f}'(\boldsymbol{x}_0)$. The minimum value of $\cos \theta$ is -1, when $\theta = \pi$. In this case $\boldsymbol{u}$ is the direction of $-\boldsymbol{f}'(\boldsymbol{x}_0)$. Hence we have the following result.

4.4 **THEOREM.** If f is differentiable at $\boldsymbol{x}_0$ and $\boldsymbol{\nabla} f(\boldsymbol{x}_0) \neq \boldsymbol{0}$, then the maximum rate of increase of f is in the direction of the gradient $\boldsymbol{\nabla} f(\boldsymbol{x}_0)$, and the maximum rate of decrease of f is in the direction of $-\boldsymbol{\nabla} f(\boldsymbol{x}_0)$. The maximum rate of increase is $|\boldsymbol{\nabla} f(\boldsymbol{x}_0)|$, and the maximum rate of decrease is $-|\boldsymbol{\nabla} f(\boldsymbol{x}_0)|$. See Figure 4.3.

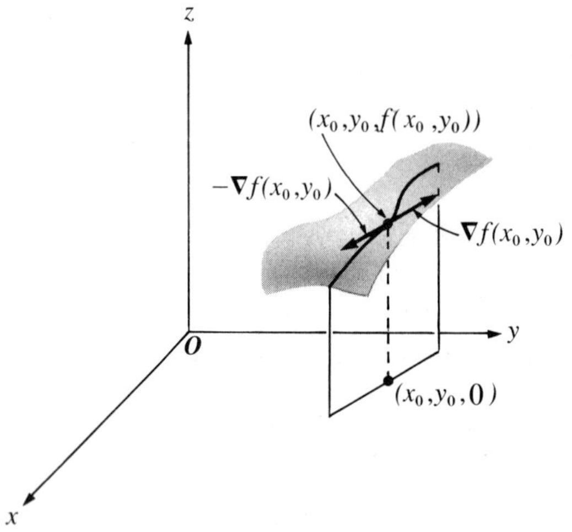

FIGURE 4.3

Proof. Our discussion above establishes everything except the maximum rate of increase and decrease. The maximum rate of increase occurs when $\theta = 0$, and so is, from (2),

$$|\boldsymbol{f}'(\boldsymbol{x}_0)| = |\boldsymbol{\nabla} f(\boldsymbol{x}_0)|$$

where $\boldsymbol{u}$ is the unit vector in the direction of $\boldsymbol{\nabla} f(\boldsymbol{x}_0)$. The maximum rate of decrease occurs when $\theta = \pi$ and so is, from (2),

$$-|\boldsymbol{f}'(\boldsymbol{x}_0)| = -|\boldsymbol{\nabla} f(\boldsymbol{x}_0)|$$

where $\boldsymbol{u}$ is the unit vector in the direction of $-\boldsymbol{\nabla} f(\boldsymbol{x}_0)$. QED

What happens if $\nabla f(\boldsymbol{x}_0)=\boldsymbol{0}$? In this case *every* $\dfrac{\partial f}{\partial \boldsymbol{u}}(\boldsymbol{x}_0)=0$ by Theorem 4.2, so there is no increase or decrease in f in *any* direction. We would expect such a point to be a candidate for f to assume an extreme value. We shall pursue this more in Section 9. (See Figure 4.4, which shows that such candidates need to be tested further.)

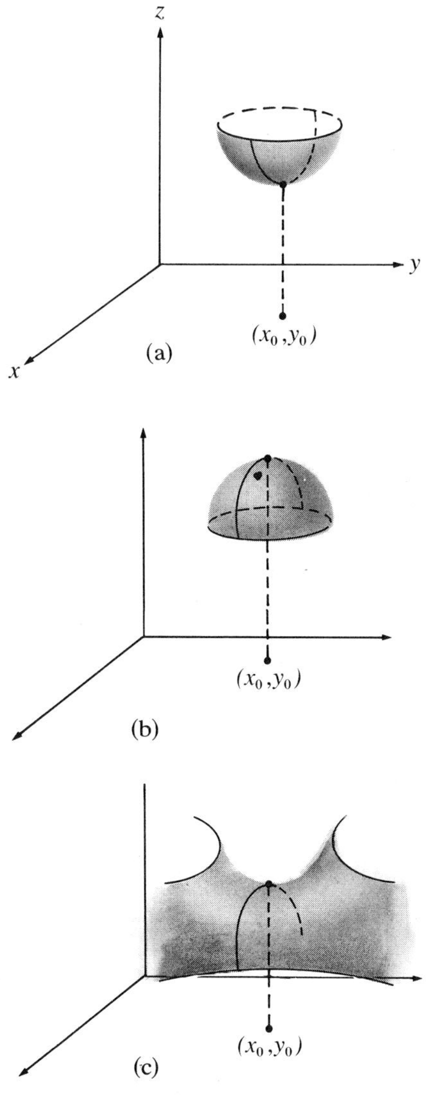

FIGURE 4.4

Theorem 4.5 is the basis for the *gradient method* of numerically maximizing a differentiable function $f: \boldsymbol{R}^n \to \boldsymbol{R}$. One chooses a point $\boldsymbol{x}_0$ and computes $\nabla f(\boldsymbol{x}_0)$. One then travels in the direction of the gradient for a short distance and reaches $\boldsymbol{x}_1$. (The distance traveled to reach $\boldsymbol{x}_1$ is calculated so that $f(\boldsymbol{x}_0+t\nabla f(\boldsymbol{x}_0))$ is maximized with respect to t.) Along the surface this is the path of steepest increase, so we expect $f(\boldsymbol{x}_1)>f(\boldsymbol{x}_0)$. Then one computes $\nabla f(\boldsymbol{x}_1)$ and travels in this direction to a point $\boldsymbol{x}_2$, where we expect $f(\boldsymbol{x}_2)>f(\boldsymbol{x}_1)$. In this way a sequence of points is generated which can be shown to converge to a point $\boldsymbol{x}_m$ where f has a relative maximum value. (In actual practice a refinement of this procedure is used, but our description conveys the underlying idea.) Similarly, one can travel in the direction of $-\nabla f(\boldsymbol{x}_0)$ to a point $\boldsymbol{y}_1$, in the direction of $-\nabla f(\boldsymbol{y}_1)$ to a point $\boldsymbol{y}_2$, etc. in search of a relative minimum for f. A practical drawback to this method is that it may take quite some time and may arrive at the point yielding the relative extremum in a roundabout fashion. It compares to hiking to the top of a mountain by following the course of a stream. Locally (i.e., at each point) the water has taken the path of swiftest descent, but in the process may have meandered around quite a bit. So while the gradient method is of use numerically, the global technique of Section 9, which avoids this problem of meandering, is preferable when it can be applied.

EXERCISES 4.4

In Exercises 1 to 8, find the directional derivative of f at the point x_0 in the given direction.

1 $f(x, y) = x^2 + y^2$, $\boldsymbol{x}_0 = (1, 2)$, $\boldsymbol{u} = (1/\sqrt{2}, 1/\sqrt{2})$

2 $f(x, y) = x^2 - y^2$, $\boldsymbol{x}_0 = (2, 1)$, $\boldsymbol{u} = (1/\sqrt{2}, 1/\sqrt{2})$

3 $f(x, y) = e^x \cos y$, $\boldsymbol{x}_0 = (0, \pi/2)$, $\boldsymbol{u} = (4/5, 3/5)$

4 $f(x, y) = e^y \sin x$, $\boldsymbol{x}_0 = (\pi, 1)$, $\boldsymbol{u} = (3/5, 4/5)$

5 $f(x, y, z) = x^2 + y^2 - z^2$, $\boldsymbol{x}_0 = (1, 2, 2)$, $\boldsymbol{u} = (1/3, 2/3, 2/3)$

6 $f(x, y, z) = 2x^2 - 4y^2 + z^2$, $\boldsymbol{x}_0 = (2, 1, 3)$, $\boldsymbol{u} = (2/3, 2/3, 1/3)$.

7 $f(x, y, z) = xyz$, $\boldsymbol{x}_0 = (1, -1, 2)$, in the direction of $\boldsymbol{v} = (1, 2, 2)$.

8 $f(x, y, z) = xye^{xyz} + yze^x$, $\boldsymbol{x}_0 = (1, 1, 1)$, $\boldsymbol{u}$ in the direction of $\boldsymbol{v} = (2, 2, 1)$.

In Exercises 9 to 12, find the direction and the rates of maximum increase and maximum decrease for f at the given point.

9 $f(x, y) = x^2 + y^2$, $\boldsymbol{x}_0 = (1, 2)$

10 $f(x, y) = x^2 - 2y^2$, $\boldsymbol{x}_0 = (3, 1)$

11 $f(x, y, z) = x^2 + y^2 + z^2$, $\boldsymbol{x}_0 = (1, 2, 1)$

12 $f(x, y, z) = 2x^2 - y^2 - 3z^2$, $\boldsymbol{x}_0 = (2, 1, 0)$

13 The *Newtonian potential* is

$$f(x, y, z) = \frac{1}{|\boldsymbol{x}|} = \frac{1}{\sqrt{x^2 + y^2 + z^2}}, \quad \text{for} \quad \boldsymbol{x} \neq \boldsymbol{0}.$$

Find the direction of maximum increase in the Newtonian potential function at the point (1, 2, 2).

14 The *logarithmic potential* is

$$g(x, y, z) = -\tfrac{1}{2}\ln(x^2 + y^2 + z^2) = \ln f(\boldsymbol{x}) \quad \text{for} \quad \boldsymbol{x} \neq \boldsymbol{0},$$

where $f(\boldsymbol{x})$ is the Newtonian potential. Find the direction of maximum increase in the logarithmic potential at the point (1, 2, 2).

15 Suppose that the temperature at (x, y, z) is given by the Newtonian potential function in Exercise 13. Assuming that heat always flows in the direction of maximum decrease in temperature, describe the resulting heat flow across a sphere of radius 3.

16 Repeat Exercise 15, assuming that the temperature is given by the logarithmic potential function of Exercise 14.

17 A skier starts his run at the point (2, 1, 8) of a mountain whose equation is $z = 14 - x^2 - 2y^2$, where x, y, and z are measured in thousands of feet. If he wants to take the path of steepest descent, then along what vector in the xy-plane does he travel?

18 In Exercise 17, suppose the skier starts at $(\frac{9}{4}, \frac{5}{4}, \frac{93}{16})$. If he wants to follow the path of steepest descent, what path should he follow? (Note that if the skier stayed on his course of steepest descent in Exercise 17, then he *would* reach the point $(\frac{9}{4}, \frac{5}{4}, \frac{93}{16})$.)

5 THE CHAIN RULE

You doubtless recall that one of the most fundamental tools in elementary differential calculus is the *Chain Rule.* Let's briefly review what this says. Suppose that the real variables x, u, and z are related by

$$z = f(u) \quad \text{where} \quad u = g(x).$$

Here we assume that the range of the function g is included in the domain of f. We then have

$$z = f(g(x)) = f \circ g(x)$$

for x in the domain of g. Let x_0 be such a point, and let $u_0 = g(x_0)$. Then

$$z_0 = f(u_0) = f \circ g(x_0).$$

Suppose that f and g are both differentiable functions. Then the Chain Rule says that

$$\left.\frac{dz}{dx}\right|_{x=x_0} = \left(\left.\frac{dz}{du}\right|_{u=u_0}\right) \cdot \left(\left.\frac{du}{dx}\right|_{x=x_0}\right),$$

i.e.,

(1) $$\frac{d(f \circ g)}{dx}(x_0) = \frac{df}{du}(g(x_0)) \cdot \frac{dg}{dx}(x_0).$$

In this section we want to derive a Chain Rule analogous to (1) for real valued functions of n real variables. It turns out that this is based on the case $z = f \circ \boldsymbol{g}(t)$ where $f : \boldsymbol{R}^n \to \boldsymbol{R}$, and $\boldsymbol{g} : \boldsymbol{R} \to \boldsymbol{R}^n$ is the type of function studied in Chapter 2. The basic tool needed to obtain the Chain Rule in this case is the following analogue for functions of several variables of the Mean Value Theorem of elementary calculus.

5.1 MEAN VALUE THEOREM. Let $f : \boldsymbol{R}^n \to \boldsymbol{R}$ be a differentiable function defined on an open set U containing the line segment joining $\boldsymbol{a}$ to $\boldsymbol{b}$ in $\boldsymbol{R}^n$. Then there is a point

$$\boldsymbol{c} = (1 - t_1)\boldsymbol{a} + t_1\boldsymbol{b} \qquad \text{where} \quad 0 < t_1 < 1$$

on this segment and strictly between $\boldsymbol{a}$ and $\boldsymbol{b}$ such that

$$f(\boldsymbol{b}) - f(\boldsymbol{a}) = \boldsymbol{f}'(\boldsymbol{c}) \cdot (\boldsymbol{b} - \boldsymbol{a}).$$

See Figure 5.1 (p. 180).

(Note that if $n = 1$, Theorem 5.1 becomes the Mean Value Theorem of elementary calculus since the segment joining a to b in that case is the closed interval $[a, b]$.)

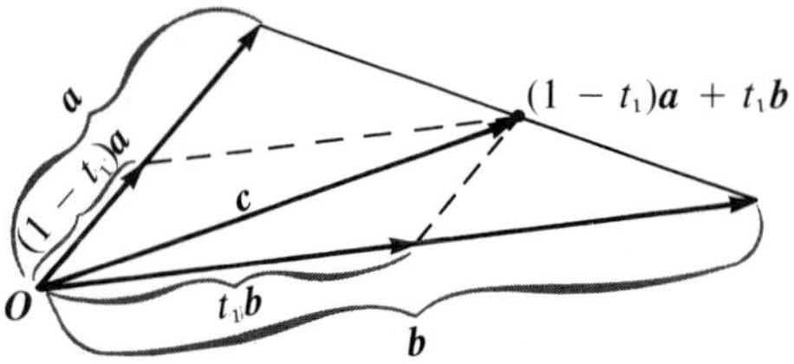

FIGURE 5.1

Proof. Recall from Proposition 1.3.8 that the segment joining $\boldsymbol{a}$ to $\boldsymbol{b}$ consists of the endpoints of vectors

$$(1-t)\boldsymbol{a}+t\boldsymbol{b}=\boldsymbol{a}+t(\boldsymbol{b}-\boldsymbol{a})$$

where $t\in[0,1]$. We will use the ordinary Mean Value Theorem to give the proof. For this we set

$$g(t)=f(\boldsymbol{a}+t(\boldsymbol{b}-\boldsymbol{a})), \qquad t\in[0,1].$$

If we can show that g is differentiable on $[0,1]$, then by the Mean Value Theorem of elementary calculus

$$g(1)-g(0)=g'(t_1)(1-0)=g'(t_1)$$

for some $t_1\in(0,1)$. This gives

(2) $$f(\boldsymbol{b})-f(\boldsymbol{a})=g'(t_1).$$

To complete the proof we need to show that $g'(t_1)$ exists for any t_1 and

$$g'(t_1)=\boldsymbol{f}'(\boldsymbol{c})\cdot(\boldsymbol{b}-\boldsymbol{a})$$

where

$$\boldsymbol{c}=(1-t_1)\boldsymbol{a}+t_1\boldsymbol{b}=\boldsymbol{a}+t_1(\boldsymbol{b}-\boldsymbol{a}).$$

Let

$$\boldsymbol{x}=\boldsymbol{a}+t(\boldsymbol{b}-\boldsymbol{a}) \quad \text{for} \quad t\in[0,1].$$

Observe that

$$\boldsymbol{x}-\boldsymbol{c}=(t-t_1)(\boldsymbol{b}-\boldsymbol{a}).$$

We now proceed to compute

$$g'(t_1)=\lim_{t\to t_1}\frac{g(t)-g(t_1)}{t-t_1}.$$

We have

$$\begin{aligned}\frac{g(t)-g(t_1)}{t-t_1} &= \frac{f(\boldsymbol{x})-f(\boldsymbol{c})}{t-t_1} \\ &= \frac{f'(\boldsymbol{c})\cdot(\boldsymbol{x}-\boldsymbol{c})+e(\boldsymbol{x})}{t-t_1} \\ &= \frac{f'(\boldsymbol{c})\cdot(t-t_1)(\boldsymbol{b}-\boldsymbol{a})+e(\boldsymbol{x})}{t-t_1} \\ &= f'(\boldsymbol{c})\cdot(\boldsymbol{b}-\boldsymbol{a})+\frac{e(\boldsymbol{x})}{t-t_1}.\end{aligned}$$

Now as $t \to t_1$, $\boldsymbol{x} \to \boldsymbol{c}$. Then by (2) of Definition 3.3

$$\frac{e(\boldsymbol{x})}{|\boldsymbol{x}-\boldsymbol{c}|} \to 0$$

since f is differentiable at $\boldsymbol{c}$. Thus

$$\frac{e(\boldsymbol{x})}{|t-t_1|\,|\boldsymbol{b}-\boldsymbol{a}|} \to 0,$$

and so

$$\frac{e(\boldsymbol{x})}{t-t_1} \to 0.$$

We therefore obtain

$$g'(t_1) = \lim_{t\to t_1} \frac{g(t)-g(t_1)}{t-t_1} = f'(\boldsymbol{c})\cdot(\boldsymbol{b}-\boldsymbol{a})$$

where $\boldsymbol{c} = (1-t_1)\boldsymbol{a}+t_1\boldsymbol{b}$ is on the line segment joining $\boldsymbol{a}$ to $\boldsymbol{b}$, and is strictly between $\boldsymbol{a}$ and $\boldsymbol{b}$ since t_1 is in the open interval (0, 1). QED

5.2 EXAMPLE. If $f: \boldsymbol{R}^2 \to \boldsymbol{R}$ is given by

$$f(x, y) = \frac{x^2}{4}+\frac{y^2}{9},$$

then find $\boldsymbol{c}$ on the line segment joining (0, 0) to (2, 3) such that

$$f'(\boldsymbol{c})\cdot(2, 3) = f(2, 3)-f(0, 0).$$

Solution. We have $f(2, 3) = 2$; $f(0, 0) = 0$; and $f'(x, y) = \left(\frac{x}{2}, \frac{2y}{9}\right)$. We want to find $\boldsymbol{c} = (c_1, c_2)$ such that

(*)
$$\left(\frac{c_1}{2}, \frac{2c_2}{9}\right)\cdot(2, 3) = 2,$$
$$c_1+\tfrac{2}{3}c_2 = 2.$$

Since (c_1, c_2) is to be on the line segment strictly between (0, 0) and (2, 3), we also have

$$\begin{aligned}(c_1, c_2) &= (0, 0) + t(2, 3), \qquad t \in (0, 1)\\ &= (2t, 3t).\end{aligned}$$

Thus

$$2t + \tfrac{2}{3}(3t) = 2, \qquad 4t = 2, \qquad t = \tfrac{1}{2}.$$

Hence $\mathbf{c} = (1, \frac{3}{2})$ works. (You can readily check this if you substitute in (*).)

As we suggested, our main use for Theorem 5.1 will lie in establishing the Chain Rule. The following special case is the key step in deriving the general Chain Rule.

5.3 CHAIN RULE. Suppose that $f: \boldsymbol{R}^n \to \boldsymbol{R}$ and $\mathbf{g}: \boldsymbol{R} \to \boldsymbol{R}^n$. Let $t_0 \in \boldsymbol{R}$, and let $\boldsymbol{x}_0 = \mathbf{g}(t_0)$ be a point such that f is differentiable and $\boldsymbol{f}' = \nabla f = \left(\frac{\partial f}{\partial x_1}, \frac{\partial f}{\partial x_2}, \ldots, \frac{\partial f}{\partial x_n}\right)$ is continuous on an open set U containing $\boldsymbol{x}_0$. Put $z = f \circ \mathbf{g}(t)$. If $\mathbf{g}$ is differentiable at t_0, then $f \circ \mathbf{g}: \boldsymbol{R} \to \boldsymbol{R}$ is differentiable at t_0 and

$$\begin{aligned}\frac{d(f \circ \mathbf{g})}{dt}(t_0) &= \frac{dz}{dt}(t_0) = \boldsymbol{f}'(\boldsymbol{x}_0) \cdot \mathbf{g}'(t_0)\\ &= \nabla f(\boldsymbol{x}_0) \cdot \frac{d\mathbf{g}}{dt}(t_0).\end{aligned} \tag{3}$$

Proof. By Theorem 5.1,

$$\begin{aligned}\frac{d(f \circ \mathbf{g})}{dt}(t_0) &= \lim_{t \to t_0} \frac{f(\mathbf{g}(t)) - f(\mathbf{g}(t_0))}{t - t_0}\\ &= \lim_{t \to t_0} \frac{\boldsymbol{f}'(\mathbf{c}(t)) \cdot (\mathbf{g}(t) - \mathbf{g}(t_0))}{t - t_0}\end{aligned}$$

for some $\mathbf{c}(t) = (1 - t_1)\mathbf{g}(t_0) + t_1\mathbf{g}(t)$ where $t_1 \in (0, 1)$, on the line segment joining $\mathbf{g}(t_0)$ to $\mathbf{g}(t)$. See Figure 5.2. Note that as $t \to t_0$, $\mathbf{g}(t) \to \mathbf{g}(t_0)$ since $\mathbf{g}$ is differentiable (hence continuous) at t_0. Thus, $\mathbf{c}(t) \to \mathbf{g}(t_0)$ as $t \to t_0$ since $\mathbf{c}(t)$ lies on the segment between $\mathbf{g}(t_0)$ and $\mathbf{g}(t)$. Since $\boldsymbol{f}'$ is continuous on this segment, as $t \to t_0$ we have

$$\boldsymbol{f}'(\mathbf{c}(t)) \to \boldsymbol{f}'(\mathbf{g}(t_0)) = \boldsymbol{f}'(\boldsymbol{x}_0).$$

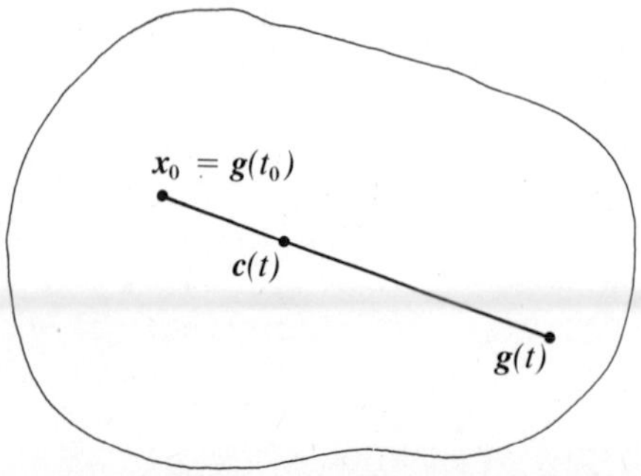

FIGURE 5.2

Hence

$$\frac{d(f \circ \mathbf{g})}{dt}(t_0) = f'(\mathbf{x}_0) \cdot \lim_{t \to t_0} \frac{\mathbf{g}(t) - \mathbf{g}(t_0)}{t - t_0}$$

$$= f'(\mathbf{x}_0) \cdot \frac{d\mathbf{g}}{dt}(t_0). \qquad \text{QED}$$

Note the perfect analogy between Equation (3) and (1), the Chain Rule of elementary calculus! To get $\frac{dz}{dt} = \frac{d(f \circ g)}{dt}$, *we take the total derivative of f with respect to the intermediate variable* $\mathbf{x}$, *take the derivative of* $\mathbf{g}$ *with respect to t, and then multiply* (by taking the dot product of these two vectors).

5.4 **EXAMPLE.** If $z = e^{xy}$ where $x = 2t^3 + 1$ and $y = t^2 + 3t$, then find $\frac{dz}{dt}$.

Solution. We have $z = e^{xy} = f(\mathbf{x})$, where $\mathbf{x}(t) = (2t^3 + 1, t^2 + 3t)$. Applying formula (3), we get

$$\frac{dz}{dt} = f'(\mathbf{x}) \cdot \dot{\mathbf{x}}(t) = (ye^{xy}, xe^{xy}) \cdot (6t^2, 2t + 3)$$

$$= 6t^2ye^{xy} + (2t + 3)xe^{xy}.$$

We can, if we like, obtain $\frac{dz}{dt}$ as a function of t alone by substituting $x = 2t^3 + 1$ and $y = t^2 + 3t$. This gives us

$$\frac{dz}{dt} = (6t^2(t^2 + 3t) + (2t + 3)(2t^3 + 1))e^{(2t^3+1)(t^2+3t)}.$$

In theory we could have obtained this by *first* substituting $x = 2t^3 + 1$ and $y = t^2 + 3t$ into the formula for z, and *then* differentiating with respect to t. *In practice* this is usually far more work than applying the Chain Rule.

To give the more general Chain Rule for functions $f: \mathbf{R}^n \to \mathbf{R}$ of the form $z = f(\mathbf{x})$ where $\mathbf{x} = \mathbf{x}(s, t, u, \ldots)$ we need to introduce the notion of a *vector* partial derivative.

5.5 **DEFINITION.** If $\mathbf{g}: \mathbf{R}^m \to \mathbf{R}^n$ is given by $\mathbf{x} = \mathbf{g}(\mathbf{t}) = (g_1(\mathbf{t}), g_2(\mathbf{t}), \ldots, g_n(\mathbf{t}))$ for $\mathbf{t} = (t_1, \ldots, t_m) \in \mathbf{R}^m$, then the **vector partial derivatives** of $\mathbf{g}$ are

$$\frac{\partial \mathbf{g}}{\partial t_i} = \left(\frac{\partial g_1}{\partial t_i}, \frac{\partial g_2}{\partial t_i}, \ldots, \frac{\partial g_n}{\partial t_i}\right) \quad \text{for} \quad i = 1, \ldots, m.$$

This is a simple process. To build the vector partial derivative of $\mathbf{g}$, we take the partial derivative of each coordinate function g_i, which is real valued. Then we put these partial derivatives together into a vector.

5.6 **EXAMPLE.** If

$$\mathbf{x}(s, t, u) = (x(s, t, u), y(s, t, u), z(s, t, u)) = (s^2 + t^2 + u^2, stu, s^2 + t^2u),$$

then find

$$\frac{\partial \boldsymbol{x}}{\partial s}, \quad \frac{\partial \boldsymbol{x}}{\partial t}, \quad \text{and} \quad \frac{\partial \boldsymbol{x}}{\partial u}.$$

Solution. From Definition 5.5,

$$\begin{aligned}
\frac{\partial \boldsymbol{x}}{\partial s} &= \left(\frac{\partial}{\partial s}(s^2+t^2+u^2), \frac{\partial}{\partial s}(stu), \frac{\partial}{\partial s}(s^2+t^2u)\right)\\
&= (2s, tu, 2s)\\
\frac{\partial \boldsymbol{x}}{\partial t} &= \left(\frac{\partial}{\partial t}(s^2+t^2+u^2), \frac{\partial}{\partial t}(stu), \frac{\partial}{\partial t}(s^2+t^2u)\right)\\
&= (2t, su, 2tu)\\
\frac{\partial \boldsymbol{x}}{\partial u} &= \left(\frac{\partial}{\partial u}(s^2+t^2+u^2), \frac{\partial}{\partial u}(stu), \frac{\partial}{\partial u}(s^2+t^2u)\right)\\
&= (2u, st, t^2).
\end{aligned}$$

In introducing this idea we are actually beginning the differential calculus of functions $\mathbf{g}: \boldsymbol{R}^m \to \boldsymbol{R}^n$, which is the subject of Chapter 6. At this point, we won't need to pursue the development of this any further, for we now are in a position to state the General Chain Rule for functions $f: \boldsymbol{R}^n \to \boldsymbol{R}$.

5.7 GENERAL CHAIN RULE FOR PARTIAL DERIVATIVES. Suppose that $f: \boldsymbol{R}^n \to \boldsymbol{R}$ is given by $w = f(\boldsymbol{x})$ and that $\mathbf{g}: \boldsymbol{R}^m \to \boldsymbol{R}^n$ is given by

$$\boldsymbol{x} = \mathbf{g}(\boldsymbol{t}) = \mathbf{g}(t_1, \ldots, t_m).$$

Suppose that f is differentiable at $\boldsymbol{x}_0$ with continuous partial derivatives. Suppose that all $\dfrac{\partial g_i}{\partial t_j}$ exist at $\boldsymbol{t}_0$ for $i = 1, 2, \ldots, n$ and $j = 1, 2, \ldots, m$. Let $\boldsymbol{x}_0 = \mathbf{g}(\boldsymbol{t}_0)$. Then

$$\frac{\partial (f \circ \mathbf{g})}{\partial t_j}(\boldsymbol{t}_0) = \boldsymbol{f}'(\boldsymbol{x}_0) \cdot \frac{\partial \mathbf{g}}{\partial t_j}(\boldsymbol{t}_0) = \boldsymbol{\nabla} f(\boldsymbol{x}_0) \cdot \frac{\partial \boldsymbol{x}}{\partial t_j}(\boldsymbol{t}_0). \tag{4}$$

In other words, *to get the partial derivative of $f \circ \mathbf{g}$ with respect to its j-th ultimate variable t_j, take the dot product of the total derivative of f with respect to the intermediate variable $\boldsymbol{x}$ and the partial derivative of $\boldsymbol{x}$ with respect to t_j.*

Proof.

$$\begin{aligned}
\frac{\partial (f \circ \mathbf{g})}{\partial t_j}(\boldsymbol{t}_0) &= \frac{\partial}{\partial t_j}[f(\mathbf{g}(t_1, t_2, \ldots, t_m))(\boldsymbol{t}_0)]\\
&= \boldsymbol{f}'(\boldsymbol{x}_0) \cdot \frac{\partial}{\partial t_j}[\mathbf{g}(\boldsymbol{t}_0)] \quad \text{by Chain Rule 5.3}\\
&= \boldsymbol{\nabla} f(\boldsymbol{x}_0) \cdot \frac{\partial \mathbf{g}}{\partial t_j}(\boldsymbol{t}_0).
\end{aligned}$$

QED

Note that if, in addition, each $\dfrac{\partial g_i}{\partial t_j}$ is continuous at $\boldsymbol{t}_0$, then each $\dfrac{\partial (f \circ \mathbf{g})}{\partial t_j}$ is continuous, since it is the sum of the products of continuous functions $\frac{\partial f}{\partial x_i}$ and $\frac{\partial g_i}{\partial t_j}$. For we have

$$f'(\boldsymbol{x}) = \left(\frac{\partial f}{\partial x_1}, \ldots, \frac{\partial f}{\partial x_i}, \ldots, \frac{\partial f}{\partial x_n}\right)$$

and

$$\frac{\partial \boldsymbol{g}}{\partial t_j} = \left(\frac{\partial x_1}{\partial t_j}, \ldots, \frac{\partial x_i}{\partial t_j}, \ldots, \frac{\partial x_n}{\partial t_j}\right)$$

so that

$$f'(\boldsymbol{x}) \cdot \frac{\partial \boldsymbol{g}}{\partial t_j} = \frac{\partial f}{\partial x_1}\frac{\partial x_1}{\partial t_j} + \cdots + \frac{\partial f}{\partial x_i}\frac{\partial x_i}{\partial t_j} + \cdots + \frac{\partial f}{\partial x_n}\frac{\partial x_n}{\partial t_j}.$$

Since each $\frac{\partial(f \circ \boldsymbol{g})}{\partial t_j}$ is continuous, we know that $f \circ \boldsymbol{g}$ is differentiable by Theorem 3.10, so its total derivative is just the result of putting together the $\frac{\partial(f \circ \boldsymbol{g})}{\partial t_j}$ for $j = 1, 2, \ldots, m$. Thus we have proved the following theorem.

5.8 GENERAL CHAIN RULE FOR TOTAL DERIVATIVES. If in Theorem 5.7 all the $\frac{\partial g_i}{\partial t_j}$ are continuous at $\boldsymbol{t}_0$, then $f \circ \boldsymbol{g}: \boldsymbol{R}^m \to \boldsymbol{R}$ defined by $w = f(\boldsymbol{g}(t))$ is differentiable at $\boldsymbol{t}_0$ and its total derivative is given by

(5)
$$f \circ \boldsymbol{g}'(\boldsymbol{t}_0) = \left(\frac{\partial(f \circ \boldsymbol{g})}{\partial t_1}(\boldsymbol{t}_0), \frac{\partial(f \circ \boldsymbol{g})}{\partial t_2}(\boldsymbol{t}_0), \ldots, \frac{\partial(f \circ \boldsymbol{g})}{\partial t_m}(\boldsymbol{t}_0)\right)$$
$$= \left(\nabla f(\boldsymbol{x}_0) \cdot \frac{\partial \boldsymbol{g}}{\partial t_1}(\boldsymbol{t}_0), \nabla f(\boldsymbol{x}_0) \cdot \frac{\partial \boldsymbol{g}}{\partial t_2}(\boldsymbol{t}_0), \ldots, \nabla f(\boldsymbol{x}_0) \cdot \frac{\partial \boldsymbol{g}}{\partial t_m}(\boldsymbol{t}_0)\right).$$

In Chapter 6, we will learn how to interpret (5) as the product of two total derivatives, so that it will be a perfect analogue of Theorem 5.3 and the Chain Rule of elementary calculus. For now, we give an example illustrating its use.

5.9 EXAMPLE. Given $w = f(\boldsymbol{x}) = x^2 + y^2 + z^2$, where $\boldsymbol{x} = (x, y, z) = (s^2 + t^2 + u^2, stu, s^2 + t^2u)$, find the total derivative of $w = f(\boldsymbol{x}(s, t, u))$.

Solution. We use Theorem 5.8 and Example 5.6, where we already computed $\frac{\partial \boldsymbol{x}}{\partial s}, \frac{\partial \boldsymbol{x}}{\partial t}, \frac{\partial \boldsymbol{x}}{\partial u}$. By Theorem 5.8, $\boldsymbol{w}'(\boldsymbol{t}) = \left(\frac{\partial w}{\partial s}, \frac{\partial w}{\partial t}, \frac{\partial w}{\partial u}\right)$ where

$$\frac{\partial w}{\partial s} = f'(\boldsymbol{x}) \cdot \frac{\partial \boldsymbol{x}}{\partial s} = (2x, 2y, 2z) \cdot (2s, tu, 2s)$$
$$= 4xs + 2ytu + 4zs.$$

$$\frac{\partial w}{\partial t} = f'(\boldsymbol{x}) \cdot \frac{\partial \boldsymbol{x}}{\partial t} = (2x, 2y, 2z) \cdot (2t, su, 2tu)$$
$$= 4xt + 2ysu + 4ztu.$$

$$\frac{\partial w}{\partial u} = f'(\boldsymbol{x}) \cdot \frac{\partial \boldsymbol{x}}{\partial u} = (2x, 2y, 2z) \cdot (2u, st, t^2)$$
$$= 4xu + 2yst + 2zt^2.$$

Thus

$$\mathbf{w}'(t) = (4xs + 2ytu + 4zs, 4xt + 2ysu + 4ztu, 4xu + 2yst + 2zt^2)$$

where, if we should want to, we could express this entirely in terms of the variables s, t, and u of $\boldsymbol{t}$ by substituting $x = s^2 + t^2 + u^2$, $y = stu$, and $z = s^2 + t^2u$.

We can use the chain rule to get a very simple means of finding the tangent plane to a *level surface* $f(x, y, z) = c$, where we assume that f has continuous partial derivatives. Suppose that $\boldsymbol{x} = \boldsymbol{x}(t) = (x(t), y(t), z(t))$ is a curve on such a level surface and passes through $\boldsymbol{x}_0 = \boldsymbol{x}(t_0)$. Then we have $f(\boldsymbol{x}(t)) = c$. Let's differentiate this with respect to t at t_0 and use the chain rule of Theorem 5.3. We get $\boldsymbol{f}'(\boldsymbol{x}(t_0)) \cdot \dot{\boldsymbol{x}}(t_0) = 0$, which says that the total derivative $\boldsymbol{f}'(\boldsymbol{x}(t_0)) = \boldsymbol{\nabla} f(\boldsymbol{x}(t_0))$ either is $\mathbf{0}$ or is perpendicular to the tangent vector of *any* curve in the surface which passes through $\boldsymbol{x}_0$. See Figure 5.3. So if $\boldsymbol{\nabla} f(\boldsymbol{x}_0) \neq \mathbf{0}$, then it is perpendicular

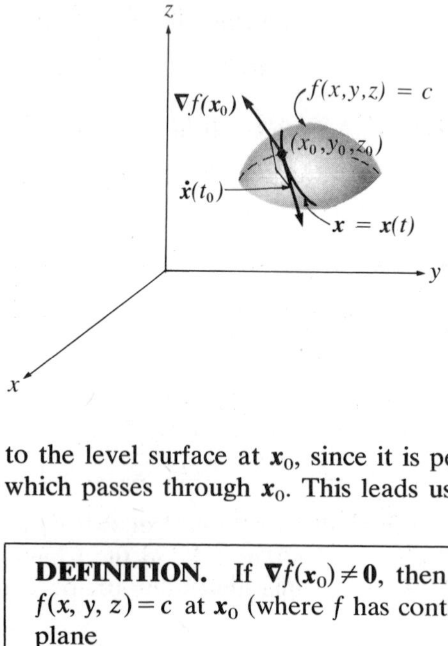

FIGURE 5.3

to the level surface at $\boldsymbol{x}_0$, since it is perpendicular to *every* curve on the surface which passes through $\boldsymbol{x}_0$. This leads us to the following definition.

5.10 **DEFINITION.** If $\boldsymbol{\nabla} f(\boldsymbol{x}_0) \neq \mathbf{0}$, then the **tangent plane to the level surface** $f(x, y, z) = c$ at $\boldsymbol{x}_0$ (where f has continuous partial derivatives near $\boldsymbol{x}_0$) is the plane

$$\boldsymbol{\nabla} f(\boldsymbol{x}_0) \cdot (\boldsymbol{x} - \boldsymbol{x}_0) = 0.$$

(Note that $\boldsymbol{\nabla} f(\boldsymbol{x}_0)$ is a normal vector to the level surface $f(\boldsymbol{x}) = c$.)

5.11 **EXAMPLE.** Find the equation of the tangent plane to the level surface $xyz = 2$ at the point $(1, 1, 2)$.

Solution. Here $f(x, y, z) = xyz$. We find $\boldsymbol{\nabla} f(\boldsymbol{x}) = (yz, xz, xy)$, so $\boldsymbol{\nabla} f(1, 1, 2) = (2, 2, 1)$. The tangent plane then has the equation

$$\begin{aligned} (2, 2, 1) \cdot (\boldsymbol{x} - (1, 1, 2)) &= 0 \\ (2, 2, 1) \cdot (x - 1, y - 1, z - 2) &= 0, \\ 2(x - 1) + 2(y - 1) + z - 2 &= 0 \\ 2x + 2y + z &= 6. \end{aligned}$$

EXERCISES 4.5

In Exercises 1 to 6, find a general formula for $\frac{dw}{dt}$ by using the Chain Rule 5.3. Evaluate at the point given.

1 $w=f(x, y)=xy^2$ where $\boldsymbol{x}=(x, y)=(\cos t, \sin t)$, $\boldsymbol{x}_0=(0, 1)$

2 $w=\sqrt{x^2+y^2}$ where $\boldsymbol{x}=(t, t^2)$; $\boldsymbol{x}_0=(-2, 4)$

3 $w=\text{Tan}^{-1}(x^2+y^2+z^2)$ where $\boldsymbol{x}=(t, t^2, t^3)$; $\boldsymbol{x}_0=(1, 1, 1)$

4 $w=e^{x2-y2-z2}$ where $\boldsymbol{x}=(t, 2t, t^2)$; $\boldsymbol{x}_0=(1, 2, 1)$

5 $w=\ln\sqrt{x^2+y^2+z^2}$ where $\boldsymbol{x}=(\cos t, \sin t, t)$; $\boldsymbol{x}_0=(0, 1, \pi/2)$

6 $w=\sqrt{x^2+y^2+z^2}$ where $\boldsymbol{x}=(t, t^2, t^3)$; $\boldsymbol{x}_0=(-1, 1, -1)$

The next two exercises illustrate Theorem 5.1.

7 Suppose that $f:\boldsymbol{R}^2\to\boldsymbol{R}$ is given by $f(x, y)=x^2+y^2$. Find a point $\mathbf{c}$ on the segment joining (0, 0) to (2, 1) such that $\boldsymbol{f}'(\mathbf{c})\cdot(2, 1)=5$.

8 Suppose that $f:\boldsymbol{R}^3\to\boldsymbol{R}$ is given by $f(x, y, z)=x^2+y^2+z^2$. Find a point $\mathbf{c}$ on the segment joining (2, 1, 1) to (3, 2, −1) such that $f(3, 2, -1)-f(2, 1, 1)=\boldsymbol{f}'(\mathbf{c})\cdot(1, 1, -2)$.

In Exercises 9 to 14, find $\frac{\partial w}{\partial s}$ and $\frac{\partial w}{\partial t}$ using Theorem 5.7.

9 $w=x^2+y^2$, $x=s\cos t$, $y=s\sin t$

10 $w=x^2y^2$, $x=s^2+st+2$, $y=3t^2+2st-1$

11 $w=x^2yz^2$, $x=ts$, $y=\frac{t}{s}$, $z=\frac{s}{t}$

12 $w=x^2e^{yz}+z^2e^{xy}$, $x=s+t$, $y=s^2-t^2$, $z=s^2+t^2$

13 $w=\ln\sqrt{x^2+y^2+z^2}$, $x=\cos st$, $y=\sin st$, $z=st$

14 $w=x^2+y^2+z^2$, $x=s^2+st^2$, $y=2st$, $z=s^3t^2$

In Exercises 15 to 22, find the total derivative of w in general and at the indicated point using Theorem 5.8.

15 $w=x^2-y^2$, $(x, y)=(s^2t+u, s^2u+t)$; at $(s_0, t_0, u_0)=(1, 2, -2)$

16 $w=\text{Tan}^{-1}\frac{y}{x}$, $(x, y)=(s^2+t^2+u^2, stu)$; at $(s_0, t_0, u_0)=(1, 1, 1)$

17 $w=xy$, $(x, y)=(r\cos\theta, r\sin\theta)$; at $(r_0, \theta_0)=(1, \pi/2)$

18 $w=x^2y+xy$, $(x, y)=(e^{st}, s^2+t^2)$; at $(s_0, t_0)=(2, 1)$

19 $w=xyz$, $(x, y, z)=(te^s, se^t, st)$; at $(s_0, t_0)=(1, 1)$

20 $w=x^2+y^2-z^2$, $(x, y, z)=(r\cos\theta, r\sin\theta, r\theta)$; at $(s_0, \theta_0)=(1, \pi)$

21 $w=\ln\sqrt{x^2+y^2+z^2}$, $(x, y, z)=(s+t+u, stu, se^{tu})$; at $(s_0, t_0, u_0)=(1, 2, 0)$

22 $w=x^2e^{yz}$, $(x, y, z)=(st+u^2, s^2ut, s+tu)$; at $(s_0, t_0, u_0)=(-1, 1, 2)$

In Exercises 23 to 27, find the tangent plane to the given level surface at the given point.

23 $x^2y-2yz+4=0$ at (−2, 2, 3)

24 $xy^2+yz^2+z^3+x^3=64$ at (0, 0, 4)

25 $x^2-3y^2+z^2=5$ at (2, 1, −2)

26 $x^2+y^2+z^2=9$ at (−2, 1, 2)

27 $x^2-y^2-z^2=1$ at (3, 2, −2)

28 A plane is flying due east with a speed of 400 miles per hour and is climbing at the rate of 528 feet per minute. It is being tracked by an airport tower. Find how fast the plane is approaching the tower when it is 3 miles above the ground over a point exactly 4 miles due west of the tower.

29 A particle of mass m moves on the surface $z=f(x, y)$ along the curve $\boldsymbol{x}=(x(t), y(t), f(x(t), y(t)))$. Its *kinetic energy* k at time t is $\frac{1}{2}m\,|\boldsymbol{v}(t)|^2$, where $\boldsymbol{v}(t)=\dot{\boldsymbol{x}}(t)$. Show that this is given by

$$k=\tfrac{1}{2}m\left[(1+f_x^2)\left(\frac{dx}{dt}\right)^2+2f_xf_y\frac{dx}{dt}\frac{dy}{dt}+(1+f_y^2)\left(\frac{dy}{dt}\right)^2\right].$$

30 A quantity of gas is governed by the *ideal gas law* $PV=nRT$ (n, R constant). (Here recall that P is the pressure, V is the volume, and T is the temperature.) The constant factor nR is determined to be 1 liter-atmosphere/°K. At a certain time $T=300$°K and is being raised by 5°K per minute. The pressure at this time is 2 atmospheres and is decreasing at the rate of 0.01 atmospheres/second. Find $\dfrac{dV}{dt}$ at this instant of time.

31 The temperature at a point (x, y, z) is given by $T=\sqrt{x^2+y^2+z^2}$. A particle traces the path $\boldsymbol{x}(t)=(\cos t, \sin t, t)$, $0\le t\le 2\pi$. At what point of the path is the lowest temperature experienced?

32 If $z=f(x, y)$ where $x=r\cos\theta$ and $y=r\sin\theta$, then

(a) find $\dfrac{\partial z}{\partial r}$ and $\dfrac{\partial z}{\partial \theta}$, and

(b) show that $\left(\dfrac{\partial z}{\partial x}\right)^2+\left(\dfrac{\partial z}{\partial y}\right)^2=\left(\dfrac{\partial z}{\partial r}\right)^2+\dfrac{1}{r^2}\left(\dfrac{\partial z}{\partial \theta}\right)^2.$

33 A region $U\subseteq \boldsymbol{R}^m$ is called *convex* if whenever $\boldsymbol{x}_1$ and $\boldsymbol{x}_2$ are in U, then the entire line segment joining $\boldsymbol{x}_1$ and $\boldsymbol{x}_2$ is contained in U. Suppose that $f:\boldsymbol{R}^n\to\boldsymbol{R}$ is differentiable on a convex open set U and $f'(\boldsymbol{x})\equiv\boldsymbol{0}$ on U. Then show that f must be constant on U.

34 Use Exercise 33 to show that two functions which have the same total derivative on an open convex set $U\subseteq\boldsymbol{R}^n$ differ by a constant on U.

6 IMPLICIT FUNCTIONS

Implicit differentiation is one of the tools of elementary calculus that you have learned as an application of the chain rule. You, like many elementary calculus students, may have found it a bit mystifying both in its significance and in its application. If you did, then you will probably find this section of special interest. The reason is that with the perspective you have by now acquired on differentiation, it is much easier to grasp the significance of implicit differentiation than it might have been when you first encountered it. And, happily enough, the technique of implicit differentiation becomes entirely straightforward once you know how to compute partial derivatives.

First, let's briefly remind ourselves of what implicit differentiation *is*. Given a functional equation $F(x, y)=0$, you *assume* that it defines y as a differentiable function of x near some point x_0. What does this mean? It means that there is a differentiable function $f:\boldsymbol{R}\to\boldsymbol{R}$ such that $F(x, f(x))=0$ for all x close to x_0. You want to calculate $f'(x_0)$, and the technique of implicit differentiation does this for you. Now at this point you may be trying to remember *why* you want to calculate $f'(x_0)$. The reason is simple: If you know $f'(x_0)$, then you can *approximate* f near x_0 by the tangent line to the graph of $y=f(x)$! And this might well be the closest you could ever come to being able to express y as an explicit function of x. For example, suppose that you are given the functional equation

$$y^7-11x^5y^6-x^{15}+11=0.$$

How in the world can you express y as a function of x? Yet if you are interested in the behavior of this function near the point (1, 1), then you can study that via the tangent approximation (cf. Section 3) $f(x) \approx 1 + f'(1)(x-1)$. (See Example 6.1 below for the calculation of $f'(1)$.)

Now that we recall a *purpose* in finding $\frac{dy}{dx}$ for an implicitly given functional relationship $F(x, y) = 0$, let's see if we can come up with the promised simple technique for finding $\frac{dy}{dx}$.

Given a functional equation $F(x, y) = 0$, let's assume for the moment that it defines y as a function of x: $y = f(x)$ for x in an open interval containing x_0. Assume further that F_x and F_y are continuous. Then we can write $F(x, f(x)) = 0$, and can differentiate this functional equation using Theorem 5.3 with $n = 2$.

$$\frac{d}{dx} F(x, f(x)) = \frac{d}{dx}(0) = 0.$$

To put things into the form of Theorem 5.3, we let $F(x, f(x))$ play the role of $f(x, y)$, and let $\mathbf{g}(t) = (x, y) = (t, f(t))$. Then by Theorem 5.3,

$$\begin{aligned} \frac{d}{dx}(F(x, f(x)) &= \frac{d}{dt} F(x(t), y(t)) = \frac{d}{dt}[F \circ \mathbf{g}(t)] \\ &= \nabla F(x, y) \cdot \left(\frac{dx}{dt}, \frac{dy}{dt}\right) \\ &= \left(\frac{\partial F}{\partial x}, \frac{\partial F}{\partial y}\right) \cdot \left(1, \frac{dy}{dt}\right) \\ &= \frac{\partial F}{\partial x} + \frac{\partial F}{\partial y}\frac{dy}{dt} = 0. \end{aligned}$$

Hence,

$$\frac{\partial F}{\partial y}\frac{dy}{dt} = -\frac{\partial F}{\partial x}.$$

So, if $\frac{\partial F}{\partial y} \neq 0$, we have

$$\frac{dy}{dt} = -\frac{\frac{\partial F}{\partial x}}{\frac{\partial F}{\partial y}}.$$

But $x = t$, so we have, if $\frac{\partial F}{\partial y} \neq 0$,

$$(1) \qquad \frac{dy}{dx} = -\frac{\frac{\partial F}{\partial x}}{\frac{\partial F}{\partial y}} = -\frac{F_x}{F_y}.$$

Equation (1) gives *precisely* the same result as the elementary calculus technique, and, as promised, is certainly a simple formula.

6.1 **EXAMPLE.** If $y^7 - 11y^6x^5 - x^{15} + 11 = 0$ defines y as a differentiable function of x, then find $\frac{dy}{dx}$. Evaluate $\left.\frac{dy}{dx}\right|_{x=1}$. When $x = 1.002$, what is an approximate value of y?

Solution. If we put $F(x, y) = y^7 - 11x^5y^6 - x^{15} + 11$, then

$$\frac{\partial F}{\partial x} = -55x^4y^6 - 15x^{14} \quad \text{and} \quad \frac{\partial F}{\partial y} = 7y^6 - 66x^5y^5.$$

So whenever $\frac{\partial F}{\partial y} \neq 0$, we get from Equation (1):

$$\begin{aligned}\frac{dy}{dx} &= -\frac{-55x^4y^6 - 15x^{14}}{7y^6 - 66x^5y^5}\\ &= \frac{55x^4y^6 + 15x^{14}}{7y^6 - 66x^5y^5}.\end{aligned}$$

(If you apply elementary calculus techniques, then you will obtain exactly the same answer, as you can check for yourself with a pencil and paper.) We thus have

$$\left.\frac{dy}{dx}\right|_{x_0=1} = \frac{70}{-59} \approx -1.186.$$

For $x = 1.002$,

$$\begin{aligned}f(x) &\approx f(1) + f'(1)(x-1)\\ &\approx 1 + (-1.186)(1.002 - 1.000) = 1 + (-1.186)(0.002)\\ &\approx 0.998.\end{aligned}$$

You might consider what means other than implicit differentiation could allow you to compute the values of y near $x = 1$.

Now, the assumption that $\frac{\partial F}{\partial y} \neq 0$ is not by any means incidental in all this. Consider the simple case of the circle $x^2 + y^2 = 16$, for instance. Here $F(x, y) = x^2 + y^2$ and so $F_x = 2x$, $F_y = 2y$. Thus $F_y = 0$ when $y = 0$, i.e., at the two points $(4, 0)$ and $(-4, 0)$. It is at these points that the whole technique of implicit differentiation breaks down. To begin with, the graph of $x^2 + y^2 = 16$ has vertical tangents at these points, so there is *no derivative* $\frac{dy}{dx}$. Moreover, in any open set U containing $x = 4$ or $x = -4$ (see Figure 6.1), there is no *single* function f defined by $F(x, y) = 0$. Above the x-axis, $y = \sqrt{16 - x^2}$, while below the x-axis $y = -\sqrt{16 - x^2}$. Finally, if we tried to use Equation (1) we would be stymied, since it would involve dividing by 0 at $x = \pm 4$.

It turns out that the assumption that $\frac{\partial F}{\partial y} \neq 0$ is enough to rule out all of the chaos encountered in the preceding paragraph. The following theorem makes this precise. We omit the proof, which is somewhat intricate.

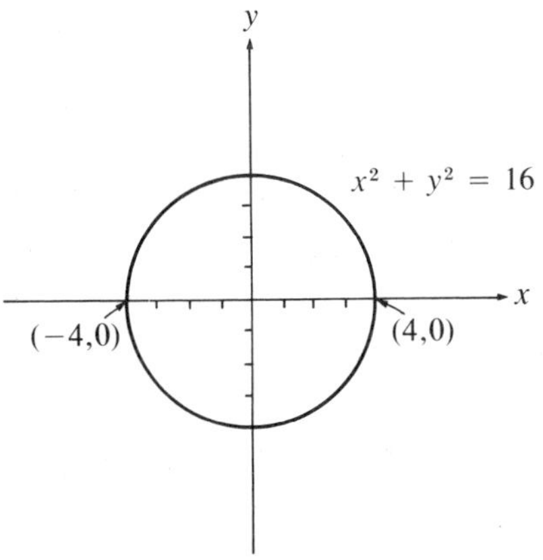

FIGURE 6.1

6.2 IMPLICIT FUNCTION THEOREM. Let $F: \boldsymbol{R}^2 \to \boldsymbol{R}$ be defined on an open set U containing $\boldsymbol{x}_0 = (x_0, y_0)$. Suppose that $F(\boldsymbol{x}_0) = 0$, that $F_y(\boldsymbol{x}_0) \neq 0$, and that F_x and F_y are continuous on U. Then there is an open interval I containing x_0 and there is a function $f: I \to \boldsymbol{R}$ satisfying the following conditions:

(a) For $x \in I$, $y = f(x)$ is such that $(x, y) \in U$;

(b) $F(x, f(x)) = 0$ for $x \in I$;

(c) $f'(x) = \dfrac{dy}{dx} = -\dfrac{F_x}{F_y}$ on I.

Returning to the circle $x^2 + y^2 = 16$, we see that for $\boldsymbol{x}_0 \neq (\pm 4, 0)$ we can safely assert

$$\frac{dy}{dx}(\boldsymbol{x}_0) = -\frac{F_x(\boldsymbol{x}_0)}{F_y(\boldsymbol{x}_0)} = -\frac{x_0}{y_0}.$$

Now, there is no reason to restrict ourselves to the case of only *two* variables x and y in Theorem 6.2. A theorem exactly analogous to Theorem 6.2 holds true for a function $F: \boldsymbol{R}^n \to \boldsymbol{R}$. We now state this theorem precisely, again leaving the proof for advanced calculus.

6.3 IMPLICIT FUNCTION THEOREM. Let $F: \boldsymbol{R}^n \to \boldsymbol{R}$ be defined on an open set U containing $\boldsymbol{x}_0 = (x_{01}, x_{02}, \ldots, x_{0n})$. Suppose that $F(\boldsymbol{x}_0) = 0$, $F_{x_i}(\boldsymbol{x}_0) = \dfrac{\partial F}{\partial x_i}(\boldsymbol{x}_0) \neq 0$ for some i, and all the partial derivatives $\dfrac{\partial F}{\partial x_j}$ are continuous on U for $j = 1, 2, \ldots, n$. Then there is an open set $V \subseteq \boldsymbol{R}^{n-1}$ containing $(x_{01}, \ldots, x_{0,i-1}, x_{0,i+1}, \ldots, x_{0n})$ and there is a function $f: \boldsymbol{R}^{n-1} \to \boldsymbol{R}$ such that:

(a) For $\boldsymbol{x} = (x_1, \ldots, x_{i-1}, x_{i+1}, \ldots, x_n) \in V$, we have $x_i = f(\boldsymbol{x})$

(b) $F(x_1, \ldots, x_{i-1}, f(x_1, \ldots, x_{i-1}, x_{i+1}, \ldots, x_n), x_{i+1}, \ldots, x_n) = 0$ for $\boldsymbol{x} \in V$

(c) $\dfrac{\partial f}{\partial x_j} = -\dfrac{\dfrac{\partial F}{\partial x_j}}{\dfrac{\partial F}{\partial x_i}} = -\dfrac{F_{x_j}}{F_{x_i}}$ on V, for $j = 1, \ldots, i-1, i+1, \ldots, n$.

The formula (c) can be remembered by the following mnemonic. Since f defines x_i as a function of the other variables, we can think of $\dfrac{\partial f}{\partial x_j}$ as $\dfrac{\partial x_i}{\partial x_j}$. If $\dfrac{\partial F}{\partial x_j}$ and

$\frac{\partial F}{\partial x_j}$ had an interpretation as quotients, then we could cancel the "∂F" parts on the right side of (c) and get $\frac{\partial x_i}{\partial x_j}$, as desired! The foregoing in part explains the origin of the notation $\frac{\partial F}{\partial x_j}$ historically, although since $\frac{\partial F}{\partial x_j}$ *isn't* interpretable as a quotient, the rationale for our mnemonic has to be classified as fortuitous nonsense. One final word: don't forget the *minus sign* in (c), which the canceling scheme will never produce for you!

6.4 EXAMPLE. Consider $x^2+4y^2-3z^2=6$ and $\boldsymbol{x}_0=(3, 0, 1)$. (a) Does the given equation define y as a function of x and z in an open set containing $(x_0, z_0)=(3, 1)$? If so, then find $\frac{\partial y}{\partial x}$ and $\frac{\partial y}{\partial z}$. (b) Does it define z as a function of x and y in an open set containing $(x_0, y_0)=(3, 0)$? If so, then find $\frac{\partial z}{\partial x}$ and $\frac{\partial z}{\partial y}$.

Solution. (a) Let $F(x, y, z)=x^2+4y^2-3z^2-6$. Since $F_y=8y$ and $F_z=-6z$, we have $F_y(\boldsymbol{x}_0)=0$ and $F_z(\boldsymbol{x}_0)=-6\neq 0$. Theorem 6.3 does not provide an answer to whether y is defined as a function of x and z, since it applies *only* if $F_y(\boldsymbol{x}_0)$ *is* different from 0. From our consideration of the circle $x^2+y^2=16$, however, we suspect that the given equation *fails* to define y as a unique function of x and z near the point $\boldsymbol{x}_0$. This is confirmed by the fact that if we move slightly toward the front from (x_0, z_0) (i.e., allow x to increase slightly), there is no y value defined by the given equation. To see this, note that any such y would have to satisfy $y=\sqrt{6+3z^2-x^2}$. However, at $\boldsymbol{x}_0$ we have $6+3z_0^2=x_0^2$. If x increases slightly, the expression under the radical becomes negative, so the equation cannot define y. Since an open set containing (x_0, z_0) *would* contain points slightly to the front of (x_0, z_0), we see that there can be no such open set U on which y is defined as a function of x and z.

(b) Since $F_z(\boldsymbol{x}_0)\neq 0$, we can apply Theorem 6.3 to conclude that the given equation *does* define $z=f(x, y)$ on an open set U containing $(x_0, y_0)=(3, 0)$. Moreover, on U we have

$$\frac{\partial z}{\partial x}=-\frac{F_x}{F_z}=-\frac{2x}{-6z}=\frac{x}{3z},$$

and

$$\frac{\partial z}{\partial y}=-\frac{F_y}{F_z}=-\frac{8y}{-6z}=\frac{4y}{3z}.$$

6.5 EXAMPLE. Find in two ways the equation of the tangent plane to the hyperboloid $x^2+4y^2-3z^2=6$ at the point $\boldsymbol{x}_0=(3, 0, 1)$.

Solution. (a) We can use the technique of Example 5.11. Putting $G(x, y, z)=x^2+4y^2-3z^2$, we have $\nabla G=(2x, 8y, -6z)=(6, 0, -6)$ at $\boldsymbol{x}_0$. So an equation of the tangent plane is

$$\begin{aligned}(6, 0, -6)\cdot(x-3, y-0, z-1)&=0,\\ 6x-18-6z+6&=0,\\ x-z&=2.\end{aligned}$$

(b) We can use Example 6.4. At $\boldsymbol{x}_0$,

$$\frac{\partial z}{\partial x} = \frac{x}{3z} = \frac{3}{3} = 1$$

and

$$\frac{\partial z}{\partial y} = \frac{4y}{3z} = 0.$$

So an equation of the tangent plane is, by (4) of Theorem 1.4,

$$z - 1 = 1(x - 3) + 0(y - 0),$$
$$2 = x - z.$$

See Figure 6.2.

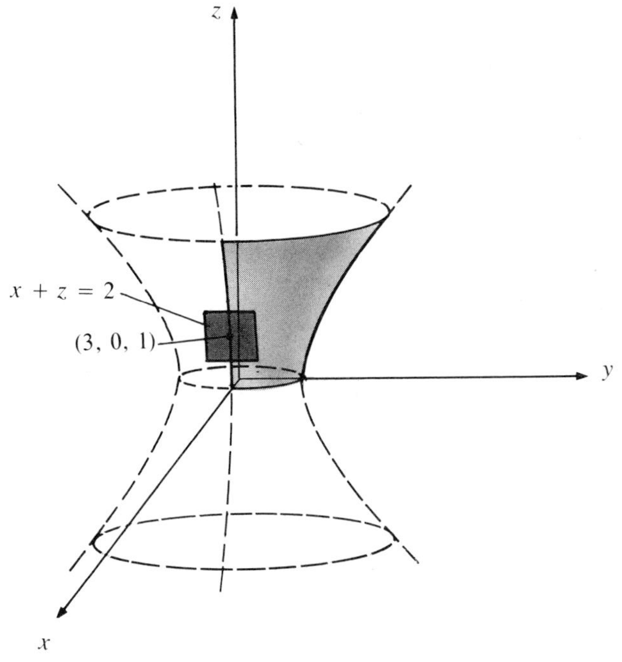

FIGURE 6.2

EXERCISES 4.6

In Exercises 1 to 4, find a formula for $\frac{dy}{dx}$ and state where it is valid. Then use the tangent approximation (Section 3) to approximate y for the given value of x.

1 $7x^2y^6 - 5xy^5 + 3y^3 - 2 = 3$; $x = 1.05$. (Note: (1, 1) satisfies the given equation.)

2 $xy^3 - x^2y + y^2 - x^5 - 19 = 0$; $x = 2.10$. (Note: (2, 3) satisfies the given equation.)

3 $x \cos y - x^2e^{xy} + y^2 \ln x - 3x^5 + 3 = 0$; $x = 0.95$. (Note: (1, 0) satisfies the given equation.)

4 $e^{x^2y^4} - x^4 + y^5 - \sin y^3 + 5x^2 - 5 = 0$; $x = 1.98$. (Note: (2, 0) satisfies the given equation.)

In Exercises 5 and 6, find formulas for $\frac{\partial z}{\partial x}$ and $\frac{\partial z}{\partial y}$ and state where they are valid. Then use the tangent approximation to approximate z for the given values of x and y.

5 $x^2+y^2+z^2=9$; $x=2.02$, $y=0.97$. (Note: (2, 1, 2) satisfies the given equation.)

6 $z^3-(x^2+y^2)z-3=0$; $x=1.98$, $y=2.03$. (Note: (2, 2, 3) satisfies the given equation.)

In Exercises 7 to 12, find formulas for $\frac{\partial z}{\partial x}$ and $\frac{\partial z}{\partial y}$ and state where they are valid.

7 $x^2 \cos xyz + y^2 \sin xyz = z^2$

8 $x+y+z-\sin xyz=0$

9 $ye^z+xz-x^2-y^2=0$

10 $z^2e^{xy}-5x^2\frac{y}{z}=7$

11 $x^3e^{y+z}-y\sin(x-z)=0$

12 $y^2ze^{xy}-x\cos(y^2-zx)=0$

In Exercises 13 to 16, find the equation of the tangent plane to the given surface at the given point. Sketch.

13 $x^2+y^2-z^2=1$ at $(3, 1, -3)$

14 $x^2-y^2-z^2=3$ at $(4, 3, 2)$

15 $4x^2+9y^2+4z^2=36$ at $(1, 1, -\sqrt{23}/2)$

16 $x^2+y^2+z^2=49$ at $(2, 3, 6)$

17 Show that the surfaces $x^2+y^2+4z=0$ and $x^2+y^2+z^2-6y+7=0$ are tangent at the point $(0, 2, -1)$.

18 Show that the surfaces $x^2+y^2+z^2=8$ and $xz=4$ are tangent at the point $(2, 0, 2)$.

19 Suppose that two surfaces $F(x, y, z)=0$ and $G(x, y, z)=0$ intersect in a curve. Show that if $\nabla F \times \nabla G \neq \mathbf{0}$, then it is a tangent vector to the curve of intersection at each point of intersection.

20 Find a tangent vector to the curve of intersection of $x^2+y^2-z=8$ and $x-y^2+z^2=-2$ at $(2, -2, 0)$.

21 Find a tangent vector to the curve of intersection of $3x^2+2y^2+z^2=49$ and $x^2+y^2-2z^2=10$ at $(3, -3, 2)$.

In numerical analysis, our technique of using the tangent line to approximate a function $y=f(x)$ defined by an equation $F(x, y)=0$ is used to obtain a *piecewise linear* approximation. We illustrate this in Exercises 22 to 24. (For simplicity we do only two steps, but on a computer one can readily generate a complete approximate graph of $y=f(x)$ in this way.)

22 Let $F(x, y)=x^2+2xy+3y^2-6$. The graph of $F(x, y)=0$ contains (1, 1). Show that an approximate value of $y=f(x)$ for $x=1.50$ is $y\approx 0.75$ by using the tangent approximation.

23 Refer to Exercise 22. Show that the approximate slope of the curve $y=f(x)$ when $x=1.50$ is -0.60.

24 Refer to Exercise 23. Use the line through $(1.50, 0.75)$ with slope -0.60 to find an approximate value of $y=f(x)$ when $x=2$.

7 HIGHER DERIVATIVES

In elementary calculus you will recall that the *second derivative* f'' of $f:\mathbf{R}\to\mathbf{R}$ is a useful tool in studying the behavior of f. In particular, f'' is essential to the study of the concavity of the graph of f, and the second derivative test is often the best way to decide if a *critical point* x_0 (i.e., a point such that $f'(x_0)=0$) gives a relative extremum at x_0. It is then natural to inquire whether we can define the notion of a *second total derivative* (and even higher order total derivatives) of a function $f:\mathbf{R}^n\to\mathbf{R}$. It turns out that we can, but a full treatment of higher total derivatives must await a study of the differential calculus of functions $f:\mathbf{R}^n\to\mathbf{R}^m$ in Chapter 6. As in elementary calculus, the most useful higher total derivative is the *second* total derivative, which is used in developing an analogue for the second derivative test for relative extrema (Section 9). In this section we introduce the second total derivative, and from it we can discern what the building blocks for higher order total derivatives must be: higher order partial derivatives, which we introduce briefly.

What reasonable interpretation can we give to the notion of a second total derivative of $f:\mathbf{R}^n \to \mathbf{R}$? Let's look at the case $n=2$ for an idea. Let D be the set of points in $\mathbf{R}^2$ where $f:\mathbf{R}^2 \to \mathbf{R}$ is differentiable. Then we can define the *total derivative function* $f':\mathbf{R}^2 \to \mathbf{R}^2$ by

$$f'(\mathbf{x}) = \nabla f(\mathbf{x}) = \left(\frac{\partial f}{\partial x}(\mathbf{x}), \frac{\partial f}{\partial y}(\mathbf{x})\right), \quad \text{for} \quad \mathbf{x} = (x, y) \in D.$$

A natural notation is

$$f' = \nabla f = \left(\frac{\partial f}{\partial x}, \frac{\partial f}{\partial y}\right).$$

Suppose that both partial derivative functions $\frac{\partial f}{\partial x}$ and $\frac{\partial f}{\partial y}$ are themselves differentiable on D. Then we can compute the total derivative of each of these by applying the gradient. Thus the total derivative of $\frac{\partial f}{\partial x}$ is

$$\nabla\left(\frac{\partial f}{\partial x}\right) = \left(\frac{\partial}{\partial x}\left(\frac{\partial f}{\partial x}\right), \frac{\partial}{\partial y}\left(\frac{\partial f}{\partial x}\right)\right),$$

and the total derivative of $\frac{\partial f}{\partial y}$ is

$$\nabla\left(\frac{\partial f}{\partial y}\right) = \left(\frac{\partial}{\partial x}\left(\frac{\partial f}{\partial y}\right), \frac{\partial}{\partial y}\left(\frac{\partial f}{\partial y}\right)\right).$$

At this point we pause to introduce names and notation for the coordinates of the vectors $\nabla\left(\frac{\partial f}{\partial x}\right)$ and $\nabla\left(\frac{\partial f}{\partial y}\right)$.

7.1 **DEFINITION.** If $f:\mathbf{R}^n \to \mathbf{R}$ is differentiable and if the partial derivative function $\frac{\partial f}{\partial x_i}:\mathbf{R}^n \to \mathbf{R}$ has a partial derivative with respect to x_j, then we write

$$\frac{\partial^2 f}{\partial x_j\,\partial x_i} = \frac{\partial}{\partial x_j}\left(\frac{\partial f}{\partial x_i}\right)$$

and call $\frac{\partial^2 f}{\partial x_j \partial x_i}$ the **second partial derivative of f with respect to x_i and x_j in that order.**

In our discussion above, then, we would write

$$\nabla\left(\frac{\partial f}{\partial x}\right) = \left(\frac{\partial^2 f}{\partial x^2}, \frac{\partial^2 f}{\partial y\,\partial x}\right),$$

and

$$\nabla\left(\frac{\partial f}{\partial y}\right)=\left(\frac{\partial^2 f}{\partial x\,\partial y},\frac{\partial^2 f}{\partial y^2}\right).$$

Another notation in use for $\dfrac{\partial^2 f}{\partial x_j\,\partial x_i}$ is $f_{x_i x_j}$. Note the *reversed* order of reading! In the ∂-notation we put the second variable with respect to which we differentiate on the *left*. In the f-subscript notation we put that second variable on the *right*! This appears at first glance to be intolerably confusing, and might well be so except for one saving fact. As we shall see below, we generally get the *same* function whether we compute $\dfrac{\partial^2 f}{\partial x_i\,\partial x_j}$ or $\dfrac{\partial^2 f}{\partial x_j\,\partial x_i}$. So we will follow the now well-established but inconsistent notation that has evolved: $f_{x_i x_j}=\dfrac{\partial^2 f}{\partial x_j\,\partial x_i}$. Variants of these notations that you may encounter elsewhere are $D_{x_j}D_{x_i}$ or $D_{ji}f$ for $\dfrac{\partial^2 f}{\partial x_j\,\partial x_i}$ and f_{ij} for $f_{x_i x_j}$. We will avoid these out of fear of compounding the inconsistency of notation!

7.2 **EXAMPLE.** If $f(x, y)=x^2 \sin xy - y^2 e^{xy}$, then find all second partial derivatives.

Solution. First,

$$\frac{\partial f}{\partial x}=f_x=2x\sin xy+x^2y\cos xy-y^3e^{xy}$$

and

$$\frac{\partial f}{\partial y}=f_y=x^3\cos xy-2ye^{xy}-xy^2e^{xy}.$$

Now we differentiate f_x and f_y.

$$\begin{aligned}f_{xx}=\frac{\partial^2 f}{\partial x^2}&=2\sin xy+2xy\cos xy+2xy\cos xy-x^2y^2\sin xy-y^4e^{xy}\\&=2\sin xy+4xy\cos xy-x^2y^2\sin xy-y^4e^{xy}.\end{aligned}$$

$$\begin{aligned}f_{xy}=\frac{\partial^2 f}{\partial y\,\partial x}&=2x^2\cos xy+x^2\cos xy-x^3y\sin xy-3y^2e^{xy}-xy^3e^{xy}\\&=3x^2\cos xy-x^3y\sin xy-3y^2e^{xy}-xy^3e^{xy}.\end{aligned}$$

$$\begin{aligned}f_{yx}=\frac{\partial^2 f}{\partial x\,\partial y}&=3x^2\cos xy-x^3y\sin xy-2y^2e^{xy}-y^2e^{xy}-xy^3e^{xy}\\&=3x^2\cos xy-x^3y\sin xy-3y^2e^{xy}-xy^3e^{xy}\end{aligned}$$

$$\begin{aligned}f_{yy}=\frac{\partial^2 f}{\partial y^2}&=-x^4\sin xy-2e^{xy}-2xye^{xy}-2xye^{xy}-x^2y^2e^{xy}\\&=-x^4\sin xy-2e^{xy}-4xye^{xy}-x^2y^2e^{xy}.\end{aligned}$$

Notice that $f_{xy}=f_{yx}$, as we promised would usually happen.

Returning to our discussion leading up to Definition 7.1, we would now like to put

$$\nabla \frac{\partial f}{\partial x} = (f_{xx}, f_{xy}) \quad \text{and} \quad \nabla \frac{\partial f}{\partial y} = (f_{yx}, f_{yy})$$

together to get the second total derivative of f. We recall that putting together the partial derivatives of f to get f' is all right so long as all the partial derivatives are continuous (Theorem 3.10). So we require that all second partial derivatives be continuous in order to build f'' up from them.

7.3 **DEFINITION.** Suppose that $f: \boldsymbol{R}^n \to \boldsymbol{R}$ is differentiable on a set $U \subseteq \boldsymbol{R}^n$. Suppose also that all $\dfrac{\partial^2 f}{\partial x_i \, \partial x_j}$ exist and are continuous on U. Then the **second total derivative** f'' is the n-by-n **Hessian matrix** H_f [named after the German mathematician Ludwig O. Hesse (1811–1874)] given on U by

$$H_f = f'' = \begin{bmatrix} \nabla \dfrac{\partial f}{\partial x_1} \\ \nabla \dfrac{\partial f}{\partial x_2} \\ \vdots \\ \nabla \dfrac{\partial f}{\partial x_n} \end{bmatrix} = \begin{bmatrix} f_{x_1x_1} & f_{x_1x_2} & \cdots & f_{x_1x_n} \\ f_{x_2x_1} & f_{x_2x_2} & \cdots & f_{x_2x_n} \\ \vdots & \vdots & \ddots & \vdots \\ f_{x_nx_1} & f_{x_nx_2} & \cdots & f_{x_nx_n} \end{bmatrix}$$

Thus f'' is the matrix whose i-th row is the (total) derivative of the i-th coordinate $\dfrac{\partial f}{\partial x_i}$ of the (total) derivative ∇f of f.

Note that we can form the Hessian matrix H_f whenever all the second partial derivatives $\dfrac{\partial^2 f}{\partial x_i \, \partial x_j}$ *exist* on U, whether they are continuous or not. As Example 3.8 would lead you to suspect, it can happen that H_f exists but is not the derivative of $f': \boldsymbol{R}^n \to \boldsymbol{R}^n$ when some $\dfrac{\partial^2 f}{\partial x_i \, \partial x_j}$ are not continuous. A more precise explanation must await a discussion of differentiability of functions $f: \boldsymbol{R}^n \to \boldsymbol{R}^m$ in Chapter 6, but as in the case of Theorem 3.10, continuity of the entries of H_f is enough to guarantee that H_f gives $(f')'$. (See Theorem 6.5.5 and Exercise 21, Exercises 6.5.) To avoid complications, we will confine ourselves to functions f whose second partial derivatives *are* continuous.

7.4 **EXAMPLE.** If $f(x, y) = x^2 - 4y^2 - 3xy$, then find f' and f''.

Solution. By inspection, $f' = \left(\dfrac{\partial f}{\partial x}, \dfrac{\partial f}{\partial y}\right) = (2x - 3y, -8y - 3x)$. Then

$$f'' = \begin{pmatrix} \nabla(2x - 3y) \\ \nabla(-8y - 3x) \end{pmatrix} = \begin{pmatrix} \dfrac{\partial}{\partial x}(2x - 3y) & \dfrac{\partial}{\partial y}(2x - 3y) \\ \dfrac{\partial}{\partial x}(-8y - 3x) & \dfrac{\partial}{\partial y}(-8y - 3x) \end{pmatrix}$$

$$= \begin{pmatrix} 2 & -3 \\ -3 & -8 \end{pmatrix}.$$

While Example 7.4 may look rather puny in comparison to Definition 7.3, it is representative of the use to which we will be putting the second derivative in Sections 8 and 9. Again we notice that $\frac{\partial^2 f}{\partial x\,\partial y} = \frac{\partial^2 f}{\partial y\,\partial x}$ $(=-3)$ in this example. We now give a theorem that guarantees this for most functions we deal with.

7.5 **THEOREM.** Suppose that $f: \boldsymbol{R}^n \to \boldsymbol{R}$ is differentiable on an open set U containing $\boldsymbol{x}_0$. Suppose that $\frac{\partial^2 f}{\partial x_i\,\partial x_j}$ and $\frac{\partial^2 f}{\partial x_j\,\partial x_i}$ exist on U and are continuous at $\boldsymbol{x}_0$. Then

$$\frac{\partial^2 f}{\partial x_i\,\partial x_j}(\boldsymbol{x}_0) = \frac{\partial^2 f}{\partial x_j\,\partial x_i}(\boldsymbol{x}_0).$$

Proof. There are only two variables that vary in this theorem, x_i and x_j. So it is sufficient to prove the theorem for $n=2$, i.e., to show that $\frac{\partial^2 f}{\partial x\,\partial y} = \frac{\partial^2 f}{\partial y\,\partial x}$ at $\boldsymbol{x}_0 = (x_0, y_0)$. We prove the theorem by applying the Mean Value Theorem of elementary calculus four times to the double difference

$$\Delta(x_1, y_1) = f(x_1, y_1) - f(x_1, y_0) - (f(x_0, y_1) - f(x_0, y_0))$$

where we choose x_1 and y_1 so that the rectangle R in Figure 7.1 lies in U. For the moment let us fix y_1. Then we have

$$\Delta(x_1, y_1) = g(x_1) - g(x_0)$$

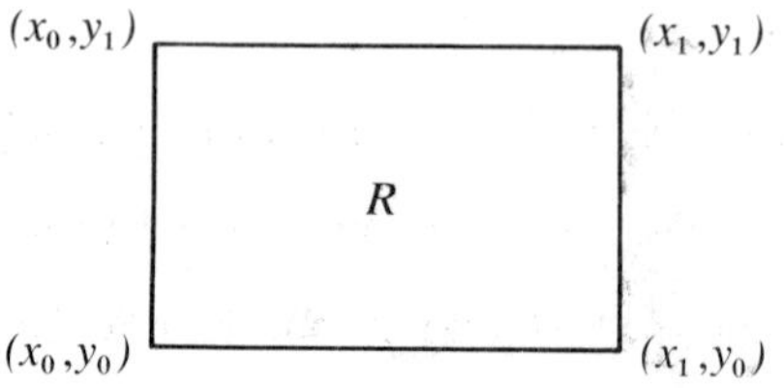

FIGURE 7.1

where $g(x) = f(x, y_1) - f(x, y_0)$ for x between x_0 and x_1. Note that g satisfies the hypotheses of the Mean Value Theorem on the closed interval $[x_0, x_1]$, so we have

(1) $$\Delta(x_1, y_1) = g'(\bar{x})(x_1 - x_0)$$

$$= \left(\frac{\partial f}{\partial x}(\bar{x}, y_1) - \frac{\partial f}{\partial x}(\bar{x}, y_0)\right)(x_1 - x_0)$$

where $\bar{x}$ is between x_1 and x_0. Now $\frac{\partial f}{\partial x}$ is continuous and differentiable on R, so we can apply the Mean Value Theorem to the function $\frac{\partial f}{\partial x}(\bar{x}, y)$ where y varies between y_0 and y_1. Then (1) gives

(2) $$\Delta(x_1, y_1) = \frac{\partial^2 f}{\partial y\,\partial x}(\bar{x}, \bar{y})(x_1 - x_0)(y_1 - y_0)$$

where $\bar{y}$ is between y_0 and y_1.

Now fix x_1 and observe that we can rearrange $\Delta(x_1, y_1)$ slightly:

$$\begin{aligned}\Delta(x_1, y_1) &= f(x_1, y_1) - f(x_0, y_1) - (f(x_1, y_0) - f(x_0, y_0)) \\ &= h(y_1) - h(y_0)\end{aligned}$$

where $h(y) = f(x_1, y) - f(x_0, y)$ for y between y_0 and y_1. Here h is a function of y alone, so from the Mean Value Theorem we get

(3)
$$\begin{aligned}\Delta(x_1, y_1) &= h'(\bar{Y})(y_1 - y_0) \\ &= \left(\frac{\partial f}{\partial y}(x_1, \bar{Y}) - \frac{\partial f}{\partial y}(x_0, \bar{Y})\right)(y_1 - y_0),\end{aligned}$$

where $\bar{Y}$ is between y_0 and y_1. Again $\dfrac{\partial f}{\partial y}$ is continuous and differentiable on R, so we can apply the Mean Value Theorem a final time:

(4)
$$\Delta(x_1, y_1) = \frac{\partial^2 f}{\partial x\, \partial y}(\bar{X}, \bar{Y})(x_1 - x_0)(y_1 - y_0)$$

where $\bar{X}$ is between x_0 and x_1. Now equate (2) and (4) and cancel the term $(x_1 - x_0)(y_1 - y_0)$:

(5)
$$\frac{\partial^2 f}{\partial y\, \partial x}(\bar{x}, \bar{y}) = \frac{\partial^2 f}{\partial x\, \partial y}(\bar{X}, \bar{Y})$$

where $(\bar{x}, \bar{y})$ and $(\bar{X}, \bar{Y})$ are both in R. Now, let $(x_1, y_1) \to (x_0, y_0)$. Then $(\bar{x}, \bar{y}) \to (x_0, y_0)$ and $(\bar{X}, \bar{Y}) \to (x_0, y_0)$, since R collapses to the point (x_0, y_0). Because $\dfrac{\partial^2 f}{\partial y\, \partial x}$ is continuous on $R \subseteq U$,

$$\frac{\partial^2 f}{\partial y\, \partial x}(\bar{x}, \bar{y}) \to \frac{\partial^2 f}{\partial y\, \partial x}(x_0, y_0)$$

and

$$\frac{\partial^2 f}{\partial x\, \partial y}(\bar{X}, \bar{Y}) \to \frac{\partial^2 f}{\partial x\, \partial y}(x_0, y_0).$$

Thus,

$$\begin{aligned}\frac{\partial^2 f}{\partial y\, \partial x}(x_0, y_0) &= \lim_{x_1 \to x_0} \frac{\partial^2 f}{\partial y\, \partial x}(\bar{x}, \bar{y}) \\ &= \lim_{x_1 \to x_0} \frac{\partial^2 f}{\partial x\, \partial y}(\bar{X}, \bar{Y}) = \frac{\partial^2 f}{\partial y\, \partial x}(x_0, y_0),\end{aligned}$$

by (5). QED

Theorem 7.5 is frequently expressed by saying that the second derivative matrix $f'' = H_f$ is **symmetric**. This just means that the entry in row i, column j (here, $f_{x_i x_j}$) is the same as the entry in row j, column i (here, $f_{x_j x_i}$) for all i and j. If a matrix A is symmetric, then its entries below the diagonal entries a_{ii} are the same as those in the *symmetric* positions above the diagonal: $a_{12} = a_{21}$, $a_{52} = a_{25}$, and, in general, $a_{ij} = a_{ji}$ for $i \neq j$. Symmetric matrices will be of considerable importance to us in Chapter 7. (See Theorems 7.4.6 and 7.4.8.)

Once we have computed the second partial derivatives of f, it is clear that we can differentiate these to generate third partial derivatives. These are the highest order partial derivatives we will treat, although we could certainly continue differentiating as many times as we liked.

7.6 DEFINITION. If $f: \boldsymbol{R}^n \to \boldsymbol{R}$ has a second total derivative at $\boldsymbol{x}_0$, then we define

$$\frac{\partial^3 f}{\partial x^3} = f_{xxx} = \frac{\partial}{\partial x}\left(\frac{\partial^2 f}{\partial x^2}\right)$$

$$\frac{\partial^3 f}{\partial y\, \partial x^2} = f_{xxy} = \frac{\partial}{\partial y}\left(\frac{\partial^2 f}{\partial x^2}\right)$$

$$\frac{\partial^3 f}{\partial x\, \partial y^2} = f_{yyx} = \frac{\partial}{\partial x}\left(\frac{\partial^2 f}{\partial y^2}\right)$$

$$\frac{\partial^3 f}{\partial y^3} = f_{yyy} = \frac{\partial}{\partial y}\left(\frac{\partial^2 f}{\partial y^2}\right).$$

We could also define more third partial derivatives, but (for most functions we meet) these will, by Theorem 7.5, coincide with the four we have defined. For instance,

$$f_{xyy} = (f_{xy})_y = (f_{yx})_y = (f_y)_{xy} = (f_y)_{yx} = f_{yyx}.$$

For most functions $f: \boldsymbol{R}^2 \to \boldsymbol{R}$ we deal with, there are likewise at most $n+1$ distinct n-th partial derivatives, namely

$$\frac{\partial^n f}{\partial x^n}, \frac{\partial^n f}{\partial y\, \partial x^{n-1}}, \frac{\partial^n f}{\partial y^2\, \partial x^{n-2}}, \ldots, \frac{\partial^n f}{\partial y^n}.$$

7.7 EXAMPLE. Find the third partial derivatives of $f(x, y) = x^3 y^3$.

Solution. The first and second partial derivatives of f are $f_x = 3x^2y^3$, $f_y = 3x^3y^2$, $f_{xx} = 6xy^3$, $f_{xy} = f_{yx} = 9x^2y^2$, and $f_{yy} = 6x^3y$. Thus, the third partial derivatives are $f_{xxx} = 6y^3$, $f_{xxy} = 18xy^2$, $f_{yyx} = 18x^2y$, and $f_{yyy} = 6x^3$.

We close the section with an indication of the relevance of higher derivatives to fields outside mathematics. In *thermodynamics* there are five basic variables: T (temperature), V (volume), P (pressure), E (energy), and S (entropy). Physicists consider that *any* two of these may be regarded as independent, with the remaining three variables viewed as functions of these two. All partial derivatives of all orders are assumed to be continuous. (In differentiating, one always indicates which is the other independent variable by a device like $\left(\frac{\partial S}{\partial T}\right)_V$, which means that T and V are the independent variables.) The *first and second laws of thermodynamics* can be expressed as

(I) $$T\left(\frac{\partial S}{\partial T}\right)_V = \left(\frac{\partial E}{\partial T}\right)_V$$

(II) $$T\left(\frac{\partial S}{\partial V}\right)_T = \left(\frac{\partial E}{\partial V}\right)_T + P$$

From these laws many relationships can be deduced.

7.8 EXAMPLE. Show that $\left(\frac{\partial P}{\partial T}\right)_V = \left(\frac{\partial S}{\partial V}\right)_T$.

Proof. We differentiate (I) with respect to V and (II) with respect to T. This gives

(a) $$T\frac{\partial^2 S}{\partial V\,\partial T} = \frac{\partial^2 E}{\partial V\,\partial T},$$

and

(b) $$\left(\frac{\partial S}{\partial V}\right)_T + T\frac{\partial^2 S}{\partial T\,\partial V} = \frac{\partial^2 E}{\partial T\,\partial V} + \left(\frac{\partial P}{\partial T}\right)_V.$$

We now substitute (a) into (b):

$$\left(\frac{\partial S}{\partial V}\right)_T + T\frac{\partial^2 S}{\partial T\,\partial V} = T\frac{\partial^2 S}{\partial V\,\partial T} + \left(\frac{\partial P}{\partial T}\right)_V.$$

Since all partials of every order are assumed continuous, we can cancel

$$T\frac{\partial^2 S}{\partial T\,\partial V} = T\frac{\partial^2 S}{\partial V\,\partial T}$$

by Theorem 7.5. This leaves $\left(\frac{\partial S}{\partial V}\right)_T = \left(\frac{\partial P}{\partial T}\right)_V$. QED

EXERCISES 4.7

In Exercises 1 to 4, find f''.

1 $f(x, y) = 3x^2 - 2y^2 - 5xy + 7$

2 $f(x, y) = x^3y^2 - 3x^2 - 4y^2 + xy - 2$.

3 $f(x, y) = x\cos(x + y)$

4 $f(x, y) = y\ln(x^2 + y^2)$

In Exercises 5 to 8, find all first, second, and third partial derivatives.

5 $f(x, y) = x^2 \sin xy$

6 $f(x, y) = \ln(x + y)$

7 $f(x, y) = \ln(x^2 + y^2)$

8 $f(x, y) = y^2 e^{xy}$

In Exercises 9 to 12, find f''.

9 $f(x, y, z)=x^2-y^2+3z^2-5$

10 $f(x, y, z)=x^2+2y^2+7z^2+5$

11 $f(x, y, z)=x^2+2y^2+5z^2-8xy-4yz-5xz+1$

12 $f(x, y, z)=3x^2-y^2+z^2-5xy-3yz-xz+5$

Exercises 13 to 17 are concerned with *Laplace's equation,* which is of importance in electricity, optics, thermodynamics and other applied areas. (*Assume all second partial derivatives are continuous.*)

13 Show that if $f(x, y)=\ln(x^2+y^2)$, then

$$\frac{\partial^2 f}{\partial x^2}+\frac{\partial^2 f}{\partial y^2}=0 \quad \text{(2-dimensional Laplace equation).}$$

14 Show that if $f(x, y)=\text{Tan}^{-1}\dfrac{2xy}{x^2-y^2}$, then f satisfies the 2-dimensional Laplace equation.

15 Show that if $f(x, y, z)=\dfrac{1}{|(x, y, z)|}=(x^2+y^2+z^2)^{-1/2}$, then f satisfies $\dfrac{\partial^2 f}{\partial x^2}+\dfrac{\partial^2 f}{\partial y^2}+\dfrac{\partial^2 f}{\partial z^2}=0$ (the 3-dimensional Laplace equation).

16 If $z=f(x, y)$ where $x=r\cos\theta$ and $y=r\sin\theta$, then show that $\dfrac{\partial^2 z}{\partial x^2}+\dfrac{\partial^2 z}{\partial y^2}=\dfrac{\partial^2 z}{\partial r^2}+\dfrac{1}{r^2}\dfrac{\partial^2 z}{\partial \theta^2}+\dfrac{1}{r}\dfrac{\partial z}{\partial r}$.

17 If $w=f(x, y, z)$ where $x=r\cos\theta$, $y=r\sin\theta$, and $z=z$, then show that

$$\frac{\partial^2 w}{\partial x^2}+\frac{\partial^2 w}{\partial y^2}+\frac{\partial^2 w}{\partial z^2}=\frac{\partial^2 w}{\partial r^2}+\frac{1}{r^2}\frac{\partial^2 w}{\partial \theta^2}+\frac{1}{r}\frac{\partial w}{\partial r}+\frac{\partial^2 w}{\partial z^2}.$$

18 Suppose that a thin flexible membrane is stretched over a portion of the xy-plane and is made to vibrate. Let $z=f(x, y, t)$ be the distance from the xy-plane to the membrane at time t. The 2-dimensional normalized **wave equation** is

$$\frac{\partial^2 z}{\partial x^2}+\frac{\partial^2 z}{\partial y^2}-\frac{\partial^2 z}{\partial t^2}=0.$$

Show that

$$z(x, y, t)=e^{x-t}-3e^{y+t}+7xy-t+5$$

satisfies the wave equation.

19 Refer to Exercise 18. Show that $z(x, y, t)=(3x-4y+5t)^{1/2}$ satisfies the 2-dimensional wave equation.

20 Refer to Example 7.8. Suppose that T and P are chosen as independent variables. Then in Example 7.8, P is a function of T and V. Also, V is now a function of T and P. So P is ultimately a function of P and T. Show that

(a) $1=\left(\dfrac{\partial P}{\partial T}\right)_V\left(\dfrac{\partial T}{\partial P}\right)_T+\left(\dfrac{\partial P}{\partial V}\right)_T\left(\dfrac{\partial V}{\partial P}\right)_T$

(b) $1=\left(\dfrac{\partial P}{\partial V}\right)_T\left(\dfrac{\partial V}{\partial P}\right)_T$

21 Refer to Example 7.8. Suppose that T and P are chosen as independent variables. Then in Example 7.8, P is a function of T and V, where T and V are functions of T and P. Then P is ultimately a function of T and P.

(a) Show that $\left(\frac{\partial P}{\partial T}\right)_V = -\left(\frac{\partial P}{\partial V}\right)_T\left(\frac{\partial V}{\partial T}\right)_P$ by computing $\left(\frac{\partial P}{\partial T}\right)_P$ (=0).

(b) Regard S as a function first of T and V (Example 7.8), and then ultimately as a function of T and P. Use part (a) and the results of Example 7.8 and Exercise 20(b) to show that $\left(\frac{\partial S}{\partial P}\right)_T = -\left(\frac{\partial V}{\partial T}\right)_P$.

8 TAYLOR POLYNOMIALS

The object of this section and the next one is to develop for functions $f:\boldsymbol{R}^2 \to \boldsymbol{R}$ an analogue of a technique you learned in elementary calculus for finding relative maxima and minima of functions $f:\boldsymbol{R} \to \boldsymbol{R}$. The tool we need to carry this out is the *second degree Taylor polynomial* $p_2(x, y)$ *for* f near a point $\boldsymbol{x}_0 = (x_0, y_0) \in \boldsymbol{R}^2$. Before plunging into the development of Taylor polynomials for functions of two variables, let's take a moment to see how, in the one-variable case, Taylor polynomials relate to extreme values. (You may want to flip back to Section 11.6 to refresh your memory about one-variable Taylor polynomials as you read this section.)

Suppose that $f:\boldsymbol{R} \to \boldsymbol{R}$ is a function such that f' and f'' exist and are continuous in an open interval I about x_0. Then we can approximate f on I by the second degree Taylor polynomial

(1) $$p_2(x) = f(x_0) + f'(x_0)(x - x_0) + \frac{1}{2!} f''(x_0)(x - x_0)^2.$$

In general, this is a *closer* approximation to f than the tangent approximation

(2) $$f(x_0) + f'(x_0)(x - x_0),$$

since it approaches $f(x_0)$ faster than $(x - x_0)^2$ approaches 0 as $x \to x_0$. It thus affords quite a close approximation to f near x_0. (See Theorem 11.6.4.)

If we are interested in maximizing or minimizing $f:\boldsymbol{R} \to \boldsymbol{R}$, what do we do? We find all points where $f'(x) = 0$, and then test these critical points to see if they give a relative maximum or minimum for f. One very useful test is the *second derivative test*, which, you probably recall, goes as follows for a critical point x_0. If $f''(x_0) < 0$, then f has a relative maximum value at x_0; if $f''(x_0) > 0$, then f has a relative minimum at x_0. This can be easily deduced from the second degree Taylor polynomial as follows. We saw in the preceding paragraphs that (under the assumptions given)

$$f(x) \approx f(x_0) + f'(x_0)(x - x_0) + \frac{1}{2!} f''(x_0)(x - x_0)^2$$

Now at a critical point x_0, $f'(x_0) = 0$, so this approximation becomes

(3) $$f(x) \approx f(x_0) + \frac{1}{2!} f''(x_0)(x - x_0)^2.$$

Consider the second degree term

(4) $$\frac{1}{2!} f''(x_0)(x - x_0)^2.$$

Since $(x-x_0)^2$ is a square, the sign of this term will be the sign of $f''(x_0)$. Thus (4) will be positive if $f''(x_0)>0$, and negative if $f''(x_0)<0$. Now we can see the behavior of f near x_0 from (3). If $f''(x_0)>0$, then $f(x)>f(x_0)$ for x near x_0, since $f(x)$ is approximately $f(x_0)$ plus the *positive* quantity (4). (See Figure 8.1(a).) This just says that $f(x_0)$ is a relative minimum if $f''(x_0)>0$.

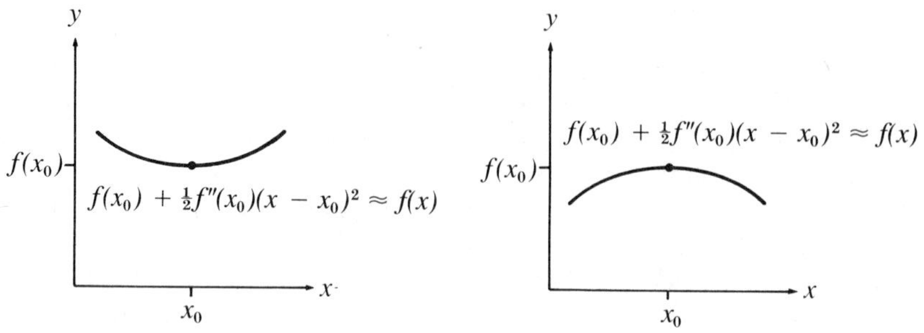

(a) (b)

FIGURE 8.1

Similarly, if $f''(x_0)<0$, then $f(x)<f(x_0)$ near x_0, since $f(x)$ is approximately $f(x_0)$ plus the *negative* quantity (4). (See Figure 8.1(b).) Hence $f(x_0)$ is a relative maximum if $f''(x)<0$. We want to develop a second derivative test for $f:\boldsymbol{R}^2\to\boldsymbol{R}$, and for this we will need to have available the second degree Taylor polynomial for a function of two real variables.

What would the second degree Taylor polynomial $p_2(x, y)$ for $f:\boldsymbol{R}^2\to\boldsymbol{R}$ be? Well, we first assume that f' and f'' exist in an open set U containing $\boldsymbol{x}_0=(x_0, y_0)$, and we assume that all second partial derivatives are continuous on U. We want to define $p_2(\boldsymbol{x})$ to be the natural analogue of (1). There is certainly no obstacle in the first two terms of (1). These just comprise the tangent approximation to f (i.e., the *first degree Taylor polynomial* for f) near $\boldsymbol{x}_0$, which we have seen in Section 3 corresponds to

(5) $$f(\boldsymbol{x}_0)+\boldsymbol{f}'(\boldsymbol{x}_0)\cdot(\boldsymbol{x}-\boldsymbol{x}_0),$$

for $f:\boldsymbol{R}^2\to\boldsymbol{R}$. But what about an analogue for (4)? Well, $f''(\boldsymbol{x}_0)$ has been defined as the matrix

$$H_f=\begin{pmatrix} f_{xx}(\boldsymbol{x}_0) & f_{xy}(\boldsymbol{x}_0)\\ f_{yx}(\boldsymbol{x}_0) & f_{yy}(\boldsymbol{x}_0)\end{pmatrix}$$

So we need to give some meaning to

(6) $$f''(\boldsymbol{x}_0)(\boldsymbol{x}-\boldsymbol{x}_0)^2.$$

To see how that can be done, we return to (4) in the $f:\boldsymbol{R}\to\boldsymbol{R}$ case, and we note that

$$f''(x_0)(x-x_0)^2=[f''(x_0)(x-x_0)](x-x_0).$$

We can give meaning to a product of this sort for $f:\boldsymbol{R}^2\to\boldsymbol{R}$ by using the following notion of multiplication of vectors by matrices.

8.1 **DEFINITION.** If

$$A = \begin{pmatrix} a_{11} & a_{12} \\ a_{21} & a_{22} \end{pmatrix} = \begin{pmatrix} \boldsymbol{r}_1 \\ \boldsymbol{r}_2 \end{pmatrix} \quad \text{and} \quad \boldsymbol{x} = \begin{pmatrix} x \\ y \end{pmatrix},$$

then we define the **product** of A and $\boldsymbol{x}$ in that order to be the vector in $\boldsymbol{R}^2$ given by

(7) $$A\boldsymbol{x} = \begin{pmatrix} \boldsymbol{r}_1 \cdot \boldsymbol{x} \\ \boldsymbol{r}_2 \cdot \boldsymbol{x} \end{pmatrix}$$

In this product, we view A as the result of taking two row vectors, $\boldsymbol{r}_1 = (a_{11}, a_{12})$ and $\boldsymbol{r}_2 = (a_{21}, a_{22})$, and putting them together. We compute A times $\boldsymbol{x}$ by computing the dot products $\boldsymbol{r}_1 \cdot \boldsymbol{x}$ and $\boldsymbol{r}_2 \cdot \boldsymbol{x}$, and then putting the results together as a vector. In such multiplications, all vectors are written as *columns* to conform to the requirement for general matrix multiplication. (See Section 6.3.)

8.2 **EXAMPLE.** If $A = \begin{pmatrix} 1 & 2 \\ 3 & 4 \end{pmatrix}$ and $\boldsymbol{x} = (-5, 3)$, then find $A\boldsymbol{x}$.

Solution

$$A\boldsymbol{x} = \begin{pmatrix} 1 & 2 \\ 3 & 4 \end{pmatrix}\begin{pmatrix} -5 \\ 3 \end{pmatrix} = \begin{pmatrix} 1(-5)+2\cdot 3 \\ 3(-5)+4\cdot 3 \end{pmatrix}$$
$$= \begin{pmatrix} 1 \\ -3 \end{pmatrix}.$$

(Note that we wrote $(-5, 3)$ as a column vector to conform with Definition 8.1.)

Now we can give the analogue for the term (4) in $p_2(x)$, and so define $p_2(\boldsymbol{x})$.

8.3 **DEFINITION.** Let $f: \boldsymbol{R}^2 \to \boldsymbol{R}$ be such that $\boldsymbol{f}'$ and f'' exist and all second partial derivatives of f are continuous on an open set U containing $\boldsymbol{x}_0$. Then the **second degree Taylor polynomial of f about $\boldsymbol{x}_0$** is

(8) $$p_2(\boldsymbol{x}) = f(\boldsymbol{x}_0) + \boldsymbol{f}'(\boldsymbol{x}_0) \cdot (\boldsymbol{x} - \boldsymbol{x}_0) + \tfrac{1}{2}[f''(\boldsymbol{x}_0)(\boldsymbol{x} - \boldsymbol{x}_0)] \cdot (\boldsymbol{x} - \boldsymbol{x}_0).$$

Notice that we have made the second degree term the result of multiplying $\boldsymbol{x} - \boldsymbol{x}_0$ by the 2-by-2 matrix $f''(\boldsymbol{x}_0)$ (in the sense of Definition 8.1), and then taking the dot product of that vector with $\boldsymbol{x} - \boldsymbol{x}_0$.

8.4 **EXAMPLE.** Find the second degree Taylor polynomial of f about $(2, 0)$ if $f(x, y) = xe^y$.

Solution. Here

$$\boldsymbol{f}'(\boldsymbol{x}) = \left(\frac{\partial f}{\partial x}, \frac{\partial f}{\partial y}\right) = (e^y, xe^y)$$

so that

$$f''(\mathbf{x}) = \begin{pmatrix} 0 & e^y \\ e^y & xe^y \end{pmatrix}.$$

We have $f(2, 0) = 2e^0 = 2$, $\mathbf{f}'(2, 0) = (e^0, 2e^0) = (1, 2)$, and

$$f''(2, 0) = \begin{pmatrix} 0 & e^0 \\ e^0 & 2e^0 \end{pmatrix} = \begin{pmatrix} 0 & 1 \\ 1 & 2 \end{pmatrix}.$$

So we obtain from (8)

$$\begin{aligned} p_2(x, y) &= 2 + (1, 2) \cdot (x-2, y-0) + \tfrac{1}{2}\left[\begin{pmatrix} 0 & 1 \\ 1 & 2 \end{pmatrix}\begin{pmatrix} x-2 \\ y-0 \end{pmatrix}\right] \cdot \begin{pmatrix} x-2 \\ y-0 \end{pmatrix} \\ &= 2 + x - 2 + 2y + \tfrac{1}{2}\begin{pmatrix} y \\ x-2+2y \end{pmatrix} \cdot \begin{pmatrix} x-2 \\ y \end{pmatrix} \\ &= x + 2y + \tfrac{1}{2}xy - y + \tfrac{1}{2}xy - y + y^2 \\ &= x + xy + y^2. \end{aligned}$$

We can easily give a formula for (8) purely in terms of partial derivatives.

8.5 **PROPOSITION.** Under the assumptions of Definition 8.3,

(9)
$$\begin{aligned} p_2(x, y) = f(\mathbf{x}_0) &+ \frac{\partial f}{\partial x}(\mathbf{x}_0)(x - x_0) + \frac{\partial f}{\partial y}(\mathbf{x}_0)(y - y_0) \\ &+ \frac{1}{2!}\left[\frac{\partial^2 f}{\partial x^2}(\mathbf{x}_0)(x - x_0)^2 + 2\frac{\partial^2 f}{\partial x\,\partial y}(\mathbf{x}_0)(x - x_0)(y - y_0)\right. \\ &\left.+ \frac{\partial^2 f}{\partial y^2}(\mathbf{x}_0)(y - y_0)^2\right], \end{aligned}$$

where $\mathbf{x}_0 = (x_0, y_0)$.

Proof. We just need to calculate each summand in (8). First,

(10)
$$f(\mathbf{x}_0) = f(x_0, y_0).$$

Also,

(11)
$$\begin{aligned} \mathbf{f}'(\mathbf{x}_0) \cdot (\mathbf{x} - \mathbf{x}_0) &= \left(\frac{\partial f}{\partial x}(\mathbf{x}_0), \frac{\partial f}{\partial y}(\mathbf{x}_0)\right) \cdot (x - x_0, y - y_0) \\ &= \frac{\partial f}{\partial x}(\mathbf{x}_0)(x - x_0) + \frac{\partial f}{\partial y}(\mathbf{x}_0)(y - y_0). \end{aligned}$$

$$\begin{aligned} f''(\mathbf{x}_0)(\mathbf{x} - \mathbf{x}_0) &= \begin{pmatrix} \dfrac{\partial^2 f}{\partial x^2}(\mathbf{x}_0) & \dfrac{\partial^2 f}{\partial y\,\partial x}(\mathbf{x}_0) \\ \dfrac{\partial^2 f}{\partial x\,\partial y}(\mathbf{x}_0) & \dfrac{\partial^2 f}{\partial y^2}(\mathbf{x}_0) \end{pmatrix} \begin{pmatrix} x - x_0 \\ y - y_0 \end{pmatrix} \\ &= \begin{pmatrix} \dfrac{\partial^2 f}{\partial x^2}(\mathbf{x}_0)(x - x_0) + \dfrac{\partial^2 f}{\partial x\,\partial y}(\mathbf{x}_0)(y - y_0) \\ \dfrac{\partial^2 f}{\partial x\,\partial y}(\mathbf{x}_0)(x - x_0) + \dfrac{\partial^2 f}{\partial y^2}(\mathbf{x}_0)(y - y_0) \end{pmatrix} \end{aligned}$$

where we used Theorem 7.5:

$$\frac{\partial^2 f}{\partial y\,\partial x}(\boldsymbol{x}_0)=\frac{\partial^2 f}{\partial x\,\partial y}(\boldsymbol{x}_0).$$

So

$$(12)\qquad [f''(\boldsymbol{x}_0)(\boldsymbol{x}-\boldsymbol{x}_0)]\cdot(\boldsymbol{x}-\boldsymbol{x}_0)=\begin{pmatrix}\dfrac{\partial^2 f}{\partial x^2}(\boldsymbol{x}_0)(x-x_0)+\dfrac{\partial^2 f}{\partial x\,\partial y}(\boldsymbol{x}_0)(y-y_0)\\ \dfrac{\partial^2 f}{\partial x\,\partial y}(\boldsymbol{x}_0)(x-x_0)+\dfrac{\partial^2 f}{\partial y^2}(\boldsymbol{x}_0)(y-y_0)\end{pmatrix}\cdot\begin{pmatrix}x-x_0\\ y-y_0\end{pmatrix}$$

$$=\frac{\partial^2 f}{\partial x^2}(\boldsymbol{x}_0)(x-x_0)^2+2\frac{\partial^2 f}{\partial x\,\partial y}(\boldsymbol{x}_0)(x-x_0)(y-y_0)+\frac{\partial^2 f}{\partial y^2}(\boldsymbol{x}_0)(y-y_0)^2$$

If we add (10), (11), and (12), then we obtain (9). QED

In working problems, you can use whichever of (8) or (9) strikes you as the more convenient formula.

Now that we have defined the second degree Taylor polynomial of f about $\boldsymbol{x}_0$, we proceed to the question of how close an approximation it provides to $f(\boldsymbol{x})$ near $\boldsymbol{x}_0$. Our next theorem states that this approximation is in general better than the tangent approximation to $f(\boldsymbol{x})$ near $\boldsymbol{x}_0$. It says, in fact, that $p_2(x, y)$ approaches $f(x, y)$ as $\boldsymbol{x}$ approaches $\boldsymbol{x}_0$ at a rate *faster* than the rate at which $|\boldsymbol{x}-\boldsymbol{x}_0|^2\to 0$ (cf. Theorem 11.6.5). The proof, which is a bit subtle, is omitted.

8.6 THEOREM. Suppose that $\boldsymbol{x}_0$, f, and U are as in Definition 8.3. Let $p_2(x, y)$ be the second degree Taylor polynomial of f about $\boldsymbol{x}_0$. Let $r_2(\boldsymbol{x})=f(\boldsymbol{x})-p_2(\boldsymbol{x})$. Then

$$\lim_{\boldsymbol{x}\to\boldsymbol{x}_0}\frac{r_2(\boldsymbol{x})}{|\boldsymbol{x}-\boldsymbol{x}_0|^2}=0.$$

That is, as $\boldsymbol{x}\to\boldsymbol{x}_0$, $p_2(\boldsymbol{x})\to f(\boldsymbol{x}_0)$ faster than $|\boldsymbol{x}-\boldsymbol{x}_0|^2\to 0$.

Theorem 8.6 says that the error we make in approximating $f(x, y)$ by its second degree Taylor polynomial $p_2(x, y)$ about (x_0, y_0) is of a *smaller order of magnitude* than $|\boldsymbol{x}-\boldsymbol{x}_0|^2$. In case $f(x, y)$ itself is a second degree polynomial, this fact, and our experience with the single variable theory, suggests that $f(x, y)$ should *coincide* with $p_2(x, y)$. Our next result confirms that expectation.

8.7 PROPOSITION. Suppose that $f(x, y)$ is a second degree polynomial and $\boldsymbol{x}_0=(x_0, y_0)\in\boldsymbol{R}^2$. Then $p_2(x, y)=f(x, y)$.

Proof. First, $p_2(x, y)$ exists since $f(x, y)$ has continuous second partial derivatives on $\boldsymbol{R}^2$. To compute the second degree Taylor polynomial about (x_0, y_0), we first write

$$f(x, y)=a_0+a_1x+a_2y+a_3x^2+a_4xy+a_5y^2$$

for constants $a_i\in\boldsymbol{R}$. Then

$$\boldsymbol{\nabla} f(x, y)=(a_1+2a_3x+a_4y,\ a_2+a_4x+2a_5y)$$

and

$$f'' = \begin{pmatrix} 2a_3 & a_4 \\ a_4 & 2a_5 \end{pmatrix}.$$

Case 1. $(x_0, y_0) = (0, 0)$. Then we have

$$\begin{aligned} p_2(x, y) &= f(0, 0) + \mathbf{f}'(0, 0) \cdot (x, y) + \tfrac{1}{2}\left[f''(0, 0)\begin{pmatrix} x \\ y \end{pmatrix}\right] \cdot \begin{pmatrix} x \\ y \end{pmatrix} \\ &= a_0 + (a_1, a_2) \cdot (x, y) + \tfrac{1}{2}\left[\begin{pmatrix} 2a_3 & a_4 \\ a_4 & 2a_5 \end{pmatrix}\begin{pmatrix} x \\ y \end{pmatrix}\right] \cdot \begin{pmatrix} x \\ y \end{pmatrix} \\ &= a_0 + a_1x + a_2y + \tfrac{1}{2}\begin{pmatrix} 2a_3x + a_4y \\ a_4x + 2a_5y \end{pmatrix} \cdot \begin{pmatrix} x \\ y \end{pmatrix} \\ &= a_0 + a_1x + a_2y + \tfrac{1}{2}(2a_3x^2 + a_4xy + a_4xy + 2a_5y^2) \\ &= a_0 + a_1x + a_2y + a_3x^2 + a_4xy + a_5y^2 \\ &= f(x, y) \end{aligned}$$

as desired.

Case 2. (x_0, y_0) arbitrary. We reduce to Case 1 by letting $u = x - x_0$ and $v = y - y_0$. Then

$$f(x, y) = f(u + x_0, v + y_0) = g(u, v)$$

where $g(u, v)$ is a polynomial of degree two in u and v. When $(x, y) = (x_0, y_0)$, we have $(u_0, v_0) = (0, 0)$; hence from Case 1,

$$P_2(u, v) = g(u, v) = f(x, y)$$

where $P_2(u, v)$ is the second degree Taylor polynomial of $g(u, v)$ about $(0, 0)$. We then have

$$\begin{aligned} f(x, y) &= P_2(u, v) \\ &= g(0, 0) + \mathbf{g}'(0, 0) \cdot (u, v) + \tfrac{1}{2}\left[g''(0, 0)\begin{pmatrix} u \\ v \end{pmatrix}\right] \cdot \begin{pmatrix} u \\ v \end{pmatrix} \\ &= f(x_0, y_0) + \mathbf{f}'(x_0, y_0) \cdot (x - x_0, y - y_0) \\ &\quad + \tfrac{1}{2}\left[f''(x_0, y_0)\begin{pmatrix} x - x_0 \\ y - y_0 \end{pmatrix}\right] \cdot \begin{pmatrix} x - x_0 \\ y - y_0 \end{pmatrix} \\ &= p_2(x, y). \end{aligned}$$

QED

Theorem 8.6 states that the second degree Taylor polynomial $p_2(\mathbf{x})$ affords a *closer* approximation to $f(\mathbf{x})$ near $\mathbf{x}_0$ than the tangent approximation $f(\mathbf{x}_0) + \mathbf{f}'(\mathbf{x}_0) \cdot (\mathbf{x} - \mathbf{x}_0)$. Our next example illustrates this greater degree of accuracy. We return to Example 3.11 to approximate $c(P, V)$ near the point $(P_0, V_0) = (100.00, 4.00)$.

8.8 EXAMPLE. In using Boyle's Law, $PV = c$, P is measured as 100.00 pounds per square inch (accurate to within 2%) and V is measured as 4.00 cubic feet (accurate to within 0.5%). This gives $c = 400.00$. Determine approximate bounds on the true value of c using the second degree Taylor polynomial.

Solution. As in Example 3.11, P lies in $[98, 102]$ and V in $[3.98, 4.02]$. By Proposition 8.7, the second degree Taylor polynomial for PV should *be* PV. This is easily checked:

$$\frac{\partial c}{\partial P} = V \quad \text{and} \quad \frac{\partial c}{\partial V} = P,$$

so

$$\frac{\partial^2 c}{\partial V\, \partial P} = 1, \quad \text{and} \quad \frac{\partial^2 c}{\partial P^2} = \frac{\partial^2 c}{\partial V^2} = 0.$$

Hence,

$$\begin{aligned} p_2(P, V) &= P_0 V_0 + (V_0, P_0) \cdot (P - P_0, V - V_0) + \tfrac{1}{2} 2(P - P_0)(V - V_0) \\ &= 400 + (4, 100) \cdot (P - 100, V - 4) + (P - 100)(V - 4) \\ &= 400 + 4(P - 100) + 100(V - 4) + (P - 100)(V - 4) \\ &= PV \end{aligned}$$

The second degree Taylor polynomial in the second-to-last line above is better suited to the present problem than is PV itself, however. For we have

$$\begin{aligned} |400 - p_2(P, V)| &= |4(P - 100) + 100(V - 4) + (P - 100)(V - 4)| \\ &\le 4\,|P - 100| + 100\,|V - 4| + |P - 100|\,|V - 4| \\ &\le 4(2) + 100(0.02) + 2(0.02) \\ &= 8 + 2 + 0.04 \\ &= 10.04. \end{aligned}$$

So the true value of c lies in the interval $[400 - 10.04, 400 + 10.04] = [389.96, 410.04]$. (The increase in the accuracy of our estimate of error may be seen by referring to Example 3.11, where we estimated c to be in the interval $[390, 410]$.)

In the preceding example, since PV *equaled* $p_2(P, V)$, we got *exact* rather than *approximate* bounds for c. In other examples (see Exercises 29–33 below), where $f(x, y) \neq p_2(x, y)$, you should follow the procedure of Example 8.8 to obtain approximate bounds for $f(x_0, y_0)$ calculated from approximate measurements.

Example 8.8 may remind you of the remarks we made just before Example 3.11 concerning the desirability of having an upper bound for the error $e(\boldsymbol{x})$ when $f(\boldsymbol{x})$ is approximated by the tangent approximation (which we call the *first degree Taylor polynomial*), $p_1(\boldsymbol{x}) = f(\boldsymbol{x}_0) + \boldsymbol{f}'(\boldsymbol{x}_0) \cdot (\boldsymbol{x} - \boldsymbol{x}_0)$. It turns out that upper bounds analogous to the one in Corollary 11.6.6 can be obtained for $e(\boldsymbol{x})$ when we approximate f by either $p_1(\boldsymbol{x})$ or $p_2(\boldsymbol{x})$. We leave a complete discussion for advanced calculus, confining ourselves to stating the following result.

8.9 THEOREM. (a) Suppose that f_{xx}, $f_{xy} = f_{yx}$, and f_{yy} are all continuous on a closed ball B about $\boldsymbol{x}_0$. Let

$$e(\boldsymbol{x}) = f(\boldsymbol{x}) - \boldsymbol{f}'(\boldsymbol{x}_0) \cdot (\boldsymbol{x} - \boldsymbol{x}_0) = f(\boldsymbol{x}) - p_1(\boldsymbol{x}).$$

If M is an upper bound on B for $|f_{xx}(x)|$, $|f_{xy}(x)|$ and $|f_{yy}(x)|$, then

$$|e(\boldsymbol{x})| \leq M\,|\boldsymbol{x}-\boldsymbol{x}_0|^2.$$

(b) Suppose that f_{xxx}, f_{xxy}, f_{xyy}, and f_{yyy} are continuous on a closed ball B about $\boldsymbol{x}_0$. Let $f(\boldsymbol{x}) = p_2(\boldsymbol{x}) + r_2(\boldsymbol{x})$, where $p_2(\boldsymbol{x})$ is the second degree Taylor polynomial of f near $\boldsymbol{x}_0$. If M is an upper bound on B for $|f_{xxx}(\boldsymbol{x})|$, $|f_{xxy}(\boldsymbol{x})|$, $|f_{xyy}(\boldsymbol{x})|$, and $|f_{yyy}(\boldsymbol{x})|$, then

$$|r_2(\boldsymbol{x})| \leq \tfrac{1}{2}M\,|\boldsymbol{x}-\boldsymbol{x}_0|^3.$$

We remark in closing that all the results of this section hold equally true for functions $f:\boldsymbol{R}^3 \to \boldsymbol{R}$, and in fact for functions $f:\boldsymbol{R}^n \to \boldsymbol{R}$. In particular, the analogue of (7) can be used to define the product of a 3-by-3 matrix

$$A = \begin{pmatrix} \boldsymbol{a}_1 \\ \boldsymbol{a}_2 \\ \boldsymbol{a}_3 \end{pmatrix}$$

and a vector

$$\boldsymbol{x} = \begin{pmatrix} x \\ y \\ z \end{pmatrix}$$

in $\boldsymbol{R}^3$. (See Exercises 7 to 12 below.) That gives

$$A\boldsymbol{x} = \begin{pmatrix} \boldsymbol{a}_1 \cdot \boldsymbol{x} \\ \boldsymbol{a}_2 \cdot \boldsymbol{x} \\ \boldsymbol{a}_3 \cdot \boldsymbol{x} \end{pmatrix}$$

Then (8) defines $p_2(\boldsymbol{x})$ if $f:\boldsymbol{R}^3 \to \boldsymbol{R}$. (See Exercises 25–28, 31–32, and 34–35.)

EXERCISES 4.8

In Exercises 1 to 12, compute the matrix-vector products.

1 $\begin{pmatrix} 1 & -1 \\ -1 & 2 \end{pmatrix}\begin{pmatrix} x \\ y \end{pmatrix}$ **2** $\begin{pmatrix} -3 & 1 \\ 1 & 0 \end{pmatrix}\begin{pmatrix} x \\ y \end{pmatrix}$ **3** $\begin{pmatrix} 1 & -4 \\ -4 & 5 \end{pmatrix}\begin{pmatrix} x-x_0 \\ y-y_0 \end{pmatrix}$

4 $\begin{pmatrix} 0 & -2 \\ -2 & 3 \end{pmatrix}\begin{pmatrix} x-x_0 \\ y-y_0 \end{pmatrix}$ **5** $\begin{pmatrix} -5 & 1 \\ 1 & -3 \end{pmatrix}\begin{pmatrix} -2 \\ 2 \end{pmatrix}$ **6** $\begin{pmatrix} -1 & 3 \\ 3 & 2 \end{pmatrix}\begin{pmatrix} 1 \\ -1 \end{pmatrix}$

7 $\begin{pmatrix} 1 & -1 & 3 \\ -1 & -2 & 1 \\ 3 & 1 & 2 \end{pmatrix}\begin{pmatrix} x \\ y \\ z \end{pmatrix}$ **8** $\begin{pmatrix} 1 & 5 & -1 \\ 5 & -1 & 2 \\ -1 & 2 & 0 \end{pmatrix}\begin{pmatrix} x \\ y \\ z \end{pmatrix}$ **9** $\begin{pmatrix} 1 & -1 & 2 \\ -1 & 0 & -3 \\ 2 & -3 & 1 \end{pmatrix}\begin{pmatrix} x-x_0 \\ y-y_0 \\ z-z_0 \end{pmatrix}$

10 $\begin{pmatrix} -1 & 0 & 2 \\ 0 & 3 & -2 \\ 2 & -2 & 5 \end{pmatrix}\begin{pmatrix} x-x_0 \\ y-y_0 \\ z-z_0 \end{pmatrix}$ **11** $\begin{pmatrix} 2 & -1 & 0 \\ -1 & 3 & 1 \\ 0 & 1 & 1 \end{pmatrix}\begin{pmatrix} -1 \\ 2 \\ 3 \end{pmatrix}$ **12** $\begin{pmatrix} 2 & -2 & 1 \\ -2 & 0 & -1 \\ 1 & -1 & 3 \end{pmatrix}\begin{pmatrix} -2 \\ -1 \\ 3 \end{pmatrix}$

In Exercises 13 to 28, compute when possible the second degree Taylor polynomials of f about the given point using either Equation (8) or (9).

13 $\sin(xy)$; $(0, 0)$ (*Hint:* Use your knowledge of Taylor polynomials for $\sin t$.)

14 $\ln(2x+y)$; $(0, 1)$

15 $\sqrt{1-x^2-y^2}$; $(1, 0)$

16 $\sqrt{25-x^2-y^2}$; $(3, 4)$

17 $xy+y^2$; $(1, 1)$

18 $2xy+x+2$; $(1, 1)$

19 $\sqrt{1+x^2+y^2}$; $(0, 0)$

20 $\sqrt{1-x^2-y^2}$; $(0, 0)$

21 y^2e^x; $(0, 0)$

22 $x \tan y$; $(0, 0)$

23 $\sin x \cos y$; $(0, 0)$

24 $x \cos y$; $(0, 0)$

25 $\sin(x+y+z)$; $(\pi/6, \pi/6, \pi/6)$

26 $\cos(x+y+z)$; $(\pi/6, \pi/6, \pi/6)$

27 $\sqrt{x^2+y^2+z^2}$; $(0, 0, 0)$

28 $\sqrt{1-x^2-2y+z}$; $(1, 0, 0)$

In Exercises 29 to 33, use the second degree Taylor polynomial to approximate the quantity in the given problem. Compare with your earlier answers using the tangent approximation.

29 Problem 11, Exercises 4.3

30 Problem 13, Exercises 4.3

31 Problem 15, Exercises 4.3

32 Problem 17, Exercises 4.3

33 Problem 19, Exercises 4.3

34 Find a formula analogous to Equation (9) for $p_2(x, y, z)$ for a function $f: \mathbf{R}^3 \to \mathbf{R}$. (*Hint:* Use Equation (8) and Definition 7.3 with $n=3$.)

9 EXTREME VALUES

We already have outlined the main point of this section in the introduction to Section 8. We want to develop machinery analogous to what you learned in elementary calculus for finding relative maxima and minima of a function $f: \mathbf{R}^n \to \mathbf{R}$. As in elementary calculus, such problems often arise in applied work. (See Exercises 11 to 24 below.) We give a detailed discussion only for the case $n=2$, but will also give the facts for the $n=3$ case. First, let's make sure we have a precise understanding of what we are talking about.

9.1 **DEFINITION.** An **absolute maximum** (respectively, **minimum**) of a function $f: \mathbf{R}^n \to \mathbf{R}$ is a value $f(\mathbf{x}_0)$ such that $f(\mathbf{x}) \leq f(\mathbf{x}_0)$ (respectively, $f(\mathbf{x}) \geq f(\mathbf{x}_0)$) for all $\mathbf{x}$ in the domain of f. A **local** or **relative maximum** (respectively, **minimum**) is a value $f(\mathbf{x}_0)$ such that $f(\mathbf{x}) \leq f(\mathbf{x}_0)$ (respectively, $f(\mathbf{x}) \geq f(\mathbf{x}_0)$) for all $\mathbf{x}$ in some open ball $B(\mathbf{x}_0, \delta)$. The term **extremum** or **extreme value** means "maximum or minimum."

9.2 **EXAMPLE.** Discuss the extreme values of $f(x, y) = 6-3x^2-y^2$.

Solution. Note that $f(x, y) = 6-(3x^2+y^2)$, and $3x^2+y^2 \geq 0$ for all (x, y). Thus the absolute maximum of f is 6, occurring when as little as possible is subtracted from 6, i.e., at $(x, y) = (0, 0)$. As x increases (or y increases) in absolute value, f decreases, so there is no local or absolute minimum. The sole extreme value then is 6. See Figure 9.1 (p. 212).

We can see that for the function f above, $\mathbf{f}'(0, 0) = \mathbf{0}$. This is what we expect at a local extreme point of a differentiable function, by analogy with elementary calculus. It is easy to prove that this is a general phenomenon.

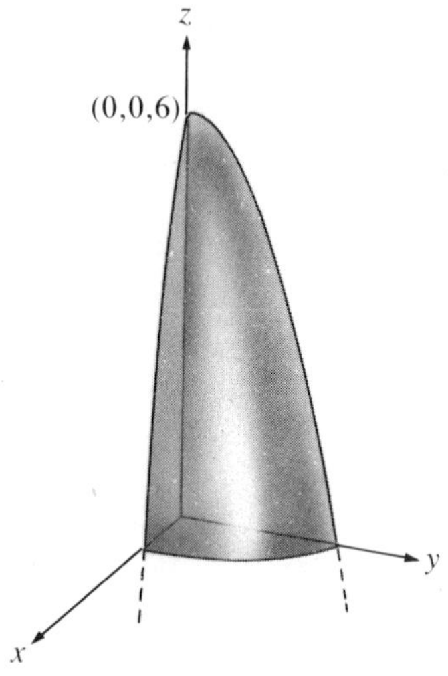

FIGURE 9.1

9.3 **THEOREM.** Suppose that $f: \boldsymbol{R}^n \to \boldsymbol{R}$ is differentiable on an open set containing $\boldsymbol{x}_0$, and suppose that f has a local extremum at $\boldsymbol{x}_0$. Then $\boldsymbol{f}'(\boldsymbol{x}_0) = \boldsymbol{0}$.

Proof. If f has a local extremum at $\boldsymbol{x}_0$, then so does each of the n functions $\hat{f}_i$ that result from freezing $n-1$ of the variables and allowing only one to vary:

$$\hat{f}_i(x) = f(x_{01}, \ldots, x_{0,i-1}, x, x_{0,i+1}, \ldots, x_{0n})$$

Since f is differentiable, so is each $\hat{f}_i$. Therefore, from elementary calculus, for each $i = 1, 2, \ldots, n$ we have

$$\frac{d\hat{f}_i}{dx}(x_{0i}) = 0.$$

That is, for $1 \leq i \leq n$, $\dfrac{\partial f}{\partial x_i}(\boldsymbol{x}_0) = 0$. Thus $\boldsymbol{\nabla} f(\boldsymbol{x}_0) = \boldsymbol{0}$. Then $\boldsymbol{f}'(\boldsymbol{x}_0) = \boldsymbol{\nabla} f(\boldsymbol{x}_0) = \boldsymbol{0}$.

QED

As in the one-variable case, then, a *necessary condition* for f to have a local extremum at $\boldsymbol{x}_0$ is $\boldsymbol{f}'(\boldsymbol{x}_0) = \boldsymbol{0}$. It is natural to call such a point by the same name used in elementary calculus.

9.4 **DEFINITION.** If $f: \boldsymbol{R}^n \to \boldsymbol{R}$ is differentiable, then a **critical** (or **stationary**) **point** of f is a point $\boldsymbol{x}_0$ of $\boldsymbol{R}^n$ such that $\boldsymbol{f}'(\boldsymbol{x}_0) = \boldsymbol{0}$.

As in elementary calculus, these are only *candidates* for local extreme points of a differentiable function.

9.5 **EXAMPLE.** Discuss the extreme values of $f: \boldsymbol{R}^2 \to \boldsymbol{R}$ given by $f(x, y) = xy$.

Solution. There are clearly no absolute extrema: xy becomes arbitrarily huge as x and y increase, and there is no lower bound for xy either. The only possible extrema, then, are local extrema. To find candidates for these, we note $\nabla f = (y, x) = \mathbf{0}$ when $x = y = 0$, i.e., at $\boldsymbol{x}_0 = \mathbf{0}$. This sole critical point *fails* to provide a local extremum, however. If x and y are both very small positive or negative numbers, then $f(x, y) > 0$. But if x is very small positive, and y is very small negative (or vice versa), then $f(x, y) < 0$.

Observe that at a critical point $\boldsymbol{x}_0$ of a function $f: \boldsymbol{R}^2 \to \boldsymbol{R}$, the tangent plane to the graph of $z = f(x, y)$ is *horizontal*, since its equation is $z - z_0 = 0(x - x_0) + 0(y - y_0)$, i.e., $z = z_0$. A point like $(0, 0)$ in Example 9.5 corresponds to an *inflection* point in elementary calculus. In optimization theory, the term *saddle point* is used for a critical point $\boldsymbol{x}_0$ that gives neither a relative maximum nor a relative minimum. See Figure 9.2, where we have drawn a sketch of the graph of

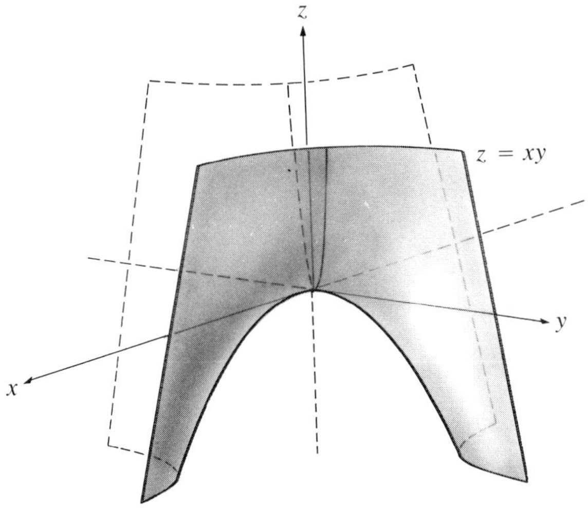

FIGURE 9.2

$z = xy$. Notice that near $(0, 0)$ the surface resembles a saddle. If f has a saddle point at $(x_0, y_0, f(x_0, y_0))$, however, its graph needn't resemble a saddle near that point. Figure 9.3 shows that the graph of the function $f(x, y) = x^3$ has a horizontal tangent plane $z = 0$ at $(0, 0, 0)$, but is not saddle-shaped near there.

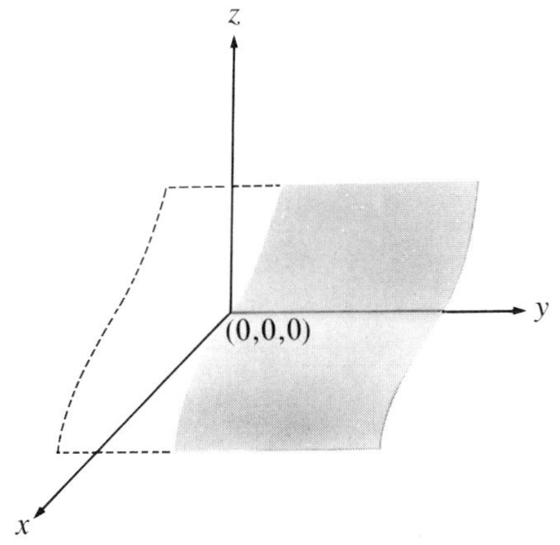

FIGURE 9.3

In light of behavior like that in Example 9.5, it is desirable to have a test to run on critical points $\boldsymbol{x}_0$ that will tell us whether the candidate $\boldsymbol{x}_0$ is an extreme point or a saddle point. A reasonable first guess would be that there should be a test corresponding to the *first derivative test* of elementary calculus: if x_0 is a critical point of $f:\boldsymbol{R}\to\boldsymbol{R}$ and f' changes sign from negative to positive (respectively, positive to negative) at x_0, then x_0 is a local minimum (respectively, maximum) point. An obstacle arises here, however. What does $\boldsymbol{x}<\boldsymbol{x}_0$ *mean* in $\boldsymbol{R}^n$? We have never tried to *order* $\boldsymbol{R}^n$. There are a number of schemes we could try, but none of them is satisfactory from the point of view of permitting us to develop a feasible first derivative test. (See, for example, Exercise 29 below.) This brings us to the *second derivative test* already discussed in Section 8. Since we have been successful in developing second degree Taylor polynomials in $\boldsymbol{R}^2$ (and even $\boldsymbol{R}^3$), the development of an analogue to the second derivative test seems promising.

If $\boldsymbol{x}_0$ is a critical point, then $f'(\boldsymbol{x}_0)=\boldsymbol{0}$, so the second degree Taylor polynomial approximation to f near $\boldsymbol{x}_0$ is

(1) $$f(\boldsymbol{x})\approx f(\boldsymbol{x}_0)+\tfrac{1}{2}[f''(\boldsymbol{x}_0)(\boldsymbol{x}-\boldsymbol{x}_0)]\cdot(\boldsymbol{x}-\boldsymbol{x}_0).$$

All we need is a means of deciding whether the second degree term

(2) $$[f''(\boldsymbol{x}_0)(\boldsymbol{x}-\boldsymbol{x}_0)]\cdot(\boldsymbol{x}-\boldsymbol{x}_0)$$

is always positive, always negative, or neither in a deleted open set containing $\boldsymbol{x}_0$. Let's state this as a theorem.

9.6 THEOREM. Suppose that the Hessian matrix $f''=H_f$ exists in an open set U containing the critical point $\boldsymbol{x}_0$, and suppose that all its entries are continuous on U. Then:
(a) If (2) is positive for all $\boldsymbol{x}\neq\boldsymbol{x}_0$ in U, then $f(\boldsymbol{x}_0)$ is a local minimum.
(b) If (2) is negative for all $\boldsymbol{x}\neq\boldsymbol{x}_0$ in U, then $f(\boldsymbol{x}_0)$ is a local maximum.
(c) If (2) is sometimes positive and sometimes negative on U, then $\boldsymbol{x}_0$ is a saddle point.

We won't give a formal proof of this, but it should be quite believable in light of Equation (1). After all, the right side of (1) closely (by Theorem 8.6) approximates f near $\boldsymbol{x}_0$. If (2) is positive on U except at $\boldsymbol{x}_0$, then the right side of (1) is always at least as big as $f(\boldsymbol{x}_0)$ for $\boldsymbol{x}\in U$, so (a) must follow. If (2) is negative on U except at $\boldsymbol{x}_0$, then the right side of (1) cannot be larger than $f(\boldsymbol{x}_0)$ for $\boldsymbol{x}\in U$, so (b) must follow. Finally, if (2) is sometimes positive and sometimes negative on U, then the right side of (1) is sometimes bigger than $f(\boldsymbol{x}_0)$ and sometimes smaller than $f(\boldsymbol{x}_0)$, so the same should be true of f, and hence (c) should follow. The full rigorous proof consists of making these very plausible statements precise and verifying their truth. We leave this for advanced calculus.

The only link remaining to be forged now is a practical procedure for determining which of the alternatives (a), (b), or (c) actually *holds* at a critical point $\boldsymbol{x}_0$. We incorporate this into the following theorem, which we state as the second derivative test for $f:\boldsymbol{R}^2\to\boldsymbol{R}$.

9.7 THEOREM. SECOND DERIVATIVE TEST. Let $f:\boldsymbol{R}^2\to\boldsymbol{R}$ be differentiable with continuous second partial derivatives in an open set U containing the critical point $\boldsymbol{x}_0$. Put

$$D=\det f''(\boldsymbol{x}_0)=f_{xx}(\boldsymbol{x}_0)f_{yy}(\boldsymbol{x}_0)-f_{xy}^2(\boldsymbol{x}_0).$$

(a) If $f_{xx}(\boldsymbol{x}_0)>0$ and $D>0$, then $f(\boldsymbol{x}_0)$ is a local minumum;
(b) If $f_{xx}(\boldsymbol{x}_0)<0$ and $D>0$, then $f(\boldsymbol{x}_0)$ is a local maximum;
(c) If $D<0$, then $\boldsymbol{x}_0$ is a saddle point.

Proof. By Theorem 9.6, it is enough to study (2). By Equation (12) of Proposition 8.5, we can express (2) as

(3) $$f_{xx}(\boldsymbol{x}_0)(x-x_0)^2+2f_{xy}(\boldsymbol{x}_0)(x-x_0)(y-y_0)+f_{yy}(\boldsymbol{x}_0)(y-y_0)^2.$$

We propose to complete the square in expression (3), assuming $f_{xx}(\boldsymbol{x}_0)\neq 0$. As you recall, one has to factor out the coefficient of $(x-x_0)^2$ first from the first two terms. This gives us

$$f_{xx}(\boldsymbol{x}_0)\left[(x-x_0)^2+\frac{2f_{xy}(\boldsymbol{x}_0)}{f_{xx}(\boldsymbol{x}_0)}(x-x_0)(y-y_0)\right]+f_{yy}(\boldsymbol{x}_0)(y-y_0)^2.$$

Now we take half the coefficient of $x-x_0$, i.e., we take

$$\frac{f_{xy}(\boldsymbol{x}_0)}{f_{xx}(\boldsymbol{x}_0)}(y-y_0),$$

square it, and add the result to the bracket. To maintain an expression equal to (2), we compensate by subtracting the same quantity outside the bracket. We get

$$f_{xx}(\boldsymbol{x}_0)\left[(x-x_0)^2+\frac{2f_{xy}(\boldsymbol{x}_0)}{f_{xx}(\boldsymbol{x}_0)}(x-x_0)(y-y_0)+\frac{f_{xy}^2(\boldsymbol{x}_0)}{f_{xx}^2(\boldsymbol{x}_0)}(y-y_0)^2\right]$$
$$+f_{yy}(\boldsymbol{x}_0)(y-y_0)^2-\frac{f_{xy}^2(\boldsymbol{x}_0)}{f_{xx}(\boldsymbol{x}_0)}(y-y_0)^2.$$

This becomes

$$f_{xx}(\boldsymbol{x}_0)\left[(x-x_0)+\frac{f_{xy}(\boldsymbol{x}_0)}{f_{xx}(\boldsymbol{x}_0)}(y-y_0)\right]^2+\frac{1}{f_{xx}(\boldsymbol{x}_0)}[f_{xx}(\boldsymbol{x}_0)f_{yy}(\boldsymbol{x}_0)-f_{xy}^2(\boldsymbol{x}_0)](y-y_0)^2,$$

which we can write as

(4) $$f_{xx}(\boldsymbol{x}_0)\left[(x-x_0)+\frac{f_{xy}(\boldsymbol{x}_0)}{f_{xx}(\boldsymbol{x}_0)}(y-y_0)\right]^2+\frac{1}{f_{xx}(\boldsymbol{x}_0)}(y-y_0)^2D.$$

Now if $f_{xx}(\boldsymbol{x}_0)>0$, the first term in (4) is positive for $\boldsymbol{x}\neq\boldsymbol{x}_0$. If also $D>0$, then the second term is also positive, i.e., (2) is positive. Thus (a) follows from Theorem 9.6(a).

If $f_{xx}(\boldsymbol{x}_0)<0$ and $D>0$, then both terms in (4) are negative for $\boldsymbol{x}\neq\boldsymbol{x}_0$, so (b) follows from Theorem 9.6(b).

If $D<0$, then there are several cases. If $f_{xx}(\boldsymbol{x}_0)>0$, then we find that (4) is positive if $y=y_0$ and x is close to x_0, because (4) reduces to

$$f_{xx}(\boldsymbol{x}_0)(x-x_0)^2.$$

Suppose, however, that we take y close to y_0 and

$$x=x_0-\frac{f_{xy}(\boldsymbol{x}_0)}{f_{xx}(\boldsymbol{x}_0)}(y-y_0).$$

The first term in (4) is then 0 and the second is negative since $D<0$. In this case, then, (4) is negative. Hence $\boldsymbol{x}_0$ is a saddle point by Theorem 9.6(c). A similar argument applies if $f_{xx}(\boldsymbol{x}_0)<0$. If $f_{xx}(\boldsymbol{x}_0)=0$ but $f_{yy}(\boldsymbol{x}_0)\neq 0$, then we can factor out $f_{yy}(\boldsymbol{x}_0)$ in (3) and complete the square as above; we then reason as we just have to conclude that $\boldsymbol{x}_0$ is a saddle point. (See Exercise 26 below.) If both $f_{xx}(\boldsymbol{x}_0)$ and $f_{yy}(\boldsymbol{x}_0)$ are 0, then (3) reduces to $2f_{xy}(\boldsymbol{x}_0)(x-x_0)(y-y_0)$. Whatever sign $f_{xy}(\boldsymbol{x}_0)$ has, we can make the expression sometimes positive and sometimes negative by taking x slightly greater or smaller than x_0, and y slightly greater than y_0. Then by Theorem 9.6(c), $\boldsymbol{x}_0$ is a saddle point. Hence (c) is proved. Thus Theorem 9.7 is fully proved.

QED

Theorem 9.7 is used in the same way as the second derivative test of elementary calculus. After finding all the critical points of a function $f:\boldsymbol{R}^2\to\boldsymbol{R}$, we test them in Theorem 9.7.

9.8 **EXAMPLE.** Find and classify all candidates for extreme points of f given by $f(x, y)=3x+12y-x^3-y^3$.

Solution. First we compute $\boldsymbol{f}'(\boldsymbol{x})=(3-3x^2,\ 12-3y^2)$. Next we find the critical points by setting $\boldsymbol{f}'(\boldsymbol{x})=\boldsymbol{0}$ and solving:

$$\begin{cases} 3-3x^2=0, & x^2=1, & x=\pm 1 \\ 12-3y^2=0, & y^2=4, & y=\pm 2. \end{cases}$$

So we have $(1,-2)$, $(1,2)$, $(-1,2)$, and $(-1,-2)$ as critical points. To classify these, we need the second partial derivatives:

$$f_{xx}=\frac{\partial}{\partial x}(3-3x^2)=-6x$$

$$f_{xy}=0$$

$$f_{yy}=\frac{\partial}{\partial y}(12-3y^2)=-6y$$

Then

$$\det f''(x, y)=\det\begin{pmatrix}-6x & 0\\ 0 & -6y\end{pmatrix}=36xy.$$

At $(1,2)$ we have $f_{xx}<0$, $\det f''(x, y)=72>0$. So we have a local maximum at $(1,2)$ by Theorem 9.7(b). Here $f(1,2)=3+24-1-8=18$. At $(1,-2)$ and $(-1,2)$ we have $\det f''(x, y)=-72<0$, so these give saddle points by Theorem 9.7(c). At $(-1,-2)$ we have $f_{xx}>0$, $\det f''(x, y)=72>0$, so we have a local minimum at $(-1,-2)$ by Theorem 9.7(a). Here $f(-1,-2)=-3-24+1+8=-18$.

The second derivative test embodied in Theorem 9.7 is quite widely, but not universally, applicable. What cases are excluded?

Notice that we have said nothing about the case $D=\det f''(x, y)=0$. This corresponds to the second derivative being 0 in the one-variable case, a circumstance in which the one-variable second derivative test fails to give information. So it is not surprising that we make no assertions in this case. At first glance it may also seem we have excluded $f_{xx}(\boldsymbol{x}_0)=0$, but we really haven't. If $f_{xx}(\boldsymbol{x}_0)=0$, then $\det f''(x, y)=-f_{xy}^2(\boldsymbol{x}_0)$ will usually be negative, so part (c) of Theorem 9.7 will apply. If $\det f''(x, y)=0$, however, then this is a case already excluded. When $\det f''(x, y)=0$, anything can happen!

9.9 EXAMPLE. Find and classify all critical points of
(i) $f(x, y) = x^2 - 2xy + y^2$
(ii) $f(x, y) = x^3 - 3xy^2 + y^2$.

Solution. (i) Here $\mathbf{f}'(x, y) = (2x - 2y, -2x + 2y) = \mathbf{0}$ when $y = x$. So we have an entire plane of critical points, with $f_{xx} = 2$, $f_{yy} = 2$, and $f_{xy} = -2$. Then

$$\det f''(x, y) = \det\begin{pmatrix} 2 & -2 \\ -2 & 2 \end{pmatrix} = 4 - 4 = 0.$$

So the test gives no information. We can see, however, that $f(x, y) = (x - y)^2$, so all the critical points $y = x$ yield absolute minima.

(ii) Here $\mathbf{f}'(x, y) = (3x^2 - 3y^2, -6xy + 2y) = \mathbf{0}$ when

$$\begin{cases} x^2 - y^2 = (x - y)(x + y) = 0 \\ -6xy + 2y = 2y(-3x + 1) = 0. \end{cases}$$

Thus $y = \pm x$ and ($y = 0$ or $x = \frac{1}{3}$). That is, $(0, 0)$, $(\frac{1}{3}, \frac{1}{3})$, and $(\frac{1}{3}, -\frac{1}{3})$ are the critical points. We also have

$$f_{xx} = 6x, \qquad f_{yy} = -6x + 2, \qquad f_{xy} = -6y.$$

At $(\frac{1}{3}, \pm\frac{1}{3})$,

$$f_{xx} = 2, \qquad f_{yy} = 0, \qquad f_{xy} = -6(\pm\tfrac{1}{3}) = \mp 2.$$

Therefore,

$$D = \det f''(\tfrac{1}{3}, \pm\tfrac{1}{3}) = 2 \cdot 0 - 4 = -4 < 0.$$

Theorem 9.7(c) then tells us that $(\frac{1}{3}, \frac{1}{3})$ and $(\frac{1}{3}, -\frac{1}{3})$ are saddle points. At $(0, 0)$

$$f_{xx} = 0, \qquad f_{yy} = 2, \qquad f_{xy} = 0.$$

Thus $D = \det f''(0, 0) = 0$. No conclusion follows from Theorem 9.7, but we see that if $y = 0$, the function $f(x, y) = x^3$ has no local extremum at $x = 0$. Therefore, $(0, 0)$ is a saddle point.

There is a second derivative test for functions $f: \mathbf{R}^n \to \mathbf{R}$, but we leave it for advanced calculus. It is best stated in terms of the theory of quadratic forms. To give you an idea of the complexity involved when quadratic forms are avoided, we state without proof the form of the general second derivative test for $f: \mathbf{R}^3 \to \mathbf{R}$.

9.10 THEOREM. Suppose that $f: \mathbf{R}^3 \to \mathbf{R}$ is differentiable on an open set U containing the critical point $\mathbf{x}_0$. Suppose that all second partial derivatives are continuous on U. For

$$f''(x) = H_f = \begin{pmatrix} f_{xx} & f_{xy} & f_{xz} \\ f_{yx} & f_{yy} & f_{yz} \\ f_{zx} & f_{zy} & f_{zz} \end{pmatrix},$$

we have

(a) if $f_{xx} > 0$,

$$\det\begin{pmatrix} f_{xx} & f_{xy} \\ f_{yx} & f_{yy} \end{pmatrix} > 0,$$

and $\det f'' > 0$ at $\boldsymbol{x}_0$, then $f(\boldsymbol{x}_0)$ is a local minimum;

(b) if $f_{xx} < 0$,

$$\det\begin{pmatrix} f_{xx} & f_{xy} \\ f_{yx} & f_{yy} \end{pmatrix} > 0,$$

and $\det f'' < 0$ at $\boldsymbol{x}_0$, then $f(\boldsymbol{x}_0)$ is a local maximum.

(Notice that there is no simple criterion for a saddle point in this case.)

EXERCISES 4.9

In Exercises 1 to 10, find and classify all critical points of the given functions.

1 $f(x, y) = 4x^2 - xy + y^2$

2 $f(x, y) = 6x - 4y - x^2 - 2y^2$

3 $f(x, y) = x^2 - xy - y^2 + 5y - 1$

4 $f(x, y) = x^3 + y^3 + 3x^2 - 3y^2 - 8$

5 $f(x, y) = x^4 + y^2 - 8x^2 - 6y + 16$

6 $f(x, y) = 2x^4 + y^4 - 2x^2 - 2y^2 + 3$

7 $f(x, y, z) = x^2 + 3y^2 + 4z^2 - 2xy - 2yz + 2xz + 3$ (*Hint:* Use Theorem 9.10.)

8 $f(x, y, z) = -x^2 - 2y^2 - z^2 + yz + 5$

9 $f(x, y, z) = 2x^2 + y^2 + 2z^2 + 2xy + 2yz + 2xz + x - 3z - 5$

10 $f(x, y, z) = -2x^2 - y^2 - 3z^2 + 2xy - 2xz + 1$

11 A box for oranges is to be made in the shape of a rectangular parallelepiped with no lid from 12 square feet of cardboard. Find the dimensions and volume of the largest such box.

12 A tool box is to be constructed from 24 square feet of lumber. It will be in the shape of a rectangular parallelepiped (including a lid). Find the dimensions for maximum volume.

13 A manufacturer wants to make boxes (without lids) of 4 cubic feet volume so as to minimize the amount of cardboard used. What dimensions should he make his boxes?

14 In Exercise 13, suppose that the boxes are made of two kinds of cardboard: the bottom is to be three times as thick (and as expensive) as the sides. Find the dimensions of the most economical box of 4 cubic feet capacity that can be made. (Assume that the sides cost \$1.00 per square foot.)

15 In the 1930's a demand function z for beef in the United States was developed, which was of the form $z = a + bx + cy + dw$ where x is the price of beef, y is a measure of the price of pork, lamb, and poultry, and w is a measure of the income of the consuming public. Here a, c, and d are positive constants, and b is a negative constant. Show that there are no points (x_0, y_0, z_0, w_0) where z achieves a local extreme value.

16 Heat loss through each wall of a structure is proportional to the surface area of the wall. Insulation is added to drive down the proportionality constant. Suppose that a house is being designed with a volume of 20,000 cubic feet. The house, of length x, width y, and height z feet, is calculated to have daily heat loss of

$$f(x, y) = 20xy + 10yz + 10xz$$

at 20°F outside temperature and 20°C inside temperature. Find the dimensions of the house for minimum daily heat loss at these temperatures. Would you find a house of these dimensions attractive?

17 A firm produces two products, A and B. The daily demand function for A is $x = 14 - 2p + 2q$, and the daily demand function for B is $y = 16 + 2p - 6q$, where p and q are the respective unit prices of A and B. It costs \$3.00 to produce a unit of A and \$2.00 to produce a unit of B.
(a) Write a formula for the company's profit from producing A and B.
(b) Find the prices it should charge to maximize its profits. How many units should be produced per day for maximum profit?

18 Refer to Exercise 17. Suppose that the demand functions are

$$x = 16 - 2p + 2q$$
$$y = 8 + 2p - 4q$$

and the production costs are to be the same as in Exercise 17. Find the prices to be charged for maximum profits.

The method of least squares is a technique for finding the straight line $y = mx + b$ which "best fits" experimentally collected data. (See Figure 9.4.) In Exercises 19 to 22 we discuss this method.

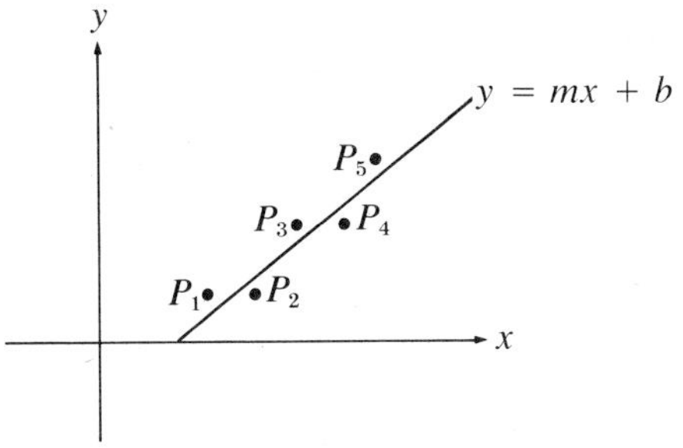

FIGURE 9.4

19 Suppose that the data are plotted as $P_i(x_i, y_i)$, $1 \le i \le n$. The method minimizes the sum $z(m, b)$ of the squares of the errors $|mx_i + b - y_i|$. Write a formula for $z(m, b)$ and show that

$$\frac{\partial z}{\partial m} = 2 \sum_{i=1}^{n} x_i(mx_i + b - y_i),$$

$$\frac{\partial z}{\partial b} = 2 \sum_{i=1}^{n} (mx_i + b - y_i).$$

20 Show that a critical point for $z(m, b)$ is given by

$$m = \frac{n \sum_{i=1}^{n} x_i y_i - \sum_{i=1}^{n} x_i \sum_{i=1}^{n} y_i}{n \sum_{i=1}^{n} x_i^2 - \left(\sum_{i=1}^{n} x_i\right)^2}$$

$$b = \frac{\left(\sum_{i=1}^{n} y_i\right)\left(\sum_{i=1}^{n} x_i^2\right) - \left(\sum_{i=1}^{n} x_i\right)\left(\sum_{i=1}^{n} x_i y_i\right)}{n \sum_{i=1}^{n} x_i^2 - \left(\sum_{i=1}^{n} x_i\right)^2}$$

21 Use the formulas of Exercise 20 to find the line best fitting the points (0, 1), (2, 3), (3, 6), and (4, 8).

22 Repeat Exercise 21 for the points $(-2, 1)$, $(0, 3)$, $(1, 6)$, and $(2, 8)$.

23 In mechanics, a *conservative force field* is one given by a formula $\boldsymbol{F} = \nabla p(x, y)$, where $-p(x, y)$ is the potential energy at the point (x, y). A *stable equilibrium point* is a point (x_0, y_0) where p has a local minimum. An *unstable equilibrium point* is a point (x_0, y_0) such that either $\boldsymbol{p}'(x_0, y_0)$ fails to exist or $\boldsymbol{p}'(x_0, y_0) = \boldsymbol{0}$, but p does not have a local minimum at (x_0, y_0). Suppose that a particle moves in the xy-plane with potential energy $-\sqrt{x^2+y^2}$. Find all equilibrium points and determine whether there is any stable equilibrium point.

24 Refer to Exercise 23. Suppose that a particle moves in space with potential energy $-\sqrt{x^2+y^2+z^2}$. Find all equilibrium points and determine whether there is a stable equilibrium point.

25 If x and y are nonnegative, then show that the minimum value of

$$f(x, y) = x + y - 2\sqrt{xy}$$

is 0. That is, show that the geometric mean of x and y is less than or equal to the arithmetic mean of x and y.

26 Prove Theorem 9.7(c) for the case $f_{xx}(\boldsymbol{x}_0) = 0$, $f_{yy}(\boldsymbol{x}_0) \neq 0$.

We can actually draw slightly more information from the second derivative test than is stated in Theorem 9.7 itself. This is the point of Exercises 27 and 28.

27 Suppose that f and $\boldsymbol{x}_0$ are as in Theorem 9.7. If $D = 0$ and $f_{xx}(\boldsymbol{x}_0) > 0$, then show that $f(\boldsymbol{x}_0)$ is a local minimum of the function $z = f(x, y_0)$ of x alone. What is the analogous result for $f(x_0, y)$?

28 Suppose that f and $\boldsymbol{x}_0$ are as in Theorem 9.7. If $D = 0$ and $f_{xx}(\boldsymbol{x}_0) < 0$, then show that $f(\boldsymbol{x}_0)$ is a local maximum of the function $z = f(x, y_0)$ of x alone. What is the analogous result for $f(x_0, y)$?

29 Suppose that we decide to define *positivity* for a vector $\boldsymbol{x} = (x, y)$ in $\boldsymbol{R}^2$ by saying that

$$(5) \quad \boldsymbol{x} > \boldsymbol{0}$$

means $x > 0$ and $y > 0$. Formulate an analogue for $f: \boldsymbol{R}^2 \to \boldsymbol{R}$ of the first derivative test of elementary calculus using (5), but then show by an example that your analogue fails to be correct.

10 LAGRANGE MULTIPLIERS

The technique developed in the preceding section is applicable to a wide class of extreme value problems, including many arising in applications (cf. Problems 11 to 24 in Exercises 4.9). However, it is not directly applicable to *constrained extreme value problems*, in which the variables x and y (and possibly z) are not free to assume any real values, but rather are restricted by some side condition (called a *constraint*).

For example, parcel post regulations stipulate that the *height* plus *girth* of a package cannot exceed 100 inches (see Figure 10.1), where the height is defined as the length of the longest edge. Suppose that we ask, "What is the box of largest volume that can be made which is mailable?" If we denote height by z, then we are asking to maximize $V = xyz$ subject to the constraint that $2x + 2y + z = 100$. We could eliminate z by using the constraint equation in the form $z = 100 - 2x - 2y$, and then substitute this into the formula for V. (This approach was actually used in Problems 11 to 14 in Exercises 4.9). In this section we develop a method originated by the French mathematician Joseph L. Lagrange (1736–1813)

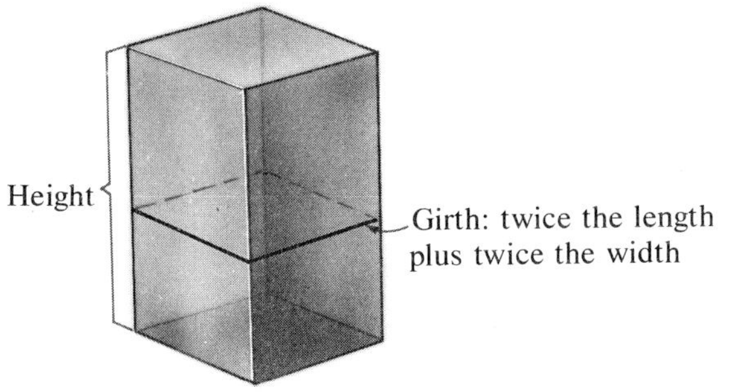

FIGURE 10.1

that makes this unnecessary. In the present example, this may not seem terribly vital, since z is so easily eliminated from the constraint equation. But consider the problem of maximizing or minimizing the electrical potential $f(x, y) = 1/\sqrt{x^2+y^2}$ over the edge of a television screen with equation $2x^4+3y^4=32$ (see Figure 10.2). Here the complexity of the problem will increase markedly if we try to eliminate y from the constraint equation. Thus, Lagrange's method seems worth pursuing.

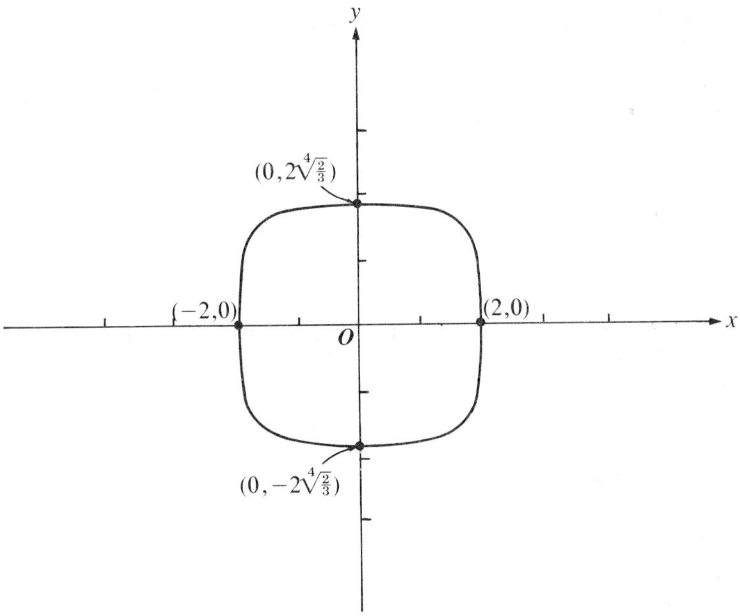

FIGURE 10.2

So let's now consider the general problem of minimizing a differentiable function $f: \mathbf{R}^2 \to \mathbf{R}$ subject to the constraint $g(x, y) = 0$. For example, $f(x, y)$ might be $1/\sqrt{x^2+y^2}$ and $g(x, y)$ might be $2x^4+3y^4-32$. To get an idea of where Lagrange's method came from, let's look at this from a geometric point of view. We are seeking the point (or points) on $g(x, y)=0$ where f takes its minimum value on that curve. Now for *any* value k, f assumes the value k precisely on the level curve $f(x, y) = k$. In our television example above, those level curves are circles $x^2+y^2=\dfrac{1}{k^2}$ that radiate inward toward the origin as k increases. Eventually they reach the curve $g(x, y)=0$ as in Figure 10.3 (p. 222). The first points of intersection $P_0(x_0, y_0)$ are the points on $g(x, y)=0$ where $f(x, y)=k_0$. No other points on $g(x, y)=0$ give so small a value to f, for any other point P_1 on

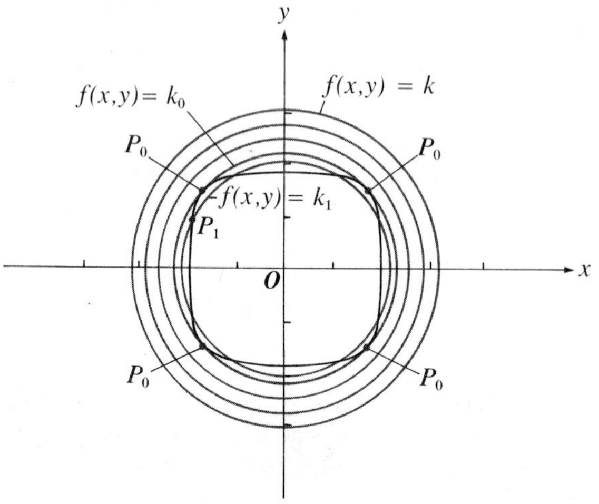

FIGURE 10.3

$g(x, y)=0$ is on a level curve of smaller radius $\dfrac{1}{k_1}$ corresponding to a *larger* value $k_1 > k_0$ (see Figure 10.3). We have now solved our problem geometrically.

What remains is to devise a means of finding $P_0=(x_0, y_0)$. Suppose then that P_0 is a point where $f:\boldsymbol{R}^2 \to \boldsymbol{R}$ achieves an extreme value k_0 on the curve $g(x, y)=0$. Then P_0 is a point of intersection of $f(x, y)=k_0$ and $g(x, y)=0$. To find such a point, we might solve for y or x in terms of the other variable in $g(x, y)=0$, and then substitute into $f(x, y)=k_0$. The Implicit Function Theorem 6.2 tells us that we can do this if

$$\nabla g(x_0, y_0)=\left(\frac{\partial g}{\partial x}(x_0, y_0), \frac{\partial g}{\partial y}(x_0, y_0)\right) \neq \mathbf{0}$$

and if ∇g has continuous entries. For in that case we have either $y=y(x)$ near x_0 or $x=x(y)$ near y_0. Suppose for definiteness that we have $y=y(x)$. Then differentiation of $g(x, y(x))=0$ using Chain Rule 5.3 gives us

$$\frac{dg}{dx}=\nabla g(x, y)\cdot(1, y'(x))=0.$$

Hence,

$$\nabla g(x_0, y_0)\cdot(1, y'(x_0))=0.$$

Thus since $\nabla g(x_0, y_0)\neq \mathbf{0}$, it is perpendicular to $(1, y'(x_0))$. Also, since f has an extreme value at (x_0, y_0), we have

$$\frac{df}{dx}(x_0, y_0)=\nabla f(x_0, y_0)\cdot(1, y'(x_0))=0.$$

If $\nabla f(x_0, y_0)\neq \mathbf{0}$, then it is also perpendicular to $(1, y'(x_0))$. Therefore, $\nabla f(x_0, y_0)$ is parallel to $\nabla g(x_0, y_0)$, and we have

(1) $$\nabla f(x_0, y_0)=\lambda \nabla g(x_0, y_0)$$

for some $\lambda \in \boldsymbol{R}$. Moreover, even if $\nabla f(x_0, y_0) = \mathbf{0}$, Equation (1) still holds since we could use $\lambda = 0$.

A similar analysis applies to the case of three variables. If we want to find extreme values of $f(x, y, z) = 0$ on a surface $g(x, y, z) = 0$, we assume that ∇f and ∇g are continuous and $\nabla g(\boldsymbol{x}_0) \neq 0$ at an extreme point $\boldsymbol{x}_0$. If, for instance, the third component $g_z(\boldsymbol{x}_0) \neq 0$, then Theorem 6.3 guarantees that $g(x, y, z) = 0$ defines $z = z(x, y)$ and then $f(x, y, z) = f(x, y, z(x, y))$. At an extreme point we have, from Theorem 9.3,

$$\frac{\partial f}{\partial x} = 0 = \frac{\partial f}{\partial y},$$

so by Chain Rule 5.7

$$\nabla f(\boldsymbol{x}_0) \cdot \left(1, 0, \frac{\partial z}{\partial x}(\boldsymbol{x}_0)\right) = 0 = \nabla f(\boldsymbol{x}_0) \cdot \left(0, 1, \frac{\partial z}{\partial y}(\boldsymbol{x}_0)\right).$$

Also, since $g(x, y, z(x, y)) = 0$, we similarly obtain

$$\frac{\partial g}{\partial x}(\boldsymbol{x}_0) = \nabla g(\boldsymbol{x}_0) \cdot \left(1, 0, \frac{\partial z}{\partial x}(\boldsymbol{x}_0)\right) = 0$$

$$= \nabla g(\boldsymbol{x}_0) \cdot \left(0, 1, \frac{\partial z}{\partial x}(\boldsymbol{x}_0)\right) = \frac{\partial g}{\partial y}(\boldsymbol{x}_0).$$

Hence if $\nabla f(\boldsymbol{x}_0) \neq \mathbf{0}$, then $\nabla f(\boldsymbol{x}_0)$ and $\nabla g(\boldsymbol{x}_0)$ are perpendicular to $(1, 0, z_x(\boldsymbol{x}_0))$ and $(0, 1, z_y(\boldsymbol{x}_0))$. The latter vectors determine the tangent plane to $g(x, y, z) = 0$ (cf. p. 150) at (x_0, y_0, z_0). Thus $\nabla f(\boldsymbol{x}_0)$ and $\nabla g(\boldsymbol{x}_0)$ are both perpendicular to this tangent plane, and hence are parallel. We therefore have

(1) $$\nabla f(\boldsymbol{x}_0) = \lambda \nabla g(\boldsymbol{x}_0)$$

for some $\lambda \in \boldsymbol{R}$. Again, (1) continues to hold even if $\nabla f(\boldsymbol{x}_0) = \mathbf{0}$ since $\lambda = 0$ is permissible. Similar reasoning applies if it is $g_x(\boldsymbol{x}_0)$ or $g_y(\boldsymbol{x}_0)$ that is a nonzero coordinate of $\nabla g(\boldsymbol{x}_0)$. In this way, the following result is proved.

10.1 THEOREM. Suppose that $f: \boldsymbol{R}^n \to \boldsymbol{R}$ ($n = 2$ or 3) is differentiable with continuous partial derivatives. Suppose that f has an extreme value on $g(\boldsymbol{x}) = 0$ at the point $\boldsymbol{x}_0 \in \boldsymbol{R}^n$. Suppose also that g has continuous partial derivatives and that $\nabla g(\boldsymbol{x}_0) \neq \mathbf{0}$. Then for some $\lambda \in \boldsymbol{R}$

(1) $$\nabla f(x_0) = \lambda \nabla g(\boldsymbol{x}_0).$$

Theorem 10.1 actually holds for *any* n, but we leave the more general discussion for advanced calculus. From Theorem 10.1 we can give Lagrange's method for finding all candidates $\boldsymbol{x}_0$ for points of $g(\boldsymbol{x}) = 0$ where $f(\boldsymbol{x})$ achieves a local extreme value. (The constant λ is called a *Lagrange multiplier.*)

10.2 **LAGRANGE MULTIPLIER METHOD.** To find all candidates $\boldsymbol{x}_0$ for extreme values of $f:\boldsymbol{R}^n \to \boldsymbol{R}$ ($n=2$ or 3), subject to the constraint $g(\boldsymbol{x})=0$, solve the system of equations

$$\begin{cases} \nabla f(\boldsymbol{x}) = \lambda \nabla g(\boldsymbol{x}) \\ g(\boldsymbol{x}) = 0 \end{cases} \tag{2}$$

for $\boldsymbol{x}$ and λ. Each solution $\boldsymbol{x}$ is a candidate for an extreme value point.

10.3 **EXAMPLE.** Find the dimensions of the box of maximum volume whose height plus girth is 100 inches.

Solution. As on p. 220, we have to maximize $V=xyz$ subject to the constraint $g(x, y, z)=2x+2y+z-100=0$. Here $\nabla V=(yz, xz, xy)$ and $\nabla g=(2, 2, 1)$. Applying 10.2, we have the system

$$\begin{cases} (yz, xz, xy)=\lambda(2, 2, 1) \\ 2x+2y+z-100=0. \end{cases}$$

These give the scalar equations

$$yz=2\lambda, \qquad xz=2\lambda, \qquad xy=\lambda, \qquad 2x+2y+z-100=0.$$

Subtraction of the second equation from the first gives

$$(y-x)z=0.$$

Since $z \neq 0$ for a box, we get $y=x$. The third equation then gives $x^2=\lambda$, which, when substituted into the second equation, yields

$$z=\frac{2\lambda}{x}=\frac{2x^2}{x}=2x.$$

We can now substitute for y and z in the final equation to obtain

$$\begin{aligned} 2x+2x+2x &= 100, \\ 6x &= 100, \\ x &= \tfrac{100}{6}=\tfrac{50}{3}. \end{aligned}$$

Hence, $y=\frac{50}{3}$ and $z=\frac{100}{3}$. Therefore, the dimensions of the mailable box of largest volume are $\frac{50}{3}$ by $\frac{50}{3}$ by $\frac{100}{3}$ inches.

You may object that we did not *test* our candidate $(\frac{50}{3}, \frac{50}{3}, \frac{100}{3})$ to verify that it truly gives the maximum of V on the plane $2x+2y+z=100$. The reason is that we found only *one* candidate for a solution to the given problem, which, on physical grounds, *has* a solution. Thus our candidate must *be* that solution. If this kind of interplay between mathematical and nonmathematical reasoning strikes you as too lacking in rigor, then you might consult an advanced calculus text for a second derivative test for Lagrange multipliers involving quadratic forms. (Or, you could work the problem by eliminating z and using Theorem 9.7.)

10.4 EXAMPLE. Find the extreme values of $f(x, y) = 1/\sqrt{x^2+y^2}$ on the curve $2x^4+3y^4=32$.

Solution. Here $g(x, y) = 2x^4+3y^4-32$. So $\nabla g = (8x^3, 12y^3)$. We also have $f(x, y) = (x^2+y^2)^{-1/2}$, so we get

$$\nabla f = \left(\frac{-x}{(x^2+y^2)^{3/2}}, \frac{-y}{(x^2+y^2)^{3/2}}\right).$$

Applying 10.2, we get

$$(3) \qquad \begin{cases} \left(\dfrac{-x}{(x^2+y^2)^{3/2}}, \dfrac{-y}{(x^2+y^2)^{3/2}}\right) = \lambda(8x^3, 12y^3) \\ 2x^4+3y^4=32 \end{cases}$$

That is,

$$\begin{cases} \dfrac{-x}{(x^2+y^2)^{3/2}} = 8\lambda x^3, \\ \dfrac{-y}{(x^2+y^2)^{3/2}} = 12\lambda y^3, \\ 2x^4+3y^4=32. \end{cases}$$

In the top equation, $x=0$ is a possible root. By the bottom equation, the corresponding y satisfies $y^4=\frac{32}{3}$, so $y=\pm\sqrt[4]{\frac{32}{3}}=\pm 2\sqrt[4]{\frac{2}{3}}$. In the middle equation, $y=0$ is a possible root, and from the bottom equation the corresponding x is ± 2. For other candidates, $x \neq 0$ and $y \neq 0$. Then we can cancel x in the first equation and y in the second equation. This gives

$$\begin{cases} \dfrac{-1}{(x^2+y^2)^{3/2}} = 8\lambda x^2, \\ \dfrac{-1}{(x^2+y^2)^{3/2}} = 12\lambda y^2. \end{cases}$$

So $8\lambda x^2 = 12\lambda y^2$. Now, λ isn't zero here; if it were zero, then x and y would also be zero by the first equation in (3). Hence we can cancel λ and get $x^2=\frac{12}{8}y^2=\frac{3}{2}y^2$. Substituting this into $2x^4+3y^4=32$, we get

$$2(\tfrac{9}{4}y^4)+3y^4=32,$$

$$9y^4+6y^4=64,$$

$$15y^4=64,$$

$$y^4=\tfrac{64}{15}$$

$$y=\pm\sqrt[4]{\frac{64}{15}}=\pm\frac{2\sqrt{2}}{\sqrt[4]{15}}.$$

Then

$$x^2 = \frac{3}{2} y^2 = \frac{3}{2} \cdot \frac{8}{\sqrt{15}} = \frac{12}{\sqrt{15}}, \quad \text{so} \quad x = \pm \frac{2\sqrt{3}}{\sqrt[4]{15}}.$$

Our candidates for extreme values of f are then $(\pm 2, 0)$, $(0, \pm 2\sqrt[4]{\frac{2}{3}})$, and $\frac{1}{\sqrt[4]{15}}(\pm 2\sqrt{3}, \pm 2\sqrt{2})$. To test these points, we compute f at each of them.

$$f(\pm 2, 0) = \frac{1}{\sqrt{4+0}} = \tfrac{1}{2}.$$

$$f(0, \pm 2\sqrt[4]{\tfrac{2}{3}}) = \frac{1}{\sqrt{0+4\sqrt{\frac{2}{3}}}} = \tfrac{1}{2}\sqrt[4]{\tfrac{3}{2}} > \tfrac{1}{2}\sqrt[4]{1} = f(\pm 2, 0).$$

$$f\left(\frac{1}{\sqrt[4]{15}}(\pm 2\sqrt{3}, \pm 2\sqrt{2})\right) = \frac{1}{\frac{1}{\sqrt[4]{15}}\sqrt{12+8}} = \sqrt{\frac{\sqrt{15}}{20}} = \frac{1}{2}\sqrt{\frac{\sqrt{15}}{5}}$$

$$= \tfrac{1}{2}\sqrt[4]{\tfrac{15}{25}} = \tfrac{1}{2}\sqrt[4]{\tfrac{3}{5}} < \tfrac{1}{2} = f(\pm 2, 0).$$

Thus, the minimum value of f on the curve $2x^4 + 3y^4 = 32$ is $\frac{1}{2}\sqrt[4]{\frac{3}{5}}$, occurring at the four points $\frac{1}{\sqrt[4]{15}}(\pm 2\sqrt{3}, \pm 2\sqrt{2})$. The maximum value of f on the same curve is $\frac{1}{2}\sqrt[4]{\frac{3}{2}}$, occurring at the two points $(0, \pm 2\sqrt[4]{\frac{2}{3}})$. We have thus solved the electrical potential problem posed on p. 221.

The method of Lagrange multipliers works so well for extreme value problems involving *one* constraint that you may be wondering whether this sort of method can be utilized for extreme value problems involving *several* constraints. Suppose, then, that we wish to find the extreme values of a function f subject to m constraints $g_1(\boldsymbol{x}) = 0$, $g_2(\boldsymbol{x}) = 0, \ldots, g_m(\boldsymbol{x}) = 0$. Is there a method like that of 10.2 for this problem? There is, and let's see if we can discover it. First, let's examine 10.2 more closely. There we have a single Lagrange multiplier λ multiplying the single gradient $\nabla g(\boldsymbol{x})$. What analogue do we know for such a multiplication when we have m λ's and m gradients? We might try the "dot product" of $\boldsymbol{\lambda} = (\lambda_1, \lambda_2, \ldots, \lambda_m)$ and $(\nabla g_1(\boldsymbol{x}), \nabla g_2(\boldsymbol{x}), \ldots, \nabla g_m(\boldsymbol{x}))$. Now in Equations (2) the condition $\nabla f(\boldsymbol{x}) = \lambda \nabla g(\boldsymbol{x})$ can be written $\nabla(f(\boldsymbol{x}) - \lambda g(\boldsymbol{x})) = 0$, so our candidates for extrema are the critical points of the function $f - \lambda g$. In the case of several constraints, then, it would seem that we should find the critical points of $f - \boldsymbol{\lambda} \cdot \boldsymbol{g}$, where $\boldsymbol{\lambda} = (\lambda_1, \lambda_2, \ldots, \lambda_m)$ and $\boldsymbol{g}(\boldsymbol{x}) = (g_1(\boldsymbol{x}), g_2(\boldsymbol{x}), \ldots, g_m(\boldsymbol{x}))$. This is exactly right.

10.5 **THEOREM.** Suppose that $n = 2$ or 3, and $f: \boldsymbol{R}^n \to \boldsymbol{R}$ is differentiable. Suppose that the equations $g_1(\boldsymbol{x}) = 0$, $g_2(\boldsymbol{x}) = 0, \ldots, g_m(\boldsymbol{x}) = 0$ (where $g_i: \boldsymbol{R}^n \to \boldsymbol{R}$ has continuous partial derivatives) define a set S in $\boldsymbol{R}^n$. If f achieves an extreme value on S at a point $\boldsymbol{x}_0$ where the gradient vectors $\nabla g_1(\boldsymbol{x}_0), \nabla g_2(\boldsymbol{x}_0), \ldots, \nabla g_m(\boldsymbol{x}_0)$ are linearly independent (see Section 1.7), then for some $\boldsymbol{\lambda} = (\lambda_1, \ldots, \lambda_m) \in \boldsymbol{R}^m$, $\boldsymbol{x}_0$ is a critical point for the function $f - \boldsymbol{\lambda} \cdot \boldsymbol{g}$, where $\boldsymbol{g} = (g_1, g_2, \ldots, g_m)$. Thus

(4) $$\nabla f(\boldsymbol{x}_0) - \lambda_1 \nabla g_1(\boldsymbol{x}_0) - \lambda_2 \nabla g_2(\boldsymbol{x}_0) - \ldots - \lambda_m \nabla g_m(\boldsymbol{x}_0) = \boldsymbol{0}.$$

The proof, which we leave for advanced calculus, involves more linear algebra and analysis than we have developed.

We can now state the analogue of 10.2 for several constraints.

10.6 **LAGRANGE'S MULTIPLIER METHOD FOR m CONSTRAINTS.** To find all candidates for the extreme values of $f:\mathbf{R}^n \to \mathbf{R}$ ($n=2$ or 3) subject to m constraints $g_1(\boldsymbol{x})=0,\ g_2(\boldsymbol{x})=0, \ldots, g_m(\boldsymbol{x})=0$, solve the system of equations

(5)
$$\begin{cases} \nabla f(\boldsymbol{x}) = \lambda_1 \nabla g_1(\boldsymbol{x}) + \lambda_2 \nabla g_2(\boldsymbol{x}) + \ldots + \lambda_m \nabla g_m(\boldsymbol{x}) \\ g_1(\boldsymbol{x})=0,\ g_2(\boldsymbol{x})=0, \ldots, g_m(\boldsymbol{x})=0, \end{cases}$$

for $\boldsymbol{x}$. Each solution $\boldsymbol{x}$ is a candidate for an extreme value point.

10.7 **EXAMPLE.** Find the extreme values of $f(x, y, z) = xy + xz$ on the curve of intersection of the right circular cylinder $x^2 + y^2 = 1$ and the hyperbolic cylinder $xz = 1$. (See Figure 10.4.)

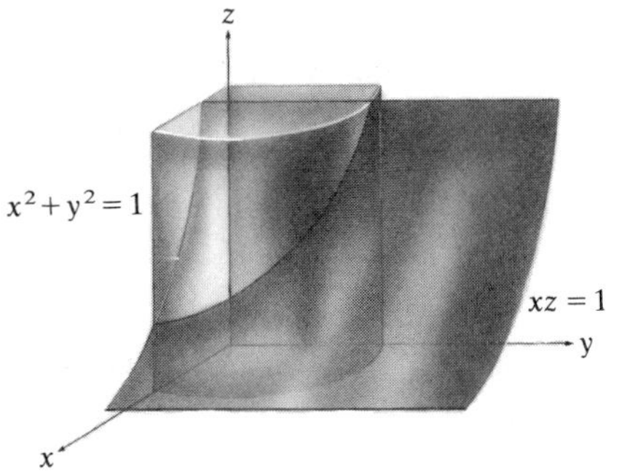

FIGURE 10.4

Solution. Here $f(x, y, z) = xy + xz$ and the two constraints are $g_1(x, y, z) = x^2 + y^2 - 1$ and $g_2(x, y, z) = xz - 1$. Since $\nabla f(x, y, z) = (y + z, x, x)$, the system (5) to be solved is

$$\begin{cases} (y+z, x, x) = \lambda_1(2x, 2y, 0) + \lambda_2(z, 0, x) \\ xz - 1 = 0, \qquad x^2 + y^2 - 1 = 0. \end{cases}$$

We thus have

$$\begin{cases} y + z = 2\lambda_1 x + \lambda_2 z & (6) \\ x = 2\lambda_1 y & (7) \\ x = \lambda_2 x & (8) \\ xz = 1 & (9) \\ x^2 + y^2 = 1. & (10) \end{cases}$$

From (9), x cannot be 0. So (8) gives $\lambda_2 = 1$, and we have $z = 1/x$ from (9). If we put this information together with (7) into (6), then we obtain

$$y + \frac{1}{x} = 2\lambda_1 x + \frac{1}{x},$$
$$y - 4\lambda_1^2 y = 0,$$
$$y(1 - 4\lambda_1^2) = 0.$$

Hence, either $y = 0$ or $\lambda_1^2 = \frac{1}{4}$, i.e., $\lambda_1 = \pm\frac{1}{2}$. But $y = 0$ would give, from (7), that $x = 0$, which we saw is impossible by (9). Hence, $\lambda_1 = \pm\frac{1}{2}$, and so from (7) $x = \pm y$, and then from (10), $2x^2=1$, i.e., $x=\pm\sqrt{\frac{1}{2}}$. Then from (9), $z=\dfrac{1}{x}=\pm\sqrt{2}$. Since we have $y=\pm x$, our candidates for extreme points are $(\sqrt{\frac{1}{2}}, \pm\sqrt{\frac{1}{2}}, \sqrt{2})$ and $(-\sqrt{\frac{1}{2}}, \pm\sqrt{\frac{1}{2}}, -\sqrt{2})$. We calculate f at these points:

$$f(\sqrt{\tfrac{1}{2}}, \sqrt{\tfrac{1}{2}}, \sqrt{2})=\tfrac{1}{2}+1=\tfrac{3}{2}$$
$$f(-\sqrt{\tfrac{1}{2}}, -\sqrt{\tfrac{1}{2}}, -\sqrt{2})=\tfrac{1}{2}+1=\tfrac{3}{2}$$
$$f(\sqrt{\tfrac{1}{2}}, -\sqrt{\tfrac{1}{2}}, \sqrt{2})=-\tfrac{1}{2}+1=\tfrac{1}{2}$$
$$f(-\sqrt{\tfrac{1}{2}}, \sqrt{\tfrac{1}{2}}, -\sqrt{2})=-\tfrac{1}{2}+1=\tfrac{1}{2}.$$

So the maximum value is $\frac{3}{2}$ and the minimum value is $\frac{1}{2}$.

EXERCISES 4.10

In Exercises 1 to 6, find the extreme values of the given function subject to the given constraint.

1 $f(x, y)=x+2y;\ x^2+y^2=1$

2 $f(x, y)=x+y;\ 9x^2+4y^2=36$

3 $f(x, y)=xy;\ x^2+y^2=1$

4 $f(x, y)=xy;\ x^2+y^2=4$

5 $f(x, y, z)=x-y+z;\ x^2+y^2+z^2=1$

6 $f(x, y, z)=2x-y+z;\ x^2+y^2+z^2=4$

In Exercises 7 to 10, solve the indicated problem by the method of Lagrange multipliers.

7 Problem 11, Exercises 4.9

8 Problem 12, Exercises 4.9

9 Problem 13, Exercises 4.9

10 Problem 14, Exercises 4.9

In Problems 11 to 14, solve the given problem using Method 10.6.

11 Find the extreme values of $f(x, y, z)=2x+y+2z$ on the curve of intersection of $x^2+y^2-4=0$ and $x+z=2$.

12 Find the extreme values of $f(x, y, z)=xz+yz$ on the curve of intersection of $x^2+z^2=2$ and $yz=2$.

13 Find the distance from (1, 1, 1) to the line of intersection of the planes $2x+y-z=1$ and $x-y+z=2$. (*Hint:* Let (x, y, z) be a point on the line and minimize the square of its distance from (1, 1, 1).)

14 Find the distance from (1, 1, 0) to the line of intersection of the planes $x-2y+z=1$ and $2x-y-z=2$.

15 What point on the sphere $x^2+y^2+z^2=4$ is farthest from the point $(1, -1, 1)$?

16 What point on the sphere $x^2+y^2+z^2=4$ is farthest from the point (0, 1, 1)?

17 Find the volume of the largest rectangular parallelepiped, with sides parallel to the coordinate planes, that can be inscribed in the ellipsoid $x^2+\dfrac{y^2}{9}+\dfrac{z^2}{4}=1$.

18 Find the volume of the largest rectangular parallelepiped, with sides parallel to the coordinate planes, that can be inscribed in the ellipsoid $\dfrac{x^2}{4}+\dfrac{y^2}{9}+\dfrac{z^2}{4}=1$.

19 Find the maximum and minimum distances from the origin to the ellipsoid $x^2+\dfrac{y^2}{9}+\dfrac{z^2}{4}=1$. (*Hint:* Find the extreme values of the square of the distance.)

20 Find the maximum and minimum distances from the origin to the ellipsoid $4x^2+9y^2+36z^2=1$.

21 What is the rectangle of minimum perimeter and area 1?

22 What is the rectangular parallelepiped of largest volume that can be inscribed in a sphere of radius a?

23 In the production model

$$p(x, y) = ax^{1-e}y^e$$

mentioned after Example 1.7, suppose that $e = \frac{1}{4}$, $a = 6$, labor costs \$10.00 per employee, and capital costs \$20.00 per unit. Assume that \$3000.00 is budgeted for production costs. How many employees and how many units of capital should there be for maximum production?

24 In Exercise 23, verify that for x and y at their optimal levels x_0 and y_0, the ratio of marginal productivity of labor to marginal cost of capital is the same as the ratio of unit labor cost to unit capital costs. (This is a basic law of economics.)

25 In Exercise 23, show that, at the optimal levels x_0 and y_0 of x and y, the expenditure of one additional dollar allows the production of λ_0 additional units of production, where λ_0 is the optimal value of λ, i.e., the value corresponding to $x = x_0$ and $y = y_0$.

REVIEW EXERCISES 4.11

1 Find a unit normal vector and the scalar equation of the tangent plane to the surface given by $z = \sqrt{25 - x^2 - y^2}$ at the point $(2, \sqrt{5}, 4)$.

2 A firm that markets lunch meat uses the demand function

$$d(p, x, y, z, v, w) = a - 10p + 3x + 2y + 2z + 3w - v$$

where a is a constant, p is the firm's price, x, y, and z are prices of competing firms' lunch meats, w is the amount of meat products used, and v is the amount of cereal filler used (all measured in appropriate units). Discuss the marginal demand for the lunch meat and its sensitivity to price and ingredient fluctuation.

3 Classify each of the following as linear or nonlinear functions, and explain.

(a) $f(x) = \pi \tan x$
(b) $f(x) = \pi x$
(c) $f(x) = \pi x - 4$
(d) $f(x, y, z) = ex - 3y + 7z$
(e) $f(x, y, z) = zx - 3y + 7z$
(f) $f(x, y) = 3e^x + 4y$
(g) $f(x, y) = 3x - 4y - 7$
(h) $f(x, y) = -2y$

4 The following problem was posed to a calculus class. A delivery truck traveling roundtrip from A to B averages 20 miles per hour on the leg from A to B. What speed must it average on the return leg so as to have averaged 40 miles per hour for the entire trip? A bright student suggested that 60 miles per hour would do on the return leg. A vote showed the majority of the class in agreement. Are you? Why or why not? What was the key assumption in the students' analysis of the problem? Is average speed a linear function of *any* constituent variable?

5 Find the total derivative of f at $(1, \frac{1}{2}\pi, 3)$ if $f(x, y, z) = x^2yz - z \cos xy$. Why can you be sure that f is differentiable at $(1, \frac{1}{2}\pi, 3)$?

6 A lead chamber used to contain radioactive wastes has inner dimensions 3 meters by 4 meters by 5 meters. Its sides are all 5 centimeters thick. Use the tangent approximation to compute the approximate volume of lead in the container.

7 Find the directional derivative of f at $\mathbf{x}_0 = (1, 2, 3)$ in the direction of $\mathbf{v} = (1, 2, 2)$ if $f(x, y, z) = 8xyz - 2x^2yz - xy^2z + 2xyz^2$. What are the direction and rate of maximum increase of f at $\mathbf{x}_0$?

8 If $w = \sqrt{x^2 + y^2}$, where $\mathbf{x}(t) = (x(t), y(t)) = (t, 2t^2)$, then find $\dfrac{dw}{dt}$ at $\mathbf{x}_0 = (-1, 2)$.

9 If $f(x, y, z) = xyz$, then find a point $\mathbf{c}$ on the segment joining $(1, 1, 1)$ to $(-1, 3, 2)$ such that $f(-1, 3, 2) - f(1, 1, 1) = \mathbf{f}'(\mathbf{c}) \cdot (-2, 2, 1)$.

10 If $w = x^2 - y^2 + 2z^2$, where $x = s^2 - t^2$, $y = 2st$, and $z = \frac{s}{t}$, then find $\frac{\partial w}{\partial s}$ and $\frac{\partial w}{\partial t}$. What is the total derivative of w at the point $(s_0, t_0) = (1, 1)$?

11 Find the equation of the tangent plane to the surface $2yz^3 - x^2y = 4z$ at the point $(1, 4, 1)$.

12 Suppose that $T(x, y, z) = \sqrt{x^2 + y^2 + z^2}$, where $x = r\cos\theta$ and $y = r\sin\theta$. Find $\frac{\partial T}{\partial r}$ and $\frac{\partial T}{\partial \theta}$.

13 If $x^2y^3 - 3x^4 + 2y^3 = 0$, then find a formula for $\frac{dy}{dx}$ and state where it is valid. Use the tangent approximation to approximate y when $x = 0.98$. (Note that $(1, 1)$ satisfies the given equation.)

14 If $x^2 - 4y^2 + z^2 = 1$, then find formulas for $\frac{\partial z}{\partial x}$ and $\frac{\partial z}{\partial y}$ and state where they are valid. Then use the tangent approximation to approximate z when $x = 2.03$ and $y = 1.01$. (Note that $(2, 1, 1)$ satisfies the given equation.)

15 If $f(x, y) = x^2y^3 + 4x^2y^2 + xy - x^2$, then find f''.

16 Find the second degree Taylor polynomial of f given by $f(x, y) = ye^x$ near the point $(x_0, y_0) = (1, 1)$.

17 Find and classify all critical points of
(a) $f(x, y) = x^3 + 3xy^2 - 3x^2 - 3y^2 + 7$
(b) $f(x, y) = x^4 + 3x^2y^2 + y^4$
(c) $f(x, y) = 2x^2 - xy - 3y^2 - 3x + 7y$

18 A crate with a lid is to be made from 18 square feet of lumber. It is to have the shape of a rectangular parallelepiped. Find the dimensions and volume of the largest such crate.

19 Find the extreme values of $f(x, y) = x - 3y$ on the ellipse $4x^2 + y^2 = 16$.

20 Find the extreme values of $f(x, y, z) = xy + yz$ on the curve of intersection of $x^2 + y^2 = 4$ and $yz = 4$.

5 MULTIPLE INTEGRATION

0 INTRODUCTION

Having developed differential calculus for functions $f: \boldsymbol{R}^n \to \boldsymbol{R}$, we are naturally led to the question of how to integrate such functions. We confine ourselves to the cases $n = 2$ and $n = 3$, leaving the discussion for general n to advanced calculus. Just as our theme in Chapter 4 was to try to extend the notions of elementary differential calculus by analogy, so in this chapter we approach integration from the standpoint of developing a suitable analogue for functions $f: \boldsymbol{R}^2 \to \boldsymbol{R}$ of the definite integral $\int_a^b f(x)\,dx$ of a function $f: \boldsymbol{R} \to \boldsymbol{R}$.

In the first section we introduce the notion of a *double integral* of a function $f: \boldsymbol{R}^2 \to \boldsymbol{R}$ over a *rectangle* in $\boldsymbol{R}^2$ by imitating the definition of the definite integral of a function $f: \boldsymbol{R} \to \boldsymbol{R}$ over a closed interval. In Section 2, we then develop a tool (analogous to the Fundamental Theorem of Integral Calculus) for evaluating double integrals by *antidifferentiation.* This tool turns out to be powerful enough to allow us to handle double integrals over sets in the plane that are more complicated than rectangles. Section 3 is devoted to presenting this. Section 4 is an optional brief refresher course in *polar coordinates* for the plane, which are then employed in Section 5 to evaluate double integrals over regions (like circular disks) whose polar coordinate description is simpler than the rectangular coordinate description. In Section 6 we turn to integration of functions $f: \boldsymbol{R}^3 \to \boldsymbol{R}$ and develop the *triple integral* over rectangular boxes and more general sets. In Sections 7 and 8, we consider evaluation of triple integrals using *alternative coordinate systems* for $\boldsymbol{R}^3$ that resemble polar coordinates for $\boldsymbol{R}^2$. We devote Section 9 to a number of applications of double and triple integration to geometry and physics. Section 10 is concerned with a classical technique for differentiating functions defined by double integrals. This goes all the way back to the origins of calculus and corresponds to an important formula you learned in elementary calculus. In the final section, we take up *improper multiple integrals*, which often arise in applied work.

1 THE DOUBLE INTEGRAL

In seeking to develop an analogue for the definite integral of a function $f: \boldsymbol{R} \to \boldsymbol{R}$, we are led to take a close look at the definite integral. First, recall that the definite integral is defined over a *closed interval* $[a, b]$. What kind of set in the plane $\boldsymbol{R}^2$ corresponds to a closed interval? The idea we have used previously of putting together two of the corresponding entities from $\boldsymbol{R}^1$ provides us with such a set. Suppose that we take a closed interval $I_1 = [a, b]$ along the x-axis copy of $\boldsymbol{R}$ and a closed interval $I_2 = [c, d]$ along the y-axis copy of $\boldsymbol{R}$ in $\boldsymbol{R}^2$. These two closed intervals define a *closed rectangle* $R = I_1 \times I_2 = \{(x, y) \mid x \in I_1, y \in I_2\}$ in $\boldsymbol{R}^2$. See Figure 1.1.

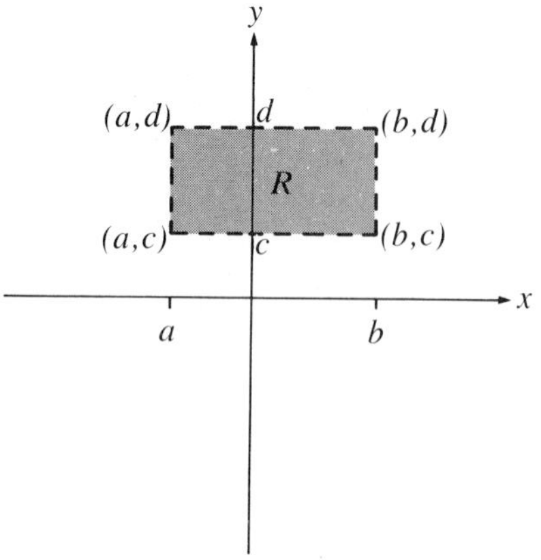

FIGURE 1.1

Now, how do we define the definite integral over a closed interval in $\boldsymbol{R}$? Here there is no universal answer. Depending on what textbook you learned elementary calculus from, you may answer in terms of upper and lower sums, Riemann sums, step functions, or perhaps something still different. We have to decide on *something* to imitate, however, so let's look at the Riemann sum approach to the definite integral [named for the 19th century German mathematician G. F. Bernhard Riemann (1826–1866)], which almost everyone at least *sees* in elementary calculus. Recall that we partition the interval $[a, b]$ into subintervals

$$P: a = x_0 < x_1 < x_2 < \ldots < x_{i-1} < x_i < \ldots < x_{m-1} < x_m = b,$$

and select a point $z_i \in [x_{i-1}, x_i]$, the i-th subinterval. We set $\Delta x_i = x_i - x_{i-1}$, the length of the i-th subinterval, and form the *Riemann sum*

(1) $$R(f, P, z) = \sum_{i=1}^{m} f(z_i)\,\Delta x_i.$$

The *definite integral* of f over $[a, b]$ is then

$$I = \int_a^b f(x)\,dx = \lim_{|P| \to 0} R(f, P, z),$$

if this limit exists, where $|P|$ is the largest Δx_i, $i = 1, 2, \ldots, m$, and the limit statement means the following. Given any $\varepsilon > 0$, there is a $\delta > 0$ such that for any partition P with $|P| < \delta$, the inequality $|R(f, P, z) - I)| < \varepsilon$ holds for all Riemann sums (i.e., for all choices of z_i) corresponding to the partition P. (Note that as $|P| \to 0$, $m \to \infty$.) We now proceed to functions $f: \boldsymbol{R}^2 \to \boldsymbol{R}$.

What corresponds to a partition of $[a, b]$ for a rectangle $R = [a, b] \times [c, d]$ in $\boldsymbol{R}^2$? Well, suppose we put together two partitions, P_1 for $I_1 = [a, b]$ and P_2 for $I_2 = [c, d]$. If P_1 is like P above and

$$P_2: c = y_0 < y_1 < y_2 < \ldots < y_{j-1} < y_j < \ldots < y_n = d,$$

then a **grid** $G = P_1 \times P_2$ of $R = I_1 \times I_2$ is a collection of *subrectangles* $R_{ij} = [x_{i-1}, x_i] \times [y_{j-1}, y_j]$ defined by vertical lines $x = x_i$, $i = 1, 2, \ldots, m$, and horizontal

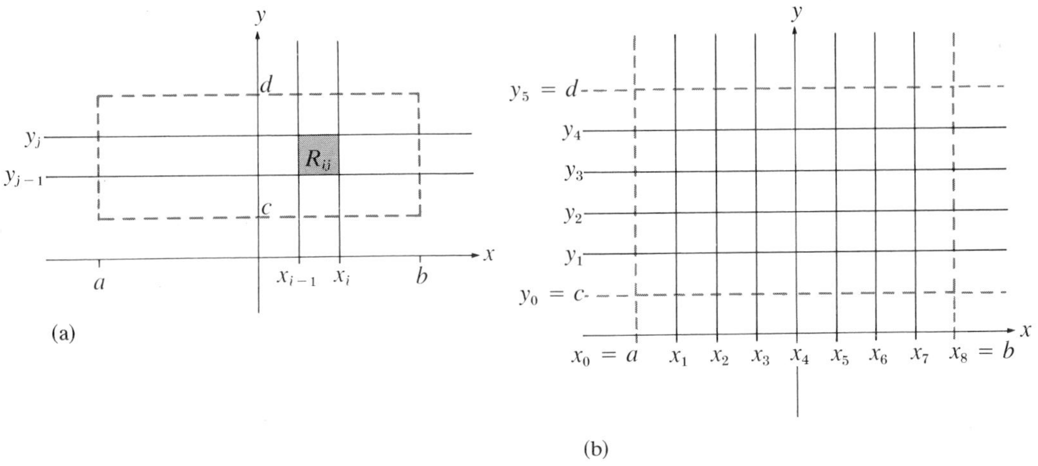

FIGURE 1.2

lines $y = y_j$, $j = 1, 2, \ldots, n$. See Figure 1.2, where a typical R_{ij} is shown in (a), and the entire grid is shown in (b) for $m = 8$ and $n = 5$.

Here

$$\begin{aligned} R_{ij} &= [x_{i-1}, x_i] \times [y_{j-1}, y_j] \\ &= \{(x, y) \mid x \in [x_{i-1}, x_i],\ y \in [y_{j-i}, y_j]\}. \end{aligned}$$

Let's pick a point $\mathbf{z}_{ij} = (\bar{x}_i, \bar{y}_j) \in R_{ij}$ for each i between 1 and m and each j between 1 and n. Then we can form the double Riemann sum by analogy with (1). In (1) we have the sum of products of $f(z_i)$ times Δx_i, where Δx_i is the *length* (i.e., *one*-dimensional measure of extent) of the subinterval containing z_i. Here, then, we should form the sum of products of $f(\mathbf{z}_{ij})$ times ΔA_{ij}, where ΔA_{ij} is the *area* (i.e., *two*-dimensional measure of extent) of the subrectangle R_{ij} containing $\mathbf{z}_{ij}$. Hence a **double Riemann sum** for $f: \mathbf{R}^2 \to \mathbf{R}$ corresponding to the grid G is

(2)
$$\begin{aligned} R(f, G, \mathbf{z}) &= \sum_{i=1}^{m} \sum_{j=1}^{n} f(\mathbf{z}_{ij})\, \Delta A_{ij} \\ &= \sum_{i=1}^{m} \sum_{j=1}^{n} f(\mathbf{z}_{ij})\, \Delta x_i\, \Delta y_j, \end{aligned}$$

where $\Delta x_i = x_i - x_{i-1}$ and $\Delta y_j = y_j - y_{j-1}$.

We can then define the double integral of f to be the limit of these Riemann sums (when the limit exists) as soon as we have a measure, like $|P|$, for the *fineness* of the grid G. We can readily supply such a measure by defining the **mesh** $|G|$ of G to be the maximum of $\{\Delta x_i, \Delta y_j\}$ for $i = 1, \ldots, m$ and $j = 1, \ldots, n$. We are now ready to give the definition.

1.1 **DEFINITION.** Let $f: \mathbf{R}^2 \to \mathbf{R}$ be defined on the rectangle $R = [a, b] \times [c, d]$. Then the **double integral of** f over R is the limit as $|G| \to 0$ of the double Riemann sums of f over R (if this limit exists). That is,

(3)
$$I = \iint_R f(\mathbf{x})\, dA = \lim_{|G| \to 0} R(f, G, \mathbf{z})$$

in the sense that given any $\varepsilon > 0$, there is a $\delta > 0$ such that $|I - R(f, G, z)| < \varepsilon$ for all Riemann sums of f corresponding to any grid G with $|G| < \delta$. If the double integral exists, we call f **integrable** over R.

Note in (3) that instead of $\iint_R f(\boldsymbol{x})\,dA$, which seems quite analogous to $\int_a^b f(x)\,dx$, we could use the notation

$$\iint_R f(x, y)\,dx\,dy, \tag{4}$$

where $\boldsymbol{x} = (x, y)$. This notation amounts to the same thing, if we think of dx and dy as representing limiting forms of Δx_i and Δy_j. *Since the subregion R_{ij} is a rectangle*, its area ΔA_{ij} is $\Delta x_i\,\Delta y_j$. So in the limiting forms, we can think of dA as given by $dx\,dy$. Later on (Section 5) we will see that (3) is the *more basic* notation since we can continue to use it even when we have *nonrectangular* subregions R_{ij}. The form (4), however, applies *only* when $dA = dx\,dy$, i.e., only when our subregions are rectangular. Because of its wider applicability, we'll tend to favor (3) over (4).

1.2 EXAMPLE. If $f: \boldsymbol{R}^2 \to \boldsymbol{R}$ is a constant function, $f(x) = c$ for all $\boldsymbol{x} \in \boldsymbol{R}^2$, then show that

$$\iint_R f(\boldsymbol{x})\,dA = cA(R),$$

where $A(R)$ is the area of the rectangle R.

Proof. Let $G = P_1 \times P_2$ be a grid of R. Then we have

$$\begin{aligned} R(f, G, \boldsymbol{z}) &= \sum_{i=1}^{m} \sum_{j=1}^{n} f(\boldsymbol{z}_{ij})\,\Delta A_{ij} \\ &= \sum_{i=1}^{m} \sum_{j=1}^{n} cA(R_{ij}) \\ &= c \sum_{i=1}^{m} \sum_{j=1}^{n} A(R_{ij}) = cA(R). \end{aligned}$$

Hence,

$$\lim_{|G| \to 0} R(f, G, \boldsymbol{z}) = cA(R).$$

QED

We can draw the following important conclusion immediately from Example 1.2 by simply letting $c = 1$.

1.3 COROLLARY. $\iint_R 1\,dA = A(R)$.

This result has the notational appeal of asserting that the double integral of dA over a rectangle R is simply the area A of R. This can be thought of as a generalization of

$$\int_a^b dx = b - a$$

where $b - a$ is the length of the interval $[a, b]$.

You may remember the following two properties of the definite integral. Suppose that f and g are integrable functions over the interval $[a, b]$. Then

(5) $$\int_a^b (f+g)(x)\,dx = \int_a^b f(x)\,dx + \int_a^b g(x)\,dx,$$

and

(6) $$\int_a^b kf(x)\,dx = k\int_a^b f(x)\,dx, \quad \text{if} \quad k \in \boldsymbol{R}.$$

These say that the definite integral from a to b is a linear function on the real vector space of integrable functions $f:[a, b] \to \boldsymbol{R}$. The double integral over a rectangle has the same linearity property.

First, we expect that the set $\mathscr{I}$ of functions $f:\boldsymbol{R}^2 \to \boldsymbol{R}$ that are defined and integrable on a rectangle R is a real vector space. Recall that the set $\mathscr{F}$ of *all* functions $f:\boldsymbol{R}^2 \to \boldsymbol{R}$ with domain R is a real vector space by Exercise 27, Exercises 1.1. Note that the zero function, $f(\boldsymbol{x}) \equiv 0$ for $\boldsymbol{x} \in R$, is integrable over R and that all the required arithmetic laws (commutative, associative, etc.) hold for $\mathscr{I}$ since they hold for the larger set $\mathscr{F}$. To show that $\mathscr{I}$ is a real vector space, we now need only show that it is closed under addition and multiplication by real scalars. That is, we need to show that if f and g are integrable, then so are $f+g$ and kf for any $k \in \boldsymbol{R}$. That verification will complete the proof of our next result.

1.4 THEOREM. The double integral over R is a linear function on the real vector space $\mathscr{I}$.

Proof. Let f and g be in $\mathscr{I}$. Then we will show that $f+g \in \mathscr{I}$, that $kf \in \mathscr{I}$ for any $k \in \boldsymbol{R}$, and that

$$\iint_R (f+g)(\boldsymbol{x})\,dA = \iint_R f(\boldsymbol{x})\,dA + \iint_R g(\boldsymbol{x})\,dA,$$

$$\iint_R kf(\boldsymbol{x})\,dA = k\iint_R f(\boldsymbol{x})\,dA.$$

First, any Riemann sum for $f+g$ has the form

(7) $$\begin{aligned}\sum_{i=1}^{m}\sum_{j=1}^{n} (f+g)(\boldsymbol{z}_{ij})\,\Delta A_{ij} &= \sum_{i=1}^{m}\sum_{j=1}^{n} [f(\boldsymbol{z}_{ij})+g(\boldsymbol{z}_{ij})]\,\Delta A_{ij}\\ &= \sum_{i=1}^{m}\sum_{j=1}^{n} f(\boldsymbol{z}_{ij})\,\Delta A_{ij} + \sum_{i=1}^{m}\sum_{j=1}^{n} g(\boldsymbol{z}_{ij})\,\Delta A_{ij}.\end{aligned}$$

Similarly, any Riemann sum for kf has the form

(8) $$\begin{aligned}\sum_{i=1}^{m}\sum_{j=1}^{n} (kf)(\boldsymbol{z}_{ij})\,\Delta A_{ij} &= \sum_{i=1}^{m}\sum_{j=1}^{n} kf(\boldsymbol{z}_{ij})\,\Delta A_{ij}\\ &= k\sum_{i=1}^{m}\sum_{j=1}^{n} f(\boldsymbol{z}_{ij})\,\Delta A_{ij}.\end{aligned}$$

Now take limits in (7) and (8) as $|G| \to 0$ and use Theorem 3.2.3, which still holds here (Exercise 21 below). We get

$$\lim_{|G|\to 0} R(f+g, G, \mathbf{z}) = \iint_R f(\mathbf{x})\,dA + \iint_R g(\mathbf{x})\,dA$$

and

$$\lim_{|G|\to 0} R(kf, G, \mathbf{z}) = k\iint_R f(\mathbf{x})\,dA$$

Hence, $\iint_R (f+g)(\mathbf{x})\,dA$ and $\iint_R kf(\mathbf{x})\,dA$ exist, and equal the precise expressions required for the double integral over R to be a linear function. QED

There are a number of other properties of the double integral, analogous to properties you saw in elementary calculus for the definite integral, that can be proved similarly. (See Exercises 13 to 16 below, for example.)

What sorts of functions are integrable over a rectangle? Almost all we will deal with, as the following theorem of advanced calculus guarantees.

1.5 **THEOREM.** If $f:\mathbf{R}^2\to\mathbf{R}$ is continuous, then f is integrable over any rectangle R in its domain.

This is reassuring, but it doesn't tell us how to *evaluate* any double integrals. In the next section we shall give a general scheme based on antidifferentiation. For now we give a special case, of interest in its own right.

1.6 **THEOREM.** Let $R=[a, b]\times[c, d]$, and let $h:\mathbf{R}^2\to\mathbf{R}$ be a continuous function of the form $h(x, y)=f(x)g(y)$. Then

$$\iint_R h(x, y)\,dA = \left(\int_a^b f(x)\,dx\right)\left(\int_c^d g(y)\,dy\right).$$

(Thus, the double integral of the *product* of two functions $f(x)$ and $g(y)$ of one real variable is the *product of the integrals*.)

Proof. Let P_1 be a partition of $[a, b]$ into n subintervals, each of length $\Delta x_i = \dfrac{b-a}{n}$; and let P_2 be a partition of $[c, d]$ into n subintervals, each of length $\Delta y_j = \dfrac{d-c}{n}$. Let $H(x)=\int_c^d h(x, y)\,dy$ for $x\in[a, b]$. Then we have

$$H(x) = \int_c^d f(x)g(y)\,dy = f(x)\int_c^d g(y)\,dy$$

$$= f(x)\sum_{j=1}^{n}\int_{y_{j-1}}^{y_j} g(y)\,dy$$

(9) $$= f(x)\sum_{j=1}^{n} g(\bar{y}_j)\,\Delta y_j,$$

applying the Mean Value Theorem for Integrals: $\int_{y_{j-1}}^{y_j} g(y)\,dy = g(\bar{y}_j)(y_j - y_{j-1})$ for some $\bar{y}_j$ between y_{j-1} and y_j. Then we get

$$\begin{aligned}
\left(\int_a^b f(x)\,dx\right)\left(\int_c^d g(y)\,dy\right) &= \int_a^b H(x)\,dx \\
&= \lim_{|P_1|\to 0} \sum_{i=1}^{n} H(\bar{x}_i)\,\Delta x_i \\
&= \lim_{n\to\infty} \sum_{i=1}^{n} H(\bar{x}_i)\,\Delta x_i \\
&= \lim_{n\to\infty} \sum_{i=1}^{n} f(\bar{x}_i)\left[\sum_{j=1}^{n} g(\bar{y}_j)\,\Delta y_j\right]\Delta x_i \text{ by (9),} \\
&= \lim_{n\to\infty} \sum_{i=1}^{n}\sum_{j=1}^{n} f(\bar{x}_i)g(\bar{y}_j)\,\Delta x_i\,\Delta y_j, \\
&\qquad \text{multiplying each term in the bracket by } f(\bar{x}_i)\,\Delta x_i, \\
&= \lim_{n\to\infty} \sum_{i=1}^{n}\sum_{j=1}^{n} h(\bar{x}_i, \bar{y}_j)\,\Delta x_i\,\Delta y_j \\
&= \lim_{|G|\to 0} R(h, G, z) \text{ where } G \text{ is the} \\
&\qquad \text{grid of } R \text{ defined by } P_1 \text{ and } P_2, \\
&= \iint_R h(x, y)\,dA. \qquad \text{QED}
\end{aligned}$$

The proof of Theorem 1.6 was carried out by first *freezing* x and then considering $\int_c^d h(x, y)\,dy$. It turns out that this process, which can be thought of as the inverse of partial differentiation, is the basis for the evaluation of double integrals in general. We pursue that in the next section, but for now we can illustrate how easy Theorem 1.6 is to use.

1.7 **EXAMPLE.** Evaluate $\iint_R e^x \cos y\,dA$ if $R = [0, 2]\times\left[\frac{\pi}{6}, \frac{\pi}{2}\right]$.

Solution. By Theorem 1.6,

$$\begin{aligned}
\iint_R e^x \cos y\,dA &= \left(\int_0^2 e^x\,dx\right)\left(\int_{\pi/6}^{\pi/2} \cos y\,dy\right) \\
&= e^x\Big]_0^2 \sin y\Big]_{\pi/6}^{\pi/2} \\
&= (e^2 - e^0)\left(\sin\frac{\pi}{2} - \sin\frac{\pi}{6}\right) \\
&= (e^2-1)(1-\tfrac{1}{2}) = \frac{e^2-1}{2}.
\end{aligned}$$

Notice that when we write $R = [a, b]\times[c, d]$, the first interval is always on the x-axis, and the second is always on the y-axis.

We close the section with a sometimes useful result, which seems quite plausible intuitively.

1.8 THEOREM. Suppose that the rectangle $R = R_1 \cup R_2$, where R_1 and R_2 are rectangles that intersect only along a common edge. If f is integrable over R_1 and R_2, then it is integrable over R and

$$\iint_R f(\mathbf{x})\,dA = \iint_{R_1} f(\mathbf{x})\,dA + \iint_{R_2} f(\mathbf{x})\,dA.$$

See Figure 1.3.

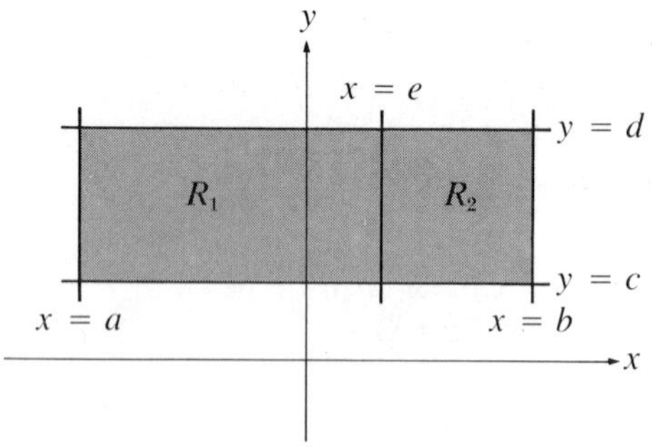

FIGURE 1.3

Proof. Suppose that R is as in Figure 1.3. Then any grid G of R induces grids G_1 of R_1 and G_2 of R_2 obtained by partitioning $[a, e]$ and $[e, b]$ by lines $x = x_i$ and $y = y_j$ (and $x = e$ if not already present in G). As $|G| \to 0$, we find that $|G_1| \to 0$ and $|G_2| \to 0$. Then

$$\begin{aligned}\iint_{R_1} f(\mathbf{x})\,dA + \iint_{R_2} f(\mathbf{x})\,dA &= \lim_{|G_1|\to 0} R(f, G_1, \mathbf{z}) + \lim_{|G_2|\to 0} R(f, G_2, \mathbf{z}) \\ &= \lim_{|G|\to 0} R(f, G, \mathbf{z}) \\ &= \iint_R f(\mathbf{x})\,dA.\end{aligned}$$

QED

EXERCISES 5.1

In Exercises 1–12, use Theorems 1.4 and 1.6 to evaluate the given double integral.

1 $\displaystyle\iint_R e^x \cos y\,dA, \quad R = [0, 1] \times \left[\frac{\pi}{2}, \pi\right]$

2 $\displaystyle\iint_R e^{x+y}\,dA, \quad R = [0, 1] \times [0, 2]$

3 $\displaystyle\iint_R \sin x \cos y\,dA, \quad R = \left[0, \frac{\pi}{2}\right] \times \left[0, \frac{\pi}{2}\right]$

4 $\displaystyle\iint_R \sec x \tan y\,dA, \quad R = \left[0, \frac{\pi}{4}\right] \times \left[0, \frac{\pi}{4}\right]$

5 $\displaystyle\iint_R (\sin y + \sin y \tan^2 x)\,dA, \quad R = \left[0, \frac{\pi}{4}\right] \times \left[0, \frac{\pi}{4}\right]$

6 $\displaystyle\iint_R (\sec^2 x \sin y - \sin y)\,dA, \quad R = \left[0, \frac{\pi}{4}\right] \times \left[\frac{\pi}{6}, \frac{\pi}{3}\right]$

7 $\iint\limits_R (x^2y-3xy^2)\,dA, \quad R=[1,2]\times[-1,1]$

8 $\iint\limits_R (x^3-y^3)\,dA, \quad R=[0,1]\times[0,1]$

9 $\iint\limits_R (x^3+y^3)\,dA, \quad R=[0,1]\times[0,1]$

10 $\iint\limits_R \frac{x}{1+y^2}\,dA, \quad R=[0,2]\times[0,1]$

11 $\iint\limits_R \frac{y^2}{1+x^2}\,dA, \quad R=[0,1]\times[0,1]$

12 $\iint\limits_R y\ln xy\,dA, \quad R=[1,2]\times[2,3]$

13 If $f(x,y)\geq 0$ on R and if f is integrable over R, then show that $\iint\limits_R f(x,y)\,dA\geq 0$.

14 If $f(x,y)\leq g(x,y)$ on R and if f and g are integrable on R, then show that $\iint\limits_R f(x,y)\,dA\leq\iint\limits_R g(x,y)\,dA$. (*Hint:* Use Exercise 13.)

15 If f is continuous on R, then show that $mA(R)\leq\iint\limits_R f(x,y)\,dA\leq MA(R)$, where M is the maximum of f on R and m is the minimum of f on R. (*Hint:* Use Exercise 14.)

16 If f is continuous on R, then show that $mA(R)\leq\left|\iint\limits_R f(x,y)\,dA\right|\leq MA(R)$, where M is an upper bound and m is a lower bound for $|f(x,y)|$ on R.

17 If

$$f(x,y)=\begin{cases}1 & \text{if } y \text{ is rational}\\ x & \text{if } y \text{ is irrational}\end{cases},$$

then show that f is not integrable over $R=[0,1]\times[0,1]$ by considering Riemann sums.

18 If

$$f(x,y)=\begin{cases}2 & \text{if } x \text{ is rational}\\ y & \text{if } x \text{ is irrational}\end{cases},$$

then show that f is not integrable over $R=[0,1]\times[0,1]$.

19 Use Exercise 15 to show that if $f:\boldsymbol{R}^2\to\boldsymbol{R}$ is continuous, then

$$\lim_{h\to 0}\frac{1}{A(S_h)}\iint\limits_{S_h} f(\boldsymbol{x})\,dA=f(\boldsymbol{x}_0)$$

if S_h is the square of side h and center $\boldsymbol{x}_0$.

20 Use Exercise 15 to prove the *Mean Value Theorem for Double Integrals:* If f is continuous on $\boldsymbol{R}^2$, then

$$\iint\limits_R f(x,y)\,dA=f(x_0,y_0)A(R)$$

for some point (x_0, y_0) in R. (*Hint:* Use, but don't attempt to prove, the Intermediate Value Theorem for Continuous Functions $f:\boldsymbol{R}^2\to\boldsymbol{R}$, which states that f takes on all values between its maximum and minimum on a rectangle.)

21 Suppose that $\lim_{|G|\to 0} R(f, G, z)$ and $\lim_{|G|\to 0} R(g, G, z)$ both exist. Then show that

$$\lim_{|G|\to 0} R(f+g, G, z) = \lim_{|G|\to 0} R(f, G, z) + \lim_{|G|\to 0} R(g, G, z)$$

where the limits are taken in the sense of Definition 1.1.

The following two exercises can often save labor. (See Example 9.4.)

22 Recall that f is an *odd function* in x if $f(-x, y) = -f(x, y)$. Suppose that f is odd in x and the rectangle $R \subseteq \boldsymbol{R}^2$ is symmetric in the y-axis. Then show that $\iint_R f(x, y)\, dA = 0$.

23 If f is an odd function in y[i.e., $f(x, -y) = -f(x, y)$] and the rectangle $R \subseteq \boldsymbol{R}^2$ is symmetric in the x-axis, then show that $\iint_R f(x, y)\, dA = 0$.

2 ITERATED INTEGRALS OVER RECTANGLES

While Theorem 1.6 permits us to reduce evaluation of multiple integrals to evaluation of two definite integrals, it applies only to a *very* limited class of functions, those of the form $h(x, y) = f(x)g(y)$. What we need is a tool like the Fundamental Theorem of Integral Calculus,

$$\int_a^b f(x)\, dx = F(b) - F(a) \quad \text{where} \quad F' = f.$$

This applies to *any* continuous function $f: \boldsymbol{R} \to \boldsymbol{R}$ and permits calculation of definite integrals by antidifferentiation. The analogous tool for functions $f: \boldsymbol{R}^2 \to \boldsymbol{R}$ is the *iterated integral*, which is a type of inverse partial derivative. How did we compute partial derivatives? We *froze* the value of one variable (x or y) and then took the derivative of f viewed as a function of just the other variable (y or x). Given a continuous function $f: \boldsymbol{R}^2 \to \boldsymbol{R}$, we can reverse this procedure as follows. *If we freeze x, then f(x, y) becomes a function of y alone.* So we can integrate it over the interval $[c, d]$, obtaining $g(x) = \int_c^d f(x, y)\, dy$. We will show in Lemma 10.2 that $g(x)$ is a continuous function. So we can integrate this function over the interval $[a, b]$. The result, $\int_a^b g(x)\, dx = \int_a^b \left(\int_c^d f(x, y)\, dy\right) dx$, is the key to evaluating double integrals.

2.1 **DEFINITION.** Let $f: \boldsymbol{R}^2 \to \boldsymbol{R}$ be integrable over $R = [a, b] \times [c, d]$. Then the **iterated integrals of f** over R are

(1) $$\int_a^b dx \int_c^d f(x, y)\, dy = \int_a^b \left(\int_c^d f(x, y)\, dy\right) dx$$

and

(2) $$\int_c^d dy \int_a^b f(x, y)\, dx = \int_c^d \left(\int_a^b f(x, y)\, dx\right) dy,$$

where in $\int_c^d f(x, y)\, dy$ the variable x is frozen, and in $\int_a^b f(x, y)\, dx$ the variable y is frozen.

Other notation in *wide use* is $\int_a^b \int_c^d f(x, y)\,dy\,dx$ for $\int_a^b dx \int_c^d f(x, y)\,dy$ and $\int_c^d \int_a^b f(x, y)\,dx\,dy$ for $\int_c^d dy \int_a^b f(x, y)\,dx$. In these notations, the *innermost* differential tells you which variable varies in the first integration. In the notation we employ, the variable that varies first is indicated by the differential next to $f(x, y)$.

2.2 EXAMPLE. If $f(x, y)=x^2y-x+y$, then find both iterated integrals of f over $R=[1, 2]\times[-1, 1]$.

Solution. From (1),

$$\begin{aligned}\int_1^2 dx \int_{-1}^1 (x^2y-x+y)\,dy &= \int_1^2 dx\,[\tfrac{1}{2}x^2y^2-xy+\tfrac{1}{2}y^2]_{-1}^1\\ &= \int_1^2 [\tfrac{1}{2}x^2-x+\tfrac{1}{2}-(\tfrac{1}{2}x^2+x+\tfrac{1}{2})]\,dx\\ &= \int_1^2 -2x\,dx = -x^2\Big]_1^2 = -(4-1)\\ &= -3.\end{aligned}$$

From (2),

$$\int_{-1}^1 dy \int_1^2 (x^2y-x+y)\,dx = \int_{-1}^1 dy \left[\frac{x^3y}{3}-\frac{x^2}{2}+xy\right]_1^2$$

$$= \int_{-1}^1 \left[\frac{8}{3}y-2+2y-\left(\frac{y}{3}-\frac{1}{2}+y\right)\right] dy$$

$$= \int_{-1}^1 \left(\frac{10}{3}y-\frac{3}{2}\right) dy = \frac{5}{3}y^2-\frac{3}{2}y\Big]_{-1}^1$$

$$= \tfrac{5}{3}-\tfrac{3}{2}-\tfrac{5}{3}-\tfrac{3}{2} = -3.$$

We got the same answer in both cases, and we will see below in Theorem 2.4 that this is always going to happen for continuous functions f.

One of the first applications you saw for the definite integral was the problem of calculating areas in $\boldsymbol{R}^2$. By analogy, we expect that the double integral may be used to calculate volumes in $\boldsymbol{R}^3$. This happens to be true, and forms the basis for the heuristic justification we will give for evaluating double integrals by calculating iterated integrals. So let's consider the volume under a surface $z=f(x, y)$ lying above a rectangle $R=[a, b]\times[c, d]$ in the xy-plane. For simplicity we assume $f(x, y)\geq 0$ on R. We give an illustration of the solid S we have just described in Figure 2.1.

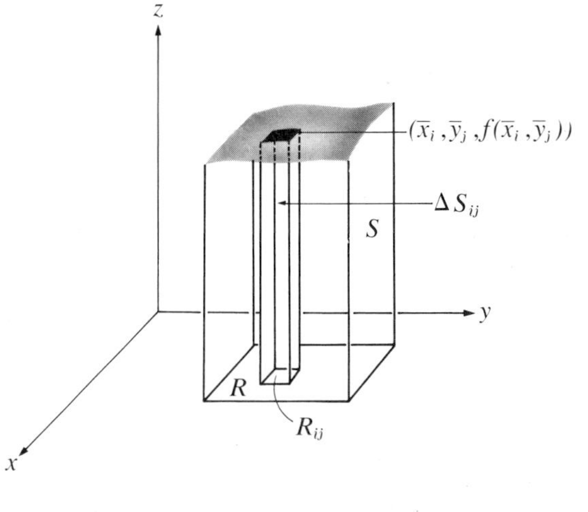

FIGURE 2.1

If we make a grid $G=P_1\times P_2$ of R, then the volume of S is the sum of the volumes of the solids ΔS_{ij} lying above the subrectangles R_{ij}, one of which is shown in Figure 2.1. If G is of fine mesh, then each such solid ΔS_{ij} is approximately a rectangular box (parallelepiped) of volume $\Delta V_{ij}=f(\mathbf{z}_{ij})\,\Delta x_i\,\Delta y_j$ for any point $\mathbf{z}_{ij}=(\bar{x}_i,\bar{y}_j)$ in R_{ij} (see Figure 2.2).

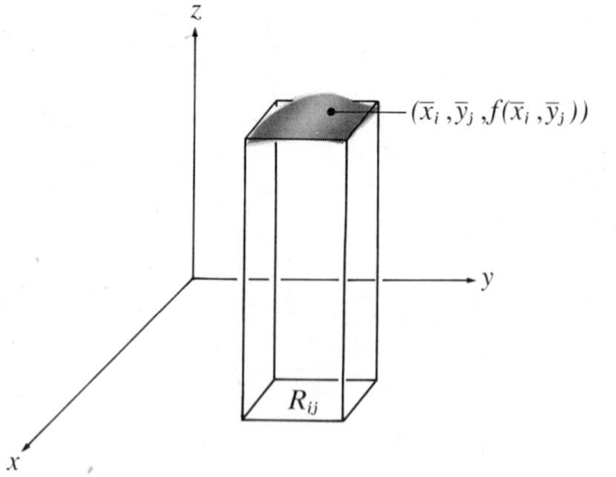

FIGURE 2.2

If S has a volume V, then V seems to be approximated by the Riemann sum

$$V\approx R(f,G,\mathbf{z})=\sum_{i=1}^{m}\sum_{j=1}^{n} f(\mathbf{z}_{ij})\,\Delta x_i\,\Delta y_j.$$

The following definition is now perfectly natural.

2.3 **DEFINITION.** Suppose that $f:\mathbf{R}^2\to\mathbf{R}$ is an integrable function over $R=[a,b]\times[c,d]$ and $f(x,y)\geq 0$ for $(x,y)\in R$. Then **the volume of the solid S lying under the graph of $z=f(x,y)$ and above R** is $V(S)=\iint_R f(x,y)\,dA$.

Now, in elementary calculus you were taught a method to evaluate the volume of a solid like S by "slicing and integrating the cross sectional area." Let's recall this method. Suppose that we slice S by planes $x=t$ parallel to the yz-plane. (See Figure 2.3.) A typical slice yields a plane cross sectional area $A(t)$

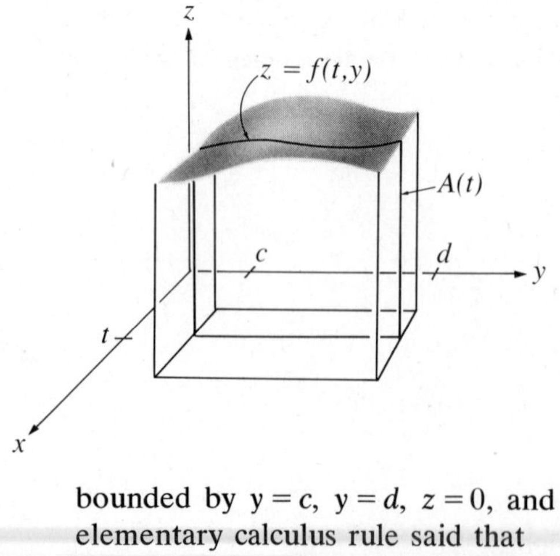

FIGURE 2.3

bounded by $y=c$, $y=d$, $z=0$, and the curve $z=f(t,y)$ in the plane $x=t$. The elementary calculus rule said that

(3) $$V(S)=\int_a^b A(t)\,dt=\int_a^b A(x)\,dx.$$

We can justify this heuristically as follows. First,

(4) $$V=\int_a^b \frac{dV(t)}{dt}\,dt,$$

where we regard $V=V(t)$ as the volume of the solid bounded by $z=0$, $z=f(x,y)$, $y=c$, $y=d$, $x=a$, and the plane $x=t$. See Figure 2.4. Since $V(t+h)-V(t)$ is the volume of an approximate box of base $A(t)$ and height h, we have

$$\frac{V(t+h)-V(t)}{h}\approx A(t)$$

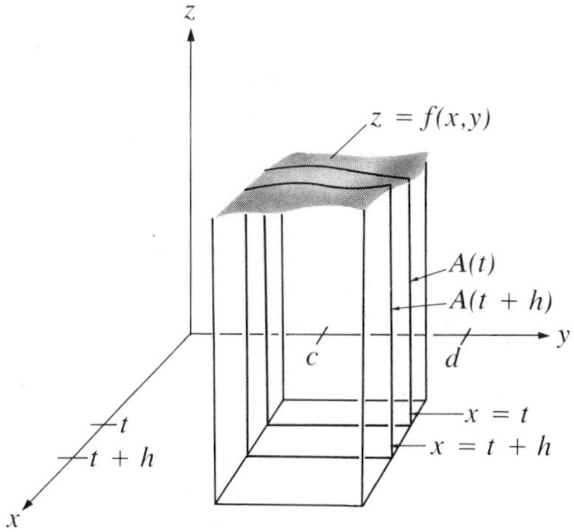

FIGURE 2.4

It can in fact be shown that $\frac{dV}{dt}=A(t)$. Then from (4) we can conclude

(3) $$V=\int_a^b A(t)\,dt=\int_a^b A(x)\,dx.$$

Now, since $A(x)$ is a plane area, we can calculate it as an integral. We have

(5) $$A(x)=\int_c^d f(x,y)\,dy$$

since this plane area is bounded by $y=c$, $y=d$, $z=0$, and $z=f(x,y)$. Combining (3) and (5), we get

$$V=\int_a^b A(x)\,dx=\int_a^b\left(\int_c^d f(x,y)\,dy\right)dx=\int_a^b dx\int_c^d f(x,y)\,dy.$$

Now, there's nothing special about plane sections $x=t$ parallel to the yz-plane. We could just as well have sliced S by planes $y=u$ parallel to the xz-plane. Had we done so, we would have gotten the formula

$$V=\int_c^d dy\int_a^b f(x,y)\,dx.$$

This discussion makes the following theorem quite believable. We leave the formal proof for advanced calculus. [The theorem is named for the Italian mathematician Guido G. Fubini (1879–1943), who made a number of fundamental contributions to the theory of integration.]

2.4 **FUBINI'S THEOREM.** Let $f: \boldsymbol{R}^2 \to \boldsymbol{R}$ be integrable over the rectangle $R = [a, b] \times [c, d]$. Then

(6)
$$\begin{aligned}\iint_R f(x, y)\, dA &= \int_a^b dx \int_c^d f(x, y)\, dy \\ &= \int_c^d dy \int_a^b f(x, y)\, dx.\end{aligned}$$

Moreover, if $f(\boldsymbol{x}) \geq 0$ for $\boldsymbol{x} \in R$, and if S is the solid region in $\boldsymbol{R}^3$ lying above R and below the graph of $z = f(x, y)$, then the volume V of S is given by either iterated integral in (6).

This theorem is of enormous importance. At one fell swoop it shows us how to evaluate a double integral as an iterated integral, and further assures us that the order of iteration we use is immaterial. (Thus, this result is a sort of inverse to Theorem 4.7.5, which said that we could interchange the order of partial differentiation for most functions f met in practice.) In applying (6), use whichever order of iteration appears easier.

2.5 **EXAMPLE.** Evaluate $\iint_R e^y \sin \frac{x}{y}\, dA$, where $R = \left[-\frac{\pi}{2}, \frac{\pi}{2}\right] \times [1, 2]$.

Solution. We can work out either

$$\int_{-\pi/2}^{\pi/2} dx \int_1^2 e^y \sin \frac{x}{y}\, dy \qquad \text{or} \qquad \int_1^2 dy \int_{-\pi/2}^{\pi/2} e^y \sin \frac{x}{y}\, dx.$$

In trying to evaluate the first integral, we would have a lot of trouble. We might try integration by parts, but even if that method worked, it would certainly take more effort than doing the second integral. So let's work with the second iterated integral.

$$\begin{aligned}\int_1^2 e^y\, dy \int_{-\pi/2}^{\pi/2} \sin \frac{x}{y}\, dx &= \int_1^2 e^y\, dy \left[-y \cos \frac{x}{y}\right]_{-\pi/2}^{\pi/2} \\ &= \int_1^2 e^y\, dy \left[-y \cos \frac{\pi/2}{y} + y \cos\left(\frac{-\pi/2}{y}\right)\right].\end{aligned}$$

But $\cos(-\theta) = \cos \theta$ for all θ, so the expression in brackets is 0. Hence, the given integral is 0.

2.6 **EXAMPLE.** Find the volume below the paraboloid $z = x^2 + y^2$ and above the unit square $S = [0, 1] \times [0, 1]$. See Figure 2.5.

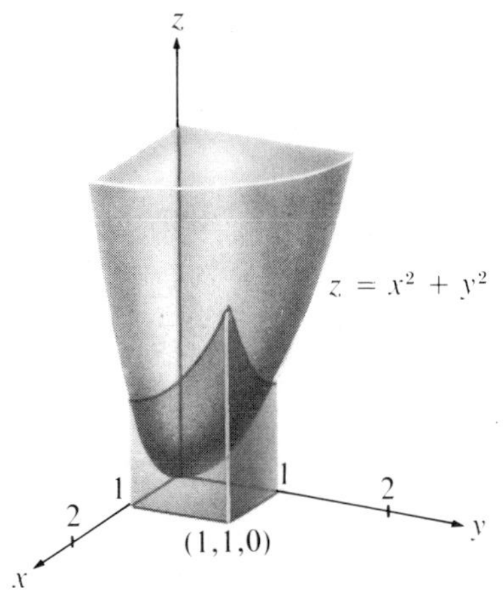

FIGURE 2.5

Solution. By Theorem 2.4,

$$
\begin{aligned}
V &= \iint_S (x^2+y^2)\,dA \\
&= \int_0^1 dx \int_0^1 (x^2+y^2)\,dy = \int_0^1 dx \left[x^2y+\frac{y^3}{3}\right]_0^1 \\
&= \int_0^1 [x^2+\tfrac{1}{3}-0-0]\,dx = \int_0^1 (x^2+\tfrac{1}{3})\,dx \\
&= \left.\frac{x^3}{3}+\frac{x}{3}\right]_0^1 = \tfrac{1}{3}+\tfrac{1}{3}=\tfrac{2}{3}.
\end{aligned}
$$

(Notice that here the order of iteration doesn't affect the complexity of the work.)

EXERCISES 5.2

In Exercises 1 to 12, evaluate the double integral.

1 $\iint_R (x^2-2xy)\,dA,\quad R=[-1,2]\times[1,4]$

2 $\iint_R (x^2-3x+2y)\,dA,\quad R=[-1,2]\times[1,4]$

3 $\iint_R (x+y)^{-2}\,dA,\quad R=[0,1]\times[1,2]$

4 $\iint_R 3x(x^2+y)^{-1}\,dA,\quad R=[0,2]\times[1,2]$

5 $\iint_R (x^2y+xy^2)\,dA,\quad R=[-1,3]\times[1,4]$

6 $\iint_R (x^3y+2x^2y^2)\,dA,\quad R=[-1,2]\times[0,2]$

7 $\iint_R \sin(x+y)\,dA,\quad R=\left[0,\frac{\pi}{2}\right]\times\left[0,\frac{\pi}{2}\right]$

8 $\iint_R \cos(x+y)\,dA,\quad R=\left[0,\frac{\pi}{2}\right]\times\left[0,\frac{\pi}{2}\right]$

9 $\iint_R y\cos xy\,dA,\quad R=[0,1]\times[0,\pi]$

10 $\iint_R x\sin xy\,dA,\quad R=[0,1]\times[0,\pi]$

11 $\iint_R \frac{1}{x+y}\,dA,\quad R=[0,1]\times[1,2]$

12 $\iint_R \frac{1}{x+y}\,dA,\quad R=[1,2]\times[0,2]$

13 Find the volume lying under the paraboloid $z=9-x^2-y^2$ and over the rectangle $R=[0,1]\times[0,2]$.

14 Find the volume lying under the paraboloid $z=16-x^2-y^2$ and above the rectangle $R=[0,2]\times[0,1]$.

15 Find the volume lying under the paraboloid $z=x^2+y^2$ and above the rectangle $R=[-1,1]\times[0,2]$.

16 Find the volume lying under the paraboloid $z=x^2+y^2$ and above the rectangle $R=[-1,2]\times[-1,1]$.

17 Find the volume lying under the surface $z=xe^{xy}$ and above the rectangle $R=[0,1]\times[1,2]$

18 Find the volume lying under the surface $z=ye^{xy}$ and above the rectangle $R=[0,1]\times[1,2]$.

19 **The trapezoidal rule for double integrals.** (a) Show that

$$\iint_R f(x,y)\,dt=\int_b^{b+\Delta y} dy \int_a^{a+\Delta x} f(x,y)\,dx$$

if $R=[a,a+\Delta x]\times[b,b+\Delta y]$.

(b) Let

$$I=\int_b^{b+\Delta y} g(y)\,dy$$

where

$$g(y)=\int_a^{a+\Delta x} f(x,y)\,dx.$$

Approximate $g(y)$ by the trapezoidal rule of elementary calculus,

$$g(y)\approx\frac{\Delta x}{2}[f(a,y)+f(a+\Delta x,y)].$$

Then do the same for I, to obtain

$$I\approx\frac{\Delta x\,\Delta y}{4}[f(a,b)+f(a+\Delta x,b)+f(a,b+\Delta y)+f(a+\Delta x,b+\Delta y)].$$

This is the trapezoidal approximation rule for a subrectangle $[a,a+\Delta x]\times[b,b+\Delta y]$. Over an entire rectangle R, one approximates $\iint_R f(x,y)\,dA$ by adding these subrectangular approximations.

20 Use the rule of Exercise 19 with $\Delta x=\Delta y=\frac{1}{4}$ to approximate $\iint_S (x+y)^2\,dA$ where $S=[0,1]\times[0,1]$. Compare this result with the exact value.

21 Use the rule of Exercise 19 to approximate $\iint_R e^{-(x^2+y^2)}\,dA$ where $R=[-1,1]\times[-1,1]$. Take $\Delta x=\Delta y=\frac{1}{4}$.

22 **Simpson's rule for double integrals.** Let $R=[a,a+2\,\Delta x]\times[b,b+2\,\Delta y]$.
(a) Show that

$$\iint_R f(x,y)\,dA=\int_b^{b+2\Delta y} h(y)\,dy$$

where

$$h(y)=\int_a^{a+2\Delta x} f(x, y)\,dx.$$

(b) Let

$$I'=\int_b^{b+2\Delta y} h(y)\,dy.$$

Approximate $h(y)$ by Simpson's rule from elementary calculus:

$$h(y)\approx\frac{\Delta x}{3}[f(a, y)+4f(a+\Delta x, y)+f(a+2\,\Delta x, y)].$$

Then do the same for I' to obtain

$$\begin{aligned}I'\approx\frac{\Delta x\,\Delta y}{9}&[f(a, b)+f(a+2\,\Delta x, b)+f(a, b+2\,\Delta y)\\&+f(a+2\,\Delta x, b+2\,\Delta y)+4f(a+\Delta x, b)\\&+4f(a+\Delta x, b+2\,\Delta y)+4f(a, b+\Delta y)\\&+4f(a+2\,\Delta x, b+\Delta y)+16f(a+\Delta x, b+\Delta y)].\end{aligned}$$

This is Simpson's approximation rule for a subrectangle. Over an entire rectangle, one approximates $\iint f(x, y)\,dA$ by adding these subrectangular approximations. A computer or calculator is needed to keep the arithmetic manageable.

23 Apply Simpson's rule to the integral of Exercise 20, using $\Delta x=\Delta y=\frac{1}{4}$.

24 Apply Simpson's rule to the integral of Exercise 21, using $\Delta x=\Delta y=\frac{1}{4}$.

3 INTEGRALS OVER GENERAL REGIONS

As you may already be thinking, not everything of interest in $\boldsymbol{R}^3$ lies over a rectangle! While we have worked out double integration over rectangles $R=[a, b]\times[c, d]$, the natural analogues of closed intervals, the class of problems to which we can apply double integration is quite limited. For instance, it seems that, just as in elementary calculus we could find the area of the set D lying above the x-axis and below the parabola $y=4-x^2$ in $\boldsymbol{R}^2$ (Figure 3.1), so we ought to be able to find the volume of the set S lying above the xy-plane and below the paraboloid $z=4-x^2-y^2$ (Figure 3.2). But at the moment we cannot find this volume because S does not lie over a rectangle R in the xy-plane. We proceed now to remedy this unsatisfactory situation.

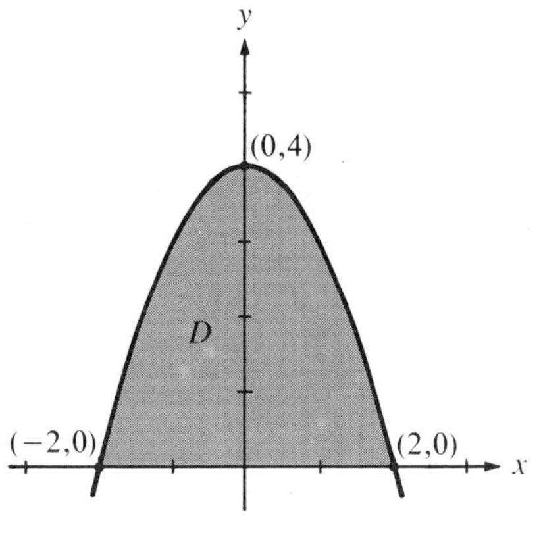

FIGURE 3.1

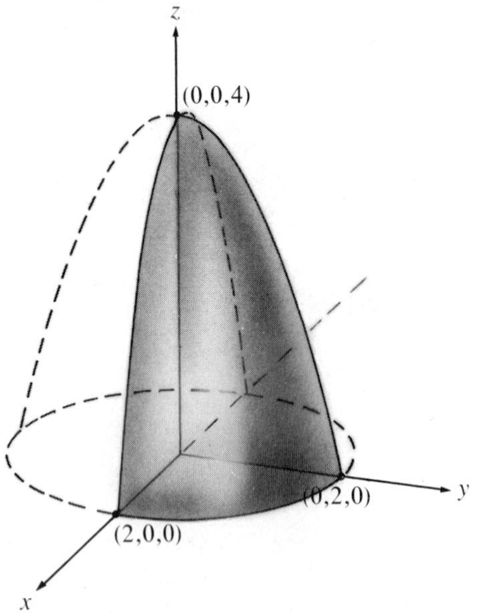

FIGURE 3.2

3.1 **DEFINITION.** Suppose that $f: \boldsymbol{R}^2 \to \boldsymbol{R}$ is defined on a closed bounded region $D \subseteq \boldsymbol{R}^2$. Let R be a rectangle completely containing D. Define $\bar{f}$ on R by

(1)
$$\bar{f}(\boldsymbol{x}) = \begin{cases} f(\boldsymbol{x}) & \text{if } \boldsymbol{x} \in D \\ 0 & \text{if } \boldsymbol{x} \notin D \end{cases}$$

for $\boldsymbol{x} \in R$. Then f is **integrable over D** if and only if $\bar{f}$ is integrable over R. If $\bar{f}$ *is* integrable over R, then we define the **double integral of f over D** by

(2)
$$\iint_D f(\boldsymbol{x})\, dA = \iint_R \bar{f}(\boldsymbol{x})\, dA.$$

While we won't prove them, the results of Sections 1 and 2 (in particular, Fubini's Theorem 2.4) hold verbatim for double integrals over closed bounded regions $D \subseteq \boldsymbol{R}^2$. Let's now take a look at how Fubini's Theorem applies to a typical case of interest. Suppose that D lies between the lines $x = a$ and $x = b$ and between the curves $y = g_1(x)$ and $y = g_2(x)$ (Figure 3.3). If we apply Fubini's

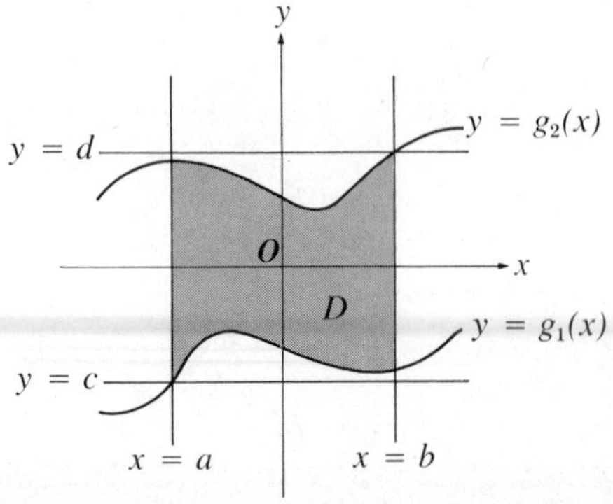

FIGURE 3.3

Theorem to evaluate $\iint_D f(\mathbf{x})\,dA$, then we obtain

$$\iint_D f(\mathbf{x})\,dA = \iint_R \bar{f}(\mathbf{x})\,dA = \int_a^b dx \int_c^d \bar{f}(x, y)\,dy$$

where $R = [a, b] \times [c, d]$ is shown in Figure 3.3. Now for each x, $\bar{f}(x, y) = 0$ if $y > g_2(x)$ or if $y < g_1(x)$. So points (x, y) in R that are above $y = g_2(x)$ or below $y = g_1(x)$ contribute *nothing* to any Riemann sum for $\bar{f}$ over R. Hence

(3)
$$\iint_D f(\mathbf{x})\,dA = \int_a^b dx \int_c^d \bar{f}(x, y)\,dy = \int_a^b dx \int_{g_1(x)}^{g_2(x)} \bar{f}(x, y)\,dy = \int_a^b dx \int_{g_1(x)}^{g_2(x)} f(x, y)\,dy.$$

Similarly, if D is a region bounded by curves $x = h_1(y)$ and $x = h_2(y)$ and lines $y = c$ and $y = d$ (Figure 3.4), then we have

(4)
$$\iint_D f(x, y)\,dA = \int_c^d dy \int_{h_1(y)}^{h_2(y)} f(x, y)\,dx.$$

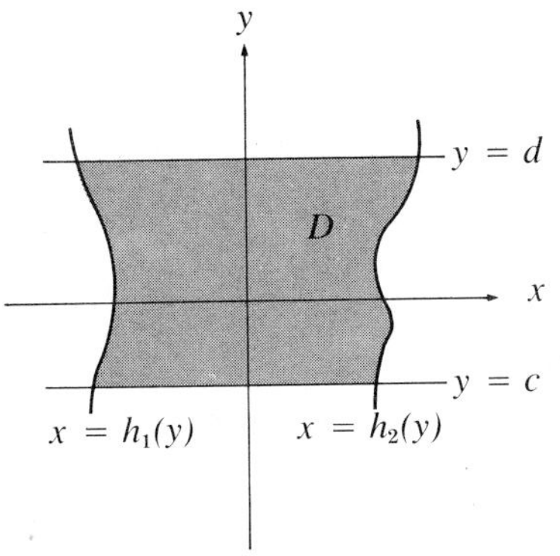

FIGURE 3.4

What do formulas (3) and (4) say? In (3), note that x is constant in the first integration $\int f(x, y)\,dy$. We find an antiderivative function $F(x, y)$ and evaluate F between $y = g_1(x)$ and $y = g_2(x)$. This gives us a function of x alone, $F(x, g_2(x)) - F(x, g_1(x))$, to be integrated between a and b to get the value of $\iint_D f(\mathbf{x})\,dA$. Similar remarks apply in (4), replacing x everywhere by y and vice versa. Before illustrating by an example, we give an extension of Definition 2.3.

3.2 **DEFINITION.** Suppose that $f: \mathbf{R}^2 \to \mathbf{R}$ is integrable over a closed bounded set $D \subseteq \mathbf{R}^2$ and $f(x, y) \geq 0$ for $(x, y) \in D$. Then the **volume of the solid S lying above D and below the graph of $z = f(x, y)$** is

$$V(S) = \iint_D f(x, y)\,dA.$$

We now *can* find the volume that, as we remarked in the opening paragraph, we *should* be able to find.

3.3 **EXAMPLE.** Find the volume lying below the paraboloid $z=4-x^2-y^2$ and above the xy-plane.

Solution. The picture is shown in Figure 3.2. Note that the paraboloid intersects the xy-plane in the circle $x^2+y^2=4$. (Let $z=0$ in the equation of the paraboloid to get this.) So the volume we want is

$$V=\iint_D (4-x^2-y^2)\,dA$$

where D is the set of points (x, y) in $\boldsymbol{R}^2$ satisfying $x^2+y^2\le 4$. By symmetry,

$$V=4\iint_{D'} (4-x^2-y^2)\,dA$$

where D' is the first quadrant portion of D (Figure 3.5). We see that D' is bounded by the curves $y=0$ and $y=\sqrt{4-x^2}$ and by the lines $x=0$ and $x=2$

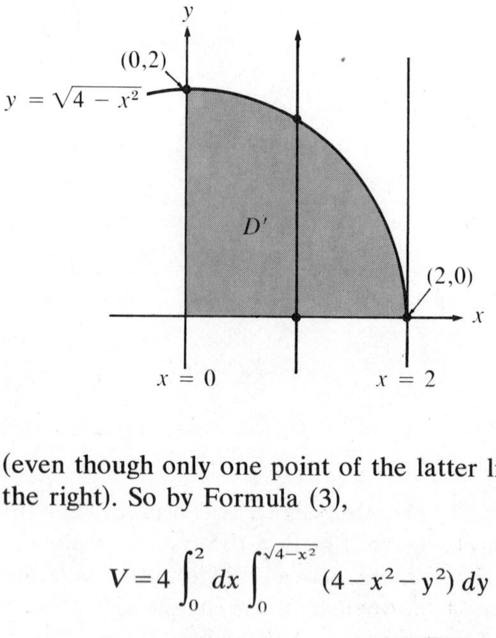

FIGURE 3.5

(even though only one point of the latter line is actually needed to bound D' on the right). So by Formula (3),

$$\begin{aligned}
V&=4\int_0^2 dx\int_0^{\sqrt{4-x^2}} (4-x^2-y^2)\,dy\\
&=4\int_0^2 dx\left[4y-x^2y-\frac{y^3}{3}\right]_0^{\sqrt{4-x^2}}\\
&=4\int_0^2 dx\left[(4-x^2)y-\frac{y^3}{3}\right]_0^{\sqrt{4-x^2}}\\
&=4\int_0^2 [(4-x^2)(4-x^2)^{1/2}-\tfrac{1}{3}(4-x^2)^{3/2}]\,dx\\
&=(4)(\tfrac{2}{3})\int_0^2 (4-x^2)^{3/2}\,dx.
\end{aligned}$$

This integral can be found in the table of integrals (Formula 26), or can be worked out using the trigonometric substitution $x=2\sin\theta$ for $-\pi/2\leq\theta\leq\pi/2$. In any event, we get

$$V=\frac{8}{3}\left[\frac{x}{4}(4-x^2)^{3/2}+\frac{3}{8}\cdot 4x(4-x^2)^{1/2}+\frac{3}{8}\cdot 16\,\mathrm{Sin}^{-1}\frac{x}{2}\right]_0^2$$

$$=\frac{8}{3}\left[0+0+6\frac{\pi}{2}-0-0-0\right]=8\pi.$$

Observe how we applied (3) in Example 3.3. We integrated first with respect to y, where y varied between $y=g_1(x)=0$ and $y=g_2(x)=\sqrt{4-x^2}$. Then we integrated the resulting $F(x)=(4-x^2)^{3/2}$ with respect to the single remaining variable x between $x=0$ and $x=2$. The directed arrow in Figure 3.5 is intended to help make the limits $g_1(x)$ and $g_2(x)$ in (3) easier to find. The equation of the *lower boundary* of D' (where the arrow first enters the region) gives the *lower limit* $y=g_1(x)$. The equation of the *upper boundary* (where the arrow emerges from the region) gives the *upper limit* $y=g_2(x)$. Notice that once the first integration with respect to y is completed, we are back to a one-variable situation, just as in elementary calculus. So there is no need to draw any arrows to determine the x limits—we already have the only arrow required, the x-axis! Our first integration "integrates out" the variable y. That brings us down to the x-axis for the second integration, where the limits are respectively the smallest and largest values of x that lie under the region of integration. In Section 6 we shall again encounter this idea of integrating out variables when we evaluate triple integrals.

The strategy discussed in the previous paragraph applies to (3), i.e., to situations in which we integrate first with respect to y. There is, as you must expect, a parallel strategy for (4) that is used when we want to integrate first with respect to x. The limits $x=h_1(y)$ and $x=h_2(y)$ are the *left* and *right boundaries* of the region of integration. These can be determined by seeing where an arrow parallel to the x-axis and directed toward the right *enters* and *leaves* the region of integration. Refer to Figure 3.4. Once x is integrated out, the limits on y are respectively the smallest and largest numerical values of y over the entire region. We will give an illustration of how this works as soon as we extend Corollary 1.3 so that we can use double integration to find areas of regions other than rectangles.

3.4 **THEOREM.** If D is a closed bounded region in the xy-plane bounded by curves $y=g_1(x)$ and $y=g_2(x)$ and by lines $x=a$ and $x=b$, then the area of D is

(5) $$A(D)=\iint_D 1\,dA.$$

The same formula holds if D is bounded by curves $x=h_1(y)$ and $x=h_2(y)$ and by lines $y=c$ and $y=d$.

Proof. If we put D inside a rectangle R, then

$$\iint_D 1\,dA=\iint_R \bar{f}(x)\,dA$$

where

$$\bar{f}(x, y) = \begin{cases} 1 & \text{if } (x, y) \in D \\ 0 & \text{if } (x, y) \notin D \end{cases}$$

Thus, since there is no contribution to any Riemann sum by points outside R, we have

$$\iint_R 1\,dA = \int_a^b dx \int_{g_1(x)}^{g_2(x)} \bar{f}(x, y)\,dy = \int_a^b dx \int_{g_1(x)}^{g_2(x)} 1\,dy$$

$$= \int_a^b [g_2(x) - g_1(x)]\,dx = A(D)$$

by elementary calculus. (See Figure 3.6.) The same reasoning applies if D is bounded by curves $x = h_1(y)$ and $x = h_2(y)$ and by lines $y = c$ and $y = d$. QED

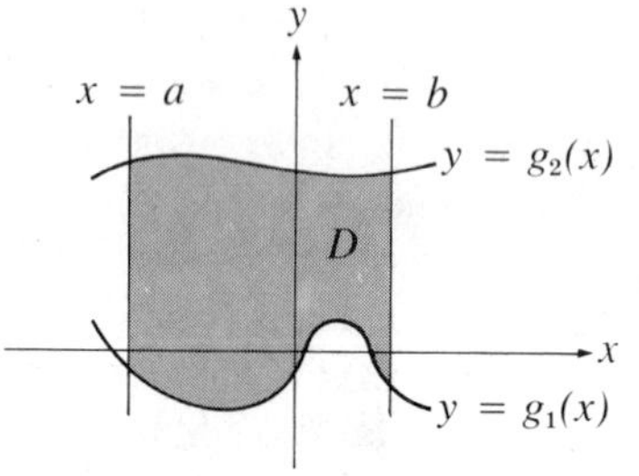

FIGURE 3.6

3.5 EXAMPLE. Find the area in the plane bounded by $x = y^2$ and $3x + 2y = 8$.

Solution. The region whose area is sought is labeled D in Figure 3.7. D is bounded by the curves $x = y^2$ and $x = \dfrac{8 - 2y}{3}$ between the lines $y = -2$ and $y = \frac{4}{3}$. So by (5) of Theorem 3.4,

$$A(D) = \iint_D 1\,dA = \int_{-2}^{4/3} dy \int_{y^2}^{(8-2y)/3} dx$$

$$= \int_{-2}^{4/3} \left(\tfrac{8}{3} - \tfrac{2}{3}y - y^2\right) dy$$

$$= \tfrac{8}{3}y - \tfrac{1}{3}y^2 - \tfrac{1}{3}y^3\Big]_{-2}^{4/3}$$

$$= \tfrac{32}{9} - \tfrac{16}{27} - \tfrac{64}{81} + \tfrac{16}{3} + \tfrac{4}{3} - \tfrac{8}{3} = \tfrac{500}{81}.$$

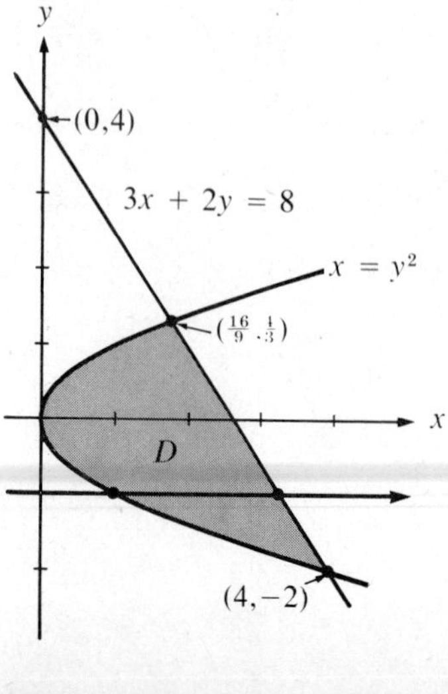

FIGURE 3.7

Notice that our directed arrow entered the region D from the left by crossing the boundary $x=y^2$, and emerged from the region on the right by crossing the boundary $x=\dfrac{8-2y}{3}$. Notice also that the problem is *much* simpler to do by using (4) instead of trying to use (3). To use (3), we would have to break D into *two* subregions D_1 and D_2 by the line $x=\frac{16}{9}$, and then integrate f over each subregion and add the results (by Theorem 1.8) to get $A(D)$. See Figure 3.8. We would have

$$A=\int_0^{16/9} dx \int_{-\sqrt{x}}^{\sqrt{x}} dy + \int_{16/9}^{4} dx \int_{-\sqrt{x}}^{(8-3x)/2} dy.$$

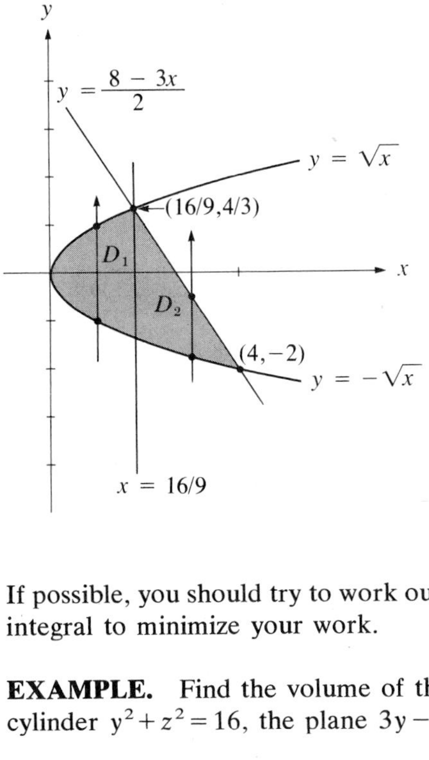

FIGURE 3.8

If possible, you should try to work out a double integral by using only one iterated integral to minimize your work.

3.6 EXAMPLE. Find the volume of the solid S in the first octant bounded by the cylinder $y^2+z^2=16$, the plane $3y-2x=0$, and the plane $x=0$.

Solution. Refer to Figure 3.9. The solid S lies below $z=\sqrt{16-y^2}$ and above the region D of the xy-plane bounded by $x=0$ (the y-axis), $y=4$, and $x=\frac{3}{2}y$ (the line $3y-2x=0$). See Figure 3.10. Thus,

$$V(S)=\iint_D \sqrt{16-y^2}\,dA=\int_0^4 dy\int_0^{3y/2}\sqrt{16-y^2}\,dx$$

$$=\int_0^4 \sqrt{16-y^2}\,dy\,[x]_0^{3y/2}=\frac{3}{2}\int_0^4 y\sqrt{16-y^2}\,dy$$

$$=\frac{3}{2}\left(-\frac{1}{2}\right)\int_0^4 -2y\sqrt{16-y^2}\,dy$$

$$=-\frac{3}{4}\frac{(16-y^2)^{3/2}}{3/2}\Bigg]_0^4=\frac{3}{4}\cdot\frac{2}{3}\cdot 16^{3/2}$$

$$=\tfrac{1}{2}\cdot 64=32.$$

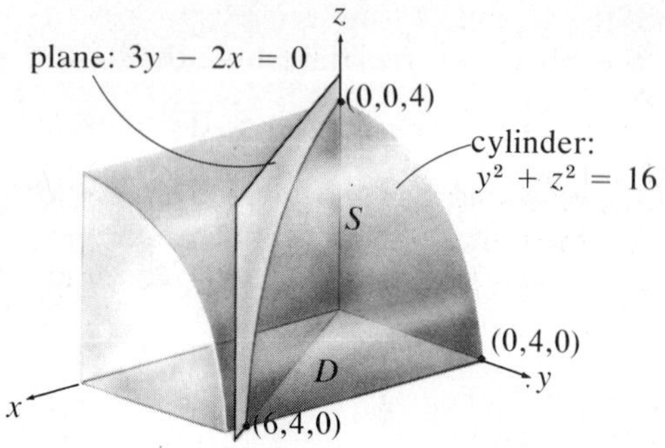

FIGURE 3.9

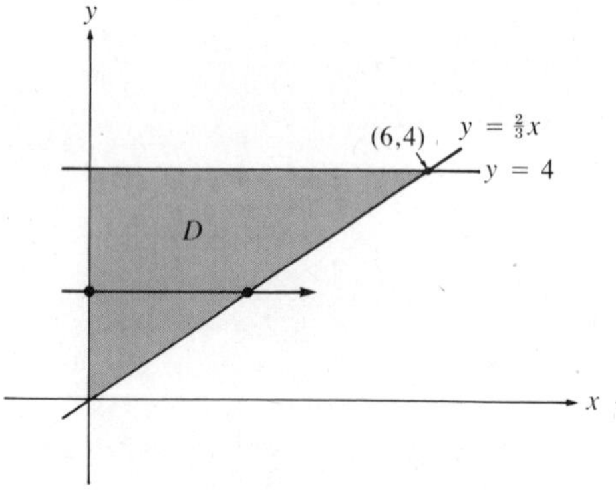

FIGURE 3.10

As you can see from Example 3.6, it often helps to draw *both* a picture of the situation in $\boldsymbol{R}^3$ and a picture of $D \subseteq \boldsymbol{R}^2$. (We integrated first with respect to x since this avoided integrating the radical until we had a factor of y to go with it and enable us to integrate it easily.)

EXERCISES 5.3

In Exercises 1 to 10, evaluate the given double integral. Draw a figure showing D in each case.

1 $\iint_D xy\,dA$, D the region bounded by the x-axis and the lines $x = 2$ and $y - 2x = 0$.

2 $\iint_D (2 - 3x + xy)\,dA$, D the region bounded by the x-axis and the lines $x = 1$ and $y - 3x = 0$.

3 $\iint_D (4 - x^2 - y)\,dA$, D the region bounded by the x-axis, the parabola $y = 4 - x^2$, and the lines $x = 0$ and $x = 2$.

4 $\iint_D (x - y^2)\,dA$, D the region bounded by the parabola $x = y^2$ and the lines $x = 1$, $y = -1$, and $y = 1$.

5 $\iint_D (3x + 2y)\,dA$, D the region bounded by the lines $x + y = 1$, $y - x = 1$, $x = 0$, and $x = 1$.

6 $\iint_D (2x - 5y)\,dA$, D the region bounded by the lines $x + y = 2$, $y - x = 2$, $x = 0$, and $x = 2$.

7 $\iint_D e^{-x-y}\,dA$, D the region bounded by the lines $y = 0$, $y = x$, $x = \frac{1}{2}$, and $x = 1$.

8 $\iint_D e^{x+y}\,dA$, D the region of Exercise 7.

9 $\iint_D (1-x^2-y^2)\,dA$, D the square with vertices $(1,0)$, $(-1,0)$, $(0,1)$, and $(0,-1)$.

10 $\iint_D (1+x^2+y^2)\,dA$, D the square of Exercise 9.

In Exercises 11 to 18 find the area of the given region in the xy-plane.

11 D the region bounded by the parabolas $y^2=12x$ and $y=\frac{2}{3}x^2$.

12 D the region bounded by the parabolas $y^2=4x$ and $y=2x^2$.

13 D the region in the first quadrant bounded by the curves $y=x^2$ and $y=x^4$.

14 D the region in the first quadrant bounded by $x=y^2$ and $x=y^4$.

15 D the region bounded by $y=e^x$, $y=0$, $x=0$, and $x=1$.

16 The region in the first quadrant bounded by the two circles $x^2+y^2=1$ and $x^2+(y-1)^2=1$.

17 The region bounded by $y=x$ and the curve $x+y^2=1$.

18 The region bounded by $y=x$ and the curve $x+y^2=4$.

In Exercises 19 to 28, find the volume of the given solid.

19 Under the plane $2x+y+z=5$ and above the rectangle $R=[0,2]\times[0,1]$.

20 Under the plane $x+y+2z=4$ and above the rectangle $R=[0,1]\times[0,2]$.

21 Under the paraboloid $z=x^2+y^2$ and above the domain D bounded by the x-axis and the lines $y=2x$ and $x=1$.

22 Under the paraboloid $z=x^2+2y^2$ and above the domain D bounded by the coordinate axes and the line $x+y=1$.

23 Under the surface $z=xy$ and above the domain D lying between $y=2x$ and $y=x^2$.

24 Under the surface $z=xy$ and above the domain lying between $y=x$ and $y=x^3$.

25 The first octant region bounded by the two cylinders $x^2+y^2=9$ and $x^2+z^2=9$.

26 The first octant region bounded by the two cylinders $x^2+y^2=1$ and $x^2+z^2=1$.

27 The first octant region bounded by the cylinder $y^2+z^2=9$, the plane $y=x$, and the yz-plane.

28 The first octant region bounded by the cylinder $y^2+z^2=4$, the plane $y=2x$, and the yz-plane.

In Exercises 29 to 34, draw a sketch of the region D over which the iterated integral is being evaluated. Express the given integral as an iterated integral (or sum of iterated integrals) in which the order of integration is reversed.

29 $\displaystyle\int_0^2 dx \int_0^{x^2} f(x,y)\,dy$

30 $\displaystyle\int_0^2 dx \int_0^{\sqrt{x}} f(x,y)\,dy$

31 $\displaystyle\int_0^1 dy \int_{e^y}^{e} f(x,y)\,dx$

32 $\displaystyle\int_1^e dy \int_0^{\ln y} f(x,y)\,dx$

33 $\displaystyle\int_{-1}^2 dx \int_{x^2}^{x+2} f(x,y)\,dy$

34 $\displaystyle\int_{-1}^2 dy \int_{y^2}^{y+2} f(x,y)\,dx$

The following two exercises will be of use in Section 9.

35 Show that the results of Exercises 22 and 23 in Exercises 5.1 extend to the region shown in Figure 3.3.

36 Repeat Exercise 35 for the region shown in Figure 3.4.

4 POLAR COORDINATES*

Regions in $\boldsymbol{R}^2$ that are symmetric about a point or that are bounded in whole or in part by circles are often more conveniently described by polar coordinates than by rectangular (Cartesian) coordinates. Given a point P in the plane, we can specify its location by giving the pair of numbers $P[r, \theta]$ where

$$r = |\overrightarrow{OP}| = \text{distance of } P \text{ from the origin,}$$

and

$$\theta = \text{angle between } \overrightarrow{OP} \text{ and } \boldsymbol{i} = (1, 0),$$

measured counterclockwise from $\boldsymbol{i}$. See Figure 4.1.

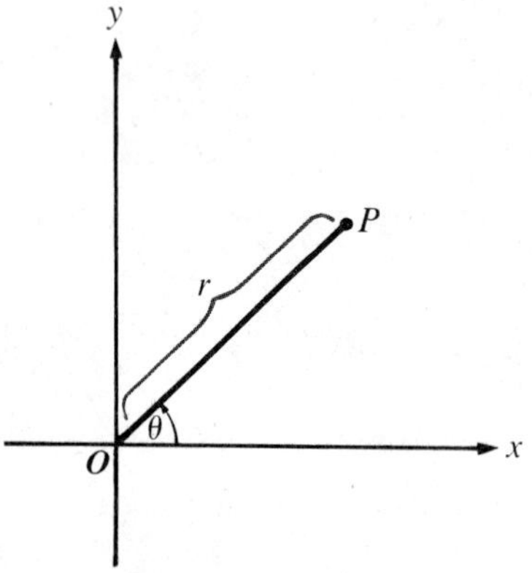

FIGURE 4.1

4.1 **DEFINITION.** Given a point P in $\boldsymbol{R}^2$ with rectangular coordinates (x, y), its **polar coordinates** $P[r, \theta]$ consist of

$$r = |\overrightarrow{OP}|$$

and

$$\theta = \text{the angle between } \overrightarrow{OP} \text{ and } \boldsymbol{i} = (1, 0),$$

measured *counterclockwise* from $\boldsymbol{i}$. If θ is negative, then we interpret the angle as being measured clockwise from $\boldsymbol{i}$. If r is negative, then $P[r, \theta]$ stands for the same point as $P[-r, \theta + \pi]$ (for which $-r > 0$). See Figure 4.2.

The origin $\boldsymbol{O}$ is called the *pole*.

Recall that θ is usually restricted to the interval $[0, 2\pi]$ and r is restricted to nonnegative real values in order to obtain a coordinatization of $\boldsymbol{R}^2$ that is one-to-one except at the pole. (Note that the pole can be represented as $\boldsymbol{O}[0, \theta]$ for any θ whatsoever.)

---

* Optional (review) section.

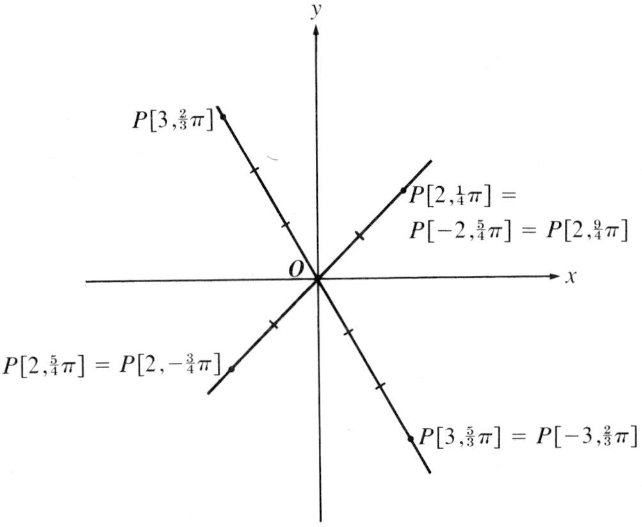

FIGURE 4.2

The next result follows immediately from Definition 4.1 and Figure 4.3.

4.2 **PROPOSITION.** If $P \in \boldsymbol{R}^2$ has rectangular coordinates $P(x, y)$ and polar coordinates $P[r, \theta]$, then

(1) $x = r\cos\theta$ (4) $\tan\theta = \dfrac{y}{x}$ if $x \neq 0$

(2) $y = r\sin\theta$ (5) $\sin\theta = \dfrac{y}{r}$ if $(x, y) \neq (0, 0)$

(3) $r = \sqrt{x^2 + y^2}$ (6) $\cos\theta = \dfrac{x}{r}$ if $(x, y) \neq (0, 0)$.

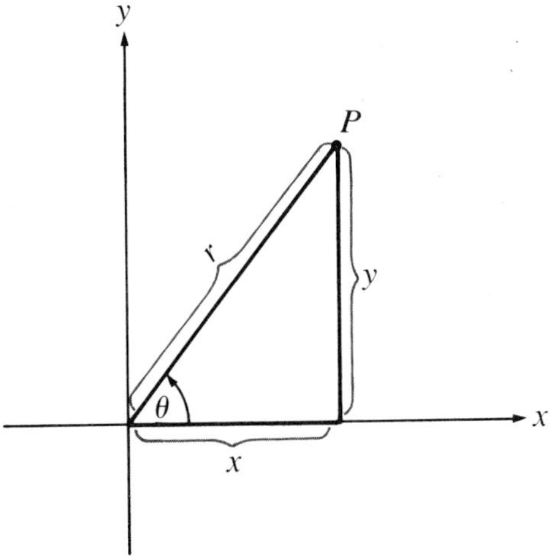

FIGURE 4.3

Rectangular coordinates derive their name from the fact that the equations $x = c$ and $y = k$ define *rectangles* with the coordinate axes (see Figure 4.4). Recall that the corresponding geometric objects in polar coordinates are *circular.* The equation $r = c$ in fact defines a circle of radius c centered at the pole, and the equation $\theta = k$ is a line through the pole with inclination k. Thus, the equations

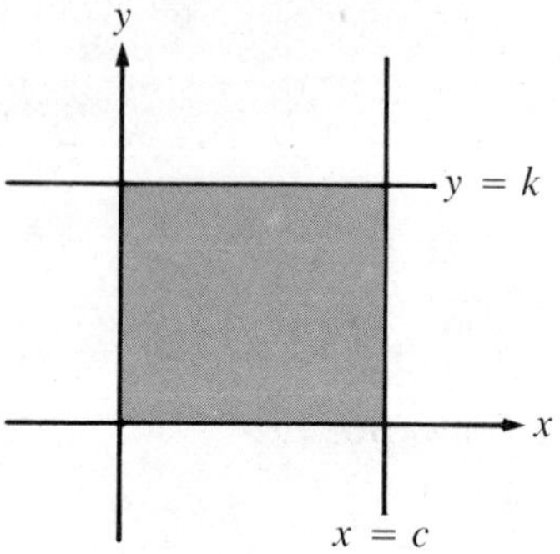

FIGURE 4.4

$r = c$, $\theta = \alpha$, and $\theta = \beta$ in general define a *circular sector* like the one shown in Figure 4.5 $\left(\text{where we have taken } c = 4,\ \alpha = 0, \text{ and } \beta = \frac{\pi}{6}\right)$. Similarly, the region

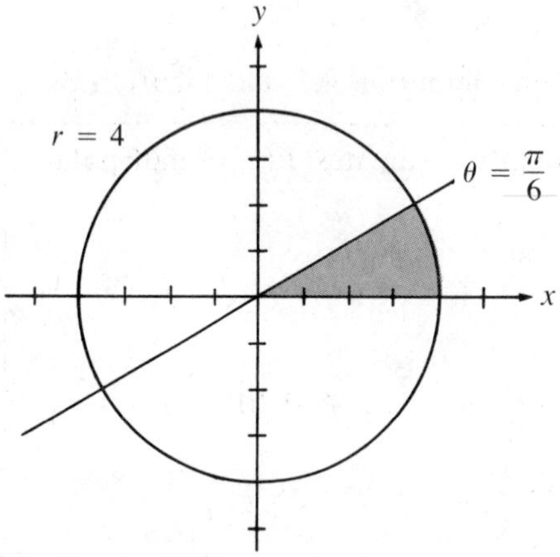

FIGURE 4.5

between $r = c_1 > 0$, $r = c_2 > 0$, $\theta = \alpha$, and $\theta = \beta$ is in general an *annular sector* as represented in Figure 4.6.

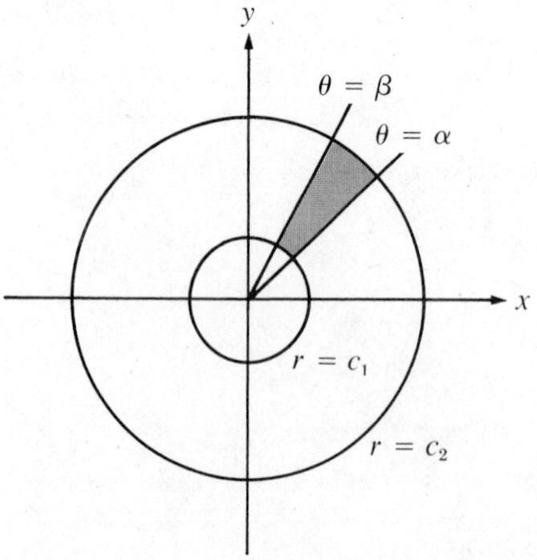

FIGURE 4.6

The next result is a simple consequence of Proposition 4.2. It is proved by substituting (5) or (6) as appropriate, multiplying the resulting equation through by r, substituting (1) or (2), and completing the square to obtain the equation in rectangular coordinates.

4.3 **PROPOSITION.** (a) The polar equation $r = a \cos \theta$, where a is positive, is the equation of the circle centered at $(\frac{1}{2}a, 0)$ with radius $\frac{1}{2}a$.

(b) The polar equation $r = a \sin \theta$, where a is positive, is the equation of the circle centered at $(0, \frac{1}{2}a)$ with radius $\frac{1}{2}a$.

Conic sections have rather nice polar equations also. Before presenting these, we review the general definition of a conic section.

4.4 **DEFINITION.** Let ℓ be a line in $\boldsymbol{R}^2$, and let F be a point not on ℓ. A **conic section** is the set of all points $P \in \boldsymbol{R}^2$ such that

$$\frac{d(P, F)}{d(P, \ell)} = e$$

where $d(P, F)$ is the distance from P to F and $d(P, \ell)$ is the perpendicular distance from P to the line ℓ. The positive number e is called the **eccentricity**, F is called a **focus**, and ℓ is a **directrix**. If $e = 1$, the conic section is a **parabola**. If $e > 1$, the conic section is a **hyperbola**. If $e < 1$, the conic section is an **ellipse**.

4.5 **THEOREM.** (a) Let C be a conic section of eccentricity e with focus F at the pole and a vertical directrix $x = d$ or $x = -d$. Then a polar equation of C is

$$r = \frac{ed}{1 \pm e \cos \theta},$$

where the plus sign is taken if the directrix is $x = d$, and the minus sign is taken if the directrix is $x = -d$.

(b) Let C be a conic section of eccentricity e with focus F at the pole and a horizontal directrix $y = d$ or $y = -d$. Then a polar equation of C is

$$r = \frac{ed}{1 \pm e \sin \theta},$$

where the plus sign is taken if the directrix is $y = d$ and the minus sign is taken if the directrix is $y = -d$.

For a proof, consult an elementary calculus text. As a simple example of the theorem, consider the locus of the polar equation

$$r = \frac{6}{1 - 2 \cos \theta}.$$

We see that this conic section has eccentricity $e = 2$, which is greater than 1. Therefore, the graph is a hyperbola with focus at the origin (pole) and directrix $x = -d$, where $de = 6$. Thus the directrix is $x = -3$. The vertices lie on the x-axis (polar axis). When $\theta = 0$, we obtain the vertex $P[-6, 0]$. When $\theta = \pi$, we get the other vertex, $P'[2, \pi] = P'(-2, 0)$. A rough sketch is shown in Figure 4.7.

We next mention three simple conditions under which a polar graph will be symmetric relative to an axis or the pole.

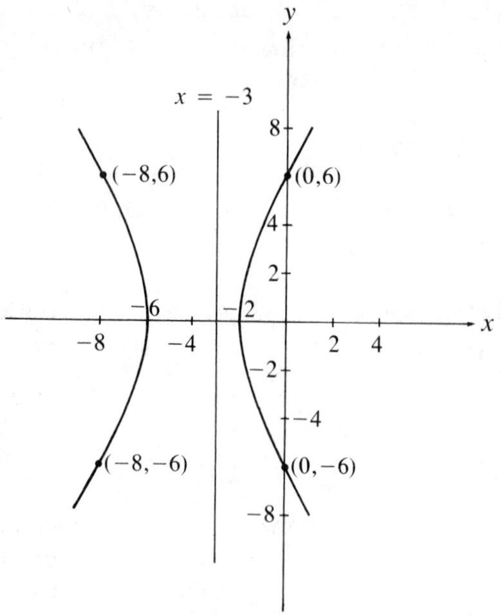

FIGURE 4.7

(a) If an equation $f(r, \theta)=0$ is unchanged when θ is replaced by $-\theta$, then the graph is **symmetric in the polar axis** $\theta=0$. For example, a conic section

$$r=\frac{de}{1\pm e\cos\theta}$$

is symmetric in the x-axis since $\cos(-\theta)=\cos\theta$.

(b) If an equation $f(r, \theta)=0$ is unchanged when θ is replaced by $\pi-\theta$, then the **graph is symmetric in the y-axis** ($\theta=\frac{1}{2}\pi$). For instance, a conic section

$$r=\frac{de}{1\pm e\sin\theta}$$

is symmetric in the y-axis since $\sin(\pi-\theta)=\sin\theta$.

(c) If an equation $f(r, \theta)=0$ is unchanged when r is replaced by $-r$, then the graph **is symmetric relative to the pole**. For example, $r^2=4\cos 2\theta$ is symmetric relative to the pole.

Using these conditions (a), (b), and (c), it is a simple matter to graph the *lemniscate* $r^2=4\cos 2\theta$ of Figure 4.8 from the points tabulated for θ in the interval $[0, \frac{1}{4}\pi]$. Likewise, in Figure 4.9 we have plotted the familiar *cardioid* $r=1+\cos\theta$ from the tabulated points corresponding to θ in the interval $[0, \pi]$.

We now consider the problem of finding the arc length of a curve given in polar coordinate form $f(r, \theta)=0$.

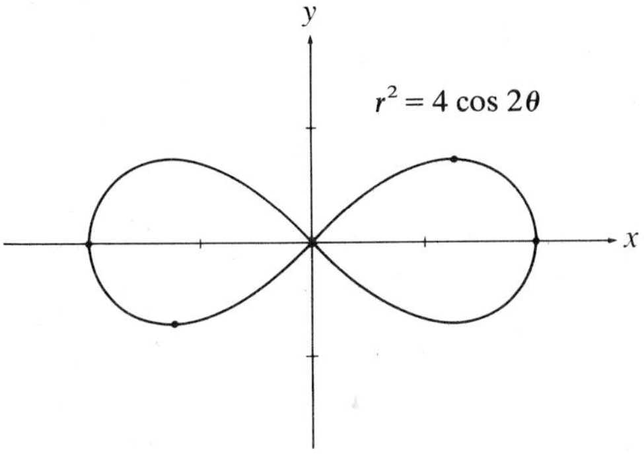

FIGURE 4.8

θ	0	$\frac{\pi}{6}$	$\frac{\pi}{4}$
r	± 2	$\pm\sqrt{2}$	0

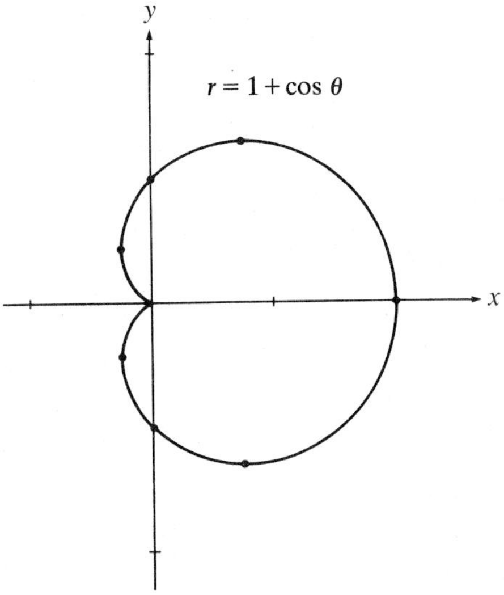

FIGURE 4.9

θ	0	$\frac{\pi}{3}$	$\frac{\pi}{2}$	$\frac{2\pi}{3}$	π
r	2	$\frac{3}{2}$	1	$\frac{1}{2}$	0

4.6 **THEOREM.** (a) If a curve is described by $f(r, \theta) = 0$, where $r = r(t)$ and $\theta = \theta(t)$ for $a \le t \le b$, then the length of arc is

$$L = \int_a^b \sqrt{\left(\frac{dr}{dt}\right)^2 + r^2\left(\frac{d\theta}{dt}\right)^2}\, dt.$$

(b) If the curve is given by an equation $r = f(\theta)$, then

$$L = \int_\alpha^\beta \sqrt{\left(\frac{dr}{d\theta}\right)^2 + r^2}\, d\theta,$$

where $\alpha = \theta(a)$ and $\beta = \theta(b)$.

Proof. We have

$$\begin{aligned}\mathbf{x}(t) &= (x(t), y(t)) \\ &= (r(t)\cos\theta(t), r(t)\sin\theta(t)).\end{aligned}$$

Using the Chain Rule (Theorem 2.1.7(6)), we get

$$\dot{\mathbf{x}}(t) = (\dot{r}(t)\cos\theta(t) - r(t)\sin\theta(t)\,\dot{\theta}(t), \dot{r}(t)\sin\theta(t) + r(t)\cos\theta(t)\,\dot{\theta}(t)).$$

So

$$\begin{aligned}|\dot{\mathbf{x}}(t)| &= (\dot{r}^2\cos^2\theta - 2r\dot{r}\dot{\theta}\sin\theta\cos\theta + r^2\dot{\theta}^2\sin^2\theta + \dot{r}^2\sin^2\theta \\ &\quad + 2r\dot{r}\dot{\theta}\sin\theta\cos\theta + r^2\dot{\theta}^2\cos^2\theta)^{1/2} \\ &= \sqrt{\dot{r}^2 + r^2\dot{\theta}^2} = \sqrt{\left(\frac{dr}{dt}\right)^2 + r^2\left(\frac{d\theta}{dt}\right)^2}.\end{aligned}$$

Thus,

$$L = \int_a^b |\dot{\mathbf{x}}(t)|\,dt = \int_a^b \sqrt{\left(\frac{dr}{dt}\right)^2 + r^2\left(\frac{d\theta}{dt}\right)^2}\,dt,$$

which proves part (a).

Now, if $r = f(\theta)$, then $\dfrac{dr}{dt} = \dfrac{dr}{d\theta}\dfrac{d\theta}{dt}$ by the Chain Rule, so we get

$$|\dot{\mathbf{x}}(t)| = \sqrt{\left(\frac{dr}{d\theta}\right)^2\left(\frac{d\theta}{dt}\right)^2 + r^2\left(\frac{d\theta}{dt}\right)^2} = \sqrt{\left(\frac{dr}{d\theta}\right)^2 + r^2}\,\frac{d\theta}{dt}.$$

So

$$\begin{aligned}L &= \int_a^b |\dot{\mathbf{x}}(t)|\,dt = \int_a^b \sqrt{\left(\frac{dr}{d\theta}\right)^2 + r^2}\,\frac{d\theta}{dt}\,dt \\ &= \int_\alpha^\beta \sqrt{\left(\frac{dr}{d\theta}\right)^2 + r^2}\,d\theta.\end{aligned}$$

QED

4.7 EXAMPLE. Find the length of arc of the cardioid $r = 1 + \cos\theta$.

Solution. First we find $\dfrac{dr}{d\theta} = -\sin\theta$. So

$$\left(\frac{dr}{d\theta}\right)^2 + r^2 = \sin^2\theta + 1 + 2\cos\theta + \cos^2\theta = 2 + 2\cos\theta.$$

Thus

$$L = \int_0^{2\pi} \sqrt{2 + 2\cos\theta}\,d\theta = \sqrt{2}\int_0^{2\pi} \sqrt{1 + \cos\theta}\,d\theta.$$

Now

$$\cos^2\tfrac{1}{2}\theta = \frac{1 + \cos\theta}{2}, \quad \text{so} \quad 2\cos^2\tfrac{1}{2}\theta = 1 + \cos\theta.$$

Hence $\sqrt{1+\cos\theta}=\sqrt{2\cos^2\frac{1}{2}\theta}=\sqrt{2}\cos\frac{1}{2}\theta$ as long as $0\le\frac{1}{2}\theta\le\frac{1}{2}\pi$, i.e., $0\le\theta\le\pi$. But by symmetry, the length L is twice the length traced as θ goes from 0 to π. So

$$L=2\sqrt{2}\int_0^{\pi}\sqrt{1+\cos\theta}\,d\theta=2\sqrt{2}\sqrt{2}\int_0^{\pi}\cos\tfrac{1}{2}\theta\,d\theta$$
$$=4\cdot 2\sin\tfrac{1}{2}\theta\Big]_0^{\pi}=8.$$

In working with $\sqrt{\left(\frac{dr}{d\theta}\right)^2+r^2}$ you must be careful that *any square root extraction done is valid over the entire interval of integration from α to β.* In the preceding example, if we had not taken this precaution and simply put $\sqrt{2}\cos\frac{1}{2}\theta$ into the expression for L from 0 to 2π, then we would have gotten $L=0$, as you should check. This absurdity results from the fact that $\sqrt{2}\cos\frac{1}{2}\theta$ is *negative* over half the interval of integration, so the positive contribution over $[0,\pi]$ is cancelled out by the negative contribution over $[\pi, 2\pi]$.

You may recall the following definition of the area of the polar region R bounded by the graph of $r=f(\theta)$ (where $f:\mathbf{R}\rightarrow\mathbf{R}$ is an integrable function) and lines $\theta=\alpha$ and $\theta=\beta$. Refer to Figure 4.10.

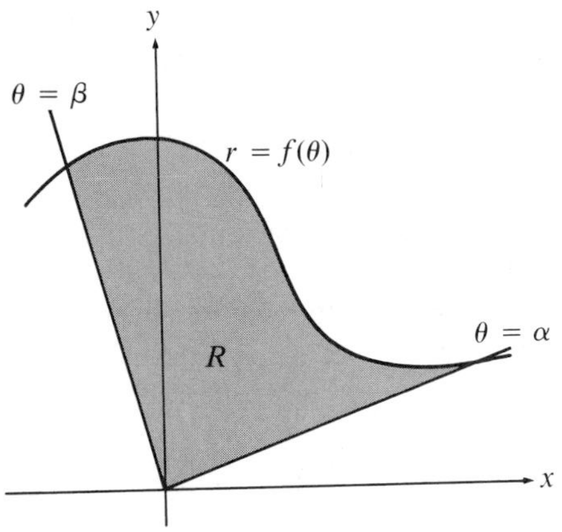

FIGURE 4.10

4.8 DEFINITION. If R is the region bounded by the graph of $r=f(\theta)$ and the lines $\theta=\alpha$ and $\theta=\beta$, then the **area** of R is

$$A(R)=\int_{\alpha}^{\beta}\tfrac{1}{2}r^2\,d\theta=\tfrac{1}{2}\int_{\alpha}^{\beta}f(\theta)^2\,d\theta.$$

An obvious question is whether this definition is *consistent* with Corollaries 1.3 and 3.4, which assert that

$$A(R)=\iint_R 1\,dA.$$

These two notions of area *do* in fact coincide, although establishing that fact involves concepts we do not yet have at our disposal. On page 269 we will informally take up this point, but a proper discussion must await the ideas of the next chapter. (See Section 6.9.)

4.9 EXAMPLE. Find the area enclosed by the cardioid $r=1+\cos\theta$.

Solution

$$A=\int_0^{2\pi}\tfrac{1}{2}(1+\cos\theta)^2\,d\theta=2\int_0^{\pi}\tfrac{1}{2}(1+\cos\theta)^2\,d\theta$$

$$=\int_0^{\pi}(1+2\cos\theta+\cos^2\theta)\,d\theta$$

$$=\int_0^{\pi}\left(1+2\cos\theta+\frac{1+\cos 2\theta}{2}\right)d\theta$$

$$=\int_0^{\pi}(\tfrac{3}{2}+2\cos\theta+\tfrac{1}{2}\cos 2\theta)\,d\theta$$

$$=\tfrac{3}{2}\theta+2\sin\theta+\tfrac{1}{4}\sin 2\theta\Big]_0^{\pi}=\frac{3\pi}{2}.$$

EXERCISES 5.4

1 Plot the points $[2, \pi/3]$, $[-2, \pi/2]$, $[-3, \pi]$, $[1, 2\pi/3]$, $[0, 7\pi/20]$.

2 Give the rectangular coordinates of the points in Exercise 1.

In Exercises 3 to 5, find a polar equation for the graph of the given rectangular equation.

3 $x^2+y^2=9$

4 $(x-2)^2+y^2=4$

5 $x^2-y^2=4$

6 Find a rectangular equation for the graph of the polar equation $\theta=\dfrac{\pi}{6}$.

In Exercises 7 to 16, draw the graph of the given polar equation. Identify the graph if it is not named.

7 $r=1+2\cos\theta$ (limaçon)

8 $r=3\cos 3\theta$ (three-petal rose)

9 $r=3\cos 2\theta$ (four-petal rose)

10 $r=3\cos 5\theta$ (five-petal rose)

11 $r=\dfrac{2}{1-\sin\theta}$

12 $r=\dfrac{4}{2+\cos\theta}$

13 $r=\dfrac{2}{1+4\cos\theta}$

14 $r=\dfrac{2}{1+4\sin\theta}$

15 $r=\theta$ (spiral of Archimedes)

16 $r=e^{\theta}$ (logarithmic spiral)

17 Find the polar equation of the ellipse with focus at the pole, directrix $x=-4$, and eccentricity $e=\frac{1}{2}$.

18 Find the polar equation of the parabola with focus at the pole and directrix $x=-3$.

19 Find the polar equation of the hyperbola having focus at the pole, directrix $x=-4$, and vertex $(-3, 0)$.

20 Find the polar equation of the hyperbola having focus at the pole, directrix $y=3$, and vertex $(0, 2)$.

In Exercises 21 to 24, find the length of arc of the given curve.

21 $r=\theta^2$, $0\le\theta\le 2\pi$

22 $r=1+\cos\theta$ between $\theta=0$ and $\theta=\pi/3$

23 $r=\sin^3 \theta/3$, entire curve

24 $r=e^{\theta/2}$ from $\theta=0$ to $\theta=2\pi$.

In Exercises 25 to 29, find the area of the given region.

25 The first quadrant region enclosed by the cardioid $r=1+\cos\theta$.

26 The region enclosed by $r^2=8\cos 2\theta$.

27 The region enclosed by one petal of $r=2\sin 3\theta$.

28 The small loop of the limaçon $r=1-2\sin\theta$.

29 The region inside the circle $r=3\sin\theta$ and outside $r=2-\sin\theta$.

5 POLAR DOUBLE INTEGRALS

Since the plane $\boldsymbol{R}^2$ can be coordinatized by polar coordinates, we can give a formula for a function $f:\boldsymbol{R}^2\to\boldsymbol{R}$ in terms of the polar coordinates $[r, \theta]$ of a point $\boldsymbol{x}\in\boldsymbol{R}^2$ if we like. If we start with a formula $z=f(x, y)$, we can substitute $x=r\cos\theta$ and $y=r\sin\theta$, thereby obtaining

$$z=f(r\cos\theta, r\sin\theta)=g(r, \theta)$$

for some function g. For instance, if $z=f(x, y)=x^2+y^2-xy$, then

$$\begin{aligned} z&=r^2\cos^2\theta+r^2\sin^2\theta-r^2\sin\theta\cos\theta\\ &=r^2(1-\sin\theta\cos\theta)\\ &=g(r, \theta)\end{aligned}$$

If we want to study this particular function on the unit disk $x^2+y^2\leq 1$, then this point of view could well be better than the rectangular coordinate description. In this section we will define the double integral of a continuous polar function $g(r, \theta)$ and will give techniques for evaluating it very much like those given in Sections 2 and 3 for double integrals in rectangular coordinates.

To begin with, we will consider the double integral of g over the polar coordinate analogue of a rectangle: a region S in the polar coordinate plane bounded by lines $\theta=\alpha$ and $\theta=\beta$ and by circular arcs $r=a$ and $r=b$. See Figure 5.1. How should $\iint_S g(r, \theta)\, dA$ be defined? Well, if we want to proceed as we did

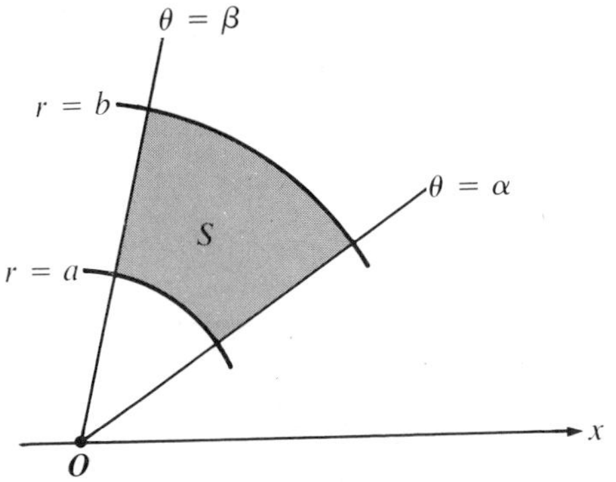

FIGURE 5.1

in Section 1, then we should make a polar grid G of S by partitioning the two intervals $[\alpha, \beta]$ and $[a, b]$. Accordingly, we let

$$P_1 : a = r_0 < r_1 < \ldots < r_{i-1} < r_i < \ldots < r_{m-1} < r_m = b$$

and

$$P_2 : \alpha = \theta_0 < \theta_1 < \theta_2 < \ldots < \theta_{j-1} < \theta_j < \ldots < \theta_{n-1} < \theta_n = \beta$$

be partitions of $[a, b]$ and $[\alpha, \beta]$. Putting them together, we get a polar grid of S into subregions S_{ij}, one of which is pictured in Figure 5.2. To form a Riemann

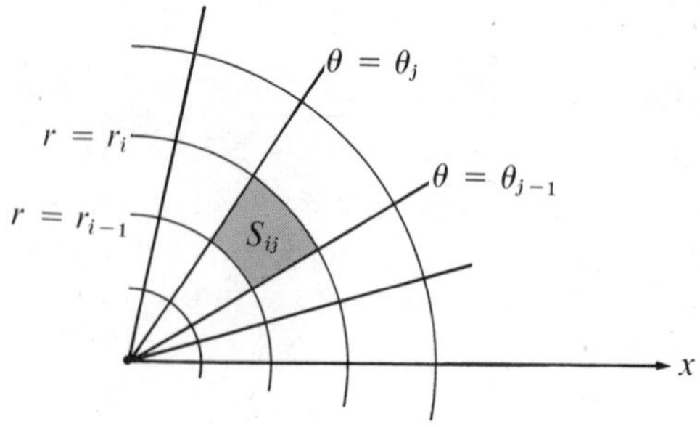

FIGURE 5.2

sum of g over S, we will pick a point $[\bar{r}_i, \bar{\theta}_j]$ in each S_{ij}, evaluate g at this point, multiply by ΔA_{ij} (the area of S_{ij}), and sum over all i and j. There is no difficulty computing $g(\bar{r}_i, \bar{\theta}_j)$, but how about ΔA_{ij}? Since S_{ij} is *not* a rectangle (we shouldn't expect it to be, since we're not working with *rectangular* coordinates any more!), we have no obvious formula (like $\Delta x_i \, \Delta y_j$ in Section 1) for ΔA_{ij}. So let's consider S_{ij} more closely.

We see that S_{ij} is a region between two arcs defining two circular sectors of the same central angle $\Delta\theta_j = \theta_j - \theta_{j-1}$. The larger sector has radius r_i, and the smaller has radius r_{i-1}. Therefore, ΔA_{ij} is the difference between the areas of these two sectors:

$$\begin{aligned}
\Delta A_{ij} &= \tfrac{1}{2} r_i^2 \, \Delta\theta_j - \tfrac{1}{2} r_{i-1}^2 \, \Delta\theta_j \\
&= \tfrac{1}{2}(r_i^2 - r_{i-1}^2) \, \Delta\theta_j \\
&= \tfrac{1}{2}(r_i + r_{i-1})(r_i - r_{i-1}) \, \Delta\theta_j \\
&= \tfrac{1}{2}(r_i + r_{i-1}) \, \Delta r_i \, \Delta\theta_j \\
&= r_i' \, \Delta r_i \, \Delta\theta_j, \quad \text{where} \quad r_i' = \frac{r_i + r_{i-1}}{2} \in [r_{i-1}, r_i] \\
&\approx \bar{r}_i \, \Delta r_i \, \Delta\theta_j, \text{ for any } \bar{r}_i \in [r_{i-1}, r_i].
\end{aligned}$$

This approximation is very close for a grid G of fine mesh, since any $\bar{r}_i \in [r_{i-1}, r_i]$ will be close to the midpoint r_i' if $[r_{i-1}, r_i]$ is very short. This approximation also

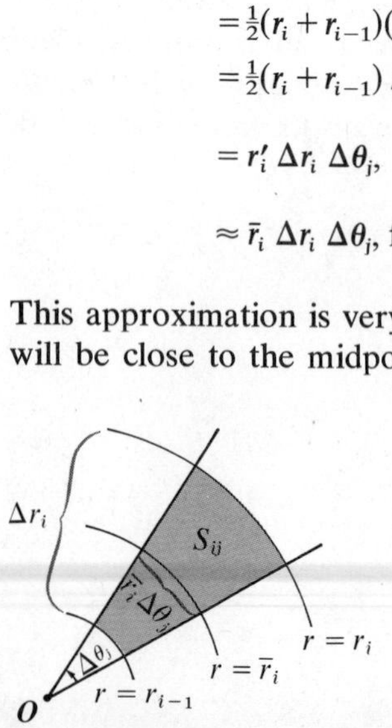

FIGURE 5.3

makes sense if we view S_{ij} as approximately a rectangle, of which the "base" is the arc $\bar{r}_i\,\Delta\theta_j$ and the height is Δr_i; hence its area is approximately the product of these. See Figure 5.3. A polar Riemann sum of g over S is now defined as

$$R(g, G, (\bar{r}_i, \bar{\theta}_j)) = \sum_{i=1}^{m}\sum_{j=1}^{n} g(\bar{r}_i, \bar{\theta}_j)\bar{r}_i\,\Delta r_i\,\Delta\theta_j$$

As we allow the mesh $|G|$ of G to approach 0, the limit of the Riemann sum can be shown to exist as a consequence of the continuity of g. It is natural to call this limit the double integral of g over S.

5.1 **DEFINITION.** Let $S \subseteq \mathbf{R}^2$ be a region in the polar coordinate plane lying between $\theta = \alpha$, $\theta = \beta$, $r = a$, and $r = b$. Let $g: \mathbf{R}^2 \to \mathbf{R}$ be continuous on S. Then the **polar double integral** of g over S is

$$\iint_S g(r, \theta)\,dA = \iint_S g(r, \theta) r\,dr\,d\theta$$

$$= \lim_{|G|\to 0} \sum_{i=1}^{m}\sum_{j=1}^{n} g(\bar{r}_i, \bar{\theta}_j)\bar{r}_i\,\Delta r_i\,\Delta\theta_j$$

where the limit is taken in the same sense as in Definition 1.1.

As in the case of Definition 4.8, we are faced with a consistency question. Is Definition 5.1 consistent with Definition 3.1 when S is a region of $\mathbf{R}^2$ described both by rectangular and by polar coordinates? It is, but a discussion of how the expressions from these two definitions can be reconciled is best given from the point of view of coordinate transformations, to be discussed in Sections 8 and 9 of the next chapter. For now, we merely emphasize that in order for consistency to exist *in polar double integrals*, we must have

$$dA = r\,dr\,d\theta, \qquad \textbf{NOT} \qquad dr\,d\theta.$$

Polar double integrals are evaluated (cf. Fubini's Theorem 2.4) by iteration, although the formal derivation of this is beyond our scope here. (See Section 6.9.)

5.2 **THEOREM.** If S and g are as in Definition 5.1, then

$$\iint_S g(r, \theta)\,dA = \int_\alpha^\beta d\theta \int_a^b g(r, \theta) r\,dr = \int_a^b r\,dr \int_\alpha^\beta g(r, \theta)\,d\theta$$

A real virtue of polar integration is that we can always switch to polar coordinates if we are confronted with a rectangular double integral that is hard to evaluate. If S is a region of $\mathbf{R}^2$ described by rectangular coordinates and T is the same region described in polar coordinates, then we will see in Section 6.9 that

$$\iint_S f(x, y)\,dA = \iint_T g(r, \theta)\,dA$$

where $g(r, \theta) = f(x(r, \theta), y(r, \theta))$ and dA on the right is $r\,dr\,d\theta$. This very reasonable formula says that if we make the substitutions $x = r\cos\theta$, $y = r\sin\theta$ in f and change xy-limits to $r\theta$-limits, then the integral's value is preserved. In physical terms, we recognize that the volume of an object is independent of the manner in which we measure it. Our next example illustrates the computational procedure.

5.3 EXAMPLE. Compute $\iint_S f(x, y)\,dA$, where $f(x, y) = 1/\sqrt{x^2+y^2}$ and S is the first quadrant region lying between the circles $x^2+y^2=1$ and $x^2+y^2=4$. See Figure 5.4.

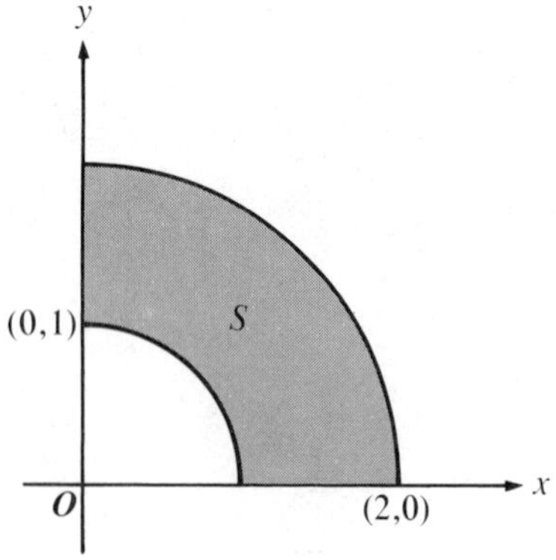

FIGURE 5.4

Solution. This looks like a job for polar coordinates, since $\sqrt{x^2+y^2}$ occurs in the formula for f and the region S lies between two circles and two lines. We have $f(r\cos\theta, r\sin\theta) = 1/r$, so we get

$$\iint_S \frac{1}{\sqrt{x^2+y^2}}\,dA = \iint_T \frac{1}{r}\,r\,dr\,d\theta = \int_0^{\pi/2} d\theta \int_1^2 dr$$

$$= \tfrac{1}{2}\pi(2-1) = \tfrac{1}{2}\pi.$$

If you're not already impressed with polar coordinates, then you might try to evaluate the given integral directly in rectangular coordinates, and compare the complexity of the solution with that of Example 5.3.

As in the rectangular case, we can extend the polar double integral of g to a more general bounded region D lying between two polar curves $r = h_1(\theta)$ and $r = h_2(\theta)$ and the lines $\theta = \alpha$ and $\theta = \beta$. See Figure 5.5. We proceed exactly as in

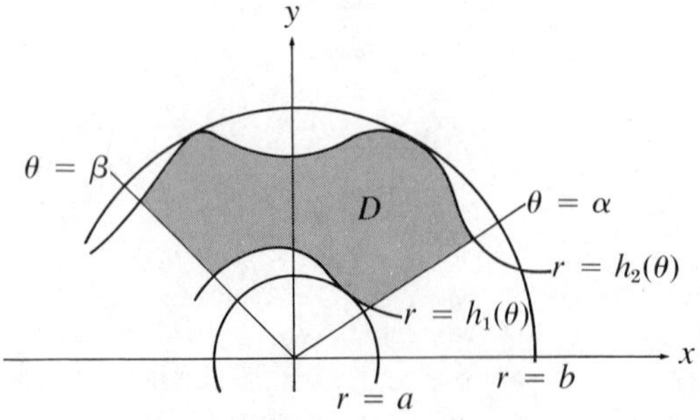

FIGURE 5.5

Definition 3.1. Since D is bounded, we can find a basic region $S \supseteq D$, where S lies between $\theta = \alpha$, $\theta = \beta$, $r = a$, and $r = b$ as in Figure 5.5. We define $\bar{g}$ on S by

(∗) $$\bar{g}(r, \theta) = \begin{cases} g(r, \theta) & \text{if } [r, \theta] \in D \\ 0 & \text{if } [r, \theta] \notin D \end{cases}$$

Then we define $\iint_D g(r, \theta)\, dA$ to be $\iint_S \bar{g}(r, \theta)\, dA$. As in the rectangular case, Fubini's Theorem allows this polar integral to be evaluated by iterated integration. Thus we have the formula

(1) $$\iint_D g(r, \theta)\, dA = \int_\alpha^\beta d\theta \int_{h_1(\theta)}^{h_2(\theta)} g(r, \theta) r\, dr.$$

With (1) available, we can give an alternative derivation for the formula in Definition 4.8 for the area of the region R in Figure 4.10, bounded by $r = f(\theta)$ and the lines $\theta = \alpha$ and $\theta = \beta$:

$$A(R) = \int_\alpha^\beta \tfrac{1}{2} f(\theta)^2\, d\theta.$$

We have from Corollary 3.4 and Definition 5.1,

$$A = \iint_R 1\, dA = \iint_R 1\, r\, dr\, d\theta = \int_\alpha^\beta d\theta \int_0^{f(\theta)} r\, dr$$

$$= \int_\alpha^\beta \left[\tfrac{1}{2} r^2\right]_0^{f(\theta)} d\theta = \int_\alpha^\beta \tfrac{1}{2} f(\theta)^2\, d\theta.$$

We next give a more concrete illustration of (1).

5.4 EXAMPLE. If $g(r, \theta) = \sec \theta$, then evaluate $\iint_D g(r, \theta)\, dA$, where D is the upper half of the disk bounded by the circle $r = 2\cos\theta$, shown in Figure 5.6.

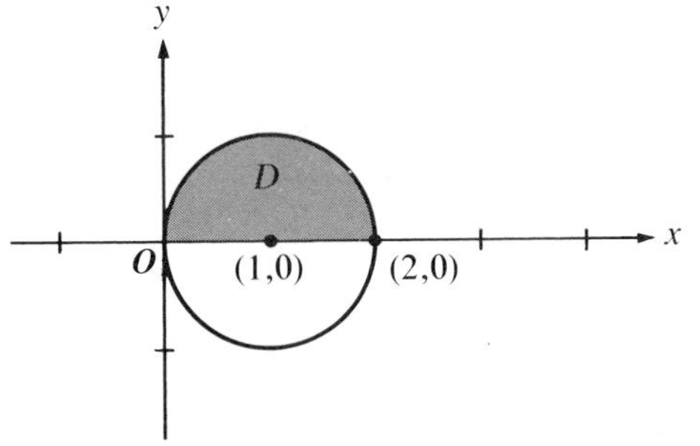

FIGURE 5.6

Solution. The region D is swept out as θ goes from 0 to $\pi/2$. So we have

$$\iint_D g(r, \theta)\, dA = \int_0^{\pi/2} d\theta \int_0^{2\cos\theta} (\sec\theta) r\, dr$$

$$= \int_0^{\pi/2} \sec\theta \left[\tfrac{1}{2} r^2\right]_0^{2\cos\theta} d\theta$$

$$= \frac{4}{2} \int_0^{\pi/2} \sec\theta \cos^2\theta\, d\theta$$

$$= 2\int_0^{\pi/2} \cos\theta\, d\theta = 2\sin\theta\Big]_0^{\pi/2}$$

$$= 2.$$

EXERCISES 5.5

In Exercises 1 to 8, use $A(S)=\iint_S 1\,dA$ to find the area of the given region S. Use polar coordinates in evaluating all integrals.

1 $S=\{(x, y) \mid x^2+y^2 \le 5\}$.

2 $S=\{(x, y) \mid x^2+y^2 \le 9\}$.

3 S is one petal of the rose $r=2\sin 3\theta$.

4 S is one petal of the rose $r=3\sin 3\theta$.

5 S is the region inside the cardioid $r=1+\cos\theta$.

6 S is the region inside the limaçon $r=2+\cos\theta$.

7 S is the region common to the circles $r=2$ and $r=4\sin\theta$.

8 S is the region common to the circles $r=3$ and $r=4\cos\theta$.

In Exercises 9 to 14, use polar integration to find the volume of the given region.

9 Under the cone $z=\sqrt{x^2+y^2}$ and above the circle $x^2+y^2=1$ in the xy-plane.

10 Under the paraboloid $z=4-x^2-y^2$ and above the circle $x^2+y^2=1$ in the xy-plane.

11 Under the paraboloid $z=x^2+y^2$ and above the annular region between $x^2+y^2=1$ and $x^2+y^2=4$ in the xy-plane.

12 The hemisphere of radius a.

13 The region between the paraboloids $z=4-3x^2-3y^2$ and $z=x^2+y^2$.

14 The region between the paraboloids $z=9-5x^2-5y^2$ and $z=4x^2+4y^2$.

15 A cylindrical drill of radius b drills through a sphere of radius a, passing straight through the center. Find the volume of the hole created.

16 A cylindrical drill of radius 1 drills through a sphere of radius 2, passing straight through the center. Find the volume of the resulting solid.

In Exercises 17 to 22, evaluate the given double integral using polar coordinates.

17 $\iint_D \frac{1}{\sqrt{x^2+y^2}}\,dA$, where D is the region bounded by the x-axis, the circle $x^2+y^2=1$, and the lines $y=x$ and $x=2$.

18 $\iint_D \frac{1}{\sqrt{x^2+y^2}}\,dA$, where D is the first quadrant region bounded by the x-axis, the line $y=x$, and the circle $x^2+y^2=4$.

19 $\iint_D \sqrt{x^2+y^2}\,dA$, where D is the upper half of the disk $r\le 2\cos\theta$.

20 $\iint_D \sqrt{x^2+y^2}\,dA$, where D is the lower half of the region $r\le 2\cos\theta$.

21 $\iint_D x^2y\,dA$, where D is the disk $x^2+y^2\le 1$.

22 $\iint_D xy^2\,dA$, where D is the disk $x^2+y^2\le 1$.

23 $\iint_D (xy-x^2-y^2)\,dA$, where D is the annular region $1\le x^2+y^2\le 9$.

24 $\iint_D (x^2+y^2-3xy)\,dA$, where D is the annular region $4\le x^2+y^2\le 16$.

6 TRIPLE INTEGRALS

If we consider a function $f: \mathbf{R}^3 \to \mathbf{R}$, then we expect that a *triple integral* of f over a region $B \subseteq \mathbf{R}^3$ in the shape of a rectangular parallelepiped ought to be definable in a manner analogous to the definition of a double integral in Section 1 for real valued functions on $\mathbf{R}^2$. Given a rectangular box

$$\begin{aligned} B &= [a, b] \times [c, d] \times [e, f] \\ &= \{(x, y, z) \mid a \le x \le b,\ c \le y \le d,\ e \le z \le f\}, \end{aligned}$$

we can chop B up into a grid G of sub-boxes B_{ijk} (see Figure 6.1) by partitioning each of the intervals $[a, b]$, $[c, d]$, and $[e, f]$ by partitions P_1, P_2, and P_3. A typical

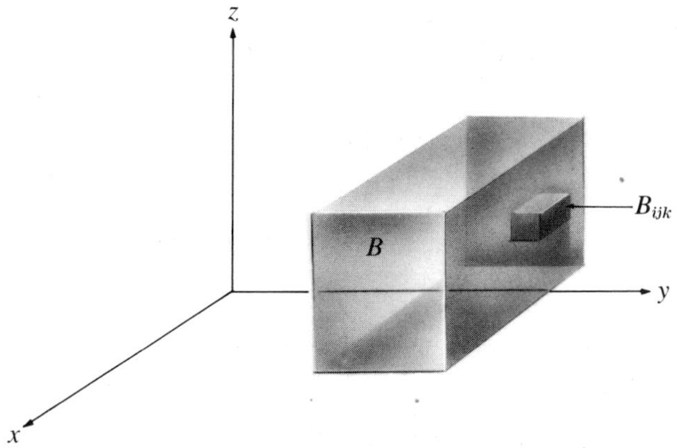

FIGURE 6.1

sub-box B_{ijk} is $[x_{i-1}, x_i] \times [y_{j-1}, y_j] \times [z_{k-1}, z_k]$. The volume of B_{ijk} is

$$\Delta V_{ijk} = \Delta x_i\, \Delta y_j\, \Delta z_k = (x_i - x_{i-1})(y_j - y_{j-1})(z_k - z_{k-1}).$$

A Riemann sum of f corresponding to G is then the triple sum

$$R(f, G, \bar{\mathbf{x}}_{ijk}) = \sum_{i=1}^{m} \sum_{j=1}^{n} \sum_{k=1}^{p} f(\bar{x}_i, \bar{y}_j, \bar{z}_k)\, \Delta V_{ijk}$$

where $\bar{\mathbf{x}}_{ijk} = (\bar{x}_i, \bar{y}_j, \bar{z}_k) \in B_{ijk}$. As we let $|P_1|$, $|P_2|$, and $|P_3| \to 0$, the size of our sub-boxes diminishes toward 0, so this Riemann sum ought to approach the triple integral we wish to define.

6.1 **DEFINITION.** The **triple integral** of $f: \mathbf{R}^3 \to \mathbf{R}$ over a rectangular parallelepiped $B \subseteq \mathbf{R}^3$ is defined as

$$\iiint\limits_B f(x, y, z)\, dV = \lim_{|G| \to 0} \sum_{i=1}^{m} \sum_{j=1}^{n} \sum_{k=1}^{p} f(\bar{x}_i, \bar{y}_j, \bar{z}_k)\, \Delta V_{ijk}.$$

This means that given $\varepsilon > 0$, there is a $\delta > 0$ such that for any grid G with $|G| < \delta$ (where $|G| = \max\limits_{i,j,k} (\Delta x_i, \Delta y_j, \Delta z_k)$),

$$\left| \iiint\limits_B f(\mathbf{x})\, dV - R(f, G, \bar{\mathbf{x}}_{ijk}) \right| < \varepsilon$$

for all Riemann sums $R(f, G, \bar{\mathbf{x}}_{ijk})$ of f over B.

The analogue of Fubini's Theorem 2.4 holds for triple integrals over rectangular boxes B, and permits us to evaluate triple integrals over such regions by simple iteration.

6.2 **EXAMPLE.** Evaluate $\iiint_B f(x, y, z)\, dV$ if B is the box $[1, 2]\times[-1, 1]\times[2, 4]$ and $f(x, y, z) = x^2z - y^2z$.

Solution. Here

$$\iiint_B f(x, y, z)\, dV = \iint_R dA \int_2^4 (x^2z - y^2z)\, dz$$

$$= \iint_R (x^2 - y^2)\, dA \int_2^4 z\, dz,$$

where R is the rectangle $[1, 2]\times[-1, 1]$,

$$= \iint_R (x^2 - y^2)\, dA[\tfrac{1}{2}z^2]_2^4 = \iint_R (x^2 - y^2)(8 - 2)\, dA$$

$$= 6\int_1^2 dx \int_{-1}^1 (x^2 - y^2)\, dy = 6\int_1^2 dx[x^2y - \tfrac{1}{3}y^3]_{-1}^1$$

$$= 6\int_1^2 (2x^2 - \tfrac{2}{3})\, dx = 12\int_1^2 (x^2 - \tfrac{1}{3})\, dx$$

$$= 12\left[\frac{x^3}{3} - \frac{x}{3}\right]_1^2 = 4(8 - 2 - 1 + 1) = 24.$$

We extend Definition 6.1 to more general regions $E \subseteq \boldsymbol{R}^3$ in the now familiar manner used in Definition 3.1 and Equation (*) of Section 5 (p. 268). If E is a bounded region of $\boldsymbol{R}^3$, then we enclose E in a box B and define $\bar{f}$ on B to coincide with f for points $\boldsymbol{x} \in E$, and to be 0 for points $\boldsymbol{x} \notin E$. Then

(2) $$\iiint_E f(\boldsymbol{x})\, dV = \iiint_B \bar{f}(\boldsymbol{x})\, dV.$$

As we expect by now, a version of Fubini's Theorem exists which asserts that the triple integral over the region E can be evaluated as an iterated integral, and we can evaluate the iterated integrals in any order that we want. Again the proof is beyond our scope here, but we do state it precisely for the case of a region E bounded above and below by surfaces $z = g(x, y)$.

6.3 **THEOREM.** Suppose that $E \subseteq \boldsymbol{R}^3$ is bounded below by the surface $z = g_1(x, y)$, and above by the surface $z = g_2(x, y)$. Then

(3) $$\iiint_E f(\boldsymbol{x})\, dV = \iint_D dA \int_{g_1(x,y)}^{g_2(x,y)} f(\boldsymbol{x})\, dz$$

where D is the perpendicular projection of E onto the xy-plane. See Figure 6.2.

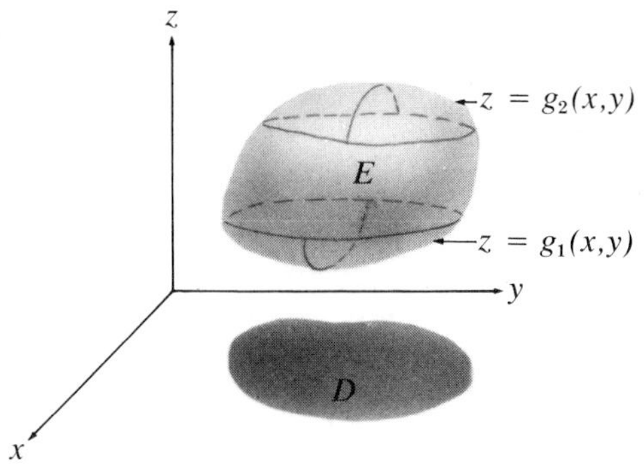

FIGURE 6.2

Notice that the right side of (3) is an ordinary double integral over D, so it can be evaluated by the methods of Section 3. (The variable z disappears when we calculate $\int_{g_1(x,y)}^{g_2(x,y)} f(x, y, z)\, dz$.)

We can integrate first with respect to y (or x) instead of with respect to z, if that is easier to do. If, for example, E lies to the right of $y = h_1(x, z)$ and to the left of $y = h_2(x, z)$, then

(4) $$\iiint_E f(\mathbf{x})\, dV = \iint_D dA \int_{h_1(x,z)}^{h_2(x,z)} f(x, y, z)\, dy$$

where D is the perpendicular projection of E onto the xz-plane. See Figure 6.3.

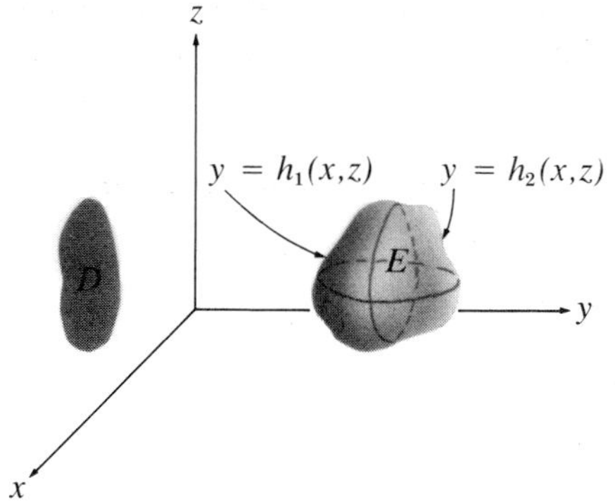

FIGURE 6.3

Similarly, if E lies in front of $x = k_1(y, z)$, and behind $x = k_2(y, z)$, then

(5) $$\iiint_E f(\mathbf{x})\, dV = \iint_D dA \int_{k_1(y,z)}^{k_2(y,z)} f(x, y, z)\, dx$$

where D is the perpendicular projection of E onto the yz-plane. (See Figure 6.4).

In practice, we use whichever of Equations (3), (4), or (5) involves the easiest calculations. If there is no difference (for example, if E is a rectangular box), then (3) is possibly the best path to follow since we obtain an xy-double integral of the exact sort we have been dealing with in the earlier parts of the chapter.

The implementation of (3), (4), and (5) requires you to correctly determine

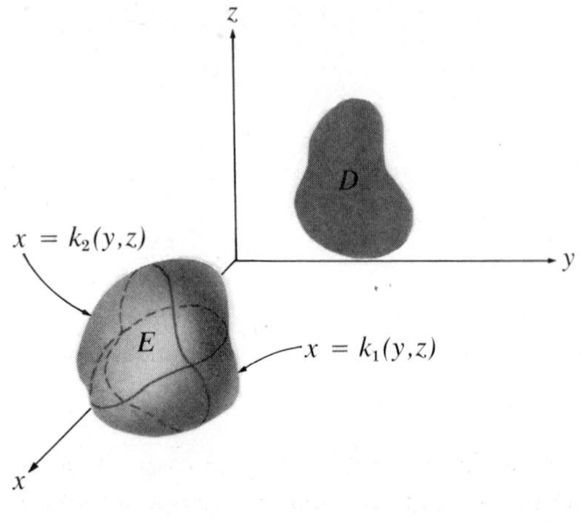

FIGURE 6.4

the limits of integration. You can do this by using the approach of Section 5.3. For instance, in (3), imagine yourself *under* the region E and shooting an arrow up through it. The equation of the first surface, say $z = g_1(x, y)$, that your arrow would puncture provides the *lower limit* in (3). Similarly, the equation of the surface from which the arrow emerges at the top provides the *upper limit* $g_2(x, y)$. In using (4), imagine yourself shooting an arrow from the *left* of the region E through it parallel to the y-axis. The *left bounding surface* gives the lower limit $h_1(x, z)$, and the *right bounding surface* gives the upper limit $h_2(x, z)$. Finally, in using (5), imagine yourself *behind* the region E shooting an arrow through it parallel to the x-axis. The *rear bounding surface* is hit first, so it gives the lower limit $k_1(y, z)$. The *front bounding surface* then gives the upper limit $k_2(y, z)$.

In all three schemes described above, once your first integration is complete you have "integrated out" the variable with respect to which the first integration was performed. This leaves you with a *double integral* over a two-dimensional region, which you evaluate as in Section 5.3. The next example shows the details in applying (5).

6.4 EXAMPLE. Evaluate $\iiint_E f(x, y, z)\, dV$, where $f(x, y, z) = 1/x$ and E is the first octant region bounded by the planes $x = 1$ and $y = z$, below by the cylinder $x = e^z$, and above by the cylinder $y^2 + z^2 = 4$. See Figure 6.5.

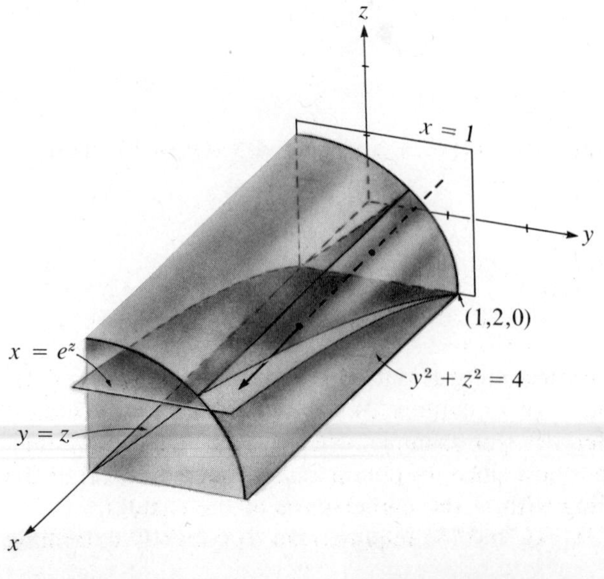

FIGURE 6.5

Solution. The region lies in front of the plane $x=1$ and behind the cylinder $x=e^z$, so that we can apply (5). Imagining an arrow shot parallel to the x-axis from the yz-plane, we get

$$\iiint_E f(x, y, z)\, dV = \iint_D dA \int_1^{e^z} \frac{1}{x}\, dx$$

$$= \iint_D \ln x \Big]_{x=1}^{x=e^z} dA$$

$$= \iint_D z\, dA,$$

where D is the region of the yz-plane shown in Figure 6.6. Evaluating the double integral over D, we obtain

$$\iint_D z\, dA = \int_0^{\sqrt{2}} z\, dz \int_z^{\sqrt{4-z^2}} dy$$

$$= \int_0^{\sqrt{2}} z(\sqrt{4-z^2}-z)\, dz$$

$$= -\tfrac{1}{2}\int_0^{\sqrt{2}} (-2z)\sqrt{4-z^2}\, dz - \int_0^{\sqrt{2}} z^2\, dz$$

$$= -\tfrac{1}{2}\cdot\tfrac{2}{3}(4-z^2)^{3/2} - \frac{z^3}{3}\Big]_0^{\sqrt{2}}$$

$$= -\frac{2\sqrt{2}}{3} + \frac{8}{3} - \frac{2\sqrt{2}}{3}$$

$$= \frac{8-4\sqrt{2}}{3}.$$

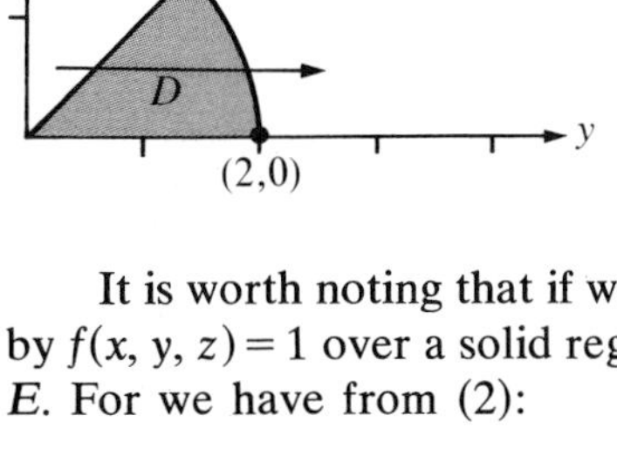

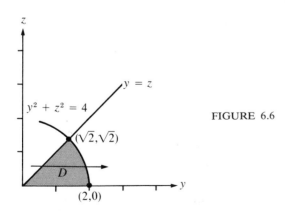

FIGURE 6.6

It is worth noting that if we compute the triple integral of the function f given by $f(x, y, z)=1$ over a solid region $E \subseteq \mathbf{R}^3$, then we simply get the *volume* $V(E)$ of E. For we have from (2):

$$\iiint_E f(\mathbf{x})\, dV = \iiint_B \bar{f}(\mathbf{x})\, dV$$

where $E \subseteq B$, and $\bar{f}(\boldsymbol{x}) = 1$ on E and $\bar{f}(\boldsymbol{x}) = 0$ on the part of B that is not in E. Moreover,

$$\iiint_B \bar{f}(\boldsymbol{x})\, dV = \lim_{|G| \to 0} R(\bar{f}, G, \bar{\boldsymbol{x}}_{ijk}) = \lim_{|G| \to 0} R(1, G, \bar{\boldsymbol{x}}_{ijk}) \quad \text{where} \quad \bar{\boldsymbol{x}}_{ijk} \in E$$

$$= \lim_{|G| \to 0} \sum_{i,j,k} \Delta V_{ijk} = V(E).$$

Hence we have proved the following analogue of Corollary 3.4.

6.5 **THEOREM.** If E is a bounded region in $\boldsymbol{R}^3$, then the volume V of E is given by

$$V = \iiint_E 1\, dV$$

where 1 denotes the function f given by $f(\boldsymbol{x}) = 1$ for all $\boldsymbol{x} \in \boldsymbol{R}^3$.

6.6 **EXAMPLE.** Find the volume of the region $E \subseteq \boldsymbol{R}^3$ bounded below by the xy-plane, above by the plane $z = x$, and by the parabolic cylinder $y^2 = 4 - 2x$.

Solution. The region is sketched in Figure 6.7. The region E lies below the

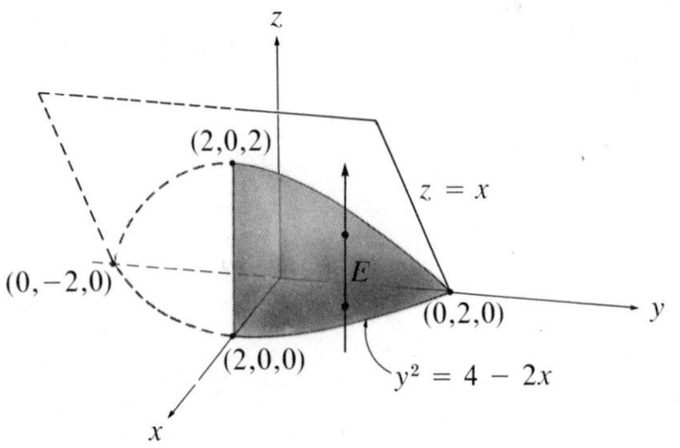

FIGURE 6.7

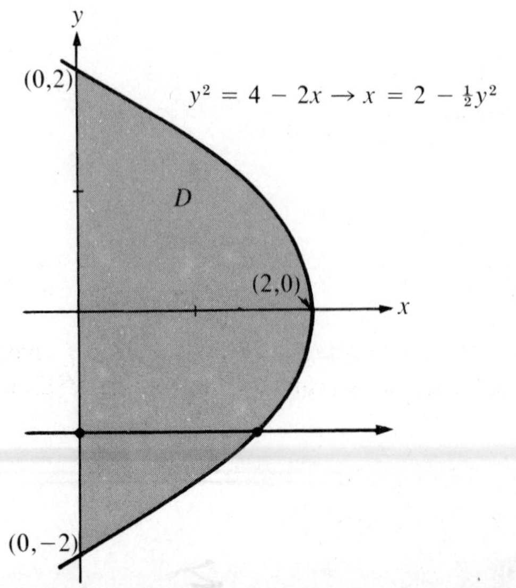

FIGURE 6.8

plane $z = x$, and above the region D of the xy-plane shown in Figure 6.8. So using (3) of Theorem 6.2 in conjunction with Theorem 6.4, we shoot our arrow upward. It pierces the floor of E at $z = 0$ and leaves through the ceiling $z = x$. We get, therefore,

$$V = \iiint_E 1\,dV = \iint_D dA \int_0^x 1\,dz = \iint_D dA[z]_0^x = \iint_D x\,dA$$

$$= \int_{-2}^{2} dy \int_0^{2-y^2/2} x\,dx = \int_{-2}^{2} dy[\tfrac{1}{2}x^2]_0^{2-y^2/2}$$

$$= \tfrac{1}{2}\int_{-2}^{2} (4 - 2y^2 + \tfrac{1}{4}y^4)\,dy = \tfrac{1}{2}\left[4y - \frac{2y^3}{3} + \tfrac{1}{20}y^5\right]_{-2}^{2}$$

$$= \frac{1}{2}\left(8 - \frac{2\cdot 8}{3} + \frac{32}{20} + 8 - \frac{2\cdot 8}{3} + \frac{32}{20}\right)$$

$$= \frac{1}{2}\left(16 - \frac{32}{3} + \frac{16}{5}\right) = 8 - \frac{16}{3} + \frac{8}{5} = \frac{120 - 80 + 24}{15}$$

$$= \frac{64}{15}.$$

EXERCISES 5.6

In Exercises 1 to 14, evaluate the given triple integral using (3), (4), or (5).

1 $\iiint_B x\,dV$, where $B = [-1, 1] \times [1, 2] \times [0, 3]$.

2 $\iiint_B y\,dV$, where $B = [0, 1] \times [-1, 1] \times [0, 2]$.

3 $\iiint_E (1 - z)\,dV$, where E lies above the triangle with vertices $(0, 0, 0)$, $(1, 0, 0)$, and $(0, 1, 0)$ and below the cylinder $z = 1 - x^2$.

4 $\iiint_E z\,dV$, where E is the pyramid with apex $(0, 0, 1)$ and with the square base having vertices $(1, 0, 0)$, $(0, 1, 0)$, $(1, 1, 0)$, and $(0, 0, 0)$.

5 $\iiint_E 24yz\,dV$, where E is the region in the first octant bounded by the cylinders $y = x^2$ and $z = 1 - y^2$.

6 $\iiint_E (x - yz)\,dV$, where E is the solid tetrahedron formed by the three coordinate planes and the plane $x + y + z = 1$.

7 $\iiint_E (xy - z)\,dV$, where E is the region of Exercise 6.

8 $\iiint_E (4x + 8y)\,dV$ where E is the solid tetrahedron formed by the three coordinate planes and the plane $6x + 3y + 2z = 6$.

9 $\iiint_E (x + y)\,dV$, where E is the region bounded by the planes $z = 0$, $x = z$, and $y = x$, and by the cylinder $y = x^2$.

10 $\iiint_E (2x - y)\,dV$, where E is the region of Exercise 9.

11 $\iiint_E x^2yz\,dV$, where E is the region bounded by the planes $y = 0$, $z = 0$, $x = 1$, and $x = 2$, and by the cylinders $y = x^2$ and $z = 1/x$.

12 $\iiint_E x^3yz\,dV$, where E is the region of Exercise 11.

13 $\iiint_E z^2\,dV$, where E is the region bounded below by the xy-plane and above by the cylinders $x^2+z=1$ and $y^2+z=1$.

14 $\iiint_E (z^2-1)\,dV$, where E is the region of Exercise 13.

In Exercises 15 to 20, find the volume of the given region *E*.

15 E is a tetrahedron formed by the three coordinate planes and the plane

$$\frac{x}{a}+\frac{y}{b}+\frac{z}{c}=1.$$

16 E is the tetrahedron formed by the three coordinate planes and $x+y+z=1$.

17 E is the first octant region bounded by $y=4-x^2$ and the planes $z=x$, $y=0$, and $z=0$.

18 E is the first octant region bounded by $z=4-x^2$ and the planes $y=x$, $y=0$, and $z=0$.

19 E is the region bounded by the paraboloids $z=8-x^2-y^2$ and $z=x^2+3y^2$.

20 E is the first octant region bounded by the planes $x=0$, $y=0$, and $z=2$ and by the surface $z=x^2+y^2$.

21 Suppose that $\lim_{|G|\to 0} R(f, G, \mathbf{z})$ and $\lim_{|G|\to 0} R(g, G, \mathbf{z})$ both exist. Then show that

$$\lim_{|G|\to 0} R(f+g, G, \mathbf{z}) = \lim_{|G|\to 0} R(f, G, \mathbf{z}) + \lim_{|G|\to 0} R(f, G, z), \text{ and}$$

$$\lim_{|G|\to 0} R(af, G, \mathbf{z}) = a \lim_{|G|\to 0} R(f, G, \mathbf{z}).$$

22 Show that the triple integral over a region E is a linear function; i.e., use Exercise 21 to show that

$$\iiint_E [f(\mathbf{x})+g(\mathbf{x})]\,dV = \iiint_E f(\mathbf{x})\,dV + \iiint_E g(\mathbf{x})\,dV,$$

$$\iiint_E (af)(\mathbf{x})\,dV = a\iiint_E f(\mathbf{x})\,dV.$$

In Exercises 23 to 28, assume that all functions are continuous.

23 If $B=B_1\cup B_2$ where B_1 and B_2 are nonoverlapping boxes, or overlap only to the extent of having a common face, then show that

$$\iiint_B f(\mathbf{x})\,dV = \iiint_{B_1} f(\mathbf{x})\,dV + \iiint_{B_2} f(\mathbf{x})\,dV.$$

24 If $f(x, y, z)\geq 0$ on E, then show that

$$\iiint_E f(x, y, z)\,dV \geq 0.$$

25 If $f(x, y, z)\leq g(x, y, z)$ on E, then show that

$$\iiint_E f(x, y, z)\,dV \leq \iiint_E g(x, y, z)\,dV.$$

26 If $E \subseteq \boldsymbol{R}^3$ is closed and bounded, then show that

$$mV(E) \leq \iiint_E f(x, y, z)\, dV \leq MV(E),$$

where M is the maximum of f on E and m is the minimum of f on E.

27 Use Exercise 26 to show that if f is continuous on an open set containing $\boldsymbol{x}_0$, then

$$\lim_{\delta \to 0} \frac{1}{V(B(\boldsymbol{x}_0, \delta))} \iiint_{B(\boldsymbol{x}_0, \delta)} f(x, y, z)\, dV = f(x_0, y_0, z_0),$$

where $B(\boldsymbol{x}_0, \delta)$ is the ball about $\boldsymbol{x}_0 = (x_0, y_0, z_0)$ of radius δ.

28 Use Exercise 26 to prove the *Mean Value Theorem for Triple Integrals*: If f is continuous on a box B, then

$$\iiint_B f(x, y, z)\, dV = f(x_0, y_0, z_0)V(B)$$

for some (x_0, y_0, z_0) in B. (See the hint for Exercise 20, Exercises 5.1.)

The next two exercises will be of use in Section 9.

29 A function f is said to be *odd* relative to x if $f(x, y, z) = -f(-x, y, z)$. If f is odd relative to x and E is symmetric in the yz-plane, then show that

$$\iiint_E f(x, y, z)\, dV = 0.$$

(*Hint:* Use the analogue of Exercise 23.)

30 If f is odd relative to y (see Exercise 29), and E is symmetric in the xz-plane, then show that

$$\iiint_E f(x, y, z)\, dV = 0.$$

(*Hint:* Use the analogue of Exercise 23.)

7 CYLINDRICAL COORDINATES

This section and the next one are in many ways the three-dimensional analogues of Sections 4 and 5. As you may already be thinking, triple integrals in Cartesian coordinates can become quite difficult to evaluate. Complicated triple integrals can often be reduced to more manageable forms by transforming from Cartesian coordinates to one of two alternative coordinate systems for $\boldsymbol{R}^3$. In this section we introduce the first of these, the cylindrical coordinate system. It is especially well suited to surfaces that are symmetric relative to the z-axis (so that cross sections made by planes $z = c$ are circles).

The cylindrical coordinate system is quite simply a hybrid produced by crossing polar coordinates (in the xy-plane) with rectangular coordinates (along the z-axis).

7.1 **DEFINITION.** The **cylindrical coordinates** of a point $P(x, y, z) \in \boldsymbol{R}^3$ are r, θ, and z, where $Q[r, \theta]$ is the polar coordinate representation of the perpendicular projection $Q(x, y, 0)$ of P onto the xy-plane. We write $P = P[r, \theta, z]$. See Figure 7.1.

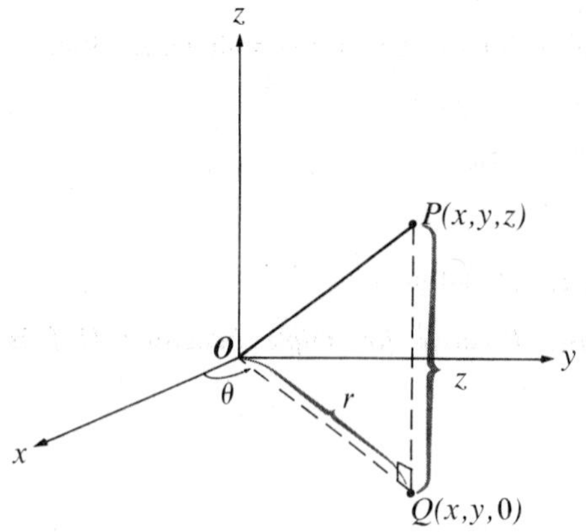

FIGURE 7.1

We thus have, directly from Proposition 4.2, the transformation equations

(1)
$$x = r\cos\theta, \qquad y = r\sin\theta, \qquad z = z$$

where

$$r = \sqrt{x^2 + y^2} \geq 0, \qquad \frac{x}{r} = \cos\theta, \qquad \frac{y}{r} = \sin\theta,$$

and

$$\frac{y}{x} = \tan\theta \quad \text{if} \quad x \neq 0.$$

As with polar coordinates for $\boldsymbol{R}^2$, there is *not* a one-to-one correspondence between points in $\boldsymbol{R}^3$ and sets of cylindrical coordinates $[r, \theta, z]$ even with the restriction $r \geq 0$. In particular, the origin can be coordinatized as $[0, \theta, 0]$ for any choice of θ.

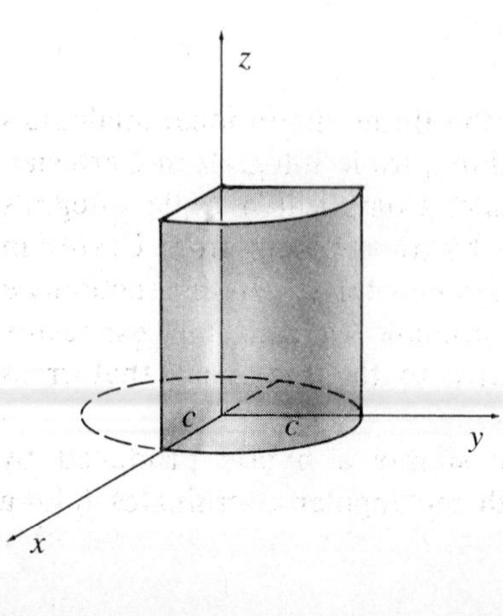

FIGURE 7.2

Let's investigate the surfaces with simple equations $r=c$, $\theta=k$, and $z=a$. The third is still a plane a units from the xy-plane—above it if $a>0$ or below it if $a<0$. The first is a right circular cylinder ($x^2+y^2=c^2$ in rectangular coordinates) perpendicular to the circle in the xy-plane of radius c and center at the origin. See Figure 7.2. The origin of the term "cylindrical coordinates" is now clear. These coordinates give the simple equation $r=c$ for right circular cylinders centered on the z-axis (which are thus analogous to the planes $x=c$ and $y=c$ in rectangular coordinates). The surface $\theta=k$ is a plane through the z-axis perpendicular to the xy-plane and making dihedral angle k with the xz-plane. See Figure 7.3. In

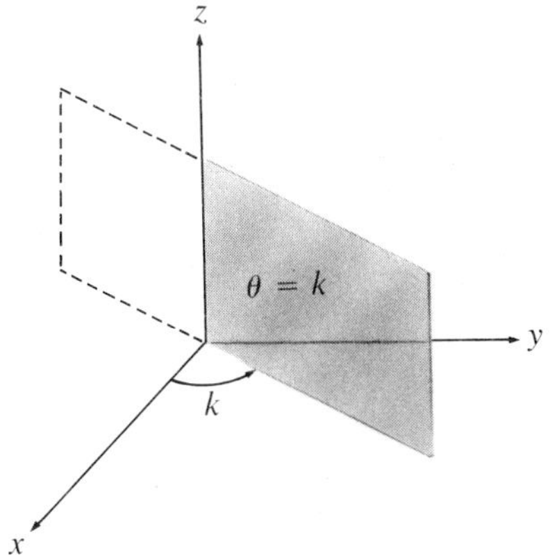

FIGURE 7.3

particular, if we restrict r to be nonnegative, then $\theta=0$ is the front half of the xz-plane and $\theta=\pi$ is its back half; $\theta=\frac{1}{2}\pi$ is the right half of the yz-plane, and $\theta=\frac{3}{2}\pi$ is its left half.

We next see how simple the cylindrical coordinate equations of some surfaces can be.

7.2 EXAMPLE. Find cylindrical coordinate equations of (a) the paraboloid $z=x^2+y^2$ and (b) the cone $z^2=x^2+y^2$.

Solution. By direct substitution, (a) $z=r^2$. Similarly, (b) $z^2=r^2$; the top nappe has the equation $z=r$ and the bottom nappe has the equation $z=-r$ if we restrict r to be nonnegative, so that $r=\sqrt{x^2+y^2}$. See Figure 7.4.

Our principal application for cylindrical coordinates will be in evaluating triple integrals. As in Section 5, a natural place to begin is a triple integral of a function $g(r, \theta, z)$ defined over a cylindrical coordinate analogue of a rectangular box. This is a region T bounded by planes $\theta=\alpha$ and $\theta=\beta$, $z=e$ and $z=f$, and right circular cylinders $r=a$ and $r=b$. We show such a region in Figure 7.5. It looks like a piece of pie out of which a bite $r=a$ has been taken. If we make a cylindrical grid G of the region, we get small subregions like the one shown in Figure 7.6. These are cylinders of height Δz_k and base area $\Delta A_{ij} \approx \bar{r}_i \, \Delta r_i \, \Delta \theta_j$ as on p. 266. So the subregion has a volume given approximately by

$$\Delta V_{ijk} \approx \bar{r}_i \, \Delta r_i \, \Delta \theta_j \, \Delta z_k.$$

We are thus led to the following definition.

7.3 DEFINITION. Let $T \subseteq \boldsymbol{R}^3$ be the region bounded by $\theta=\alpha$, $\theta=\beta$, $z=e$, $z=f$, and $r=b$, $r=a$. Let $g: \boldsymbol{R}^3 \to \boldsymbol{R}$ be continuous on T. Then the **cylindrical triple**

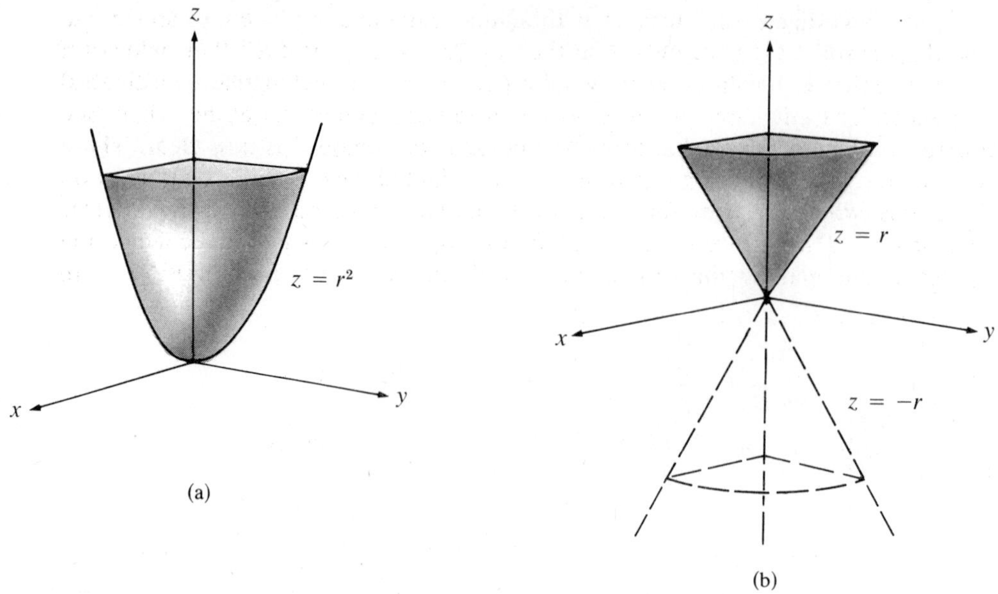

FIGURE 7.4

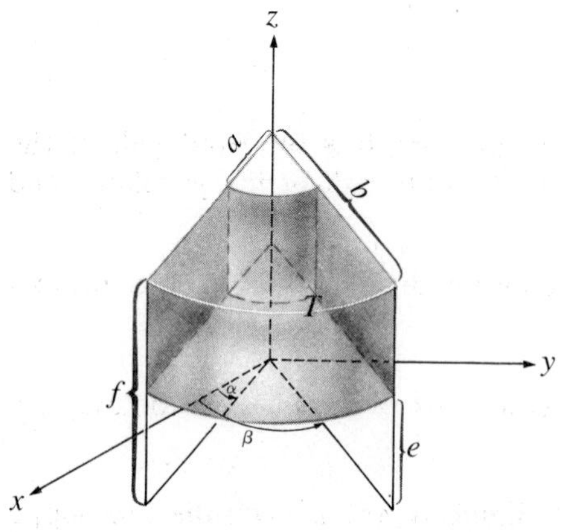

FIGURE 7.5

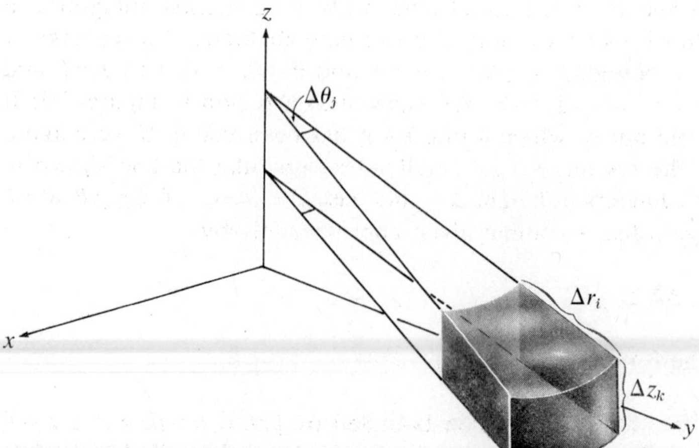

FIGURE 7.6

integral of g over T is

$$\iiint_T g(r, \theta, z)\, dV = \iiint_T g(r, \theta, z) r\, dr\, d\theta\, dz$$

$$= \lim_{|G| \to 0} \sum_{i=1}^{m} \sum_{j=1}^{n} \sum_{k=1}^{p} g(\bar{r}_i, \bar{\theta}_j, \bar{z}_k)\bar{r}_i\, \Delta r_i\, \Delta\theta_j\, \Delta z_k$$

where the limit is taken in the same sense as in Definition 6.1.

We must again defer a discussion of the consistency of Definition 7.3 and Equation (2) on p. 272 until the end of the next chapter. That consistency once more requires

$$dV = r\, dr\, d\theta\, dz, \quad \textbf{NOT} \quad dr\, d\theta\, dz.$$

Take care, as in Section 5, to remember r.

As in the rectangular case, a cylindrical integral is evaluated by iteration. We state the appropriate version of Fubini's Theorem, but leave its justification for Section 6.9.

7.4 **THEOREM.** If g and T are as in Definition 7.3, then

$$\iiint_T g(r, \theta, z)\, dV = \int_\alpha^\beta d\theta \int_a^b r\, dr \int_e^f g(r, \theta, z)\, dz$$

$$= \int_a^b r\, dr \int_\alpha^\beta d\theta \int_e^f g(r, \theta, z)\, dz$$

$$= \int_e^f dz \int_\alpha^\beta d\theta \int_a^b r\, dr, \text{ etc.}$$

7.5 **EXAMPLE.** If $g(r, \theta, z) = zr^2 \cos\theta$, then evaluate $\iiint_T g(r, \theta, z)\, dV$ if T is bounded by $\theta = 0$, $\theta = \frac{1}{2}\pi$, $z = 0$, $z = 1$, $r = 1$, and $r = 3$.

Solution. From (2) we have

$$\iiint_T zr^2 \cos\theta\, dV = \int_0^{\pi/2} \cos\theta\, d\theta \int_1^3 r^3\, dr \int_0^1 z\, dz$$

$$= \int_0^{\pi/2} \cos\theta\, d\theta \int_1^3 r^3\, dr[\tfrac{1}{2}]$$

$$= \tfrac{1}{2} \int_0^{\pi/2} \cos\theta [\tfrac{1}{4}r^4]_1^3\, d\theta$$

$$= \tfrac{1}{2} \cdot \tfrac{80}{4} \sin\theta \Big]_0^{\pi/2}$$

$$= 10$$

If U is a more general bounded region described by cylindrical coordinates, then we extend Definition 7.3 as follows. Let T be a region containing U and bounded by $\theta = \alpha$, $\theta = \beta$, $z = e$, $z = f$, $r = a$, and $r = b$. Define

$$\bar{g}(r, \theta, z) = \begin{cases} g(r, \theta, z) & \text{if } [r, \theta, z] \in U \\ 0 & \text{if } [r, \theta, z] \notin U \end{cases}$$

Then we define

$$\iiint\limits_U g(r, \theta, z)\, dV = \iiint\limits_T \bar{g}(r, \theta, z)\, dV.$$

Fubini's Theorem extends to such regions U. Rather than state all possible forms, we content ourselves with the following representative version.

7.6 **THEOREM.** If U is bounded by $\theta = \alpha$, $\theta = \beta$, $z = e$, $z = f$, and cylinders $r = h_1(\theta)$ and $r = h_2(\theta)$, then

$$\iiint\limits_U g(r, \theta, z)\, dV = \int_\alpha^\beta d\theta \int_{h_1(\theta)}^{h_2(\theta)} r\, dr \int_e^f g(r, \theta, z)\, dz.$$

As in the case of double integrals, we can change an intractable rectangular triple integral to cylindrical form. If U is a subset of $\boldsymbol{R}^3$ described by rectangular coordinates, and W is the same region described in cylindrical coordinates, then

$$\iiint\limits_U f(x, y, z)\, dV = \iiint\limits_W g(r, \theta, z)\, dV$$

where $g(r, \theta, z) = f(x(r, \theta), y(r, \theta), z)$ and on the right $dV = r\, dr\, d\theta\, dz$. Our next example shows how to employ this result, whose justification again must await the tools of Section 6.9.

7.7 **EXAMPLE.** Evaluate $\iiint\limits_U \sqrt{x^2+y^2}\, dV$, where U is the region lying above the xy-plane and below the cone $z = 4 - \sqrt{x^2+y^2}$. (See Figure 7.7, where we show the first octant portion of U.)

Solution. Transforming to cylindrical coordinates, we obtain

$$\iiint\limits_U \sqrt{x^2+y^2}\, dV = \iiint\limits_U r\, r\, dr\, d\theta\, dz = \int_0^{2\pi} d\theta \int_0^4 r^2\, dr \int_0^{4-r} dz$$

$$= \int_0^{2\pi} d\theta \int_0^4 r^2(4-r)\, dr$$

$$= \int_0^{2\pi} d\theta \int_0^4 (4r^2 - r^3)\, dr$$

$$= \int_0^{2\pi} d\theta \left[\frac{4r^3}{3} - \frac{1}{4}r^4\right]_0^4$$

$$= 2\pi\left[\frac{256}{3} - 64\right] = 2\pi\left[\frac{64}{3}\right] = \frac{128\pi}{3}.$$

Notice that evaluation of the given integral in rectangular coordinates would be much more complicated than the foregoing.

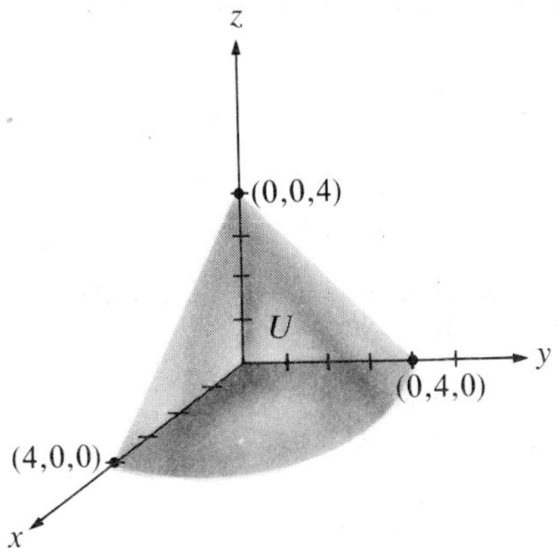

FIGURE 7.7

7.8 EXAMPLE. Find the volume of the region U in Example 7.7.

Solution. We can use Theorem 6.5.

$$V=\iiint_U 1\,dV=\iiint_U r\,dr\,d\theta\,dz=\int_0^{2\pi}d\theta\int_0^4 r\,dr\int_0^{4-r}dz$$

$$=\int_0^{2\pi}d\theta\int_0^4 r(4-r)\,dr=\int_0^{2\pi}d\theta\int_0^4(4r-r^2)\,dr$$

$$=\int_0^{2\pi}d\theta\left[2r^2-\frac{r^3}{3}\right]_0^4=2\pi\left[32-\frac{64}{3}\right]=\frac{64\pi}{3}.$$

If you happen to recall the formula for the volume of a right circular cone of base radius a and height h, $V=\frac{1}{3}\pi a^2h$, then you can check our result in Example 7.6 by setting $a=4$ and $h=4$. (If you don't happen to recall this formula, then be sure to do Exercise 9 below, where you are asked to derive it!)

EXERCISES 5.7

In Exercises 1 and 2, find cylindrical coordinates for each of the points with given rectangular coordinates.

1 (a) $(1, 2, 3)$ (b) $(0, 2, -2)$ (c) $(2, 3, -1)$ (d) $(-2, -3, 1)$ (e) $(2, -3, -1)$

2 (a) $(-1, 1, 1)$ (b) $(2, -1, 3)$ (c) $(2, 0, -3)$ (d) $(-2, 3, -1)$ (e) $(1, 2, -3)$

In Exercises 3 and 4, find rectangular coordinates for each of the points with given cylindrical coordinates.

3 (a) $\left[2, \frac{2\pi}{3}, 1\right]$

(b) $\left[2, \frac{\pi}{6}, 2\right]$

(c) $[1, \pi, 2]$

(d) $\left[2, \frac{7\pi}{6}, -1\right]$

(e) $\left[2, \frac{5\pi}{3}, -1\right]$

4 (a) $\left[-2, \frac{5\pi}{6}, 1\right]$

(b) $\left[-2, \frac{\pi}{3}, 1\right]$

(c) $[-1, \pi, 2]$

(d) $\left[-2, \frac{4\pi}{3}, -1\right]$

(e) $\left[-3, \frac{11\pi}{6}, -2\right]$

In Exercises 5 and 6, a rectangular coordinate equation of a surface is given. Obtain a cylindrical coordinate equation of the same surface.

5 (a) $x^2+y^2=16$
(b) $x^2-y^2=9$
(c) $x^2+y^2+9z^2=9$
(d) $2x^2-4y^2+4z^2=4$

6 (a) $4x^2+y^2=16$
(b) $x^2+y^2=4z$
(c) $3x^2-4y^2-12z^2=12$
(d) $3x^2+3y^2-5z^2=0$

In Exercises 7 and 8, a cylindrical coordinate equation of a surface is given. Sketch the surface and obtain a rectangular coordinate equation for it.

7 (a) $r=3$
(b) $r=4\cos\theta$
(c) $r=2\csc\theta$
(d) $r=\dfrac{6}{2-\cos\theta}$
(e) $z=2r$
(f) $r^2+z^2=9$

8 (a) $r=4\sin\theta$
(b) $r=2\sec\theta$
(c) $r=\dfrac{2}{1-\cos\theta}$
(d) $r=\dfrac{3}{2+4\cos\theta}$
(e) $z^2=4r^2$
(f) $r^2+z^2=4$

9 Use cylindrical coordinate integration to find a formula for the volume enclosed by a right circular cone of base radius a and height h.

10 Use cylindrical coordinate integration to find a formula for the volume enclosed by a sphere of radius a.

11 Find the volume of the region bounded by $x^2+y^2+z=4$ and the xy-plane.

12 Find the volume of the region bounded by $x^2+y^2+z=9$ and the xy-plane.

13 Find the volume of the region inside the sphere $x^2+y^2+z^2=9$ and outside the cone $z=3-\sqrt{x^2+y^2}$.

14 Find the volume lying under the graph of $z=\sqrt{x^2+y^2}$ and above the unit disk $\{(x, y) \mid x^2+y^2\le 1\}$.

15 Find the volume of the solid generated by revolving the region in the xz-plane bounded by $z=2x^2$, the x-axis, and the line $x=1$ about the z-axis.

16 Find the volume of the solid generated by revolving the region in the xz-plane bounded by $z=x^2$, the x-axis, and the line $x=2$ about the z-axis.

In Exercises 17 to 22, evaluate the triple integral.

17 $\iiint_U \sqrt{x^2+y^2}\,z\,dV$, where U is bounded on the left by $y=0$, below by $z=0$, and above by $z=1-\frac{1}{2}\sqrt{x^2+y^2}$.

18 $\iiint_U \sqrt{x^2+y^2}\,z\,dV$, where U is bounded on the left by $y=0$, below by $z=0$, and above by $z=1-\sqrt{x^2+y^2}$.

19 $\iiint_U xyz\,dV$, where U is the region in the first octant bounded by $x^2+y^2=a^2$, the xy-plane, and $z=4$.

20 $\iiint_U yz\,dV$, where U is bounded by the planes $y=0$, $z=0$, and $z=y$, and the cylinder $x^2+y^2=1$.

21 $\iiint_U z^4\,dV$, where U is the region bounded by the cones $z=\sqrt{x^2+y^2}$ and $z=-\sqrt{x^2+y^2}$, the cylinder $x^2+y^2=1$, in back by $x=0$, and to the left by $y=0$.

22 $\iiint_U \sqrt{x^2+y^2}\,dV$, where U is the first octant region bounded by $x^2+y^2=2x$ and $z^2=x^2+y^2$.

8 SPHERICAL COORDINATES

As we mentioned at the start of Section 7, the cylindrical coordinate system is a hybrid produced by crossing polar coordinates for $\boldsymbol{R}^2$ with Cartesian coordinates for $\boldsymbol{R}^3$. The second alternative coordinate system for $\boldsymbol{R}^3$ is, by contrast, the result of applying the idea behind polar coordinates to three dimensions. We will have more to say in this direction, but first pause to define what is meant by the *spherical coordinates* of a point $P \in \boldsymbol{R}^3$.

8.1 **DEFINITION.** The **spherical coordinates** of a point $P(x, y, z) \in \boldsymbol{R}^3$ are ρ, ϕ, and θ where $\rho = |\overrightarrow{OP}| = \sqrt{x^2+y^2+z^2}$, ϕ is the angle between $\boldsymbol{k}$ and $\overrightarrow{OP}$, and θ is the polar (or cylindrical) angle between $\boldsymbol{i}$ and the perpendicular projection $\overrightarrow{OQ}$ of $\overrightarrow{OP}$ on the xy-plane measured counterclockwise. We write $P = P\{\rho, \phi, \theta\}$. See Figure 8.1.

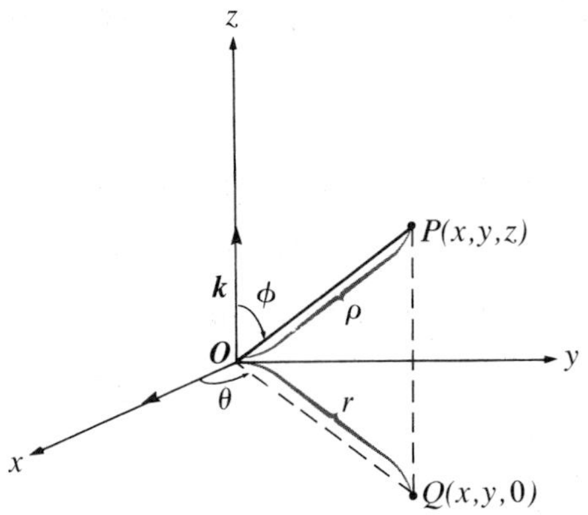

FIGURE 8.1

To obtain equations connecting the rectangular and spherical coordinates of P, it is helpful to observe that $\overrightarrow{OQ}$ is just the cylindrical coordinate r of P. (See Figure 8.1 and also Figure 7.1.) This observation leads us to the following proposition.

8.2 **PROPOSITION.** If $P \in \boldsymbol{R}^3$ has rectangular coordinates (x, y, z) and spherical coordinates $\{\rho, \phi, \theta\}$, then

$$\rho = \sqrt{x^2+y^2+z^2}$$
$$x = \rho \sin\phi \cos\theta$$
$$y = \rho \sin\phi \sin\theta$$
$$z = \rho \cos\phi$$

$$\left.\begin{aligned} \cos\theta &= \frac{x}{\sqrt{x^2+y^2}} \\ \sin\theta &= \frac{y}{\sqrt{x^2+y^2}} \\ \cos\phi &= \frac{z}{\rho} \\ \sin\phi &= \frac{\sqrt{x^2+y^2}}{\rho} \end{aligned}\right\} \quad \begin{aligned} &(\text{if } \sqrt{x^2+y^2} \neq 0 \text{ and hence} \\ &\rho = \sqrt{x^2+y^2+z^2} \neq 0) \end{aligned}$$

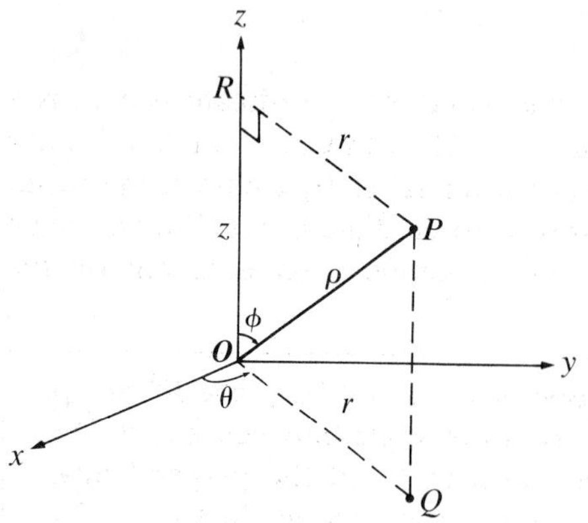

FIGURE 8.2

Proof. The equation $\rho=\sqrt{x^2+y^2+z^2}$ is part of Definition 8.1. We can obtain the next three equations from Figure 8.2. By equation (1) on p. 280, we have

(1) $$x=r\cos\theta \quad \text{and} \quad y=r\sin\theta,$$

where $r=\sqrt{x^2+y^2}$. Observe that triangle PRO has a right angle at R. Then $\sin\phi=\dfrac{r}{\rho}$ and $\cos\phi=\dfrac{z}{\rho}$. Hence $r=\rho\sin\phi$ and $z=\rho\cos\phi$. The second equation is one of those we wanted to prove. If we substitute $r=\rho\sin\phi$ into (1), we have two others:

$$x=\rho\sin\phi\cos\theta \quad \text{and} \quad y=\rho\sin\phi\sin\theta.$$

Now, as usual in polar or cylindrical coordinates,

$$\cos\theta=\frac{x}{r}=\frac{x}{\sqrt{x^2+y^2}} \quad \text{and} \quad \sin\theta=\frac{y}{r}=\frac{x}{\sqrt{x^2+y^2}}.$$

Finally, we already saw that

$$\cos\phi=\frac{z}{\rho} \quad \text{and} \quad \sin\phi=\frac{r}{\rho}=\frac{\sqrt{x^2+y^2}}{\rho}.$$

QED

We don't have a one-to-one correspondence between points in $\boldsymbol{R}^3$ and triples $\{\rho, \phi, \theta\}$ of spherical coordinates, but we can come pretty close if we limit θ to the interval $[0, 2\pi)$ and ϕ to the interval $[0, \pi]$. Most points then have a unique triple of spherical coordinates, but again the origin can be represented as $\{0, 0, \theta\}$ or $\{0, \pi, \theta\}$ for any θ.

It may be of interest to note that spherical coordinates are very nearly those used in navigation. In navigation one considers the surface of the earth to be approximately a sphere of radius $\rho_0 \approx 4000$ miles. If we set up the origin at the center of the earth, then we can locate any point on the surface by spherical coordinates $\{\rho_0, \phi, \theta\}$. In fact, since ρ_0 is nearly the same for all points, it suffices to give just $\{\phi, \theta\}$. Note that the curves $\theta=c$ for $\rho=\rho_0$ are *meridians* (great semicircles passing through the poles) and the curves $\phi=k$ are *parallels* of latitude. See Figure 8.3.

In this scheme, the *prime meridian* $\theta=0$ passes through Greenwich, England. Instead of measuring θ from 0 to 2π eastward around the earth, navigators measure θ from 0 to π east from Greenwich and from 0 to π west from

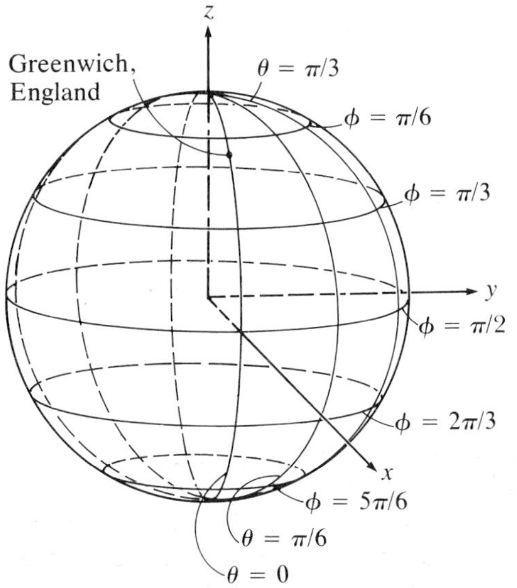

FIGURE 8.3

Greenwich. Moreover, latitude is not measured simply by ϕ but rather by $\frac{1}{2}\pi - \phi$ for localities north of the equator and by $\phi - \frac{1}{2}\pi$ for localities south of the equator. For this reason, ϕ is called the *co-latitude*. Navigators usually measure in degrees rather than radians. Thus, Moscow is located roughly by latitude 56°N and longitude 37°E, meaning $\phi \approx 34°$, $\theta \approx 37°$. Sydney, Australia is located roughly by latitude 33°S and longitude 152°E, meaning $\phi \approx 123°$, $\theta \approx 152°$. Monrovia, Liberia is located roughly by longitude 10°W and latitude 8°N, meaning $\theta \approx 350°$, $\phi \approx 82°$. Weather reports on the positions of winter storms, hurricanes, and typhoons are usually given in this way.

Let's describe, as we did for cylindrical coordinates, the simple equations of the form

"spherical coordinate = constant."

First, $\rho = a$ is clearly a sphere of radius a centered at the origin. See Figure 8.4.

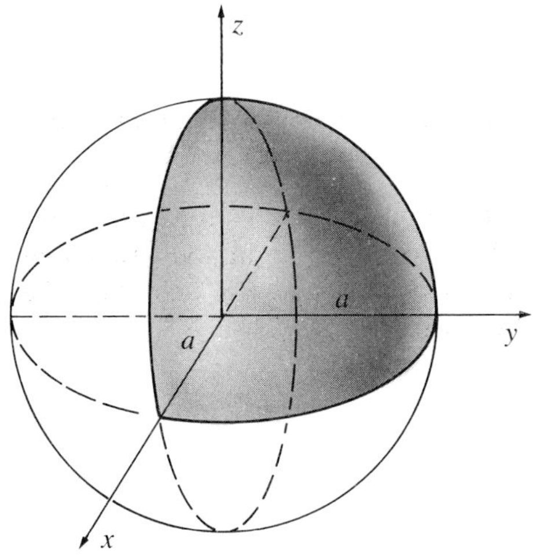

FIGURE 8.4

Its rectangular coordinate equation is $x^2 + y^2 + z^2 = a^2$. Thus, we have the exact three-dimensional analogue of a polar coordinate equation $r = a$, explaining in part why we said that the spherical coordinate system is the three-dimensional version of the polar coordinate system for $\boldsymbol{R}^2$. Since the angle θ in spherical

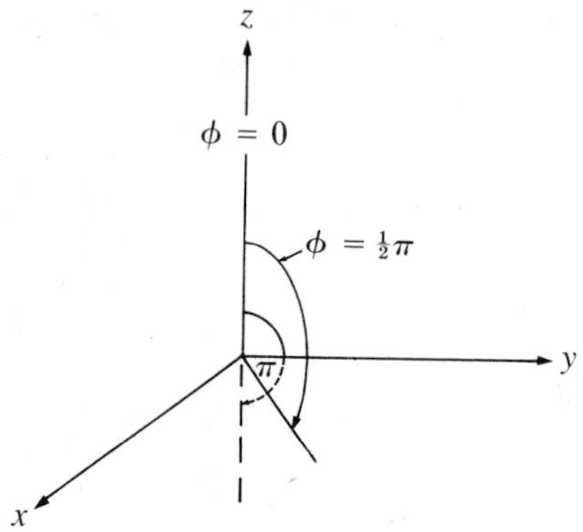

FIGURE 8.5

coordinates is the same as the angle θ in cylindrical coordinates, the graph of $\theta = c$ is still a plane through the z-axis perpendicular to the xy-plane. See Figure 7.3. Finally, we come to $\phi = k$. In this case there is some degeneracy. If $k = 0$, we get the positive z-axis. If $k = \pi$, then we get the negative z-axis. If $k = \frac{1}{2}\pi$, we get the xy-plane. (See Figure 8.5.) In all other cases we get one nappe of a cone, as shown in Figures 8.6(a) and 8.6(b). Hence, just as cylindrical coordinates were very well suited for both right circular cones and right circular cylinders, so spherical coordinates are ideal for both right circular cones and spheres.

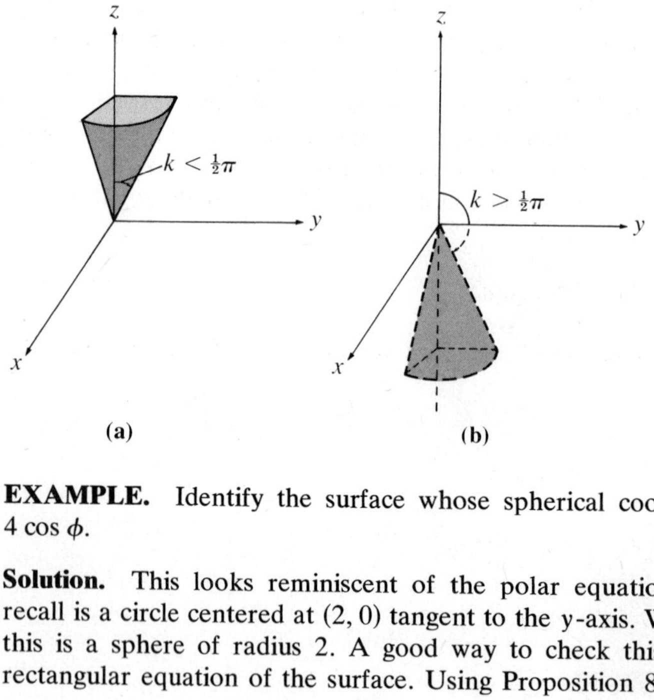

FIGURE 8.6

8.3 EXAMPLE. Identify the surface whose spherical coordinate equation is $\rho = 4\cos\phi$.

Solution. This looks reminiscent of the polar equation $r = 4\cos\theta$, which we recall is a circle centered at (2, 0) tangent to the y-axis. We might guess then that this is a sphere of radius 2. A good way to check this guess is to obtain the rectangular equation of the surface. Using Proposition 8.2, we have

$$\rho = 4\frac{z}{\rho}, \quad \text{hence} \quad \rho^2 = 4z.$$

This gives

$$x^2 + y^2 + z^2 = 4z, \qquad x^2 + y^2 + z^2 - 4z = 0, \qquad x^2 + y^2 + (z-2)^2 = 4.$$

Hence, we indeed have a sphere of radius 2, centered at (0, 0, 2) and hence tangent to the xy-plane. Our reasoning-by-analogy was exactly right. (Analogy *does* have its limits, of course. See Exercise 28 below.)

As was the case with cylindrical coordinates, our principal application for spherical coordinates will be in triple integration. We thus want to assign a meaning to

$$\iiint_U g(\rho, \phi, \theta)\, dV,$$

where the formula for the continuous function g is expressed in terms of spherical coordinates. By now you probably realize that this will be the limit of a Riemann sum

$$\sum_{i=1}^{m} \sum_{j=1}^{n} \sum_{k=1}^{p} g(\bar{\rho}_i, \bar{\phi}_j, \bar{\theta}_k)\, \Delta V_{ijk}$$

where ΔV_{ijk} is the volume of a small region U_{ijk} formed by making a spherical coordinate grid of U. In Figure 8.7 we try to picture U_{ijk}. It lies between spherical

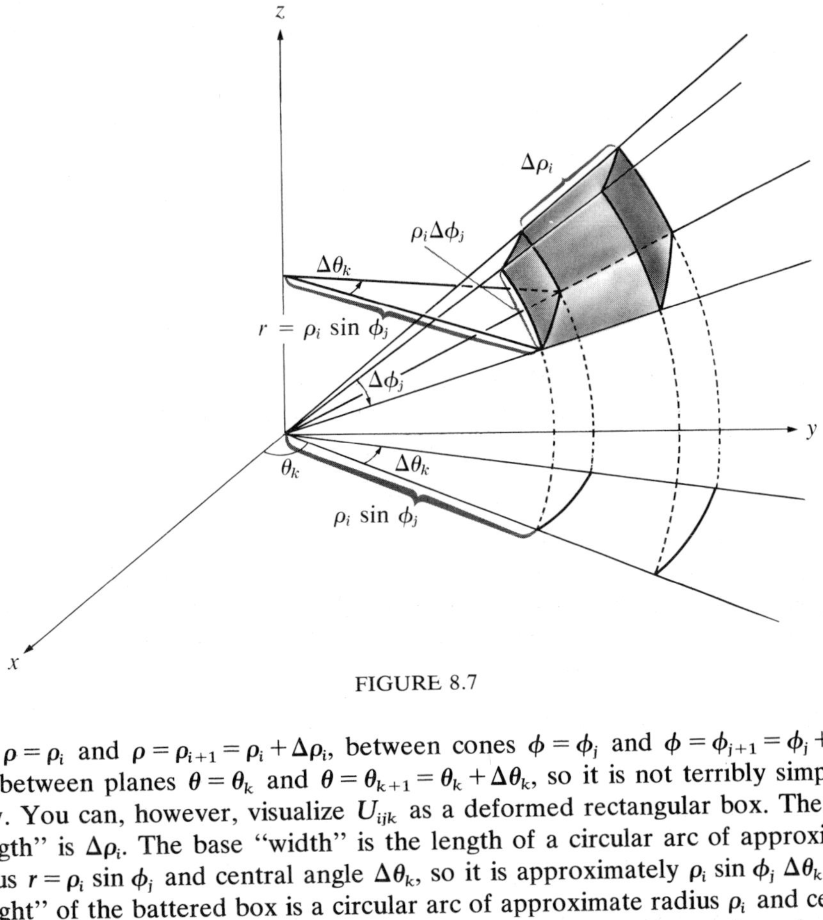

FIGURE 8.7

arcs $\rho = \rho_i$ and $\rho = \rho_{i+1} = \rho_i + \Delta\rho_i$, between cones $\phi = \phi_j$ and $\phi = \phi_{j+1} = \phi_j + \Delta\phi_j$, and between planes $\theta = \theta_k$ and $\theta = \theta_{k+1} = \theta_k + \Delta\theta_k$, so it is not terribly simple to draw. You can, however, visualize U_{ijk} as a deformed rectangular box. The base "length" is $\Delta\rho_i$. The base "width" is the length of a circular arc of approximate radius $r = \rho_i \sin \phi_j$ and central angle $\Delta\theta_k$, so it is approximately $\rho_i \sin \phi_j\, \Delta\theta_k$. The "height" of the battered box is a circular arc of approximate radius ρ_i and central angle $\Delta\phi_j$, so it is approximately $\rho_i\, \Delta\phi_j$. Thus, we can approximate the volume of U_{ijk} by

$$\Delta V_{ijk} \approx (\Delta\rho_i)(\rho_i \sin \phi_j\, \Delta\theta_k)\rho_i\, \Delta\phi_j$$

Thus,

$$\Delta V_{ijk} \approx \rho_i^2 \sin \phi_j\, \Delta\rho_i\, \Delta\phi_j\, \Delta\theta_k.$$

This won't change much if we vary ρ_i and ϕ_j slightly, so

$$\Delta V_{ijk} \approx \bar{\rho}_i^2 \sin \bar{\phi}_j \, \Delta\rho_i \, \Delta\phi_j \, \Delta\theta_k.$$

Thus, the following definition seems reasonable.

8.4 **DEFINITION.** Let $T \subseteq \boldsymbol{R}^3$ be a region on which $g: \boldsymbol{R}^3 \to \boldsymbol{R}$ is continuous. Then the **spherical triple integral** of g over T is

$$\iiint_T g(\rho, \phi, \theta)\, dV = \iiint_T g(\rho, \phi, \theta)\rho^2 \sin \phi \, d\rho \, d\phi \, d\theta$$

$$= \lim_{|G| \to 0} \sum_{i=1}^{m} \sum_{j=1}^{n} \sum_{k=1}^{p} g(\bar{\rho}_i, \bar{\phi}_j, \bar{\theta}_k)\bar{\rho}_i^2 \sin \phi_j \, \Delta\rho_i \, \Delta\phi_j \, \Delta\theta_k,$$

where the limit is taken in the same sense as in Definition 6.1. Here we imagine T to be bounded by spheres $\rho = a$ and $\rho = b$, planes $\theta = \alpha$ and $\theta = \beta$, and cones $\phi = \gamma$ and $\phi = \delta$. But, as before, we can extend the definition to an arbitrary bounded region U by enclosing U in such a basic region T and defining

$$\iiint_U g(\rho, \phi, \theta)\, dV = \iiint_T \bar{g}(\rho, \phi, \theta)\, dV$$

where $\bar{g}$ is given by

$$\bar{g}(\rho, \phi, \theta) = \begin{cases} 0 & \text{if } \{\rho, \phi, \theta\} \notin U \\ g(\rho, \phi, \theta) & \text{if } \{\rho, \phi, \theta\} \in U \end{cases}$$

Again we state that Definition 8.4 can be shown to be consistent with Definitions 6.1 and 7.3, and emphasize that *in spherical coordinates,*

$$dV = \rho^2 \sin \phi \, d\rho \, d\phi \, d\theta, \quad \textbf{NOT} \quad d\rho \, d\phi \, d\theta.$$

Details must be postponed until Section 6.9. There is also a version of Fubini's Theorem that permits spherical triple integrals to be evaluated by iteration, without regard to the order of iteration. The following version covers most cases of interest.

8.5 **THEOREM.** If g and T are as in Definition 8.4, then

$$\iiint_T g(\rho, \phi, \theta)\, dV = \int_\alpha^\beta d\theta \int_a^b \rho^2 \, d\rho \int_\gamma^\delta g(\rho, \phi, \theta) \sin \phi \, d\phi.$$

If U is bounded by planes $\theta = \alpha$ and $\theta = \beta$ where $\alpha \leq \beta$, cones $\phi = \gamma$ and $\phi = \delta$ where $\gamma \leq \delta$, and surfaces $\rho = h_1(\phi)$ and $\rho = h_2(\phi)$ where $h_1(\phi) \leq h_2(\phi)$, then

$$\iiint_U g(\rho, \phi, \theta)\, dV = \int_\alpha^\beta d\theta \int_\gamma^\delta \sin \phi \, d\phi \int_{h_1(\phi)}^{h_2(\phi)} g(\rho, \phi, \theta)\rho^2 \, d\rho.$$

8.6 EXAMPLE. Derive the formula for the volume of a sphere of radius a by triple integration.

Solution. Using Theorem 6.5, we get

$$V = \iiint_B 1 \, dV = \iiint_B \rho^2 \sin\phi \, d\rho \, d\phi \, d\theta$$

where B is the solid ball of radius a. We can set up coordinates at the center of the ball (Figure 8.8). If there is ever a job for spherical coordinates, this must be

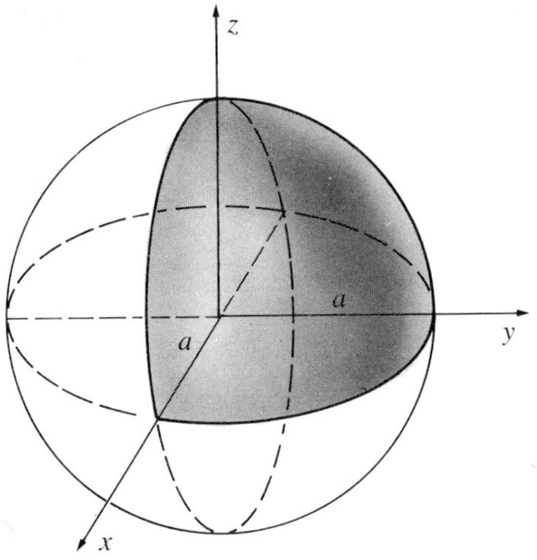

FIGURE 8.8

one, so we evaluate the integral using spherical coordinates.

$$\begin{aligned} V &= \int_0^{2\pi} d\theta \int_0^{\pi} \sin\phi \, d\phi \int_0^a \rho^2 \, d\rho \\ &= \int_0^{2\pi} d\theta \int_0^{\pi} \sin\phi \left[\frac{\rho^3}{3}\right]_0^a d\phi \\ &= \frac{a^3}{3} \int_0^{2\pi} d\theta \int_0^{\pi} \sin\phi \, d\phi = \frac{a^3}{3} \int_0^{2\pi} d\theta [-\cos\phi]_0^{\pi} \\ &= \frac{a^3}{3} \int_0^{2\pi} d\theta (1+1) = \frac{2a^3}{3} \theta \Big]_0^{2\pi} \\ &= \tfrac{2}{3}a^3(2\pi) = \tfrac{4}{3}\pi a^3. \end{aligned}$$

You will be more impressed by the usefulness of spherical coordinates if you try to do this problem using rectangular coordinates. (If you did Problem 10 of Exercises 5.7, you may be interested in comparing your solution to Example 8.6 also.)

As in the case of cylindrical coordinates, we can change an intractable rectangular triple integral $\iiint_U f(x, y, z) \, dV$ to a spherical integral. If U is described by rectangular coordinates and W is the same region described in spherical coordinates, then

$$\iiint_U f(x, y, z) \, dV = \iiint_W f(x(\rho, \phi, \theta), y(\rho, \phi, \theta), z(\rho, \phi, \theta)) \, dV$$

where on the right $dV=\rho^2 \sin\phi \, d\rho \, d\phi \, d\theta$. This will be justified in Section 6.9. Our next example illustrates how it can be used to simplify stubborn triple integrals.

8.7 EXAMPLE. Evaluate $\iiint_C \sqrt{x^2+y^2+z^2}\, dV$, where C is the "ice cream cone"

$$\left\{(x, y, z) \mid x^2+y^2+z^2 \le 1,\ x^2+y^2 \le \frac{z^2}{3},\ z \ge 0\right\}.$$

Solution. C can be described in spherical coordinates as follows. First, $z \ge 0$ corresponds to $\phi \le \frac{1}{2}\pi$, and $x^2+y^2+z^2 \le 1$ corresponds to $\rho \le 1$. The bounding surface

$$x^2+y^2=\frac{z^2}{3}$$

becomes

$$r^2=\frac{\rho^2 \cos^2\phi}{3}$$

where $r=\rho \sin\phi$, so

$$\rho^2 \sin^2\phi=\frac{\rho^2 \cos^2\phi}{3}.$$

This simplifies to $\tan^2\phi = 1/3$, i.e., $\tan\phi = 1/\sqrt{3}$ $\left(\text{since } \phi \le \frac{\pi}{2}\right)$, so $\phi=\frac{\pi}{6}$, as shown in Figure 8.9.

Our integral becomes, since $\sqrt{x^2+y^2+z^2}=\rho$,

$$\begin{aligned}
\iiint_C \rho \, dV &= \iiint_C \rho^3 \sin\phi \, d\rho \, d\phi \, d\theta \\
&= \int_0^{2\pi} d\theta \int_0^{\pi/6} \sin\phi \, d\phi \int_0^1 \rho^3 \, d\rho \\
&= \int_0^{2\pi} d\theta \int_0^{\pi/6} \sin\phi \, d\phi \left[\frac{\rho^4}{4}\right]_0^1 \\
&= \tfrac{1}{4}\int_0^{2\pi} d\theta \int_0^{\pi/6} \sin\phi \, d\phi \\
&= \tfrac{1}{4}\int_0^{2\pi} d\theta [-\cos\phi]_0^{\pi/6} \\
&= \tfrac{1}{4}\int_0^{2\pi} d\theta [-\tfrac{1}{2}\sqrt{3}+1] \\
&= \tfrac{1}{4}(1-\tfrac{1}{2}\sqrt{3})2\pi \\
&= \tfrac{1}{2}\pi(1-\tfrac{1}{2}\sqrt{3}).
\end{aligned}$$

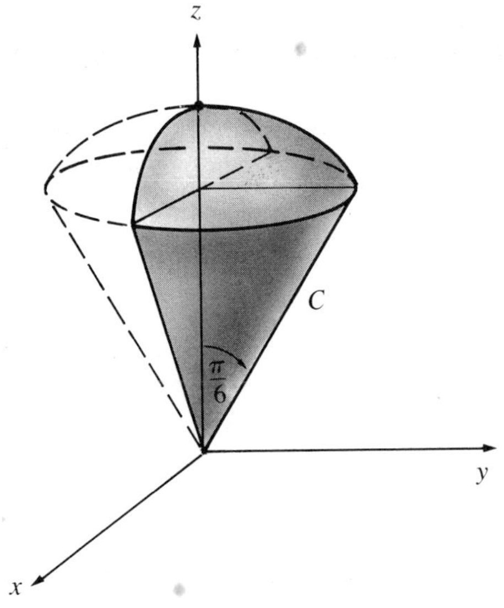

FIGURE 8.9

EXERCISES 5.8

In Exercises 1 and 2, find spherical coordinates for each of the points with the given rectangular coordinates.

1 (a) $(1, 2, 3)$ (b) $(-1, 1, 1)$ (c) $(0, 2, -2)$ (d) $(2, 0, -3)$ (e) $(-2, 3, -1)$

2 (a) $(2, -1, 3)$ (b) $(2, 3, -1)$ (c) $(-2, -3, 1)$ (d) $(2, -3, -1)$ (e) $(3, -1, -2)$

In Exercises 3 and 4, find rectangular coordinates for each of the points with the given spherical coordinates.

3 (a) $\{3, \pi/6, \pi/4\}$ (b) $\{2, \pi/2, \pi/3\}$ (c) $\{2, 3\pi/4, \pi/6\}$ (d) $\{2, 3\pi/4, 7\pi/6\}$ (e) $\{2, 3\pi/2, \pi\}$

4 (a) $\{2, \pi/3, \pi/2\}$ (b) $\{3, \pi/2, \pi/2\}$ (c) $\{2, \pi/6, 3\pi/4\}$ (d) $\{2, 3\pi/4, 5\pi/3\}$ (e) $\{2, \pi/4, \pi/6\}$

In Exercises 5 and 6, a rectangular coordinate equation of a surface is given. Obtain a spherical coordinate equation for the surface.

5 (a) $x^2+y^2+z^2=9$ (b) $x^2+y^2+z^2-6z=0$ (c) $x^2+y^2=16$

6 (a) $2x^2+2y^2+z^2-6z=0$ (b) $x^2+y^2-3z^2=0$ (c) $x^2+y^2=z$

In Exercises 7 and 8, a spherical coordinate equation of a surface is given. Sketch the surface and obtain a rectangular coordinate equation for it.

7 (a) $\rho=5$ (b) $\theta=\pi/6$ (c) $\rho=8\cos\phi$

8 (a) $\phi=\pi/4$ (b) $\rho=8\sec\phi$ (c) $\rho\sin\phi=8$

9 Use spherical coordinates to find a formula for the volume enclosed by a right circular cylinder of radius a and height h. (*Hint:* See Exercises 5(c) and 8(c).)

10 Use spherical coordinates to find a formula for the volume enclosed by a right circular cone of base radius a and height h.

11 Find the volume of the "ice cream cone" in Example 8.7.

12 Find the volume of the region described by

$$x^2+y^2+z^2\le 1,\ x^2+y^2\le 3z^2,\ z\ge 0.$$

13 Find the volume of the region lying between the spheres $x^2+y^2+z^2=1$ and $x^2+y^2+z^2=4$.

14 Find the volume of the region bounded above by the sphere $\rho=2$, below by the xy-plane, and on the sides by $x^2+y^2=1$. (*Hint:* You can save some work by using the formula from Exercise 9.)

15 Find the volume enclosed by the surface $\rho=1-\cos\phi$. Sketch it. (The region can be obtained by revolving the region bounded by the cardioid $r=1-\cos\theta$ about the polar (z) axis in the zx-plane.)

16 Find the volume enclosed by the surface $\rho=a(1-\cos\phi)$.

In Exercises 17 to 22, evaluate the triple integral.

17 $\iiint_B \sqrt{x^2+y^2+z^2}\,dV$, where B is the ball of radius 1 centered at the origin.

18 $\iiint_B z\,dV$, where B is the first octant part of the ball in Exercise 17.

19 $\displaystyle\iiint_A \frac{1}{x^2+y^2+z^2}\,dV$, where A is the region between the spheres $x^2+y^2+z^2=1$ and $x^2+y^2+z^2=4$.

20 $\displaystyle\iiint_A \frac{1}{x^2+y^2+z^2}\,dV$, where A is the region between the spheres $x^2+y^2+z^2=4$ and $x^2+y^2+z^2=9$.

21 $\iiint_B (x^2+y^2)\,dV$, where B is the region of Exercise 18.

22 $\iiint_U \sqrt{x^2+y^2+z^2}\,dV$, where U is the region enclosed by $\rho=1-\cos\phi$. (See Exercise 15.)

23 If a curve in $\mathbf{R}^3$ is parametrized in spherical coordinates by $\rho=\rho(t)$, $\phi=\phi(t)$, and $\theta=\theta(t)$, then use the Chain Rule to show that its arc length between $t=a$ and $t=b$ is

$$L=\int_a^b \sqrt{\dot{\rho}^2+(\rho^2\sin^2\phi)\dot{\theta}^2+\rho^2\dot{\phi}^2}\,dt.$$

(*Hint:* Make use of Proposition 8.2 and the arc length formula in rectangular coordinates.)

24 A submarine sails due north from the equator in the mid-Pacific to the north pole. Find the length of its journey. (THINK before resorting to Exercise 23!)

25 A submarine sails due south from a point on the equator in the Indian Ocean. Find the length it must travel to reach the south pole. (See Exercise 24.)

26 A *conical helix* is given parametrically by $\rho=t$, $\phi=\pi/4$, and $\theta=t$. (This curve winds around the right circular cone $\phi=\pi/4$ as it ascends.) Find its arc length from $t=0$ to $t=\pi$. (See Exercise 23.)

27 Find the length of arc of the curve on the surface of a right circular cone given by $\rho=t^2$, $\phi=\pi/6$, and $\theta=t$ between $t=0$ and $t=1$. (See Exercise 23.)

28 The reasoning-by-analogy used in Example 8.3 would suggest that the graph of $\rho=a\sin\phi$ should *also* be a sphere of radius $a/2$. Is it? Draw the graph. Is this related in *any* way to the graph of $r=a\sin\theta$ in polar coordinates?

29 Compute $\displaystyle\iiint_E \frac{1}{\sqrt{x^2+y^2}}\,dV$, where E is the region enclosed by the graph of $\rho=a\sin\phi$.

30 Compute $\displaystyle\iiint_E \frac{1}{x^2+y^2+z^2}\,dV$, where E is as in Exercise 29.

9 APPLICATIONS

You probably recall from your elementary calculus course that the ordinary definite integral $\int_a^b f(x)\,dx$ of a function $f: \mathbf{R} \to \mathbf{R}$ has a number of important geometrical and physical applications. You have no doubt used the definite integral to compute areas and volumes, and at least some of the physical quantities mass, center of mass, work, fluid pressure, and gravitational attraction. We have already used double and triple integrals to calculate areas and volumes, and in this section we show how multiple integration may be used to calculate a number of important physical quantities.

First we consider the *mass* of a plane *lamina*. A lamina may be thought of as a sheet of nonhomogeneous material, i.e., material so thin that we can neglect its thickness and give its (variable) density $d(x, y)$ as mass per unit area. Suppose the material occupies a bounded region $D \subseteq \mathbf{R}^2$. To see why the mass is defined as it is, let's subdivide the region D by means of a grid G (as in Figure 9.1) of a

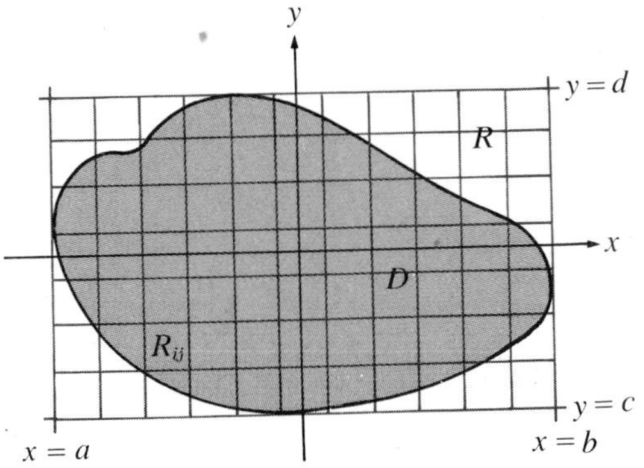

FIGURE 9.1

rectangle $R = [a, b] \times [c, d]$ which contains D. If the grid is fine, and the density function $d(x, y)$ is continuous, then on a small rectangle R_{ij} of the grid, d will be nearly constant. Hence the mass of the lamina corresponding to this subrectangle will be approximately

$$d(\bar{x}_i, \bar{y}_j)\,\Delta x_i\,\Delta y_j,$$

where $(\bar{x}_i, \bar{y}_j)$ is any point of D in R_{ij}, and, as usual,

$$\Delta x_i = x_i - x_{i-1} \quad \text{and} \quad \Delta y_j = y_j - y_{j-1}$$

are the dimensions of R_{ij}. Hence, the mass of the lamina is approximated by a Riemann sum

$$R(d, G, (\bar{x}_i, \bar{y}_j)) = \sum_{i=1}^{m} \sum_{j=1}^{n} d(\bar{x}_i, \bar{y}_j)\,\Delta x_i\,\Delta y_j.$$

This approximation stands to get better as we make our grid G finer, and so the following definition seems appropriate.

9.1 **DEFINITION.** If a lamina with a continuous density function d occupies the region $D \subseteq \mathbf{R}^2$, then the **mass** of the lamina is

$$M = \iint_D d(x, y)\,dA$$

9.2 EXAMPLE. A lamina occupies the region bounded by $y=1-x^2$ and the x-axis. Its density is given by $d(x, y)=|xy|$. Find the mass.

Solution. The region is shown in Figure 9.2. Notice that the density function is symmetric relative to the y-axis: $d(x, y)=d(-x, y)=|x|\,y$ for any point $(x, y)\in D$. So the mass is *double* the mass of the first quadrant portion D^+ of the lamina.

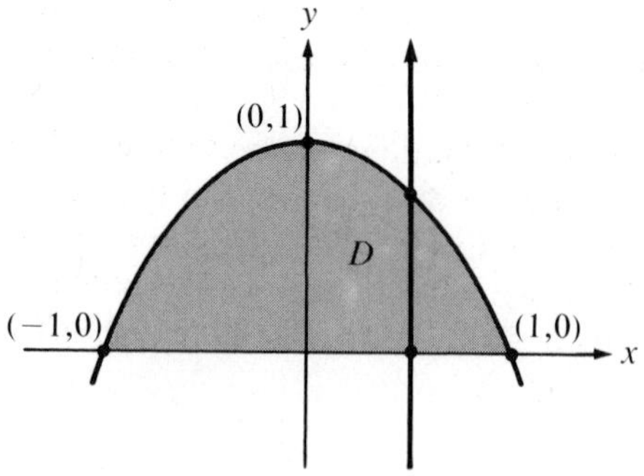

FIGURE 9.2

Hence we have

$$M=2\iint_{D^+} d(x, y)\,dA = 2\int_0^1 dx\int_0^{1-x^2} xy\,dy$$

$$=2\int_0^1 x\,dx\int_0^{1-x^2} y\,dy = 2\int_0^1 x\,dx\left[\tfrac{1}{2}y^2\right]_0^{1-x^2}$$

$$=\int_0^1 x(1-x^2)^2\,dx = -\tfrac{1}{2}\int_0^1 -2x(1-x^2)^2\,dx$$

$$=-\frac{1}{2}\frac{(1-x^2)^3}{3}\Bigg]_0^1=\frac{1}{2}\cdot\frac{1}{3}=\frac{1}{6}.$$

Now we want to pursue this point of view to find the *center of mass* of a plane lamina. The technique used is suggested by the discrete case. Suppose for the moment that, instead of a lamina, we had a discrete system of point masses at the points $(x_1, y_1), (x_2, y_2), \ldots, (x_k, y_k)$, where at point $\mathbf{x}_i=(x_i, y_i)$ there is a mass m_i. The center of mass of such a system is simply $(\bar{x}, \bar{y})$, where $\bar{x}$ is the weighted average of the x-coordinates of the point masses and $\bar{y}$ is the weighted average of the y-coordinates of the point masses. Thus,

$$\bar{x}=\frac{\sum_{i=1}^{k} m_i x_i}{\sum_{i=1}^{k} m_i} \quad\text{and}\quad \bar{y}=\frac{\sum_{i=1}^{k} m_i y_i}{\sum_{i=1}^{k} m_i}.$$

While we don't have such a system, we can *approximate* our lamina in this way. Namely, if we finely partition the domain D as in Figure 9.1, then we can imagine the entire mass of each subrectangle R_{ij} as concentrated at the point $(\bar{x}_i, \bar{y}_j)$. This gives us a discrete system close to the situation in our lamina. Then an approximation to the coordinates $(\bar{x}, \bar{y})$ of the center of mass of the lamina is given by the weighted averages of the $\bar{x}_i$ and $\bar{y}_j$ respectively. Now each subrectangular mass, we have seen above, is approximately $d(\bar{x}_i, \bar{y}_j)\,\Delta A_{ij}$ where $\Delta A_{ij}=$

$A(R_{ij})$. Thus we have

$$\bar{x} \approx \frac{\sum_{i=1}^{m} \sum_{j=1}^{n} \bar{x}_i \, d(\bar{x}_i, \bar{y}_j) \, \Delta A_{ij}}{\sum_{i=1}^{m} \sum_{j=1}^{n} d(\bar{x}_i, \bar{y}_j) \, \Delta A_{ij}}$$

and

$$\bar{y} \approx \frac{\sum_{i=1}^{m} \sum_{j=1}^{n} \bar{y}_j \, d(\bar{x}_i, \bar{y}_j) \, \Delta A_{ij}}{\sum_{i=1}^{m} \sum_{j=1}^{n} d(\bar{x}_i, \bar{y}_j) \, \Delta A_{ij}}.$$

As our grid gets finer and finer, these approximations should approach the exact coordinates of the center of mass. Thus the following definition is natural.

9.3 **DEFINITION.** If a lamina occupies the region $D \subseteq \mathbf{R}^2$ and has a continuous density function $d(x, y)$, then the **center of mass** of the lamina is the point $(\bar{x}, \bar{y})$ where

$$\bar{x} = \frac{\iint_D x \, d(x, y) \, dA}{M} \quad \text{and} \quad \bar{y} = \frac{\iint_D y \, d(x, y) \, dA}{M}.$$

Here the numerators are called the **moments of the lamina with respect to the coordinate axes:**

$$M_y = \iint_D x \, d(x, y) \, dA \text{ is the } \textit{moment with respect to the y-axis,}$$

$$M_x = \iint_D y \, d(x, y) \, dA \text{ is the } \textit{moment with respect to the x-axis.}$$

Be careful that you don't get confused by the notation. The formula for M_x has a y in the integral (*not* x), and the formula for M_y has an x in the integral (*not* y). The reason is that x measures the distance from the y-*axis* to (x, y) and y measures the distance from the x-*axis* to (x, y). Some students prefer to use the notation $M\bar{x}$ for M_y and $M\bar{y}$ for M_x because there *is* an x in the integral for $M\bar{x}$ and a y in the integral for $M\bar{y}$.

9.4 **EXAMPLE.** Find the center of mass of the lamina in Example 9.2.

Solution. We have already calculated $M = \frac{1}{6}$. So we need only M_x and M_y. We have

$$M_x = M\bar{y} = \iint_D y \, d(x, y) \, dA = 2 \iint_{D^+} xy^2 \, dA$$

$$= 2 \int_0^1 x \, dx \int_0^{1-x^2} y^2 \, dy = \tfrac{2}{3} \int_0^1 x \, dx [y^3]_0^{1-x^2}$$

$$= \tfrac{2}{3} \int_0^1 x(1-x^2)^3 \, dx = -\tfrac{1}{2} \cdot \tfrac{2}{3} \int_0^1 -2x(1-x^2)^3 \, dx$$

$$= -\frac{1}{3} \frac{(1-x^2)^4}{4} \Bigg]_0^1 = \frac{1}{12}.$$

Thus,

$$\bar{y} = \frac{M\bar{y}}{M} = \frac{\frac{1}{12}}{\frac{1}{6}} = \frac{1}{2}.$$

Similarly,

$$M_y = M\bar{x} = \iint_D x\, d(x, y)\, dA = \iint_D x\, |x|\, y\, dA.$$

Now

$$x\,|x|\,y = x^2 y \quad \text{if} \quad x \geq 0,$$

and

$$x\,|x|\,y = -x^2 y \quad \text{if} \quad x < 0,$$

so the integrand is an *odd* function of x. Since the region D is symmetric in the y-axis, we know that $M_y = \iint_D x\,|x|\,y\,dA$ will be 0 by Exercise 35 of Exercises 5.3. In case you took little note of that exercise, you might try working out the integral for M_y directly and noting how much work is needed. In any case, $(\bar{x}, \bar{y}) = (0, \frac{1}{2})$.

Observe that if the density function is a constant d, then

$$M = \iint_D d\, dA = d\, A(D),$$

$$M_x = \iint_D d\, y\, dA = d \iint_D y\, dA,$$

$$M_y = \iint_D d\, x\, dA = d \iint_D x\, dA.$$

These quantities are of geometric significance.

9.5 **DEFINITION.** If a lamina has constant density d, then its **mass** is $d\,A(D)$, where $D \subseteq \mathbf{R}^2$ is the region occupied by the lamina. The center of mass is called the **centroid** or **geometric center** of the region D.

Polar integration can often help reduce the complexity of center of mass problems.

9.6 **EXAMPLE.** Find the centroid of a lamina occupying the upper half of the unit disk $D = \{(x, y) \mid x^2 + y^2 \leq 1\}$. See Figure 9.3.

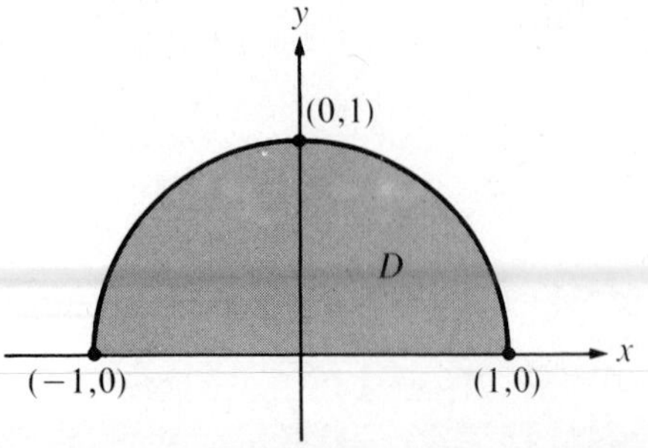

FIGURE 9.3

Solution. If d is the density, then $M = d\,A(D) = d\dfrac{\pi}{2}$. By the symmetry of D in the y-axis, it is clear that $\bar{x} = 0$. From Definition 9.3,

$$M_x = M\bar{y} = \iint_D y\,d\,dA = d\iint_D r\sin\theta\, r\,dr\,d\theta$$

$$= d\int_0^1 r^2\,dr\int_0^\pi \sin\theta\,d\theta = d\int_0^1 r^2\,dr[-\cos\theta]_0^\pi$$

$$= 2d\tfrac{1}{3}r^3]_0^1 = \frac{2d}{3}.$$

Hence

$$\bar{y} = \frac{M\bar{y}}{M} = \frac{\frac{2d}{3}}{d\frac{\pi}{2}} = \frac{4}{3\pi}.$$

So the centroid is $(\bar{x}, \bar{y}) = (0, 4/3\pi)$.

A striking geometric application of the centroid that goes back about 1700 years to the Greek mathematician Pappus of Alexandria is presented in the following theorem.

9.7 **THEOREM.** Let $D \subseteq \boldsymbol{R}^2$ be a region bounded by the graphs of two continuous functions, $y = f(x)$ and $y = g(x)$, and the lines $x = a$ and $x = b$, where we assume $0 \le f(x) \le g(x)$ for $x \in [a, b]$. See Figure 9.4. Let E be the solid of revolution obtained by revolving D about the x-axis. Then

$$V(E) = 2\pi\bar{y}A(D),$$

i.e., the *volume of E is the area of D times the circumference of the circle traversed by the centroid $(\bar{x}, \bar{y})$ of D.*

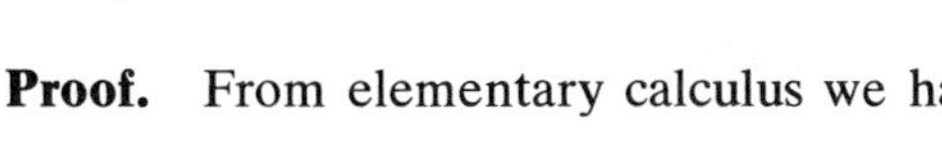

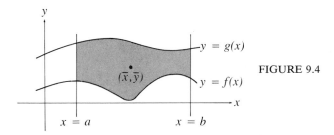

FIGURE 9.4

Proof. From elementary calculus we have

$$V(E) = \int_a^b \pi(g(x)^2 - f(x)^2)\,dx.$$

If we regard D as a lamina of constant density d, then we have

$$M_x = M\bar{y} = \iint_D d\,y\,dA = d\int_a^b dx\int_{f(x)}^{g(x)} y\,dy$$

$$= \tfrac{1}{2}d\int_a^b (g(x)^2 - f(x)^2)\,dx.$$

Since

$$\bar{y}=\frac{M\bar{y}}{M}=\frac{M\bar{y}}{A(D)\,d}=\frac{\frac{1}{2}d\int_a^b (g(x)^2-f(x)^2)\,dx}{A(D)\,d},$$

we have

$$2\pi\bar{y}A(D)=\pi\int_a^b (g(x)^2-f(x)^2)\,dx=V(E). \qquad \text{QED}$$

This classic theorem enables us to easily compute the volume of a doughnut (or tire inner tube) in the shape of a **torus,** the figure resulting when a circular disk D of radius a is revolved about a line ℓ at a distance $b>a$ from the center of D. Refer to Figures 9.5 and 9.6, where we take ℓ to be the x-axis. The centroid of D

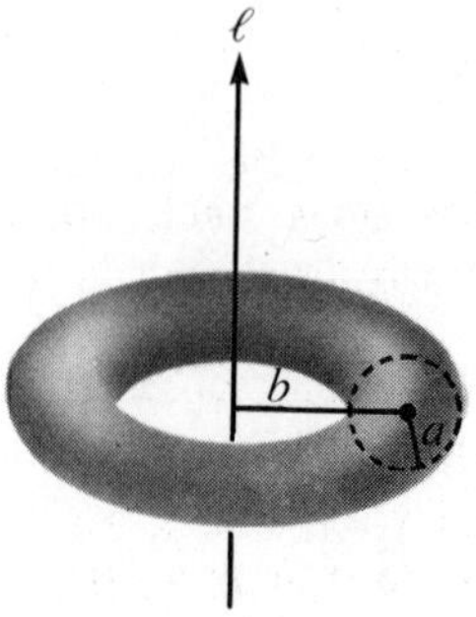

FIGURE 9.5

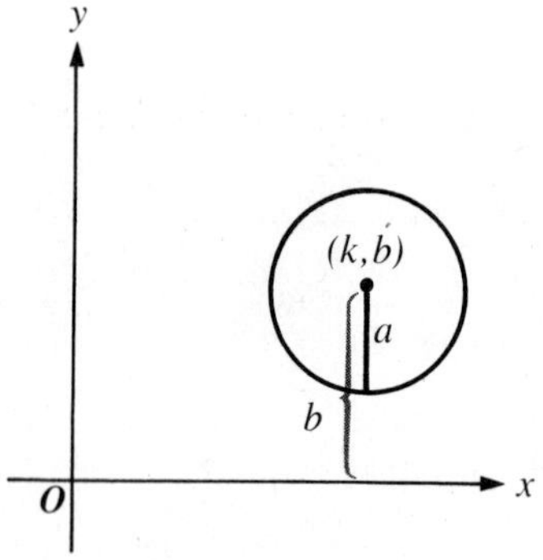

FIGURE 9.6

is its center, say (k, b). So the torus E has volume

$$V(E)=2\pi\bar{y}A(D)=2\pi b\,\pi a^2=2\pi^2a^2b.$$

You will be immediately impressed by Pappus if you try to calculate $V(E)$ by integration without using his theorem!

We can quite naturally carry over the concepts of mass and moments to regions $E\subseteq \mathbf{R}^3$. In fact, here they seem even more appropriate because now the density function $d(x, y, z)$ is in terms of mass per unit volume. Rather than repeat in three dimensions the approximations by discrete mass distributions that led up to Definitions 9.1 and 9.3, we simply give the analogous definition.

9.8 **DEFINITION.** Let a substance S occupy the region $E \subseteq \mathbf{R}^3$ and have a continuous density function $d(x, y, z)$. Then the **mass** of the substance is

$$M = \iiint_E d(x, y, z)\, dV.$$

The **center of mass** of the substance is $(\bar{x}, \bar{y}, \bar{z})$, where

$$\bar{x} = \frac{\iiint_E x\, d(x, y, z)\, dV}{M}, \qquad \bar{y} = \frac{\iiint_E y\, d(x, y, z)\, dV}{M},$$

$$\bar{z} = \frac{\iiint_E z\, d(x, y, z)\, dV}{M}.$$

The numerators of $\bar{x}$, $\bar{y}$, and $\bar{z}$ respectively are called the **moments of S with respect to the yz-, xz-, and xy-planes** respectively. The notations M_{yz} or $M\bar{x}$, M_{xz} or $M\bar{y}$, and M_{xy} or $M\bar{z}$ are used for these moments.

It appears that a considerable amount of energy needs to be expended to find center of mass, but fortunately the symmetry present in most commonly encountered solids often helps to reduce the work. We mention two helpful rules worth bearing in mind.

(1) *If the region E is symmetric in the xy-plane and the density function d satisfies $d(x, y, -z) = d(x, y, z)$, then $\bar{z} = 0$.*

(2) *If E is symmetric in the z-axis* (i.e., whenever (x, y, z) is in E, then so is $(-x, -y, z)$) *and the density function satisfies $d(-x, -y, z) = d(x, y, z)$, then $\bar{x} = \bar{y} = 0$.*

Rule (1) follows from Exercise 29 of Exercises 5.6:

$$M_{xy} = M\bar{z} = \frac{\iiint_E z\, d(x, y, z)\, dV}{M} = 0$$

because $zd(x, y, z)$ is an odd function in z; at $(x, y, -z)$, its value is $-zd(x, y, z)$, which is the negative of its value at (x, y, z). Similarly, (2) follows. Note that *Rules (1) and (2) apply equally well to regions E that are symmetric in the other axes or coordinate planes if the density functions obey the corresponding rules.*

Application of these rules, coupled with the judicious use of cylindrical or spherical coordinates, can often keep the work involved in computing centers of mass within reasonable bounds.

9.9 **EXAMPLE.** A substance S is in the shape of half a spherical ball of radius 2. Its density at a point P is proportional to the distance of P from the center of the base. Find the center of mass.

Solution. Here we can represent the solid (see Figure 9.7) as occupying the region

$$E = \{(x, y, z) \mid x^2 + y^2 + z^2 \leq 4,\ z \geq 0\}.$$

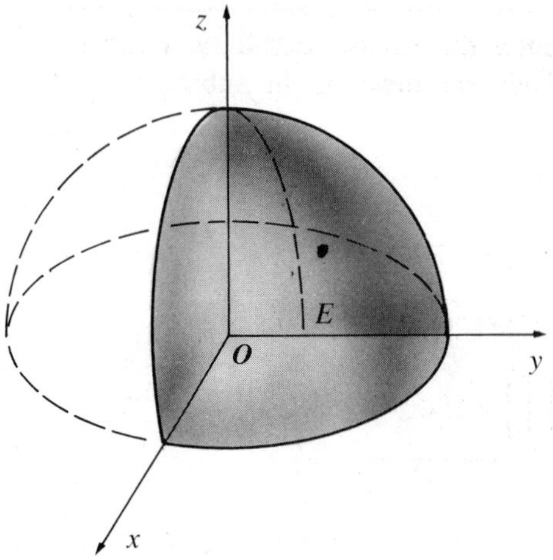

FIGURE 9.7

The center of the base is then (0, 0, 0), so $d(x, y, z) = k\sqrt{x^2+y^2+z^2}$. Now the region E is symmetric in the z-axis and $d(x, y, z) = d(-x, -y, z)$, so Rule (2) tells us that $\bar{x} = \bar{y} = 0$. We still need to determine $\bar{z}$. First,

$$M = \iiint_E k\sqrt{x^2+y^2+z^2}\, dV.$$

This looks like a job for spherical coordinates. Since $\sqrt{x^2+y^2+z^2} = \rho$, we get

$$\begin{aligned}
M &= k\iiint_E \rho\, dV = k\int_0^{2\pi} d\theta \int_0^{\pi/2} d\phi \int_0^2 \rho\rho^2 \sin\phi\, d\rho \\
&= k\int_0^{2\pi} d\theta \int_0^{\pi/2} \sin\phi\, d\phi \int_0^2 \rho^3\, d\rho \\
&= k\int_0^{2\pi} d\theta \int_0^{\pi/2} \sin\phi [\tfrac{1}{4}\rho^4]_0^2\, d\phi \\
&= 4k\int_0^{2\pi} d\theta \int_0^{\pi/2} \sin\phi\, d\phi = 4k\int_0^{2\pi} d\theta [-\cos\phi]_0^{\pi/2} \\
&= 4k\int_0^{2\pi} d\theta = 8\pi k.
\end{aligned}$$

$$\begin{aligned}
M_{xy} = M\bar{z} &= k\iiint_E \rho\cos\phi\, dV \\
&= k\int_0^{2\pi} d\theta \int_0^{\pi/2} \sin\phi\cos\phi\, d\phi \int_0^2 \rho^4\, d\rho \\
&= k\int_0^{2\pi} d\theta \int_0^{\pi/2} \sin\phi\cos\phi [\tfrac{1}{5}\rho^5]_0^2\, d\phi \\
&= \tfrac{32}{5}k\int_0^{2\pi} d\theta [\tfrac{1}{2}\sin^2\phi]_0^{\pi/2} \\
&= \tfrac{16}{5}k\int_0^{2\pi} d\theta = \tfrac{32}{5}\pi k.
\end{aligned}$$

Hence $\bar{z} = \dfrac{M\bar{z}}{M} = \dfrac{\frac{32}{5}\pi k}{8\pi k} = \dfrac{4}{5}$. Thus the center of mass is $(0, 0, \frac{4}{5})$.

The final physical quantities we discuss in this section of applications are of considerable importance in studying dynamical systems. First we need to recall that the **kinetic energy** of a point mass m moving with velocity $\boldsymbol{v}$ is $K=\frac{1}{2}m\,|\boldsymbol{v}|^2$. Suppose next that a solid S rotates with constant angular velocity ω about an axis ℓ. (Here ω is a scalar quantity.) See Figure 9.8, where we show the path described by a typical point P in the solid. If this point P is at a distance r from the axis ℓ of rotation, then its linear speed is

$$|\boldsymbol{v}| = \omega r.$$

A small part of S surrounding P of mass ΔM, say, then has kinetic energy of approximately $\frac{1}{2}\Delta M\,|\boldsymbol{v}|^2=\frac{1}{2}\omega^2 r^2\,\Delta M$. Since a grid on S reduces S to many such small parts, we see that the total kinetic energy K of S is approximated by

$$\sum_{i,j,k} \tfrac{1}{2}\omega^2[r(x_i, y_j, z_k)]^2\,\Delta M_{ijk},$$

where

$$\Delta M_{ijk} = d(\bar{x}_i, \bar{y}_j, \bar{z}_k)\,\Delta V_{ijk}.$$

Hence we are led once more to the following natural definition.

9.10 **DEFINITION.** If a solid S of density $d(x, y, z)$ occupying a region $E\subseteq \boldsymbol{R}^3$ rotates about an axis ℓ with constant angular velocity ω, then the **total kinetic energy** of S is

$$K=\tfrac{1}{2}\omega^2 \iiint_E r^2\,dM=\tfrac{1}{2}\omega^2 \iiint_E r^2\,d(x, y, z)\,dV,$$

where $r=r(x, y, z)$ is the distance from $P(x, y, z)$ to the axis ℓ of rotation. We have

$$K=\tfrac{1}{2}\omega^2 I_\ell$$

where

$$I_\ell=\iiint_E r^2\,d(x, y, z)\,dV$$

is the **moment of inertia** of S about ℓ. The **radius of gyration** R of S about ℓ is defined by

$$R^2=\frac{I_\ell}{M},$$

where M is the mass of S. (This represents the distance from ℓ to a point P where the entire mass M of S could be concentrated to yield the same moment of inertia I_ℓ and the same kinetic energy.)

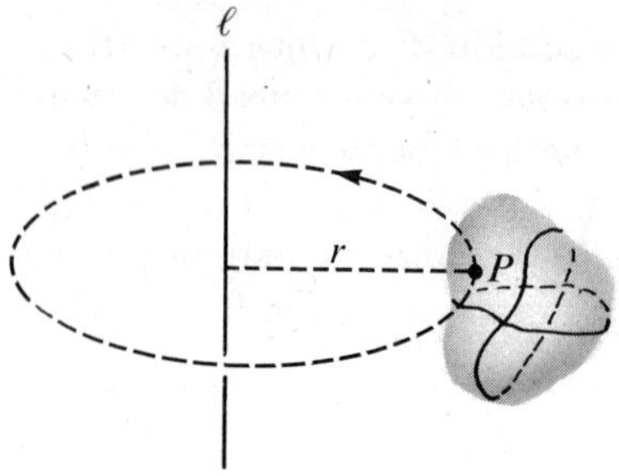

FIGURE 9.8

In practice, ℓ is usually the z-axis, x-axis, or y-axis. In these cases, we get the following formulas for I_ℓ:

$$I_x = \iiint_E (y^2+z^2)\, d(x, y, z)\, dV$$

$$I_y = \iiint_E (x^2+z^2)\, d(x, y, z)\, dV$$

$$I_z = \iiint_E (x^2+y^2)\, d(x, y, z)\, dV.$$

9.11 EXAMPLE. A bar of gold bullion of constant density δ occupies the box $B=[0, 3]\times[0, 1]\times[0, 1]$. Find its moment of inertia and radius of gyration about the z-axis.

Solution. Here $M=3\delta$ since $V=3\cdot 1\cdot 1=3$.

$$I_z = \iiint_B (x^2+y^2)\, \delta\, dV = \delta \int_0^3 dx \int_0^1 dy \int_0^1 (x^2+y^2)\, dz$$

$$= \delta \int_0^3 dx \int_0^1 (x^2+y^2)\, dy = \delta \int_0^3 [x^2y+\tfrac{1}{3}y^3]_0^1\, dx$$

$$= \delta \int_0^3 (x^2+\tfrac{1}{3})\, dx = \delta[\tfrac{27}{3}+\tfrac{3}{3}] = 10\delta.$$

Then $R^2=\dfrac{10\delta}{3\delta}=\dfrac{10}{3}$, so $R=\sqrt{10/3}$.

EXERCISES 5.9

In Exercises 1 to 8, find the mass and center of mass of a lamina occupying the given region $D\subseteq R^2$ and having the given density function.

1 The triangle with vertices $(0, 0)$, $(a, 0)$, (b, c), constant density d.

2 The semidisk $\{(x, y) \mid x^2+y^2\le a^2, y\ge 0\}$, constant density d.

3 The rectangle $[1, 2]\times[1, 3]$ where $d(x, y)=xy$.

4 The triangular region with vertices $(0, 0)$, $(0, 3)$, $(4, 0)$, where $d(x, y)=xy$.

5 The region bounded by $y^2=x$ and $y=x^2$, where $d(x, y)=ky$.

6 The first quadrant region enclosed by the circle $x^2+y^2=4$, where $d(x, y)=\sqrt{x^2+y^2}$.

7 The first quadrant region bounded by the circles $x^2+y^2=1$ and $x^2+y^2=4$, constant density d.

8 The region enclosed by the cardioid $r=2(1+\cos\theta)$, constant density d.

9 Find the volume of the torus formed by revolving the circle $(x-2)^2+(y-3)^2=9$ about the line $x=6$.

10 Find the volume of the torus formed by revolving the circle $(x+1)^2+(y-2)^2=4$ about the line $y=5$.

11 Find the volume of the solid formed when the region of Exercise 1 is revolved about the line $x=-a$.

12 Find the volume of the solid formed when the region of Exercise 2 is revolved about the line $x=-a$.

In Exercises 13 to 18, find the mass and center of mass of a substance S occupying the given region $E\subseteq R^3$, and having the given density function.

13 The rectangular box $[0, 1]\times[0, 1]\times[0, 1]$, where $d(x, y, z)=z$.

14 The rectangular box $[0, 2]\times[0, 1]\times[0, 1]$, where $d(x, y, z)=y$.

15 The region in Example 9.9 but of constant density.

16 The region described by $x^2+y^2+z^2\leq 9$ for $z\geq 0$, where $d(x, y, z)=k\sqrt{x^2+y^2+z^2}$.

17 The solid right circular cone in Figure 9.9, constant density.

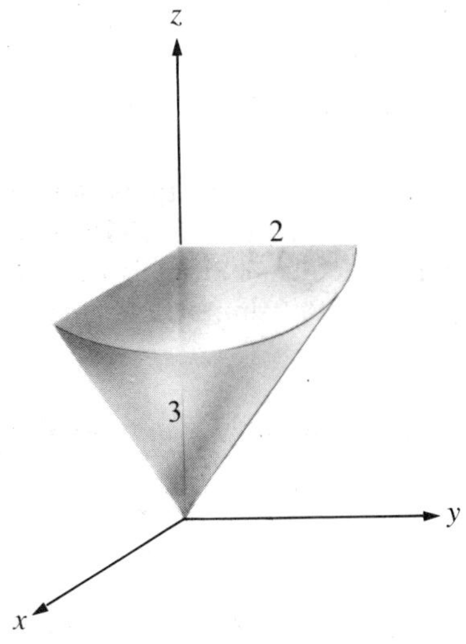

FIGURE 9.9

18 The first octant region enclosed by the sphere $x^2+y^2+z^2=4$, constant density.

In Exercises 19 to 22, find the moment of inertia and radius of gyration about the axis given.

19 The region of Example 9.11, x-axis.

20 The ball $x^2+y^2+z^2\leq a^2$, constant density, z-axis.

21 The ball $x^2+y^2+z^2\leq 9$, $d(x, y, z)=k\sqrt{x^2+y^2+z^2}$, z-axis.

22 The ball $x^2+y^2+z^2\leq a^2$, $d(x, y, z)=k\sqrt{x^2+y^2+z^2}$, z-axis.

23 The ball in Exercise 21 rotates about the z-axis with angular velocity $\omega=2$. Find the kinetic energy.

24 The ball in Exercise 21 rotates about the x-axis with angular velocity $\omega=2$. Find the kinetic energy.

25 A 12-inch long-playing record of constant density d and thickness h revolves at $33\frac{1}{3}$ revolutions per minute. Use Definition 9.10 to express its kinetic energy as a triple integral and evaluate it using cylindrical coordinates.

26 Suppose that a wire is parametrized by arc length (Definition 2.2.8): $\boldsymbol{x} = \boldsymbol{x}(s)$, $a \le s \le b$, and has density $d(s)$. Its mass is defined as $M = \int_a^b d(s)\, ds$, and its center of mass by

$$(\bar{x}, \bar{y}, \bar{z}) = \frac{1}{M}\left(\int_a^b x(s)\, d(s)\, ds, \int_a^b y(s)\, d(s)\, ds, \int_a^b z(s)\, d(s)\, ds\right).$$

Find the center of mass of a wire of constant density in the shape $x^2 + y^2 = a^2$, $y \ge 0$.

27 Find the center of mass of the first quadrant portion of the wire described in Exercise 26.

28 Another theorem of Pappus says that if a smooth curve $\boldsymbol{x} = \boldsymbol{x}(s)$ lying to the right of the y-axis is revolved about the y-axis, then the surface of revolution formed has surface area $A = 2\pi\bar{x}L$, where $\bar{x}$ is the x-coordinate of the center of mass and L is the length of the curve. Use Exercise 26 to prove this. (*Hint:* At a typical point $\boldsymbol{x}(s)$, a short portion of the curve of length ΔS_i is approximately linear. When revolved it traces an approximate frustum of a cone with area $\Delta A_i \approx 2\pi x\, \Delta S_i$.)

29 Apply Exercise 28 to calculate:
(a) The surface area of a sphere obtained by revolving the graph of $x = \sqrt{a^2 - y^2}$ in the xy-plane about the y-axis.
(b) The surface area of the torus (p. 302).

30 Apply Exercise 28 to calculate the surface area of the right circular cone $\phi = \pi/4$.

10 LEIBNIZ'S RULE

This section derives its name from a rule for differentiation of functions defined by integrals. This rule was first given by Gottfried W. Leibniz (1646–1716) who, with Newton, was one of the co-founders of modern calculus. Leibniz's rule can be thought of as the two-variable analogue of the following important rule for differentiating functions defined by indefinite integrals.

10.1 THEOREM. Let f be continuous on the interval $[a, b]$ and let $F:[a, b] \to \boldsymbol{R}$ be defined by $F(x) = \int_a^x f(t)\, dt$. Then F is differentiable and

$$F'(x) = f(x).$$

This important fact is crucial to establishing the Fundamental Theorem of Integral Calculus: For continuous f',

$$\int_a^b f'(x)\, dx = f(b) - f(a).$$

Since most of intermediate calculus seems to consist in establishing several-variable analogues for important parts of elementary calculus, it is natural to ask if this fundamental differentiation-of-integrals property has an analogue for functions $f: \boldsymbol{R}^2 \to \boldsymbol{R}$. It has, and we now set about establishing that analogue. We begin with a lemma, which was first mentioned in Section 2.

10.2 LEMMA. Suppose that $g: \boldsymbol{R}^2 \to \boldsymbol{R}$ is continuous on $R = [a, b] \times [c, d]$. For $x \in [a, b]$, define $G(x) = \int_c^d g(x, y)\, dy$. Then G is continuous on $[a, b]$.

Proof. Let $x_0 \in [a, b]$. We need to show that $\lim_{x \to x_0} [G(x) - G(x_0)] = 0$. We have

$$G(x) - G(x_0) = \int_c^d g(x, y)\, dy - \int_c^d g(x_0, y)\, dy$$
$$= \int_c^d [g(x, y) - g(x_0, y)]\, dy.$$

Since $g(\ , y)$ is continuous on $[a, b]$, we can make $g(x, y) - g(x_0, y)$ as small as we please by taking x close enough to x_0. Then $G(x) - G(x_0)$ will approach 0 as $x \to x_0$, as required. To be precise, given $\varepsilon > 0$, it can be shown[1] that there is a single $\delta > 0$ such that, for all $y \in [c, d]$,

$$|g(x, y) - g(x_0, y)| < \frac{\varepsilon}{d - c} \quad \text{when} \quad |x - x_0| < \delta.$$

So

$$-\frac{\varepsilon}{d - c} < g(x, y) - g(x_0, y) < \frac{\varepsilon}{d - c} \quad \text{if} \quad |x - x_0| < \delta.$$

Then

$$G(x) - G(x_0) = \int_c^d [g(x, y) - g(x_0, y)]\, dy$$
$$< \int_c^d \frac{\varepsilon}{d - c}\, dy = \frac{\varepsilon}{d - c}(d - c) = \varepsilon.$$

And similarly,

$$G(x) - G(x_0) = \int_c^d [g(x, y) - g(x_0, y)]\, dy$$
$$> \int_c^d -\frac{\varepsilon}{d - c}\, dy = -\frac{\varepsilon}{d - c}(d - c) = -\varepsilon.$$

So if $|x - x_0| < \delta$, then $|G(x) - G(x_0)| < \varepsilon$. Hence $\lim_{x \to x_0} [G(x) - G(x_0)] = 0$. QED

We are now ready to state the simplest case of Leibniz's rule. In this first version, we integrate over a rectangle. This first Leibniz Rule is frequently useful in mathematical physics.

10.3 **THEOREM (LEIBNIZ).** Let $f: \boldsymbol{R}^2 \to \boldsymbol{R}$ be continuous and have a continuous partial derivative f_x on a rectangle $R = [a, b] \times [c, d]$. For $x \in [a, b]$, define $F(x) = \int_c^d f(x, y)\, dy$. Then F is differentiable on $[a, b]$, and

$$F'(x) = \int_c^d \frac{\partial f(x, y)}{\partial x}\, dy.$$

Proof. We will use Theorem 10.1, Lemma 10.2, Fubini's Theorem 2.4, and the

1. See, for example, R. C. Buck, *Advanced Calculus*, 3rd Ed., McGraw-Hill, New York, 1978, p. 85.

Fundamental Theorem of Integral Calculus. First we remark that, by Lemma 10.2, the function $G:[a, b]\to R$ defined by

$$G(x)=\int_c^d \frac{\partial f(x, y)}{\partial x}\,dy$$

is continuous, since by hypothesis $\dfrac{\partial f(x, y)}{\partial x}=f_x$ is continuous on R. Thus Theorem 10.1 tells us that

$$\int_c^d \frac{\partial f(x, y)}{\partial x}\,dx = G(x)=\frac{d}{dx}\int_a^x G(t)\,dt$$

$$=\frac{d}{dx}\left(\int_a^x dt\int_c^d \frac{\partial f(t, y)}{\partial t}\,dy\right)$$

$$=\frac{d}{dx}\left(\int_c^d dy\int_a^x \frac{\partial f(t, y)}{\partial t}\,dt\right)$$ by Theorem 2.4 applied to $R_x=[a, x]\times[c, d]$ (see Figure 10.1)

$$=\frac{d}{dx}\left(\int_c^d dy[f(t, y)]_a^x\right)$$ by the Fundamental Theorem of Integral Calculus, y being frozen in the integration relative to t.

$$=\frac{d}{dx}\left(\int_c^d [f(x, y)-f(a, y)]\,dy\right)$$

$$=\frac{d}{dx}\left(\int_c^d f(x, y)-\int_c^d f(a, y)\,dy\right)$$

$$=\frac{d}{dx}(F(x)-F(a))=F'(x).$$ QED

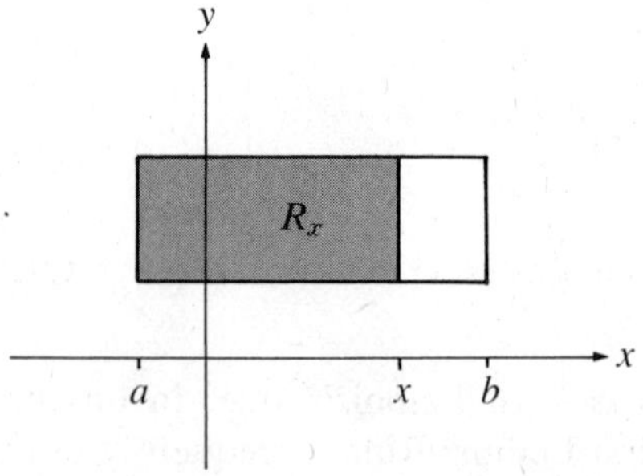

FIGURE 10.1

Before proceeding, we pause to give a simple illustration of Theorem 10.3.

10.4 EXAMPLE. Verify Theorem 10.3 in case $f(x, y)=x^2y+x^3$, for $c=3$ and $d=4$.

Solution. $F(x)=\int_3^4 f(x, y)\,dy=\int_3^4 (x^2y+x^3)\,dy$

$$=\frac{x^2y^2}{2}+x^3y\Big]_{y=3}^{y=4}$$

$$=8x^2+4x^3-\tfrac{9}{2}x^2-3x^3$$

$$=\tfrac{7}{2}x^2+x^3.$$

Thus, $F'(x)=7x+3x^2$. This should be the same as

$$\int_3^4 \frac{\partial f}{\partial x}\,dy=\int_3^4 (2xy+3x^2)\,dy=[xy^2+3x^2y]_{y=3}^{y=4}$$
$$=16x+12x^2-9x-9x^2=7x+3x^2.$$

So Theorem 10.3 works in this case.

There is nothing *special* about x. Hence, the following result is obtained from Theorem 10.3 by simply interchanging the roles of x and y.

10.5 **THEOREM (LEIBNIZ).** Let $f: \boldsymbol{R}^2 \to \boldsymbol{R}$ be continuous and have a continuous partial derivative f_y on $R=[a, b]\times[c, d]$. Define $F(y)=\int_a^b f(x, y)\,dx$ for each $y\in[c, d]$. Then F is differentiable on $[c, d]$ and

$$\frac{dF}{dy}=\int_a^b \frac{\partial f(x, y)}{\partial y}\,dx.$$

We now give an example of Theorem 10.5 that is a little less trivial than Example 10.4. In this next example it can be shown that there is no simpler formula for $F(y)$ than the defining integral, and yet Leibniz's Rule allows us to get a simple formula for $\dfrac{dF}{dy}$ *directly* in terms of y!

10.6 **EXAMPLE.** Find $\displaystyle\frac{d}{dy}\int_\pi^{2\pi} \frac{\cos xy}{x}\,dx.$

Solution.
$$\frac{d}{dy}\int_\pi^{2\pi} \frac{\cos xy}{x}\,dx=\int_\pi^{2\pi} \frac{\partial}{\partial y}\left(\frac{\cos xy}{x}\right)dx$$
$$=\int_\pi^{2\pi} \frac{-x \sin xy}{x}\,dx$$
$$=\int_\pi^{2\pi} -\sin xy\,dx$$
$$=\frac{\cos xy}{y}\Big]_{x=\pi}^{x=2\pi}$$
$$=\frac{\cos 2\pi y}{y}-\frac{\cos \pi y}{y}.$$

We now are ready to consider the more general analogue of 10.1 for regions that are not necessarily rectangular.

10.7 **LEIBNIZ'S RULE.** Suppose that f and f_x are continuous on $R=[a, b]\times[c, d]$. Suppose also that $h_1(x)$ and $h_2(x)$ have continuous derivatives on $[a, b]$, with $c\le h_1(x)\le d$ and $c\le h_2(x)\le d$ for all x in $[a, b]$. Define

$$F(x)=\int_{h_1(x)}^{h_2(x)} f(x, y)\,dy$$

for $x\in[a, b]$. Then F is differentiable on $[a, b]$ and

$$F'(x)=\int_{h_1(x)}^{h_2(x)} \frac{\partial f(x, y)}{\partial x}\,dy \quad + \quad f(x, h_2(x))h_2'(x) \quad - \quad f(x, h_1(x))h_1'(x).$$

Proof. Let $G(u, v, w)=\int_v^w f(u, y)\,dy$. Then we have $F(x)=G(x, h_1(x), h_2(x))$. We can therefore apply the Chain Rule (Theorem 4.5.3) to get

$$F'(x)=\nabla G\cdot\frac{d}{dx}(x, h_1(x), h_2(x))$$

$$=\frac{\partial G}{\partial u}+\frac{\partial G}{\partial v}\frac{dh_1}{dx}+\frac{\partial G}{\partial w}\frac{dh_2}{dx}$$

$$=\int_{h_1(x)}^{h_2(x)}\frac{\partial f(x, y)}{\partial x}\,dy+\frac{\partial}{\partial v}\left(\int_v^{h_2(x)} f(x, y)\,dy\right)h_1'(x)+\frac{\partial}{\partial w}\left(\int_{h_1(x)}^{w} f(x, y)\,dy\right)h_2'(x)$$

by Theorem 10.3

$$=\int_{h_1(x)}^{h_2(x)}\frac{\partial f}{\partial x}\,dy-\frac{\partial}{\partial v}\left(\int_{h_2(x)}^{v} f(x, y)\,dy\right)h_1'(x)+f(x, h_2(x))h_2'(x)$$

by Theorem 10.1

$$=\int_{h_1(x)}^{h_2(x)}\frac{\partial f(x, y)}{\partial x}\,dy-f(x, h_1(x))h_1'(x)+f(x, h_2(x))h_2'(x)$$

by Theorem 10.1 again. QED

As before, we have the dual result for differentiating with respect to y.

10.8 **LEIBNIZ'S RULE.** Suppose that f and f_y are continuous on $R=[a, b]\times[c, d]$. Suppose also that $g_1(y)$ and $g_2(y)$ have continuous derivatives on $[c, d]$, with $a\le g_1(y)\le b$ and $a\le g_2(y)\le b$ for all $y\in[c, d]$. Define

$$F(y)=\int_{g_1(y)}^{g_2(y)} f(x, y)\,dx.$$

Then F is differentiable on $[c, d]$ and

$$\frac{dF}{dy}=\int_{g_1(y)}^{g_2(y)}\frac{\partial f(x, y)}{\partial y}\,dx + f(g_2(y), y)g_2'(y) - f(g_1(y), y)g_1'(y).$$

10.9 **EXAMPLE.** If $F(x)=\int_{x^2}^{e^{x2}}\frac{1}{y}\cos(xy)\,dy$, then find $F'(x)$. (Assume that y is never 0.)

Solution. The integrand is continuous since $y\neq 0$, so 10.7 applies to give:

$$F'(x)=\int_{x^2}^{e^{x2}}\frac{\partial}{\partial x}\left(\frac{1}{y}\cos xy\right)dy+2xe^{x^2}\frac{1}{e^{x2}}\cos(xe^{x^2})-2x\frac{1}{x^2}\cos x^3$$

$$=\int_{x^2}^{e^{x2}}(-\sin xy)\,dy+2x\cos(xe^{x^2})-\frac{2}{x}\cos x^3$$

$$=\frac{1}{x}\cos xy\Big]_{y=x^2}^{y=e^{x2}}+2x\cos(xe^{x^2})-\frac{2}{x}\cos x^3$$

$$=\frac{1}{x}\cos xe^{x^2}-\frac{1}{x}\cos x^3+2x\cos xe^{x^2}-\frac{2}{x}\cos x^3$$

$$=\left(\frac{1}{x}+2x\right)\cos xe^{x^2}-\frac{3}{x}\cos x^3.$$

You have probably noticed that Theorems 10.7 and 10.1 look rather different. You therefore may be thinking that this is one instance in which the several-variable analogue of an elementary calculus fact is very much different from that elementary calculus fact. Before you become too convinced of this, however, consider the following.

10.10 EXAMPLE. Assuming the hypotheses of Theorem 10.7, find $F'(x)$ if $F(x) = \int_a^x f(x, y)\,dy$. What happens if $f(x, y) = g(y)$ is a function of y alone?

Solution. We have from 10.7, with $h_2(x) = x$ and $h_1(x) = a$,

$$F'(x) = \int_a^x \frac{\partial f}{\partial x}\,dy + f(x, x)1 - f(x, a)0$$

$$= \int_a^x \frac{\partial f}{\partial x}\,dy + f(x, x).$$

If $f(x, y) = g(y)$, then $\dfrac{\partial f}{\partial x} = 0$, so we get

$$F'(x) = \frac{d}{dx}\int_a^x g(y)\,dy = f(x, x) = g(x),$$

which is *precisely* 10.1.

So Leibniz's Rule is a *generalization* of 10.1, since it gives this elementary calculus result as a special case.

EXERCISES 5.10

In Exercises 1 to 4, verify Theorems 10.3 and 10.5 in the manner of Example 10.4.

1 $f(x, y) = xy + xy^2 + 3$; $R = [0, 1] \times [1, 2]$

2 $f(x, y) = xy + x^2y - 1$; $R = [0, 1] \times [1, 2]$

3 $f(x, y) = x + y - y^2$; $R = [-1, 1] \times [0, 1]$

4 $f(x, y) = x - y + x^2$; $R = [-1, 1] \times [0, 1]$

In Exercises 5 to 8, verify whichever of Theorems 10.3 or 10.5 applies.

5 $F(y) = \int_0^1 (x + y)^5\,dx$

6 $F(x) = \int_0^1 e^{-xy}\,dy$ (*Hint:* Use integration by parts or a table to evaluate the integral of the partial derivative.)

7 $F(y) = \int_0^1 \mathrm{Tan}^{-1}\dfrac{x}{y}\,dx \quad (y > 2/\pi)$

8 $F(x) = \int_0^1 \mathrm{Tan}^{-1}\dfrac{y}{x}\,dy \quad (x > 2/\pi)$

In Exercises 9 to 18, use Theorems 10.3, 10.5, 10.7 or 10.8 to differentiate the given function.

9 $F(y) = \int_\pi^{2\pi} \dfrac{\sin xy}{x}\,dx.$

10 $F(x) = \int_{\pi/2}^{\pi} \dfrac{\cos xy}{y}\,dy.$

11 $F(x)=\int_1^2 y^x\,dy$; assume $x>-1$.

12 $F(y)=\int_1^2 x^y\,dx$; assume $y>-1$.

13 $F(x)=\int_{x^2}^{1} \frac{1}{y}\sin xy\,dy$; assume $y\neq 0$

14 $F(y)=\int_{e^{2y}}^{y^2} \frac{\sin xy}{x}\,dx$; assume $x\neq 0$.

15 $F(y)=\int_{y^2}^{y} \frac{\sin xy}{x}\,dx$; assume $x\neq 0$.

16 $F(y)=\int_{y^2}^{y} \frac{\cos xy}{y}\,dy$; assume $y\neq 0$.

17 $F(x)=\int_{x}^{x^2} \frac{e^{xy}}{y}\,dy$; assume $y\neq 0$.

18 $F(y)=\int_{y}^{y^3} \frac{e^{-xy}}{x}\,dx$; assume $x\neq 0$.

19 Show that Theorem 10.7 is a generalization of Theorem 10.3 by deducing 10.3 from 10.7 in a suitable situation.

20 Show that Theorem 10.8 is a generalization of Theorem 10.5 by deducing 10.5 from 10.8 in a suitable situation.

21 Use Leibniz's Rule 10.5 to show that for $\varepsilon>0$,

$$\int_\varepsilon^1 x^a \ln x\,dx=-\frac{1}{(a+1)^2}[1+\varepsilon^{a+1}(a+1)\ln\varepsilon-\varepsilon^{y+1}].$$

22 Take limits as $\varepsilon\to 0+$ in the result of Exercise 21 to show that

$$\int_0^1 x^\alpha \ln x\,dx=-\frac{1}{(\alpha+1)^2}.$$

[Recall that $\int_0^1 x^\alpha \ln x\,dx$ is an *improper* integral (cf. Section 11) whose value is defined to be $\lim_{\varepsilon\to 0+}\int_\varepsilon^1 x^\alpha \ln x\,dx$ if this limit exists.]

23 Use the fact that

$$\int_0^a \frac{dx}{x^2+a^2}=\frac{\pi}{4a}$$

and Theorem 10.8 to show that

$$\int_0^a \frac{dx}{(x^2+a^2)^2}=\frac{1}{4a^3}+\frac{\pi}{8a^3}.$$

24 Use the result of Exercise 23 and Theorem 10.8 to find

$$\int_0^a \frac{dx}{(x^2+a^2)^3}.$$

Versions of Leibniz's Rule exist for functions of more than one variable defined by integrals. The following two exercises give two examples. Assume that f satisfies the hypotheses of Theorems 10.3 and 10.5.

25 If $F(x,y)=\int_a^x f(t,y)\,dt$, then show that $F_x=f(x,y)$ and $F_y=\int_a^x f_y(t,y)\,dt$.

26 If $F(x,y,z)=\int_y^z f(x,t)\,dt$, then show that $F_x=\int_y^z f_x(x,t)\,dt$, $F_y=-f(x,y)$, and $F_z=f(x,z)$.

11 IMPROPER INTEGRALS

A very important concept in statistics, physics, chemistry, engineering, and mathematical probability is the notion of improper integral. You probably studied this concept in elementary calculus. If you did, then the first part of this section will be review and you can use it mainly for reference. If you *didn't* study improper integrals in elementary calculus, then you will need to study the first part of the section very carefully before proceeding to improper multiple integrals.

Recall that if f is continuous on $[a, b]$ and $f(x) \geq 0$ on $[a, b]$, then $\int_a^b f(x)\,dx$ gives the area of the region S under the graph of f between $x = a$ and $x = b$, as shown in Figure 11.1. Here $S \subseteq \boldsymbol{R}^2$ is a bounded region. A reasonable question is

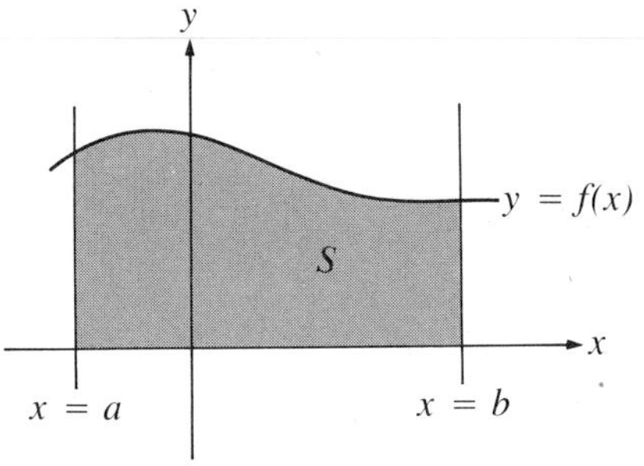

FIGURE 11.1

whether we can ever assign a finite area to an *unbounded* region $S \subseteq \boldsymbol{R}^2$. In particular, if S is the region lying under the graph of a continuous function f for $x \geq a$ (or $x \leq a$), can this area ever be finite? (See Figure 11.2). There is a natural sense in which to answer this question. Namely, we will say that the area is finite if $\int_a^b f(x)\,dx$ (respectively, $\int_c^a f(x)\,dx$) approaches some limit A as $b \to +\infty$ (respectively, $c \to -\infty$).

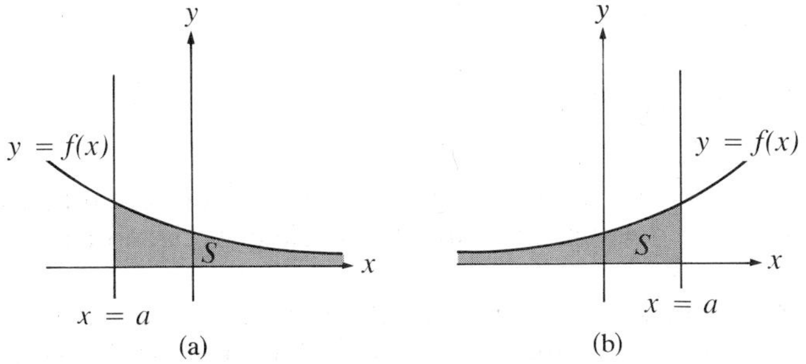

FIGURE 11.2

11.1 DEFINITION. If f is continuous on $\boldsymbol{R}$, then

$$\int_a^{+\infty} f(x)\,dx = \lim_{b \to +\infty} \int_a^b f(x)\,dx \quad \text{and} \quad \int_{-\infty}^{a} f(x)\,dx = \lim_{c \to -\infty} \int_c^a f(x)\,dx$$

if these limits exist as real numbers. Also,

$$\int_{-\infty}^{+\infty} f(x)\,dx = \int_{-\infty}^{0} f(x)\,dx + \int_0^{+\infty} f(x)\,dx$$

provided *both* of the improper integrals in the sum exist.

11.2 EXAMPLE. Consider $y = \dfrac{1}{x}$ for $x \geq 1$. (a) Is the area of the region S lying under the graph of this function finite? (b) If S is revolved about the x-axis to generate a solid of revolution T, does T have finite volume? See Figure 11.3.

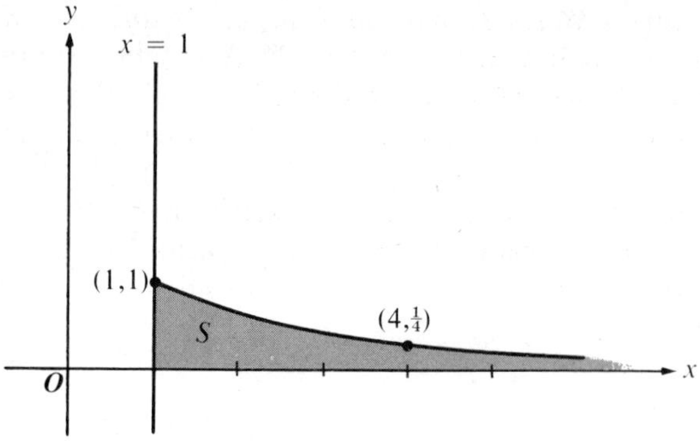

FIGURE 11.3

Solution. (a) The area of S is given by the value of the improper integral

$$\int_1^{+\infty} \frac{1}{x}\,dx = \lim_{b\to+\infty} \int_1^b x^{-1}\,dx = \lim_{b\to+\infty} \ln x \Big]_1^b$$

$$= \lim_{b\to+\infty} \ln b = +\infty.$$

So the area is *not finite.*

(b) The volume of T is given by the improper integral

$$\int_1^{+\infty} \pi\left(\frac{1}{x}\right)^2 dx = \pi \lim_{b\to+\infty} \int_1^b x^{-2}\,dx$$

$$= \pi \lim_{b\to+\infty} \left[-\frac{1}{x}\right]_1^b$$

$$= \pi \lim_{b\to+\infty} \left(-\frac{1}{b}+1\right) = \pi.$$

So the volume of T *is finite,* π cubic units. This may strike you as odd, but it helps to bring out the distinction between planar areas and three-dimensional volumes.

In addition to improper integrals defined over infinite intervals, it is sometimes necessary to consider integrals of functions $f:\boldsymbol{R}\to\boldsymbol{R}$ over finite intervals $[a, b]$ which contain a point d where f has an infinite discontinuity.

11.3 DEFINITION. If $d\in[a, b]$, $f:\boldsymbol{R}\to\boldsymbol{R}$ is continuous at all points of $[a, b]$ except at d, and

$$\lim_{x\to d+} f(x) = \pm\infty \quad \text{or} \quad \lim_{x\to d-} f(x) = \pm\infty,$$

then we define the following **improper integrals:**

$$\int_a^d f(x)\,dx = \lim_{r\to d-} \int_a^r f(x)\,dx \quad \text{if this limit exists;}$$

$$\int_d^b f(x)\,dx = \lim_{r\to d+} \int_r^b f(x)\,dx \quad \text{if this limit exists;}$$

$$\int_a^b f(x)\,dx = \int_a^d f(x)\,dx + \int_d^b f(x)\,dx \quad \text{if } \textit{both} \text{ improper integrals exist.}$$

A classic test of whether an elementary calculus student looks before he or she leaps is given by posing $\int_{-1}^{1} x^{-2}\,dx$ on a final examination. If the student doesn't notice that 0 is an infinite discontinuity for $f(x)=x^{-2}=1/x^2$, then he or she might try to evaluate this integral as $-\frac{1}{x}\Big]_{-1}^{1}=-1/1+1/-1=-2$. This answer is absurd, of course, since $f(x)>0$ on $[-1, 1]$, so $\int_{-1}^{1} f(x)\,dx$ *can't* be negative!

11.4 **EXAMPLE.** Does $\int_{-1}^{1} x^{-2}\,dx$ exist?

Solution. According to Definition 11.3, we have to investigate $\int_{-1}^{0} x^{-2}\,dx$ and $\int_{0}^{1} x^{-2}\,dx$. Using Definition 11.3, we get

$$\int_{-1}^{0} x^{-2}\,dx = \lim_{r\to 0-}\int_{-1}^{r} x^{-2}\,dx = \lim_{r\to 0-}\left[-\frac{1}{x}\right]_{-1}^{r}$$
$$= \lim_{r\to 0-}\left[-\frac{1}{r}+\frac{1}{-1}\right]=+\infty.$$

Since *both* $\int_{-1}^{0} x^{-2}\,dx$ and $\int_{0}^{1} x^{-2}\,dx$ have to exist in order for $\int_{-1}^{1} x^{-2}\,dx$ to exist, we conclude that this improper integral *doesn't* exist.

An improper integral of either type 11.1 or 11.3 that exists is called **convergent.** An improper integral of either type that fails to exist is called **divergent.**

Now we want to consider improper double and triple integrals. First suppose that $f:\mathbf{R}^n\to\mathbf{R}$ $(n=2$ or $3)$ is bounded and $D\subseteq\mathbf{R}^n$ is an unbounded region. We want to define $\iint_D f(\mathbf{x})\,dA$ or $\iiint_D f(\mathbf{x})\,dV$ as the case may be $(n=2$ or $n=3)$. How can we do this? If we look back at Definition 11.1, we see that in the single-variable case we took the limit of the values of the ordinary definite integral over larger and larger intervals $[a, b]$. So it seems natural to do the same sort of thing here.

11.5 **DEFINITION.** Let $f:\mathbf{R}^n\to\mathbf{R}$ be continuous $(n=2$ or $3)$, and let $D\subseteq\mathbf{R}^n$ be an unbounded set. The **improper multiple integral** of f over D is defined as the limit as $k\to\infty$ of the ordinary multiple integral of f over D_k, where $\{D_k\}$ is a family of bounded subsets of D, each of nonzero area (or volume), such that $D_k\subseteq D_{k+1}$, $\bigcup_{k=1}^{\infty} D_k=D$, and for any bounded subset B of D we have $B\subseteq D_k$ for some k. Thus

$$\iint_D f(\mathbf{x})\,dA=\lim_{k\to\infty}\iint_{D_k} f(\mathbf{x})\,dA$$

and

$$\iiint_D f(\mathbf{x})\,dV=\lim_{k\to\infty}\iiint_{D_k} f(\mathbf{x})\,dV$$

if these limits exist.

In Figure 11.4 we try to suggest how a typical family $\{D_k\}$ in $\mathbf{R}^2$ might look.

At first glance this definition may appear wholly satisfactory, but before we start using it, we have to prove that it is *independent of how we choose the* D_k. Otherwise, if for some choice of D_k the limits in Definition 11.5 exist, and for

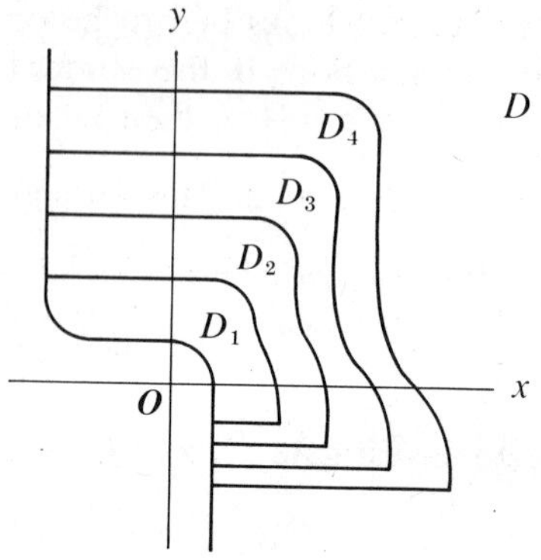

FIGURE 11.4

some different choice of D_k the limits don't exist, then we have not given a definition at all, but have loosed chaos upon ourselves. We consider the case in which f is always nonnegative or always nonpositive on D. In general we can break D up into regions where f is either always nonnegative or always nonpositive. We leave the full discussion for advanced calculus.

11.6 THEOREM. Suppose that $f:\mathbf{R}^n \to \mathbf{R}$ ($n=2$ or 3) is nonnegative (or nonpositive) on the unbounded region D. If for some particular choice of D_k in Definition 11.5 the limit $\lim_{k\to\infty} \iint_{D_k} f(\mathbf{x})\, dA$ (respectively, $\lim_{k\to\infty} \iiint_{D_k} f(\mathbf{x})\, dV$) exists, then for *any* choice of D_k in 11.5, this limit exists and has the same value L.

Proof. Suppose that $f(\mathbf{x})\geq 0$ on $D\subseteq \mathbf{R}^2$. Let $\{C_k\}$ be any other family of bounded subsets of D chosen as in Definition 11.5. Then for any ℓ, D_ℓ is a bounded subset of D. Hence there is some m such that $D_\ell \subseteq C_m$. Since C_m is a bounded subset of D, there is also some index p such that $C_m \subseteq D_p$. We then have $D_\ell \subseteq C_m \subseteq D_p$ and therefore

(1) $$\iint_{D_\ell} f(\mathbf{x})\, dA \leq \iint_{C_m} f(\mathbf{x})\, dA \leq \iint_{D_p} f(\mathbf{x})\, dA$$

by Exercise 14, Exercises 5.13. Since

$$\iint_{D_p} f(\mathbf{x})\, dA \leq L = \lim_{k\to\infty} \iint_{D_k} f(\mathbf{x})\, dA,$$

we see that the increasing sequence of real numbers $\iint_{C_m} f(\mathbf{x})\, dA$ is bounded above by L, so $\lim_{m\to\infty} \iint_{C_m} f(\mathbf{x})\, dA$ exists $\leq L$ by Theorem 11.1.14. But

$$L = \lim_{\ell\to\infty} \iint_{D_\ell} f(\mathbf{x})\, dA \leq \lim_{m\to\infty} \iint_{C_m} f(\mathbf{x})\, dA \leq L$$

from (1). Thus $\lim_{m\to\infty} \iint_{C_m} f(\mathbf{x})\, dA = L$. The same argument applies if $f(\mathbf{x})\leq 0$ or $D\subseteq \mathbf{R}^3$.

QED

We can now feel free to choose whatever sequence of D_k in Definition 11.5 is best suited to a particular example.

11.7 EXAMPLE. Determine whether

$$\iint_D \frac{1}{x^2(1+y^2)}\,dA$$

is convergent or divergent if $D=[1, +\infty)\times[0, +\infty)$.

Solution. Here we have

$$\iint_D \frac{1}{x^2(1+y^2)}\,dA = \lim_{n\to+\infty} \iint_{D_n} \frac{1}{x^2(1+y^2)}\,dA,$$

where $D_n=[1, n]\times[0, n]$. See Figure 11.5. Over D_n we can iterate, and this gives

$$\begin{aligned}\iint_D \frac{1}{x^2(1+y^2)}\,dA &= \lim_{n\to+\infty} \int_1^n \frac{1}{x^2}\,dx \int_0^n \frac{1}{1+y^2}\,dy \\ &= \lim_{n\to+\infty} \left[-\frac{1}{x}\right]_1^n [\mathrm{Tan}^{-1}\, y]_0^n \\ &= \lim_{n\to+\infty} \left[-\frac{1}{n}+1\right][\mathrm{Tan}^{-1}\, n]=1\cdot\tfrac{1}{2}\pi=\tfrac{1}{2}\pi.\end{aligned}$$

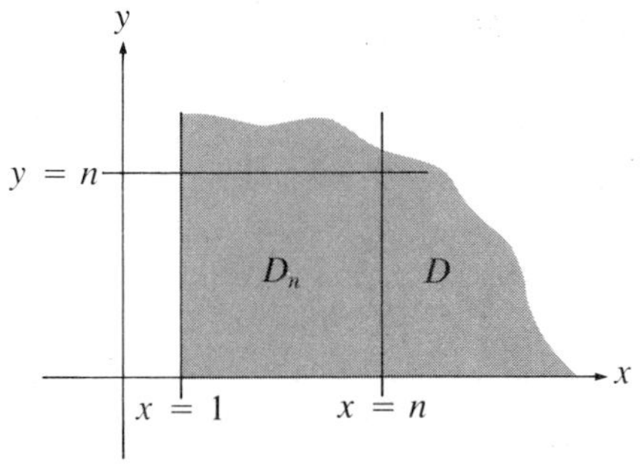

FIGURE 11.5

Improper integrals play a very significant role in probability and statistics and many scientific applications. We will now briefly discuss one important area where they occur—*quantum theory*. This turns out to be in part probabilistic, so we begin by mentioning some basic ideas of probability.

Loosely speaking, a **probability density function** on a subset $S\subseteq \mathbf{R}^3$ is a continuous nonnegative function $p:\mathbf{R}^3\to\mathbf{R}$ such that $\iiint_S p(\mathbf{x})\,dV=1$. If an experiment is performed whose set of outcomes is in S, then the **probability** that the outcome lies in a subset $T\subseteq S$ is

$$\Pr[\text{Outcome in } T]=\iiint_T p(\mathbf{x})\,dV.$$

In quantum theory, electrons (like light) are assumed to have the dual nature of waves and particles. The basic mathematical object in quantum theory is the *wave function* $\Psi(\mathbf{x}(t))$. (If you are familiar with waves, then you can think of this as the amplitude of the electron-wave at time t; but you don't have to give it an independent meaning to follow the ensuing discussion.) The theory[1] interprets $\Psi^2(\boldsymbol{x}(t))$ as the probability density function for finding the electron at a point $\boldsymbol{x}=(x, y, z)\in \boldsymbol{R}^3$ at time t. Since the electron-particle E must be *somewhere* in $\boldsymbol{R}^3$ at any time t,

$$\Pr[E \text{ in } \boldsymbol{R}^3]=1=\iiint_{\boldsymbol{R}^3} \Psi^2(\boldsymbol{x})\, dV.$$

The *atomic orbital wave function* for the hydrogen atom is given by

(2) $$\Psi(x, y, z)=\frac{c^{3/2}}{\sqrt{\pi}}\, e^{-c\sqrt{x^2+y^2+z^2}},$$

where $(x, y, z)=(x(t), y(t), z(t))$ and $c>0$ is a constant ($c=Z/a_0$) where Z is the nuclear charge and a_0 is the Bohr radius, if you are familiar with nuclear physics).

11.8 EXAMPLE. Show that Ψ^2, where Ψ is given by (2), is a probability density function on $\boldsymbol{R}^3$.

Solution. Since $\Psi^2\geq 0$ on $\boldsymbol{R}^3$, we need only verify that $\iiint_{\boldsymbol{R}^3} \Psi^2(\boldsymbol{x})\, dV=1$. We have, using spherical coordinates, and for D_b a ball of radius b,

$$\begin{aligned}
\iiint_{\boldsymbol{R}^3} \Psi^2(\boldsymbol{x})\, dV &= \frac{c^3}{\pi}\iiint_{\boldsymbol{R}^3} e^{-2c\rho}\rho^2 \sin\phi\, d\rho\, d\phi\, d\theta \\
&= \frac{c^3}{\pi}\int_0^{2\pi} d\theta \int_0^{\pi} \sin\phi\, d\phi \int_0^{+\infty} \rho^2 e^{-2c\rho}\, d\rho \\
&= \frac{c^3}{\pi}\cdot 2\pi[-\cos\phi]_0^{\pi} \int_0^{+\infty} \rho^2 e^{-2c\rho}\, d\rho \\
&= 2c^3[2]\int_0^{+\infty} \rho^2 e^{-2c\rho}\, d\rho \\
&= 4c^3 \int_0^{+\infty} \rho^2 e^{-2c\rho}\, d\rho \\
&= 4c^3 \lim_{b\to+\infty} \int_0^{b} \rho^2 e^{-2c\rho}\, d\rho
\end{aligned}$$

We can evaluate this last integral by parts or use Formulas #18 (with $a=-2c$ and $n=2$) and #17 in the Table of Integrals. In either case,

$$\begin{aligned}
\int_0^b \rho^2 e^{-2c\rho}\, d\rho &= -\frac{1}{2c}\rho^2 e^{-2c\rho} + \frac{2}{2c}\frac{e^{-2c\rho}}{4c^2}(-2c\rho-1)\Big]_0^b \\
&= -\frac{1}{2c}\frac{b^2}{e^{2cb}} + \frac{1}{4c^3}\frac{1}{e^{2cb}}(-2cb-1)+\frac{1}{4c^3}.
\end{aligned}$$

1. Due in large part to Werner Heisenberg (1901–1976), the 1932 Nobel laureate in physics for his "uncertainty principle."

Thus,

$$\iiint_{\mathbf{R}^3} \Psi^2(\mathbf{x})\, dV = 4c^3 \lim_{b\to+\infty}\left[-\frac{1}{2c}\frac{b^2}{e^{2cb}} + \frac{1}{4c^3}\left(\frac{-2cb}{e^{2cb}} - \frac{1}{e^{2cb}} + 1\right)\right]$$
$$= 4c^3\left[0 + \frac{1}{4c^3}(0-0+1)\right]$$
$$= 1,$$

as desired. Here we used the fact that

$$\lim_{b\to+\infty}\frac{b^2}{e^{2cb}} = \lim_{b\to+\infty}\frac{b}{ce^{2cb}} = 0,$$

as follows easily from using L'Hôpital's Rule on the functions $\frac{x^2}{e^{2cx}}$ and $\frac{x}{e^{2cx}}$ (see Example 11.1.6.

We now turn to improper integrals over a bounded region D which contains an infinite discontinuity point $\mathbf{x}_0$ for the function $f:\mathbf{R}^n \to \mathbf{R}$ ($n=2$ or 3). We proceed as we did in Definition 11.5. We consider a family of subregions $D_k \subseteq D$ such that $D_k \subseteq D_{k+1}$, and for any subset $E \subseteq D$ on which f is bounded, $E \subseteq D_m$ for some m. For example, if $\mathbf{x}_0 = \mathbf{0}$ and $D = B(\mathbf{0}, 1)$ (the ball about $\mathbf{0}$ of radius 1), then for D_k we could take $\left\{\mathbf{x} \,\middle|\, \frac{1}{k} \le |\mathbf{x}| \le 1\right\}$. Then $\bigcup_{k=1}^{\infty} D_k = D = B(\mathbf{0}, 1)$, and *any* subset E of D which omits $\mathbf{0}$ is eventually contained in some D_m. See Figure 11.6.

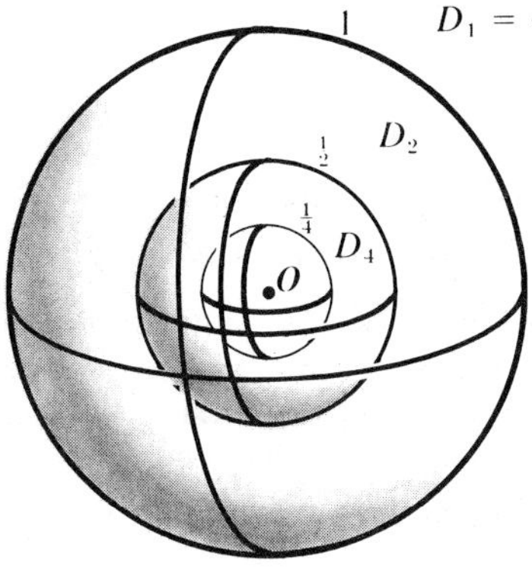

FIGURE 11.6

11.9 **DEFINITION.** Let $D \subseteq \mathbf{R}^n$ ($n=2$ or 3) and let $f:\mathbf{R}^n \to \mathbf{R}$ be continuous on D except at $\mathbf{x}_0$, where $\lim_{\mathbf{x}\to\mathbf{x}_0} f(\mathbf{x}) = \infty$. The **improper integral** of f over D is

$$\iint_D f(\mathbf{x})\, dA = \lim_{k\to\infty}\iint_{D_k} f(\mathbf{x})\, dA$$

if this limit exists, or

$$\iiint_D f(\mathbf{x})\, dV = \lim_{k\to\infty}\iiint_{D_k} f(\mathbf{x})\, dV$$

if this limit exists. Here $\{D_k\}$ is a family of increasing subsets of D such that any $E \subseteq D$ on which f is bounded is contained in some D_m, and $\bigcup_{k=1}^{\infty} D_k = D$.

We could again prove a theorem like 11.6, but instead ask you to believe that we have given a legitimate definition so that we can concentrate on showing you how to apply Definition 11.9 in practice.

11.10 EXAMPLE. Determine whether the triple integral of the Newtonian potential function

$$f(x, y, z) = \frac{1}{\sqrt{x^2+y^2+z^2}}$$

is convergent over the unit ball $x^2+y^2+z^2 \leq 1$.

Solution. This is clearly a job for spherical coordinates. We have

$$\iiint_B f(\mathbf{x})\, dV = \lim_{n\to\infty} \iiint_{B_n} f(\mathbf{x})\, dV, \quad \text{where} \quad B_n = \left\{\mathbf{x} \;\middle|\; \frac{1}{n} \leq |\mathbf{x}| \leq 1\right\}$$

$$= \lim_{n\to\infty} \iiint_{B_n} \frac{1}{\rho} \rho^2 \sin\phi \, d\rho\, d\phi\, d\theta$$

$$= \lim_{n\to\infty} \int_0^{2\pi} d\theta \int_0^{\pi} \sin\phi\, d\phi \int_{1/n}^{1} \rho\, d\rho$$

$$= \lim_{n\to\infty} (2\pi)(2)\left[\tfrac{1}{2}\rho^2\right]_{1/n}^{1}$$

$$= \lim_{n\to\infty} 2\pi\left[1 - \left(\frac{1}{n}\right)^2\right] = 2\pi.$$

So the integral does converge, to 2π. Notice that this integral is actually transformed to a *proper* triple integral $\iiint_B \rho \sin\phi\, d\rho\, d\phi\, d\theta$ if we change to spherical coordinates and allow ourselves to cancel $1/\rho$ with ρ^2. This cancellation is permissible on every B_n, and it can be shown that it is then permissible on B. We won't take the trouble to do so here, since this situation is not of that general a degree of occurrence. If you do meet an improper integral of this type, however, you won't go wrong if you carry out the cancellation and then evaluate it as a proper integral.

Our final example shows that occasionally we can evaluate improper multiple integrals without Definition 11.9.

11.11 EXAMPLE. Determine whether

$$\iint_D \frac{1}{\sqrt{xy}}\, dA$$

converges, where $D = [0, 1] \times [0, 1]$.

Solution. Definition 11.9 does *not* apply to

$$f(x, y) = \frac{1}{\sqrt{xy}}$$

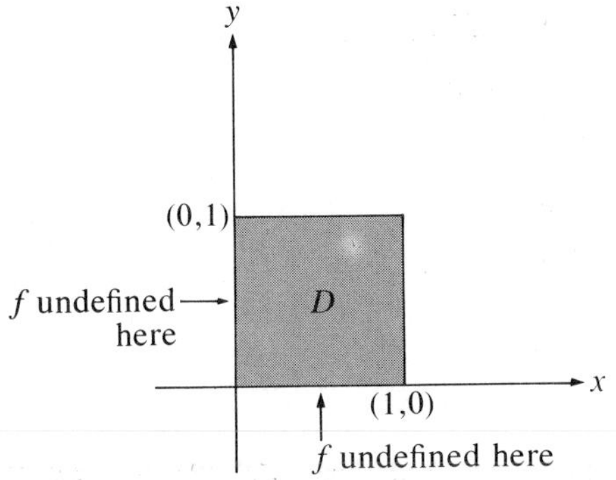

FIGURE 11.7

since this function fails to be continuous on both segments of the coordinate axes between 0 and 1. (See Figure 11.7.) Observe, however, that

$$\iint_D \frac{1}{\sqrt{xy}}\,dA = \iint_D \frac{1}{\sqrt{x}}\frac{1}{\sqrt{y}}\,dx\,dy$$

$$= \int_0^1 \frac{1}{\sqrt{x}}\,dx \int_0^1 \frac{1}{\sqrt{y}}\,dy \quad \text{by Theorem 1.6,}$$

$$= \left(\int_0^1 \frac{1}{\sqrt{x}}\,dx\right)^2.$$

So we need only investigate

$$\int_0^1 \frac{1}{\sqrt{x}}\,dx = \int_0^1 x^{-1/2}\,dx$$

$$= \lim_{a\to 0+} \int_a^1 x^{-1/2}\,dx$$

$$= \lim_{a\to 0+} 2x^{1/2}\Big]_a^1$$

$$= \lim_{a\to 0+} 2[1-a^{1/2}] = 2.$$

Hence

$$\iint_D \frac{1}{\sqrt{xy}}\,dA$$

converges to $2^2 = 4$.

EXERCISES 5.11

In Exercises 1 to 10, determine whether the given improper definite integral is convergent or divergent. If convergent, then find its value.

1 $\displaystyle\int_0^2 \frac{dx}{x+4x^{-1}}$

2 $\displaystyle\int_0^2 \frac{dx}{\sqrt{4-x^2}}$

3 $\int_0^3 \frac{dx}{(x-1)^{2/3}}$

4 $\int_0^4 \frac{dx}{(x-2)^2}$

5 $\int_0^2 \frac{3\,dx}{x-1}$

6 $\int_1^{+\infty} e^{-x}\,dx$

7 $\int_0^{+\infty} xe^{-x}\,dx$

8 $\int_{-\infty}^{+\infty} \frac{dx}{x^2+1}$

9 $\int_1^{+\infty} \frac{2}{x+1}\,dx$

10 $\int_2^{+\infty} \frac{dx}{x(\ln x)}$

In Exercises 11 to 22, determine whether the given improper multiple integral is convergent or divergent. If convergent, find its value.

11 $\iint_D \frac{1}{x^2y^2}\,dA,\qquad D=[1,+\infty)\times[1,+\infty).$

12 $\iint_D \frac{1}{(x-1)^2y^2}\,dA,\qquad D=[2,+\infty)\times[2,+\infty).$

13 $\iint_D \frac{1}{\sqrt{x^2+y^2}}\,dA,\qquad D=\{(x,y)\mid x^2+y^2\le 1\}.$

14 $\iint_D \frac{1}{x^2+y^2}\,dA,$ $\quad D$ as in Exercise 13.

15 $\iint_D \frac{1}{xy^2}\,dA,$ $\quad D$ as in Exercise 11.

16 $\int_{\mathbf{R}} e^{-x^2}\,dx$ (*Hint:* Use the trick of Example 11.11 in reverse.)

17 $\iiint_{\mathbf{R}^3} \frac{1}{\sqrt{x^2+y^2+z^2}}\,dV$

18 $\iiint_{\mathbf{R}^3} \frac{1}{x^2+y^2+z^2}\,dV$

19 $\iiint_{\mathbf{R}^3} \frac{1}{x^2yz^2}\,dV$

20 $\iiint_D \frac{1}{x^2y^2z^2}\,dV,$ where $D=\{(x,y,z)\mid x\ge 1, y\ge 1, z\ge 1\}$

21 $\iiint_B \frac{z}{(x^2+y^2+z^2)^{3/2}}\,dV,$ B the unit ball.

22 $\iiint_B \frac{z}{\sqrt{x^2+y^2+z^2}}\,dV,$ B the unit ball.

23 The **error function** erf is defined by

$$\operatorname{erf}(x)=\frac{2}{\sqrt{\pi}}\int_0^x e^{-t^2}\,dt,\qquad x\ge 0.$$

This function is of great importance in statistics and applied mathematics. Show that $\lim_{x\to+\infty}\operatorname{erf}(x)=1.$

24 The **normal probability density** in $\boldsymbol{R}^1$ is

$$n(x)=\frac{1}{\sqrt{2\pi}}e^{-x^2/2} \quad \text{for} \quad x\geq 0.$$

Show that this is a probability density on $[0, +\infty)$.

25 The **symmetric probability density** in $\boldsymbol{R}^2$ is

$$n(x, y)=\frac{1}{2\pi\sigma^2}e^{-(x^2+y^2)/2\sigma^2}.$$

Show that this is a probability density on $\boldsymbol{R}^2$. (Here σ is a constant.)

26 Show that

$$n(\boldsymbol{x})=\frac{2}{\pi\sigma^2}e^{-(x^2+y^2)/2\sigma^2}$$

defines a probability density on the first octant. (Here σ is a constant.)

In Exercises 27 to 30, refer to Example 11.8.

The hydrogen *molecule* consists of *two* hydrogen atoms, to which we may assign wave functions Ψ_a and Ψ_b, where a and b stand for the two nuclei. The following two problems are basic to the theory of the chemical bond worked out in 1927 by W. Heitler and F. London. (See, for example, J. Murrell, S. Kettle, and J. Tedder, *Valence Theory*, John Wiley, New York, 1965, Chapter 10.)

27 (a) Show that

$$\int_0^\infty x^n e^{-ax}\,dx=\frac{n!}{a^{n+1}}$$

using mathematical induction on n and Formulas 17 and 18 in the Table of Integrals.

(b) In quantum theory (see Example 11.8), atomic orbital wave functions are solutions to the **Schrödinger wave equation** [named for the Austrian physicist Erwin Schrödinger (1887–1961)] that are classified, in part, by principal quantum numbers n. In Example 11.8 we considered the case $n=1$. For $n=2$, the wave function is

$$\Psi_2(x, y, z)=\frac{1}{4\sqrt{2}}\frac{c^{3/2}}{\sqrt{\pi}}(1-c\rho)e^{-c\rho/2}.$$

Show that Ψ_2^2 is a probability density function on $\boldsymbol{R}^3$. (Use (a).)

28 Establish the *orthogonality condition*

$$\iiint_{\boldsymbol{R}^3}\Psi_1(\boldsymbol{x})\Psi_2(\boldsymbol{x})\,dV=0,$$

where $\Psi_1(\boldsymbol{x})$ is the $\Psi(\boldsymbol{x})$ in Example 11.8 and $\Psi_2(\boldsymbol{x})$ is as in Exercise 27.

29 If we label the electrons in the hydrogen molecule as 1 and 2, then one possible natural wave function for the molecule is the normalized product $\Psi_{a1}\Psi_{b2}=k\left(\frac{c^3}{\pi}e^{-2c\rho}\right)$ of the corresponding atomic orbital wave functions from Example 11.8. Show that k should be chosen as $\frac{2\sqrt{2\pi}}{c^{3/2}}$ in order for the square of $\Psi_{a1}\Psi_{b2}$ (or, symmetrically, $\Psi_{a2}\Psi_{b1}$) to be a probability density function.

30 Quantum theory views $\Psi_{a1}\Psi_{b2}$ (and $\Psi_{a2}\Psi_{b1}$) as a model too rigid for hydrogen atoms bound in a molecule. The more refined model allows either nucleus a or nucleus b to randomly capture either electron 1 or electron 2. The wave function proposed to quantify this refined model is

$$\Psi_+ = \tfrac{1}{2}(\Psi_{a1}\Psi_{b2} + \Psi_{a2}\Psi_{b1}).$$

Show that Ψ_+^2 is a probability density function on $\boldsymbol{R}^3$. (Use Exercises 27 and 28.)

REVIEW EXERCISES 5.12

In Exercises 1 to 11, evaluate the given double or triple integral.

1 $\iint_R e^x \sin y \, dA$, $R = [0, 2]\times[\pi/3, \pi/2]$.

2 $\iint_R e^x \sin \frac{y}{x} \, dA$, $R = [1, 2]\times[-\pi/2, \pi/2]$.

3 $\iint_D (x + y^2) \, dA$, D the region bounded by the parabola $y = x^2$ and the lines $x = -1$, $x = 1$, $y = 1$.

4 $\iint_D (2x - 3y + xy) \, dA$, D the region bounded by the y-axis, the parabola $x = 4 - y^2$, and the lines $y = 0$, $y = 2$.

5 $\iint_D (x^2 + y^2)^{3/2} \, dA$, D the region in the first quadrant bounded by the y-axis, the line $y = \frac{1}{\sqrt{3}} x$, and the circle $x^2 + y^2 = 4$.

6 $\iint_D \frac{1}{\sqrt{x^2 + y^2}} \, dA$, where D is the top half of the disk $(x-1)^2 + y^2 = 1$.

7 $\iiint_E xy \, dV$, where E is the first octant region bounded by the cylinders $y = x^2$ and $z = 1 - y^2$.

8 $\iiint_E (x + 2y) \, dV$, where E is enclosed by the tetrahedron formed by the three coordinate planes and the plane $6x + 3y + 2z = 6$.

9 $\iiint_U xyz \, dV$, where U is the first octant region bounded by the xy-plane, the plane $z = 3$, and the cylinder $x^2 + y^2 = a^2$.

10 $\iiint_U z^2 \, dV$, where U is the region bounded by the cones $z = \sqrt{x^2 + y^2}$ and $z = -\sqrt{x^2 + y^2}$, and the cylinder $x^2 + y^2 = 4$.

11 $\iiint_B \sqrt{x^2 + y^2 + z^2} \, dV$, where B is the ball of radius 2 centered at the origin.

In Exercises 12 to 18, find the volume of the given region in $\boldsymbol{R}^3$.

12 Under the paraboloid $z = x^2 + y^2$ and above the rectangle $[0, 1]\times[0, 2]$.

13 In the first octant bounded by the cylinder $y^2 + z^2 = 16$ and the planes $y = x$ and $x = 0$.

14 Under the cone $z = \sqrt{x^2 + y^2}$ and above the region between the circles $x^2 + y^2 = 4$ and $x^2 + y^2 = 16$ in the xy-plane.

15 Between the paraboloids $z = 2 - 3x^2 - 3y^2$ and $z = -2 + x^2 + y^2$.

16 Inside the sphere $x^2 + y^2 + z^2 = 16$ and outside the cone $z = 4 - \sqrt{x^2 + y^2}$.

17 Above the xy-plane, below the sphere $x^2 + y^2 + z^2 = 9$, and inside the cylinder $x^2 + y^2 = 4$.

18 The solid formed when the triangle with vertices (0, 0), (3, 0), and (2, 4) is revolved about the line $x = -3$.

19 Write $\int_0^1 dy \int_{e^y}^{e} f(x, y)\, dx$ as an iterated integral (or sum of iterated integrals) in which the integration is the first performed with respect to y.

20 Reverse the order of integration in $\int_1^e dy \int_0^{\ln y} f(x, y)\, dx$.

21 Find by double integration the area bounded by the graphs of $y = e^x$, $x = 0$, $y = 0$, and $x = -1$.

22 Find by double integration the area of one petal of the rose $r = \cos 3\theta$.

In Exercises 23 to 25, find the center of mass of the given region.

23 The triangular region with vertices (0, 0), (0, 4), and (3, 0) if $d(x, y) = xy$.

24 The region in the first quadrant enclosed by $x^2 + y^2 = 9$ if $d(x, y) = \sqrt{x^2 + y^2}$.

25 (a) The region above the xy-plane and below $x^2 + y^2 + z^2 \leq 16$ if $d(x, y, z) = \sqrt{x^2 + y^2 + z^2}$.
(b) Find the moment of inertia and radius of gyration relative to the z-axis of this region.
(c) This region rotates about the z-axis with angular velocity $\omega = 3$. What is the kinetic energy?

In Exercises 26 and 27, find the derivative of the given function.

26 (a) $F(x) = \int_0^1 e^{xy}\, dx$ (b) $F(y) = \int_{\pi/6}^{\pi/2} \frac{\cos xy}{x}\, dx$

27 (a) $F(y) = \int_1^{y^2} \frac{\sin xy}{x}\, dx \quad (x \neq 0)$ (b) $F(x) = \int_{e^{x3}}^{x^3} \frac{\sin xy}{y}\, dy \quad (y \neq 0)$

In Exercises 28 to 30, evaluate the given integrals.

28 $\iint_D \frac{1}{\sqrt{x^2 + y^2}}\, dA$, $D = \{(x, y) \mid x^2 + y^2 \leq 9\}$ **29** $\iint_D \frac{1}{(x^2 + 1)y^2}\, dA$, $D = [1, +\infty) \times [1, +\infty)$

30 $\int_0^{+\infty} e^{-x^2}\, dx$

31 In a simple model for the flow of the Gulf Stream, the speed is $v(x, z) = e^{-x^2 - z^2}$ where v is in miles per hour, x is measured in tens of miles off the east coast of North America, and z is measured in Gulf Stream units (each about 4 miles) below the surface of the water. Assume that the water occupies the region $x \geq 0$, $z \geq 0$. Find the amount of water crossing the plane $y = 0$ each hour.

6 VECTOR DIFFERENTIATION

0 INTRODUCTION

Having developed the differential and integral calculus of functions $f: \boldsymbol{R}^n \to \boldsymbol{R}$, we are now ready to consider the calculus of vector valued functions on $\boldsymbol{R}^n$. These are functions $\boldsymbol{F}: \boldsymbol{R}^n \to \boldsymbol{R}^m$; recall that this means that $\boldsymbol{F}$ has domain $D \subseteq \boldsymbol{R}^n$ and range $R \subseteq \boldsymbol{R}^m$. In most concrete examples which we will consider, m and n will be 2 or 3. The differential calculus of these functions is the subject of this chapter; their integral calculus is the subject of Chapter 7.

The differential calculus of such functions is developed by fusing together the differential calculus you learned in Chapter 4 for m separate functions $f: \boldsymbol{R}^n \to \boldsymbol{R}$. Hence the basic ideas in this chapter are very simple extensions of the ideas of Chapter 4. We begin by discussing *limits* and *continuity* of vector valued functions on $\boldsymbol{R}^n$, and considering how the notion of *differentiability* from Chapter 4 should apply to our new context. We find once more that linear functions $\boldsymbol{L}: \boldsymbol{R}^n \to \boldsymbol{R}^m$ are the key ingredient in the concept of differentiability. We then proceed to study such linear functions, called *linear transformations* in this case. Just as we saw that a linear function $\ell: \boldsymbol{R}^n \to \boldsymbol{R}$ could be realized concretely by a fixed vector $\boldsymbol{m}$ in $\boldsymbol{R}^n$, so we will see that linear transformations $\boldsymbol{L}: \boldsymbol{R}^n \to \boldsymbol{R}^m$ can be represented by m fixed vectors in $\boldsymbol{R}^n$, each one written under the preceding to form an m-by-n *matrix.* This idea and its computational implications are treated in most of Sections 2, 3, and 4. After we have enough machinery, we proceed to study differentiation. In Sections 5, 6, and 7, topics corresponding to those of Sections 3, 5, and 6 of Chapter 4 are presented.

In the concluding two sections of the chapter, we show how the notion of differentiation of functions $\boldsymbol{F}: \boldsymbol{R}^n \to \boldsymbol{R}^m$ can be used to give a comprehensive development of *change of variables* in double and triple integrals. This at once provides a uniform approach both for the techniques of Sections 5, 7, and 8 of Chapter 5 and also for a still wider class of integrals obtained from Cartesian integrals by change of coordinates.

1 LIMITS, CONTINUITY, AND A PREVIEW OF DIFFERENTIATION

The fusing process referred to in the introduction requires first of all that for a function $\boldsymbol{F}:\boldsymbol{R}^n \to \boldsymbol{R}^m$ we can produce m real-valued functions on $\boldsymbol{R}^n$. This is easy to do. For example, suppose we have $\boldsymbol{F}:\boldsymbol{R}^3 \to \boldsymbol{R}^2$ given by

$$\boldsymbol{F}(x, y, z) = (x^2+y^2+z^2, xyz) \in \boldsymbol{R}^2.$$

Then the coordinates of the image vector in $\boldsymbol{R}^2$ provide two real-valued functions on $\boldsymbol{R}^3$. Namely, we have

$$f_1(x, y, z) = x^2+y^2+z^2, \quad \text{and} \quad f_2(x, y, z) = xyz.$$

It is natural to call these functions the *coordinate functions* of $\boldsymbol{F}$.

1.1 **DEFINITION.** If $\boldsymbol{F}:\boldsymbol{R}^n \to \boldsymbol{R}^m$ is given by $\boldsymbol{F}(\boldsymbol{x}) = \boldsymbol{F}(x_1, x_2, \ldots, x_n) = (y_1, y_2, \ldots, y_m) = (f_1(\boldsymbol{x}), f_2(\boldsymbol{x}), \ldots, f_m(\boldsymbol{x}))$ for $\boldsymbol{x}$ in the domain D of $\boldsymbol{F}$, then the functions $f_i:\boldsymbol{R}^n \to \boldsymbol{R}$ are called the **coordinate functions** of $\boldsymbol{F}$.

These functions will play a major role in developing the calculus of functions $\boldsymbol{F}:\boldsymbol{R}^n \to \boldsymbol{R}^m$. We can begin the calculus of vector valued functions on $\boldsymbol{R}^n$ by giving the expected definition of the limit of $\boldsymbol{F}(\boldsymbol{x})$ as $\boldsymbol{x} \to \boldsymbol{x}_0$.

1.2 **DEFINITION.** Let $\boldsymbol{F}:\boldsymbol{R}^n \to \boldsymbol{R}^m$ be defined on some deleted open ball $B(\boldsymbol{x}_0, r) - \{\boldsymbol{x}_0\}$ surrounding $\boldsymbol{x}_0$. Then $\lim_{\boldsymbol{x}\to\boldsymbol{x}_0} \boldsymbol{F}(\boldsymbol{x}) = \boldsymbol{\ell}$ means that for every $\varepsilon > 0$ there is a $\delta > 0$ such that

$$|\boldsymbol{F}(\boldsymbol{x}) - \boldsymbol{\ell}| < \varepsilon \quad \text{for all } \boldsymbol{x} \text{ satisfying } 0 < |\boldsymbol{x} - \boldsymbol{x}_0| < \delta.$$

This just says that when $\boldsymbol{x}$ is not equal to $\boldsymbol{x}_0$ but within distance δ of $\boldsymbol{x}_0$, then $\boldsymbol{F}(\boldsymbol{x})$ is within distance ε of $\boldsymbol{\ell}$.

The following result reduces the calculation of limits of a function $\boldsymbol{F}:\boldsymbol{R}^n \to \boldsymbol{R}^m$ to calculation of limits of the coordinate functions f_i of $\boldsymbol{F}$.

1.3 **THEOREM.** $\lim_{\boldsymbol{x}\to\boldsymbol{x}_0} \boldsymbol{F}(\boldsymbol{x}) = \boldsymbol{\ell} = (\ell_1, \ell_2, \ldots, \ell_m)$ holds if and only if $\lim_{\boldsymbol{x}\to\boldsymbol{x}_0} f_i(\boldsymbol{x}) = \ell_i$ for each coordinate function f_i of $\boldsymbol{F}$.

Proof. (a) Suppose that $\lim_{\boldsymbol{x}\to\boldsymbol{x}_0} \boldsymbol{F}(\boldsymbol{x}) = \boldsymbol{\ell}$. Then for any $\varepsilon > 0$, there is a $\delta > 0$ such that if $0 < |\boldsymbol{x} - \boldsymbol{x}_0| < \delta$, then $|\boldsymbol{F}(\boldsymbol{x}) - \boldsymbol{\ell}| < \varepsilon$. Thus for $0 < |\boldsymbol{x} - \boldsymbol{x}_0| < \delta$ we have

$$\sqrt{\sum_{i=1}^{m} (f_i(\boldsymbol{x}) - \ell_i)^2} < \varepsilon.$$

Hence for $0 < |\boldsymbol{x} - \boldsymbol{x}_0| < \delta$, and for each coordinate function f_i,

$$|f_i(\boldsymbol{x}) - \ell_i| \le \sqrt{\sum_{j=1}^{m} (f_j(\boldsymbol{x}) - \ell_j)^2} < \varepsilon.$$

Thus $\lim_{\boldsymbol{x}\to\boldsymbol{x}_0} f_i(\boldsymbol{x}) = \ell_i$ as desired.

(b) Suppose that $\lim_{\boldsymbol{x}\to\boldsymbol{x}_0} f_i(\boldsymbol{x}) = \ell_i$ for each i. Then given $\varepsilon > 0$, we can find δ_i so that

$$|f_i(\boldsymbol{x})-\ell_i|<\frac{\varepsilon}{\sqrt{m}} \quad \text{when} \quad 0<|\boldsymbol{x}-\boldsymbol{x}_0|<\delta_i.$$

Hence if $\delta=\min(\delta_1, \delta_2, \ldots, \delta_m)$, we have for $0<|\boldsymbol{x}-\boldsymbol{x}_0|<\delta$,

$$|f_i(\boldsymbol{x})-\ell_i|<\frac{\varepsilon}{\sqrt{m}}, \quad \text{for each} \quad i=1, 2, \ldots, m.$$

Then for $0<|\boldsymbol{x}-\boldsymbol{x}_0|<\delta$,

$$|\boldsymbol{F}(\boldsymbol{x})-\boldsymbol{\ell}|=\sqrt{\sum_{i=1}^{m}(f_i(\boldsymbol{x})-\ell_i)^2}\leq\sqrt{m\max(f_i(\boldsymbol{x})-\ell_i)^2}<\sqrt{m\frac{\varepsilon^2}{m}}=\varepsilon,$$

where $\max(f_i(\boldsymbol{x})-\ell_i)^2$ is the largest of

$$(f_1(\boldsymbol{x})-\ell_1)^2, (f_2(\boldsymbol{x})-\ell_2)^2, \ldots, (f_m(\boldsymbol{x})-\ell_m)^2.$$

Thus $\lim_{\boldsymbol{x}\to\boldsymbol{x}_0}\boldsymbol{F}(\boldsymbol{x})=\boldsymbol{\ell}$ as desired. QED

1.4 EXAMPLE. Find $\lim_{\boldsymbol{x}\to\boldsymbol{x}_0}\boldsymbol{F}(\boldsymbol{x})$ if $\boldsymbol{F}:\boldsymbol{R}^2\to\boldsymbol{R}^3$ is given by

$$\boldsymbol{F}(\boldsymbol{x})=\boldsymbol{F}(x, y)=\left(\frac{x^4-y^4}{x^2+y^2}, x^2-y^2, 3e^{x+y}\right)$$

and $\boldsymbol{x}_0=\boldsymbol{0}=(0, 0)$.

Solution. $$\lim_{(x,y)\to(0,0)}\frac{x^4-y^4}{x^2+y^2}=\lim_{(x,y)\to(0,0)}(x^2-y^2)=0,$$

and

$$\lim_{(x,y)\to(0,0)}3e^{x+y}=3.$$

So $\lim_{\boldsymbol{x}\to\boldsymbol{0}}\boldsymbol{F}(\boldsymbol{x})=(0, 0, 3)$.

The notion of continuity for a function $\boldsymbol{F}:\boldsymbol{R}^n\to\boldsymbol{R}^m$ is just what you expect it to be.

1.5 DEFINITION. A function $\boldsymbol{F}:\boldsymbol{R}^n\to\boldsymbol{R}^m$ is **continuous** at $\boldsymbol{x}_0\in\boldsymbol{R}^n$ if $\lim_{\boldsymbol{x}\to\boldsymbol{x}_0}\boldsymbol{F}(\boldsymbol{x})=\boldsymbol{F}(\boldsymbol{x}_0)$.

We remark that, in view of Theorem 1.3, continuity of $\boldsymbol{F}$ at $\boldsymbol{x}_0$ is equivalent to the continuity of *all* of its coordinate functions f_i at $\boldsymbol{x}_0$. We observe that the function $\boldsymbol{F}$ in Example 1.4 is not continuous at $\boldsymbol{x}_0=\boldsymbol{0}$ since its first coordinate function is not defined there. But the function $\boldsymbol{F}$ *is* continuous at all other points $\boldsymbol{x}_0\in\boldsymbol{R}^2$, and becomes continuous at $\boldsymbol{0}$ if we extend its domain by defining $f_1(\boldsymbol{0})=0$.

The next result, analogous to Theorem 3.2.8, illustrates how the results of Section 3.2 extend to vector valued functions on $\boldsymbol{R}^n$.

1.6 **THEOREM.** If $\boldsymbol{G}:\boldsymbol{R}^p \to \boldsymbol{R}^n$ is continuous at $\boldsymbol{x}_0 \in \boldsymbol{R}^p$ and $\boldsymbol{F}:\boldsymbol{R}^n \to \boldsymbol{R}^m$ is continuous at $\mathbf{y}_0 = \boldsymbol{G}(\boldsymbol{x}_0) \in \boldsymbol{R}^n$, then $\boldsymbol{F} \circ \boldsymbol{G}:\boldsymbol{R}^p \to \boldsymbol{R}^m$ is continuous at $\boldsymbol{x}_0$.

Proof. We know that as $\boldsymbol{x} \to \boldsymbol{x}_0$, then $\mathbf{y} = \boldsymbol{G}(\boldsymbol{x}) \to \boldsymbol{G}(\boldsymbol{x}_0) = \mathbf{y}_0$. We also know that as $\mathbf{y} \to \mathbf{y}_0$, $\boldsymbol{F}(\mathbf{y}) \to \boldsymbol{F}(\mathbf{y}_0)$. Thus

$$\begin{aligned}\lim_{\boldsymbol{x}\to\boldsymbol{x}_0} \boldsymbol{F} \circ \boldsymbol{G}(\boldsymbol{x}) &= \lim_{\boldsymbol{x}\to\boldsymbol{x}_0} \boldsymbol{F}(\boldsymbol{G}(\boldsymbol{x})) \\ &= \lim_{\mathbf{y}\to\mathbf{y}_0} \boldsymbol{F}(\mathbf{y}) = \boldsymbol{F}(\mathbf{y}_0) \\ &= \boldsymbol{F}(\boldsymbol{G}(\boldsymbol{x}_0)) = \boldsymbol{F} \circ \boldsymbol{G}(\boldsymbol{x}_0).\end{aligned}$$

QED

(In Exercise 11 below you are asked to justify the formal calculations made in taking the limits as we did in the proof.)

It has already begun to appear that the calculus of functions $\boldsymbol{F}:\boldsymbol{R}^n \to \boldsymbol{R}^m$ is strikingly similar to the calculus of functions $f:\boldsymbol{R}^n \to \boldsymbol{R}$. Indeed, by virtue of Theorem 1.3, it seems that we can actually *reduce* the calculus of a function $\boldsymbol{F}:\boldsymbol{R}^n \to \boldsymbol{R}^m$ to the calculus of its coordinate functions $f_i:\boldsymbol{R}^n \to \boldsymbol{R}$. Hence we would expect the calculus of $\boldsymbol{F}$ to be obtained by simply combining somehow the calculus of its coordinate functions. This turns out to be essentially what happens. So the obvious question at this point is, then, *how* do we combine the calculus of the f_i to get that of $\boldsymbol{F}$? We have seen in Theorem 1.3 that this is very simple in the case of limits and continuity. That brings us to differentiation. What should it mean for a function $\boldsymbol{F}:\boldsymbol{R}^n \to \boldsymbol{R}^m$ to be differentiable at $\boldsymbol{x}_0 \in \boldsymbol{R}^n$? Let's see if we can carry Definition 4.3.3 over to such a function $\boldsymbol{F}$.

There should be a *linear* function $\boldsymbol{L}:\boldsymbol{R}^n \to \boldsymbol{R}^m$ that we can use to approximate $\boldsymbol{F}$ near $\boldsymbol{x}_0$. Thus we should have

(1) $$\boldsymbol{F}(\boldsymbol{x}) = \boldsymbol{F}(\boldsymbol{x}_0) + \boldsymbol{L}(\boldsymbol{x} - \boldsymbol{x}_0) + \boldsymbol{E}(\boldsymbol{x})$$

where

$$\lim_{\boldsymbol{x}\to\boldsymbol{x}_0} \frac{\boldsymbol{E}(\boldsymbol{x})}{|\boldsymbol{x} - \boldsymbol{x}_0|} = \mathbf{0}.$$

Here $\boldsymbol{E}(\boldsymbol{x})$ is the error we make in using $\boldsymbol{F}(\boldsymbol{x}_0) + \boldsymbol{L}(\boldsymbol{x} - \boldsymbol{x}_0)$ to approximate $\boldsymbol{F}(\boldsymbol{x})$, and this should approach $\mathbf{0}$ as $\boldsymbol{x} \to \boldsymbol{x}_0$ faster than $|\boldsymbol{x} - \boldsymbol{x}_0| \to 0$.

On the basis of our experience in Chapter 4 (cf. Theorem 4.3.4) and in Theorem 1.3, we expect that the value of the linear function $\boldsymbol{L}$ at $\boldsymbol{x} - \boldsymbol{x}_0$ should be calculable by putting together the dot products

$$\nabla f_i(\boldsymbol{x}_0) \cdot (\boldsymbol{x} - \boldsymbol{x}_0), \quad \text{for} \quad i = 1, 2, \ldots, m.$$

(This would indeed give us a vector in $\boldsymbol{R}^m$, which is what $\boldsymbol{L}(\boldsymbol{x} - \boldsymbol{x}_0)$ should be.) Again this is essentially correct, but before we can explain why, we are going to have to study linear functions $\boldsymbol{L}:\boldsymbol{R}^n \to \boldsymbol{R}^m$ just as we did in Section 4.2 for the case $m = 1$. It turns out that there is a richer theory of linear functions in the $m > 1$ case than there was for the $m = 1$ case. This will occupy our attention for the next few sections. For now, we can certainly define what is meant by a linear function $\boldsymbol{L}:\boldsymbol{R}^n \to \boldsymbol{R}^m$. Definition 4.2.3 carries over without difficulty. (Out of longstanding usage, these functions are called *linear transformations* when $m > 1$.)

1.7 **DEFINITION.** A function $\boldsymbol{L}:\boldsymbol{R}^n \to \boldsymbol{R}^m$ is a **linear transformation** if $\boldsymbol{L}(\boldsymbol{x} + \mathbf{y}) = \boldsymbol{L}(\boldsymbol{x}) + \boldsymbol{L}(\mathbf{y})$, and $\boldsymbol{L}(a\boldsymbol{x}) = a\boldsymbol{L}(\boldsymbol{x})$ for all $\boldsymbol{x}, \mathbf{y} \in \boldsymbol{R}^n$ and all $a \in \boldsymbol{R}$.

Thus, these are precisely the vector valued functions on $\boldsymbol{R}^n$ which preserve the vector space structure built upon addition and multiplication by scalars. It turns out that linear transformations are constructed by putting together a sequence of linear functions $\ell_i:\boldsymbol{R}^n \to \boldsymbol{R}$.

1.8 **THEOREM.** $\mathbf{L}:\boldsymbol{R}^n \to \boldsymbol{R}^m$ is a linear transformation if and only if each of its coordinate functions $\ell_i:\boldsymbol{R}^n \to \boldsymbol{R}$ is a linear function.

Proof. (a) Suppose that $\mathbf{L}$ is a linear transformation. Then $\mathbf{L}(\mathbf{x}+\mathbf{y})=\mathbf{L}(\mathbf{x})+\mathbf{L}(\mathbf{y})$, so we have

$$\begin{aligned}\mathbf{L}(\mathbf{x}+\mathbf{y}) &= (\ell_1(\mathbf{x}+\mathbf{y}),\ldots,\ell_i(\mathbf{x}+\mathbf{y}),\ldots,\ell_m(\mathbf{x}+\mathbf{y}))\\ &= \mathbf{L}(\mathbf{x})+\mathbf{L}(\mathbf{y})\\ &= (\ell_1(\mathbf{x}),\ldots,\ell_i(\mathbf{x}),\ldots,\ell_m(\mathbf{x}))+(\ell_1(\mathbf{y}),\ldots,\ell_i(\mathbf{y}),\ldots,\ell_m(\mathbf{y})).\end{aligned}$$

Thus for each i, the i-th coordinates of $\mathbf{L}(\mathbf{x}+\mathbf{y})$ and $\mathbf{L}(\mathbf{x})+\mathbf{L}(\mathbf{y})$ coincide. This means that

$$\ell_i(\mathbf{x}+\mathbf{y})=\ell_i(\mathbf{x})+\ell_i(\mathbf{y}).$$

Similarly, $\mathbf{L}(a\mathbf{x})=a\mathbf{L}(\mathbf{x})$ gives, for each i,

$$\ell_i(a\mathbf{x})=a\ell_i(\mathbf{x}).$$

Thus, each $\ell_i:\boldsymbol{R}^n \to \boldsymbol{R}$ is a linear function as claimed.

(b) If each $\ell_i:\boldsymbol{R}^n \to \boldsymbol{R}$ is a linear function, then we have

$$\begin{aligned}\mathbf{L}(\mathbf{x}+\mathbf{y}) &= (\ell_1(\mathbf{x}+\mathbf{y}),\ldots,\ell_i(\mathbf{x}+\mathbf{y}),\ldots,\ell_m(\mathbf{x}+\mathbf{y}))\\ &= (\ell_1(\mathbf{x})+\ell_1(\mathbf{y}),\ldots,\ell_i(\mathbf{x})+\ell_i(\mathbf{y}),\ldots,\ell_m(\mathbf{x})+\ell_m(\mathbf{y}))\\ &= (\ell_1(\mathbf{x}),\ldots,\ell_i(\mathbf{x}),\ldots,\ell_m(\mathbf{x}))+(\ell_1(\mathbf{y}),\ldots,\ell_i(\mathbf{y}),\ldots,\ell_m(\mathbf{y}))\\ &= \mathbf{L}(\mathbf{x})+\mathbf{L}(\mathbf{y}).\end{aligned}$$

Similarly,

$$\begin{aligned}\mathbf{L}(a\mathbf{x}) &= (\ell_1(a\mathbf{x}),\ldots,\ell_i(a\mathbf{x}),\ldots,\ell_m(a\mathbf{x}))\\ &= (a\ell_1(\mathbf{x}),\ldots,a\ell_i(\mathbf{x}),\ldots,a\ell_m(\mathbf{x}))\\ &= a(\ell_1(\mathbf{x}),\ldots,\ell_i(\mathbf{x}),\ldots,\ell_m(\mathbf{x}))\\ &= a\mathbf{L}(\mathbf{x}).\end{aligned}$$

Thus, if each ℓ_i is linear, then so is $\mathbf{L}$, as claimed. QED

We now can tell at a glance whether a function $\boldsymbol{F}:\boldsymbol{R}^n \to \boldsymbol{R}^m$ is a linear transformation, since we know how to tell if its coordinate functions $\ell_i:\boldsymbol{R}^n \to \boldsymbol{R}$ are linear. (See Theorem 4.2.4.)

1.9 **EXAMPLE.** Decide whether the following functions $\boldsymbol{F}:\boldsymbol{R}^3 \to \boldsymbol{R}^2$ are linear. (a) $\boldsymbol{F}(x, y, z)=(x+y-1, 3x+2y+5z)$; (b) $\boldsymbol{G}(x, y, z)=(2x, -6y+4z)$.

Solution. (a) $\boldsymbol{F}$ is *not* linear, because its first coordinate function f_1 is not linear:

$$f_1(x, y, z)=x+y-1\neq \boldsymbol{m}\cdot(x, y, z) \text{ for any vector } \boldsymbol{m}=(m_1, m_2, m_3)\in \boldsymbol{R}^3.$$

(b) $\boldsymbol{G}$ *is* linear. We have $g_1(x, y, z) = (2, 0, 0) \cdot (x, y, z)$ and $g_2(x, y, z) = (0, -6, 4) \cdot (x, y, z)$, so both g_1 and g_2 are linear functions by Theorem 4.2.4.

We can exploit Theorem 1.8 to obtain the analogue of Theorem 4.2.6.

1.10 **THEOREM.** If $\boldsymbol{L}: \boldsymbol{R}^n \to \boldsymbol{R}^m$ is a linear transformation, then $\boldsymbol{L}$ is continuous at all points $\boldsymbol{x}_0$ of $\boldsymbol{R}^n$.

Proof. If $\boldsymbol{L}$ is linear, then each of its coordinate functions $\ell_i : \boldsymbol{R}^n \to \boldsymbol{R}$ is linear. Hence by Theorem 4.2.6, each of the ℓ_i is continuous on all of $\boldsymbol{R}^n$. Thus $\lim_{\boldsymbol{x} \to \boldsymbol{x}_0} \ell_i(\boldsymbol{x}) = \ell_i(\boldsymbol{x}_0)$ for any $\boldsymbol{x}_0 \in \boldsymbol{R}$. Then by Theorem 1.3, $\lim_{\boldsymbol{x} \to \boldsymbol{x}_0} \boldsymbol{L}(\boldsymbol{x}) = \boldsymbol{L}(\boldsymbol{x}_0)$, so $\boldsymbol{L}$ is continuous at any $\boldsymbol{x}_0 \in \boldsymbol{R}$. QED

We will exploit Theorem 1.8 still more in the next section to obtain a concrete representation for a linear transformation $\boldsymbol{L}: \boldsymbol{R}^n \to \boldsymbol{R}^m$ by an m-by-n matrix.

EXERCISES 6.1

In Exercises 1 to 8, find the limit of the given function at the given point. Is the function continuous at the given point?

1 $\boldsymbol{F}(x, y) = \left(3x^2y, -\text{Tan}^{-1} e^{x-y}, \sin \frac{\pi xy}{6}\right)$; $\boldsymbol{x}_0 = (1, 1)$.

2 $\boldsymbol{F}(x, y) = \left(\tan \pi(xy), e^{x^2y}, \frac{x^2+1}{y^2+1}\right)$; $\boldsymbol{x}_0 = (2, 3)$.

3 $\boldsymbol{F}(x, y, z) = \left(\frac{x^4 - y^4}{x^2 - y^2}, yze^{xy}, x^2yz, x^2 - y^2z^2\right)$; $\boldsymbol{x}_0 = (2, 2, 1)$.

4 $\boldsymbol{F}(x, y, z) = \left(x^2 \frac{\sin(x^2 + y^2 + z^2)}{x^2 + y^2 + z^2}, \sqrt{x^2y - z}\right)$; $\boldsymbol{x}_0 = (2, 1, 0)$.

5 $\boldsymbol{F}(x, y, z) = \left(\frac{3x^2 + y}{y - 2z}, e^{5x-y}, xy + z^3\right)$; $\boldsymbol{x}_0 = (1, 2, 1)$.

6 $\boldsymbol{F}(x, y, z) = \left(\sin \pi(x^2 - yz), \cos \pi(y^2 - xz), \frac{\ln xyz}{x^2 + y^2}\right)$; $\boldsymbol{x}_0 = (1, 2, 0)$.

7 $\boldsymbol{F}(x, y, z) = \left(\frac{1 - \cos xyz}{xyz}, e^{xy-z}, \sin \frac{\pi x}{y + z}\right)$; $\boldsymbol{x}_0 = (1, 2, 0)$.

8 $\boldsymbol{F}(x, y, z, w) = \left(\frac{\sin xyzw}{xyzw}, x^2y^2zw, x + yzw\right)$; $\boldsymbol{x}_0 = (0, 1, 2, 3)$.

9 Show that $\boldsymbol{F}(x, y, z) = \left(x^2 + y^2 + z^2, \sin \frac{x^2yz}{x^2 + y^2 + z^2 + 1}\right)$ is continuous on all of $\boldsymbol{R}^3$.

10 Show that $\boldsymbol{F}(x, y, z) = (e^{\sqrt{x^2+y^2+z^2}}, (\sin xy)e^{\cos xy})$ is continuous on all of $\boldsymbol{R}^3$.

11 Prove Theorem 1.6 using Definition 1.2.

12 Prove that if $\boldsymbol{F}: \boldsymbol{R}^n \to \boldsymbol{R}^m$ and $\boldsymbol{G}: \boldsymbol{R}^n \to \boldsymbol{R}^m$ are continuous at $\boldsymbol{x}_0$, then so is $\boldsymbol{F} + \boldsymbol{G}$.

In Exercises 13 to 20, decide whether the given function is linear. Explain.

13 $\boldsymbol{F}(x, y, z) = (x - y + z, 3x - \frac{1}{4}z, 2x + y + 6z)$.

14 $\boldsymbol{F}(x, y, z) = (3x + 4yz, -5x + 6z, 2x - 3y + z)$.

15 $F(x, y)=(-3x-1, 2y)$.

16 $F(x, y)=(4x, 5y, 4x+5y)$.

17 $F(x, y, z)=(4x-3y, 3z+4)$.

18 $F(x, y, z)=\left(\frac{5x}{3}+\frac{z}{2}, \frac{3y+4z}{5}\right)$.

19 $F(x, y, z, w)=(2x-y+z+3w, -x+3y-w, x-2z-w)$.

20 $F(x, y, z, w)=(x-2y+z-w, x-yz+3w, 2x-3y+z)$.

21 (a) If F and G are linear transformations from R^n to R^m, then show that $F+G$ is also a linear transformation from R^n to R^m, where $F+G$ is defined by $(F+G)(x)=F(x)+G(x)$.
(b) If $F: R^n \to R^m$ is a linear transformation and $a \in R$, then show that aF is also a linear transformation, where aF is defined by $(aF)(x)=aF(x)$.

22 Let $\mathscr{L}(R^n, R^m)$ be the set of all linear transformations from R^n into R^m. Show that $\mathscr{L}(R^n, R^m)$ is a real vector space.

23 If $G: R^p \to R^n$ and $F: R^n \to R^m$ are linear transformations, then show that $F \circ G: R^p \to R^m$ is also a linear transformation. (**This exercise is needed in Section 3 (Theorem 3.8).**)

24 As for real-valued functions $f: R^n \to R$, a vector-valued function $F: R^n \to R^m$ is called **affine** if $F(x)=L(x)+b$ for some fixed $b \in R^m$. Here $L: R^n \to R^m$ is linear. Show that F is affine if and only if all its coordinate functions are affine functions.

25 Prove that every affine function $F: R^n \to R^m$ is continuous on all of R^n.

2 REPRESENTATION OF A LINEAR TRANSFORMATION BY A MATRIX

In the preceding section we defined the concept of a linear transformation $L: R^n \to R^m$ and we saw (Theorem 1.8) that L is linear if and only if each of its coordinate functions $\ell_i: R^n \to R$ is a linear function. Now suppose that we are given a linear transformation $L: R^n \to R^m$. Since the ℓ_i are all linear functions, we have for each i,

(1) $$\ell_i(x)=m_i \cdot x$$

for some vector $m_i \in R^n$ by Theorem 4.2.4. As you probably recall, m_i is the vector given by

(2) $$\begin{aligned} m_i &= (\ell_i(1, 0, 0, \dots, 0), \ell_i(0, 1, 0, \dots, 0), \\ &\quad \ell_i(0, 0, 1, \dots, 0), \dots, \ell_i(0, 0, 0, \dots, 1)) \\ &= (\ell_i(e_1), \ell_i(e_2), \ell_i(e_3), \dots, \ell_i(e_n)), \end{aligned}$$

where, as usual, e_j stands for the standard basis vector of R^n which has a 1 in the j-th entry and 0's in every other entry. Now, since m_i *completely determines* ℓ_i, *we can represent* ℓ_i *by the vector* m_i. This suggests a natural way to represent the linear transformation L. Since L is made up of its m coordinate functions $\ell_1, \ell_2, \dots, \ell_m$, we can represent L by the *matrix* that results from putting the representing vectors m_i for ℓ_i together in the following sense.

2.1 **DEFINITION.** If $\boldsymbol{L}:\boldsymbol{R}^n \to \boldsymbol{R}^m$ is a linear transformation with coordinate functions $\ell_1, \ell_2, \ldots, \ell_m$, then the **matrix** $M_{\boldsymbol{L}}$ **of** $\boldsymbol{L}$ **relative to the standard bases** $(\boldsymbol{e}_1, \ldots, \boldsymbol{e}_n)$ for $\boldsymbol{R}^n$ and $(\boldsymbol{e}_1, \ldots, \boldsymbol{e}_m)$ for $\boldsymbol{R}^m$ is the m-by-n matrix given by

(3)
$$M_{\boldsymbol{L}} = \begin{pmatrix} \boldsymbol{m}_1 \\ \boldsymbol{m}_2 \\ \cdot \\ \cdot \\ \cdot \\ \boldsymbol{m}_m \end{pmatrix} = \begin{pmatrix} \ell_1(\boldsymbol{e}_1) & \ell_1(\boldsymbol{e}_2) & \cdots & \ell_1(\boldsymbol{e}_n) \\ \ell_2(\boldsymbol{e}_1) & \ell_2(\boldsymbol{e}_2) & \cdots & \ell_2(\boldsymbol{e}_n) \\ \cdot & & & \\ \cdot & & & \\ \cdot & & & \\ \ell_m(\boldsymbol{e}_1) & \ell_m(\boldsymbol{e}_2) & \cdots & \ell_m(\boldsymbol{e}_n) \end{pmatrix}$$

We form the matrix by simply arraying in the natural order the row vectors $\boldsymbol{m}_i$ that represent the linear coordinate functions ℓ_i of $\boldsymbol{L}$. Thus $M_{\boldsymbol{L}}$ has m rows and, since each $\boldsymbol{m}_i \in \boldsymbol{R}^n$, it has n columns. So it is an m-by-n matrix.

2.2 **EXAMPLE.** Give the matrices of the following linear transformations relative to the standard bases.
(a) $\boldsymbol{L}(x, y, z) = (3x - 5y, -x + 5y - z, 3x - y + z)$
(b) $\boldsymbol{T}(x, y, z) = (2x - y - z, 3x + 4z)$.

Solution. (a) We simply use (3). Here $\ell_1(x, y, z) = 3x - 5y = (3, -5, 0) \cdot (x, y, z)$; $\ell_2(x, y, z) = -x + 5y - z = (-1, 5, -1) \cdot (x, y, z)$; and $\ell_3(x, y, z) = 3x - y + z = (3, -1, 1) \cdot (x, y, z)$. Hence $\boldsymbol{m}_1 = (3, -5, 0)$, $\boldsymbol{m}_2 = (-1, 5, -1)$, and $\boldsymbol{m}_3 = (3, -1, 1)$. Thus,

$$M_{\boldsymbol{L}} = \begin{pmatrix} \boldsymbol{m}_1 \\ \boldsymbol{m}_2 \\ \boldsymbol{m}_3 \end{pmatrix} = \begin{pmatrix} 3 & -5 & 0 \\ -1 & 5 & -1 \\ 3 & -1 & 1 \end{pmatrix}.$$

(b) Here $t_1(x, y, z) = 2x - y - z = (2, -1, -1) \cdot (x, y, z)$ and $t_2(x, y, z) = 3x + 4z = (3, 0, 4) \cdot (x, y, z)$. Thus, $\boldsymbol{m}_1 = (2, -1, -1)$ and $\boldsymbol{m}_2 = (3, 0, 4)$. We then have

$$M_{\boldsymbol{T}} = \begin{pmatrix} \boldsymbol{m}_1 \\ \boldsymbol{m}_2 \end{pmatrix} = \begin{pmatrix} 2 & -1 & -1 \\ 3 & 0 & 4 \end{pmatrix}.$$

Consider for a moment the *columns of* $M_{\boldsymbol{L}}$. Notice that the first column is

$$\begin{pmatrix} \ell_1(\boldsymbol{e}_1) \\ \ell_2(\boldsymbol{e}_1) \\ \cdot \\ \cdot \\ \cdot \\ \ell_m(\boldsymbol{e}_1) \end{pmatrix} = \boldsymbol{L}(\boldsymbol{e}_1) \in \boldsymbol{R}^m$$

where the vector $\boldsymbol{L}(\boldsymbol{e}_1)$ is written as a column vector. Now observe that the same phenomenon is true of *every* column: the j-th column is $\boldsymbol{L}(\boldsymbol{e}_j)$.

Hence we can write M_L as a matrix of columns in the form

(4) $$M_L = (L(e_1), L(e_2), \ldots, L(e_n)).$$

This gives us an alternative procedure for finding the matrix of a linear transformation $L: R^n \to R^m$. We can compute the value of L on each standard basis vector $e_1, e_2, \ldots, e_n$, and then put the results into the *columns* of M_L.

2.3 **EXAMPLE.** Find the matrix of each of the linear transformations in Example 2.2 using (4) instead of (3).

Solution. Here $e_1 = i = (1, 0, 0)$, $e_2 = j = (0, 1, 0)$, and $e_3 = k = (0, 0, 1)$.

(a) We have

$$L(1, 0, 0) = (3, -1, 3) = \begin{pmatrix} 3 \\ -1 \\ 3 \end{pmatrix}$$

as a column vector,

$$L(0, 1, 0) = (-5, 5, -1) = \begin{pmatrix} -5 \\ 5 \\ -1 \end{pmatrix}$$

as a column vector, and

$$L(0, 0, 1) = (0, -1, 1) = \begin{pmatrix} 0 \\ -1 \\ 1 \end{pmatrix}$$

as a column vector. Thus, from (4),

$$M_L = \begin{pmatrix} 3 & -5 & 0 \\ -1 & 5 & -1 \\ 3 & -1 & 1 \end{pmatrix}.$$

(b) We have

$$T(1, 0, 0) = (2, 3) = \begin{pmatrix} 2 \\ 3 \end{pmatrix}$$

as a column vector,

$$T(0, 1, 0) = (-1, 0) = \begin{pmatrix} -1 \\ 0 \end{pmatrix}$$

as a column vector, and

$$T(0, 0, 1) = (-1, 4) = \begin{pmatrix} -1 \\ 4 \end{pmatrix}$$

as a column vector. Hence, from (4)

$$M_T = \begin{pmatrix} 2 & -1 & -1 \\ 3 & 0 & 4 \end{pmatrix}.$$

Whether you use (3) or (4) to construct M_L is really up to you. At our stage, (3) may seem more natural, but in linear algebra (4) turns out to be the more natural scheme to follow.

We remarked before that we could represent the linear function $\ell_i : R^n \to R$ by the vector $m_i = (\ell_i(e_1), \ell_i(e_2), \ldots, \ell_i(e_n))$ since ℓ_i is completely determined by m_i; that is, $\ell_i(x) = m_i \cdot x$ for all $x \in R^n$. Now, the same is true of representing L by its matrix M_L relative to the standard bases. For what is $L(x)$ for any $x \in R^n$? It is given by

$$\begin{aligned} \boldsymbol{L}(\boldsymbol{x}) &= (\ell_1(\boldsymbol{x}), \ell_2(\boldsymbol{x}), \ldots, \ell_m(\boldsymbol{x})) \\ &= (\boldsymbol{m}_1 \cdot \boldsymbol{x}, \boldsymbol{m}_2 \cdot \boldsymbol{x}, \ldots, \boldsymbol{m}_m \cdot \boldsymbol{x}) \end{aligned}$$

Here the $\boldsymbol{m}_i$'s are just the *rows* of $M_{\boldsymbol{L}}$. So once we have $M_{\boldsymbol{L}}$, we can compute the value of $\boldsymbol{L}$ on any vector $\boldsymbol{x} \in \boldsymbol{R}^n$. We state this important fact as a theorem.

2.4 **THEOREM.** If $\boldsymbol{L}: \boldsymbol{R}^n \to \boldsymbol{R}^m$ is a linear transformation and $M_{\boldsymbol{L}}$ is its matrix relative to the standard bases, then for any $\boldsymbol{x} \in \boldsymbol{R}^n$,

(5) $$\boldsymbol{L}(\boldsymbol{x}) = \begin{pmatrix} \mathbf{m}_1 \cdot \boldsymbol{x} \\ \mathbf{m}_2 \cdot \boldsymbol{x} \\ \cdot \\ \cdot \\ \cdot \\ \mathbf{m}_m \cdot \boldsymbol{x} \end{pmatrix} \in \boldsymbol{R}^m.$$

We have written $\boldsymbol{L}(\boldsymbol{x})$ and $\boldsymbol{x}$ as column vectors, which gives (5) a familiar appearance. In fact, if you look back at Definition 4.8.1, you will see that, in case $m = n = 2$, (5) simply reduces to the *product* of $M_{\boldsymbol{L}}$ and $\boldsymbol{x}$ as defined in Section 4.8. In general, if we compute the dot product of each *row* of $M_{\boldsymbol{L}}$ with $\boldsymbol{x}$, and put the results into a column vector, then we get $\boldsymbol{L}(\boldsymbol{x})$, the value of the linear transformation $\boldsymbol{L}: \boldsymbol{R}^n \to \boldsymbol{R}^m$ at $\boldsymbol{x} \in \boldsymbol{R}^n$. The following definition is thus a direct extension of Definition 4.8.1 and a useful dividend of our scheme to represent $\boldsymbol{L}$ by its matrix $M_{\boldsymbol{L}}$ relative to the standard bases.

2.5 **DEFINITION.** If

$$A = \begin{pmatrix} a_{11} & a_{12} & \cdots & a_{1n} \\ a_{21} & a_{22} & \cdots & a_{2n} \\ \cdot & \cdot & \cdot & \cdot \\ \cdot & \cdot & \cdot & \cdot \\ \cdot & \cdot & \cdot & \cdot \\ a_{m1} & a_{m2} & \cdots & a_{mn} \end{pmatrix} = \begin{pmatrix} \boldsymbol{r}_1 \\ \boldsymbol{r}_2 \\ \cdot \\ \cdot \\ \cdot \\ \boldsymbol{r}_m \end{pmatrix}$$

is an m-by-n matrix and

$$\boldsymbol{x} = \begin{pmatrix} x_1 \\ x_2 \\ \cdot \\ \cdot \\ \cdot \\ x_n \end{pmatrix} \in \boldsymbol{R}^n,$$

then the **product of A and $\boldsymbol{x}$ in that order** is given by

(6) $$A\boldsymbol{x} = \begin{pmatrix} \boldsymbol{r}_1 \cdot \boldsymbol{x} \\ \boldsymbol{r}_2 \cdot \boldsymbol{x} \\ \cdot \\ \cdot \\ \cdot \\ \boldsymbol{r}_m \cdot \boldsymbol{x} \end{pmatrix} \in \boldsymbol{R}^m.$$

In (6) one writes $\boldsymbol{x}$ and $A\boldsymbol{x}$ as column vectors to conform with the requirements for general matrix multiplication to be defined in the next section. Therorem 2.4 then takes the following form.

2.4 **RESTATED.** If $\boldsymbol{L}:\boldsymbol{R}^n \to \boldsymbol{R}^m$ is a linear transformation and $M_{\boldsymbol{L}}$ is its matrix relative to the standard bases, then for any $\boldsymbol{x} \in \boldsymbol{R}^n$ we have

(7)
$$\boldsymbol{L}(\boldsymbol{x}) = M_{\boldsymbol{L}}\boldsymbol{x} = \begin{pmatrix} \boldsymbol{m}_1 \\ \boldsymbol{m}_2 \\ \cdot \\ \cdot \\ \cdot \\ \boldsymbol{m}_m \end{pmatrix} \begin{pmatrix} x_1 \\ x_2 \\ \cdot \\ \cdot \\ \cdot \\ x_n \end{pmatrix} = \begin{pmatrix} \boldsymbol{m}_1 \cdot \boldsymbol{x} \\ \boldsymbol{m}_2 \cdot \boldsymbol{x} \\ \cdot \\ \cdot \\ \cdot \\ \boldsymbol{m}_m \cdot \boldsymbol{x} \end{pmatrix} \in \boldsymbol{R}^m.$$

2.6 **EXAMPLE.** Use Theorem 2.4 to compute $\boldsymbol{L}(3, -1, 1)$ and $\boldsymbol{T}(3, -1, 1)$, where $\boldsymbol{L}$ and $\boldsymbol{T}$ are the linear transformations of Examples 2.2 and 2.3.

Solution. From either example, we have

$$M_{\boldsymbol{L}} = \begin{pmatrix} 3 & -5 & 0 \\ -1 & 5 & -1 \\ 3 & -1 & 1 \end{pmatrix}.$$

Thus by (7)

$$\boldsymbol{L}(3, -1, 1) = \begin{pmatrix} 3 & -5 & 0 \\ -1 & 5 & -1 \\ 3 & -1 & 1 \end{pmatrix} \begin{pmatrix} 3 \\ -1 \\ 1 \end{pmatrix} = \begin{pmatrix} 3 \cdot 3 + (-5)(-1) + 0(1) \\ -1 \cdot 3 + 5(-1) + (-1)1 \\ 3 \cdot 3 + (-1)(-1) + 1 \cdot 1 \end{pmatrix}$$
$$= \begin{pmatrix} 9+5+0 \\ -3-5-1 \\ 9+1+1 \end{pmatrix} = \begin{pmatrix} 14 \\ -9 \\ 11 \end{pmatrix} \in \boldsymbol{R}^3.$$

Similarly, we have

$$M_{\boldsymbol{T}} = \begin{pmatrix} 2 & -1 & -1 \\ 3 & 0 & 4 \end{pmatrix},$$

so

$$\boldsymbol{T}(3, -1, 1) = \begin{pmatrix} 2 & -1 & -1 \\ 3 & 0 & 4 \end{pmatrix} \begin{pmatrix} 3 \\ -1 \\ 1 \end{pmatrix} = \begin{pmatrix} 2 \cdot 3 + (-1)(-1) + (-1)1 \\ 3 \cdot 3 + \ 0(-1) \ + 4 \cdot 1 \end{pmatrix}$$
$$= \begin{pmatrix} 6+1-1 \\ 9+0+4 \end{pmatrix} = \begin{pmatrix} 6 \\ 13 \end{pmatrix} \in \boldsymbol{R}^2.$$

The next example suggests the usefulness of matrix-vector multiplication in organizing data.

2.7 **EXAMPLE.** Three orange growers, G_1, G_2, and G_3, store their picked oranges in a marketing cooperative's facilities during the three-month picking season. Storage costs are discounted after the first month (M_1), reflecting the lower overheads associated with long-term accounts. The growers' storage requirements in hundreds of boxes of oranges are as follows:

	M_1	M_2	M_3
G_1	20	35	25
G_2	20	15	12
G_3	15	25	20

The storage costs are 30¢ per box for the first month, 20¢ per box for the second month, and 10¢ per box for the third month. Find each grower's total storage costs for the season.

Solution. We can represent the monthly storage costs in dollars per box by the vector

$$\boldsymbol{c} = \begin{pmatrix} 0.3 \\ 0.2 \\ 0.1 \end{pmatrix}.$$

If we put

$$A = \begin{pmatrix} 2000 & 3500 & 2500 \\ 2000 & 1500 & 1200 \\ 1500 & 2500 & 2000 \end{pmatrix},$$

then the rows of this matrix represent the monthly storage requirements of each grower. The total money expended by each grower is thus the dot product of his row in A with the vector $\boldsymbol{c}$. So we can calculate the total storage costs in dollars for each grower by computing $A\boldsymbol{c}$. This gives

$$A\boldsymbol{c} = \begin{pmatrix} 2000 & 3500 & 2500 \\ 2000 & 1500 & 1200 \\ 1500 & 2500 & 2000 \end{pmatrix} \begin{pmatrix} 0.3 \\ 0.2 \\ 0.1 \end{pmatrix} = \begin{pmatrix} 1550 \\ 1020 \\ 1150 \end{pmatrix}.$$

Thus Grower 1 spends \$1550, Grower 2 spends \$1020, and Grower 3 spends \$1150.

We conclude the section with a result which says that the phenomenon in Theorem 2.4 *characterizes* the matrix of a linear transformation $\boldsymbol{L}: \boldsymbol{R}^n \to \boldsymbol{R}^m$ relative to the standard bases. This result will be of use in the next section.

2.8 THEOREM. Let $\boldsymbol{L}: \boldsymbol{R}^n \to \boldsymbol{R}^m$ be a linear transformation. If A is *any* m-by-n matrix with the property that $\boldsymbol{L}(\boldsymbol{x}) = A\boldsymbol{x}$ for $\boldsymbol{x} \in \boldsymbol{R}^n$, then $A = M_{\boldsymbol{L}}$. That is, $M_{\boldsymbol{L}}$ is the *unique* matrix M with the property that $\boldsymbol{L}(\boldsymbol{x})$ can be evaluated as the product of M and $\boldsymbol{x}$ in that order.

Proof. Suppose that

$$A = \begin{pmatrix} a_{11} & a_{12} & \cdots & a_{1n} \\ a_{21} & a_{22} & \cdots & a_{2n} \\ \cdot & \cdot & \cdot & \cdot \\ \cdot & \cdot & \cdot & \cdot \\ \cdot & \cdot & \cdot & \cdot \\ a_{m1} & a_{m2} & \cdots & a_{mn} \end{pmatrix}.$$

We have by hypothesis that $\boldsymbol{L}(\boldsymbol{e}_1) = A\boldsymbol{e}_1$. We can calculate

$$A\boldsymbol{e}_1 = \begin{pmatrix} a_{11} & a_{12} & \cdots & a_{1n} \\ a_{21} & a_{22} & \cdots & a_{2n} \\ \vdots & \vdots & & \vdots \\ a_{m1} & a_{m2} & \cdots & a_{mn} \end{pmatrix} \begin{pmatrix} 1 \\ 0 \\ 0 \\ \vdots \\ 0 \end{pmatrix} = \begin{pmatrix} a_{11} \\ a_{21} \\ \vdots \\ a_{m1} \end{pmatrix}$$

Hence the first column of A is $\boldsymbol{L}(\boldsymbol{e}_1)$. Similarly, for

$$\boldsymbol{e}_j = \begin{pmatrix} 0 \\ \vdots \\ 0 \\ 1 \\ 0 \\ \vdots \\ 0 \end{pmatrix}$$

where the 1 occurs as the j-th entry, we have $\boldsymbol{L}(\boldsymbol{e}_j) = A\boldsymbol{e}_j$, and we can calculate

$$A\boldsymbol{e}_j = \begin{pmatrix} a_{11} & \cdots & a_{1j} & \cdots & a_{1n} \\ a_{21} & \cdots & a_{2j} & \cdots & a_{2n} \\ \vdots & & \vdots & & \vdots \\ a_{m1} & \cdots & a_{mj} & \cdots & a_{mn} \end{pmatrix} \begin{pmatrix} 0 \\ \vdots \\ 0 \\ 1 \\ 0 \\ \vdots \\ 0 \end{pmatrix} = \begin{pmatrix} a_{1j} \\ a_{2j} \\ \vdots \\ a_{mj} \end{pmatrix}$$

So the j-th column of A is $\boldsymbol{L}(\boldsymbol{e}_j)$ for any $j = 1, 2, \ldots, n$. Thus, $A = (\boldsymbol{L}(\boldsymbol{e}_1), \boldsymbol{L}(\boldsymbol{e}_2), \ldots, \boldsymbol{L}(\boldsymbol{e}_n)) = M_{\boldsymbol{L}}$ by (4). QED

EXERCISES 6.2

In Exercises 1 to 20, give the matrix of the linear transformation $\boldsymbol{L}$.

1 $\boldsymbol{L}(x, y, z) = (2x, 3y, -z)$

2 $\boldsymbol{L}(x, y, z) = (2z, -\frac{1}{2}x, 4y)$

3 $\boldsymbol{L}(x, y, z) = (-2x + y - 5z, \frac{1}{2}x - \frac{1}{4}y + z, 3x - y - z)$

4 $\boldsymbol{L}(x, y, z) = (3x - z, 2y - 4x + z, 2x - 3y)$

5 $\boldsymbol{L}(x, y, z) = (3x - y, 2x + 3z)$.

6 $L(x, y) = (2x, 3y - x)$

7 $L(x, y) = (-x + y, 2y, 3x + y)$

8 $L(x, y) = (x - \frac{1}{2}y, 2y - x, 5x)$

9 $L(x, y, z, w) = (2x - 3y + z + w, x + y - 2z - 5w, \frac{1}{2}x - \frac{3}{2}z + \frac{1}{3}w)$

10 $L(x, y, z) = (\frac{1}{3}x - y + 3z, 2x - z, x - 3y + \frac{1}{2}z, -\frac{3}{2}x + \frac{1}{2}y - z)$

11 $L(\boldsymbol{i}) = (1, -1, 2), L(\boldsymbol{j}) = (-1, 1, 0), L(\boldsymbol{k}) = (2, 1, 1)$

12 $L(\boldsymbol{i}) = (0, -1, 1), L(\boldsymbol{j}) = (2, 0, 5), L(\boldsymbol{k}) = (-1, 2, 1)$

13 $L(\boldsymbol{i}) = (1, 3), L(\boldsymbol{j}) = (-1, 1), L(\boldsymbol{k}) = (-1, 1)$

14 $L(\boldsymbol{i}) = (1, 2), L(\boldsymbol{j}) = (-2, 1), L(\boldsymbol{k}) = (3, 7)$

15 $L(1, 0) = (1, -\frac{1}{2}), L(0, 1) = (-2, \frac{1}{4})$

16 $L(1, 0) = (1, \frac{1}{4}), L(0, 1) = (-\frac{1}{2}, 7)$

17 $L(1, 0) = (1, -1, 0), L(0, 1) = (0, -3, 0)$

18 $L(1, 0) = (1, 2, 3), L(0, 1) = (4, 5, 6)$

19 $L(1, 0, 0) = (2, -1, 1, 3),\ L(0, 1, 0) = (-\frac{1}{2}, 0, 1, 1),\ L(0, 0, 1) = (0, \frac{1}{3}, -\frac{1}{2}, 6)$

20 $L(1, 0, 0, 0) = (2, 1, -1),\ L(0, 1, 0, 0) = (-\frac{1}{2}, 0, 1),\ L(0, 0, 1, 0) = (\frac{3}{2}, 1, 0), L(0, 0, 0, 1) = (-2, -1, 1)$.

In Exercises 21 to 30, compute *L* of the given vector for the indicated *L*.

21 $(1, -1, 3)$; L of Exercise 3.

22 $(1, -1, 1)$; L of Exercise 4.

23 $(2, \frac{1}{2}, 1)$; L of Exercise 11.

24 $(3, -1, 1)$; L of Exercise 10.

25 $(2, -1)$; L of Exercise 7.

26 $(-3, 1)$; L of Exercise 8.

27 $(\frac{1}{2}, \frac{1}{3})$; L of Exercise 17.

28 $(-1, 3)$; L of Exercise 18.

29 $(-1, 1, 2, 1)$; L of Exercise 9.

30 $(1, 2, 3)$; L of Exercise 19.

31 In Example 2.7, suppose that a second storage facility is available which charges 25¢ per box for each of the first two months, and 10¢ per box for the third month. Should any of the growers *change* to this second facility?

32 Show that the system of linear equations

$$\begin{cases} a_{11}x + a_{12}y + a_{13}z = b_1 \\ a_{21}x + a_{22}y + a_{23}z = b_2 \\ a_{31}x + a_{32}y + a_{33}z = b_3 \end{cases}$$

can be written as an equation between two vectors, one of which is a matrix-vector product. (This was first observed by the English mathematician Sir Arthur Cayley in 1858.) **We will use this result in Section 4.**

33 Write the following system of first order differential equations as a matrix-vector product (cf. Exercise 32).

$$\begin{cases} \dfrac{dx_1}{dt} = \quad x_2 - x_3 \\ \dfrac{dx_2}{dt} = x_1 + x_2 \\ \dfrac{dx_3}{dt} = x_1 \quad + 2x_3 \end{cases}$$

34 Write the following system of first order differential equations as a matrix-vector product (cf. Exercise 32).

$$\begin{cases} \dfrac{dx_1}{dt} = 3x_1 - x_2 \\ \dfrac{dx_2}{dt} = -x_1 \qquad + x_3 \\ \dfrac{dx_3}{dt} = 2x_1 + x_2 - 2x_3 \end{cases}$$

35 A company has two factories, each producing two kinds of radios, as given by the following table of daily output.

	Factory 1	Factory 2
Radio A	2000	1000
Radio B	1500	2000

Factory 1 works 5 days per week. Factory 2 works 6 days per week. Find the total weekly production of each radio by the company.

36 A company produces three kinds of watches at each of two factories, as given by the following table of daily output.

	Factory 1	Factory 2
W_1	100	100
W_2	50	200
W_3	150	100

Factory 1 works 6 days per week. Factory 2 is on a 5 day per week schedule. Find the total weekly production of each kind of watch by the company.

37 Let A be an m-by-n matrix. Suppose that the function $\boldsymbol{F}: \boldsymbol{R}^n \to \boldsymbol{R}^m$ is defined by

$$\boldsymbol{F}(\boldsymbol{x}) = A\boldsymbol{x} \quad \text{for} \quad \boldsymbol{x} \in \boldsymbol{R}^n,$$

where the product $A\boldsymbol{x}$ is taken in the sense of Definition 2.5. Then show that $\boldsymbol{F}$ is a linear transformation.

3 MATRIX ARITHMETIC

We have seen how to represent a linear transformation $\boldsymbol{L}: \boldsymbol{R}^n \to \boldsymbol{R}^m$ by an m-by-n matrix $M_{\boldsymbol{L}}$. In this section we want to develop the arithmetic of matrices, which we will need to work with linear transformations. To start, suppose that we have two linear transformations $\boldsymbol{L}_1$ and $\boldsymbol{L}_2$ from $\boldsymbol{R}^n$ into $\boldsymbol{R}^m$ with representing matrices M_1 and M_2. What matrix M should represent the *sum* $\boldsymbol{L}_1 + \boldsymbol{L}_2$ defined by

$$(\boldsymbol{L}_1 + \boldsymbol{L}_2)(\boldsymbol{x}) = \boldsymbol{L}_1(\boldsymbol{x}) + \boldsymbol{L}_2(\boldsymbol{x})?$$

Well, from Equation (4) of the last section, we have

$$\begin{aligned} M_{\boldsymbol{L}_1+\boldsymbol{L}_2} &= ((\boldsymbol{L}_1 + \boldsymbol{L}_2)(\boldsymbol{e}_1), \ldots, (\boldsymbol{L}_1 + \boldsymbol{L}_2)(\boldsymbol{e}_n)) \\ &= (\boldsymbol{L}_1(\boldsymbol{e}_1) + \boldsymbol{L}_2(\boldsymbol{e}_1), \ldots, \boldsymbol{L}_1(\boldsymbol{e}_n) + \boldsymbol{L}_2(\boldsymbol{e}_n)) \end{aligned}$$

since by definition $(\boldsymbol{L}_1 + \boldsymbol{L}_2)(\boldsymbol{e}_j) = \boldsymbol{L}_1(\boldsymbol{e}_j) + \boldsymbol{L}_2(\boldsymbol{e}_j)$. So each column of M is obtained by adding the corresponding columns of M_1 and M_2. Thus for each row i and each column j, the (i, j)-entry of M is the sum of the (i, j)-entries from M_1 and M_2. Accordingly, the following definition is the natural one.

3.1 **DEFINITION.** If A and B are m-by-n matrices with $A=(a_{ij})$ and $B=(b_{ij})$, then $A+B$ is the m-by-n matrix whose (i, j) entry is $a_{ij}+b_{ij}$, for $1 \leq i \leq m$ and $1 \leq j \leq n$.

A similar analysis of the matrix of $a\boldsymbol{L}$ for any real number a leads us to give the following definition.

3.2 **DEFINITION.** If A is an m-by-n matrix and $a \in \boldsymbol{R}$, then aA is the m-by-n matrix whose (i, j)-entry is aa_{ij}. In particular, the (i, j)-entry of $-A$ is $-a_{ij}$. We also define $A-B$ to mean $A+(-B)$.

3.3 **EXAMPLE.** Compute $2A-3B$ if

$$A=\begin{pmatrix} 2 & 1 & -1 \\ 0 & 1 & 1 \end{pmatrix}, \qquad B=\begin{pmatrix} 1 & 2 & -3 \\ -1 & 1 & 0 \end{pmatrix}.$$

Solution.
$$2A-3B=\begin{pmatrix} 4 & 2 & -2 \\ 0 & 2 & 2 \end{pmatrix}-\begin{pmatrix} 3 & 6 & -9 \\ -3 & 3 & 0 \end{pmatrix}$$
$$=\begin{pmatrix} 1 & -4 & 7 \\ 3 & -1 & 2 \end{pmatrix}.$$

What about the *product* of two matrices A and B? We have already considered this in one case. In Definition 2.5 we gave the definition of the product AB in case A is m-by-n and B is n-by-1, i.e., in case

$$B=\boldsymbol{b}=\begin{pmatrix} b_1 \\ b_2 \\ \cdot \\ \cdot \\ \cdot \\ b_n \end{pmatrix} \in \boldsymbol{R}^n.$$

What we do in this case is to write

$$A=\begin{pmatrix} \boldsymbol{r}_1 \\ \boldsymbol{r}_2 \\ \cdot \\ \cdot \\ \cdot \\ \boldsymbol{r}_m \end{pmatrix}.$$

Then the product AB is the m-by-1 matrix

$$AB=\begin{pmatrix} \boldsymbol{r}_1 \cdot \boldsymbol{b} \\ \boldsymbol{r}_2 \cdot \boldsymbol{b} \\ \cdot \\ \cdot \\ \cdot \\ \boldsymbol{r}_m \cdot \boldsymbol{b} \end{pmatrix}.$$

The idea here is that the i-th entry of the product is obtained by taking the dot product of the i-th row vector of A with the column vector $\boldsymbol{b} = B$. Now, in the general case, how can we extend this idea? The natural way to proceed is to observe that if B is not a simple column vector, it nevertheless is made up of several (say p) columns arranged one after the other. Symbolically we have

$$B = \begin{pmatrix} b_{11} & b_{12} & \cdots & b_{1j} & \cdots & b_{1p} \\ b_{21} & b_{22} & \cdots & b_{2j} & \cdots & b_{2p} \\ \cdot & \cdot & & \cdot & & \cdot \\ \cdot & \cdot & & \cdot & & \cdot \\ \cdot & \cdot & & \cdot & & \cdot \\ b_{n1} & b_{n2} & \cdots & b_{nj} & \cdots & b_{np} \end{pmatrix} = (\boldsymbol{c}_1, \boldsymbol{c}_2, \ldots, \boldsymbol{c}_j, \ldots, \boldsymbol{c}_p)$$

where $\boldsymbol{c}_j$ is the j-th column vector. To give the product AB meaning, we define it to be the result of taking the products $A\boldsymbol{c}_1, A\boldsymbol{c}_2, \ldots, A\boldsymbol{c}_p$ as in Definition 2.5, and then making these vectors $A\boldsymbol{c}_j$ the columns of the m-by-p product matrix AB.

3.4 **DEFINITION.** Suppose that A is an m-by-n matrix and B is an n-by-p matrix. The **product** of A and B in that order is

(1) $$AB = (A\boldsymbol{c}_1, A\boldsymbol{c}_2, \ldots, A\boldsymbol{c}_j, \ldots, A\boldsymbol{c}_p)$$

where $B = (\boldsymbol{c}_1, \boldsymbol{c}_2, \ldots, \boldsymbol{c}_j, \ldots, \boldsymbol{c}_p)$ as a matrix of columns. Thus, AB is an m-by-p matrix.

You recall that each $A\boldsymbol{c}_j$ is computed by taking the dot products $\boldsymbol{r}_1 \cdot \boldsymbol{c}_j, \boldsymbol{r}_2 \cdot \boldsymbol{c}_j, \ldots, \boldsymbol{r}_i \cdot \boldsymbol{c}_j, \ldots, \boldsymbol{r}_m \cdot \boldsymbol{c}_j$ of the successive rows of A with $\boldsymbol{c}_j$ and arranging the m real numbers so obtained in a column. That is,

$$A\boldsymbol{c}_j = \begin{pmatrix} \boldsymbol{r}_1 \cdot \boldsymbol{c}_j \\ \cdot \\ \cdot \\ \cdot \\ \boldsymbol{r}_i \cdot \boldsymbol{c}_j \\ \cdot \\ \cdot \\ \cdot \\ \boldsymbol{r}_m \cdot \boldsymbol{c}_j \end{pmatrix}.$$

So we have the following entry-by-entry representation for AB:

(2) $$AB = \begin{pmatrix} \boldsymbol{r}_1 \cdot \boldsymbol{c}_1 & \boldsymbol{r}_1 \cdot \boldsymbol{c}_2 & \cdots & \boldsymbol{r}_1 \cdot \boldsymbol{c}_j & \cdots & \boldsymbol{r}_1 \cdot \boldsymbol{c}_p \\ \boldsymbol{r}_2 \cdot \boldsymbol{c}_1 & \boldsymbol{r}_2 \cdot \boldsymbol{c}_2 & \cdots & \boldsymbol{r}_2 \cdot \boldsymbol{c}_j & \cdots & \boldsymbol{r}_2 \cdot \boldsymbol{c}_p \\ \cdot & \cdot & & \cdot & & \cdot \\ \cdot & \cdot & & \cdot & & \cdot \\ \cdot & \cdot & & \cdot & & \cdot \\ \boldsymbol{r}_i \cdot \boldsymbol{c}_1 & \boldsymbol{r}_i \cdot \boldsymbol{c}_2 & \cdots & \boldsymbol{r}_i \cdot \boldsymbol{c}_j & \cdots & \boldsymbol{r}_i \cdot \boldsymbol{c}_p \\ \cdot & \cdot & & \cdot & & \cdot \\ \cdot & \cdot & & \cdot & & \cdot \\ \cdot & \cdot & & \cdot & & \cdot \\ \boldsymbol{r}_m \cdot \boldsymbol{c}_1 & \boldsymbol{r}_m \cdot \boldsymbol{c}_2 & \cdots & \boldsymbol{r}_m \cdot \boldsymbol{c}_j & \cdots & \boldsymbol{r}_m \cdot \boldsymbol{c}_p \end{pmatrix}$$

This shows that the typical (i, j)-entry of AB is

$$\boldsymbol{r}_i \cdot \boldsymbol{c}_j = (r_{i1}, r_{i2}, \ldots, r_{in}) \cdot \begin{pmatrix} c_{1j} \\ c_{2j} \\ \cdot \\ \cdot \\ \cdot \\ c_{nj} \end{pmatrix}$$

$$= r_{i1}c_{1j} + r_{i2}c_{2j} + \cdots + r_{in}c_{nj}$$

$$= \sum_{k=1}^{n} r_{ik}c_{kj}.$$

Hence one sometimes abbreviates (2) in the form $AB = (\sum_{k=1}^{n} r_{ik}c_{kj})$, where the term in parentheses represents the entry in row i, column j of AB. For computational purposes, you can simply remember either Equation (1) or (2).

Take particular note of the fact that *the product AB is defined only when A has the same number (n) of columns as B has rows*, since this is the only time the dot products $\boldsymbol{r}_i \cdot \boldsymbol{c}_j$ will be meaningful.

3.5 **EXAMPLE.** Compute AB if possible.

(a) $A = \begin{pmatrix} 3 & 1 & -1 \\ 2 & 1 & 1 \end{pmatrix}$, $\quad B = \begin{pmatrix} 1 & 2 & 3 \\ 1 & 2 & -1 \end{pmatrix}$

(b) $A = \begin{pmatrix} 3 & 1 & -1 \\ 2 & 1 & 1 \end{pmatrix}$, $\quad B = \begin{pmatrix} 1 & 4 \\ -1 & 1 \\ 3 & 1 \end{pmatrix}$

(c) $A = (1, 2)$, $\quad B = \begin{pmatrix} 1 & 2 & 1 \\ -1 & 1 & 3 \end{pmatrix}$.

Solution. (a) Here A is 2-by-③ and B is ②-by-3. The circled numbers fail to match, so we cannot compute AB: the number (3) of columns of A does *not* equal the number (2) of rows of B.

(b) Here A is 2-by-③ and B is ③-by-2. So we can compute the product. Using (2) we get

$$AB = \begin{pmatrix} 3(1)+1(-1)+(-1)3 & 3(4)+1(1)-1(1) \\ 2(1)+1(-1)+1(3) & 2(4)+1(1)+1(1) \end{pmatrix}$$

$$= \begin{pmatrix} 3-1-3 & 12+1-1 \\ 2-1+3 & 8+1+1 \end{pmatrix} = \begin{pmatrix} -1 & 12 \\ 4 & 10 \end{pmatrix}.$$

(c) Here A is 1-by-② and B is ②-by-3, so we can compute AB. We get

$$AB = (1-2, 2+2, 1+6) = (-1, 4, 7).$$

We can now add, subtract, and multiply matrices as long as their sizes are compatible with such operations. (The question of division must be deferred until the next section.) You may be wondering whether addition and multiplication of matrices obey the same familiar rules that addition and multiplication of real numbers obey. In the case of addition, the answer is pretty much "yes." In fact, addition and scalar multiplication of matrices behave exactly like addition and scalar multiplication of vectors.

3.6 THEOREM. If A, B, C are m-by-n matrices, and a and b are any real numbers, then

(i) $A+(B+C)=(A+B)+C$;
(ii) $A+B=B+A$;
(iii) $A+0=A$, where 0 is the m-by-n matrix with 0 in each position;
(iv) $-A+A=0$
(v) $a(A+B)=aA+aB$;
(vi) $(a+b)A=aA+bA$;
(vii) $(ab)A=a(bA)$
(viii) $1A=A$ and $0A=0$.

Hence the set of all m-by-n matrices is a real vector space.

Partial Proof. The reasoning is almost identical to that used in proving Theorem 1.1.11, since we have defined addition and multiplication by scalars entry-wise. For example, to prove (v), note that the (i, j)-entry of $a(A+B)$ is $a(a_{ij}+b_{ij})= aa_{ij}+ab_{ij}$, which is the same as the (i, j)-entry of $aA+aB$, for any i and j. Hence $a(A+B)=aA+aB$. The remaining assertions can be proved in the same way.

Multiplication of two vectors is quite a different story, however. The commutative law, $AB=BA$, fails utterly. In the first place, AB might be defined (e.g., if A is 2-by-3 and B is 3-by-1) while BA might *not* be defined. But this isn't the only difficulty. Even if *both* AB and BA are defined, they need not be equal! Consider for example

$$A=\begin{pmatrix}0 & 1\\ 0 & 0\end{pmatrix}$$

and

$$B=\begin{pmatrix}0 & 0\\ 0 & 1\end{pmatrix}.$$

Then

$$AB=\begin{pmatrix}0 & 1\\ 0 & 0\end{pmatrix} \quad \text{while} \quad BA=\begin{pmatrix}0 & 0\\ 0 & 0\end{pmatrix}.$$

So $AB \neq BA$! A second look reveals further bizarre behavior. Notice that neither B nor A is the zero matrix 0, and yet $BA=0$. Again, such a thing never happens when multiplying real numbers. Furthermore, we will see in the next section that, unlike the situation in the real number system, even if B is not the zero matrix, we still may not be able to define $A \div B$ even if the sizes of A and B are compatible! By now you may be viewing matrix multiplication with a jaundiced eye. Before you reject it as a totally chaotic concept unworthy of the name multiplication, we will show that it *does* have at least one important arithmetical property: the associative law $(AB)C=A(BC)$ holds when the matrices are of compatible sizes. While this may not seem like much, for our purposes it turns out to be crucial.

3.7 THEOREM. If A is m-by-n, B is n-by-p, and C is p-by-q, then

$$(AB)C=A(BC). \tag{3}$$

Proof. Each matrix in (3) is m-by-q, so we need only check that the corresponding entries of $(AB)C$ and $A(BC)$ are equal. The (i, j)-entry of $(AB)C$ is the dot product of row i of AB with column j of C. Row i of AB is the result of computing the successive dot products of $\boldsymbol{a}_i = (a_{i1}, a_{i2}, \ldots, a_{in})$, the i-th row of A, with each column

$$\begin{pmatrix} b_{11} \\ b_{21} \\ \cdot \\ \cdot \\ \cdot \\ b_{n1} \end{pmatrix}, \begin{pmatrix} b_{12} \\ b_{22} \\ \cdot \\ \cdot \\ \cdot \\ b_{n2} \end{pmatrix}, \ldots, \begin{pmatrix} b_{1p} \\ b_{2p} \\ \cdot \\ \cdot \\ \cdot \\ b_{np} \end{pmatrix}.$$

of B. Thus row i of AB is

$$\left(\sum_{k=1}^{n} a_{ik}b_{k1}, \sum_{k=1}^{n} a_{ik}b_{k2}, \ldots, \sum_{i=1}^{n} a_{ik}b_{kp}\right).$$

Since the j-th column of C is

$$\mathbf{c}_j = \begin{pmatrix} c_{1j} \\ c_{2j} \\ \cdot \\ \cdot \\ \cdot \\ c_{pj} \end{pmatrix},$$

the (i, j)-entry of $(AB)C$ is

$$\sum_{k=1}^{n} a_{ik}b_{k1}c_{1j} + \sum_{k=1}^{n} a_{ik}b_{k2}c_{2j} + \cdots + \sum_{k=1}^{n} a_{ik}b_{kp}c_{pj}$$

i.e.,

(4) $$\sum_{\ell=1}^{p}\left(\sum_{k=1}^{n} a_{ik}b_{k\ell}\right)c_{\ell j} = \sum_{k=1}^{n}\sum_{\ell=1}^{p} a_{ik}b_{k\ell}c_{\ell j}.$$

Next we compute the (i, j)-entry of $A(BC)$. This is the dot product of $\boldsymbol{a}_i$ with column j of BC. The latter column is the result of computing the successive dot products of each row of B,

$$(b_{11}, b_{12}, \ldots, b_{1p}), (b_{21}, b_{22}, \ldots, b_{2p}), \ldots, (b_{n1}, b_{n2}, \ldots, b_{np}),$$

with $\mathbf{c}_j$. Thus column j of BC is

$$\begin{pmatrix} \sum_{\ell=1}^{p} b_{1\ell}c_{\ell j} \\ \sum_{\ell=1}^{p} b_{2\ell}c_{\ell j} \\ \cdot \\ \cdot \\ \cdot \\ \sum_{\ell=1}^{p} b_{n\ell}c_{\ell j} \end{pmatrix}.$$

Therefore, the (i, j)-entry of $A(BC)$ is

$$a_{i1}\sum_{\ell=1}^{p} b_{1\ell}c_{\ell j} + a_{i2}\sum_{\ell=1}^{p} b_{2\ell}c_{\ell j} + \cdots + a_{in}\sum_{\ell=1}^{p} b_{n\ell}c_{\ell j},$$

i.e.,

(5) $$\sum_{k=1}^{n} a_{ik}\left(\sum_{\ell=1}^{p} b_{k\ell}c_{\ell j}\right) = \sum_{k=1}^{n}\sum_{\ell=1}^{p} a_{ik}b_{k\ell}c_{\ell j}.$$

Since (4) and (5) coincide, we have shown that the (i, j)-entries of $(AB)C$ and $A(BC)$ are equal. QED

We now are in a position to establish a very important result on the matrix which represents the composite $\mathbf{L}_2 \circ \mathbf{L}_1$ of two linear transformations $\mathbf{L}_1 : \mathbf{R}^p \to \mathbf{R}^n$, $\mathbf{L}_2 : \mathbf{R}^n \to \mathbf{R}^m$.

3.8 **THEOREM.** Suppose that $\mathbf{L}_1 : \mathbf{R}^p \to \mathbf{R}^n$ and $\mathbf{L}_2 : \mathbf{R}^n \to \mathbf{R}^m$ are linear transformations with matrices M_1 and M_2, respectively, relative to the standard bases. Then $\mathbf{L}_2 \circ \mathbf{L}_1 : \mathbf{R}^p \to \mathbf{R}^m$ is a linear transformation whose matrix relative to the standard bases is the matrix product M_2M_1.

Proof. We know that $\mathbf{L}_2 \circ \mathbf{L}_1$ is a linear transformation from Exercise 23 of Exercises 6.1. Also, we know that M_1 is an n-by-p matrix, and M_2 is a m-by-n matrix, by Definition 2.1, so that M_2M_1 is m-by-p, the size of the matrix of $\mathbf{L}_2 \circ \mathbf{L}_1 : \mathbf{R}^p \to \mathbf{R}^m$. Now we can compute, for $\mathbf{x} \in \mathbf{R}^p$, regarded as a p-by-1 matrix,

$$\begin{aligned}\mathbf{L}_2 \circ \mathbf{L}_1(\mathbf{x}) = \mathbf{L}_2(\mathbf{L}_1(\mathbf{x})) = \mathbf{L}_2(M_1\mathbf{x}) &\quad \text{by Theorem 2.4,}\\ = M_2(M_1\mathbf{x}) &\quad \text{by Theorem 2.4 again,}\\ = (M_2M_1)\mathbf{x} &\quad \text{by Theorem 3.7.}\end{aligned}$$

Since M_2M_1 is *one* matrix such that $\mathbf{L}_2 \circ \mathbf{L}_1(\mathbf{x}) = (M_2M_1)\mathbf{x}$, we conclude from Theorem 2.8 that M_2M_1 is *the* matrix of $\mathbf{L}_2 \circ \mathbf{L}_1$ relative to the standard bases. QED

EXERCISES 6.3

In Exercises 1 to 6, compute (a) $3A - 2B$, (b) AB, (c) BA if possible.

1 $A = \begin{pmatrix} 2 & 1 & -1 \\ 3 & 0 & 1 \end{pmatrix}$, $B = \begin{pmatrix} 2 & -2 & 1 \\ -1 & 1 & 2 \end{pmatrix}$

2 $A = \begin{pmatrix} 1 & 5 \\ -1 & 1 \end{pmatrix}$, $B = \begin{pmatrix} 1 & 0 \\ 1 & 1 \end{pmatrix}$

3 $A = \begin{pmatrix} 1 & 2 & 0 \\ -1 & 1 & 1 \\ 0 & 1 & 7 \end{pmatrix}$, $B = \begin{pmatrix} 1 & -1 & 0 \\ 2 & 1 & 1 \end{pmatrix}$.

4 $A = \begin{pmatrix} -1 & 1 & 3 \\ 2 & 1 & 0 \end{pmatrix}$, $B = \begin{pmatrix} 1 & 0 & -4 \\ 0 & 2 & -1 \\ 3 & -1 & 2 \end{pmatrix}$.

5 $A = \begin{pmatrix} 2 & -1 & 0 \\ -1 & 2 & 0 \\ 0 & 1 & 3 \end{pmatrix}$, $B = \begin{pmatrix} 1 & -1 & 1 \\ -3 & 1 & 0 \\ 2 & -1 & -3 \end{pmatrix}$.

6 $A=\begin{pmatrix}1 & -1 & 0 & 1\\ 2 & 0 & -1 & 1\\ 3 & -1 & 2 & 2\end{pmatrix}$, $B=\begin{pmatrix}0 & 1 & -1\\ -1 & 2 & 1\\ 1 & 0 & 2\\ -1 & 0 & -3\end{pmatrix}$.

In Exercises 7 to 12, find the matrix of the linear transformation $L_2 \circ L_1$.

7 $L_1(x, y, z) = (x-2y+z, -2x+z, -x+y-3z)$
$L_2(x, y, z) = (x-2y-3z, 4x-y+z)$.

8 $L_1(x, y) = (x-y, x+y, 2x-3y)$
$L_2(x, y, z) = (2x+y-z, -x-2y+z)$.

9 $L_1(x, y) = (-x+y, -2x+3y)$
$L_2(x, y) = (3x+2y, 2x-y)$.

10 $L_1(x, y, z) = (x-y, y-z, z-x)$
$L_2(x, y, z) = (2x-y+z, x-2y+z, x+y-3z)$.

11 $L_1(x, y, z, w) = (2x-y+3z+w, y-3z-w)$
$L_2(x, y) = (x+2y, 3x-y, -x+2y)$.

12 $L_1(x, y) = (x+y, -x+2y, 3y)$
$L_2(x, y, z) = (x+z, x+y, y+z, x+y-z)$.

In Exercises 13 to 16, use Theorems 3.8 and 2.4 to find the given vector.

13 $L_2 \circ L_1(1, 5, -2)$ if $L_1(\boldsymbol{i}) = (2, -1, 1)$, $L_1(\boldsymbol{j}) = (3, 1, 2)$, $L_1(\boldsymbol{k}) = (-1, 0, 2)$; $L_2(\boldsymbol{i}) = (2, -1)$, $L_2(\boldsymbol{j}) = (-1, 2)$, $L_2(\boldsymbol{k}) = (3, -1)$.

14 $L_2 \circ L_1(-1, 3, 3)$ if $L_1(\boldsymbol{i}) = (1, -1, 0)$, $L_1(\boldsymbol{j}) = (1, 0, -1)$, $L_1(\boldsymbol{k}) = (0, -1, 1)$, $L_2(\boldsymbol{i}) = (3, 4)$, $L_2(\boldsymbol{j}) = (-1, 3)$, $L_2(\boldsymbol{k}) = (2, 1)$.

15 $L_2 \circ L_1(1, 2, 3)$ if L_1 and L_2 are as in Exercise 7.

16 $L_2 \circ L_1(1, -1, 1)$ if L_1 and L_2 are as in Exercise 10.

17 Verify Theorem 3.7 for

$$A=\begin{pmatrix}2 & 1 & 1\\ -1 & 2 & 0\end{pmatrix},\quad B=\begin{pmatrix}1 & -1 & 0\\ 0 & 4 & 2\\ 1 & 0 & 3\end{pmatrix},\quad C=\begin{pmatrix}3 & -1\\ 1 & -2\\ 0 & 3\end{pmatrix}.$$

18 Verify Theorem 3.7 for

$$A=\begin{pmatrix}-1 & 2 & 3\\ 0 & -2 & 1\end{pmatrix},\quad B=\begin{pmatrix}2 & -1 & 1\\ 0 & 3 & 1\\ 1 & -1 & 2\end{pmatrix},\quad C=\begin{pmatrix}1 & -3\\ 2 & 1\\ -1 & 3\end{pmatrix}.$$

19 Prove parts (i) and (vi) of Theorem 3.6.

20 Prove parts (iii), (iv), and (vii) of Theorem 3.6.

21 (a) What is the matrix M_L of the linear transformation $L: \boldsymbol{R}^3 \to \boldsymbol{R}^3$ given by $L(x, y, z) = (cx, cy, cz)$ for some scalar $c \in \boldsymbol{R}$?
(b) Show that the matrix M_L in (a) commutes with every 3-by-3 matrix A; that is, $AM_L = M_L A$.

22 Show that if a 3-by-3 matrix S commutes with every 3-by-3 matrix A (i.e., $SA = AS$), then

$$S=\begin{pmatrix}c & 0 & 0\\ 0 & c & 0\\ 0 & 0 & c\end{pmatrix}$$

for some scalar $c \in \boldsymbol{R}$. (*Hint:* Consider SE_{ij}, where E_{ij} has 1 as (i, j)-entry and 0 in every other position.)

The next two exercises will be helpful for the next section.

23 The three-by-three *identity matrix* is

$$I_3 = \begin{pmatrix} 1 & 0 & 0 \\ 0 & 1 & 0 \\ 0 & 0 & 1 \end{pmatrix}.$$

Show that $I_3A = AI_3 = A$ for every 3-by-3 matrix A.

24 The two-by-two *identity matrix* is

$$I_2 = \begin{pmatrix} 1 & 0 \\ 0 & 1 \end{pmatrix}.$$

Show that $I_2A = AI_2$ for every 2-by-2 matrix A.

25 Let R_{12} be the elementary row operator that interchanges rows 1 and 2 of a 3-by-3 matrix. Show that $R_{12}(A) = E_{12}A$, where $E_{12} = R_{12}(I_3)$.

26 Generalize Exercise 25: Show that if R_{ij} is the elementary row operator that interchanges rows i and j of an m-by-n matrix A, then $R_{ij}(A) = E_{ij}A$, where $E_{ij} = R_{ij}(I_m)$. Here I_m is the m-by-m identity matrix having a 1 in each diagonal position (i, i) and a 0 in each off-diagonal position (i, j) for $i \neq j$.

27 Let $R_i(c)$ be the elementary row operator that multiplies row i of a 3-by-3 matrix by $c \neq 0$. Show that $R_i(c)(A) = E_i(c)A$, where $E_i(c) = R_i(c)(I_3)$.

28 Generalize Exercise 27: Show that if $R_i(c)$ is the elementary row operator that multiplies row i of an m-by-n matrix A by $c \neq 0$, then $R_i(c)(A) = E_i(c)A$, where $E_i(c) = R_i(c)(I_m)$, and I_m is as in Exercise 26.

29 Let $R_{23}(c)$ be the elementary row operator that adds $c\boldsymbol{r}_3$ to $\boldsymbol{r}_2$ for

$$A = \begin{pmatrix} \boldsymbol{r}_1 \\ \boldsymbol{r}_2 \\ \boldsymbol{r}_3 \end{pmatrix},$$

a 3-by-3 matrix. Show that $R_{23}(c)(A) = E_{23}(c)A$, where $E_{23}(c) = R_{23}(c)(I_3)$.

30 Generalize Exercise 29: Let $R_{ij}(c)$ be the elementary row operator that adds $c\boldsymbol{r}_j$ to $\boldsymbol{r}_i$ of the m-by-n matrix A whose rows are $\boldsymbol{r}_1, \boldsymbol{r}_2, \ldots, \boldsymbol{r}_m$. Show that $R_{ij}(c)(A) = E_{ij}(c)A$, where $E_{ij}(c) = R_{ij}(c)(I_m)$ and I_m is as in Exercise 26.

31 Prove the *left-distributive* law for matrix multiplication: If A is m-by-n and B and C are n-by-p, then $A(B+C) = AB + AC$.

32 Prove the *right-distributive* law for matrix multiplication: If A and B are m-by-n, and C is n-by-p, then $(A+B)C = AC + BC$.

33 The *Pauli-spin* matrices in atomic physics are

$$S_x = \begin{pmatrix} 0 & 1 \\ 1 & 0 \end{pmatrix}, \qquad S_y = \begin{pmatrix} 0 & -i \\ i & 0 \end{pmatrix}, \qquad S_z = \begin{pmatrix} 1 & 0 \\ 0 & -1 \end{pmatrix},$$

where $i^2 = -1$. Show that $S_x^2 = S_y^2 = S_z^2 = I_2$.

34 Refer to Exercise 33. Show that $S_xS_y = -S_yS_x = iS_z$, $S_yS_z = -S_zS_y = iS_x$, and $S_zS_x = -S_xS_z = iS_y$.

4 MATRIX INVERSION

We have learned how to add, subtract, and multiply matrices. It is natural now to ask about division. If you recall the situation in the arithmetic of the rational numbers or complex numbers (cf. Section 8.5), then you will recognize our approach to matrix division. How does one divide a rational number p/q by a rational number r/s? By *inverting* r/s *if possible* (i.e., if $r \neq 0$), *and then multiplying:*

$$\frac{p}{q} \div \frac{r}{s} = \frac{p}{q} \cdot \left(\frac{r}{s}\right)^{-1} = \frac{p}{q} \cdot \frac{s}{r}.$$

The same approach is used in complex arithmetic:

$$z_1 \div z_2 = z_1 \cdot (z_2)^{-1} = z_1 \cdot \frac{1}{z_2}$$

if we can invert z_2. Thus it seems reasonable to define division of a matrix M_1 by another matrix M_2 to mean the *product* of M_1 and the *inverse* M_2^{-1} of M_2.

This leaves us with the task of defining M_2^{-1} as the analogue of $\left(\frac{r}{s}\right)^{-1}$ for $\frac{r}{s}$ a rational number. Recall that $\left(\frac{r}{s}\right)^{-1}$ is characterized by the property

$$\left(\frac{r}{s}\right)\left(\frac{r}{s}\right)^{-1} = 1 = \left(\frac{r}{s}\right)^{-1}\left(\frac{r}{s}\right).$$

Since matrix multiplication is not in general commutative, it seems that we should define M_2^{-1} by requiring that *both* $M_2M_2^{-1}$ and $M_2^{-1}M_2$ give the matrix analogue of the rational number 1. To begin with, in order to ensure that both $M_2M_2^{-1}$ and $M_2^{-1}M_2$ be *defined*, we restrict our attention to the case of *square* (i.e., n-by-n for some n) matrices. Next, we have to define the n-by-n matrix which is analogous to the number 1 (cf. Exercises 23–24, Exercises 6.3).

4.1 **DEFINITION.** The ***n*-by-*n* identity matrix** is

$$I_n = \begin{pmatrix} 1 & 0 & 0 & \cdots & 0 \\ 0 & 1 & 0 & \cdots & 0 \\ 0 & 0 & 1 & & 0 \\ \cdot & \cdot & \cdot & \cdot & \cdot \\ \cdot & \cdot & \cdot & \cdot & \cdot \\ \cdot & \cdot & \cdot & \cdot & \cdot \\ 0 & 0 & 0 & \cdots & 1 \end{pmatrix}$$

We must proceed to show that for any n-by-n matrix A,

(*) $$AI_n = I_nA = A,$$

i.e., that I_n *does* behave like 1 with respect to multiplication. Both I_nA and AI_n are n-by-n matrices, so to check (*) we just have to verify that a_{ij} is the

(i, j)-entry of both AI_n and I_nA. But this follows immediately:

$$\underbrace{(a_{i1} \cdots a_{ij} \cdots a_{in})}_{\text{row } i \text{ of } A} \cdot \begin{pmatrix} 0 \\ \cdot \\ \cdot \\ \cdot \\ 0 \\ 1 \\ 0 \\ \cdot \\ \cdot \\ \cdot \\ 0 \end{pmatrix} = a_{ij} = (0, \ldots 0, 1, 0, \ldots 0) \cdot \underbrace{\begin{pmatrix} a_{1j} \\ \cdot \\ \cdot \\ \cdot \\ a_{ij} \\ \cdot \\ \cdot \\ \cdot \\ a_{nj} \end{pmatrix}}_{\text{column } j \text{ of } A}$$

where the j-th column of I_n $(= \boldsymbol{e}_j)$ and the i-th row of I_n $(= \boldsymbol{e}_i)$ are shown. Thus (*) is proved.

Now we can define the inverse of an n-by-n matrix A.

4.2 **DEFINITION.** If A is an n-by-n matrix, then its **inverse matrix** is the n-by-n matrix A^{-1} such that

$$A^{-1}A = AA^{-1} = I_n,$$

if such a matrix A^{-1} exists.

In the definition we have referred to *the* n-by-n matrix whose product with A is I_n. This is justified by the following uniqueness result, which guarantees that there is at most one such matrix.

4.3 **PROPOSITION.** If the n-by-n matrix A has an inverse matrix and if C is *any* n-by-n matrix such that $AC = I_n$, then $C = A^{-1}$.

Proof. Multiply each side of the equation $AC = I_n$ on the left side by A^{-1}. This gives, in view of Theorem 3.7 and the defining property of I_n,

$$\begin{aligned} A^{-1}(AC) &= A^{-1}I_n \\ (A^{-1}A)C &= A^{-1} \\ I_nC &= A^{-1} \\ C &= A^{-1} \end{aligned}$$

QED

Now that we know that an n-by-n matrix A has at most one inverse, how can we determine whether a given A is invertible, and, if it is, how can we find A^{-1}? Well, first note that $I_n = (\boldsymbol{e}_1, \boldsymbol{e}_2, \ldots, \boldsymbol{e}_n)$. So we are seeking a matrix $A^{-1} = (\boldsymbol{x}_1, \boldsymbol{x}_2, \ldots, \boldsymbol{x}_n)$ such that AA^{-1} has $\boldsymbol{e}_j$ as its j-th column for any j. What is the j-th column of AA^{-1}? If

$$A = \begin{pmatrix} \boldsymbol{r}_1 \\ \boldsymbol{r}_2 \\ \cdot \\ \cdot \\ \cdot \\ \boldsymbol{r}_n \end{pmatrix},$$

then the j-th column of AA^{-1} is just

$$A\boldsymbol{x}_j = \begin{pmatrix} \boldsymbol{r}_1 \cdot \boldsymbol{x}_j \\ \boldsymbol{r}_2 \cdot \boldsymbol{x}_j \\ \cdot \\ \cdot \\ \cdot \\ \boldsymbol{r}_n \cdot \boldsymbol{x}_j \end{pmatrix}$$

by (6) of Definition 2.5. Hence we can determine the j-th column of A^{-1} by solving the system of linear equations

$$\begin{cases} \boldsymbol{r}_1 \cdot \boldsymbol{x}_j = 0 \\ \quad \cdot \qquad \cdot \\ \quad \cdot \qquad \cdot \\ \quad \cdot \qquad \cdot \\ \boldsymbol{r}_j \cdot \boldsymbol{x}_j = 1 \\ \quad \cdot \qquad \cdot \\ \quad \cdot \qquad \cdot \\ \quad \cdot \qquad \cdot \\ \boldsymbol{r}_n \cdot \boldsymbol{x}_j = 0 \end{cases}$$

Using Exercise 32, Exercises 6.2, we can write this system more compactly as

(1) $$A\boldsymbol{x}_j = \boldsymbol{e}_j, \quad j = 1, 2, \ldots, n.$$

Moreover, we know how to determine whether (1) has a solution and, if it has, how to find that solution. We can use the technique of *Gaussian elimination* from Section 1.6.

The only obstacle to carrying out the foregoing scheme seems to lie in the necessity of solving n *separate* systems of the form (1) (for $j = 1, j = 2, \ldots, j = n$) in order to determine all the columns of A^{-1}. This in fact is what we will do, but we will solve *all n systems together.* We can do that by taking advantage of the fact that in each system (1), the matrix of coefficients is A. In fact, the only thing that changes from one system to the next is the column of constants $\boldsymbol{e}_j$. So we can write all n systems (1) in the matrix form of p. 50 ff.,

(2) $$(A \,\vdots\, \boldsymbol{e}_1\, \boldsymbol{e}_2 \ldots \boldsymbol{e}_n) = (A \,\vdots\, I_n).$$

We can then proceed to solve the systems represented in (2) by using elementary row operations on the array (2) in an effort to reduce it to $(I_n \,\vdots\, C)$. Our next result assures us that, if A is invertible, then we will be successful in that effort.

4.4 **PROPOSITION.** If the n-by-n matrix A has an inverse, then A can be reduced to I_n by a sequence of elementary row operations.

Proof. First we note that if A has an inverse, then the homogeneous system of linear equations

(3) $$A\boldsymbol{x} = \boldsymbol{0}$$

has only the trivial solution $\boldsymbol{x} = \boldsymbol{0}$. For we can multiply both sides of (3) by A^{-1} to obtain

$$\begin{aligned} A^{-1}(A\boldsymbol{x}) &= A^{-1}\boldsymbol{0} \\ (A^{-1}A)\boldsymbol{x} &= \boldsymbol{0} \\ I_n\boldsymbol{x} &= \boldsymbol{0} \\ \boldsymbol{x} &= \boldsymbol{0}. \end{aligned}$$

Therefore, if we solve (3) by Gaussian elimination, we will obtain the solution

$$\begin{aligned} x_1 \qquad\qquad &= 0 \\ x_2 \qquad &= 0 \\ &\ \ \vdots \\ x_n &= 0. \end{aligned}$$

This means that by elementary row operations, the matrix $(A \vdots 0)$ can be reduced to $(I_n \vdots 0)$. Hence A can be reduced to I_n. QED

We can draw the following immediate conclusion from Proposition 4.4.

4.5 **COROLLARY.** If the n-by-n matrix A cannot be reduced to I_n by means of elementary row operations, then A is not invertible.

Thus if we *can't* reduce (2) to the form

(4) $$(I_n \vdots C),$$

then we can conclude that A doesn't have an inverse.

What if A *is* invertible? Then we *can* reduce (2) to (4). We can then conclude that $C = A^{-1}$. Why? Because for each j, the j-th column to the right of the dotted line is the solution $\boldsymbol{x}_j$ to the system of equations (1). This means then that the matrix C is one matrix which satisfies

(5) $$AC = I_n.$$

But since A is invertible, $C = A^{-1}$ by Proposition 4.3.

To sum up, here is our method: *Try to reduce the array* $(A \vdots I_n)$ *to* $(I_n \vdots C)$ *by performing elementary row operations. If we can't do so, then* A *isn't invertible. If we can, then* $C = A^{-1}$.

4.6 **EXAMPLE.** Find the inverse of the following matrix, if it has an inverse:

$$A = \begin{pmatrix} 1 & -1 & 1 \\ -1 & 2 & -1 \\ 2 & -1 & 1 \end{pmatrix}.$$

Solution. We reduce $(I_n \vdots A)$ to $(A^{-1} \vdots I_n)$ by performing the indicated elementary row operations.

$$\left(\begin{array}{rrr:rrr} 1 & 0 & 0 & 1 & -1 & 1 \\ 0 & 1 & 0 & -1 & 2 & -1 \\ 0 & 0 & 1 & 2 & -1 & 1 \end{array}\right) \xrightarrow[\text{Add } -2R_1 \text{ to } R_3]{\text{Add } R_1 \text{ to } R_2}$$

$$\left(\begin{array}{rrr:rrr} 1 & 0 & 0 & 1 & -1 & 1 \\ 1 & 1 & 0 & 0 & 1 & 0 \\ -2 & 0 & 1 & 0 & 1 & -1 \end{array}\right) \xrightarrow[\text{Add } -R_2 \text{ to } R_3]{\text{Add } R_2 \text{ to } R_1}$$

$$\left(\begin{array}{rrr:rrr} 2 & 1 & 0 & 1 & 0 & 1 \\ 1 & 1 & 0 & 0 & 1 & 0 \\ -3 & -1 & 1 & 0 & 0 & -1 \end{array}\right) \xrightarrow[\text{Add } -R_3 \text{ to } R_1]{-R_3}$$

$$\left(\begin{array}{rrr:rrr} -1 & 0 & 1 & 1 & 0 & 0 \\ 1 & 1 & 0 & 0 & 1 & 0 \\ 3 & 1 & -1 & 0 & 0 & 1 \end{array}\right).$$

Thus,

$$A^{-1} = \begin{pmatrix} -1 & 0 & 1 \\ 1 & 1 & 0 \\ 3 & 1 & -1 \end{pmatrix}.$$

As a check, you can verify that $A^{-1}A = I_3$.

In general, even if A is a matrix of integers, A^{-1} need not be, as the following example illustrates.

4.7 EXAMPLE. Find the inverse of the following matrix, if it has an inverse.

$$A = \begin{pmatrix} 2 & -1 \\ 1 & 1 \end{pmatrix}.$$

Solution. We proceed as in Example 4.6.

$$\left(\begin{array}{rr:rr} 1 & 0 & 2 & -1 \\ 0 & 1 & 1 & 1 \end{array}\right) \xrightarrow[R_2 \text{ and } R_1]{\text{Interchange}} \left(\begin{array}{rr:rr} 0 & 1 & 1 & 1 \\ 1 & 0 & 2 & -1 \end{array}\right) \xrightarrow[\text{to } R_2]{\text{Add } -2R_1}$$

$$\left(\begin{array}{rr:rr} 0 & 1 & 1 & 1 \\ 1 & -2 & 0 & -3 \end{array}\right) \xrightarrow{-R_2/3} \left(\begin{array}{rr:rr} 0 & 1 & 1 & 1 \\ -\frac{1}{3} & \frac{2}{3} & 0 & 1 \end{array}\right) \xrightarrow[\text{to } R_1]{\text{Add } -R_2}$$

$$\left(\begin{array}{rr:rr} \frac{1}{3} & \frac{1}{3} & 1 & 0 \\ -\frac{1}{3} & \frac{2}{3} & 0 & 1 \end{array}\right).$$

Hence,

$$A^{-1} = \begin{pmatrix} \frac{1}{3} & \frac{1}{3} \\ -\frac{1}{3} & \frac{2}{3} \end{pmatrix} = \frac{1}{3}\begin{pmatrix} 1 & 1 \\ -1 & 2 \end{pmatrix}.$$

4.8 EXAMPLE. Find the inverse of the following matrix, if it has an inverse.

$$A = \begin{pmatrix} 3 & 1 & 5 \\ 2 & 4 & 1 \\ -4 & 2 & -9 \end{pmatrix}.$$

Solution. Our procedure gives

$$\left(\begin{array}{ccc:ccc} 1 & 0 & 0 & 3 & 1 & 5 \\ 0 & 1 & 0 & 2 & 4 & 1 \\ 0 & 0 & 1 & -4 & 2 & -9 \end{array}\right) \xrightarrow[\text{to } R_1]{\text{Add } -R_2}$$

$$\left(\begin{array}{ccc:ccc} 1 & -1 & 0 & 1 & -3 & 4 \\ 0 & 1 & 0 & 2 & 4 & 1 \\ 0 & 0 & 1 & -4 & 2 & -9 \end{array}\right) \xrightarrow[\text{Add } 4R_1 \text{ to } R_3]{\text{Add } -2R_1 \text{ to } R_2}$$

$$\left(\begin{array}{ccc:ccc} 1 & -1 & 0 & 1 & -3 & 4 \\ -2 & 3 & 0 & 0 & 10 & -7 \\ 4 & -4 & 1 & 0 & -10 & 7 \end{array}\right) \xrightarrow{\text{Add } R_2 \text{ to } R_3}$$

$$\left(\begin{array}{ccc:ccc} 1 & -1 & 0 & 1 & -3 & 4 \\ -2 & 3 & 0 & 0 & 10 & -7 \\ 2 & -1 & 1 & 0 & 0 & 0 \end{array}\right).$$

Having obtained a row of 0's in our reduction process on A, we see that we *cannot* reduce it to I_3. Thus, A is *not* an invertible matrix.

The elementary row operation scheme involves considerable arithmetic, even if A is 2-by-2. Time can therefore often be saved in the 2-by-2 case by using a *formula* for A^{-1} which applies whenever $\det A \neq 0$. (We will see in Theorem 7.3 below that this is a necessary and sufficient condition for A to be invertible.) We proceed now to give such a formula.

4.9 THEOREM. If

$$A = \begin{pmatrix} a & b \\ c & d \end{pmatrix}$$

and $\det A \neq 0$, then

(6) $$A^{-1} = \frac{1}{\det A} \begin{pmatrix} d & -b \\ -c & a \end{pmatrix}.$$

(The matrix

$$\begin{pmatrix} d & -b \\ -c & a \end{pmatrix}$$

is called the **adjoint matrix** adj A of A.)

Proof. We need to verify that $AC = CA = I_2$, where C is the right side of (6). We show that $AC = I_2$, and leave $CA = I_2$ for Exercise 15 below. We have

$$AC = \begin{pmatrix} a & b \\ c & d \end{pmatrix} \frac{1}{ad-bc} \begin{pmatrix} d & -b \\ -c & a \end{pmatrix}$$

$$= \frac{1}{ad-bc} \begin{pmatrix} a & b \\ c & d \end{pmatrix} \begin{pmatrix} d & -b \\ -c & a \end{pmatrix}$$

$$= \frac{1}{ad-bc} \begin{pmatrix} ad-bc & 0 \\ 0 & -bc+ad \end{pmatrix} = \begin{pmatrix} 1 & 0 \\ 0 & 1 \end{pmatrix} = I_2.$$

In the same way, $CA = I_2$, so $C = A^{-1}$. QED

If we apply (6) to the matrix in Example 4.7, then since

$$\det\begin{pmatrix} 2 & -1 \\ 1 & 1 \end{pmatrix} = 2+1 = 3,$$

we have

$$A^{-1} = \frac{1}{3}\begin{pmatrix} 1 & 1 \\ -1 & 2 \end{pmatrix}.$$

Notice how *much* easier this is than the work done in Example 4.7.

It turns out that there is an analogue of (6) for 3-by-3 and, in fact, for n-by-n matrices. But it quickly becomes more work than our reduction scheme. The problem is that adj A is a complicated matrix to calculate if $n>2$. So we will content ourselves with giving the three-dimensional analogue of (6). In this case

$$\operatorname{adj} A = \begin{pmatrix} m_{11} & -m_{21} & m_{31} \\ -m_{12} & m_{22} & -m_{32} \\ m_{13} & -m_{23} & m_{33} \end{pmatrix} = ((-1)^{i+j} m_{ij})^t$$

where m_{ij} is the (i, j) *minor*, the determinant of the matrix A_{ij} which results when row i and column j are deleted from A. Thus, to calculate adj A for the matrix of Example 4.6 (as you need to do in Exercise 18 below), you have to evaluate *nine different* 2-by-2 determinants, roughly the same amount of arithmetic done in Example 4.6. In some cases though, one of which we will see in Section 10.3, the following can save labor.

4.10 **THEOREM.** If A is a 3-by-3 matrix such that $\det A \neq 0$, then

(7) $$A^{-1} = \frac{1}{\det A} \operatorname{adj} A.$$

You have the opportunity to apply (7) in Exercises 18 and 19 below, and to compare the amount of work involved in using (7) with the amount involved in using elementary row operations. The proof of (7) consists of solving (1) for each j by Cramer's Rule. (See Exercise 31 below.)

EXERCISES 6.4

In Exercises 1 to 14, find the inverse of the given matrix, if it has an inverse, by reducing the array $(A \vdots I_n)$ to $(I_n \vdots A^{-1})$.

1 $A = \begin{pmatrix} 2 & -3 \\ -1 & 2 \end{pmatrix}$

2 $A = \begin{pmatrix} -2 & 3 \\ 1 & -2 \end{pmatrix}$

3 $A = \begin{pmatrix} 3 & 1 \\ -1 & 1 \end{pmatrix}$

4 $A = \begin{pmatrix} 3 & -2 \\ -2 & 2 \end{pmatrix}$

5 $A = \begin{pmatrix} 1 & 2 \\ 3 & 6 \end{pmatrix}$

6 $A = \begin{pmatrix} -2 & 1 \\ -6 & 3 \end{pmatrix}$

7 $A = \begin{pmatrix} 1 & 2 & 3 \\ 2 & 5 & 3 \\ 1 & 0 & 8 \end{pmatrix}$

8 $A = \begin{pmatrix} 1 & -1 & 2 \\ 2 & 1 & -1 \\ 3 & 2 & -2 \end{pmatrix}$

9 $A = \begin{pmatrix} 1 & 0 & 1 \\ 2 & -1 & 1 \\ -2 & -2 & 1 \end{pmatrix}$

10 $A = \begin{pmatrix} 1 & 1 & 1 \\ 1 & -1 & 2 \\ 1 & 1 & 4 \end{pmatrix}$

11 $A = \begin{pmatrix} 1 & 6 & 4 \\ 2 & 4 & -1 \\ -1 & 2 & 5 \end{pmatrix}$

12 $A = \begin{pmatrix} 1 & -2 & -4 \\ 2 & -3 & -6 \\ -3 & 1 & 2 \end{pmatrix}$

13 $A = \begin{pmatrix} 2 & -1 & 0 & 0 \\ -1 & 2 & -1 & 0 \\ 0 & -1 & 2 & -1 \\ 0 & 0 & -1 & 2 \end{pmatrix}$

14 $A = \begin{pmatrix} 2 & -2 & 0 & 0 \\ -1 & 2 & -1 & 0 \\ 0 & -1 & 2 & -1 \\ 0 & 0 & -1 & 2 \end{pmatrix}$

15 Complete the proof of Theorem 4.9 by showing that $CA = I_2$.

16 Use (6) to invert the matrix in Exercise 2.

17 Use (6) to invert the matrix in Exercise 3.

18 Use (7) to invert the matrix in Example 4.6. Is this more or less work than done in the example?

19 Use (7) to invert the matrix in Exercise 9. Is this more or less work than done in using the reduction scheme?

20 Show that if A and B are invertible n-by-n matrices, then AB is invertible. (*Hint:* Consider $B^{-1}A^{-1}$.)

21 Under what conditions is a diagonal matrix

$$\begin{pmatrix} d_1 & 0 & 0 \\ 0 & d_2 & 0 \\ 0 & 0 & d_3 \end{pmatrix}$$

invertible?

22 Show that the matrix

$$A = \begin{pmatrix} \cos\theta & \sin\theta & 0 \\ -\sin\theta & \cos\theta & 0 \\ 0 & 0 & 1 \end{pmatrix}$$

is invertible and find A^{-1}.

23 Show that the matrix

$$A = \begin{pmatrix} 1 & 0 & 0 \\ 0 & \cos\theta & -\sin\theta \\ 0 & \sin\theta & \cos\theta \end{pmatrix}$$

is invertible and find A^{-1}.

24 If A is an invertible 3-by-3 matrix, then show that the system of linear equations $A\boldsymbol{x}=\boldsymbol{b}$ has a unique solution in $\boldsymbol{R}^3$.

25 Show that the homogeneous system of equations $A\boldsymbol{x}=\boldsymbol{0}$ has a unique solution if and only if A is an invertible n-by-n matrix. (**This is needed in Section 9.3 and in Section 10.2.**) (*Hint*: Use Exercise 24 for one part, the proof of Proposition 4.4 for the other.)

26 Use the results of Exercises 9 and 24 to show that the system

$$\begin{aligned} x \quad\;\; + z &= 5 \\ 2x - \;\; y + z &= e \\ -2x - 2y + z &= \pi \end{aligned}$$

has a unique solution in $\boldsymbol{R}^3$. Find that solution.

27 Use the results of Exercises 9 and 24 to show that the system

$$\begin{aligned} x + \quad\;\; z &= \;\;7 \\ 2x - \;\; y + z &= \;\;5 \\ -2x - 2y + z &= -2 \end{aligned}$$

has a unique solution in $\boldsymbol{R}^3$. Find that solution.

28 Show that the upper triangular matrix

$$A = \begin{pmatrix} a_{11} & a_{12} & a_{13} \\ 0 & a_{22} & a_{23} \\ 0 & 0 & a_{33} \end{pmatrix}$$

has an inverse if $a_{11}a_{22}a_{33} \neq 0$. Show that the inverse is also upper triangular.

29 Let $\boldsymbol{T}:\boldsymbol{R}^n \to \boldsymbol{R}^n$ be a linear transformation whose matrix relative to the standard bases is A. If A is invertible, then show that $\boldsymbol{T}$ has an inverse function $\boldsymbol{T}^{-1}:\boldsymbol{R}^n \to \boldsymbol{R}^n$ such that $\boldsymbol{T}^{-1}(\boldsymbol{T}(\boldsymbol{x})) = \boldsymbol{T}(\boldsymbol{T}^{-1}(\boldsymbol{x})) = \boldsymbol{x}$ for every $\boldsymbol{x} \in \boldsymbol{R}^n$. Show that $\boldsymbol{T}^{-1}$ is also a linear transformation.

30 Let $\boldsymbol{T}:\boldsymbol{R}^n \to \boldsymbol{R}^n$ be a linear transformation that has an inverse function $\boldsymbol{T}^{-1}:\boldsymbol{R}^n \to \boldsymbol{R}^n$ such that $\boldsymbol{T}^{-1}(\boldsymbol{T}(\boldsymbol{x})) = \boldsymbol{T}(\boldsymbol{T}^{-1}(\boldsymbol{x})) = \boldsymbol{x}$ for every $\boldsymbol{x} \in \boldsymbol{R}^n$. Show that $\boldsymbol{T}^{-1}$ is also a linear transformation whose matrix relative to the standard bases for $\boldsymbol{R}^n$ is the inverse of the matrix A of $\boldsymbol{T}$ relative to the standard bases.

31 Use Cramer's Rule to show that the j-th column of A^{-1} is given by (7) for (a) $j=1$; (b) $j=2$; (c) $j=3$.

5 DIFFERENTIABLE FUNCTIONS

In Section 1 (p. 331) we discussed how to define the notion of a differentiable function $\boldsymbol{F}:\boldsymbol{R}^n \to \boldsymbol{R}^m$. We simply transport the property used in defining a differentiable function $f:\boldsymbol{R}^n \to \boldsymbol{R}$ in Chapter 4 (Definition 4.3.3) to the present context.

5.1 **DEFINITION.** Suppose that $\boldsymbol{F}:\boldsymbol{R}^n \to \boldsymbol{R}^m$ is defined on an open set U containing $\boldsymbol{x}_0 \in \boldsymbol{R}^n$. Then $\boldsymbol{F}$ is **differentiable** at $\boldsymbol{x}_0$ if there is a linear transformation $\boldsymbol{L}:\boldsymbol{R}^n \to \boldsymbol{R}^m$ such that for $\boldsymbol{x} \in U$

(1) $$\boldsymbol{F}(\boldsymbol{x}) = \boldsymbol{F}(\boldsymbol{x}_0) + \boldsymbol{L}(\boldsymbol{x}-\boldsymbol{x}_0) + \boldsymbol{E}(\boldsymbol{x}),$$

where

$$\lim_{\boldsymbol{x}\to\boldsymbol{x}_0} \frac{\boldsymbol{E}(\boldsymbol{x})}{|\boldsymbol{x}-\boldsymbol{x}_0|} = \boldsymbol{0}.$$

You will notice that Definition 5.1 is the exact analogue of Definition 4.3.3 for functions $\boldsymbol{F}:\boldsymbol{R}^n \to \boldsymbol{R}^m$. We will then refer to

$$\boldsymbol{F}(\boldsymbol{x}_0) + \boldsymbol{L}(\boldsymbol{x}-\boldsymbol{x}_0)$$

as the *tangent approximation* of $\boldsymbol{F}$ near $\boldsymbol{x}_0$. From our experience in Chapter 4 we expect there to be a simple formula for $\boldsymbol{L}(\boldsymbol{x}-\boldsymbol{x}_0)$. In fact, we know from Theorem 2.4 that $\boldsymbol{L}(\boldsymbol{x}-\boldsymbol{x}_0) = M_{\boldsymbol{L}}(\boldsymbol{x}-\boldsymbol{x}_0)$, where $M_{\boldsymbol{L}}$ is the matrix of $\boldsymbol{L}$ relative to the standard bases. What do you suppose the entries of $M_{\boldsymbol{L}}$ are? From Theorem 4.3.4, we expect the entries to be partial derivatives. Moreover,

$$\boldsymbol{L}(\boldsymbol{x}) = \begin{pmatrix} \ell_1(\boldsymbol{x}) \\ \ell_2(\boldsymbol{x}) \\ \cdot \\ \cdot \\ \cdot \\ \ell_m(\boldsymbol{x}) \end{pmatrix}$$

where $\ell_i:\boldsymbol{R}^n \to \boldsymbol{R}$ are the coordinate functions of $\boldsymbol{L}$. So in view of Theorem 4.3.4, natural candidates for the rows of $M_{\boldsymbol{L}}$ would be the gradients $\nabla f_i(\boldsymbol{x}_0)$, where the $f_i:\boldsymbol{R}^n \to \boldsymbol{R}$ are the coordinate functions of $\boldsymbol{F}$. That is, we expect to form $M_{\boldsymbol{L}}$ by differentiating the coordinate functions of $\boldsymbol{F}$, and arraying the resulting total derivatives $\nabla f_i(\boldsymbol{x}_0)$ in rows. This expectation is exactly right.

5.2 **THEOREM.** Suppose that $\boldsymbol{F}:\boldsymbol{R}^n \to \boldsymbol{R}^m$ is differentiable at $\boldsymbol{x}_0$. Then (1) becomes

(2) $$\boldsymbol{F}(\boldsymbol{x}) = \boldsymbol{F}(\boldsymbol{x}_0) + M_{\boldsymbol{L}}(\boldsymbol{x}-\boldsymbol{x}_0) + \boldsymbol{E}(\boldsymbol{x}),$$

where

$$M_{\boldsymbol{L}} = \begin{pmatrix} \nabla f_1(\boldsymbol{x}_0) \\ \nabla f_2(\boldsymbol{x}_0) \\ \cdot \\ \cdot \\ \cdot \\ \nabla f_m(\boldsymbol{x}_0) \end{pmatrix},$$

and $f_i:\boldsymbol{R}^n \to \boldsymbol{R}$ are the coordinate functions of $\boldsymbol{F}$. Thus $M_{\boldsymbol{L}}$ is the m-by-n matrix whose entry in row i, column j is $\dfrac{\partial f_i}{\partial x_j}(\boldsymbol{x}_0)$.

Proof. As remarked above, Theorem 2.4 tells us that (1) takes the form (2) where $M_{\boldsymbol{L}}$ is the matrix of $\boldsymbol{L}$ relative to the standard bases. If we then write

$$M_{\boldsymbol{L}} = \begin{pmatrix} \boldsymbol{r}_1 \\ \cdot \\ \cdot \\ \cdot \\ \boldsymbol{r}_i \\ \cdot \\ \cdot \\ \cdot \\ \boldsymbol{r}_m \end{pmatrix},$$

our task is reduced to showing that $\boldsymbol{r}_i = \nabla f_i(\boldsymbol{x}_0)$. This we can show if we study the i-th coordinates of the vectors in (2). For we can write (2) as

$$(3) \qquad \begin{pmatrix} f_1(\boldsymbol{x}) \\ \cdot \\ \cdot \\ \cdot \\ f_i(\boldsymbol{x}) \\ \cdot \\ \cdot \\ \cdot \\ f_m(\boldsymbol{x}) \end{pmatrix} = \begin{pmatrix} f_1(\boldsymbol{x}_0) \\ \cdot \\ \cdot \\ \cdot \\ f_i(\boldsymbol{x}_0) \\ \cdot \\ \cdot \\ \cdot \\ f_m(\boldsymbol{x}_0) \end{pmatrix} + \begin{pmatrix} \boldsymbol{r}_1 \cdot (\boldsymbol{x}-\boldsymbol{x}_0) \\ \cdot \\ \cdot \\ \cdot \\ \boldsymbol{r}_i \cdot (\boldsymbol{x}-\boldsymbol{x}_0) \\ \cdot \\ \cdot \\ \cdot \\ \boldsymbol{r}_m \cdot (\boldsymbol{x}-\boldsymbol{x}_0) \end{pmatrix} + \begin{pmatrix} \boldsymbol{e}_1(\boldsymbol{x}) \\ \cdot \\ \cdot \\ \cdot \\ \boldsymbol{e}_i(\boldsymbol{x}) \\ \cdot \\ \cdot \\ \cdot \\ \boldsymbol{e}_m(\boldsymbol{x}) \end{pmatrix}$$

If we equate the i-th coordinates of the vectors in (3), we get

$$(4) \qquad f_i(\boldsymbol{x}) = f_i(\boldsymbol{x}_0) + \boldsymbol{r}_i \cdot (\boldsymbol{x}-\boldsymbol{x}_0) + \boldsymbol{e}_i(\boldsymbol{x}).$$

Moreover, since

$$\lim_{\boldsymbol{x}\to\boldsymbol{x}_0} \frac{E(\boldsymbol{x})}{|\boldsymbol{x}-\boldsymbol{x}_0|} = \boldsymbol{0} = \begin{pmatrix} 0 \\ \cdot \\ \cdot \\ \cdot \\ 0 \\ \cdot \\ \cdot \\ \cdot \\ 0 \end{pmatrix},$$

we have

$$\lim_{\boldsymbol{x}\to\boldsymbol{x}_0} \frac{e_i(\boldsymbol{x})}{|\boldsymbol{x}-\boldsymbol{x}_0|} = 0.$$

Hence, by Definition 4.3.3, each coordinate function $f_i : \boldsymbol{R}^n \to \boldsymbol{R}$ is differentiable at $\boldsymbol{x}_0$, and $\boldsymbol{r}_i$ is its total derivative at $\boldsymbol{x}_0$ in view of (4). Thus, $\boldsymbol{r}_i = \nabla f_i(\boldsymbol{x}_0)$ by Theorem 4.3.4. QED

5.3 **DEFINITION.** If $\boldsymbol{F}:\boldsymbol{R}^n \to \boldsymbol{R}^m$ has coordinate functions $f_1, f_2, \ldots, f_m$ which have partial derivatives with respect to each x_i at $\boldsymbol{x}_0$, then the m-by-n matrix $J_{\boldsymbol{F}}(\boldsymbol{x}_0)$ whose (i, j)-entry is $\dfrac{\partial f_i}{\partial x_j}(\boldsymbol{x}_0)$ is called the **Jacobian matrix of $\boldsymbol{F}$ at $\boldsymbol{x}_0$** [in honor of Carl G. J. Jacobi (1804–1851), a German mathematician who made many important contributions to analysis]. Thus,

$$J_{\boldsymbol{F}}(\boldsymbol{x}_0)=\begin{pmatrix} \dfrac{\partial f_1}{\partial x_1}(\boldsymbol{x}_0) & \cdots & \dfrac{\partial f_1}{\partial x_j}(\boldsymbol{x}_0) & \cdots & \dfrac{\partial f_1}{\partial x_n}(\boldsymbol{x}_0) \\ \vdots & & \vdots & & \vdots \\ \dfrac{\partial f_i}{\partial x_1}(\boldsymbol{x}_0) & \cdots & \dfrac{\partial f_i}{\partial x_j}(\boldsymbol{x}_0) & \cdots & \dfrac{\partial f_i}{\partial x_n}(\boldsymbol{x}_0) \\ \vdots & & \vdots & & \vdots \\ \dfrac{\partial f_m}{\partial x_1}(\boldsymbol{x}_0) & \cdots & \dfrac{\partial f_m}{\partial x_j}(\boldsymbol{x}_0) & \cdots & \dfrac{\partial f_m}{\partial x_n}(\boldsymbol{x}_0) \end{pmatrix}.$$

If $\boldsymbol{F}$ is differentiable at $\boldsymbol{x}_0$, then $J_{\boldsymbol{F}}$ (or, equivalently, the corresponding linear transformation $\boldsymbol{L}:\boldsymbol{R}^n \to \boldsymbol{R}^m$) is called the **total derivative** of $\boldsymbol{F}$ at $\boldsymbol{x}_0$, sometimes denoted $\boldsymbol{F}'(\boldsymbol{x}_0)$. Note that $M_{\boldsymbol{L}} = J_{\boldsymbol{F}}(\boldsymbol{x}_0) = \boldsymbol{F}'(\boldsymbol{x}_0)$ in this case.

5.4 **EXAMPLE.** Suppose that $\boldsymbol{F}:\boldsymbol{R}^3 \to \boldsymbol{R}^2$ is given by

$$\boldsymbol{F}(x, y, z) = \begin{pmatrix} x^2+e^y+z^3 \\ \sin xy + yze^x \end{pmatrix}.$$

Find $J_{\boldsymbol{F}}(\boldsymbol{x})$ for any $\boldsymbol{x} \in \boldsymbol{R}^3$. In particular, what is $J_{\boldsymbol{F}}(1, \pi, -1)$?

Solution. Here the coordinate functions of $\boldsymbol{F}$ are $f_1(x, y, z) = x^2+e^y+z^3$ and $f_2(x, y, z) = \sin xy + yze^x$. So $\nabla f_1(\boldsymbol{x}) = (2x, e^y, 3z^2)$ and $\nabla f_2(\boldsymbol{x}) = (y \cos xy + yze^x,$ $x \cos xy + ze^x, ye^x)$. Thus,

$$J_{\boldsymbol{F}}(\boldsymbol{x}) = \begin{pmatrix} \nabla f_1(\boldsymbol{x}) \\ \nabla f_2(\boldsymbol{x}) \end{pmatrix}$$

$$= \begin{pmatrix} 2x & e^y & 3z^2 \\ y \cos xy + yze^x & x \cos xy + ze^x & ye^x \end{pmatrix}.$$

If $\boldsymbol{x}_0 = (1, \pi, -1)$, then

$$J_{\boldsymbol{F}}(\boldsymbol{x}_0) = \begin{pmatrix} 2 & e^\pi & 3 \\ -\pi(1+e) & -(1+e) & \pi e \end{pmatrix}$$

In the preceding example, we notice that all the partial derivatives of all the coordinate functions of $\boldsymbol{F}$ in $J_{\boldsymbol{F}}(\boldsymbol{x})$ are continuous. So, recalling Theorem 4.3.10, we expect that in fact $\boldsymbol{F}$ is differentiable *everywhere* with total derivative $J_{\boldsymbol{F}}$. This expectation is easy to confirm.

5.5 THEOREM. If $\boldsymbol{F}:\boldsymbol{R}^n \to \boldsymbol{R}^m$ has Jacobian matrix $J_{\boldsymbol{F}}(\boldsymbol{x}_0)$ at $\boldsymbol{x}_0 \in \boldsymbol{R}^n$ and all the $\dfrac{\partial f_i}{\partial x_j}$ are defined near and continuous at $\boldsymbol{x}_0$, then $\boldsymbol{F}$ is differentiable at $\boldsymbol{x}_0$ and so

$$J_{\boldsymbol{F}}(\boldsymbol{x}_0) = \boldsymbol{F}'(\boldsymbol{x}_0) = M_{\boldsymbol{L}}.$$

Proof. If every entry $\dfrac{\partial f_i}{\partial x_j}$ is continuous at $\boldsymbol{x}_0$, then each coordinate function $f_i:\boldsymbol{R}^n \to \boldsymbol{R}$ is differentiable at $\boldsymbol{x}_0$ and has total derivative $\nabla f_i(\boldsymbol{x}_0)$ by Theorem 4.3.10. Hence, for each i there is a linear function $\ell_i:\boldsymbol{R}^n \to \boldsymbol{R}$ such that for $\boldsymbol{x}$ near $\boldsymbol{x}_0$ we have

(5) $$f_i(\boldsymbol{x}) = f_i(\boldsymbol{x}_0) + \ell_i(\boldsymbol{x}-\boldsymbol{x}_0) + e_i(\boldsymbol{x}),$$

where

(6) $$\lim_{\boldsymbol{x}\to\boldsymbol{x}_0} \frac{e_i(\boldsymbol{x})}{|\boldsymbol{x}-\boldsymbol{x}_0|} = 0.$$

Putting the i-th coordinates together, $i = 1, 2, \ldots, m$, we get from (5)

$$\boldsymbol{F}(\boldsymbol{x}) = \boldsymbol{F}(\boldsymbol{x}_0) + \boldsymbol{L}(\boldsymbol{x}-\boldsymbol{x}_0) + \boldsymbol{E}(\boldsymbol{x})$$

where $\boldsymbol{L}$ has coordinate functions $\ell_1, \ell_2, \ldots, \ell_m$. Hence $\boldsymbol{L}$ is a linear transformation by Theorem 1.8. Moreover, (6) tells us that

$$\lim_{\boldsymbol{x}\to\boldsymbol{x}_0} \frac{\boldsymbol{E}(\boldsymbol{x})}{|\boldsymbol{x}-\boldsymbol{x}_0|} = \boldsymbol{0}$$

since each coordinate has limit 0. Then by Definition 5.1, F is differentiable at $\boldsymbol{x}_0$. Hence by Theorem 5.2 and Definition 5.3, $\boldsymbol{F}'(\boldsymbol{x}_0) = J_{\boldsymbol{F}}(\boldsymbol{x}_0)$. QED

Thus most functions $\boldsymbol{F}:\boldsymbol{R}^n \to \boldsymbol{R}^m$ which we have occasion to consider will be differentiable on their entire domains. We only need to check that all the partial derivatives $\dfrac{\partial f_i}{\partial x_j}$ $(j = 1, 2, \ldots, n)$ of all the coordinate functions $f_1, f_2, \ldots, f_m$ are continuous. Once this is done, we can calculate the total derivative $\boldsymbol{F}'(\boldsymbol{x}_0)$ at any point $\boldsymbol{x}_0$ by simply evaluating $J_{\boldsymbol{F}}$ at $\boldsymbol{x}_0$. For instance, in Example 5.4 we see that $\boldsymbol{F}$ is differentiable on all of $\boldsymbol{R}^3$ and its total derivative at $\boldsymbol{x}_0 = (1, \pi, -1)$ is

$$\begin{pmatrix} 2 & e^{\pi} & 3 \\ -\pi(1+e) & -(1+e) & \pi e \end{pmatrix}.$$

Finally, it is worth noting that the definition of differentiation and total derivative given here for functions $\boldsymbol{F}:\boldsymbol{R}^n \to \boldsymbol{R}^m$ actually *includes* the notions of differentiation considered both in Chapter 4 and in Chapter 2. If $m = 1$, we are in the situation of Chapter 4 and then Definition 5.1 *becomes* Definition 4.3.3. This assures us that the reasoning by analogy we used to produce Definition 5.1 was on target. Also if $n = 1$, then we are in the situation of Chapter 2, because we have a function $\boldsymbol{f}:\boldsymbol{R}\rightarrow\boldsymbol{R}^m$ given by

$$\boldsymbol{f}(t)=\begin{pmatrix} f_1(t) \\ \cdot \\ \cdot \\ \cdot \\ f_m(t) \end{pmatrix}.$$

We know from Theorem 5.2 that if $\boldsymbol{f}$ is differentiable at t_0 in the sense of Definition 5.1, then the total derivative of $\boldsymbol{f}$ is

$$\boldsymbol{f}'(t_0)=\begin{pmatrix} \frac{\partial f_1}{\partial t}(t_0) \\ \cdot \\ \cdot \\ \cdot \\ \frac{\partial f_m}{\partial t}(t_0) \end{pmatrix}=\begin{pmatrix} \frac{df_1}{dt}(t_0) \\ \cdot \\ \cdot \\ \cdot \\ \frac{df_m}{dt}(t_0) \end{pmatrix}$$

since each f_i is a function of just the single variable t. We can just as well write the column vector as a row vector. If we do, we have

$$\boldsymbol{f}'(t_0)=\left(\frac{df_1}{dt}(t_0), \ldots, \frac{df_m}{dt}(t_0)\right).$$

This is *precisely* the meaning we gave to $\boldsymbol{f}'(t_0)$ in Definition 2.1.6! In particular, if $m=n=1$, we get

$$f'(t_0)=\left(\frac{df}{dt}(t_0)\right)$$

and arrive back at the elementary calculus concept of derivative as a special case. Hence the notion of differentiation given here includes every previous concept of differentiation you have encountered in your study of calculus. It is then the most general notion of differentiation you have seen formulated. If you major in mathematics, then you may see this notion extended even further, but always by extending the idea of (1) which says that, near $\boldsymbol{x}_0$, $\boldsymbol{F}$ can be closely approximated by the affine function $\boldsymbol{F}(\boldsymbol{x}_0)+\boldsymbol{L}(\boldsymbol{x}-\boldsymbol{x}_0)$ for some linear function $\boldsymbol{L}$.

EXERCISES 6.5

In Exercises 1 to 10, use Theorem 5.5 to show that $\boldsymbol{F}$ is differentiable, and find its total derivative $\boldsymbol{F}'(\boldsymbol{x}_0)$ at the given point $\boldsymbol{x}_0$.

1 $\boldsymbol{F}(x, y)=\begin{pmatrix} x^2+y^2 \\ x^2-y^2 \end{pmatrix}$; $\boldsymbol{x}_0=(1, 2)$

2 $\boldsymbol{F}(x, y)=\begin{pmatrix} x^2+5xy-y^2 \\ 3xy^2+x^2y \end{pmatrix}$; $\boldsymbol{x}_0=(1, 1)$

3 $\boldsymbol{F}(x, y)=\begin{pmatrix} x+y \\ x-y \\ xy \end{pmatrix}$; $\boldsymbol{x}_0=(1, 2)$

4 $\boldsymbol{F}(x, y, z)=\begin{pmatrix} xe^{yz}-ye^{xz} \\ \ln(x^2+y^2-z^2) \end{pmatrix}$; $\boldsymbol{x}_0=(1, 1, 0)$

5 $\boldsymbol{F}(x, y, z)=\begin{pmatrix}\ln(x^2+y^2+z^2)\\ 3e^{xyz}\end{pmatrix}$; $\boldsymbol{x}_0=(1, 0, 1)$

6 $\boldsymbol{F}(x, y, z)=\begin{pmatrix}\sqrt{3x^2-y^2}+2xye^z\\ \cos \pi xyz-\ln(x^2+z^2)\\ \tan \pi xyz-\sin \pi xyz\end{pmatrix}$; $\boldsymbol{x}_0=(1, 1, 1)$

7 $\boldsymbol{F}(x, y, z)=\begin{pmatrix}x \cos \pi yz-yz \sin \pi x\\ \sqrt{x^2+4y^2+e^{z^2}}\\ \tan \pi xy-2xyz^2\end{pmatrix}$; $\boldsymbol{x}_0=(1, 1, 0)$

8 $\boldsymbol{F}(x, y, z)=\begin{pmatrix}x \cos y+y \cos z\\ x \sin y+y \sin z\\ xyz\end{pmatrix}$; $\boldsymbol{x}_0=(1, \pi, 1)$

9 $\boldsymbol{F}(x, y, z)=\begin{pmatrix}x^2y-z\\ x^2-z^2\\ \cos \pi xyz\end{pmatrix}$; $\boldsymbol{x}_0=(1, -1, \frac{1}{2})$

10 $\boldsymbol{F}(x, y, z)=\begin{pmatrix}x\\ y\end{pmatrix}$; $\boldsymbol{x}_0=(x_0, y_0, z_0)$

11 If $\boldsymbol{L}:\boldsymbol{R}^n \to \boldsymbol{R}^m$ is a linear transformation, then show that $\boldsymbol{L}$ is differentiable at all points of $\boldsymbol{R}^n$. What is $\boldsymbol{L}'$?

12 Let $\boldsymbol{A}:\boldsymbol{R}^n \to \boldsymbol{R}^m$ be an **affine transformation**, i.e., $\boldsymbol{A}(\boldsymbol{x})=\boldsymbol{L}(\boldsymbol{x})+\boldsymbol{a}$, where $\boldsymbol{L}:\boldsymbol{R}^n \to \boldsymbol{R}^m$ is linear and $\boldsymbol{a} \in \boldsymbol{R}^m$ is fixed. Show that $\boldsymbol{A}$ is differentiable at all points of $\boldsymbol{R}^n$. What is $\boldsymbol{A}'$?

13 Prove that if $\boldsymbol{F}:\boldsymbol{R}^n \to \boldsymbol{R}^m$ is differentiable at $\boldsymbol{x}_0 \in \boldsymbol{R}^n$, then $\boldsymbol{F}$ is continuous at $\boldsymbol{x}_0$. (*Hint:* See Theorem 4.3.7.)

14 Prove that if any one of the coordinate functions f_i of $\boldsymbol{F}:\boldsymbol{R}^n \to \boldsymbol{R}^m$ fails to be differentiable at $\boldsymbol{x}_0$, then $\boldsymbol{F}$ cannot be differentiable at $\boldsymbol{x}_0$.

In Exercises 15 to 18, show that the given function is not differentiable at the given point. (*Use Exercises 13 and 14 as needed.*)

15 $\boldsymbol{F}(x, y)=\begin{pmatrix}\dfrac{xy}{x^2+y^2}\\ \cos xy\end{pmatrix}$ for $(x, y)\neq(0, 0)$; $\boldsymbol{F}(0, 0)=\begin{pmatrix}0\\ 1\end{pmatrix}$; $\boldsymbol{x}_0=(0, 0)$.

16 $\boldsymbol{F}(x, y)=\begin{pmatrix}\dfrac{x^2y+xy^2}{x^2+y^2}\\ e^{x^2}+y^2\end{pmatrix}$ for $(x, y)\neq(0, 0)$; $\boldsymbol{F}(0, 0)=\begin{pmatrix}0\\ 1\end{pmatrix}$; $\boldsymbol{x}_0=(0, 0)$.

17 $\boldsymbol{F}(x, y)=\begin{pmatrix}\dfrac{xy}{x^2-y^2}\\ \sin xy\end{pmatrix}$ for $y\neq \pm x$; $\boldsymbol{F}(x, \pm x)=\begin{pmatrix}0\\ \pm\sin x^2\end{pmatrix}$; $\boldsymbol{x}_0=(0, 0)$.

18 $\boldsymbol{F}(x, y)=\begin{pmatrix}|x|\\ 3x+y\end{pmatrix}$; $\boldsymbol{x}_0=(0, 0)$.

19 Use the tangent approximation to $\boldsymbol{F}$ (p. 360) to approximate $\boldsymbol{F}(2.10, 1.60, -0.90)$ if

$$\boldsymbol{F}(x, y, z)=\begin{pmatrix}x \cos y-z^3\\ xz+x \sin y\\ xyz\end{pmatrix}. \quad [\text{Use } \boldsymbol{x}_0=(2, \tfrac{1}{2}\pi, -1).]$$

20 Use the tangent approximation to $\boldsymbol{F}$ to approximate $\boldsymbol{F}(1.01, 1.03, 0.99)$ if

$$\boldsymbol{F}(x, y, z)=\begin{pmatrix}xyz\\ x^2y-z^3\\ xy-yz\end{pmatrix}.$$

21 If $f:\boldsymbol{R}^n \to \boldsymbol{R}$ is differentiable on all of $\boldsymbol{R}^n$, then $\boldsymbol{f}'(\boldsymbol{x}_0)\in \boldsymbol{R}^n$ exists for every $\boldsymbol{x}_0 \in \boldsymbol{R}^n$. Thus we can regard $\boldsymbol{f}':\boldsymbol{R}^n \to \boldsymbol{R}^n$. Show that if all entries of $J_{f'}$ exist near and are continuous at $\boldsymbol{x}_0$, then $J_{f'}(\boldsymbol{x}_0)$ is the *Hessian matrix* $H_f(\boldsymbol{x}_0)$ defined in Section 4.7. Thus, Definition 4.7.3 is a special case of Definition 5.3.

22 Refer to Exercise 21. To define the *third total derivative* f''' of $f:\boldsymbol{R}^n \to \boldsymbol{R}$, we regard f''

as a function from $\boldsymbol{R}^n$ to $\boldsymbol{R}^{n^2}$ as follows. For each $\boldsymbol{x}_0 \in \boldsymbol{R}^n$, $f''(\boldsymbol{x}_0)$ is the n^2-tuple $(\boldsymbol{r}_1, \boldsymbol{r}_2, \ldots, \boldsymbol{r}_n)$ where the $\boldsymbol{r}_i$ are the rows of $H_f(\boldsymbol{x}_0)$, the Hessian matrix. For example, if $n=2$, then we would write

$$f''(\boldsymbol{x}_0) = (f_{xx}(\boldsymbol{x}_0), f_{xy}(\boldsymbol{x}_0), f_{yx}(\boldsymbol{x}_0), f_{yy}(\boldsymbol{x}_0)).$$

If $f'' : \boldsymbol{R}^n \to \boldsymbol{R}^{n^2}$ has a total derivative at $\boldsymbol{x}_0$, then that derivative is called the third total derivative of f at $\boldsymbol{x}_0$. If $f : \boldsymbol{R}^2 \to \boldsymbol{R}$ has continuous third partial derivatives on $\boldsymbol{R}^2$, then write out the 4-by-2 third total derivative matrix f''' of f at any $\boldsymbol{x}_0 \in \boldsymbol{R}^2$. Show that

$$f''' = \begin{pmatrix} \nabla f_{xx} \\ \nabla f_{xy} \\ \nabla f_{yx} \\ \nabla f_{yy} \end{pmatrix},$$

so f''' is obtainable by taking the total derivative of each entry in H_f, just as H_f was obtainable by taking the total derivative of each entry of

$$\nabla f = \begin{pmatrix} f_x \\ f_y \end{pmatrix} \in \boldsymbol{R}^2.$$

A similar scheme could be used to define a k-th total derivative $f^{(k)}$ for any positive integer k.)

23 Suppose that $\boldsymbol{F} : \boldsymbol{R}^2 \to \boldsymbol{R}^2$ is such that for every $\boldsymbol{x}_0 \in \boldsymbol{R}^2$ we have $\boldsymbol{F}'(\boldsymbol{x}_0) = \boldsymbol{0}$. Show that $\boldsymbol{F}(x, y) = \boldsymbol{a}$, where $\boldsymbol{a}$ is constant.

24 Suppose that $\boldsymbol{F} : \boldsymbol{R}^2 \to \boldsymbol{R}^2$ and $\boldsymbol{G} : \boldsymbol{R}^2 \to \boldsymbol{R}^2$ are such that $\boldsymbol{F}'(\boldsymbol{x}_0) = \boldsymbol{G}'(\boldsymbol{x}_0)$ for all $\boldsymbol{x}_0 \in \boldsymbol{R}^2$. Then show that $\boldsymbol{F}(\boldsymbol{x}_0) = \boldsymbol{G}(\boldsymbol{x}_0) + \boldsymbol{a}$ for a constant vector $\boldsymbol{a}$.

25 Suppose that $\boldsymbol{F} : \boldsymbol{R}^n \to \boldsymbol{R}^m$ is such that for every $\boldsymbol{x}_0 \in \boldsymbol{R}^n$ we have $\boldsymbol{F}'(\boldsymbol{x}_0) = A$, a matrix of constants. Show that $\boldsymbol{F}$ is an affine function.

26 If $\boldsymbol{F} : \boldsymbol{R}^n \to \boldsymbol{R}^m$ is such that for every $\boldsymbol{x}_0 \in \boldsymbol{R}^n$ we have $\boldsymbol{F}'(\boldsymbol{x}_0) = A$, a matrix of constants, and $\boldsymbol{F}(\boldsymbol{0}) = \boldsymbol{0}$, then show that $\boldsymbol{F}$ is a linear transformation.

6 THE CHAIN RULE

We turn now to the problem of differentiating a composite $\boldsymbol{F} \circ \boldsymbol{G}$ of two functions $\boldsymbol{F} : \boldsymbol{R}^n \to \boldsymbol{R}^m$ and $\boldsymbol{G} : \boldsymbol{R}^p \to \boldsymbol{R}^n$. We see that $\boldsymbol{F} \circ \boldsymbol{G} : \boldsymbol{R}^p \to \boldsymbol{R}^m$ and so, if it is differentiable, its total derivative will be the m-by-p matrix $J_{\boldsymbol{F} \circ \boldsymbol{G}}$. The Chain Rule we are about to give states that, for most functions we meet, $J_{\boldsymbol{F} \circ \boldsymbol{G}}$ is the *matrix product* $J_{\boldsymbol{F}} J_{\boldsymbol{G}}$ of the m-by-n total derivative of $\boldsymbol{F}$ and the n-by-p total derivative of $\boldsymbol{G}$. This is the crowning extension of the Chain Rule of elementary calculus, where $m = n = p = 1$. Recall that in that case, if we write $z = F \circ G(t) = F(x)$ where $x = G(t)$, then

(1) $$\frac{d(F \circ G)}{dt} = \frac{dz}{dt} = \frac{dz}{dx}\frac{dx}{dt} = F'(x)G'(t).$$

That is, in elementary calculus, the derivative of the composite of two differentiable functions is the *product* of the derivatives. Looking at this from our present point of view, we can interpret all derivatives as 1-by-1 matrices. Then (1) asserts that $J_{F \circ G}(t) = J_F(G(t)) J_G(t)$.

Similarly, the Chain Rule in Section 4.5 can be interpreted as asserting that $J_{F \circ G} = J_F J_G$. Consider first the Chain Rule 4.5.3. In our present notation, we have

$G : \mathbf{R}^1 \to \mathbf{R}^n$ and $F : \mathbf{R}^n \to \mathbf{R}^1$, where all the partial derivatives $\frac{\partial F}{\partial x_i}$ are continuous. If we put $z = F(G(t))$, then Equation (3) of 4.5.3 becomes

$$J_{F \circ G}(t) = \frac{dz}{dt} = \mathbf{F}'(\mathbf{x}) \cdot \dot{\mathbf{G}}(t) = J_F(\mathbf{G}(t)) J_{\mathbf{G}}(t).$$

Here we can regard $\mathbf{F}'(\mathbf{x}) = \nabla F(\mathbf{x})$ as a 1-by-n matrix and

$$\dot{\mathbf{G}}(t) = \begin{pmatrix} \frac{dg_1}{dt} \\ \cdot \\ \cdot \\ \cdot \\ \frac{dg_n}{dt} \end{pmatrix}$$

as an n-by-1 matrix (cf. p. 364). The dot product of the vectors $\mathbf{F}'(\mathbf{x})$ and $\dot{\mathbf{G}}(t)$ can be naturally interpreted as the matrix product $J_F(\mathbf{x}) J_{\mathbf{G}}(\mathbf{t})$, where $\mathbf{x} = \mathbf{G}(\mathbf{t})$.

Next, suppose that we are in the situation of Chain Rule 4.5.8. Then $F : \mathbf{R}^n \to \mathbf{R}^1$ and $\mathbf{G} : \mathbf{R}^m \to \mathbf{R}^n$ have continuous partial derivatives. If we put $\mathbf{x} = \mathbf{G}(\mathbf{t})$ and $w = F(\mathbf{G}(\mathbf{t}))$, then Equation (5) of 4.5.8 becomes

(2) $$J_{F \circ \mathbf{G}}(\mathbf{t}) = \mathbf{w}'(\mathbf{t}) = \nabla F(\mathbf{x}) \left(\frac{\partial \mathbf{G}}{\partial t_1}, \ldots, \frac{\partial \mathbf{G}}{\partial t_m} \right) = J_F(\mathbf{x}) J_{\mathbf{G}}(\mathbf{t}).$$

(Here

$$\mathbf{x} = \mathbf{G}(\mathbf{t}) = \begin{pmatrix} g_1(\mathbf{t}) \\ g_2(\mathbf{t}) \\ \cdot \\ \cdot \\ \cdot \\ g_n(\mathbf{t}) \end{pmatrix},$$

so the vector partial derivatives $\frac{\partial \mathbf{G}}{\partial t_j}$ are the columns of $J_{\mathbf{G}}$.)

Thus we have seen that in every case that we previously considered, the Chain Rule can be interpreted as stating

$$J_{F \circ \mathbf{G}}(\mathbf{t}) = J_F(\mathbf{G}(\mathbf{t})) J_{\mathbf{G}}(\mathbf{t}).$$

We now show that this continues to hold in the setting of this chapter.

6.1 **GENERAL CHAIN RULE FOR JACOBIANS.** Suppose that $\mathbf{F} : \mathbf{R}^n \to \mathbf{R}^m$ is differentiable at $\mathbf{x}_0 = \mathbf{G}(\mathbf{t}_0) \in \mathbf{R}^n$ and $\mathbf{G} : \mathbf{R}^p \to \mathbf{R}^n$ is differentiable at $\mathbf{t}_0 \in \mathbf{R}^p$. If every entry of $J_{\mathbf{F}}(\mathbf{x}_0)$ is continuous, then

(3) $$J_{\mathbf{F} \circ \mathbf{G}}(\mathbf{t}_0) = J_{\mathbf{F}}(\mathbf{x}_0) J_{\mathbf{G}}(\mathbf{t}_0),$$

i.e., *the Jacobian of the composite* $\mathbf{F} \circ \mathbf{G}$ *at* $\mathbf{t}_0$ *is the product of the Jacobians* $J_{\mathbf{F}}(G(\mathbf{t}_0))$ *and* $J_{\mathbf{G}}(\mathbf{t}_0)$.

Proof. Observe that each side of Equation (3) is an m-by-p matrix. So we need only show that the (i, j)-entries of each of these matrices are the same. Consider first the (i, j)-entry of $J_{F\circ G}(t_0)$. This is

$$\frac{\partial(F\circ G)_i}{\partial t_j}(t_0)$$

where $(F\circ G)_i$ stands for the i-th coordinate function of $F\circ G$. Let's see what $(F\circ G)_i$ *is* in terms of the coordinate functions of F and G. Note that

$$(F\circ G)(t) = F(G(t)) = F(x) = \begin{pmatrix} f_1(x) \\ f_2(x) \\ \cdot \\ \cdot \\ \cdot \\ f_m(x) \end{pmatrix}$$

where $x = G(t) \in R^n$. Thus,

$$(F\circ G)_i(t) = f_i(x) = f_i(G(t)) = (f_i\circ G)(t).$$

Hence,

$$(F\circ G)_i = f_i\circ G.$$

Now $G: R^p \to R^n$ and $f_i: R^n \to R$, so we can use Equation (2) above or Equation (4) of Chain Rule 4.5.7 to compute $\dfrac{\partial(F\circ G)_i}{\partial t_j}(t_0)$. We get

$$\begin{aligned}\frac{\partial(F\circ G)_i}{\partial t_j}(t_0) &= \frac{\partial(f_i\circ G)}{\partial t_j}(t_0)\\ &= \nabla f_i(x_0)\cdot\frac{\partial G}{\partial t_j}(t_0)\\ &= [\text{row } i \text{ of } J_F(x_0)]\cdot[\text{column } j \text{ of } J_G(t_0)]\\ &= \text{the } (i, j)\text{-entry of } J_F(x_0)J_G(t_0).\end{aligned}$$

Thus the (i, j)-entries of $J_{F\circ G}(t_0)$ and $J_F(x_0)J_G(t_0)$ are the same. Hence, $J_{F\circ G}(t_0) = J_F(x_0)J_G(t_0)$. QED

6.2 COROLLARY. If the hypotheses of 6.1 hold and also each entry of J_G is continuous at t_0, then $F\circ G$ is differentiable at t_0.

Proof. The entries of $J_{F\circ G}(t_0)$ are, by Equation (3), sums of products of entries from $J_F(x_0)$ and $J_G(t_0)$. Since products and sums of continuous functions are continuous, all entries of $J_{F\circ G}(t_0)$ are then continuous. So by Theorem 5.5, $F\circ G$ is differentiable. QED

Corollary 6.2 and (3) tell us that if J_F has continuous entries at $G(t_0)$ and J_G has continuous entries at t_0, then the *total derivative of the composite* $F\circ G$ *is the matrix product of the total derivatives* $J_F(G(t_0))$ and $J_G(t_0)$. That is,

(*) $$M_{L_{F\circ G}} = M_{L_F}M_{L_G} = M_{L_F\circ L_G}.$$

In fact, (*) holds more generally whenever $\boldsymbol{F}$ is differentiable at $\boldsymbol{G}(\boldsymbol{t}_0)$ and $\boldsymbol{G}$ is differentiable at $\boldsymbol{t}_0$, but we leave the general discussion for advanced calculus.

Aside from its esthetic appeal as an extension of all previous chain rules you have studied, Chain Rule 6.1 is a general enough formulation to allow you to solve any problem which requires calculation of some $\dfrac{\partial z_i}{\partial u_j}$, where $\boldsymbol{z}=\boldsymbol{F}(\boldsymbol{x})$ and $\boldsymbol{x}=\boldsymbol{G}(\boldsymbol{u})$. In such a case, $\dfrac{\partial z_i}{\partial u_j}$ is just the (i, j)-entry of $J_{\boldsymbol{F}\circ\boldsymbol{G}}$, so is the dot product of row i of $J_{\boldsymbol{F}}$ with column j of $J_{\boldsymbol{G}}$.

6.3 EXAMPLE. Suppose that

$$\begin{cases} u=x^2+y^2 \\ v=x^2-y^2 \\ w=x^2y^2, \end{cases}$$

and that $z=u^3+v^3+w^3$ and $q=uvw$. Find $\dfrac{\partial z}{dy}$ and $\dfrac{\partial q}{\partial x}$.

Solution. We can write

$$(u, v, w)=\boldsymbol{G}(x, y)=(x^2+y^2, x^2-y^2, x^2y^2)$$

and

$$(z, q)=\boldsymbol{F}(u, v, w)=(u^3+v^3+w^3, uvw).$$

Then we have $\boldsymbol{G}:\boldsymbol{R}^2\to\boldsymbol{R}^3$ and $\boldsymbol{F}:\boldsymbol{R}^3\to\boldsymbol{R}^2$. In this form, we are asked to find the (1, 2) and (2, 1) entries of $J_{\boldsymbol{F}\circ\boldsymbol{G}}$. We have

$$J_{\boldsymbol{G}}=\begin{pmatrix} 2x & 2y \\ 2x & -2y \\ 2xy^2 & 2x^2y \end{pmatrix}$$

and

$$J_{\boldsymbol{F}}=\begin{pmatrix} 3u^2 & 3v^2 & 3w^2 \\ vw & uw & uv \end{pmatrix}.$$

Then

$$\begin{aligned}\frac{\partial z}{\partial y} &= (1, 2) \text{ entry of } J_{\boldsymbol{F}}J_{\boldsymbol{G}}=[\text{row 1 of } J_{\boldsymbol{F}}]\cdot[\text{column 2 of } J_{\boldsymbol{G}}] \\ &= (3u^2\ 3v^2\ 3w^2)\cdot\begin{pmatrix} 2y \\ -2y \\ 2x^2y \end{pmatrix} \\ &= 6u^2y-6v^2y+6w^2x^2y \end{aligned}$$

$$\begin{aligned}\frac{\partial q}{\partial x} &= (2, 1) \text{ entry of } J_{\boldsymbol{F}}J_{\boldsymbol{G}}=(vw\ \ uw\ \ uv)\cdot\begin{pmatrix} 2x \\ 2x \\ 2xy^2 \end{pmatrix} \\ &= 2xvw+2xuw+2xy^2uv. \end{aligned}$$

If we needed to express our answers just in terms of x and y, then we would substitute $u=x^2+y^2$, $v=x^2-y^2$, and $w=x^2y^2$ into the expressions we got for $\dfrac{\partial z}{\partial y}$ and $\dfrac{\partial q}{\partial x}$. This is usually more work than it is worth, and is unnecessary for purposes of computation, as the following example illustrates.

6.4 **EXAMPLE.** If $\boldsymbol{G}(x, y)=(x^2+y^2, x^2-y^2, x^2y^2)$ and $\boldsymbol{F}(u, v, w)=(u^3+v^3+w^3, uvw)$, then find $J_{\boldsymbol{F}\circ\boldsymbol{G}}(1, 1)$.

Solution. These are the very same $\boldsymbol{F}$ and $\boldsymbol{G}$ used in the preceding example. The only new features are that a specific point $(1, 1)$ is given and the entire total derivative $J_{\boldsymbol{F}\circ\boldsymbol{G}}(1, 1)$ is asked for, rather than just a couple of its entries. We then have to apply Equation (3) of Chain Rule 6.1, where $\boldsymbol{t}_0=(1, 1)$. Here $\boldsymbol{G}(\boldsymbol{t}_0)=\boldsymbol{G}(1, 1)=(2, 0, 1)$, so

$$J_{\boldsymbol{F}\circ\boldsymbol{G}}(1, 1)=J_{\boldsymbol{F}}(\boldsymbol{G}(1, 1))J_{\boldsymbol{G}}(1, 1)=J_{\boldsymbol{F}}(2, 0, 1)J_{\boldsymbol{G}}(1, 1).$$

From Example 6.3,

$$J_{\boldsymbol{G}}(1, 1)=\begin{pmatrix} 2 & 2 \\ 2 & -2 \\ 2 & 2 \end{pmatrix} \quad \text{and} \quad J_{\boldsymbol{F}}(2, 0, 1)=\begin{pmatrix} 12 & 0 & 3 \\ 0 & 2 & 0 \end{pmatrix}.$$

Thus

$$J_{\boldsymbol{F}\circ\boldsymbol{G}}(1, 1)=\begin{pmatrix} 12 & 0 & 3 \\ 0 & 2 & 0 \end{pmatrix}\begin{pmatrix} 2 & 2 \\ 2 & -2 \\ 2 & 2 \end{pmatrix}=\begin{pmatrix} 30 & 30 \\ 4 & -4 \end{pmatrix}.$$

Observe, as a check, that if we substitute 1 for x, 1 for y, 2 for u, 0 for v, and 1 for w in Example 6.3, then we obtain $\dfrac{\partial z}{\partial y}=30$ and $\dfrac{\partial q}{\partial x}=4$, which agree with the $(1, 2)$ and $(2, 1)$ entries we calculated here for $J_{\boldsymbol{F}\circ\boldsymbol{G}}(1, 1)$.

Be *sure in applying Equation* (3) *of* 6.1 *that you evaluate* $J_{\boldsymbol{F}}$ *at* $\boldsymbol{x}_0=\boldsymbol{G}(\boldsymbol{t}_0)\in \boldsymbol{R}^n$ *and* $J_{\boldsymbol{G}}$ *at* $\boldsymbol{t}_0$. A common mistake is to evaluate both $J_{\boldsymbol{F}}$ and $J_{\boldsymbol{G}}$ at $\boldsymbol{t}_0$ in problems where this is possible (e.g., if $\boldsymbol{F}:\boldsymbol{R}^3\to\boldsymbol{R}^3$ and $\boldsymbol{G}:\boldsymbol{R}^3\to\boldsymbol{R}^3$). You can avoid such a pitfall by committing Equation (3) to memory.

In other texts, you may find the Chain Rule given as a chain of explicit equations for the entries of $J_{\boldsymbol{F}\circ\boldsymbol{G}}$. Such formulations usually have an unwieldy bulk to them, so we have avoided presenting them. If the need arises, they can be obtained directly from 6.1 in a straightforward way. The following example illustrates the procedure.

6.5 **EXAMPLE.** Suppose that $x=x(u, v, w)$, $y=y(u, v, w)$, and $z=z(u, v, w)$. If $q=f(x, y, z)$, then state the version of the Chain Rule that gives $\dfrac{\partial q}{\partial u}, \dfrac{\partial q}{\partial v}, \dfrac{\partial q}{\partial w}$.

Solution. Here we have $q=f(\boldsymbol{x})=f(\mathbf{g}(u, v, w))$, where $\boldsymbol{x}=\mathbf{g}(u, v, w)$.

$$J_f(\boldsymbol{x})=\nabla f(\boldsymbol{x})=\left(\frac{\partial f}{\partial x}, \frac{\partial f}{\partial y}, \frac{\partial f}{\partial z}\right).$$

$$J_{\mathbf{g}}(u, v, w)=\begin{pmatrix} \dfrac{\partial x}{\partial u} & \dfrac{\partial x}{\partial v} & \dfrac{\partial x}{\partial w} \\[2ex] \dfrac{\partial y}{\partial u} & \dfrac{\partial y}{\partial v} & \dfrac{\partial y}{\partial w} \\[2ex] \dfrac{\partial z}{\partial u} & \dfrac{\partial z}{\partial v} & \dfrac{\partial z}{\partial w} \end{pmatrix}$$

Then, applying Equation (3), we get

(4)
$$\begin{aligned} J_{f\circ\mathbf{g}}(u, v, w) &= J_f(\boldsymbol{x})J_{\mathbf{g}}(u, v, w) \\ &= \left(\frac{\partial f}{\partial x}\frac{\partial x}{\partial u}+\frac{\partial f}{\partial y}\frac{\partial y}{\partial u}+\frac{\partial f}{\partial z}\frac{\partial z}{\partial u}, \frac{\partial f}{\partial x}\frac{\partial x}{\partial v}+\frac{\partial f}{\partial y}\frac{\partial y}{\partial v}+\frac{\partial f}{\partial z}\frac{\partial z}{\partial v},\right. \\ &\quad \left.\frac{\partial f}{\partial x}\frac{\partial x}{\partial w}+\frac{\partial f}{\partial y}\frac{\partial y}{\partial w}+\frac{\partial f}{\partial z}\frac{\partial z}{\partial w}\right). \end{aligned}$$

Thus, in coordinate form, we can state the chain rule for this situation as

(5)
$$\begin{cases} \dfrac{\partial q}{\partial u}=\dfrac{\partial f}{\partial x}\dfrac{\partial x}{\partial u}+\dfrac{\partial f}{\partial y}\dfrac{\partial y}{\partial u}+\dfrac{\partial f}{\partial z}\dfrac{\partial z}{\partial u} \\[2ex] \dfrac{\partial q}{\partial v}=\dfrac{\partial f}{\partial x}\dfrac{\partial x}{\partial v}+\dfrac{\partial f}{\partial y}\dfrac{\partial y}{\partial v}+\dfrac{\partial f}{\partial z}\dfrac{\partial z}{\partial v} \\[2ex] \dfrac{\partial q}{\partial w}=\dfrac{\partial f}{\partial x}\dfrac{\partial x}{\partial w}+\dfrac{\partial f}{\partial y}\dfrac{\partial y}{\partial w}+\dfrac{\partial f}{\partial z}\dfrac{\partial z}{\partial w}. \end{cases}$$

As you can see from Equation (4), the form assumed by Equation (3), is considerably less involved than Equations (5). Since it is easier to keep track of less involved formulas, you will probably find it best always to apply Equation (3) to a given problem, and then derive whatever coordinate information you may need as we did in the foregoing example. This keeps formalism to a minimum, and avoids certain problems that can arise with Chain Rules in the form of Equations (5). (See Exercise 30 below.)

EXERCISES 6.6

In Exercises 1 to 12, find the Jacobian matrix $J_{F\circ G}$ at the given point for the given functions F and G.

1 $\boldsymbol{F}(u, v)=(uv, u+v)$; $\boldsymbol{G}(x, y)=(x^2+y^2, x^2-y^2)$, $(x_0, y_0)=(1, 1)$.

2 $\boldsymbol{F}(u, v, w)=uvw$; $\boldsymbol{G}(x, y, z)=(x \cos y, ye^x, z)$; $(x_0, y_0, z_0)=(1, \pi, 2)$.

3 $\boldsymbol{F}(x, y)=(x+y, \sqrt{y})$; $\boldsymbol{G}(t)=(1+t, t^2)$; $t_0=-1$.

4 $\boldsymbol{F}(x, y)=(x-y, xy)$; $\boldsymbol{G}(t)=(t^2-2t, t^3+2t)$; $t_0=1$.

5 $\boldsymbol{F}(u, v)=(3+u+v, uv, u-2v)$; $\boldsymbol{G}(x, y)=(x^2-y^2, 2xy)$; $\boldsymbol{x}_0=(1, 1)$.

6 $\boldsymbol{F}(u, v)=(u^2+v^2, u^2-v^2, 5uv)$; $\boldsymbol{G}(x, y)=(x+3xy-y^2, x^2-4y)$; $\boldsymbol{x}_0=(-1, 1)$.

7 $\boldsymbol{F}(x, y)=(e^{x+y^2}, e^{x^2+y})$; $\boldsymbol{G}(r, s, t)=(r+s^2+t^2, s+t^2+r^3)$; $(r_0, s_0, t_0)=(-1, 0, 0)$.

8 $\boldsymbol{F}(x, y)=(x^3+y^3, x^2y^2-xy)$; $\boldsymbol{G}(r, s, t)=(r^2+s^2+t^2, r+s^2+t^3)$; $(r_0, s_0, t_0)=(-1, 1, 0)$.

9 $\boldsymbol{F}(u, v, w)=(uv, vw)$; $\boldsymbol{G}(x, y)=(xy, x+y, x-y)$; $(x_0, y_0)=(2, -1)$.

10 $\boldsymbol{F}(u, v, w)=(u^2v^2, v^2+w^2)$; $\boldsymbol{G}(x, y)=(x^2+y^2, x^2-y^2, xy)$; $(x_0, y_0)=(-1, 1)$.

11 $\boldsymbol{F}(u, v)=(u^2, u^2+v^2, v^2)$; $\boldsymbol{G}(x, y, z)=(\sqrt{x^2+y^2+z^2}, \sqrt{x^2-y^2})$; $\boldsymbol{x}_0=(1, 0, -1)$.

12 $\boldsymbol{F}(u, v, w)=(uv-uw+v^2, u^3-v^3+w^3, u^2v^2w)$; $\boldsymbol{G}(x, y, z)=(x^2+y^2-z^2, xyz-x^2-y^2-z^2, xy+yz+xz)$; $\boldsymbol{x}_0=(1, -1, 0)$.

13 If $u=x^2+y^2$, $v=x^2-y^2$, $w=x^2y^2$, $p=u^3+v^3+w^3$, and $q=uvw$, then use the tangent approximation to approximate p when $x=1.03$ and $y=1.01$.

14 If $u=x^3y+xy^3$, $v=x^3+y^3$, $w=x^2y^2+x^3y^3$, $p=u^2+v^2+w^2$, and $q=\sqrt{uvw}$, then use the tangent approximation to approximate p when $x=1.01$ and $y=0.97$.

In Exercises 15 to 20, find the indicated partial derivatives.

15 If $\begin{cases} u=x^2+y^2 \\ v=x^2-y^2 \end{cases}$ and $\begin{cases} w=uv \\ z=u+v \end{cases}$, then find $\dfrac{\partial w}{\partial x}$ and $\dfrac{\partial z}{\partial y}$.

16 If $\begin{cases} u=x^2-3y^2+xy \\ v=2xy-3x^2y+y^3 \end{cases}$ and $\begin{cases} w=u^2+v \\ z=u-v \end{cases}$, then find $\dfrac{\partial w}{\partial y}$ and $\dfrac{\partial z}{\partial x}$.

17 If $\begin{cases} u=x^2-y^2 \\ v=2xy \end{cases}$ and $\begin{cases} p=3+u+v \\ q=uv \\ r=u-2v \end{cases}$, then find $\dfrac{\partial p}{\partial x}$, $\dfrac{\partial q}{\partial y}$, and $\dfrac{\partial r}{\partial x}$.

18 If $\begin{cases} u=3x^2+xy \\ v=y^2-4xy \end{cases}$ and $\begin{cases} p=3u+v^2-uv \\ q=4u-v+uv \\ r=2u^2-3v \end{cases}$, then find $\dfrac{\partial p}{\partial x}$, $\dfrac{\partial q}{\partial y}$, and $\dfrac{\partial r}{\partial y}$.

19 If $\begin{cases} x=r+s^2+t^2 \\ y=r^3+s+t^2 \end{cases}$ and $\begin{cases} u=e^{x+y^2} \\ v=e^{x^2+y} \end{cases}$ then find $\dfrac{\partial u}{\partial r}$, $\dfrac{\partial v}{\partial s}$, and $\dfrac{\partial u}{\partial t}$.

20 If $\begin{cases} x=r^2+3st-rs \\ y=rs+rt+st \end{cases}$ and $\begin{cases} u=x^3+y^3-x^2y \\ v=x^2y^2-xy+x^3y \end{cases}$ then find $\dfrac{\partial u}{\partial r}$, $\dfrac{\partial v}{\partial s}$, and $\dfrac{\partial v}{\partial t}$.

21 If $w=G(x, y, z)$ where $x=x(s, t)$, $y=y(s, t)$, and $z=z(s, t)$, then write down the Chain Rule formulas for $\dfrac{\partial w}{\partial s}$ and $\dfrac{\partial w}{\partial t}$.

22 If $w=g(x, y, z)$ where $x=x(s, t, u)$, $y=y(s, t, u)$, and $z=z(s, t, u)$, then write down the Chain Rule formulas for $\dfrac{\partial w}{\partial s}$, $\dfrac{\partial w}{\partial t}$ and $\dfrac{\partial w}{\partial u}$.

23 If $w=f(x, y, z)$ and cylindrical coordinates $x=r\cos\theta$, $y=r\sin\theta$, and $z=z$ are introduced, then $w=f(r\cos\theta, r\sin\theta, z)=g(r, \theta, z)$. Express $\dfrac{\partial g}{\partial r}$, $\dfrac{\partial g}{\partial \theta}$, and $\dfrac{\partial g}{\partial z}$ in terms of $\dfrac{\partial f}{\partial x}$, $\dfrac{\partial f}{\partial y}$, $\dfrac{\partial f}{\partial z}$, r, and θ.

24 If $w=f(x, y, z)$ and spherical coordinates $x=\rho\sin\phi\cos\theta$, $y=\rho\sin\phi\sin\theta$, and $z=\rho\cos\phi$ are introduced, then $w=f(\rho\sin\phi\cos\theta, \rho\sin\phi\sin\theta, \rho\cos\phi)=g(\rho, \phi, \theta)$. Express $\dfrac{\partial g}{\partial \rho}$, $\dfrac{\partial g}{\partial \phi}$, and $\dfrac{\partial g}{\partial \theta}$ in terms of $\dfrac{\partial f}{\partial x}$, $\dfrac{\partial f}{\partial y}$, $\dfrac{\partial f}{\partial z}$, ρ, ϕ, and θ.

25 If $f:\boldsymbol{R}^2\to\boldsymbol{R}$ is given by $f(x, y)=g(y+ax)+h(y-ax)$ for some constant a, then show that f satisfies the **wave equation** $\dfrac{\partial^2 f}{\partial x^2}=a^2\dfrac{\partial^2 f}{\partial y^2}$.

26 If $f:\boldsymbol{R}\to\boldsymbol{R}$, and if $g:\boldsymbol{R}^2\to R$ is given by $g(x, y)=f(x/y)$ for $y\neq 0$, then show that $x\dfrac{\partial g}{\partial x}+y\dfrac{\partial g}{\partial y}=0$.

27 If $f:\boldsymbol{R}^2\to\boldsymbol{R}$, and if $g:\boldsymbol{R}^2\to\boldsymbol{R}$ is given by $g(x, y)=f(u, v)$ where $u=x+y$ and $v=x-y$, then show that $\left(\dfrac{\partial f}{\partial u}\right)^2-\left(\dfrac{\partial f}{\partial v}\right)^2-\dfrac{\partial g}{\partial x}\dfrac{\partial g}{\partial y}=0$.

28 A function $f:\boldsymbol{R}^n\to\boldsymbol{R}$ is called **homogeneous of degree *k*** if $f(t\boldsymbol{x})=t^kf(\boldsymbol{x})$. (Thus a *linear* function is homogeneous of degree 1.) Prove the following result due to Euler. If $f:\boldsymbol{R}^n\to\boldsymbol{R}$ is homogeneous of degree k, then

$$\nabla f(\boldsymbol{x})\cdot\boldsymbol{x}=kf(\boldsymbol{x}).$$

29 Prove the converse of Euler's result in Exercise 28. If $f:\boldsymbol{R}^n\to\boldsymbol{R}$ is differentiable on an open set U and for all $\boldsymbol{x}\in U$

$$\nabla f(\boldsymbol{x})\cdot\boldsymbol{x}=kf(\boldsymbol{x})$$

holds, then f is homogeneous of degree k. (*Hint*: Consider the function $g(t)=t^{-k}f(t\boldsymbol{x})$.)

30 Care is essential in working with Chain Rules in coordinate form. Suppose that $q=f(x, y, z)$ where $z=z(x, y)$. Consider the following "application" of Chain Rules (5), where we let $u=x$, $v=y$, and $w=z(x, y)$:

$$\frac{\partial q}{\partial x}=\frac{\partial q}{\partial x}\frac{\partial x}{\partial x}+\frac{\partial q}{\partial y}\frac{\partial y}{\partial x}+\frac{\partial q}{\partial z}\frac{\partial z}{\partial x}.$$

Clearly $\dfrac{\partial x}{\partial x}=1$ and $\dfrac{\partial y}{\partial x}=0$, so we have

$$\frac{\partial q}{\partial x}=\frac{\partial q}{\partial x}+\frac{\partial q}{\partial z}\frac{\partial z}{\partial x}.$$

Cancelling $\dfrac{\partial q}{\partial x}$, we obtain

$$0=\frac{\partial q}{\partial z}\frac{\partial z}{\partial x}.$$

If in particular $q=17x+3y+z$ and $z=x+5$, we get $\dfrac{\partial q}{\partial z}=1$ and $\dfrac{\partial z}{\partial x}=1$. Thus $0=\dfrac{\partial q}{\partial z}\dfrac{\partial z}{\partial x}=1\cdot 1=1$. Since $0\neq 1$, something is wrong! Explain what.

7 IMPLICIT FUNCTIONS

In Section 4.6 we considered the functional equation $F(\boldsymbol{x})=0$ for a function $\boldsymbol{F}:\boldsymbol{R}^n\to\boldsymbol{R}$ and asked when it would define one of the variables x_i as a differentiable function of the other variables. In particular, if $n=3$ and $F(x_0, y_0, z_0)=0$, we saw that $F(x, y, z)=0$ defined z as a function of x and y near (x_0, y_0) provided that F_x, F_y, and F_z are continuous near (x_0, y_0, z_0) and $F_z\neq 0$. In this case we have

$$\frac{\partial z}{\partial x}=-\frac{F_x}{F_z}\quad\text{and}\quad\frac{\partial z}{\partial y}=-\frac{F_y}{F_z}$$

near (x_0, y_0). Hence, even if we couldn't solve explicitly for $z=f(x, y)$, we could still compute approximations for z near (x_0, y_0) from the tangent approximation formula

$$z \approx f(\boldsymbol{x}_0)+\nabla f(\boldsymbol{x}_0)\cdot(\boldsymbol{x}-\boldsymbol{x}_0)=z_0+\frac{\partial f}{\partial x}(x_0, y_0)(x-x_0)+\frac{\partial f}{\partial y}(x_0, y_0)(y-y_0).$$

Now we are considering functions $\boldsymbol{F}:\boldsymbol{R}^n \to \boldsymbol{R}^m$, and might inquire whether an analogous theory of implicit functions exists here. Supposing that $n>m$, we can write $n=k+m$, and then write the functional equation $\boldsymbol{F}(\boldsymbol{x})=\boldsymbol{0}$ in the form $\boldsymbol{F}(x_1, \ldots, x_k, u_1, \ldots, u_m)=\boldsymbol{0}$. In this form we can reasonably ask if this functional equation defines $\boldsymbol{u}=(u_1, \ldots, u_m)\in \boldsymbol{R}^m$ as a differentiable function of $(x_1, \ldots, x_k)\in \boldsymbol{R}^k$. It turns out that a direct analogue of Theorem 4.6.3 holds in this case. To see how it arises, let's take a close look at an example with $n=4$ and $m=2$.

Given $\boldsymbol{F}(x, y, u, v)=(0, 0)$, we have, in terms of the coordinate functions F_1 and F_2 of $\boldsymbol{F}$,

(1)
$$\begin{cases} F_1(x, y, u, v)=0 \\ F_2(x, y, u, v)=0. \end{cases}$$

If $(x_0, y_0, u_0, v_0)\in \boldsymbol{R}^4$ is a point where $\boldsymbol{F}(x_0, y_0, u_0, v_0)=\boldsymbol{0}$, i.e., a point satisfying (1), then we are interested in whether $\boldsymbol{u}=(u, v)$ is defined as a differentiable function $\boldsymbol{f}(\boldsymbol{x})$ for $\boldsymbol{x}=(x, y)$ near (x_0, y_0). We proceed as we did in Section 4.6, p. 189. That is, assume for the moment that (1) *does* define $\boldsymbol{u}=\boldsymbol{f}(\boldsymbol{x})$ as a differentiable function of x and y. If the equations in (1) are linear, then we can easily solve for u and v as functions of x and y as in Section 1.6. In general, we can only *approximate* u and v linearly as functions of x and y. To do this, we need the partial derivatives of u and v with respect to x and y, i.e., we need the entries of $J_{\boldsymbol{f}}$. As in Section 4.6, the way to find these entries of $J_{\boldsymbol{f}}$ is by using the Chain Rule. We have

$$\boldsymbol{0}=\boldsymbol{F}(\boldsymbol{x}, \boldsymbol{u})=\boldsymbol{F}(\boldsymbol{x}, \boldsymbol{f}(\boldsymbol{x}))=\boldsymbol{F}\circ \boldsymbol{G}(\boldsymbol{x}),$$

where $\boldsymbol{G}(\boldsymbol{x})=(\boldsymbol{x}, \boldsymbol{f}(\boldsymbol{x}))\in \boldsymbol{R}^4$. Then by Chain Rule 6.1,

$$0=J_{\boldsymbol{F}\circ\boldsymbol{G}}(\boldsymbol{x})=J_{\boldsymbol{F}}(\boldsymbol{G}(\boldsymbol{x}))J_{\boldsymbol{G}}(\boldsymbol{x})$$

$$=\begin{pmatrix} \dfrac{\partial F_1}{\partial x} & \dfrac{\partial F_1}{\partial y} & \dfrac{\partial F_1}{\partial u} & \dfrac{\partial F_1}{\partial v} \\ \dfrac{\partial F_2}{\partial x} & \dfrac{\partial F_2}{\partial y} & \dfrac{\partial F_2}{\partial u} & \dfrac{\partial F_2}{\partial v} \end{pmatrix}\begin{pmatrix} \dfrac{\partial x}{\partial x} & \dfrac{\partial x}{\partial y} \\ \dfrac{\partial y}{\partial x} & \dfrac{\partial y}{\partial y} \\ \dfrac{\partial u}{\partial x} & \dfrac{\partial u}{\partial y} \\ \dfrac{\partial v}{\partial x} & \dfrac{\partial v}{\partial y} \end{pmatrix}$$

$$=\begin{pmatrix} \dfrac{\partial F_1}{\partial x} & \dfrac{\partial F_1}{\partial y} & \dfrac{\partial F_1}{\partial u} & \dfrac{\partial F_1}{\partial v} \\ \dfrac{\partial F_2}{\partial x} & \dfrac{\partial F_2}{\partial y} & \dfrac{\partial F_2}{\partial u} & \dfrac{\partial F_2}{\partial v} \end{pmatrix}\begin{pmatrix} 1 & 0 \\ 0 & 1 \\ \dfrac{\partial u}{\partial x} & \dfrac{\partial u}{\partial y} \\ \dfrac{\partial v}{\partial x} & \dfrac{\partial v}{\partial y} \end{pmatrix}$$

$$= \begin{pmatrix} \frac{\partial F_1}{\partial x} + \frac{\partial F_1}{\partial u}\frac{\partial u}{\partial x} + \frac{\partial F_1}{\partial v}\frac{\partial v}{\partial x} & \frac{\partial F_1}{\partial y} + \frac{\partial F_1}{\partial u}\frac{\partial u}{\partial y} + \frac{\partial F_1}{\partial v}\frac{\partial v}{\partial y} \\ \frac{\partial F_2}{\partial x} + \frac{\partial F_2}{\partial u}\frac{\partial u}{\partial x} + \frac{\partial F_2}{\partial v}\frac{\partial v}{\partial x} & \frac{\partial F_2}{\partial y} + \frac{\partial F_2}{\partial u}\frac{\partial u}{\partial y} + \frac{\partial F_2}{\partial v}\frac{\partial v}{\partial y} \end{pmatrix}$$

$$= \begin{pmatrix} \frac{\partial F_1}{\partial x} & \frac{\partial F_1}{\partial y} \\ \frac{\partial F_2}{\partial x} & \frac{\partial F_2}{\partial y} \end{pmatrix} + \begin{pmatrix} \frac{\partial F_1}{\partial u}\frac{\partial u}{\partial x} + \frac{\partial F_1}{\partial v}\frac{\partial v}{\partial x} & \frac{\partial F_1}{\partial u}\frac{\partial u}{\partial y} + \frac{\partial F_1}{\partial v}\frac{\partial v}{\partial y} \\ \frac{\partial F_2}{\partial u}\frac{\partial u}{\partial x} + \frac{\partial F_2}{\partial v}\frac{\partial v}{\partial x} & \frac{\partial F_2}{\partial u}\frac{\partial u}{\partial y} + \frac{\partial F_2}{\partial v}\frac{\partial v}{\partial y} \end{pmatrix},$$

Finally, this reduces to

$$(2) \qquad -\begin{pmatrix} \frac{\partial F_1}{\partial x} & \frac{\partial F_1}{\partial y} \\ \frac{\partial F_2}{\partial x} & \frac{\partial F_2}{\partial y} \end{pmatrix} = \begin{pmatrix} \frac{\partial F_1}{\partial u} & \frac{\partial F_1}{\partial v} \\ \frac{\partial F_2}{\partial u} & \frac{\partial F_2}{\partial v} \end{pmatrix} \begin{pmatrix} \frac{\partial u}{\partial x} & \frac{\partial u}{\partial y} \\ \frac{\partial v}{\partial x} & \frac{\partial v}{\partial y} \end{pmatrix}.$$

If the matrix

$$\begin{pmatrix} \frac{\partial F_1}{\partial u} & \frac{\partial F_1}{\partial v} \\ \frac{\partial F_2}{\partial u} & \frac{\partial F_2}{\partial v} \end{pmatrix}$$

is invertible, then we can solve (2) to obtain

$$(3) \qquad J_f = \begin{pmatrix} \frac{\partial u}{\partial x} & \frac{\partial u}{\partial y} \\ \frac{\partial v}{\partial x} & \frac{\partial v}{\partial y} \end{pmatrix} = -\begin{pmatrix} \frac{\partial F_1}{\partial u} & \frac{\partial F_1}{\partial v} \\ \frac{\partial F_2}{\partial u} & \frac{\partial F_2}{\partial v} \end{pmatrix}^{-1} \begin{pmatrix} \frac{\partial F_1}{\partial x} & \frac{\partial F_1}{\partial y} \\ \frac{\partial F_2}{\partial x} & \frac{\partial F_2}{\partial y} \end{pmatrix}$$

Now while (3) is a satisfactory *theoretical* solution for J_f, it has the *computational* drawback of requiring us to invert the matrix

$$\begin{pmatrix} \frac{\partial F_1}{\partial u} & \frac{\partial F_1}{\partial v} \\ \frac{\partial F_2}{\partial u} & \frac{\partial F_2}{\partial v} \end{pmatrix}$$

in order to write down explicit formulas for the entries of J_f. This can be a bit of a chore, and no simple shortcut is available to avoid it. We can, however, obtain *explicit formulas* for the entries of J_f by using (6) in Theorem 4.9 to compute the inverse. We obtain in this way

$$(4) \qquad \begin{pmatrix} \frac{\partial F_1}{\partial u} & \frac{\partial F_1}{\partial v} \\ \frac{\partial F_2}{\partial u} & \frac{\partial F_2}{\partial v} \end{pmatrix}^{-1} = \frac{1}{\det \begin{pmatrix} \frac{\partial F_1}{\partial u} & \frac{\partial F_1}{\partial v} \\ \frac{\partial F_2}{\partial u} & \frac{\partial F_2}{\partial v} \end{pmatrix}} \begin{pmatrix} \frac{\partial F_2}{\partial v} & -\frac{\partial F_1}{\partial v} \\ -\frac{\partial F_2}{\partial u} & \frac{\partial F_1}{\partial u} \end{pmatrix}.$$

Substituting (4) into (3), we get

$$J_f = -\frac{1}{\det\begin{pmatrix} \frac{\partial F_1}{\partial u} & \frac{\partial F_1}{\partial v} \\ \frac{\partial F_2}{\partial u} & \frac{\partial F_2}{\partial v} \end{pmatrix}} \begin{pmatrix} \frac{\partial F_2}{\partial v} & -\frac{\partial F_1}{\partial v} \\ -\frac{\partial F_2}{\partial u} & \frac{\partial F_1}{\partial u} \end{pmatrix} \begin{pmatrix} \frac{\partial F_1}{\partial x} & \frac{\partial F_1}{\partial y} \\ \frac{\partial F_2}{\partial x} & \frac{\partial F_2}{\partial y} \end{pmatrix}$$

Thus,

(5)
$$J_f = -\frac{1}{\det\begin{pmatrix} \frac{\partial F_1}{\partial u} & \frac{\partial F_1}{\partial v} \\ \frac{\partial F_2}{\partial u} & \frac{\partial F_2}{\partial v} \end{pmatrix}} \begin{pmatrix} \frac{\partial F_1}{\partial x}\frac{\partial F_2}{\partial v} - \frac{\partial F_1}{\partial v}\frac{\partial F_2}{\partial x} & \frac{\partial F_1}{\partial y}\frac{\partial F_2}{\partial v} - \frac{\partial F_1}{\partial v}\frac{\partial F_2}{\partial y} \\ \frac{\partial F_1}{\partial u}\frac{\partial F_2}{\partial x} - \frac{\partial F_1}{\partial x}\frac{\partial F_2}{\partial u} & \frac{\partial F_1}{\partial u}\frac{\partial F_2}{\partial y} - \frac{\partial F_1}{\partial y}\frac{\partial F_2}{\partial u} \end{pmatrix}.$$

At this point we introduce some handy notation.

7.1 **DEFINITION.** The **Jacobian determinant** of F_1 and F_2 with respect to u and v is

$$\frac{\partial(F_1, F_2)}{\partial(u, v)} = \det\begin{pmatrix} \frac{\partial F_1}{\partial u} & \frac{\partial F_1}{\partial v} \\ \frac{\partial F_2}{\partial u} & \frac{\partial F_2}{\partial v} \end{pmatrix}.$$

With this notation, (5) becomes

(6)
$$J_f = \begin{pmatrix} \frac{\partial u}{\partial x} & \frac{\partial u}{\partial y} \\ \frac{\partial v}{\partial x} & \frac{\partial v}{\partial y} \end{pmatrix} = -\frac{1}{\frac{\partial(F_1, F_2)}{\partial(u, v)}} \begin{pmatrix} \frac{\partial(F_1, F_2)}{\partial(x, v)} & \frac{\partial(F_1, F_2)}{\partial(y, v)} \\ \frac{\partial(F_1, F_2)}{\partial(u, x)} & \frac{\partial(F_1, F_2)}{\partial(u, y)} \end{pmatrix}.$$

Thus each member of J_f has denominator $\dfrac{\partial(F_1, F_2)}{\partial(u, v)}$. In its numerator is

$$-\frac{\partial(F_1, F_2)}{\partial(\_\_, \_\_)},$$

where the blanks are filled in so that if symbolic cancellation were available, then it would give the entry of J_f sought. For example,

(7)
$$\boxed{\frac{\partial u}{\partial x} = -\frac{\dfrac{\partial(F_1, F_2)}{\partial(x, v)}}{\dfrac{\partial(F_1, F_2)}{\partial(u, v)}},}$$

since in this case if we were allowed to cancel, we would cancel the $\partial(F_1, F_2)$'s in the numerators and the v's in the denominators to get

$$\frac{\dfrac{1}{\partial(x,\ )}}{\dfrac{1}{\partial(u,\ )}} \to \frac{\partial u}{\partial x}.$$

(Compare the remarks on pp. 191–192.) Similarly,

$$(8) \qquad \frac{\partial u}{\partial y} = -\frac{\dfrac{\partial(F_1, F_2)}{\partial(y, v)}}{\dfrac{\partial(F_1, F_2)}{\partial(u, v)}}, \quad \frac{\partial v}{\partial x} = -\frac{\dfrac{\partial(F_1, F_2)}{\partial(u, x)}}{\dfrac{\partial(F_1, F_2)}{\partial(u, v)}}, \quad \text{and} \quad \frac{\partial v}{\partial y} = -\frac{\dfrac{\partial(F_1, F_2)}{\partial(u, y)}}{\dfrac{\partial(F_1, F_2)}{\partial(u, v)}}.$$

The foregoing discussion is justified by the following extension of Theorem 4.6.3, whose proof we leave for advanced calculus. To facilitate stating this theorem we introduce special notation. If $\boldsymbol{F}:\boldsymbol{R}^n \to \boldsymbol{R}^m$, where $n = k + m$, then we write vectors in $\boldsymbol{R}^n$ in the form $(\boldsymbol{x}, \boldsymbol{u}) = (x_1, \ldots, x_k, u_1, \ldots, u_m)$. We also *partition* the m-by-n Jacobian matrix $J_{\boldsymbol{F}}$ as follows:

$$J_{\boldsymbol{F}} = \left(\begin{array}{ccc|ccc} \dfrac{\partial F_1}{\partial x_1} & \cdots & \dfrac{\partial F_1}{\partial x_k} & \dfrac{\partial F_1}{\partial u_1} & \cdots & \dfrac{\partial F_1}{\partial u_m} \\ \vdots & & \vdots & \vdots & & \vdots \\ \dfrac{\partial F_m}{\partial x_1} & \cdots & \dfrac{\partial F_m}{\partial x_k} & \dfrac{\partial F_m}{\partial u_1} & \cdots & \dfrac{\partial F_m}{\partial u_m} \end{array}\right) = (\boldsymbol{F}_{\boldsymbol{x}}, \boldsymbol{F}_{\boldsymbol{u}}).$$

Thus $\boldsymbol{F}_{\boldsymbol{u}}$, the second matrix in the pair, is an m-by-m matrix.

7.2 IMPLICIT FUNCTION THEOREM. Let $\boldsymbol{F}:\boldsymbol{R}^n \to \boldsymbol{R}^m$, where $n = k + m$, have continuous partial derivatives $\dfrac{\partial F_i}{\partial x_j}$ and $\dfrac{\partial F_i}{\partial u_j}$ on an open set U containing $(\boldsymbol{x}_0, \boldsymbol{u}_0) \in \boldsymbol{R}^n$. Suppose that $\boldsymbol{F}(\boldsymbol{x}_0, \boldsymbol{u}_0) = \boldsymbol{0}$ and that $\boldsymbol{F}_{\boldsymbol{u}}(\boldsymbol{u}_0)$ is an invertible m-by-m matrix. Then there is an open set $V \subseteq \boldsymbol{R}^k$ containing $\boldsymbol{x}_0$ and a unique function $\boldsymbol{f}: V \to \boldsymbol{R}^m$ such that

(a) $\boldsymbol{f}(\boldsymbol{x}_0) = \boldsymbol{u}_0$,

(b) $\boldsymbol{F}(\boldsymbol{x}, \boldsymbol{f}(\boldsymbol{x})) = \boldsymbol{0}$ for $\boldsymbol{x} \in V$,

(c) $\boldsymbol{F}_{\boldsymbol{u}}(\boldsymbol{f}(\boldsymbol{x}))$ is invertible for $\boldsymbol{x} \in V$,

(d) $\boldsymbol{f}$ has continuous partial derivatives $\dfrac{\partial u_i}{\partial x_j}$ on V, and

(e) $J_{\boldsymbol{f}}(\boldsymbol{x}) = -[\boldsymbol{F}_{\boldsymbol{u}}(\boldsymbol{f}(\boldsymbol{x}))]^{-1}\boldsymbol{F}_{\boldsymbol{x}}(\boldsymbol{x})$ for $\boldsymbol{x} \in V$.

Returning to our discussion above, we see that if the matrix

$$\boldsymbol{F}_{\boldsymbol{u}} = \begin{pmatrix} \dfrac{\partial F_1}{\partial u} & \dfrac{\partial F_1}{\partial v} \\ \dfrac{\partial F_2}{\partial u} & \dfrac{\partial F_2}{\partial v} \end{pmatrix}$$

is invertible at $\boldsymbol{u}_0$, then our derivation of Formulas (3) and (6) is valid, by Theorem 7.2(e). The following basic theorem of linear algebra provides a simple test for determining whether the matrix $\boldsymbol{F}_{\boldsymbol{u}}(\boldsymbol{u}_0)$ is invertible. We leave the proof for a course in linear algebra. We shall use the result mainly in the cases $m=2$ and $m=3$. (The "if" part was proved in Theorem 4.9, for the case $m=2$.)

7.3 **THEOREM.** An m-by-m matrix B is invertible if and only if $\det B \neq 0$.

7.4 **EXAMPLE.** Let

$$\begin{cases} F_1(x, y, u, v) = x^2y + xy^2 + u^2 - v^2 \\ F_2(x, y, u, v) = e^{x+y} - v. \end{cases}$$

Show that $\boldsymbol{F}(0, 0, 1, 1) = \boldsymbol{0}$, and that $\boldsymbol{F}(x, y, u, v) = \boldsymbol{0}$, where

$$\boldsymbol{F} = \begin{pmatrix} F_1 \\ F_2 \end{pmatrix}$$

defines (u, v) as a differentiable function of x and y near $(x_0, y_0) = (0, 0)$. Use the tangent approximation to approximate u at $(x, y) = (-0.10, 0.05)$.

Solution. Here

$$\boldsymbol{F}_{\boldsymbol{u}} = \begin{pmatrix} 2u & -2v \\ 0 & -1 \end{pmatrix} = \begin{pmatrix} 2 & -2 \\ 0 & -1 \end{pmatrix}$$

at $\boldsymbol{u}_0 = (u_0, v_0) = (1, 1)$. Since $\det \boldsymbol{F}_{\boldsymbol{u}}(\boldsymbol{u}_0) = -2 \neq 0$, $\boldsymbol{F}_{\boldsymbol{u}}(\boldsymbol{u}_0)$ is invertible by Theorem 7.3. Also,

$$\boldsymbol{F}(0, 0, 1, 1) = (0 + 0 + 1 - 1, e^0 - 1) = (0, 0).$$

Thus by Theorem 7.2, on an open set V containing $(x_0, y_0) = (0, 0)$, the functional equation $\boldsymbol{F}(x, y, u, v) = \boldsymbol{0}$ defines $\boldsymbol{u} = (u, v)$ as a differentiable function of x and y. Since $\boldsymbol{F}(0, 0, 1, 1) = (0, 0)$, we have $u(0, 0) = 1$. Then

$$u(-0.10, 0.05) \approx u(0, 0) + \nabla u(0, 0) \cdot (-0.10, 0.05).$$

Thus,

$$u(-0.10, 0.05) \approx 1 - 0.10 \frac{\partial u}{\partial x}(0, 0) + 0.05 \frac{\partial u}{\partial y}(0, 0).$$

We have from (6), or from (7) and (8),

$$\frac{\partial u}{\partial x}(0, 0) = -\frac{\dfrac{\partial(F_1, F_2)}{\partial(x, v)}}{\dfrac{\partial(F_1, F_2)}{\partial(u, v)}}(0, 0, 1, 1)$$

$$= -\frac{\det\begin{pmatrix} 2xy + y^2 & -2v \\ e^{x+y} & -1 \end{pmatrix}}{\det\begin{pmatrix} 2u & -2v \\ 0 & -1 \end{pmatrix}}(0, 0, 1, 1,) = -\frac{2}{-2} = 1.$$

$$\frac{\partial u}{\partial y}(0,0) = -\frac{\dfrac{\partial(F_1,F_2)}{\partial(y,v)}}{\dfrac{\partial(F_1,F_2)}{\partial(u,v)}}(0,0,1,1)$$

$$= -\frac{\det\begin{pmatrix} x^2+2xy & -2v \\ e^{x+y} & -1 \end{pmatrix}}{\det\begin{pmatrix} 2u & -2v \\ 0 & -1 \end{pmatrix}}(0,0,1,1) = -\frac{2}{-2} = 1.$$

So $u(-0.10, 0.05) \approx 1 - 0.10 + 0.05 = 0.95$.

Even though in this example we can actually solve explicitly for $v(=e^{x+y})$ and then u in terms of x and y, the expression for u is complicated enough that our method is actually the *easiest* way to compute $u(-0.10, 0.05)$. (The exact value of $u(-0.10, 0.05)$, to two decimal places, is actually 0.95.)

We could derive formulas for part (e) of Theorem 7.2 corresponding to formulas (7) and (8) if we wanted to take the trouble. Instead, we simply assert that these can be formed in the same way we formed (7) and (8). If you consider the same sort of symbolic cancellation and remember the minus sign, then you can't go wrong. Thus, from $\mathbf{F}(x_1, \ldots, x_k, u_1, \ldots, u_m) = \mathbf{0}$, we write the coordinate functions

$$\begin{cases} F_1(x_1, \ldots, x_k, u_1, \ldots, u_m) = 0 \\ F_2(x_1, \ldots, x_k, u_1, \ldots, u_m) = 0 \\ \quad \vdots \qquad\qquad \vdots \qquad\qquad \vdots \\ F_m(x_1, \ldots, x_k, u_1, \ldots, u_m) = 0 \end{cases}$$

If

$$\frac{\partial(F_1, \ldots, F_m)}{\partial(u_1, \ldots, u_m)}(\mathbf{u}_0) = \det\begin{pmatrix} \dfrac{\partial F_1}{\partial u_1} & \cdots & \dfrac{\partial F_1}{\partial u_m} \\ \vdots & & \vdots \\ \dfrac{\partial F_m}{\partial u_1} & \cdots & \dfrac{\partial F_m}{\partial u_m} \end{pmatrix}(\mathbf{u}_0)$$

is not 0, then $\mathbf{u}$ is defined as a differentiable function of $\mathbf{x}$, and

(9)
$$\frac{\partial u_i}{\partial x_j} = -\frac{\dfrac{\partial(F_1, \ldots, F_i, \ldots, F_j, \ldots, F_m)}{\partial(u_1, \ldots, x_j, \ldots, u_j, \ldots, u_m)}}{\dfrac{\partial(F_1, \ldots, F_i, \ldots, F_j, \ldots, F_m)}{\partial(u_1, \ldots, u_i, \ldots, u_j, \ldots, u_m)}}$$

7.5 EXAMPLE. Let $F_1(x, y, z, u, v) = x^2 + y^2 - z - u^3 + 2v$, and $F_2(x, y, z, u, v) = 2xy - 2yz - 4u + v^3$. Show that $\mathbf{F}(1, 1, -1, -1, -2) = \mathbf{0}$, and then that $\mathbf{F}(x, y, z, u, v) = \mathbf{0}$, where F_1 and F_2 are the coordinate functions of $\mathbf{F}$, defines (u, v) as a differentiable function of (x, y, z). Find $\dfrac{\partial u}{\partial y}$ near $(1, 1, -1)$.

Solution. By direct substitution,

$$\boldsymbol{F}(1, 1, -1, -1, -2) = \mathbf{0}.$$

We calculate

$$\frac{\partial(F_1, F_2)}{\partial(u, v)} = \det \boldsymbol{F}_u = \det\begin{pmatrix} -3u^2 & 2 \\ -4 & 3v^2 \end{pmatrix},$$

so at $(1, 1, -1, -1, -2)$ we get

$$\frac{\partial(F_1, F_2)}{\partial(u, v)} = \det\begin{pmatrix} -3 & 2 \\ -4 & 12 \end{pmatrix} = -36 + 8 \neq 0.$$

So by Theorems 7.2 and 7.3, (u, v) is a differentiable function of x, y, and z. We have from (9)

$$\frac{\partial u}{\partial y} = -\frac{\dfrac{\partial(F_1, F_2)}{\partial(y, v)}}{\dfrac{\partial(F_1, F_2)}{\partial(u, v)}} = -\frac{\det\begin{pmatrix} 2y & 2 \\ 2x - 2z & 3v^2 \end{pmatrix}}{\det\begin{pmatrix} -3u^2 & 2 \\ -4 & 3v^2 \end{pmatrix}}$$

$$= -\frac{6yv^2 - 4x + 4z}{-9u^2v^2 + 8} \quad \text{near} \quad (1, 1, -1).$$

EXERCISES 6.7

In Exercises 1 to 12, show that the given coordinate functions of $\boldsymbol{F}: \boldsymbol{R}^m \to \boldsymbol{R}^n$ define u, v (and w) as functions of x, y (and z) near the given point $P_0 \in \boldsymbol{R}^m$. Find the indicated partial derivatives of u, v, (and w), using Theorem 7.2(e) or Equation (9).

1 $F_1(x, y, u, v) = x + 2y - u + v = 0$,
$F_2(x, y, u, v) = -2x + y + 2u + 2v = 0$,
$P_0 = (x_0, y_0, u_0, v_0) = (1, 4, 4, -5)$;
$\dfrac{\partial u}{\partial x}, \dfrac{\partial u}{\partial y}, \dfrac{\partial v}{\partial x}, \dfrac{\partial v}{\partial y}$ at (x, y, u, v).

2 $F_1(x, y, u, v) = 3x + 2y - 3u + 2v = 0$,
$F_2(x, y, u, v) = 5x - 2y + u - 2v = 0$,
$P_0 = (x_0, y_0, u_0, v_0) = (2, 1, 8, 8)$;
$\dfrac{\partial u}{\partial x}, \dfrac{\partial u}{\partial y}, \dfrac{\partial v}{\partial x}, \dfrac{\partial v}{\partial y}$.

3 $F_1(x, y, z, u, v) = -x + 2y - 5z + 3u - 5v = 0$,
$F_2(x, y, z, u, v) = x + 4y + 2z + 2u - 3v = 0$,
$P_0 = (1, 1, -1, 3, 3)$; $\dfrac{\partial u}{\partial x}, \dfrac{\partial u}{\partial y}, \dfrac{\partial u}{\partial z}$.

4 $F_1(x, y, z, u, v) = 2x - y + 5z + u + v = 0$,
$F_2(x, y, z, u, v) = -x + 2y + 3z - 2u + 3v = 0$,
$P_0 = (1, 2, -1, 3, 2)$; $\dfrac{\partial v}{\partial x}, \dfrac{\partial v}{\partial y}, \dfrac{\partial v}{\partial z}$.

5 $F_1(x, y, u, v) = x^2 - y^2 - u^3 + v^2 + 4 = 0$,
$F_2(x, y, u, v) = 2xy + y^2 - 2u^2 + 3v^4 + 8 = 0$,
$P_0 = (2, -1, 2, 1)$; $\dfrac{\partial u}{\partial x}, \dfrac{\partial u}{\partial y}, \dfrac{\partial v}{\partial x}, \dfrac{\partial v}{\partial y}$.

6 $F_1(x, y, u, v) = x^2 - y^2 + uv - v^2 + 3 = 0$,
$F_2(x, y, u, v) = x + y^2 + u^2 + uv - 2 = 0$,
$P_0 = (2, 1, -1, 2)$; $\dfrac{\partial u}{\partial x}, \dfrac{\partial u}{\partial y}, \dfrac{\partial v}{\partial x}, \dfrac{\partial v}{\partial y}$ at P_0.

7 $F_1(x, y, z, u, v) = x^2 + y^2 + z^2 - u^2 - v^2 = 1$,
$F_2(x, y, z, u, v) = x^2 - 2xy + y^2 + u + v = 0$,
$P_0 = (1, 1, -1, 1, -1)$; $\dfrac{\partial u}{\partial x}, \dfrac{\partial u}{\partial y}, \dfrac{\partial u}{\partial z}$ at P_0.

8 $F_1(x, y, z, u, v) = xv + yu + z + u^2 = 0$,
$F_2(x, y, z, u, v) = xyz + u + v + 1 = 0$,
$P_0 = (1, 1, -1, 1, -1) = (x_0, y_0, z_0, u_0, v_0)$;
$\dfrac{\partial u}{\partial x}, \dfrac{\partial u}{\partial y}, \dfrac{\partial u}{\partial z}$ at P_0.

9 $F_1(x, y, u, v, w) = x - y + 2u + v + 2w - 1 = 0$,
$F_2(x, y, u, v, w) = -x + 2y + uv + w - 1 = 0$,
$F_3(x, y, u, v, w) = x^2 + y + uw + vw = 0$,
$P_0 = (x_0, y_0, u_0, v_0, w_0) = (1, 1, 1, 1, -1)$;
$\dfrac{\partial u}{\partial y}, \dfrac{\partial v}{\partial x}, \dfrac{\partial w}{\partial y}$ at P_0.

10 $F_1(x, y, u, v, w) = x^2 + 2y^2 - 3u^2 + 4uv - v^2 + w^2 - 1 = 0$,
$F_2(x, y, u, v, w) = x + 3y - 4xy + 4u^2 - 2v^2 + w^2 - 5 = 0$,
$F_3(x, y, u, v, w) = x^2 - y^2 + 4u^2 + 2v - 3w^2 - 1 = 0$;
$P_0 = (x_0, y_0, u_0, v_0, w_0) = (1, 1, -1, 0, 1)$;
$\dfrac{\partial u}{\partial x}, \dfrac{\partial v}{\partial y}, \dfrac{\partial w}{\partial x}$ at P_0.

11 $F_1(x, y, z, u, v, w) = e^x \cos y + e^z \cos u + e^v \cos w + x - 3 = 0$,
$F_2 = (x, y, z, u, v, w) = e^x \sin y + e^z \sin u + e^v \cos w - 1 = 0$,
$F_3(x, y, z, u, v, w) = e^x \tan y + e^z \tan u + e^v \tan w + z = 0$,
$P_0 = (x_0, y_0, z_0, u_0, v_0, w_0) = (0, 0, 0, 0, 0, 0)$;
$\dfrac{\partial u}{\partial x}, \dfrac{\partial u}{\partial y}, \dfrac{\partial v}{\partial x}, \dfrac{\partial v}{\partial y}, \dfrac{\partial w}{\partial x}$ at P_0.

12 F_1, F_2, F_3, P_0 of Exercise 11; $\dfrac{\partial u}{\partial x}, \dfrac{\partial u}{\partial z}, \dfrac{\partial v}{\partial z}, \dfrac{\partial w}{\partial y}, \dfrac{\partial w}{\partial z}$ at P_0.

In Exercises 13 to 16, show that the given equations define u and v as functions of x and y near the point $P_0 = (x_0, y_0, u_0, v_0)$. Find $\dfrac{\partial u}{\partial x}$ and $\dfrac{\partial u}{\partial y}$ at P_0, and the equation of the tangent plane to the surface $u = u(x, y)$ at (x_0, y_0). Use this to approximate u at the given point (x, y).

13 $u^2 - v^2 - 3x^3 + 3y - 2 = 0$,
$u + v - 2x^2 - y^2 - 1 = 0$,
$P_0 = (-1, 1, 3/2, 5/2)$, $(x, y) = (-0.80, 0.80)$.

14 $uv - v^2 + x^2 - y^2 + 3 = 0$,
$u^2 + uv + x + y^2 - 2 = 0$,
$P_0 = (2, 1, -1, 2)$, $(x, y) = (2.05, 0.85)$. (See Exercise 6.)

15 $-u^3 + v^2 + x^2 - y^2 + 4 = 0$,
$-2u^2 + 3v^4 + 2xy + y^2 + 8 = 0$,
$P_0 = (2, -1, 2, 1)$, $(x, y) = (1.98, -1.01)$. (See Exercise 5.)

16 $u^3 - v^2 + 2x - y = 1$,
$u^2 + v^3 + x + y = 4$,
$P_0 = (1, 1, 1, 1)$, $(x, y) = (1.13, 0.82)$.

In Exercises 17 to 20, show that the given equations define y and z as a function of x in the

neighborhood of the given point P_0, Then find an equation for the tangent line at P_0 to the curve of intersection of the surfaces defined by the given equations.

17 $x^2+2y^2-z^2=2,\ 2x-y+z-1=0,$ $P_0=(2,1,-2)$.

18 $x-y+z^3=1,\ 2x+y^3-z=2,$ $P_0=(1,1,1)$.

19 $x^3+y^3+z^3-3xyz-14=0,$ $x^2+y^2+z^2-6=0,\ P_0=(2,-1,1)$.

20 $2x^3y+yx^2+z^2=0,\ x+y+z-1=0,$ $P_0=(-1,1,1)$.

21 If

$$\begin{cases} u^2+v^2-3x^2-y^2=0 \\ uv+xy \qquad\qquad\quad =0 \end{cases}$$

define u and v as functions of (x, y) near a point P_0, then give formulas for u_x, u_y, v_x, and v_y.

22 If

$$\begin{cases} u^2+xu-uv+y^2=0 \\ uv+v^2-xy \qquad =0 \end{cases}$$

define u and v as functions of x and y, then give formulas for u_x, u_y, v_x, and v_y.

23 If

$$\begin{cases} x^2+y^2+u^2+v^2 \ =1 \\ -x^2+y^2+u^2+2v^2=1 \end{cases}$$

define u and v as functions of (x, y), then find formulas for $\dfrac{\partial u}{\partial x}$ and $\dfrac{\partial^2 u}{\partial x^2}$.

24 In Exercise 23, find formulas for $\dfrac{\partial u}{\partial y}$ and $\dfrac{\partial^2 u}{\partial y^2}$.

25 In Exercise 23, find formulas for $\dfrac{\partial v}{\partial x}$ and $\dfrac{\partial^2 v}{\partial x^2}$.

26 In Exercise 23, find formulas for $\dfrac{\partial v}{\partial y}$ and $\dfrac{\partial^2 v}{\partial y^2}$.

27 In Exercise 23, find a formula for $\dfrac{\partial^2 u}{\partial y\,\partial x}$.

28 In Exercise 23, find a formula for $\dfrac{\partial^2 v}{\partial x\,\partial y}$.

29 The system

$$\begin{cases} x=u^2-v^2 \\ y=2uv \end{cases}$$

defines x and y explicitly as functions of (u, v), and may also define u and v implicitly as functions of (x, y). Assuming that this is the case, find $\dfrac{\partial u}{\partial x}, \dfrac{\partial u}{\partial y}, \dfrac{\partial v}{\partial x}$ and $\dfrac{\partial v}{\partial y}$. Then show that

$$\frac{\partial(u, v)}{\partial(x, y)}\frac{\partial(x, y)}{\partial(u, v)}=1.$$

(So symbolic cancellation works again!)

30 Repeat Exercise 29 for the system

$$\begin{cases} x=u^3-v^3 \\ y=u^2+uv. \end{cases}$$

8 TRANSFORMATION OF COORDINATES

We have already experienced a number of situations in which a coordinatization of $\boldsymbol{R}^n$ ($n=2$ or 3) different from the standard Cartesian coordinate system was appropriate. In particular, in Sections 4, 5, 7, and 8 of Chapter 5, we used to good advantage polar coordinates in $\boldsymbol{R}^2$ and cylindrical and spherical coordinates in $\boldsymbol{R}^3$. We found that when new coordinates were introduced, integrals of functions $f(\boldsymbol{x})$ changed their appearance, and often became simpler if we chose the proper coordinate system. In this section we will discuss change of coordinates from a more general point of view, and in the next section apply our results to the general problem of changing coordinates in multiple integrals.

To describe change of coordinates algebraically, we consider two copies of $\boldsymbol{R}^n$, and view a change of coordinates as given by a mapping from the first copy of $\boldsymbol{R}^n$ onto the second copy. The first copy we will coordinatize by n-tuples $(u_1, u_2, \ldots, u_n) \in \boldsymbol{R}^n$, where the u_i's represent the *new* coordinates of a point $\boldsymbol{x} \in \boldsymbol{R}^n$. The second copy of $\boldsymbol{R}^n$ will be coordinatized by the usual *Cartesian* coordinates $(x_1, \ldots, x_n)$. Since the new coordinatization puts new coordinates on the points of $\boldsymbol{R}^n$, we view the coordinatization as a map $\boldsymbol{T}: \boldsymbol{R}^n \to \boldsymbol{R}^n$ with $\boldsymbol{T}(\boldsymbol{u}) = \boldsymbol{x}$.

8.1 DEFINITION. A **coordinate transformation** of $\boldsymbol{R}^n$ is a function $\boldsymbol{T}: \boldsymbol{R}^n \to \boldsymbol{R}^n$, written $\boldsymbol{x} = \boldsymbol{T}(\boldsymbol{u})$. The coordinates $x_1, \ldots, x_n$ of $\boldsymbol{x}$ are called the **old** (or **standard** or **Cartesian**) coordinates, while the coordinates $u_1, u_2, \ldots, u_n$ of $\boldsymbol{u}$ are called the **new** (or **curvilinear**) coordinates.

8.2 EXAMPLE (POLAR COORDINATES). We have $(x, y) = \boldsymbol{T}(r, \theta) = (r\cos\theta, r\sin\theta)$. Given a point (r, θ) in the first copy of $\boldsymbol{R}^2$, its image is the point $\boldsymbol{x} = (x, y)$ whose distance from the origin is r and such that the angle from $\boldsymbol{i}$ to $\boldsymbol{x}$ in the counterclockwise sense is θ. See Figure 8.1. With this point of view, we can

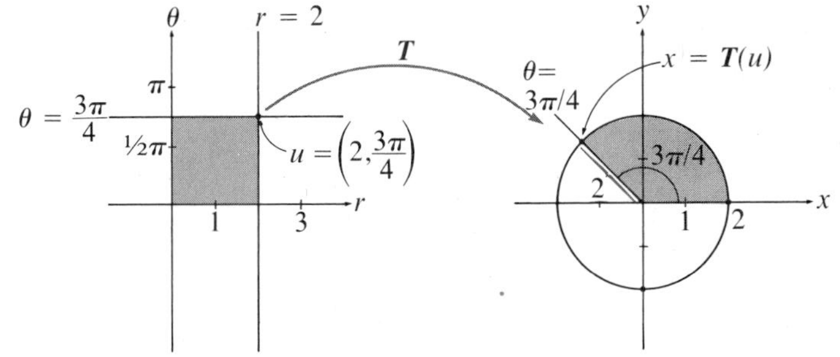

FIGURE 8.1

now describe *geometric* effects of $\boldsymbol{T}$ as well as algebraic effects. For example, $\boldsymbol{T}$ maps a line like $r=2$ in the (r, θ)-copy of $\boldsymbol{R}^2$ onto a circle in the (x, y)-copy of $\boldsymbol{R}^2$ centered at the origin and having a radius of 2. It maps a line like $\theta = 3\pi/4$ onto a ray (or an entire line if we allow $r<0$).

The term *curvilinear* arises from the fact that *lines* $u_i = c$ are usually transformed to *curves* by a coordinate transformation $\boldsymbol{T}$, as in Example 8.2.

Intuitively speaking, an alternative coordinate system for $\boldsymbol{R}^n$ ought to be such that we can freely pass back and forth between it and the standard coordinate system. In Definition 8.1 we have required only that we be able to go *one* way: from $\boldsymbol{u}$ to $\boldsymbol{x}=\boldsymbol{T}(\boldsymbol{u})$. When can we also go back from $\boldsymbol{x}$ to $\boldsymbol{u}$? Precisely when at each point $\boldsymbol{u}$ the transformation $\boldsymbol{T}:\boldsymbol{R}^n \to \boldsymbol{R}^n$ has an *inverse* map $\boldsymbol{S}=\boldsymbol{T}^{-1}$ such that $\boldsymbol{T}\circ\boldsymbol{S}=\boldsymbol{S}\circ\boldsymbol{T}=\boldsymbol{I}$, the identity map. Let's see why this is so. First suppose that we can go back via a map $\boldsymbol{S}$. Then $\boldsymbol{u}=\boldsymbol{S}(\boldsymbol{x})$ gives

$$\boldsymbol{x}=\boldsymbol{T}(\boldsymbol{u})=\boldsymbol{T}(\boldsymbol{S}(\boldsymbol{x}))=(\boldsymbol{T}\circ\boldsymbol{S})(\boldsymbol{x})$$

and

$$\boldsymbol{u}=\boldsymbol{S}(\boldsymbol{x})=\boldsymbol{S}(\boldsymbol{T}(\boldsymbol{u}))=(\boldsymbol{S}\circ\boldsymbol{T})(\boldsymbol{u}).$$

Thus $\boldsymbol{S}=\boldsymbol{T}^{-1}$. On the other hand, if $\boldsymbol{T}^{-1}$ exists, then we can apply it to $\boldsymbol{x}=\boldsymbol{T}(\boldsymbol{u})$ to get

$$\boldsymbol{T}^{-1}(\boldsymbol{x})=\boldsymbol{u}.$$

Thus we can then go back from $\boldsymbol{x}$ to $\boldsymbol{u}$.

This is the reason that we usually adopt the conventions $r\geq 0, 0\leq\theta<2\pi$ when working with polar coordinates. With these conventions, we can solve the equations

$$x=r\cos\theta$$
$$y=r\sin\theta,$$

at every point except the origin, for r and θ, obtaining

$$r=\sqrt{x^2+y^2},$$

$$\theta=\begin{cases}\tan^{-1}\dfrac{y}{x} & \text{if} \quad x\neq 0\\[2ex] \dfrac{\pi}{2} & \text{if} \quad x=0,\ y>0\\[2ex] \dfrac{3\pi}{2} & \text{if} \quad x=0,\ y<0.\end{cases}$$

We can give a general criterion for the existence of $\boldsymbol{T}^{-1}$, but the proof is left for advanced calculus.

8.3 INVERSE TRANSFORMATION THEOREM. Let $\boldsymbol{T}:\boldsymbol{R}^n \to \boldsymbol{R}^n$ be a coordinate transformation, $\boldsymbol{T}(\boldsymbol{u})=\boldsymbol{x}$. Suppose that on an open set U containing $\boldsymbol{u}_0$, all entries of $J_{\boldsymbol{T}}$ are continuous, and that $J_{\boldsymbol{T}}(\boldsymbol{u}_0)$ is an invertible n-by-n matrix. Then there is an open set V containing $\boldsymbol{x}_0$ and a unique differentiable function $\boldsymbol{S}:\boldsymbol{R}^n \to \boldsymbol{R}^n$ such that

(a) $\boldsymbol{S}(\boldsymbol{x}_0)=\boldsymbol{u}_0$;
(b) $\boldsymbol{S}(\boldsymbol{T}(\boldsymbol{u}))=\boldsymbol{u}$ for $\boldsymbol{u}\in U$;
(c) $\boldsymbol{T}(\boldsymbol{S}(\boldsymbol{x}))=\boldsymbol{x}$ for $\boldsymbol{x}\in V$;
(d) $J_{\boldsymbol{S}}(\boldsymbol{x}_0)=J_{\boldsymbol{T}}(\boldsymbol{u}_0)^{-1}$.

8.4 EXAMPLE. For the polar coordinate transformation $\boldsymbol{T}(r, \theta) = (x, y) = (r\cos\theta, r\sin\theta)$, we have

$$J_{\boldsymbol{T}} = \begin{pmatrix} \dfrac{\partial x}{\partial r} & \dfrac{\partial x}{\partial \theta} \\ \dfrac{\partial y}{\partial r} & \dfrac{\partial y}{\partial \theta} \end{pmatrix} = \begin{pmatrix} \cos\theta & -r\sin\theta \\ \sin\theta & r\cos\theta \end{pmatrix}.$$

Then

$$\begin{aligned} \det J_{\boldsymbol{T}} &= \frac{\partial(x, y)}{\partial(r, \theta)} = \det\begin{pmatrix} \cos\theta & -r\sin\theta \\ \sin\theta & r\cos\theta \end{pmatrix} \\ &= r\cos^2\theta + r\sin^2\theta = r. \end{aligned}$$

So near any $\boldsymbol{u}_0 = (r_0, \theta_0)$ where $r_0 \neq 0$, we know, in view of Theorem 7.3, that $\boldsymbol{T}$ has an inverse transformation $\boldsymbol{T}^{-1}$. Moreover, from part (d) of Theorem 8.3,

(1) $$\boxed{J_{\boldsymbol{T}^{-1}} = (J_{\boldsymbol{T}})^{-1}.}$$

The relation (1) should remind you of the formula for finding the derivative of an inverse function in elementary calculus. Recall that if s and t are monotonic differentiable inverse functions with $y = s(x)$ equivalent to $x = t(y)$, then

$$\frac{dt}{dy} = \frac{1}{\dfrac{ds}{dx}}.$$

That is,

(2) $$\frac{dx}{dy} = \frac{1}{\dfrac{dy}{dx}}.$$

We can state a very appealing consequence of Theorem 8.3(d) using the following basic result of linear algebra, whose proof we once more leave for a linear algebra course.

8.5 THEOREM. If A and B are n-by-n matrices, then $\det(AB) = (\det A)(\det B)$.

8.6 COROLLARY. If A is an invertible n-by-n matrix, then $\det A^{-1} = \dfrac{1}{\det A}$.

Proof. In this case we have $I_n = AA^{-1}$, so

$$1 = \det I_n = \det(AA^{-1}) = (\det A)(\det A^{-1})$$

by Theorem 8.5. Thus $\det A^{-1} = \dfrac{1}{\det A}$. QED

8.7 **COROLLARY.** If $T: R^n \to R^n$ is a differentiable coordinate transformation with

$$x = (x_1, \ldots, x_n) = T(u) = T(u_1, \ldots, u_n),$$

and if T has a differentiable inverse, then $\det J_{T^{-1}} = \dfrac{1}{\det J_T}$. That is,

(3)
$$\frac{\partial(u_1, \ldots, u_n)}{\partial(x_1, \ldots, x_n)} = \frac{1}{\dfrac{\partial(x_1, \ldots, x_n)}{\partial(u_1, \ldots, u_n)}}.$$

Note the close symbolic resemblance between equations (2) and (3), again reflective of the close analogy that exists between the single-variable and several-variable theories.

We can apply Corollary 8.7 to the polar coordinate transformation T in Example 8.4 to conclude immediately that

(4)
$$\det J_{T^{-1}} = \frac{\partial(r, \theta)}{\partial(x, y)} = \frac{1}{r}$$

when $r \neq 0$. Thus Corollary 8.7 enables us to compute the Jacobian determinant of the inverse of a coordinate transformation T *without* actually finding the defining formulas for the inverse transformation T^{-1}. We will see in the next section that Corollary 8.7 can sometimes save us considerable labor in cases where what we really need is the Jacobian determinant of a transformation. Here, for example, to compute $J_{T^{-1}}$ from the formulas $r = \sqrt{x^2 + y^2}$ and $\theta = \tan^{-1} y/x$ would be a lot more work than applying Corollary 8.7. The next result shows how determinants can give important geometric information about a transformation.

8.8 **THEOREM.** Suppose that $T: R^2 \to R^2$ is a linear transformation of coordinates. Let R be the rectangle defined by $a \leq u \leq b$, $c \leq v \leq d$. Then $T(R)$ is a (possibly degenerate) parallelogram whose area is given by

(5)
$$A(T(R)) = |\det M_T|\, A(R).$$

Proof. There is no loss in assuming that $a = c = 0$, for if this isn't the case initially, we can, before proceeding, translate axes in the uv-plane, making a new origin at (a, c). See Figure 8.2. Such a translation of axes has no effect upon R or

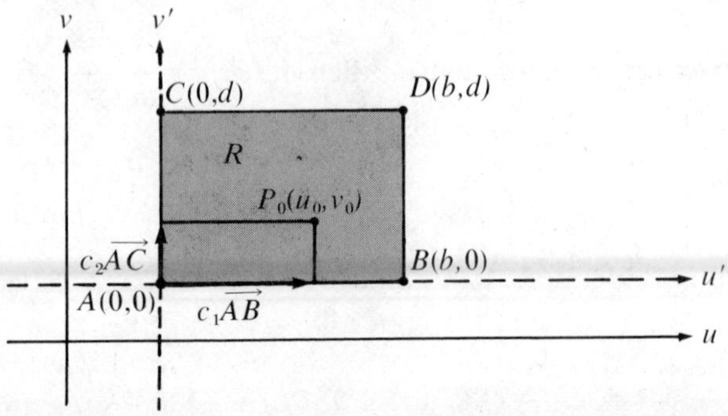

FIGURE 8.2

its area. Now suppose that

$$M_{\boldsymbol{T}} = \begin{pmatrix} a_{11} & a_{12} \\ a_{21} & a_{22} \end{pmatrix}$$

is the matrix of $\boldsymbol{T}$ relative to the standard bases. Then we can compute the images of the vertices A, B, C, and D of R using Theorem 2.4.

(i) $$\boldsymbol{T}(A) = \boldsymbol{T}\begin{pmatrix} 0 \\ 0 \end{pmatrix} = \begin{pmatrix} a_{11} & a_{12} \\ a_{21} & a_{22} \end{pmatrix}\begin{pmatrix} 0 \\ 0 \end{pmatrix} = \begin{pmatrix} 0 \\ 0 \end{pmatrix}.$$

(ii) $$\boldsymbol{T}(B) = \boldsymbol{T}\begin{pmatrix} b \\ 0 \end{pmatrix} = \begin{pmatrix} a_{11} & a_{12} \\ a_{21} & a_{22} \end{pmatrix}\begin{pmatrix} b \\ 0 \end{pmatrix} = b\begin{pmatrix} a_{11} \\ a_{21} \end{pmatrix}.$$

(iii) $$\boldsymbol{T}(C) = \boldsymbol{T}\begin{pmatrix} 0 \\ d \end{pmatrix} = \begin{pmatrix} a_{11} & a_{12} \\ a_{21} & a_{22} \end{pmatrix}\begin{pmatrix} 0 \\ d \end{pmatrix} = d\begin{pmatrix} a_{12} \\ a_{22} \end{pmatrix}.$$

(iv) $$\boldsymbol{T}(D) = \boldsymbol{T}\left(\begin{pmatrix} b \\ 0 \end{pmatrix} + \begin{pmatrix} 0 \\ d \end{pmatrix}\right) = \boldsymbol{T}\begin{pmatrix} b \\ 0 \end{pmatrix} + \boldsymbol{T}\begin{pmatrix} 0 \\ d \end{pmatrix} = b\begin{pmatrix} a_{11} \\ a_{21} \end{pmatrix} + d\begin{pmatrix} a_{12} \\ a_{22} \end{pmatrix}.$$

By the parallelogram law, $\boldsymbol{T}(D)$ is the fourth vertex of a parallelogram in the xy-plane whose other vertices are $(0, 0)$, $\boldsymbol{T}(B)$, and $\boldsymbol{T}(C)$. See Figure 8.3.

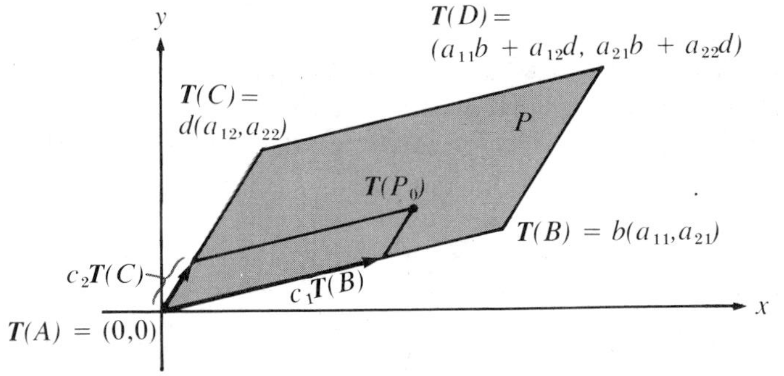

FIGURE 8.3

Moreover, if $P_0(u_0, v_0)$ is *any* point inside R or on its boundary, we see from Figure 8.2 that

$$\begin{aligned} (u_0, v_0) &= c_1\overrightarrow{AB} + c_2\overrightarrow{AC} \text{ for suitable } c_1 \text{ and } c_2 \text{ in the interval } [0, 1] \\ &= c_1\begin{pmatrix} b \\ 0 \end{pmatrix} + c_2\begin{pmatrix} 0 \\ d \end{pmatrix}. \end{aligned}$$

Therefore, the parallelogram law again tells us that

$$\boldsymbol{T}\begin{pmatrix} u_0 \\ v_0 \end{pmatrix} = c_1\boldsymbol{T}\begin{pmatrix} b \\ 0 \end{pmatrix} + c_2\boldsymbol{T}\begin{pmatrix} 0 \\ d \end{pmatrix} = c_1\boldsymbol{T}(B) + c_2\boldsymbol{T}(C)$$

is the fourth vertex of a parallelogram in the xy-plane whose other vertices are $(0, 0)$, $c_1\boldsymbol{T}(B)$, and $c_2\boldsymbol{T}(C)$. Hence, $\boldsymbol{T}$ maps R onto the parallelogram labeled P in Figure 8.3. (Note that we cannot assert that the four vertices of P are *distinct* points in the plane, so that P may be degenerate, i.e., a line segment or even a single point, and have area zero.) We can compute the area of P using Corollary

1.5.9. We get

$$\begin{aligned}
A(\boldsymbol{T}(R)) &= |[\boldsymbol{T}(B)-\boldsymbol{T}(A)]\times[\boldsymbol{T}(C)-\boldsymbol{T}(A)]| \\
&= |b(a_{11}, a_{21})\times d(a_{12}, a_{22})| \\
&= |bd|\,|(a_{11}, a_{21}, 0)\times(a_{12}, a_{22}, 0)| \\
&= |bd|\,|(0, 0, a_{11}a_{22}-a_{12}a_{21})| \\
&= |bd|\,|a_{11}a_{22}-a_{12}a_{21}| \\
&= A(R)\,|\det M_{\boldsymbol{T}}|.
\end{aligned}$$

QED

If $\boldsymbol{T}:\boldsymbol{R}^2\to\boldsymbol{R}^2$ is an *affine* transformation of the form $\boldsymbol{T}(u, v)=\boldsymbol{L}(u, v)+(a, b)$ where $\boldsymbol{L}$ is linear and $(a, b)\in\boldsymbol{R}^2$ is a fixed vector, then its effect on areas is exactly the same as the effect of $\boldsymbol{L}$ on areas, since translation by the fixed vector (a, b) does not affect areas in any way. Thus, the next result follows immediately from Theorem 8.8.

8.9 COROLLARY. If $\boldsymbol{T}:\boldsymbol{R}^2\to\boldsymbol{R}^2$ is an affine transformation, $\boldsymbol{T}(\boldsymbol{u})=M\boldsymbol{u}+\boldsymbol{b}$, where M is an invertible matrix, then $\boldsymbol{T}$ maps a rectangle R to a parallelogram $\boldsymbol{T}(R)$ whose area is given by (5).

The analogue of Theorem 8.8 holds for the linear transformations $\boldsymbol{T}:\boldsymbol{R}^3\to\boldsymbol{R}^3$. (In fact, it holds for $\boldsymbol{T}:\boldsymbol{R}^n\to\boldsymbol{R}^n$ for general n, but we leave that for later courses.)

8.10 THEOREM. Suppose that $\boldsymbol{T}:\boldsymbol{R}^3\to\boldsymbol{R}^3$ is a linear transformation of coordinates. If P is a rectangular parallelepiped defined by $a\le u\le b$, $c\le v\le d$, and $e\le w\le f$, then $\boldsymbol{T}(P)$ is a (possibly degenerate) parallelepiped with volume

(6) $$V(\boldsymbol{T}(P))=|\det M_{\boldsymbol{T}}|\;V(P).$$

Outline of Proof. The proof exactly parallels that of Theorem 8.8. If $M_{\boldsymbol{T}}=\begin{pmatrix} a_{11} & a_{12} & a_{13} \\ a_{21} & a_{22} & a_{23} \\ a_{31} & a_{32} & a_{33}\end{pmatrix}$, then we can show by calculation that the images of the planes $u=a$ and $u=b$, for example, are parallel (or, in the degenerate case, coincident) planes. Similarly, we can show the same result for the images of the planes $v=c$ and $v=d$, and for the images of $w=c$ and $w=d$. Thus, $\boldsymbol{T}(P)$ will be a parallelepiped, possibility degenerate. Also, using Theorem 1.8.1, we can compute its volume, which turns out to be $|\det M_{\boldsymbol{T}}|\;V(P)$. (Note that if $\boldsymbol{T}$ is invertible, then $\det M_{\boldsymbol{T}}\neq 0$ by Theorem 7.3. Thus, $\boldsymbol{T}(P)$ is a non-degenerate parallelepiped.)

QED

You are asked to fill in the various parts of this outline of the proof in Exercises 29 to 33 below. The analogue of Corollary 8.9 follows immediately.

8.11 COROLLARY. If $\boldsymbol{T}:\boldsymbol{R}^3\to\boldsymbol{R}^3$ is an affine transformation. $\boldsymbol{T}(\boldsymbol{u})=M\boldsymbol{u}+\boldsymbol{b}$, where M is an invertible 3-by-3 matrix, then $\boldsymbol{T}$ maps a rectangular parallelepiped P to a parallelepiped $\boldsymbol{T}(P)$ whose volume is given by (6).

8.12 EXAMPLE. Find the area of the parallelogram P bounded by the lines $2x-y=1$, $2x-y=3$, $x+y=-2$, and $x+y=0$.

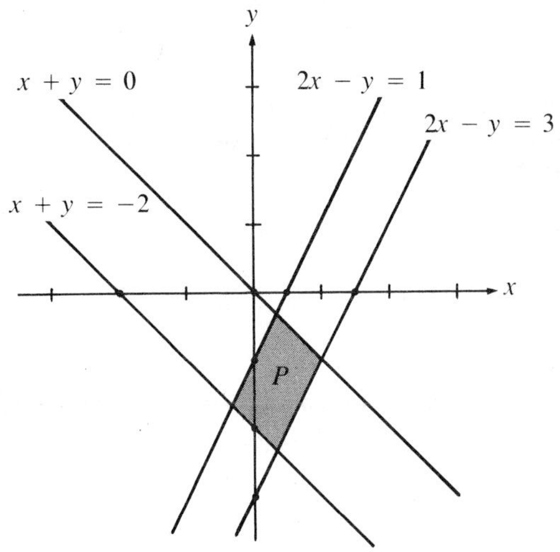

FIGURE 8.4

Solution. Refer to Figure 8.4. Here it seems natural to let $u = 2x - y$ and $v = x + y$. This defines a linear transformation $\boldsymbol{S}: \boldsymbol{R}^2 \to \boldsymbol{R}^2$ given by

$$\begin{pmatrix} u \\ v \end{pmatrix} = \begin{pmatrix} 2 & -1 \\ 1 & 1 \end{pmatrix} \begin{pmatrix} x \\ y \end{pmatrix}.$$

We are thus asked to find the area of the portion P of the xy-plane which corresponds to the rectangle $R = \boldsymbol{S}(P)$ in the uv-plane given by $1 \le u \le 3$ and $-2 \le v \le 0$. See Figure 8.5. We have $P = \boldsymbol{S}^{-1}(R) = \boldsymbol{T}(R)$. Since

$$M_{\boldsymbol{S}} = \begin{pmatrix} 2 & -1 \\ 1 & 1 \end{pmatrix},$$

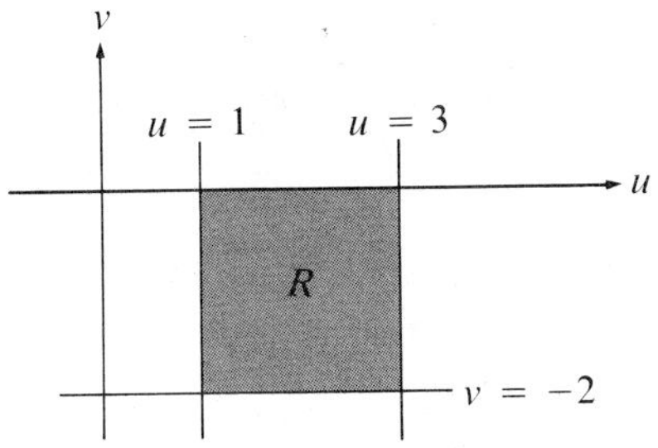

FIGURE 8.5

we calculate $\det M_{\boldsymbol{S}} = 2 + 1 = 3$. Then by Corollary 8.6, $\det M_{\boldsymbol{T}} = \frac{1}{3}$. So from Theorem 8.8 we have

$$\begin{aligned} A(P) &= |\det M_{\boldsymbol{T}}|\, A(R) \\ &= \tfrac{1}{3}(3-1)(0-(-2)) \\ &= \tfrac{1}{3}2 \cdot 2 \\ &= \tfrac{4}{3}. \end{aligned}$$

Observe that the only alternative we have to the solution just given would be to use Corollary 1.5.9. That would entail *considerably* more work, since we would first need to find the coordinates of three of the vertices of P.

EXERCISES 6.8

In Exercises 1 to 4, compute the Jacobian matrix and determinant of the given coordinate transformation T, and find the Jacobian determinant of T^{-1} when it exists. Draw images of the lines $u=1$, $u=-1$, $v=1$, and $v=-1$ in the xy-plane.

1 $T(u, v)=(3u-2v, u+v)=(x, y)$

2 $T(u, v)=(3u-v, -u+\frac{1}{3}v)=(x, y)$

3 $\begin{cases} x=u^2+v^2 \\ y=u^2-v^2 \end{cases}$

4 $\begin{cases} x=u \\ y=v+u^2 \end{cases}$

In Exercises 5 to 12, compute the Jacobian matrix and determinant of the given coordinate transformation T, and find the Jacobian determinant of T^{-1} when it exists.

5 $T(u, v)=(2u-4v, -u+2v)$

6 $T(u, v)=(-u+2v, \frac{1}{2}u-v)$

7 $T(u, v, w)=(-u+2v+w, 2u-v-2w, u+v-w)=(x, y, z)$

8 $T(u, v, w)=(u-3v+2w, 2u+v-3w, 3u-2v-w)=(x, y, z)$

9 (*Cylindrical coordinates*) $T(r, \theta, z)=(r\cos\theta, r\sin\theta, z)=(x, y, z)$

10 $T(u, v, w)=(u\sin v, u\cos v, -w)=(x, y, z)$

11 (*Spherical coordinates*) $T(\rho, \phi, \theta)=(\rho\sin\phi\cos\theta, \rho\sin\phi\sin\theta, \rho\cos\phi)=(x, y, z)$

12 $\begin{cases} x=u^2+v^2+w^2 \\ y=u^2+v^2-w^2 \\ z=uvw \end{cases}$

In Exercises 13 to 16, find the area or volume of the given parallelogram in R^2 or parallelepiped in R^3.

13 Bounded by the lines $x-y=-1$, $x-y=1$, $x+y=0$, $x+y=2$.

14 Bounded by the lines $3x+y=1$, $3x+y=6$, $x+2y=2$, $x+2y=7$.

15 Bounded by the planes $x-z=-1$, $x-z=1$, $y+z=0$, $y+z=2$, $x+z=1$, $x+z=3$.

16 Bounded by the planes $2x-y+z=0$, $2x-y+z=3$, $x+2y-z=2$, $x+2y-z=5$, $-x+y+z=3$, $-x+y+z=6$.

17 Suppose that

$$T(u, v)=(x, y)=(f(u, v), g(u, v)),$$

where f and g have continuous partial derivatives with respect to u and v. If $\frac{\partial(x, y)}{\partial(u, v)}\neq 0$, then show that

$$\frac{\partial(x, y)}{\partial(u, v)}\frac{\partial(u, v)}{\partial(x, y)}=1$$

by direct calculation, assuming T^{-1} exists.

18 Suppose that

$$T(u, v, w) = (x, y, z) = (f(u, v, w), g(u, v, w), h(u, v, w))$$

where f, g, and h have continuous partial derivatives with respect to u, v, and w. If $\dfrac{\partial(x, y, z)}{\partial(u, v, w)} \neq 0$, then show that

$$\frac{\partial(x, y, z)}{\partial(u, v, w)} \frac{\partial(u, v, w)}{\partial(x, y, z)} = 1$$

by direct calculation, assuming T^{-1} exists.

19 Show that T given by $T(u, v) = (u^3, v) = (x, y)$ has an inverse transformation T^{-1} everywhere (including points $(u_0, 0)$) even though J_T fails to be invertible on the u-axis. (Thus, the condition that $J_T(\mathbf{u}_0)$ be invertible in Theorem 8.3 is *sufficient* but *not necessary* for T^{-1} to exist.)

20 Show that T given by $T(u, v, w) = (u^3, v^3, w) = (x, y, z)$ has an inverse transformation T^{-1} everywhere, even though J_T fails to be invertible at points on the u- or v-axis. (See the remark in Exercise 19.)

21 Show that T given by $T(u, v) = (u^2, v) = (x, y)$ has no inverse on any open set containing the v-axis. Show that the v-axis comprises *exactly* the set of points in $\mathbf{R}^2$ where J_T fails to be invertible.

22 Show that T given by $T(u, v, w) = (u^2, v^2, w) = (x, y, z)$ has no inverse on any open set containing points on either the u-axis or the v-axis. Where is J_T not invertible?

23 Let $T: \mathbf{R}^2 \to \mathbf{R}^2$, $T(u, v) = (x, y)$ be a differentiable coordinate transformation. Show that the first column of $J_T(u_0, v_0)$ is a tangent vector to the curve in the xy-plane which is the image of $v = v_0$ in the uv-plane. This is the curve $\mathbf{x} = \mathbf{x}(u, v_0)$ which results when we let u vary near (u_0, v_0).

24 Let $T: \mathbf{R}^3 \to \mathbf{R}^3$ be a differentiable coordinate transformation. Show that each column of $J_T(u_0, v_0, w_0)$ is a tangent vector to a curvilinear curve which results when u or v or w is allowed to vary near (u_0, v_0, w_0).

25 Show that for the polar coordinate transformation $T: \mathbf{R}^2 \to \mathbf{R}^2$ of Examples 8.2 and 8.4, the columns of J_T are orthogonal vectors.

26 Show that for the cylindrical coordinate transformation of Exercise 9, the columns of J_T are mutually orthogonal vectors.

27 An n-by-n matrix A is called **orthogonal** in case its columns are mutually orthogonal and every column has length 1. In case $n = 2$ or 3, show that an orthogonal matrix A has inverse A^t, the transpose matrix which results when the rows of A are put into the columns of A^t. For example,

$$\begin{pmatrix} a_{11} & a_{12} & a_{13} \\ a_{21} & a_{22} & a_{23} \\ a_{31} & a_{32} & a_{33} \end{pmatrix}^t = \begin{pmatrix} a_{11} & a_{21} & a_{31} \\ a_{12} & a_{22} & a_{32} \\ a_{13} & a_{23} & a_{33} \end{pmatrix}.$$

28 Use Exercise 27 to find the inverse of the "normalized" Jacobian matrix for polar coordinates,

$$\begin{pmatrix} \cos\theta & -\sin\theta \\ \sin\theta & \cos\theta \end{pmatrix}.$$

In Exercises 29 to 33, refer to Theorem 8.10.

29 Show that the images of the planes $u = a$ and $u = b$ under $\boldsymbol{T}$ are parallel planes (*Hint:* The vector equation of $u = a$ is $(u, v, w) = (a, v, w)$).

30 Show that the images of the planes $v = c$ and $v = d$ under $\boldsymbol{T}$ are parallel planes.

31 Show that the images of the planes $w = e$ and $w = f$ are parallel planes.

32 Show that the volume of $\boldsymbol{T}(P)$ is $|\det M_{\boldsymbol{T}}|\, V(P)$.

33 Show that $\boldsymbol{T}(P)$ is a nondegenerate parallelepiped if $\boldsymbol{T}$ is an invertible linear transformation.

9 CHANGE OF VARIABLES IN MULTIPLE INTEGRALS

With the aid of the ideas in the preceding section, we can now give a uniform approach to changing variables in double and triple integrals. You will see that the general formula we obtain includes the special formulas we developed in Sections 5, 7, and 8 of Chapter 5 for the cases of changing the Cartesian variables x, y, and z to polar, cylindrical, or spherical coordinate variables. Moreover, the general formula is a direct analogue of the change-of-variable theorem for definite integrals in elementary calculus. Let's first recall the situation in elementary calculus.

Given an intractable integral to evaluate, in which part of the integrand is the derivative of some other part $g(x)$, we let $u = g(x)$. Then $du = g'(x)\,dx$. Then the integral $\int_c^d f(g(x))g'(x)\,dx$ becomes $\int_a^b f(u)\,du$, where $a = g(c)$ and $b = g(d)$.

From the point of view of Section 8, we are changing coordinates in $\boldsymbol{R} = \boldsymbol{R}^1$ by letting $u = g(x)$. Observe, however, that this is the *reverse* of our convention of changing coordinates by putting new coordinates on $\boldsymbol{R}^n$ via a transformation $\boldsymbol{T}$ where $\boldsymbol{T}(\boldsymbol{u}) = \boldsymbol{x}$. To put the elementary calculus change-of-variable formula in tune with this convention, we prefer to write the formula in the last paragraph as

$$\int_c^d f(g(u))g'(u)\,du = \int_a^b f(x)\,dx. \tag{1}$$

(This is certainly the same formula—we merely interchanged the dummy variables x and u.) One usually assumes here that g' is continuous on $[c, d]$ and that f is continuous on $[a, b]$. As you have doubtless guessed, it is formula (1) which we will generalize here to the case of functions $f: \boldsymbol{R}^n \to \boldsymbol{R}$ ($n = 2$ or 3) and coordinate transformations $\boldsymbol{g}$ (or $\boldsymbol{T}$): $\boldsymbol{R}^n \to \boldsymbol{R}^n$.

So suppose that we have a differentiable coordinate transformation $\boldsymbol{T}: \boldsymbol{R}^n \to \boldsymbol{R}^n$ ($n = 2$ or 3) given as $\boldsymbol{T}(\boldsymbol{u}) = \boldsymbol{x}$, where $\boldsymbol{u} = (u_1, u_2, \ldots, u_n)$ and $\boldsymbol{x} = (x_1, x_2, \ldots, x_n)$. Since $\boldsymbol{T}$ is differentiable, near any point $\boldsymbol{u}_0$ we can *approximate* $\boldsymbol{T}$ by its tangent approximation.

$$\boldsymbol{T}(\boldsymbol{u}) \approx \boldsymbol{T}(\boldsymbol{u}_0) + J_{\boldsymbol{T}}(\boldsymbol{u}_0)(\boldsymbol{u} - \boldsymbol{u}_0).$$

The effect of $\boldsymbol{T}$ on volumes ΔV_{ijk} of elementary rectangular parallelepipeds in $\boldsymbol{R}^3$ (or areas ΔA_{ij} of elementary subrectangles in $\boldsymbol{R}^2$) near $\boldsymbol{u}_0$ is thus approximately the effect of the tangent approximation. By Corollaries 8.9 and 8.11, the tangent approximation simply multiplies these volumes or areas by $|\det J_{\boldsymbol{T}}(\boldsymbol{u}_0)|$, the magnitude of the Jacobian determinant of $\boldsymbol{T}$ at $\boldsymbol{u}_0$. It turns out that this local reasoning can be extended to the entire region of integration. The precise result is the following.

9.1 **CHANGE OF VARIABLES THEOREM.** Let U be an open subset of $\boldsymbol{R}^n$ ($n=2$ or 3), and let $\boldsymbol{T}$ be a differentiable coordinate transformation given by $\boldsymbol{x}=\boldsymbol{T}(\boldsymbol{u})$ such that the entries of $J_{\boldsymbol{T}}$ are continuous and $\det J_{\boldsymbol{T}}$ is never 0 on U. Suppose that $E \subseteq U$ and that $f:\boldsymbol{R}^n \to \boldsymbol{R}$ is integrable over $\boldsymbol{T}(E)$. Then $f \circ \boldsymbol{T}$ is integrable over E, and we have:

($n=2$):

(2)
$$\iint_{\boldsymbol{T}(E)} f(\boldsymbol{x})\, dA = \iint_{E} f \circ \boldsymbol{T}(\boldsymbol{u})\, |\det J_{\boldsymbol{T}}|\, dA$$
$$= \iint_{E} f(x(\boldsymbol{u}), y(\boldsymbol{u})) \left|\frac{\partial(x, y)}{\partial(u, v)}\right| du\, dv$$

($n=3$):

(3)
$$\iiint_{\boldsymbol{T}(E)} f(\boldsymbol{x})\, dV = \iiint_{E} f \circ \boldsymbol{T}(\boldsymbol{u})\, |\det J_{\boldsymbol{T}}|\, dV$$
$$= \iiint_{E} f(x(\boldsymbol{u}), y(\boldsymbol{u}), z(\boldsymbol{u})) \left|\frac{\partial(x, y, z)}{\partial(u, v, w)}\right| du\, dv\, dw.$$

Idea of the Proof. Suppose that we are in the $n=2$ case. The double integrals in (2) are both limits of Riemann sums. Suppose that we make a grid G of E by rectangles R_{ij}, and choose $\bar{\boldsymbol{u}}_{ij} \in R_{ij}$. Then by the reasoning of the preceding paragraph, $\boldsymbol{T}(E)$ is chopped into subregions $\boldsymbol{T}(R_{ij})$, which are approximately the parallelograms produced by the tangent approximations

$$\boldsymbol{T}(\bar{\boldsymbol{u}}_{ij}) + J_{\boldsymbol{T}}(\bar{\boldsymbol{u}}_{ij})(\boldsymbol{u} - \bar{\boldsymbol{u}}_{ij}).$$

See Figure 9.1. Thus,

(4)
$$A(\boldsymbol{T}(R_{ij})) \approx A(J_{\boldsymbol{T}}(R_{ij})) = |\det J_{\boldsymbol{T}}(\bar{\boldsymbol{u}}_{ij})|\, A(R_{ij})$$

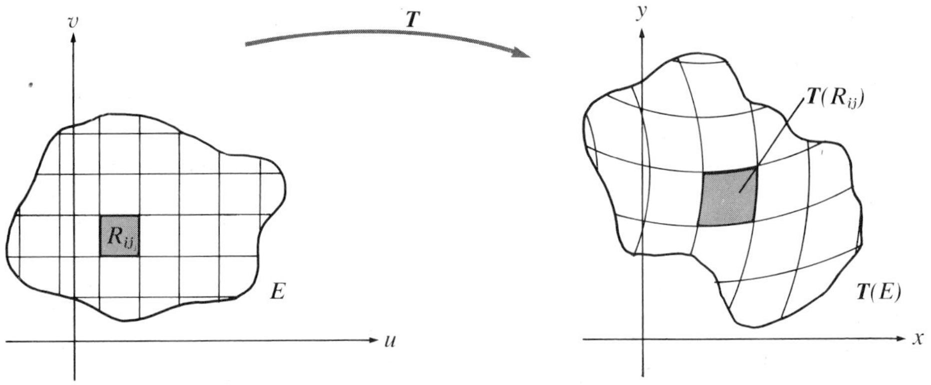

FIGURE 9.1

by Corollary 8.9. Now we can approximate $\iint_{T(E)} f(\boldsymbol{x})\,dA$ by the sum

$$\sum_{i,j} f(x(\bar{\boldsymbol{u}}_{ij}), y(\bar{\boldsymbol{u}}_{ij}))A(\boldsymbol{T}(R_{ij})),$$

which by (4) is approximately

$$\sum_{i,j} f(x(\bar{\boldsymbol{u}}_{ij}), y(\bar{\boldsymbol{u}}_{ij}))\,|\det J_{\boldsymbol{T}}(\bar{\boldsymbol{u}}_{ij})|\,A(R_{ij})$$

$$= \sum_{i,j} f(x(\bar{\boldsymbol{u}}_{ij}), y(\bar{\boldsymbol{u}}_{ij}))\left|\frac{\partial(x, y)}{\partial(u, v)}(\bar{\boldsymbol{u}}_{ij})\right|\Delta u_i\,\Delta v_j.$$

The last sum is a Riemann sum $R(f\circ \boldsymbol{T}, G, \bar{\boldsymbol{u}}_{ij})$ over our grid G of rectangles of length Δu_i and height Δv_j. To complete the proof, we must show that our approximation of $\iint_{T(E)} f(\boldsymbol{x})\,dA$ by this quantity is accurate enough that $R(f\circ \boldsymbol{T}, G, \bar{\boldsymbol{u}}_{ij})$ actually approaches $\iint_{T(E)} f(\boldsymbol{x})\,dA$ as $m(G)\to 0$. If this is so, then $f\circ \boldsymbol{T}$ will be integrable over E, and

$$\iint_{T(E)} f(\boldsymbol{x})\,dx\,dy = \iint_E f\circ \boldsymbol{T}(\boldsymbol{u})\,|\det J_{\boldsymbol{T}}|\,du\,dv.$$

That will prove (2). Formula (3) is proved by the same reasoning in three dimensions. The rigorous proof that our approximation scheme actually *does* work requires a subtler version of Theorem 8.8 and a delicate study of the errors in our approximations, and so we leave the details for advanced calculus.

Note that (2) and (3) are exact analogues of (1). Our assumptions on $\boldsymbol{T}$ can be shown to imply that either $\det J_{\boldsymbol{T}}>0$ on E or $\det J_{\boldsymbol{T}}<0$ on E, so the absolute value signs merely serve to ensure that the area factor will be positive. We are thus multiplying $f(\boldsymbol{T}(\boldsymbol{u}))$ by the positive real number which is our measure of the size of the total derivative of $\boldsymbol{T}$ at $\boldsymbol{u}$. If in (1) we insist that all definite integrals have lower limit *smaller* than upper limit, then we would need to enclose $g'(u)$ in absolute value signs also. For instance, the substitution $x=1-u^2$ transforms the integral $\int_0^1 \sqrt{x}\,dx$ to

$$\int_0^1 \sqrt{1-u^2}\,|-2u|\,du = \int_1^0 \sqrt{1-u^2}(-2u)\,du.$$

Thus in (1) we actually multiply $f(g(u))$ by the *size* of $g'(u)$.

We will now present a string of examples illustrating Theorem 9.1. First we consider the familiar alternative coordinate systems of the last chapter. Then we consider applications of Theorem 9.1 in more general situations.

9.2 EXAMPLE. Obtain the polar coordinate integral expression $\iint_E f(r\cos\theta, r\sin\theta)r\,dr\,d\theta$ from Theorem 9.1.

Solution. Here $f:\boldsymbol{R}^2\to \boldsymbol{R}$ is a function of x and y, and polar coordinates are introduced by $x=r\cos\theta$ and $y=r\sin\theta$. Then $\boldsymbol{T}(r, \theta)=(x, y)=(r\cos\theta, r\sin\theta)$.

From Example 8.4,

$$\det J_T = \frac{\partial(x, y)}{\partial(r, \theta)} = r.$$

If, as usual, we require $r > 0$, then we get from (2) of Theorem 9.1

$$\iint_{T(E)} f(x, y)\, dA = \iint_{E} f(r\cos\theta, r\sin\theta) r\, dr\, d\theta,$$

as desired.

9.3 EXAMPLE. Obtain the cylindrical coordinate integral expression

$$\iiint_{E} f(r\cos\theta, r\sin\theta, z) r\, dr\, d\theta\, dz$$

from Theorem 9.1.

Solution. Here $f: \boldsymbol{R}^3 \to \boldsymbol{R}$ is a function of x, y, and z, and new coordinates are introduced by $x = r\cos\theta$, $y = r\sin\theta$, and $z = z$. Then $\boldsymbol{T}(r, \theta, z) = (x, y, z) = (r\cos\theta, r\sin\theta, z)$. We have

$$\frac{\partial(x, y, z)}{\partial(r, \theta, z)} = \det\begin{pmatrix} \cos\theta & -r\sin\theta & 0 \\ \sin\theta & r\cos\theta & 0 \\ 0 & 0 & 1 \end{pmatrix}$$

$$= r\cos^2\theta + r\sin^2\theta = r.$$

Thus, if we again require $r > 0$, then from (3) of Theorem 9.1,

$$\iiint_{T(E)} f(x, y, z)\, dV = \iiint_{E} f(r\cos\theta, r\sin\theta, z) r\, dr\, d\theta\, dz,$$

as desired.

9.4 EXAMPLE. Obtain the spherical coordinate integral expression

$$\iiint_{E} f(\rho\sin\phi\cos\theta, \rho\sin\phi\sin\theta, \rho\cos\phi)\rho^2\sin\phi\, d\rho\, d\phi\, d\theta$$

from Theorem 9.1.

Solution. Here $f: \boldsymbol{R}^3 \to \boldsymbol{R}$ is a function of x, y, and z, and new coordinates are introduced by $x = \rho\sin\phi\cos\theta$, $y = \rho\sin\phi\sin\theta$, and $z = \rho\cos\phi$, where $0 \le \phi \le \pi$,

$0 \le \rho$, and $0 \le \theta < 2\pi$. We have

$$\frac{\partial(x, y, z)}{\partial(\rho, \phi, \theta)} = \det\begin{pmatrix} \sin\phi\cos\theta & \rho\cos\phi\cos\theta & -\rho\sin\phi\sin\theta \\ \sin\phi\sin\theta & \rho\cos\phi\sin\theta & \rho\sin\phi\cos\theta \\ \cos\phi & -\rho\sin\phi & 0 \end{pmatrix}$$

$$= \rho^2\sin\phi\cos^2\phi\cos^2\theta + \rho^2\sin^3\phi\sin^2\theta + \rho^2\sin\phi\cos^2\phi\sin^2\theta + \rho^2\sin^3\phi\cos^2\theta$$

$$= \rho^2\sin\phi\cos^2\phi + \rho^2\sin^3\phi$$

$$= \rho^2\sin\phi(\cos^2\phi + \sin^2\phi) = \rho^2\sin\phi.$$

Thus from (3) of Theorem 9.1, if we restrict $\phi \ne 0$, $\phi \ne \pi$, then $\rho^2\sin\phi > 0$ since $0 < \phi < \pi$, so

$$\iiint_{T(E)} f(x, y, z)\, dV$$

$$= \iiint_{E} f(\rho\sin\phi\cos\theta, \rho\sin\phi\sin\theta, \rho\cos\phi)\rho^2\sin\phi\, d\rho\, d\phi\, d\theta$$

as desired.

The preceding three examples establish the formulas $dA = r\,dr\,d\theta$, $dV = r\,dr\,d\theta\,dz$, and $dV = \rho^2\sin\phi\,d\rho\,d\phi\,d\theta$ for polar, cylindrical, and spherical coordinates respectively as consequences of Theorem 9.1. Previously we arrived at them (in Sections 5, 7, and 8 of Chapter 5) through geometric arguments on approximations of areas in $\boldsymbol{R}^2$ and volumes in $\boldsymbol{R}^3$. These approximations are now fully justified by Theorem 9.1, as are the change of variable techniques and consistency claims in Chapter 5 (pp. 263, 283, 284, 292, etc.).

The impact of Theorem 9.1 is actually far greater than the foregoing three examples suggest. By providing a uniform method for transforming integrals under a change of coordinates, it allows us to handle a far wider class of integrals than we could previously evaluate readily.

9.5 EXAMPLE. Evaluate $\iint_P (2x^2 + xy - y^2)\, dA$ where $P = \{(x, y) \mid 1 \le 2x - y \le 3, -2 \le x + y \le 0\}$.

Solution. Note that P is the region of Example 8.12. Refer to Figures 8.4 and 8.5. While we could actually evaluate the integral in terms of Cartesian coordinates, it would involve decomposing P into subregions and then working out some messy polynomial integrations. A much simpler approach is to make the change of variables $\boldsymbol{x} = \boldsymbol{T}(\boldsymbol{u}) = \boldsymbol{S}^{-1}(\boldsymbol{u})$ where $\boldsymbol{S}$ is defined by

(5) $$\begin{cases} u = 2x - y \\ v = x + y. \end{cases}$$

Then $P = \boldsymbol{T}(R)$, where R is the rectangle $[1, 3] \times [-2, 0]$ in the uv-plane. Observe that the integrand

(*) $$2x^2 + xy - y^2 = (2x + y)(x - y) = uv,$$

so we don't even have to solve (5) for x and y in terms of u and v to transform our integral. (Of course, we *could* do so if we didn't notice (*).) From (5) we have

$$\frac{\partial(u, v)}{\partial(x, y)} = \det\begin{pmatrix} 2 & -1 \\ 1 & 1 \end{pmatrix} = 3.$$

Thus by Corollary 8.7, $\dfrac{\partial(x, y)}{\partial(u, v)} = \dfrac{1}{3}$. Theorem 9.1(2) then gives

$$\begin{aligned}
\iint_P (2x^2 + xy - y^2)\, dA &= \iint_{T(R)} (2x^2 + xy - y^2)\, dx\, dy \\
&= \iint_R uv\tfrac{1}{3}\, du\, dv \\
&= \tfrac{1}{3} \int_1^3 u\, du \int_{-2}^0 v\, dv \\
&= \tfrac{1}{3}\left[\tfrac{1}{2}u^2\right]_1^3 \left[\tfrac{1}{2}v^2\right]_{-2}^0 \\
&= \tfrac{1}{3}\tfrac{1}{4}[9-1][0-4] \\
&= -\tfrac{8}{3}.
\end{aligned}$$

9.6 EXAMPLE. Evaluate $\iiint_D z^2\, dV$ where D is the interior of the ellipsoid

$$\frac{x^2}{4} + \frac{y^2}{16} + \frac{z^2}{9} = 1.$$

Solution. We can make a linear change of coordinates which will replace D by the interior of a sphere, and then we can use spherical coordinates. First, if we let $(x, y, z) = \boldsymbol{T}(u, v, w)$ where

$$(6) \qquad \begin{cases} u = \tfrac{1}{2}x \\ v = \tfrac{1}{4}y, \\ w = \tfrac{1}{3}z \end{cases}$$

then $D = \boldsymbol{T}(E)$ where E is the unit ball $u^2 + v^2 + w^2 \leq 1$. (See Figure 9.2.) Here we

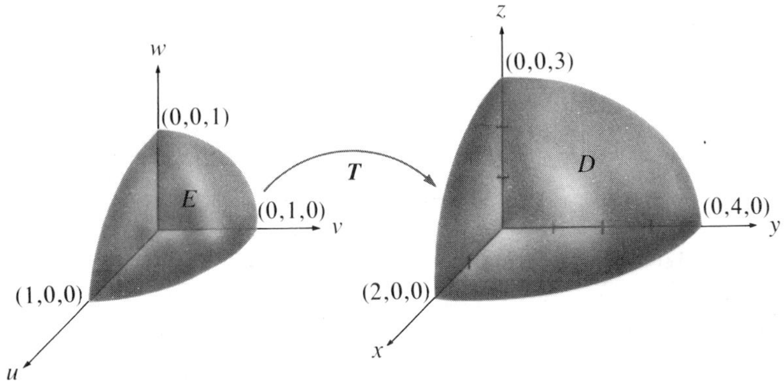

FIGURE 9.2

can solve for x, y, and z in (6), getting

(7) $$\begin{cases} x = 2u \\ y = 4v \\ z = 3w \end{cases}$$

Then

$$\frac{\partial(x, y, z)}{\partial(u, v, w)} = \det\begin{pmatrix} 2 & 0 & 0 \\ 0 & 4 & 0 \\ 0 & 0 & 3 \end{pmatrix} = 24.$$

So by Theorem 9.1(3) we have

(8) $$\iiint\limits_{T(E)} z^2\, dV = \iiint\limits_{E} (3w)^2 24\, du\, dv\, dw = 216 \iiint\limits_{E} w^2\, du\, dv\, dw.$$

We can now introduce spherical coordinates. Let

$$u = \rho \sin\phi \cos\theta$$
$$v = \rho \sin\phi \sin\theta$$
$$w = \rho \cos\phi$$

Here $(u, v, w) = \boldsymbol{S}(\rho, \phi, \theta)$, where $\boldsymbol{S}$ is the spherical coordinate transformation. We have from Example 9.4

$$\frac{\partial(u, v, w)}{\partial(\rho, \phi, \theta)} = \rho^2 \sin\phi.$$

So from Theorem 9.1(3), (8) becomes

$$\begin{aligned}
\iiint\limits_{T(E)} z^2\, dV &= 216 \iiint\limits_{E} w^2\, du\, dv\, dw = 216 \iiint\limits_{S^{-1}(E)} \rho^4 \sin\phi \cos^2\phi\, d\rho\, d\phi\, d\theta \\
&= 216 \int_0^{2\pi} d\theta \int_0^{\pi} \sin\phi \cos^2\phi\, d\phi \int_0^1 \rho^4\, d\rho \\
&= 216(2\pi)\Big[-\tfrac{1}{3}\cos^3\phi\Big]_0^{\pi} \Big[\frac{\rho^5}{5}\Big]_0^1 \\
&= 432\pi\big[\tfrac{1}{3} + \tfrac{1}{3}\big]\tfrac{1}{5} \\
&= \frac{864\pi}{15} = \frac{288\pi}{5}.
\end{aligned}$$

EXERCISES 6.9

In Exercises 1 to 14, evaluate the given double or triple integrals.

1 $\iint\limits_{P} 3x\, dx\, dy$, $P = \{(x, y) \mid 1 \le 2x - y \le 3, -2 \le x + y \le 0\}$

2 $\iint_P x^2\,dA,\ P=\{(x,y)\mid -1\le x-y\le 1,\ 0\le x+y\le 2\}$

3 $\iint_P (3x^2+7xy+2y^2)\,dA,\ P=\{(x,y)\mid 1\le 3x+y\le 3,\ 2\le x+2y\le 5\}$

4 $\iint_P (3x^2+5xy-2y^2)\,dA,\ P=\{(x,y)\mid -1\le 3x-y\le 2,\ -3\le x+2y\le -1\}$

5 $\iint_D x\,dA,\ D=\{(x,y)\mid 1\le x(1-y)\le 2,\ 1\le xy\le 3\}$

6 $\iint_D (x^4-y^4)\,dA,\ D=\{(x,y)\mid 0\le x^2-y^2\le 2,\ 0\le xy\le 1\}$

7 $\iint_D (x^2+y^2)^{3/2}\,dA,\ D=\{(x,y)\mid 1\le x^2+y^2\le 4\}$

8 $\iiint_P (y-z)\,dV,\ P=\{(x,y,z)\mid -1\le x-z\le 1,\ 0\le y+z\le 2,\ 1\le x+z\le 3\}$

9 $\iiint_P (xy-yz+xz-z^2)\,dV$, P as in Exercise 8.

10 $\iiint_P (x^2-z^2)\,dV$, P as in Exercise 8.

11 $\iiint_P (2x^2-2y^2-z^2+3xy-xz+3yz)\,dV$,
$P=\{(x,y,z)\mid 1\le 2x-y+z\le 3,\ 0\le x+2y-z\le 1,\ 0\le -x+y+z\le 2\}$

12 $\iiint_D (12x^2yz+8xyz-4yz-6x^2y-4xz+1)\,dV$,
$D=\{(x,y,z)\mid 0\le x-y^2+z^2\le 1,\ 0\le x^2-y^2+z\le 2,\ 1\le x^3-y+z^2\le 2\}$

13 $\iiint_D \frac{z}{\sqrt{x^2+y^2}}\,dV$, D the region between $x^2+y^2=4$, $x^2+y^2=9$, the xy-plane, and $z=2$.

14 $\iiint_D \frac{1}{x^2+y^2+z^2}\,dV$, D the region between the spheres $x^2+y^2+z^2=1$ and $x^2+y^2+z^2=9$.

15 Establish the formula $A=\pi ab$ for the area enclosed by the ellipse $\frac{x^2}{a^2}+\frac{y^2}{b^2}=1$.

16 Introduce an appropriate change of variable to find the area enclosed by the ellipse $x^2+4xy+5y^2=4$. (*Hint:* Complete the square.)

17 Establish the formula $V=\frac{4\pi}{3}abc$ for the region enclosed by the ellipsoid $\frac{x^2}{a^2}+\frac{y^2}{b^2}+\frac{z^2}{c^2}=1$.

18 Introduce an appropriate change of variable to find the volume enclosed by the ellipsoid $x^2+8y^2+6z^2+4xy-2xz+4yz=9$. (*Hint:* Complete the square to obtain $(ax+by+cz)^2$ plus other terms for suitable a, b, and c.)

19 Evaluate $\iiint\limits_D z^2\,dv$, where D is the interior of the ellipsoid

$$\frac{x^2}{9}+\frac{y^2}{4}+\frac{z^2}{4}=1.$$

20 Evaluate $\iint\limits_D e^{(y-x)/(y+x)}\,dA$, where D is the triangle bounded by the two coordinate axes and the line $x+y=2$.

21 Show that the change of coordinates $\boldsymbol{T}:\boldsymbol{R}^3\to\boldsymbol{R}^3$ preserves volumes if $\boldsymbol{T}(u,v,w)=(x,y,z)$ where

$$u=x$$
$$v=ax+y$$
$$w=bx+cy+z \quad \text{where } a, b, c \text{ are any constants.}$$

22 Repeat Exercise 21 for the transformation $\boldsymbol{T}$ given by $\boldsymbol{T}(u,v,w)=(x,y,z)$ where

$$u=x+ay+bz$$
$$v=y+cz$$
$$w=z.$$

23 Show that $\iiint\limits_D xy\,dV=0$ if $D=\{(x,y,z)\mid 0\le z\le x+2,\ 0\le x^2+4y^2\le 4\}$

24 Find $\iiint\limits_D x\,dV$ if D is the region of Exercise 23.

25 Outline the proof of Theorem 9.1 in case $n=3$.

REVIEW EXERCISES 6.10

1 Find $\lim_{\boldsymbol{x}\to\boldsymbol{x}_0}\boldsymbol{F}(\boldsymbol{x})$ and determine whether $\boldsymbol{F}$ is continuous at $\boldsymbol{x}_0$.

(a) $\boldsymbol{F}(x,y)=\left(\operatorname{Tan}^{-1}\dfrac{x}{y},\ \sqrt{x^2+y^2},\ \dfrac{y^3}{x^2}\right)$, $\boldsymbol{x}_0=(1,1)$.

(b) $\boldsymbol{F}(x,y,z)=\left(\dfrac{1-\cos yz}{yz},\ \sqrt{x^2+y^2+z^2}\right)$, $\boldsymbol{x}_0=(2,2,0)$.

In Exercises 2 to 4, determine whether the given function is linear. If it is, give M_F.

2 (a) $\boldsymbol{F}(x,y)=(-\pi x-3y,\ x+ey)$
(b) $\boldsymbol{F}(x,y,z)=(-\frac{1}{2}x-y-z,\ 2x-y+2z,\ -x-2y)$

3 (a) $\boldsymbol{F}(x,y)=(2x-y,\ 2x-xy)$
(b) $\boldsymbol{F}(x,y,z)=(2x+ey-z,\ -\pi x+y-z,\ xy-2z)$

4 $\boldsymbol{F}$ is the reflection in the plane through the origin perpendicular to $\boldsymbol{v}_0=(1,1,1)$:

$$\boldsymbol{F}(\boldsymbol{x})=\boldsymbol{x}-\frac{2\boldsymbol{x}\cdot\boldsymbol{v}_0}{|\boldsymbol{v}_0|^2}\boldsymbol{v}_0.$$

5 Compute $\boldsymbol{L}(\boldsymbol{x}_0)$ for the linear transformation $\boldsymbol{L}$ and vector $\boldsymbol{x}_0$:
(a) $\boldsymbol{L}(1,0)=(1,2)$, $\boldsymbol{L}(0,1)=(-1,1)$; $\boldsymbol{x}_0=(\frac{1}{2},-2)$.
(b) $\boldsymbol{L}(1,0,0)=(-1,\frac{1}{2},2)$, $\boldsymbol{L}(0,1,0)=(0,1,-2)$, $\boldsymbol{L}(0,0,1)=(3,-1,1)$; $\boldsymbol{x}_0=(-1,2,-2)$.

6 Compute, where possible, $2A-3B$, AB, and BA.

(a) $A=\begin{pmatrix} 0 & 1 & 2 \\ -1 & 1 & 0 \\ 1 & 5 & -2 \end{pmatrix}$, $B=\begin{pmatrix} 0 & 1 & 3 \\ -1 & 1 & 4 \end{pmatrix}$

(b) $A=\begin{pmatrix} -1 & 3 & 2 \\ 0 & 2 & -3 \\ 1 & -3 & 4 \end{pmatrix}$, $B=\begin{pmatrix} -1 & 1 & -1 \\ 2 & 0 & -1 \\ -1 & 1 & 0 \end{pmatrix}$

7 Find $M_{\mathbf{L}_1\circ\mathbf{L}_2}$ and $\mathbf{L}_1\circ\mathbf{L}_2(-2, 1)$.

(a) $\mathbf{L}_1(x, y, z)=(x-y+2z, -x-3y+4z)$
$\mathbf{L}_2(x, y) \quad =(3x-y, 2x+3y, -x+2y)$

(b) $\mathbf{L}_1(x, y) \quad =(-x+y, 2x-y, 3y)$
$\mathbf{L}_2(x, y) \quad =(x-y, x-2y)$

8 If $\mathbf{L}_1(x, y, z)=(-x+2y-z, 3x-y-z, x+y)$ and
$\mathbf{L}_2(x, y, z)=(2y-3z, x-3y+z, -x+2y-z)$,
then find $\mathbf{L}_2\circ\mathbf{L}_1(3, -1, 1)$.

In Exercises 9 to 13, determine whether A is invertible and find A^{-1} if it exists.

9 $A=\begin{pmatrix} 2 & -3 \\ -1 & \frac{3}{2} \end{pmatrix}$

10 $A=\begin{pmatrix} 5 & 2 \\ 2 & 1 \end{pmatrix}$

11 $A=\begin{pmatrix} 2 & -1 & 0 \\ -1 & 2 & -1 \\ 0 & -1 & 2 \end{pmatrix}$

12 $A=\begin{pmatrix} 1 & -1 & 2 \\ 0 & -4 & 5 \\ 2 & 2 & -1 \end{pmatrix}$

13 $A=\begin{pmatrix} 2 & -2 & 0 & 0 \\ -1 & 2 & -1 & 0 \\ 0 & -1 & 2 & -1 \\ 0 & 0 & -2 & 2 \end{pmatrix}$

14 Decide whether $\boldsymbol{F}$ is differentiable at $\boldsymbol{x}_0$ and, if it is, find its total derivative at $\boldsymbol{x}_0$.

(a) $\boldsymbol{F}(x, y)=\begin{pmatrix} \sqrt{x-y} \\ x^2+xy-xy^3 \end{pmatrix}$; $\boldsymbol{x}_0=(1, 1)$

(b) $\boldsymbol{F}(x, y, z)=\begin{pmatrix} \sqrt{x^2+y^2+z^2} \\ 3xyz-x^2z-yz^2-2 \end{pmatrix}$; $\boldsymbol{x}_0=(1, 2, 2).,$

In Exercises 15 to 16, find $J_{\boldsymbol{F}\circ\boldsymbol{G}}$ at the given point.

15 $\boldsymbol{F}(u, v, w)=(u^2v, w^2)$, $\boldsymbol{G}(x, y, z)=(xye^z, x^2\cos yz, x+y+z)$ at $(x_0, y_0, z_0)=(-1, \pi, 1)$.

16 $\boldsymbol{F}(u, v)=(uv, u^2+v^2, u^2-v^2)$, $\boldsymbol{G}(x, y, z)=(\sqrt{x^2+y^2+z^2}, xyz)$ at $(x_0, y_0, z_0)=(1, 1, \sqrt{2})$.

17 If

$$\begin{cases} p=u^2-v^2 \\ q=2uv-u^2v-v^2 \\ r=uv^2-u^2 \end{cases} \quad \text{where} \quad \begin{cases} u=x^2-y^2 \\ v=xy^2-y^3 \end{cases}$$

then find $\dfrac{\partial p}{\partial x}$, $\dfrac{\partial q}{\partial y}$ and $\dfrac{\partial r}{\partial x}$ when $x=1$ and $y=2$.

18 Find the tangent line to the curve of intersection of the surfaces $x^2+y^2-z^2=1$ and $x-2y+z+1=0$ at $(1, 2, 2)$.

19 Show that

$$\begin{cases} u^2-v^2-2xy+x^2+3=0 \\ 2u-v-x^2-xy-y^2+6=0 \end{cases}$$

define u and v as functions of x and y near $(x_0, y_0, u_0, v_0) = (1, 2, 1, 1)$. Find $\frac{\partial u}{\partial x}$ and $\frac{\partial u}{\partial y}$, and the equation of the tangent plane to the surface $z = u(x, y)$ at (x_0, y_0). Approximate $u(0.98, 2.02)$.

20 If

$$\mathbf{T}: \begin{cases} x = u^2 - v^2 \\ y = 2uv \end{cases},$$

then find $J_{\mathbf{T}}$, $\det J_{\mathbf{T}}$, and $\det J_{\mathbf{T}^{-1}}$, if it exists.

21 Find the area of the parallelogram bounded by the lines $2x - y = 1$, $2x - y = 6$, $x + 2y = -1$, and $x + 2y = 4$.

22 Find the volume of the parallelepiped bounded by the planes $x + z = 0$, $x + z = 3$, $y - z = -2$, $y - z = -1$, $y + z = 1$, and $y + z = 5$.

In Exercises 23 to 27, evaluate the given integral.

23 $\iint_D y^2 \, dA$, $D = \{(x, y) \mid -1 \le x - y \le 1, 0 \le x + y \le 2\}$.

24 $\iint_D (x^2 - y^2) \, dA$, $D = \{(x, y) \mid 0 \le x^2 + y^2 \le 2, 0 \le xy \le 1\}$.

25 $\iiint_D (x^2 - z^2) \, dV$, $D = \{(x, y, z) \mid -1 \le x - z \le 1, 0 \le y + z \le 2, 1 \le x + z \le 3\}$.

26 $\iiint_D \frac{z^2}{(x^2 + y^2)^{3/2}} \, dV$, D the region between $x^2 + y^2 = 4$, $x^2 + y^2 = 9$, the xy-plane, and $z = 3$.

27 $\iiint_D z^2 \, dV$, where D is the interior of $\frac{x^2}{4} + \frac{y^2}{4} + \frac{z^2}{9} = 1$.

7 VECTOR INTEGRATION

0 INTRODUCTION

In Chapter 5 we developed integration for functions $f: \mathbf{R}^n \to \mathbf{R}$. In Chapter 6 we studied functions $\mathbf{F}: \mathbf{R}^n \to \mathbf{R}^m$ and developed their differential calculus. So it is natural now to consider integration of functions $\mathbf{F}: \mathbf{R}^n \to \mathbf{R}^m$. A comprehensive treatment of this requires the notion of a differential form which we leave for advanced calculus. We give only an introduction to the basic ideas for the cases $(n, m) = (2, 2)$, $(2, 3)$ and $(3, 3)$. This is the part of vector integration which is most commonly used in applications, and it will prepare you for the more general point of view of advanced calculus, should you study that subject, by supplying you with concrete examples of the more abstract theory.

We begin with the important notion of the *line integral* of a function $\mathbf{F}: \mathbf{R}^n \to \mathbf{R}^n$. This, you will see, is an extension of the ordinary definite integral over closed intervals to more general one-dimensional paths, namely curves in $\mathbf{R}^n$. In the second section we discuss *Green's Theorem*, a very useful analogue for double integrals and line integrals of the Fundamental Theorem on Evaluation of Definite Integrals. We then develop in Section 4 another analogue of this basic theorem of one-variable integration, a particularly appealing one because of its very close resemblance to the Fundamental Theorem. This resemblance is brought out using the *gradient operator*, which is studied in Section 3 along with two other important vector differential operators, the *divergence* and *curl*. In Sections 5 and 6 we introduce the *surface integral* of a function $\mathbf{F}: \mathbf{R}^3 \to \mathbf{R}^3$ and find a formula for the area of a surface in $\mathbf{R}^3$ as a special kind of surface integral. The concluding two sections are devoted to results on surface integrals analogous to Green's Theorem. These are useful not only because they can often simplify evaluation of surface integrals, but also because of their important applications in physics, chemistry, and engineering, a few of which we discuss in some detail.

1 LINE INTEGRALS

If we consider the ordinary definite integral $\int_a^b f(x)\,dx$ of a function $f: \mathbf{R} \to \mathbf{R}$ over an interval $[a, b]$, then we are led to formulate an analogous notion for functions

$f: \boldsymbol{R}^n \to \boldsymbol{R}$. In Chapter 5 we developed the multiple integral of f over a set $D \subseteq \boldsymbol{R}^n$. Our point of view was to define the integral of f over a region D having the *same dimension*, n, as the domain of f. In the case $n=1$, then, our procedure reduced to the ordinary definite integral of f over an interval $[a, b]$. We are now going to define a new type of integral of a function $\boldsymbol{F}: \boldsymbol{R}^n \to \boldsymbol{R}^n$ over a *one-dimensional* subset γ of $\boldsymbol{R}^n$ which is more general than a simple interval $[a, b]$ on one of the coordinate axes of $\boldsymbol{R}^n$. We will do this in such a way that when $n=1$ and $\gamma=[a, b]$, then we again get the ordinary definite integral $\int_a^b f(x)\,dx$.

A one-dimensional subset of $\boldsymbol{R}^n$ is a *curve* in $\boldsymbol{R}^n$, and in the notation of Chapter 2 we represent such a path γ as $\boldsymbol{x}=\boldsymbol{x}(t)=\boldsymbol{g}(t)$ where $\boldsymbol{g}: \boldsymbol{R} \to \boldsymbol{R}^n$ is a continuous function defined on some interval $[a, b] \subseteq \boldsymbol{R}$. We will require that the curve γ be **piecewise smooth**. This means that $[a, b]$ can be partitioned into a finite number of subintervals $[a_i, b_i]$ on each of which γ is *smooth* in the sense that $\boldsymbol{g}'$ exists and is continuous on $[a_i, b_i]$. (Recall that if γ is smooth over $[a_i, b_i]$, then it has an arc length over $[a_i, b_i]$. See Section 2.2.) Thus the curve γ might look something like that shown in Figure 1.1.

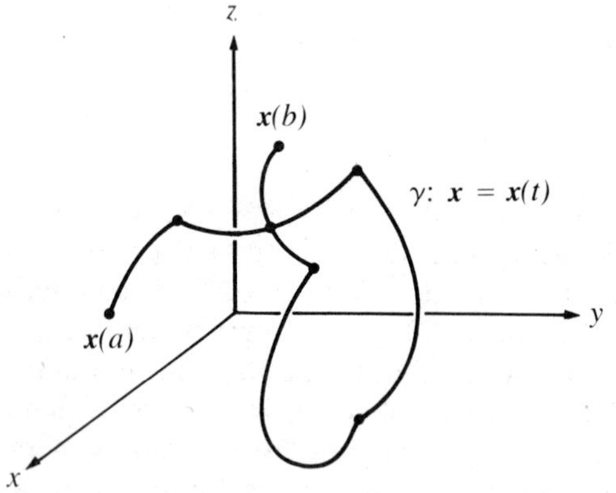

FIGURE 1.1

We will also usually restrict attention to *simple* curves γ which do not cross themselves (as the γ in Figure 1.1 does). Such curves can be parametrized on $[a, b]$ so that $x(t_1)=x(t_2)$ can hold only when $t_1=t_2$ or when $t_1=a$ and $t_2=b$. In the latter case, γ is a **simple closed curve** like the one shown in Figure 1.2.

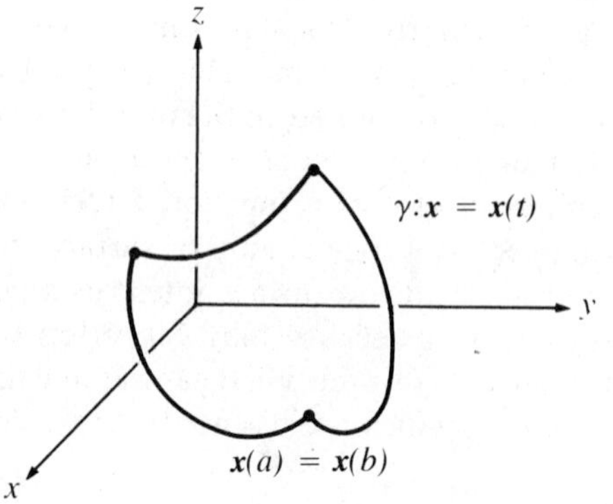

FIGURE 1.2

With this restriction, we can in a natural way put an **orientation** on the curve γ. Namely, we will say that $\boldsymbol{x}(t_1)$ **precedes** $\boldsymbol{x}(t_2)$ if $t_1<t_2$. We will often attach arrowheads to the curve to mark the positive direction, i.e., the direction corresponding to increasing t. See Figure 1.3.

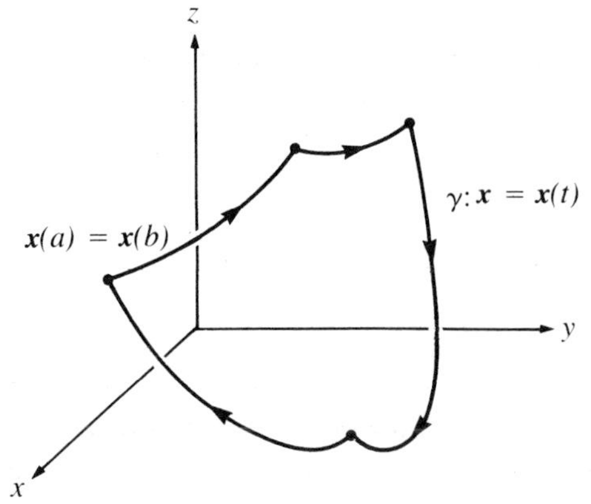

FIGURE 1.3

At this point, you may want to turn back to Chapter 2 to refresh your memory on parametrization of curves. The next example illustrates two general principles. *First,* whenever you have a curve whose Cartesian coordinate equation can be reduced to the sum of two squares equated to 1, set one quantity being squared equal to $\cos t$ and the other equal to $\sin t$. (See Exercise 26 below.) *Second,* if a curve γ is parametrized by $\boldsymbol{x} = \boldsymbol{x}(t)$, $t \in [a, b]$, then the orientation of γ is *reversed* by using the parametrization $\boldsymbol{x} = \boldsymbol{x}(a + b - t)$, $t \in [a, b]$. (See Figure 1.4.)

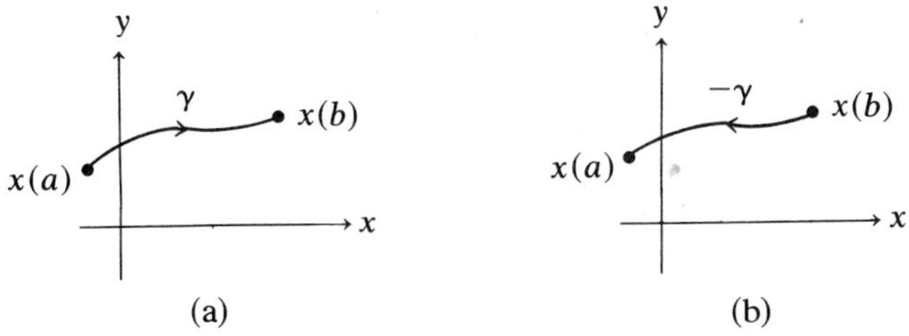

FIGURE 1.4

Observe that we start when $t = a$ at $\boldsymbol{x}(b)$, and we end when $t = b$ at $\boldsymbol{x}(a)$ in this parametrization. We call the resulting oriented curve $-\gamma$ and refer to it as the **negative** or **opposite** of the original curve γ.

1.1 **EXAMPLE.** Parametrize the simple closed curve shown in Figure 1.5.

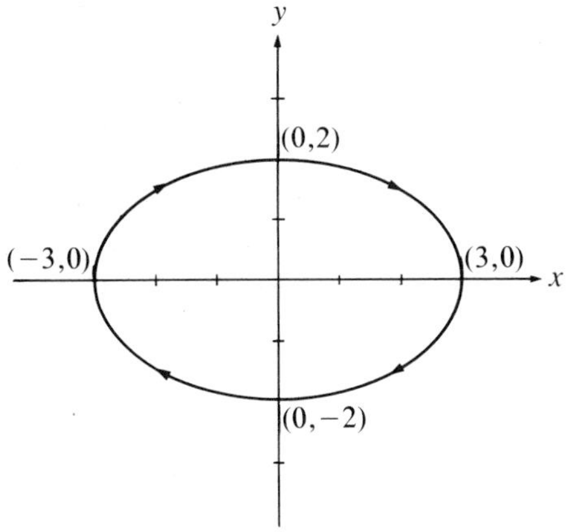

FIGURE 1.5

Solution. The curve is the ellipse $\dfrac{x^2}{9} + \dfrac{y^2}{4} = 1$ traversed in the *clockwise* sense. The usual parametrization, $\boldsymbol{x} = \boldsymbol{x}(t) = (3 \cos t,\ 2 \sin t)$ for $0 \le t \le 2\pi$ gives this curve, but *not* with this orientation, since as t increases one travels along the curve in the *counterclockwise* sense. We can, however, obtain the clockwise orientation by replacing t by $2\pi - t$ in the usual parametrization. This gives the parametrization

$$\begin{aligned} \boldsymbol{x} &= (3\cos(2\pi - t), 2\sin(2\pi - t)) \\ &= (3\cos t, -2\sin t), \quad t \in [0, 2\pi]. \end{aligned}$$

Now we can describe the *line integral* of a continuous vector function $\boldsymbol{F}: \boldsymbol{R}^n \to \boldsymbol{R}^n$ over a curve $\gamma: \boldsymbol{x} = \boldsymbol{x}(t)$ between two points $\boldsymbol{x}(a)$ and $\boldsymbol{x}(b)$. We approach this via one of the line integral's important physical interpretations, *work*. Recall that if a constant force $\boldsymbol{F}: \boldsymbol{R}^3 \to \boldsymbol{R}^3$ acts to move a particle along the line segment joining $\boldsymbol{x}(a)$ to $\boldsymbol{x}(b)$, then the **work** W done by the force is defined to be the product of the distance $|\boldsymbol{x}(b) - \boldsymbol{x}(a)|$ times the coordinate of $\boldsymbol{F}$ acting in the direction of the motion. See Figure 1.6. The coordinate of $\boldsymbol{F}$ in the direction of

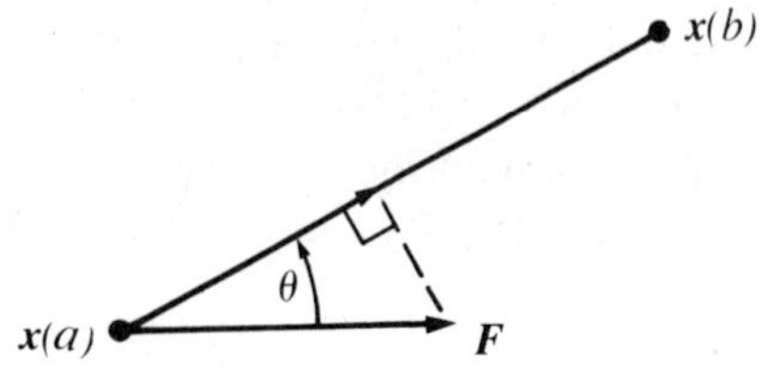

FIGURE 1.6

the motion is $|\boldsymbol{F}| \cos\theta$. (See Definition 1.2.16.) Thus,

(1)
$$\begin{aligned} W &= |\boldsymbol{F}|\,|\boldsymbol{x}(b) - \boldsymbol{x}(a)| \cos\theta \\ &= \boldsymbol{F} \cdot (\boldsymbol{x}(b) - \boldsymbol{x}(a)). \end{aligned}$$

The more typical situation arises when $\boldsymbol{F}$ is no longer constant and the path of motion is no longer a line but rather a curve $\gamma: \boldsymbol{x} = \boldsymbol{x}(t)$, extending from $\boldsymbol{x}(a)$ to $\boldsymbol{x}(b)$. See Figure 1.7.

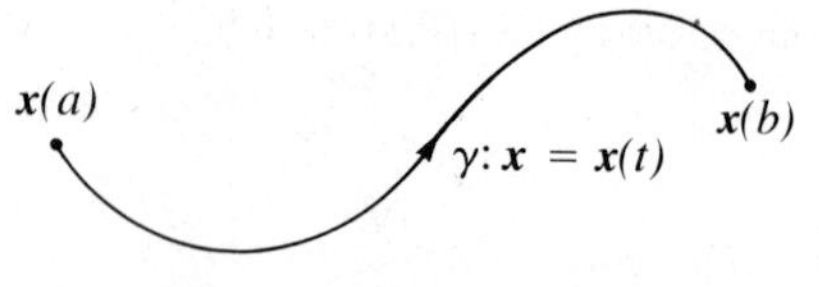

FIGURE 1.7

Suppose that the curve is smooth and simple. If we partition $[a, b]$ by $P: a = t_0 < t_1 < \ldots < t_{i-1} < t_i < \ldots < t_k = b$, then intuitively it seems that the amount of work ΔW_i done on the arc between $\boldsymbol{x}(t_{i-1})$ and $\boldsymbol{x}(t_i)$ will be approximately the work done by the constant force $\boldsymbol{F}(\boldsymbol{x}(\bar{t}_i))$ for some $\bar{t}_i \in [t_{i-1}, t_i]$ in moving the particle along the straight line segment joining $\boldsymbol{x}(t_{i-1})$ to $\boldsymbol{x}(t_i)$. See Figure 1.8.

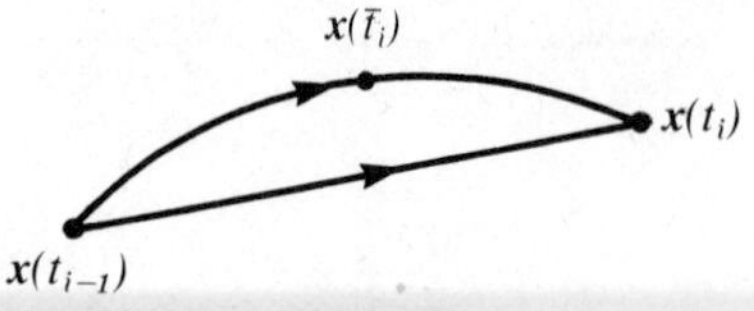

FIGURE 1.8

Thus, from (1),

(2)
$$\Delta W_i \approx \boldsymbol{F}(\boldsymbol{x}(\bar{t}_i)) \cdot (\boldsymbol{x}(t_i) - \boldsymbol{x}(t_{i-1})).$$

Since γ is smooth, we know that

(3) $$\boldsymbol{x}(t_i) - \boldsymbol{x}(t_{i-1}) \approx \dot{\boldsymbol{x}}(t_{i-1})(t_i - t_{i-1})$$

from the differentiability of $\boldsymbol{x} = \boldsymbol{x}(t)$ at t_{i-1}. Now, since $\dot{\boldsymbol{x}}$ is continuous, $\dot{\boldsymbol{x}}(t_{i-1}) \approx \dot{\boldsymbol{x}}(\bar{t}_i)$. Hence, (2) becomes

(4) $$\Delta W_i \approx \boldsymbol{F}(\boldsymbol{x}(\bar{t}_i)) \cdot \dot{\boldsymbol{x}}(\bar{t}_i)\, \Delta t_i$$

where $\Delta t_i = t_i - t_{i-1}$. The *total work* done by $\boldsymbol{F}$ in moving the particle from $\boldsymbol{x}(a)$ to $\boldsymbol{x}(b)$ along γ is thus approximately the Riemann sum

$$\sum_{i=1}^{k} \boldsymbol{F}(\boldsymbol{x}(\bar{t}_i)) \cdot \dot{\boldsymbol{x}}(\bar{t}_i)\, \Delta t_i$$

for the real valued function $\boldsymbol{F}(\boldsymbol{x}(t)) \cdot \dot{\boldsymbol{x}}(t)$. For this reason, physicists *define* the work to be the limit of these Riemann sums, if the limit exists.

1.2 **DEFINITION.** The **work** done by a force $\boldsymbol{F}: \boldsymbol{R}^n \to \boldsymbol{R}^n$ in moving a particle along a smooth simple curve $\gamma: \boldsymbol{x} = \boldsymbol{x}(t)$ in $\boldsymbol{R}^n$ from $\boldsymbol{x}(a)$ to $\boldsymbol{x}(b)$ is

(5) $$W = \int_a^b \boldsymbol{F}(\boldsymbol{x}(t)) \cdot \dot{\boldsymbol{x}}(t)\, dt,$$

if this integral exists. (In view of our assumptions about the curve, this integral *will* exist if $\boldsymbol{F}$ is, for example, continuous, as was assumed in the preceding discussion.)

Now let's examine the integral in (5) more closely. It is the limit of sums of terms

$$\begin{aligned}\boldsymbol{F}(\boldsymbol{x}(\bar{t}_i) \cdot \dot{\boldsymbol{x}}(\bar{t}_i))\, \Delta t_i &= \boldsymbol{F}(\boldsymbol{x}(\bar{t}_i)) \cdot \frac{\dot{\boldsymbol{x}}(\bar{t}_i)}{|\dot{\boldsymbol{x}}(\bar{t}_i)|}\, |\dot{\boldsymbol{x}}(\bar{t}_i)|\, \Delta t_i \\ &= (\boldsymbol{F}(\boldsymbol{x}(\bar{t}_i)) \cdot \boldsymbol{T}(\bar{t}_i))\, \Delta s_i,\end{aligned}$$

where $\boldsymbol{T}(\bar{t}_i)$ is the unit tangent vector at $\boldsymbol{x}(\bar{t}_i)$ and Δs_i approximates the arc length of γ over the subinterval $[t_{i-1}, t_i]$. Observe that

(6) $$\boldsymbol{F}(\boldsymbol{x}(\bar{t}_i)) \cdot \boldsymbol{T}(\bar{t}_i)$$

is the coordinate of $\boldsymbol{F}(\boldsymbol{x}(\bar{t}_i))$ in the direction of the unit tangent, hence in the direction of the path $\boldsymbol{x} = \boldsymbol{x}(t)$. Then the integral in (5) is the limit of the sum of the values of $\boldsymbol{F}(\boldsymbol{x}(t))$ in the direction of the curve times numbers Δs_i representing short distances along the curve. Thus we can regard the integral in (5) as the analogue of $\int_a^b f(x)\, dx$ in elementary calculus, for $\int_a^b f(x)\, dx$ is likewise the limit of the sum of the values $f(\bar{x}_i)$ of f along the interval $[a, b]$ times numbers Δx_i representing short distances along this interval. We thus are led to the following definition.

1.3 **DEFINITION.** If $\boldsymbol{F}: \boldsymbol{R}^n \to \boldsymbol{R}^n$ is continuous and $\gamma: \boldsymbol{x} = \boldsymbol{x}(t)$ for $a \le t \le b$ is a smooth simple curve in $\boldsymbol{R}^n$, then the **line integral** (or *contour integral*) of $\boldsymbol{F}$ over γ is

(7) $$\int_\gamma \boldsymbol{F} \cdot d\boldsymbol{x} = \int_a^b \boldsymbol{F}(\boldsymbol{x}(t)) \cdot \dot{\boldsymbol{x}}(t)\, dt.$$

There are a number of different notations in use for $\int_\gamma \boldsymbol{F}\cdot d\boldsymbol{x}$, some of which it will be convenient to use from time to time. One can rewrite the integrand on the right of (7) as

$$\int_a^b \boldsymbol{F}(\boldsymbol{x}(t))\cdot\frac{\dot{\boldsymbol{x}}(t)}{|\dot{\boldsymbol{x}}(t)|}|\dot{\boldsymbol{x}}(t)|\,dt=\int_a^b \boldsymbol{F}(\boldsymbol{x}(t))\cdot\boldsymbol{T}(t)\,ds$$

so that an alternate notation to (7) is

(8) $$\int_\gamma \boldsymbol{F}\cdot\boldsymbol{T}\,ds=\int_a^b \boldsymbol{F}(\boldsymbol{x}(t))\cdot\boldsymbol{T}(t)\,ds.$$

If γ is parametrized by its arc length (see Definition 2.2.8), then $|\dot{\boldsymbol{x}}(t)|=1$, so $\boldsymbol{T}(t)=\dot{\boldsymbol{x}}(t)$ and $ds=dt$. In this case, then, (7) and (8) coincide.

We can also write (7) in coordinate form. We develop this only for the cases $n=2$ and 3. Suppose that

$$\boldsymbol{F}(\boldsymbol{x})=P(\boldsymbol{x})\boldsymbol{i}+Q(\boldsymbol{x})\boldsymbol{j}+R(\boldsymbol{x})\boldsymbol{k},$$

where $R(\boldsymbol{x})=0$ if $n=2$. We can write

$$\boldsymbol{F}(\boldsymbol{x})=(P(\boldsymbol{x}),Q(\boldsymbol{x}),R(\boldsymbol{x})),$$
$$\dot{\boldsymbol{x}}(t)=\left(\frac{dx}{dt},\frac{dy}{dt},\frac{dz}{dt}\right).$$

Then

$$\boldsymbol{F}(\boldsymbol{x})\cdot\dot{\boldsymbol{x}}(t)=P(\boldsymbol{x})\frac{dx}{dt}+Q(\boldsymbol{x})\frac{dy}{dt}+R(\boldsymbol{x})\frac{dz}{dt}.$$

So (7) becomes

$$\int_\gamma \boldsymbol{F}\cdot d\boldsymbol{x}=\int_a^b\left(P(x,y,z)\frac{dx}{dt}+Q(x,y,z)\frac{dy}{dt}+R(x,y,z)\frac{dz}{dt}\right)dt,$$

or, equivalently,

(9) $$\int_\gamma \boldsymbol{F}\cdot d\boldsymbol{x}=\int_\gamma P(x,y,z)\,dx+Q(x,y,z)\,dy+R(x,y,z)\,dz.$$

In evaluating line integrals, it is usually best to use (7), even if another notation is initially given.

1.4 EXAMPLE. Compute $\int_\gamma 2xy\,dx+(x^2+y^2)\,dy$, where γ is the parabola $y^2=x$ between $(0,0)$ and $(1,-1)$.

Solution. We show the curve in Figure 1.9.

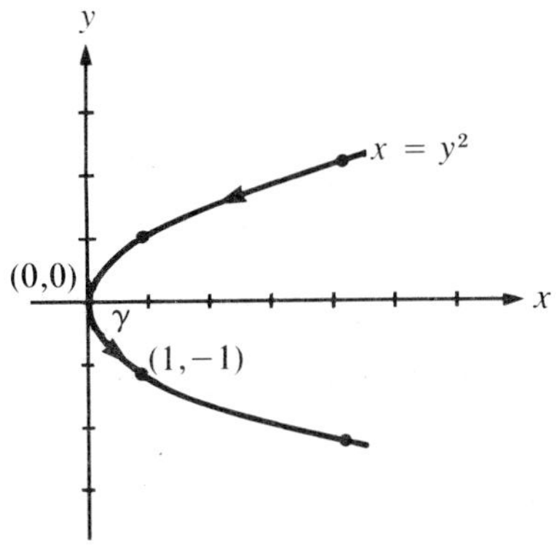

FIGURE 1.9

To reduce to the form (7), we have to parametrize γ. Since on the *lower* half of the parabola $x = y^2$ we have $y = -\sqrt{x}$, we let $x = t^2$ and $y = -t$, $t \in [0, 1]$. That is, $\boldsymbol{x} = \boldsymbol{x}(t) = (t^2, -t)$. Note that $\boldsymbol{x}(0) = (0, 0)$ and $\boldsymbol{x}(1) = (1, -1)$. We then have

$$d\boldsymbol{x} = \left(\frac{dx}{dt}, \frac{dy}{dt}\right) dt = (2t, -1)\, dt = (2t\, dt, -dt).$$

Our integral thus becomes

$$\begin{aligned}\int_\gamma 2xy\, dx + (x^2 + y^2)\, dy &= \int_0^1 2t^2(-t)(2t\, dt) + (t^4 + t^2)(-dt)\\ &= \int_0^1 (-4t^4 - t^4 - t^2)\, dt\\ &= \int_0^1 (-5t^4 - t^2)\, dt\\ &= -t^5 - \frac{t^3}{3}\Big]_0^1 = -4/3.\end{aligned}$$

We extend line integrals to piecewise smooth paths by adding the line integrals over the smooth pieces.

1.5 **DEFINITION.** If $\gamma = \gamma_1 \cup \gamma_2 \cup \ldots \cup \gamma_k$ is piecewise smooth where each γ_i is smooth, then

$$\int_\gamma \boldsymbol{F} \cdot d\boldsymbol{x} = \sum_{i=1}^{k} \int_{\gamma_i} \boldsymbol{F} \cdot d\boldsymbol{x}.$$

1.6 **EXAMPLE.** Evaluate $\int_\gamma \boldsymbol{F} \cdot d\boldsymbol{x}$ if $\boldsymbol{F}(x, y) = (x + y, x - y)$ and γ is the square shown in Figure 1.10.

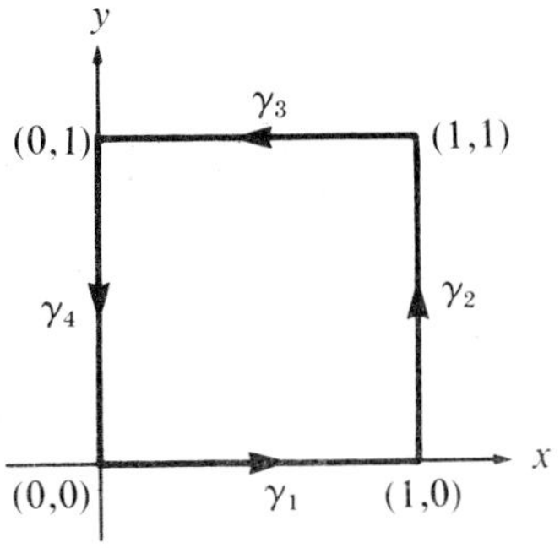

FIGURE 1.10

Solution. We have $\gamma = \gamma_1 \cup \gamma_2 \cup \gamma_3 \cup \gamma_4$, where we can parametrize each γ_i as follows.

$$\gamma_1 : \boldsymbol{x}(t) = (t, 0), \quad 0 \le t \le 1, \quad \boldsymbol{F}(\boldsymbol{x}(t)) = (t, t), \quad d\boldsymbol{x} = (1, 0)\, dt = (dt, 0).$$

$$\gamma_2 : \boldsymbol{x}(t) = (1, t), \quad 0 \le t \le 1, \quad \boldsymbol{F}(\boldsymbol{x}(t)) = (t+1, 1-t), \quad d\boldsymbol{x} = (0, 1)\, dt = (0, dt).$$

$$\gamma_3 : \boldsymbol{x}(t) = (1-t, 1), \quad 0 \le t \le 1, \quad \boldsymbol{F}(\boldsymbol{x}(t)) = (2-t, -t), \quad d\boldsymbol{x} = (-1, 0)\, dt = (-dt, 0).$$

$$\gamma_4 : \boldsymbol{x}(t) = (0, 1-t), \quad 0 \le t \le 1, \quad \boldsymbol{F}(\boldsymbol{x}(t)) = (1-t, t-1), \quad d\boldsymbol{x} = (0, -1)\, dt = (0, -dt).$$

So we have

$$\begin{aligned}\int_\gamma \boldsymbol{F}(\boldsymbol{x}(t)) \cdot d\boldsymbol{x}(t) &= \int_0^1 (t, t) \cdot (dt, 0) + \int_0^1 (t+1, 1-t) \cdot (0, dt) \\ &\quad + \int_0^1 (2-t, -t) \cdot (-dt, 0) + \int_0^1 (1-t, t-1) \cdot (0, -dt) \\ &= \int_0^1 t\, dt + \int_0^1 (1-t)\, dt + \int_0^1 (t-2)\, dt + \int_0^1 (1-t)\, dt \\ &= \int_0^1 0\, dt = 0.\end{aligned}$$

Our next result is a very basic one which will be of great importance throughout the chapter. It says that if the orientation of γ is reversed, then the line integral of $\boldsymbol{F}$ over γ changes sign.

1.7 **THEOREM** $\displaystyle\int_{-\gamma} \boldsymbol{F} \cdot d\boldsymbol{x} = -\int_\gamma \boldsymbol{F} \cdot d\boldsymbol{x}.$

Proof. Suppose that γ is given by $\boldsymbol{x} = \boldsymbol{x}(t)$ for $t \in [a, b]$. Then $-\gamma$ is given by $\boldsymbol{x} = \boldsymbol{x}(a+b-t)$ for $t \in [a, b]$. So,

(10) $$\int_{-\gamma} \boldsymbol{F} \cdot d\boldsymbol{x} = \int_a^b \boldsymbol{F}(\boldsymbol{x}(a+b-t)) \cdot \dot{\boldsymbol{x}}(a+b-t)\, dt.$$

To work this integral out, we introduce the change of variable

$$u = a + b - t.$$

Then

$$du = -dt, \quad \text{so} \quad dt = -du.$$

Also, when $t = a$, then $u = b$, and when $t = b$, we get $u = a$. By the Chain Rule (Theorem 2.1.7(6)),

$$\begin{aligned}\dot{\boldsymbol{x}}(a+b-t) &= \frac{d\boldsymbol{x}(u)}{dt} = \frac{d\boldsymbol{x}(u)}{du}\frac{du}{dt} \\ &= \frac{d\boldsymbol{x}(u)}{du}(-1) = -\frac{d\boldsymbol{x}(u)}{du}\end{aligned}$$

Thus (10) becomes

$$\int_{-\gamma} \boldsymbol{F} \cdot d\boldsymbol{x} = \int_b^a \boldsymbol{F}(\boldsymbol{x}(u)) \cdot \left(-\frac{d\boldsymbol{x}(u)}{du}\right)(-du)$$

$$= \int_b^a \boldsymbol{F}(\boldsymbol{x}(u)) \cdot \frac{d\boldsymbol{x}(u)}{du}\, du$$

$$= -\int_a^b \boldsymbol{F}(\boldsymbol{x}(u)) \cdot \frac{d\boldsymbol{x}(u)}{du}\, du = -\int_{\gamma} \boldsymbol{F} \cdot d\boldsymbol{x}$$

since in the next-to-last definite integral u is a dummy variable. QED

What if we change the parametrization of γ but don't change the orientation? That is, we change from $\boldsymbol{x} = \boldsymbol{x}(t)$ for $t \in [a, b]$ to $\boldsymbol{x} = \mathbf{y}(u)$ for $u \in [c, d]$, where $t = g(u)$ has continuous derivative and the orientation is preserved. We can show that the line integral is unchanged. Thus, line integrals are *independent of the parametrization of γ as long as the orientation is not changed.*

1.8 **THEOREM.** Suppose that γ is parametrized by $\boldsymbol{x} = \boldsymbol{x}(t)$ for $t \in [a, b]$, and also by $\boldsymbol{x} = \mathbf{y}(u)$ for $u \in [c, d]$, where $t = g(u)$ with g' continuous. If the orientation of γ is preserved (i.e., $a = g(c)$ and $b = g(d)$), then the value of $\int_\gamma \boldsymbol{F} \cdot d\boldsymbol{x}$ is the same whether computed as

$$\int_a^b \boldsymbol{F}(\boldsymbol{x}(t)) \cdot \dot{\boldsymbol{x}}(t)\, dt$$

or as

$$\int_c^d \boldsymbol{F}(\mathbf{y}(u)) \cdot \frac{d\mathbf{y}}{du}\, du.$$

Proof. We have $\mathbf{y}(u) = \boldsymbol{x} = \boldsymbol{x}(t) = \boldsymbol{x}(g(u))$. So by the Chain Rule again,

$$\frac{d\mathbf{y}(u)}{du} = \frac{d\boldsymbol{x}}{dt}\frac{dt}{du}.$$

We also have $a = g(c)$ and $b = g(d)$. Hence,

$$\int_a^b \boldsymbol{F}(\boldsymbol{x}(t)) \cdot \dot{\boldsymbol{x}}(t)\, dt = \int_c^d \boldsymbol{F}(\boldsymbol{x}(g(u))) \cdot \frac{d\boldsymbol{x}}{dt}\frac{dt}{du}\, du$$

$$= \int_c^d \boldsymbol{F}(\mathbf{y}(u)) \cdot \frac{d\mathbf{y}}{du}\, du.$$ QED

So in working out a line integral, we are free to choose any parametrization for γ consistent with its orientation. The moral of this is: parametrize γ as simply as possible!

EXERCISES 7.1

In Exercises 1 to 14, evaluate the given line integral.

1 $\int_\gamma \boldsymbol{F} \cdot d\boldsymbol{x}$ where $\boldsymbol{F}(x, y) = (x^2y, x^2 - y)$ and γ is the curve $\boldsymbol{x}(t) = (t, 1-t)$, $0 \le t \le 1$.

2 $\int_\gamma \boldsymbol{F} \cdot \boldsymbol{T}\, ds$ where $\boldsymbol{F}(x, y) = (x + 2y, x^2 - y^2)$ and γ is the triangular path from $(0, 0)$ to $(1, 0)$ to $(1, 1)$ to $(0, 0)$.

3 $\int_\gamma \boldsymbol{F} \cdot d\boldsymbol{x}$ where $\boldsymbol{F}(x, y) = (\sqrt{x^2+y^2}, \sqrt{1-x^2})$ and γ is the upper half of the unit circle $x^2+y^2=1$, traversed from $(1, 0)$ to $(-1, 0)$.

4 $\int_\gamma \boldsymbol{F} \cdot d\boldsymbol{x}$ where $\boldsymbol{F}(x, y) = (\sqrt{1-y^2}, \sqrt{1-x^2})$, and γ is the curve of Exercise 3.

5 $\int_\gamma xy\, dx + (y^2+1)\, dy$ where γ is the curve $y^2 = x$ from $(0, 0)$ to $(1, 1)$.

6 $\int_\gamma (x^2-y^2)\, dx + 2xy\, dy$ where γ is the curve $y = x^2$ from $(-1, 1)$ to $(2, 4)$.

7 $\displaystyle\int_\gamma \frac{-y}{x^2+y^2}\, dx + \frac{x}{x^2+y^2}\, dy$ where γ is the circle $x^2+y^2=9$ traversed counterclockwise from $(3, 0)$. **(This exercise is referred to in Section 4.)**

8 $\displaystyle\int_\gamma \frac{x+y}{x^2+y^2}\, dx - \frac{x-y}{x^2+y^2}\, dy$ where γ is the path in Exercise 7.

9 $\int_\gamma \boldsymbol{F} \cdot d\boldsymbol{x}$ where $\boldsymbol{F}(x, y, z) = (y, z, x)$ and γ is the line segment joining $(0, 0, 0)$ to $(2, 4, -1)$.

10 $\int_\gamma \boldsymbol{F} \cdot \boldsymbol{T}\, ds$ where $\boldsymbol{F}(x, y, z) = (x^2, y^2, z^2)$ and γ is the curve $\boldsymbol{x}(t) = (t, t^2, t^3)$, $0 \le t \le 1$.

11 $\int_\gamma x\, dx - y\, dy + z\, dz$ where γ is the helix

$$\boldsymbol{x}(t) = \left(\cos t, \sin t, \frac{t}{\pi}\right), \qquad 0 \le t \le 2\pi.$$

12 $\int_\gamma \boldsymbol{F} \cdot d\boldsymbol{x}$ where $\boldsymbol{F}(x_1, x_2, x_3, x_4) = (x_1 - x_2, x_2 - x_3, x_4 - x_1, x_1x_2x_3x_4)$ and γ is the curve $\boldsymbol{x}(t) = (t, t^2, t^3, t^4)$, $0 \le t \le 1$.

13 $\int_\gamma (xz+y^2)\, dx + (yz-x^2)\, dy + (xy-z^2)\, dz$ where γ is the line segment from $(1, 2, 3)$ to $(4, 3, 5)$ followed by the segment from $(4, 3, 5)$ to $(1, 1, 1)$.

14 $\int_\gamma y\, dx + z\, dy + x\, dz$ where γ is the curve of intersection of the plane $x+y=2$ and the sphere $(x-1)^2+(y-1)^2+z^2=2$, traversed in the counterclockwise sense viewed from the point $(2, 2, 0)$.

15 Find the work done by a force $\boldsymbol{F}(x, y) = (2-y, x)$ in moving a particle along one arch of the **cycloid** given by $\boldsymbol{x}(t) = (t - \sin t, 1 - \cos t)$, $0 \le t \le 2\pi$.

16 Find the work done by a force $\boldsymbol{F}(x, y) = (1-y, x)$ in moving a particle along one arch of the cycloid $\boldsymbol{x}(t) = 2(t - \sin t, 1 - \cos t)$, $0 \le t \le 2\pi$.

17 The repelling force between a charged particle P at the origin and an oppositely charged particle Q at (x, y) is

$$\boldsymbol{F}(x, y) = \frac{1}{(x^2+y^2)^{3/2}}(x, y).$$

Find the work done by $\boldsymbol{F}$ as it moves Q along the line segment from $(1, 0)$ to $(-1, 2)$.

18 Refer to Exercise 17. Suppose that $\boldsymbol{F}$ moves Q along the polygonal path from $(1, 0)$ to $(1, 1)$ to $(-1, 2)$. What is the work done?

19 The **line integral of a scalar valued function** $f: \boldsymbol{R}^n \to \boldsymbol{R}$ over a curve γ is defined by

$$\int_\gamma f(\boldsymbol{x})\, ds = \int_\gamma f(\boldsymbol{x}(t))\, |\dot{\boldsymbol{x}}(t)|\, dt.$$

(Recall that $ds = |\dot{\boldsymbol{x}}(t)|\, dt$.) If $f(\boldsymbol{x}(t))$ gives the density of a wire in the shape of γ at the point $\boldsymbol{x}(t)$, then the **mass** of the wire is defined to be $\int_\gamma f(\boldsymbol{x})\, ds$. Formulate a definition of the *center of mass* of the wire.

20 Formulate a definition of the *moment of inertia* of a wire in the shape of $\boldsymbol{x} = \boldsymbol{x}(t)$ relative to an axis ℓ at distance $d(x, y, z)$ from (x, y, z) on γ. (Assume that the wire has density $f(x, y, z)$.)

21 Use Exercise 19 to compute the mass of a helical wire $\boldsymbol{x}(t)=(\cos t, \sin t, t)$, $0 \leq t \leq 2\pi$, if its density is given by $\delta(x, y, z)=x^2+y^2+z^2$.

22 Find the center of mass of the wire in Exercise 21.

23 Find the moment of inertia of the wire in Exercise 21 about the z-axis.

The following exercise is needed for Example 4.4.

24 A force $\boldsymbol{F}$ acts on a particle to move it from $\boldsymbol{x}(t_0)$ to $\boldsymbol{x}(t_1)$ along the path $\boldsymbol{x}=\boldsymbol{x}(t)$. Show that the work done is the change in kinetic energy, i.e.,

$$W=\tfrac{1}{2}m\,|\boldsymbol{v}(t_1)|^2-\tfrac{1}{2}m\,|\boldsymbol{v}(t_0)|^2.$$

(This is one form of the *Law of Conservation of Energy*.) [*Hint:* Recall that $\boldsymbol{F}=m\ddot{\boldsymbol{x}}(t)$ by Newton's law, and compute $\dfrac{d}{dt}(|\boldsymbol{v}(t)|^2)$.]

25 If γ is a smooth, but not necessarily simple, closed curve which does not pass through the origin, then the *winding number* of γ is defined as $\dfrac{1}{2\pi}\displaystyle\int_\gamma \boldsymbol{F}\cdot d\boldsymbol{x}$ where

$$\boldsymbol{F}(x, y)=\left(\frac{-y}{x^2+y^2}, \frac{x}{x^2+y^2}\right).$$

It can be shown (in advanced calculus, for instance) that the winding number is always an integer which gives the number of times γ winds around the origin. If it is positive, then γ encircles the origin in a counterclockwise sense. If it is negative, then γ encircles the origin in a clockwise sense. Compute the winding numbers of

$$\gamma_1: \boldsymbol{x}(t)=(2\cos t, 2\sin t), \qquad 0 \leq t \leq 6\pi$$

and

$$\gamma_2: \boldsymbol{x}(t)=(\cos t, -\sin t), \qquad 0 \leq t \leq 2\pi$$

26 Use the general principles on p. 405 to parametrize the *hypocycloid of four cusps*

$$x^{2/3}+y^{2/3}=a^{2/3}$$

in (a) the counterclockwise sense; (b) the clockwise sense.

27 If $\boldsymbol{F}(x, y)=\left(\dfrac{-1}{\sqrt[3]{y}}, \dfrac{1}{\sqrt[3]{x}}\right)$, then compute $\int_\gamma \boldsymbol{F}\cdot d\boldsymbol{x}$ over the hypocycloid $x^{2/3}+y^{2/3}=1$ parametrized in the counterclockwise sense.

2 GREEN'S THEOREM

This section takes its title from a theorem named in honor of the English mathematician George Green (1793–1841), who first used it in 1828. The theorem was actually discovered somewhat earlier by Gauss and Lagrange, but became widely known because of Green's applications to electricity and magnetism, fluid flow, and other parts of physics whose modern treatments make heavy use of it.

We will suggest some of the important applications of the theorem later, but can at the outset suggest its mathematical appeal. This stems from its relationship to the Fundamental Integration Theorem of elementary calculus. That fundamental theorem can be stated in the form

(1) $$\int_a^b f'(x)\,dx=f(b)-f(a),$$

if $f: \boldsymbol{R} \to \boldsymbol{R}$ is a smooth function on $[a, b]$. What does this say *qualitatively*? Simply that the integral of the derivative of f over $[a, b]$ is completely determined by the values of f on the *boundary* of $[a, b]$, i.e., on the two-point set $\{a, b\}$. We have already seen that an important analogue of (1) holds for double and triple integrals, namely Fubini's Theorem. That theorem states that a double or triple integral of a function $f: \boldsymbol{R}^n \to \boldsymbol{R}$ over a region $D \subseteq \boldsymbol{R}^n$ can be evaluated by partial antidifferentiation. You will recall that the partially antidifferentiated functions also are evaluated just on the *boundary* of D. For example, if D lies between surfaces $z = g_1(x, y)$ and $z = g_2(x, y)$, cylinders $y = h_1(x)$ and $y = h_2(x)$, and planes $x = a$ and $x = b$, then

$$\iiint_D f(x, y, z)\, dV = \int_a^b dx \int_{h_1(x)}^{h_2(x)} dy \int_{g_1(x,y)}^{g_2(x,y)} f(x, y, z)\, dz.$$

Now Green's Theorem is a similar sort of result for a function $\boldsymbol{F}: \boldsymbol{R}^2 \to \boldsymbol{R}^2$, the first of several analogues of (1) which we will see in this chapter. Roughly, Green's Theorem says that the integral of a certain kind of derivative of $\boldsymbol{F}$ over D can be evaluated by computing the line integral of $\boldsymbol{F}$ over the *boundary* γ of D. This has the happy effect of at times allowing complicated double integrals to be evaluated by computing simple line integrals, and at other times permitting complicated line integrals to be evaluated by computing simple double integrals. Before we can realize such dividends, however, we first have to formulate the theorem precisely. A formulation in full generality involves a subtler discussion of bounding curves γ of regions $D \subseteq \boldsymbol{R}^2$ than we have the means to give. So we will be content to state the theorem in a form general enough to apply to nearly any case encountered in practice. We will first state it for rather simple regions and then describe how it can be extended to more complex regions. We begin with a definition.

2.1 **DEFINITION.** A region $D \subseteq \boldsymbol{R}^2$ is **simple** if it is the part of $\boldsymbol{R}^2$ enclosed by a piecewise smooth simple closed curve γ that intersects any vertical or horizontal line $x = a$ or $y = c$ either in at most two points A and B or in one line segment AB.

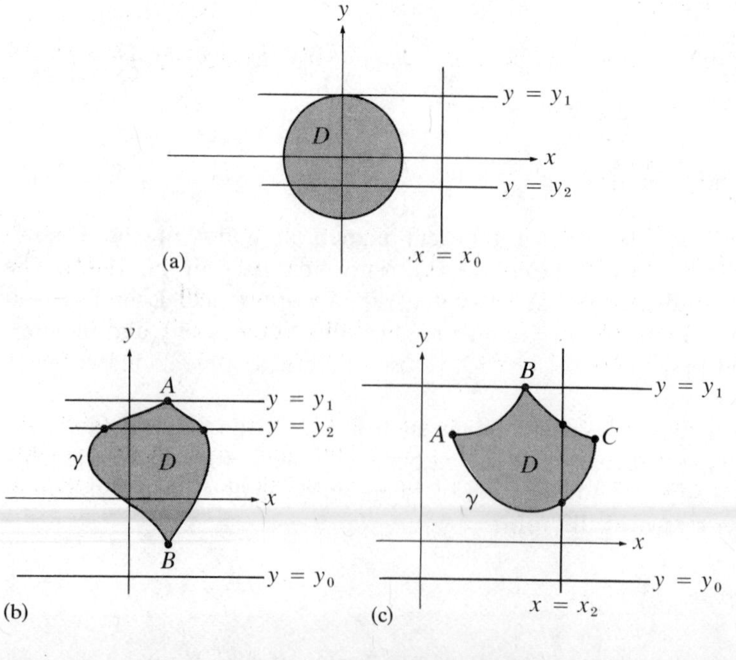

FIGURE 2.1

As the name suggests, simple regions are geometrically uncomplicated. Figure 2.1 shows three examples of simple regions.

The region in Figure 2.1(a) is bounded by a *smooth* curve, the graph of $x^2+y^2=1$. The regions in Figure 2.1(b) and (c) are bounded by *piecewise smooth* curves γ. Points where $\dot{\mathbf{x}}(t)$ fails to be continuous are labeled. In all three cases, vertical and horizontal lines intersect γ in 0, 1, or 2 points. Figure 2.2 shows

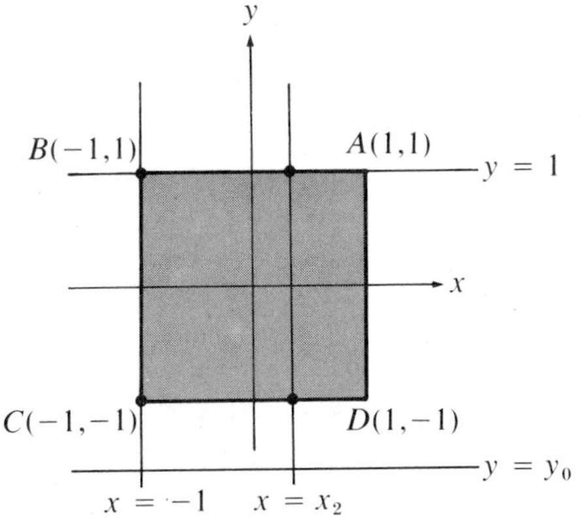

FIGURE 2.2

another type of simple region, in this case a square. Notice that some lines parallel to a coordinate axis intersect γ in at most two points, while others meet γ in entire line segments.

Figure 2.3 shows a non-simple region. While *some* lines (like $x=a$ and $y=c$) intersect γ in at most two points, there are others (like $y=d$) that intersect γ in *four* points.

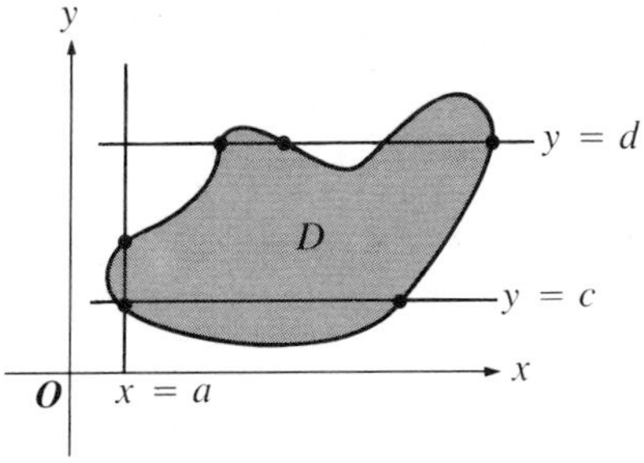

FIGURE 2.3

We try to picture a general simple region in Figure 2.4. The regions D and R in Figure 2.5 fail to be simple because the line segments shown in Figure 2.6 intersect the boundaries in more than two points or a single line segment. In each

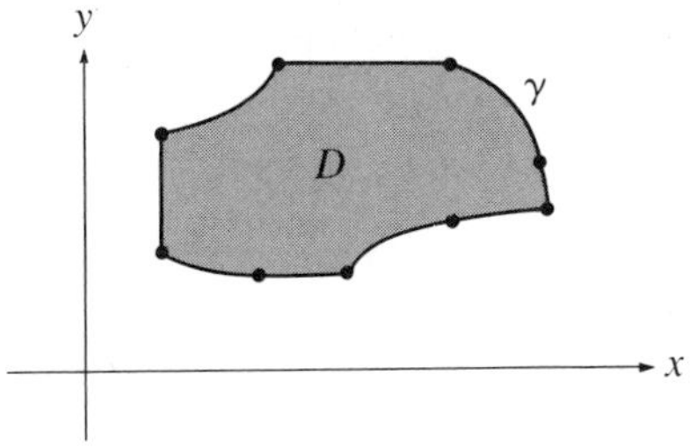

FIGURE 2.4

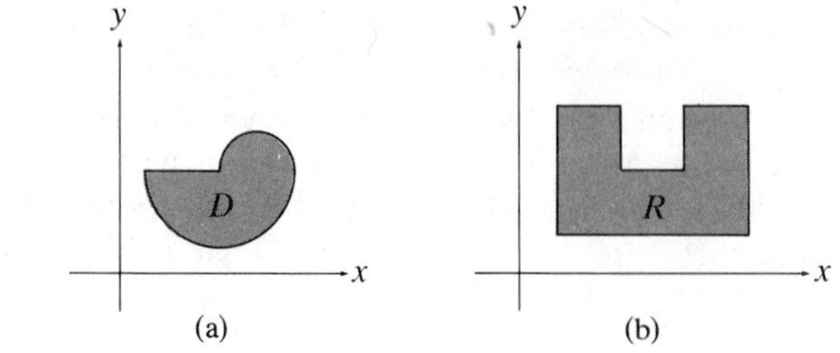

FIGURE 2.5

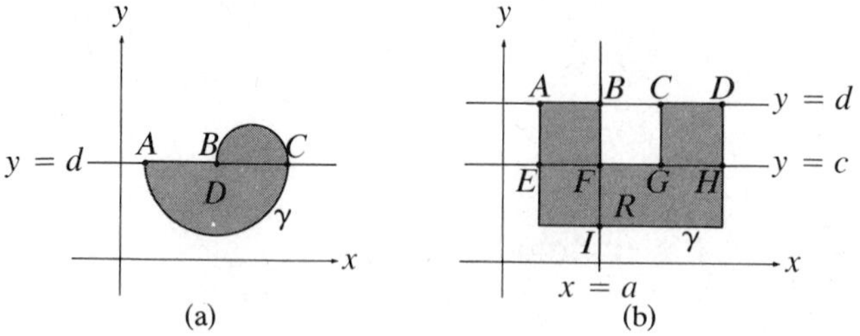

FIGURE 2.6

case, though, we could decompose D and R into subregions that *are* simple, and we show this in Figure 2.7.

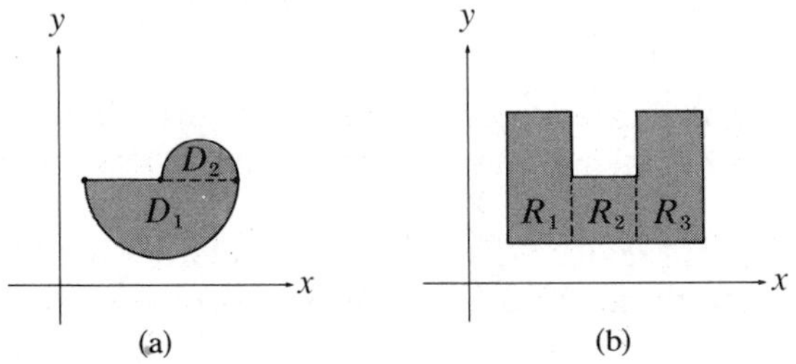

FIGURE 2.7

Our first version of Green's Theorem will apply only to simple regions. But then we will be able to extend it to regions like D in Figure 2.5, which can be decomposed into the union of simple regions. We can now state our first version of Green's Theorem.

2.2 **GREEN'S THEOREM.** Let $\boldsymbol{F}: \boldsymbol{R}^2 \to \boldsymbol{R}^2$ be given by

$$\boldsymbol{F}(x, y) = P(x, y)\boldsymbol{i} + Q(x, y)\boldsymbol{j},$$

where $P, Q: \boldsymbol{R}^2 \to \boldsymbol{R}$ have continuous first partial derivatives. Suppose that D is a simple region with boundary γ. Then

(2)
$$\iint_D \left(\frac{\partial Q}{\partial x} - \frac{\partial P}{\partial y}\right) dA = \oint_\gamma P\,dx + Q\,dy = \oint_\gamma \boldsymbol{F} \cdot d\boldsymbol{x}.$$

Here $\oint$ means that γ is parametrized by $\boldsymbol{x} = \boldsymbol{x}(t)$, $a \le t \le b$, in such a way that as t increases from a to b, γ is traversed once in a *counterclockwise* sense, i.e., so that D lies to the *left* as the boundary is traversed.

Proof. We will show separately that

(3)
$$\oint_\gamma P(x, y)\,dx = -\iint_D \frac{\partial P}{\partial y}\,dA$$

and

(4) $$\oint_\gamma Q(x, y)\, dy = \iint_D \frac{\partial Q}{\partial x}\, dA.$$

Here by $\oint_\gamma P(x, y)\, dx$ and $\oint_\gamma Q(x, y)\, dy$ we mean respectively

$$\int_a^b (P(\boldsymbol{x}(t)), 0) \cdot \left(\frac{dx}{dt}, \frac{dy}{dt}\right) dt \quad \text{and} \quad \int_a^b (0, Q(\boldsymbol{x}(t))) \cdot \left(\frac{dx}{dt}, \frac{dy}{dt}\right) dt$$

if γ is parametrized by $\boldsymbol{x} = \boldsymbol{x}(t)$ for $t \in [a, b]$. Then (2) will follow simply by adding (3) and (4).

First we establish (3). Referring to Figure 2.8, we see that γ consists of the two vertical segments BA and EF together with the oriented arcs $\widehat{AE}$ and $\widehat{FB}$. On

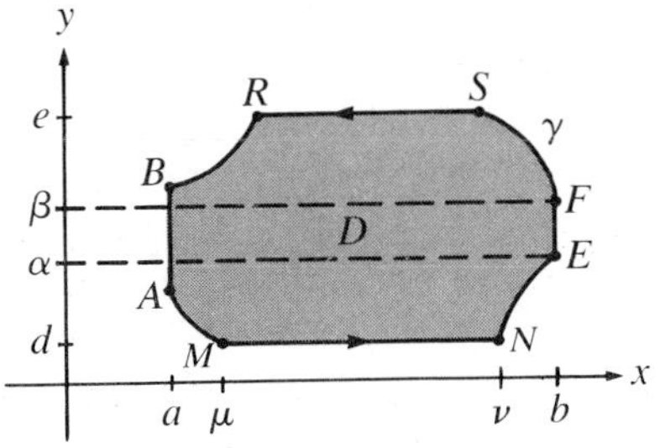

FIGURE 2.8

the segment EF we can parametrize by

$$\boldsymbol{x}(t) = (x, y) = (b, t), \qquad \alpha \le t \le \beta.$$

Then, $\dfrac{dx}{dt} = 0$, so $dx = 0$. We therefore get

$$\int_{EF} P(x, y)\, dx = 0.$$

Similarly, $\int_{BA} P(x, y)\, dx = 0$. Thus from Definition 1.5,

(5) $$\oint_\gamma P(x, y)\, dx = \int_{\widehat{AE}} P(x, y)\, dx + \int_{\widehat{FB}} P(x, y)\, dx.$$

Since a vertical line crosses each of $\widehat{AE}$ and $\widehat{BF}$ *exactly once*, we see that these arcs are the graphs of *functions* $y = g_1(x)$ and $y = g_2(x)$ for $x \in [a, b]$. So they can be parametrized as curves γ_i for $i = 1, 2$, by $\boldsymbol{x} = (t, g_i(t))$ with $t \in [a, b]$. Thus $x = t$, $y = g_i(t)$. It is natural then to use x $(= t)$ as parameter instead of t. Since γ_2 is traced out as x goes from a to b, γ_2 joins B to F. Since γ traverses the boundary of D in the counterclockwise sense, γ proceeds from F to B. Thus we have from (5) and Theorem 1.7,

$$\begin{aligned}\oint_{\gamma} P(x, y)\,dx &= \int_{\gamma_1} P(x, y)\,dx + \int_{-\gamma_2} P(x, y)\,dx \\ &= \int_{\gamma_1} P(x, y)\,dx - \int_{\gamma_2} P(x, y)\,dx \\ &= \int_a^b P(x, g_1(x))\,dx - \int_a^b P(x, g_2(x))\,dx \\ &= -\int_a^b [P(x, g_2(x)) - P(x, g_1(x))]\,dx \\ &= -\int_a^b P(x, y)\Big]_{y=g_1(x)}^{y=g_2(x)}\,dx \\ &= -\int_a^b dx \int_{g_1(x)}^{g_2(x)} \frac{\partial P(x, y)}{\partial y}\,dy \\ &= -\iint_D \frac{\partial P}{\partial y}\,dA \quad \text{by Fubini's Theorem.}\end{aligned}$$

Thus (3) is proved. The approach to (4) is quite similar and is left as Exercise 32 below. Addition of (3) and (4) gives us (2). QED

As we suggested before stating Theorem 2.2, (2) can be shown to hold also for any region D that is a finite union $D_1 \cup D_2 \cup \ldots \cup D_k$ of nonoverlapping simple regions. Consider, for example, the region D in Figure 2.9. We have

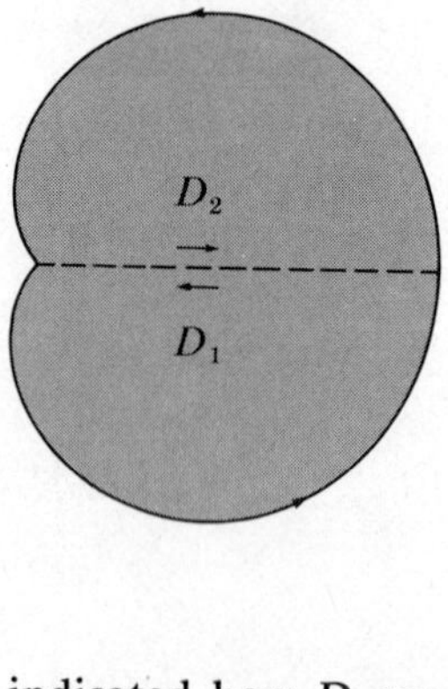

FIGURE 2.9

indicated how D can be subdivided into two simple subregions D_1 and D_2. On each of these, (2) holds by Theorem 2.2. Thus we have

(6) $$\begin{aligned}\iint_D \left(\frac{\partial Q}{\partial x} - \frac{\partial P}{\partial y}\right) dA &= \iint_{D_1} \left(\frac{\partial Q}{\partial x} - \frac{\partial P}{\partial y}\right) dA + \iint_{D_2} \left(\frac{\partial Q}{\partial x} - \frac{\partial P}{\partial y}\right) dA \\ &= \oint_{\partial D_1} P\,dx + Q\,dy + \oint_{\partial D_2} P\,dx + Q\,dy\end{aligned}$$

where ∂D_i stands for the boundary of D_i. Notice that each part of ∂D_i that is not part of $\gamma = \partial D$ is traversed twice in *opposite* directions in the evaluation of the line integral in (6). (See Figure 2.9.) Hence the net contribution to the line integral in (6) from the common boundary of D_1 and D_2 is 0, by Theorem 1.7. Thus, the only nonzero contributions in (6) come from sub-arcs of γ. The same reasoning applies if $D = D_1 \cup D_2 \cup \ldots \cup D_k$, for $k > 2$. Thus (6) becomes

(7) $$\iint_D \left(\frac{\partial Q}{\partial x} - \frac{\partial P}{\partial y}\right) dA = \sum_{i=1}^{k} \oint_{\partial D_i} P\,dx + Q\,dy$$
$$= \oint_\gamma P\,dx + Q\,dy.$$

We have thus extended (2) to regions D that can be decomposed into finite

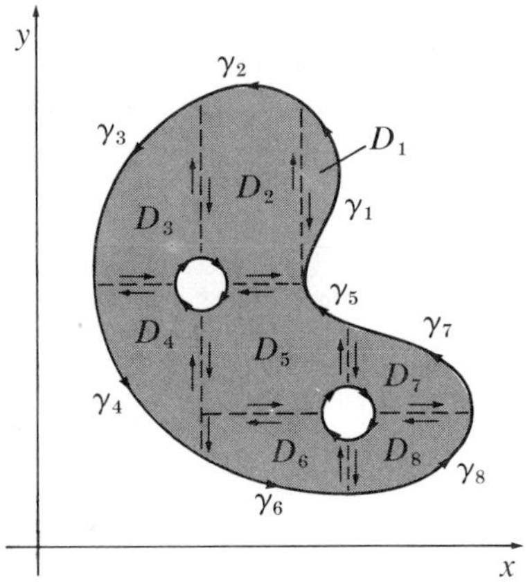

FIGURE 2.10

unions of simple regions. (See, for example, Figure 2.10.) The formal statement is the following.

2.3 **GREEN'S THEOREM.** Let $\boldsymbol{F}$ be as in Theorem 2.2. Suppose that D is decomposable into the union of a finite number of non-overlapping simple regions D_i with piecewise smooth counterclockwise oriented simple closed curves γ_i as boundaries. Then

(2) $$\iint_D \left(\frac{\partial Q}{\partial x} - \frac{\partial P}{\partial y}\right) dA = \int_\gamma P\,dx + Q\,dy,$$

where $\gamma = \partial D = \gamma_1 \cup \gamma_2 \cup \ldots \cup \gamma_k$.

As suggested above, Theorem 2.3 is sufficiently general for most purposes. We mention that you can think of $\dfrac{\partial Q}{\partial x} - \dfrac{\partial P}{\partial y}$ as a kind of derivative of the vector field $\boldsymbol{F}(x, y) = P(x, y)\boldsymbol{i} + Q(x, y)\boldsymbol{j}$. (This will be made more precise in the next section, when we define the vector differential operator curl.) In this interpretation the resemblance of (2) to (1) is easily seen: (2) now states that the double integral over $D \subseteq \boldsymbol{R}^2$ of a type of derivative of $\boldsymbol{F}$ is determined by the values of $\boldsymbol{F}$ on $\gamma = \partial D$. In fact, the exact value is the line integral of $\boldsymbol{F}$ over γ. The following examples illustrate how useful (2) can be.

2.4 EXAMPLE. If $\boldsymbol{F}(x, y)=(x^2+y^2)\boldsymbol{i}+(2xy)\boldsymbol{j}$, then compute $\oint_\gamma \boldsymbol{F}\cdot d\boldsymbol{x}$, where γ is the boundary of the unit square shown in Figure 2.11.

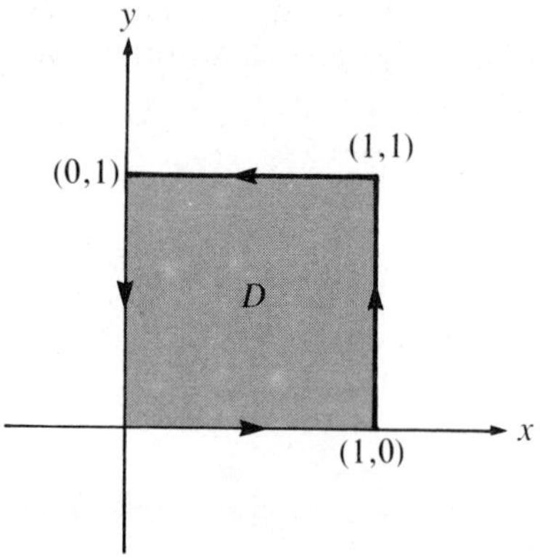

FIGURE 2.11

Solution. We *could* compute the line integral as in the preceding section, but this would require us to parametrize the four sides of D separately, work out each line integral, and add. If we use (2), though, we have to compute only one double integral:

$$\oint_\gamma \boldsymbol{F}\cdot d\boldsymbol{x}=\iint_D\left[\frac{\partial(2xy)}{\partial x}-\frac{\partial}{\partial y}(x^2+y^2)\right]dA$$

$$=\iint_D(2y-2y)\,dA=0.$$

While it is unreasonable to expect every application of (2) to turn out *this* simply, Example 2.4 should convince you that (2) can often save a lot of labor in evaluating line integrals. Sometimes we can use (2) to save labor in evaluating double integrals, too. A useful tool is the following result, which is a simple corollary of (2).

2.5 **THEOREM.** If D is a region satisfying the hypotheses of Theorem 2.3, then its area is given by

$$A(D)=\oint_{\partial D} x\,dy=-\oint_{\partial D} y\,dx=\tfrac{1}{2}\oint_{\partial D} -y\,dx+x\,dy.$$

Proof.

$$\oint_{\partial D} x\,dy=\oint_{\partial D} 0\,dx+x\,dy$$

$$=\iint_D\left(\frac{\partial x}{\partial x}-\frac{\partial 0}{\partial y}\right)dA=\iint_D dA=A(D),$$

and

$$-\oint_{\partial D} y\,dx = \oint_{\partial D} (-y)\,dx + 0\,dy$$

$$= \iint_D \left(\frac{\partial 0}{\partial x} - \frac{\partial(-y)}{\partial y}\right) dA = \iint_D dA = A(D),$$

where in each case the conversion from the line integral to the double integral is justified by Theorem 2.3. Finally,

$$\tfrac{1}{2}\oint_{\partial D} -y\,dx + x\,dy = \tfrac{1}{2}\oint_{\partial D} -y\,dx + \tfrac{1}{2}\oint_{\partial D} x\,dy$$

$$= \tfrac{1}{2}A(D) + \tfrac{1}{2}A(D) = A(D).$$

QED

2.6 **EXAMPLE.** Find the area enclosed by the ellipse

$$\frac{x^2}{16} + \frac{y^2}{9} = 1.$$

Solution. To do this problem by double integration, we would have to change coordinates (see Section 6.9) or evaluate a complicated definite integral. Alternatively, we can use Theorem 2.5. We get $A(D) = \int_\gamma x\,dy$, where γ is the ellipse

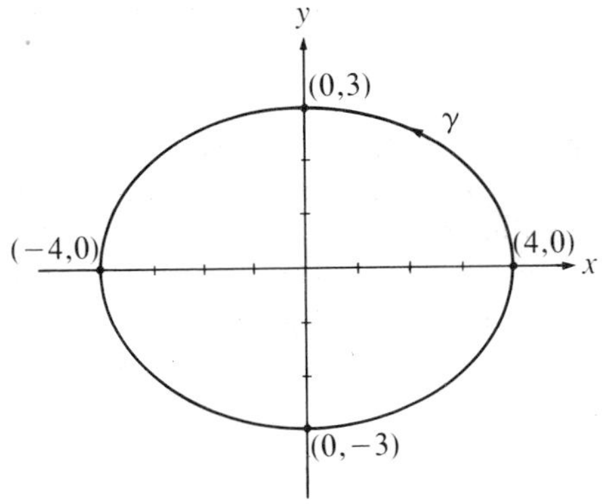

FIGURE 2.12

shown in Figure 2.12. We parametrize γ by $x = 4\cos t$ and $y = 3\sin t$ for $0 \le t \le 2\pi$. Then $dy = 3\cos t\,dt$, so

$$A(D) = \oint_\gamma x\,dy = \int_0^{2\pi} (4\cos t)3\cos t\,dt$$

$$= 12\int_0^{2\pi} \cos^2 t\,dt = 12(\tfrac{1}{2})\int_0^{2\pi} (1 + \cos 2t)\,dt$$

$$= 6[t + \tfrac{1}{2}\sin 2t]_0^{2\pi} = 12\pi.$$

We next give an example illustrating how we can sometimes use Green's Theorem in the form of Theorem 2.3 to work out the line integral of a function

$\boldsymbol{F}: \boldsymbol{R}^2 \rightarrow \boldsymbol{R}^2$ over the boundary of a region D even if $\boldsymbol{F}$ is not continuous on the entire region.

2.7 EXAMPLE. If

$$\boldsymbol{F}(x, y) = \left(-\frac{y}{x^2+y^2}, \frac{x}{x^2+y^2}\right) = (P(x, y), Q(x, y)),$$

then compute $\oint_\gamma \boldsymbol{F} \cdot d\boldsymbol{x}$ where γ is the ellipse of Example 2.6.

Solution. The given line integral becomes quite complicated if we try to evaluate it directly. Instead, we use the following trick based on Green's Theorem. This is suggested by the fact that $\dfrac{\partial Q}{\partial x} - \dfrac{\partial P}{\partial y} = 0$, as you can check by direct calculation. However, we *can't* apply Green's Theorem directly to the interior of the ellipse, since $\boldsymbol{F}$ is not continuous on the entire interior, in particular at the origin. To avoid this difficulty, let D be the region shown in Figure 2.13 *between* γ and the

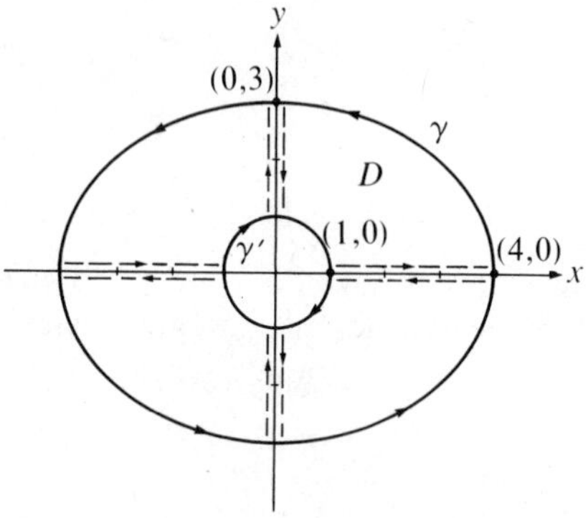

FIGURE 2.13

circle $\gamma': x^2+y^2=1$, oriented clockwise (so that D is on the *left* as we traverse the circle). Then γ' is parametrized by $x = \cos t$ and $y = -\sin t$ for $t \in [0, 2\pi]$. Now $\gamma \cup \gamma'$ forms the boundary of D, oriented so that D is on the left as $\gamma \cup \gamma'$ is traversed. D can be decomposed into the union of the four simple subregions of D located in the four quadrants. Then (8) and Theorem 2.3 give

$$\oint_{\gamma \cup \gamma'} \boldsymbol{F} \cdot d\boldsymbol{x} = \iint_D \left(\frac{\partial Q}{\partial x} - \frac{\partial P}{\partial y}\right) dA = 0.$$

Therefore, since

$$\oint_{\gamma \cup \gamma'} \boldsymbol{F} \cdot d\boldsymbol{x} = \oint_\gamma \boldsymbol{F} \cdot d\boldsymbol{x} + \oint_{\gamma'} \boldsymbol{F} \cdot d\boldsymbol{x} = 0,$$

we obtain

$$\oint_\gamma \boldsymbol{F} \cdot d\boldsymbol{x} = -\oint_{\gamma'} \boldsymbol{F} \cdot d\boldsymbol{x} = \oint_{-\gamma'} \boldsymbol{F} \cdot d\boldsymbol{x},$$

where $-\gamma'$ is parametrized in the usual way by

$$\boldsymbol{x}(t) = (\cos t, \sin t), \ t \in [0, 2\pi].$$

Then

$$\dot{\boldsymbol{x}}(t)\,dt = (-\sin t, \cos t)\,dt.$$

On $-\gamma'$ we have $x^2+y^2=1$, so

$$\boldsymbol{F}(x(t), y(t)) = (-\sin t, \cos t).$$

Thus

$$\begin{aligned}\oint_{\gamma} \boldsymbol{F}\cdot d\boldsymbol{x} &= \oint_{-\gamma'} \boldsymbol{F}(\boldsymbol{x}(t))\cdot \dot{\boldsymbol{x}}(t)\,dt \\ &= \oint_{-\gamma'} (-\sin t, \cos t)\cdot(-\sin t, \cos t)\,dt \\ &= \int_0^{2\pi} (\sin^2 t+\cos^2 t)\,dt = \int_0^{2\pi} 1\,dt \\ &= 2\pi.\end{aligned}$$

We close by suggesting briefly the physical interpretations (2) possesses. (We will have more to say about these in Sections 3, 4, 7, and 8.) First, $\int_\gamma \boldsymbol{F}\cdot d\boldsymbol{x}$ has already been interpreted as the work done by a force $\boldsymbol{F}=(P, Q)$ in moving a particle around γ. If

$$\frac{\partial Q}{\partial x}-\frac{\partial P}{\partial y}=0$$

inside the entire (simple) region D enclosed by γ and if $\dfrac{\partial Q}{\partial x}$ and $\dfrac{\partial P}{\partial y}$ are continuous throughout D, then by Theorem 2.2 this work will be 0.

If $\boldsymbol{F}=(P, Q)$ is the electric field intensity produced by a steady magnetic field and $\boldsymbol{F}=\nabla f=\left(\dfrac{\partial f}{\partial x}, \dfrac{\partial f}{\partial y}\right)$ for some scalar function $f:\boldsymbol{R}^2\to\boldsymbol{R}$, then it can be shown that

$$\frac{\partial Q}{\partial x}-\frac{\partial P}{\partial y}=0$$

(cf. Section 3). Then the *electromotive force* or *potential difference* produced on any simple closed curve γ in $\boldsymbol{R}^2$ is

$$\begin{aligned}\text{emf} &= \oint_{\gamma} \boldsymbol{F}\cdot d\boldsymbol{x} && \text{by definition of emf} \\ &= 0 && \text{by Green's Theorem.}\end{aligned}$$

Thus in such a situation no current flows in the circuit γ.

A distinct but analogous physical interpretation of Green's Theorem occurs in fluid dynamics. Here we interpret $\boldsymbol{F}$ as the velocity field of a flowing fluid in $\boldsymbol{R}^2$. In this case $\int_\gamma \boldsymbol{F}\cdot d\boldsymbol{x}$ is called the **circulation** of $\boldsymbol{F}$ about γ. The terminology is natural in that

$$\oint_{\gamma} \boldsymbol{F} \cdot d\boldsymbol{x} = \int_{\gamma} \boldsymbol{F} \cdot \boldsymbol{T}\, ds$$

measures the strength of $\boldsymbol{F}$ in the direction of γ, a strength which tends to produce circulation of the fluid about γ. If this line integral is nonzero, a *vortex* is said to exist in the region D enclosed by γ. But if $\dfrac{\partial Q}{\partial x} = \dfrac{\partial P}{\partial y}$, and these partial derivatives are continuous, then by (2)

$$\int_{\gamma} \boldsymbol{F} \cdot d\boldsymbol{x} = \iint_{D} \left(\frac{\partial Q}{\partial x} - \frac{\partial P}{\partial y}\right) dA = 0.$$

In this case, there is no tendency to create a vortex anywhere.

EXERCISES 7.2

In Exercises 1 to 14, use Green's Theorem to evaluate the given integrals.

1 $\oint_{\gamma} 2y\, dx + 3x\, dy$, where γ is the polygonal path from $(-1, 2)$ to $(3, 2)$ to $(3, 5)$ to $(-1, 5)$ to $(-1, 2)$

2 $\oint_{\gamma} (x^2 - y^3)\, dx + (x^3 + y^2)\, dy$, where γ is the unit circle $x^2 + y^2 = 1$ traversed counterclockwise.

3 $\oint_{\gamma} \boldsymbol{F} \cdot d\boldsymbol{x}$, where $\boldsymbol{F}$ is as in Exercise 2, but γ is the circle $x^2 + y^2 = 9$ traversed *clockwise.*

4 $\oint_{\gamma} -ye^x\, dx + xe^y\, dy$, where γ is the polygonal path from $(1, 1)$ to $(-1, 1)$ to $(-1, -1)$ to $(1, -1)$ to $(1, 1)$.

5 $\oint_{\gamma} y^2 e^x\, dx + 2ye^x\, dy$, where γ is the path of Exercise 1.

6 $\int_{\gamma} (e^x + y - 2x^2)\, dx + (7x - \sin y)\, dy$, where γ is the triangular path from $(0, 0)$ to $(0, 1)$ to $(1, 0)$ to $(0, 0)$.

7 $\int_{\gamma} (x^2 + y)\, dx + (2x - y)\, dy$, where $\gamma = \gamma_1 \cup \gamma_2$. Here γ_1 is the polygonal path of Exercise 1 and γ_2 is the unit circle parametrized in the clockwise sense.

8 $\int_{\gamma} (x^3 + \sin x + y)\, dx + (2x - \sin y \cos y)\, dy$, where γ is as in Exercise 7.

9 $\oint_{\gamma} (\sin y - x^2 y)\, dx + (x \cos y + xy^2)\, dy$, where γ is the curve of Exercise 2.

10 $\oint_{\gamma} (x - xy)\, dx + (y^3 + 1)\, dy$, where γ is the polynomial path from $(1, 0)$ to $(2, 0)$ to $(2, 1)$ to $(1, 1)$ to $(1, 0)$.

11 $\oint_{\gamma} \dfrac{-y}{x^2 + y^2}\, dx + \dfrac{x}{x^2 + y^2}\, dy$, where γ is the ellipse $4x^2 + y^2 = 16$, traversed counterclockwise.

12 $\oint_\gamma \frac{-y}{x^2+y^2}\,dx + \frac{x}{x^2+y^2}\,dy$, where γ is the ellipse $5x^2+y^2=25$, traversed *clockwise.*

13 $\int_\gamma \frac{x}{x^2+y^2}\,dx + \frac{y}{x^2+y^2}\,dy$, where $\gamma=\gamma_1\cup\gamma_2$. Here γ_1 is the parabolic path from $(-1, 0)$ to $(2, 3)$ along $y=x^2-1$, and γ_2 is the line segment from $(2, 3)$ to $(-1, 0)$.

14 $\int_\gamma \frac{x}{x^2+y^2}\,dx + \frac{y}{x^2+y^2}\,dy$, where $\gamma=\gamma_1\cup\gamma_2$. Here γ_1 is the parabolic path from $(-2, 0)$ to $(3, 5)$ along $y=x^2-4$, and γ_2 is the line segment $(3, 5)$ to $(-2, 0)$.

In Exercises 15 to 20, find the area of the given region.

15 The region D enclosed by the ellipse $4x^2+y^2=16$.

16 The region D enclosed by the hypocycloid $x^{2/3}+y^{2/3}=1$. (See Exercises 26–27, Exercises 7.1.)

17 The region D enclosed by the hypocycloid $x^{2/3}+y^{2/3}=4$.

18 The triangle with vertices $(1, 2)$, $(3, 2)$, and $(3, 5)$.

19 The annular region between $x^2+y^2=16$ and $x^2+y^2=9$. (*Hint: Think* before resorting to Theorem 2.5!)

20 The annular region between $x^2+y^2=4$ and $x^2+y^2=16$.

21 Show that if $\boldsymbol{F}(x, y)=(y^2e^x, 2ye^x)$ is the velocity field of a fluid flow, then there is no tendency for a vortex around the origin.

22 Repeat Exercise 21 for $\boldsymbol{F}(x, y)=(x^3+y^3+y\sin x, 3xy^2-\cos x)$.

23 If D is a region satisfying the hypotheses of Theorem 2.3, then find a formula for $\bar{x}$ and $\bar{y}$, the coordinates of the centroid of D, as a line integral.

24 Generalize Example 2.7. Namely, let

$$\boldsymbol{F}(\boldsymbol{x})=(P(\boldsymbol{x}), Q(\boldsymbol{x})) \quad \text{where} \quad \frac{\partial P}{\partial y}=\frac{\partial Q}{\partial x} \quad \text{for} \quad \boldsymbol{x}\neq\boldsymbol{0}.$$

Suppose that γ_1 and γ_2 are smooth simple closed curves with the same orientation, which enclose regions D_i such that $\boldsymbol{0}\in D_1\subseteq D_2$. Then show that

$$\oint_{\gamma_1} \boldsymbol{F}\cdot d\boldsymbol{x} = \oint_{\gamma_2} \boldsymbol{F}\cdot d\boldsymbol{x}.$$

25 Generalize Exercise 11. Namely, prove that if

$$\boldsymbol{F}(x)=\left(-\frac{y}{x^2+y^2}, \frac{x}{x^2+y^2}\right),$$

then for any smooth simple closed curve γ

$$\oint_\gamma \boldsymbol{F}\cdot d\boldsymbol{x} = \begin{cases} 2\pi & \text{if } \boldsymbol{0} \text{ is enclosed by } \gamma \\ 0 & \text{if } \boldsymbol{0} \text{ is not enclosed by } \gamma. \end{cases}$$

26 If γ is a smooth curve parametrized by

$$\boldsymbol{x}=\boldsymbol{x}(t)=(x(t), y(t)) \quad \text{for} \quad t\in[a, b],$$

then show that

$$\boldsymbol{N}(t)=\frac{1}{|\dot{\boldsymbol{x}}(t)|}\left(\frac{dy}{dt}, -\frac{dx}{dt}\right)$$

is a unit normal vector to γ at every point.

27 Use Exercise 26 to show that

$$\oint_\gamma -Q\,dx + P\,dy = \oint_\gamma \boldsymbol{F} \cdot \boldsymbol{N}\,ds.$$

In case $\boldsymbol{F} = (P, Q)$ is the velocity field of a fluid flow in $\boldsymbol{R}^2$, this integral is called the **flux** across γ. Explain how it measures the amount of fluid flowing across γ. **(This is referred to in Sections 6 and 8.)**

28 The **divergence** of $\boldsymbol{F} = (P, Q)$ is defined to be

$$\operatorname{div} \boldsymbol{F} = \frac{\partial P}{\partial x} + \frac{\partial Q}{\partial y}.$$

Use Exercise 27 and Theorem 2.3 to show that, under suitable hypotheses,

$$\iint_D \operatorname{div} \boldsymbol{F}\,dA = \int_\gamma \boldsymbol{F} \cdot \boldsymbol{N}\,ds.$$

(In Section 8 we will see how Green's Theorem in this form can be extended to three-dimensional vector fields $\boldsymbol{F}: \boldsymbol{R}^3 \to \boldsymbol{R}^3$. This exercise is also referred to in Section 3.)

The next three exercises suggest how Theorem 2.6 can be used to prove Theorem 6.9.1 in the case of a differentiable transformation of coordinates

$$\boldsymbol{T}(u, v) = (x, y) = (x(u, v), y(u, v))$$

in the plane. Suppose in Exercises 29 to 31 that $\dfrac{\partial(x, y)}{\partial(u, v)} > 0$ **on the region** $E = \boldsymbol{T}(D) \subseteq \boldsymbol{R}^2$.

29 Show that

$$A(E) = \int_\gamma x(u, v)\frac{\partial y}{\partial u}\,du + x(u, v)\frac{\partial y}{\partial v}\,dv,$$

where $\gamma = \partial E$. (*Hint:* Parametrize γ by $u = u(t)$ and $v = v(t)$ where $t \in [a, b]$. This will provide a parametrization of ∂E.)

30 Show that $\dfrac{\partial(x, y)}{\partial(u, v)} = \dfrac{\partial}{\partial u}\left(x\dfrac{\partial y}{\partial v}\right) - \dfrac{\partial}{\partial v}\left(x\dfrac{\partial y}{\partial u}\right)$.

31 Apply Green's Theorem to the line integral in Exercise 29 to show that

$$\iint_E dx\,dy = \iint_D \frac{\partial(x, y)}{\partial(u, v)}\,du\,dv.$$

(This approach can provide a rigorous proof of the formula

$$\iint_E f(x, y)\,dx\,dy = \iint_D f(x(u, v), y(u, v))\frac{\partial(x, y)}{\partial(u, v)}\,du\,dv$$

Exercise 31 shows that the *local* area distortion under $\boldsymbol{T}$ by the factor $\det J_{\boldsymbol{T}}(u_0, v_0)$ extends to the *global* area of D, i.e., $A(E) = \iint_D |\det J_{\boldsymbol{T}}|\,dA$.)

32 Use the approach of the text in proving (3) to prove (4).

3 VECTOR DIFFERENTIAL OPERATORS

In the preceding section we referred to the expression

$$\frac{\partial Q}{\partial x}-\frac{\partial P}{\partial y}$$

as a *sort of derivative* of the vector function $\boldsymbol{F}:\boldsymbol{R}^2 \rightarrow \boldsymbol{R}^2$ given by

$$\boldsymbol{F}(x, y)=(P(x, y), Q(x, y)).$$

In this section we will make this precise by introducing two vector differential operators called the *divergence* and *curl* operators. We will see that $\frac{\partial Q}{\partial x}-\frac{\partial P}{\partial y}$ is the result of applying the curl operator to

$$\boldsymbol{F}(x, y)=P(x, y)\boldsymbol{i}+Q(x, y)\boldsymbol{j}.$$

This will enable us to reformulate Green's Theorem in terms of the differentiation operator curl. At the end of this section, we will see that such a formulation strikingly displays the analogy between Green's Theorem and the Fundamental Theorem for Definite Integrals,

$$\int_a^b f'(x)\, dx=f(b)-f(a).$$

It also points the way to related results in Sections 4, 7, and 8.

Both the divergence and curl operators are closely related to the *gradient*. Recall that if $f:\boldsymbol{R}^3 \rightarrow \boldsymbol{R}$ has partial derivatives with respect to x, y, and z, then its gradient is

$$\nabla f=\left(\frac{\partial f}{\partial x}, \frac{\partial f}{\partial y}, \frac{\partial f}{\partial z}\right)=\frac{\partial f}{\partial x}\boldsymbol{i}+\frac{\partial f}{\partial y}\boldsymbol{j}+\frac{\partial f}{\partial z}\boldsymbol{k}.$$

To define divergence and curl, it is convenient to assign a meaning to ∇ alone. We do this as follows.

3.1 **DEFINITION.** The differential operator **del** (or **gradient**) is

$$\nabla=\left(\frac{\partial}{\partial x_1}, \frac{\partial}{\partial x_2}, \frac{\partial}{\partial x_3}, \ldots, \frac{\partial}{\partial x_n}\right),$$

where ∇ acts on functions $f:\boldsymbol{R}^n \rightarrow \boldsymbol{R}$ by the rule

$$\nabla(f)=\nabla f=\left(\frac{\partial f}{\partial x_1}, \frac{\partial f}{\partial x_2}, \ldots, \frac{\partial f}{\partial x_n}\right).$$

(Thus ∇ can act only on real valued functions that have partial derivatives with respect to each variable x_i.)

We can easily derive the following formal properties of the gradient operator ∇, which bring out its analogy to the differential operator $\frac{d}{dx}$ in elementary calculus.

3.2 **THEOREM.** If f and g are differentiable functions $f, g: \boldsymbol{R}^n \to \boldsymbol{R}$, then:
(a) *Linearity:* $\nabla(f+g) = \nabla f + \nabla g$
$\nabla(af) = a\,\nabla f$ for all $a \in \boldsymbol{R}$.
(b) $\nabla(f(\boldsymbol{x})g(\boldsymbol{x})) = f(\boldsymbol{x})\,\nabla g(\boldsymbol{x}) + [\nabla f(\boldsymbol{x})]g(\boldsymbol{x})$

(c) $\nabla\left(\dfrac{f(\boldsymbol{x})}{g(\boldsymbol{x})}\right) = \dfrac{g(\boldsymbol{x})\,\nabla f(\boldsymbol{x}) - f(\boldsymbol{x})\,\nabla g(\boldsymbol{x})}{g(\boldsymbol{x})g(\boldsymbol{x})}$ if $g(\boldsymbol{x}) \neq 0$.

(d) If $h: \boldsymbol{R} \to \boldsymbol{R}$, and we put $t = f(\boldsymbol{x})$, then

$$\nabla[h \circ f(\boldsymbol{x})] = h'(f(\boldsymbol{x}))\,\nabla f(\boldsymbol{x});$$

that is,

$$\nabla(h(t)) = \frac{dh}{dt}\nabla t.$$

Partial Proof. (a) These follow from the linearity of partial differentiation relative to each x_i. For instance,

$$\frac{\partial}{\partial x_i}(f+g) = \frac{\partial f}{\partial x_i} + \frac{\partial g}{\partial x_i},$$

so all the coordinates of $\nabla(f+g)$ and $\nabla f + \nabla g$ agree. A similar proof can be given $\nabla(af)$ and $a\,\nabla f$. We leave (b) and (c) as Exercises 9 and 10 below.

(d) This follows on applying the Chain Rule to the composite function $h \circ f: \boldsymbol{R}^n \to \boldsymbol{R}^1$. Its total derivative at $\boldsymbol{x}$ is (by Chain Rule 6.6.1)

$$\begin{aligned}
\nabla(h \circ f)(\boldsymbol{x}) &= \nabla(h(f(\boldsymbol{x}))) \\
&= J_h(f(\boldsymbol{x}))J_f(\boldsymbol{x}) \\
&= \frac{dh}{dt}(f(\boldsymbol{x}))\,\nabla f(\boldsymbol{x}) \\
&= h'(f(\boldsymbol{x}))\,\nabla f(\boldsymbol{x}).
\end{aligned}$$

QED

Once we have a meaning for the vector differential operator ∇, we can start performing vector operations with it. This leads to the divergence operator, which has many important applications in mechanics, fluid dynamics, electricity, and magnetism.

3.3 **DEFINITION.** Suppose that $\boldsymbol{F}: \boldsymbol{R}^n \to \boldsymbol{R}^n$ is a differentiable vector function. Then the **divergence** of $\boldsymbol{F}$ is

$$\operatorname{div} \boldsymbol{F} = \nabla \cdot \boldsymbol{F} = \frac{\partial f_1}{\partial x_1} + \frac{\partial f_2}{\partial x_2} + \dots + \frac{\partial f_n}{\partial x_n},$$

where $f_1, f_2, \dots, f_n$ are the coordinate functions of $\boldsymbol{F}$.

Observe that in the light of Definition 3.1, the notation $\nabla \cdot \boldsymbol{F}$ makes sense. We have in the $n = 3$ case, for instance,

$$\nabla = \left(\frac{\partial}{\partial x}, \frac{\partial}{\partial y}, \frac{\partial}{\partial z}\right) = \frac{\partial}{\partial x}\boldsymbol{i} + \frac{\partial}{\partial y}\boldsymbol{j} + \frac{\partial}{\partial z}\boldsymbol{k}$$

and

$$F = (f_1, f_2, f_3) = f_1\boldsymbol{i} + f_2\boldsymbol{j} + f_3\boldsymbol{k}.$$

So if we formally compute the dot product $\nabla \cdot F$, we obtain

$$\begin{aligned}\nabla \cdot F &= \frac{\partial}{\partial x}(f_1) + \frac{\partial}{\partial y}(f_2) + \frac{\partial}{\partial z}(f_3) \\ &= \frac{\partial f_1}{\partial x} + \frac{\partial f_2}{\partial y} + \frac{\partial f_3}{\partial z}\end{aligned}$$

We can easily give meaning to the divergence of $F : \boldsymbol{R}^2 \to \boldsymbol{R}^2$ if we adopt this formal point of view. Namely, if

$$F(x, y) = (P(x, y), Q(x, y)),$$

then

$$F = P\boldsymbol{i} + Q\boldsymbol{j} + 0\boldsymbol{k}.$$

Hence,

$$\nabla \cdot F = \frac{\partial P}{\partial x} + \frac{\partial Q}{\partial y}.$$

The term "divergence" arises from fluid dynamics, where div F measures the tendency of fluid to flow outward. If you worked Exercise 28 in Exercises 7.2, then you have already seen something of this for a two-dimensional flow. Let's consider now a three-dimensional flow.

Suppose that a fluid flow has differentiable velocity

$$\boldsymbol{v}(x, y, z) = v_1(x, y, z)\boldsymbol{i} + v_2(x, y, z)\boldsymbol{j} + v_3(x, y, z)\boldsymbol{k}.$$

Consider a small rectangular box ΔB with edges Δx, Δy, and Δz. See Figure 3.1.

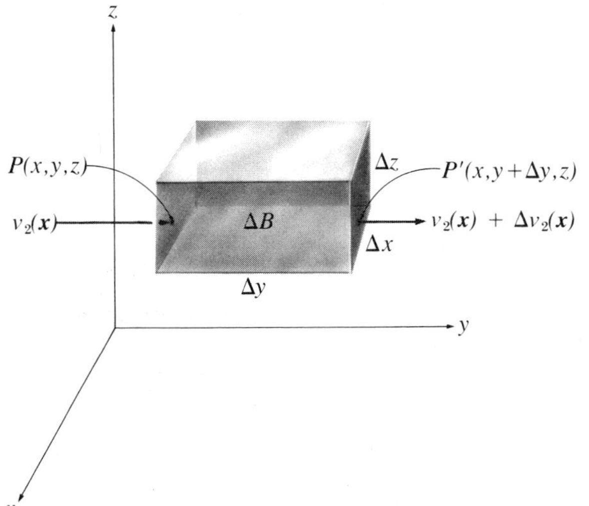

FIGURE 3.1

In a short time span Δt, how much fluid enters the left wall of ΔB? The component of the flow moving left to right is approximately $v_2(\boldsymbol{x})\boldsymbol{j}$, where $\boldsymbol{x}$ is a point on the left wall. At each point of the left wall, then, the amount of liquid flowing across that point during time Δt is approximately $v_2(\boldsymbol{x})\,\Delta t$ linear units. The *total amount* of fluid flowing across the entire wall during time Δt is then approximately

(1) $$v_2(\boldsymbol{x})\,\Delta x\,\Delta z\,\Delta t \quad \text{cubic units.}$$

(For example, if $v_2 = 3$ feet per second, then about 3 linear feet per second pass each unit area of the left wall. So $3\Delta x\,\Delta z$ cubic feet cross the wall per second.)

Now, how much fluid leaves the right wall during Δt? Well, at the right wall the velocity at point $P'(x, y+\Delta y, z)$ is

$$v_2(x, y+\Delta y, z) \approx v_2(x, y, z) + \frac{\partial v_2(x, y, z)}{\partial y}\,|(x, y+\Delta y, z) - (x, y, z)|$$

by the usual tangent approximation for v_2 near (x, y, z). Thus

$$v_2(x, y+\Delta y, z) \approx v_2(x, y, z) + \frac{\partial v_2(\boldsymbol{x})}{\partial y}\,\Delta y.$$

So the total amount of fluid flowing across the right wall during the time Δt is approximately

(2) $$\left(v_2(\boldsymbol{x}) + \frac{\partial v_2(\boldsymbol{x})}{\partial y}\,\Delta y\right)\Delta x\,\Delta z\,\Delta t.$$

Comparing (1) and (2), we see that the net *outflow* of fluid from ΔB in the y-direction is approximately

(3) $$\frac{\partial v_2(\boldsymbol{x})}{\partial y}\,\Delta x\,\Delta y\,\Delta z\,\Delta t.$$

Identical reasoning on the front and back walls, and on the top and bottom, will show that the net outflow from ΔB in the x-direction is approximately

(4) $$\frac{\partial v_1}{\partial x}(\boldsymbol{x})\,\Delta x\,\Delta y\,\Delta z\,\Delta t,$$

and that in the z-direction is approximately

(5) $$\frac{\partial v_3}{\partial z}(\boldsymbol{x})\,\Delta x\,\Delta y\,\Delta z\,\Delta t.$$

Hence the total net outflow from ΔB per unit time is approximated by adding (3), (4) and (5) and dividing by Δt. This gives

$$\left(\frac{\partial v_1}{\partial x} + \frac{\partial v_2}{\partial y} + \frac{\partial v_3}{\partial z}\right)\Delta x\,\Delta y\,\Delta z.$$

Thus the total net outflow from ΔB per unit of volume per unit time is approximated by

(6) $$\operatorname{div} v(\boldsymbol{x}) = \frac{\partial v_1}{\partial x}(\boldsymbol{x}) + \frac{\partial v_2}{\partial y}(\boldsymbol{x}) + \frac{\partial v_3}{\partial z}(\boldsymbol{x}).$$

As Δx, Δy, Δz, and Δt approach 0, this approximation seems to measure the *outflow per unit volume per unit time* more and more accurately. For this reason, $\operatorname{div} \boldsymbol{v}(\boldsymbol{x})$ is defined in fluid dynamics as the rate of flow outward from $\boldsymbol{x}$ per unit time. If $\operatorname{div} \boldsymbol{v}(\boldsymbol{x}_0) > 0$, then one says that there is a *source* for the flow at $\boldsymbol{x}_0$. If $\operatorname{div} \boldsymbol{v}(\boldsymbol{x}_0) < 0$, then there is a *sink* at $\boldsymbol{x}_0$.

The key properties of the divergence operator are given in the following result.

3.4 THEOREM. Suppose that $\boldsymbol{F}, \boldsymbol{G} : \boldsymbol{R}^n \to \boldsymbol{R}^n$ and $g : \boldsymbol{R}^n \to \boldsymbol{R}$ are differentiable. Then

(a) *Linearity:* $\operatorname{div}(\boldsymbol{F} + \boldsymbol{G}) = \operatorname{div} \boldsymbol{F} + \operatorname{div} \boldsymbol{G}$
$\operatorname{div}(c\boldsymbol{F}) = c \operatorname{div} \boldsymbol{F}$ for any $c \in \boldsymbol{R}$.

(b) $\operatorname{div}(g(\boldsymbol{x})\boldsymbol{G}(\boldsymbol{x})) = g(\boldsymbol{x}) \operatorname{div} \boldsymbol{G}(\boldsymbol{x}) + \nabla g(\boldsymbol{x}) \cdot \boldsymbol{G}(\boldsymbol{x})$.

Proof. Part (a) is straightforward, so is left as Exercise 11 below. We prove (b) in case $n = 3$. (The same reasoning applies in any dimension.) We have

$$\begin{aligned}
\operatorname{div}(g(\boldsymbol{x})\boldsymbol{G}(\boldsymbol{x})) &= \frac{\partial}{\partial x}(g(\boldsymbol{x})g_1(\boldsymbol{x})) + \frac{\partial}{\partial y}(g(\boldsymbol{x})g_2(\boldsymbol{x})) + \frac{\partial}{\partial z}(g(\boldsymbol{x})g_3(\boldsymbol{x})) \\
&= \frac{\partial g(\boldsymbol{x})}{\partial x} g_1(\boldsymbol{x}) + g(\boldsymbol{x}) \frac{\partial g_1(\boldsymbol{x})}{\partial x} + \frac{\partial g(\boldsymbol{x})}{\partial y} g_2(\boldsymbol{x}) \\
&\quad + g(\boldsymbol{x}) \frac{\partial g_2(\boldsymbol{x})}{\partial y} + \frac{\partial g(\boldsymbol{x})}{\partial z} g_3(\boldsymbol{x}) + g(\boldsymbol{x}) \frac{\partial g_3(\boldsymbol{x})}{\partial z} \\
&= \frac{\partial g(\boldsymbol{x})}{\partial x} g_1(\boldsymbol{x}) + \frac{\partial g(\boldsymbol{x})}{\partial y} g_2(\boldsymbol{x}) + \frac{\partial g(\boldsymbol{x})}{\partial z} g_3(\boldsymbol{x}) \\
&\quad + g(\boldsymbol{x}) \frac{\partial g_1(\boldsymbol{x})}{\partial x} + g(\boldsymbol{x}) \frac{\partial g_2(\boldsymbol{x})}{\partial y} + g(\boldsymbol{x}) \frac{\partial g_3(\boldsymbol{x})}{\partial z} \\
&= \left(\frac{\partial g}{\partial x}, \frac{\partial g}{\partial y}, \frac{\partial g}{\partial z}\right) \cdot (g_1(\boldsymbol{x}), g_2(\boldsymbol{x}), g_3(\boldsymbol{x})) + g(\boldsymbol{x}) \operatorname{div} \boldsymbol{G}(\boldsymbol{x}) \\
&= g(\boldsymbol{x}) \operatorname{div} \boldsymbol{G}(\boldsymbol{x}) + \nabla g(\boldsymbol{x}) \cdot \boldsymbol{G}(\boldsymbol{x}).
\end{aligned}$$

QED

The next example suggests something of the usefulness of Theorem 3.4 in applications.

3.5 EXAMPLE. Inverse square law fields. Suppose that for nonzero $\boldsymbol{x} \in \boldsymbol{R}^3$,

$$\boldsymbol{F}(\boldsymbol{x}) = k \frac{1}{|\boldsymbol{x}|^2} \boldsymbol{u},$$

where k is constant and $\boldsymbol{u}$ is the unit vector giving the direction of $\boldsymbol{x}$, i.e.,

$$\boldsymbol{u} = \frac{1}{|\boldsymbol{x}|} \boldsymbol{x}.$$

Find $\operatorname{div} \boldsymbol{F}(\boldsymbol{x})$.

Solution. We have

$$F(x) = k\frac{1}{|x|^3}x = k\ |x|^{-3}\ x.$$

So if we put

$$g(x) = k\ |x|^{-3} = \frac{k}{(x^2+y^2+z^2)^{3/2}}, \qquad G(x) = x$$

then we can apply Theorem 3.4(b). We get

$$\begin{aligned} \operatorname{div} F(x) &= k\ |x|^{-3} \operatorname{div} x + \nabla(k\ |x|^{-3}) \cdot x \\ &= k\ |x|^{-3}(\operatorname{div} x) + k\ \nabla(|x|^{-3}) \cdot x. \end{aligned}$$

We next apply Theorem 3.2(d) to the second term, with $f(x) = |x|$ and $h(t) = t^{-3}$. This, together with the observation

$$\operatorname{div} x = \operatorname{div}(x, y, z) = \frac{\partial x}{\partial x} + \frac{\partial y}{\partial y} + \frac{\partial z}{\partial z},$$

gives

(7) $$\begin{aligned} \operatorname{div} F(x) &= k\ |x|^{-3}\ (1+1+1) + k(-3)\ |x|^{-4}\ \nabla(|x|) \cdot x \\ &= 3k\ |x|^{-3} - 3k\ |x|^{-4}\ \nabla(|x|) \cdot x. \end{aligned}$$

Now

$$\begin{aligned} \nabla(|x|) &= \nabla(\sqrt{x^2+y^2+z^2}) \\ &= \left(\frac{x}{\sqrt{x^2+y^2+z^2}}, \frac{y}{\sqrt{x^2+y^2+z^2}}, \frac{z}{\sqrt{x^2+y^2+z^2}}\right) \\ &= \frac{1}{|x|}x. \end{aligned}$$

Thus

$$\nabla(|x|) \cdot x = \frac{1}{|x|}x \cdot x = |x|.$$

So from (7),

$$\operatorname{div} F(x) = 3k\ |x|^{-3} - 3k\ |x|^{-4}\ |x| = 0.$$

Inverse square law fields play a major role in electromagnetic theory and gravitational theory. For instance, the electric force on a unit charge at x produced by a unit charge of the same sign at $\mathbf{0}$ is

$$E(x) = k\frac{1}{|x|^2}u.$$

The gravitational field of the earth at a point $\boldsymbol{x}$ is given by

$$\boldsymbol{G}(\boldsymbol{x}) = -gR^2 \frac{1}{|\boldsymbol{x}|^2} \boldsymbol{u},$$

where R is the radius of the earth and g is the acceleration due to gravity at the surface of the earth. (The origin is taken at the center of the earth.) Thus, Example 3.5 applies to both the electrical and gravitational situations.

3.6 **DEFINITION.** A vector field $\boldsymbol{F}:\boldsymbol{R}^n \to \boldsymbol{R}^n$ is called **conservative** if $\boldsymbol{F}(\boldsymbol{x}) = \boldsymbol{\nabla} p(\boldsymbol{x})$ for some function $p:\boldsymbol{R}^n \to \boldsymbol{R}$. In this case, p is called a **potential function** for $\boldsymbol{F}$.

We will see in the next section that such fields conserve total (=potential+kinetic) energy, or, equivalently, do the same amount of work in moving a particle from a point $\boldsymbol{x}_0$ to $\boldsymbol{x}_1$ no matter what piecewise smooth path from $\boldsymbol{x}_0$ to $\boldsymbol{x}_1$ is taken. For the present, we can show that inverse square law fields are conservative.

3.7 **EXAMPLE.** Show that the inverse square law field $\boldsymbol{F}$ of Example 3.5 is conservative, with the potential function $p(\boldsymbol{x}) = -k\dfrac{1}{|\boldsymbol{x}|}$.

Proof. We need only verify that $\boldsymbol{\nabla} p(x) = \boldsymbol{F}(\boldsymbol{x})$. Here

$$\begin{aligned}\boldsymbol{\nabla} p(\boldsymbol{x}) &= -k\, \boldsymbol{\nabla}\left(\frac{1}{|\boldsymbol{x}|}\right) = -k\, \boldsymbol{\nabla}(|\boldsymbol{x}|^{-1}) \\ &= k\, |\boldsymbol{x}|^{-2}\, \boldsymbol{\nabla}(|\boldsymbol{x}|) \cdot \boldsymbol{x} \quad \text{by Theorem 3.2(d)} \\ &= k \frac{1}{|\boldsymbol{x}|^2} \frac{1}{|\boldsymbol{x}|} \boldsymbol{x} \\ &= \boldsymbol{F}(\boldsymbol{x}).\end{aligned}$$

QED

Since in Example 3.5 we showed that div $\boldsymbol{F}(\boldsymbol{x}) = 0$, we have from Example 3.7 that for a potential function p of a conservative vector field $\boldsymbol{F}$

$$\operatorname{div} \boldsymbol{\nabla} p(\boldsymbol{x}) = 0 = \boldsymbol{\nabla} \cdot \boldsymbol{\nabla} p(\boldsymbol{x}).$$

Thus

$$\boldsymbol{\nabla} \cdot \left(\frac{\partial p}{\partial x}, \frac{\partial p}{\partial y}, \frac{\partial p}{\partial z}\right) = 0.$$

This is the same as

$$\frac{\partial^2 p}{\partial x^2} + \frac{\partial^2 p}{\partial y^2} + \frac{\partial^2 p}{\partial z^2} = 0.$$

The latter equation occurs so often that it carries a special name, that of the French mathematician Pierre S. Laplace (1749–1827), who played a major role in the development of mechanics.

3.8 **DEFINITION.** The operator ∇^2 or Δ given by

$$\nabla^2 = \boldsymbol{\nabla} \cdot \boldsymbol{\nabla} = \frac{\partial^2}{\partial x^2} + \frac{\partial^2}{\partial y^2} + \frac{\partial^2}{\partial z^2}$$

is called the **Laplacian operator**. The equation

$$\nabla^2 p(\boldsymbol{x}) = 0$$

is called **Laplace's equation**. A function p satisfying Laplace's equation is called a **harmonic function**.

Many important functions arising in physics, including electrical potential, pressure, and temperature functions, are harmonic functions. Their study forms an important part of modern analysis. A significant problem in mathematical physics is to find solutions of Laplace's equation that satisfy various conditions.

We come now to the final differential operator to be discussed in this section, the curl operator.

3.9 **DEFINITION.** If $\boldsymbol{F}:\boldsymbol{R}^3 \to \boldsymbol{R}^3$ is a differentiable vector field, then its **curl** is

$$\textbf{curl}\ \boldsymbol{F} = \boldsymbol{\nabla} \times \boldsymbol{F} = \left(\frac{\partial}{\partial x}, \frac{\partial}{\partial y}, \frac{\partial}{\partial z}\right) \times (f_1, f_2, f_3)$$

$$= \left(\frac{\partial f_3}{\partial y} - \frac{\partial f_2}{\partial z}, \frac{\partial f_1}{\partial z} - \frac{\partial f_3}{\partial x}, \frac{\partial f_2}{\partial x} - \frac{\partial f_1}{\partial y}\right).$$

(Note that the curl operator is defined *only* in the case of dimension 3.)

3.10 **EXAMPLE.** If

$$\boldsymbol{F}(x, y, z) = (3x^2yz - 5x^2yz^2, e^{x^2+y^2+z^2}, \sin x + \cos y - \tan z),$$

then find **curl** $\boldsymbol{F}$.

Solution.

$$\textbf{curl}\ \boldsymbol{F} = \boldsymbol{\nabla} \times \boldsymbol{F}$$
$$= (-\sin y - 2ze^{x^2+y^2+z^2}, 3x^2y - 10x^2yz - \cos x, 2xe^{x^2+y^2+z^2} - 3x^2z + 5x^2z^2).$$

Corresponding to Theorems 3.2 and 3.4, we have the following collection of properties for the curl operator.

3.11 **THEOREM.** Let $\boldsymbol{F}, \boldsymbol{G}:\boldsymbol{R}^3 \to \boldsymbol{R}^3$ and $f:\boldsymbol{R}^3 \to \boldsymbol{R}$ be differentiable. Then

(a) *Linearity*: **curl** $(\boldsymbol{F}+\boldsymbol{G}) =$ **curl** $\boldsymbol{F} +$ **curl** $\boldsymbol{G}$

curl $(a\boldsymbol{F}) = a$ **curl** $\boldsymbol{F}$ for any $a \in \boldsymbol{R}$.

(b) **curl** $(f(\boldsymbol{x})\boldsymbol{F}(\boldsymbol{x})) = f(\boldsymbol{x})$ **curl** $\boldsymbol{F}(\boldsymbol{x}) + \boldsymbol{\nabla} f(\boldsymbol{x}) \times \boldsymbol{F}(\boldsymbol{x})$.

(c) $\operatorname{div}(\boldsymbol{F} \times \boldsymbol{G}) = \boldsymbol{G} \cdot$ **curl** $\boldsymbol{F} - \boldsymbol{F} \cdot$ **curl** $\boldsymbol{G}$

(d) If the second partial derivatives of f are continuous, then **curl** $\boldsymbol{\nabla} f = \boldsymbol{0}$.

(e) If the second partial derivatives of $\boldsymbol{F}$ are continuous, then div(**curl** $\boldsymbol{F}$) $= 0$.

Partial Proof. The proof of part (a) is very easy, so is left as an exercise for you (Exercise 12 below). The proofs of (b) and (c) are likewise easy to give if the left side is expanded using Definitions 3.3 and 3.9, and the rule for the derivative of a product of two real valued functions is used (Exercises 13 to 15 below). As for (d), we have

$$\begin{aligned}\textbf{curl}\,\nabla f = \nabla\times\nabla f &= \left(\frac{\partial}{\partial x}, \frac{\partial}{\partial y}, \frac{\partial}{\partial z}\right)\times\left(\frac{\partial f}{\partial x}, \frac{\partial f}{\partial y}, \frac{\partial f}{\partial z}\right)\\ &= \left(\frac{\partial^2 f}{\partial y\,\partial z}-\frac{\partial^2 f}{\partial z\,\partial y}, \frac{\partial^2 f}{\partial z\,\partial x}-\frac{\partial^2 f}{\partial x\,\partial z}, \frac{\partial^2 f}{\partial x\,\partial y}-\frac{\partial^2 f}{\partial y\,\partial x}\right)\\ &= (0, 0, 0) \text{ by Theorem 6.7.5,}\\ &= \mathbf{0}.\end{aligned}$$

To show (e), we write

$$\begin{aligned}\text{div}\,\textbf{curl}\,\boldsymbol{F} &= \frac{\partial}{\partial x}\left(\frac{\partial f_3}{\partial y}-\frac{\partial f_2}{\partial z}\right)+\frac{\partial}{\partial y}\left(\frac{\partial f_1}{\partial z}-\frac{\partial f_3}{\partial x}\right)+\frac{\partial}{\partial z}\left(\frac{\partial f_2}{\partial x}-\frac{\partial f_1}{\partial y}\right)\\ &= \frac{\partial^2 f_3}{\partial x\,\partial y}-\frac{\partial^2 f_2}{\partial x\,\partial z}+\frac{\partial^2 f_1}{\partial y\,\partial z}-\frac{\partial^2 f_3}{\partial y\,\partial x}+\frac{\partial^2 f_2}{\partial z\,\partial x}-\frac{\partial^2 f_1}{\partial z\,\partial y}\\ &= 0, \text{ again by Theorem 6.7.5.}\end{aligned}$$

QED

We close this section by returning to Green's Theorem and the concept of circulation discussed at the end of Section 2. Suppose that $\boldsymbol{F}:\boldsymbol{R}^2\to\boldsymbol{R}^2$ is given by

$$\boldsymbol{F}(x, y) = (P(x, y), Q(x, y)).$$

Then we can regard the function as $\boldsymbol{F}:\boldsymbol{R}^3\to\boldsymbol{R}^3$ by writing

$$\boldsymbol{F}(x, y, z) = (P(x, y), Q(x, y), 0).$$

Then application of the curl operator gives

$$\begin{aligned}\textbf{curl}\,\boldsymbol{F}(x, y, z) &= \left(\frac{\partial 0}{\partial y}-\frac{\partial Q}{\partial z}, \frac{\partial P}{\partial z}-\frac{\partial 0}{\partial x}, \frac{\partial Q}{\partial x}-\frac{\partial P}{\partial y}\right)\\ &= \left(0, 0, \frac{\partial Q}{\partial x}-\frac{\partial P}{\partial y}\right).\end{aligned}$$

Now, assuming that the partial derivatives of $\boldsymbol{F}$ are continuous, we can put Green's Theorem 2.3 into the form

(8)
$$\begin{aligned}\int_\gamma \boldsymbol{F}\cdot d\boldsymbol{x} &= \iint_D \left(\frac{\partial Q}{\partial x}-\frac{\partial P}{\partial y}\right) dA\\ &= \iint_D \textbf{curl}\,\boldsymbol{F}\cdot\boldsymbol{k}\, dA\end{aligned}$$

where $\gamma = \partial D$. Thus, the circulation of $\boldsymbol{F}$ around γ is the double integral of the coordinate of **curl** $\boldsymbol{F}$ in the direction perpendicular to D. This suggests the origin of the term "curl." It is a measure of the tendency for a flow with velocity field $\boldsymbol{F}$ to curl around the region D enclosed by γ. A vector field $\boldsymbol{F}$ is called **irrotational** if **curl** $\boldsymbol{F} = \mathbf{0}$. This stems from the fact that if $\boldsymbol{F}$ is irrotational, then the circulation is

0 around every piecewise smooth simple closed curve γ. So there is no tendency to create a vortex anywhere.

We will see in Section 7 that an analogue of Green's Theorem in the form (8) holds for fields $\boldsymbol{F}:\boldsymbol{R}^3 \to \boldsymbol{R}^3$. This form also displays the analogy between Green's Theorem and the Fundamental Theorem

$$\int_a^b f'(x)\,dx = f(b) - f(a).$$

For in (8), $\textbf{curl } \boldsymbol{F} \cdot \boldsymbol{k} = \pm|\textbf{curl } \boldsymbol{F}|$. So the double integral of the magnitude (i.e., the scalar measure) of the differential operator curl applied to $\boldsymbol{F}$ is determined by the values of $\boldsymbol{F}$ on γ. In fact, it is the line integral of $\boldsymbol{F}$ over γ.

EXERCISES 7.3

In Exercises 1 to 4, compute the divergence of the given vector field $\boldsymbol{F}$ at $\boldsymbol{x}$ and the given $\boldsymbol{x}_0$.

1 $\boldsymbol{F}(\boldsymbol{x}) = x\boldsymbol{i} + (xz+3)\boldsymbol{j} + (yz+x)\boldsymbol{k}$; $\boldsymbol{x}_0 = (2, -2, 3)$.

2 $\boldsymbol{F}(\boldsymbol{x}) = (x_1^2 + x_2^2 + x_3^2 + x_4^2,\ x_1x_2x_3x_4,\ x_1x_2 - x_3x_4,\ x_1x_4 - x_2x_3)$; $\boldsymbol{x}_0 = (1, -1, 2, 3)$.

3 $\boldsymbol{F}(\boldsymbol{x}) = xy^2z^2\boldsymbol{i} + z^2 \sin y\boldsymbol{j} + x^2e^y\boldsymbol{k}$; $\boldsymbol{x}_0 = (1, \pi, -2)$.

4 $\boldsymbol{F}(\boldsymbol{x}) = \dfrac{x+z}{x^2+y^2+z^2}\boldsymbol{i} + \dfrac{y-x}{x^2+y^2+z^2}\boldsymbol{j} + \dfrac{z-y}{x^2+y^2+z^2}\boldsymbol{k}$; $\boldsymbol{x}_0 = (1, 0, -2)$.

In Exercises 5 to 8, compute the curl of the given field $\boldsymbol{F}$ at $\boldsymbol{x}$ and at $\boldsymbol{x}_0$.

5 $\boldsymbol{F}(\boldsymbol{x}) = 3x^2\boldsymbol{i} + xy\boldsymbol{j} + z\boldsymbol{k}$; $\boldsymbol{x}_0 = (-3, 4, 2)$.

6 The $\boldsymbol{F}$ and $\boldsymbol{x}_0$ of Exercise 1.

7 $\boldsymbol{F}(\boldsymbol{x}) = e^x \sin y \cos z\boldsymbol{i} + e^x \cos y \cos z\boldsymbol{j} - e^x \sin y \sin z\boldsymbol{k}$; $\boldsymbol{x}_0 = (-2, \pi/2, 5\pi/6)$.

8 $\boldsymbol{F}(\boldsymbol{x}) = (ye^{x^2z} + 2x^2yze^{x^2z},\ xe^{x^2z},\ x^3ye^{x^2z})$; $\boldsymbol{x}_0 = (1, 2, 3)$.

In Exercises 9 to 13, prove the indicated parts of the given theorems in this section.

9 Theorem 3.2(b).

10 Theorem 3.2(c).

11 Theorem 3.4(a).

12 Theorem 3.11(a).

13 Theorem 3.11(b).

14 Prove that if all second partial derivatives of the coordinate functions f_1, f_2, and f_3 of $\boldsymbol{F}$ are continuous, then

$$\textbf{curl}(\textbf{curl } \boldsymbol{F}) = \boldsymbol{\nabla}(\text{div } \boldsymbol{F}) - \nabla^2\boldsymbol{F},$$

where

$$\nabla^2\boldsymbol{F} = (\nabla^2 f_1)\boldsymbol{i} + (\nabla^2 f_2)\boldsymbol{j} + (\nabla^2 f_3)\boldsymbol{k}.$$

15 Prove Theorem 3.11(c).

16 Compute $\textbf{curl}(\textbf{curl } \boldsymbol{F})$ using Exercise 14, if

$$\boldsymbol{F}(x, y, z) = (xz + y^2, xy + z^2, x^2 + yz).$$

17 Verify the formula for $\text{div}(\boldsymbol{F} \times \boldsymbol{G})$ in Exercise 15, if

$$\boldsymbol{F} = (x^2, y^2, z^2) \quad \text{and} \quad \boldsymbol{G} = (y^2, z^2, x^2).$$

18 The **trace** of an n-by-n matrix A is defined to be the sum of the main diagonal entries

$$\mathrm{Tr}(A) = \sum_{i=1}^{n} a_{ii} = a_{11} + a_{22} + \ldots + a_{nn}.$$

What matrix with which we are familiar has div $\boldsymbol{F}$ for its trace?

19 Refer to Exercise 18. Express $\nabla^2 f$ as the trace of an appropriate matrix.

In Exercises 20 to 23, show that the functions are harmonic.

20 $f(x, y, z) = x^2 + y^2 - 2z^2$.

21 $f(x, y, z) = 3x^2 + 2y^2 - 5z^2$.

22 $f(x, y, z) = \sqrt{x^2 + y^2 + z^2}$, $(x, y, z) \neq \boldsymbol{0}$.

23 $f(x, y) = \ln \sqrt{x^2 + y^2}$, $(x, y) \neq (0, 0)$.

24 Show that $\boldsymbol{\nabla}(|\boldsymbol{x}|) = \dfrac{1}{|\boldsymbol{x}|}\boldsymbol{x}$, if $\boldsymbol{x} \neq \boldsymbol{0}$.

25 From Exercise 24, conclude without further computing what $\mathbf{curl}\left(\dfrac{1}{|\boldsymbol{x}|}\boldsymbol{x}\right)$ is.

Exercises 26 and 27 make use of Exercise 28 in Exercises 7.2, which showed that Green's Theorem can be put in the form

(9) $$\iint_D \mathrm{div}\, \boldsymbol{F}\, dA = \oint_\gamma \boldsymbol{F} \cdot \boldsymbol{N}\, ds,$$

where $\boldsymbol{N}(\boldsymbol{x})$ is a unit vector orthogonal to $\boldsymbol{F}(\boldsymbol{x})$.

26 Prove *Green's first identity*, an analogue of integration by parts. If $f: \boldsymbol{R}^2 \to \boldsymbol{R}$ is differentiable on a simple region D and all second partial derivatives of $g: \boldsymbol{R}^2 \to \boldsymbol{R}$ are continuous on D, then

(10) $$\iint_D f(\boldsymbol{x})\, \nabla^2 g(\boldsymbol{x})\, dA = \int_{\partial D} f(\boldsymbol{x})(\boldsymbol{\nabla} g \cdot \boldsymbol{N})\, ds - \iint_D \boldsymbol{\nabla} f \cdot \boldsymbol{\nabla} g\, dA$$

$$= \int_{\partial D} f(\boldsymbol{x}) \frac{\partial g}{\partial \boldsymbol{N}}\, ds - \iint_D \boldsymbol{\nabla} f \cdot \boldsymbol{\nabla} g\, dA.$$

27 Obtain *Green's second identity* from Exercise 26. Assume that all second partial derivatives of f and g are continuous on D. Then

(11) $$\iint_D [f(\boldsymbol{x})\, \nabla^2 g(\boldsymbol{x}) - g(\boldsymbol{x})\, \nabla^2 f(\boldsymbol{x})]\, dA = \int_{\partial D} f(\boldsymbol{x})(\boldsymbol{\nabla} g \cdot \boldsymbol{N})\, ds - \int_{\partial D} g(\boldsymbol{x})(\boldsymbol{\nabla} f \cdot \boldsymbol{N})\, ds$$

$$= \int_{\partial D} \left(f \frac{\partial g}{\partial \boldsymbol{N}} - g \frac{\partial f}{\partial \boldsymbol{N}}\right) ds.$$

What can you conclude if f and g are harmonic?

28 A fluid with velocity field $\boldsymbol{F}$ is called **incompressible** if div $\boldsymbol{F}(\boldsymbol{x}) = 0$. A fluid flow is called **solenoidal** if $\boldsymbol{F}(x) = \mathbf{curl}\, \boldsymbol{G}(\boldsymbol{x})$ for some $\boldsymbol{G}$. What relationship can you conclude exists between incompressible and solenoidal flows? (Assume appropriate differentiability hypotheses.)

In Exercises 29 to 31, refer to Exercises 27 and 28 of Exercises 7.2.

29 Suppose that a flowing fluid in the plane $\boldsymbol{R}^2$ has density $\delta(x, y, t)$. Show that $\oint_\gamma \delta(x, y, t)\boldsymbol{F}\cdot\boldsymbol{N}\,ds$ is the rate of loss of mass per unit time from the disk D enclosed by a circle γ of radius a centered at a point $\boldsymbol{x}_0 \in \boldsymbol{R}^2$.

30 Use Exercise 29 to show that

$$\lim_{a\to 0}\frac{1}{\pi a^2}\oint_\gamma \delta(x, y, t)\boldsymbol{F}\cdot\boldsymbol{N}\,ds$$

gives the rate of decrease of the density per unit time per unit area at $\boldsymbol{x}_0$. $\left(\text{This is then } -\dfrac{\partial\delta}{\partial t}.\right)$ Show that this limit is the divergence of the function $\delta\boldsymbol{F}$ at $\boldsymbol{x}_0$, and hence obtain the two-dimensional *continuity equation of hydrodynamics,*

$$\frac{\partial\delta}{\partial t}+\operatorname{div}(\delta\boldsymbol{F})=0.$$

If $\delta(x, y, t)$ is a constant, then conclude that the fluid is incompressible. [*Hint:* Use the Mean Value Theorem (Exercise 20, Exercises 5.1) for double integrals.]

We make the terminology "curl" even more compelling in the next two exercises, which are set in the xy-plane.

31 Suppose that γ is a circle enclosing the disk D with center $\boldsymbol{x}_0$ and radius a. Then use the Mean Value Theorem for double integrals cited in Exercise 30 to show that

$$\oint_\gamma \boldsymbol{F}\cdot d\boldsymbol{x}=(\textbf{curl }\boldsymbol{F}(x_1, y_1)\cdot\boldsymbol{k})A(D)$$

for some $(x_1, y_1)\in D$.

32 Conclude from Exercise 31 that the z-coordinate of **curl** $\boldsymbol{F}$ is the limit of the circulation per unit area at $\boldsymbol{x}_0$ if **curl** $\boldsymbol{F}$ is continuous on an open set containing $\boldsymbol{x}_0$.

33 Use Exercises 31 and 32 to compute **curl** $\boldsymbol{F}(\boldsymbol{0})$ if $\boldsymbol{F}(x, y)=-y\boldsymbol{i}+x\boldsymbol{j}$. (Find the limit of $1/\pi a^2$ times the circulation around the circle γ of radius a and center $\boldsymbol{0}$.)

34 A field $\boldsymbol{F}:\boldsymbol{R}^3\to\boldsymbol{R}^3$ is called **central** if $\boldsymbol{F}(\boldsymbol{x})=f(|\boldsymbol{x}|)\boldsymbol{x}$ for some $f:\boldsymbol{R}\to\boldsymbol{R}$. Show that at all points $\boldsymbol{x}_0\neq\boldsymbol{0}$, the curl of a central field is $\boldsymbol{0}$.

35 (Laplacian in polar coordinates) Suppose that $z=f(x, y)$ and we introduce polar coordinates $x=r\cos\theta$ and $y=r\sin\theta$. Then $z=f(r\cos\theta, r\sin\theta)=g(r,\theta)$. Show that

$$\nabla^2 z=\frac{\partial^2 g}{\partial r^2}+\frac{1}{r^2}\frac{\partial^2 g}{\partial\theta^2}+\frac{1}{r}\frac{\partial g}{\partial r}.$$

36 (Laplacian in cylindrical coordinates) Suppose that $w=f(x, y, z)$ and we introduce cylindrical coordinates $x=r\cos\theta$, $y=r\sin\theta$, and $z=z$. Then $w=f(r\cos\theta, r\sin\theta, z)=g(r,\theta,z)$. Show that

$$\nabla^2 f(x, y, z)=\frac{1}{r}\frac{\partial g}{\partial r}+\frac{\partial^2 g}{\partial r^2}+\frac{1}{r^2}\frac{\partial^2 g}{\partial\theta^2}+\frac{\partial^2 g}{\partial z^2}.$$

37 (Laplacian in spherical coordinates) Suppose that $w=f(x, y, z)$ and we introduce spherical coordinates $x=\rho \sin \phi \cos \theta$, $y=\rho \sin \phi \sin \theta$, and $z=\rho \cos \phi$. Then

$$w=f(\rho \sin \phi \cos \theta, \rho \sin \phi \sin \theta, \rho \cos \phi)=g(\rho, \phi, \theta).$$

Show that

$$\nabla^2 w=\frac{\partial^2 g}{\partial \rho^2}+\frac{2}{\rho}\frac{\partial g}{\partial \rho}+\frac{1}{\rho^2}\frac{\partial^2 g}{\partial \phi^2}+\frac{\cos \phi}{\rho^2 \sin \phi}\frac{\partial g}{\partial \phi}+\frac{1}{\rho^2 \sin \phi}\frac{\partial^2 g}{\partial \theta^2}.$$

4 INDEPENDENCE OF PATH

This section is about a very pleasing analogue of the Fundamental Theorem of Integral Calculus and its ramifications. If $f:\mathbf{R}^n \to \mathbf{R}$ is differentiable at $\mathbf{x}_0 \in \mathbf{R}^n$, then we know that its total derivative is $\nabla f(\mathbf{x}_0)$. Now, we can regard $\nabla f:\mathbf{R}^n \to \mathbf{R}^n$ as a vector field whose value at $\mathbf{x}_0$ is $\nabla f(\mathbf{x}_0)$. Suppose that γ is a piecewise smooth path in $\mathbf{R}^n$ from $\mathbf{x}_1$ to $\mathbf{x}_2$. We can form the line integral

$$\int_\gamma \nabla f \cdot d\mathbf{x},$$

in analogy to the definite integral $\int_a^b f'(x)\,dx$. Thus we might hope that the value of this line integral would be $f(\mathbf{x}_2)-f(\mathbf{x}_1)$, the difference between the values of f at the end and the beginning of γ. This hope is confirmed by the following fundamental theorem on evaluation of line integrals.

4.1 **THEOREM.** Suppose that $f:\mathbf{R}^n \to \mathbf{R}$ is differentiable, with continuous partial derivatives on a region D containing $\mathbf{x}_1$ and $\mathbf{x}_2$. Then for any piecewise smooth path γ in D from $\mathbf{x}_1$ to $\mathbf{x}_2$,

$$\int_\gamma \nabla f \cdot d\mathbf{x}=f(\mathbf{x}_2)-f(\mathbf{x}_1).$$

Proof. Since γ is a union of smooth paths, the line integral over γ is (by Definition 1.5) the sum of the line integrals over the smooth paths. So it is enough to show that the result holds on a smooth path. Suppose, then, that we parametrize γ by $\mathbf{x}=\mathbf{x}(t)$, $t_1 \le t \le t_2$. Thus $\mathbf{x}_1=\mathbf{x}(t_1)$ and $\mathbf{x}_2=\mathbf{x}(t_2)$, and we have

$$\begin{aligned}\int_\gamma \nabla f \cdot d\mathbf{x} &= \int_{t_1}^{t_2} \nabla f(\mathbf{x}) \cdot \dot{\mathbf{x}}(t)\,dt \\ &= \int_{t_1}^{t_2} \frac{df(\mathbf{x}(t))}{dt}\,dt \quad \text{by Chain Rule 4.5.3} \\ &= f(\mathbf{x}(t))\Big]_{t_1}^{t_2} \quad \text{by the Fundamental Theorem of Integral Calculus} \\ &= f(\mathbf{x}(t_2))-f(\mathbf{x}(t_1)) \\ &= f(\mathbf{x}_2)-f(\mathbf{x}_1).\end{aligned}$$

QED

4.2 **EXAMPLE.** If

$$f(x, y)=\frac{x}{\sqrt{x^2+y^2}},$$

then compute $\int_\gamma \nabla f \cdot d\boldsymbol{x}$, where γ is the arc of the circle $x^2+y^2=9$ joining $\boldsymbol{x}_1=(3,0)$ to $\boldsymbol{x}_2=(-1,2\sqrt{2})$ in the counterclockwise sense (Figure 4.1).

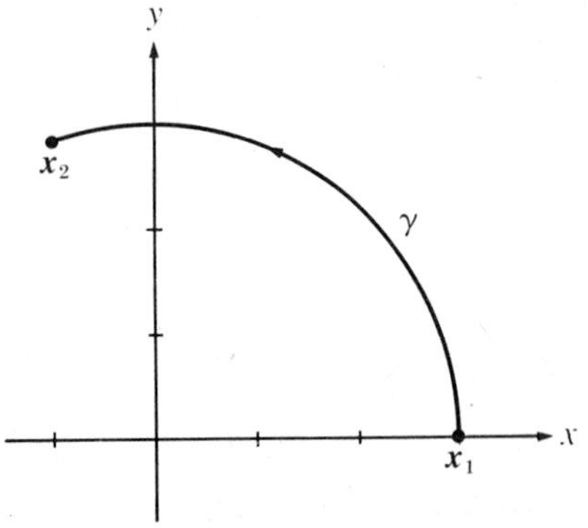

FIGURE 4.1

Solution. We don't even have to compute ∇f or parametrize γ! We simply use Theorem 4.1, and have

$$\begin{aligned}\int_\gamma \nabla f \cdot d\boldsymbol{x} &= f(-1, 2\sqrt{2}) - f(3,0) \\ &= \frac{-1}{3} - \frac{3}{3} \\ &= -\frac{4}{3}.\end{aligned}$$

Theorem 4.1 is thus already seen to be a very useful result in reducing labor. We can see that the line integral of ∇f over a path γ depends only on the initial and terminal points of γ, and *not* on the nature of γ as long as it is at least piecewise smooth. On *every* piecewise smooth path from $\boldsymbol{x}_2$ to $\boldsymbol{x}_1$ we get $\int_\gamma \nabla f \cdot d\boldsymbol{x} = f(\boldsymbol{x}_2) - f(\boldsymbol{x}_1)$. We say then that the line integral of ∇f between two points $\boldsymbol{x}_1$ and $\boldsymbol{x}_2$ in D is **independent of path**. We can illustrate something of the impact of this by considering the following two results.

4.3 **THEOREM.** The line integral $\int_\gamma \boldsymbol{F} \cdot d\boldsymbol{x}$ is independent of path in D if and only if $\int_\gamma \boldsymbol{F} \cdot d\boldsymbol{x} = 0$ for every piecewise smooth closed curve γ in D.

Proof. If the line integral is independent of path, then choose any piecewise smooth closed curve γ in D. If $\boldsymbol{x}_1$ and $\boldsymbol{x}_2$ are any two points on γ, then γ is decomposed into two piecewise smooth subpaths γ_1 from $\boldsymbol{x}_1$ to $\boldsymbol{x}_2$ and γ_2 from $\boldsymbol{x}_2$

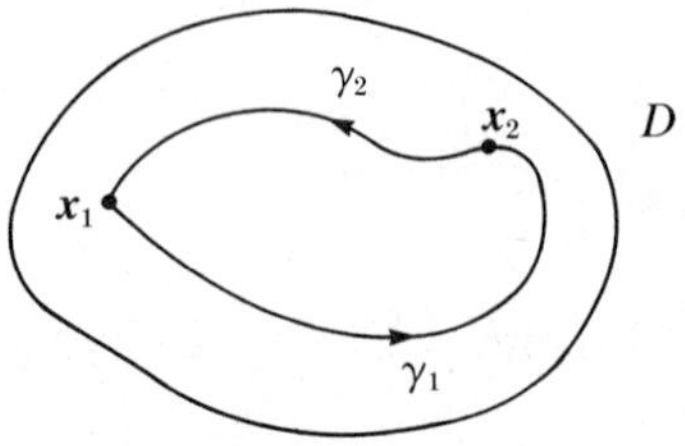

FIGURE 4.2

to $\boldsymbol{x}_1$ in opposite senses. See Figure 4.2. Therefore, γ_1 and $-\gamma_2$ are two paths connecting $\boldsymbol{x}_1$ to $\boldsymbol{x}_2$ in the same sense. So

$$\int_{\gamma_1} \boldsymbol{F} \cdot d\boldsymbol{x} = \int_{-\gamma_2} \boldsymbol{F} \cdot d\boldsymbol{x}.$$

Thus,

$$\begin{aligned}\int_{\gamma} \boldsymbol{F} \cdot d\boldsymbol{x} &= \int_{\gamma_1} \boldsymbol{F} \cdot d\boldsymbol{x} + \int_{\gamma_2} \boldsymbol{F} \cdot d\boldsymbol{x} \quad \text{by Definition 1.5,} \\ &= \int_{\gamma_1} \boldsymbol{F} \cdot d\boldsymbol{x} - \int_{-\gamma_2} \boldsymbol{F} \cdot d\boldsymbol{x} \quad \text{by Theorem 1.7,} \\ &= 0.\end{aligned}$$

Conversely, if $\int_{\gamma} \boldsymbol{F} \cdot d\boldsymbol{x} = 0$ for any piecewise smooth closed curve γ in D, then we claim that $\int_{\gamma_1} \boldsymbol{F} \cdot d\boldsymbol{x} = \int_{\gamma_2} \boldsymbol{F} \cdot d\boldsymbol{x}$ for any piecewise smooth curves γ_1 and γ_2 joining $\boldsymbol{x}_1$ to $\boldsymbol{x}_2$ in the same sense. For $\gamma = \gamma_1 \cup (-\gamma_2)$ is a piecewise smooth

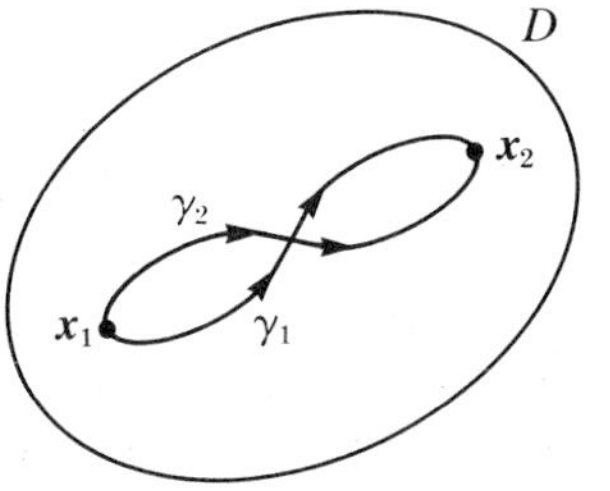

FIGURE 4.3

closed path. See Figure 4.3. Then by hypothesis, $\int_{\gamma} \boldsymbol{F} \cdot d\boldsymbol{x} = 0$. But again by Theorem 1.7, we have

$$0 = \int_{\gamma} \boldsymbol{F} \cdot d\boldsymbol{x} = \int_{\gamma_1} \boldsymbol{F} \cdot d\boldsymbol{x} - \int_{\gamma_2} \boldsymbol{F} \cdot d\boldsymbol{x},$$

so

$$\int_{\gamma_1} \boldsymbol{F} \cdot d\boldsymbol{x} = \int_{\gamma_2} \boldsymbol{F} \cdot d\boldsymbol{x}.$$

QED

We can now justify the terminology "conservative vector field" introduced in Definition 3.6. Recall that $\boldsymbol{F}: \boldsymbol{R}^n \to \boldsymbol{R}^n$ is called *conservative* if $\boldsymbol{F}(\boldsymbol{x}) = \nabla f(\boldsymbol{x})$ for some $f: \boldsymbol{R}^n \to \boldsymbol{R}$.

4.4 EXAMPLE. Conservation of Energy. Suppose that $\boldsymbol{F}: \boldsymbol{R}^n \to \boldsymbol{R}^n$ is a conservative vector field on D, and that $\boldsymbol{F}(\boldsymbol{x}) = \nabla f(\boldsymbol{x})$ for some $f: \boldsymbol{R}^n \to \boldsymbol{R}$. Then

$$p(\boldsymbol{x}) = -f(\boldsymbol{x})$$

is called the **potential energy** of the field. Show that if $\boldsymbol{F}$ moves a particle from $\boldsymbol{x}_1$ in D to $\boldsymbol{x}_2$ in D along any piecewise smooth path γ in D, parametrized by $\boldsymbol{x} = \boldsymbol{x}(t)$, then the *total energy* $T(\boldsymbol{x})$ remains the same, where

$$T(\boldsymbol{x}) = p(\boldsymbol{x}) + \tfrac{1}{2}m\,|\boldsymbol{v}(t)|^2$$

is the sum of the potential energy and the kinetic energy.

Solution. We first recall that the work done in moving the particle from $x_1 = x(t_1)$ to $x_2 = x(t_2)$ is the change

$$\tfrac{1}{2}m\,|v(t_2)|^2 - \tfrac{1}{2}m\,|v(t_1)|^2$$

in kinetic energy, by Exercise 24 of Exercises 7.1. On the other hand, we can use Theorem 4.1 to compute W:

$$\begin{aligned} W &= \int_\gamma F \cdot dx = \int_\gamma \nabla f \cdot dx = -\int_\gamma \nabla p \cdot dx \\ &= -[p(x_2) - p(x_1)] = p(x_1) - p(x_2). \end{aligned}$$

Equating the two expressions for W, we obtain

$$\tfrac{1}{2}m\,|v(t_2)|^2 + p(x_2) = \tfrac{1}{2}m\,|v(t_1)|^2 + p(x_1).$$

Thus,

$$T(x_2) = T(x_1).$$

QED

This result is called the *Law of Conservation of Energy*, and it is of fundamental importance in physics. You might be wondering if the line integral of some function $F: R^n \to R^n$ can be independent of path in a region D and arise in some way *other* than as in Theorem 4.1. That is, could F be something other than the gradient of a potential function? The answer afforded by the next theorem is "no." The *only* functions that have line integrals independent of path in D are conservative vector fields, as long as D is a reasonably uncomplicated region. The technical name for the sort of region D specified in Theorem 4.5 is *arcwise connected.* If you study topology, then you will consider this concept in more detail.

4.5 **THEOREM.** Let $D \subseteq R^n$ be open, with the property that any two points x_1 and x_2 in D can be joined by a piecewise smooth curve wholly in D. If $F: R^n \to R^n$ is continuous and such that its line integral is independent of path in D, then F is a conservative vector field. In fact,

$$F = \nabla f, \quad \text{for} \quad f(x) = \int_{x_1}^{x} F \cdot dx,$$

where the line integral is taken over any piecewise smooth path γ in D joining an arbitrarily chosen but fixed point x_1 to x.

Proof. We have to show for each $x_0 \in D$ that $\nabla f(x_0) = F(x_0)$ if $f(x) = \int_{x_1}^{x} F \cdot dx$. Since D is open, we know that there exists $B(x_0, \delta) \subseteq D$ for some $\delta > 0$. Then if $|h| < \delta$, we have $x_0 + he_i \in D$ where $e_i = (0, \ldots, 0, 1, 0, \ldots, 0)$ is the i-th standard basis vector for R^n. We can join x_1 to x_0 in D by a piecewise smooth path and

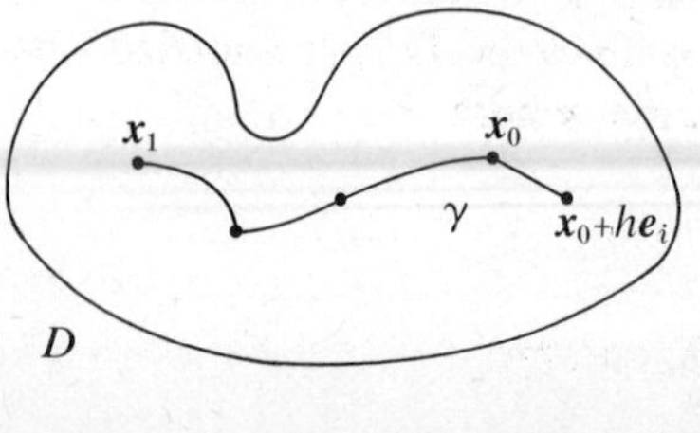

FIGURE 4.4

then join $\boldsymbol{x}_0$ to $\boldsymbol{x}_0 + h\boldsymbol{e}_i$ along the line segment $\boldsymbol{x} = \boldsymbol{x}_0 + t\boldsymbol{e}_i$, with t varying between 0 and h. See Figure 4.4, where the entire path from $\boldsymbol{x}_1$ to $\boldsymbol{x}_0 + h\boldsymbol{e}_i$ is labeled γ.

Now, by the independence of path, we have

$$\begin{aligned} f(\boldsymbol{x}_0 + h\boldsymbol{e}_i) - f(\boldsymbol{x}_0) &= \int_{\boldsymbol{x}_1}^{\boldsymbol{x}_0 + h\boldsymbol{e}_i} \boldsymbol{F} \cdot d\boldsymbol{x} - \int_{\boldsymbol{x}_1}^{\boldsymbol{x}_0} \boldsymbol{F} \cdot d\boldsymbol{x} \\ &= \int_{\boldsymbol{x}_1}^{\boldsymbol{x}_0 + h\boldsymbol{e}_i} \boldsymbol{F} \cdot d\boldsymbol{x} + \int_{\boldsymbol{x}_0}^{\boldsymbol{x}_1} \boldsymbol{F} \cdot d\boldsymbol{x} \\ &= \int_{\boldsymbol{x}_0}^{\boldsymbol{x}_0 + h\boldsymbol{e}_i} \boldsymbol{F} \cdot d\boldsymbol{x} \\ &= \int_0^h F(\boldsymbol{x}_0 + t\boldsymbol{e}_i) \cdot \boldsymbol{e}_i \, dt \end{aligned}$$

since on the path $\boldsymbol{x} = \boldsymbol{x}_0 + t\boldsymbol{e}_i$ we have $\dot{\boldsymbol{x}}(t) = \boldsymbol{e}_i$. Now we can compute

$$\begin{aligned} \frac{\partial f}{\partial x_i}(\boldsymbol{x}_0) &= \lim_{h \to 0} \frac{f(\boldsymbol{x}_0 + h\boldsymbol{e}_i) - f(\boldsymbol{x}_0)}{h} \\ &= \lim_{h \to 0} \frac{1}{h} \int_0^h \boldsymbol{F}(\boldsymbol{x}_0 + t\boldsymbol{e}_i) \cdot \boldsymbol{e}_i \, dt. \end{aligned}$$

By the Mean Value Theorem for definite integrals,

$$\begin{aligned} \int_0^h \boldsymbol{F}(\boldsymbol{x}_0 + t\boldsymbol{e}_i) \cdot \boldsymbol{e}_i \, dt &= \boldsymbol{F}(\boldsymbol{x}_0 + \bar{h}\boldsymbol{e}_i) \cdot \boldsymbol{e}_i (h - 0) \\ &= h\boldsymbol{F}(\boldsymbol{x}_0 + \bar{h}\boldsymbol{e}_i) \cdot \boldsymbol{e}_i \quad \text{for some } \bar{h} \text{ between 0 and } h. \end{aligned}$$

Thus,

$$\begin{aligned} \frac{\partial f}{\partial x_i}(\boldsymbol{x}_0) &= \lim_{h \to 0} \frac{1}{h} \cdot h\boldsymbol{F}(\boldsymbol{x}_0 + \bar{h}\boldsymbol{e}_i) \cdot \boldsymbol{e}_i \\ &= \boldsymbol{F}(\boldsymbol{x}_0) \cdot \boldsymbol{e}_i, \quad \text{since } \boldsymbol{F} \text{ is continuous and as } h \to 0, \text{ so does } \bar{h} \\ &= f_i(\boldsymbol{x}_0) \end{aligned}$$

where f_i is the i-th coordinate function of $\boldsymbol{F}$. Therefore,

$$\begin{aligned} \boldsymbol{F}(\boldsymbol{x}_0) &= (f_1(\boldsymbol{x}_0), \ldots, f_n(\boldsymbol{x}_0)) = \left(\frac{\partial f}{\partial x_1}(\boldsymbol{x}_0), \ldots, \frac{\partial f}{\partial x_n}(\boldsymbol{x}_0)\right) \\ &= \boldsymbol{\nabla} f(\boldsymbol{x}_0). \end{aligned}$$

QED

Let's see what we have established so far. Theorems 4.1 and 4.5 say that $\boldsymbol{F}$ is a conservative vector field on an open arcwise connected region D if and only if $\int_\gamma \boldsymbol{F} \cdot d\boldsymbol{x}$ is independent of path in D. This is in turn equivalent, by Theorem 4.3, to $\int_\gamma \boldsymbol{F} \cdot d\boldsymbol{x} = 0$ for any piecewise smooth closed curve γ in D.

A still unanswered question is the following. Given a vector field $\boldsymbol{F}: \boldsymbol{R}^n \to \boldsymbol{R}^n$, *how* can we determine whether $\boldsymbol{F}$ is conservative, i.e., whether $\int_\gamma \boldsymbol{F} \cdot d\boldsymbol{x}$ is independent of path in reasonable domains D? Consider the case $n = 3$, for example. We know that $\boldsymbol{F}$ will be conservative if and only if $\boldsymbol{F} = \boldsymbol{\nabla} p$ for some

$p: \boldsymbol{R}^3 \to \boldsymbol{R}$. If this holds, then we have

$$\boldsymbol{F}(\boldsymbol{x}) = \left(\frac{\partial p}{\partial x}, \frac{\partial p}{\partial y}, \frac{\partial p}{\partial z}\right) = (f_1(\boldsymbol{x}), f_2(\boldsymbol{x}), f_3(\boldsymbol{x})).$$

Supposing that $\boldsymbol{F}$ is differentiable, we get

(1) $$J_{\boldsymbol{F}} = \begin{pmatrix} p_{xx} & p_{xy} & p_{xz} \\ p_{yx} & p_{yy} & p_{yz} \\ p_{zx} & p_{zy} & p_{zz} \end{pmatrix} = \begin{pmatrix} \frac{\partial f_1}{\partial x} & \frac{\partial f_1}{\partial y} & \frac{\partial f_1}{\partial z} \\ \frac{\partial f_2}{\partial x} & \frac{\partial f_2}{\partial y} & \frac{\partial f_2}{\partial z} \\ \frac{\partial f_3}{\partial x} & \frac{\partial f_3}{\partial y} & \frac{\partial f_3}{\partial z} \end{pmatrix}$$

Recall that if the entries p_{ij} of $J_{\boldsymbol{F}}$ are continuous, then by Theorem 4.7.5 $J_{\boldsymbol{F}}$ will be a *symmetric* matrix, i.e.,

$$\frac{\partial f_1}{\partial y} = \frac{\partial f_2}{\partial x}, \qquad \frac{\partial f_1}{\partial z} = \frac{\partial f_3}{\partial x}, \quad \text{and} \quad \frac{\partial f_2}{\partial z} = \frac{\partial f_3}{\partial y}.$$

This extends easily.

4.6 **THEOREM.** If $\boldsymbol{F}(\boldsymbol{x}) = (f_1(\boldsymbol{x}), f_2(\boldsymbol{x}), \ldots, f_n(\boldsymbol{x}))$ is conservative and $f_1, f_2, \ldots, f_n$ have continuous partial derivatives on a set $D \subseteq \boldsymbol{R}^n$, then $J_{\boldsymbol{F}}(\boldsymbol{x})$ is a symmetric matrix, i.e.,

$$\frac{\partial f_i}{\partial x_j} = \frac{\partial f_j}{\partial x_i} \quad \text{for} \quad i, j = 1, 2, \ldots, n.$$

Proof. Here, since

$$\boldsymbol{F}(\boldsymbol{x}) = \nabla p(\boldsymbol{x}) = \left(\frac{\partial p}{\partial x_1}, \frac{\partial p}{\partial x_2}, \ldots, \frac{\partial p}{\partial x_n}\right)$$

for some $p: \boldsymbol{R}^n \to \boldsymbol{R}$, we have

$$f_i = \frac{\partial p}{\partial x_i} \quad \text{and} \quad f_j = \frac{\partial p}{\partial x_j}.$$

Then

$$\frac{\partial f_i}{\partial x_j} = \frac{\partial^2 p}{\partial x_j\, \partial x_i} = \frac{\partial^2 p}{\partial x_i\, \partial x_j} = \frac{\partial f_j}{\partial x_i}$$

by Theorem 4.7.5. QED

The condition that $J_{\boldsymbol{F}}(\boldsymbol{x})$ is symmetric is thus a *necessary* condition for $\boldsymbol{F}$ to be a conservative vector field. However, this is not a *sufficient* condition for $\boldsymbol{F}$ to be conservative on an arcwise connected set $D \subseteq \boldsymbol{R}^n$. This fact is illustrated by our next example.

4.7 EXAMPLE. Show that the vector field

$$\boldsymbol{F}(x, y)=\left(-\frac{y}{x^2+y^2}, \frac{x}{x^2+y^2}\right)$$

has a symmetric Jacobian matrix and yet is not a conservative vector field on the annular region D lying between the two circles $x^2+y^2=1$ and $x^2+y^2=16$ (see Figure 4.5).

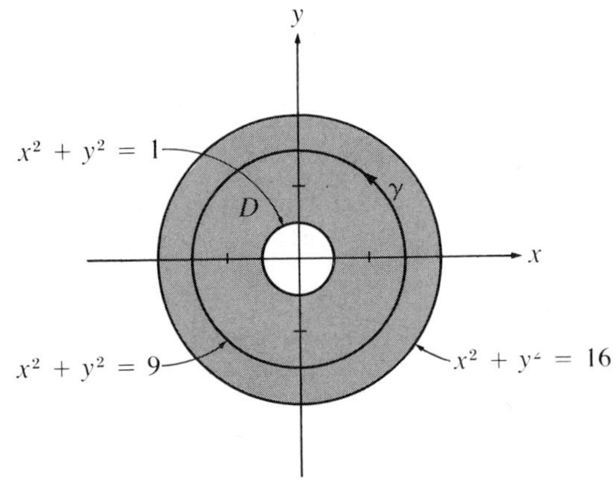

FIGURE 4.5

Solution. Here

$$f_1(\boldsymbol{x})=-\frac{y}{x^2+y^2} \quad \text{and} \quad f_2(\boldsymbol{x})=\frac{x}{x^2+y^2}.$$

A routine calculation gives us

$$\frac{\partial f_1}{\partial y}=\frac{\partial f_2}{\partial x}=\frac{y^2-x^2}{(x^2+y^2)^2}$$

on D. Thus,

$$J_{\boldsymbol{F}}=\begin{pmatrix}\partial f_1/\partial x & \partial f_1/\partial y\\ \partial f_2/\partial x & \partial f_2/\partial y\end{pmatrix}$$

is symmetric on D. But Example 2.7 showed that $\int_\gamma \boldsymbol{F}\cdot d\boldsymbol{x}=2\pi\neq 0$ where γ is the circle $x^2+y^2=9$ traversed in the counterclockwise sense. So, by Theorem 4.3, $\int_\gamma \boldsymbol{F}\cdot d\boldsymbol{x}$ is *not* independent of path on D. Hence, by Theorem 4.1, $\boldsymbol{F}$ is not a conservative vector field.

It was found that for *some* regions D, the fact that $J_{\boldsymbol{F}}(\boldsymbol{x})$ is symmetric *was* enough to guarantee that $\boldsymbol{F}$ would be conservative on D. One of the major problems in the nineteenth century was to characterize the *most general* such region. (Note that arcwise connected regions are *too* general, since D in Example 4.7 is arcwise connected.) It was eventually discovered that the most general sets D for which a symmetric $J_{\boldsymbol{F}}(\boldsymbol{x})$ is enough to guarantee $\boldsymbol{F}$ conservative are **simply connected** regions. We will use this term for regions D such that every planar simple closed curve γ in D bounds a planar surface S completely enclosed in D.[1]

1. The term "simply connected" is usually applied to a somewhat wider class of regions in advanced calculus and topology, but we do not have the tools to treat the more general notion.

For example, rectangles and disks in $\mathbf{R}^2$ and balls, the interiors of spheres, cubes, tetrahedra, ellipsoids, and similar figures in $\mathbf{R}^3$ are simply connected. The region D of Example 4.7 is *not* simply connected because the simple closed curve γ shown in Figure 4.5 encloses a disk S that is *not* completely contained in D. Using Green's Theorem, we can now prove the following partial converse of Theorem 4.6 for dimension 2. (This result also holds in three dimensions, and we will use it, deferring the proof to Section 7, Theorem 7.4. Higher dimensions must wait for advanced calculus.)

4.8 **THEOREM.** Suppose that for $\boldsymbol{F}:\mathbf{R}^2 \to \mathbf{R}^2$, $J_{\boldsymbol{F}}(\boldsymbol{x})$ has continuous entries and is symmetric in a simply connected region $D \subseteq \mathbf{R}^2$. Then $\boldsymbol{F}$ is a conservative vector field on D.

Proof. Let γ be any simple closed curve in D. We will show that $\int_\gamma \boldsymbol{F} \cdot d\boldsymbol{x} = 0$. Then the result will follow from Theorem 4.3. Writing $\boldsymbol{F}(\boldsymbol{x}) = (P(\boldsymbol{x}), Q(\boldsymbol{x}))$, we have from Green's Theorem,

(2) $$\int_\gamma \boldsymbol{F} \cdot d\boldsymbol{x} = \iint_S \left(\frac{\partial Q}{\partial x} - \frac{\partial P}{\partial y}\right) dA$$

where S is the region enclosed by γ. ($S \subseteq D$ since D is simply connected.) But since

$$J_{\boldsymbol{F}}(\boldsymbol{x}) = \begin{pmatrix} \partial P/\partial x & \partial P/\partial y \\ \partial Q/\partial x & \partial Q/\partial y \end{pmatrix}$$

is symmetric, $\dfrac{\partial P}{\partial y} = \dfrac{\partial Q}{\partial x}$. Thus from (2)

$$\int_\gamma \boldsymbol{F} \cdot d\boldsymbol{x} = 0.$$

QED

All right, suppose that we confine ourselves to nice regions D that are simply connected. If we have $J_{\boldsymbol{F}}(\boldsymbol{x})$ symmetric, then we *know* that $\boldsymbol{F}:\mathbf{R}^n \to \mathbf{R}^n$ is a conservative vector field (for $n = 2$ and 3). So $\boldsymbol{F} = \nabla p$ for some $p:\mathbf{R}^n \to \mathbf{R}$. How can we find p? We simply solve the partial differential equations

(3) $$\frac{\partial p}{\partial x} = f_1, \quad \frac{\partial p}{\partial y} = f_2, \quad \text{and, if present} \quad \frac{\partial p}{\partial z} = f_3.$$

We *partially integrate* (3). That is, we freeze y and z in $\dfrac{\partial p}{\partial x} = f_1$ and integrate with respect to x. Similarly, we integrate $\dfrac{\partial p}{\partial y} = f_2$ and $\dfrac{\partial p}{\partial z} = f_3$ with respect to y and z, respectively. We will obtain

(4) $$p(x, y, z) = \int f_1(x, y, z)\, dx + c_1(y, z),$$

(5) $$p(x, y, z) = \int f_2(x, y, z)\, dy + c_2(x, z),$$

(6) $$p(x, y, z) = \int f_3(x, y, z)\, dz + c_3(x, y),$$

where $c_1(y, z)$, $c_2(x, z)$, and $c_3(x, y)$ are "constants" of integration, i.e., functions not involving the variable with respect to which we integrated. We have to determine c_1, c_2, and c_3 so that (4), (5), and (6) coincide. This we can usually do by inspection, as the following example illustrates.

4.9 **EXAMPLE.** Determine whether $\boldsymbol{F}(x, y, z)=(2xy, x^2+z^2, 2yz)$ is conservative on the domain D enclosed by the ellipsoid $x^2+2y^2+3z^2=27$. If it is, find p so that $\boldsymbol{F}=\nabla p$ on D.

Solution. D is simply connected. We have

$$J_{\boldsymbol{F}}(\boldsymbol{x})=\begin{pmatrix} 2y & 2x & 0 \\ 2x & 0 & 2z \\ 0 & 2z & 2y \end{pmatrix}$$

Since $J_{\boldsymbol{F}}(\boldsymbol{x})$ is symmetric, we know that $\boldsymbol{F}=\nabla p=\left(\frac{\partial p}{\partial x}, \frac{\partial p}{\partial y}, \frac{\partial p}{\partial z}\right)$ for some $p:\boldsymbol{R}^3 \to \boldsymbol{R}$. We have

(7) $$\frac{\partial p}{\partial x}=2xy, \qquad \frac{\partial p}{\partial y}=x^2+z^2, \qquad \frac{\partial p}{\partial z}=2yz.$$

Integration of (7) with respect to x, y, and z gives

(8) $$p(x, y, z)=x^2y+c_1(y, z)=x^2y+z^2y+c_2(x, z)=yz^2+c_3(x, y).$$

Now, for the equalities in (8) to hold simultaneously, we can take $c_1(y, z)=yz^2$ and $c_3(x, y)=x^2y$. We can also take $c_2(x, z)=0$, since no term involving x and z turned up in our integrations in (8). Thus,

$$p(x, y, z)=x^2y+z^2y$$

will do. (As a check, a simple calculation verifies $\nabla p=\boldsymbol{F}$.)

It turns out that first checking $J_{\boldsymbol{F}}(\boldsymbol{x})$ for symmetry can actually be omitted. We can simply try to find a potential function p by integrating (3). If $\boldsymbol{F}$ is *not* a conservative vector field, then it will be impossible to reconcile (4), (5), and (6) as we did in Example 4.9. The following example illustrates that situation.

4.10 **EXAMPLE.** Determine whether $\boldsymbol{F}(x, y)=(x^2y, x^2-y^2)$ is a conservative vector field on the unit disk enclosed by $x^2+y^2=1$.

Solution. Let's set $\boldsymbol{F}=\nabla p$ and try to find p. We have

(9) $$f_1(x, y)=x^2y=\frac{\partial p}{\partial x}, \quad \text{and} \qquad f_2(x, y)=x^2-y^2=\frac{\partial p}{\partial y}.$$

Integrating (9), we come to

(10) $$p(x, y)=\frac{x^3y}{3}+c_1(y) \qquad p(x, y)=x^2y-\frac{y^3}{3}+c_2(x).$$

Now, it is clear that no function p can be simultaneously of the two forms in (10), for the first form is $\frac{x^3y}{3}$ plus a function of y alone. That can *never* equal the second form, $x^2y-\frac{y^3}{3}$ plus a function of x alone! So $\boldsymbol{F}$ can't be a conservative vector field.

Of course, we *could* have given a quicker solution by computing

$$J_{\boldsymbol{F}}(\boldsymbol{x})=\begin{pmatrix}2xy & x^2\\ 2x & -2y\end{pmatrix}$$

and noting that it is not symmetric. Thus $\boldsymbol{F}$ can't be a conservative vector field by Theorem 4.6. In practice, computing $J_{\boldsymbol{F}}(\boldsymbol{x})$ will save work when $\boldsymbol{F}$ is *not* conservative, and not computing $J_{\boldsymbol{F}}(\boldsymbol{x})$ will save work when $\boldsymbol{F}$ *is* conservative! So it's pretty much up to you (by guesswork, coin flipping, or whatever) whether to check $J_{\boldsymbol{F}}(\boldsymbol{x})$ for symmetry.

EXERCISES 7.4

In Exercises 1 to 10, compute the given line integrals by finding a potential function for $\boldsymbol{F}$ and using Theorem 4.1.

1 $\int_\gamma 2xy\,dx+(x^2+3y)\,dy$, where γ is the polygonal path from $(0,0)$ to $(0,1)$ to $(1,1)$.

2 $\int_\gamma y^2\,dx+2xy\,dy$, where γ is the path from $(0,0)$ to $(2,4)$ along the parabola $y=x^2$.

3 $\int_\gamma \boldsymbol{F}\cdot d\boldsymbol{x}$, where $\boldsymbol{F}(x,y)=\left(-\frac{y}{x^2+y^2},\frac{x}{x^2+y^2}\right)$ and γ is any path from $(1,0)$ to $(1,1)$ in the counterclockwise sense.

4 $\int_\gamma \boldsymbol{F}\cdot d\boldsymbol{x}$, where $\boldsymbol{F}$ is as in Exercise 3 and γ is the straight line segment from $(1,0)$ to $(1,1)$.

5 $\oint_\gamma (2xy^3+5\cos x)\,dx+(3x^2y^2-4e^y)\,dy$, where γ is the unit circle $x^2+y^2=1$.

6 $\oint_\gamma (3x^2\sin y+\cos^2 y)\,dx-x(\sin 2y-x^2\cos y)\,dy$, where γ is the square with vertices $(-1,-1)$, $(1,-1)$, $(1,1)$, and $(-1,1)$.

7 $\int_\gamma (6x^2y^2-3yz^2)\,dx+(4x^3y-3xz^2)\,dy-6xyz\,dz$, where γ is the polygonal path from $(-1,0,1)$ to $(2,0,1)$ to $(2,1,1)$ to $(2,1,-2)$.

8 $\int_\gamma \boldsymbol{F}\cdot d\boldsymbol{x}$, where $\boldsymbol{F}(\boldsymbol{x})=(x^2+yz,\,y^2+xz,\,z^2+xy)$ and γ is the curve $\boldsymbol{x}(t)=(t,t^2,t^3)$ from $(0,0,0)$ to $(1,1,1)$.

9 $\int_\gamma ze^{xz}\sin yz\,dx+ze^{xz}\cos yz\,dy+e^{xz}(x\sin yz+y\cos yz)\,dz$, where γ is any path from $(-1,1,\pi/2)$ to $(3,-1,3\pi/2)$.

10 $\int_\gamma \boldsymbol{F}\cdot d\boldsymbol{x}$, where $\boldsymbol{F}(\boldsymbol{x})=(2xy^2z^2w^2,\,2x^2yz^2w^2,\,2x^2y^2zw^2,\,2x^2y^2z^2w)$ and γ is the curve $\boldsymbol{x}(t)=(t,t^2,t^3,t^4)$ from $(1,1,1,1)$ to $(2,4,8,16)$ in $\boldsymbol{R}^4$.

In Exercises 11 to 18, decide whether the given vector field is conservative on the set D. If it is, then find a potential function.

11 $\boldsymbol{F}(\boldsymbol{x})=(3x^2+y,\,e^y+x)$, D the unit disk $x^2+y^2\le 1$.

12 $\boldsymbol{F}(\boldsymbol{x})=(3y^2+6xy,\,3x^2+6y)$, D as in Exercise 11.

13 $\boldsymbol{F}(x,y)=\left(\frac{x}{\sqrt{x^2+y^2}},\frac{y}{\sqrt{x^2+y^2}}\right)$, D the annular region between the circles $x^2+y^2=1$ and $x^2+y^2=4$.

14 $\boldsymbol{F}(x, y) = \left(\frac{y}{x^2+y^2}, -\frac{x}{x^2+y^2}\right)$, D as in Exercise 13.

15 $\boldsymbol{F}(\boldsymbol{x}) = (2xyz + z^2 - 2y^2 + 1, x^2z - 4xy, x^2y + 2xz - 2)$, D the unit ball $x^2+y^2+z^2 \leq 1$.

16 $\boldsymbol{F}(\boldsymbol{x}) = (2xy^2, x^2z^3, 3x^2y^2z^2)$, D as in Exercise 15.

17 $\boldsymbol{F}(x, y, z) = (3x^2 - 3yz, 3y^2 - 3xz, -3xy)$, D the region between the spheres $x^2+y^2+z^2 = 1$ and $x^2+y^2+z^2 = 9$.

18 $\boldsymbol{F}(\boldsymbol{x}) = \left(\frac{x}{\sqrt{x^2+y^2+z^2}}, \frac{y}{\sqrt{x^2+y^2+z^2}}, \frac{z}{\sqrt{x^2+y^2+z^2}}\right)$, D the region between the two cylinders $x^2+y^2 = 1$ and $x^2+y^2 = 9$.

19 A force $\boldsymbol{F}$ of friction has constant magnitude and direction always *opposite* to the direction of motion. Show that such a force is not conservative by calculating the line integral of $\boldsymbol{F}$ over an arbitrary path $\boldsymbol{x} = \boldsymbol{x}(t)$ in $\boldsymbol{R}^n$.

20 Suppose that $\boldsymbol{F} = \boldsymbol{\nabla} p$ is conservative. Show that at each $\boldsymbol{x}_0$ the force $\boldsymbol{F}(\boldsymbol{x}_0)$ is orthogonal to the equipotential surface $p(\boldsymbol{x}) = p(\boldsymbol{x}_0)$ through that point.

21 If $\boldsymbol{F}$ is the inverse square law field of Example 3.5, then show that the work done in moving a particle from $\boldsymbol{x}_0$ to a point $\boldsymbol{x}_1$ that is very far from the origin is approximately $k/|\boldsymbol{x}_0|$.

22 Using

$$\boldsymbol{G}(\boldsymbol{x}) = -gR^2 \frac{1}{|\boldsymbol{x}|^2} \frac{\boldsymbol{x}}{|\boldsymbol{x}|}$$

to represent the gravitational field of the earth at a point $\boldsymbol{x}$, find the amount of work needed to move a particle from $\boldsymbol{x}_0$ on the earth's surface completely out of the gravitational influence of the earth (cf. Exercise 21).

23 Near the surface of the earth the gravitational field $\boldsymbol{G}$ acts on a particle of mass m approximately as the constant field

$$\boldsymbol{F}(x, y, z) = (0, 0, -mg),$$

where g is the acceleration due to gravity at the surface of the earth. Find the potential function p that is 0 at the origin. (The origin is at the center of the earth.)

24 Refer to Exercise 23. Find the path followed by a particle of mass m that has velocity $\boldsymbol{v} = (v_1, v_2, v_3)$ at the center of the earth and is acted on only by $\boldsymbol{F}$.

25 In Exercise 24, show that the total energy is constant at every point of the path $\boldsymbol{x} = \boldsymbol{x}(t)$ of the particle.

26 Show that the central field of Exercise 34 of Exercises 7.3 is conservative. (*Hint:* If $p(r) = \int_0^r yf(y)\,dy$, then $\boldsymbol{F}(\boldsymbol{x}) = \boldsymbol{\nabla} p(|\boldsymbol{x}|)$.)

5 SURFACES AND SURFACE AREA

In order to describe two additional analogues of the Fundamental Theorem of Integral Calculus, we need to develop a new type of double integral called a *surface integral.* This will bear roughly the same relation to the ordinary double integral as the line integral bears to the ordinary definite integral. Instead of integrating a function $f: \boldsymbol{R}^2 \rightarrow \boldsymbol{R}$ over a region $D \subseteq \boldsymbol{R}^2$, we will be integrating a vecto field $\boldsymbol{F}: \boldsymbol{R}^3 \rightarrow \boldsymbol{R}^3$ over a surface $S \subseteq \boldsymbol{R}^3$.

Our first step is to develop a *vector representation* for surfaces $S \subseteq \boldsymbol{R}^3$ analogous to the vector representation of curves γ in $\boldsymbol{R}^n$ developed in Chapter 2.

Let's remind ourselves of how we parametrize curves. We think of a vector function $\boldsymbol{x}=\boldsymbol{x}(t)$ as a function that deforms a closed interval $I=[a, b]$ in $\boldsymbol{R}^1$ into a curve in $\boldsymbol{R}^n$. As t varies over I, the curve γ is traced out in $\boldsymbol{R}^n$. See Figure 5.1. Is

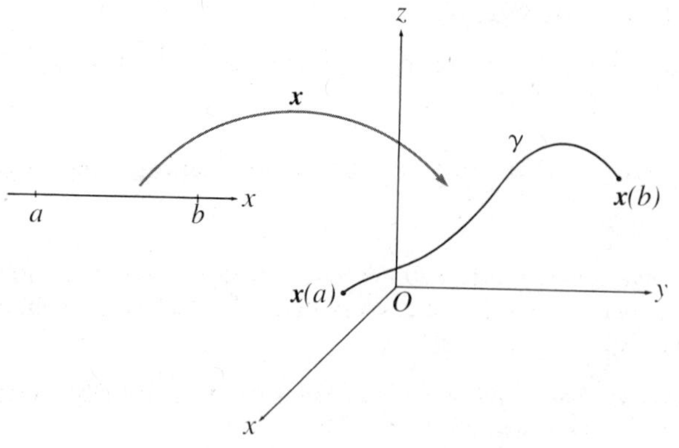

FIGURE 5.1

there a corresponding way of thinking of surfaces in $\boldsymbol{R}^3$? There is, if we think of a surface S in $\boldsymbol{R}^3$ as being obtained by *deforming some two-dimensional set* $D\subseteq\boldsymbol{R}^2$. See Figure 5.2. Thus, for example, if D is a rectangle, we can think of it as being

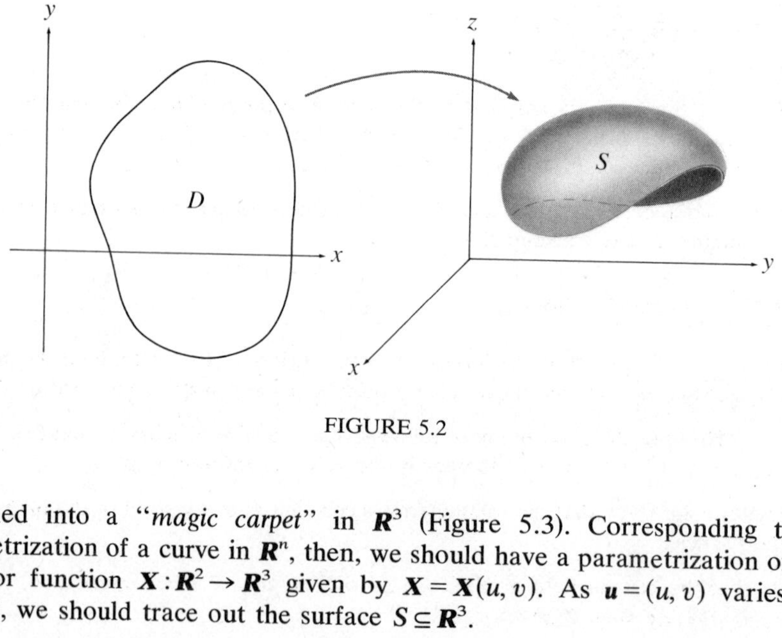

FIGURE 5.2

deformed into a "*magic carpet*" in $\boldsymbol{R}^3$ (Figure 5.3). Corresponding to our parametrization of a curve in $\boldsymbol{R}^n$, then, we should have a parametrization of S by a vector function $\boldsymbol{X}:\boldsymbol{R}^2\to\boldsymbol{R}^3$ given by $\boldsymbol{X}=\boldsymbol{X}(u, v)$. As $\boldsymbol{u}=(u, v)$ varies over $D\subseteq\boldsymbol{R}^2$, we should trace out the surface $S\subseteq\boldsymbol{R}^3$.

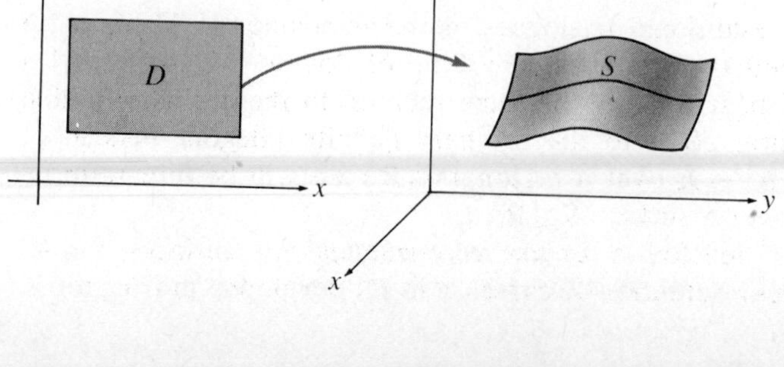

FIGURE 5.3

5.1 **DEFINITION.** A **parametrization** of a **surface patch** $S \subseteq \boldsymbol{R}^3$ is a function $\boldsymbol{X}: \boldsymbol{R}^2 \to \boldsymbol{R}^3$ such that for some set $D \subseteq \boldsymbol{R}^2$,

$$S = \boldsymbol{X}(D) = \{(x, y, z) \in \boldsymbol{R}^3 \mid (x, y, z) = \boldsymbol{X}(u, v) \quad \text{for some} \quad (u, v) \in D\}.$$

A **parametrized surface** is a union of parametrized surface patches.

As we recall from Chapter 1, a *plane* can be represented as the graph of $ax + by + cz = d$, where not all of a, b, c are 0. If, say, $a \neq 0$, then we can solve for x,

$$x = \frac{d}{a} - \frac{b}{a} y - \frac{c}{a} z.$$

Then we can parametrize the plane as a surface patch by letting $y = u$, $z = v$, and $x = \frac{d}{a} - \frac{b}{a} u - \frac{c}{a} v$. This gives us $\boldsymbol{X}(u, v) = \left(\frac{d}{a} - \frac{b}{a} u - \frac{c}{a} v, u, v\right)$.

A similar approach works for a surface given explicitly as the graph of a function $f: \boldsymbol{R}^2 \to \boldsymbol{R}$. If we have $z = f(x, y)$, then we parametrize by the same approach we used to parametrize a curve $y = f(x)$ in $\boldsymbol{R}^2$. Namely, we let

$$x = u, \qquad y = v, \qquad z = f(u, v), \quad \text{for} \quad (u, v) \in D = \text{Domain } f.$$

In this way we obtain

$$\boldsymbol{X}(u, v) = (u, v, f(u, v)).$$

We can handle a much wider class of surfaces than those that are graphs of functions $z = f(x, y)$, however. Our next example provides us with two illustrations of this. The second part is also of interest, since it shows how some familiar surfaces cannot be parametrized as single patches, but instead must be parametrized as the *union* of several surface patches.

5.2 **EXAMPLE.** Parametrize (a) a sphere of radius a centered at $\mathbf{0}$; (b) a right circular cylinder of radius a and height h, with the z-axis as its axis and including a top and bottom (Figure 5.4).

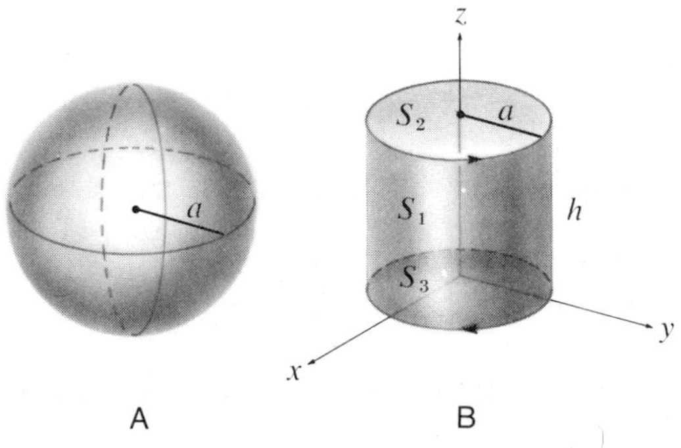

FIGURE 5.4

Solution. (a) We parametrize using spherical coordinates on the sphere, which has equation $\rho = a$. We let

$$x = a \sin v \cos u, \qquad y = a \sin v \sin u, \qquad z = a \cos v.$$

Thus

$$\mathbf{X}(u, v) = (a \sin v \cos u, a \sin v \sin u, a \cos v)$$

where $0 \leq u \leq 2\pi$ and $0 \leq v \leq \pi$. Here D is the rectangle $[0, 2\pi] \times [0, \pi]$.

(b) We have to parametrize this cylinder as three patches: top, curved cylindrical portion, and bottom. For the cylindrical patch S_1, we can use cylindrical coordinates with $r = a$. Thus let $\mathbf{X}_1(u, v) = (x, y, z)$ where

$$x = a \cos u, \qquad y = a \sin u, \qquad z = v,$$

for $u \in [0, 2\pi]$ and $v \in [0, h]$. For the top (S_2) and bottom (S_3) we use polar coordinates in the planes $z = 0$ and $z = h$. Thus, for the top, let $\mathbf{X}_2(u, v) = (x, y, z)$, where

$$x = u \cos v, \qquad y = u \sin v, \qquad z = h,$$

for $u \in [0, a]$ and $v \in [0, 2\pi]$. For the bottom, let $\mathbf{X}_3(u, v) = (x, y, z)$ where

$$x = u \cos(-v), \qquad y = u \sin(-v), \qquad z = 0,$$

for $u \in [0, a]$ and $v \in [0, 2\pi]$.

(We *could* have used v instead of $-v$ in parametrizing S_3, of course. But we will see after Definition 6.8 that our choice is a good one.)

In developing line integrals, we essentially confined our attention to smooth curves γ. Likewise, we will concentrate our attention on *smooth surfaces* in defining surface integrals. Intuitively, a smooth surface patch is a magic carpet that is comfortable to ride: there are no sharp points. We were able to describe smooth curves mathematically by the requirement that the tangent vector $\dot{\mathbf{x}}(t)$ should vary continuously. We can then describe a smooth surface mathematically by requiring the *tangent plane* to vary continuously. And how do we describe the tangent plane to a parametrically defined surface S? A natural approach is the following. At each point $\mathbf{X}(u_0, v_0)$ on S we have curves (Figure 5.5) $\mathbf{x}_1 = \mathbf{X}(u, v_0)$

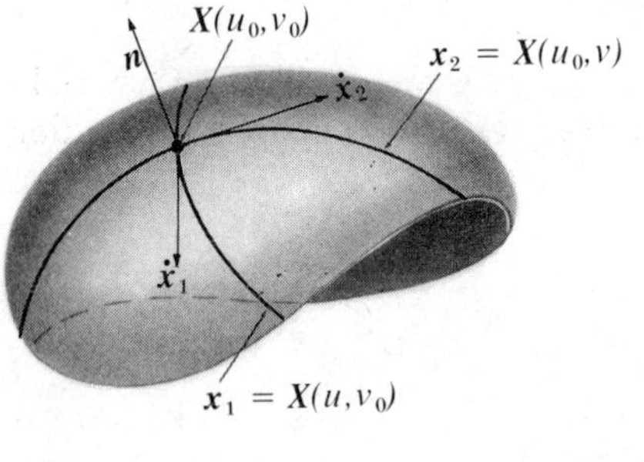

FIGURE 5.5

and $\mathbf{x}_2 = \mathbf{X}(u_0, v)$ passing through that point. Tangent vectors to these curves are given by

$$\dot{\mathbf{x}}_1 = \mathbf{X}_u(u_0, v_0) = \frac{\partial \mathbf{X}}{\partial u}(u_0, v_0) \quad \text{and} \quad \dot{\mathbf{x}}_2 = \mathbf{X}_v(u_0, v_0) = \frac{\partial \mathbf{X}}{\partial v}(u_0, v_0).$$

If these vectors are not collinear, then they determine the plane with a normal vector

(1) $$\mathbf{n}(u_0, v_0) = (\mathbf{X}_u \times \mathbf{X}_v)(u_0, v_0).$$

If now $\mathbf{X}_u$ and $\mathbf{X}_v$ are continuous, that is, if all entries of $J_{\mathbf{X}}$ are continuous, then $\mathbf{n}$ will vary continuously as u and v vary. So the tangent plane will vary continuously. Hence we give the following definition.

5.3 **DEFINITION.** A surface patch $S \subseteq \boldsymbol{R}^2$ is **smooth** if it has a parametrization $\boldsymbol{X}: \boldsymbol{R}^2 \to \boldsymbol{R}^3$ with $\boldsymbol{X} = \boldsymbol{X}(u, v)$ for $(u, v) \in D \subseteq \boldsymbol{R}^2$ such that all entries of $J_{\boldsymbol{X}}$ are continuous on D.

In Example 5.2, observe that the parametrizations show that the sphere and the top, bottom, and sides of the cylinder are smooth surface patches.

In what follows, we assume that $\boldsymbol{X}_u$ and $\boldsymbol{X}_v$ will almost never be collinear at points on a smooth surface patch S, i.e., that $\boldsymbol{X}_u \times \boldsymbol{X}_v \neq \boldsymbol{0}$ will be a normal vector to S at almost all points of S.

We turn now to the problem of defining the *area* of a smooth surface patch S, parametrized by $\boldsymbol{X}: D \to \boldsymbol{R}^3$. The idea is an old, familiar one. We make a grid of $D \subseteq \boldsymbol{R}^2$ by rectangles. Consider a subrectangle R_{ij} of D that is mapped by $\boldsymbol{X}$ onto a small magic sub-carpet of S (Figure 5.6).

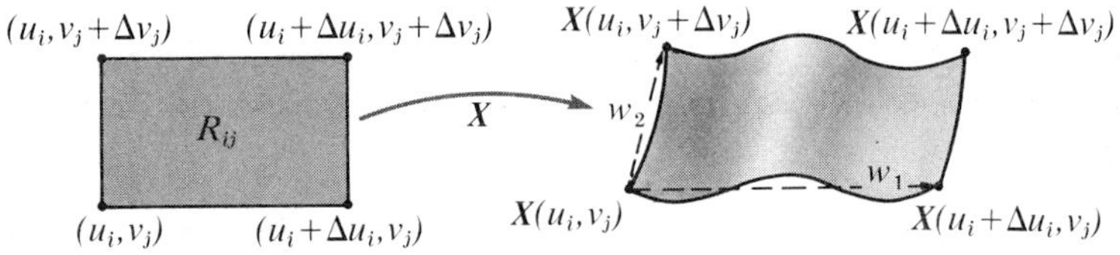

FIGURE 5.6

Now, intuitively, the area of this magic sub-carpet should be approximately the same as the area $|\boldsymbol{w}_1 \times \boldsymbol{w}_2|$ of the parallelogram determined by

$$\boldsymbol{w}_1 = \boldsymbol{X}(u_i + \Delta u_i, v_j) - \boldsymbol{X}(u_i, v_j)$$

and

$$\boldsymbol{w}_2 = \boldsymbol{X}(u_i, v_j + \Delta v_j) - \boldsymbol{X}(u_i, v_j).$$

Moreover, all the entries of $J_{\boldsymbol{X}}$ are continuous, so $\boldsymbol{X}$ is differentiable at (u_i, v_j) with total derivative

$$J_{\boldsymbol{X}}(u_i, v_j) = \begin{pmatrix} \frac{\partial x}{\partial u} & \frac{\partial x}{\partial v} \\ \frac{\partial y}{\partial u} & \frac{\partial y}{\partial v} \\ \frac{\partial z}{\partial u} & \frac{\partial z}{\partial v} \end{pmatrix} (u_i, v_j) \tag{2}$$

by Theorem 6.5.5. Thus, by the definition of $\boldsymbol{X}$ differentiable at (u_i, v_j), we can approximate $\boldsymbol{w}_1$ and $\boldsymbol{w}_2$ by the following tangent approximations:

$$\boldsymbol{w}_1 \approx J_{\boldsymbol{X}}(u_i, v_j)\left[\begin{pmatrix} u_i + \Delta u_i \\ v_j \end{pmatrix} - \begin{pmatrix} u_i \\ v_j \end{pmatrix}\right] = J_{\boldsymbol{X}}(u_i, v_j)\begin{pmatrix} \Delta u_i \\ 0 \end{pmatrix}$$

$$\approx \begin{pmatrix} \frac{\partial x}{\partial u}(u_i, v_j)\,\Delta u_i \\ \frac{\partial y}{\partial u}(u_i, v_j)\,\Delta u_i \\ \frac{\partial z}{\partial u}(u_i, v_j)\,\Delta u_i \end{pmatrix} \quad \text{from (2)}$$

$$\approx \Delta u_i \boldsymbol{X}_u(u_i, v_j).$$

and similarly

$$\mathbf{w}_2 \approx \Delta v_j \mathbf{X}_v(u_i, v_j).$$

Hence the area of the magic sub-carpet is approximately

$$\begin{aligned}|\mathbf{w}_1 \times \mathbf{w}_2| &\approx |\Delta u_i \mathbf{X}_u(u_i, v_j) \times \Delta v_j \mathbf{X}_v(u_i, v_j)| \\ &\approx |(\mathbf{X}_u \times \mathbf{X}_v)(u_i, v_j)|\, \Delta u_i\, \Delta v_j.\end{aligned}$$

An approximation to the total area $A(S)$ of S is obtained by adding the foregoing approximations over all subrectangles R_{ij}. We get

(3) $$A(S) \approx \sum_i \sum_j |(\mathbf{X}_u \times \mathbf{X}_v)(u_i, v_j)|\, \Delta u_i\, \Delta v_j.$$

We now define $A(S)$ to be the limit of the approximation (3) as our grid becomes finer and finer. (It can be shown that this is essentially independent of the parametrization $\mathbf{X}$ used for S. See Exercises 25 and 26 below.)

5.4 **DEFINITION.** If $S \subseteq \mathbf{R}^3$ is a smooth surface patch parametrized by $\mathbf{X}: \mathbf{R}^2 \to \mathbf{R}^3$, where $\mathbf{X}$ maps $D \subseteq \mathbf{R}^2$ onto S, then the **area** of S is

(4) $$A(S) = \iint_D |\mathbf{X}_u \times \mathbf{X}_v|\, du\, dv,$$

whenever this integral exists.

5.5 **EXAMPLE.** Use Definition 5.4 to compute the surface area of a sphere of radius a.

Solution. We parametrized the sphere in Example 5.2 as $\mathbf{X}(u, v) = (a \sin v \cos u, a \sin v \sin u, a \cos v)$ for $(u, v) \in D = [0, 2\pi] \times [0, \pi]$. We then have

$$\begin{aligned}\mathbf{X}_u &= (-a \sin v \sin u, a \sin v \cos u, 0), \\ \mathbf{X}_v &= (a \cos v \cos u, a \cos v \sin u, -a \sin v).\end{aligned}$$

Thus,

$$\begin{aligned}\mathbf{X}_u \times \mathbf{X}_v &= -a^2(\sin^2 v \cos u, \sin^2 v \sin u, \sin v \cos v \sin^2 u + \sin v \cos v \cos^2 u) \\ &= -a^2(\sin^2 v \cos u, \sin^2 v \sin u, \sin v \cos v).\end{aligned}$$

Hence,

$$\begin{aligned}|\mathbf{X}_u \times \mathbf{X}_v| &= a^2\sqrt{\sin^4 v \cos^2 u + \sin^4 v \sin^2 u + \sin^2 v \cos^2 v} \\ &= a^2\sqrt{\sin^4 v + \sin^2 v(1 - \sin^2 v)} \\ &= a^2\sqrt{\sin^2 v} \\ &= a^2 \sin v \quad \text{since} \quad v \in [0, \pi].\end{aligned}$$

(Note that $\mathbf{X}_u \times \mathbf{X}_v \neq \mathbf{0}$ except at the two points $v = 0$ and $v = \pi$, the poles.) We now get from (4)

$$A(S)=\iint_D a^2 \sin v\, du\, dv = a^2\int_0^{2\pi} du \int_0^{\pi} \sin v\, dv$$
$$=2\pi a^2[-\cos v]_0^{\pi}=4\pi a^2.$$

(You may have encountered this formula before.)

When a surface patch $S \subseteq \boldsymbol{R}^3$ is given explicitly as the graph of a function $f:\boldsymbol{R}^2 \to \boldsymbol{R}$, then (4) assumes a slightly simpler form, one that will remind you of the arc length formula of elementary calculus.

5.6 COROLLARY. If S is the graph of $z=f(x, y)$ for $(x, y)\in D\subseteq \boldsymbol{R}^2$, then the area of S is given by

(5) $$A(S)=\iint_D \sqrt{1+|\nabla f|^2}\, dx\, dy.$$

Proof. We use the parametrization mentioned just before Example 5.2,

$$\boldsymbol{X}(u, v)=(u, v, f(u, v)) \quad \text{for} \quad (u, v)\in D.$$

We have

$$\boldsymbol{X}_u=\left(1, 0, \frac{\partial f}{\partial u}\right) \quad \text{and} \quad \boldsymbol{X}_v=\left(0, 1, \frac{\partial f}{\partial v}\right).$$

Thus,

$$\boldsymbol{X}_u \times \boldsymbol{X}_v=\left(-\frac{\partial f}{\partial u}, -\frac{\partial f}{\partial v}, 1\right).$$

Hence

$$|\boldsymbol{X}_u \times \boldsymbol{X}_v|=\sqrt{\left(\frac{\partial f}{\partial u}\right)^2+\left(\frac{\partial f}{\partial v}\right)^2+1}$$
$$=\sqrt{\left(\frac{\partial f}{\partial x}\right)^2+\left(\frac{\partial f}{\partial y}\right)^2+1}$$
$$=\sqrt{|\nabla f|^2+1}.$$

QED

If f is differentiable, then its total derivative is ∇f, so that (5) is the exact analogue for $f:\boldsymbol{R}^2\to \boldsymbol{R}$ of the arc length formula

$$s=\int_a^b \sqrt{1+[f'(x)]^2}\, dx$$

from elementary calculus.

5.7 EXAMPLE. Find the area of the surface $z=x^{3/2}+2\sqrt{2}y$ that lies above the rectangle $[0, 1]\times[1, 2]$.

Solution. We are given

$$f(x, y) = x^{3/2} + 2\sqrt{2}\, y.$$

Thus,

$$\nabla f(x, y) = (\tfrac{3}{2}x^{1/2}, 2\sqrt{2}),$$
$$|\nabla f(x, y)|^2 = \tfrac{9}{4}x + 8,$$
$$\sqrt{1 + |\nabla f(x, y)|^2} = \sqrt{\tfrac{9}{4}x + 9} = \tfrac{3}{2}\sqrt{x + 4}.$$

We get from (5)

$$A(S) = \iint_D \tfrac{3}{2}\sqrt{x+4}\, dx\, dy = \tfrac{3}{2}\int_0^1 \sqrt{x+4}\, dx \int_1^2 dy$$
$$= \frac{3}{2}\frac{(x+4)^{3/2}}{3/2}\bigg]_0^1 = 5^{3/2} - 4^{3/2}$$
$$= 5\sqrt{5} - 8.$$

EXERCISES 7.5

In Exercises 1 to 10, parametrize the given surface.

1 The paraboloid $x^2 + y^2 - 2z = 0$ above the square $[1, 2] \times [2, 5]$.

2 The cone $z = \sqrt{x^2 + y^2}$ lying above the disk $x^2 + y^2 \le 4$.

3 The cone $z = 4 - \sqrt{x^2 + y^2}$ lying above the disk $x^2 + y^2 \le 16$.

4 The part of the plane $z = 2x + 2y$ that lies inside the cylinder $x^2 + y^2 = 1$.

5 The part of the plane $z = 3x - 2y$ that lies inside the cylinder $x^2 + y^2 = 4$.

6 The part of the cylinder $x^2 + y^2 = 4$ between the planes $z = 0$ and $z = x + 2$.

7 The part S of the sphere $x^2 + y^2 + z^2 = 1$ that lies above the cone $z = \sqrt{x^2 + y^2}$.

8 The portion of the paraboloid $z = x^2 + y^2$ lying between the planes $z = 1$ and $z = 4$.

9 The torus T obtained by revolving a circle in the xz-plane with center $(b, 0, 0)$ and radius $a < b$ about the z-axis (Figure 5.7).

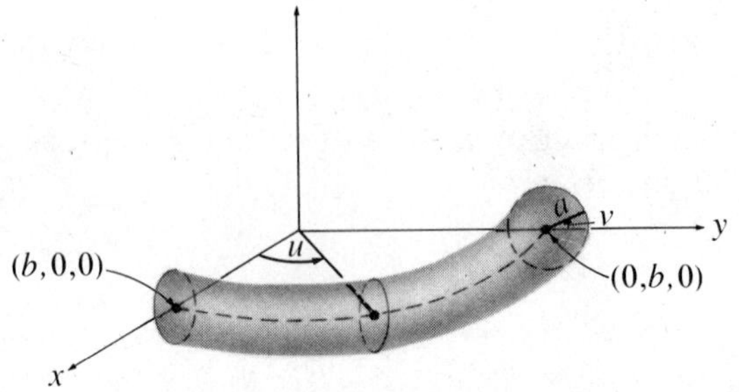

FIGURE 5.7

10 The part S of the sphere $x^2 + y^2 + z^2 = 4$ inside the cylinder $x^2 + y^2 = 2y$. (Recall that the cylinder is $r = 2 \sin \theta$ in cylindrical coordinates.)

In Exercises 11 to 20, use (4) of Definition 5.4 or (5) of Corollary 5.6 to compute the area of the given surface.

11 The portion of the saddle $z=x^2-y^2$ that lies inside the cylinder $x^2+y^2=1$.

12 The portion of the saddle $z=y^2-x^2$ that lies inside the cylinder $x^2+y^2=4$.

13 The portion of the paraboloid $z=x^2+y^2$ between the planes $z=1$ and $z=4$.

14 The portion of the paraboloid $z=x^2+y^2$ under the plane $z=4$.

15 The **spiral ramp** (or **helicoid**) given by $\mathbf{X}(u, v)=(u \cos v, u \sin v, v)$ for $(u, v) \in D=[0, 1]\times[0, \pi]$. Sketch this surface.

16 The spiral ramp of Exercise 15, where $D=[0, \sqrt{2}]\times[0, \pi]$.

17 The surface S of Exercise 5.

18 The surface of intersection of the cylinders $x^2+y^2=1$ and $x^2+z^2=1$.

19 The torus T of Exercise 9.

20 The surface S of Exercise 10.

21 Find the area of the lateral surface S_1 in Example 5.2(b).

22 Find the area of the lateral surface S of a right circular cone of radius a and height h. Show that your answer reduces to $\pi a \ell$, where ℓ is the *slant height* $\sqrt{a^2+h^2}$ shown in Figure 5.8.

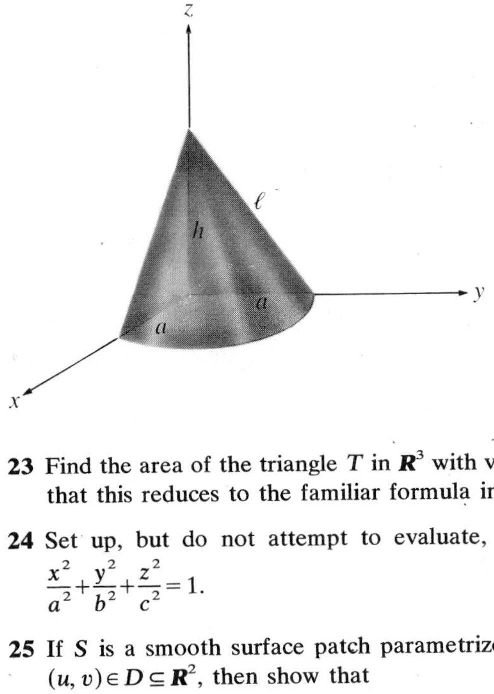

FIGURE 5.8

23 Find the area of the triangle T in $\mathbf{R}^3$ with vertices $(a, 0, 0)$, $(0, b, 0)$, and $(0, 0, c)$. Show that this reduces to the familiar formula in $\mathbf{R}^2$ if a or b or c is 0.

24 Set up, but do not attempt to evaluate, the integral for the area of the ellipsoid $\frac{x^2}{a^2}+\frac{y^2}{b^2}+\frac{z^2}{c^2}=1$.

25 If S is a smooth surface patch parametrized by $\mathbf{X}(u, v)=(x(u, v), y(u, v), z(u, v))$ for $(u, v)\in D\subseteq \mathbf{R}^2$, then show that

$$|\mathbf{X}_u \times \mathbf{X}_v|=\sqrt{\frac{\partial(y, z)}{\partial(u, v)}^2+\frac{\partial(z, x)}{\partial(u, v)}^2+\frac{\partial(x, y)}{\partial(u, v)}^2}.$$

26 Let $\mathbf{X}=\mathbf{X}(u, v)$ for $(u, v)\in D\subseteq \mathbf{R}^2$ and $\mathbf{X}=\mathbf{Y}(s, t)$ for $(s, t)\in E\subseteq \mathbf{R}^2$ be two parametrizations for a smooth surface patch $S\subseteq \mathbf{R}^3$. These are called *equivalent* if there is an invertible transformation of coordinates $\mathbf{T}:\mathbf{R}^2\to\mathbf{R}^2$ such that $\mathbf{T}(E)=D$, where T is differentiable with all entries of J_T continuous and

$$\det J_T=\frac{\partial(u, v)}{\partial(s, t)}\neq 0$$

on D. Use Exercise 25 and Theorem 6.9.1 to show that $A(S)$ as computed from (4) will be the same for both parametrizations $\boldsymbol{X}$ and $\boldsymbol{Y}$. Thus, the area of S is *independent* of the parametrization of the surface patch.

27 Suppose that a curve C is the graph in the xz-plane of $z=f(x)$, where $0 \le a \le x \le b$. Let S be the surface of revolution obtained by revolving C about the z-axis. Obtain a parametrization of S.

28 Use Exercise 27 to show that $A(S)=2\pi\int_a^b u\sqrt{1+f'(u)^2}\,du$.

29 Prove the following **Theorem of Pappus**. The area of the surface of revolution in Exercise 27 is $A(S)=2\pi\bar{x}L$, where $\bar{x}$ is the x-coordinate of the center of gravity of the curve C (see Exercise 19 of Exercises 7.1) and L is the length of C.

30 Use the Theorem of Pappus in Exercise 29 to rework Exercise 19.

6 SURFACE INTEGRALS

We are now ready to introduce the final kind of integral we shall discuss, the *surface integral* of a continuous function f over a smooth surface patch $S \subseteq \boldsymbol{R}^3$. We first define the surface integral for scalar functions $f:\boldsymbol{R}^3 \to \boldsymbol{R}$, and then define it for vector functions $\boldsymbol{F}:\boldsymbol{R}^3 \to \boldsymbol{R}^3$. In the latter case we will recognize the surface integral as the natural analogue of the line integral of a two-dimensional vector function.

Our first task, then, is to define the surface integral of a continuous function $f:\boldsymbol{R}^3 \to \boldsymbol{R}$ over a smooth surface patch S. We do this by analogy with the notion of the line integral of a scalar function $f:\boldsymbol{R}^n \to \boldsymbol{R}$ over a curve γ in $\boldsymbol{R}^n$. (See Exercises 19 to 23 of Exercises 7.1.) Recall that the latter integral is defined by

$$\int_\gamma f(\boldsymbol{x})\,ds = \int_a^b f(\boldsymbol{x}(t))\,|\dot{\boldsymbol{x}}(t)|\,dt$$

if γ is parametrized by $\boldsymbol{x}=\boldsymbol{x}(t)$ for $t \in [a, b]$.

Let $\boldsymbol{X}:D \to \boldsymbol{R}^3$ be a parametrization of the smooth surface patch S, where $D \subseteq \boldsymbol{R}^2$. We make a grid of D by rectangles as in the preceding section, and define the surface integral of f over S to be the limit of the Riemann sums

$$\sum_{i=1}^{m}\sum_{j=1}^{n} f(\boldsymbol{X}(u_i, v_j))\,|\boldsymbol{X}_u \times \boldsymbol{X}_v(u_i, v_j)|\,\Delta u_i\,\Delta v_j$$

as the mesh of our grid tends to zero. These Riemann sums are obtained by adding the products of the values of f at points $\boldsymbol{X}(u_i, v_j)$ on the surface with

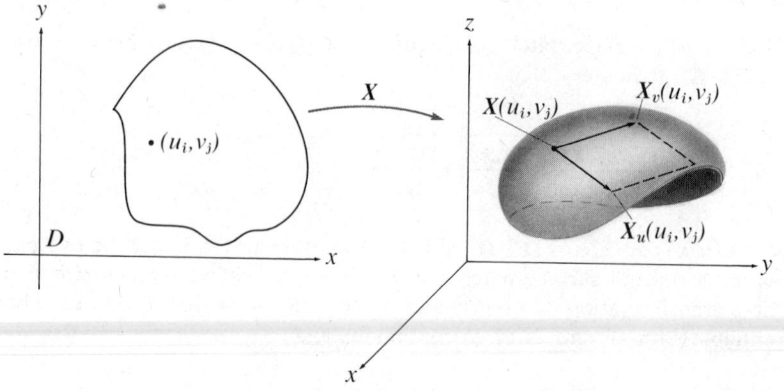

FIGURE 6.1

elementary area approximations for small pieces of S surrounding the points. Refer to Figure 6.1. Since f, $\mathbf{X}_u$, and $\mathbf{X}_v$ are all continuous, the Riemann sums will approach the ordinary double integral of the real-valued function

$$f(\mathbf{X}(u, v))\,|\mathbf{X}_u \times \mathbf{X}_v|$$

over the domain D in $\mathbf{R}^2$. We are thus led to the following definition.

6.1 **DEFINITION.** Let $f:\mathbf{R}^3 \to \mathbf{R}$ be continuous on a smooth surface patch $S \subseteq \mathbf{R}^3$ that is parametrized by $\mathbf{X}: D \to \mathbf{R}^3$ for $D \subseteq \mathbf{R}^2$. Then the **surface integral** of f over S is

(1) $$\iint_S f(\mathbf{x})\,dA = \iint_D f(\mathbf{X}(u, v))\,|\mathbf{X}_u \times \mathbf{X}_v|\,du\,dv,$$

whenever this integral exists.

6.2 **EXAMPLE.** Compute the surface integral of $f(x, y, z) = 3z^2 + 4$ over the sphere in Example 5.5.

Solution. In Example 5.5 we computed $|\mathbf{X}_u \times \mathbf{X}_v| = a^2 \sin v$. Then (1) gives

$$\begin{aligned}\iint_{\mathbf{X}(D)} f(\mathbf{x})\,dA &= \iint_D (3a^2\cos^2 v + 4)a^2 \sin v\,du\,dv \\ &= a^2\int_0^{2\pi} du \int_0^{\pi} 3a^2\cos^2 v \sin v\,dv + 4\iint_D a^2 \sin v\,du\,dv \\ &= 2\pi a^2\left[3a^2\left(-\frac{\cos^3 v}{3}\right)\right]_0^{\pi} + 4(4\pi a^2) \\ &= 2\pi a^2[2a^2] + 16\pi a^2 \\ &= 4\pi a^2(a^2 + 4).\end{aligned}$$

We next want to define the surface integral of a continuous vector valued function $\mathbf{F}:\mathbf{R}^3 \to \mathbf{R}^3$ over a smooth surface patch S. Just as the *line integral* was defined with *work* in mind, so the *surface integral* will be defined with the physical concept of *flux* in mind. (See Exercise 27 of Exercises 7.2.) For this, we need the notion of a *standard* unit normal vector to S (cf. formula (1), p. 452).

6.3 **DEFINITION.** If $S \subseteq \mathbf{R}^3$ is a smooth surface patch parametrized by $\mathbf{X}:\mathbf{R}^2 \to \mathbf{R}^3$, then the **standard unit normal** $\mathbf{N}$ to S at a point $\mathbf{X}(u, v)$ where $\mathbf{X}_u \times \mathbf{X}_v \neq \mathbf{0}$ is

(2) $$\mathbf{N} = \frac{\mathbf{X}_u \times \mathbf{X}_v}{|\mathbf{X}_u \times \mathbf{X}_v|}(u, v).$$

Note that the negative of $\mathbf{N}$ is another unit normal vector pointing in the opposite direction, and that no matter how S is parametrized, $(\mathbf{X}_u, \mathbf{X}_v, \mathbf{N})$ is always a *right-handed system*. See Figure 6.2.

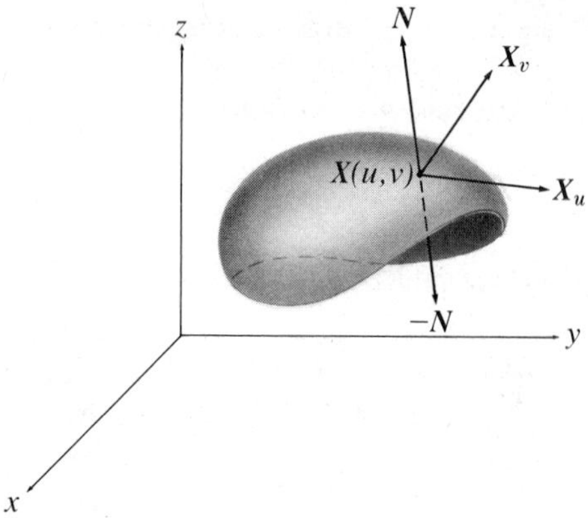

FIGURE 6.2

We can now define the surface integral of a continuous vector field

6.4 **DEFINITION.** Let $\boldsymbol{F}:\boldsymbol{R}^3 \to \boldsymbol{R}^3$ be a continuous vector field on a smooth surface patch $S \subseteq \boldsymbol{R}^3$. Let $\boldsymbol{X}:D \to \boldsymbol{R}^3$ parametrize S. Then the **surface integral** of $\boldsymbol{F}$ over S is

(3) $$\iint_S \boldsymbol{F}\cdot d\boldsymbol{S} = \iint_{\boldsymbol{X}(D)} \boldsymbol{F}\cdot\boldsymbol{N}\, dA = \iint_D \boldsymbol{F}\cdot\boldsymbol{N}\,|\boldsymbol{X}_u \times \boldsymbol{X}_v|\, du\, dv,$$

whenever this integral exists.

Note from (2) that we have

$$\boldsymbol{F}\cdot\boldsymbol{N}\,|\boldsymbol{X}_u \times \boldsymbol{X}_v|\, du\, dv = \boldsymbol{F}\cdot\frac{\boldsymbol{X}_u \times \boldsymbol{X}_v}{|\boldsymbol{X}_u \times \boldsymbol{X}_v|}\,|\boldsymbol{X}_u \times \boldsymbol{X}_v|\, du\, dv$$
$$= \boldsymbol{F}\cdot(\boldsymbol{X}_u \times \boldsymbol{X}_v)\, du\, dv.$$

We thus can rewrite (3) as

(4) $$\iint_S \boldsymbol{F}\cdot d\boldsymbol{S} = \iint_D \boldsymbol{F}\cdot(\boldsymbol{X}_u \times \boldsymbol{X}_v)\, du\, dv.$$

The notation $\iint_S \boldsymbol{F}\cdot d\boldsymbol{S}$ is intended to parallel the notation $\int_\gamma \boldsymbol{F}\cdot d\boldsymbol{x}$ for a line integral. We will find that surface integrals of vector functions over surfaces correspond closely to line integrals of vector functions over curves.

6.5 **EXAMPLE.** Suppose that

$$\boldsymbol{F}(x, y, z) = \left(\frac{x}{\sqrt{x^2+y^2+z^2}}, \frac{y}{\sqrt{x^2+y^2+z^2}}, \frac{z}{\sqrt{x^2+y^2+z^2}}\right).$$

Find the surface integral of $\boldsymbol{F}$ over the sphere of Example 5.5.

Solution. We have $\sqrt{x^2+y^2+z^2} = a$ and

$$x = a \sin v \cos u, \qquad y = a \sin v \sin u, \qquad z = a \cos v$$

on the surface. Hence, on the sphere,

$$\boldsymbol{F}(x, y, z) = (\sin v \cos u, \sin v \sin u, \cos v).$$

From Example 5.5,

$$\mathbf{X}_u \times \mathbf{X}_v = -a^2(\sin^2 v \cos u, \sin^2 v \sin u, \sin v \cos v).$$

Then from (4),

$$\begin{aligned}
\iint_S \boldsymbol{F} \cdot d\boldsymbol{S} &= \iint_D \boldsymbol{F} \cdot (\mathbf{X}_u \times \mathbf{X}_v)\, du\, dv \\
&= -a^2 \iint_D (\sin^3 v \cos^2 u + \sin^3 v \sin^2 u + \sin v \cos^2 v)\, du\, dv \\
&= -a^2 \iint_D (\sin^3 v + \sin v \cos^2 v)\, du\, dv \\
&= -a^2 \iint_D \sin v(\sin^2 v + \cos^2 v)\, du\, dv \\
&= -a^2 \iint_D \sin v\, du\, dv = -4\pi a^2, \quad \text{as in Example 5.5.}
\end{aligned}$$

An *important notation* for the surface integral of a vector function $\boldsymbol{F}$ arises if we write $\boldsymbol{F}$ in coordinate form as $\boldsymbol{F}(\boldsymbol{x}) = \boldsymbol{F}(x, y, z) = (P(x, y, z), Q(x, y, z), R(x, y, z))$. Then it is natural to also write $\mathbf{X}: \boldsymbol{R}^2 \to \boldsymbol{R}^3$ in terms of its coordinate functions as

$$\mathbf{X}(u, v) = (x(u, v), y(u, v), z(u, v)).$$

Then

$$\mathbf{X}_u = \left(\frac{\partial x}{\partial u}, \frac{\partial y}{\partial u}, \frac{\partial z}{\partial u}\right),$$

and

$$\mathbf{X}_v = \left(\frac{\partial x}{\partial v}, \frac{\partial y}{\partial v}, \frac{\partial z}{\partial v}\right).$$

Thus

$$\begin{aligned}
\mathbf{X}_u \times \mathbf{X}_v &= \left(\frac{\partial y}{\partial u}\frac{\partial z}{\partial v} - \frac{\partial y}{\partial v}\frac{\partial z}{\partial u}, \frac{\partial z}{\partial u}\frac{\partial x}{\partial v} - \frac{\partial z}{\partial v}\frac{\partial x}{\partial u}, \frac{\partial x}{\partial u}\frac{\partial y}{\partial v} - \frac{\partial x}{\partial v}\frac{\partial y}{\partial u}\right) \\
&= \left(\frac{\partial(y, z)}{\partial(u, v)}, \frac{\partial(z, x)}{\partial(u, v)}, \frac{\partial(x, y)}{(u, v)}\right)
\end{aligned}$$

Then (4) becomes

$$\iint_S \boldsymbol{F}\cdot d\boldsymbol{S} = \iint_D \boldsymbol{F}\cdot(\mathbf{X}_u \times \mathbf{X}_v)\, du\, dv$$

$$= \iint_D P(\boldsymbol{x})\frac{\partial(y, z)}{\partial(u, v)}\, du\, dv + Q(\boldsymbol{x})\frac{\partial(z, x)}{\partial(u, v)}\, du\, dv + R(\boldsymbol{x})\frac{\partial(x, y)}{\partial(u, v)}\, du\, dv.$$

Imagining that symbolic cancellation between $\partial(u, v)$ and $du\, dv$ is possible, we arrive at the notation

(5) $$\iint_S \boldsymbol{F}\cdot d\boldsymbol{S} = \iint_D P(\boldsymbol{x})\, dy \wedge dz + Q(\boldsymbol{x})\, dz \wedge dx + R(\boldsymbol{x})\, dx \wedge dy.$$

We introduce the *wedge* $\wedge$ to remind ourselves that, *unlike* the case for the $dx\, dy$ in a double integral, the *order* of dx and dy *matters* here. After all, $dx \wedge dy$ really came from

$$\frac{\partial(x, y)}{\partial(u, v)} = \det\begin{pmatrix} \dfrac{\partial x}{\partial u} & \dfrac{\partial x}{\partial v} \\ \dfrac{\partial y}{\partial u} & \dfrac{\partial y}{\partial v} \end{pmatrix}$$

via symbolic cancellation. Note that

$$\frac{\partial(y, x)}{\partial(u, v)} = -\frac{\partial(x, y)}{\partial(u, v)},$$

since the sign of a determinant is *reversed* if we interchange the order of its rows. The notation (5) leads to the important concept of *differential forms*, which you may study if you take advanced calculus.

We now can give an indication of the importance of surface integrals in physics. In the study of fluid flows (and also electrical, magnetic, gravitational, and other fields), the *flux* is a measure of the amount of fluid per unit area flowing across a given smooth surface S per unit time. Let $\boldsymbol{F}: \boldsymbol{R}^3 \to \boldsymbol{R}^3$ be the velocity field of the flow. We assume that $\boldsymbol{F}$ is continuous on S. Consider now a small patch of S near a typical point $\mathbf{X}(u_i, v_j)$. Since S is smooth, this patch is nearly planar (i.e., flat). By the continuity of $\boldsymbol{F}$, the velocity at all points of the patch is close to $\boldsymbol{F}(\mathbf{X}(u_i, v_j))$. Thus, the coordinate of the velocity normal to the patch is approximately $\boldsymbol{F}(\mathbf{X}(u_i, v_j)) \cdot \boldsymbol{N}(u_i, v_j)$. This measures the rate at which the fluid is flowing across the patch. In a small time period Δt, the mass of the fluid that crosses the patch is therefore approximately

$$\delta \boldsymbol{F}(\mathbf{X}(u_i, v_j)) \cdot \boldsymbol{N}(u_i, v_j)\, \Delta S_{ij}\, \Delta t,$$

where δ is the density and ΔS_{ij} is the area of the patch. (Refer to Figure 6.3, where we show the small cylinder filled by the fluid during the period Δt.) Hence the *total mass of the fluid crossing the entire surface* S during time Δt will be

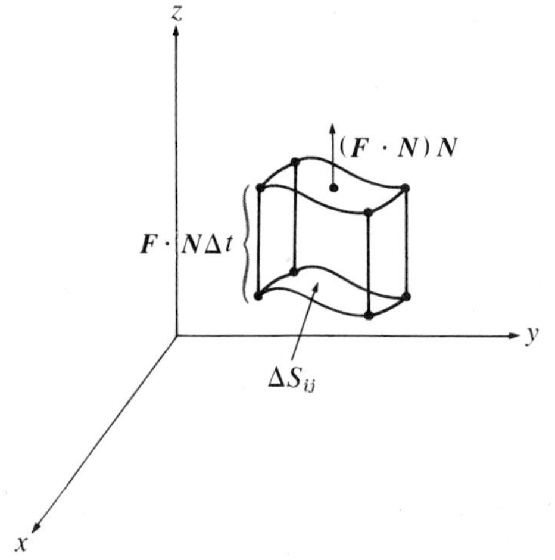

FIGURE 6.3

approximately

$$\delta\,\Delta t \sum_{i,j} \boldsymbol{F}(\boldsymbol{X}(u_i, v_j)) \cdot \boldsymbol{N}(u_i, v_j)\,\Delta S_{ij}.$$

The following definition is thus appropriate.

6.6 **DEFINITION.** The **flux** across S of a fluid flowing with continuous velocity field $\boldsymbol{F}: \boldsymbol{R}^3 \to \boldsymbol{R}^3$ is

$$\Phi(S) = \iint_S \boldsymbol{F} \cdot d\boldsymbol{S}.$$

For instance, if the $\boldsymbol{F}$ of Example 6.5 is the velocity field of a fluid flow with a sink at the origin, then the flux across any sphere of radius a is $-4\pi a^2$.

You will recall that in the case of line integrals we were able to define the line integral of a continuous vector function $\boldsymbol{F}: \boldsymbol{R}^n \to \boldsymbol{R}^n$ over a *piecewise* smooth curve $\gamma = \gamma_1 \cup \gamma_2 \cup \ldots \cup \gamma_k$ (for smooth γ_i) to be the sum of the $\int_{\gamma_i} \boldsymbol{F} \cdot d\boldsymbol{x}$. By a similar approach we can extend surface integrals to most, *but not all*, surfaces obtained by gluing together smooth surface patches S_i along common boundary curves $\boldsymbol{X}_i(\gamma_i)$, where $\boldsymbol{X}_i : D_i \to \boldsymbol{R}^3$ parametrizes S_i and $\gamma_i = \partial(D_i)$. See Figure 6.4.

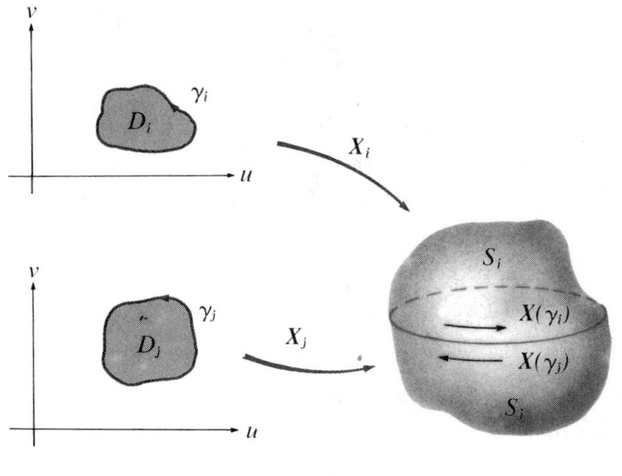

FIGURE 6.4

6.7 **DEFINITION.** A surface $S \subseteq \boldsymbol{R}^3$ is **piecewise smooth** if $S = S_1 \cup S_2 \cup \ldots \cup S_k$ for smooth surface patches S_i such that for $i \neq j$, either $S_i \cap S_j$ is a common boundary curve for both S_i and S_j or else $S_i \cap S_j = \varnothing$.

In Figure 5.4(b), the cylinder is represented as $S_1 \cup S_2 \cup S_3$, where $S_1 \cap S_2$ is the boundary of both the top disk S_2 and of S_1; $S_2 \cap S_3 = \varnothing$; and $S_1 \cap S_3$ is the common boundary of both S_1 and S_3. As you are well aware if you have ever stepped on the top of a can of soup, the cylinder is not smooth. But it *is* piecewise smooth.

In order to satisfactorily extend surface integrals to piecewise smooth surfaces, we need to require the surfaces to be *orientable.* Intuitively, this simply means that it is possible to parametrize the patches S_i in such a way that the normal vector $\mathbf{X}_u \times \mathbf{X}_v$ does not reverse its orientation as we move from one surface patch S_i to an adjacent patch S_j. We would, for example, be able to orient the cylinder in Figure 5.4(b) by parametrizing it so that $\mathbf{X}_u \times \mathbf{X}_v$ always points outward. (In Section 8, we would say that $S_1 \cup S_2 \cup S_3$ is the *positively oriented boundary* of the solid it encloses in that case.) It turns out that the next definition assures such orientability, although we we will not show that in general. (Instead, we will illustrate it in the case of the cylinder.)

6.8 **DEFINITION.** A piecewise smooth surface $S = S_1 \cup S_2 \cup \ldots \cup S_k$ is **orientable** if there are parametrizations $\mathbf{X}_i$ of the S_i such that any two common boundary curves $\mathbf{X}(\gamma_i)$ and $\mathbf{X}(\gamma_j)$ of adjoining patches S_i and S_j are traced in opposite directions on S as γ_i and γ_j are traced in the counterclockwise sense around D_i and D_j. See Figure 6.4.

Returning again to Example 5.2(b), we can see from Figure 6.5 not only that

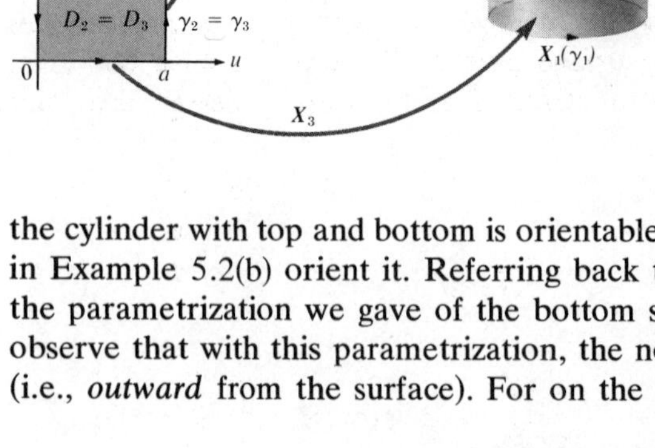

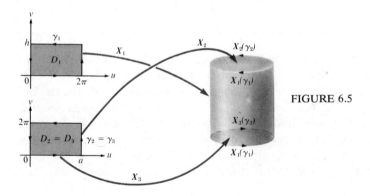

FIGURE 6.5

the cylinder with top and bottom is orientable, but that the parametrizations given in Example 5.2(b) orient it. Referring back to that example, you will recall that the parametrization we gave of the bottom seemed at the time a trifle odd. But observe that with this parametrization, the normal vector to S_3 points downward (i.e., *outward* from the surface). For on the bottom disk S_3 we parametrized by

$$\mathbf{X}(u, v) = (u \cos v, -u \sin v, 0) \quad \text{for} \quad (u, v) \in D_3 = [0, a] \times [0, 2\pi].$$

Therefore,

$$\mathbf{X}_u = (\cos v, -\sin v, 0), \qquad \mathbf{X}_v = (-u \sin v, -u \cos v, 0),$$
$$\mathbf{X}_u \times \mathbf{X}_v = (0, 0, -u) = -u\mathbf{k}.$$

On the top, we parametrized S_2 by

$$\mathbf{X}(u, v) = (u \cos v, u \sin v, h) \quad \text{for} \quad (u, v) \in D_2 = [0, a] \times [0, 2\pi].$$

Hence in this case we obtain

$$\mathbf{X}_u = (\cos v, \sin v, 0), \qquad \mathbf{X}_v = (-u \sin v, u \cos v, 0),$$
$$\mathbf{X}_u \times \mathbf{X}_v = (0, 0, u) = u\boldsymbol{k},$$

which points upward since u is positive. Again, this is outward from the surface. Finally, we parametrized the cylindrical middle portion S_1 by

$$\mathbf{X}(u, v) = (a \cos u, a \sin u, v) \quad \text{for} \quad (u, v) \in D_1 = [0, 2\pi] \times [0, h].$$

We thus have

$$\mathbf{X}_u = (-a \sin u, a \cos u, 0), \qquad \mathbf{X}_v = (0, 0, 1),$$
$$\mathbf{X}_u \times \mathbf{X}_v = (a \cos u, a \sin u, 0),$$

which again points outward from the surface.

Note also that had we used the apparently more natural parametrization of S_3 given by

$$\mathbf{X}_4(u, v) = (u \cos v, u \sin v, 0) \quad \text{for} \quad (u, v) \in [0, a] \times [0, 2\pi],$$

then we would *not* have had an oriented surface. The reason is that $\mathbf{X}_4(\gamma_3)$ would traverse the bottom circle in the *same* direction that $\mathbf{X}_1(\gamma_3)$ traversed the bottom circle. Moreover, the normal vector $\mathbf{X}_u \times \mathbf{X}_v$ would, as in the case of S_2, have been $(0, 0, u) = u\boldsymbol{k}$, which points *inward* unlike the other two normals.

In the case of piecewise smooth orientable surfaces S, we can extend the notion of surface integral in Definition 6.4.

6.9 **DEFINITION.** If S is a piecewise smooth orientable surface in $\boldsymbol{R}^3$, $S = S_1 \cup S_2 \cup \ldots \cup S_k$ where the S_i's are parametrized in accord with Definition 6.8, then the **surface integral** of a continuous vector field $\boldsymbol{F}: \boldsymbol{R}^3 \to \boldsymbol{R}^3$ over S is defined by

(6) $$\iint_S \boldsymbol{F} \cdot d\boldsymbol{S} = \sum_{i=1}^{k} \iint_{S_i} \boldsymbol{F} \cdot d\boldsymbol{S}_i.$$

Definition 6.9 is seen to bear the same relation to Definition 6.4 that Definition 1.5 bears to Definition 1.3.

6.10 **EXAMPLE.** Compute the surface integral of

$$\boldsymbol{F}(x, y, z) = (-y, x + y, z)$$

over the solid cylinder

$$S = \{(x, y, z) \mid x^2 + y^2 = 1, 0 \le z \le 1\} \cup \{(x, y, z) \mid 0 \le x^2 + y^2 \le 1, z = 0, z = 1\}.$$

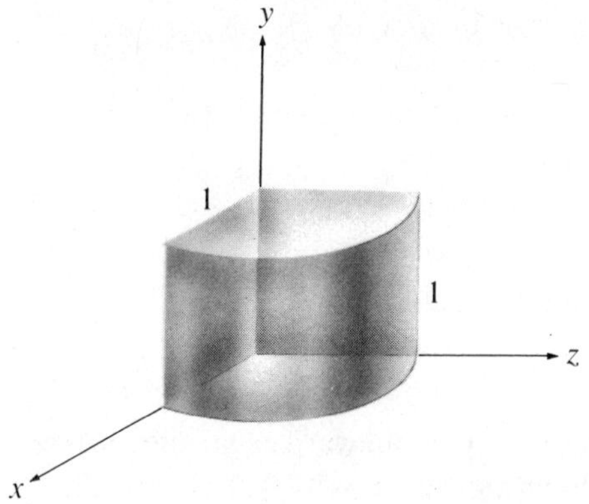

FIGURE 6.6

Solution. S is a cylinder including top $z=1$ and bottom $z=0$. See Figure 6.6, where the first octant portion is shown. We can use the parametrization of Example 5.2(b) with $a=1$ and $h=1$. From the discussion following Definition 6.8 we have:

(i) On the curved portion S_1, $\mathbf{X}(u, v)=(\cos u, \sin u, v)$ for (u, v) in $[0, 2\pi]\times[0, 1]$. Thus, on S_1,

$$\mathbf{F}(x, y, z)=(-\sin u, \sin u+\cos u, v).$$

We also calculated $\mathbf{X}_u \times \mathbf{X}_v=(\cos u, \sin u, 0)$.

(ii) On the top S_2, $\mathbf{X}(u, v)=(u \cos v, u \sin v, 1)$ for (u, v) in $[0, 1]\times[0, 2\pi]$. Therefore, on S_2,

$$\mathbf{F}(x, y, z)=(-u \sin v, u \cos v+u \sin v, 1).$$

We also calculated $\mathbf{X}_u \times \mathbf{X}_v=(0, 0, u)$.

(iii) On the bottom S_3, $\mathbf{X}(u, v)=(u \cos v, -u \sin v, 0)$ for (u, v) in $[0, 1]\times[0, 2\pi]$. Then on S_3,

$$\mathbf{F}(x, y, z)=(u \sin v, u \cos v-u \sin v, 0).$$

We also calculated $\mathbf{X}_u \times \mathbf{X}_v=(0, 0, -u)$.

From (i), (ii), and (iii), and Definition 6.9, we obtain

$$\iint_S \mathbf{F}\cdot d\mathbf{S}=\iint_{S_1} \mathbf{F}\cdot d\mathbf{S}+\iint_{S_2} \mathbf{F}\cdot d\mathbf{S}+\iint_{S_3} \mathbf{F}\cdot d\mathbf{S}$$

$$=\iint_{D_1}(-\sin u, \sin u+\cos u, v)\cdot(\cos u, \sin u, 0)\, du\, dv$$

$$+\iint_{D_2}(-u \sin v, u \cos v+u \sin v, 1)\cdot(0, 0, u)\, du\, dv$$

$$+\iint_{D_3}(u \sin v, u \cos v-u \sin v, 0)\cdot(0, 0, -u)\, du\, dv$$

$$=\iint_{D_1} \sin^2 u\, du\, dv+\iint_{D_2} u\, du\, dv+0$$

$$= \int_0^1 dv \int_0^{2\pi} \frac{1-\cos 2u}{2}\, du + \int_0^{2\pi} dv \int_0^1 u\, du$$
$$= 1[\tfrac{1}{2}u - \tfrac{1}{4}\sin 2u]_0^{2\pi} + 2\pi[\tfrac{1}{2}u^2]_0^1$$
$$= \pi + \pi = 2\pi.$$

Our remarks about Example 5.2(b) may give the impression that *any* piecewise smooth surface can be oriented if we just take the trouble to parametrize it properly. But this is not the case. The celebrated **Möbius band** [named for the German mathematician Augustus F. Möbius (1790–1868), who was a student of Gauss] is *not* orientable. The band is made by deforming the rectangle $R = R_1 \cup R_2$ as shown in Figure 6.7. One *twists* the rectangle and then glues edge

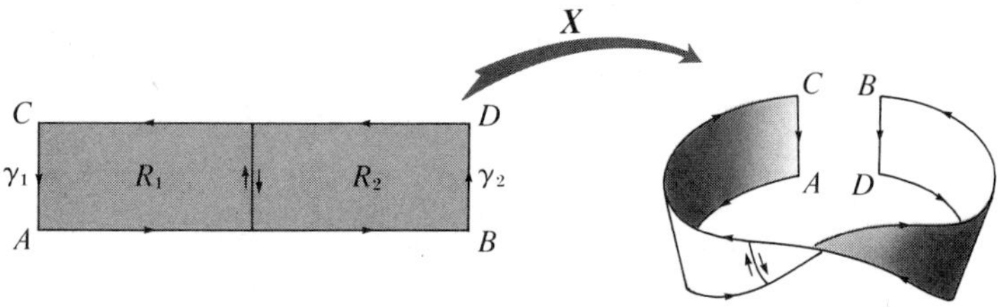

FIGURE 6.7

AC to edge DB. It is fun (and easy) to construct the band from a long strip of paper and then study it first hand. (See Exercise 26.) Figure 6.7 shows that the common boundary at the glued spot where R_1 and R_2 come together is traversed in the *same* sense on both γ_1 and γ_2. Unlike the oriented cylinder in Example 5.2(b), this non-orientable surface has only *one* side. Suppose you begin painting the band on "one side" of AC, and never lift your brush from the surface. Then, after passing through the twist, you will arrive at BD on the "reverse side"! So you can continue, and thus paint the entire band with one stroke! (Have a friend hold a pencil perpendicular to AC while you traverse the band by holding another pencil perpendicular to the band. If you begin with your pencil pointing in the same direction as your friend's, then you will arrive back at BD with your pencil pointing the *opposite* way!) This says that $\mathbf{N}$ *reverses* its orientation near the point where you began to traverse the strip. For after a small displacement, $\mathbf{N}$ points the opposite way! Such surfaces are studied further in topology, but we will stick to orientable surfaces aside from this curious one-sided surface.

EXERCISES 7.6

In Exercises 1 to 18, compute the surface integral of the given function over the given surface.

1 $\iint_S (x^2+y^2)z\, dA$, where S is the top hemisphere of $x^2+y^2+z^2=1$.

2 $\iint_S (x^2+y^2)z^2\, dA$, where S is as in Exercise 1.

3 $\iint_S z\, dA$, where S is the part of the cylinder $x^2+y^2=4$ between the planes $z=0$ and $z=x+2$.

4 $\iint_S z^2\, dA$, where S is as in Exercise 3.

5 $\iint_S (x^2+y^2)\, dA$, where S is the part of the sphere $x^2+y^2+z^2=1$ lying above the cone $z=\sqrt{x^2+y^2}$.

6 $\iint_S x\,dA$, where S is the triangle with vertices $(1, 0, 0)$, $(0, 1, 0)$, $(0, 0, 1)$.

7 $\iint_S x^2\,dA$, where $S = S_1 \cup S_2 \cup S_3$ is the cylinder of Example 5.2(b). (Use Definition 6.9.)

8 $\iint_S x\,dA$, where S is the surface of Exercise 3 complete with top and bottom. (Use Definition 6.9.)

9 $\iint_S \boldsymbol{F} \cdot d\boldsymbol{S}$, where $\boldsymbol{F}(x, y, z) = (x, y, z)$ and S is the sphere $x^2 + y^2 + z^2 = 1$.

10 $\iint_S \boldsymbol{F} \cdot d\boldsymbol{S}$, where $\boldsymbol{F}(x, y, z) = (xz, yz, x^2)$ and S is as in Exercise 9.

11 $\iint_S \boldsymbol{F} \cdot d\boldsymbol{S}$, where $\boldsymbol{F}(x, y, z) = (y, y - x, e^x)$ and S is the part of the paraboloid $z = x^2 + y^2$ lying above the rectangle $[0, 1] \times [0, 3]$.

12 $\iint_S \boldsymbol{F} \cdot d\boldsymbol{S}$, where $\boldsymbol{F}(x, y, z) = (x^2, y^2, z^2)$ and S is the portion of $z = \sqrt{x^2 + y^2}$ between the planes $z = 1$ and $z = 2$.

13 $\iint_S \boldsymbol{F} \cdot \boldsymbol{N}\,dS$, where $\boldsymbol{F}(x, y, z) = (x^2, y^2, z^2)$ and S is the cone $z = \sqrt{x^2 + y^2}$ between $z = 0$ and $z = 3$.

14 $\iint_S \boldsymbol{F} \cdot d\boldsymbol{S}$, where $\boldsymbol{F}(x, y, z) = (x, y, z)$ and S is the cylinder $x^2 + y^2 = 4$ between $z = 0$ and $z = x + 2$.

15 $\iint_S \boldsymbol{F} \cdot d\boldsymbol{S}$, where $\boldsymbol{F}(x, y, z) = (x, y, z)$ and S is the cylinder $x^2 + y^2 = 1$, $0 \le z \le 1$ with bottom but no top. (Use Definition 6.9.)

16 $\iint_S \boldsymbol{F} \cdot \boldsymbol{N}\,dS$, where $\boldsymbol{F}(x, y, z) = (2x, -3y, z)$ and S is the part of the cylinder $x^2 + y^2 = 1$ between $z = 0$ and $z = x + 2$.

17 $\iint_S \boldsymbol{F} \cdot d\boldsymbol{S}$, where $\boldsymbol{F}(x, y, z) = (x + y, y, x + z)$ and S is the surface of Example 6.10.

18 $\iint_S \boldsymbol{F} \cdot d\boldsymbol{S}$, where $\boldsymbol{F}(x, y, z) = (y, y - x, y + z)$ and S is the surface of Example 6.10.

19 The **electrostatic potential** V at $\boldsymbol{0}$ due to a charge distribution of charge density $\delta(x, y, z)$ on S is

$$V(\boldsymbol{0}) = \iint_S \frac{\delta(\boldsymbol{x})}{|\boldsymbol{x}|}\,dS.$$

If $\delta(x, y, z) = \delta$, a constant, then find $V(\boldsymbol{0})$ for S the sphere $x^2 + y^2 + z^2 = a^2$.

20 Repeat Exercise 19 for the portion of $z = \sqrt{x^2 + y^2}$ lying between $z = 1$ and $z = 3$.

21 If a fluid flows with velocity $\boldsymbol{F}(x, y, z) = \dfrac{\boldsymbol{x}}{|\boldsymbol{x}|^3}$, then find the flux across the sphere $x^2 + y^2 + z^2 = 1$. (Note that $\boldsymbol{F}$ is an inverse square law field. The next exercise says that our answer won't change if we consider bigger spheres.)

22 If a fluid flows with velocity $\boldsymbol{F}(x, y, z) = \dfrac{k}{|\boldsymbol{x}^2|} \dfrac{\boldsymbol{x}}{|\boldsymbol{x}|}$, then show that the flux across any sphere of radius a centered at $\boldsymbol{0}$ is the same as that across any other concentric sphere.

23 In the study of heat flows, one supposes that the temperature $T(x, y, z)$ has a continuous gradient $\boldsymbol{\nabla} T(\boldsymbol{x})$. The negative gradient gives the magnitude and direction of the rate of flow of the heat at $\boldsymbol{x}$. If $T(x, y, z) = x^2 + y^2$, then find the flux across the cylinder $x^2 + y^2 = 1$ between $z = 0$ and $z = 1$.

24 Refer to Exercise 23. Find the flux across the sphere $x^2 + y^2 + z^2 = 1$ if $T(x, y, z) = \sqrt{x^2 + y^2 + z^2}$.

25 Is the surface integral in Definition 6.4 independent of parametrization?

26 Construct two paper Möbius bands.
(a) Using a pair of scissors, pierce one band midway between the edges, and cut along the entire surface parallel to the edges. What do you obtain?
(b) Repeat (a) for the resulting surface. What happens?
(c) Try making two parallel cuts on the second band one third of the width of the band from each edge. What happens?

7 THEOREM OF STOKES

This section takes its name from the Irish mathematician George Gabriel Stokes (1819–1903), who in 1854 published a theorem that can be regarded as an extension of Green's Theorem 2.3 from two to three dimensions. This theorem of Stokes pointed the way to a far-reaching generalized version of Green's Theorem in n dimensions, which you will learn if you study advanced calculus. It is also of great importance in engineering and physics.

Recall that in Section 3 (p. 435) we saw the following: If $\boldsymbol{F}:\boldsymbol{R}^2 \to \boldsymbol{R}^2$ is a vector field whose coordinate functions have continuous first partial derivatives on a simple region $D \subseteq \boldsymbol{R}^2$ with boundary γ traversed counterclockwise, then Green's Theorem can be put in the form

(1) $$\iint_D (\textbf{curl}\,\boldsymbol{F} \cdot \boldsymbol{k})\, dA = \oint_\gamma \boldsymbol{F} \cdot d\boldsymbol{x}.$$

How can we extend this to three dimensions? By considering for $\boldsymbol{F}:\boldsymbol{R}^3 \to \boldsymbol{R}^3$ the *surface integral* of $\textbf{curl}\,\boldsymbol{F} \cdot \boldsymbol{N}$ over a smooth surface patch S. Recall that S is parametrized by $\boldsymbol{X}:\boldsymbol{R}^2 \to \boldsymbol{R}^3$ so that $S = \boldsymbol{X}(D)$ for $D \subseteq \boldsymbol{R}^2$. (See Figure 7.1.)

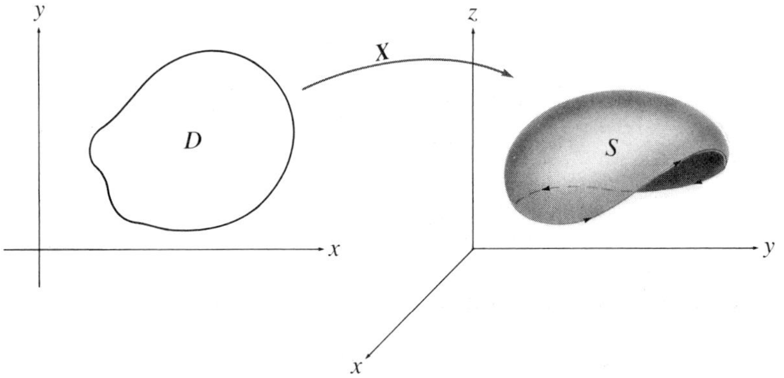

FIGURE 7.1

As γ is traversed in the counterclockwise sense, the boundary $\partial S = \boldsymbol{X}(\gamma)$ of S is traversed in what we will call the **positive sense**. Then the analogue of (1) is

(2) $$\iint_S \textbf{curl}\,\boldsymbol{F} \cdot \boldsymbol{N}\, dA = \oint_{\partial S} \boldsymbol{F} \cdot d\boldsymbol{x}$$

where $\oint$ means that ∂S is traversed in the positive sense. Notice that (2) asserts that the value of the surface integral $\iint_S \textbf{curl}\,\boldsymbol{F} \cdot d\boldsymbol{S}$ of the quantity $\boldsymbol{\nabla} \times \boldsymbol{F}$ (which we again can think of as a *kind of derivative* of $\boldsymbol{F}$) over S is completely determined by the values of $\boldsymbol{F}$ on the *boundary* of S. So (2) is one more analogue of the Fundamental Theorem on Evaluation of Definite Integrals in elementary calculus. We leave a proof of (2) in full generality for advanced calculus. We can, however, establish the following version, which will be adequate for most computational purposes.

7.1 **THEOREM OF STOKES.** Suppose that $S \subseteq \boldsymbol{R}^3$ is a smooth surface patch parametrized by

$$\mathbf{X}(u, v) = (x(u, v), y(u, v), z(u, v)),$$

where x, y, and z have continuous partial derivatives on a simple region $D \subseteq \boldsymbol{R}^2$ bounded by a piecewise smooth positively oriented simple closed curve γ. Suppose that $\boldsymbol{F}: \boldsymbol{R}^3 \to \boldsymbol{R}^3$ is given by

$$\boldsymbol{F}(x, y, z) = (P(x, y, z), Q(x, y, z), R(x, y, z)),$$

where P, Q, and R have continuous partial derivatives on S. Then

$$\iint_S \mathbf{curl}\, \boldsymbol{F} \cdot d\boldsymbol{S} = \oint_{\partial S} \boldsymbol{F} \cdot d\boldsymbol{x}. \tag{2}$$

Proof. The proof is easiest to give if we adopt the notation (5) of Section 6 (p. 462). In this notation, since

$$\mathbf{curl}\, \boldsymbol{F} = \left(\frac{\partial R}{\partial y} - \frac{\partial Q}{\partial z}, \frac{\partial P}{\partial z} - \frac{\partial R}{\partial x}, \frac{\partial Q}{\partial x} - \frac{\partial P}{\partial y}\right),$$

we have

$$\iint_S \mathbf{curl}\, \boldsymbol{F} \cdot d\boldsymbol{S} = \iint_S \left(\frac{\partial R}{\partial y} - \frac{\partial Q}{\partial z}\right) dy \wedge dz + \left(\frac{\partial P}{\partial z} - \frac{\partial R}{\partial x}\right) dz \wedge dx$$
$$+ \left(\frac{\partial Q}{\partial x} - \frac{\partial P}{\partial y}\right) dx \wedge dy.$$

We need to prove that this is the same as

$$\oint_{\partial S} \boldsymbol{F} \cdot d\boldsymbol{x} = \int_{\partial S} P\, dx + Q\, dy + R\, dz.$$

This will follow if we can prove the following three equalities:

$$\int_{\partial S} P\, dx = \iint_S -\frac{\partial P}{\partial y}\, dx \wedge dy + \frac{\partial P}{\partial z}\, dz \wedge dx, \tag{3}$$

$$\int_{\partial S} Q\, dy = \iint_S -\frac{\partial Q}{\partial z}\, dy \wedge dz + \frac{\partial Q}{\partial x}\, dx \wedge dy, \tag{4}$$

(5) $$\int_{\partial S} R\,dz = \iint_S -\frac{\partial R}{\partial x}\,dz \wedge dx + \frac{\partial R}{\partial y}\,dy \wedge dz.$$

Each of these is proved by applying Green's Theorem. We will prove (3) and leave the similar proofs of (4) and (5) as Exercises 24 and 25 below. Suppose that γ is parametrized by

$$\boldsymbol{x}(t) = (x, y) = (u(t), v(t)) \quad \text{for} \quad t \in [a, b].$$

Then ∂S is parametrized by

$$\mathbf{X}(\boldsymbol{x}(t)) = (x(\boldsymbol{x}(t)), y(\boldsymbol{x}(t)), z(\boldsymbol{x}(t))).$$

We have then

$$\oint_{\partial S} P\,dx = \int_a^b (P(\mathbf{X}(\boldsymbol{x}(t)), 0, 0) \cdot \dot{\mathbf{X}}(\boldsymbol{x}(t))\,dt$$

$$= \int_a^b P(\mathbf{X}(u(t), v(t)))\frac{dx(\boldsymbol{x}(t))}{dt}\,dt$$

$$= \int_a^b P(\mathbf{X}(u(t), v(t)))\left(\frac{\partial x}{\partial u}\frac{du}{dt} + \frac{\partial x}{\partial v}\frac{dv}{dt}\right)dt$$

by Chain Rule 4.5.3

$$= \oint_\gamma (P \circ \mathbf{X})(u, v)\frac{\partial x}{\partial u}\,du + (P \circ \mathbf{X})(u, v)\frac{\partial x}{\partial v}\,dv$$

(6) $$= \iint_D \left[\frac{\partial}{\partial u}\left(P \circ \mathbf{X}(u, v)\frac{\partial x}{\partial v}\right) - \frac{\partial}{\partial v}\left(P \circ \mathbf{X}(u, v)\frac{\partial x}{\partial u}\right)\right] du\,dv$$

by Green's Theorem. [Note that $(x, y) = (u(t), v(t))$, so $(dx, dy) = \left(\frac{du}{dt}\,dt, \frac{dv}{dt}\,dt\right) = (du, dv)$.]

We are now reduced to calculating the bracket in the double integral. By the product rule for derivatives and the Chain Rule, we have

(7) $$\frac{\partial}{\partial u}\left(P \circ \mathbf{X}(u, v)\frac{\partial x}{\partial v}\right) = \frac{\partial}{\partial u}(P \circ \mathbf{X})\frac{\partial x}{\partial v} + (P \circ \mathbf{X})\frac{\partial^2 x}{\partial u\,\partial v}$$

$$= \left(\frac{\partial P}{\partial x}\frac{\partial x}{\partial u} + \frac{\partial P}{\partial y}\frac{\partial y}{\partial u} + \frac{\partial P}{\partial z}\frac{\partial z}{\partial u}\right)\frac{\partial x}{\partial v} + (P \circ \mathbf{X})\frac{\partial^2 x}{\partial u\,\partial v},$$

(8) $$\frac{\partial}{\partial v}(P \circ \mathbf{X}(u, v))\frac{\partial x}{\partial u} = \frac{\partial}{\partial v}(P \circ \mathbf{X})\frac{\partial x}{\partial u} + (P \circ \mathbf{X})\frac{\partial^2 x}{\partial v\,\partial u}$$

$$= \left(\frac{\partial P}{\partial x}\frac{\partial x}{\partial v} + \frac{\partial P}{\partial y}\frac{\partial y}{\partial v} + \frac{\partial P}{\partial z}\frac{\partial z}{\partial v}\right)\frac{\partial x}{\partial u} + (P \circ \mathbf{X})\frac{\partial^2 x}{\partial v\,\partial u}.$$

Our hypotheses say $\dfrac{\partial^2 x}{\partial u\,\partial v}$ and $\dfrac{\partial^2 x}{\partial v\,\partial u}$ are continuous, so they are equal by Theorem 4.7.5. Then subtraction of (8) from (7) gives us

$$\frac{\partial}{\partial u}\left(P\circ\mathbf{X}(u,v)\frac{\partial x}{\partial v}\right)-\frac{\partial}{\partial v}\left(P\circ\mathbf{X}(u,v)\frac{\partial x}{\partial u}\right)=\frac{\partial P}{\partial y}\left(\frac{\partial y}{\partial u}\frac{\partial x}{\partial v}-\frac{\partial y}{\partial v}\frac{\partial x}{\partial u}\right)$$
$$+\frac{\partial P}{\partial z}\left(\frac{\partial z}{\partial u}\frac{\partial x}{\partial v}-\frac{\partial z}{\partial v}\frac{\partial x}{\partial u}\right)$$
$$=-\frac{\partial P}{\partial y}\frac{\partial(x,y)}{\partial(u,v)}+\frac{\partial P}{\partial z}\frac{\partial(z,x)}{\partial(u,v)}.$$

So, substituting above in (6), we obtain

$$\oint_{\partial S} P\,dx=\iint_D\left[-\frac{\partial P}{\partial y}\frac{\partial(x,y)}{\partial(u,v)}+\frac{\partial P}{\partial z}\frac{\partial(z,x)}{\partial(u,v)}\right]du\,dv$$
$$=\iint_D -\frac{\partial P}{\partial y}\,dx\wedge dy+\frac{\partial P}{\partial z}\,dz\wedge dx.$$

Hence (3) is proved. The same approach establishes (4) and (5). Then (2) follows upon adding (3), (4), and (5). QED

As you might expect, Theorem 7.1 can be extended to piecewise smooth orientable surfaces S. From Definition 6.9, if $S=S_1\cup S_2\cup\ldots\cup S_k$ for smooth patches S_i, then

(9) $$\iint_S \mathbf{curl}\,\boldsymbol{F}\cdot d\boldsymbol{S}=\sum_{i=1}^{k}\iint_{S_i}\mathbf{curl}\,\boldsymbol{F}\cdot d\boldsymbol{S}$$
$$=\sum_{i=1}^{k}\oint_{\partial S_i}\boldsymbol{F}\cdot d\boldsymbol{x}\quad\text{by Theorem 7.1.}$$

Now recall from Definition 6.8 that any piece of common boundary of two of the S_i will be traversed twice in *opposite* directions in the line integral (Figure 7.2), so

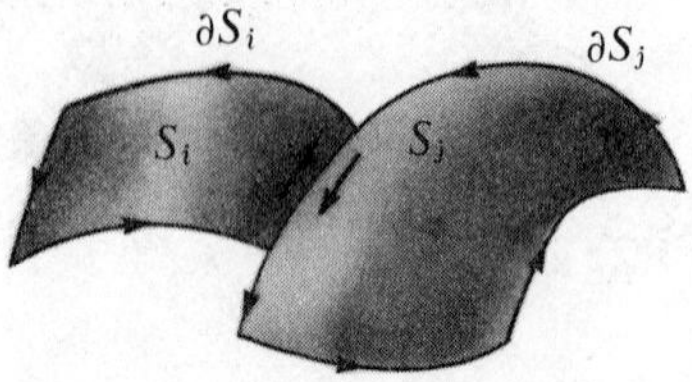

FIGURE 7.2

the line integrals over these two pieces will cancel by Theorem 1.8. So the line integral in (9) is actually

$$\oint_{\partial S}\boldsymbol{F}\cdot d\boldsymbol{x}.$$

Thus,

$$\iint_S \textbf{curl}\, \boldsymbol{F} \cdot d\boldsymbol{S} = \oint_{\partial S} \boldsymbol{F} \cdot d\boldsymbol{x}$$

holds if S is piecewise smooth and orientable. We state this as an extension of Theorem 7.1.

7.2 **THEOREM.** If S is a piecewise smooth orientable surface, and $\boldsymbol{F}$ is as in Theorem 7.1, then

(2)
$$\iint_S \textbf{curl}\, \boldsymbol{F} \cdot \boldsymbol{N}\, dA = \oint_{\partial S} \boldsymbol{F} \cdot d\boldsymbol{x}.$$

The remarks on p. 423 and p. 435 about circulation in a fluid flow in $\boldsymbol{F}^2$ apply equally well to fluid flows in $\boldsymbol{R}^3$. As before, $\oint_{\partial S} \boldsymbol{F} \cdot d\boldsymbol{x}$ is the circulation around ∂S. For a flow in $\boldsymbol{R}^3$ whose velocity field $\boldsymbol{F}$ satisfies the hypotheses of Theorems 7.1 or 7.2, we can conclude that the circulation around ∂S is given by $\iint_S \textbf{curl}\, \boldsymbol{F} \cdot \boldsymbol{N}\, dA$. In particular, if S_r is a closed disk of radius r centered at $\boldsymbol{x}_0$, then the circulation around ∂S_r is $\iint_{S_r} \textbf{curl}\, \boldsymbol{F} \cdot \boldsymbol{N}\, dA$ where $\boldsymbol{N}$ is perpendicular to the disk. (See Figure 7.3.) If r is small, then the surface integral will be approximately

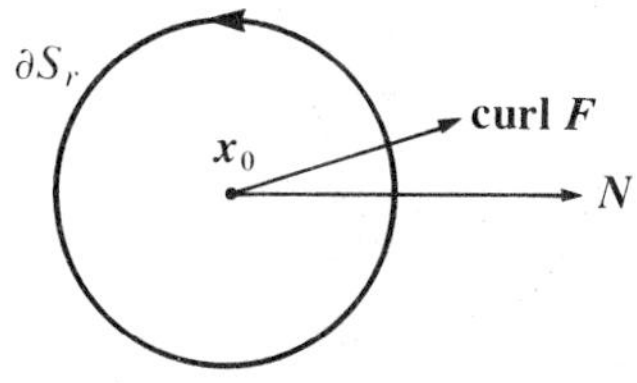

FIGURE 7.3

$(\textbf{curl}\, \boldsymbol{F}(\boldsymbol{x}_0) \cdot \boldsymbol{N}) A(S_r)$. Hence, among all axes $\boldsymbol{N}$ of rotation, $\textbf{curl}\, \boldsymbol{F}(\boldsymbol{x}_0)$ is the one that maximizes the circulation around small circles centered at $\boldsymbol{x}_0$ and perpendicular to $\boldsymbol{N}$. Physically, a paddle wheel inserted into the flow at $\boldsymbol{x}_0$ will revolve most rapidly counterclockwise if it is placed with its axle in the direction of $\textbf{curl}\, \boldsymbol{F}(\boldsymbol{x}_0)$. See Figure 7.4. If $\textbf{curl}\, \boldsymbol{F}(\boldsymbol{x}_0) = \boldsymbol{0}$, then the wheel will not revolve at all; this is the origin of the term "irrotational" for flows such that $\textbf{curl}\, \boldsymbol{F} = \boldsymbol{0}$.

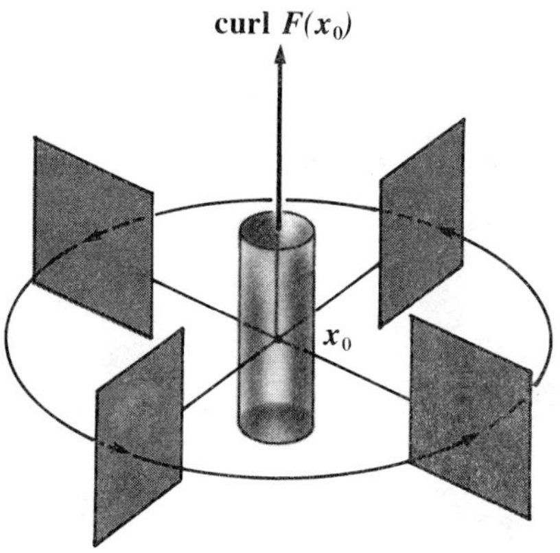

FIGURE 7.4

Theorems 7.1 and 7.2 can save considerable labor in evaluating surface integrals of curls. The next example illustrates this.

7.3 EXAMPLE. Evaluate $\iint_{S_r} \mathbf{curl}\, \boldsymbol{F} \cdot \boldsymbol{N}\, dA$ if

$$\boldsymbol{F}(x, y, z) = z\boldsymbol{i} + x\boldsymbol{j} + y\boldsymbol{k}$$

and S is the surface $z = x^2 + y^2$ lying below the plane $z = 1$.

Solution. The surface can be parametrized by

$$\boldsymbol{X}(u, v) = (x, y, z) = (u, v, u^2 + v^2),$$

where D is the disk $0 \le u^2 + v^2 \le 1$. According to Theorem 7.1,

$$\iint_S \mathbf{curl}\, \boldsymbol{F} \cdot \boldsymbol{N}\, dA = \oint_{\partial S} \boldsymbol{F} \cdot d\boldsymbol{x} = \oint_{\partial S} (z, x, y) \cdot d\boldsymbol{x} \tag{2}$$

where ∂S is the circle $\gamma: x^2 + y^2 = 1$, $z = 1$ traversed counterclockwise. (See Figure 7.5.)

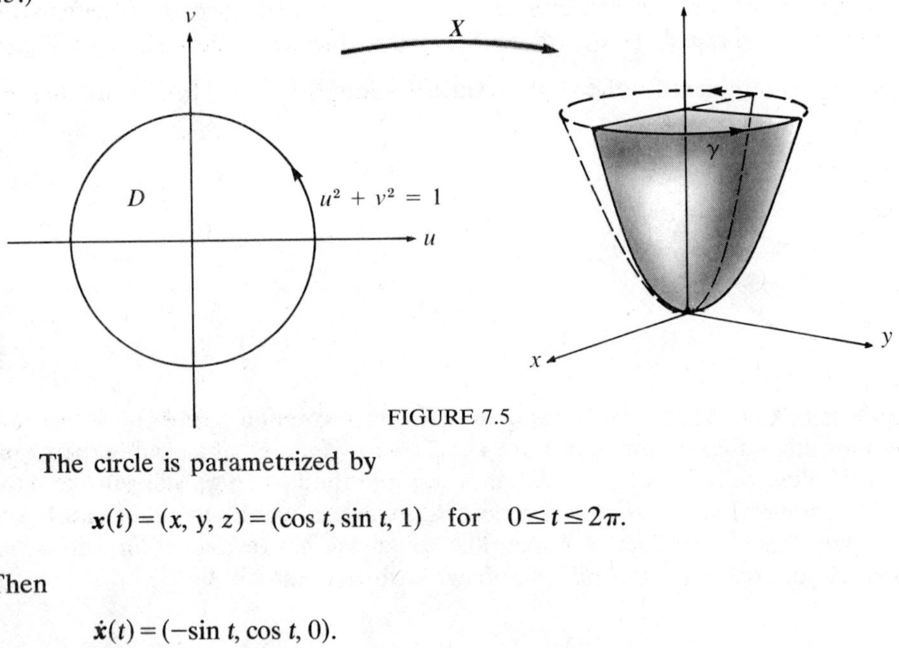

FIGURE 7.5

The circle is parametrized by

$$\boldsymbol{x}(t) = (x, y, z) = (\cos t, \sin t, 1) \quad \text{for} \quad 0 \le t \le 2\pi.$$

Then

$$\dot{\boldsymbol{x}}(t) = (-\sin t, \cos t, 0).$$

Hence from (2),

$$\begin{aligned}\iint_S \mathbf{curl}\, \boldsymbol{F} \cdot \boldsymbol{N}\, dA &= \int_0^{2\pi} (1, \cos t, \sin t) \cdot (-\sin t, \cos t, 0)\, dt \\ &= \int_0^{2\pi} (-\sin t + \cos^2 t)\, dt \\ &= \cos t\Big]_0^{2\pi} + \int_0^{2\pi} \frac{1 + \cos 2t}{2}\, dt \\ &= 0 + \tfrac{1}{2}[t + \tfrac{1}{2}\sin 2t]_0^{2\pi} \\ &= \tfrac{1}{2} \cdot 2\pi = \pi.\end{aligned}$$

A second look at Example 7.3 reveals that we have an analogue of independence-of-path for line integrals. Namely, the surface integral in Example 7.3 is *independent of surface* in the following sense. If S is *any* surface so parametrized that its positively oriented boundary is γ, then

$$\iint_S \mathbf{curl}\, \boldsymbol{F} \cdot \boldsymbol{N}\, dA = \pi.$$

That is *exactly* what Stokes's Theorem says. The surface integral depends *only* on γ, not on the surface S which γ bounds!

Speaking of independence of path, we can now establish the three-dimensional analogue of Theorem 4.8, as promised on p. 446.

7.4 **THEOREM.** Suppose that $\boldsymbol{F}: \boldsymbol{R}^3 \to \boldsymbol{R}^3$, and that $J_{\boldsymbol{F}}(\boldsymbol{x})$ has continuous entries and is symmetric on a simply connected region $D \subseteq \boldsymbol{R}^3$. Then $\boldsymbol{F}$ is a conservative vector field on D.

Proof. Let γ be any planar simple closed curve in D. By the definition of a simply connected region, γ bounds a surface patch $S \subseteq D$. Then by (2) we have

$$\oint_\gamma \boldsymbol{F} \cdot d\boldsymbol{x} = \pm \iint_S (\mathbf{curl}\, \boldsymbol{F} \cdot \boldsymbol{N})\, dA. \tag{10}$$

(The plus or minus sign depends on how S is parametrized. The plus sign applies if γ is the positively oriented boundary of S.) Now if $\boldsymbol{F} = (f_1, f_2, f_3)$, then

$$\mathbf{curl}\, \boldsymbol{F} = \left(\frac{\partial f_3}{\partial y} - \frac{\partial f_2}{\partial z}, \frac{\partial f_1}{\partial z} - \frac{\partial f_3}{\partial x}, \frac{\partial f_2}{\partial x} - \frac{\partial f_1}{\partial y}\right) = \mathbf{0},$$

since

$$J_{\boldsymbol{F}} = \begin{pmatrix} \frac{\partial f_1}{\partial x} & \frac{\partial f_1}{\partial y} & \frac{\partial f_1}{\partial z} \\ \frac{\partial f_2}{\partial x} & \frac{\partial f_2}{\partial y} & \frac{\partial f_2}{\partial z} \\ \frac{\partial f_3}{\partial x} & \frac{\partial f_3}{\partial y} & \frac{\partial f_3}{\partial z} \end{pmatrix}$$

is symmetric. Now (10) says that

$$\oint_\gamma \boldsymbol{F} \cdot d\boldsymbol{x} = \pm \iint_S \mathbf{0} \cdot \boldsymbol{N}\, dA = 0.$$

So by Theorem 4.3, $\boldsymbol{F}$ is conservative. QED

We can often save work by using the approach of the foregoing proof in computational examples. When given a line integral $\oint_\gamma \boldsymbol{F} \cdot d\boldsymbol{x}$ to calculate over a closed path γ, first compute $\mathbf{curl}\, \boldsymbol{F}$. If $\mathbf{curl}\, \boldsymbol{F} = \mathbf{0}$ (for instance, if $\boldsymbol{F}$ is conservative and J_F has continuous entries—cf. Theorem 3.11(d)), then (2) gives

$$\oint_\gamma \boldsymbol{F} \cdot d\boldsymbol{x} = \iint_S \boldsymbol{0} \cdot d\boldsymbol{S} = 0$$

for any smooth surface S which γ bounds. There is usually no quicker way of computing the answer in this case. Even if **curl** $\boldsymbol{F}$ *isn't* $\boldsymbol{0}$, it still might be easier to use (2) than to compute the line integral directly (cf. Example 7.6 below). So the effort expended in computing **curl** $\boldsymbol{F}$ is justified by the potential gain.

The importance of Theorems 7.1 and 7.2 in electrical theory is easy to indicate. The **current density vector** $\boldsymbol{J}$ may be thought of as giving the magnitude and direction of a flow of electric charges whose motion constitutes an electrical current. The total current I flowing across a surface S is defined to be

(11) $$I = \iint_S \boldsymbol{J} \cdot d\boldsymbol{S}.$$

The **magnetic field intensity vector** $\boldsymbol{H}$ of a steady magnetic field is related to $\boldsymbol{J}$ by *Maxwell's Second Law* (named for the English physicist James Clerk Maxwell (1831–1879)),

(12) $$\boldsymbol{J} = \textbf{curl}\, \boldsymbol{H}$$

Using (12) we can derive a basic law given by Andre M. Ampère (1775–1836).

7.5 EXAMPLE. Establish *Ampère's circuital law*,

(13) $$\oint_\gamma \boldsymbol{H} \cdot d\boldsymbol{x} = I$$

where γ is a closed path traversed counterclockwise.

Solution. Let S be any piecewise smooth oriented surface for which γ is the positively oriented boundary. Then the Theorem of Stokes gives

$$\begin{aligned}\oint_\gamma \boldsymbol{H} \cdot d\boldsymbol{x} &= \iint_S \textbf{curl}\, \boldsymbol{H} \cdot d\boldsymbol{S} \\ &= \iint_S \boldsymbol{J} \cdot d\boldsymbol{S} \quad \text{by (12)} \\ &= I \quad \text{by (11)}.\end{aligned}$$

[Actually, (12) and (13) are equivalent. One can also derive (12) from (13).[1]]

While the Theorem of Stokes is ordinarily a tool for reducing computation of surface integrals to line integrals, which are usually easier to evaluate, our final

1. See, for example, W. H. Hayt, Jr., *Engineering Electromagnetics*, 3rd Edition, McGraw-Hill, New York, 1974, p. 255.

example shows that it can be worthwhile to bear the theorem in mind when confronted with a line integral.

7.6 EXAMPLE. Compute the circulation of $\boldsymbol{F}(x, y, z) = (y, z, x)$ around the circle γ of intersection of the sphere $x^2+y^2+z^2=1$ with the plane $y=z$ oriented as shown in Figure 7.6.

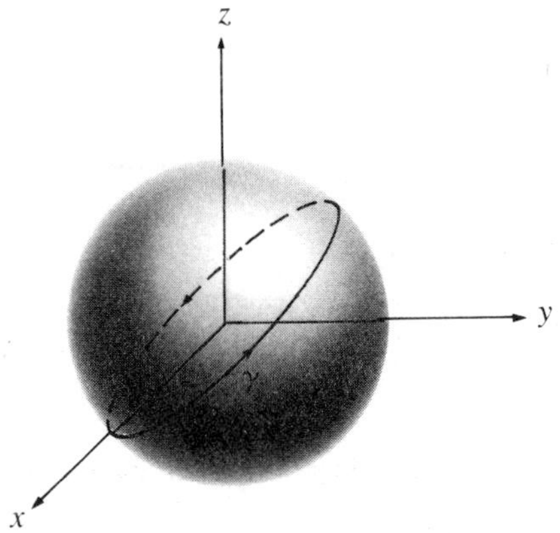

FIGURE 7.6

Solution. Observe that γ bounds a disk S, which we can parametrize by

$$\mathbf{X} = (x, y, z) = \left(u \cos v, \frac{1}{\sqrt{2}} u \sin v, \frac{1}{\sqrt{2}} u \sin v\right)$$

for $u \in [0, 1]$ and $v \in [0, 2\pi]$. (We use this parametrization so that $y=z$ and $x^2+y^2+z^2=1$.) Then we obtain

$$\mathbf{X}_u = \left(\cos v, \frac{1}{\sqrt{2}} \sin v, \frac{1}{\sqrt{2}} \sin v\right)$$

$$\mathbf{X}_v = \left(-u \sin v, \frac{1}{\sqrt{2}} u \cos v, \frac{1}{\sqrt{2}} u \cos v\right),$$

so

$$\mathbf{X}_u \times \mathbf{X}_v = \left(0, \frac{1}{\sqrt{2}}, \frac{1}{\sqrt{2}}\right) u.$$

We also have **curl** $\boldsymbol{F} = (-1, -1, -1)$. So by Theorem 6.1,

$$\oint_\gamma \boldsymbol{F} \cdot d\boldsymbol{x} = \iint_S \textbf{curl } \boldsymbol{F} \cdot \mathbf{N}\, dA = \iint_S (-1, -1, -1) \cdot \left(0, -\frac{1}{\sqrt{2}}, \frac{1}{\sqrt{2}}\right) u\, du\, dv$$

$$= \iint_S 0\, du\, dv = 0.$$

EXERCISES 7.7

In Exercises 1 to 16, apply the Theorem of Stokes to evaluate the given surface or line integral.

1 $\displaystyle\oint_\gamma 2xy^2z\, dx + 2x^2yz\, dy + (x^2y^2 - 2z)\, dz$, where γ is the curve given by $\boldsymbol{x}(t) = (\cos t, \sin t, \sin t)$ for $0 \le t \le 2\pi$.

2 $\oint_\gamma (6x^2y^2-3yz^2)\,dx+(4x^3y-3xz^2)\,dy-6xyz\,dz$, where γ is the curve of intersection of the surface $z=x^2+y^2$ with the plane $z=2+x+y$.

3 $\iint_S \mathbf{curl}\,\boldsymbol{F}\cdot d\boldsymbol{S}$, where S is the portion of the paraboloid $z=4-x^2-y^2$ above the xy-plane and $\boldsymbol{F}(x, y, z)=(-y+z, x+z, -x-y)$.

4 $\iint_S \mathbf{curl}\,\boldsymbol{F}\cdot d\boldsymbol{S}$, where $\boldsymbol{F}(x, y, z)=(y-z, yz, -xz)$ and S is the portion of the unit cube $[0, 1]\times[0, 1]\times[0, 1]$ above the xy-plane, with outward pointing normal.

5 $\oint_\gamma \boldsymbol{F}\cdot d\boldsymbol{x}$, where

$$\boldsymbol{F}=\left(\frac{x}{\sqrt{x^2+y^2+z^2}}, \frac{y}{\sqrt{x^2+y^2+z^2}}, \frac{z}{\sqrt{x^2+y^2+z^2}}\right)$$

and γ is the curve of intersection of the sphere $(x-5)^2+(y-2)^2+(z-3)^2=1$ with the plane $3x-5y-z=2$.

6 $\oint_\gamma \boldsymbol{F}\cdot d\boldsymbol{x}$, where $\boldsymbol{F}$ is as in Exercise 5 and γ is the curve of intersection of the sphere $(x-3)^2+(y-3)^2+(z-2)^2=1$ with the plane $2x-y-z=1$.

7 $\iint_S \mathbf{Curl}\,\boldsymbol{F}\cdot\mathbf{N}\,dA$, where $\boldsymbol{F}(x, y, z)=(y, 2z, 3x)$ and S is the hemisphere $z=\sqrt{1-x^2-y^2}$.

8 $\iint_S \mathbf{curl}\,\boldsymbol{F}\cdot\mathbf{N}\,dA$, where $\boldsymbol{F}(x, y, z)=(y^2, x, -xz)$ and S is the part of $z=1-x^2-y^2$ lying above the xy-plane.

9 $\oint_\gamma \boldsymbol{F}\cdot d\boldsymbol{x}$, where $\boldsymbol{F}(x, y, z)=(z, 0, 0)$ and γ is the curve of intersection of the plane $z=x+y$ with the cylinder $x^2+y^2=1$, traversed counterclockwise in the plane $z=x+y$ when viewed from above.

10 $\oint_\gamma z\,dx+x\,dy+y\,dz$, where γ is the circle of intersection of the plane $x+y+z=0$ and $x^2+y^2+z^2=1$, traversed counterclockwise as viewed from above the plane $x+y+z=0$.

11 $\iint_S \mathbf{curl}\,\boldsymbol{F}\cdot d\boldsymbol{S}$, where $\boldsymbol{F}(x, y, z)=(y, z, x)$ and S is the portion of the cylinder $x^2+y^2=1$ cut off by the plane $z=x+3$ and the xy-plane.

12 $\iint_S \mathbf{curl}\,\boldsymbol{F}\cdot d\boldsymbol{S}$, where $\boldsymbol{F}(x, y, z)=(y^2, xy, -2xz)$ and S is the hemisphere $z=\sqrt{1-x^2-y^2}$.

13 $\int_{\partial S} \boldsymbol{F}\cdot d\boldsymbol{x}$, where $\boldsymbol{F}(x, y, z)=(e^x\sin y, e^x\cos y-z, y)$ and $\partial S=\gamma=\gamma_1\cup\gamma_2$ where S is the part of $z=\sqrt{x^2+y^2}$ between $z=1$ and $z=2$.

14 $\int_{\partial S} \boldsymbol{F}\cdot d\boldsymbol{x}$, where $\boldsymbol{F}(x, y, z)=(z, 0, -x)$ and S is the part of the cylinder $r=2+\cos\theta$ above the xy-plane and below the cone $z=\sqrt{x^2+y^2}$.

15 $\oint_\gamma \boldsymbol{F}\cdot d\boldsymbol{x}$, where $\boldsymbol{F}(x, y, z)=(y-z, z-x, x-y)$ and γ is the intersection of the cylinder $x^2+y^2=1$ and the plane $x+z=1$, traversed counterclockwise when viewed from above.

16 $\oint_\gamma \boldsymbol{F}\cdot d\boldsymbol{x}$, where $\boldsymbol{F}(x, y, z)=(y^2+z^2, x^2+z^2, x^2+y^2)$ and γ is the intersection of the hemisphere $z=\sqrt{4-(x-2)^2-y^2}$ and the cylinder $x^2+y^2=2x$.

17 Use Theorem 7.1 to show that for any vector field $\boldsymbol{F}:\boldsymbol{R}^3\to\boldsymbol{R}^3$ such that $J_{\boldsymbol{F}}$ has continuous entries, $\iint_S \mathbf{curl}\,\boldsymbol{F}\cdot d\boldsymbol{S}=0$ if S is the sphere $x^2+y^2+z^2=a^2$. (*Hint:* Cut the sphere into two hemispheres along a great circle C. Apply Theorem 7.1 to each hemisphere and note that the two line integrals over C cancel.)

18 Repeat Exercise 17 for the ellipsoid $\frac{x^2}{a^2}+\frac{y^2}{b^2}+\frac{z^2}{c^2}=1$.

19 Suppose that a fluid flow has velocity vector $\boldsymbol{F}(x, y, z)=(z, x, y)$. Let D_r be a disk centered at $\boldsymbol{x}_0=(x_0, y_0, z_0)$ of radius r with unit normal vector $\boldsymbol{N}=(a, b, c)$. Calculate the circulation of $\boldsymbol{F}$ around ∂D_r, the positively oriented boundary of D_r.

20 In Exercise 19, maximize your answer by maximizing the expression $a+b+c$ subject to the constraint $a^2+b^2+c^2=1$ (cf. Section 4.10). Is the resulting

$$\boldsymbol{N}=\frac{1}{|\mathbf{curl}\ \boldsymbol{F}(\boldsymbol{x}_0)|}\mathbf{curl}\ \boldsymbol{F}(\boldsymbol{x}_0)?$$

(It *should* be, according to the discussion following Theorem 7.2. See page 473.)

21 Let $\boldsymbol{F}(x, y, z)=\left(-\frac{y}{x^2+y^2}, \frac{x}{x^2+y^2}, z\right)$.

(a) Show that $\mathbf{curl}\ \boldsymbol{F}(\boldsymbol{x})=\mathbf{0}$ for all $\boldsymbol{x}$ not lying in the z-axis.

(b) Let γ be the boundary of the disk $x^2+z^2=1$, $y=1$ traversed counterclockwise as viewed from the right around the y-axis. Show that $\oint_\gamma \boldsymbol{F}\cdot d\boldsymbol{x}=0$.

22 Refer to Exercise 21. Let γ be the boundary of the disk $x^2+y^2=1$, $z=0$, traversed counterclockwise as viewed from above. Can we conclude from Theorem 7.1 that $\int_\gamma \boldsymbol{F}\cdot d\boldsymbol{x}=0$? Why or why not?

23 Refer to Example 7.5. In the case of time-varying magnetic fields, Maxwell's Second Equation has the form

$$\mathbf{curl}\ \boldsymbol{H}=\frac{\partial \boldsymbol{D}}{\partial t}+\boldsymbol{J}$$

where $\boldsymbol{D}$ is the *electric flux density* (or *displacement density*) vector field. Show that in this situation, Ampère's circuital law becomes

$$\oint_\gamma \boldsymbol{H}\cdot d\boldsymbol{x}=I+\iint_S \frac{\partial \boldsymbol{D}}{\partial t}\cdot d\boldsymbol{S}.$$

24 Derive Equation (4) in the proof of Theorem 7.1.

25 Derive Equation (5) in the proof of Theorem 7.1.

8 THEOREM OF GAUSS

There is one more integration theorem which states that the integral of a certain kind of derivative of a function $\boldsymbol{F}$ over a region is determined by the values of $\boldsymbol{F}$ on the boundary of the region. This theorem was established by Gauss in connection with his work on electrical theory, and, together with the theorems of Green and Stokes, plays a prominent role in that theory.

Previously (Theorems 4.1 and 7.1), we have given integration theorems of this sort where the "certain kinds" of derivatives referred to were ∇f and $\nabla\times\boldsymbol{F}$. It is only natural, then, to suspect that there should be a theorem of the same type in which the "certain kind" of derivative is the divergence $\nabla\cdot\boldsymbol{F}$ of $\boldsymbol{F}$. This is indeed the case, and the main theorem of this section is also frequently called the *Divergence Theorem.* If you worked Exercise 28 of Exercises 7.2, then you can easily guess the form this theorem will take. That exercise showed how Green's Theorem in the plane could be formulated as

(1) $$\iint_D \operatorname{div} \boldsymbol{F}\, dA = \oint_{\partial D} \boldsymbol{F} \cdot \boldsymbol{N}\, ds$$

for a vector field $\boldsymbol{F} = (P, Q) : \boldsymbol{R}^2 \to \boldsymbol{R}^2$ such that P and Q have continuous partial derivatives on D, and ∂D is traversed counterclockwise. Here $\boldsymbol{N}$ is a unit normal to γ. We can write down an analogue of (1) for $\boldsymbol{F} : \boldsymbol{R}^3 \to \boldsymbol{R}^3$ as follows. Suppose that $D \subseteq \boldsymbol{R}^3$ is a solid region whose boundary is a **positively oriented** smooth surface patch S. This means that S is parametrized by $\boldsymbol{X} : E \to \boldsymbol{R}^3$ for some $E \subseteq \boldsymbol{R}^2$ such that at each point of S, the standard unit normal vector

$$\boldsymbol{N} = \frac{\boldsymbol{X}_u \times \boldsymbol{X}_v}{|\boldsymbol{X}_u \times \boldsymbol{X}_v|}$$

points *outward* from D. See Figure 8.1.

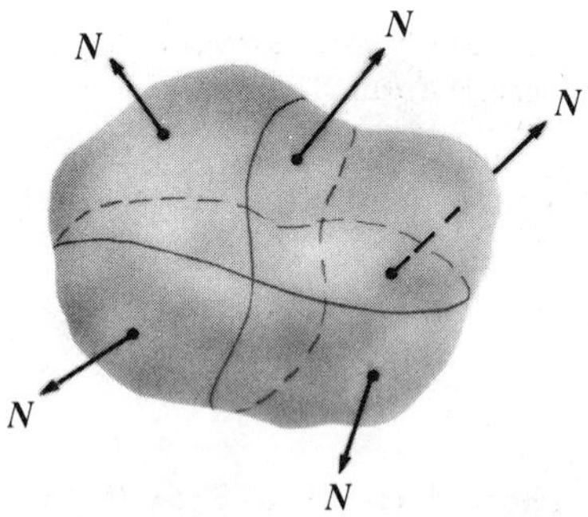

FIGURE 8.1

If $\boldsymbol{F} = (P, Q, R) : \boldsymbol{R}^3 \to \boldsymbol{R}^3$, where P, Q, and R have continuous partial derivatives on D, then the analogue of (1) is

(2) $$\iiint_D \operatorname{div} \boldsymbol{F}\, dV = \iint_S \boldsymbol{F} \cdot \boldsymbol{N}\, dA.$$

The Theorem of Gauss says that (2) is correct. It thus may be regarded as a final variation on this chapter's theme that the integral of a derivative of a function $\boldsymbol{F}$ over a region is determined by the values of $\boldsymbol{F}$ on the boundary of the region, provided that $\boldsymbol{F}$, the region, and the boundary satisfy appropriate smoothness and orientation hypotheses. In the present case we will require D to be a *simple region* in $\boldsymbol{R}^3$ (compare Definition 2.1).

8.1 **DEFINITION.** A region $D \subseteq \boldsymbol{R}^3$ is **simple** if it is the portion of $\boldsymbol{R}^3$ enclosed by a piecewise smooth orientable surface S that intersects any line parallel to a coordinate axis either in at most two points or in one line segment.

Thus, balls and solid parallelepipeds are simple regions. The region between two concentric spheres or inside a torus is not simple, however. See Figure 8.2.

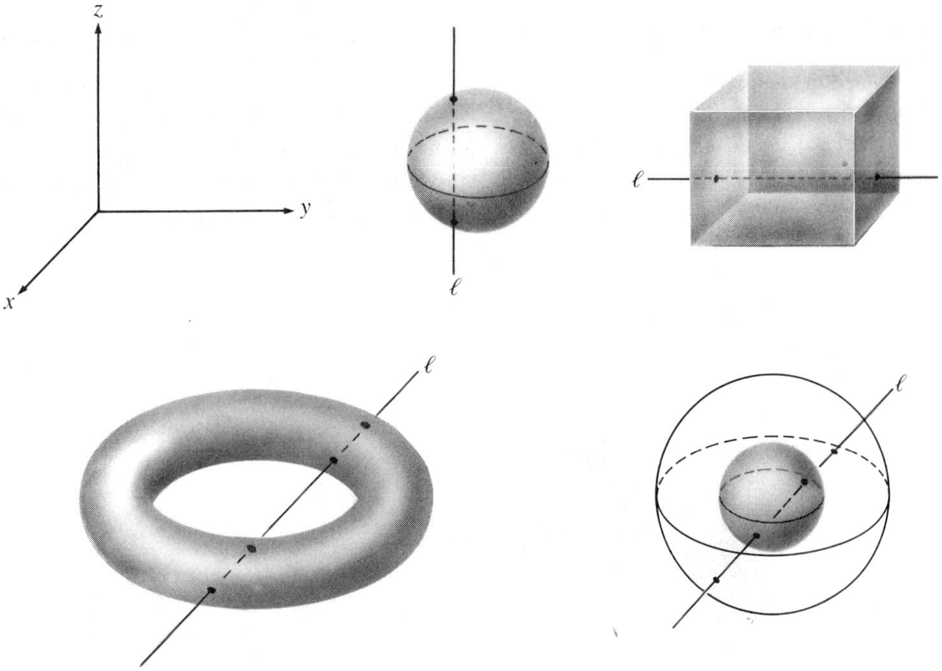

FIGURE 8.2

8.2 **THEOREM OF GAUSS (DIVERGENCE THEOREM).** Let $D \subseteq \boldsymbol{R}^3$ be a simple region with positively oriented piecewise smooth boundary S. Suppose that $\boldsymbol{F} = (P, Q, R): \boldsymbol{R}^3 \to \boldsymbol{R}^3$ is such that P, Q, and R have continuous partial derivatives on D and S. Then, using notation (5) (p. 462),

(2)
$$\iiint_D \operatorname{div} \boldsymbol{F}\, dV = \iint_S \boldsymbol{F} \cdot \boldsymbol{N}\, dA$$

$$= \iint_S P\, dy \wedge dz + Q\, dz \wedge dx + P\, dx \wedge dy.$$

Proof. The proof is similar in approach to the proof of Theorem 7.1. If we can show

(3)
$$\iiint_D \frac{\partial P}{\partial x}\, dV = \iint_S P\, dy \wedge dz,$$

(4)
$$\iiint_D \frac{\partial Q}{\partial y}\, dV = \iint_S Q\, dz \wedge dx,$$

(5)
$$\iiint_D \frac{\partial R}{\partial z}\, dV = \iint_S R\, dx \wedge dy,$$

then (2) will follow by adding (3), (4), and (5), since $\frac{\partial P}{\partial x}+\frac{\partial Q}{\partial y}+\frac{\partial R}{\partial z}=\operatorname{div}\boldsymbol{F}$. Each of (3), (4), (5) is shown in the same way, so we will establish (3) and leave (4) and (5) as Exercises 15 and 16 below. Since D is a simple region, its boundary S consists of a surface patch S_1, the graph of $x=g_1(y,z)$, S_2, the graph of $x=g_2(y,z)$, and possibly a cylindrical surface patch S_3 generated by a line moving parallel to the x-axis. See Figure 8.3. The surface integral in (3) is then the sum of the surface

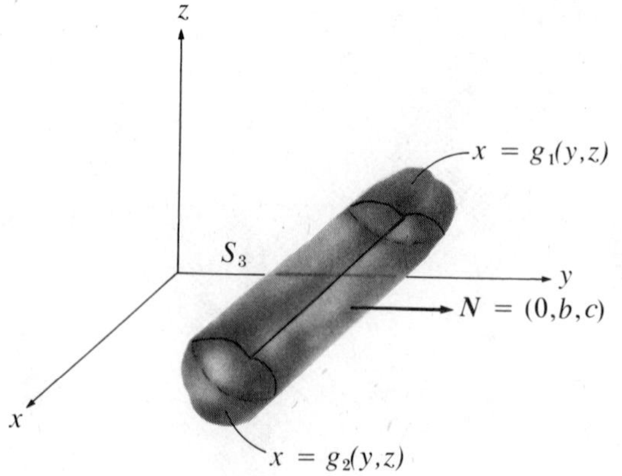

FIGURE 8.3

integrals over each subsurface,

$$\iint_S P\,dy\wedge dz=\iint_{S_1} P\,dy\wedge dz+\iint_{S_2} P\,dy\wedge dz+\iint_{S_3} P\,dy\wedge dz.$$

Let's consider $\iint_{S_3} P\,dy\wedge dz$ first. Here the normal $\boldsymbol{N}$ has zero as x-coordinate, since S_3 is parallel to the x-axis, so

$$\iint_{S_3} P\,dy\wedge dz=\iint_{S_3}(P,0,0)\cdot\boldsymbol{N}\,dA=0.$$

For the other surfaces, we can naturally parametrize by

$$\begin{aligned}&S_1:\boldsymbol{X}(u,v)=(g_1(u,v),u,v), && (u,v)\in E\subseteq\boldsymbol{R}^2\\ &S_2:\boldsymbol{X}(u,v)=(g_2(u,v),u,v), && (u,v)\in E\subseteq\boldsymbol{R}^2\end{aligned}\tag{6}$$

where E is the projection of D on the yz-plane. However, we must use a parametrization that orients S *positively;* i.e., the normal $\boldsymbol{N}$ must point *outward* from D. Thus, for S_1 the first coordinate should be negative, while for S_2 it should be positive. Note that

$$\boldsymbol{X}_u=\left(\frac{\partial g_i}{\partial u},1,0\right),$$

$$\boldsymbol{X}_v=\left(\frac{\partial g_i}{\partial v},0,1\right),$$

so

$$\boldsymbol{X}_u\times\boldsymbol{X}_v=\left(1,-\frac{\partial g_i}{\partial u},-\frac{\partial g_i}{\partial v}\right).$$

Hence the parametrization for S_2 is all right, but it must be *reversed* for S_1. That is, if we use (6), we have

$$\iint_{S_1} P\,dy \wedge dz = -\iint_E (P, 0, 0) \cdot (\mathbf{X}_u \times \mathbf{X}_v)\,du\,dv$$
$$= -\iint_E P(g_1(u, v), u, v)\,du\,dv.$$

So we obtain

$$\iint_S P\,dy \wedge dz = \iint_{S_1} P\,dy \wedge dz + \iint_{S_2} P\,dy \wedge dz$$
$$= -\iint_E P(g_1(u, v), u, v)\,du\,dv + \iint_E P(g_2(u, v), u, v)\,du\,dv$$
$$= -\iint_E P(g_1(y, z), y, z)\,dy\,dz + \iint_E P(g_2(y, z), y, z)\,dy\,dz$$
$$= \iint_E [P(g_2(y, z), y, z) - P(g_1(y, z), y, z)]\,dy\,dz$$
$$= \iint_E \left[\int_{g_1(y,z)}^{g_2(y,z)} \frac{\partial P}{\partial x}\,dx\right] dy\,dz \quad \text{by Fubini's Theorem}$$
$$= \iiint_D \frac{\partial P}{\partial x}\,dx\,dy\,dz.$$

This proves (3). The proofs of (4) and (5) are similar. Then (2) follows by adding (3), (4), and (5). QED

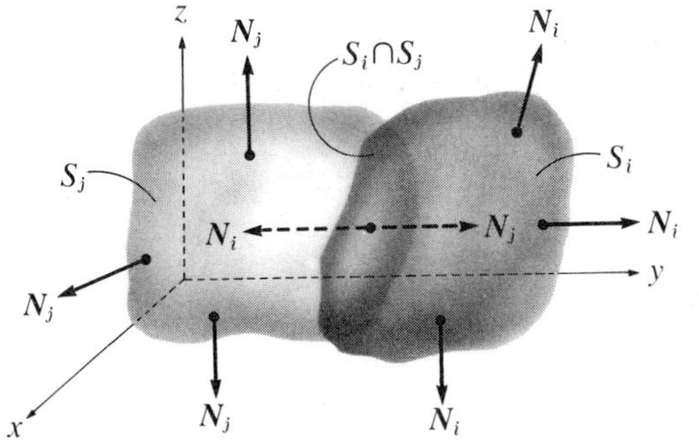

FIGURE 8.4

Just as we were able to extend Green's Theorem 2.2 for simple regions in $\boldsymbol{R}^2$ to Theorem 2.3 for regions that could be decomposed into a finite union of simple regions, so we can extend Theorem 8.2 to more general regions. For this, we consider a region D that is the union $D_1 \cup D_2 \cup \ldots \cup D_k$ of simple regions. On any common boundary surface $S_i \cap S_j$ of two subregions D_i and D_j, the outward pointing normals will be oppositely directed from each other as shown in Figure 8.4. Then the corresponding surface integrals

$$\iint_{S_i} \boldsymbol{F} \cdot \mathbf{N}_i\,dS \quad \text{and} \quad \iint_{S_j} \boldsymbol{F} \cdot \mathbf{N}_j\,dS$$

will be negatives of each other, since $N_j = -N_i$. Thus, these surface integrals will cancel each other if we compute $\iint_{\cup S_i} \boldsymbol{F} \cdot \boldsymbol{N}\, dA$. Hence we have

$$\iiint_D \operatorname{div} \boldsymbol{F}\, dV = \sum_{i=1}^{k} \iiint_{D_i} \operatorname{div} \boldsymbol{F}\, dV$$

$$= \sum_{i=1}^{k} \iint_{S_i} \boldsymbol{F} \cdot \boldsymbol{N}_i\, dA, \quad \text{by Theorem 8.2}$$

$$= \iint_S \boldsymbol{F} \cdot \boldsymbol{N}\, dA$$

where S is the union of the *non-coincident* portions of the boundaries of the regions D_i. Thus S is precisely the boundary of $D = D_1 \cup D_2 \cup \ldots \cup D_k$. We can state this formally as follows.

8.3 **THEOREM OF GAUSS.** If D is a finite union of simple regions D_i, each with a positively oriented piecewise smooth boundary S_i, and $\boldsymbol{F}: \boldsymbol{R}^3 \to \boldsymbol{R}^3$ is as in Theorem 8.2, then

(2) $$\iiint_D \operatorname{div} \boldsymbol{F}\, dV = \iint_S \boldsymbol{F} \cdot \boldsymbol{N}\, dA,$$

where $S = \partial D$ is the union of the portions of the S_i not common to any two D_i, oriented consistently with the positive orientation of the S_i.

Now, while (2) tells us that a triple integral over a solid D of the derivative-like quantity $\operatorname{div} \boldsymbol{F} = \boldsymbol{\nabla} \cdot \boldsymbol{F}$ is determined by the values of $\boldsymbol{F}$ over the two-dimensional bounding surface S of D, there is usually not much reduction of complexity if we try to evaluate the surface integral $\iint_S \boldsymbol{F} \cdot \boldsymbol{N}\, dA$ instead of the triple integral $\iiint_D \operatorname{div} \boldsymbol{F}\, dV$. Indeed, one of the prime uses of the Divergence Theorem is to afford us a simple means of evaluating surface integrals. With our present tools, it is often simpler to evaluate $\iiint_S \operatorname{div} \boldsymbol{F}\, dV$ than to compute $\iint_S \boldsymbol{F} \cdot \boldsymbol{N}\, dA$.

8.4 **EXAMPLE.** If $\boldsymbol{F}(x, y, z) = (x^3, y^3, -e^z)$, then compute $\iint_S \boldsymbol{F} \cdot \boldsymbol{N}\, dA$ if S is the cylinder (including top and bottom) $x^2 + y^2 = 1$, $0 \le z \le 1$ with outward pointing normal. See Figure 8.5 for the first octant portion.

Solution. To do this as a surface integral, we would need to parametrize separately the bottom, top, and cylindrical side of S as in Example 5.2(b). Then to evaluate the surface integral, we would have to evaluate it over each portion of S, and add the results. This seems like a job better handled by the Theorem of Gauss. We have

$$\operatorname{div} \boldsymbol{F} = 3x^2 + 3y^2 - e^z,$$

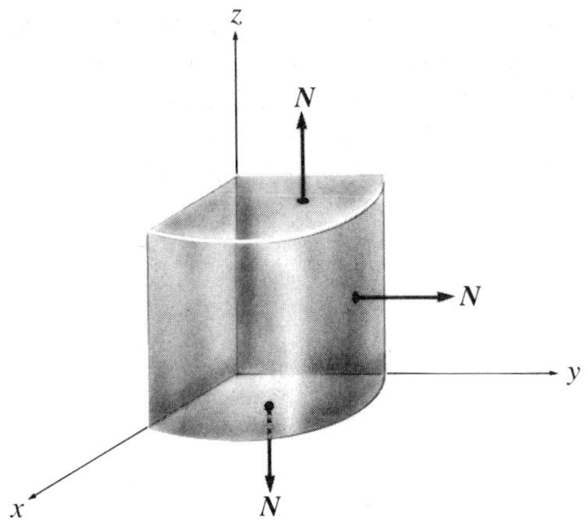

FIGURE 8.5

so by (2)

$$\iint_S \boldsymbol{F} \cdot \boldsymbol{N}\, dA = \iiint_D (3x^2+3y^2-e^z)\, dV$$

where D is the cylindrical solid enclosed by S. To evaluate the triple integral, we use cylindrical coordinates, obtaining

$$\iiint_D [3(x^2+y^2)-e^z]\, dx\, dy\, dz = \iiint_D (3r^2-e^z)\, r\, dr\, d\theta\, dz$$

$$= \int_0^1 r\, dr \int_0^{2\pi} d\theta \int_0^1 (3r^2-e^z)\, dz$$

$$= \int_0^1 r\, dr \int_0^{2\pi} d\theta [3r^2 z - e^z]_0^1$$

$$= \int_0^1 r\, dr \int_0^{2\pi} d\theta (3r^2 - e + 1)$$

$$= 2\pi \int_0^1 (3r^3 - re + r)\, dr$$

$$= 2\pi[\tfrac{3}{4}r^4 - \tfrac{1}{2}r^2 e + \tfrac{1}{2}r^2]_0^1$$

$$= 2\pi[\tfrac{3}{4} - \tfrac{1}{2}e + \tfrac{1}{2}]$$

$$= \pi(\tfrac{5}{2} - e).$$

A host of important physical applications of Gauss's Theorem exist. The next example illustrates how they arise. The technique employed will remind you of Example 2.8.

8.5 EXAMPLE. If $\boldsymbol{F}$ is an inverse square law field (Example 3.5), then show that the flux across the boundary S of any simple region $D \subseteq \boldsymbol{R}^3$ that contains the origin is constant.

Solution. Recall that $\boldsymbol{F}$ is given by

$$\boldsymbol{F}(\boldsymbol{x}) = k \frac{1}{|\boldsymbol{x}|^2} \frac{\boldsymbol{x}}{|\boldsymbol{x}|}.$$

We can't then apply Theorem 8.2 immediately, because D contains the origin, where $\boldsymbol{F}$ is not even defined, let alone differentiable, so the hypotheses of Theorem 8.2 are not met. To get around this, we consider the region E inside S and outside a sphere S_ε centered at $\mathbf{0}$ and of radius ε so small that $S_\varepsilon \subseteq D$. See Figure 8.6. On E all the hypotheses of Theorem 8.3 *are* met, and by Example

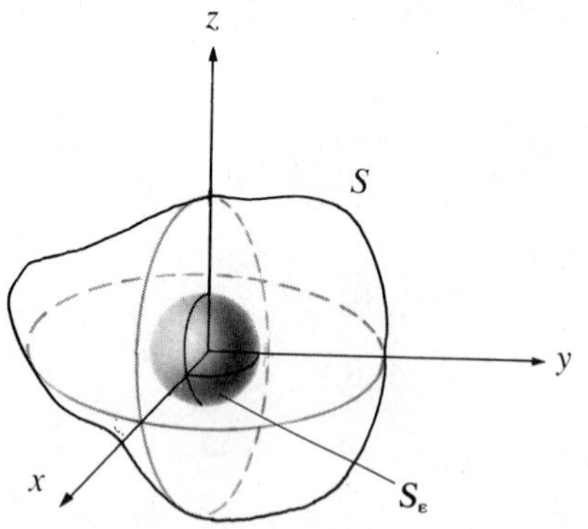

FIGURE 8.6

3.5, $\operatorname{div} \boldsymbol{F} = 0$ on all of E. Now E is bounded by $\partial E = S \cup S_\varepsilon$. If we orient these surfaces so that the normal $\boldsymbol{N}$ points toward $\mathbf{0}$ on S_ε and away from $\mathbf{0}$ on S, then $\boldsymbol{N}$ always points *outward* from E. So we can apply Theorem 8.3 and get

$$\iint_{\partial E} \boldsymbol{F} \cdot \boldsymbol{N}\, dA = \iiint_{E} \operatorname{div} \boldsymbol{F}\, dV = 0.$$

But then

$$\begin{aligned}
\iint_{\partial E} \boldsymbol{F} \cdot \boldsymbol{N}\, dA &= \iint_{S \cup S_\varepsilon} \boldsymbol{F} \cdot \boldsymbol{N}\, dA \\
&= \iint_{S} \boldsymbol{F} \cdot \boldsymbol{N}\, dA + \iint_{S_\varepsilon} \boldsymbol{F} \cdot \boldsymbol{N}\, dA \\
&= 0.
\end{aligned}$$

Hence the desired flux is

$$\Phi = \iint_{S} \boldsymbol{F} \cdot \boldsymbol{N}\, dA = -\iint_{S_\varepsilon} \boldsymbol{F} \cdot \boldsymbol{N}\, dA$$

where S_ε has inward pointing normal $\boldsymbol{N} = -\dfrac{\boldsymbol{x}}{|\boldsymbol{x}|}$. We thus have

$$\begin{aligned}
\Phi &= -k \iint_{S_\varepsilon} \frac{\boldsymbol{x}}{|\boldsymbol{x}|^3} \cdot \left(-\frac{\boldsymbol{x}}{|\boldsymbol{x}|}\right) dA \\
&= k \iint_{S_\varepsilon} \frac{1}{|\boldsymbol{x}|^2}\, dS = \frac{1}{\varepsilon^2} k \iint_{S_\varepsilon} dA \\
&= \frac{1}{\varepsilon^2} k A(S_\varepsilon) = \frac{1}{\varepsilon^2} k 4\pi\varepsilon^2 \\
&= 4\pi k
\end{aligned}$$

(using the result of Example 5.5). QED

We have already mentioned that inverse square law fields describe a number of important physical phenomena in gravitation, magnetism, electricity, and other areas of science. The result of Example 8.5 bears on such applications. As an illustration, let's return to electricity theory. You may have studied *Coulomb's Law* (named for the French army engineer and scientist Charles A. Coulomb (1736–1806)). Coulomb determined experimentally that the force between two charged particles in a vacuum is inversely proportional to the square of the distance between them. That is,

$$|\boldsymbol{F}| = k\frac{q_1 q_2}{r^2},$$

for some constant k. This constant of proportionality is written as $\dfrac{1}{4\pi\varepsilon_0}$ where ε_0 is a constant called the *permittivity* of free space. We can see from Example 8.5 why the factor 4π was introduced. If a charge q_0 is at the origin, then the force on a similarly charged particle at $\boldsymbol{x}$ carrying a charge a q_1 is

$$\boldsymbol{F} = \frac{1}{4\pi\varepsilon_0}\frac{|\boldsymbol{x}|}{|\boldsymbol{x}|^3} q_0 q_1.$$

The flux Φ across any closed surface is then, by Example 8.5, $q_0 q_1$ times the reciprocal of the permittivity. The *electric flux density* $\boldsymbol{D} = \varepsilon_0 \boldsymbol{E}$, where $\boldsymbol{E} = \dfrac{1}{q_1}\boldsymbol{F}$, is introduced, and the *electric flux* across S is then defined as $\Psi = \iint_S \boldsymbol{D} \cdot d\boldsymbol{S}$. Gauss was able to formulate the following fundamental law, which bears his name, by the simple application of Theorem 8.2 given in Example 8.5.

8.6 EXAMPLE. Obtain *Gauss's Law*: If a charge q_0 at $\boldsymbol{0}$ produces an electric flux density $\boldsymbol{D}$, then the electric flux passing through any piecewise smooth closed surface S containing $\boldsymbol{0}$ is q_0.

Solution. Let q_1 be the total charge on S. The electric flux is then

$$\begin{aligned}
\Psi = \iint_S \boldsymbol{D} \cdot d\boldsymbol{S} = \iint_S \varepsilon_0 \boldsymbol{E} \cdot d\boldsymbol{S} &= \varepsilon_0 \iint_S \frac{\boldsymbol{F}}{q_1} \cdot d\boldsymbol{S} \\
&= \varepsilon_0 \iint_S \frac{q_0}{4\pi\varepsilon_0}\frac{\boldsymbol{x}}{|\boldsymbol{x}|^3} \cdot d\boldsymbol{S} \\
&= \frac{1}{4\pi} q_0 \iint_S \frac{\boldsymbol{x}}{|\boldsymbol{x}|^3} \cdot d\boldsymbol{S} \\
&= \frac{1}{4\pi} q_0 4\pi \quad \text{by Example 8.5} \\
&= q_0.
\end{aligned}$$

QED

Electricity is by no means the only application in which Gauss's Theorem plays an important role. Some further applications will be found in Exercises 25 to 32 below.

Exercises 26 to 28 are aimed at deriving the *heat equation*, which is fundamental in thermodynamics and describes many important diffusion processes. If you do these exercises, then you will need to use the following facts from physics about heat flow.

(i) The amount of heat required to raise the temperature of a small body ΔB having constant density δ by T degrees is proportional to T and the mass m of the body. So it is

$$c\,\delta T V(\Delta B),$$

where c is a constant called the *specific heat* of the material.

(ii) The rate at which heat crosses a small patch of area ΔA on a surface S is

$$-k\mathbf{N}\cdot\nabla T\,\Delta A,$$

where $\mathbf{N}$ is the standard unit normal (Definition 6.3), k is a constant called the *coefficient of thermal conductivity*, and $T=T(x, y, z, t)$ is the temperature at $\mathbf{x}=(x, y, z)$ at time t. It is assumed that ∇T has entries whose partial derivatives are continuous on $\mathbf{R}^3$. (The minus sign reflects the fact that heat flows *away* from hot sources toward colder bodies.)

(iii) The total thermal energy (i.e., heat) that enters a body B during time Δt is the amount of heat absorbed by B during this time interval. This is the principle of *Conservation of Thermal Energy*.

Exercises 29 to 32 are aimed at deriving the celebrated Principle of Archimedes (287–212 BC) for floating bodies, the discovery of which, according to legend, prompted Archimedes to leave his bath and run naked through the streets of ancient Syracuse, shouting "Eureka!"—"I have found it!" If you do these exercises, then you will need to use the following definition of vector surface integrals and vector triple integrals of vector functions.

8.7 **DEFINITION.** Suppose that $\mathbf{F}:\mathbf{R}^3\to\mathbf{R}^3$ given by $\mathbf{F}(\mathbf{x})=(f_1(\mathbf{x}), f_2(\mathbf{x}), f_3(\mathbf{x}))$ is continuous on a simple region $D\subseteq\mathbf{R}^3$ with positively oriented boundary S. Then the **vector surface integral** of $\mathbf{F}$ over S is

$$\iint_S \mathbf{F}\,dA=\left(\iint_S f_1(\mathbf{x})\,dA, \iint_S f_2(\mathbf{x})\,dA, \iint_S f_3(\mathbf{x})\,dA\right),$$

where $\iint_S f_i(\mathbf{x})\,dA$ is as in Definition 6.1. The **vector triple integral** of $\mathbf{F}$ over D is

D is

$$\iiint_D \mathbf{F}(\mathbf{x})\,dV=\left(\iiint_D f_1(\mathbf{x})\,dV, \iiint_D f_2(\mathbf{x})\,dV, \iiint_D f_3(\mathbf{x})\,dV\right).$$

EXERCISES 7.8

In Exercises 1 to 14, evaluate the surface integral $\iint_S \mathbf{F}\cdot\mathbf{N}\,dA$ for the given $\mathbf{F}$ and S by using the Theorem of Gauss.

1 $\mathbf{F}(x, y, z)=2x\mathbf{i}+3y\mathbf{j}-4z\mathbf{k}$, and S is the sphere $x^2+y^2+z^2=4$ with outward pointing normal.

2 $\mathbf{F}(x, y, z)=(2x, x^2-xz^2, x^2y-y^3)$, and S is the sphere $x^2+y^2+z^2=1$ with inward pointing normal.

3 $\boldsymbol{F}(x, y, z) = \left(\dfrac{x}{(x^2+y^2+z^2)^{3/2}}, \dfrac{y}{(x^2+y^2+z^2)^{3/2}}, \dfrac{z}{(x^2+y^2+z^2)^{3/2}}\right)$, and S is the surface that bounds the solid box $[-1, 3]\times[2, 5]\times[-1, 1]$ with outward pointing normal.

4 $\boldsymbol{F}(x, y, z)$ of Exercise 3, and S is the ellipsoid $\dfrac{x^2}{9}+\dfrac{y^2}{4}+\dfrac{z^2}{16}=1$.

5 $\boldsymbol{F}(x, y, z) = (3x^2+z^2, xy-z^3, z+x^2-y^2)$, and S is the tetrahedral surface formed by the three coordinate planes and the plane $x+y+z=1$, with outward pointing normal.

6 $\boldsymbol{F}(x, y, z) = (x^3+z^2, y^3+x^2, z^3+y^2)$, and S is the sphere $x^2+y^2+z^2=1$, with outward pointing normal.

7 $\boldsymbol{F}(x, y, z) = (x^3+y^3+z^3, x^2y+\frac{1}{3}y^3+z^2, (x^2+y^2)z)$, and S is the cylindrical surface $x^2+y^2=1$ for $0\leq z\leq 1$, including top and bottom.

8 $\boldsymbol{F}$ as in Exercise 7, and S is the cone $z=1-\sqrt{x^2+y^2}$ (including the floor in the xy-plane) with outward pointing normal.

9 $\boldsymbol{F}(x, y, z) = (x^2-yz, y^2-xz, z^2-xy)$, and S is the surface of Exercise 3.

10 $\boldsymbol{F}(x, y, z) = (x^2-\cos yz, y^2+\sin 2xz, z^2+\tan 5xy)$, and S is the surface of Exercise 3.

11 $\boldsymbol{F}(x, y, z) = (x^2+ye^z, y^2+ze^x, z^2+xe^y)$, and S is the part of the cylinder $x^2+y^2=1$ between the xy-plane and $z=x+1$ (including top and bottom) with outward pointing normal.

12 $\boldsymbol{F}(x, y, z) = (x^2+y^2+z^2, y^2+2x^3z^3, z^2-x^2y^3e^{xy})$, and S is as in Exercise 11.

13 $\boldsymbol{F}(x, y, z) = (x^2+yz, y-xe^z, z-x^2\sqrt{1-y^2})$, and S is the paraboloid $z=x^2+y^2$ between $z=0$ and $z=2x$ (including the top), with outward pointing normal.

14 $\boldsymbol{F}(x, y, z) = (x^2-y^3z, y-x^2+z^2, z-x^2+y^2)$, and S is as in Exercise 11.

15 Establish (4) in the proof of Theorem 8.2.

16 Establish (5) in the proof of Theorem 8.2.

17 Prove *Green's First Identity* in three dimensions: If $f:\boldsymbol{R}^3\to\boldsymbol{R}$ is differentiable on a simple region D and all second partial derivatives of g are continuous on D, then

$$\iiint_D f(\boldsymbol{x})\,\nabla^2 g(\boldsymbol{x})\,dV = \iint_S (f(\boldsymbol{x})\,\boldsymbol{\nabla} g)\cdot\boldsymbol{N}\,dA - \iiint_D \boldsymbol{\nabla} f\cdot\boldsymbol{\nabla} g\,dV,$$

where S is the positively oriented boundary of D (cf. Exercise 26, Exercises 7.3).

18 Obtain *Green's Second Identity* from Exercise 17. Assuming the same hypotheses,

$$\iiint_D (f(\boldsymbol{x})\,\nabla^2 g(\boldsymbol{x}) - g(\boldsymbol{x})\,\nabla^2 f(\boldsymbol{x}))\,dV = \iint_S [f(\boldsymbol{x})(\boldsymbol{\nabla} g\cdot\boldsymbol{N}) - g(\boldsymbol{x})(\boldsymbol{\nabla} f\cdot\boldsymbol{N})]\,dA$$

$$= \iint_S \left(f\frac{\partial g}{\partial \boldsymbol{N}} - g\frac{\partial f}{\partial \boldsymbol{N}}\right) dA$$

(cf. Exercise 27, Exercises 7.3).

19 If $u:\boldsymbol{R}^3\to\boldsymbol{R}$ and $\boldsymbol{v}:\boldsymbol{R}^3\to\boldsymbol{R}^3$ have Jacobian matrices with continuous entries and D is a simple region with positively oriented boundary S, then establish the following *integration-by-parts* formula:

$$\iiint_D u\,\operatorname{div}\boldsymbol{v}\,dV = \iint_S u\boldsymbol{v}\cdot\boldsymbol{N}\,dA - \iiint_D \boldsymbol{v}\cdot\boldsymbol{\nabla} u\,dV.$$

20 If f is harmonic (Definition 3.8) on D in Exercise 18, then show that $\iint_S \frac{\partial f}{\partial \mathbf{N}}\, dA = 0$.

21 Suppose that f is harmonic on D and $f(\mathbf{x}) = 0$ on $S = \partial D$. Then use Exercise 17 to show that f is identically zero on all of D.

22 Let $g, h: \mathbf{R}^3 \to \mathbf{R}$. The relations $\nabla^2 f(\mathbf{x}) = g(\mathbf{x})$ for $\mathbf{x} \in D \subseteq \mathbf{R}^3$ and $f(\mathbf{x}) = h(\mathbf{x})$ for $\mathbf{x} \in S = \partial D$ are called **Poisson's equation,** after the French mathematician and physicist Siméon D. Poisson (1781–1840). Show that if g and h are given, then there can be at most one function f that solves Poisson's equation on a simple region D. (Use Exercise 21. The equation $f(\mathbf{x}) = h(\mathbf{x})$ on S is called a *boundary condition.*)

23 If $\mathbf{F}: \mathbf{R}^3 \to \mathbf{R}^3$ is the velocity field of a fluid flow, where the fluid has density $\delta(x, y, z)$, then show that $\iint_S \delta(x, y, z)\mathbf{F} \cdot \mathbf{N}\, dA$ is the net loss of mass per unit time from the ball $B(\mathbf{x}_0, a)$. Here S is the sphere of radius a centered at $\mathbf{x}_0$.

24 Use Exercise 23 to show that $\lim_{a \to 0} \frac{1}{\frac{4}{3}\pi a^3} \iint_S \delta(x, y, z, t)\mathbf{F} \cdot \mathbf{N}\, dA$ is the rate $\frac{\partial \delta}{\partial t}$ of decrease of the density per unit time per unit volume at $\mathbf{x}_0$.

25 Show that the limit in Exercise 24 is the divergence of $\delta\mathbf{F}$ at $\mathbf{x}_0$ if $\mathbf{F}$ and δ have continuous partial derivatives. Then derive the *continuity equation* of fluid mechanics, $\frac{\partial \delta}{\partial t} + \operatorname{div}(\delta\mathbf{F}) = 0$.

26 (a) Show that (iii) above (p. 488) can be formulated as

$$\left(\iint_S k\mathbf{N} \cdot \nabla T\, dA\right) \Delta t = \iiint_B c\delta\, \Delta T\, dV,$$

where S is the positively oriented boundary of B.

(b) Conclude that

$$\iint_S k\mathbf{N} \cdot \nabla T\, dA = \iiint_B c\delta \frac{\partial T(\mathbf{x}, t)}{\partial t}\, dV.$$

27 Use Theorem 8.2 and Exercise 26(b) to show that

$$\iiint_B \left[\nabla \cdot (k\, \nabla T) - \frac{\partial}{\partial t}(c\delta T)\right] dV = 0.$$

28 Use Exercise 27 to show that

$$\text{(iv)} \qquad \frac{\partial T}{\partial t} = \frac{k}{\delta c} \nabla^2 T.$$

The constant $k/\delta c$ is called the *coefficient of thermometric conductivity*. Equation (iv) is called the **heat equation**.

29 Suppose that $f: \mathbf{R}^3 \to \mathbf{R}$ has continuous partial derivatives on a simple region D with piecewise smooth positively oriented boundary S.

(a) If $\mathbf{a} \in \mathbf{R}^3$ is fixed, then show that

$$\mathbf{a} \cdot \left(\iint_S f(\mathbf{x})\mathbf{N}\, dA - \iiint_D \nabla f\, dV\right) = 0.$$

(*Hint:* Use Theorems 8.2 and 3.4(c) with $\mathbf{G}(\mathbf{x}) = \mathbf{a}$.)

(b) Since $\boldsymbol{a}$ is arbitrary, conclude that

$$\iint_S f(\boldsymbol{x})\boldsymbol{N}\,dA = \iiint_D \nabla f\,dV.$$

30 Denote by $\boldsymbol{B}$ the force of buoyancy exerted on a floating body by a surrounding liquid. Let $p(\boldsymbol{x})$ be the fluid pressure on the body at $\boldsymbol{x}$. By considering the force on a small area of the surface S of the body near $\boldsymbol{x}$, justify the definition given in physics,

$$\boldsymbol{B} = -\iint_S p(\boldsymbol{x})\boldsymbol{N}\,dA.$$

31 The *hydrostatic law* states that

$$\nabla p(\boldsymbol{x}) = \delta(\boldsymbol{x})\mathbf{g},$$

where $\delta(\boldsymbol{x})$ is the density of the liquid at $\boldsymbol{x}$ and $\mathbf{g}$ is the force exerted by gravity. Use this and Theorem 8.2 to show that

$$\boldsymbol{B} = -\iiint_V \delta(\boldsymbol{x})\mathbf{g}\,dV.$$

32 Let $\boldsymbol{W} = m\mathbf{g}$ be the weight vector of the liquid displaced by the body. Show that $\boldsymbol{W} = \mathbf{g}\iiint_B \delta(\boldsymbol{x})\,dV$, and hence derive the **Principle of Archimedes,**

$$\boldsymbol{W} + \boldsymbol{B} = \mathbf{0}.$$

33 (a) Apply Theorem 8.2 to $\boldsymbol{F}\times\boldsymbol{a}$ to show that

$$\boldsymbol{a}\cdot\left(\iint_S \boldsymbol{N}\times\boldsymbol{F}\,dA - \iiint_D \mathbf{curl}\,\boldsymbol{F}\,dV\right) = 0,$$

where $\boldsymbol{a}\in\boldsymbol{R}^3$ is fixed. (*Hint:* Use Theorem 3.11(c), with $\boldsymbol{G}(\boldsymbol{x}) = \boldsymbol{a}$.)

(b) Show that $\iint_S \boldsymbol{N}\times\boldsymbol{F}\,dA = \iiint_D \mathbf{curl}\,\boldsymbol{F}\,dV$.

REVIEW EXERCISES 7.9

In Exercises 1 to 10, evaluate the given line integral.

1 $\int_\gamma \boldsymbol{F}\cdot d\boldsymbol{x}$ where $\boldsymbol{F}(x, y) = (xy^2, x+y^2)$ and γ is $\boldsymbol{x}(t) = (\sin t, 1-\sin t)$ for $0\le t\le \frac{1}{2}\pi$.

2 $\int_\gamma \boldsymbol{F}\cdot\boldsymbol{T}\,ds$ where $\boldsymbol{F}(x, y) = (xy, y^2+1)$ and γ is the triangular path from $(0, 0)$ to $(1, 0)$ to $(1, 1)$.

3 $\int_\gamma (2x^2-y^3)\,dx + (x^3-4y^2+y)\,dy$ where γ is the circle $x^2+y^2=4$ traversed clockwise.

4 $\int_\gamma -2ye^x\,dx + xe^y\,dy$ where γ is the polygonal path from $(1, 1)$ to $(-1, 1)$ to $(-1, -1)$ to $(1, -1)$ to $(1, 1)$.

5 $\int_\gamma (-\cos y - y^3)\,dx + (x\sin y - 3xy^2)\,dy$ where γ is the ellipse $\dfrac{x^2}{4}+\dfrac{y^2}{16}=1$ traversed counterclockwise.

6 $\int_{\partial S} \boldsymbol{F}\cdot d\boldsymbol{x}$ where $\boldsymbol{F}(x, y, z) = (e^x\sin y, e^x\cos y - z, y)$ and S is the part of the cylinder $x^2+y^2=1$ between $z=1$ and $z=2$.

7 $\int_{\partial S} \boldsymbol{F} \cdot d\boldsymbol{x}$ where $\boldsymbol{F}(x, y, z) = (2z, 0, x)$ and S is the part of the cylinder $r = 2 + \cos\theta$ above the xy-plane and below the cone $z = \sqrt{x^2+y^2}$.

8 $\int_\gamma \boldsymbol{F} \cdot d\boldsymbol{x}$ where $\boldsymbol{F}(x, y, z) = (2y - z, 2z - x, 2x - y)$ and γ is the intersection of the cylinder $x^2 + y^2 = 4$ and the plane $x + z = 4$ traversed counterclockwise when viewed from above.

9 $\int_\gamma 3x^2y^2z\,dx + 2x^3yz\,dy + x^3y^2\,dz$ where γ is the polygonal path from $(0, 0, 0)$ to $(2, -1, 0)$ to $(2, 0, 0)$ to $(-1, 2, 1)$.

10 $\int_\gamma y \sin z\,dx + x \sin z\,dy + xy \cos z\,dz$ where γ is the path $\boldsymbol{x}(t) = (t, t^2, t^3)$ from $(1, 1, 1)$ to $(2, 4, 8)$.

11 Show that $\operatorname{div} \dfrac{\boldsymbol{x}}{|\boldsymbol{x}|^3} = 0$.

12 If $\boldsymbol{F}(\boldsymbol{x}) = (x^2 + y^2 - xyz, xz^2 \cos y, z^2 e^{x^2+y^2})$, then compute $\operatorname{div} \boldsymbol{F}$ and **curl** $\boldsymbol{F}$ in general and at $\boldsymbol{x}_0 = (-1, \frac{1}{2}\pi, 3)$.

13 Is $\boldsymbol{F}(x, y) = \left(-\dfrac{x}{(x^2+y^2)^{3/2}}, -\dfrac{y}{(x^2+y^2)^{3/2}}\right)$ conservative on the annular region between $x^2 + y^2 = 1$ and $x^2 + y^2 = 4$?

14 Is $\boldsymbol{F}(x, y, z) = (y - 3z + 2x^2, x + 2z, 2y - 3x + 4xz)$ conservative on the unit ball?

15 Is $\boldsymbol{F}(x, y, z) = (2x^2 + 8xy^2, 3x^3y - 3xy, 4y^2z - 2x^3z)$ conservative on the unit ball?

16 Is $\boldsymbol{F}(x, y) = (x^2 - 3xy, y^2 + 3xy)$ conservative on the unit circle?

17 (a) Parametrize the part S of the plane $z = x + y$ that lies inside the cylinder $x^2 + y^2 = 1$. (b) Find the area of S.

18 (a) Parametrize the part S of the sphere $x^2 + y^2 + z^2 = 4$ that lies above the cone $z = \sqrt{x^2 + y^2}$. (b) Find the area of S.

In Exercises 19 to 25, compute the given surface integral.

19 $\iint_S (y + z)\,dA$ where S is the surface of Exercise 17,

20 $\iint_S y^2\,dA$ where S is the cylinder $x^2 + y^2 = 1$ between $z = 0$ and $z = 1$, including top and bottom.

21 $\iint_S \boldsymbol{F} \cdot d\boldsymbol{S}$ where $\boldsymbol{F}(x, y, z) = (x, -2y, 1)$ and S is the triangle with vertices $(1, 0, 0)$, $(0, 1, 0)$, and $(0, 0, 1)$.

22 $\iint_S \boldsymbol{F} \cdot \boldsymbol{N}\,dA$ where $\boldsymbol{F}(x, y, z) = (x + 2, -2y - 2, 2)$ and S is the triangle with vertices $(1, 0, 0)$, $(0, 1, 0)$, and $(0, 0, 1)$.

23 $\iint_S \boldsymbol{F} \cdot d\boldsymbol{S}$ where $\boldsymbol{F}(x, y, z) = (x^3 + y^3 + e^{z^2}, x^2y + \frac{1}{3}y^3 + z^3, \frac{1}{2}x^2z + \frac{1}{2}y^2z - x^3y^3)$ and S is the cylindrical surface $x^2 + y^2 = 1$ for $0 \le z \le 1$, including top and bottom.

24 $\iint_S \boldsymbol{F} \cdot d\boldsymbol{S}$ where $\boldsymbol{F}(x, y, z) = (x^3 + y^2z, x^2 + y^3 + x^2z^2, x^2y^3 + z^3)$ and S is the sphere $x^2 + y^2 + z^2 = 9$.

25 $\iint_S P\,dy \wedge dz + Q\,dz \wedge dx + R\,dx \wedge dy$ where $P(\boldsymbol{x}) = x + y^2 + z^2$, $Q(\boldsymbol{x}) = x^2 - 2y + z^3$, and $R(\boldsymbol{x}) = x^2 + y^2 + z$, and S is the ellipsoid $\dfrac{x^2}{9} + \dfrac{y^2}{36} + \dfrac{z^2}{4} = 1$.

8 ELEMENTARY DIFFERENTIAL EQUATIONS

0 INTRODUCTION

Historically, mathematics has developed in response to a number of different influences. One of these, which we have seen illustrated in our study of multivariable calculus, can be described as the esthetic goal of extending existing mathematical systems, such as single variable calculus, to newer, more general or abstract settings. In carrying out such extensions, problems arise whose resolution often requires the development of new mathematics or the use of other already existing areas of mathematics. We have seen, for instance, the role played by linear algebra in building the calculus of several variables.

Another strong influence in the development of mathematics has been, and continues to be, the need for mathematics in quantifying other fields. The area of mathematics known as differential equations has received tremendous impetus from such other fields as physics and engineering, and today continues to be one of the most widely applied parts of mathematics. By a *differential equation* we mean an equation involving one or more variables and derivatives of one or more of these variables. In the present chapter and the next two, we give an introduction to differential equations as a mathematical subject, and try to indicate its usefulness in quantifying other fields of study by presenting several illustrations of the role differential equations play in what is called *mathematical modeling.* Before getting down to the details of this, let's consider for a moment the way mathematics interacts with other fields in their quantification. In doing this we can perhaps obtain something of an overview of the externally stimulated growth of mathematics we just spoke of.

To begin with, researchers in external fields study what mathematicians like to call the "real world," as opposed to the *idealized* mathematical world consisting of abstract notions like functions and points. Laws are desired to explain *observed data* collected by these researchers. Such laws take the form of *mathematical statements* (equations, theorems, or the like) about a mathematical "model" of the real phenomena being studied. This mathematical model is some abstract mathematical system that we hope contains all the essential variables and relationships present in the real world setting. By then carrying out purely mathematical work on the model, mathematicians can obtain results that, when interpreted back in the real world setting, provide predictions about what *behavior* will be observed there. *Comparison* of the predictions with the observed results of controlled experiments then measures the degree to which the mathematical model accurately captures the essence of the real world situation. If the predictions correlate almost perfectly with the experimental results, then the model will be adopted and its results will be called *laws* of the external field. If the predictions correlate poorly with the experimental results, then a revised or refined model may be constructed, subject perhaps to still further refinement after additional experimentation. The process is shown schematically in Figure 0.1.

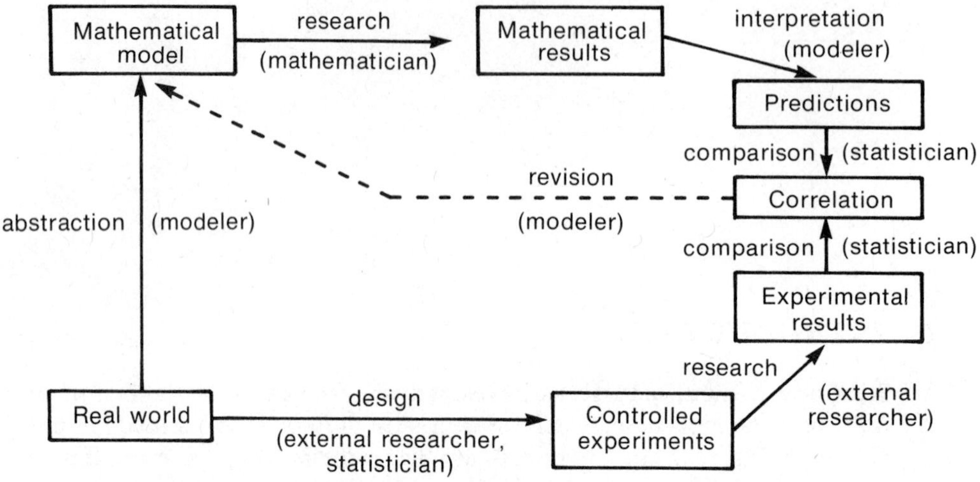

FIGURE 0.1

As you can see, a number of skilled professionals may be involved in the overall process, especially when the phenomena under study or the mathematical models are complex. In many cases, though, the external researcher assumes several (or all) of the roles, and hence needs to know a good deal about both mathematics and statistics in order to be as effective as possible. In a large number of situations the model is based on one or more differential equations, and the research needed to obtain mathematical results consists of obtaining an exact or approximate solution of the differential equation(s).

A simple example involves the motion of a body near the earth's surface falling freely from rest. If we decide that air resistance is not important, then we can model this phenomenon mathematically by the differential equation

$$a = \frac{d^2s}{dt^2} = g$$

where s is the number of feet (or meters) the body has fallen after t seconds, a is the magnitude of the acceleration of the body at that time, and g is the magnitude of the acceleration due to gravity (positive direction being downward). This differential equation can be solved by simple integration, giving first

$$\frac{ds}{dt} = gt + c_1.$$

Now, since the body falls from rest, $\frac{ds}{dt} = 0$ when $t = 0$. Hence substitution of $t = 0$

gives $c_1 = 0$. Integrating again, we have

$$s = \tfrac{1}{2}gt^2 + c_2.$$

If we measure distances from the initial point of the body, then $s = 0$ when $t = 0$. Hence $c_2 = 0$. We have then the formula

$$s = \tfrac{1}{2}gt^2. \tag{1}$$

To test the accuracy of this law, we could drop some bodies from rest and measure how far they fall during various time intervals. If we dropped stones or bricks, then we would probably be pretty pleased with our model's predictions. If we dropped paper or feathers, however, then our experimental results would probably not correlate very well with the predictions based on Equation (1). We might then be led to refine our model to take account of air resistance. (See Section 2 below.)

While mathematical models, especially in contemporary applied mathematics, need not always involve differential equations, still it is probably fair to say that historically differential equations have been the most effective tools of the mathematical modeler. In studying other fields, especially the physical, engineering, and life sciences, you will see many, many mathematical models that employ differential equations. Thus the material of Chapters 8, 9, and 10 constitutes an introduction to one of the principal parts of applied mathematics.

We begin this chapter by considering the simplest kinds of differential equations, and methods for obtaining their solutions. The first four sections take up the most commonly encountered types of differential equations, which involve only first derivatives. Applications to a number of other fields are presented in Sections 2 and 8, and elsewhere. Sections 6 and 7 consider an important class of differential equations that involve both first and second derivatives. These are certain kinds of *linear* differential equations, and their solution requires consideration of functions defined on the set $\boldsymbol{C}$ of complex numbers. The necessary groundwork is laid in Section 5. The chapter concludes with a section on methods of solution involving power series and numerical approximation of solutions that cannot be found by the methods of earlier sections.

You may notice that the word "linear" occurs in the titles of Sections 3, 6, and 7. You will see that, as in multivariable calculus, the concept of linearity plays a very important role in differential equations. In fact, it is linear differential equations that have a well-developed theory utilizing many ideas of linear algebra. Nonlinear differential equations are much harder to deal with than linear ones, and, reminiscent of the case in differentiation of functions of several variables, nonlinear equations are often approached from the point of view of being approximately linear. One often replaces a model involving a nonlinear differential equation by an approximating model involving a linear differential equation, for instance. The emphasis in these three chapters will consequently be on linear differential equations. If you take subsequent courses on differential equations, then you may go further into the study of nonlinear equations.

1 SEPARABLE EQUATIONS

In the introduction we said that a differential equation is an equation involving variables and derivatives of these variables, and we considered the simple example $\dfrac{d^2s}{dt^2} = g$. To study differential equations systematically, we need to consider in turn various types of differential equations. In this text, for example, we will consider solving only what are called *ordinary differential equations.* These are of the general form

(1) $$F\left(x, y, \frac{dy}{dx}, \frac{d^2y}{dx^2}, \ldots, \frac{d^ny}{dx^n}\right) = 0.$$

Here the variable y is considered to be a function of the *single* variable x, and a *solution* of (1) means an explicit formula $y = f(x)$ for a function f satisfying (1). We call (1) an ordinary differential equation of *order n* because the order of the highest derivative present in (1) is n. In this chapter we will initially consider only equations of the first order; in Sections 6 and 7 we will study second order equations. In the next chapter we will allow n to be an arbitrary positive integer.

Besides ordinary differential equations, we have already encountered examples of *partial differential equations* such as Laplace's equation,

$$\frac{\partial^2 u}{\partial x^2} + \frac{\partial^2 u}{\partial y^2} + \frac{\partial^2 u}{\partial z^2} = 0,$$

and the heat equation (Exercises 7.8). We leave a discussion of methods of solution for partial differential equations for a later course.

In this section we will study perhaps the simplest kind of first order differential equations, examples of which you have already met in elementary calculus.

1.1 **DEFINITION.** A first order differential equation is called **separable** if it can be reduced to the form

(2) $$g(y)\frac{dy}{dx} = h(x).$$

The term *separable* is appropriate here, since viewing $\frac{dy}{dx}$ as the quotient of the differentials dy and dx allows us to put (2) in the form

(3) $$g(y)dy = h(x)\,dx,$$

where the variables x and y are indeed separated by the equal sign. Now suppose that g and h are continuous, so that (3) can be integrated. If G is an antiderivative of g and H is an antiderivative of h, then we have, on integrating (3),

(4) $$G(y) = H(x) + c.$$

For any particular value of the constant c of integration, this already gives y *implicitly* as a function of x near any point (x_0, y_0) where $g(y_0) \neq 0$. More precisely, if we let $F(x, y) = G(y) - H(x) - c$, then $\frac{\partial F}{\partial y} = G'(y) = g(y)$. Therefore, the Implicit Function Theorem 4.6.2 guarantees that we can solve (4) for y as a unique function of x near (x_0, y_0), provided that $\frac{\partial F}{\partial y}(x_0, y_0) = g(y_0) \neq 0$. In what follows, then, we shall usually assume $g(y) \neq 0$ on any region where we seek to solve (2).

The foregoing discussion shows that any possible solution of (2) must be implicitly defined by (4). To check that (2) really *has* solutions, we can verify that any function $y = y(x)$ defined by (4) actually satisfies (2). This we can do easily by differentiating (4) implicitly with respect to x. We get

$$G'(y)\frac{dy}{dx} = H'(x),$$

that is,

(2) $$g(y)\frac{dy}{dx} = h(x).$$

Thus, the functions defined by (4) as c takes on all possible real values actually *are* solutions of (2). It is appropriate now to introduce some traditional terminology.

1.2 DEFINITION. The **general solution** of a differential equation (1) is the set of all functions $y = y(x)$ that satisfy the given equation.

We can now sum up our discussion of separable equations in the following theorem.

1.3 THEOREM. A separable differential Equation (2) such that g and h are continuous and $g(y) \neq 0$ has as general solution the set of all functions $y = y(x)$ defined by (4) for all possible values of $c \in \boldsymbol{R}$.

1.4 EXAMPLE. Find all differentiable functions $y = y(x)$ whose graphs have the property that at any point not on the y-axis the slope is equal to half the ratio of the y-coordinate to the x-coordinate.

Solution. We have to translate the given geometrical condition into a differential equation. At any point of the graph of $y = y(x)$, the slope is of course $\frac{dy}{dx}$. Our condition is then

(5) $$\frac{dy}{dx} = \frac{1}{2}\frac{y}{x}, \qquad x \neq 0.$$

We can put this equation into the form (2) if we impose the restriction $y \neq 0$. [Note that the constant function $y = 0$ *is* a solution to (5), so at the end of our work we will need to check that it is included in our general solution.] Then multiplication of both sides of (5) by $2/y$ gives

$$\frac{2}{y}\frac{dy}{dx} = \frac{1}{x},$$

$$\frac{2}{y}\,dy = \frac{1}{x}\,dx.$$

Hence,

$$2\int \frac{1}{y}\,dy = \int \frac{1}{x}\,dx,$$

$$2\ln|y| = \ln|x| + k,$$

$$\ln y^2 = \ln|x| + k.$$

We can put this implicit equation for y into a geometrically more recognizable form if we write the arbitrary constant k as $\ln \bar{c}$ for $\bar{c}>0$. Then

$$\begin{aligned}\ln y^2 &= \ln |x| + \ln \bar{c}\\ &= \ln \bar{c}\,|x|.\end{aligned}$$

Hence,

$$y^2 = \bar{c}\,|x|,$$

that is,

$$y^2 = \pm \bar{c}x.$$

As $\bar{c}$ takes on all positive values, $\pm\bar{c}$ takes on all real values except 0. We noted above, though, that we had to include the solution $y=0$ (which corresponds to $\bar{c}=0$) in our general solution. We can thus write the general solution in the form

$$\{y = y(x) \mid y^2 = cx,\ c \in \boldsymbol{R}\}.$$

Geometrically we have the x-axis together with a family of parabolas, some of which are drawn in Figure 1.1.

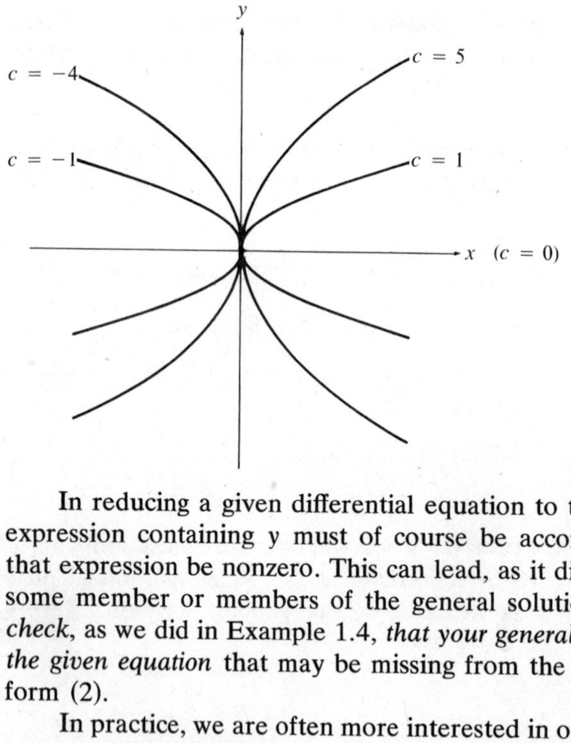

FIGURE 1.1

In reducing a given differential equation to the form (2), any division by an expression containing y must of course be accompanied by the restriction that that expression be nonzero. This can lead, as it did in Example 1.4, to the loss of some member or members of the general solution. You must therefore *always check*, as we did in Example 1.4, *that your general solution contains all solutions of the given equation* that may be missing from the general solution of the reduced form (2).

In practice, we are often more interested in one *particular* solution to (1) than we are in the complete general solution. You will recall, for instance, in our discussion of the equation

(6) $$\frac{d^2s}{dt^2} = g$$

in the introduction, that we did not simply integrate

$$\frac{ds}{dt} = gt + c_1$$

to obtain the general solution

$$s=\tfrac{1}{2}gt^2+c_1t+c_2.$$

Rather, we used the conditions $\frac{ds}{dt}=0$ and $s=0$ at $t=0$ to obtain the single solution $s=\frac{1}{2}gt^2$ that reflected the conditions present in the real world situation being modeled by (6). Even though our independent variable x in (1) does not always represent time, a condition of the form $y=y_0$ when $x=x_0$ is called an *initial condition.*

1.5 **DEFINITION.** A first order differential equation $F(x, y, y')=0$ with an initial condition $y=y_0$ when $x=x_0$ is called an **initial value problem.**

With reference to (2), an initial condition has the effect of picking out one particular solution to the given equation from among the infinitely many members of the general solution. This occurs because an initial condition determines a particular value of the constant c of integration in (4). In fact, we have in (4),

$$G(y_0)=H(x_0)+c,$$
$$c=G(y_0)-H(x_0).$$

Thus (4) takes the form

$$G(y)=H(x)+G(y_0)-H(x_0)$$

which defines y as a *unique* function of x near any point where $G'(y)=g(y)$ is not zero. We thus have the following existence-uniqueness theorem for separable first order initial value problems.

1.6 **THEOREM.** The separable first order initial value problem

(7) $$g(y)\frac{dy}{dx}=h(x), \qquad y(x_0)=y_0$$

where g and h are continuous, has a unique solution $y=y(x)$ on some interval containing x_0 if $g(y_0)\neq 0$.

1.7 **EXAMPLE.** Find the function $y=y(x)$ whose graph passes through $(4, -2)$ such that the slope of the graph at any point not on the y-axis is half the ratio of the y-coordinate to the x-coordinate.

Solution. This is just an initial value problem for Example 1.4. In that example we found that all solutions to the unconstrained problem had the form

$$y^2=cx$$

for c a real number. If $y(4)=-2$, then we have

$$(-2)^2=4c,$$
$$4=4c,$$
$$c=1.$$

Thus the desired curve is the parabola whose equation is $y^2 = x$. The corresponding function is given by $y = -\sqrt{x}$, since $(4, -2)$ lies below the x-axis.

Sometimes in solving a separable equation we will not obtain from (4) an explicit formula $y = y(x)$ because it is too much work, or perhaps impossible, to do so. This shouldn't be too disturbing in light of the techniques we have learned for handling implicit functions.

1.8 EXAMPLE. Solve the initial value problem $(y + e^{-y})\dfrac{dy}{dx} = x^2$, $y(1) = 1$.

Solution. We have $(y + e^{-y})dy = x^2 dx$, so after a simple integration we get

(8) $$\tfrac{1}{2}y^2 - e^{-y} = \tfrac{1}{3}x^3 + c.$$

When $x = 1$ we have $y = 1$, so substitution of this information leads to $\frac{1}{2} - e^{-1} = \frac{1}{3} + c$. Hence, $c = \dfrac{1}{6} - \dfrac{1}{e}$. Putting this into (8) we have,

(9) $$\tfrac{1}{2}y^2 - e^{-y} = \frac{1}{3}x^3 + \frac{1}{6} - \frac{1}{e}.$$

There is no apparent way to solve (9) for y as a function of x near $(1, 1)$. We can, however, be sure that (9) *does* define a unique solution by the Implicit Function Theorem, since if we put

$$F(x, y) = \tfrac{1}{2}y^2 - e^{-y} - \frac{1}{3}x^3 - \frac{1}{6} + \frac{1}{e},$$

then $\dfrac{\partial F}{\partial y} = y + e^{-y} \neq 0$ near $y = 1$.

EXERCISES 8.1

In Exercises 1 to 8, find the general solution of the given first order differential equation. If convenient, express the solutions in explicit form.

1 $\dfrac{dy}{dx} = x^3 e^{-2y}$

2 $(x-4)y^4 - x^3(y^2-3)\dfrac{dy}{dx} = 0.$

3 $2(y-1)\dfrac{dy}{dx} - 3x^2 - 4x = 2.$

4 $(x+2)y^2 + x^2(y-1)\dfrac{dy}{dx} = 0.$

5 $\dfrac{dy}{dx} = \dfrac{2xy - 2x}{x^2+1}.$

6 $2(y-1)\dfrac{dy}{dx} - e^x + x^2 = 1.$

7 $\dfrac{dy}{dx} = 5y - 1.$

8 $\dfrac{dy}{dx} = \dfrac{xy + 2x}{x^2+1}.$

In Exercises 9 to 14, solve the given initial value problem.

9 $\dfrac{dy}{dx} = \dfrac{\cos x}{3y^2 + e^y}$, $y(0) = 2.$

10 $x \sin y + (x^2+1) \cos y \dfrac{dy}{dx} = 0$, $y(1) = \frac{1}{2}\pi.$

11 Exercise 2, where $y(1) = 2.$

12 Exercise 3, where $y(0) = -1.$

13 $(y+1)\, dx - y(x+2)\, dy = 0$, $y(0) = 1.$

14 $\sqrt{1+x^2}\, dy - xy^3\, dx = 0$, $y(0) = 2.$

15 Show that the equation

$$\frac{dy}{dx} = F(ax + by + c) \quad \text{where} \quad b \neq 0$$

becomes separable if we change variables by letting $v = ax + by + c$ become the new dependent variable.

16 Show that in Exercise 15 we can use the substitution $v = ax + by + d$ for *any* constant d.

17 Solve $y' = (y + x + 5)^2$ by the method of Exercise 15.

18 Solve $y' = (y + 9x - 3)^2$ by the method of Exercise 15.

19 (a) Find the general equation for the family of curves whose slope at each point $P(x, y)$ is the product of the x and y coordinates.
(b) Which of these curves goes through $(-7, 1)$?

20 (a) Find the general equation for the family of curves whose slope at each point $P(x, y)$ is twice the product of the x-coordinate and the square of the y-coordinate.
(b) Which of these curves goes through $(5, 0)$?

21 Find the general equation for the family of curves whose slope at each point $P(x, y)$ is twice the slope of the line through P and the origin $\mathbf{0}$. Which of these goes through $(4, 0)$?

22 Find the general equation for the family of curves whose slope at each point $P(x, y)$ such that $x \neq 0$ is 3 times the slope of the line through P and the origin $\mathbf{0}$. Which of these goes through $(-3, 0)$?

23 Find the general equation for the family of curves whose slope at each point $P(x, y)$ not on the x-axis is twice the ratio of the x-coordinate to the y-coordinate. Which of these passes through $(2, 0)$?

24 Find the general equation for the family of curves whose slope at each point $P(x, y)$ not on the x-axis is the negative ratio of the x-coordinate to the y-coordinate. Which of these passes through $(2, 0)$?

2 APPLICATIONS

In this section we consider a number of examples of how differential equations are used to model situations that arise in fields other than mathematics.

We have already considered the notion of bodies falling freely from rest. In Section 0 we modeled that situation as follows. Let s be the number of units in the distance the body has fallen after t seconds. Let a be the magnitude of the acceleration of the body, in appropriate units of distance per second per second, at time t. (We are thus measuring both distance and time from the start of the motion.) Then we proposed to model the motion by the differential equation

(1) $$a = \frac{d^2s}{dt^2} = g,$$

where g is the magnitude of the acceleration due to gravity. This led to the formula

(2) $$s = \tfrac{1}{2} gt^2$$

for the distance fallen after t seconds. You may well have seen this formula before and worked with it. But, as we pointed out in Section 0, it really is not universally applicable. The reason is not hard to see, either. In using (1) to model the motion, we are essentially asserting that the *only* physical factor influencing the motion is gravity. One obvious factor being suppressed in this model is air resistance acting on the body as it falls. Our experience is that air resistance can be significant for certain types of bodies. For bodies falling from rest, experiments suggest that air resistance is proportional to the speed of the falling body. If we denote resistance by R, then

(3) $$R=-kv$$

for a positive constant k that depends on the shape and mass of the body and possibly other factors. (The resisting force is directed upward since it impedes motion, hence the negative sign.) If we incorporate (3) into our model, then we have from $F=ma$,

$$m\frac{d^2s}{dt^2}=mg-kv,$$

i.e.,

(4) $$m\frac{dv}{dt}=mg-kv.$$

2.1 EXAMPLE. Find formulas for v and s at time t, using (4) to model the motion of a body falling from rest. (Assume $k\neq 0$.)

Solution. We have from (4)

$$m\,dv=(mg-kv)\,dt,$$

$$m\frac{dv}{mg-kv}=dt,$$

$$-\frac{m}{k}\frac{-k\,dv}{mg-kv}=dt.$$

In this form we can integrate to obtain

(5) $$-\frac{m}{k}\ln(mg-kv)=t+c.$$

We don't need absolute value signs since $mg-kv$ will, on physical grounds, be positive from (4). Since the body falls from rest, $v=0$ when $t=0$. Hence

$$-\frac{m}{k}\ln mg=c.$$

Substituting back into (5), we have

$$-\frac{m}{k}\ln(mg-kv)=t-\frac{m}{k}\ln mg,$$

$$-\frac{m}{k}(\ln(mg-kv)-\ln mg)=t,$$

$$\ln\frac{mg-kv}{mg}=-\frac{kt}{m},$$

$$\frac{mg-kv}{mg}=e^{-kt/m}$$

$$mg-kv=mge^{-kt/m},$$

$$kv=mg(1-e^{-kt/m}),$$

(6) $$v=\frac{mg}{k}(1-e^{-kt/m}).$$

This is one of the quantities sought. Since $v=\frac{ds}{dt}$, we can get the other by integrating (6). This gives

(7) $$s+c=\frac{mg}{k}\left(t-\left(-\frac{m}{k}\right)e^{-kt/m}\right).$$

Again, since the body falls from its initial position, we have $s=0$ when $t=0$. Thus

$$c=\frac{mg}{k}\left(\frac{m}{k}\right)=\frac{m^2g}{k^2}.$$

Substituting into (7) we obtain

(8) $$s=\frac{mg}{k}t-\frac{m^2g}{k^2}(1-e^{-kt/m}).$$

From our model, we have obtained formulas for s and v that could be put to experimental test. Notice that these formulas differ sharply from those obtained in Section 0 using (1) to model the motion. Most noticeable is the difference in the nature of the speed. Using (1) we got $v=gt$, so the speed is predicted to increase without bound as t increases. But if we let $t\to\infty$ in (6) we get

$$\lim_{t\to\infty} v=\lim_{t\to\infty}\frac{mg}{k}(1-e^{-kt/m})=\frac{mg}{k}.$$

Thus in this model, the speed tends to a limiting value that depends upon the mass and the coefficient k of resistance.

Which model is "right"? Such a question is really meaningless. Both models give predictions about the behavior of v, and all we can really ask is which set of predictions correlates better with experimentation. In some cases the model based on (1) is perfectly adequate. In others, the model based on (4) is preferable. In still other cases, involving high speed motion (with an initial velocity), air resistance seems to be better approximated by $-kv^2$. Thus the choice of model to use may be governed not by mathematical factors but by purely physical considerations having direct effects on the phenomena being studied. Again we emphasize that applied mathematics deals with the real world, where the absolute black-or-white, right-or-wrong criteria of pure mathematics may not be entirely appropriate. Most often, no model is *absolutely* accurate. Rather, it is a question of which model gives the best *approximation* to the observed behavior.

A wide range of important phenomena involving growth and decay are modeled by use of the exponential function. Let $y(t)$ be the amount of a growing (or decaying) quantity present at time t.

2.2 **DEFINITION.** A quantity $y=y(t)$ has an **exponential rate of growth or decay** if for some constant k

(9)
$$\frac{dy}{dt}=ky,$$

that is, if at any time t the rate of change of the amount of the quantity is proportional to the amount present at that time. If $k>0$, then we say that y *grows exponentially*. If $k<0$, then y *decays exponentially*.

We can easily solve (9) for y, and the resulting formula will show the origin of the term "exponential growth." We have

$$dy=ky\,dt,$$

$$\frac{dy}{y}=k\,dt \qquad \text{(assuming } y\neq 0\text{)},$$

(10)
$$\ln y=kt+c \qquad \text{(assuming } y>0\text{)}.$$

Assuming that y is measured to be y_0 when $t=0$, we have

$$\ln y_0=c.$$

Thus (10) becomes

$$\ln y-\ln y_0=kt$$

$$\ln\frac{y}{y_0}=kt,$$

$$\frac{y}{y_0}=e^{kt},$$

(11)
$$y=y_0e^{kt}.$$

Formulas (9) and (11) occur so often in applications that they are given the name *Law of Exponential Growth* (or *Decay*). Radioactive elements decay according to this law, investments earning continuously compounded interest grow according to it, and colonies of bacteria grow for a time at a rate given by this law. (Of course, populations grow *discretely*, not continuously, but the growth of large populations can be usefully treated as a continuous process.)

Recently, the use of (9) to model the earth's population has led to serious concern. Notice that if we let t increase indefinitely in (11), then y grows without bound if $k>0$, since

$$\lim_{t\to\infty} y=\lim_{t\to\infty} y_0e^{kt}=\infty.$$

Since the resources of the earth, in particular its capacity to provide food for human consumption, are finite, we see that a constant growth rate $k>0$ is *impossible to sustain* indefinitely. There has thus developed a movement to voluntarily reduce k to zero ("zero population growth") before such catastrophes as widespread famine, disease, or thermonuclear warfare occur and reduce k to zero or even below zero.

The contemporary discipline of *demography* studies human population growth. Since the exponential growth law with $k>0$ cannot provide a long-term model for the growth of any population (human or otherwise), demographic modelers have sought to develop better population growth models. Observation of bacteria and insect colonies, and even colonies of rats, suggests that for large values of y, death rates increase rapidly and reproduction decreases rapidly as living conditions, food supplies, and other resources deteriorate. This has led to the following model, which is called the *logistic growth* model.

2.3 EXAMPLE. Suppose that a population changes according to the law

(12) $$\frac{dy}{dt} = ky - \ell y^2,$$

where $k>0$ and $\ell>0$ are constants. Discuss the behavior of y.

Discussion. We can think of the term ℓy^2 as a damper on growth in this population. In a cannibalistic or warlike or disease- or famine-ridden population, it represents the death rate. In contemporary society, it might reflect decisions on the part of human beings not to reproduce in light of crowding or declining quality of life. We now proceed to solve (12) for $y = y(t)$. We first separate the variables. This gives

(13) $$\frac{dy}{y(k-\ell y)} = dt.$$

To integrate, we need to decompose the left-hand side into partial fractions.

$$\frac{1}{y(k-\ell y)} = \frac{A}{y} + \frac{B}{k-\ell y},$$

$$1 = A(k-\ell y) + By$$

Letting $y=0$, we get $A=1/k$. Letting $y=k/\ell$, we have $B=\ell/k$. So (13) becomes

$$\frac{1}{k}\frac{dy}{y} + \frac{\ell}{k}\frac{dy}{k-\ell y} = dt.$$

Integrating, we get

$$\frac{1}{k}\ln y + \frac{\ell}{k}\left(-\frac{1}{\ell}\right)\ln|k-\ell y| = t+c,$$

$$\frac{1}{k}\ln\frac{y}{|k-\ell y|} = t+c,$$

$$\ln\frac{y}{|k-\ell y|} = k(t+c),$$

$$\frac{y}{k-\ell y} = \pm e^{k(t+c)} = \pm e^{kt}e^{kc} = Ke^{kt},$$

where $K = \pm e^{kc}$. Thus we have

$$y = kKe^{kt} - y\ell Ke^{kt},$$
$$y(1 + \ell Ke^{kt}) = kKe^{kt},$$
$$y = \frac{kKe^{kt}}{1 + \ell Ke^{kt}}.$$

What happens now as $t \to \infty$? We have

$$\lim_{t\to\infty} y = \lim_{t\to\infty} \frac{kK}{e^{-kt} + \ell K} = \frac{k}{\ell}.$$

Thus, the population tends toward a *stable level* given by the ratio k/ℓ of the constants in (12).

Certainly the predicted long-term behavior of the population in this model is more reasonable than that of the exponential model based on (9). While you might think that this more sophisticated logistic growth model is only of recent origin, it was in fact first introduced in 1846 by the Belgian sociologist Pierre-François Verhulst (1804–1849) to model Belgium's population. Verhulst obtained a stable population estimate of about 9,400,000, less than 3% under Belgium's actual population of 9,650,000 in its 1970 census!

Verhulst's work unfortunately attracted little attention, and lay forgotten for many years. But in 1920 the logistic growth model was rediscovered by the American biologists Raymond Pearl (1879–1940) and Louis J. Reed (1886–1966). Following a study of United States census figures from 1790 through 1910, Pearl and Reed introduced (12) with $k = 3.13395 \times 10^{-2}$ and $\ell = 1.58864 \times 10^{-10}$ as a model for the population of the United States. This model also proved to be quite accurate. It predicted a stable population $k/\ell = 197{,}272{,}500$, whereas the 1970 population was 203,184,772.

As mentioned earlier, radioactive substances decay according to the exponential laws (9) and (11). The 1960 Nobel Prize for Chemistry was awarded to Willard F. Libby (1908–   ) of the University of California, Los Angeles, for work leading to the important archeological tool *carbon 14 dating.* Libby developed a technique for accurately measuring the amount of radioactive carbon 14 present in ancient relics. Living plants and animals assimilate this isotope from the atmosphere, where it is produced as a result of cosmic radiation. Upon death, the carbon 14 decay is no longer offset by assimilation. The isotope has a half life of 5730 years, meaning that after 5730 years, half the original amount of carbon 14 still is present in the remains of, for instance, a once living tree or animal. This fact permits us to use (11) to determine when the ancient tree or animal lived.

2.4 **EXAMPLE.** (a) Determine the value of the coefficient k of decay for carbon 14. (b) If human hair is found that contains 20% of the concentration of carbon 14 found in a living human, then when was the hair part of a living body?

Solution. (a) From (11) we have $y = y_0 e^{kt}$. When $t = 5730$, then $y = \frac{1}{2}y_0$. So,

$$\tfrac{1}{2}y_0 = y_0 e^{5730k},$$
$$\tfrac{1}{2} = e^{5730k},$$
$$\ln\tfrac{1}{2} = 5730\,k$$
$$k = \frac{\ln\frac{1}{2}}{5730} = \frac{-\ln 2}{5730} \approx -0.00012097.$$

(b) From (11) we have $\dfrac{y}{y_0}=e^{kt}$. Hence

$$\ln\frac{y}{y_0}=kt,$$

$$t=\frac{1}{k}\ln\frac{y}{y_0}.$$

Here $y=\frac{1}{5}y_0$, so that

$$t=\frac{1}{k}\ln\frac{1}{5}=-\frac{1}{k}\ln 5=5730\frac{\ln 5}{\ln 2}\approx 13{,}305.$$

Thus the ancient man or woman lived about 13,300 years ago.

A phenomenon closely related to exponential growth and decay is the cooling (or heating) experienced by a body placed in a surrounding cooler (or hotter) medium. This phenomenon is modeled by **Newton's law of cooling.** We can state this law as follows. Let $T(t)$ be the number of degrees in the temperature of a body at time t. Let C be the number of degrees in the (assumed constant) temperature of the surrounding medium. Then the rate at which the body cools is given by

(14) $$\boxed{\frac{dT}{dt}=k(T-C).}$$

That is, the rate of change of the body's temperature at any instant is proportional to the difference between the body's temperature at that instant and the temperature of the surrounding medium.

2.5 EXAMPLE. A cup of black coffee is poured from a pot whose contents are maintained at 95°C. The cup is put on a table in a room whose temperature is kept at 20°C. After one minute the coffee in the cup has cooled to 90°C. How long will it take the coffee to reach a drinkable temperature of 65°C?

Solution. We solve (14) by separating the variables. Since $C=20$, (14) becomes $\dfrac{dT}{T-20}=k\,dt$. Therefore,

(15) $$\ln|T-20|=kt+c_1.$$

We are told that $T=95$ when $t=0$. Thus, $\ln 75=c_1$. Then we have, on substituting for c_1 in (15) and using $-\ln 75=\ln\frac{1}{75}$,

(16) $$\ln\frac{T-20}{75}=kt.$$

We also are told that $T=90$ when $t=1$. Hence we obtain

$$\ln\tfrac{70}{75}=k,\quad \text{i.e.,}\quad k=\ln\tfrac{14}{15}.$$

So (16) can be put into the form

$$t=\frac{\ln\dfrac{T-20}{75}}{\ln\frac{14}{15}}.$$

When $T=65$, we have then

$$t=\frac{\ln\frac{45}{75}}{\ln\frac{14}{15}}=\frac{\ln\frac{3}{5}}{\ln\frac{14}{15}}\approx 7.4.$$

So it takes approximately $7\frac{1}{2}$ minutes for the cup of coffee to reach a drinkable temperature.

Many important chemical reactions involve the interaction of single molecules of two substances S and T to form a molecule of new substance U. This is denoted schematically by

$$S+T\rightarrow U.$$

For example, if S is sodium and T is chlorine, then U is common table salt. The **law of mass action** states that the rate of production of U at any instant is proportional to the amounts of S and T present at that instant.

2.6 **EXAMPLE.** Suppose that equal amounts a of S and T are placed in a reacting chamber. Find a formula for the amount of compound U produced after t seconds, if after 10 seconds the amount of U present is $\frac{1}{2}a$.

Solution. Let $y(t)$ be the number of units of U present after t seconds. Then the amounts of S and T present at that time are both given by $a-y(t)$, since for every molecule of U formed, one molecule of S and one molecule of T are used up. The law of mass action then tells us that

(17) $$\frac{dy}{dt}=k(a-y)(a-y).$$

We can solve (17) easily. Separating the variables, we get

$$\frac{dy}{(a-y)^2}=k\,dt,$$

$$\int (a-y)^{-2}\,dy=\int k\,dt,$$

$$(a-y)^{-1}=kt+c,$$

$$\frac{1}{a-y}=kt+c.$$

Now when $t=0$, we have $y=0$, so that $\dfrac{1}{a}=c$. This gives us

(18) $$\frac{1}{a-y}=kt+\frac{1}{a}.$$

When $t=10$, we have $y=\frac{1}{2}a$. Thus,

$$\frac{2}{a}=10k+\frac{1}{a}, \quad \text{i.e.,} \quad k=\frac{1}{10a}.$$

From (18), then,

$$\frac{1}{a-y}=\frac{t}{10a}+\frac{1}{a}=\frac{t+10}{10a},$$

$$10a=(t+10)(a-y)=at+10a-y(t+10),$$

$$y(t+10)=at,$$

$$y=\frac{at}{t+10}.$$

Notice that the "law" of mass action applies only up to a point. We can write our formula for y as

$$y=a\left(\frac{t}{t+10}\right).$$

In this form, we see that the amount of U produced *approaches* a ultimately but never reaches a in a finite time. Thus, while the law may give an accurate description of the reaction process during its *early* (and most active) stage, this model is too crude to describe the entire reaction. This affords us one more illustration of how every modeling process must be assessed both for accuracy of approximation and for the extent of its applicability.

We close this section with an example that introduces the *orthogonal trajectories* of a family $\mathcal{F}$ of curves in $\mathbf{R}^2$. These are curves γ that intersect every member of $\mathcal{F}$ at right angles. Such mutually orthogonal families of curves arise frequently in applications. [For example, in the theories of heat, electricity, magnetism, and gravitation, the force field vectors are tangent vectors to a family of orthogonal curves to the level (equipotential) curves.]

2.7 EXAMPLE. Find the family of all orthogonal trajectories to the family of curves $x^2+y^2=c^2$.

Solution. At each point $P(x, y)$ on the curve $x^2+y^2=c^2$ (other than the points where $y=0$), we have $\frac{dy}{dx}=-\frac{x}{y}$. Since the slope of an orthogonal curve is the negative reciprocal of the slope of the curve,

(19) $$\frac{dy}{dx}=\frac{y}{x}$$

holds at each point (x, y) such that $x\neq 0$ of an orthogonal trajectory $y=y(x)$.

Separation of the variables in (19) brings us to

$$\frac{dy}{y} = \frac{dx}{x},$$
$$\ln|y| = \ln|x| + c,$$
$$\ln|y| - \ln|x| = c$$
$$\ln\frac{|y|}{|x|} = c$$
$$\frac{y}{x} = \pm e^c = k,$$
$$y = kx.$$

In this solution we required $x \neq 0$. So we must add that line (the y-axis) to our solution. Thus the family of orthogonal trajectories consists precisely of all lines through the origin. See Figure 2.1.

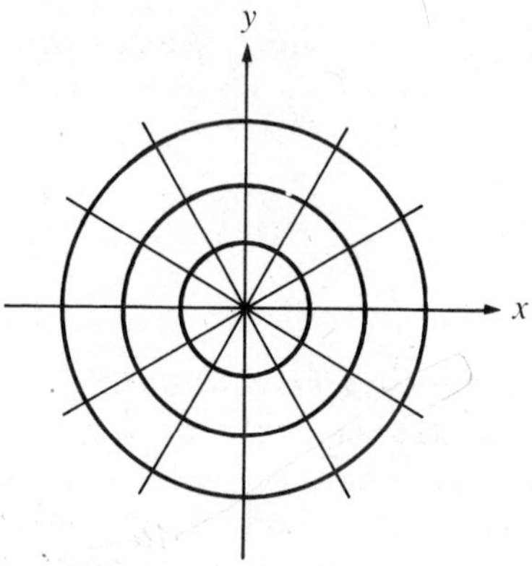

FIGURE 2.1

EXERCISES 8.2

In Exercises 1 to 6, find the family of orthogonal trajectories to the given family of curves.

1 $x^2 - y^2 = c^2$

2 $xy = c$

3 $x^2 + y^2 = cx$

4 $y = cx^2$

5 $y^2 = 2cx + c^2$

6 $x^2 + 2y^2 = c^2$

7 Some banks now pay interest *continuously*, that is, according to the law (9). If a bank pays 5.25% interest compounded continuously, then after one year how much interest has accumulated in an account opened with \$1000, in which no subsequent deposits or withdrawals are made? Thus, what annually compounded interest rate is equivalent to a continuous rate of 5.25%?

8 Find how long is required for a \$1000 certificate of deposit to double in value if it earns a continuously compounded rate of interest of (a) 5.25%, (b) 7.50%.

9 Assume in Example 2.1 that air resistance is proportional to the *square* of the speed. Find formulas for v and s at time t. What is $\lim_{t\to\infty} v$ in this case?

10 Suppose that a ball is thrown upward with initial speed v_0. If we neglect air resistance, then how far will the ball rise? If air resistance is as in Example 2.1, how far will the ball rise?

11 In launching rockets from the earth's surface, the force due to gravity can no longer be assumed constant. Instead, *Newton's law of gravitational attraction* is used. Assume that the earth is a ball of radius $R \approx 4000$ miles and mass M. If the rocket is x miles from the center of the earth and has mass m, then this law states that $F = k\dfrac{mM}{x^2}$.

(a) Find k by using the fact that when $x = R$, then $F = -mg$. (Note that the positive direction is now taken as *upward*.)

(b) Solve the resulting differential equation

$$m\frac{dv}{dt} = \frac{-mgR^2}{x^2}$$

for v by using

$$\frac{dv}{dt} = \frac{dv}{dx}\frac{dx}{dt} = \frac{dv}{dx}v.$$

Assume that the rocket has initial velocity v_0.

(c) How large should v_0 be in order for the rocket to *escape* the earth's gravity? (That is, in order for v to always be positive no matter how large x becomes.)

12 A sky diver falls from rest toward the earth. The diver and equipment weigh 192 pounds. Assume that the air resistance in pounds is $\frac{1}{2}v$ before the parachute opens (v measured in feet per second), and $\frac{3}{4}v^2$ after the parachute opens. Find formulas for v before and after the parachute opens. Assume that the parachute opens 12 seconds after free-fall begins. Find the limiting value of v after the parachute opens. What is the limiting value if the parachute fails to open?

13 The population of the United States in 1900 was about 76,000,000, and in 1950 it was about 150,700,000. Assuming that the population growth rate was exponential, obtain projections for the population from 1910 to 1970, and compare with the following census figures (rounded to the nearest 100,000).

1910	92,000,000	1950	150,700,000
1920	105,800,000	1960	179,300,000
1930	122,800,000	1970	204,000,000
1940	131,700,000		

14 In the population model of Example 2.3, denote the limiting population k/ℓ by L. Show that this gives

$$\frac{y}{L-y} = \ell K e^{kt}.$$

Show that if we measure time from the instant when $y = \frac{1}{2}L$, then the formula simplifies to

$$y = \frac{L}{1+e^{-kt}}. \tag{20}$$

(In the Pearl-Reed model, the time $t = 0$ was taken as April 1, 1914.)

15 Use formula (20) from Exercise 14, with time measured from the time of the 1940 census figure given in Exercise 13. Determine k so that the population predicted by the model for 1960 agrees with the census figure. Now compute the predicted 1970, 1980, and 2000 U.S. populations. Compute your percent error for the 1970 prediction. (What is the limiting population in this model?)

16 (a) The population of the earth was about $2\frac{1}{2}$ billion (2.5×10^9) in 1950, and about 4 billion in 1975. Assuming exponential growth, obtain projections for the population of the earth in 2000 and 2050. Assume that 25 billion is the absolute maximum number of people the earth can support. When will the limiting population be achieved if the growth rate remains constant?

(b) Assume that the growth rate is reduced to half of the 1950–to–1975 rate. Obtain new estimates for the quantities sought in part (a).

17 A piece of charcoal found in a cave contains 30% of the amount of carbon 14 found in trees now living. Determine when an ancient caveman cut down a tree to build a fire that produced the charcoal.

18 A fossil is found containing 10% of the carbon 14 found in living shellfish. When did the shellfish that became the fossil die?

19 A by-product of atmospheric nuclear tests is the isotope strontium 90, which research has linked to bone cancer in humans. The half life of strontium 90 is 28 years. How much strontium 90 will remain in the year 2000 from that released by a nuclear explosion in 1950?

20 A heated office building is maintained at 21°C. At 2:00 P.M., a murdered executive is found by a secretary who immediately calls the police. The coroner arrives at 3:00 P.M., and finds that the executive's body temperature is 31°C. One hour later, it has dropped to 29°C. Assuming that the executive's body temperature was 37°C when he was killed, establish the time of death.

21 A hospitalized criminal suffering from a severe case of influenza has been running a body temperature of 40°C. He is found shot to death in his bed at 11:00 P.M. At that time his body temperature is 37°C. One hour later the body temperature is 35°C. Find the time of death, if the hospital room is maintained at 20°C.

22 A metal foundry has a cooling bath maintained at 10°C. Hot metal at 250°C is plunged into the bath, and after 30 seconds it has cooled to 150°C. How long will it take for the metal to cool to the handling temperature of 60°C?

23 Suppose in Example 2.6 that *unequal* amounts a of S and b of T are placed in the reacting chamber. Find a formula for the amount of U present after t seconds, using the law of mass action.

24 Suppose in Example 2.6 that 2 molecules of S combine with one molecule of T to form each molecule of U. If initial amounts a of S and $b=\frac{1}{2}a$ of T are placed in a reaction chamber, find a formula for the amount of U present after t seconds. Use the law of mass action.

25 (a) A simple model for the spread of epidemics (or rumors) is based on the differential equation

$$\frac{dy}{dt}=ky(1-y),$$

where y is the fraction of the population infected at time t. Thus, the rate of spread is jointly proportional to the number of infected people and the number of susceptible but not yet infected people. Assuming that $y=y_0>0$ when $t=0$, find a formula for y at time t.

(b) What happens in (a) as $t\to\infty$? Is a quarantine useful in this situation? Comment on your impression of the accuracy of this model.

26 A spherical mothball evaporates at a rate proportional to its surface area. Using the fact that $V=\frac{4}{3}\pi r^3$ and $S=4\pi r^2$ are the respective formulas for the volume and surface area of a mothball of radius r, give a formula for $\dfrac{dV}{dt}$. Solve the resulting differential equation to get a formula for V. If the ball initially has volume $\frac{1}{8}$ cubic inch, and after two weeks has volume $\frac{1}{27}$ cubic inch, then how long will it take for the mothball to evaporate entirely?

27 The *Domar growth model* (named for the Polish-born economist Evsey D. Domar (1914–) of the Massachusetts Institute of Technology) assumes that a certain fixed percentage k of a country's gross national product $y(t)$ is invested to produce growth of the total capital $x(t)$. The rate of growth of the capital is measured by this investment. It is further assumed that $y(t)$ is a fixed percentage ℓ of the total capital $x(t)$. Show that in this model $x(t)$ grows exponentially.

28 In 1930, the following model was proposed by the American psychologist Louis L. Thurstone (1887–1955), a pioneer in learning theory and intelligence measurement. First, a person's knowledge is specified at time t by a function $k(t)$, where $0 \le k(t) \le 1$. (Think of $k(t)$ as the percentage score that the person would achieve at time t on an objective test in the particular field of study.) Next, every act a person performs is either helpful or harmful to learning. Let $s(t)$ be the frequency with which acts contributing toward learning are performed, and $f(t)$ the frequency with which harmful acts are performed. Then Thurstone asserted that

$$k(t) = \frac{s(t)}{s(t) + f(t)}.$$

He also proposed that the rates of change of $s(t)$ and $f(t)$ should be measured by the formulas

$$\frac{ds}{dt} = Kk(t) \quad \text{and} \quad \frac{df}{dt} = -K(1 - k(t)),$$

where K is a positive constant.

(a) Show that from this the Chain rule gives $\dfrac{ds}{df} = -\dfrac{s}{f}$.

(b) Solve the equation in (a) to obtain $sf = D$, where the constant D measures the difficulty of the material.

(c) Show that then $k(t) = \dfrac{s^2}{s^2 + D}$.

(d) Show that $\dfrac{dt}{ds} = \dfrac{1}{K} + \dfrac{D}{Ks^2}$.

(e) Separate the variables in (d) and integrate to get an equation defining s implicitly as a function of t, hence by (c) giving $k(t)$ as a function of t.

3 FIRST ORDER LINEAR DIFFERENTIAL EQUATIONS[1]

In Section 1.6 we considered the problem of solving a system of linear equations

$$\begin{cases} a_{11}x_1 + a_{12}x_2 + \ldots + a_{1n}x_n = b_1 \\ a_{21}x_1 + a_{22}x_2 + \ldots + a_{2n}x_n = b_2 \\ \quad \vdots \qquad\quad \vdots \qquad\qquad\quad \vdots \qquad\; \vdots \\ a_{m1}x_1 + a_{m2}x_2 + \ldots + a_{mn}x_n = b_m. \end{cases}$$

We can put this into the form of a matrix equation

(1) $$A\mathbf{x} = \mathbf{b},$$

where A is the m-by-n matrix of coefficients a_{ij}, $\mathbf{x} \in \mathbf{R}^n$, and $\mathbf{b} \in \mathbf{R}^m$. Recall from

1. If you haven't yet studied Section 6.2, then begin reading this section with Equation (7), p. 516.

Section 6.2 that the function $\boldsymbol{T}:\boldsymbol{R}^n \to \boldsymbol{R}^m$ defined by

(2) $$\boldsymbol{T}(\boldsymbol{x}) = A\boldsymbol{x}$$

is a *linear transformation*, that is,

$$\boldsymbol{T}(\boldsymbol{x}+\boldsymbol{y}) = \boldsymbol{T}(\boldsymbol{x})+\boldsymbol{T}(\boldsymbol{y})$$
$$\boldsymbol{T}(a\boldsymbol{x}) = a\boldsymbol{T}(\boldsymbol{x})$$

for all $\boldsymbol{x}$ and $\boldsymbol{y} \in \boldsymbol{R}^n$ and $a \in \boldsymbol{R}$. Thus we can reformulate (1) as follows. In solving a system (1) of m linear equations in n unknowns, we seek the set of all $\boldsymbol{x} \in \boldsymbol{R}^n$ that the linear transformation $\boldsymbol{T}:\boldsymbol{R}^n \to \boldsymbol{R}^m$ determined by (2) maps onto a given vector $\boldsymbol{b} \in \boldsymbol{R}^m$. That is, we seek to solve

(3) $$\boldsymbol{T}(\boldsymbol{x}) = \boldsymbol{b}$$

for $\boldsymbol{x} \in \boldsymbol{R}^n$.

In this section we will consider a problem of the form (3) where $\boldsymbol{T}$ is a particular linear transformation on a real vector space of *functions* instead of a linear transformation on $\boldsymbol{R}^n$. Our $\boldsymbol{T}$ will be what is called a *linear differential operator* of order 1. That is, $\boldsymbol{T}$ will involve $\dfrac{dy}{dx}$. The equation of the form (3) will be called a **first order linear differential equation.** Later on we will consider second order linear differential equations (Sections 6 and 7) and n-th order differential equations (Chapter 9) for arbitrary n.

To bring the analogy with Section 1.6 into focus, we first introduce the analogue of the real vector space $\boldsymbol{R}^n$.

3.1 **PROPOSITION.** The set $C^\infty(\boldsymbol{R})$ of all infinitely differentiable functions $f:\boldsymbol{R}\to\boldsymbol{R}$ with domain $\boldsymbol{R}$ is a real vector space relative to the operations of addition and scalar multiplication defined by

(4) $$(f+g)(x) = f(x)+g(x)$$
$$(af)(x) = af(x) \qquad \text{for} \quad a \in \boldsymbol{R}.$$

Proof. First note that $C^\infty(\boldsymbol{R}) \subseteq \mathscr{F}$, where $\mathscr{F}$ is the set of *all* functions $f:\boldsymbol{R}\to\boldsymbol{R}$ with domain $\boldsymbol{R}$. Since $\mathscr{F}$ is a real vector space by Exercise 27 of Exercises 1.1, all the required algebraic identities hold for $C^\infty(\boldsymbol{R})$. The proof consists then in merely noting that, by elementary calculus, the identically 0 function (given by $0(x)=0$ for all $x \in \boldsymbol{R}$), $f+g$, and af are all in $C^\infty(\boldsymbol{R})$ for all f and $g \in C^\infty(\boldsymbol{R})$ and all $a \in \boldsymbol{R}$. Then by Definition 1.1.14, $C^\infty(\boldsymbol{R})$ is a real vector space.

3.2 **DEFINITION.** A **linear operator** on $C^\infty(\boldsymbol{R})$ is a linear transformation $T:C^\infty(\boldsymbol{R}) \to C^\infty(\boldsymbol{R})$.

3.3 **EXAMPLE.** The operator L defined by

(5) $$L(f)(x) = f'(x)+p(x)f(x)$$

for a fixed $p(x) \in C^\infty(\boldsymbol{R})$ is a linear operator on $C^\infty(\boldsymbol{R})$. (L is called a **linear differential operator** since it is linear and involves differentiation.)

Proof. First observe that if $f(x)$ is in $C^\infty(\boldsymbol{R})$, then so are $f'(x)$ and $p(x)f(x)$. Hence, so is $L(f)(x)=f'(x)+p(x)f(x)$. To verify that L is a linear operator, we have to verify that L preserves sums and scalar multiplication in $C^\infty(\boldsymbol{R})$. If $f_1(x)$ and $f_2(x)\in C^\infty(\boldsymbol{R})$, then we have

$$\begin{aligned}L(f_1+f_2)(x)&=\frac{d}{dx}(f_1(x)+f_2(x))+p(x)(f_1(x)+f_2(x))\\&=\frac{df_1}{dx}+\frac{df_2}{dx}+p(x)f_1(x)+p(x)f_2(x)\\&=\frac{df_1}{dx}+p(x)f_1(x)+\frac{df_2}{dx}+p(x)f_2(x)\\&=L(f_1)(x)+L(f_2)(x).\end{aligned}$$

Thus L preserves sums. If $a\in\boldsymbol{R}$ and $f(x)\in C^\infty(\boldsymbol{R})$, then we have

$$\begin{aligned}L(af)(x)&=\frac{d}{dx}(af(x))+p(x)(af(x))\\&=a\frac{df}{dx}+ap(x)f(x)\\&=af'(x)+ap(x)f(x)\\&=a(L(f)(x)).\end{aligned}$$

Then L preserves scalar multiplication. QED

3.4 **DEFINITION.** The general **first order linear differential equation** is

(6) $$L(y)(x)=\frac{dy}{dx}+p(x)y(x)=q(x)$$

where $p(x)$ and $q(x)\in C^\infty(\boldsymbol{R})$ and $y=y(x)$ is a function in $C^\infty(\boldsymbol{R})$ to be determined.

In Table 3.1 we show schematically the analogy between the familiar vector spaces $\boldsymbol{R}^n$ and systems of linear equations (1), on one hand, and the vector space $C^\infty(\boldsymbol{R})$ and the general first order linear differential equation (6) on the other.

Table 3.1

	Section 1.6	**Section 8.3**
Space:	real vector space $\boldsymbol{R}^n$	real vector space $C^\infty(\boldsymbol{R})$
Vectors:	n-tuples of real numbers	infinitely differentiable functions
Addition:	$\boldsymbol{x}+\boldsymbol{y}=(x_1+y_1,\ldots,x_n+y_n)$	$(f+g)(x)=f(x)+g(x)$
Scalar multiplication:	$a\boldsymbol{x}=(ax_1,\ldots,ax_n)$	$(af)(x)=af(x)$
Structure preserving mappings (linear transformations):	$\boldsymbol{T}(\boldsymbol{x}+\boldsymbol{y})=\boldsymbol{T}(\boldsymbol{x})+\boldsymbol{T}(\boldsymbol{y})$ $\boldsymbol{T}(a\boldsymbol{x})=a\boldsymbol{T}(\boldsymbol{x})$	$L(f+g)=L(f)+L(g)$ $L(af)=aL(f)$
Equation to be solved:	$\boldsymbol{T}(\boldsymbol{x})=\boldsymbol{b}$, where $\boldsymbol{T}(\boldsymbol{x})=A\boldsymbol{x}$	$L(y)(x)=q(x)$, where $L(y)=y'+p(x)y$

It is our purpose in this section to solve (6). In view of (5), we can state the general first order linear differential equation (6) as

(7) $$\frac{dy}{dx}+p(x)y(x)=q(x).$$

A simple example of a first order linear differential equation is offered by Newton's law of cooling, Equation (14) on p. 507. That law in fact can be written as

(8) $$\frac{dT}{dt}-kT=-kC.$$

We see that (8) is of the form (7), with t as independent variable in place of x, with $p(t)=-k$ (a constant function), and with $q(t)=-kC$ (another constant function).

The analogy between systems of linear equations in Section 1.6 and Equation (7) is a very close one. You will recall (Theorem 1.6.7) that the general solution to (1) was the set $\{\boldsymbol{x}_0+\boldsymbol{x}_p \mid A\boldsymbol{x}_0=\boldsymbol{0}\}$ where $\boldsymbol{x}_p$ was one particular solution of (1) and $\boldsymbol{x}_0$ ranged over the set of all solutions of the associated *homogeneous linear equation*

$$A\boldsymbol{x}=\boldsymbol{0}.$$

The general solution of (7) is similarly describable, and the proof of this fact corresponds very closely to the proof of Theorem 1.6.7.

3.5 **THEOREM.** The general solution to (7) is the set of all functions $y(x)$ of the form $y_0(x)+y_p(x)$, where $y_p(x)$ is one fixed particular solution of (7) and $y_0(x)$ is any solution of the *associated homogeneous linear differential equation*

(9) $$L(y)(x)=y'(x)+p(x)y(x)=0.$$

Proof. First, every function of the form $y(x)=y_0(x)+y_p(x)$ *is* a solution of (7), since

$$\begin{aligned}L(y)(x)&=L(y_0)(x)+L(y_p)(x)\\&=\frac{d}{dx}(y_0(x)+y_p(x))+p(x)(y_0(x)+y_p(x))\\&=y_0'(x)+y_p'(x)+p(x)y_0(x)+p(x)y_p(x)\\&=y_0'(x)+p(x)y_0(x)+y_p'(x)+p(x)y_p(x)\\&=0+q(x)=q(x).\end{aligned}$$

Moreover, suppose that $y(x)$ is *any* solution to (7). Then consider $y(x)-y_p(x)$. We have

$$\begin{aligned}L(y-y_p)(x)&=\frac{d}{dx}(y(x)-y_p(x))+p(x)(y(x)-y_p(x))\\&=y'(x)-y_p'(x)+p(x)y(x)-p(x)y_p(x)\\&=y'(x)+p(x)y(x)-(y_p'(x)+p(x)y_p(x))\\&=q(x)-q(x)=0.\end{aligned}$$

Thus $y(x)-y_p(x)$ is a solution to the associated homogeneous linear Equation (9). Hence $y(x)-y_p(x)=y_0(x)$ for some solution $y_0(x)$ to (9). But then $y(x)=y_0(x)+y_p(x)$. Hence every solution $y(x)$ to (7) must have the required form. We previously checked that every $y(x)$ of this form is a solution to (7). So the general solution consists precisely of the set of all functions of the prescribed form.

QED

Theorem 3.5 gives us a nice *theoretical* characterization of the set of all solutions of (7), but we still face the *practical* problem of finding a formula that will yield all the solutions of (7). It seems natural to approach this in the two stages suggested by Theorem 3.5, and that in fact is precisely what is done in general to solve linear differential equations (cf. Sections 6 and 7 of this chapter and Sections 2 and 3 of the next chapter). But in the case of first order linear differential equations, it turns out that we can carry out both steps simultaneously. The trick lies in multiplying

(7) $$\frac{dy}{dx}+p(x)y=q(x)$$

through by the nonzero quantity $e^{\int p(x)dx}$. This gives us

(10) $$\frac{dy}{dx}e^{\int p(x)dx}+p(x)ye^{\int p(x)dx}=q(x)e^{\int p(x)dx}.$$

The left-hand side is (as you should verify by differentiation) exactly the derivative of the function $y(x)e^{\int p(x)dx}$. Thus if we take antiderivatives in (10), we obtain

(11) $$\begin{aligned} ye^{\int p(x)dx} &= \int q(x)e^{\int p(x)dx}\,dx \\ &= Q(x)+C, \end{aligned}$$

where $Q(x)$ is some fixed antiderivative of $q(x)e^{\int p(x)dx}$ and C is an arbitrary constant. Multiplication of (11) through by $e^{-\int p(x)dx}$ then gives

(12) $$\begin{aligned} y &= Q(x)e^{-\int p(x)dx}+Ce^{-\int p(x)dx} \\ &= y_p(x)+y_0(x). \end{aligned}$$

CLAIM: The function $y_p(x)=Q(x)e^{-\int p(x)dx}$ is a particular solution of (7) and $y_0(x)=Ce^{-\int p(x)dx}$ is the general solution of (9).

This claim is easy to verify. First we have

$$\begin{aligned} \frac{dy_p(x)}{dx} &= Q'(x)e^{-\int p(x)dx}+Q(x)\frac{d}{dx}(e^{-\int p(x)dx}) \\ &= q(x)e^{\int p(x)dx}e^{-\int p(x)dx}+Q(x)e^{-\int p(x)dx}(-p(x)) \\ &= q(x)+y_p(x)(-p(x)), \end{aligned}$$

$$\frac{dy_p(x)}{dx}+p(x)y_p(x)=q(x).$$

Hence $y_p(x)$ is a solution of (7). Next, observe that (9) is a separable equation that can, if $y \neq 0$, be put in the form

$$\frac{dy}{y} = -p(x)\,dx.$$

Integration gives us

$$\ln|y| = -\int p(x)\,dx + k = -\int p(x)\,dx + \ln c,$$

$$\ln\frac{|y|}{c} = -\int p(x)\,dx,$$

$$\frac{|y|}{c} = e^{-\int p(x)dx},$$

$$y = Ce^{-\int p(x)dx} = y_0(x),$$

where $C = \pm c$. Thus every solution of (9) must have the form $Ce^{-\int p(x)dx}$. (And if we allow C to be zero, we get the solution $y = 0$ in this form also.) It is easy to (and you *should*) verify that every such function actually *is* a solution to (9). Thus, $y_0(x)$ is the general solution of (9), so our claim is proved. In establishing the above claim, we have established most of the following result.

3.6 **THEOREM.** The general solution of the first order linear differential equation

(7) $$\frac{dy}{dx} + p(x)y = q(x)$$

is given by

(12) $$y(x) = e^{-\int p(x)dx}\int q(x)e^{\int p(x)dx}\,dx + Ce^{-\int p(x)dx}.$$

There is a unique solution of the form (12) that satisfies the initial condition $y(x_0) = y_1$.

Proof. The discussion just given has proved everything except the final assertion. To establish that, we write (12) as

(13) $$y(x) = y_p(x) + Ce^{-P(x)},$$

where $\dfrac{dP(x)}{dx} = p(x)$. The condition $y(x_0) = y_1$ allows us to solve (13) for a unique C:

$$y_1 = y_p(x_0) + Ce^{-P(x_0)},$$

$$C = (y_1 - y_p(x_0))e^{P(x_0)}.$$

Thus there is a unique function of the form (12) that satisfies the initial condition $y(x_0) = y_1$. QED

In solving a first order linear differential equation in the form (7), you can either use the formula (12) or, as many students prefer, multiply through by the *integrating factor* $e^{\int p(x)dx}$ and use the fact that the left-hand side of the resulting equation is the derivative of $ye^{\int p(x)dx}$. Then taking antiderivatives will bring you to (12) without having committed that formula to memory. The next example illustrates how either approach results in the same solution.

3.7 EXAMPLE. Find the general solution of $(x^2+1)\dfrac{dy}{dx}-2xy=x^3+x=x(x^2+1)$.

Solution. We can put this into the form (7) by dividing through by the nonzero quantity x^2+1. This brings us to

(14) $$\frac{dy}{dx}-\frac{2x}{x^2+1}y=x.$$

We proceed to solve (14) in both of the ways suggested above.

(a) We have $p(x)=-\dfrac{2x}{x^2+1}$, so

$$\begin{aligned}\int p(x)\,dx &= -\ln|x^2+1|\\ &= \ln\frac{1}{x^2+1}\end{aligned}$$

since x^2+1 is always positive. So we multiply (14) through by the integrating factor

$$e^{\int p(x)dx}=e^{\ln(1/x^2+1)}=\frac{1}{x^2+1}.$$

This brings us to

$$\frac{1}{x^2+1}\frac{dy}{dx}-\frac{2x}{(x^2+1)^2}y=\frac{x}{x^2+1}.$$

Now, recalling (10) and (11), we have

$$\frac{d}{dx}\left(\frac{1}{x^2+1}y\right)=\frac{1}{2}\frac{2x}{x^2+1}.$$

So we can integrate to obtain

$$\begin{aligned}\frac{1}{x^2+1}y &= \tfrac{1}{2}\ln(x^2+1)+C,\\ &= \ln\sqrt{x^2+1}+C,\\ y &= (x^2+1)\ln\sqrt{x^2+1}+C(x^2+1).\end{aligned}$$

Here we have the general solution in a form from which we can read off $y=C(x^2+1)$ as the general solution of the associated homogeneous linear equation.

(b) We can use (12). As above, we calculate

$$\int p(x)\,dx = -\ln(x^2+1), \quad \text{so} \quad e^{-\int p(x)dx} = e^{\ln(x^2+1)} = x^2+1.$$

So (12) gives us

$$\begin{aligned} y &= C(x^2+1)+(x^2+1)\int xe^{-\ln(x^2+1)}\,dx \\ &= C(x^2+1)+(x^2+1)\int \frac{x}{x^2+1}\,dx \\ &= \bar{C}(x^2+1)+(x^2+1)\tfrac{1}{2}\ln(x^2+1) \\ &= \bar{C}(x^2+1)+(x^2+1)\ln\sqrt{x^2+1}. \end{aligned}$$

Note that the constant of integration from $\int \frac{x\,dx}{x^2+1}$ can be consolidated with C, since it is multiplied by the factor x^2+1.

First order linear differential equations are of wide application, as suggested by the following example, which is based on a recent attempt to improve the Thurstone learning model (Exercise 28 of Exercises 8.2).

3.8 **EXAMPLE.** The following learning model was proposed in 1960 by William K. Estes (1919–) of Indiana University, and later of Rockefeller University. As in Thurstone's model, knowledge is measured by a function $k(t)$, where $0 \le k(t) \le 1$, measuring the probability at time t that the person will answer a randomly selected test question correctly. Learning means changing (hopefully, *increasing*) $k(t)$. To try to capture the observed phenomenon that individuals learn at different rates, Estes modeled the change in $k(t)$ as follows. If a *helpful* act (such as studying, coming to class, or doing homework) is performed, then $k(t)$ changes to $k(t)+c(1-k(t))$, where $c \in [0, 1]$ is the individual's *coefficient of learning* and might be measured by something like I.Q. If a *negative* act is performed (such as cutting class or failing to do homework), then $k(t)$ changes to $k(t)-ck(t)$. If λ (where $0 \le \lambda \le 1$) measures the percentage of the available time that the student performs helpful acts, then we wish to find a formula for his knowledge at time t.

Before working out the solution, let's convince ourselves that the constant c is at least a reasonable measure of a person's aptitude for learning. Observe that if $c=0$, then the person *cannot* learn at all. No matter what he does, his knowledge never changes. On the other hand, if c is close to 1, then positive activity quickly increases $k(t)$ to nearly 1, so that the student is a superior learner.

Solution. We have

$$\frac{dk}{dt} = \lambda c(1-k(t)) + (1-\lambda)(-ck(t))$$

since 100 λ% of the available time, the student is increasing $k(t)$ by the factor $c(1-k(t))$, and the remainder of the time $k(t)$ is decreasing by the factor $-ck(t)$. Expanding and simplifying, we get

$$\frac{dk}{dt} = \lambda c - \lambda ck(t) - ck(t) + \lambda ck(t), \qquad \frac{dk}{dt} + ck(t) = \lambda c,$$

which is first order linear with integrating factor $e^{\int c\,dt} = e^{ct}$. Multiplying through by e^{ct}, we obtain

$$e^{ct}\frac{dk}{dt} + e^{ct}ck(t) = \lambda ce^{ct}$$

$$\frac{d}{dt}(e^{ct}k(t)) = \lambda ce^{ct},$$

$$e^{ct}k(t) = \lambda e^{ct} + C$$

$$k(t) = \lambda + Ce^{-ct}.$$

Thus this model predicts that a person's knowledge is a function not only of his or her native intelligence, but also of how much effort he or she expends in activity related to learning. That seems to correlate rather well with most students' experience. (For more on this model, see Exercises 26 and 27 below.)

EXERCISES 8.3

In Exercises 1 to 4, solve the homogeneous linear differential equation.

1 $\dfrac{dy}{dx} + y \sin x = 0.$

2 $x\dfrac{dy}{dx} + y = 0.$

3 $\cos x\dfrac{dy}{dx} + y \sin x = 0.$

4 $\left(\dfrac{1}{x} - x\right)\dfrac{dy}{dx} + y = 0$, where $y = 3$ when $x = 0$.

In Exercises 5 to 15, solve the given first order linear differential equation.

5 $\dfrac{dy}{dx} - y = e^{3x}.$

6 $x\dfrac{dy}{dx} - 2y = x^3 + x.$

7 $x\dfrac{dy}{dx} - 3y = x^2 + 1.$

8 $\dfrac{dy}{dx} - y = x^2.$

9 $x^2\,dy - \sin 3x\,dx = -2xy\,dx.$

10 $\cos x\dfrac{dy}{dx} + y \sin x = 1$

11 $(1-x)\dfrac{dy}{dx} + y = x^2 - 2x + 1$ where $y(0) = 1.$

12 $\dfrac{dy}{dx} - 3y = e^{3x} \sin x$ where $y(0) = 1.$

13 $x\dfrac{dy}{dx} - 2y = x^5$ where $y(1) = -6$

14 $\dfrac{dy}{dx} - \dfrac{2x}{x^2+1}y = 1$ where $y(\pi) = 0.$

15 $(x+1)\dfrac{dy}{dx} - y = x$ where $y(0) = 1.$

16 The **Bernoulli equation** [first solved by Jakob Bernoulli (1654–1705) in 1695] is

(15) $$\frac{dy}{dx} + p(x)y = q(x)y^n,$$

where $n \neq 0$ and $n \neq 1$. (If $n=0$ or 1, then we have a first order linear equation.) Show that if (15) is multiplied through by y^{-n} and the substitution $z = y^{1-n}$ is made, then a first order linear equation

$$\frac{dz}{dx} + (1-n)p(x)z = (1-n)q(x)$$

results. (This method of solution was published by Leibniz in 1696.)

In Exercises 17 to 20, use the method of Exercise 16 to solve the given equation.

17 $x\frac{dy}{dx} + 3y = x^3y^2$.

18 $x\frac{dy}{dx} - (3x+6)y = -9xe^{-x}y^{4/3}$.

19 $\frac{dy}{dx} + y = xy^3$ where $y(0) = 2$.

20 $x^2\frac{dy}{dx} + 2xy = y^3$.

21 The simple electric circuit shown in Figure 3.1 is described by a first order linear differential equation. When the switch at S is closed, current flows. In 1847 the German physicist Gustav R. Kirchhoff (1824–1887) formulated a mathematical model for electric circuits. Now known as *Kirchhoff's Voltage Law*, it states that the sum of the voltage drops around the circuit at each instant is equal to the electromotive force (EMF) $E(t)$ at that instant. In Figure 3.1, let R be the resistance in ohms, L the

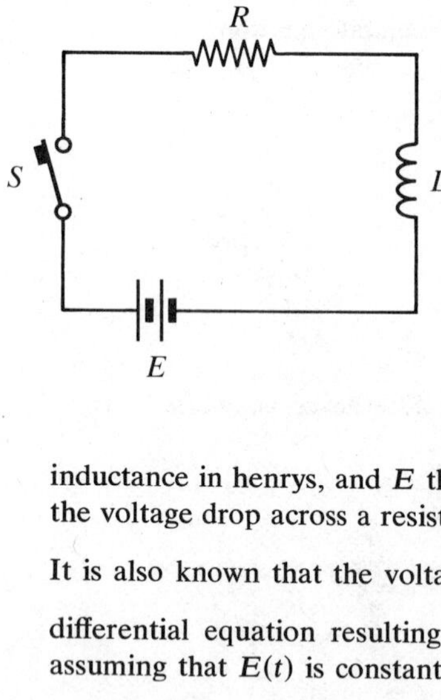

FIGURE 3.1

inductance in henrys, and E the EMF in volts. From experimentation it is known that the voltage drop across a resistor is $Ri(t)$ where $i(t)$ is the current in amperes at time t. It is also known that the voltage drop across an inductor is $L\frac{di}{dt}$. (a) Write down the differential equation resulting from Kirchhoff's Law. (b) Find the general solution, assuming that $E(t)$ is constant.

22 In Exercise 21, suppose that $E = 12$, $R = 16$, and $L = 0.02$. Find a formula for $i(t)$ and, by calculating $\lim_{t\to\infty} i(t)$, find the steady-state current flow in the circuit.

23 Brine made up of 25 grams of salt per liter of water is flowing into a tank at the rate of 3 liters per minute. The tank initially held 1 kilogram of salt dissolved in 10 liters of water. As the brine enters, it is mixed thoroughly with the contents of the tank and the mixture is drawn off at the rate of 3 liters per minute. Find a formula for the amount $s(t)$ of salt in the tank after t minutes. What is the long-term amount of salt in the tank? (*Hint:* $\frac{ds}{dt}$ is the rate at which the salt enters the tank minus the rate at which it leaves the tank.)

24 A tank initially holding 50 gallons of pure water receives a flow of brine containing 2 pounds of salt per gallon at the rate of 3 gallons per minute. The well-stirred mixture is drained off at 3 gallons per minute. Find the long-term concentration of salt in the tank.

25 A room contains 1000 cubic feet of air. Cigarette smokers enter the room and light up, introducing smoke (which contains 4% carbon monoxide) at the rate of $\frac{1}{10}$ cubic foot per minute. This diffuses uniformly throughout the room. An air conditioner removes the air and pollutants from the room at the rate of $\frac{1}{5}$ cubic foot per minute. Find a formula for the amount of carbon monoxide in the room after t minutes. If a concentration of carbon monoxide of 10 parts per million is harmful to human health, then how long a time is required for the contents of the room to become dangerous?

26 Refer to Example 3.8. Some political scientists have proposed using the Estes learning model for public opinion. Here one interprets "helpful act" as information received that is favorable to a certain product (or governmental policy), and "harmful act" as negative information received. Also, one interprets $k(t)$ as the probability that the person will *respond favorably* to a pollster's question about the product (or policy). The constant c can be interpreted as a stubborness factor (or a measure of sales resistance or receptiveness). Under this model, what is the long-term tendency for $k(t)$? Explain how this leads to the practice of saturation advertising for a product, or, in the case of countries with a controlled press, to reporting of "good news" (favorable to the regime) only.

27 Appraise the accuracy of the Estes model for (a) learning, and (b) opinion change. In (b), take into account the issue of *credibility*. Is this factor ignored in the Estes model or is it accounted for? Can you suggest a model which might better weight this factor?

28 The *Solow growth model* was developed by the American economist Robert M. Solow (1924—) of the Massachusetts Institute of Technology during the 1960's. It assumes that the gross national product $y(t)$ is a function $f(x(t), z(t))$ of the total capital $x(t)$ and the size $z(t)$ of the labor force. It is further assumed that for every positive constant c,

$$f(cx(t), cz(t)) = cf(x(t), z(t)).$$

Finally, it is assumed that labor grows exponentially at rate k and capital grows at a rate ℓ proportional to the gross national product. If $w(t) = x(t)/z(t)$, then show that w satisfies the linear differential equation

$$\frac{dw}{dt} + kw = \ell g(w),$$

where $g(w) = f(w, 1)$.

29 A simple version of the *law of supply and demand* in economics is

$$\frac{dp}{dt} = k(d(t) - s(t)), \tag{16}$$

where $p(t)$ is the price of a given product at time t, and $d(t)$ and $s(t)$ are respectively the demand and supply at time t. If the supply is seasonal (for instance, if the product is agricultural), then a simple model for $s(t)$ is given by

$$s(t) = c(1 - \cos at),$$

where c and a are positive constants. A linear model for demand as a function of price is

$$d(t) = b - \ell p(t),$$

where again b and ℓ are positive constants, and $0<p(t)<\frac{b}{\ell}$. Put these supply and demand expressions into (16) and solve the resulting first order linear differential equation for $p(t)$. Discuss the long-term behavior of $p(t)$. (*Hint:* For large t, make the substitution $\theta = \operatorname{Tan}^{-1}\frac{k\ell}{a}$.)

4 HOMOGENEOUS EQUATIONS; EXACT EQUATIONS

In this section we consider two special types of first order differential equations. One of these is *almost* separable, and its solution can be reduced to the solution of a separable equation. The other can be looked upon as the old problem (cf. Section 7.4) of finding a potential function for a two-dimensional conservative vector field, formulated in the language of differential equations.

While the first type of equation we consider is almost separable, it also bears a relationship to linearity and, unfortunately perhaps, carries the name "homogeneous." This term seems to be too well entrenched to change, so you will have to take care to distinguish "homogeneous" from the "homogeneous linear" differential equations encountered in Sections 3 and 6, and in Chapter 9. The term "homogeneous" in this section is attached to *functions* (rather than linear equations) of the following sort.

4.1 **DEFINITION.** A function $f:\boldsymbol{R}^2 \to \boldsymbol{R}$ is **homogeneous of degree *k*** if, for any a in $\boldsymbol{R}$ such that (ax, ay) is in the domain of f,

$$f(ax, ay) = a^k f(x, y)$$

for all $x, y \in \boldsymbol{R}$.

Notice that any *linear* function $f:\boldsymbol{R}^2 \to \boldsymbol{R}$ is homogeneous of degree 1. But there are many nonlinear homogeneous functions. The polynomial function

$$p(x, y) = x^2y^2 + 5x^3y - y^4$$

is homogeneous of degree 4. And you can easily verify that any polynomial function which is a sum of monomials of degree k is homogeneous of degree k. There are still more homogeneous functions. For instance,

$$f(x, y) = \sin\frac{xy}{x^2+y^2}$$

is homogeneous of degree 0. But

$$f(x, y) = \sin\frac{xy}{x^2+y^2} + x$$

is not homogeneous of any degree.

4.2 **DEFINITION.** A first order differential equation

(1) $$P(x, y)+Q(x, y)\frac{dy}{dx}=0$$

is called a **homogeneous (nonlinear) differential equation** if both P and Q are homogeneous of the same degree k.

In 1691, Leibniz discovered the following technique for reducing the solution of (1) to the solution of a separable first order equation.

4.3 **THEOREM.** A homogeneous Equation (1) reduces to a separable equation if the change of variables

(2) $$v(x)=\frac{y(x)}{x}$$

is made.

Proof. We can write (2) as $y(x)=xv(x)$. Differentiating this, we obtain

$$\frac{dy}{dx}=v(x)+x\frac{dv}{dx}.$$

Substituting into (1), we transform it to

$$P(x, xv)+Q(x, xv)\left(v+x\frac{dv}{dx}\right)=0,$$

$$x^kP(1, v)+x^kQ(1, v)v+x^{k+1}Q(1, v)\frac{dv}{dx}=0.$$

We can now multiply through by x^{-k} to get

(3) $$P(1, v)+Q(1, v)v=-xQ(1, v)\frac{dv}{dx}.$$

Now we can separate the variables in (3); under the assumption $P(1, v)+Q(1, v)v\neq 0$, we obtain

(4) $$\frac{1}{x}\,dx=-\frac{Q(1, v)}{P(1, v)+Q(1, v)v}\,dv.$$

We have then reduced (1) to the separable Equation (4). QED

Notice that once (4) is solved for $v=v(x)$, we have y automatically since $y=xv(x)$.

4.4 EXAMPLE. Solve $x^2+y^2-2x^2\dfrac{dy}{dx}=0.$

Solution. Here, notice that $P(x, y)=x^2+y^2$ and $Q(x, y)=-2x^2$ are both homogeneous of degree 2. So we use (2) and let $y=xv(x)$. Then,

$$\frac{dy}{dx}=v(x)+x\frac{dv}{dx}.$$

The given equation becomes

$$x^2+x^2v^2-2x^2\left(v+x\frac{dv}{dx}\right)=0,$$

$$x^2+x^2v^2-2x^2v-2x^3\frac{dv}{dx}=0,$$

$$x^2(1-2v+v^2)=2x^3\frac{dv}{dx}$$

$$\frac{1}{2}\frac{1}{x}\,dx=\frac{dv}{(1-v)^2}$$

$$=-(1-v)^{-2}(-dv).$$

Integrating, we obtain

$$\tfrac{1}{2}\ln|x|+c=-\frac{(1-v)^{-1}}{-1}=\frac{1}{1-v}.$$

This can be solved for v explicitly as follows.

$$1-v=\frac{1}{\ln\sqrt{|x|}+c}$$

$$v=1-\frac{1}{\ln\sqrt{|x|}+c}.$$

Since $y=xv$, we finally have as solution

$$y=x-\frac{x}{\ln\sqrt{|x|}+c}.$$

As you might expect, the separation of variables leading up to (4) may sometimes give an expression that is quite a chore (or even impossible) to antidifferentiate. Nevertheless, the substitution (2) *does* provide a formula for $y=y(x)$ as an integral. So we can work satisfactorily with y using numerical integration techniques if we can't antidifferentiate.

The second type of first order differential equations we treat in this section are *exact equations*.

4.5 **DEFINITION.** A first order differential equation is **exact** if it has the form

(5) $$P(x, y)+Q(x, y)\frac{dy}{dx}=0$$

where $P(x, y)=\dfrac{\partial f}{\partial x}$ and $Q(x, y)=\dfrac{\partial f}{\partial y}$ for a function $f: \boldsymbol{R}^2 \to \boldsymbol{R}$.

Observe that if $y=y(x)$ is a solution to (5), then

$$\frac{\partial f}{\partial x}+\frac{\partial f}{\partial y}\frac{dy}{dx}=0.$$

so that, by Chain Rule 4.5.3, we have

(6) $$\frac{df(x, y)}{dx}=0.$$

We then can solve (5) for y as an implicit function of x immediately, obtaining

(7) $$f(x, y)=c,$$

where c is an arbitrary constant. Exact equations are thus trivial to solve, once we recognize them as being exact. We can state this for reference in the form of the following theorem, which goes back to the French mathematician Alexis C. Clairaut (1713–1765) in 1743.

4.6 **THEOREM.** An exact differential equation

(5) $$P(x, y)+Q(x, y)\frac{dy}{dx}=0$$

has general solution given by

(6) $$f(x, y)=c.$$

How do we recognize a given equation (5) as being exact? If we suppose that $P(x, y)$ and $Q(x, y)$ have continuous partial derivatives, then the next result follows easily from Sections 4.7 and 7.4.

4.7 **THEOREM.** Suppose that P and Q are continuous and have continuous partial derivatives on a simply connected domain $D \subseteq \boldsymbol{R}^2$. Then a given first order differential equation

(5) $$P(x, y)+Q(x, y)\frac{dy}{dx}=0$$

is exact if and only if

(6) $$\frac{\partial P}{\partial y}=\frac{\partial Q}{\partial x}.$$

Proof. If (5) is exact, then for some $f: \boldsymbol{R}^2 \to \boldsymbol{R}$,

$$P(x, y) = \frac{\partial f}{\partial x} \quad \text{and} \quad Q(x, y) = \frac{\partial f}{\partial y}.$$

Then

$$\frac{\partial P}{\partial y} = \frac{\partial^2 f}{\partial y\, \partial x} = \frac{\partial^2 f}{\partial x\, \partial y} = \frac{\partial Q}{\partial x}$$

as desired, by Theorem 4.7.5. On the other hand, if (6) holds on D, then we conclude that $(P(x, y), Q(x, y)) = \nabla(x, y)$ for some $f: \boldsymbol{R}^2 \to \boldsymbol{R}$ by Theorem 7.4.8. Thus, $P(x, y) = \dfrac{\partial f}{\partial x}$ and $Q(x, y) = \dfrac{\partial f}{\partial y}$, so that (5) is exact. QED

Theorem 4.7 essentially reduces the solution of exact differential equations to the determination of the potential function $f(x, y)$ of the conservative vector field $(P(x, y), Q(x, y))$. Thus, no new procedures are needed. We merely employ the approach of Section 7.4.

4.8 **EXAMPLE.** Solve the differential equation

$$3x^2 + 2y \sin 2x + (2 \sin^2 x + 3y^2) \frac{dy}{dx} = 0.$$

Solution. Here

$$P(x, y) = 3x^2 + 2y \sin 2x \quad \text{and} \quad Q(x, y) = 2 \sin^2 x + 3y^2$$

are clearly continuous on all of $\boldsymbol{R}^2$, which is certainly simply connected. (Here we are assuming that $y = y(x)$ is differentiable on $\boldsymbol{R}^2$.) Applying Theorem 4.7, we get

$$\frac{\partial P}{\partial y} = 2 \sin 2x = 4 \sin x \cos x = \frac{\partial Q}{\partial x},$$

so the given equation is exact. We can find its solution $f(x, y) = c$ by setting

$$\frac{\partial f}{\partial x} = 3x^2 + 2y \sin 2x \quad \text{and} \quad \frac{\partial f}{\partial y} = 2 \sin^2 x + 3y^2,$$

and partially integrating. From $\dfrac{\partial f}{\partial x}$ we obtain

(7) $$f(x, y) = x^3 - y \cos 2x + g(y)$$

for some function $g(y)$ involving only y and constants. From $\dfrac{\partial f}{\partial y}$ we get

(8) $$f(x, y) = 2y \sin^2 x + y^3 + h(x)$$

where $h(x)$ involves only x and constants. At first glance, it appears difficult to reconcile (7) and (8), but we can use the identity $\cos 2x = 1 - 2 \sin^2 x$ in (7) to put it in the form

(9) $$f(x, y) = x^3 - y + 2y\sin^2 x + g(y).$$

Comparing (8) and (9), we see that we must have

$$y^3 = -y + g(y) \quad \text{and} \quad h(x) = x^3.$$

Thus the solution we seek is given from (8) by

$$f(x, y) = x^3 + y^3 + 2y\sin^2 x = c.$$

We could also write the solution in the form (7) as

$$f(x, y) = x^3 - y\cos 2x + y + y^3 = c.$$

If you compare the solutions of Example 4.8 and Example 7.4.9, then you will see that they boil down to the same basic idea.

Some first order equations (5) that are not exact can be reduced to exact equations if we multiply through by a function $M(x, y)$. Such a function is called an **integrating factor.** (Compare the discussion leading up to Theorem 3.6.) How can we find an integrating factor? This is often not at all an easy matter, but we will indicate one approach that is sometimes useful. If $M(x, y)$ is an integrating factor for (5), then

(10) $$M(x, y)P(x, y) + M(x, y)Q(x, y)\frac{dy}{dx} = 0$$

must be exact. Assuming that M, P, and Q are continuous and have continuous partial derivatives on some simply connected domain D, we know from Theorem 4.7 that

$$\frac{\partial(MP)}{\partial y} = \frac{\partial(MQ)}{\partial x},$$

$$M\frac{\partial P}{\partial y} + \frac{\partial M}{\partial y}P = M\frac{\partial Q}{\partial x} + \frac{\partial M}{\partial x}Q,$$

(11) $$M\left(\frac{\partial Q}{\partial x} - \frac{\partial P}{\partial y}\right) = P\frac{\partial M}{\partial y} - Q\frac{\partial M}{\partial x}.$$

Now (11) is generally *much* harder to try to solve for M than (5) is to solve for y, so we have to make some sort of simplifying assumption. One such assumption is that our multiplying factor $M(x, y)$ *is a function of* x alone. In that case $\frac{\partial M}{\partial y} = 0$, so (11) becomes

$$\frac{dM}{dx} = \frac{-M\left(\frac{\partial Q}{\partial x} - \frac{\partial P}{\partial y}\right)}{Q},$$

(12) $$\frac{dM}{M} = -\frac{1}{Q}\left(\frac{\partial Q}{\partial x} - \frac{\partial P}{\partial y}\right)dx.$$

If we are lucky enough that the right side of (12) involves only x and we can find an antiderivative $A(x)$ for it, then we see that $\ln M = A(x)$, so that

$$M = e^{A(x)} \tag{13}$$

will be an integrating factor for (5). Similar analysis applies if M is assumed to be a function of y alone (Exercise 21 below) to show that

$$M = e^{B(y)} \tag{14}$$

will be an integrating factor of (5) if $B(y)$ is an antiderivative of

$$\frac{dM}{M} = \frac{1}{P}\left(\frac{\partial Q}{\partial x} - \frac{\partial P}{\partial y}\right) dy, \tag{15}$$

provided (15) involves only y.

EXERCISES 8.4

In Exercises 1 to 18, solve the given differential equation.

1 $x^2 + y^2 - xy\dfrac{dy}{dx} = 0.$

2 $(xy - y^2)\,dx - x^2\,dy = 0.$

3 $y + \sqrt{x^2 + y^2} - x\dfrac{dy}{dx} = 0$ where $y(1) = 0$.

4 $x^2 + y^2 + 2xy\dfrac{dy}{dx} = 0$, where $y(0) = 1$.

5 $\dfrac{dy}{dx} = 2\dfrac{y^3}{x^3} + \dfrac{y}{x}$ where $y(1) = 2$

6 $x^2\dfrac{dy}{dx} = 2y^2 + xy$, where $y(1) = -1$.

7 $x\dfrac{dy}{dx} - y = xe^{y/x}$, where $y(2) = 0$.

8 $xy\dfrac{dy}{dx} = \frac{1}{2}x^2 + \frac{3}{2}y^2$, where $y(2) = 5$.

9 $y + (x - \sin y)\dfrac{dy}{dx} = 0$, where $y(1) = \pi$.

10 $2xy + (x^2 + \cos y)\dfrac{dy}{dx} = 0$, where $y(1) = 0$.

11 $(3x^2 + 4xy)\,dx + (2x^2 + 2y)\,dy = 0.$

12 $(y\cos x + 2xe^y) + (\sin x + x^2e^y + 2)\dfrac{dy}{dx} = 0.$

13 $(3xy + y^2) + (x^2 + xy)\dfrac{dy}{dx} = 0$

[*Hint:* Use Equation (12).]

14 $x^2\dfrac{dy}{dx} + 2xy = y.$

15 $x\dfrac{dy}{dx} + y = y\sin y\dfrac{dy}{dx}.$

16 $e^x\sin y + (e^x\cos y + 2y)\dfrac{dy}{dx} = 0.$

17 $2x^2y + (x^3 + x\cos y)\dfrac{dy}{dx} = 0$, where $y(1) = \pi$. [*Hint:* Use Equation (12).]

18 $(2x^2 + y) + x(xy - 1)\dfrac{dy}{dx} = 0.$

19 Find an integrating factor for $(xy^2 + 4x^2y) + (3x^2y + 4x^3)\dfrac{dy}{dx} = 0$ and solve.

20 Repeat Exercise 19 for $2y^3 - 3xy + x(3x - y^2)\dfrac{dy}{dx} = 0$. (Look for an integrating factor of the form x^my^n.)

21 Show that if $M(y)$ is an integrating factor for (5), then $M = e^{B(y)}$ where $B(y)$ is an antiderivative of

$$\frac{1}{P}\left(\frac{\partial Q}{\partial x} - \frac{\partial P}{\partial y}\right).$$

22 If $a_1b_2 - a_2b_1 \neq 0$, then show that the substitutions

$$x = X + h,\ y = Y + k$$

reduce the first order equation

$$(a_1x + b_1y + c) + (a_2x + b_2y + c_2)\frac{dy}{dx} = 0 \tag{16}$$

to a homogeneous equation in X and Y, where (h, k) is any solution to

$$\begin{cases} a_1h + b_1k + c_1 = 0 \\ a_2h + b_2k + c_2 = 0. \end{cases}$$

23 If $a_1b_2 - a_2b_1 = 0$, then show that the substitution $z = a_1x + b_1y$ reduces the Equation (16) to a separable equation.

24 Solve $2x - y - 4 + (2y - x + 5)\dfrac{dy}{dx} = 0$ by the method of Exercise 22.

25 Solve $x + 2y + 3 + (2x + 4y - 1)\dfrac{dy}{dx} = 0$ by the method of Exercise 23.

26 Show that if $P(x, y) + Q(x, y)\dfrac{dy}{dx} = 0$ is homogeneous, then an integrating factor is

$$M(x, y) = \frac{1}{xP(x, y) + yQ(x, y)}$$

if the denominator is not zero.

27 Suppose that $P(x, y) + Q(x, y)\dfrac{dy}{dx} = 0$ is both homogeneous and exact and, further, $xP(x, y) + yQ(x, y)$ is not a constant. Then show that the general solution is $xP(x, y) + yQ(x, y) = c$.

28 The *Clairaut equation* is

$$y = px + f(p) \tag{17}$$

where $p = \dfrac{dy}{dx}$ and f is differentiable. Show that differentiation of (17) gives the equation

$$\frac{dp}{dx}(x + f'(p)) = 0. \tag{18}$$

29 For the particular Clairaut equation $y = px + p^2$, solve (18) by equating each factor to 0. Show that $\dfrac{dp}{dx} = 0$ leads to the one-parameter family of solutions $y = cx + c^2$, and $x + f'(p) = 0$ leads to the single solution $y = -\frac{1}{4}x^2$. Describe the geometric relationship between the single solution and the one-parameter family.

30 Repeat Exercise 29 for the equation

$$y = px - \tfrac{1}{3}p^3.$$

5 COMPLEX NUMBERS AND EXPONENTIALS

In contrast to the first order equations considered so far, second and higher order linear differential equations are most conveniently solved if we have available the complex exponential function. This section provides the basic ideas needed, beginning with the system of complex numbers.[1]

In your previous experience with mathematics, you have no doubt observed that successively larger systems of numbers were introduced in order to solve various sorts of polynomial equations. For instance, the natural number system $\boldsymbol{N}$ was inadequate to solve equations $ax = b$, where $a, b \in \boldsymbol{N}$. Thus the system $\boldsymbol{F}$ of positive fractions was introduced. This system proved inadequate for solving equations $ax + b = 0$, where $a, b \in \boldsymbol{F}$, so the system $\boldsymbol{Q}$ of rational numbers was introduced. But $\boldsymbol{Q}$ wasn't rich enough to extract square roots, i.e., to solve an equation like $x^2 - 2 = 0$. Thus the system $\boldsymbol{R}$ of real numbers was introduced. But some quadratic equations like

(1) $$x^2 + 1 = 0,$$

still remain unsolvable in $\boldsymbol{R}$. As you recall from high school, this leads to the introduction of the system $\boldsymbol{C}$ of complex numbers. The following scheme provides us not only with the solutions of (1) but also the solutions of *any* polynomial equation of any degree.

5.1 **DEFINITION.** The set $\boldsymbol{C}$ of **complex numbers** consists of all ordered pairs (a, b) of real numbers added, subtracted, and multiplied as follows.

$$(a, b) \pm (c, d) = (a \pm c, b \pm d),$$
$$(a, b)(c, d) = (ac - bd, ad + bc).$$

The complex number $(0, 1)$ is denoted i. A complex number $(a, 0)$ is identified with the real number a. A complex number $(0, b)$ is called a *pure imaginary* number. The **real part** of (a, b) is a. The **imaginary part** of (a, b) is b.

There are a number of things to note here. First,

$$i^2 = ii = (0, 1)(0, 1) = (0 - 1, 0 + 0) = (-1, 0).$$

Since we identify $(-1, 0)$ with $-1 \in \boldsymbol{R}$, we have

(2) $$i^2 = -1,$$

so that we can indeed solve (1), and we see that its roots are $x = i$ and $x = -i$.

1. The first part of this section, up to p. 538 where the discussion of complex exponentials begins, is included for purposes of reference and review.

Second, for any real number a,

$$(a, 0)(c, d) = (ac - 0d, ad + 0c) = (ac, ad)$$
$$= a(c, d),$$

so that $\boldsymbol{C}$ is a *real vector space*. Addition is ordinary vector addition as in $\boldsymbol{R}^2$, and since we identify $a \in \boldsymbol{R}$ with $(a, 0)$, we see that multiplication by real scalars is carried out just as in $\boldsymbol{R}^2$ also. This permits us to represent complex numbers *graphically* as points (or vectors) in $\boldsymbol{R}^2$. This scheme goes back to the Swiss

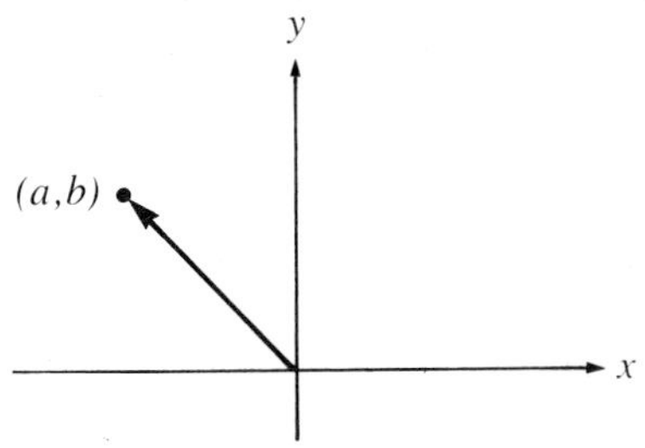

FIGURE 5.1

mathematician Jean R. Argand (1768–1822) in 1806, and is illustrated in Figure 5.1. Using the standard basis vectors $(1, 0)$ and $(0, 1)$ as usual, we can write

$$\begin{aligned}(a, b) &= (a, 0) + (0, b)\\ &= a(1, 0) + b(0, 1)\\ &= a1 + bi\\ &= a + bi, \qquad (3)\end{aligned}$$

since $(0, 1) = i$ and we identify $(1, 0)$ with $1 \in \boldsymbol{R}$. The latter is a natural identification since $(1, 0)(a, b) = (a, b)$ for all $(a, b) \in \boldsymbol{C}$. The *traditional* notation for complex numbers is the notation (3) and we shall use it for the most part, rather than (a, b). Since $(a, b) = (c, d)$ if and only if $a = c$ and $b = d$, we have

$$a + bi = c + di$$

if and only if $a = c$ and $b = d$. Finally, note that our identification scheme $a \leftrightarrow (a, 0)$ is consistent with the arithmetic of $\boldsymbol{R}$ and $\boldsymbol{C}$, since

$$a + b \leftrightarrow (a, 0) + (b, 0) = (a + b, 0)$$

and

$$ab \leftrightarrow (a, 0)(b, 0) = (ab - 0, 0 + 0) = (ab, 0).$$

For all practical purposes, then, we can view $\boldsymbol{R}$ and its arithmetic as a *subsystem* of $\boldsymbol{C}$ and its arithmetic.

To divide, we need the **complex conjugate** $\overline{a + bi} = a - bi$. Writing z for an arbitrary complex number $a + bi$, we have the following properties for the conjugation operation.

5.2 **COROLLARY.** (a) **Linearity:** $\overline{z_1+z_2}=\bar{z}_1+\bar{z}_2$, $\overline{az}=a\bar{z}$ for $a\in \boldsymbol{R}$. (b) $\overline{z_1z_2}=\bar{z}_1\bar{z}_2$. (c) $z\bar{z}$ is a nonnegative real number.

Partial Proof. Parts (a) and (b) are Exercise 1 below. For (c), we have

$$\begin{aligned} z\bar{z} &= (a+bi)(a-bi) = a^2+b^2+(-ab+ab)i \\ &= a^2+b^2, \end{aligned}$$

so $z\bar{z}$ is real and nonnegative. QED

Corollary 5.2 permits us to invert any nonzero complex number $z=a+bi$. Since $z\bar{z}=a^2+b^2$, we see that

$$z^{-1}=\frac{1}{a^2+b^2}\,\bar{z}=\frac{a-bi}{a^2+b^2}$$

has the property that $zz^{-1}=1$. If $z_2\neq 0$, we can thus define

$$z_1\div z_2=z_1z_2^{-1}.$$

The *quadratic formula* provides a complete solution for the general quadratic equation

$$ax^2+bx+c=0, \qquad a, b, c\in \boldsymbol{R}, \qquad a\neq 0.$$

The roots are given by

$$x_1=\frac{-b+\sqrt{b^2-4ac}}{2a}, \qquad x_2=\frac{-b-\sqrt{b^2-4ac}}{2a}.$$

If $b^2-4ac=0$, then $x_1=x_2$, so the equation has equal real roots. If $b^2-4ac>0$, then x_1 and x_2 are unequal and real. If $b^2-4ac<0$, then x_1 and x_2 are conjugate complex numbers.

Considerably more is true. If we want to solve polynomial equations, then we do not ever have to enlarge $\boldsymbol{C}$ as we enlarged $\boldsymbol{N}$, $\boldsymbol{F}$, $\boldsymbol{Q}$, and $\boldsymbol{R}$. This fact, known as the "Fundamental Theorem of Algebra," was proved by Gauss in his Ph.D. dissertation in 1799. There are many proofs known, but they involve more advanced ideas than we can treat here, so we leave the proof for later courses.

5.3 **THEOREM.** Every polynomial equation

(4) $$a_nx^n+a_{n-1}x^{n-1}+\ldots+a_2x^2+a_1x+a_0=0$$

with real or complex coefficients a_i has n roots in $\boldsymbol{C}$ (some of which may be repeated roots).

5.4 **COROLLARY.** If $z\in \boldsymbol{C}$ is a root of (4), and all $a_i\in \boldsymbol{R}$, then $\bar{z}$ is also a root of (4).

Proof. If $a_nz^n+\ldots+a_1z+a_0=0$, then taking conjugates of each side and using

Corollary 5.2 (a) and (b), we have

$$\overline{a_n z^n + \ldots + a_1 z + a_0} = \bar{0} = 0,$$
$$\overline{a_n z^n} + \ldots + \overline{a_1 z} + \overline{a_0} = 0,$$
$$a_n \overline{z^n} + \ldots + a_1 \bar{z} + a_0 = 0,$$
$$a_n \bar{z}^n + \ldots + a_1 \bar{z} + a_0 = 0.$$

Thus $\bar{z}$ is a root of (4). QED

Part (c) of Corollary 5.2 leads to the following important notion.

5.5 **DEFINITION.** The **absolute value** or **modulus** of the complex number $z = a + bi$ is

$$|z| = \sqrt{z\bar{z}} = \sqrt{a^2 + b^2}.$$

We can now give the *polar form* of $z = a + bi$.

5.6 **DEFINITION.** If $z = x + yi$, then the **polar form** of z is $z = r(\cos\theta + i\sin\theta)$, where $[r, \theta]$ are polar coordinates for $(x, y) \in \boldsymbol{R}^2$, with $r = |z|$. The angle θ is called the **argument** of z.

Thus, given $z = x + yi$, we take $r = |z|$; and for θ we can take the angle from the x-axis to the vector $(x, y) \in \boldsymbol{R}^2$ measured counterclockwise. See Figure 5.2.

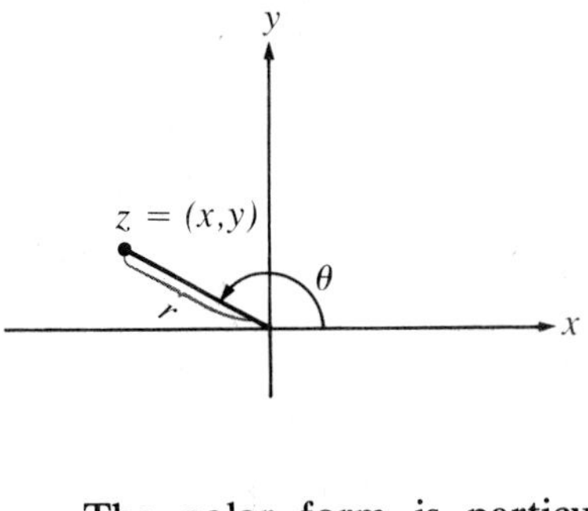

FIGURE 5.2

The polar form is particularly well suited to multiplication of complex numbers, as shown by the following result of the English mathematician Abraham De Moivre (1667–1754).

5.7 **DE MOIVRE'S THEOREM.** (a) If

$$z_1 = r_1(\cos\theta_1 + i\sin\theta_1) \quad \text{and} \quad z_2 = r_2(\cos\theta_2 + i\sin\theta_2),$$

then

$$z_1 z_2 = r_1 r_2[\cos(\theta_1 + \theta_2) + i\sin(\theta_1 + \theta_2)].$$

(b) If $z = r(\cos\theta + i\sin\theta)$, then

$$z^n = r^n(\cos n\theta + i\sin n\theta).$$

You are asked to prove this result in Exercises 7 and 8 below.

5.8 **EXAMPLE.** (a) Describe the geometric effect of multiplying $1-\sqrt{3}\,i$ by i. (b) Compute $(1-i)^5$.

Solution. (a) We can express $1-\sqrt{3}\,i=(1,-\sqrt{3})$ in polar form as $2(\cos 5\pi/3+i\sin 5\pi/3)$, and $i=(0,1)$ as $\cos\frac{1}{2}\pi+i\sin\frac{1}{2}\pi$. If we multiply $1-\sqrt{3}\,i$ by i, then according to Theorem 5.7(a), we obtain a product whose argument is $\dfrac{5\pi}{3}+\dfrac{\pi}{2}=\dfrac{13\pi}{6}\left(\text{or }\dfrac{\pi}{6}\right)$ and whose absolute value is $2\cdot 1=2$. We can think of $1-\sqrt{3}i$ as being rotated counterclockwise through the angle $\frac{1}{2}\pi$. See Figure 5.3.

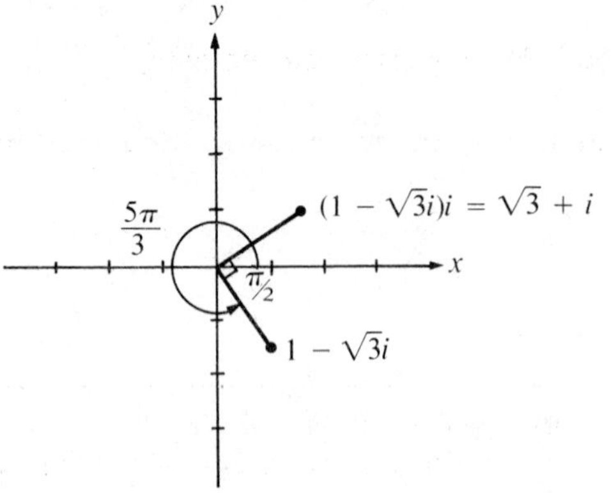

FIGURE 5.3

(b) In polar form we have

$$1-i=(1,-1)=\sqrt{2}\left(\cos\frac{7\pi}{4}+i\sin\frac{7\pi}{4}\right).$$

Thus by Theorem 5.7(b),

$$\begin{aligned}(1-i)^5&=(\sqrt{2})^5\left(\cos\frac{35\pi}{4}+i\sin\frac{35\pi}{4}\right)\\&=4\sqrt{2}\left(\cos\frac{3\pi}{4}+i\sin\frac{3\pi}{4}\right)\\&=4\sqrt{2}\left(-\frac{1}{\sqrt{2}}+i\frac{1}{\sqrt{2}}\right)\\&=-4+4i.\end{aligned}$$

We can use De Moivre's Theorem to find the n-th roots of a given complex number. These are determined by the n-th roots of 1.

5.9 **THEOREM.** (a) For each positive integer n, the equation $z^n=1$ has exactly n distinct roots given by $\omega, \omega^2, \omega^3, \ldots, \omega^{n-1}, \omega^n=1$, where

(5) $$\omega=\cos\frac{2\pi}{n}+i\sin\frac{2\pi}{n}$$

is called a **primitive n-th root of 1**.

(b) Every nonzero complex number $z=r(\cos\theta+i\sin\theta)$ has exactly n distinct n-th roots in $\mathbf{C}$ given by $\sigma, \sigma\omega, \sigma\omega^2, \ldots, \sigma\omega^{n-1}$, where

(6) $$\sigma=r^{1/n}\left(\cos\frac{\theta}{n}+i\sin\frac{\theta}{n}\right)$$

is called a **primitive n-th root of z.**

Proof. (a) It is clear from Theorem 5.7(b) that $\omega^n = \cos 2\pi + i \sin 2\pi = 1$. Also, for each integer k $(1 \le k \le n)$, we have $\omega^k = \cos \frac{2\pi k}{n} + i \sin \frac{2\pi k}{n}$. Thus, $(\omega^k)^n = \cos 2\pi k + i \sin 2\pi k = 1$. Thus, all of the listed powers of ω are n-th roots of 1. They are all distinct, since they have different arguments

$$\frac{2\pi}{n}, \frac{4\pi}{n}, \frac{6\pi}{n}, \ldots, \frac{(n-1)2\pi}{n}, 2\pi.$$

By Theorem 5.3, the polynomial equation $x^n - 1 = 0$ has n roots in $\boldsymbol{C}$, so we have *all* of the n-th roots of 1 in the set $\{\omega, \omega^2, \omega^3, \ldots, \omega^{n-1}, \omega^n\}$.

(b) First observe that each $\sigma\omega^k$ is an n-th root of z, since

$$\begin{aligned}(\sigma\omega^k) &= \sigma^n(\omega^k)^n \\ &= \sigma^n 1 \quad \text{(by part (a))} \\ &= (r^{1/n})^n\left(\cos\frac{n\theta}{n} + i \sin\frac{n\theta}{n}\right) \\ &= r(\cos\theta + i\sin\theta) = z.\end{aligned}$$

Moreover, we have *all* of the n-th roots of z, since if ϕ is any n-th root of z, then $\phi^n = z$. But then

$$\left(\frac{\phi}{\sigma}\right)^n = \frac{\phi^n}{\sigma^n} = \frac{z}{z} = 1.$$

Hence ϕ/σ is an n-th root of 1. By part (a), then, $\phi/\sigma = \omega^k$ for some k, $1 \le k \le n$. Hence $\phi = \sigma\omega^k$ as desired. QED

5.10 **EXAMPLE.** (a) Find the fourth roots of 1. (b) Find the fourth roots of $-8 + 8\sqrt{3}i$.

Solution. (a) We have from (5)

$$\begin{aligned}\omega &= \cos 2\pi/4 + i \sin 2\pi/4 = \cos \tfrac{1}{2}\pi + i \sin \tfrac{1}{2}\pi = i, \\ \omega^2 &= \cos 4\pi/4 + i \sin 4\pi/4 = i^2 \qquad = -1, \\ \omega^3 &= \omega\omega^2 \qquad = (i)(-1) \qquad = -i, \\ \omega^4 &= \omega\omega^3 \qquad = i(-i) = -i^2 \qquad = 1.\end{aligned}$$

So the fourth roots of 1 are i, $-i$, 1, and -1.

(b) Let

$$\begin{aligned}z = -8 + 8\sqrt{3}i &= 16\left(-\frac{1}{2} + \frac{\sqrt{3}}{2} i\right) \\ &= 16\left(\cos\frac{2\pi}{3} + i \sin\frac{2\pi}{3}\right).\end{aligned}$$

Then from (6)

$$\begin{aligned}\sigma &= 16^{1/4}\left(\cos\frac{2\pi}{12}+i\sin\frac{2\pi}{12}\right)\\ &= 2\left(\cos\frac{\pi}{6}+i\sin\frac{\pi}{6}\right)=2\left(\frac{\sqrt{3}}{2}+\frac{1}{2}i\right)\\ &= \sqrt{3}+i,\\ \sigma\omega &= (\sqrt{3}+i)i=-1+\sqrt{3}i,\\ \sigma\omega^2 &= (\sigma\omega)\omega=(-1+\sqrt{3}i)i=-\sqrt{3}-i,\\ \sigma\omega^3 &= (\sigma\omega^2)\omega=(-\sqrt{3}-i)i=1-\sqrt{3}i.\end{aligned}$$

So the fourth roots of $-8+8\sqrt{3}i$ are $\sqrt{3}+i$, $-\sqrt{3}-i$, $-1+\sqrt{3}i$, and $1-\sqrt{3}i$.

In Figure 5.4 we show the roots found in parts (a) and (b). Note that in each

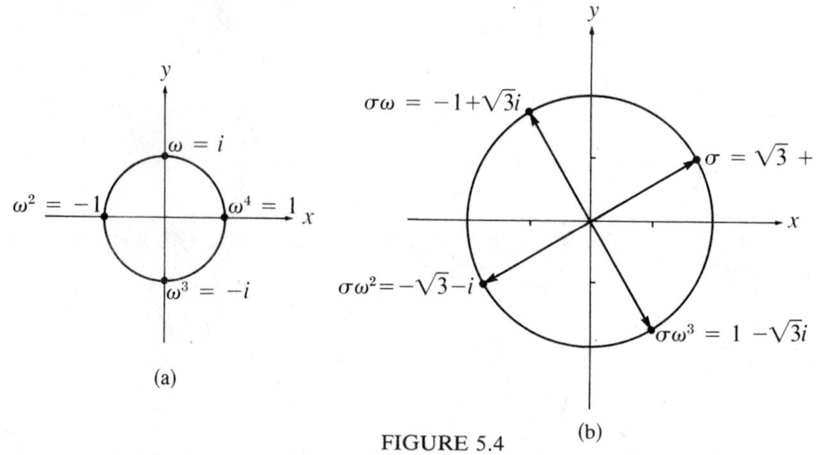

FIGURE 5.4

case the roots are equally spaced around the circle centered at the origin with radius $r^{1/n}$; as the powers of ω increase, we hop around this circle in the counterclockwise sense.

We now come to the main business of this section, the definition of e^z for a complex variable $z=x+iy$. What meaning can we attach to this? Well, if the usual laws of exponents are to hold, then we must have

$$e^{x+iy}=e^xe^{iy}.$$

The factor e^x presents no difficulty. We already know what that is. So we are faced with defining e^{iy}. Recall (cf. Section 11.8, Example 11.8.6) that for all *real* numbers y we have

(7) $$e^y=1+y+\frac{y^2}{2!}+\frac{y^3}{3!}+\ldots+\frac{y^n}{n!}+\ldots$$

Let's see what would happen if we use (7) with iy in place of y. We would obtain

$$e^{iy}=1+iy+\frac{(iy)^2}{2!}+\frac{(iy)^3}{3!}+\frac{(iy)^4}{4!}+\frac{(iy)^5}{5!}$$
$$+\frac{(iy)^6}{6!}+\ldots+\frac{(iy)^{2n}}{(2n)!}+\frac{(iy)^{2n+1}}{(2n+1)!}+\ldots$$

$$= 1 + iy - \frac{1}{2!}y^2 - i\frac{1}{3!}y^3 + \frac{1}{4!}y^4 + i\frac{1}{5!}y^5 - \frac{1}{6!}y^6$$

$$+ \ldots + \frac{(-1)^n}{(2n)!}y^{2n} + i\frac{(-1)^n}{(2n+1)!}y^{2n+1} + \ldots$$

$$= \left(1 - \frac{1}{2!}y^2 + \frac{1}{4!}y^4 - \frac{1}{6!}y^6 + \ldots + \frac{(-1)^n}{(2n)!}y^{2n} + \ldots\right)$$

$$+ i\left(y - \frac{1}{3!}y^3 + \frac{1}{5!}y^5 - \ldots + \frac{(-1)^n}{(2n+1)!}y^{2n+1} + \ldots\right)$$

$$= \cos y + i \sin y.$$

We can thus avoid all reference to infinite series by giving the following definition.

5.11 **DEFINITION.** $e^{iy} = \cos y + i \sin y,$

$$e^{x+iy} = e^x(\cos y + i \sin y).$$

Note that e^{iy} is a unit vector and

$$e^{2\pi i/n} = \cos\frac{2\pi}{n} + i \sin\frac{2\pi}{n} = \omega,$$

a primitive n-th root of 1. So *all* n-th roots of 1 are of the form $e^{2\pi i r/n}$ for $1 \le r \le n$.

5.12 **COROLLARY.** If $z_1 z_2 \in \mathbf{C}$, then $e^{z_1}e^{z_2} = e^{z_1+z_2}$.

Proof. Let $z_1 = x_1 + iy_1$, and $z_2 = x_2 + iy_2$. Then

$$e^{z_1}e^{z_2} = e^{x_1}(\cos y_1 + i \sin y_1)e^{x_2}(\cos y_2 + i \sin y_2)$$

$$= e^{x_1}e^{x_2}(\cos y_1 + i \sin y_1)(\cos y_2 + i \sin y_2)$$

$$= e^{x_1+x_2}(\cos(y_1 + y_2) + i \sin(y_1 + y_2))$$

by Theorem 5.7(a)

$$= e^{z_1+z_2}.$$

QED

Notice that the real and imaginary parts of e^{iy} are the cosine and sine functions, respectively. Writing Re z for the real part of z and Im z for the imaginary part, we have then

(8) $$\text{Re}(e^{iy}) = \cos y$$

(9) $$\text{Im}(e^{iy}) = \sin y.$$

We now extend the calculus of real values functions of a real variable x to complex valued functions of x in the natural way.

5.13 **DEFINITION.** If $f(x)=g(x)+ih(x)$ is a complex valued function of the real variable x, where $g(x)=\text{Re}\, f(x)$ and $h(x)=\text{Im}\, f(x)$, then we differentiate and integrate f according to the rules

(10) $$\frac{d}{dx}(f(x))=\frac{dg}{dx}+i\frac{dh}{dx},$$

(11) $$\int f(x)\,dx=\int g(x)\,dx+i\int h(x)\,dx.$$

5.14 **COROLLARY.** If $\alpha \in \boldsymbol{C}$, then

(12) $$\frac{d}{dx}(e^{\alpha x})=\alpha e^{\alpha x}$$

and if $\alpha \neq 0$,

(13) $$\int e^{\alpha x}\,dx=\frac{1}{\alpha}e^{\alpha x}+c.$$

for all $c \in \boldsymbol{C}$.

Proof. Clearly (13) follows once (12) is established. For (12) we use Definitions 5.11 and 5.13 as follows. Suppose that $\alpha=a+bi$, where $a, b \in \boldsymbol{R}$. Then

$$\begin{aligned}
\frac{d}{dx}e^{\alpha x} &= \frac{d}{dx}e^{(a+bi)x}\\
&= \frac{d}{dx}e^{ax+bix}\\
&= \frac{d}{dx}(e^{ax}(\cos bx+i\sin bx))\\
&= \frac{d}{dx}(e^{ax}\cos bx)+i\frac{d}{dx}(e^{ax}\sin bx)\\
&= ae^{ax}\cos bx-be^{ax}\sin bx+i(ae^{ax}\sin bx+be^{ax}\cos bx)\\
&= ae^{ax}(\cos bx+i\sin bx)+ibe^{ax}(\cos bx+i\sin bx), \text{ using (2)}\\
&= (a+bi)e^{ax}(\cos bx+i\sin bx)\\
&= \alpha e^{ax+ibx}=\alpha e^{\alpha x}.
\end{aligned}$$

QED

EXERCISES 8.5

1 (a) Prove Corollary 5.2(a).
(b) Prove Corollary 5.2(b).

2 (a) Show that $z+\bar{z}=2\,\text{Re}(z)$.
(b) Show that $z-\bar{z}=2i\,\text{Im}(z)$.

3 If $z_1=1-i$ and $z_2=3+2i$, then compute

$$z_1+z_2, \quad z_1z_2, \quad \text{and} \quad z_1 \div z_2.$$

4 (a) Prove that multiplication of complex numbers is associative and commutative.
(b) Show that the distributive law holds for $\boldsymbol{C}$.

5 Find all roots of the polynomial equation $x^4+3x^2+2=0$.

6 Find all roots of the polynomial equation $x^4+5x^2+6=0$.

7 Prove Theorem 5.7(a).

8 Prove Theorem 5.7(b).

9 Describe the geometric effect of multiplying $2-3i$ by $1+i$.

10 Compute $(1+i)^8$.

11 Compute $(\sqrt{3}-i)^4$.

12 Find all five fifth roots of 1.

13 Find the cube roots of $z=\frac{27}{2}(1+i\sqrt{3})$.

14 Find the cube roots of $z=32(\sqrt{3}-i)$.

15 Find the sixth roots of -1.

16 Find the sixth roots of 64.

17 If $\omega \neq 1$ is an n-th root of 1, then show that ω is a root of the *cyclotomic polynomial* $1+x+x^2+\ldots+x^{n-1}=0$.

18 A **primitive n-th root of 1** is an n-th root μ of 1 such that every n-th root of 1 is some positive integral power of μ. Find all the primitive eighth roots of 1.

19 Establish *Euler's equation*, $e^{i\pi}=-1$.
What is e^i? What is $e^{2\pi i}$?

20 Compute $e^{\pi i/3}$ and $e^{\ln 2+\pi i/4}$.

21 Show from Definition 5.11 that

$$\cos\theta=\frac{e^{i\theta}+e^{-i\theta}}{2}=\mathrm{Re}(e^{i\theta}).$$

22 Show from Definition 5.11 that

$$\sin\theta=\frac{e^{i\theta}-e^{-i\theta}}{2i}=\mathrm{Im}(e^{i\theta}).$$

23 Find the first three derivatives of $e^x(\cos 2x+i\sin 2x)$.

24 Find the first three derivatives of $e^{2x}(\cos x+i\sin x)$.

25 Find the eighth derivative of $e^{x/\sqrt{2}}\left(\cos\dfrac{x}{\sqrt{2}}+i\sin\dfrac{x}{\sqrt{2}}\right)$.

26 Find the twelfth derivative of $e^{\sqrt{3}x/2}\left(\cos\dfrac{x}{2}+i\sin\dfrac{x}{2}\right)$.

27 By computing $\int e^{\alpha x}\,dx$ for $\alpha=a+ib$, obtain formulas for $\int e^{ax}\cos bx\,dx$ and $\int e^{ax}\sin bx\,dx$. **(This is used in Section 7, p. 553.)**

28 Using the identities in Exercises 21 and 22, show that $\int_0^{2\pi}\sin mx\cos nx\,dx=0$ for any positive integers m and n.

29 Assuming that m and n are unequal positive integers, show that $\int_0^{2\pi}\sin mx\sin nx\,dx=0$. What happens if $m=n$?

30 Compute $\int_0^{2\pi}\cos mx\cos nx\,dx$ for m and n positive integers.

6 SECOND ORDER HOMOGENEOUS LINEAR DIFFERENTIAL EQUATIONS WITH CONSTANT COEFFICIENTS

Now that we have learned how to solve a fairly broad range of first order differential equations, we turn to second order equations that take the form

$$F\left(x, y, \frac{dy}{dx}, \frac{d^2y}{dx^2}\right)=0.$$

Apart from some special types considered in Exercises 23 to 29 below, we will confine ourselves to linear equations. The development very closely parallels that in Section 3. Our main results, Theorems 6.8 and 6.9, go back to the Swiss mathematician Johann Bernoulli (1667–1748), the brother of Jakob, and were obtained prior to 1700.

6.1 **DEFINITION.** The **general second order linear differential equation** is

(1) $$L(y)(x)=s(x),$$

where

(2) $$L(y)=\frac{d^2y}{dx^2}+p(x)\frac{dy}{dx}+q(x)y$$

for functions $p(x)$, $q(x)$, $s(x)\in C^\infty(\boldsymbol{R})$.

The terminology results from the fact that L is again a linear operator on $C^\infty(\boldsymbol{R})$.

6.2 **LEMMA.** The transformation $L:C^\infty(\boldsymbol{R})\to C^\infty(\boldsymbol{R})$ given by (2) is a linear operator.

The proof proceeds just as the proof of Example 3.3 and is left as Exercise 1 below. We thus once more have an analogy between (1) and the system

$$A\boldsymbol{x}=\boldsymbol{b}$$

of linear equations corresponding to the linear transformation $\boldsymbol{T}$ defined by $\boldsymbol{T}(\boldsymbol{x})=A\boldsymbol{x}$. As you expect, then, we have the following analogue of Theorem 3.5.

6.3 **THEOREM.** Every solution to (1) has the form

(3) $$y(x)=y_0(x)+y_p(x),$$

where $y_p(x)$ is one particular solution to (1) and $y_0(x)$ is any solution to the associated homogeneous linear differential equation

(4) $$L(y)(x)=0.$$

The proof is just like the proof of Theorem 3.5, so is left as Exercise 2 below. Thus, we can solve (1) by performing two steps. *First*, find the general solution of the associated homogeneous linear differential Equation (4). *Then*, find one particular solution $y_p(x)$ to (1). Our goal in this section is to carry out the first step, for the important special class of linear differential Equations (1) for which $p(x)=p$ and $q(x)=q$ are *constants*. Thus we can write (2) as

$$L(y)=\frac{d^2y}{dx^2}+p\frac{dy}{dx}+qy,$$

where $p, q \in \boldsymbol{R}$. To solve (4) in this case it is helpful to write the linear operator L in terms of the fundamental differentiation operator.

6.4 **DEFINITION.** The **linear differential operators** D and D^2 are given by

$$D = \frac{d}{dx}$$

$$D^2 = \frac{d^2}{dx^2}.$$

With this notation we can write (2) in the form

$$\begin{aligned} L(y) &= D^2y + pDy + qy \\ &= (D^2 + pD + q)(y). \end{aligned} \tag{5}$$

The expression $D^2 + pD + q$ is called a second degree **polynomial differential operator** in D. These sorts of operators behave just like ordinary polynomials when it comes to their arithmetic. In particular, we have the following very important composition rule.

6.5 **LEMMA.** $(D - r_1)\circ(D - r_2) = D^2 - (r_1 + r_2)D + r_1r_2$.

Proof. We merely have to verify that each side has the same effect on a function $y(x) \in C^\infty(\boldsymbol{R})$. We have

$$\begin{aligned} (D - r_1)\circ(D - r_2)(y) &= (D - r_1)((D - r_2)(y)) \\ &= (D - r_1)\left(\frac{dy}{dx} - r_2y\right) \\ &= D\left(\frac{dy}{dx} - r_2y\right) - r_1\left(\frac{dy}{dx} - r_2y\right) \\ &= \frac{d^2y}{dx^2} - r_2\frac{dy}{dx} - r_1\frac{dy}{dx} + r_1r_2y \\ &= \frac{d^2y}{dx^2} - (r_1 + r_2)\frac{dy}{dx} + r_1r_2y \\ &= D^2y - (r_1 + r_2)Dy + r_1r_2y \\ &= (D^2 - (r_1 + r_2)D + r_1r_2)(y). \end{aligned}$$

QED

In view of this lemma, we can factor any second order polynomial differential operator $D^2 + pD + q$ as

$$D^2 + pD + q = (D - r_1)\circ(D - r_2),$$

where r_1 and r_2 are the roots of the corresponding polynomial $r^2 + pr + q$.

6.6 **DEFINITION.** The **auxiliary polynomial** to the polynomial operator $D^2 + pD + q$ is the polynomial $r^2 + pr + q$. The **auxiliary equation** to the homogeneous linear differential equation

(4) $$(D^2 + pD + q)(y) = 0$$

is

(6) $$r^2 + pr + q = 0.$$

We are going to solve (4) by using the auxiliary Equation (6). Since its roots may be complex, we will need the following result to give the general real solution to (4).

6.7 **LEMMA.** The complex function

$$y(x) = f(x) + ig(x)$$

is a solution of (4), if and only if the real and imaginary parts, $\mathrm{Re}(y(x)) = f(x)$ and $\mathrm{Im}(y(x)) = g(x)$, are real solutions of (4).

Proof. If $y(x) = f(x) + ig(x)$ is a solution of (4), then we have

$$Dy = \frac{dy}{dx} = f'(x) + ig'(x),$$

$$D^2 y = \frac{d^2 y}{dx^2} = f''(x) + ig''(x).$$

Thus

$$f''(x) + ig''(x) + pf'(x) + ipg'(x) + qf(x) + iqg(x) = 0,$$
$$f''(x) + pf'(x) + qf(x) + i(g''(x) + pg'(x) + qg(x)) = 0 + 0i,$$

so

$$f''(x) + pf'(x) + qf(x) = 0 \quad \text{and} \quad g''(x) + pg'(x) + qg(x) = 0.$$

Thus $f(x)$ and $g(x)$ are both solutions of (4). On the other hand, if $f(x)$ and $g(x)$ are real solutions of (4), then $y(x) = f(x) + ig(x)$ is a complex solution of (4), since we have

$$\begin{aligned} L(y)(x) &= (D^2 + pD + q)(f(x) + ig(x)) \\ &= f''(x) + ig''(x) + pf'(x) + ipg'(x) + qf(x) + iqg(x) \\ &= f''(x) + pf'(x) + qf(x) + i(g''(x) + pg'(x) + qg(x)) \\ &= 0 + i0 = 0. \end{aligned}$$

QED

With this background, we can now solve (4). Our method, essentially due to the French mathematician Jean d'Alembert (1717–1783) and the Italian mathematician Jacopo Riccati (1676–1754), will be to reduce the solution of (4) to the solution of *two first order linear* differential equations, which we can then easily solve using Theorem 3.6. To start with, we write (4) as $(D^2 + pD + q)y = 0$.

If r_1 and r_2 are the roots of the auxiliary equation $r^2+pr+q=0$, then we have from Lemma 6.5

$$(D-r_1)\circ(D-r_2)(y)=0. \tag{7}$$

Now the trick! Let

$$z=(D-r_2)(y). \tag{8}$$

If we substitute (8) into (7), then it becomes

$$(D-r_1)(z)=0,$$

$$\frac{dz}{dx}-r_1 z=0.$$

This is a *first order homogeneous* linear differential equation whose solution is given by

$$z(x)=ke^{r_1 x}.$$

Now we can substitute back into (8), getting

$$ke^{r_1(x)}=(D-r_2)(y).$$

This can be put in the form

$$\frac{dy}{dx}-r_2 y=ke^{r_1(x)}, \tag{9}$$

which we recognize as the general first order linear differential equation. It thus has integrating factor $e^{-r_2 x}$. If we multiply (9) through by this factor, we obtain, as usual,

$$\frac{d}{dx}(e^{-r_2 x}y)=ke^{(r_1-r_2)x}. \tag{10}$$

At this point we have to distinguish two cases.

(a) Suppose that $r_1 \neq r_2$. Then we integrate (10) to obtain

$$e^{-r_2 x}y=\frac{k}{r_1-r_2}e^{(r_1-r_2)x}+c_2,$$

$$y=\frac{k}{r_1-r_2}e^{r_1 x}+c_2 e^{r_2 x},$$

$$y=c_1 e^{r_1 x}+c_2 e^{r_2 x}, \tag{11}$$

where $c_1=\dfrac{k}{r_1-r_2}$

(b) Suppose that $r_1=r_2$. Then (10) is really

$$\frac{d}{dx}(e^{-r_2 x}y)=k.$$

So integration gives $e^{-r_2x}y = kx + c_1$, or, equivalently,

(12) $$y = c_1e^{r_2x} + c_2xe^{r_2x},$$

where we can write c_2 in place of k for notational symmetry.

It is easy to verify that any function of the form (11) is a solution to (7) in case $r_1 \neq r_2$ and that any solution of the form (12) is solution to (7) in case $r_1 = r_2$ (Exercise 3 below). Thus we have actually found the *general real solution* to (7) provided r_1 and r_2 are real numbers. Let's state this formally before taking up the case in which r_1 and r_2 are complex numbers.

6.8 **THEOREM.** If the auxiliary equation for

(4) $$(D^2 + pD + q)(y) = 0$$

has real roots r_1 and r_2, then the general solution of (4) is given by

(11) $$y = c_1e^{r_1x} + c_2e^{r_2x} \quad \text{if} \quad r_1 \neq r_2,$$

and is given by

(12) $$y = c_1e^{rx} + c_2xe^{rx} \quad \text{if} \quad r_1 = r_2 = r.$$

What if r_1 and r_2 are complex roots? Then

$$r_1 = \frac{-p + \sqrt{p^2 - 4q}}{2} = a + bi,$$

$$r_2 = \frac{-p - \sqrt{p^2 - 4q}}{2} = a - bi$$

are conjugate complex numbers. In particular, $r_1 \neq r_2$. Then by (13) of Corollary 5.14, the integration we performed deriving (11) is valid. Thus

(11′) $$y = k_1e^{(a+bi)x} + k_2e^{(a-bi)x}$$

is the only possible form for a *complex solution* to (7). It is easy to check (Exercise 4 below) that any complex function of the form (11′) really *is* a solution, so (11′) gives the *general complex solution* to (7). We can rewrite (11′) using Definition 5.11 to get

$$\begin{aligned} y &= k_1e^{ax}(\cos bx + i \sin bx) + k_2e^{ax}(\cos(-bx) + i \sin(-bx)) \\ &= k_1e^{ax}(\cos bx + i \sin bx) + k_2e^{ax}(\cos bx - i \sin bx) \\ &= (k_1 + k_2)e^{ax} \cos bx + i(k_1 - k_2)e^{ax} \sin bx, \end{aligned}$$

or, equivalently,

(13) $$y = c_1e^{ax} \cos bx + ic_2e^{ax} \sin bx$$

where $c_1 = k_1 + k_2$, and $c_2 = k_1 - k_2$. From Lemma 6.7, the real and imaginary parts $c_1e^{ax} \cos bx$ and $c_2e^{ax} \sin bx$ are real solutions to (7). These in fact *generate* the general real solution.

6.9 **THEOREM.** If the auxiliary equation for

(4) $$(D^2+pD+q)(y)=0$$

has complex roots $r_1=a+bi$ and $r_2=a-bi$, then the general real solution of (4) is given by

(14) $$y=c_1e^{ax}\cos bx+c_2e^{ax}\sin bx.$$

Proof. Our discussion leading up to the theorem shows that any real function of the form (14) is a real solution. All that remains is to show that *every* real solution to (4) must have the form (14). For that, if $y(x)=f(x)$ is any real solution, then $y_a(x)=f(x)+if(x)$ is a complex solution. Then $y_a(x)$ must have the form (13). Hence we have

$$y_a(x)=K_1e^{ax}\cos bx+K_2e^{ax}\sin bx,$$

where $K_1=d_1+id_2$ and $K_2=e_1+ie_2$ are complex constants, Thus,

$$\begin{aligned} y_a(x) &= d_1e^{ax}\cos bx+id_2e^{ax}\cos bx+ie_1e^{ax}\sin bx-e_2e^{ax}\sin bx \\ &= d_1e^{ax}\cos bx-e_2e^{ax}\sin bx+i(d_2e^{ax}\cos bx+e_1e^{ax}\sin bx) \\ &= f(x)+if(x). \end{aligned}$$

Hence $f(x)$ must *simultaneously* have the forms

$$d_1e^{ax}\cos bx-e_2e^{ax}\sin bx$$

and

$$d_2e^{ax}\cos bx+e_1e^{ax}\sin bx.$$

Then subtraction of these expressions gives

$$(d_1-d_2)e^{ax}\cos bx-(e_1+e_2)e^{ax}\sin bx=0.$$

Since $e^{ax}\neq 0$,

(15) $$(d_1-d_2)\cos bx-(e_1+e_2)\sin bx=0$$

holds for all x. Differentiation of (15) gives

$$-b(d_1-d_2)\sin bx-b(e_1+e_2)\cos bx=0.$$

Now $b\neq 0$ since r_1 and r_2 are not real. Therefore,

(16) $$(d_1-d_2)\sin bx-(e_1+e_2)\cos bx=0.$$

Differentiation of (16) and division by b give

(17) $$(d_1-d_2)\cos bx+(e_1+e_2)\sin bx=0.$$

Addition of (15) and (17) gives

$$2(d_1-d_2)\cos bx=0,$$

for all x. Thus $d_1=d_2$. Then (15) gives $e_1=-e_2$. Hence $f(x)$ has the required form (14), where $c_1=d_1=d_2$, and $c_2=e_1=-e_2$. QED

Using Formulas (11), (12), and (14), it is a purely mechanical exercise to find the general solution of the second order homogeneous linear differential Equation (4) with constant coefficients. *It is thus worthwhile to commit Theorems 6.8 and 6.9 to memory.* The ease of application of these theorems is brought out by the next three examples.

6.10 EXAMPLE. Find the general solution of

$$3\frac{d^2y}{dx^2}-15\frac{dy}{dx}-42y=0.$$

Solution. First, we put this in the standard form (4) by dividing through by 3 and writing D^2y for $\frac{d^2y}{dx}$ and Dy for $\frac{dy}{dx}$. We have then $(D^2-5D-14)y=0$. The auxiliary equation is $r^2-5r-14=0$, which factors as $(r-7)(r+2)=0$, so $r_1=7$ and $r_2=-2$. Then from (11), the general solution is

$$y=c_1e^{7x}+c_2e^{-2x}.$$

6.11 EXAMPLE. Find the general solution of

$$5\frac{d^2y}{dx^2}-40\frac{dy}{dx}+80y=0.$$

Solution. Proceeding as in Example 6.10, we divide through by 5 to get $(D^2-8D+16)y=0$. The auxiliary equation is $r^2-8r+16=(r-4)^2$, which has the single repeated root 4. So by (12), the general solution is

$$y=c_1e^{4x}+c_2xe^{4x}.$$

6.12 EXAMPLE. Find the general real solution of $y''-6y'+10y=0$.

Solution. In standard form (4) we have

$$(D^2-6D+10)y=0.$$

The auxiliary equation is $r^2-6r+10=0$, whose roots are

$$r_1=\frac{6+\sqrt{36-40}}{2}=3+i$$

and

$$r_2=\frac{6-\sqrt{36-40}}{2}=3-i.$$

Thus in Theorem 6.9, $a=3$ and $b=1$. The general real solution is then by (14)

$$y = c_1 e^{3x} \cos x + c_2 e^{3x} \sin x.$$

The solutions $e^{r_1 x}$ and $e^{r_2 x}$ in (11) are called *basic* solutions. The same term is applied to e^{rx} and xe^{rx} in (12), and $e^{ax} \cos bx$ and $e^{ax} \sin bx$ in (14). In Chapter 9 we will see that the set of all solutions of (4) forms a real vector space and that the basic solutions form a basis (in the sense of Section 1.7) for this space.

For a second order linear differential Equation (4), an **initial condition** is a condition of the form

$$y(x_0) = y_0, \qquad y'(x_0) = y_1.$$

In addition to this sort of condition, we can consider a **boundary condition** in which y or y' is specified at two points, say $y(x_0) = y_0$ and $y(x_1) = y_1$. Initial value problems always have a unique solution, but boundary value problems may or may not have unique solutions.

6.13 EXAMPLE. Solve the initial value problem

$$3\frac{d^2y}{dx^2} - 15\frac{dy}{dx} - 42y = 0, \qquad y(0) = 1,\ y'(0) = 6.$$

Solution. The general solution is, from Example 6.10,

$$y = c_1 e^{7x} + c_2 e^{-2x}.$$

Then,

$$y' = 7c_1 e^{7x} - 2c_2 e^{-2x}.$$

Our initial conditions give

(15) $$1 = c_1 + c_2$$

(16) $$6 = 7c_1 - 2c_2.$$

Adding twice Equation (15) to Equation (16), we obtain

$$8 = 9c_1, \qquad c_1 = \tfrac{8}{9}.$$

Then from (15), $c_2 = \frac{1}{9}$. So the solution to the given initial value problem is

$$y = \tfrac{8}{9}e^{7x} + \tfrac{1}{9}e^{-2x}.$$

6.14 EXAMPLE. Solve, if possible, the boundary value problems

(a) $y'' - 6y' + 10y = 0, \qquad y(0) = 1,\ y(\pi/2) = 1.$

(b) $y'' - 6y' + 10y = 0, \qquad y(0) = 1,\ y(\pi) = 0.$

Solution. (a) From Example 6.12, the general solution is

$$y = c_1 e^{3x} \cos x + c_2 e^{3x} \sin x.$$

We have from the boundary conditions

$$1 = c_1$$
$$1 = c_2 e^{3\pi/2}.$$

Thus the desired solution is

$$y = e^{3x} \cos x + e^{3(x-\pi/2)} \sin x.$$

(b) Here the boundary conditions give

$$1 = c_1$$
$$0 = -c_1 e^{3\pi}.$$

Thus we have the impossible situation $c_1 = 1$ and $c_1 = 0$. Hence this boundary value problem has no solution.

EXERCISES 8.6

1 Prove Lemma 6.2.

2 Prove Theorem 6.3.

3 (a) Verify that any function of the form (11) is a solution of (4) in case $r_1 \neq r_2$.
(b) Verify that any function of the form (12) is a solution to (4) in case $r_1 = r_2 = r$.

4 Verify that any function of the form (11′) is a solution to (4) in case $r_1 = a + bi$ and $r_2 = a - bi$.

In Exercises 5 to 12, find the general solution of the given homogeneous linear differential equation.

5 $3\dfrac{d^2y}{dx^2} + 15\dfrac{dy}{dx} + 18y = 0.$

6 $2\dfrac{d^2y}{dx^2} + 12\dfrac{dy}{dx} + 10y = 0.$

7 $3y'' + 18y' + 27y = 0.$

8 $-2y'' + 8y' - 8y = 0.$

9 $\dfrac{d^2x}{dt^2} + 2\dfrac{dx}{dt} + 5x = 0.$

10 $\dfrac{d^2x}{dt^2} + 4\dfrac{dx}{dt} + 13x = 0.$

11 $y'' + 9y = 0.$

12 $y'' + 4y = 0.$

In Exercises 13 to 18, solve the given initial value problems. (*Use the results of earlier exercises as needed.*)

13 $y'' + 9y = 0, \quad y(0) = 1, \quad y'(0) = -3.$

14 $y'' + 4y = 0, \quad y(0) = 1, \quad y'(0) = 0.$

15 $y'' + 9y = 0, \quad y(\pi/6) = 1, \quad y'(\pi/6) = 2.$

16 $y'' + 4y = 0, \quad y(\pi/4) = -1, \quad y'(\pi/4) = 0.$

17 $3y'' + 18y' + 27y = 0, \quad y(0) = 1, \quad y'(0) = 1.$

18 $-2y'' + 8y' - 8y = 0, \quad y(0) = 1, \quad y'(0) = 1.$

In Exercises 19 to 22, solve the given boundary value problems if possible.

19 $y'' + 9y = 0, \quad y(0) = 1, \quad y(\pi) = 0.$

20 $y'' + 4y = 0, \quad y(0) = 1, \quad y(\pi) = 0.$

21 $y'' + 9y = 0, \quad y(0) = 1, \quad y(\pi/2) = 1.$

22 $y'' + 4y = 0, \quad y(0) = 1, \quad y(\pi/2) = 1.$

Exercises 23 to 29 are devoted to some special types of nonlinear second order differential equations for which special techniques exist.

23 (a) Show that the substitution $v = y'$ reduces the second order differential equation $F(x, y', y'') = 0$ to a first order differential equation.

(b) Show that the substitution $v = \dfrac{dy}{dx}$ reduces the second order differential equation $F(y, y', y'') = 0$ to a first order equation.

$$\left(\textit{Hint:}\quad \frac{d^2y}{dx^2} = \frac{dv}{dx} = v\frac{dv}{dy}.\right)$$

24 Solve $x\dfrac{d^2y}{dx^2} = \dfrac{dy}{dx} + 2\sqrt{x^2 + \left(\dfrac{dy}{dx}\right)^2}$.

25 Solve $y\dfrac{d^2y}{dx^2} + \left(\dfrac{dy}{dx}\right)^2 = \dfrac{dy}{dx}$.

26 Solve $y\dfrac{d^2y}{dx^2} - \left(\dfrac{dy}{dx}\right)^2 = 0$.

27 Solve $y\dfrac{d^2y}{dx^2} = \left(\dfrac{dy}{dx}\right)^2 + 2\dfrac{dy}{dx}$.

28 Solve $y'' - 3y^2 = 0, \quad y(0) = 2, \quad y'(0) = 4.$

29 Solve $y'y'' - x = 0, \quad y(1) = 2, \quad y'(1) = 1.$

7 NONHOMOGENEOUS SECOND ORDER LINEAR EQUATIONS WITH CONSTANT COEFFICIENTS

Returning now to the general second order linear differential equation with constant coefficients

$$L(y)(x) = (D^2 + pD + q)y(x) = s(x), \tag{1}$$

we need to find only one particular solution $y = y_p(x)$ to give the general solution, in view of Theorem 6.3. For Theorems 6.8 and 6.9 provide us with the general solution to the associated homogeneous linear differential equation

$$L(y)(x) = 0. \tag{2}$$

The most natural thing to do is to try to use the reduction of order trick (p. 545) again, since we can easily solve first order nonhomogeneous differential equations by the methods of Section 3. Let's see where this leads us.

Suppose that r_1 and r_2 are the roots of the auxiliary equation $r^2 + pr + q = 0$. Then we can write (1) as

$$(D - r_1)\circ(D - r_2)(y) = s(x),$$

and as before let $z = (D - r_2)(y)$. We then have

$$(D - r_1)(z) = s(x),$$

$$\frac{dz}{dx} - r_1 z = s(x).$$

This is first order linear with integrating factor e^{-r_1x}. Multiplying through by this integrating factor, we obtain

$$\frac{d}{dx}(e^{-r_1x}z) = e^{-r_1x}s(x),$$

$$e^{-r_1x}z = \int e^{-r_1x}s(x)\,dx + k_1,$$

$$z = e^{r_1x}\int e^{-r_1x}s(x)\,dx + k_1e^{r_1x}$$

Now substituting

$$(D-r_2)(y) = \frac{dy}{dx} - r_2y$$

for z, we have

$$\frac{dy}{dx} - r_2y = e^{r_1x}\int e^{-r_1x}s(x)\,dx + k_1e^{r_1x}.$$

This again is first order linear with integrating factor e^{-r_2x}. Multiplying through by this integrating factor, we have

(3) $$\frac{d}{dx}(e^{-r_2x}y) = e^{(r_1-r_2)x}\int e^{-r_1x}s(x)\,dx + k_1e^{(r_1-r_2)x}.$$

If $r_1 \neq r_2$, then we get

$$e^{-r_2x}y = \int e^{(r_1-r_2)x}\left(\int e^{-r_1x}s(x)\,dx\right)dx + \frac{k_1}{r_1-r_2}e^{(r_1-r_2)x} + c_2$$

(4) $$y = e^{r_2x}\int e^{(r_1-r_2)x}\left(\int e^{-r_1x}s(x)\,dx\right)dx + c_1e^{r_1x} + c_2e^{r_2x}.$$

This is the general solution of (1), since we recognize the last two terms as making up the general solution of (2). On the other hand, if $r_1 = r_2 = r$, then (3) has the form

$$\frac{d}{dx}(e^{-rx}y) = \int e^{-rx}s(x)\,dx + c_1$$

Hence,

$$e^{-rx}y = \int\left(\int e^{-rx}s(x)\,dx\right)dx + c_1x + c_2,$$

(5) $$y = e^{rx}\int\left(\int e^{-rx}s(x)\,dx\right)dx + c_1xe^{rx} + c_2e^{rx}.$$

Again we recognize that the last two terms constitute the general solution of (2). The integrals in (4) and (5) thus give us particular solutions of (1), and in finding them we can ignore constants of integration like k, c_1, and c_2 above, because these constants are associated with the *homogeneous* problem (2) whose solutions we can write down more easily by using Theorems 6.8 and 6.9.

7.1 EXAMPLE. Find the general solution of

$$3\frac{d^2y}{dx^2}-15\frac{dy}{dx}-42y=3e^{7x}\sin x.$$

Solution. The associated homogeneous problem is the one we solved in Example 6.10 with general solution $y_0=c_1e^{7x}+c_2e^{-2x}$. To apply the solution technique above, we have to divide through by 3, getting

$$(D^2-5D-14)(y)=e^{7x}\sin x,$$
$$(D-7)\circ(D+2)(y)=e^{7x}\sin x.$$

Then we let $z=(D+2)y$, so we have

$$(D-7)z=e^{7x}\sin x,$$
$$\frac{dz}{dx}-7z=e^{7x}\sin x.$$

Multiplying through by the integrating factor e^{-7x}, we obtain

$$\frac{d}{dx}(ze^{-7x})=\sin x.$$

Integrating and suppressing the constant of integration in accord with our remarks above, we get

$$ze^{-7x}=-\cos x,$$
$$z=-e^{7x}\cos x,$$
$$\frac{dy}{dx}+2y=-e^{7x}\cos x.$$

This first order linear differential equation has integrating factor e^{2x}. Multiplying through by that, we get

$$\frac{d}{dx}(e^{2x}y)=-e^{9x}\cos x.$$

Using integration by parts, Formula 30 of the table of integrals, or Exercise 27 of Exercises 8.5, we have

$$e^{2x}y=-\tfrac{1}{82}e^{9x}\sin x-\tfrac{9}{82}e^{9x}\cos x,$$
$$y=-\tfrac{1}{82}e^{7x}\sin x-\tfrac{9}{82}e^{7x}\cos x.$$

With this particular solution, we can now write down the general solution as

$$y=-\frac{e^{7x}}{82}(\sin x+9\cos x)+c_1e^{7x}+c_2e^{-2x}.$$

As you can see from Example 7.1, and can well imagine from Formulas (4) and (5), the integration necessary to find a particular solution using the reduction of order trick can be difficult. An additional drawback to using (4) and (5) is the rather complicated nature of these formulas, which makes them hard to remember. Because of these impediments, an alternative method to using (4) and (5) has been developed. This is known as the **method of undetermined coefficients,** and is basically a catalogue of the results arising from using (4) and (5) with various commonly encountered functions $s(x)$ in (1). This method gives the general form of a particular solution $y_p(x)$ to be expected for (1) in the case of several general types of $s(x)$. The exact form of $y_p(x)$ can then be found by finding $y_p'(x)$ and $y_p''(x)$, and substituting into (1). We first formulate the method, then illustrate its use in the problem of Example 7.1, and finally show how to use it in a wider variety of problems.

7.2 **METHOD OF UNDETERMINED COEFFICIENTS.** If $s(x)$ in (1) has one of the forms given in Table 7.1, then a particular solution $y_p(x)$ exists having the form indicated **unless** *such a* $y_p(x)$ *happens to be a solution of the associated homogeneous problem* (2). If the indicated $y_p(x)$ *is* a solution to (2), then there is a particular solution of the form $x^m y_p(x)$, where m is the least positive integer such that $x^m y_p(x)$ is *not* a solution to (2). (Since (1) is second order, $m \le 2$.)

Table 7.1

$s(x)$	$y_p(x)$
$a_0 + a_1x + a_2x^2 + \ldots + a_nx^n$	$A_0 + A_1x + A_2x^2 + \ldots + A_nx^n$
ae^{kx}	Ae^{kx}
$e^{kx}(a_0 + a_1x + \ldots + a_nx^n)$	$e^{kx}(A_0 + A_1x + \ldots + A_nx^n)$
$a \cos cx + b \sin cx$	$A \cos cx + B \sin cx$
$e^{kx}(a \cos cx + b \sin cx)$	$e^{kx}(A \cos cx + B \sin cx)$
$e^{kx}(a_0 + a_1x + \ldots + a_kx^k)\cos cx$ $+ e^{kx}(b_0 + b_1x + \ldots + b_kx^k)\sin cx$	$e^{kx}(A_0 + A_1x + \ldots + A_kx^k)\cos cx$ $+ e^{kx}(B_0 + B_1x \ldots + B_kx^k)\sin cx$

We won't prove that the $y_p(x)$ mentioned, or $x^m y_p(x)$ as the case may be, actually *must* be a particular solution. As we suggested, this could be accomplished by a case-by-case study of formulas (4) and (5). It will be easier to see where the various $y_p(x)$ come from, however, after the methods of Chapter 9 are available. (See Section 9.3.) We will now just illustrate how to use the method. First, let's return to Example 7.1.

7.3 **EXAMPLE.** Solve Example 7.1,

$$3\frac{d^2y}{dx^2} - 15\frac{dy}{dx} - 42y = 3e^{7x} \sin x,$$

using the method of undetermined coefficients.

Solution. First we divide through by 3. Then we have $s(x) = e^{7x} \sin x$. This is of the form $e^{kx}(a \cos cx + b \sin cx)$, where $k = 7$, $a = 0$, $b = 1$, and $c = 1$. So our $y_p(x)$

has, from Table 7.1, the form

$$y_p(x) = e^{7x}(A \cos x + B \sin x).$$

(**Note:** BOTH the sine and cosine terms *must* appear in $y_p(x)$ even if $s(x)$ involves only the sine or only the cosine.) Here the coefficients A and B must be determined from the fact that $y_p(x)$ is a solution to the given equation. (Hence the name "method of undetermined coefficients.") We have

$$\begin{aligned} y_p'(x) &= 7e^{7x}(A \cos x + B \sin x) + e^{7x}(-A \sin x + B \cos x) \\ &= e^{7x}(7A + B)\cos x + e^{7x}(-A + 7B)\sin x. \\ y_p''(x) &= 7e^{7x}(7A + B)\cos x - e^{7x}(7A + B)\sin x \\ &\quad + 7e^{7x}(-A + 7B)\sin x + e^{7x}(-A + 7B)\cos x \\ &= e^{7x}(48A + 14B)\cos x + e^{7x}(-14A + 48B)\sin x. \end{aligned}$$

Thus,

$$\begin{aligned} y'' - 5y' - 14y &= e^{7x}(48A + 14B)\cos x + e^{7x}(-14A + 48B)\sin x \\ &\quad + e^{7x}(-35A - 5B)\cos x + e^{7x}(5A - 35B)\sin x \\ &\quad + e^{7x}(-14A)\cos x + e^{7x}(-14B)\sin x \\ &= e^{7x}(-A + 9B)\cos x + e^{7x}(-9A - B)\sin x \\ &= e^{7x} \sin x. \end{aligned}$$

Equating coefficients of $e^{7x} \sin x$ and $e^{7x} \cos x$ in this equation, we get

(6) $\quad \begin{cases} -A + 9B = 0 \end{cases}$

(7) $\quad -9A - B = 1$

Multiplying (6) by -9 and adding to (7), we find

$$-82B = 1, \qquad B = -\tfrac{1}{82}.$$

From (6), then, $A = 9B = -\frac{9}{82}$. Hence,

$$\begin{aligned} y_p(x) &= e^{7x}(-\tfrac{9}{82} \cos x - \tfrac{1}{82} \sin x) \\ &= -\frac{e^{7x}}{82}(9 \cos x + \sin x). \end{aligned}$$

So, as before, we find that the general solution is

$$y = -\frac{e^{7x}}{82}(9 \cos x + \sin x) + c_1 e^{7x} + c_2 e^{-2x}.$$

7.4 EXAMPLE. Solve $\dfrac{d^2y}{dx^2} - 5\dfrac{dy}{dx} - 14y = 5e^{7x}$.

Solution. Here $s(x) = 5e^{7x}$ is a solution to the associated homogeneous problem (see Example 6.10, 7.1, or 7.3), since it is of the form $c_1 e^{7x} + c_2 e^{-2x}$ (with $c_1 = 5$, $c_2 = 0$). So Method 7.2 tells us to take $y_p(x) = Axe^{7x}$ *instead of the* Ae^{7x}

appearing in Table 7.1. We have

$$y_p'(x) = Ae^{7x} + 7xAe^{7x}$$
$$y_p''(x) = 7Ae^{7x} + 7Ae^{7x} + 49xAe^{7x}$$
$$= 14Ae^{7x} + 49xAe^{7x}.$$

Thus substitution in the given equation gives us

$$y'' - 5y' - 14y = 14Ae^{7x} + 49xAe^{7x} - 5Ae^{7x} - 35xAe^{7x} - 14Axe^{7x}$$
$$= 9Ae^{7x} = 5e^{7x}.$$

Hence $A = \frac{5}{9}$. Thus $y_p(x) = \frac{5}{9}xe^{7x}$. Therefore, the general solution is

$$y(x) = \tfrac{5}{9}xe^{7x} + c_1e^{7x} + c_2e^{-2x}.$$

Apart from being conceptually simpler (no substitutions, integrating factors, or integration are required), the method of undetermined coefficients, when it applies, is usually quicker than using the approach leading up to Formulas (4) and (5). It also has the esthetic appeal of reducing the solution of a *linear* differential equation to the solution of a *system of linear equations* like (6) and (7).

The following simple result makes the method of undetermined coefficients of wider applicability than might be initially supposed.

7.5 **THEOREM.** The second order linear differential equation

(8) $$L(y) = \frac{d^2y}{dx^2} + p\frac{dy}{dx} + qy = s(x) + t(x)$$

has a particular solution $y_{p_1}(x) + y_{p_2}(x)$ where $y_{p_1}(x)$ is a particular solution to $L(y) = s(x)$ and $y_{p_2}(x)$ is a particular solution to $L(y) = t(x)$.

Proof. This is a simple consequence of the fact that L is a linear operator (Lemma 6.2). For we have

$$L(y_{p_1}(x) + y_{p_2}(x)) = L(y_{p_1}(x)) + L(y_{p_2}(x))$$
$$= s(x) + t(x).$$

QED

We weren't kidding: that *was* simple! It is also a helpful tool in reducing complexity. To find a particular solution of a given linear Equation (8), we can just find separate particular solutions of the equations

$$L(y) = s(x) \quad \text{and} \quad L(y) = t(x),$$

and then put them together by adding to obtain the desired particular solution of (8). This scheme goes by the name "method of superposition" in the literature of differential equations.

7.6 **EXAMPLE.** Find the general solution of

$$\frac{d^2y}{dx^2} - 5\frac{dy}{dx} - 14y = e^{7x}\sin x + 14x^2 - 2.$$

Solution. We have the general solution of the associated homogeneous problem, and a particular solution to

$$\frac{d^2y}{dx^2}-5\frac{dy}{dx}-14y=e^{7x}\sin x$$

from Examples 7.1 and 7.3. So all we need is a particular solution of

(9) $$\frac{d^2y}{dx^2}-5\frac{dy}{dx}-14y=14x^2-2.$$

According to Table 7.1, since $14x^2-2$ is a polynomial of degree 2, (9) will have a particular solution of the form

$$y_p(x)=A_0+A_1x+A_2x^2.$$

Then
$$y_p'(x)=\quad A_1\ +2A_2x,$$

$$y_p''(x)=\quad 2A_2,$$

and so, substituting in the given equation, we have

$$\begin{aligned}\frac{d^2y}{dx^2}-5\frac{dy}{dx}-14y&=2A_2-5A_1-10A_2x-14A_0-14A_1x-14A_2x^2\\&=(-14A_0-5A_1+2A_2)-(10A_2+14A_1)x-14A_2x^2.\\&=14x^2-2.\end{aligned}$$

Equating coefficients of x, x^2, and constants, we get

$$\begin{aligned}&-14A_2=14, \qquad A_2=-1;\\&-10A_2-14A_1=0,\\&A_1=-\tfrac{10}{14}A_2=\tfrac{5}{7},\end{aligned}$$

and

$$\begin{aligned}&-14A_0-5A_1+2A_2=-2,\\&-14A_0=-2+\tfrac{25}{7}+2=\tfrac{25}{7},\\&A_0=-\tfrac{25}{98}.\end{aligned}$$

Hence a particular solution to (9) is

$$y_p(x)=-\tfrac{25}{98}+\tfrac{5}{7}x-x^2.$$

The general solution of the given problem is then given by

$$y=-\tfrac{25}{98}+\tfrac{5}{7}x-x^2-\frac{e^{7x}}{82}(9\cos x+\sin x)+c_1e^{7x}+c_2e^{-2x}.$$

While the method of undetermined coefficients is of fairly wide applicability, *it is not universally applicable.* If $s(x)$ does not have one of the forms in Table 7.1 and is not a sum of such forms, you may have to resort to the technique of Example 7.1.

7.7 EXAMPLE. Find the general solution of

$$\frac{d^2y}{dx^2}-8\frac{dy}{dx}+16y=xe^{-x^2+4x}$$

Solution. Table 7.1 has no suggested $y_p(x)$ when $s(x)$ involves e to a power of x other than the first power. So we write

$$[(D-4)\circ(D-4)](y)=xe^{-x^2+4x}$$

and let $z=(D-4)(y)$. We then have

$$(D-4)(z)=xe^{-x^2+4x}$$

$$\frac{dz}{dx}-4z=xe^{-x^2+4x}$$

This first order linear equation has the integrating factor e^{-4x}. Multiplying through by that, we get

$$\frac{d}{dx}(ze^{-4x})=xe^{-x^2},$$

so

$$ze^{-4x}=-\tfrac{1}{2}e^{-x^2},$$

$$z=-\tfrac{1}{2}e^{-x^2+4x}.$$

Since $z=(D-4)(y)$, we have

$$\frac{dy}{dx}-4y=-\tfrac{1}{2}e^{-x^2+4x}.$$

Again we use the integrating factor e^{-4x}, getting

$$\frac{d}{dx}(ye^{-4x})=-\tfrac{1}{2}e^{-x^2}$$

(10) $$ye^{-4x}=-\tfrac{1}{2}\int e^{-x^2}\,dx.$$

Even though e^{-x^2} does not have an antiderivative expressible in terms of other elementary functions, the **error function** $erf(x)$ is defined by

$$\text{erf}(x)=\frac{2}{\sqrt{\pi}}\int_0^x e^{-t^2}\,dt,$$

and tables of this function exist just as for $\exp(x)$, $\ln x$, and others. So, choosing the right constant of integration enables us to write (10) as

$$ye^{-4x}=-\tfrac{1}{4}\sqrt{\pi}\,\text{erf}(x).$$

Hence,

$$y_p(x)=-\tfrac{1}{4}\sqrt{\pi}e^{4x}\,\text{erf}(x).$$

Thus the given differential equation has general solution

$$y=-\tfrac{1}{4}\sqrt{\pi}e^{4x}\,\mathrm{erf}(x)+c_1e^{4x}+c_2xe^{4x},$$

using the result of Example 6.11.

EXERCISES 8.7

In Exercises 1 to 8, find a particular solution of the given differential equation.

1 $\dfrac{d^2y}{dx^2}+3\dfrac{dy}{dx}-y=x^2-1.$

2 $\dfrac{d^2y}{dx^2}+2\dfrac{dy}{dx}-3y=2x^2-x+1.$

3 $\dfrac{d^2y}{dx^2}-3\dfrac{dy}{dx}-4y=2\sin x.$

4 $\dfrac{d^2y}{dx^2}+2\dfrac{dy}{dx}-5y=\cos 2x.$

5 $\dfrac{d^2y}{dx^2}+\dfrac{dy}{dx}-6y=2e^{-x}.$

6 $\dfrac{d^2y}{dx^2}+3\dfrac{dy}{dx}-5y=3e^{2x}.$

7 $\dfrac{d^2y}{dx^2}+\dfrac{dy}{dx}-2y=xe^{x}.$

8 $\dfrac{d^2y}{dx^2}-4y=3e^{2x}.$

In Exercises 9 to 22, find the general solution of the given differential equation.

9 $\dfrac{d^2y}{dx^2}-5\dfrac{dy}{dx}+6y=xe^{x}.$

10 $\dfrac{d^2y}{dx^2}+3\dfrac{dy}{dx}-y=\cos x.$

11 $\dfrac{d^2y}{dx^2}+2\dfrac{dy}{dx}+y=3e^{-x}.$

12 $\dfrac{d^2y}{dx^2}+6\dfrac{dy}{dx}+9y=2e^{-3x}.$

13 $\dfrac{d^2y}{dx^2}-2\dfrac{dy}{dx}-3y=2e^{x}-10\sin x.$

14 $\dfrac{d^2y}{dx^2}-5\dfrac{dy}{dx}-6y=2e^{x}+\cos x.$

15 $\dfrac{d^2y}{dx^2}-2\dfrac{dy}{dx}-3y=2e^{x}-10\sin x,$ where $y(0)=2,\ y'(0)=4.$

16 $\dfrac{d^2y}{dx^2}-5\dfrac{dy}{dx}-6y=2e^{x}\cos x,$ where $y(0)=0,\ y'(0)=1.$

17 $y''-2y'+y=e^{x},$ where $y(0)=1,\ y'(0)=1.$

18 $y''-6y'+9y=e^{3x},$ where $y(0)=1,\ y'(0)=1.$

19 $y''-3y'+2y=2x^2+e^{x}+2xe^{x}+4e^{3x}.$

20 $y''+2y'+2y=x\cos 2x+\sin 2x.$

21 $\dfrac{d^2y}{dx^2}-2\dfrac{dy}{dx}+y=2xe^{-x2+x}.$

22 $\dfrac{d^2y}{dx^2}-6\dfrac{dy}{dx}+9y=xe^{-x2+3x}.$

23 Find the general solution to $\dfrac{d^2y}{dx^2}+9y=\cos 2x$ by finding a complex solution to $y''+9y=e^{2ix}$ using the method of undetermined (complex) coefficients and then taking its real part.

24 Use the approach of Exercise 23 to find the general solution of $y''+y'-2y=4\sin 2x.$

25 By letting $x=e^{t}$, reduce

$$x^2\frac{d^2y}{dx^2}-2x\frac{dy}{dx}+2y=6\ln x$$

to a second order linear equation with constant coefficients. Find the general solution on $(0,+\infty)$.

26 Use the method of Exercise 25 to solve

$$x^2\frac{d^2y}{dx^2}+2x\frac{dy}{dx}-2y=6x$$

on $(0, +\infty)$. (The equations in Exercises 25 and 26 are called **Cauchy–Euler** equations.)

8 APPLICATIONS OF SECOND ORDER LINEAR EQUATIONS

Some important applications of second order linear equations to physics and engineering grow out of the study of oscillatory motion and electrical circuits. We will discuss the first of these in some detail, and then see that the models for both phenomena are so closely parallel that a great deal of information can be obtained with very little additional work.

We will start with the simplest kind of vibrating system, the one pictured in Figure 8.1. Here we have a mass m attached to a fixed spring. If the mass is

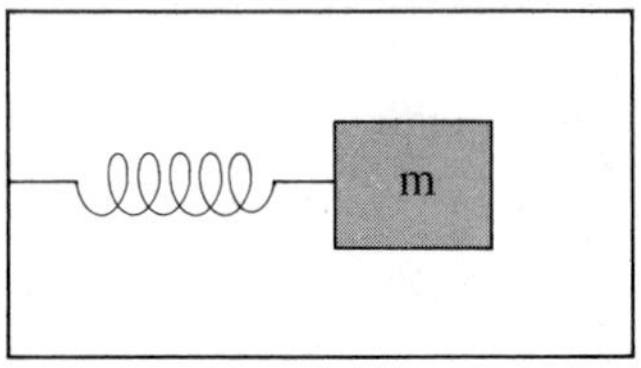

FIGURE 8.1

moved from its initial position and then released, it will oscillate back and forth about its equilibrium position. To analyze the motion, we need only Newton's law $\boldsymbol{F}=m\boldsymbol{a}$ and *Hooke's law*, which was formulated by the English physicist Robert Hooke (1635–1703) in 1676. This law says that the spring exerts a force on the mass proportional to the displacement $\boldsymbol{x}$ from the equilibrium position and directed oppositely to the motion. That is,

$$\boldsymbol{F}=-h\boldsymbol{x},$$

where $\boldsymbol{x}$ is the directed distance from the equilibrium position to the mass at time t and h is a positive constant called the *spring constant.* Since the motion and forces are all one-dimensional, we will consider them as real variables, and write the differential equation of the system as

(1) $$m\frac{d^2x}{dt^2}=-hx,$$

since $\frac{d^2x}{dt^2}$ gives the acceleration at time t.

8.1 THEOREM. Under the model (1), the motion of the mass in Figure 8.1 is oscillatory about the equilibrium position.

Proof. All we need to do is solve (1) for $x = x(t)$. First we put it in the standard form

$$\frac{d^2x}{dt^2} + \frac{h}{m} x = 0,$$

$$\left(D^2 + \frac{h}{m}\right)x = 0.$$

This second order homogeneous linear equation has an auxiliary equation whose roots are $\pm\sqrt{h/m}\,i$. Thus the general solution to (1) is

(2) $$x(t) = c_1 \cos\sqrt{h/m}\,t + c_2 \sin\sqrt{h/m}\,t.$$

In this form we can't visualize the motion very well, but the following trick produces a formula that is easy to interpret. Let $a = \sqrt{c_1^2 + c_2^2}$. Then observe that $\left(\frac{c_1}{a}, \frac{c_2}{a}\right)$ is on the unit circle $x^2 + y^2 = 1$. Hence by basic trigonometry, there is an angle ϕ_0 such that $\cos\phi_0 = \frac{c_1}{a}$ and $\sin\phi_0 = \frac{c_2}{a}$. (See Figure 8.2.)

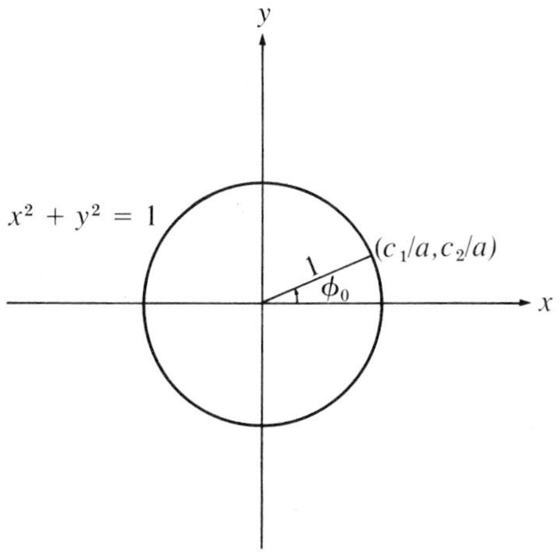

FIGURE 8.2

We can thus rewrite (2) as

$$x(t) = a\left(\frac{c_1}{a}\cos\sqrt{h/m}\,t + \frac{c_2}{a}\sin\sqrt{h/m}\,t\right)$$
$$= a(\cos\phi_0 \cos\sqrt{h/m}\,t + \sin\phi_0 \sin\sqrt{h/m}\,t),$$

(3) $$\boxed{x(t) = a\cos(\sqrt{h/m}\,t - \phi_0).}$$

The motion is now seen to be *periodic*. We graph $x(t)$ versus t in Figure 8.3. The

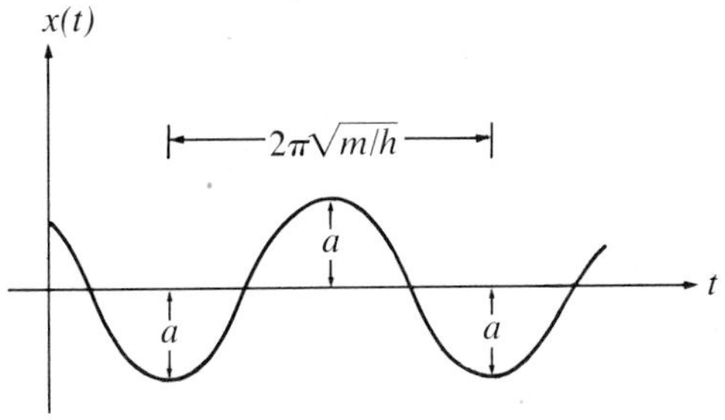

FIGURE 8.3

mass clearly oscillates between two extremes a units to the left and right of the equilibrium position.

QED

8.2 **DEFINITION.** The motion described by (3) is called **simple harmonic motion.** The **amplitude** is a, the maximum displacement from the equilibrium position. The **frequency** is $\omega = \sqrt{h/m}$, measured in radians per second. The **period** is $2\pi/\omega$, the time required for one complete cycle. The **phase angle** is ϕ_0.

With the notation of this definition, we can put (3) into the simpler form

(4) $$x(t) = a\cos(\omega t - \phi_0).$$

Notice that we could express the displacement in terms of the sine function by simply using a different phase angle $\phi_1 = \phi_0 - \frac{1}{2}\pi$. Then,

$$\begin{aligned} x(t) &= a\cos(\omega t - \phi_0) \\ &= a\cos(\phi_0 - \omega t), \quad \text{since} \quad \cos\theta = \cos(-\theta), \\ &= a\sin(\tfrac{1}{2}\pi - (\phi_0 - \omega t)) \\ &= a\sin(\omega t - \phi_1). \end{aligned}$$

In practice the model (1) often fails to give an accurate representation of oscillatory motion encountered in real-world situations. We have neglected air resistance, friction, and any other kind of damping force that might exist, for example. We thus are led to consider the slightly more sophisticated vibrating system pictured in Figure 8.4. We have added a "dashpot" D, which damps the

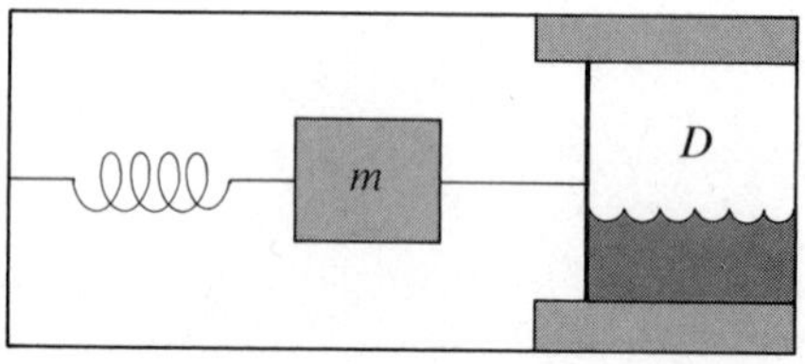

FIGURE 8.4

motion, and can be thought of as representing air or fluid resistance to the motion of m. Consider, for example, the vertical motion of the front of an automobile after it strikes a bump in the road. The oscillatory motion is damped by the front shock absorbers, and the motion dies down rather than continuing indefinitely as in (4).

The resisting force to the rightward motion of the mass is often observed to be proportional to the velocity $\dot{x}(t) = \dfrac{dx}{dt}$ of the mass. We thus can model this system by the equation

$$m\frac{d^2x}{dt^2} = -hx - k\frac{dx}{dt}$$

where k is a positive constant. In standard form, this second order linear homogeneous equation is

(5) $$\frac{d^2x}{dt^2} + \frac{k}{m}\frac{dx}{dt} + \frac{h}{m}x = 0.$$

By the quadratic formula the solution depends on the sign of

$$\left(\frac{k}{m}\right)^2 - 4\frac{h}{m} = \frac{k^2-4mh}{m^2},$$

and hence on the sign of k^2-4mh. As we might expect, then, the relative sizes of the spring constant h, the damping constant k, and the mass m determine the nature of the motion.

8.3 THEOREM. Under the model (5), the system in Figure 8.4 tends to return to its equilibrium position.

Rather than proving this directly, we will give a discussion of the motion that will tell us considerably more than the theorem asserts about the motion of the system in each of the cases $k^2-4mh>0$, $k^2-4mh<0$, and $k^2-4mh=0$. In operator form (5) is

$$\left(D^2+\frac{k}{m}D+\frac{h}{m}\right)(x)=0.$$

The auxiliary polynomial has roots

$$r_1 = \frac{1}{2}\left(-\frac{k}{m}+\sqrt{\frac{k^2-4mh}{m^2}}\right) = \frac{1}{2m}(-k+\sqrt{k^2-4mh}),$$

$$r_2 = \frac{1}{2m}(-k-\sqrt{k^2-4mh}).$$

(a) If $k^2-4mh>0$, then $k>2\sqrt{mh}$, so the damping constant is more than double the geometric mean of m and h. In this case one says that the motion is *overdamped.* Clearly $r_2<r_1$, and since $k>\sqrt{k^2-4mh}$ we have

$$-k+\sqrt{k^2-4mh}<0.$$

Thus $r_2<r_1<0$. We have as the general solution to (5)

(6) $$x(t)=c_1e^{r_1t}+c_2e^{r_2t}.$$

Hence,

$$\lim_{t\to\infty} x(t) = c_1\lim_{t\to\infty} e^{r_1t}+c_2\lim_{t\to\infty} e^{r_2t}=0+0=0,$$

since r_2 and r_1 are negative. So Theorem 8.3 follows in this case. The motion is *not* oscillatory, as can be seen from (6), and dies out rather rapidly if the r_i are not extremely small. Depending on c_1 and c_2, k, m, and h, the mass may pass its equilibrium position once. Two typical graphs of (6) are shown in Figure 8.5.

(b) If $k^2-4mh<0$, then $k<2\sqrt{mh}$. In this case the motion is said to be *underdamped.* The roots r_1 and r_2 are complex,

$$r_1=\frac{-k}{2m}+i\frac{\sqrt{4mh-k^2}}{2m}, \qquad r_2=\frac{-k}{2m}-i\frac{\sqrt{4mh-k^2}}{2m}.$$

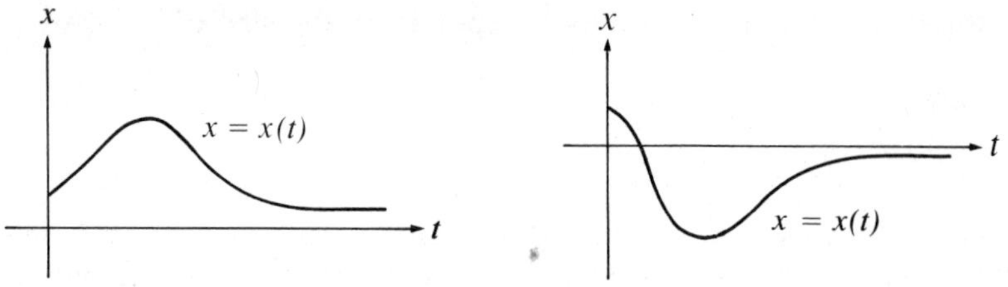

FIGURE 8.5

Hence we have as the general solution of (5)

$$x(t)=e^{-kt/2m}\left(c_1\cos\frac{\sqrt{4mh-k^2}}{2m}t+c_2\sin\frac{\sqrt{4mh-k^2}}{2m}t\right).$$

If, as in Theorem 8.1, we let $a=\sqrt{c_1^2+c_2^2}$, and choose ϕ_0 so that $\cos\phi_0=\dfrac{c_1}{a}$ and $\sin\phi_0=\dfrac{c_2}{a}$, then

(7) $$x(t)=ae^{-kt/2m}\cos\left(\frac{\sqrt{4mh-k^2}}{2m}t-\phi_0\right).$$

In this case then we get *damped vibration*: the graph of $x(t)$ resembles a cosine curve whose amplitude gradually decreases toward 0, because we have

$$\begin{aligned}\lim_{t\to\infty}|x(t)|&=|a|\lim_{t\to\infty}\left(e^{-kt/2m}\left|\cos\left(\frac{\sqrt{4mh-k^2}}{2m}t-\phi_0\right)\right|\right)\\&\le|a|\lim_{t\to\infty}e^{-kt/2m}1=0.\end{aligned}$$

So Theorem 8.3 holds here. The mass does oscillate this time about its equilibrium position, as shown in Figure 8.6, which represents a typical graph of (7). We put

$$\omega_1=\frac{\sqrt{4mh-k^2}}{2m},\quad\text{so}\quad x(t)=ae^{-kt/2m}\cos(\omega_1t-\phi_0).$$

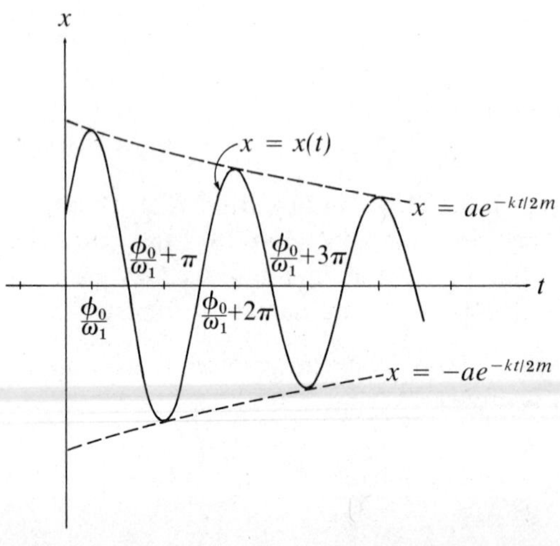

FIGURE 8.6

(c) If $k^2 - 4mh = 0$, then $k = 2\sqrt{mh}$. In this case the motion is said to be *critically damped.* This is an unstable situation in which a small change in either k, m, or h will put us into case (a) or case (b). We have $r_1 = r_2 = -k/2m$, so the general solution of (5) is

(8) $$x(t) = c_1 e^{-kt/2m} + c_2 t e^{-kt/2m}.$$

There is no oscillation, and as $t \to \infty$, $x(t)$ again approaches 0. (Use L'Hôpital's Rule to see that $c_2 t e^{-kt/2m} = c_2 \dfrac{t}{e^{kt/2m}}$ goes to 0 as $t \to \infty$.) Two typical graphs of (8) are shown in Figure 8.7.

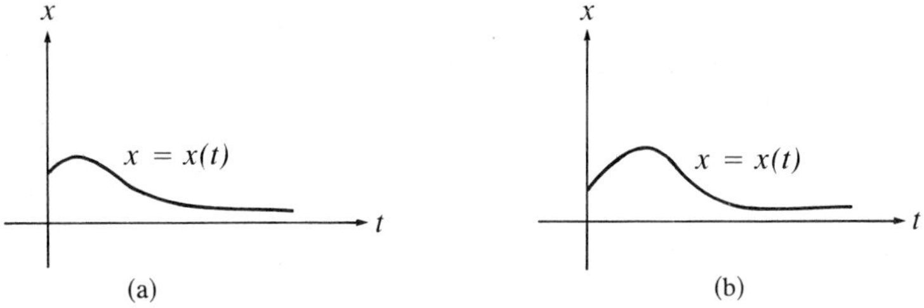

FIGURE 8.7

8.4 EXAMPLE. In the system in Figure 8.4, suppose that a force $F = 13$ newtons is needed to move the 1 kg mass 1 meter to the right. Suppose that the damping constant is 4 kg/sec. If the spring is compressed $\frac{1}{4}$ meter to the left and the mass is released with zero initial velocity, then find a formula for the displacement $x(t)$ of the mass from its equilibrium position at time t. What kind of motion does the mass exhibit?

Solution. We are given that $k = 4$ and $m = 1$, and that a force of 13 newtons exerted on the spring moves the mass 1 meter to the right. By Hooke's Law then, $13 = h \cdot 1$, so $h = 13$. Then (5) becomes

$$\ddot{x}(t) + 4\dot{x}(t) + 13x = 0.$$

The auxiliary equation is $r^2 + 4r + 13 = 0$, which has roots

$$r = \frac{-4 \pm \sqrt{16 - 52}}{2} = -2 \pm 3i.$$

Then

(*) $$x(t) = e^{-2t}(c_1 \cos 3t + c_2 \sin 3t).$$

The motion is thus *underdamped.* We know that $x(0) = -\frac{1}{4}$ and $\dot{x}(0) = 0$. So we have $-\frac{1}{4} = c_1$. Differentiating (*), we obtain

$$\dot{x}(t) = -2e^{-2t}(c_1 \cos 3t + c_2 \sin 3t) + e^{-2t}(-3c_1 \sin 3t + 3c_2 \cos 3t).$$

Thus when $t = 0$ we have

$$0 = -2c_1 + 3c_2 = \tfrac{1}{2} + 3c_2.$$

Thus $c_2 = -\frac{1}{6}$. So our solution to this initial value problem is

$$x(t) = -\tfrac{1}{2}e^{-2t}\left(\tfrac{1}{2}\cos 3t + \tfrac{1}{3}\sin 3t\right).$$

The motion we have discussed so far is called **free motion** because of the absence of any external force on the system. Let us now suppose that our system in Figure 8.4 is also acted on by an external force $F(t)$. In place of the homogeneous equation (5), we have then the nonhomogeneous equation

(9) $$\boxed{m\frac{d^2x}{dt^2} + k\frac{dx}{dt} + hx = F(t)}$$

as the differential equation of the system. In this case the motion is called **forced vibration** or **oscillation.** A common type of external force $F(t)$ is $F_0 \cos \omega_0 t$, where $\omega_0 \neq \omega = \sqrt{h/m}$. We can analyze the undamped system modeled by (9) with $k = 0$ in this case rather easily.

8.5 THEOREM. Under the model (9) with $k = 0$ and $F(t) = F_0 \cos \omega_0 t$, where $\omega_0 \neq \omega$, the motion of the system is described by

$$x(t) = c_1 \cos \omega t + c_2 \sin \omega t + \frac{2F_0}{m(\omega^2 - \omega_0^2)} \sin \tfrac{1}{2}(\omega - \omega_0)t \, \sin \tfrac{1}{2}(\omega + \omega_0)t.$$

if the system starts from rest.

Proof. In this case, (9) reduces to

(10) $$m\frac{d^2x}{dt^2} + hx = F_0 \cos \omega_0 t.$$

The homogeneous part of the solution was found in Theorem 8.1 to be given by

(2) $$x_0(t) = c_1 \cos \omega t + c_2 \sin \omega t.$$

So all we need is a particular solution $x_p(t)$ of (10). From the last section, we know there is a particular solution of the form

$$x_p(t) = A \cos \omega_0 t + B \sin \omega_0 t.$$

Differentiation gives

(11) $$\dot{x}_p(t) = -A\omega_0 \sin \omega_0 t + B\omega_0 \cos \omega_0 t,$$

(12) $$\ddot{x}_p(t) = -A\omega_0^2 \cos \omega_0 t - B\omega_0^2 \sin \omega_0 t = -\omega_0^2 x_p(t).$$

Thus, substituting into (10), we get

$$-m\omega_0^2 x(t) + hx(t) = F_0 \cos \omega_0 t, \; (h - m\omega_0^2)(A \cos \omega_0 t + B \sin \omega_0 t) = F_0 \cos \omega_0 t.$$

Hence,

$$A = \frac{F_0}{h - m\omega_0^2} = \frac{F_0}{m\omega^2 - m\omega_0^2} \quad \text{since} \quad \omega = \sqrt{h/m}$$

$$= \frac{F_0}{m(\omega^2 - \omega_0^2)},$$

$$B = 0.$$

So the general solution of (10) is

$$x(t) = c_1 \cos \omega t + c_2 \sin \omega t + \frac{F_0}{m(\omega^2 - \omega_0^2)} \cos \omega_0 t. \tag{13}$$

If the system starts from rest, then $x(0) = \dot{x}(0) = 0$, so we have

$$x(0) = 0 = c_1 + 0 + \frac{F_0}{m(\omega^2 - \omega_0^2)}, \qquad c_1 = -\frac{F_0}{m(\omega^2 - \omega_0^2)};$$

$$\dot{x}(0) = 0 = 0 + \omega c_2 - 0, \qquad c_2 = 0.$$

Thus (13) becomes

$$x(t) = \frac{F_0}{m(\omega^2 - \omega_0^2)} (\cos \omega_0 t - \cos \omega t),$$

$$x(t) = \frac{2F_0}{m(\omega^2 - \omega_0^2)} \sin \tfrac{1}{2}(\omega - \omega_0)t \sin \tfrac{1}{2}(\omega + \omega_0)t, \tag{14}$$

where we used the trigonometric identities

$$\cos(\theta_1 + \theta_2) = \cos \theta_1 \cos \theta_2 - \sin \theta_1 \sin \theta_2 \tag{15}$$

$$\cos(\theta_1 - \theta_2) = \cos \theta_1 \cos \theta_2 + \sin \theta_1 \sin \theta_2, \tag{16}$$

with $\theta_1 = \frac{1}{2}(\omega + \omega_0)t$ and $\theta_2 = \frac{1}{2}(\omega - \omega_0)t$. [We subtracted (15) from (16), as you should verify.] QED

Note that if ω_0 is close to ω, then $\omega + \omega_0$ is much larger than $\omega - \omega_0$, so $\sin \frac{1}{2}(\omega + \omega_0)t$ oscillates with small period $4\pi/(\omega + \omega_0)$. The resulting motion is rapidly oscillating with frequency $\frac{1}{2}(\omega + \omega_0)$, and with a slowly changing amplitude influenced by the factor $\sin \frac{1}{2}(\omega - \omega_0)$. (AM radio broadcasting uses such an ω_0, hence the name "amplitude modulation.") In Figure 8.8 we show a graph of the motion described by (14), which is said to exhibit *beats*.

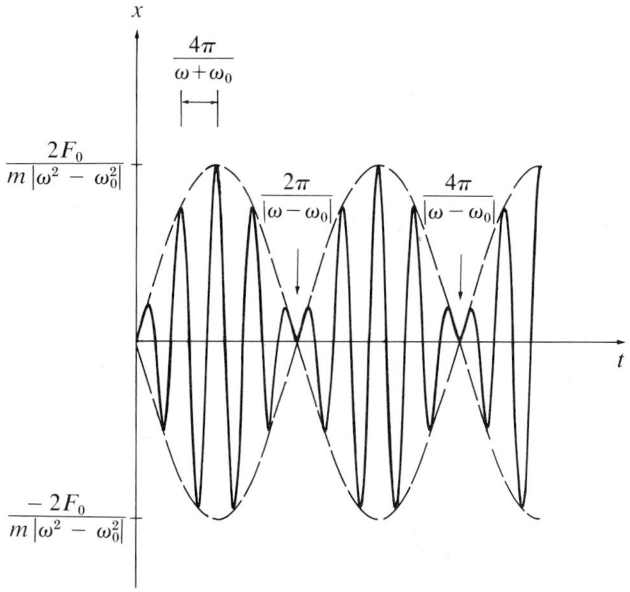

FIGURE 8.8

In case $\omega = \omega_0$, the analysis of Theorem 8.5 changes radically. Since $A \cos \omega_0 t + B \sin \omega_0 t$ is then a *homogeneous* solution, the particular solution $x_p(t)$ must (by Method 7.2) have the form $tA \cos \omega_0 t + tB \sin \omega_0 t$; thus, the motion oscillates *without bound* as $t \to \infty$, a phenomenon known as **resonance**. In practical terms, the spring in Figure 8.4 will be stretched to the point of breaking or losing its elasticity, or the chamber walls at one end or the other will be broken down. You can consider this case in more detail in Exercise 3 below.

Let's return now to the general (damped) case in (9), with external force $F(t) = F_0 \cos \omega_0 t$, where we no longer restrict ω_0 to be different from $\omega = \sqrt{h/m}$.

8.6 THEOREM. Under the model (9) with $F(t) = F_0 \cos \omega_0 t$, the motion of the system approaches a steady oscillation about the equilibrium point.

Proof. In the course of proving Theorem 8.3, we found that the homogeneous part of the solution to (9) [i.e., the solution to (5)] had one of the forms (6), (7), or (8). In all three cases, $\lim_{t \to \infty} x_0(t) = 0$, so we see that the motion will ultimately be determined by the nature of a particular solution $x_p(t)$ to (9). For this reason, $x_0(t)$ is called a **transient solution** and $x_p(t)$ a **steady state solution**. We know that the particular solution to (9) has the form

(17) $$x_p(t) = A \cos \omega_0 t + B \sin \omega_0 t,$$

so the motion does approach a steady oscillation, since we can reduce (17) to the form $x_p(t) = a \cos(\omega_0 t - \phi_0)$ by introducing an appropriate phase angle ϕ_0 as in Theorem 8.1.
QED

Let's take a closer look at $x_p(t)$. The formulas (11) and (12) give $\dot{x}_p(t)$ and $\ddot{x}_p(t)$. Substitution of these into (9) gives

(18) $$(h - m\omega_0^2)(A \cos \omega_0 t + B \sin \omega_0 t) + k(-A\omega_0 \sin \omega_0 t + B\omega_0 \cos \omega_0 t) = F_0 \cos \omega_0 t.$$

Recalling that $\omega = \sqrt{h/m}$, we can write $h - m\omega_0^2 = m(\omega^2 - \omega_0^2)$. Equating coefficients of $\cos \omega_0 t$ and $\sin \omega_0 t$ in (18),

(19) $$\begin{cases} m(\omega^2 - \omega_0^2)A + k\omega_0 B = F_0, \\ -k\omega_0 A + m(\omega^2 - \omega_0^2)B = 0. \end{cases}$$

This system has (Exercise 2 below) the solution

(20) $$A = \frac{m(\omega^2 - \omega_0^2)F_0}{m^2(\omega^2 - \omega_0^2)^2 + k^2\omega_0^2}, \quad B = \frac{k\omega_0 F_0}{m^2(\omega^2 - \omega_0)^2 + k^2\omega_0^2}.$$

So (17) becomes

(17′) $$x_p(t) = \frac{F_0}{m^2(\omega^2 - \omega_0^2)^2 + k^2\omega_0^2}(m(\omega^2 - \omega_0^2)\cos \omega_0 t + k\omega_0 \sin \omega_0 t).$$

Now if we let

$$c = \sqrt{m^2(\omega^2 - \omega_0^2)^2 + k^2\omega_0^2},$$

then (cf. Theorem 8.1) we can choose ϕ_0 so that

$$\cos \phi_0 = \frac{m(\omega^2 - \omega_0^2)}{c}, \qquad \sin \phi_0 = \frac{k\omega_0}{c}.$$

We thus have

$$x_p(t) = \frac{F_0}{c}(\cos \phi_0 \cos \omega_0 t + \sin \phi_0 \sin \omega_0 t),$$

(21) $$x_p(t) = \frac{F_0}{c}\cos(\omega_0 t - \phi_0).$$

Here c is never 0, even if $\omega_0 = \omega$. Hence, unlike the undamped case, the motion here is always bounded. It is easy to show (Exercise 4 below) that c takes on its minimum when $\omega_0^2 = \omega^2 - \frac{1}{2}\frac{k^2}{m^2}$. In this case the amplitude of $x_p(t)$ is maximized.

Seismographs are constructed as vibrating systems of frequency ω such that $\sqrt{\omega^2 - \frac{1}{2}\frac{k^2}{m^2}}$ is located in the middle of the range of frequencies associated with earthquakes. They thus can detect even slight earthquakes by responding strongly to frequencies near ω_0.

We turn now to a different area, electrical circuits, whose mathematical models are strikingly similar to those of the vibrating systems just considered. The type of circuit we consider is shown in Figure 8.9. It has an external power source

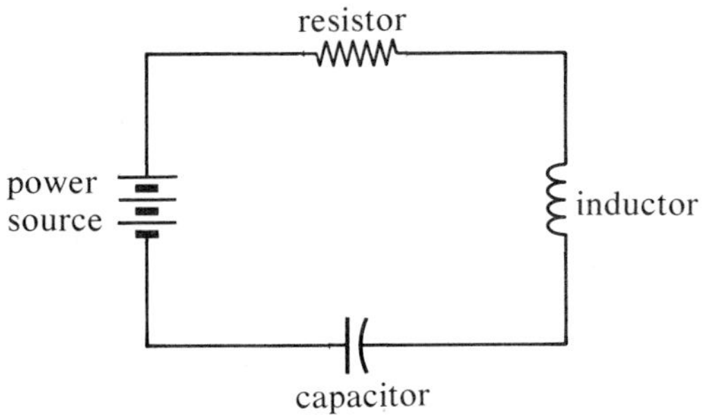

FIGURE 8.9

(EMF), a resistor, an inductor, and a capacitor (or "condenser"). The power source has a voltage $E(t)$, the resistor a constant resistance R measured in ohms, the inductor a fixed inductance L measured in henrys, and the capacitor a fixed capacitance C measured in farads. (The quantity $1/C$ is called the *elastance*.) Let $q(t)$ be the charge in coulombs on the capacitor at time t, and let $i(t) = \frac{dq}{dt}$ be the current flowing in the circuit at time t.

We use *Kirchhoff's voltage law*, which says that the sum of the voltage drops around the circuit at any instant is equal to the impressed voltage $E(t)$. From laboratory measurements, the voltage drops across the resistor, inductor, and capacitor are given respectively by $Ri(t)$, $L\frac{di}{dt}$, and $\frac{1}{C}q(t)$. Thus Kirchhoff's law gives the differential equation

(22) $$L\frac{di}{dt} + Ri(t) + \frac{1}{C}q(t) = E(t).$$

Since $i = \dfrac{dq}{dt}$, this can be written

(23) $$L\frac{d^2q}{dt^2} + R\frac{dq}{dt} + \frac{1}{C}q(t) = E(t).$$

From a mathematical standpoint, (23) is *indistinguishable* from the equation (9) of a damped vibrating system. We have, in fact, the following table of correspondences.

Table 8.1

Vibrating System	**Electrical Circuit**
Differential Equation (9)	Differential Equation (23)
Displacement $x(t)$	Charge $q(t)$
Velocity $\dfrac{dx}{dt}$	Current $\dfrac{dq}{dt} = i(t)$
Mass m	Inductance L
Damping constant k	Resistance R
Spring constant h	Elastance $1/C$
External force $F(t)$	Electromotive force $E(t)$
Frequency $\omega = \sqrt{h/m}$	Frequency $\omega = \sqrt{1/CL}$

If we wish to consider just current and avoid working directly with $q(t)$, then we can do so by differentiating (22). We get the second order linear equation

(24) $$L\frac{d^2i}{dt^2} + R\frac{di}{dt} + \frac{1}{C}i = \dot{E}(t),$$

which again has the same basic form as (9).

Since we have already analyzed (9) thoroughly, we have in effect a complete solution of (23) or (24). All we need to do is translate our earlier results by means of Table 8.1. Suppose, for example, that $E(t) = \dfrac{1}{\omega_0}E_0 \sin \omega_0 t$. Then $\dot{E}(t) = E_0 \cos \omega_0 t$ and Theorem 8.6 applies. There is then a *transient current* of the form (6), (7), or (8) and a *steady state current* $i_p(t)$ in the circuit given by the analogue of (17′) or (21). Thus,

(25) $$i_p(t) = \frac{E_0}{I}(L(\omega^2 - \omega_0^2)\cos \omega_0 t + R\omega_0 \sin \omega_0 t)$$

(26) $$= \frac{E_0}{I}\cos(\omega_0 t - \phi_0),$$

where $I = \sqrt{L^2(\omega^2 - \omega_0^2)^2 + R^2\omega_0^2}$ is called the *impedance* of the circuit.

The remarks following Equation (21) suggest that receivers for signals transmitted along electrical circuits should have frequencies ω such that $\sqrt{\omega_0^2-\frac{1}{2}(R/L)^2}$ is close to the mean signal frequency.

8.7 **EXAMPLE.** In the circuit of Figure 8.9, suppose that $E(t)=400\sin 200t$, $R=40$, $L=0.25$, and $C=4.0\times 10^{-4}$. Determine the frequency ω, phase angle ϕ_0, impedance, and steady state current.

Solution. We have $\dot{E}(t)=8.0\times 10^4\cos 200t$, so $E_0=8.0\times 10^4$. Our Equation (24) thus is

$$0.25\frac{d^2i}{dt^2}+40\frac{di}{dt}+(0.25\times 10^4)i=8.0\times 10^4\cos 200t.$$

In standard form, this is

$$\frac{d^2i}{dt^2}+160\frac{di}{dt}+10^4i=32.0\times 10^4\cos 200t.$$

The auxiliary equation is

$$r^2+160r+10^4=0$$

whose roots are given by

$$r=\tfrac{1}{2}(-160\pm\sqrt{25600-40000})=-80\pm\tfrac{1}{2}\sqrt{-14400}=-80\pm 60i.$$

Thus the transient current has the form

$$i(t)=e^{-80t}(c_1\cos 60t+c_2\sin 60t).$$

From the preceding discussion, the steady state current is given by (25) or (26). Since $E_0=8.0\times 10^4$, $\omega^2=1/CL=1.0\times 10^4$, $L^2=6.25\times 10^{-2}$, $R^2=1.6\times 10^3$, and $\omega_0=200$, we have

$$\begin{aligned} I&=\sqrt{6.25\times 10^{-2}(10^4-4\times 10^4)^2+1.6\times 10^3(4\times 10^4)}\\ &=\sqrt{120.25\times 10^6}\\ &\approx 10.966\times 10^3. \end{aligned}$$

So the impedance is approximately 1.1×10^4. We have $\omega=1.0\times 10^2$, and the phase angle ϕ_0 is given by

$$\cos\phi_0=\frac{L(\omega^2-\omega_0^2)}{I}=\frac{-7500}{\sqrt{120.25\times 10^6}}\approx -0.68394,$$

$$\sin\phi_0=\frac{R\omega_0}{I}=\frac{8000}{\sqrt{120.25\times 10^6}}\approx 0.72954.$$

So $\phi_0\approx 2.324$ radians (about 133°). We can conveniently give the steady state current in the form (26) now.

$$\begin{aligned} i_p(t)&=\frac{E_0}{I}\cos(\omega_0 t-\phi_0)\\ &\approx 7.3\cos(200t-2.3) \end{aligned}$$

where we round off to the two significant figures of accuracy of the measurements of R, L, and C.

If we wanted an explicit formula for $i(t)$ at any instant, we would usually find it more convenient to work with the form (23) than with (24).

8.8 EXAMPLE. Suppose in Example 8.7 that $E(t)=100\sin 200t$, $q(0)=0$, and $i(0)=0$. Find a formula for $i(t)$ at any time t.

Solution. Our equation in form (23) is

$$0.25\frac{d^2q}{dt^2}+40\frac{dq}{dt}+(0.25\times 10^4)q=100\sin 200t.$$

In standard form we have

(27) $$\frac{d^2q}{dt^2}+1.6\times 10^2\frac{dq}{dt}+1.0\times 10^4 q=400\sin 200t.$$

The homogeneous part of the solution is

$$q(t)=e^{-80t}(c_1\cos 60t+c_2\sin 60t).$$

To find a particular solution, let $q_1(t)=A\cos 200t+B\sin 200t$. Then

$$q_1'(t)=-200A\sin 200t+200B\cos 200t,$$
$$q''(t)=-40{,}000A\cos 200t-40{,}000B\sin 200t$$
$$=-40{,}000q_1(t).$$

Substituting into (27) we have

$$-40{,}000(A\cos 200t+B\sin 200t)+32{,}000(-A\sin 200t+B\cos 200t)$$
$$+10{,}000(A\cos 200t+B\sin 200t)=400\sin 200t.$$

Equating coefficients,

$$(-40{,}000+10{,}000)A+32{,}000B=0,$$
$$-32{,}000A+(-40{,}000+10{,}000)B=400.$$

Dividing by 2000,

$$-15A+16B=0$$
$$-16A-15B=\tfrac{1}{5}.$$

This system has solution $A=-16/2405$, $B=-3/481$. Thus the general solution to (27) is

$$q(t)=e^{-80t}(c_1\cos 60t+c_2\sin 60t)-\tfrac{16}{2405}\cos 200t-\tfrac{3}{481}\sin 200t.$$

When $t=0$, $q=0$. Hence $0=c_1-\frac{16}{2405}$, so $c_1=\frac{16}{2405}$. Next,

$$\begin{aligned} i(t)=\frac{dq}{dt}&=-80e^{-80t}(c_1\cos 60t+c_2\sin 60t)\\ &\quad+60e^{-80t}(-c_1\sin 60t+c_2\cos 60t)\\ &\quad+\tfrac{3200}{2405}\sin 200t-\tfrac{600}{481}\cos 200t.\\ &=e^{-80t}[(-80c_1+60c_2)\cos 60t+(-60c_1-80c_2)\sin 60t]\\ &\quad+\tfrac{3200}{2405}\sin 200t-\tfrac{600}{481}\cos 200t. \end{aligned}$$

Since $i(0)=0$, we have

$$\begin{aligned} 0&=-80c_1+60c_2-\tfrac{600}{481}\\ c_2&=\tfrac{1}{60}(\tfrac{600}{481}+80c_1)=\tfrac{10}{481}+\tfrac{4}{3}c_1\\ &=\tfrac{10}{481}+\tfrac{64}{7215}=\tfrac{214}{7215}. \end{aligned}$$

Hence our desired formula for $i(t)$ is

$$\begin{aligned} i(t)&=e^{-80t}(\tfrac{600}{481}\cos 60t-\tfrac{1112}{481}\sin 60t)+\tfrac{3200}{2405}\sin 200t-\tfrac{600}{481}\cos 200t\\ &\approx e^{-80t}(1.2\cos 60t-2.3\sin 60t)+1.3\sin 200t-1.2\cos 200t. \end{aligned}$$

EXERCISES 8.8

1 Show that Equation (4) also models the motion of a point revolving with constant angular velocity ω counterclockwise about the circle $x^2+y^2=a^2$, which starts at an initial position (x_0, y_0) such that $\dfrac{x_0}{a}=\cos\phi_0$.

2 Show that the system (19) has solution (20).

3 (a) Suppose in Theorem 8.5 that $\omega_0=\omega$. Find a particular solution to (10).
(b) Show that if $\omega_0=\omega$, then the general solution $x(t)$ of (10) tends to infinity as $t\to\infty$, thus establishing the assertion made just before Theorem 8.6.

4 Show that

$$c=\sqrt{m^2(\omega^2-\omega_0^2)^2+k^2\omega_0^2}$$

has its minimum, and so $x_p(t)$ in (17) has its maximum amplitude, when $\omega_0^2=\omega^2-\dfrac{1}{2}\dfrac{k^2}{m^2}$.

5 A mass of weight 16 pounds hangs at rest from an eight foot long spring attached to the ceiling. The mass is pulled down six inches and then released (with no initial velocity). Find a formula for $x(t)$, the number of feet from the ceiling to the mass, at time t. (Use $g=32$ ft/sec^2.)

6 (a) A mass with weight 8 pounds is placed on the lower end of a spring suspended from the ceiling. The weight pulls the spring 6 inches down and comes to rest in its equilibrium position. The weight is then pulled down 3 inches farther and released with initial velocity $v_0=1$ ft/sec downward. Find a formula in form (2) for the displacement $x(t)$ of the mass from its equilibrium position at time t. (Use $g=32$ ft/sec^2.)
(b) Express $x(t)$ in the form (3) and determine the amplitude, period, and frequency of the motion.

7 In the system of Figure 8.4, a force of 32 newtons is required to move the 1 kg mass 2 meters to the right. The mass is moved $\frac{1}{2}$ meter to the right and released. Assume that the damping constant is 8 kg/sec. Find $x(t)$ at time t. What type of motion does the system exhibit?

8 Repeat Exercise 7 if the damping constant is 10 kg/sec.

9 Consider the undamped system in Figure 8.4 where $m = 1$ kg and a force of 4 newtons moves the mass 1 meter to the right. Suppose that a force $F = 5 \cos t$ is applied to the system. Find a formula for $x(t)$, assuming that the system starts from rest in its equilibrium position.

10 In the system of Figure 8.4, suppose that $m = \frac{1}{2}$ kg. A force of 20 newtons is required to move the mass 2 meters to the right. If the damping constant is 2 and the system begins from rest in its equilibrium position, then find a formula for the displacement $x(t)$ at time t if an external force $F = 5 \cos 2t$ is applied to the system. What is the resonant frequency?

11 A circuit consists of a capacitor and an inductor. Show that the charge $q(t)$ on the capacitor is periodic. What is the frequency? What is the nature of the current?

12 If an electromotive force $E = E_0 \cos \omega t$ (where $\omega = 1/\sqrt{CL}$) is added to the circuit in Exercise 11, then what happens to the charge as $t \to \infty$? What about the current?

13 If a circuit consists of a resistor, capacitor, and inductor (but no impressed electromotive force), then show that as $t \to \infty$ the charge decreases to 0.

14 Discuss the long-term behavior of the charge and current in Exercise 13 if $E(t) = E_0 \cos \omega_0 t$, where $\omega_0 \neq \omega$.

15 A circuit consists of a resistor, inductor, and capacitor with $R = 6.0 \times 10^2$, $C = 0.5 \times 10^{-5}$, and $L = 4.0 \times 10^{-1}$. If $q(0) = 1.0 \times 10^{-6}$ and $i(0) = 0$, then find $q(t)$ and $i(t)$.

16 Repeat Exercise 15 if $R = 2 \times 10^2$, $C = 5.0 \times 10^{-6}$, and $L = 1.0$.

17 In the circuit of Figure 8.9, suppose that $R = 16$, $L = 2.0 \times 10^{-2}$, $C = 2.0 \times 10^{-4}$, and $E = 12$. If $q(0) = 4.8 \times 10^{-3}$ and $i(0) = 0$, then find formulas for $q(t)$ and $i(t)$.

18 Suppose in the circuit of Figure 8.9 that $E(t) = 110 \sin 120\pi t$, $R = 5.0 \times 10^2$, $C = 5.0 \times 10^{-6}$, and $L = 3.0$. Find the steady state current.

19 Suppose in the circuit of Figure 8.9 that $E(t) = 100 \sin 60t$, $R = 2$, $L = 0.1$, and $C = \frac{1}{260}$. If $q(0) = i(0) = 0$ then find $q(t)$ and $i(t)$. What is the steady state current?

20 The **catenary** (described by Leibniz in 1691 and derived fully by Jakob Bernoulli shortly thereafter) is the curve assumed by a flexible cable of uniformly distributed weight that hangs from two supports B and C as in Figure 8.10. Let $P(x, y)$ be a typical

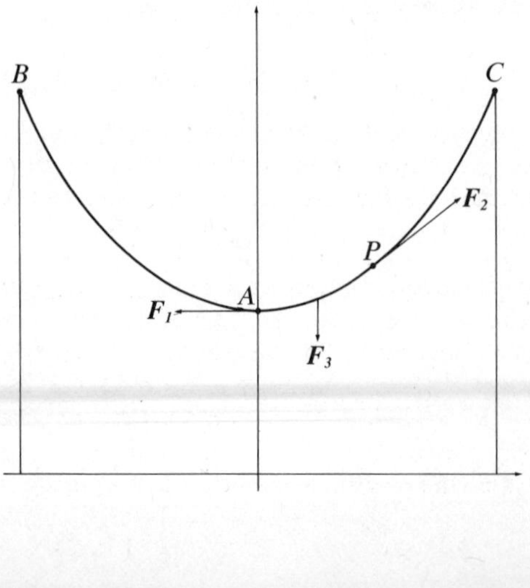

FIGURE 8.10

point on the curve, and $A(0, a)$ the lowest point. The forces on arc AP are:
(a) a downward force $\boldsymbol{F}_3$ of magnitude ws, where w is the weight of one unit of length of the cable and s is the length of the arc along AP;
(b) a tangential force $\boldsymbol{F}_2$ acting at P due to the support at C;
(c) a horizontal force $\boldsymbol{F}_1$ at A due to the support at B.

Since the cable is in equilibrium, the sums of the horizontal and vertical components of these forces are zero. Show then that

$$|\boldsymbol{F}_2| \sin \theta = ws$$
$$|\boldsymbol{F}_2| \cos \theta = |\boldsymbol{F}_1|$$

Divide these to obtain the differential equation

(28) $$\frac{dy}{dx} = \frac{1}{b} s$$

where $b = |\boldsymbol{F}_1|/w$. Differentiate to obtain the nonlinear second order differential equation

(29) $$\frac{d^2y}{dx^2} = \frac{1}{b}\sqrt{1 + (dy/dx)^2}.$$

21 In Exercise 20, use the substitution $v = \frac{dy}{dx}$ suggested in Exercise 23(a), Exercises 8.6, and Formula 27 of the Table of Integrals to obtain $y = b \cosh \frac{x}{b} + (a - b)$. (If the origin is b units below A, then this gives

$$y = a \cosh \frac{x}{a} = \tfrac{1}{2} a (e^{x/a} + e^{-x/a}),$$

the standard-form equation of the catenary.)

9 POWER SERIES AND NUMERICAL SOLUTIONS[1]

Thus far the only second order linear differential equations

(1) $$y'' + p(x)y' + q(x)y = s(x)$$

for which we have a solution method are those with constant coefficients $p(x) = p$ and $q(x) = q$. While a large number of important applications involve this restricted class of equations, there are other important applications in which the more general Equation (1) occurs. According to Theorem 6.3, since (1) is a linear equation, all solutions take the form $y_0(x) + y_p(x)$ where $y_0(x)$ is any solution to the associated homogeneous linear equation

(2) $$y'' + p(x)y' + q(x)y = 0$$

and $y_p(x)$ is one particular solution to (1). We will give a method of finding $y_p(x)$ in Section 9.3. In this section, we present an introduction to two very important methods of finding the general solution of the associated homogeneous linear differential equation (2).

For the first method, we shall restrict attention to equations (2) such that both $p(x)$ and $q(x)$ are *analytic* at $x_0 \in \boldsymbol{R}$. By this we mean that the Taylor series

1. Sections 6 to 9 of Chapter 11 are prerequisite to the power series method.

expansions for $p(x)$ and $q(x)$ converge to these functions on some interval I about x_0. Such a point x_0 is called in differential equations an *ordinary point.* Power series techniques often are still useful near *singular points* where $p(x)$ and $q(x)$ are not analytic, but we must leave a discussion of more general cases to a later course in differential equations. Our approach here will be to assume that the general solution to (2) is an analytic function

(3) $$y(x) = \sum_{n=0}^{\infty} a_n x^n,$$

where the coefficients are real numbers. We shall then determine the coefficients by a process very much like that used in Section 7. Theoretical justification of this technique rests on the following result of the German mathematician I. Lazarus Fuchs (1833–1902), whose proof we leave for a course in differential equations.

9.1 **THEOREM.** If x_0 is an ordinary point of (2), then the general solution of (2) is an analytic function $y(x)$ whose Taylor series converges to $y(x)$ on an interval $|x-x_0|<r$. This interval is at least as large as the common interval I on which the Taylor series for $p(x)$ and $q(x)$ converge to these functions.

One can actually say more. It turns out that the series for $y(x)$ can be written as

(4) $$y(x) = c_1 y_1(x) + c_2 y_2(x),$$

where c_1 and c_2 are arbitrary constants and $y_1(x)$ and $y_2(x)$ are analytic functions that are linearly independent (in the sense of Section 1.7) in the real vector space $C^\infty(\boldsymbol{R})$; that is, $ay_1(x)+by_2(x)=0$, where $a, b \in \boldsymbol{R}$, occurs only when $a=b=0$. Thus the general solution to (2) has the same general appearance as in the constant coefficient case when

$$y(x) = c_1 e^{r_1 x} + c_2 e^{r_2 x},$$

or

$$y(x) = c_1 e^{rx} + c_2 x e^{rx},$$

or

$$y(x) = c_1 e^{ax} \cos bx + c_2 e^{ax} \sin bx.$$

In order to find the general solution (4) for (2), we write

(3) $$\boxed{y(x) = \sum_{n=0}^{\infty} a_n x^n.}$$

Then by Theorem 11.9.6, we know that inside the interval of convergence of (3) we have

(5) $$\boxed{y'(x) = \sum_{n=1}^{\infty} n a_n x^{n-1}.}$$

Similarly, inside the interval of convergence,

(6) $$y''(x) = \sum_{n=2}^{\infty} n(n-1)a_n x^{n-2}.$$

If we substitute (3), (5), and (6) into (2), then we have

$$\sum_{n=2}^{\infty} n(n-1)a_n x^{n-2} + \sum_{n=1}^{\infty} na_n p(x)x^{n-1} + \sum_{n=0}^{\infty} a_n q(x)x^n = 0$$

If we equate the coefficients of x^n (for $n = 0, 1, 2, \ldots$) to 0, then we obtain a series of relations on the coefficients $a_0, a_1, a_2, a_3, \ldots$. We can solve these relations in the sense of expressing the later coefficients in terms of the constants a_0 and a_1, which then are the c_1 and c_2 in (4). Some rewriting of summation indices is usually needed to carry this out. We illustrate the procedure in the following examples. In the first, we choose a simple equation with constant coefficients which we could easily solve explicitly by the method of Section 6, so as to demonstrate the reasonableness of Theorem 9.1 and the power series method just outlined.

9.2 EXAMPLE. Use the power series scheme to solve $y'' + y = 0$.

Solution. We assume that

$$y(x) = \sum_{n=0}^{\infty} a_n x^n.$$

Then we have (3) and (6) to substitute into the given equation. We get

$$\sum_{n=2}^{\infty} n(n-1)a_n x^{n-2} + \sum_{n=0}^{\infty} a_n x^n = 0.$$

The natural impulse is to add these series by adding corresponding coefficients of the powers of x. The representations present, however, are expressed in terms of *different powers* of x. To get series that we can more readily combine, we change our index of summation in the first series to m. So we have

$$\sum_{m=2}^{\infty} m(m-1)a_m x^{m-2} + \sum_{n=0}^{\infty} a_n x^n = 0.$$

Now it seems appropriate to let $m-2=n$. Then $m-1=n+1$, and $m=n+2$. Note that when $m=2$, $n=0$. Thus we have

$$\sum_{n=0}^{\infty} (n+2)(n+1)a_{n+2}x^n + \sum_{n=0}^{\infty} a_n x^n = 0,$$

(7) $$\sum_{n=0}^{\infty} [(n+2)(n+1)a_{n+2} + a_n]x^n = 0.$$

Thus the power series on the left is the power series for the identically zero function. By Theorem 11.9.7, we know that the only power series for that function is

$$0 + 0x + 0x^2 + 0x^3 + \ldots + 0x^n + \ldots$$

So every single coefficient in (7) has to be 0. Hence,

(8) $$a_{n+2} = -\frac{1}{(n+2)(n+1)}\, a_n.$$

When $n=0$, this says $a_2 = -\frac{1}{2\cdot 1}\, a_0$.

When $n=1$, this says $a_3 = -\frac{1}{3\cdot 2}\, a_1$.

When $n=2$, this says $a_4 = -\frac{1}{4\cdot 3}\, a_2 = \frac{1}{4!}\, a_0$.

When $n=3$, this says $a_5 = -\frac{1}{5\cdot 4}\, a_3 = \frac{1}{5!}\, a_1$.

When $n=4$, this says $a_6 = -\frac{1}{6\cdot 5}\, a_4 = -\frac{1}{6!}\, a_0$.

When $n=5$, this says $a_7 = -\frac{1}{7\cdot 6}\, a_5 = -\frac{1}{7!}\, a_1$, etc.

From (3) then, our general solution takes the form

$$\begin{aligned} y(x) &= a_0 + a_1 x - \frac{1}{2!}\, a_0 x^2 - \frac{1}{3!}\, a_1 x^3 + \frac{1}{4!}\, a_0 x^4 + \frac{1}{5!}\, a_1 x^5 \\ &\quad - \frac{1}{6!}\, a_0 x^6 - \frac{1}{7!}\, a_1 x^7 + \dots \\ &= a_0\left(1 - \frac{1}{2!}x^2 + \frac{1}{4!}x^4 - \frac{1}{6!}x^6 + - \dots\right) \\ &\quad + a_1\left(x - \frac{1}{3!}x^3 + \frac{1}{5!}x^5 - \frac{1}{7!}x^7 + - \dots\right) \\ &= a_0 \cos x + a_1 \sin x, \end{aligned}$$

since we recognize the series in parentheses as the Taylor series for $\sin x$ and $\cos x$, which converge for all x.

As a simple check, we can put the given constant coefficient equation in the standard form $(D^2+1)y=0$. The auxiliary equation has roots $\pm i$. So the general solution is $y(x) = c_1 \cos x + c_2 \sin x$, in agreement with our calculated result using the power series method.

An equation like (8) that determines later coefficients in terms of earlier ones is called a *recurrence relation*.

Naturally, the usefulness of the power series method lies *not* in affording a more involved way of solving constant coefficient equations, but rather in providing a method to attack *variable* coefficient Equations (2). Thus our next example is more illustrative of the practical application of the method.

9.3 EXAMPLE. Find a power series representation for the solution of

$$\frac{d^2y}{dx^2} + x\frac{dy}{dx} + (2x^2+1)y = 0.$$

Solution. Here $p(x) = x$ and $q(x) = 2x^2+1$. Thus both $p(x)$ and $q(x)$ are analytic at 0. Letting $y(x)$ be defined by (3), we have, from (5),

$$x\frac{dy}{dx}=\sum_{n=1}^{\infty} na_nx^n,$$

and from (3),

$$(2x^2+1)y=\sum_{n=0}^{\infty} 2a_nx^{n+2}+\sum_{n=0}^{\infty} a_nx^n.$$

Substituting into the given equation, we get from these and (6),

$$\sum_{n=2}^{\infty} n(n-1)a_nx^{n-2}+\sum_{n=1}^{\infty} na_nx^n+\sum_{n=0}^{\infty} 2a_nx^{n+2}+\sum_{n=0}^{\infty} a_nx^n=0,$$

(9) $$\sum_{n=2}^{\infty} n(n-1)a_nx^{n-2}+a_0+\sum_{n=1}^{\infty} (n+1)a_nx^n+\sum_{n=0}^{\infty} 2a_nx^{n+2}=0.$$

We are again faced with the need to combine these summations on differing powers of x. In the first sum we change to index of summation m and let $m-2=n$. In the third sum we change to index of summation k and let $k+2=n$. Then (9) becomes

$$\sum_{m=2}^{\infty} m(m-1)a_mx^{m-2}+a_0+\sum_{n=1}^{\infty} (n+1)a_nx^n+\sum_{k=0}^{\infty} 2a_kx^{k+2}=0,$$

$$\sum_{n=0}^{\infty} (n+2)(n+1)a_{n+2}x^n+a_0+\sum_{n=1}^{\infty} (n+1)a_nx^n+\sum_{n=2}^{\infty} 2a_{n-2}x^n=0.$$

We can combine terms for $n\geq 2$, since the last sum begins only at $n=2$. Collecting the earlier terms, we have

$$(2)(1)a_2+(3)(2)a_3x+a_0+2a_1x+\sum_{n=2}^{\infty}[(n+2)(n+1)a_{n+2}+(n+1)a_n+2a_{n-2}]x^n=0,$$

$$(a_0+2a_2)+2(a_1+3a_3)x+\sum_{n=2}^{\infty}[(n+2)(n+1)a_{n+2}+(n+1)a_n+2a_{n-2}]x^n=0.$$

Thus

$$a_0+2a_2=0,\qquad a_2=-\tfrac{1}{2}a_0;$$

$$a_1+3a_3=0,\qquad a_3=-\tfrac{1}{3}a_1;$$

and for $n\geq 2$,

$$a_{n+2}=-\frac{1}{(n+2)(n+1)}(2a_{n-2}+(n+1)a_n).$$

For $n=2$ this gives

$$a_4=-\frac{1}{4\cdot 3}(2a_0-\tfrac{3}{2}a_0)=-\tfrac{1}{24}a_0.$$

For $n=3$ this gives

$$a_5=-\frac{1}{5\cdot 4}(2a_1-4(\tfrac{1}{3})a_1)=-\tfrac{1}{20}(\tfrac{2}{3}a_1)=-\tfrac{1}{30}a_1.$$

For $n=4$, this gives

$$a_6=-\frac{1}{6\cdot 5}(2(-\tfrac{1}{2}a_0)+5(-\tfrac{1}{24})a_0)=\tfrac{29}{720}a_0.$$

For $n=5$, this gives

$$a_7=-\frac{1}{7\cdot 6}(2(-\tfrac{1}{3}a_1)+6(-\tfrac{1}{30}a_1))=\tfrac{13}{630}a_1.$$

Thus (3) becomes

$$\begin{aligned} y(x)&=a_0+a_1x-\tfrac{1}{2}a_0x^2-\tfrac{1}{3}a_1x^3-\tfrac{1}{24}a_0x^4-\tfrac{1}{30}a_1x^5+\tfrac{29}{720}a_0x^6+\tfrac{13}{630}a_1x^7+\ldots \\ &=a_0(1-\tfrac{1}{2}x^2-\tfrac{1}{24}x^4+\tfrac{29}{720}x^6+\ldots)+a_1(x-\tfrac{1}{3}x^3-\tfrac{1}{30}x^5+\tfrac{13}{630}x^7+\ldots). \\ &=a_0y_1(x)+a_1y_2(x). \end{aligned}$$

In this example, $y_1(x)$ and $y_2(x)$ are not recognizable as familiar functions. The reason for this is that they are *not* elementary functions! For small values of x, however, we can compute them to any desired accuracy by taking enough terms of the series, by Theorem 11.4.5.

If we were interested in values of $y(x)$ near $x_0\neq 0$, we would, instead of (3), use

(10) $$y(x)=\sum_{n=0}^{\infty}a_n(x-x_0)^n,$$

the Taylor series expansion near x_0. The next example illustrates that if we change variables by letting $t=x-x_0$, then the calculations are just as in the preceding examples.

9.4 EXAMPLE. Find a power series representation for the solution of

$$\frac{d^2y}{dx^2}+(x-1)\frac{dy}{dx}+y=0$$

in powers of $x-1$.

Solution. Let $t=x-1$. Then $\frac{dy}{dx}=\frac{dy}{dt}\frac{dt}{dx}=\frac{dy}{dt}$ and $\frac{d^2y}{dx^2}=\frac{d^2y}{dt^2}\frac{dt}{dx}=\frac{d^2y}{dt^2}$. Thus our original equation is equivalent to

$$\frac{d^2y}{dt^2}+t\frac{dy}{dt}+y=0.$$

We want a power series solution in powers of t now. So suppose that $y=\sum_{n=0}^{\infty}a_nt^n$. Then

$$t\frac{dy}{dt}=\sum_{n=1}^{\infty}na_nt^n$$

and

$$\frac{d^2y}{dt^2}=\sum_{n=2}^{\infty}n(n-1)a_nt^{n-2}.$$

Substituting into our given equation, we get

$$\sum_{n=2}^{\infty} n(n-1)a_n t^{n-2} + \sum_{n=1}^{\infty} na_n t^n + \sum_{n=0}^{\infty} a_n t^n = 0,$$

$$\sum_{n=2}^{\infty} n(n-1)a_n t^{n-2} + \sum_{n=1}^{\infty} (n+1)a_n t^n + a_0 = 0.$$

In the first summation we switch to summation index m and then let $m-2=n$; so $m=n+2$ and $m-1=n+1$. We get

$$\sum_{n=0}^{\infty} (n+1)(n+2)a_{n+2} t^n + \sum_{n=1}^{\infty} (n+1)a_n t^n + a_0 = 0,$$

$$\sum_{n=1}^{\infty} [(n+1)(n+2)a_{n+2} + (n+1)a_n] t^n + a_0 + (1 \cdot 2)a_2 = 0.$$

We have then $a_2 = -\frac{1}{2}a_0$, and for $n \geq 1$, $a_{n+2} = -\dfrac{1}{n+2}\, a_n$. Thus

$$a_3 = -\tfrac{1}{3}a_1,$$

$$a_4 = -\tfrac{1}{4}a_2 = \tfrac{1}{8}a_0,$$

$$a_5 = -\tfrac{1}{5}a_3 = \frac{1}{3 \cdot 5}\, a_1,$$

$$a_6 = -\tfrac{1}{6}a_4 = -\frac{1}{8 \cdot 6}\, a_0$$

We therefore have

$$\begin{aligned} y(t) &= a_0 + a_1 t - \tfrac{1}{2}a_0 t^2 - \tfrac{1}{3}a_1 t^3 + \tfrac{1}{8}a_0 t^4 + \frac{1}{3 \cdot 5}\, a_1 t^5 - \frac{1}{8 \cdot 6}\, a_0 t^6 + \dots \\ &= a_0\left(1 - \tfrac{1}{2}t^2 + \frac{1}{2^2 \cdot 2!}\, t^4 - \frac{1}{2^3 \cdot 3!}\, t^6 + - \dots\right) \\ &\quad + a_1\left(t - \tfrac{1}{3}t^3 + \frac{1}{3 \cdot 5}\, t^5 - + \dots\right) \\ &= a_0 \sum_{i=0}^{\infty} \frac{(-1)^i t^{2i}}{2^i i!} + a_1 \sum_{j=0}^{\infty} \frac{(-1)^j t^{2j+1}}{1 \cdot 3 \cdot 5 \dots (2j+1)}. \end{aligned}$$

Hence,

$$\begin{aligned} y(x) &= a_0 \sum_{i=0}^{\infty} \frac{(-1)^i (x-1)^{2i}}{2^i i!} + a_1 \sum_{j=0}^{\infty} \frac{(-1)^j (x-1)^{2j+1}}{1 \cdot 3 \cdot 5 \dots (2j+1)} \\ &= a_0 y_1(x) + a_1 y_2(x). \end{aligned}$$

Again, we do not recognize $y_1(x)$ and $y_2(x)$ as elementary functions, but they are defined by convergent alternating series, so they can be approximated to any desired degree of accuracy. See Theorem 11.4.5.

At this point, you may have the idea that power series afford a universal solution method for differential equations. The problem with this method is that if you want to calculate values of the series solution $y = y(x) = \sum_{n=0}^{\infty} a_n (x - x_0)^n$ for certain values of x, then you are faced with the task of finding the sums of several infinite series of constants. Unless these happen to be geometric series, that is a formidable task for which the methods of Chapter 11 are of little help. The best way to proceed is to numerically approximate $y(x_1)$ for values x_1 of interest by taking a finite partial sum $\sum_{n=0}^{k} a_n(x_1-x_0)^n$.

An often shorter approach is to use a *numerical solution method* for the given differential equation in the first place. Such methods generate a list of *approximate* values of the solution $y=y(x)$ at a number of points $x_1, x_2, \ldots, x_n$. Such a list is reminiscent of the tables of values of transcendental functions (or values from a pocket calculator) that provide a large part of our knowledge about functions like sine, exponentials, and logarithms. In recent years the development of electronic calculators and small computers has enormously increased the practical importance of numerical solution methods, some of which have been known theoretically (but difficult to apply in concrete examples) for years. You will learn more about these if you take later courses in differential equations and numerical analysis. In the rest of this section we will just give you an introduction to the idea of numerical procedures for solving differential equations.

Euler developed the following *linear* approximation scheme for the unique solution $y=y(x)$ to the first order initial value problem

(11) $$\frac{dy}{dx}=f(x, y), \qquad y(x_0)=y_0.$$

This method is also called **Euler's method** or the **tangent-line** method. We know x_0 and y_0, so we know the slope $y'(x_0)=f(x_0, y_0)$ of the tangent line to the solution $y=y(x)$ at (x_0, y_0). If x_1 is close to x_0, then we can use the tangent approximation (cf. Sections 4.2 and 4.3) to approximate y at x_1. We have

(12) $$\begin{aligned} y(x_1) &\approx y_0+y'(x_0)(x_1-x_0) \\ &\approx y_0+f(x_0, y_0)(x_1-x_0)=y_1. \end{aligned}$$

Once we have the approximation y_1 for $y(x_1)$, we can approximate $y'(x_1)$ by

$$y'(x_1)\approx f(x_1, y_1).$$

This approximates the slope of the tangent line to $y=y(x)$ at $(x_1, y(x_1))$. Then if x_2 is near x_1, we use the tangent approximation

$$\begin{aligned} y(x_2) &\approx y_1+y'(x_1)(x_2-x_1) \\ &\approx y_1+f(x_1, y_1)(x_2-x_1)=y_2. \end{aligned}$$

We can now continue this scheme, generating in general the approximation

(13) $$\boxed{y(x_{n+1})\approx y_n+f(x_n, y_n)(x_{n+1}-x_n)=y_{n+1}.}$$

If in (13) we take steps of uniform length $h=x_{i+1}-x_i$, then we have

(14) $$y(x_{n+1})\approx y_n+hf(x_n, y_n)=y_{n+1}.$$

To illustrate the method, we consider a very simple first order linear equation that we can readily solve for $y=y(x)$ explicitly. This will allow us to study the accuracy of our Euler approximation.

9.5 EXAMPLE. Find approximate values of y at x_0, x_1, x_2, x_3, x_4, and x_5 if

$$\frac{dy}{dx} = x - 2y, \qquad y(x_0) = 1,$$
$$x_0 = 0, \qquad x_{i+1} = x_i + h, \qquad h = 0.1.$$

Solution. We have $y(x_0) = y_0 = 1$. From (12),

$$y_1 = y_0 + (x_0 - 2y_0)(0.1) = 1 + (-2)(0.1) = 0.8000.$$

From (13) or (14) with $n = 1$, 2, 3, and 4:

$$\begin{aligned}
y_2 &= y_1 + (x_1 - 2y_1)(0.1)\\
&= 0.8000 + (0.1000 - 1.6000)(0.1)\\
&= 0.8000 - 0.1500 = 0.6500.\\
y_3 &= y_2 + (x_2 - 2y_2)(0.1)\\
&= 0.6500 + (0.2000 - 1.300)(0.1)\\
&= 0.6500 - 0.1100 = 0.5400,\\
y_4 &= y_3 + (x_3 - 2y_3)(0.1)\\
&= 0.5400 + (0.3000 - 1.0800)(0.1)\\
&= 0.5400 - 0.0780 = 0.4620\\
y_5 &= y_4 + (x_4 - 2y_4)(0.1)\\
&= 0.4620 + (0.4000 - 0.9240)(0.1)\\
&= 0.4620 - 0.0524 = 0.4096.
\end{aligned}$$

We can easily solve this first order linear equation by the method of Section 3, since it has integrating factor e^{2x}. We get

$$\frac{d}{dx}(e^{2x}y) = xe^{2x}$$
$$y = \tfrac{1}{2}x - \tfrac{1}{4} + ce^{-2x}.$$

When $x = x_0 = 0$, $y = 1$, so we get $c = \frac{5}{4}$. Thus the exact solution is $y = \frac{5}{4}e^{-2x} + \frac{1}{2}x - \frac{1}{4}$. In Figure 9.1 we illustrate graphically how our points $P_i(x_i, y_i)$

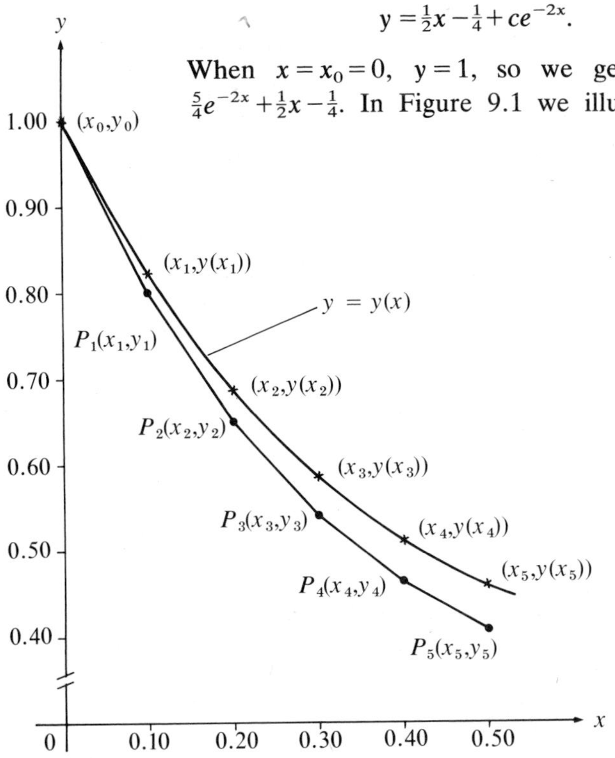

FIGURE 9.1

compare with the corresponding points $(x_i, y(x_i))$ on the graph of $y=y(x)$. Connecting these points with the line segments shown produces an approximate graph of $y=y(x)$. In Table 9.1 we show the approximate values of y_i in comparison with the values of $y(x_i)$ to four decimal places. We show also the percent error we have if we use y_i to approximate $y(x_i)$.

Table 9.1

x_n	$y(x_n)=\frac{5}{4}e^{-2x_n}+\frac{1}{2}x_n-\frac{1}{4}$	y_n	$\lvert y_n-y(x_n)\rvert$	% error
0	1.0000	1.0000	0	0.00
0.1	0.8234	0.8000	0.0234	2.84
0.2	0.6879	0.6500	0.0379	5.51
0.3	0.5860	0.5400	0.0460	7.85
0.4	0.5117	0.4620	0.0497	9.71
0.5	0.4598	0.4096	0.0502	10.92

As you can see, the error is sizeable. The error can be reduced if a smaller step size h is used (see Exercise 28 below). But the Euler method is basically not an extremely accurate method. For this reason, a *refinement* of Euler's method or more sophisticated techniques of numerical analysis are employed in practice. An improved version of Euler's approach consists in replacing $f(x_n, y_n)$ in (14) by the *average* of $f(x_n, y_n)$ and $f(x_{n+1}, y_{n+1})$ as computed using y_{n+1} in (14). If we call the new points so generated $\bar{y}_1, \bar{y}_2, \ldots, \bar{y}_n, \ldots$, then we have

(15)
$$\begin{aligned}\bar{y}_{n+1} &= \bar{y}_n+\tfrac{1}{2}(x_{n+1}-x_n)[f(x_n, \bar{y}_n)+f(x_{n+1}, y_{n+1})]\\ &\quad \text{where } y_{n+1} \text{ is computed from } \bar{y}_n \text{ by (14)}\\ &= \bar{y}_n+\tfrac{1}{2}(x_{n+1}-x_n)[f(x_n, \bar{y}_n)+f(x_{n+1}, \bar{y}_n+(x_{n+1}-x_n)f(x_n, \bar{y}_n))]\end{aligned}$$

Observe that (15) requires a good deal more calculation than (14). In particular, it requires *two* evaluations of $f(x, y)$ at each step instead of one. If you have a hand-held calculator with a memory, (15) is a feasible scheme to use, however, at least when f is not too complicated. A programmable hand-held calculator makes (15) easy and quick to implement.

9.6 EXAMPLE. Repeat Example 9.5 using the improved Euler method (15).

Solution. We still have $y(x_0)=\bar{y}_0=1$. Continuing from here, we first find y_{n+1} at each stage, and then we use (15) to find $\bar{y}_{n+1}$.

$$\begin{aligned}\bar{y}_1 &= \bar{y}_0+\tfrac{1}{2}(0.1)(x_0-2\bar{y}_0+x_1-2y_1), \quad \text{where} \quad y_1=0.8000\\ &= 1.000+(0.05)(-2.0000-1.5000)=0.8250.\\ y_2 &= \bar{y}_1+0.1(x_1-2\bar{y}_1)\\ &= 0.8250+0.1(0.1000-1.6500)=0.6700.\\ \bar{y}_2 &= \bar{y}_1+\tfrac{1}{2}(0.1)(x_1-2\bar{y}_1+x_2-2y_2)\\ &= 0.8250+0.05(-1.5500-1.1400)=0.6905.\\ y_3 &= \bar{y}_2+(0.1)(x_2-2\bar{y}_2)\\ &= 0.6905+0.1(0.2000-1.3810)=0.5724.\end{aligned}$$

$$\bar{y}_3 = \bar{y}_2 + \tfrac{1}{2}(0.1)(x_2 - 2\bar{y}_2 + x_3 - 2y_3)$$
$$= 0.6905 + (0.05)(-1.1810 - 0.8448) = 0.5892.$$
$$y_4 = \bar{y}_3 + (0.1)(x_3 - 2\bar{y}_3)$$
$$= 0.5892 + 0.1(-0.8784) = 0.5014.$$
$$\bar{y}_4 = \bar{y}_3 + \tfrac{1}{2}(0.1)(x_3 - 2\bar{y}_3 + x_4 - 2y_4)$$
$$= 0.5892 + (0.05)(-0.8784 - 0.6028) = 0.5151.$$
$$y_5 = \bar{y}_4 + (0.1)(x_4 - 2\bar{y}_4)$$
$$= 0.5151 + 0.1(-0.6302) = 0.4521.$$
$$\bar{y}_5 = \bar{y}_4 + \tfrac{1}{2}(0.1)(x_4 - 2\bar{y}_4 + x_5 - 2y_5)$$
$$= 0.5151 + (0.05)(-0.6302 - 0.4042) = 0.4634.$$

Table 9.2 shows the marked improvement in accuracy that the improved Euler method yields in this example, and is suggestive of the gain in accuracy usually afforded by this method.

Table 9.2

x_n	$y(x_n)$	$\bar{y}_n$	$\lvert\bar{y}_n - y(x_n)\rvert$	% error
0	1.0000	1.0000	0	0.00
0.1	0.8234	0.8250	0.0016	0.19
0.2	0.6879	0.6905	0.0026	0.38
0.3	0.5860	0.5892	0.0032	0.55
0.4	0.5117	0.5151	0.0034	0.66
0.5	0.4598	0.4634	0.0036	0.78

While the error continues to increase as we get farther away from x_0, the percent error remains less than 1% throughout. For many computational purposes this is adequate accuracy.

EXERCISES 8.9

In Exercises 1 to 14, find the general solution of the given differential equation in power series form. In Exercises 1 to 4, compare your solution with the explicit solution obtained using the methods of the preceding sections.

1 $\dfrac{d^2y}{dx^2} - 4y = 0$, about $x_0 = 0$.

2 $\dfrac{d^2y}{dx^2} + 4\dfrac{dy}{dx} + 4y = 0$, about $x_0 = 0$.

3 $\dfrac{d^2y}{dx^2} + 4y = 0$, about $x_0 = 0$.

4 $\dfrac{d^2y}{dx^2} + 9y = 0$ about $x_0 = 0$.

5 $\dfrac{d^2y}{dx^2} + x\dfrac{dy}{dx} + 2y = 0$, about $x_0 = 0$.

6 $\dfrac{d^2y}{dx^2} - xy' - y = 0$, about $x_0 = 0$.

7 $y'' + xy' + (x^2 + 2)y = 0$, about $x_0 = 0$.

8 $y'' + xy' + (3x + 2)y = 0$, about $x = 0$.

9 $(2x + 1)y'' + y' + 2y = 0$, near $x_0 = 0$.

10 $(1 + x)y'' - y = 0$, near $x_0 = 0$.

11 $(x^2 + 1)y'' + x\dfrac{dy}{dx} + xy = 0$, about $x_0 = 0$.

12 $y'' - 2(x - 1)y' - y = 0$, about $x_0 = 1$.

13 $y''-xy=0$, about $x_0=1$. [This equation is known as the *Airy equation*, after the English mathematician and astronomer George Airy (1801–1892).]

14 $y''-2xy+\lambda y=0$, about $x_0=0$. [This equation is known as the *Hermite equation*, after the French mathematician Charles Hermite (1822–1901). If λ is an even integer, then one of the basic solutions is a polynomial that gives rise to the *Hermite polynomial.*]

15 $(1-x^2)y''-2xy'+\lambda(\lambda+1)y=0$ about $x_0=0$. [This is called the *Legendre equation*, after the French mathematician Adrien Legendre (1752–1833). If λ is a positive integer, then one of the basic solutions is a polynomial that gives rise to the *Legendre polynomial.*]

16 $(1-x^2)y''-xy'+\lambda y=0$ about $x_0=0$. [This is called the *Tchebysheff equation*, after the Russian mathematician Pafnuti L. Tchebysheff (1821–1894). If λ is the square of a positive integer, then one of the basic solutions is a polynomial that gives rise to the *Tchebysheff polynomial.*]

In Exercises 17 to 20, find the power series solution about $x_0=0$ for the given initial value problem.

17 $y''+xy'+2y=0$, $y(0)=0$, $y'(0)=1$.

18 $y''-xy'-y=0$, $y(0)=0$, $y'(0)=1$.

19 $(x^2+1)y''+xy'+xy=0$, $y(0)=2$, $y'(0)=3$.

20 $(x^2-1)y''+3xy'+xy=0$, $y(0)=4$, $y'(0)=6$.

In Exercises 21 to 24, find approximate values for the solution $y(x)$ to the given initial value problem at the points listed. Use Euler's method. In Exercises 21 and 22, find explicit formulas for the solution and compare with your Euler approximations. Carry your calculations to three places (four if you have a hand-held calculator).

21 $\dfrac{dy}{dx}-4y=1-x$, $y(0)=1$ at $x=0, 0.1, 0.2, 0.3, 0.4, 0.5$.

22 $\dfrac{dy}{dx}=2x+y$, $y(0)=1$ at $x=0, 0.1, 0.2, 0.3, 0.4, 0.5$.

23 $y\dfrac{dy}{dx}=e^{-x^2}$, $y(0)=1$ at $x=0, 0.1, 0.2, 0.3, 0.4, 0.5$.

24 $\dfrac{dy}{dx}=x^2y$, $y(0)=1$ at $x=0, 0.1, 0.2, 0.3, 0.4, 0.5$.

25 (a) Repeat Exercise 21 using the improved Euler method.
(b) Compare with the exact solution values.

26 (a) Repeat Exercise 22 using the improved Euler method.
(b) Compare with the exact solution values.

27 Work Exercise 21 using the Euler method and $h=0.05$ to approximate $y(x)$ for $x=0.1$ and $x=0.2$. Compare with the results of Exercise 25.

28 Work Example 9.5 using the Euler method and $h=0.05$ to approximate $y(x)$ for $x=0.1$ and $x=0.2$. Compare with the results of Example 9.6.

29 Euler's method extends to second order initial value problems

$$y''=f(x, y, y'),\ y(x_0)=y_0,\ y'(x_0)=v_0$$

as follows. Let $v=y'$. Then the given equation is equivalent to the system of two first order equations

$$\begin{cases} v=y' \\ v'=f(x, y, v) \end{cases}$$

with initial values $y(x_0)=y_0$, $v(x_0)=v_0$. Show that Euler's approach leads to

$$\begin{cases} y_1 = y_0 + (x_1 - x_0)v_0 \\ v_1 = v_0 + (x_1 - x_0)f(x_0, y_0, v_0) \end{cases}$$

as second stage approximations.

30 In Exercise 29, show that the general formula is

$$\begin{cases} y_{n+1} = y_n + hy_n \\ v_{n+1} = v_n + hf(x_n, y_n, v_n) \end{cases}$$

where $h = x_{n+1} - x_n$.

31 Apply the technique of Exercises 29 and 30 to approximate the solution $y = y(x)$ to

$$\frac{d^2y}{dx^2} + \frac{2x}{x^2-1}\frac{dy}{dx} - \frac{6}{x^2-1}y = 0,$$

where $y(0) = 1$ and $y'(0) = 0$, at $x = 0, 0.1, 0.2, 0.3, 0.4, 0.5$. Carry calculations to three decimal places.

32 Repeat Exercise 31 for

$$\frac{d^2y}{dx^2} + \frac{1}{x}\frac{dy}{dx} - x^2 - 2y = 0, \qquad y(1) = -2, \qquad y'(1) = 3.$$

33 In Exercise 9, use Theorem 9.1 to show that the power series solution converges at least on the interval $(-\frac{1}{2}, \frac{1}{2})$.

34 In Exercise 10, use Theorem 9.1 to show that the power series solution converges at least on the interval $(-1, \cdot 1)$.

35 Solve $y'' + xy' + 2y = x$ by assuming a power series particular solution $y_p(x)$ (cf. Exercise 5).

36 Solve $y'' - xy' - y = x$ by assuming a power series particular solution $y_p(x)$.

REVIEW EXERCISES 8.10

In Exercises 1 to 15, solve the given differential equation.

1 $\dfrac{dy}{dx} = y^3e^{-2x}$.

2 $x \cos y - (x^2+1)\sin y \dfrac{dy}{dx} = 0$, $y(1) = 0$.

3 $\sin x \dfrac{dy}{dx} + y \cos x = 0$.

4 $(x^2+1)\dfrac{dy}{dx} - 4xy = x$, $y(0) = 1$.

5 $x\dfrac{dy}{dx} - y = x^4y^3$, $y(1) = 2$.

6 $(3x^2 - 2y^2)\,dx + (1 - 4xy)\,dy = 0$.

7 $(x^2 - 3y^2)\,dx + 2xy\,dy = 0$.

8 $\dfrac{d^2y}{dx^2} - 3\dfrac{dy}{dx} + 2y = 0$.

9 $\dfrac{d^2y}{dx^2} - 6\dfrac{dy}{dx} + 9y = 0$.

10 $\dfrac{d^2y}{dx^2} + \dfrac{dy}{dx} + 2y = 0$.

11 $\dfrac{d^2y}{dx^2} - 3\dfrac{dy}{dx} + 2y = 0$, $y(0) = 1$, $y'(0) = 3$. (See Exercise 8.)

12 $\dfrac{d^2y}{dx^2}+\dfrac{dy}{dx}-6y=xe^{-x}$.

13 $\dfrac{d^2y}{dx^2}+3\dfrac{dy}{dx}-y=\sin 2x$.

14 $\dfrac{d^2y}{dx^2}+5\dfrac{dy}{dx}-6y=2e^{3x}$.

15 $\dfrac{d^2y}{dx^2}+5\dfrac{dy}{dx}-6y=3e^{2x}$.

16 Find a power series about $x_0=0$ for $y=y(x)$ if $\dfrac{d^2y}{dx^2}-9y=0$. Check by solving the equation for y.

17 Find a power series about $x_0=0$ for $y=y(x)$ if

$$(x^2+2)\frac{d^2y}{dx^2}+5x\frac{dy}{dx}+4y=0.$$

18 Find a power series about $x_0=1$ for $y=y(x)$ if $x^2y''+xy'+y=0$.

19 Use Euler's method to find approximate values of $y=y(x)$ at $x=0$, 0.1, 0.2, 0.3, 0.4, and 0.5 if

$$\frac{dy}{dx}=xy^2, \qquad y(0)=1.$$

20 Redo Exercise 19 using the improved Euler method.

21 Suppose that the Indians who sold Manhattan for \$24 in 1624 had invested their money in a savings account paying interest at the continuously compounded rate of 5%. How much would their account have contained in 1974, assuming no additional deposits and no withdrawals?

22 A cup of soup is made by pouring water at 100°C over soup mix. The resulting mixture has an initial temperature of 98°C. It is placed in a room maintained at 20°C. After two minutes, its temperature is down to 85°C. When will it reach a temperature of 59°C?

23 An industrial plant starts dumping a pollutant into an initially unpolluted lake at the rate of 10 gallons per day. Each day, the lake loses 1000 gallons of its contents of ten million gallons by evaporation and receives 990 gallons of pure water from a small stream. Find a formula for the number of gallons of pollutant in the lake after t days. If a concentration of 10 parts pollutant per million parts of fluid is enough to kill fish, then how long will it take for fish in the lake to begin dying?

24 In the undamped system of Figure 8.4, suppose that $m=1$ kg and a force of 32 newtons is required to move the mass two meters to the right. Suppose that a force $F=5\cos t$ is applied to the system. Find a formula for $x(t)$, assuming that the system starts from its equilibrium position at rest.

25 In the system of Figure 8.4, suppose that the damping constant is 2 and $m=\frac{1}{2}$ kg. A force of 13 newtons moves the mass 2 meters to the right. If the system begins from rest in its equilibrium position, then find a formula for $x(t)$ at time t if an external force $F=5\cos 2t$ is applied. Find the resonant frequency.

26 A circuit consists of a resistor, inductor, and capacitor with $R=1.0\times10^2$, $C=4.0\times10^{-6}$, and $L=1.0\times10^{-1}$. If $q(0)=1.0\times10^{-6}$ and $i(0)=0$, then find $q(t)$ and $i(t)$.

27 Suppose that in the circuit of Figure 8.9, $E(t)=110\sin 120\pi t$, $R=3.0\times10^3$, $C=2.5\times10^{-6}$, and $L=10$. Find the steady state current.

28 (a) If $z_1=2+3i$ and $z_2=-1+2i$, then compute z_1+z_2, z_1z_2, and $z_1\div z_2$.
(b) Describe the geometric effect of multiplying $-3+2i$ by $\sqrt{3}-i$.
(c) Compute $(-1+\sqrt{3}i)^4$.
(d) Find the six complex sixth roots of 1.
(e) Find the sixth roots of -64.

29 (a) Compute $e^{\pi i/6}$ and $e^{\ln 3+\pi i/2}$.
(b) Find the sixth derivative of $e^{x/4}(\cos\frac{1}{4}x+i\sin\frac{1}{4}x)$.

9 LINEAR DIFFERENTIAL EQUATIONS

0 INTRODUCTION

As suggested in Section 8.0, the class of linear differential equations has the best developed theory and techniques of solution. The present chapter is devoted to presenting that theory and deriving the solution techniques from it. As in Sections 6 and 7 of the previous chapter, the emphasis is on linear differential equations with constant coefficients. This time, however, we consider equations having arbitrary order n.

Not surprisingly, perhaps, in view of the linearity of the equations, we need some additional linear algebra to build the theory. This is supplied in Section 1, after we have set the scene for it by discussing n-th order linear differential operators and equations.

In Section 2 we proceed to develop solution techniques, which are natural extensions of those in Section 8.6, for solving homogeneous linear differential equations of order n. In Section 3 we extend the method of undetermined coefficients of Section 8.7 to find a particular solution of an n-th order nonhomogeneous linear equation, and present a new method for finding a particular solution that applies to a much wider class of equations than the method of undetermined coefficients. In the final section we introduce the Laplace transform as an aid in solving initial value problems directly.

1 LINEAR DIFFERENTIAL OPERATORS AND EQUATIONS OF ORDER n[1]

The notion of linear differential operators of orders 1 and 2 given in Sections 8.3 and 8.6 extends immediately to order n.

1. Section 1.7 is needed for the part of this section that follows Example 1.7.

1.1 **DEFINITION.** The **general n-th order linear differential operator** $L: C^\infty(\mathbf{R}) \to C^\infty(\mathbf{R})$ is defined by

(1) $$L(y) = \frac{d^n y}{dx^n} + p_{n-1}(x)\frac{d^{n-1}y}{dx^{n-1}} + \ldots + p_1(x)\frac{dy}{dx} + p_0(x)y$$

where $p_0(x)$, $p_1(x)$, $p_2(x), \ldots, p_{n-1}(x)$ are functions in $C^\infty(\mathbf{R})$. The **general n-th order linear differential equation** is

(2) $$L(y)(x) = s(x)$$

where L is given by (1) and $s(x) \in C^\infty(\mathbf{R})$.

Our purpose in this chapter is to develop methods of solution for (2) that parallel the ones given in Sections 8.6 and 8.7 for the case $n = 2$. First, we can characterize the set of all solutions of (2) in the same way we did in the second order case.

1.2 **THEOREM.** The set S of all solutions of (2) is given by

$$S = \{y_0(x) + y_p(x) \mid y_0(x) \text{ a solution of } L(y) = 0\},$$

where $y_p(x)$ is one particular solution to (2).

Proof. First we observe that L given by (1) is a linear transformation on the real vector space $C^\infty(\mathbf{R})$ into itself. Certainly $L(y) \in C^\infty(\mathbf{R})$ if $y \in C^\infty(\mathbf{R})$. As for linearity, we have

$$\begin{aligned}
L(y_1 + y_2) &= \frac{d^n(y_1+y_2)}{dx^n} + p_{n-1}(x)\frac{d^{n-1}(y_1+y_2)}{dx^{n-1}} + \ldots \\
&\quad + p_1(x)\frac{d(y_1+y_2)}{dx} + p_0(x)(y_1+y_2) \\
&= \frac{d^n y_1}{dx^n} + \frac{d^n y_2}{dx^n} + p_{n-1}(x)\left(\frac{d^{n-1}y_1}{dx^{n-1}} + \frac{d^{n-1}y_2}{dx^{n-1}}\right) \\
&\quad + \ldots + p_1(x)\left(\frac{dy_1}{dx} + \frac{dy_2}{dx}\right) + p_0(x)(y_1+y_2) \\
&= \frac{d^n y_1}{dx^n} + p_{n-1}(x)\frac{d^{n-1}y_1}{dx^{n-1}} + \ldots + p_1(x)\frac{dy_1}{dx} + p_0(x)y_1 \\
&\quad + \frac{d^n y_2}{dx^n} + p_{n-1}(x)\frac{d^{n-1}y_2}{dx^{n-1}} + \ldots + p_1(x)\frac{dy_2}{dx} + p_0(x)y_2 \\
&= L(y_1) + L(y_2), \\
L(cy) &= \frac{d^n(cy)}{dx^n} + p_{n-1}(x)\frac{d^{n-1}(cy)}{dx^{n-1}} + \ldots + p_1(x)\frac{d(cy)}{dx} + p_0(x)(cy) \\
&= c\frac{d^n y}{dx^n} + cp_{n-1}(x)\frac{d^{n-1}y}{dx^{n-1}} + \ldots + cp_1(x)\frac{dy}{dx} + cp_0(x)y \\
&= cL(y).
\end{aligned}$$

So now we can argue just as in Therorems 8.3.5 and 8.6.3 to show that S has the form claimed. QED

Thus, just as in the first and second order cases, we see that the linearity of L reduces the solution of (2) to the following two steps.

(a) Find the complete solution of the associated homogeneous problem $L(y)=0$.

(b) Find one particular solution of (2).

Section 2 considers (a) and Section 3 considers (b) for constant coefficient equations (2), those corresponding to polynomial operators L given by

(3) $$L(y)=\frac{d^n y}{dx^n}+p_{n-1}\frac{d^{n-1}y}{dx^{n-1}}+\ldots+p_1\frac{dy}{dx}+p_0 y$$

where $p_0, p_1, \ldots, p_{n-1} \in \boldsymbol{R}$.

To carry out Step (a), we need to take advantage of the fact that the set of all solutions to the associated homogeneous n-th order linear problem

(4) $$L(y)=0$$

is a real vector space.

1.3 **THEOREM.** The set $Ker\,L$ of all solutions to (4) is a real vector space called the **kernel** of the linear transformation $L: C^\infty(\boldsymbol{R}) \to C^\infty(\boldsymbol{R})$.

Proof. Since $Ker\,L \subseteq C^\infty(\boldsymbol{R})$, all required identities for a real vector space must hold for all functions in $Ker\,L$, since they hold for all functions in the real vector space $C^\infty(\boldsymbol{R})$ by Proposition 8.3.1. So all we need to do to see that $Ker\,L$ is a real vector space is to show that $Ker\,L$ is closed under the vector operations of addition and multiplication by real scalars c. But since L is a linear transformation, this is immediate. If $y_1, y_2 \in Ker\,L$, then $y_1+y_2 \in Ker\,L$ since

$$L(y_1+y_2)=L(y_1)+L(y_2)=0+0=0.$$

Likewise, if $c \in \boldsymbol{R}$ and $y \in Ker\,L$, then $cy \in Ker\,L$ since

$$L(cy)=cL(y)=c0=0.$$ QED

To describe $Ker\,L$ efficiently, we need the following existence-uniqueness theorem, whose proof is beyond our scope here.

1.4 **THEOREM (EXISTENCE-UNIQUENESS THEOREM FOR N-TH ORDER LINEAR INITIAL VALUE PROBLEM).** If $(y_0, y_1, \ldots, y_{n-1}) \in \boldsymbol{R}^n$, then there is one and only one function $y(x) \in C^\infty(\boldsymbol{R})$ that is a solution to (4) and satisfies the initial conditions

$$y(x_0)=y_0,\ y'(x_0)=y_1,\ y''(x_0)=y_2, \ldots, y^{(n-1)}(x_0)=y_{n-1}.$$

We can use this theorem to set up a linear transformation from the real vector space $Ker\,L$ onto the real vector space $\boldsymbol{R}^n$ as follows.

1.5 **LEMMA.** Let $x_0 \in \boldsymbol{R}$. Then the map $T_{x_0}: Ker\,L \to \boldsymbol{R}^n$ defined by

$$T_{x_0}(y)=\left(y(x_0), \frac{dy}{dx}(x_0), \frac{d^2y}{dx^2}(x_0), \ldots, \frac{d^{n-1}y}{dx^{n-1}}(x_0)\right)$$

is a linear transformation that maps *onto* $\boldsymbol{R}^n$ (that is, the range of T_{x_0} is Ran $T_{x_0} = \boldsymbol{R}^n$) and is a *1-1 mapping* (that is, $T_{x_0}(y_1) = T_{x_0}(y_2)$ can happen only if y_1 and y_2 are the same function: $y_1(x) = y_2(x)$ for all x.)

Proof. First, we must show that T_{x_0} is a linear transformation. This follows from the linearity of the differential operators $D, D^2, \ldots, D^{n-1}$, for if $y_1(x), y_2(x) \in$ *Ker L*, then

$$\begin{aligned}
T_{x_0}(y_1+y_2) &= \left((y_1+y_2)(x_0), \right. \\
&\qquad \left. \frac{d(y_1+y_2)}{dx}(x_0), \frac{d^2(y_1+y_2)}{dx^2}(x_0), \ldots, \frac{d^{n-1}(y_1+y_2)}{dx^{n-1}}(x_0)\right) \\
&= \left(y_1(x_0)+y_2(x_0), \left(\frac{dy_1}{dx}+\frac{dy_2}{dx}\right)(x_0), \left(\frac{d^2y_1}{dx^2}+\frac{d^2y_2}{dx^2}\right)(x_0), \right. \\
&\qquad \left. \ldots, \left(\frac{d^{n-1}y_1}{dx^{n-1}}+\frac{d^{n-1}y_2}{dx^{n-1}}\right)(x_0)\right) \\
&= \left(y_1(x_0)+y_2(x_0), \frac{dy_1}{dx}(x_0)+\frac{dy_2}{dx}(x_0), \frac{d^2y_1}{dx^2}(x_0)+\frac{d^2y_2}{dx^2}(x_0), \right. \\
&\qquad \left. \ldots, \frac{d^{n-1}y_1}{dx^{n-1}}(x_0)+\frac{d^{n-1}y_2}{dx^{n-1}}(x_0)\right) \\
&= \left(y_1(x_0), \frac{dy_1}{dx}(x_0), \frac{d^2y_1}{dx^2}(x_0), \ldots, \frac{d^{n-1}y_1}{dx^{n-1}}(x_0)\right) \\
&\quad + \left(y_2(x_0), \frac{dy_2}{dx}(x_0), \frac{d^2y_2}{dx^2}(x_0), \ldots, \frac{d^{n-1}y_2}{dx^{n-1}}(x_0)\right) \\
&= T_{x_0}(y_1)+T_{x_0}(y_2).
\end{aligned}$$

Similarly, if $c \in \boldsymbol{R}$ and $y \in$ *Ker L*, then

$$\begin{aligned}
T_{x_0}(cy) &= \left(cy(x_0), \frac{d(cy)}{dx}(x_0), \frac{d^2(cy)}{dx^2}(x_0), \ldots, \frac{d^{n-1}(cy)}{dx^{n-1}}(x_0)\right) \\
&= \left(cy(x_0), c\frac{dy}{dx}(x_0), c\frac{d^2y}{dx^2}(x_0), \ldots, c\frac{d^{n-1}y}{dx^{n-1}}(x_0)\right) \\
&= c\left(y(x_0), \frac{dy}{dx}(x_0), \frac{d^2y}{dx^2}(x_0), \ldots, \frac{d^{n-1}y}{dx^{n-1}}(x_0)\right) \\
&= cT_{x_0}(y).
\end{aligned}$$

Next, we have to show that T_{x_0} is 1–1 and onto $\boldsymbol{R}^n$. But this follows from Theorem 1.4. Let $(z_0, z_1, \ldots, z_{n-1})$ be *any* vector in $\boldsymbol{R}^n$. According to Theorem 1.4, there is one and only one function $y(x) \in$ *Ker L* such that

$$T_{x_0}(y) = (z_0, z_1, \ldots, z_{n-1}).$$

Since there *is* one such function $y(x)$, we see that T_{x_0} is onto. Since there is *only one* such function $y(x)$, we see that T_{x_0} is 1–1, for if

$$T_{x_0}(y_1) = T_{x_0}(y_2) = (z_0, z_1, \ldots, z_{n-1}),$$

then by uniqueness in Theorem 1.4, $y_1(x) = y_2(x)$. QED

A linear transformation T that is 1–1 and onto connects two real vector spaces, which are *indistinguishable* algebraically. This is true because the elements of the two spaces match up exactly under the (1–1) onto map T,

$$\boldsymbol{v} \leftrightarrow T(\boldsymbol{v}),$$

and since T preserves the operations of addition and multiplication by scalars, the alter egos $T(\boldsymbol{v})$ of the vectors $\boldsymbol{v}$ in the first space combine exactly as do the vectors $\boldsymbol{v}$ in the first space. Algebraic structures whose forms are identical carry the term *isomorphic*, derived from the Greek words for "same" or "equal" (cf. isosceles triangles) and "form" (cf. metamorphosis).

1.6 **DEFINITION.** Two real vector spaces V and W are called **isomorphic** if there is a 1–1 onto linear transformation $T: V \to W$. Such a T is called an **isomorphism**.

Thus T_{x_0} in Lemma 1.5 is an isomorphism. Any invertible linear transformation $T: V \to W$ is an isomorphism.

1.7 **EXAMPLE.** Show that if $T: V \to W$ is a linear transformation which has an inverse map $T^{-1}: W \to V$ such that $T \circ T^{-1} = I_W$ and $T^{-1} \circ T = I_V$, then T and T^{-1} are isomorphisms.

Proof. T must be 1–1 and onto since T^{-1} exists. For if $T(\boldsymbol{v}_1) = T(\boldsymbol{v}_2)$, then we can apply T^{-1} to get

$$\begin{aligned} T^{-1} \circ T(\boldsymbol{v}_1) &= T^{-1} \circ T(\boldsymbol{v}_2), \\ I_V(\boldsymbol{v}_1) &= I_V(\boldsymbol{v}_2), \\ \boldsymbol{v}_1 &= \boldsymbol{v}_2. \end{aligned}$$

Also, if $\boldsymbol{w} \in W$, then let $\boldsymbol{v} = T^{-1}(\boldsymbol{w})$. We have then $T(\boldsymbol{v}) = T(T^{-1}(\boldsymbol{w})) = T \circ T^{-1}(\boldsymbol{w}) = I_W(\boldsymbol{w}) = \boldsymbol{w}$. So T is an isomorphism. Then T^{-1} is 1–1 and onto because T^{-1} has an inverse (namely, T). It remains only to show that T^{-1} is a linear transformation. To this end, let $\boldsymbol{w}_1, \boldsymbol{w}_2 \in W$. We have $\boldsymbol{w}_1 = T(\boldsymbol{v}_1)$ and $\boldsymbol{w}_2 = T(\boldsymbol{v}_2)$ for unique $\boldsymbol{v}_1, \boldsymbol{v}_2 \in V$ since T is onto and 1–1. Then $\boldsymbol{v}_1 = T^{-1}(\boldsymbol{w}_1)$ and $\boldsymbol{v}_2 = T^{-1}(\boldsymbol{w}_2)$. Therefore,

$$\begin{aligned} T^{-1}(\boldsymbol{w}_1 + \boldsymbol{w}_2) &= T^{-1}(T(\boldsymbol{v}_1) + T(\boldsymbol{v}_2)) \\ &= T^{-1}(T\boldsymbol{v}_1 + \boldsymbol{v}_2)) \text{ since } T \text{ is linear} \\ &= T^{-1} \circ T(\boldsymbol{v}_1 + \boldsymbol{v}_2) \\ &= I_V(\boldsymbol{v}_1 + \boldsymbol{v}_2) \\ &= \boldsymbol{v}_1 + \boldsymbol{v}_2 \\ &= T^{-1}(\boldsymbol{w}_1) + T^{-1}(\boldsymbol{w}_2). \end{aligned}$$

Similarly, if $\boldsymbol{w} \in W$ and $c \in \boldsymbol{R}$, then we have $\boldsymbol{w} = T(\boldsymbol{v})$ for unique $\boldsymbol{v} \in V$ such that

$v = T^{-1}(w)$, and

$$\begin{aligned} T^{-1}(cw) &= T^{-1}(cT(v)) \\ &= T^{-1}(T(cv)) \text{ since } T \text{ is linear} \\ &= T^{-1} \circ T(cv) \\ &= I_V(cv) = cv = cT^{-1}(w). \end{aligned}$$

So T^{-1} is a linear transformation. QED

We can now determine the dimension of *Ker L*, the solution space of (3). Recall from Section 1.7 that a real vector space V has *dimension* n in case V contains a set of n linearly independent vectors and also every set of more than n vectors in V is linearly dependent. Since isomorphic vector spaces are indistinguishable, the next result is not surprising.

1.8 **LEMMA.** If V and W are isomorphic real vector spaces and $\dim V = n$, then $\dim W = n$ also.

Proof. Suppose that $\{v_1, \ldots, v_n\}$ is a linearly independent set of vectors in V. Then we claim that $\{T(v_1), \ldots, T(v_n)\}$ is a linearly independent set of vectors in W. We can show this by showing that

(5) $$c_1T(v_1) + c_2T(v_2) + \ldots + c_nT(v_n) = \mathbf{0}$$

can happen only if $c_1 = c_2 = \ldots = c_n = 0$ (Corollary 1.7.3). But if (5) holds, then

$$T(c_1v_1) + T(c_2v_2) + \ldots + T(c_nv_n) = \mathbf{0},$$
$$T(c_1v_1 + c_2v_2 + \ldots + c_nv_n) = \mathbf{0},$$

by the linearity of T. Applying T^{-1} we have

(6) $$c_1v_1 + c_2v_2 + \ldots + c_nv_n = T^{-1}(\mathbf{0}) = \mathbf{0}$$

since $T(\mathbf{0}) = \mathbf{0}$ (Exercise 1 below) and T is 1–1. But since $\{v_1, \ldots, v_n\}$ is linearly independent in V, (6) can happen only if $c_1 = c_2 = \ldots = c_n = 0$, as desired.

It remains only to show that any set $\{w_1, \ldots, w_k\}$, for $k > n$, of vectors in W that has more than n elements must be linearly dependent. Note that each $w_i = T(v_i)$ for unique $v_i \in V$, since T is 1–1 and onto. If now $\{w_1, w_2, \ldots, w_k\}$ were linearly independent in W, then by the reasoning used above with T^{-1} in place of T,

$$\{T^{-1}(w_1), T^{-1}(w_2), \ldots, T^{-1}(w_k)\} = \{v_1, v_2, \ldots, v_k\}$$

would be linearly independent in V. But that is impossible, since $k > n = \dim V$. So $\{w_1, \ldots, w_k\}$ must be linearly dependent in W. QED

1.9 **THEOREM.** The solution space *Ker L* of (4) has dimension n.

Proof. By Lemma 1.5, $T_{x_0}: Ker\,L \to \mathbf{R}^n$ is a 1–1 onto linear transformation. So by Definition 1.6, *Ker L* and $\mathbf{R}^n$ are isomorphic. But from Section 1.7 we know that $\dim \mathbf{R}^n = n$. Hence by Lemma 1.8, $\dim Ker\,L = n$ also. QED

Now we see why the general solution of (4) when $n=2$ and L is given by (3) had the forms it did in Section 8.6. The solution space of (4) is *two-dimensional* when $n=2$, so it has a basis of two elements. Theorems 8.6.8 and 8.6.9 seem to suggest that, depending on the nature of the coefficients in (3), this basis might take the form

$$\{e^{r_1x}, e^{r_2x}\}, \{e^{rx}, xe^{rx}\} \quad \text{or} \quad \{e^{ax}\cos bx, e^{ax}\sin bx\}.$$

To see that this suggestion is in fact correct, we need the next result of linear algebra.

1.10 **THEOREM.** Suppose that the real vector space V has dimension n. Suppose that V is spanned by a set

$$S=\{\mathbf{s}_1, \mathbf{s}_2, \ldots, \mathbf{s}_n\}$$

of n vectors (i.e., every vector $\boldsymbol{v}$ in S is a linear combination $a_1\mathbf{s}_1+a_2\mathbf{s}_2+\ldots+a_n\mathbf{s}_n$, where $a_i \in \boldsymbol{R}$, of vectors in S). Then S is a linearly independent set. (Hence S is a basis for V, since it contains $n=\dim V$ vectors. See Definition 1.7.9.)

Proof. We shall show that S cannot be linearly dependent. We do this by showing that if S *were* linearly dependent, then there could not be *any* linearly independent set in V with n elements, a contradiction to $\dim V=n$. So suppose that S is linearly dependent. Then some element is a linear combination of the others. Without loss of generality, we can assume that $\mathbf{s}_n=\sum_{i=1}^{n-1} b_i\mathbf{s}_i$ is a linear combination of the earlier $\mathbf{s}_i$. But then $S-\{\mathbf{s}_n\}=\{\mathbf{s}_1, \mathbf{s}_2, \ldots, \mathbf{s}_{n-1}\}$ spans V also, since for any $\boldsymbol{v}\in V$,

$$\begin{aligned}\boldsymbol{v} &= \sum_{i=1}^{n} a_i\mathbf{s}_i = \sum_{i=1}^{n-1} a_i\mathbf{s}_i + a_n\mathbf{s}_n \\ &= \sum_{i=1}^{n-1} a_i\mathbf{s}_i + a_n \sum_{i=1}^{n-1} b_i\mathbf{s}_i \\ &= \sum_{i=1}^{n-1} (a_i + a_nb_i)\mathbf{s}_i\end{aligned}$$

is a linear combination of $\mathbf{s}_1, \mathbf{s}_2, \ldots, \mathbf{s}_{n-1}$. Then we claim that *any* set $\{\boldsymbol{v}_1, \boldsymbol{v}_2, \ldots, \boldsymbol{v}_n\}$ of n vectors in V would be linearly dependent. We shall show in fact that

(7) $$\sum_{j=1}^{n} c_j\boldsymbol{v}_j = \mathbf{0}$$

has a nontrivial solution $\boldsymbol{c}=(c_1, c_2, \ldots, c_n)\in\boldsymbol{R}^n$. First, since $\{\mathbf{s}_1, \mathbf{s}_2, \ldots, \mathbf{s}_{n-1}\}$ spans V, we can write for $j=1, 2, \ldots, n$,

(8) $$\boldsymbol{v}_j = \sum_{i=1}^{n-1} a_{ij}\mathbf{s}_i$$

Then, substituting (8) into (7), we have

$$\sum_{j=1}^{n} c_j\left(\sum_{i=1}^{n-1} a_{ij}\mathbf{s}_i\right)=\mathbf{0},$$

$$\sum_{j=1}^{n}\sum_{i=1}^{n-1} a_{ij}c_j\mathbf{s}_i=\mathbf{0},$$

(9) $$\sum_{i=1}^{n-1}\left(\sum_{j=1}^{n} a_{ij}c_j\right)\mathbf{s}_i=\mathbf{0}.$$

Now (9) will hold if we can find $c_1, c_2, \ldots, c_n \in \boldsymbol{R}$ so that $\sum_{j=1}^{n} a_{ij}c_j=0$ for $i=1,\ldots,n-1$. We can write this system of equations as

(10) $$\begin{cases} a_{11}c_1 + a_{12}c_2 + \ldots + a_{1n}c_n = 0, \\ a_{21}c_1 + a_{22}c_2 + \ldots + a_{2n}c_n = 0 \\ \quad\vdots \\ a_{n-1,1}c_1 + a_{n-1,2}c_2 + \ldots + a_{n-1,n}c_n = 0 \end{cases}$$

Since (10) is a system of $n-1$ homogeneous equations in n unknowns, it has a nontrivial solution $\boldsymbol{c}=(c_1, c_2, \ldots, c_n)\in\boldsymbol{R}^n$ by Theorem 1.7.7. Thus, not all the $c_i=0$ in (7). So any set $\{\boldsymbol{v}_1, \ldots, \boldsymbol{v}_n\}$ of n vectors in V is linearly dependent. This contradicts dim $V=n$. Thus S *cannot* be linearly dependent. So S is linearly independent. QED

1.11 **COROLLARY.** The solution space *Ker L* of the second order homogeneous linear differential Equation (4) with L given by (3) and $n=2$ has basis

$$B_1=\{e^{r_1x}, e^{r_2x}\},\ B_2=\{e^{rx}, xe^{rx}\},\quad\text{or}\quad B_3=\{e^{ax}\cos bx, e^{ax}\sin bx\}$$

for some $r, r_1, r_2, a, b \in \boldsymbol{R}$.

Proof. By Theorems 8.6.8 and 8.6.9, one of these sets B_i spans *Ker L*, which has dimension 2 by Theorem 1.9. Then by Theorem 1.10, B_i is a linearly independent set having $2=\dim$ *Ker L* elements, so by definition B_i is a basis for *Ker L*. QED

In view of Theorem 1.10, we can solve (4) in the following way: once we can find a set B of n different functions $\{y_1(x), y_2(x), \ldots, y_n(x)\}$ that are solutions of (4) and that span the solution space *Ker L*, we can then assert that they form a basis for *Ker L*. The general solution of (4) is therefore given by

$$y(x)=c_1y_1(x)+c_2y_2(x)+\ldots+c_ny_n(x).$$

Moreover, this is the most *economical* representation for the general solution of (4), since the set B forms a basis for *Ker L*, so we couldn't possibly span *Ker L* with fewer than n functions. (See Exercise 8 below.)

EXERCISES 9.1

1 If $T: V\rightarrow W$ is any linear transformation, then show that $T(\mathbf{0})=\mathbf{0}$.

2 If $T: V \to W$ is a linear transformation, then show that the range Ran $T = \{\boldsymbol{w} \in W \mid \boldsymbol{w} = T(\boldsymbol{v}) \text{ for some } \boldsymbol{v} \in V\}$ is a real vector space.

3 Show that the vector space $\boldsymbol{R}_n[x]$ of all real polynomials in x of degree at most n is isomorphic to $\boldsymbol{R}^{n+1}$. [*Hint:* Define an isomorphism by mapping $a_0 + a_1x + \ldots + a_nx^n$ onto $(a_0, a_1, \ldots, a_n)$.]

4 If V is any real vector space of dimension n, then show that V is isomorphic to $\boldsymbol{R}^n$. [*Hint:* Choose a basis $B = \{\boldsymbol{v}_1, \boldsymbol{v}_2, \ldots, \boldsymbol{v}_n\}$ of V and define an isomorphism by mapping $\boldsymbol{v} = \sum_{i=1}^n a_i\boldsymbol{v}_i$ onto $(a_1, a_2, \ldots, a_n)$.]

5 Show that if $B = \{v_1, v_2, \ldots, v_n\}$ is a basis for V, then B spans V.

6 Show that if a linearly independent set $S = \{\boldsymbol{v}_1, \boldsymbol{v}_2, \ldots, \boldsymbol{v}_n\}$ spans V, then S is a basis. [*Hint:* Use the argument of Theorem 1.10.]

7 If $y_1(x), y_2(x), \ldots, y_n(x)$ are linearly independent solutions of (4), then show that every solution $y(x)$ of (4) can be *uniquely* expressed as $y(x) = c_1y_1(x) + c_2y_2(x) + \ldots + c_ny_n(x)$ for $c_i \in \boldsymbol{R}$.

8 Prove the last statement of the section. That is, prove that if L is given by (1), then *Ker L* cannot be spanned by fewer than n functions.

9 Prove directly that $\{e^{r_1x}, e^{r_2x}\}$ is a linearly independent set in $C^\infty(\boldsymbol{R})$ if $r_1 \neq r_2$. [*Hint:* Multiply $c_1e^{r_1x} + c_2e^{r_2x} = 0$ through by e^{-r_2x} and differentiate.]

10 Prove directly that $\{e^{rx}, xe^{rx}\}$ is a linearly independent set in $C^\infty(\boldsymbol{R})$.

Exercises 11 to 20 introduce a tool we will need in Section 3.

11 The **Wronskian** of the ordered set of functions $(f_1(x), f_2(x), \ldots, f_n(x))$ in $C^\infty(\boldsymbol{R})$. [named for the Polish mathematician J. M. Hoëne-Wronski (1778–1853)] is det $W(x)$, where

$$W(x) = \begin{pmatrix} f_1(x) & f_2(x) & \ldots & f_n(x) \\ f_1'(x) & f_2'(x) & \ldots & f_n'(x) \\ \cdot & \cdot & & \cdot \\ \cdot & \cdot & & \cdot \\ \cdot & \cdot & & \cdot \\ f_1^{(n-1)}(x) & f_2^{(n-1)}(x) & \ldots & f_n^{(n-1)}(x) \end{pmatrix}$$

Show that if det $W(x_0) \neq 0$ for at least one x_0 in an interval I, then

$$S = \{f_1(x), f_2(x), \ldots, f_n(x)\}$$

is a linearly independent set over I. [*Hint:* Write $c_1f_1(x) + c_2f_2(x) + \ldots + c_nf_n(x) = 0$ and differentiate $n-1$ times. Substitute $x = x_0$ and use the fact that the matrix of coefficients of the resulting system of linear equations is invertible by Theorem 6.7.3.

12 Use Exercise 11 to show that $\{e^{ax} \cos bx, e^{ax} \sin bx\}$ is a linearly independent set in $C^\infty(\boldsymbol{R})$.

13 If $y_1(x)$ and $y_2(x)$ both satisfy (4) with $n = 2$ on an interval I, then show that

$$(y_1y_2'' - y_2y_1'') + p_1(y_1y_2' - y_1'y_2) = 0.$$

14 Show that the derivative of the Wronskian of $y_1(x)$ and $y_2(x)$ is $y_1y_2'' - y_1''y_2$.

15 Let $w(x) = \det W(x)$ be the Wronskian of $y_1(x)$ and $y_2(x)$ in Exercise 13. Use Exercises 13 and 14 to show that

$$w'(x) + p_1w(x) = 0.$$

16 Use Exercise 15 to prove *Abel's identity*, named for the Norwegian mathematician

Niels H. Abel (1802–1829):

$$W(x) = c \exp\left(\int p_1(x)\, dx\right) \quad \text{for} \quad c \in \mathbf{R}.$$

17 Use Exercise 16 to prove that if $y_1(x)$ and $y_2(x)$ are solutions of (4) with $n = 2$ on an open interval I, then either $w(x) \equiv 0$ on I or else $w(x)$ is never zero on I.

18 Conclude from Exercise 16 that if $B_1 = \{y_1(x), y_2(x)\}$ and $B_2 = \{z_1(x), z_2(x)\}$ are any two bases for *Ker L*, then the Wronskians of B_1 and B_2 can differ only by a multiplicative constant.

19 If $y_1(x), \ldots, y_n(x)$ are linearly independent solutions of (4) on an interval I, then show that their Wronskian is never zero on I. [*Hint:* Suppose that the Wronskian is zero at some $x_0 \in I$. Then show that there are constants $c_1, c_2, \ldots, c_n$ not all zero such that $y(x) = \sum_{i=1}^{n} c_i y_i(x)$ and its first $(n-1)$ derivatives are 0 at x_0. Then use Theorem 1.4 to conclude that $y(x) \equiv 0$ on I.]

20 If

$$f_1(x) = e^{r_1 x}, f_2(x) = e^{r_2 x}, \ldots, f_n(x) = e^{r_n x},$$

then show that the Wronskian of $f_1(x), f_2(x), \ldots, f_n(x)$ is

$$\det W(x) = e^{(r_1 + r_2 + \ldots + r_n)x} \det V$$

where

$$V = \begin{pmatrix} 1 & 1 & \ldots & 1 \\ r_1 & r_2 & \ldots & r_n \\ r_1^2 & r_2^2 & \ldots & r_n^2 \\ \cdot & \cdot & & \cdot \\ \cdot & \cdot & & \cdot \\ \cdot & \cdot & & \cdot \\ r_1^{n-1} & r_2^{n-1} & \ldots & r_n^{n-1} \end{pmatrix}$$

is the *Vandermonde matrix* of the French mathematician Alexis T. Vandermonde (1735–1796).

21 Show that $\det V \neq 0$, where V is the Vandermonde matrix of Exercise 20. [*Hint:* Use induction to show that $\det V$ is the product of all terms of the form $r_j - r_i$ where $i < j$. Conclude that $\{f_1(x), f_2(x), \ldots, f_n(x)\}$ is linearly independent in $C^\infty(\mathbf{R})$.]

2 HOMOGENEOUS n-th ORDER LINEAR DIFFERENTIAL EQUATIONS WITH CONSTANT COEFFICIENTS

We now put to use the theory developed in the last section to solve n-th order linear differential equations $L(y) = s(x)$ where

(1) $$L = D^n + p_{n-1}D^{n-1} + \ldots + p_1 D + p_0$$

is a *polynomial operator*, i.e. the coefficients p_i are constants. Following the course suggested by Theorem 1.2, we solve the associated homogeneous problem

(2) $$L(y) = 0$$

in this section. Then in the next section we take up methods for finding a particular solution $y = y_p(x)$ to the general nonhomogeneous equation $L(y) = s(x)$.

As we would suspect, our results in this section parallel Theorems 8.6.8 and 8.6.9 for the case $n = 2$. First, we associate with each polynomial operator L given by (1) its auxiliary polynomial.

2.1 **DEFINITION.** If L is a polynomial operator given by (1), then its **auxiliary polynomial** is

(3) $$p(r) = r^n + p_{n-1}r^{n-1} + \ldots + p_1 r + p_0.$$

For example, the auxiliary polynomial corresponding to the differential equation

$$\frac{d^4y}{dx^4} - 13\frac{d^2y}{dx^2} + 36y = 0$$

is

$$r^4 - 13r^2 + 36,$$

since we can write the differential equation in operator form as $(D^4 - 13D^2 + 36)y = 0$. Because multiplication of polynomials is commutative, the next result is not surprising. It is, however, of crucial importance in solving (2).

2.2 **LEMMA.** Any two first degree polynomial operators $D - r_1$ and $D - r_2$ ($r_1, r_2 \in \boldsymbol{R}$ not necessarily distinct) commute with each other, that is,

$$(D - r_1)\circ(D - r_2) = (D - r_2)\circ(D - r_1).$$

Proof. This follows immediately from Lemma 8.6.5. From that result we know that

$$\begin{aligned}(D - r_1)\circ(D - r_2) &= D^2 - (r_1 + r_2)D + r_1 r_2 \\ &= D^2 - (r_2 + r_1)D + r_2 r_1 \\ &= (D - r_2)\circ(D - r_1)\end{aligned}$$

where the second equality follows from the commutativity of addition ($r_1 + r_2 = r_2 + r_1$) and multiplication ($r_1 r_2 = r_2 r_1$) of real numbers. QED

Throughout this book, we have stressed that mathematics is a *reasonable* subject in the sense that deeper, more general results ought to look analogous to simpler, more special results we've met before. With this in mind, it is natural to conjecture what the general solution to (2) looks like in the various cases when the auxiliary polynomial (3) has distinct or repeated real or complex roots. For instance, we expect by analogy with Theorem 8.6.8 that if (3) has distinct real roots $r_1, r_2, \ldots, r_n$, then the real vector space of solutions of (2) ought to have basis $\{e^{r_1 x}, e^{r_2 x}, \ldots, e^{r_n x}\}$, and so every solution should be expressible in the form

(4) $$y = c_1 e^{r_1 x} + c_2 e^{r_2 x} + \ldots + c_n e^{r_n x}.$$

This expectation is confirmed by our next result.

2.3 **THEOREM.** If $p(r)=0$ has distinct real roots $r_1, r_2, \ldots, r_n$, then $Ker\, L$ has basis

$$B=\{e^{r_1x}, e^{r_2x}, \ldots, e^{r_nx}\},$$

and so every solution to (2) has the form (4).

Proof. We first show that each e^{r_ix} is in $Ker\, L$. We have $L=(D-r_1)\circ\ldots\circ(D-r_i)\circ\ldots\circ(D-r_n)$. But then by Lemma 2.2,

$$\begin{aligned} L &= (D-r_1)\circ\ldots\circ(D-r_{i-1})\circ(D-r_{i+1})\circ\ldots\circ(D-r_n)\circ(D-r_i) \\ &= q(D)\circ(D-r_i), \end{aligned}$$

where $q(D)$ is a polynomial operator of degree $n-1$. We now have

$$\begin{aligned} L(e^{r_ix}) &= q(D)\circ(D-r_i)(e^{r_ix}) \\ &= q(D)\circ(r_ie^{r_ix}-r_ie^{r_ix}) \\ &= 0. \end{aligned}$$

Thus, $e^{r_ix}\in Ker\, L$, so $B\subseteq Ker\, L$ as claimed. Second, we show that B is a linearly independent set. Then since B has n ($=\dim Ker\, L$) elements, B is a basis for $Ker\, L$. We proceed by mathematical induction on n. When $n=1$, B is obviously linearly independent. Suppose that this holds for $n=k$ now, and let $n=k+1$. We need to show that

(5) $$c_1e^{r_1x}+\ldots+c_ke^{r_kx}+c_{k+1}e^{r_{k+1}x}=0$$

can hold only if $c_1=c_2=\ldots=c_{k+1}=0$. If we multiply (5) through by $e^{-r_{k+1}x}$, we obtain

(6) $$c_1e^{(r_1-r_{k+1})x}+\ldots+c_ke^{(r_k-r_{k+1})x}+c_{k+1}=0.$$

Differentiation of (6) gives

(7) $$d_1e^{(r_1-r_{k+1})x}+\ldots+d_ke^{(r_k-r_{k+1})x}=0$$

where $d_i=c_i(r_i-r_{k+1})$. Since $\{r_1-r_{k+1}, \ldots, r_k-r_{k+1}\}$ is a set of k distinct real numbers, our induction hypothesis tells us that $\{e^{(r_1-r_{k+1})x}, \ldots, e^{(r_k-r_{k+1})x}\}$ is linearly independent. Then in (7) each d_i must be zero. Now $r_i-r_{k+1}\neq 0$, because r_{k+1} is distinct from each r_i for $i\leq k$. Therefore, for $i=1, 2, \ldots, k$ we have $c_i=0$. Then (5) becomes

$$c_{k+1}e^{r_{k+1}x}=0.$$

Since $e^{r_{k+1}x}$ is never 0, we must have $c_{k+1}=0$. We have now shown that B is linearly independent for $n=k+1$ if it is linearly independent for $n=k$. Hence B is linearly independent for all n. QED

2.4 **EXAMPLE.** Solve $\dfrac{d^4y}{dx^4}-13\dfrac{d^2y}{dx^2}+36y=0.$

Solution. The auxiliary polynomial, as we saw following Definition 2.1, is

$$\begin{aligned} r^4 - 13r^2 + 36 &= (r^2 - 9)(r^2 - 4) \\ &= (r-3)(r+3)(r-2)(r+2). \end{aligned}$$

So here $r_1 = 3$, $r_2 = -3$, $r_3 = 2$, and $r_4 = -2$ are distinct real numbers. By Theorem 2.3, the general solution is given by all functions y of the form

$$y = c_1 e^{3x} + c_2 e^{-3x} + c_3 e^{2x} + c_4 e^{-2x}, \quad \text{for} \quad c_i \in \boldsymbol{R} \quad \text{and} \quad i = 1, 2, 3, 4.$$

Now suppose that some real root of (3) is repeated k times. By analogy with the case $k = 2$ in Theorem 8.6.8, we expect that our basis B of $Ker\, L$ will contain elements $e^{rx}, xe^{rx}, x^2e^{rx}, \ldots, x^{k-1}e^{rx}$. We proceed now to confirm this expectation. First we show that all such functions actually do belong to $Ker\, L$.

2.5 LEMMA. Let $p(r)$ be the auxiliary polynomial of L in (2). If the equation $p(r) = 0$ has a *k-fold repeated root* r, then $e^{rx}, xe^{rx}, \ldots, x^{k-1}e^{rx}$ are all in $Ker(D-r)^k$, and so are solutions of (2).

Proof. We proceed by induction on k. The case $k = 1$ has already been verified at the start of the proof of Theorem 2.3. Suppose then that the result is true for $k = m$, and let $k = m + 1$. We are assuming then that $e^{rx}, xe^{rx}, \ldots, x^{m-1}e^{rx} \in Ker(D-r)^m$. Then they are all in $Ker(D-r)^{m+1}$ since $(D-r)^{m+1} = (D-r)\circ(D-r)^m$, so anything mapped to 0 by $(D-r)^m$ will be mapped to zero by $(D-r)^{m+1}$. We need only show then that $x^m e^{rx} \in Ker(D-r)^{m+1}$. We have

$$\begin{aligned} (D-r)^{m+1}(x^m e^{rx}) &= (D-r)^m (D-r)(x^m e^{rx}) \\ &= (D-r)^m (mx^{m-1}e^{rx} + rx^m e^{rx} - rx^m e^{rx}) \\ &= (D-r)^m (mx^{m-1}e^{rx}) \\ &= m(D-r)^m (x^{m-1}e^{rx}) \\ &= m0 = 0, \end{aligned}$$

as desired. Thus $\{e^{rx}, xe^{rx}, \ldots, x^{k-1}e^{rx}\} \subseteq Ker(D-r)^k$ for any positive integer k. Then each $x^i e^{rx}$ is a solution to (2), since if $p(s)$ has the factorization $q(s)(s-r)^k$, then we get

$$\begin{aligned} L(x^i e^{rx}) &= q(D)(D-r)^k (x^i e^{rx}) \\ &= q(D)(0) = 0. \end{aligned}$$

QED

Now that we know that all the expected functions are solutions, we need to show that they span $Ker\, L$. We do this first for the case when there is only one repeated root r of (3).

2.6 LEMMA. If $p(r) = 0$ has distinct roots $r_1, r_2, \ldots, r_{n-k}$ and a k-fold repeated root r, then $Ker\, L$ has basis

$$B = \{e^{r_1 x}, e^{r_2 x}, \ldots, e^{r_{n-k} x}, e^{rx}, xe^{rx}, x^2 e^{rx}, \ldots, x^{k-1}e^{rx}\}.$$

So every solution of (2) is a linear combination of these elements.

Proof. We've checked that every $x^i e^{rx}$ is a solution to (2). The verification that each $e^{r_j x}$ is a solution of (2) for $j = 1, \ldots, n-k$ is just like the first part of the proof of Theorem 2.3. So we need only to show that B spans *Ker L*. For since B has n elements, it will then be a basis for *Ker L* by Theorem 1.10. We use the approach of Theorem 8.6.8. First we let

$$z_1 = (D - r_2)\circ \ldots \circ (D - r_{n-k}) \circ (D - r)^k (y).$$

Then (2) becomes

$$(D - r_1) z_1 = 0$$

whose solution is, by Theorem 8.3.6,

$$z_1 = C_1' e^{r_1 x}.$$

Continuing, we let $z_2 = (D - r_3)\circ \ldots \circ (D - r_{n-k}) \circ (D - r)^k (y)$, and just as in Theorem 8.6.8 we obtain

$$z_2 = \bar{c}_1 e^{r_1 x} + \bar{c}_2 e^{r_2 x}.$$

Continuing in this way for each $i \leq n - k$, we eventually arrive at

$$z_{n-k} = C_1 e^{r_1 x} + \ldots + C_{n-k} e^{r_{n-k} x},$$

where $z_{n-k} = (D - r)^k (y)$. Then we let $w_1 = (D - r)^{k-1}(y)$. We have

$$(D - r)(w_1) = C_1 e^{r_1 x} + \ldots + C_{n-k} e^{r_{n-k} x}.$$

This first order linear equation in w_1 has integrating factor e^{-rx}, so we obtain

$$e^{-rx} \frac{dw_1}{dx} - re^{-rx} w_1 = e^{-rx}(C_1 e^{r_1 x} + \ldots + C_{n-k} e^{r_{n-k} x}),$$

$$\frac{d}{dx}(e^{-rx} w_1) = C_1 e^{r_1 x - rx} + \ldots + C_{n-k} e^{r_{n-k} x - rx},$$

$$\begin{aligned} w_1 &= (D - r)^{k-1}(y) \\ &= c_1' e^{r_1 x} + \ldots + c_{n-k}' e^{r_{n-k} x} + c_{n-k+1}' e^{rx}. \end{aligned}$$

Next we let $w_2 = (D - r)^{k-2}(y)$. So we have

$$(D - r)(w_2) = w_1 = c_1' e^{r_1 x} + \ldots + c_{n-k}' e^{r_{n-k} x} + c_{n-k+1}' e^{rx}.$$

This first order linear equation in w_2 again has integrating factor e^{-rx}. So we get

$$\frac{d}{dx}(e^{-rx} w_2) = c_1' e^{r_1 x - rx} + \ldots + c_{n-k}' e^{r_{n-k} x - rx} + c_{n-k+1}',$$

$$e^{-rx} w_2 = c_1'' e^{r_1 x - rx} + \ldots + c_{n-k}'' e^{r_{n-k} x - rx} + c_{n-k+1}' x + c_{n-k+2}',$$

$$w_2 = c_1'' e^{r_1 x} + \ldots + c_{n-k}'' e^{r_{n-k} x} + c_{n-k+1}' x e^{rx} + c_{n-k+2}' e^{rx}.$$

Continuing in this way, we will be able to express each $w_i = (D - r)^{k-i}(y)$ as a linear combination of $e^{r_1 x}, \ldots, e^{r_{n-k} x}, e^{rx}, xe^{rx}, \ldots, x^{i-1} e^{rx}$. When $i = k$, we thus will have y as a linear combination of the elements of B. Therefore, the solution space *Ker L* of (2) is spanned by B. QED

Now, if more than one root of $p(r)=0$ is a repeated root (say s and r are repeated), then we expect that our basis B for $Ker\,L$ will contain elements

$$e^{rx}, xe^{rx}, \ldots, x^{k-1}e^{rx}, e^{sx}, xe^{sx}, \ldots, x^{\ell-1}e^{sx}$$

if r is a k-fold root and s is an ℓ-fold root. The next result confirms that expectation. The proof, whose details we omit, simply consists in continuing the argument contained in the proof of Lemma 2.6.

2.7 **THEOREM.** If the equation $p(r)=0$ has distinct roots

$$r_1, r_2, \ldots, r_m, s_1, s_2, \ldots, s_j$$

where the r_i are simple roots and each of the s_i is a k_i-fold repeated root (where $m+k_1+k_2+\ldots+k_j=n$), then $Ker\,L$ has basis

$$B=\{e^{r_1x}, \ldots, e^{r_mx}, e^{s_1x}, xe^{s_1x}, \ldots, x^{k_1-1}e^{s_1x}, \ldots, e^{s_jx}, xe^{s_jx}, \ldots, x^{k_j-1}e^{s_jx}\}. \tag{8}$$

2.8 **EXAMPLE.** Solve

$$\frac{d^5y}{dx^5}-6\frac{d^4y}{dx^4}+12\frac{d^3y}{dx^3}-8\frac{d^2y}{dx^2}=0.$$

Solution. In operator form this is

$$(D^5-6D^4+12D^3-8D^2)(y)=0.$$

The auxiliary polynomial is

$$\begin{aligned} r^5-6r^4+12r^3-8r^2 &= r^2(r^3-6r^2+12r-8) \\ &= r^2(r-2)^3. \end{aligned}$$

So the roots are $r=0$ with multiplicity two and $r=2$ with multiplicity three. Then a basis for the solution space is, by Theorem 2.7,

$$\{1=e^{0x}, x=xe^{0x}, e^{2x}, xe^{2x}, x^2e^{2x}\}.$$

So we can write the general solution in the form

$$y=c_1+c_2x+c_3e^{2x}+c_4xe^{2x}+c_5x^2e^{2x}.$$

Theorem 2.7 gives a complete solution of (2) in case the auxiliary equation $p(r)=0$ has only real roots. What if complex roots occur? In this case we need to widen our notion of vector spaces to include **complex vector spaces**. These are sets on which addition and multiplication by *complex* scalars $a+bi$, for $a, b \in \mathbf{R}$, are defined and for which the defining identities for a real vector space are satisfied. (See p. 11.)

2.9 **THEOREM.** (a) The set of all complex valued functions $y(x)=g(x)+ih(x)$ that satisfy (2) is a complex vector space denoted $(Ker\,L)_{\mathbf{C}}$.

(b) If $y(x)=g(x)+ih(x) \in (Ker\,L)_{\mathbf{C}}$, then the real part $g(x)$ and imaginary part $h(x)$ are real valued solutions of (2), so they are in $(Ker\,L)_{\mathbf{R}}$.

The proof of this is *exactly* like the proofs of Theorem 1.3 and Lemma 8.6.7. So we leave it as Exercises 25 and 26 below.

Notice that since $\boldsymbol{R} \subseteq \boldsymbol{C}$, we can regard each element of $(\mathit{Ker}\, L)_{\boldsymbol{R}}$ as an element of $(\mathit{Ker}\, L)_{\boldsymbol{C}}$, so the latter is a more extensive vector space. The Existence-Uniqueness Theorem 1.4 still holds if the initial values $y(x_0), y'(x_0), \ldots, y^{(n-1)}(x_0)$ are complex numbers, so we can prove, just as we did in Theorem 1.9, that $(\mathit{Ker}\, L)_{\boldsymbol{C}}$ is of dimension n over $\boldsymbol{C}$. Since by Corollary 8.5.14, integration of complex exponential functions follows exactly the same rules as integration of real exponential functions, we can prove that Theorem 2.7 continues to hold true for the case when $r_i, s_i \in \boldsymbol{C}$. We state this formally.

2.10 COROLLARY. If some or all of the r_i and s_i in Theorem 2.6 are complex, then the complex vector space $(\mathit{Ker}\, L)_{\boldsymbol{C}}$ has the basis $\boldsymbol{B_C}$ given by (8).

If we are content to work with complex valued functions, then Corollary 2.10 completes the solution of (2). But generally, since the coefficients p_i in (2) are real numbers, we want the *real* solution to (2). Fortunately, this is easy to extract from the complex solution. Recall from Corollary 8.5.4 that if $r = a + bi$ is a root of the auxiliary polynomial (3), then so is $\bar{r} = a - bi$. So the complex roots of (3) occur in conjugate pairs. In $\boldsymbol{B_C}$, then, we have

(9)
$$\begin{aligned} x^k e^{rx} &= x^k e^{ax}(\cos bx + i \sin bx) \\ x^k e^{\bar{r}x} &= x^k e^{ax}(\cos bx - i \sin bx). \end{aligned}$$

So from the two complex functions (9) in $\boldsymbol{B_C}$ we get, taking real and imaginary parts, the two real functions

(10)
$$\boxed{x^k e^{ax} \cos bx \quad \text{and} \quad x^k e^{ax} \sin bx,}$$

which by Theorem 2.9(b) are in $(\mathit{Ker}\, L)_{\boldsymbol{R}}$, and hence in $(\mathit{Ker}\, L)_{\boldsymbol{C}}$. If we take the real and imaginary parts of all the n functions in $\boldsymbol{B_C}$, then we obtain exactly n real valued functions in $(\mathit{Ker}\, L)_{\boldsymbol{C}}$. These functions still span $(\mathit{Ker}\, L)_{\boldsymbol{C}}$, since the complex solutions (9) are complex linear combinations of the real solutions (10). Thus, the set $\boldsymbol{B_R}$ of all these real valued functions is a basis for $(\mathit{Ker}\, L)_{\boldsymbol{C}}$, by Theorem 1.10. They are in fact, by our next result, a basis for the real solution space $(\mathit{Ker}\, L)_{\boldsymbol{R}}$.

2.11 THEOREM. The real solution space $(\mathit{Ker}\, L)_{\boldsymbol{R}}$ of (2) has basis $\boldsymbol{B_R}$ obtained by taking the real and imaginary parts of the basis elements in $\boldsymbol{B_C}$ of Corollary 2.10.

Proof. As we already observed, Theorem 2.9(b) guarantees that all elements of $\boldsymbol{B_R}$ are real solutions. We know from the preceding paragraph that $\boldsymbol{B_R}$ is a basis for $(\mathit{Ker}\, L)_{\boldsymbol{C}}$, so these n elements are linearly independent over $\boldsymbol{C}$. That is, the only way that

(11)
$$\sum_{k=1}^{n} \alpha_k \boldsymbol{b}_k = 0$$

can hold for $\alpha_k \in \boldsymbol{C}$ and $\boldsymbol{b}_k \in \boldsymbol{B_R}$ is if $\alpha_1 = \alpha_2 = \ldots = \alpha_n = 0$. But then the only way (11) can hold for $\alpha_k \in \boldsymbol{R} \subseteq \boldsymbol{C}$ and $\boldsymbol{b}_k \in \boldsymbol{B_R}$ is if $\alpha_1 = \alpha_2 = \ldots = \alpha_n = 0$. So $\boldsymbol{B_R}$ is a set of n linearly independent functions in $(\mathit{Ker}\, L)_{\boldsymbol{R}}$, which, by Theorem 1.9, has dimension n over $\boldsymbol{R}$. Hence, $\boldsymbol{B_R}$ is a basis for $(\mathit{Ker}\, L)_{\boldsymbol{R}}$. QED

The following example illustrates both Corollary 2.10 and Theorem 2.11.

2.12 EXAMPLE. Find the general real solution of

$$\frac{d^6y}{dx^6}+\frac{d^4y}{dx^4}-\frac{d^2y}{dx^2}-y=0.$$

Solution. In operator form this is

$$(D^6+D^4-D^2-1)(y)=0.$$

The auxiliary polynomial is

$$\begin{aligned}p(r)=r^6+r^4-r^2-1&=r^4(r^2+1)-(r^2+1)\\&=(r^4-1)(r^2+1)\\&=(r^2-1)(r^2+1)^2.\end{aligned}$$

The roots of $p(r)=0$ are 1, -1, i, i, $-i$, and $-i$. So the basic complex solution is given by

$$B_{\boldsymbol{C}}=\{e^x, e^{-x}, e^{ix}, e^{-ix}, xe^{ix}, xe^{-ix}\}.$$

Since $e^{ix}=\cos x+i\sin x$, the basic real solution is given by

$$B_{\boldsymbol{R}}=\{e^x, e^{-x}, \cos x, \sin x, x\cos x, x\sin x\}.$$

We can then write the general real solution in the form

$$y=c_1e^x+c_2e^{-x}+c_3\cos x+c_4x\cos x+c_5\sin x+c_6x\sin x$$

for $c_k\in\boldsymbol{R}$. This gives a typical element of $(Ker\,L)_{\boldsymbol{R}}$. By contrast, a typical element of $(Ker\,L)_{\boldsymbol{C}}$ can be expressed in terms of the elements of $B_{\boldsymbol{C}}$ in the form

$$y=c_1e^x+c_2e^{-x}+c_3e^{ix}+c_4e^{-ix}+c_5xe^{ix}+c_6xe^{-ix}\quad\text{for}\quad c_k\in\boldsymbol{C}.$$

EXERCISES 9.2

In Exercises 1 to 16, find the general (real) solution to the given differential equation. (If you don't see an obvious factorization of the auxiliary polynomial, then look for roots that are integers. Recall that such roots must be divisors of the constant term.)

1 $\dfrac{d^3y}{dx^3}-5\dfrac{d^2y}{dx^2}+6\dfrac{dy}{dx}=0.$

2 $\dfrac{d^3y}{dx^3}-4\dfrac{d^2y}{dx^2}+4\dfrac{dy}{dx}=0.$

3 $y''''-4y'''+4y''=0.$

4 $y''''-8y'''+16y''=0.$

5 $y'''+3y''+3y'+y=0.$

6 $\dfrac{d^4y}{dx^4}+2\dfrac{d^2y}{dx^2}+y=0.$

7 $\dfrac{d^4y}{dx^4}+3\dfrac{d^2y}{dx^2}-4y=0.$

8 $y'''+y''-2y=0.$

9 $\dfrac{d^3y}{dx^3}-3\dfrac{d^2y}{dx^2}-\dfrac{dy}{dx}+3y=0.$

10 $y'''-5y''+9y'-5y=0.$

11 $\frac{d^4y}{dx^4}-5\frac{d^3y}{dx^3}+6\frac{d^2y}{dx^2}+4\frac{dy}{dx}-8y=0.$

12 $\frac{d^4y}{dx^4}+2\frac{d^3y}{dx^3}-2\frac{dy}{dx}-y=0.$

13 $\frac{d^6y}{dx^6}+2\frac{d^4y}{dx^4}+\frac{d^2y}{dx^2}=0.$

14 $\frac{d^8y}{dx^8}+8\frac{d^4y}{dx^4}+16y=0.$

15 $\frac{d^4y}{dx^4}+y=0.$

16 $\frac{d^4y}{dx^4}+16y=0.$

In Exercises 17 to 24, solve the given initial value problem.

17 $\frac{d^3y}{dx^3}+\frac{d^2y}{dx^2}=0,\ y(0)=2,\ y'(0)=1,\ y''(0)=-1.$

18 $\frac{d^3y}{dx^3}+\frac{dy}{dx}=0,\ y(0)=0,\ y'(0)=1,\ y''(0)=2.$

19 $\frac{d^4y}{dx^4}-y=0,\ y(0)=0,\ y'(0)=0,\ y''(0)=1,\ y'''(0)=1.$

20 $\frac{d^4y}{dx^4}-16y=0,\ y(0)=0,\ y'(0)=0,\ y''(0)=2,\ y'''(0)=2.$

21 $y'''-6y''+11y'-6y=0,\ y(0)=0,\ y'(0)=0,\ y''(0)=2.$

22 $y'''-7y''+10y'=0,\ y(0)=1,\ y'(0)=0,\ y''(0)=2.$

23 Exercise 6 where $y(0)=1,\ y'(0)=1,\ y''(0)=0,\ y'''(0)=0.$

24 Exercise 15 where $y(0)=1,\ y''(0)=0,\ y'''(0)=0.$

25 Prove Theorem 2.8(a).

26 Prove Theorem 2.8(b).

27 Consider the two-mass vibrating system of Figure 2.1. The mass m_2 is pulled to the right and released. Let $y_2(t)$ be the distance by which m_2 is to the right of its original

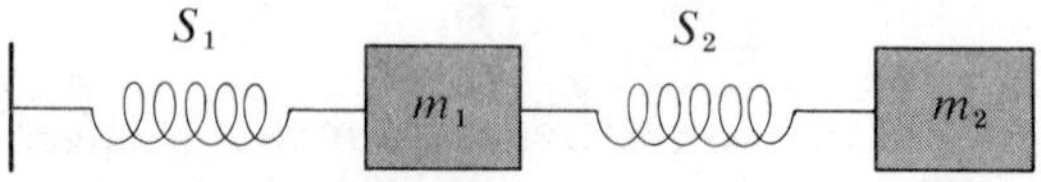

FIGURE 2.1

equilibrium position at time t, and let $y_1(t)$ be the distance by which m_1 is to the right of its original position at time t. Show, using Hooke's Law, that the forces F_2 on m_2 and F_1 on m_1 at time t are given by

$$\begin{cases} F_1=-k_1y_1(t)+k_2(y_2(t)-y_1(t))=m_1\dfrac{d^2y_1}{dt^2} \\ F_2=-k_2(y_2(t)-y_1(t))=m_2\dfrac{d^2y_2}{dt^2} \end{cases}$$

where k_1 and k_2 are the spring constants of S_1 and S_2 respectively.

28 Show that the system of equations in Exercise 27 is equivalent to the system

$$\begin{cases} m_1m_2\dfrac{d^4y_1}{dt^4}+[m_2(k_2+k_1)+m_1k_2]\dfrac{d^2y_1}{dt^2}+k_1k_2y_1=0 \\ \dfrac{m_1}{k_1}\dfrac{d^2y_1}{dt^2}+\dfrac{k_1+k_2}{k_2}y_1-y_2=0. \end{cases}$$

[*Hint:* Differentiate the first equation in Exercise 27 twice and substitute in the second equation.]

29 Show that the auxiliary equation for the first equation in Exercise 28 has four distinct pure imaginary roots, so the motion is bounded.

3 NONHOMOGENEOUS n-th ORDER LINEAR EQUATIONS

To complete the solution of the n-th order linear differential equation

(1) $$L(y)=\frac{d^n y}{dx^n}+p_{n-1}\frac{d^{n-1}y}{dx^{n-1}}+\ldots+p_1\frac{dy}{dx}+p_0 y=s(x),$$

we now just need to find one particular solution $y_p(x)$ to (1). In view of Theorem 1.2, the complete general solution to (1) will then be

$$\{y_0(x)+y_p(x) \mid y_0(x)\in Ker\, L\},$$

and we can determine the solution space $Ker\, L$ of the associated homogeneous problem $L(y)=0$ from Theorem 2.3, Theorem 2.7, or Theorem 2.11. As we would expect, the Method of Undetermined Coefficients 8.7.2 works whenever $s(x)$ has one of the forms appearing in Table 8.7.1. If you examine them, you will notice that all the $s(x)$ in Table 8.7.1 are *themselves* solutions of a homogeneous linear differential equation $M(y)=0$ for some linear differential operator M. For example, the n-th degree polynomial $s(x)=a_0+a_1x+a_2x^2+\ldots+a_nx^n$ is a solution of $M(y)=0$ where $M=D^{n+1}$; i.e., $s(x)$ is a solution of $\frac{d^{n+1}y}{dx^{n+1}}=0$. Similarly, $s(x)=ae^{kx}$ is a solution of $M(y)=\frac{dy}{dx}-ky=0$. The method of undetermined coefficients is based on the use of M suggested by the following simple result.

3.1 **LEMMA.** If $s(x)$ is a solution of the homogeneous equation $M(y)=0$ where M is a linear differential operator, then any particular solution $y_p(x)$ of (1) must be a solution of the homogeneous linear equation

(2) $$M\circ L(y)=0.$$

Proof. If $y=y_p(x)$ is a solution of (1), then $(M\circ L)(y(x))=M(L(y_p(x))=M(s(x))=0$. Thus $y_p(x)$ is a solution of $(M\circ L)(y)=0$. QED

Let's illustrate how Lemma 3.1 leads to the method of undetermined coefficients by returning to Example 8.7.1.

3.2 **EXAMPLE.** Solve

$$3\frac{d^2y}{dx^2}-15\frac{dy}{dx}-42y=3e^{7x}\sin x$$

using Lemma 3.1.

Solution. In standard form, $L(y)=s(x)$, this is

(3) $$\frac{d^2y}{dx^2}-5\frac{dy}{dx}-14y=e^{7x}\sin x.$$

The associated homogeneous problem has general solution $y = c_1e^{7x} + c_2e^{-2x}$. We recognize $e^{7x} \sin x = \operatorname{Im} e^{(7+i)x}$ as a solution to

$$[D-(7+i)]\circ[D-(7-i)](y)=0,$$
$$M(y)=(D^2-14D+50)(y)=0.$$

Thus, by Lemma 3.1, we can find a particular solution to the given equation among the solutions to the homogeneous linear equation

(4) $$(D^2-14D+50)\circ(D^2-5D-14)(y)=0.$$

The auxiliary polynomial has roots 7, -2, $7+i$, and $7-i$. So the general real solution of (4) is

$$y = c_1e^{7x} + c_2e^{-2x} + Ae^{7x} \cos x + Be^{7x} \sin x.$$

By Lemma 3.1 there is a particular solution of (3) of this form. Since the first two terms are solutions to the *homogeneous* problem associated with (3), we can take $c_1 = c_2 = 0$ and look for a $y_p(x)$ of the form $Ae^{7x} \cos x + Be^{7x} \sin x$. This is *precisely* the form for $y_p(x)$ suggested in Table 8.7.1, and is the form we used in Example 8.7.3. Proceeding exactly as we did before, we find $A = -9/82$ and $B = -1/82$. So

$$y_p(x) = -\frac{1}{82} e^{7x}(9 \cos x + \sin x).$$

Then the general solution of (3) is

$$y = c_1e^{7x} + c_2e^{-2x} - \frac{1}{82} e^{7x}(9 \cos x + \sin x).$$

We could use the method of Example 3.2 to justify each particular solution suggested in Table 8.7.1. The reasoning of Example 3.2, for instance, justifies the third line of the table. When dealing with higher-order equations, we apply the suggestions of Table 8.7.1 just as we did in Section 8.7.

3.3 EXAMPLE. Find the general solution of

$$\frac{d^3y}{dx^3} - \frac{d^2y}{dx^2} + 4\frac{dy}{dx} - 4y = 20 \sin 2x.$$

Solution. First we solve the associated homogeneous problem, which in operator form is

(5) $$(D^3-D^2+4D-4)(y)=0.$$

Here the auxiliary polynomial is

$$\begin{aligned} r^3-r^2+4r-4 &= r^2(r-1)+4(r-1) \\ &= (r^2+4)(r-1). \end{aligned}$$

The roots are $\pm 2i$ and 1, so the general solution to (5) is

$$y = c_1 \cos 2x + c_2 \sin 2x + c_3e^x.$$

Observe that $s(x) = 20 \sin 2x$ is then a solution to (5). According to the instructions for using Table 8.7.1, we should *not* seek a $y_p(x)$ of the form $A \cos 2x + B \sin 2x$ (which is a homogeneous solution, and hence not the particular solution we are seeking). Instead, we should seek a $y_p(x)$ of the form

$$y_p(x) = Ax \cos 2x + Bx \sin 2x.$$

This is so because the latter $y_p(x)$ is *not* a homogeneous solution. We have then

$$\begin{aligned}
y_p'(x) &= A \cos 2x - 2Ax \sin 2x + B \sin 2x + 2Bx \cos 2x,\\
y_p''(x) &= -2A \sin 2x - 2A \sin 2x - 4Ax \cos 2x + 2B \cos 2x\\
&\quad - 4Bx \sin 2x + 2B \cos 2x\\
&= -4A \sin 2x - 4Bx \sin 2x + 4B \cos 2x - 4Ax \cos 2x,\\
y_p'''(x) &= -8A \cos 2x - 4B \sin 2x - 8Bx \cos 2x - 8B \sin 2x\\
&\quad - 4A \cos 2x + 8Ax \sin 2x\\
&= -12A \cos 2x - 12B \sin 2x + 8Ax \sin 2x - 8Bx \cos 2x.
\end{aligned}$$

Substituting into the given equation, we have

$$\begin{aligned}
&y_p'''(x) - y_p''(x) + 4y_p'(x) - 4y_p(x)\\
&\quad = -12A \cos 2x - 12B \sin 2x + 8Ax \sin 2x - 8Bx \cos 2x\\
&\qquad - (-4A \sin 2x - 4Bx \sin 2x + 4B \cos 2x - 4Ax \cos 2x)\\
&\qquad + 4A \cos 2x - 8Ax \sin 2x + 4B \sin 2x + 8Bx \cos 2x\\
&\qquad - 4Ax \cos 2x - 4Bx \sin 2x\\
&\quad = 20 \sin 2x
\end{aligned}$$

Equating coefficients, we get

$$\begin{aligned}
&\cos 2x\colon\ -12A - 4B + 4A = 0,\ -8A - 4B = 0,\ A = -\tfrac{1}{2}B\\
&\sin 2x\colon\ -12B + 4A + 4B = 20,\ 4A - 8B = 20,\ -2B - 8B = 20,\ B = -2.
\end{aligned}$$

Thus $A = 1$. (Notice that if we equate the coefficients of $x \sin 2x$ and $x \cos 2x$ to 0, then we get the *useless* equations $8A - 4B - 8A + 4B = 0$ and $-8B + 4A + 8B - 4A = 0$.) So

$$y_p(x) = x \cos 2x - 2x \sin 2x.$$

The general solution of the given equation can thus be put in the form

$$y = c_1 \cos 2x + c_2 \sin 2x + c_3 e^x + x \cos 2x - 2x \sin 2x.$$

The method of undetermined coefficients is easy to use when it applies. Its major drawback is that it applies only to a very narrow range of Equations (1) in which $s(x)$ is a solution of some homogeneous linear differential equation with constant coefficients. A method of finding a particular solution of (1) that is of wider applicability is the method known as **variation of parameters**. This seems to have been first used by Euler in 1739 and, in essentially the form presented here, by Lagrange in 1774. The method is far more general than the method of undetermined coefficients in that it can be used even when the coefficients in (1) are *variable* functions $p_i(x)$.

To show why the technique works, we will make use of the *Wronskian*, which was introduced in Exercises 11 to 20 of Exercises 9.1. Recall that if $S = \{f_1(x), f_2(x), \ldots, f_n(x)\} \subseteq C^\infty(\mathbf{R})$, then the Wronskian of this set is $\det W_S(x)$ where

$$W_S(x) = \begin{pmatrix} f_1(x) & f_2(x) & \cdots & f_n(x) \\ f_1'(x) & f_2'(x) & \cdots & f_n'(x) \\ \vdots & \vdots & & \vdots \\ f_1^{(n-1)}(x) & f_2^{(n-1)}(x) & \cdots & f_n^{(n-1)}(x) \end{pmatrix}$$

The next result is basic to the method of variation of parameters.

3.4 **THEOREM.** If $B = \{y_1(x), y_2(x), \ldots, y_n(x)\}$ is a basis for the solution space $Ker\, L$ of $L(y) = 0$ on some interval I, then the Wronskian $\det W_B(x)$ is never 0.

Proof. We rule out the possibility that $\det W_B(x)$ could be 0 by showing that $\det W_B(x_0) = 0$ for some x_0 would imply that B was *not* a linearly independent set, contrary to the fact that B is a *basis* for $Ker\, L$. Suppose then that $\det W_B(x_0) = 0$. Then by Theorem 6.7.3, the matrix $W_B(x_0)$ would not be invertible. Hence the system of linear equations

(6) $$W_B(x_0)(\mathbf{c}) = 0$$

would have a nontrivial solution $\mathbf{c} = (c_1, c_2, \ldots, c_n) \neq \mathbf{0}$ by Exercise 25 of Exercises 6.4. Thus we would have from (6)

(7) $$\begin{cases} y_1(x_0)c_1 + y_2(x_0)c_2 + \ldots + y_n(x_0)c_n = 0 \\ y_1'(x_0)c_1 + y_2'(x_0)c_2 + \ldots + y_n'(x_0)c_n = 0 \\ \vdots \\ y_1^{(n-1)}(x_0)c_1 + y_2^{(n-1)}(x_0)c_2 + \ldots + y_n^{(n-1)}(x_0)c_n = 0. \end{cases}$$

So if we put $y(x) = c_1y_1(x) + c_2y_2(x) + \ldots + c_ny_n(x)$, then $y(x)$ is a solution to the initial value problem

$$L(y) = 0,\ y(x_0) = 0,\ y'(x_0) = 0, \ldots, y^{(n-1)}(x_0) = 0,$$

since the left-hand sides of the equations in (1) are just $y(x_0), y'(x_0), \ldots, y^{(n-1)}(x_0)$. But certainly the function $\bar{y}(x) \equiv 0$ is a solution to this same initial value problem. By the uniqueness part of Theorem 1.4, then, we must have

$$0 \equiv y(x) = c_1y_1(x) + c_2y_2(x) + \ldots + c_ny_n(x).$$

But then B could not be a linearly independent set, because not all the c_i would be 0 since $\mathbf{c} \neq \mathbf{0}$. This is impossible since B is a basis of $Ker\, L$. Hence $\det W_B(x_0) = 0$ is impossible. Thus $\det W_B(x_0)$ can never be 0 on I. QED

Now we can develop the method of variation of parameters. This method finds a particular solution of (1) that has the form

(8) $$y_p(x) = c_1(x)y_1(x) + c_2(x)y_2(x) + \ldots + c_n(x)y_n(x).$$

where $c_1(x)$, $c_2(x), \ldots, c_n(x)$ are *functions* to be determined. To make their determination feasible we will impose simplifying conditions on these "variable parameters" $c_i(x)$. From (8) we get

(9)
$$\begin{aligned} y_p'(x) &= c_1(x)y_1'(x)+\ldots+c_n(x)y_n'(x) \\ &\quad +(c_1'(x)y_1(x)+\ldots+c_n'(x)y_n(x)). \end{aligned}$$

The first condition we impose is that the expression in parentheses in (9) be 0. Then we differentiate the resulting simpler (9) to get

(10)
$$\begin{aligned} y_p''(x) &= c_1(x)y_1''(x)+\ldots+c_n(x)y_n''(x) \\ &\quad +(c_1'(x)y_1'(x)+\ldots+c_n'(x)y_n'(x)). \end{aligned}$$

Our second condition is that the expression in parentheses in (10) also be 0. Continuing in this way we come to

(11)
$$\begin{aligned} y_p^{(n-1)}(x) &= c_1(x)y_1^{(n-1)}(x)+\ldots+c_n(x)y_n^{(n-1)}(x) \cdot \\ &\quad +(c_1'(x)y_1^{(n-2)}(x)+\ldots+c_n'(x)y_n^{(n-2)}(x)), \end{aligned}$$

where we impose as the $(n-1)$-st condition that the expression in parentheses in (11) be 0. Then we finally have

(12)
$$\begin{aligned} y_p^{(n)}(x) &= c_1(x)y_1^{(n)}(x)+\ldots+c_n(x)y_n^{(n)}(x) \\ &\quad +(c_1'(x)y_1^{(n-1)}(x)+\ldots+c_n'(x)y_n^{(n-1)}(x)). \end{aligned}$$

Our n-th condition is that $y_p(x)$ be a solution to (1). Substituting (9), (10), ..., (11), (12) into (1), we have

$$\begin{aligned} &c_1(x)y_1^{(n)}+\ldots+c_n(x)y_n^{(n)}+(c_1'(x)y_1^{(n-1)} \\ &\quad +\ldots+c_n'(x)y_n^{(n-1)})+\ldots+p_1(c_1(x)y_1'+\ldots+c_n(x)y_n') \\ &\quad +p_0(c_1(x)y_1+\ldots+c_n(x)y_n)=s(x). \end{aligned}$$

Collecting terms, we have

$$\begin{aligned} &c_1'(x)y_1^{(n-1)}(x)+\ldots+c_n'(x)y_n^{(n-1)}(x) \\ &\quad +c_1(x)(y_1^{(n)}+\ldots+p_1y_1'+p_0y_1) \\ &\quad +\ldots+c_n(x)(y_n^{(n)}+\ldots+p_1y_n'+p_0y_n)=s(x) \end{aligned}$$

But each y_i is a solution of $L(y)=0$, so this reduces to

(13)
$$c_1'(x)y_1^{(n-1)}(x)+\ldots+c_n'(x)y_n^{(n-1)}(x)=s(x).$$

If we put (13) together with all our previous conditions, then we obtain the system of equations

$$\begin{cases} y_1(x)c_1'(x) \ +\ldots+\ y_n(x)c_n'(x) \ =0, \\ y_1'(x)c_1'(x) \ +\ldots+\ y_n'(x)c_n'(x) \ =0, \\ \quad \cdot \qquad\qquad \cdot \quad \cdot \qquad\qquad \cdot\ \cdot \\ \quad \cdot \qquad\qquad \cdot \quad \cdot \qquad\qquad \cdot\ \cdot \\ \quad \cdot \qquad\qquad \cdot \quad \cdot \qquad\qquad \cdot\ \cdot \\ y_1^{(n-2)}(x)c_1'(x)+\ldots+y_n^{(n-2)}(x)c_n'(x)=0, \\ y_1^{(n-1)}(x)c_1'(x)+\ldots+y_n^{(n-1)}(x)c_n'(x)=s(x) \end{cases}$$

which in matrix form we can write as

(14) $$W_B(x)\mathbf{c}'(x)=\begin{pmatrix}0\\0\\ \cdot\\ \cdot\\ \cdot\\ s(x)\end{pmatrix}$$

Since by Theorem 3.4 det $W_B(x)\neq 0$, we know from Theorem 6.7.3 again that $W_B(x)$ is an invertible matrix for all x. Thus we can solve (14) to obtain unique $c_1'(x), c_2'(x), \ldots, c_n'(x)$. If we integrate these, we obtain $c_1(x), \ldots, c_n(x)$ such that

$$y_p(x)=c_1(x)y_1(x)+\ldots+c_n(x)y_n(x)$$

is a particular solution to (1). We thus have established most of the following result.

3.5 **THEOREM.** The equation (1) has a particular solution $y_p(x)$ of the form

$$y_p(x)=c_1(x)y_1(x)+\ldots+c_n(x)y_n(x)$$

where $B=\{y_1(x),\ldots, y_n(x)\}$ is a basis for the solution space *Ker L* of the associated homogeneous problem $L(y)=0$. Moreover,

(15) $$y_p(x)=\sum_{i=1}^{n} y_i(x)\int\frac{s(x)\det W_i(x)}{\det W(x)}\,dx,$$

where $W_i(x)$ is the matrix that results when the i-th column of $W(x)$ is replaced by

$$\begin{pmatrix}0\\0\\ \cdot\\ \cdot\\ \cdot\\ 1\end{pmatrix}$$

Proof. All that remains is to obtain (15). But it follows upon solving (14) by Cramer's Rule 1.8.9. For we have

$$c_i'(x)=\frac{\det \bar{W}_i(x)}{\det W(x)}$$

where $\bar{W}_i(x)$ is obtained from $W(x)$ by replacing the i-th column by

$$\begin{pmatrix}0\\0\\ \cdot\\ \cdot\\ \cdot\\ s(x)\end{pmatrix}$$

Factoring $s(x)$ from the i-th column, we see that

$$\det \bar{W}_i(x) = s(x) \det W_i(x),$$

so

$$c_i'(x) = \frac{s(x) \det W_i(x)}{\det W(x)}.$$

Thus, (15) follows from $y_p(x) = c_1(x)y_1(x) + \ldots + c_n(x)y_n(x)$, where

$$c_i(x) = \int c_i'(x)\, dx = \int \frac{s(x) \det W_i(x)}{\det W(x)}\, dx. \qquad \text{QED}$$

It is fairly clear from (15) that whenever the method of undetermined coefficients can be applied, it will be less work to use it than to use the method of variation of parameters. The price we pay for the wider applicability of the latter method is a considerable increase in the complexity of the calculations required in (15), especially if $n > 2$. In the $n = 2$ case, however, the method of variation of parameters is a valuable complement to the method of undetermined coefficients. In general, its wide applicability makes it very important in theoretical work.

3.6 EXAMPLE. Find the general solution of

$$\frac{d^2y}{dx^2} + y = \csc x.$$

Solution. The function $s(x) = \csc x$ is one for which the method of undetermined coefficients fails to apply. So we try to use Theorem 3.5. The associated homogeneous problem,

$$(D^2 + 1)(y) = 0,$$

has solution $y_0(x) = c_1 \cos x + c_2 \sin x$. Thus for $y_p(x)$ we take $c_1(x)\cos x + c_2(x)\sin x$. Since $B = \{\cos x, \sin x\}$, we have

$$\det W_B(x) = \det\begin{pmatrix} \cos x & \sin x \\ -\sin x & \cos x \end{pmatrix} = 1.$$

We now apply (15) and get

$$\begin{aligned} y_p(x) &= \cos x \int \csc x \det\begin{pmatrix} 0 & \sin x \\ 1 & \cos x \end{pmatrix} dx \\ &\quad + \sin x \int \csc x \det\begin{pmatrix} \cos x & 0 \\ -\sin x & 1 \end{pmatrix} dx \\ &= \cos x \int \csc x(-\sin x)\, dx + \sin x \int \csc x \cos x\, dx \\ &= -\cos x \int dx + \sin x \int \frac{\cos x}{\sin x}\, dx \\ &= -x \cos x + \sin x \ln |\sin x|. \end{aligned}$$

So the general solution of the given equation is

$$y(x) = c_1 \cos x + c_2 \sin x - x \cos x + \sin x \ln |\sin x|.$$

Notice that in Example 3.6 the integrals in (15) fortuitously turned out to be easy to evaluate. As you can well imagine, even in the case $n = 2$, the integrals in (15) may be difficult or impossible to evaluate in terms of elementary functions. Nevertheless, (15) does supply an explicit formula for $y_p(x)$, which can always be approximated by numerical methods such as Simpson's rule if we need numerical values of $y_p(x)$. Thus, Theorem 3.5 is an effective general method of obtaining a particular solution of (1). We can solve *any* equation (1) completely now.

EXERCISES 9.3

In Exercises 1 to 18, find the general solution of the given differential equation. Use the method of undetermined coefficients when possible, and the method of variation of parameters otherwise.

1 $\dfrac{d^3y}{dx^3} + \dfrac{dy}{dx} = 4 \cos x$

2 $\dfrac{d^3y}{dx^3} - 4\dfrac{dy}{dx} = 8x$

3 $y'''' + y'' = 12x^2$

4 $y'''' + 4y'' = x^2$

5 $y''' - y' = x + 1$

6 $\dfrac{d^4y}{dx^4} - 16y = 16x^4 + 16x + 32$

7 $\dfrac{d^3y}{dx^3} - 4\dfrac{dy}{dx} = x + 3 \cos x + e^{-2x}$

8 $\dfrac{d^3y}{dx^3} + \dfrac{dy}{dx} = 2x^2 + 4 \sin x.$

9 $y'''' + y'' = 3x^2 - 4 \sin x - 2 \cos x$

10 $y'''' + 2y'' + y = 1 + 3 \cos 2x$

11 $y'' + y = \tan x$

12 $y'' + y = \sec^3 x$

13 $\dfrac{d^2y}{dx^2} - 3\dfrac{dy}{dx} + 2y = -\dfrac{e^{2x}}{1 + e^x}$

14 $\dfrac{d^2y}{dx^2} - y = \dfrac{2}{e^x + 1}$

15 $y'' + y = \cos^2 x$

16 $y'' + 9y = 9 \sec^2 3x$

17 $\dfrac{d^3y}{dx^3} + \dfrac{dy}{dx} = \tan x$

18 $\dfrac{d^3y}{dx^3} - 3\dfrac{d^2y}{dx^2} + 3\dfrac{dy}{dx} - y = \dfrac{2e^x}{x^2}$

19 Solve the initial value problem $y'' + y = \csc x$, $y(\frac{1}{2}\pi) = 1$, $y'(\frac{1}{2}\pi) = -1$. (*Hint:* Use the result of Example 3.6.)

20 Solve the initial value problem $y'' + y = \tan x$, $y(\frac{1}{4}\pi) = y'(\frac{1}{4}\pi) = 1$. (*Hint:* Use the result of Exercise 11.)

21 Solve the initial value problem $\dfrac{d^3y}{dx^3} + \dfrac{dy}{dx} = xe^x$, $y(0) = y'(0) = 0$, $y''(0) = 1$.

22 Solve the initial value problem $\dfrac{d^3y}{dx^3} + 4\dfrac{dy}{dx} = x^2e^x$, $y(0) = y'(0) = 1$, $y''(0) = 0$.

23 Show that the equation $L(y) = ae^{cx}$, where L is given by (1) and c is not a root of the auxiliary polynomial $P(r)$, has a particular solution

$$y_p(x) = \frac{a}{P(c)} e^{cx}.$$

24 If in Exercise 23 c is a k-fold root of the auxiliary polynomial $P(r) = (r - c)^k Q(r)$, where c is not a root of $Q(r)$, then show that $L(y) = ae^{cx}$ has a particular solution

$$y_p(x) = \frac{a}{k!Q(c)} x^k e^{cx}.$$

25 Use Exercises 23 and 24 to solve
(a) $y''' - y'' - y' + y = 9e^{2x}$;
(b) $y''' - 3y'' + 4y = 6e^{2x}$.

26 Use Exercises 23 and 24 to solve
(a) $\dfrac{d^3y}{dx^3} - 2\dfrac{d^2y}{dx^2} - 4\dfrac{dy}{dx} + 8y = 5e^{3x}$;
(b) $\dfrac{d^3y}{dx^3} - 2\dfrac{d^2y}{dx^2} - 4\dfrac{dy}{dx} + 8y = 5e^{2x}$.

4 INTRODUCTION TO THE LAPLACE TRANSFORM

We conclude this chapter with an introduction to a concept that is of far-reaching importance in applied mathematics. This is the Laplace transform, which was introduced in 1782 by Pierre Laplace. It is a tool that will permit us to solve a linear differential initial value problem *directly*. That is, we won't first have to find the general solution, and then determine values for the arbitrary constants to yield the unique solution that satisfies the given initial conditions. This approach is not only more direct, but also turns out to be ideally suited to many problems in electrical circuit theory that involve impulse functions and forcing functions with jump discontinuities. Thus, if you are studying electrical engineering, you will have occasion to use the Laplace transform extensively. Moreover, the Laplace transform turns out to be a major tool in the study of partial differential equations and integral equations that arise in a wide variety of applications. Time and space permit us only a brief introduction to this important area. This should serve to give you a foundation for further work with the Laplace transform in later courses.

We have already met many situations in which transformation can simplify a problem under study. Most of our attention has centered on *linear* transformations. The Laplace transform is such a linear transformation that maps functions from one vector space of functions to another. So far in this chapter our real vector space of functions has been $C^\infty(\boldsymbol{R})$, but this space is not nearly extensive enough to encompass all the functions that arise in applications of the Laplace transform. Instead of infinitely differentiable functions $f(x)$, we will consider functions $f(x)$ that grow less rapidly than exponential functions e^{tx} as $x \to +\infty$.

4.1 **DEFINITION.** The set $E[0, +\infty)$ of **integrable functions of exponential order** is the set of all continuous functions $f: \boldsymbol{R} \to \boldsymbol{R}$ defined on $[0, +\infty)$ such that for some $\alpha \in \boldsymbol{R}$ we have that

(*) $$\int_0^{+\infty} e^{-tx} f(x)\, dx \text{ exists as a real number for each } t \in (\alpha, +\infty).$$

Condition (*) not only says that $f(x)$ grows less rapidly than e^{tx} as $x \to +\infty$, but also requires that $f(x)$ grow *so much* less rapidly than e^{tx} that the improper integral of the quotient function $\dfrac{f(x)}{e^{tx}}$ over $[0, +\infty)$ exists.

4.2 EXAMPLE. Show that $f(x) = ax^2 + cx + d$ is a member of $E[0, +\infty)$. Use $\alpha = 0$ in Definition 4.1.

Proof. For any $t \in (0, +\infty)$ we have to evaluate $\int_0^{+\infty} e^{-tx} f(x)\, dx$. We can do so using integration by parts, or Formulas 17 and 18 in the Table of Integrals. The latter give

$$\int_0^{+\infty} e^{-tx}(ax^2+cx+d)\,dx = \lim_{b\to+\infty}\left[a\int_0^b x^2e^{-tx}\,dx + c\int_0^b xe^{-tx}\,dx + d\int_0^b e^{-tx}\,dx\right]$$

$$= \lim_{b\to+\infty}\left[a\left(-\frac{1}{t}x^2e^{-tx} + \frac{2}{t^3}e^{-tx}(-tx-1)\right)\right.$$

$$\left. + \frac{c}{t^2}e^{-tx}(-tx-1) - \frac{d}{t}e^{-tx}\right]_0^b,$$

which exists as a real number since

$$\lim_{b\to+\infty} b^2e^{-tb} = \lim_{b\to+\infty} be^{-tb} = \lim_{b\to+\infty} e^{-tb} = 0.$$

QED

Algebraically, $E[0, +\infty)$ has a familiar structure.

4.3 PROPOSITION. $E[0, +\infty)$ is a real vector space.

Proof. We know that $E[0, +\infty) \subseteq \mathscr{F}(\mathbf{R})$, which is a real vector space by Exercise 27, Exercises 1.1. Hence, all the required algebraic laws of Definition 1.1.14 hold for $E[0, +\infty)$, since they hold on $\mathscr{F}(\mathbf{R})$. All we need to show now is that if f and g are functions in $E[0, +\infty)$, then so are $f+g$ and af for any $a \in \mathbf{R}$. First, $f+g \in E[0, +\infty)$ because

$$\int_0^{+\infty} e^{-tx}(f+g)(x)\,dx = \int_0^{+\infty} [e^{-tx}f(x) + e^{-tx}g(x)]\,dx$$

$$= \int_0^{+\infty} e^{-tx}f(x)\,dx + \int_0^{+\infty} e^{-tx}g(x)\,dx$$

since both improper integrals on the right exist. Also,

$$\int_0^{+\infty} e^{-tx}(af)(x)\,dx = a\int_0^{+\infty} e^{-tx}f(x)\,dx,$$

so $af \in E[0, +\infty)$ if $f \in E[0, +\infty)$.

QED

4.4 **DEFINITION.** The **Laplace transform** is a map $\mathscr{L}: E[0, +\infty) \to \mathscr{F}(\mathbf{R})$ defined by $\mathscr{L}(f) = F$, where

(1) $$F(t) = \int_0^{+\infty} e^{-tx}f(x)\,dx$$

for $t \in (\alpha, +\infty)$. Here α is as in Definition 4.1.

Before proceeding, we emphasize the nature of the mapping $\mathscr{L}$. It acts on functions $f(x)$ from the real vector space $E[0, +\infty)$. For any such function f, $\mathscr{L}$ produces a *new* function $\mathscr{L}(f) = F$ defined on $(\alpha, +\infty)$. For each $t \in (\alpha, +\infty)$, the value of $\mathscr{L}(f)$ at t is given by the improper integral in (1), which is a well-defined real number in view of Definition 4.1. We next illustrate how we can identify $\mathscr{L}(f)$ in the case of some simple functions f.

4.5 EXAMPLE. (a) If $f(x)\equiv 1$ for $x\geq 0$, then find $\mathscr{L}(f(x))$. (b) If $f(x)=x^n$ for n a positive integer, then find $\mathscr{L}(f(x))$.

Solution. (a) From (1) we have for $t>0$,

$$\begin{aligned}\mathscr{L}(f)(t) &= \int_0^{+\infty} e^{-tx}\,dx = \lim_{b\to+\infty}\int_0^b e^{-tx}\,dx \\ &= \lim_{b\to+\infty} -\frac{1}{t}e^{-tx}\Big]_0^b = \lim_{b\to+\infty}\left[-\frac{1}{t}e^{-tb}+\frac{1}{t}\right] \\ &= \frac{1}{t}.\end{aligned}$$

Thus $\mathscr{L}(f)=F$ where $F(t)=\dfrac{1}{t}$.

(b) For $t>0$, we get

$$\begin{aligned}\mathscr{L}(f)(t) &= \int_0^{+\infty} x^n e^{-tx}\,dx \\ &= \lim_{b\to+\infty}\int_0^b x^n e^{-tx}\,dx \\ &= \lim_{b\to+\infty}\left(-\frac{1}{t}x^n e^{-tx}\Big]_0^b - \frac{n}{-t}\int_0^b x^{n-1}e^{-tx}\,dx\right)\end{aligned}$$

by Formula 18 of the Table of Integrals

$$= \lim_{b\to+\infty}\left(0-\frac{n}{t^2}x^{n-1}e^{-tx}\Big]_0^b + \frac{n(n-1)}{t^2}\int_0^b x^{n-2}e^{-tx}\,dx\right)$$

again by the same formula

$$\begin{aligned}&= \ldots = \frac{n!}{t^{n+1}}\lim_{b\to+\infty}[e^{-tx}(-tx-1)+1]_0^b \\ &= \frac{n!}{t^{n+1}}.\end{aligned}$$

Thus $\mathscr{L}(f)=F$ where $F(t)=\dfrac{n!}{t^{n+1}}$.

As promised, $\mathscr{L}$ is a linear transformation.

4.6 THEOREM. $\mathscr{L}: E[0,+\infty)\to\mathscr{F}(\mathbf{R})$ is a linear transformation.

Proof. For $t\in(\alpha,+\infty)$, we have

$$\begin{aligned}(f+g)(t) &= \int_0^{+\infty} e^{-tx}(f+g)(x)\,dx \\ &= \int_0^{+\infty} e^{-tx}(f(x)+g(x))\,dx \\ &= \int_0^{+\infty} e^{-tx}f(x)\,dx + \int_0^{+\infty} e^{-tx}g(x)\,dx \\ &= \mathscr{L}(f)(t)+\mathscr{L}(g)(t).\end{aligned}$$

We also have

$$\begin{aligned}\mathscr{L}(af)(t) &= \int_0^{+\infty} e^{-tx}(af)(x)\,dx \\ &= a\int_0^{+\infty} e^{-tx}f(x)\,dx \\ &= a\mathscr{L}(f)(t).\end{aligned}$$

Thus $\mathscr{L}$ is a linear transformation. QED

In order to use the Laplace transform $\mathscr{L}$ to solve an initial value problem of the form

(2) $$L(y) = s(x),\ y(0) = y_0,\ y'(0) = y_1, \ldots, y^{(n-1)}(0) = y_{n-1}$$

we are going to need to know how to apply $\mathscr{L}$ to an expression

(3) $$L(y) = y^{(n)} + p_{n-1}y^{(n-1)} + \ldots + p_1 y' + p_0 y$$

where the p_i are constants. Since $\mathscr{L}$ is linear, we have

(4) $$\mathscr{L}(L(y)) = \mathscr{L}(y^{(n)}) + p_{n-1}\mathscr{L}(y^{(n-1)}) + \ldots + p_1\mathscr{L}(y') + p_0\mathscr{L}(y).$$

At first glance (4) seems a bit forbidding, since it looks as if we will not only have to apply $\mathscr{L}$ to $y = y(x)$ using (1), but also need to apply $\mathscr{L}$ to $y', y'', \ldots, y^{(n-1)}$ and $y^{(n)}$. Fortunately, this is not as difficult as it looks. For the functions $y = y(x)$ that are most often met in practice, in fact, it is a simple matter to calculate $\mathscr{L}$ of y', y'', and so forth once $\mathscr{L}(y)$ is known.

4.7 **THEOREM.** If $y, y', \ldots, y^{(n)}$ all belong to $E[0, +\infty)$, then $\mathscr{L}(y^{(i)})$ for $i = 1, \ldots, n$ is given by

(5) $$\mathscr{L}(y^{(i)})(t) = -y^{(i-1)}(0) - ty^{(i-2)}(0) - \ldots - t^{i-1}y(0) + t^i\mathscr{L}(y)(t).$$

Proof. This result follows by applying

(6) $$\boxed{\mathscr{L}(y')(t) = -y(0) + t\mathscr{L}(y)(t)}$$

repeatedly. So let's first prove (6). We have from (1)

$$\mathscr{L}(y')(t) = \int_0^{+\infty} e^{-tx}y'(x)\,dx = \lim_{b\to+\infty}\int_0^b e^{-tx}y'(x)\,dx.$$

Using integration by parts (with $u = e^{-tx}$ and $dv = y'(x)\,dx$), we obtain

$$\begin{aligned}\mathscr{L}(y')(t) &= \lim_{b\to+\infty}\left\{e^{-tx}y(x)\Big]_0^b - \int_0^b -te^{-tx}y(x)\,dx\right\} \\ &= \lim_{b\to+\infty}\left\{e^{-tb}y(b) - 1y(0) + t\int_0^b e^{-tx}y(x)\,dx\right\} \\ &= \lim_{b\to+\infty} e^{-tb}y(b) - y(0) + t\lim_{b\to+\infty}\int_0^{+\infty} e^{-tx}y(x)\,dx \\ &= 0 - y(0) + t\mathscr{L}(y)(t),\end{aligned}$$

using Exercise 39 below. Thus (6) is proved. We can then use mathematical induction. Suppose that (5) holds for $i=k$; that is,

(7) $$\mathscr{L}(y^{(k)})(t)=-y^{(k-1)}(0)-ty^{(k-2)}(0)-\ldots-t^{k-1}y(0)+t^k\mathscr{L}(y)(t).$$

Then we use (6) to show that (5) must hold also for $i=k+1$. We have

$$\begin{aligned}\mathscr{L}(y^{(k+1)}(t))&=\mathscr{L}(y^{(k)\prime}(t))\\&=-y^{(k)}(0)+t\mathscr{L}(y^{(k)}(t))\quad\text{by (6)}\\&=-y^{(k)}(0)-ty^{(k-1)}(0)-t^2y^{(k-2)}(0)\\&\quad-\ldots-t^ky(0)+t^{k+1}\mathscr{L}(y)(t)\end{aligned}$$

by our induction hypothesis (7). Thus the truth of (5) for $i=k$ implies its truth for $i=k+1$. Since (5) holds for $i=1$ by (6), (5) holds for all i. QED

4.8 EXAMPLE. If $y=y(x)\in E[0,+\infty)$, then compute $\mathscr{L}\left(\dfrac{d^2y}{dx^2}+4\dfrac{dy}{dx}+4y\right)$, if $y(0)=1$ and $y'(0)=2$.

Solution. Using Theorems 4.6 and 4.7, we get

$$\begin{aligned}\mathscr{L}\left(\frac{d^2y}{dx^2}+4\frac{dy}{dx}+4y\right)&=\mathscr{L}(y'')+4\mathscr{L}(y')+4\mathscr{L}(y)\\&=-y'(0)-ty(0)+t^2\mathscr{L}(y)-4y(0)+4t\mathscr{L}(y)+4\mathscr{L}(y)\\&=-2-t-4+(t^2+4t+4)\mathscr{L}(y)\\&=(t^2+4t+4)\mathscr{L}(y)-(t+6).\end{aligned}$$

Observe that in Example 4.8, we took $\mathscr{L}$ of $(D^2+4D+4)y$ and obtained an expression in which the coefficient of $\mathscr{L}(y)$ was $t^2+4t+4=P(t)$, *where $P(r)$ is the auxiliary polynomial of D^2+4D+4.* This will *always* happen, and remembering that will give you a convenient check of your application of Theorem 4.7. (See Exercise 40 below.)

We now have a simple formula (5) for working out $\mathscr{L}(L(y))$ in (4). So we can transform (2) to the form

(8) $$\mathscr{L}(L(y))=\mathscr{L}(s)$$

where $\mathscr{L}(L(y))$ involves only the constants $y(0), y'(0), \ldots, y^{(n-1)}(0)$ given in the statement of the initial value problem, powers of t, and $\mathscr{L}(y)$. Thus (8) is an equation of the form

$$q_1(t)\mathscr{L}(y)+q_2(t)y(0)+\ldots+q_n(t)y^{(n-1)}(0)=\mathscr{L}(s),$$

where the $q_i(t)$ are polynomials in t. This is an algebraic equation that we can readily solve for $\mathscr{L}(y)$:

(9) $$\mathscr{L}(y)=\frac{\mathscr{L}(s)-q_2(t)y(0)-\ldots-q_n(t)y^{(n-1)}(0)}{q_1(t)}$$

Thus we will have the solution of (2) if we can somehow determine what function $y=y(x)$ has a Laplace transform given by the right side of (9). You may well object at this point that we do not even know that there *is* any such function

$y = y(x)$ and, if there is, that there aren't *several* such functions. For while we showed in Theorem 4.6 that $\mathscr{L}$ is a linear transformation, we did *not* show that $\mathscr{L}$ is onto (i.e., that there must be such a $y = y(x)$) and 1–1 (i.e., that there can be only one such $y = y(x)$). It can be shown that this is essentially the case, but such a proof involves notions from advanced calculus and complex function theory that we can't go into here. For our purposes, the following heuristic argument is appropriate. First, any $y = y(x)$ satisfying (9) will be a solution to the initial value problem (2) as we remarked. But by Theorem 1.4, (2) has only *one* solution. So there can be *at most* one $y = y(x)$ that satisfies (9). Second, Theorem 1.4 asserts that there *is* a solution y to (2). It doesn't assert that $y \in E[0, +\infty)$, i.e., that $\mathscr{L}(y)$ is sure to be defined, but this does seem plausible. Thus, there *ought* to be a $y = y(x)$ satisfying (9).

All right, assuming that (9) has a unique solution $y = y(x)$, how can we go about *finding* this function? The most efficient way is by consulting a *table of Laplace transforms.* Such a table (Table 4.1), when read from left to right, gives the Laplace transforms of commonly encountered functions $f(x)$. (Notice that the first three entries follow from Example 4.5.) *But it can also be read right-to-left.* That is, by looking in the right-hand column for the function $F(t)$ defined by the right side of (9), we can find the unique $y = y(x)$ such that $\mathscr{L}(y) = F$ in the left-hand column. In Table 4.1 we list only enough transforms to do the examples and exercises in this section. Much more extensive tables[1] have been compiled for use in solving initial value problems that commonly arise in engineering and science.

The following example illustrates the use of the table.

4.9 EXAMPLE. If

$$\mathscr{L}(y)(t) = \frac{t}{(t^2+4)^2} - \frac{3}{(t-3)^2},$$

then find $y = y(x)$.

Solution. We don't find $\mathscr{L}(y)(t)$ listed in the righthand column of Table 4.1. However, it is the *difference* of terms resembling entries 10 and 5, respectively. We have, in fact,

$$\begin{aligned}\mathscr{L}(y)(t) &= \frac{1}{4}\frac{4t}{(t^2+4)^2} - 3\frac{1}{(t-3)^2}\\ &= \tfrac{1}{4}\mathscr{L}(x \sin 2x) - 3\mathscr{L}(xe^{3x})\\ &= \mathscr{L}(\tfrac{1}{4}x \sin 2x - 3xe^{3x}) \text{ by Theorem 4.6.}\end{aligned}$$

Thus $y(x) = \frac{1}{4}x \sin 2x - 3xe^{3x}$.

We close the section with two examples that show how the Laplace transform can be used to solve initial value problems. First, we consider a very simple example to make sure that you understand how to proceed. Then we will see a more typical example of how to use the Laplace transform in practice.

4.10 EXAMPLE. Solve

$$\frac{dy}{dx} + y = x \cos x, \quad y(0) = 0.$$

1. See, for example, any edition of *Standard Mathematical Tables*, CRC Press, Cleveland.

Table 4.1

$f(x)$	$\mathscr{L}(f)(t)$
1. 1	$\frac{1}{t}, \quad t>0$
2. x	$\frac{1}{t^2}, \quad t>0$
3. x^n	$\frac{n!}{t^{n+1}}, \quad t>0$
4. e^{ax}	$\frac{1}{t-a}, \quad t>a$
5. xe^{ax}	$\frac{1}{(t-a)^2}, \quad t>a$
6. $x^n e^{ax}$	$\frac{n!}{(t-a)^{n+1}}, \quad t>a$
7. $\cos bx$	$\frac{t}{t^2+b^2}, \quad t>0$
8. $\sin bx$	$\frac{b}{t^2+b^2}, \quad t>0$
9. $x \cos bx$	$\frac{t^2-b^2}{(t^2+b^2)^2}, \quad t>0$
10. $x \sin bx$	$\frac{2bt}{(t^2+b^2)^2}, \quad t>0$
11. $e^{ax} \cos bx$	$\frac{t-a}{(t-a)^2+b^2}, \quad t>a$
12. $e^{ax} \sin bx$	$\frac{b}{(t-a)^2+b^2}, \quad t>a$
13. $xe^{ax} \cos bx$	$\frac{(t-a)^2-b^2}{[(t-a)^2+b^2]^2}, \quad t>a$
14. $xe^{ax} \sin bx$	$\frac{2b(t-a)}{[(t-a)^2+b^2]^2}, \quad t>a$
15. $\cosh ax$	$\frac{t}{t^2-a^2}, \quad t>a$
16. $\sinh ax$	$\frac{a}{t^2-a^2}, \quad t>a$
17. $e^{ax}-e^{bx}$	$\frac{a-b}{(t-a)(t-b)}, \quad t>\max(a, b)$
18. $\sin bx - bx \cos bx$	$\frac{2b^3}{(t^2+b^2)^2}, \quad t>0$

Solution. We apply $\mathscr{L}$ to the given equation to get

$$\begin{aligned}
\mathscr{L}(y'+y) &= \mathscr{L}(x\cos x)\\
\mathscr{L}(y')+\mathscr{L}(y) &= \frac{t^2-1}{(t^2+1)^2} \text{ by Theorem 4.6 and Table 4.1(9),}\\
-0+t\mathscr{L}(y)+\mathscr{L}(y) &= \frac{t^2-1}{(t^2+1)^2} \text{ by Theorem 4.7 since } y(0)=0,\\
(t+1)\mathscr{L}(y) &= \frac{t^2-1}{(t^2+1)^2},\\
\mathscr{L}(y) &= \frac{t-1}{(t^2+1)^2}\\
&= \frac{t}{(t^2+1)^2}-\frac{1}{(t^2+1)^2}\\
&= \frac{1}{2}\frac{2t}{(t^2+1)^2}-\frac{1}{2}\frac{2}{(t^2+1)^2}.
\end{aligned}$$

Thus $y=y(x)=\frac{1}{2}x\sin x-\frac{1}{2}\sin x+\frac{1}{2}x\cos x$ by Table 4.1 (10) and (18).

To solve the foregoing problem using the approach of Section 8.3, we would have to *first* find the general solution of $\dfrac{dy}{dx}+y=x\cos x$. This has integrating factor e^x, so we would be faced with $\dfrac{d}{dx}(e^x y)=xe^x\cos x$. Integrating the right-hand side would not be easy, and afterwards we would *still* need to determine the value of the constant that gives $y(0)=0$. Thus, the Laplace transform method is actually the *quickest and easiest* way to solve the given equation.

If you ever wondered why you had to learn the technique of *partial fractions* decomposition, the next example will give you an answer.

4.11 EXAMPLE. Solve the initial value problem

$$\frac{d^2y}{dx^2}+4\frac{dy}{dx}+4y=3e^{-2x},\ y(0)=1,\ y'(0)=2.$$

Solution. We apply $\mathscr{L}$, and use Theorems 4.6 and 4.7, Example 4.8, and entry 4 of Table 4.1. We get

$$\begin{aligned}
(t^2+4t+4)\mathscr{L}(y)-(t+6) &= 3\frac{1}{t+2},\\
(t+2)^2\mathscr{L}(y) &= \frac{3}{t+2}+t+6,\\
(t+2)^2\mathscr{L}(y) &= \frac{t^2+8t+15}{t+2}\\
\mathscr{L}(y) &= \frac{t^2+8t+15}{(t+2)^3}.
\end{aligned}$$

Again we don't find the right side in Table 4.1; but if we decompose it into partial fractions, then we will be able to use the table. We have

(10)
$$\frac{t^2+8t+15}{(t+2)^3}=\frac{A}{t+2}+\frac{B}{(t+2)^2}+\frac{C}{(t+2)^3},$$
$$t^2+8t+15=A(t+2)^2+B(t+2)+C.$$

Letting $t=-2$, we find $C=3$. Letting $t=0$ and then letting $t=-1$, we obtain

$$15=4A+2B+3, \quad \text{i.e.,} \quad 6=2A+B$$
$$8=A+B+3, \quad \text{i.e.,} \quad 5=A+B.$$

Subtraction gives $A=1$, and then $B=4$. So (10) is

$$\mathscr{L}(y)=\frac{t^2+8t+15}{(t+2)^3}=\frac{1}{t+2}+\frac{4}{(t+2)^2}+\frac{3}{(t+2)^3}$$
$$=\frac{1}{t-(-2)}+4\frac{1}{[t-(-2)]^2}+\frac{3}{2}\frac{2!}{[t-(-2)]^{2+1}}$$

In this form, entries 4, 5, and 6 of Table 4.1 give us

$$y=y(x)=e^{-2x}+4xe^{-2x}+\tfrac{3}{2}x^2e^{-2x}.$$

Again, the Laplace transform approach is quicker than our previous methods, which would require us to:

(a) solve the associated homogeneous problem $\dfrac{d^2y}{dx^2}+4\dfrac{dy}{dx}+4y=0$ to get $y=c_1e^{-2x}+c_2xe^{-2x}$;
(b) find a particular solution of the form ax^2e^{-2x} (since e^{-2x} and xe^{-2x} are solutions to the homogeneous problem) using the method of undetermined coefficients; and
(c) determine the values of the constants c_1 and c_2 that satisfy the initial conditions $y(0)=1$ and $y'(0)=2$.

EXERCISES 9.4

In Exercises 1 to 10, compute the Laplace transform of the given function $f(x)$. (This will verify some of Table 4.1.)

1 $f(x)=x$

2 $f(x)=ax+b$

3 $f(x)=e^{ax}$

4 $f(x)=xe^{ax}$

5 $f(x)=\cos bx$

6 $f(x)=\sin bx$

7 $f(x)=e^{ax}\cos bx$

8 $f(x)=e^{ax}\sin bx$

9 $f(x)=\cosh ax$

10 $f(x)=\sinh ax$

In Exercises 11 to 20, use Table 4.1 to find $y=y(x)$ from the given $\mathscr{L}(y)$.

11 $\mathscr{L}(y)=\dfrac{1}{t^{n+1}}$

12 $\mathscr{L}(y)=\dfrac{1}{(t-a)^{n+1}}$

13 $\mathscr{L}(y)=\dfrac{1}{t^2+a^2}$

14 $\mathscr{L}(y)=\dfrac{t}{(t^2+a^2)^2}$

15 $\mathscr{L}(y)=\dfrac{1}{(t^2+a^2)^2}$

16 $\mathscr{L}(y)=\dfrac{1}{(t-3)(t-4)}$

17 $\mathscr{L}(y)=\dfrac{t+1}{(t-1)^2(t^2+1)}$

18 $\mathscr{L}(y)=\dfrac{10}{(t-1)(t^2+9)}$

19 $\mathscr{L}(y)=\dfrac{1}{(t^2+6t+13)}$

20 $\mathscr{L}(y)=\dfrac{t+2}{t^2+4t+7}$

In Exercises 21 to 34, solve the given initial value problem using the Laplace transform.

21 $y'+2y=0,\ y(0)=3.$

22 $y'-2y=e^{5x},\ y(0)=3.$

23 $y'-y=x,\ y(0)=2.$

24 $y''+4y=0,\ y(0)=y'(0)=-2.$

25 $y''-2y'-8y=0,\ y(0)=3,\ y'(0)=6.$

26 $y''-y'-2y=3e^x,\ y(0)=1,\ y'(0)=0.$

27 $\dfrac{d^2y}{dx^2}+4y=3\sin 2x,\ y(0)=1,\ y'(0)=1.$

28 $y''+y=8e^{-2x}\sin x,\ y(0)=y'(0)=0.$

29 $\dfrac{d^3y}{dx^3}+2\dfrac{d^2y}{dx^2}=4,\ y(0)=y'(0)=y''(0)=0.$

30 $\dfrac{d^3y}{dx^3}-y=0,\ y(0)=1,\ y'(0)=y''(0)=0.$

31 $y'''+4y''+5y'+2y=10\cos x,\ y(0)=y'(0)=0,\ y''(0)=3.$

32 Exercise 31, where $y(0)=y'(0)=y''(0)=1.$

33 $\dfrac{d^4y}{dx^4}-4\dfrac{d^3y}{dx^3}+6\dfrac{d^2y}{dx^2}-4\dfrac{dy}{dx}+y=0,\ y(0)=y''(0)=0,\ y'(0)=y'''(0)=1.$

34 Exercise 33, where $y(0)=y'(0)=y''(0)=0,\ y'''(0)=1.$

35 Use Definition 4.4 to show that if $\mathscr{L}(f)=F$, then $\mathscr{L}(xf(x))(t)=-\dfrac{dF}{dt}.$

36 Use Exercise 35 to determine entry 4 of Table 4.1 from entry 3, and entries 9 and 10 from entries 7 and 8.

37 Use Exercise 35 to derive entries 13 and 14 of Table 4.1 from entries 11 and 12.

38 Find the Laplace transform of $x^2\sin bx$ and $x^2\cos bx$, using Exercise 35.

39 Show that if (*) in Definition 4.1 holds, then necessarily

$$\lim_{b\to+\infty} e^{-tb}f(b)=0.$$

[*Hint:* Show that if, for example, the limit is positive, then the improper integral in (*) diverges to $+\infty$.]

40 Show that when $\mathscr{L}$ is applied to (2), then the coefficient of $\mathscr{L}(y)$ in the resulting equation is the auxiliary polynomial of the linear differential operator L given by (3).

41 Show from Exercise 39 that the function $f(x)=e^{x^2}$ does *not* have a Laplace transform. [Thus, the methods of this section cannot be used to solve *all* initial value problems, e.g., $y''+2y'-3y=e^{x^2},\ y(0)=y'(0)=1$.]

REVIEW EXERCISES 9.5

1 Show that a subset $B=\{v_1,\ldots,v_n\}$ of a real vector space V is a basis if and only if B spans V and no smaller set spans V.

2 Find the Wronskian of $S=\{e^{3x}\cos 2x,\ e^{3x}\sin 2x,\ e^{2x}\}$.

3 Use Exercise 2 to show that the set S is linearly independent in $C^{\infty}(\boldsymbol{R})$.

4 Repeat Exercise 3 for $S=\{e^x, e^{-x}, \sin 2x, \cos 2x\}$.

In Exercises 5 to 26, solve the given differential equation.

5 $\dfrac{d^3y}{dx^3}-6\dfrac{d^2y}{dx^2}+9\dfrac{dy}{dx}=0.$

6 $\dfrac{d^3y}{dx^3}-7\dfrac{d^2y}{dx^2}+10\dfrac{dy}{dx}=0.$

7 $\dfrac{d^4y}{dx^4}-6\dfrac{d^3y}{dx^3}+9\dfrac{d^2y}{dx^2}=0.$

8 $\dfrac{d^4y}{dx^4}-y=0.$

9 $\dfrac{d^3y}{dx^3}+\dfrac{dy}{dx}=2\sin x.$

10 $\dfrac{d^3y}{dx^3}-9\dfrac{dy}{dx}=x.$

11 $y'''-y=x+1.$

12 $\dfrac{d^4y}{dx^4}-y=8x^2e^{-x}.$

13 $y''+y=\sin^2 x.$

14 $y''+9y=9\csc^3 3x.$

15 $\dfrac{d^3y}{dx^3}+\dfrac{dy}{dx}=\sec^3 x.$

16 $\dfrac{d^3y}{dx^3}+2\dfrac{d^2y}{dx^2}-\dfrac{dy}{dx}-2y=\dfrac{1}{e^x+1}.$

17 $y''-y'-2y=0$, where $y(0)=1$, $y'(0)=0$.

18 $y''+4y=0$, where $y(0)=y'(0)=1$.

19 $\dfrac{d^2y}{dx^2}+y=\sin 2x$, where $y(0)=0$, $y'(0)=1$.

20 $y''-2y'+2y=e^{-x}$, where $y(0)=0$, $y'(0)=1$.

21 $y''+5y'+6y=e^{-x}\cos 2x$, where $y(0)=y'(0)=0$.

22 $\dfrac{d^3y}{dx^3}+\dfrac{d^2y}{dx^2}=1$, where $y(0)=y'(0)=y''(0)=0$.

23 Compute the Laplace transform of $f(x)=x^2+1$.

24 Repeat Exercise 23 for $f(x)=e^x\cos 2x$.

10 SYSTEMS OF DIFFERENTIAL EQUATIONS

0 INTRODUCTION

Throughout this book there has been a recurring theme that linear algebra can be a remarkably useful tool in the study of the calculus of several variables and differential equations. In this chapter that theme reaches its completion as we study systems of (for the most part linear) differential equations.

We begin by developing an *elimination* scheme paralleling the technique of Gaussian elimination of Section 1.6. We see how we can naturally formulate systems of first order linear equations in *vector form* so that notationally they look like first order linear equations considered in Section 8.3. It turns out that *homogeneous* first order systems with constant coefficients can be solved in a manner analogous to that used in Sections 8.6 and 9.2. Corresponding to the roots of the auxiliary polynomial are the *eigenvalues* of the matrix A associated with the system. These are the roots of what is called the *characteristic polynomial* of A, the analogue of the auxiliary polynomial of n-th order linear equations with constant coefficients.

To handle *nonhomogeneous* linear systems, we need to develop techniques analogous to the method of undetermined coefficients of Sections 8.7 and 9.3 and the method of variation of parameters of Section 9.3. In Section 3 we present the necessary techniques, which correspond very closely to the earlier methods. In Section 4 we present a number of applications that involve systems of differential equations.

In the last section, we extend the Laplace transform of Section 9.4 to systems of linear differential equations, where it once more enables us to solve initial value problems directly.

1 ELIMINATION; REDUCTION OF n-th ORDER EQUATIONS TO FIRST ORDER SYSTEMS

This chapter is concerned with systems of differential equations. Such systems arise naturally in many applied fields. We will consider several examples in detail in Section 4, but for now present a simple example whose practical impact has become considerable in the last two decades.

1.1 EXAMPLE. Suppose that a small body (such as an artificial earth satellite or deep space probe) of mass m is in motion in space. Write a system of differential equations describing the motion.

Solution. Let $P(x(t), y(t), z(t))$ be the position of the body at time t. Let $\boldsymbol{F}=(f_1, f_2, f_3)$ be the force exerted on the body at time t. This force includes the gravitational forces exerted on the body by celestial bodies plus forces due to propulsion and/or resistance due to the outer atmosphere of earth. In general this force depends on $P(x(t), y(t), z(t))$ and the velocity of the body at time t in addition to depending on t directly. Newton's law of motion applied to each coordinate gives us the system

(1)
$$\begin{cases} m\dfrac{d^2x}{dt^2}=f_1\left(t, x, y, z, \dfrac{dx}{dt}, \dfrac{dy}{dt}, \dfrac{dz}{dt}\right) \\ m\dfrac{d^2y}{dt^2}=f_2\left(t, x, y, z, \dfrac{dx}{dt}, \dfrac{dy}{dt}, \dfrac{dz}{dt}\right) \\ m\dfrac{d^2z}{dt^2}=f_3\left(t, x, y, z, \dfrac{dx}{dt}, \dfrac{dy}{dt}, \dfrac{dz}{dt}\right). \end{cases}$$

Before proceeding, let's draw some general conclusions from this example. First, the natural notation for the independent variable is t, rather than the x that we used for the most part in the previous two chapters. This allows us to use (x, y, z) as usual for the coordinates of a moving point in space. Second, the system (1) would assume a simpler form if we wrote it as a single *vector* equation

(2)
$$m\ddot{\boldsymbol{x}}=\boldsymbol{F}(t, \boldsymbol{x}(t), \dot{\boldsymbol{x}}(t))$$

where $\boldsymbol{x}=\boldsymbol{x}(t)=(x(t), y(t)\ z(t))$. We are merely applying the *idea* behind vector functions here (first mentioned in Section 1.0). Namely, a vector formulation will often simplify a problem originally analyzed in terms of coordinates. Here (2) looks reminiscent of a single second order differential equation

(3)
$$m\frac{d^2x}{dt^2}=f\left(t, x(t), \frac{dx}{dt}\right).$$

It is natural to suspect that techniques used for equations of the form (3) might have analogues for equations of the form (2) that might help us solve (1). We pursue this idea in Section 2 and exploit it throughout the remainder of the chapter.

We have seen how systems of differential equations can arise outside mathematics. Now we show how they naturally arise *inside* mathematics.

1.2 THEOREM. An n-th order differential equation

(4)
$$\frac{d^ny}{dt^n}=F\left(t, y, \frac{dy}{dt}, \ldots, \frac{d^{n-1}y}{dt^{n-1}}\right)$$

is equivalent to a system of n first order differential equations.

Proof. The trick here is to make the substitutions defined by

(5) $$x_1 = y,\ x_2 = \frac{dy}{dt}, \ldots, x_n = \frac{d^{n-1}y}{dt^{n-1}}.$$

Then the given equation transforms to the system

(6) $$\begin{cases} \dfrac{dx_1}{dt} = x_2 \\ \dfrac{dx_2}{dt} = x_3 \\ \quad\vdots \\ \dfrac{dx_{n-1}}{dt} = x_n \\ \dfrac{dx_n}{dt} = F(t, x_1, x_2, \ldots, x_n). \end{cases}$$

Thus any solution $y = y(t)$ to the given n-th order equation provides us with a solution to (6). Conversely, if we have any solution $(x_1(t), x_2(t), \ldots, x_n(t))$ to the system (6), then the substitutions (5) clearly give us a solution of the n-th order equation (4). Thus (6) is equivalent to (4). QED

Theorem 1.2 is of great importance. For us, its main value lies in the fact that it enables us, for theoretical purposes, to consider just *first order* systems. Any higher order equations that may be involved in more general systems can be replaced by a number of first order equations using (6). So the original system is then equivalent to the resulting first order system of more equations. We can thus limit our theoretical development to the systems specified below.

1.3 DEFINITION. The **general first order system** of n differential equations is

(7) $$\begin{cases} \dfrac{dx_1}{dt} = f_1(t, x_1, x_2, \ldots, x_n) \\ \dfrac{dx_2}{dt} = f_2(t, x_1, x_2, \ldots, x_n) \\ \quad\vdots \\ \dfrac{dx_n}{dt} = f_n(t, x_1, x_2, \ldots, x_n) \end{cases}$$

Actually, for purposes of developing effective general solution techniques, (7) is still a bit too general. For most of this chapter we will restrict attention to first order *linear* systems.

1.4 **DEFINITION.** The **general first order linear system** of n differential equations is

(8)
$$\begin{cases} \dfrac{dx_1}{dt} = a_{11}(t)x_1(t) + a_{12}(t)x_2(t) + \ldots + a_{1n}(t)x_n(t) + b_1(t), \\ \dfrac{dx_2}{dt} = a_{21}(t)x_1(t) + a_{22}(t)x_2(t) + \ldots + a_{2n}(t)x_n(t) + b_2(t), \\ \quad \vdots \qquad \vdots \qquad \vdots \qquad \vdots \qquad \vdots \\ \dfrac{dx_n}{dt} = a_{n1}(t)x_1(t) + a_{n2}(t)x_2(t) + \ldots + a_{nn}(t)x_n(t) + b_n(t) \end{cases}$$

In vector-matrix form we can write (8) more compactly as

(9)
$$\frac{d\boldsymbol{x}}{dt} = A(t)\boldsymbol{x}(t) + \boldsymbol{b}(t),$$

where $A(t) = (a_{ij}(t))$ is the n-by-n matrix of coefficient functions in (8), and $\boldsymbol{b}(t) = (b_1(t), \ldots, b_n(t))$. The map $T: C^\infty(\boldsymbol{R}, \boldsymbol{R}^n) \to C^\infty(\boldsymbol{R}, \boldsymbol{R}^n)$ defined by

(10)
$$T(\boldsymbol{x}(t)) = A(t)\boldsymbol{x}(t)$$

is easily shown to be a linear transformation on the real vector space $C^\infty(\boldsymbol{R}, \boldsymbol{R}^n)$ of all infinitely differentiable functions $\boldsymbol{x}: \boldsymbol{R} \to \boldsymbol{R}^n$ (Exercise 21 below). The differential operator $D = \dfrac{d}{dt}$ is certainly a linear transformation on this space. Hence, so is $D - T$. We can then rewrite (9) as

(11)
$$(D - T)(\boldsymbol{x}(t)) = \boldsymbol{b}(t),$$

where $D - T$ is a linear transformation. Then the approach of Theorem 9.1.2 will prove the following result (Exercise 22 below).

1.5 **THEOREM.** If $\boldsymbol{x} = \boldsymbol{x}_p$ is one solution of (8), then the set S of all solutions of (8) is given by

$$S = \{\boldsymbol{x}_0(t) + \boldsymbol{x}_p(t) \mid (D - T)(\boldsymbol{x}_0(t)) = \boldsymbol{0}\}.$$

Analogues of further results in Section 9.1 also hold. The following basic existence-uniqueness theorem implies Theorem 9.1.4, in view of Theorem 1.2.

1.6 **EXISTENCE-UNIQUENESS THEOREM.** Suppose that the functions $a_{ij}(t)$, for $i, j = 1, \ldots, n$, are continuous on an interval I containing t_0. Then there is a unique solution $\boldsymbol{x} = \boldsymbol{x}(t) = (x_1(t), \ldots, x_n(t))$ to

(12)
$$\frac{d\boldsymbol{x}}{dt} = A(t)\boldsymbol{x}(t)$$

satisfying the initial conditions $\boldsymbol{x}(t) = \boldsymbol{x}_0$.

The proof of this result is beyond our scope here. We will, however, use it to establish the analogue of Theorem 9.1.9.

1.7 **THEOREM.** The set $Ker(D-T)$ of all solutions of the homogeneous system (12) is a real vector space of dimension n.

Proof. Since $D-T$ is a linear transformation, $Ker(D-T)$ is a real vector space as usual (Exercise 23 below). We now proceed as in the proof of Theorem 9.1.9. For $t_0 \in \boldsymbol{R}$, we define $F_{t_0}: Ker(D-T) \to \boldsymbol{R}^n$ as follows. Given the vector function

$$\boldsymbol{x} = \boldsymbol{x}(t) = (x_1(t), \ldots, x_n(t))$$

in $Ker(D-T)$, the mapping F_{t_0} evaluates $\boldsymbol{x}$ at t_0:

$$F_{t_0}(\boldsymbol{x}) = \boldsymbol{x}(t_0) = (x_1(t_0), \ldots, x_n(t_0)) \in \boldsymbol{R}^n.$$

Then it is easy to show that F_{t_0} is linear (Exercise 24 below). Also, F_{t_0} is 1–1 and onto by Theorem 1.6, since for any vector $\mathbf{y} = (y_1, y_2, \ldots, y_n) \in \boldsymbol{R}^n$ there is one and only one solution $\boldsymbol{x} = \boldsymbol{x}(t)$ for which $F_{t_0}(\boldsymbol{x}) = \mathbf{y}$. That is, if $F_{t_0}(\boldsymbol{x}) = F_{t_0}(\boldsymbol{z}) = \mathbf{y}$, then $\boldsymbol{x}(t) = \boldsymbol{z}(t)$ for all t. So F_{t_0} is an isomorphism of vector spaces. Hence $\dim Ker(D-T) = \dim \boldsymbol{R}^n = n$ by Lemma 9.1.8. QED

We have all the theoretical groundwork we need to solve (8) now. Paralleling Chapter 9, we will concentrate on the case when all the coefficient functions $a_{ij}(t)$ are constants. The resemblance between (8) and the systems of linear equations treated in Section 1.6 is then striking. As you might expect, there is a solution technique analogous to the method of Gaussian elimination. We simply call it the **method of elimination**. Just like the technique of Gaussian elimination, it employs three *elementary operations* on (8), which do not change its set of solutions. These three operations are:

(i) Interchange two equations in (8);
(ii) Multiply some equation through by a constant $c \neq 0$;
(iii) Add to some equation (say the j-th) the result of applying a polynomial operator $p(D)$ to some other equation (say the i-th).

Clearly, operations (i) and (ii) have no effect on the set of solutions of (8). Let's consider (iii), then. It will replace (8) by a system (8′) identical to (8) except for the j-th equation, which is

(13) $$\frac{dx_j}{dt} + p(D)\frac{dx_i}{dt} = (a_{j1} + a_{i1}p(D))x_1 + \ldots + (a_{jn} + a_{in}p(D))x_n + (b_j(t) + p(D)b_i(t)).$$

If now $\boldsymbol{x}(t) = (x_1(t), \ldots, x_n(t))$ is a solution to (8), then $\boldsymbol{x}(t)$ satisfies (13), and so $\boldsymbol{x}(t)$ is also a solution of (8′). Conversely, if $\boldsymbol{x}(t)$ is a solution of (8′), then it will also be a solution of the system (8″) that results if we apply $-p(D)$ to the i-th equation in (8′) and add to the j-th equation. But this replaces (13) by the j-th equation of (8). So $\boldsymbol{x}(t)$ is a solution of (8). Thus (8) and (8′) have *identical* solution sets. So application of the operations (i), (ii), and (iii) will lead from (8) to an equivalent system. The following example shows that, just as with Gaussian elimination, judicious choice of operations will lead us to a readily solved system.

1.8 EXAMPLE. Solve the system

(14)
$$\begin{cases} \dfrac{dx_1}{dt} = 4x_1 - 2x_2 + 2t \\ \dfrac{dx_2}{dt} = 8x_1 - 4x_2 + 1 \end{cases}$$

Solution. Since in using (iii) we will be dealing with polynomial operators, we first put (14) into operator form. We get

$$\begin{cases} (D-4)x_1 + \quad 2x_2 = 2t \\ \quad -8x_1 + (D+4)x_2 = 1 \end{cases}$$

We can eliminate x_1 if we multiply the first equation by 8 and add to it the result of applying $(D-4)$ to the second. We have

$$\begin{cases} 8(D-4)x_1 + \qquad 16x_2 = 16t \\ -8(D-4)x_1 + (D-4)(D+4)x_2 = (D-4)(1) = -4, \end{cases}$$

which when added gives

$$(D^2 - 16 + 16)x_2 = 16t - 4,$$
$$\frac{d^2x_2}{dt^2} = 16t - 4.$$

We can integrate the last equation, and obtain

$$\frac{dx_2}{dt} = 8t^2 - 4t + c_1,$$
$$x_2 = \tfrac{8}{3}t^3 - 2t^2 + c_1 t + c_2.$$

If we substitute this in the second given equation, we have

$$-8x_1 + 8t^2 - 4t + c_1 + \tfrac{32}{3}t^3 - 8t^2 + 4c_1 t + 4c_2 = 1,$$
$$x_1 = \tfrac{4}{3}t^3 - \tfrac{1}{2}t + \tfrac{1}{2}c_1 t + \tfrac{1}{8}c_1 + \tfrac{1}{2}c_2 - \tfrac{1}{8}.$$

So we obtain the solution

$$x_1(t) = \tfrac{4}{3}t^3 - \tfrac{1}{2}t + \tfrac{1}{2}c_1 t + \tfrac{1}{8}c_1 + \tfrac{1}{2}c_2 - \tfrac{1}{8}$$
$$x_2(t) = \tfrac{8}{3}t^3 - 2t^2 + c_1 t + c_2.$$

In vector form this is

(15)
$$\boldsymbol{x}(t) = \begin{pmatrix} \frac{4}{3}t^3 - \frac{1}{2}t - \frac{1}{8} \\ \frac{8}{3}t^3 - 2t^2 \end{pmatrix} + c_1 \begin{pmatrix} \frac{1}{2}t + \frac{1}{8} \\ t \end{pmatrix} + c_2 \begin{pmatrix} \frac{1}{2} \\ 1 \end{pmatrix}$$

The solution (15) is expressed according to the form given in Theorem 1.5. The first vector is, as you can easily verify, a *particular* solution $\boldsymbol{x}_p(t)$ of (14). The two vectors

$$\boldsymbol{v}_1 = \begin{pmatrix} \frac{1}{2}t + \frac{1}{8} \\ t \end{pmatrix} \quad \text{and} \quad \boldsymbol{v}_2 = \begin{pmatrix} \frac{1}{2} \\ 1 \end{pmatrix}$$

are likewise easily seen to be solutions to the associated *homogeneous* system

$$\begin{cases} \dfrac{dx_1}{dt} = 4x_1 - 2x_2 \\ \dfrac{dx_2}{dt} = 8x_1 - 4x_2 \end{cases}$$

The work in Example 1.8 shows that $\boldsymbol{v}_1$ and $\boldsymbol{v}_2$ span this solution space, which by Theorem 1.7 is 2-dimensional. So by Theorem 9.1.10, $\{\boldsymbol{v}_1, \boldsymbol{v}_2\}$ is a basis for the homogeneous solution space. So, just as in Section 1.6, the elimination method not only solves a given system of differential equations but also expresses the solution in the form of Theorem 1.5, and provides us with a basis of the solution space of the associated homogeneous problem.

A word of *caution* is in order when it comes to checking your answers to problems like (14). If we had chosen to eliminate x_2 instead of x_1, then instead of (15) we would have come to

(16) $$\boldsymbol{x}(t) = \begin{pmatrix} \frac{4}{3}t^3 \\ \frac{8}{3}t^3 - 2t^2 + t \end{pmatrix} + k_1 \begin{pmatrix} 1 \\ 2 \end{pmatrix} + k_2 \begin{pmatrix} t \\ 2t - \frac{1}{2} \end{pmatrix}.$$

This completely different *looking* description of the general solution actually describes the *same* set as (15). The description just utilizes a *different particular solution* $\boldsymbol{x}_p(t)$ and a *different basis for the homogeneous solution space.* Depending on how you carry out the method of elimination, then, your answer may or may not look anything like someone else's (e.g., the answer book's). The best check is to verify that your particular solution *is* a solution of the given system and that your candidates for a basis of the homogeneous solution space do actually *belong* to the space.

While we have described the method of elimination only for first order systems, it can also be used for systems that involve higher order derivatives *without* first using Theorem 1.2 to reduce them to equivalent first order systems involving more equations. It is usually easier to work with a *smaller number* of equations, so the method of our concluding example is recommended for such problems.

1.9 EXAMPLE. Solve the system

(17) $$\begin{cases} 2\dfrac{dx}{dt} + \dfrac{d^2y}{dt^2} = -x + 4y - 7e^{-t} \\ \dfrac{dx}{dt} - \dfrac{dy}{dt} = 2y - 3e^{-t} \end{cases}$$

Solution. Putting (17) into operator form, we get

$$\begin{cases} (2D+1)x + (D^2-4)y = -7e^{-t} \\ Dx - (D+2)y = -3e^{-t} \end{cases}$$

We can eliminate y if we apply $D-2$ to the second equation and add to the first. This gives

(18) $$\begin{aligned} (2D+1+D^2-2D)x &= -7e^{-t} + 3e^{-t} + 6e^{-t}, \\ (D^2+1)x &= 2e^{-t} \end{aligned}$$

We can solve (18) by the methods of Section 8.7 or 9.3. The roots of the auxiliary polynomial are $\pm i$, so the associated homogeneous problem has solution

$$x_0(t)=c_1\cos t+c_2\sin t.$$

The method of undetermined coefficients tells us to seek a particular solution of (18) of the form $x_p(t)=Ae^{-t}$. Then $\dot{x}_p=-Ae^{-t}$ and $\ddot{x}_p=Ae^{-t}$. So substituting into (18) we have

$$Ae^{-t}+Ae^{-t}=2e^{-t}.$$

Hence $A=1$, so (18) has general solution

(19) $$x(t)=c_1\cos t+c_2\sin t+e^{-t}.$$

If we substitute this into the second equation in (17), we have

$$-c_1\sin t+c_2\cos t-e^{-t}-\frac{dy}{dt}=2y-3e^{-t},$$

$$\frac{dy}{dt}+2y=-c_1\sin t+c_2\cos t+2e^{-t}.$$

This first order linear equation has integrating factor e^{2t}, so we get

$$\frac{d}{dt}(e^{2t}y)=-c_1e^{2t}\sin t+c_2e^{2t}\cos t+2e^{t}.$$

Using Formulas 29 and 30 in the Table of Integrals (or integration by parts), we obtain

$$e^{2t}y=c_1(\tfrac{1}{5}e^{2t}\cos t-\tfrac{2}{5}e^{2t}\sin t)+c_2(\tfrac{1}{5}e^{2t}\sin t+\tfrac{2}{5}e^{2t}\cos t)+c_3+2e^{t},$$

(20) $$y=\tfrac{1}{5}c_1(\cos t-2\sin t)+\tfrac{1}{5}c_2(2\cos t+\sin t)+c_3e^{-2t}+2e^{-t}.$$

Putting (19) and (20) in vector form, we have

$$\boldsymbol{x}=c_1\begin{pmatrix}\cos t\\ \tfrac{1}{5}\cos t-\tfrac{2}{5}\sin t\end{pmatrix}+c_2\begin{pmatrix}\sin t\\ \tfrac{2}{5}\cos t+\tfrac{1}{5}\sin t\end{pmatrix}+c_3\begin{pmatrix}0\\ e^{-2t}\end{pmatrix}+\begin{pmatrix}e^{-t}\\ 2e^{-t}\end{pmatrix}.$$

We have particular solution

$$\begin{pmatrix}e^{-t}\\ 2e^{-t}\end{pmatrix},$$

and our basis for the homogeneous solution space has *three* vectors in it even though (17) consisted of just two equations. The reason for this is that if we replaced (17) by an equivalent system of *first order* equations using Theorem 1.2, the single second order equation would be replaced by two first order equations. So (17) would lead to a system of *three* first order equations. The homogeneous solution space thus has dimension 3 by Theorem 1.7.

EXERCISES 10.1

In Exercises 1 to 16, solve the given system of differential equations.

1 $\begin{cases} \dfrac{dx}{dt} = x + 1 \\ \dfrac{dy}{dt} = -y + 2 \end{cases}$

2 $\begin{cases} \dfrac{dx}{dt} = 2x - 5 \\ \dfrac{dy}{dt} = -3y + 1 \end{cases}$

3 $\begin{cases} \dfrac{dx}{dt} = -x - 3e^{-2t} \\ \dfrac{dy}{dt} = -2x - y - 6e^{-2t} \end{cases}$

4 $\begin{cases} \dfrac{dx_1}{dt} = 2x_1 + 4x_2 + 2 \\ \dfrac{dx_2}{dt} = x_1 - x_2 + 4 \end{cases}$

5 $\begin{cases} \dfrac{dx}{dt} = 3x - 2y - e^{-t}\sin t \\ \dfrac{dy}{dt} = 4x - y + 2e^{-t}\cos t \end{cases}$

6 $\begin{cases} \dfrac{dx}{dt} = 2x - 5y - e^{t}\sin 2t \\ \dfrac{dy}{dt} = x - 2y + 2e^{t}\cos 2t \end{cases}$

7 $\begin{cases} \dfrac{dx}{dt} = 3x - 4y + e^{t} \\ \dfrac{dy}{dt} = x - y - e^{t} \end{cases}$

8 $\begin{cases} \dfrac{dx}{dt} = x - y + e^{t} \\ \dfrac{dy}{dt} = x + 3y + 2e^{t} \end{cases}$

9 $\begin{cases} \dfrac{dx}{dt} = 3x + y - 2z \\ \dfrac{dy}{dt} = -x + 2y + z \\ \dfrac{dz}{dt} = 4x + y - 3z \end{cases}$

10 $\begin{cases} \dfrac{dx}{dt} = -x + y \\ \dfrac{dy}{dt} = 2x - 2y + 2z \\ \dfrac{dz}{dt} = -y - z \end{cases}$

11 $\begin{cases} \dfrac{dx}{dt} = x - 3y + 2z \\ \dfrac{dy}{dt} = -2y + 2z \\ \dfrac{dz}{dt} = x - 5y + 2z \end{cases}$

12 $\begin{cases} \dfrac{dx}{dt} = -x + z \\ \dfrac{dy}{dt} = 2z \\ \dfrac{dz}{dt} = x - 2y - 3z \end{cases}$

13 $\begin{cases} \dfrac{dx_1}{dt} + \dfrac{dx_2}{dt} = 2x_1 + 4x_2 \\ 2\dfrac{dx_1}{dt} + \dfrac{dx_2}{dt} = 2x_1 + 6x_2 \end{cases}$

14 $\begin{cases} 2\dfrac{dx_1}{dt} + \dfrac{dx_2}{dt} = x_1 + x_2 \\ \dfrac{dx_1}{dt} + \dfrac{dx_2}{dt} = -2x_1 - x_2 \end{cases}$

15 $\begin{cases} -\dfrac{dx}{dt} + \dfrac{d^2y}{dt^2} = 2x + 4y + 4t \\ \dfrac{dx}{dt} + \dfrac{dy}{dt} = -3x - 7y \end{cases}$

16 $\begin{cases} \dfrac{d^2x_1}{dt^2} + \dfrac{dx_2}{dt} = x_1 - x_2 + 1 \\ \dfrac{dx_1}{dt} + \dfrac{d^2x_2}{dt^2} = x_1 - x_2 \end{cases}$

In Exercises 17 and 18, show that the given system is inconsistent, so that it cannot be put into the form (8).

17 $\begin{cases} \dfrac{d^2x}{dt^2} + \dfrac{dy}{dt} = x + y + \sin t \\ \dfrac{dx}{dt} = -x - y + 2e^{t} \end{cases}$

18 $\begin{cases} \dfrac{d^2x}{dt^2} + \dfrac{dy}{dt} = 4x + 2y + 5e^{t} \\ \dfrac{dx}{dt} = -2x - y + \cos t \end{cases}$

In Exercises 19 and 20, solve the initial value problem.

19 $\begin{cases} \dfrac{dx}{dt} = -4x - 6y + 9e^{-3t} \\ \dfrac{dy}{dt} = -x - 5y + e^{-3t} \end{cases}, \quad \boldsymbol{x}(0) = (-9, 4)$

20 $\begin{cases} \dfrac{d^2x}{dt^2} = x + 2y + t \\ \dfrac{d^2y}{dt^2} = 3x + 2y \end{cases}, \quad \boldsymbol{x}(0) = (0, 0), \dot{\boldsymbol{x}}(0) = (0, 1)$

21 (a) Show that $C^\infty(\boldsymbol{R}, \boldsymbol{R}^n)$ is a real vector space.
(b) Show that $T: C^\infty(\boldsymbol{R}, \boldsymbol{R}^n) \to C^\infty(\boldsymbol{R}, \boldsymbol{R}^n)$ defined by (10) is a linear transformation.

22 Prove Theorem 1.5.

23 Show that $Ker(D - T)$ in Theorem 1.7 is a real vector space, using the approach of Theorem 9.1.3.

24 Show that $F_{t_0}: Ker(D - T) \to \boldsymbol{R}^n$ in Theorem 1.7 is a linear transformation.

2 EIGENVALUES AND SOLUTION OF HOMOGENEOUS LINEAR SYSTEMS

The elimination technique presented in the last section works well when we have a two-dimensional first order linear system

(1) $$\frac{d\boldsymbol{x}}{dt} = A\boldsymbol{x}(t) + \boldsymbol{b}(t)$$

to solve. If we want to develop solution methods for n-dimensional systems (1), though, then we will have to come up with a better scheme. Unlike the situation in Section 1.6, *we don't have a matrix formulation* of the elimination method, so in higher dimensions the coordinate-wise calculations it requires will become too unwieldy to perform. This section is devoted to the development of a more efficient approach to solving (1).

Previously we have enjoyed great success in trying to develop vector methods in imitation of methods we already know for the one-dimensional analogues of n-dimensional problems. Thus it seems appropriate here to try to develop a solution method for (1) analogous to the one in Section 8.3 for the first order linear equation

(2) $$\frac{dx}{dt} = ax + b,$$

which is the one-variable version of (1). Indeed, we already have a start in Theorem 1.5, the analogue of Theorem 8.3.5 for (2). We can thus follow the approach of Section 8.6 and solve (1) in the familiar two-step sequence we have previously followed in solving linear problems. That is, we first solve the associated *homogeneous* problem

(3) $$\boxed{\frac{d\boldsymbol{x}}{dt} = A\boldsymbol{x}}$$

and then find a particular solution $\mathbf{x}_p(t)$ for (1). This section is devoted to the first step. The second step will be carried out in Section 3.

The one-dimensional analogue of (3) is

$$\frac{dx}{dt} = ax, \tag{4}$$

which we know has general solution $x(t) = ce^{at}$, where $c \in \mathbf{R}^1$ is an arbitrary constant. Now what sort of analogous solutions should be expected for (3)? To get an idea, consider a very special case of (3) in which A is a *diagonal* matrix

$$A = \text{diag}(\lambda_1, \lambda_2, \ldots, \lambda_n) \quad \text{for} \quad \lambda_i \in \mathbf{R}.$$

Then in coordinate form the system (3) is

$$\text{(3}'\text{)} \quad \begin{cases} \dfrac{dx_1}{dt} = \lambda_1 x_1 \\ \dfrac{dx_2}{dt} = \qquad \lambda_2 x_2 \\ \quad \vdots \qquad\qquad \ddots \\ \dfrac{dx_n}{dt} = \qquad\qquad \lambda_n x_n. \end{cases}$$

The system (3′) can be solved *immediately*, since each equation is of the form (4). We have, in fact, $x_i = c_i e^{\lambda_i t}$. So the general solution is given by

$$\mathbf{x} = \begin{pmatrix} c_1 \\ 0 \\ \cdot \\ \cdot \\ \cdot \\ 0 \end{pmatrix} e^{\lambda_1 t} + \begin{pmatrix} 0 \\ c_2 \\ \cdot \\ \cdot \\ \cdot \\ 0 \end{pmatrix} e^{\lambda_2 t} + \ldots + \begin{pmatrix} 0 \\ 0 \\ \cdot \\ \cdot \\ \cdot \\ c_n \end{pmatrix} e^{\lambda_n t}$$

$$= \mathbf{c}_1 e^{\lambda_1 t} + \mathbf{c}_2 e^{\lambda_2 t} + \ldots + \mathbf{c}_n e^{\lambda_n t}.$$

Thus the set $B = \{\mathbf{c}_1 e^{\lambda_1 t}, \ldots, \mathbf{c}_n e^{\lambda_n t}\}$ spans the solution space of (3), which by Theorem 1.7 is n-dimensional. By Theorem 9.1.10, then, B is a basis for the solution space.

Considering this special case, it seems reasonable to seek solutions of (3) of the form

$$\mathbf{x} = \mathbf{c}e^{\lambda t} \tag{5}$$

for certain real numbers λ and certain vectors $\mathbf{c} \in \mathbf{R}^n$. The question is, *which real numbers λ and vectors $\mathbf{c}$ will give solutions to* (3)? This is easy to answer. If we differentiate (5) and substitute into (3), we get $\dfrac{d\mathbf{x}}{dt} = \lambda \mathbf{c}e^{\lambda t}$, and then

$$\lambda \mathbf{c}e^{\lambda t} = A\mathbf{c}e^{\lambda t},$$

which if multiplied through by the nonzero quantity $e^{-\lambda t}$ becomes

(6) $$A\mathbf{c} = \lambda\mathbf{c}.$$

So in order for (5) to give a solution of (3), λ and $\mathbf{c}$ must satisfy (6). The case $\mathbf{c}=\mathbf{0}$ is of no interest, since the corresponding solution (5) is then the trivial solution $\boldsymbol{x}=\mathbf{0}$ to (3), which one needn't be very clever to find. So we should seek solutions $\mathbf{c}\neq\mathbf{0}$ and λ to (6). Such solutions are so important in mathematics that they have special names.

2.1 **DEFINITION.** If A is an n-by-n matrix of real numbers, then an **eigenvalue** of A is a number λ such that there is a nonzero vector $\mathbf{c}$ (called an **eigenvector belonging to** λ) for which (6) holds.

The first question to ask about eigenvalues and eigenvectors is, how can they be found? Our next result says that the answer lies in solving the polynomial equation

$$p(\lambda)=0$$

for λ, where $p(x)$ is the analogue (see Exercise 15 below) of the auxiliary polynomial which in Chapter 9 was associated with an n-th order linear differential equation.

2.2 **THEOREM.** The n-by-n matrix A has eigenvalue λ if and only if λ satisfies the **characteristic equation**

(7) $$\det(A-\lambda I_n)=0$$

where I_n is the n-by-n identity matrix. (The polynomial $\det(A-\lambda I_n)$ is called the **characteristic polynomial** of A.)

Proof. We know that λ is an eigenvalue if and only if (6) holds for some $\mathbf{c}\neq\mathbf{0}$. But this is equivalent to the existence of a nontrivial solution $\boldsymbol{x}=\mathbf{c}$ of the homogeneous system of n linear equations

(8) $$(A-\lambda I_n)\boldsymbol{x}=\mathbf{0},$$

since $(A-\lambda I_n)\boldsymbol{x}=A\boldsymbol{x}-\lambda\boldsymbol{x}$. By Exercise 25 of Exercises 6.4, (8) has a nontrivial solution if and only if the matrix $A-\lambda I_n$ is not invertible. By Theorem 6.7.3, this in turn is equivalent to $\det(A-\lambda I_n)=0$. QED

This theorem can be used in conjunction with the methods of Section 1.6 to find eigenvalues and corresponding eigenvectors.

2.3 **EXAMPLE.** Find the eigenvalues of

$$A=\begin{pmatrix} 2 & -9 \\ -1 & 2 \end{pmatrix},$$

and for each eigenvalue find all corresponding eigenvectors.

Solution. We have

$$\det(A-\lambda I_2)=\det\begin{pmatrix}2-\lambda & -9\\ -1 & 2-\lambda\end{pmatrix}$$
$$=(2-\lambda)^2-9=4-4\lambda+\lambda^2-9$$
$$=\lambda^2-4\lambda-5=(\lambda-5)(\lambda+1).$$

So the eigenvalues are $\lambda=5$ and $\lambda=-1$. To find corresponding eigenvectors, we solve (6) for $\boldsymbol{c}$. For $\lambda=5$ we have $A\boldsymbol{c}=5\boldsymbol{c}$,

$$\begin{pmatrix}2 & -9\\ -1 & 2\end{pmatrix}\begin{pmatrix}c_1\\ c_2\end{pmatrix}=\begin{pmatrix}5c_1\\ 5c_2\end{pmatrix},\ \begin{pmatrix}2c_1-9c_2\\ -c_1+2c_2\end{pmatrix}=\begin{pmatrix}5c_1\\ 5c_2\end{pmatrix},$$
$$\begin{pmatrix}-3c_1-9c_2\\ -c_1-3c_2\end{pmatrix}=\begin{pmatrix}0\\ 0\end{pmatrix}.$$

This system is clearly equivalent to the single equation $c_1=-3c_2$. So $\boldsymbol{c}=c_2(-3, 1)$, where c_2 is arbitrary. For $\lambda=-1$ we get

$$\begin{pmatrix}2 & -9\\ -1 & 2\end{pmatrix}\begin{pmatrix}c_1\\ c_2\end{pmatrix}=\begin{pmatrix}-c_1\\ -c_2\end{pmatrix},\qquad \begin{pmatrix}2c_1-9c_2\\ -c_1+2c_2\end{pmatrix}=\begin{pmatrix}-c_1\\ -c_2\end{pmatrix},$$
$$\begin{pmatrix}3c_1-9c_2\\ -c_1+3c_2\end{pmatrix}=\begin{pmatrix}0\\ 0\end{pmatrix}.$$

This system is equivalent to the single equation $c_1=3c_2$. So $\boldsymbol{c}=c_2(3, 1)$, where c_2 is arbitrary.

Now we can return to the solution of (3).

2.4 **THEOREM.** The function $\boldsymbol{x}(t)=\boldsymbol{c}e^{\lambda t}$ is a nontrivial solution of (3) if and only if λ is an eigenvalue of A and $\boldsymbol{c}$ is an eigenvector belonging to λ.

Proof. We showed in the discussion leading up to (6) that if $\boldsymbol{x}(t)=\boldsymbol{c}e^{\lambda t}$ is a nontrivial solution to (3), then $\boldsymbol{c}$ must be an eigenvector of A belonging to the eigenvalue λ. Conversely, if $\boldsymbol{c}$ is an eigenvector belonging to λ, then we have $A\boldsymbol{c}=\lambda\boldsymbol{c}$. Hence for $\boldsymbol{x}=\boldsymbol{c}e^{\lambda t}$,

$$\frac{d\boldsymbol{x}}{dt}=\lambda\boldsymbol{c}e^{\lambda t}=A\boldsymbol{c}e^{\lambda t}=A\boldsymbol{x}.$$

Hence $\boldsymbol{x}$ is a solution of (3). QED

In view of this result, if we can find n linearly independent solutions to (3) of the form (5), then we will have a basis for the solution space of (3). We thus will have solved (3) completely. There is, as in the case of a single n-th order linear differential equation, a *Wronskian determinant* to help us decide whether a set of n vector functions is linearly independent.

2.5 DEFINITION. If $\boldsymbol{x}_1(t), \boldsymbol{x}_2(t), \ldots, \boldsymbol{x}_n(t): \boldsymbol{R} \to \boldsymbol{R}^n$ are vector functions

$$\boldsymbol{x}_i(t) = \begin{pmatrix} x_{1i}(t) \\ x_{2i}(t) \\ \cdot \\ \cdot \\ \cdot \\ x_{ni}(t) \end{pmatrix},$$

then the **Wronskian** of these functions at t_0 is

$$\det W(t_0) = \det(\boldsymbol{x}_1(t_0), \boldsymbol{x}_2(t_0), \ldots, \boldsymbol{x}_n(t_0)).$$

For example, if

$$\boldsymbol{x}_1(t) = \begin{pmatrix} -3e^{5t} \\ e^{5t} \end{pmatrix}, \boldsymbol{x}_2(t) = \begin{pmatrix} 3e^{-t} \\ e^{-t} \end{pmatrix},$$

then the Wronskian of the two functions at $t = 0$ is

$$\det W(0) = \det\begin{pmatrix} -3 & 3 \\ 1 & 1 \end{pmatrix} = -6.$$

At a general point t_0,

$$\det W(t_0) = \det\begin{pmatrix} -3e^{5t_0} & 3e^{-t_0} \\ e^{5t_0} & e^{-t_0} \end{pmatrix} = -6e^{4t_0}.$$

We have the following analogue of Exercise 11, Exercises 9.1, as a basic criterion for linear independence.

2.6 LEMMA. Let the $\boldsymbol{x}_i(t)$ be as in Definition 2.5. If for some t_0 in an interval I we have $\det W(t_0) \neq 0$, then the set S of functions $\boldsymbol{x}_i(t)$ is linearly independent on I.

Proof. If S were linearly dependent on I, then for some choice of constants c_1, $c_2, \ldots, c_n$ that are not all zero, we would have

$$c_1\boldsymbol{x}_1(t) + \ldots + c_n\boldsymbol{x}_n(t) = \boldsymbol{0}, \quad \text{for all } t \in I.$$

That is, $W(t)\mathbf{c} = \boldsymbol{0}$, where

$$\mathbf{c} = \begin{pmatrix} c_1 \\ c_2 \\ \cdot \\ \cdot \\ \cdot \\ c_n \end{pmatrix}.$$

Then the system $W(t)\boldsymbol{x} = \boldsymbol{0}$ would have a nontrivial solution for all t in I. Hence, by Exercise 25 of Exercises 6.4, $W(t)$ would not be invertible for any $t \in I$. Then $\det W(t) = 0$ for all $t \in I$ by Theorem 6.7.3. But by hypothesis, $\det W(t_0) \neq 0$. Thus S must be linearly independent. QED

Thus, for instance, the vector functions considered after Definition 2.5 are linearly independent on any interval containing 0, since their Wronskian is nonzero at $t=0$.

We can now give a criterion for n solutions of (3) of the form (5) to be linearly independent, and hence a basis for the solution space of (3).

2.7 **THEOREM.** Suppose that the n-by-n matrix A has n linearly independent eigenvectors $\boldsymbol{c}_1, \boldsymbol{c}_2, \ldots, \boldsymbol{c}_n$ where $\boldsymbol{c}_i$ belongs to the eigenvalue λ_i. Then the set $S=\{\boldsymbol{x}_1(t), \ldots, \boldsymbol{x}_n(t)\}$, where $\boldsymbol{x}_i(t)=\boldsymbol{c}_i e^{\lambda_i t}$, is linearly independent on any interval I containing 0.

Proof. We have $\det W(0)=\det(\boldsymbol{c}_1, \boldsymbol{c}_2, \ldots, \boldsymbol{c}_n)$. We claim that $\det W(0)\neq 0$. This is proved by noting that the system of equations $y_1\boldsymbol{c}_1+y_2\boldsymbol{c}_2+\ldots+y_n\boldsymbol{c}_n=\mathbf{0}$ can have only the trivial solution

$$y_1=y_2=\ldots=y_n=0$$

by the linear independence of $\{\boldsymbol{c}_1, \ldots, \boldsymbol{c}_n\}$. In matrix form this system is

$$W(0)\mathbf{y}=\mathbf{0}.$$

So by Theorem 6.7.3 and Exercise 25 of Exercises 6.4, $\det W(0)\neq 0$. Thus, by Lemma 2.6, S is linearly independent on I. QED

How can we determine whether A has n linearly independent eigenvectors? The following result of linear algebra gives two important circumstances in which this happens.

2.8 **THEOREM.** The n-by-n matrix A has n linearly independent eigenvectors if either

(a) A has n distinct eigenvalues

or

(b) A is a symmetric matrix.

Proof. We leave (b) for a linear algebra course. It is, however, easy to prove (a) by mathematical induction. We will in fact show that:

(*) If A is a square matrix of any size (say m-by-m) and $\{\lambda_1, \lambda_2, \ldots, \lambda_n\}$ is a set of n distinct eigenvalues of A, then $\{\boldsymbol{c}_1, \boldsymbol{c}_2, \ldots, \boldsymbol{c}_n\}$ is a linearly independent set, where $\boldsymbol{c}_i$ is any eigenvector corresponding to the eigenvalue λ_i for $i=1, 2, \ldots, n$.

This is clear in case $n=1$. Suppose that it holds for $n=k$, and let $n=k+1$. Then $\lambda_1, \lambda_2, \ldots, \lambda_k, \lambda_{k+1}$ are distinct eigenvalues with corresponding eigenvectors $\boldsymbol{c}_1, \boldsymbol{c}_2, \ldots, \boldsymbol{c}_k, \boldsymbol{c}_{k+1}$. By our induction hypothesis, $\{\boldsymbol{c}_1, \boldsymbol{c}_2, \ldots, \boldsymbol{c}_k\}$ is linearly independent. Let $a_1, a_2, \ldots, a_k, a_{k+1}$ be scalars such that

(9) $$a_1\boldsymbol{c}_1+\ldots+a_k\boldsymbol{c}_k+a_{k+1}\boldsymbol{c}_{k+1}=\mathbf{0}.$$

Multiplying through by λ_{k+1} we get

(10) $$\lambda_{k+1}a_1\boldsymbol{c}_1+\ldots+\lambda_{k+1}a_k\boldsymbol{c}_k+\lambda_{k+1}a_{k+1}\boldsymbol{c}_{k+1}=\mathbf{0}.$$

If we apply A to both sides of (9), we have

(11) $$a_1\lambda_1\mathbf{c}_1 + \ldots + a_k\lambda_k\mathbf{c}_k + a_{k+1}\lambda_{k+1}\mathbf{c}_{k+1} = \mathbf{0}.$$

since $A\mathbf{c}_i = \lambda_i\mathbf{c}_i$ for each i. Subtracting (11) from (10), we have

$$a_1(\lambda_{k+1} - \lambda_1)\mathbf{c}_1 + \ldots + a_k(\lambda_{k+1} - \lambda_k)\mathbf{c}_k = \mathbf{0}.$$

Since $\{\mathbf{c}_1, \ldots, \mathbf{c}_k\}$ is linearly independent, each coefficient

$$a_i(\lambda_{k+1} - \lambda_i) = 0.$$

Then for each $i = 1, \ldots, k$ we have $a_i = 0$, since $\lambda_{k+1} - \lambda_i \neq 0$. (Remember, the λ's are *distinct*.) That means that (9) reduces to

$$a_{k+1}\mathbf{c}_{k+1} = \mathbf{0}.$$

Since $\mathbf{c}_{k+1}$ is an eigenvector, $\mathbf{c}_{k+1} \neq \mathbf{0}$. Therefore, $a_{k+1} = 0$. This means that every coefficient in (9) is 0, so that

$$\{\mathbf{c}_1, \mathbf{c}_2, \ldots, \mathbf{c}_k, \mathbf{c}_{k+1}\}$$

is a linearly independent set by Corollary 1.7.3. Thus (*) holds for $n = k+1$ whenever it holds for $n = k$. Since it holds for $n = 1$, (*) holds for every positive integer n. Thus (a) holds for all n. QED

The next three examples illustrate the importance of Theorem 2.8 in solving systems of linear differential equations. First, note that if the n-by-n matrix A has n distinct eigenvalues λ_i for $i = 1, 2, \ldots, n$, with corresponding eigenvectors $\mathbf{c}_i$, then the functions $\mathbf{x}_i(t) = \mathbf{c}_i e^{\lambda_i t}$ in Theorem 2.7 form a basis for the solution space of

(3) $$\frac{d\mathbf{x}}{dt} = A\mathbf{x}.$$

Thus the general solution of (3) has the form

$$\mathbf{x} = \mathbf{x}(t) = \sum_{i=1}^{n} \mathbf{c}_i e^{\lambda_i t},$$

where $\mathbf{c}_i$ is an eigenvector corresponding to the eigenvalue λ_i.

2.9 EXAMPLE. Solve the system

$$\begin{cases} \dfrac{dx}{dt} = 2x - 9y \\ \dfrac{dy}{dt} = -x + 2y. \end{cases}$$

Solution. In matrix-vector form (3) we have

$$A = \begin{pmatrix} 2 & -9 \\ -1 & 2 \end{pmatrix},$$

the matrix of Example 2.3. We know that the eigenvalues of A are $\lambda=5$ and $\lambda=-1$. We also found corresponding basic eigenvectors to be $\boldsymbol{c}=(-3, 1)$ for $\lambda=5$ and $\boldsymbol{c}=(3, 1)$ for $\lambda=-1$. These are clearly linearly independent, as promised by Theorem 2.8. Hence, so are the corresponding solutions of the form (5) by Theorem 2.7. Thus a basis for the solution space of the given system is $B=\{\boldsymbol{x}_1(t), \boldsymbol{x}_2(t)\}$, where

$$\boldsymbol{x}_1(t)=\begin{pmatrix}-3\\1\end{pmatrix}e^{5t}, \qquad \boldsymbol{x}_2(t)=\begin{pmatrix}3\\1\end{pmatrix}e^{-t}.$$

We can write the general solution as

$$\boldsymbol{x}=c_1\begin{pmatrix}-3\\1\end{pmatrix}e^{5t}+c_2\begin{pmatrix}3\\1\end{pmatrix}e^{-t}.$$

In solving (3), the eigenvalues are computed by solving the characteristic Equation (7), and they may well be complex numbers. If you look over Theorems 2.4 and 2.7, you will see that we *never* assumed that the λ_i had to be *real.* All the calculations used to prove those results are in fact still valid in case some or all the λ_i are *complex.* The set S in Theorem 2.7 will then be a basis for the *complex solution space* of (3). As in Section 9.2 (Theorem 9.2.10), we get a basis for the real solution space by taking real and imaginary parts of the members of S. The following example illustrates the details.

2.10 EXAMPLE. Solve the system

$$\begin{cases}\dfrac{dx}{dt}=2x+4y\\[2mm] \dfrac{dy}{dt}=-2x-2y\end{cases}$$

Solution. In the form (3),

$$A=\begin{pmatrix}2 & 4\\-2 & -2\end{pmatrix}.$$

To find the eigenvalues, we solve the characteristic equation

$$\det(A-\lambda I_2)=\det\begin{pmatrix}2-\lambda & 4\\-2 & -2-\lambda\end{pmatrix}=-4+\lambda^2+8=0,$$

$$\lambda^2+4=0, \qquad \lambda=\pm 2i.$$

To find the corresponding eigenvectors, we solve $A\boldsymbol{c}=\lambda\boldsymbol{c}$. For $\lambda=2i$ we get

$$\begin{pmatrix}2 & 4\\-2 & -2\end{pmatrix}\begin{pmatrix}c_1\\c_2\end{pmatrix}=\begin{pmatrix}2ic_1\\2ic_2\end{pmatrix}, \qquad \begin{pmatrix}2c_1+4c_2\\-2c_1-2c_2\end{pmatrix}=\begin{pmatrix}2ic_1\\2ic_2\end{pmatrix},$$

$$\begin{pmatrix}(1-i)c_1+2c_2\\-c_1-(1+i)c_2\end{pmatrix}=\begin{pmatrix}0\\0\end{pmatrix}.$$

This is equivalent to the single equation $c_2=\frac{1}{2}(i-1)c_1$, since the second equation gives

$$c_2 = -\frac{1}{1+i} c_1 = -\frac{1-i}{2} c_1 = \frac{i-1}{2} c_1.$$

So $\boldsymbol{c} = \frac{1}{2}c_1(2, i-1)$. For $\lambda = -2i$ we get

$$\begin{pmatrix} 2 & 4 \\ -2 & -2 \end{pmatrix}\begin{pmatrix} c_1 \\ c_2 \end{pmatrix} = \begin{pmatrix} -2ic_1 \\ -2ic_2 \end{pmatrix}, \quad \begin{pmatrix} 2c_1 + 4c_2 \\ -2c_1 - 2c_2 \end{pmatrix} = \begin{pmatrix} -2ic_1 \\ -2ic_2 \end{pmatrix},$$

$$\begin{pmatrix} (1+i)c_1 + 2c_2 \\ -c_1 - (1-i)c_2 \end{pmatrix} = \begin{pmatrix} 0 \\ 0 \end{pmatrix}.$$

This is equivalent to the single equation $c_2 = -\frac{1}{2}(1+i)c_1$. So $\boldsymbol{c} = \frac{1}{2}c_1(2, -1-i)$. A basis for the complex solution space is then

$$\left\{\begin{pmatrix} 2 \\ i-1 \end{pmatrix} e^{2it}, \quad \begin{pmatrix} 2 \\ -1-i \end{pmatrix} e^{-2it}\right\}$$

Since $e^{2it} = \cos 2t + i \sin 2t$, we can write

$$\begin{aligned} \begin{pmatrix} 2 \\ i-1 \end{pmatrix} e^{2it} &= \begin{pmatrix} 2\cos 2t + 2i\sin 2t \\ (i-1)\cos 2t + i(i-1)\sin 2t \end{pmatrix} \\ &= \begin{pmatrix} 2\cos 2t + 2i\sin 2t \\ -\cos 2t - \sin 2t + i(\cos 2t - \sin 2t) \end{pmatrix}. \end{aligned}$$

Taking real and imaginary parts, we have

$$\boldsymbol{x}_1(t) = \begin{pmatrix} 2\cos 2t \\ -\cos 2t - \sin 2t \end{pmatrix} \quad \text{and} \quad \boldsymbol{x}_2(t) = \begin{pmatrix} 2\sin 2t \\ \cos 2t - \sin 2t \end{pmatrix}$$

These make up the basic real solution. We can thus write the general real solution as

$$\boldsymbol{x}(t) = c_1\begin{pmatrix} 2\cos 2t \\ -\cos 2t - \sin 2t \end{pmatrix} + c_2\begin{pmatrix} 2\sin 2t \\ \cos 2t - \sin 2t \end{pmatrix}$$

What if we had taken the real and imaginary parts of the second complex basis vector instead of the first in Example 2.10? Well,

$$\begin{aligned} \begin{pmatrix} 2 \\ -1-i \end{pmatrix} e^{-2it} &= \begin{pmatrix} 2 \\ -1-i \end{pmatrix}(\cos 2t - i\sin 2t) \\ &= \begin{pmatrix} 2\cos t \\ -\cos 2t - \sin 2t \end{pmatrix} + i\begin{pmatrix} -2\sin t \\ -\cos 2t + \sin 2t \end{pmatrix} \end{aligned}$$

So we would have obtained the *slightly different basis* $\{\boldsymbol{x}_1(t), -\boldsymbol{x}_2(t)\}$ for the solution space. As a practical matter, then, we can get by with a little less computation than we carried out in Example 2.10. For each conjugate pair λ and $\bar{\lambda}$ of eigenvalues of A, it suffices to compute just *one* eigenvector, say the one corresponding to λ, if we want only the real solution to (3). Note the parallel here with Sections 8.6 and 9.2 for the case of complex roots of the auxiliary polynomial.

Even when A fails to have n distinct eigenvalues, there may still be n linearly independent eigenvectors. The next example, which illustrates Theorem 2.8(b), shows you how to apply Theorem 2.7 in such circumstances.

2.11 EXAMPLE. Solve the system

$$\begin{cases} \dfrac{dx}{dt} = 4x - 2y + 2z \\ \dfrac{dy}{dt} = -2x + y - z \\ \dfrac{dz}{dt} = 2x - y + z. \end{cases}$$

Solution. Here we have

$$A = \begin{pmatrix} 4 & -2 & 2 \\ -2 & 1 & -1 \\ 2 & -1 & 1 \end{pmatrix},$$

which is symmetric; by Theorem 2.8(b), there will be a basis of eigenvectors. To find the eigenvalues of A, we solve the characteristic equation

$$\det(A - \lambda I_3) = \det\begin{pmatrix} 4-\lambda & -2 & 2 \\ -2 & 1-\lambda & -1 \\ 2 & -1 & 1-\lambda \end{pmatrix} = 0,$$

$$\begin{aligned} (4-\lambda)(1-\lambda)^2 + 4 + 4 - 4(1-\lambda) - (4-\lambda) - 4(1-\lambda) &= 0, \\ (4-\lambda)(1-2\lambda+\lambda^2) + 8 - 4 + 4\lambda - 4 + \lambda - 4 + 4\lambda &= 0, \\ 4 - 9\lambda + 6\lambda^2 - \lambda^3 - 4 + 9\lambda &= 0, \\ \lambda^3 - 6\lambda^2 &= 0, \\ \lambda^2(\lambda - 6) &= 0. \end{aligned}$$

So the eigenvalues are $\lambda = 6$ and the double root $\lambda = 0$. Solving for the corresponding eigenvectors, we have for $\lambda = 6$

$$\begin{pmatrix} 4 & -2 & 2 \\ -2 & 1 & -1 \\ 2 & -1 & 1 \end{pmatrix}\begin{pmatrix} x \\ y \\ z \end{pmatrix} = \begin{pmatrix} 6x \\ 6y \\ 6z \end{pmatrix}, \qquad \begin{pmatrix} 4x - 2y + 2z \\ -2x + y - z \\ 2x - y + z \end{pmatrix} = \begin{pmatrix} 6x \\ 6y \\ 6z \end{pmatrix},$$

$$\begin{pmatrix} -2x - 2y + 2z \\ -2x - 5y - z \\ 2x - y - 5z \end{pmatrix} = \begin{pmatrix} 0 \\ 0 \\ 0 \end{pmatrix}, \qquad \begin{pmatrix} x + y - z \\ 2x + 5y + z \\ 2x - y - 5z \end{pmatrix} = \begin{pmatrix} 0 \\ 0 \\ 0 \end{pmatrix}.$$

We can solve this system by Gaussian elimination. In matrix form we have

$$\begin{pmatrix} 1 & 1 & -1 & 0 \\ 2 & 5 & 1 & 0 \\ 2 & -1 & -5 & 0 \end{pmatrix} \xrightarrow[\text{Add } -2R_1 \text{ to } R_3]{\text{Add } -2R_1 \text{ to } R_2} \begin{pmatrix} 1 & 1 & -1 & 0 \\ 0 & 3 & 3 & 0 \\ 0 & -3 & -3 & 0 \end{pmatrix} \xrightarrow[\frac{1}{3}R_2]{\text{Add } R_2 \text{ to } R_3}$$

$$\begin{pmatrix} 1 & 1 & -1 & 0 \\ 0 & 1 & 1 & 0 \\ 0 & 0 & 0 & 0 \end{pmatrix} \xrightarrow[\text{to } R_1]{\text{Add } -R_2} \begin{pmatrix} 1 & 0 & -2 & 0 \\ 0 & 1 & 1 & 0 \\ 0 & 0 & 0 & 0 \end{pmatrix}.$$

Thus $x = 2z$, $y = -z$, and $z = z$. So a basis for the **eigenspace** (see Exercise 19 below) of eigenvectors belonging to $\lambda = 6$ is $(2, -1, 1)$.

For $\lambda = 0$ we have

$$\begin{pmatrix} 4 & -2 & 2 \\ -2 & 1 & -1 \\ 2 & -1 & 1 \end{pmatrix} \begin{pmatrix} x \\ y \\ z \end{pmatrix} = \begin{pmatrix} 0 \\ 0 \\ 0 \end{pmatrix}.$$

In matrix form,

$$\begin{pmatrix} 2 & -1 & 1 & 0 \\ -2 & 1 & -1 & 0 \\ 2 & -1 & 1 & 0 \end{pmatrix} \xrightarrow[\text{to } R_3]{\text{Add } R_2}$$

$$\begin{pmatrix} 2 & -1 & 1 & 0 \\ -2 & 1 & -1 & 0 \\ 0 & 0 & 0 & 0 \end{pmatrix} \xrightarrow[\frac{1}{2}R_1]{\text{Add } R_1 \text{ to } R_2} \begin{pmatrix} 1 & -\frac{1}{2} & -\frac{1}{2} & 0 \\ 0 & 0 & 0 & 0 \\ 0 & 0 & 0 & 0 \end{pmatrix}.$$

Thus $x = \frac{1}{2}y + \frac{1}{2}z$, $y = y$, and $z = z$, so

$$\begin{pmatrix} x \\ y \\ z \end{pmatrix} = y\begin{pmatrix} \frac{1}{2} \\ 1 \\ 0 \end{pmatrix} + z\begin{pmatrix} \frac{1}{2} \\ 0 \\ 1 \end{pmatrix}.$$

Hence the eigenspace has basis $\{(\frac{1}{2}, 1, 0), (\frac{1}{2}, 0, 1)\}$. The given system's solution space then has basis

$$\left\{\begin{pmatrix} 2 \\ -1 \\ 1 \end{pmatrix} e^{6t}, \begin{pmatrix} \frac{1}{2} \\ 1 \\ 0 \end{pmatrix} e^{0t}, \begin{pmatrix} \frac{1}{2} \\ 0 \\ 1 \end{pmatrix} e^{0t}\right\}.$$

The general solution is therefore given by

$$\boldsymbol{x}(t) = c_1\begin{pmatrix} 2 \\ -1 \\ 1 \end{pmatrix} e^{6t} + c_2\begin{pmatrix} \frac{1}{2} \\ 1 \\ 0 \end{pmatrix} + c_3\begin{pmatrix} \frac{1}{2} \\ 0 \\ 1 \end{pmatrix}.$$

What happens in case A has a repeated eigenvalue λ of multiplicity k, but *doesn't* have as many as k linearly independent eigenvectors belonging to λ? A full discussion of this case requires more sophisticated linear algebra than we can go into. In lieu of a full discussion, we conclude the section with an example of how a basis for the solution space of (3) can still be found in this situation in the low-dimensional cases $k = 2$ and $k = 3$.

2.12 EXAMPLE. Solve the system

$$(12)\qquad \begin{cases} \dfrac{dx}{dt} = -2x + 3y \\ \dfrac{dy}{dt} = \qquad -2y. \end{cases}$$

Solution. Here

$$A = \begin{pmatrix} -2 & 3 \\ 0 & -2 \end{pmatrix}.$$

We have

$$\det(A - \lambda I_2) = \det\begin{pmatrix} -2-\lambda & 3 \\ 0 & -2-\lambda \end{pmatrix} = (\lambda + 2)^2 = 0$$

when $\lambda = -2$. If $\begin{pmatrix} x \\ y \end{pmatrix}$ is an eigenvector belonging to -2, then we have

$$\begin{pmatrix} -2 & 3 \\ 0 & -2 \end{pmatrix}\begin{pmatrix} x \\ y \end{pmatrix} = \begin{pmatrix} -2x \\ -2y \end{pmatrix}, \qquad \begin{pmatrix} -2x + 3y \\ -2y \end{pmatrix} = \begin{pmatrix} -2x \\ -2y \end{pmatrix}.$$

This gives $3y = 0$, $y = y$. So all eigenvectors have the form $\mathbf{c}_1 = c(1, 0)$. From $(1, 0)$ we get a basic solution

$$\mathbf{x}_1(t) = \begin{pmatrix} 1 \\ 0 \end{pmatrix} e^{-2t} = \begin{pmatrix} e^{-2t} \\ 0 \end{pmatrix}$$

for the given system. But its solution space is *two*-dimensional, so we need a second solution $\mathbf{x}_2(t)$ that is linearly independent of $\mathbf{x}_1(t)$. Analogy with the case of repeated roots of the auxiliary equation would suggest that we look for a solution of the form $\mathbf{c}_2 te^{-2t}$ for some $\mathbf{c}_2 \in \mathbf{R}^2$. Unfortunately, it turns out that *there is no nontrivial solution of this form,* as you are asked to show in Exercise 16 below. What then? Suppose we hunt for a solution of the slightly more complicated form

$$(13)\qquad \mathbf{x}_2(t) = \mathbf{c}_2 te^{-2t} + \mathbf{c}_3 e^{-2t}$$

for some

$$\mathbf{c}_2 = \begin{pmatrix} c_{21} \\ c_{22} \end{pmatrix} \quad \text{and} \quad \mathbf{c}_3 = \begin{pmatrix} c_{31} \\ c_{32} \end{pmatrix}$$

in $\mathbf{R}^2$. For $\mathbf{x}_2(t)$ to be a solution of (12), we must have

$$\frac{d\mathbf{x}_2}{dt} = A\mathbf{x}_2,$$

that is,

$$\mathbf{c}_2 e^{-2t} - 2\mathbf{c}_2 te^{-2t} - 2\mathbf{c}_3 e^{-2t} = (A\mathbf{c}_2)te^{-2t} + (A\mathbf{c}_3)e^{-2t}.$$

Equating the coefficients of te^{-2t}, we obtain

$$-2\mathbf{c}_2 = A\mathbf{c}_2,$$

that is, $(A+2I_2)\mathbf{c}_2 = 0$. This means that $\mathbf{c}_2$ is an eigenvector belonging to the eigenvalue $\lambda = -2$, so from above we have

$$\mathbf{c}_2 = \begin{pmatrix} c_{21} \\ c_{22} \end{pmatrix} = c_2 \begin{pmatrix} 1 \\ 0 \end{pmatrix}.$$

Equating the coefficients of e^{-2t}, we have

$$\mathbf{c}_2 - 2\mathbf{c}_3 = A\mathbf{c}_3,$$

that is, $(A+2I_2)\mathbf{c}_3 = \mathbf{c}_2$, which gives

$$\text{(14)} \qquad \begin{cases} c_2 - 2c_{31} = -2c_{31} + 3c_{32} \\ - 2c_{32} = \phantom{-2c_{31}+3} -2c_{32} \end{cases}$$

The second equation in (14) gives no information, but the first says that $c_{32} = \frac{1}{3}c_2$. There is no restriction on c_{31}; i.e., $c_{31} = c_3$ is arbitrary. Hence

$$\mathbf{c}_3 = \begin{pmatrix} c_{31} \\ c_{32} \end{pmatrix} = \begin{pmatrix} c_3 \\ \frac{1}{3}c_2 \end{pmatrix}.$$

Therefore, (13) has led us to

$$\mathbf{x}(t) = c_2 \begin{pmatrix} 1 \\ 0 \end{pmatrix} te^{-2t} + c_3 \begin{pmatrix} 1 \\ 0 \end{pmatrix} e^{-2t} + c_2 \begin{pmatrix} 0 \\ \frac{1}{3} \end{pmatrix} e^{-2t}.$$

This suggests that (12) has general solution

$$\begin{aligned} \mathbf{x}(t) &= \bar{c}_1 \begin{pmatrix} 1 \\ 0 \end{pmatrix} e^{-2t} + c_2 \begin{pmatrix} 1 \\ 0 \end{pmatrix} te^{-2t} + c_3 \begin{pmatrix} 1 \\ 0 \end{pmatrix} e^{-2t} + c_2 \begin{pmatrix} 0 \\ \frac{1}{3} \end{pmatrix} e^{-2t} \\ &= c_1 \begin{pmatrix} 1 \\ 0 \end{pmatrix} e^{-2t} + c_2 \left[\begin{pmatrix} 1 \\ 0 \end{pmatrix} te^{-2t} + \begin{pmatrix} 0 \\ \frac{1}{3} \end{pmatrix} e^{-2t} \right], \end{aligned}$$

where $c_1 = \bar{c}_1 + c_3$. It is easy to confirm this if we let

$$\mathbf{x}_2(t) = \begin{pmatrix} 1 \\ 0 \end{pmatrix} te^{-2t} + \begin{pmatrix} 0 \\ \frac{1}{3} \end{pmatrix} e^{-2t},$$

for the Wronskian of $B = \{\mathbf{x}_1(t), \mathbf{x}_2(t)\}$ is then

$$\det W_B = \det \begin{pmatrix} e^{-2t} & te^{-2t} \\ 0 & \frac{1}{3}e^{-2t} \end{pmatrix} = \tfrac{1}{3}e^{-4t} \neq 0.$$

Therefore B is linearly independent by Lemma 2.6, so it is a basis for the solution space of (12).

As this example illustrates, if λ is an eigenvalue of A of multiplicity two that produces only one linearly independent eigenvector $\mathbf{c}_1$, then besides the basic solution $\mathbf{x}_1(t)=\mathbf{c}_1e^{\lambda t}$ of Theorem 2.7 for (3), an additional basic solution of the form

(15) $$\mathbf{x}_2(t)=\mathbf{c}_1te^{\lambda t}+\mathbf{c}_2e^{\lambda t}$$

where $(A-\lambda I_n)\mathbf{c}_2=\mathbf{c}_1$ should be sought.

If λ is an eigenvalue of multiplicity three that produces only one linearly independent vector, then an analysis similar to the foregoing suggests that besides the basic solution $\mathbf{c}_1e^{\lambda t}$, additional basic solutions $\mathbf{x}_2(t)$ of the form (15) and also $\mathbf{x}_3(t)$ given by

(16) $$\mathbf{x}_3(t)=\frac{1}{2!}\mathbf{c}_1t^2e^{\lambda t}+\mathbf{c}_2te^{\lambda t}+\mathbf{c}_3e^{\lambda t},$$

where $(A-\lambda I_n)\mathbf{c}_3=\mathbf{c}_2$ and $(A-\lambda I_n)\mathbf{c}_2=\mathbf{c}_1$, should be sought.

EXERCISES 10.2

In Exercises 1 to 10, solve the given first order system.

1 $\begin{cases}\dfrac{dx}{dt}=3x+8y\\ \dfrac{dy}{dt}=x+y\end{cases}$

2 $\begin{cases}\dfrac{dx}{dt}=x-y\\ \dfrac{dy}{dt}=5x-3y\end{cases}$

3 $\begin{cases}\dfrac{dx}{dt}=x-y\\ \dfrac{dy}{dt}=x+2y+z\\ \dfrac{dz}{dt}=-2x+y-z\end{cases}$

4 $\begin{cases}\dfrac{dx}{dt}=3x+2y+4z\\ \dfrac{dy}{dt}=2x+2z\\ \dfrac{dz}{dt}=4x+2y+3z\end{cases}$

5 $\begin{cases}\dfrac{dx}{dt}=y+z\\ \dfrac{dy}{dt}=x+z\\ \dfrac{dz}{dt}=x+y\end{cases}$

6 $\begin{cases}\dfrac{dx}{dt}=2x-5y\\ \dfrac{dy}{dt}=x-2y-3z\\ \dfrac{dz}{dt}=y+2z\end{cases}$

7 $\begin{cases}\dfrac{dx}{dt}=x-y\\ \dfrac{dy}{dt}=x+3y\end{cases}$

8 $\begin{cases}\dfrac{dx}{dt}=4x-y\\ \dfrac{dy}{dt}=x+2y\end{cases}$

9 $\begin{cases}\dfrac{dx}{dt}=x+y+z\\ \dfrac{dy}{dt}=2x+y-z\\ \dfrac{dz}{dt}=-y+z\end{cases}$

10 $\begin{cases}\dfrac{dx}{dt}=x+y+z\\ \dfrac{dy}{dt}=2x+y-z\\ \dfrac{dz}{dt}=-3x+2y+4z\end{cases}$

In Exercises 11 to 14, solve the given initial value problem.

11 $\begin{cases} \dfrac{dx}{dt} = -2x + y \\ \dfrac{dy}{dt} = -5x + 4y, \quad \boldsymbol{x}(0) = (1, 3) \end{cases}$

12 $\begin{cases} \dfrac{dx}{dt} = x - 5y \\ \dfrac{dy}{dt} = x - 3y, \quad \boldsymbol{x}(0) = (1, 1) \end{cases}$

13 $\begin{cases} \dfrac{dx}{dt} = x + y + 2z \\ \dfrac{dy}{dt} = 2y + 2z \\ \dfrac{dz}{dt} = -x + y + 3z, \quad \boldsymbol{x}(0) = (2, 0, 1) \end{cases}$

14 $\begin{cases} \dfrac{dx}{dt} = 3x - z \\ \dfrac{dy}{dt} = -2x + 2y + z, \quad \boldsymbol{x}(0) = (-1, 2, 8) \\ \dfrac{dz}{dt} = 8x - 3z \end{cases}$

15 Let $p(D)(y) = 0$ be a homogeneous n-th order linear differential equation. Suppose that we transform it into a system of the form (3) using the approach of Theorem 1.2. Then show that the characteristic polynomial $\det(A - \lambda I_n)$ of this system is identical to the auxiliary polynomial $p(r)$ of the original n-th order homogeneous problem $p(D)(y) = 0$.

16 Verify that Example 2.12 has no nontrivial solution of the form $\mathbf{c}_2 te^{-2t}$.

17 Show that if $\boldsymbol{x}_1(t), \boldsymbol{x}_2(t), \ldots, \boldsymbol{x}_n(t)$ are linearly independent solutions of the system (3) on an interval I, then their Wronskian determinant is never 0 on I. (*Hint:* Imitate the proof of Theorem 9.3.4.) **This exercise will be used in Section 3.**

18 In Example 2.10, show that the basic real solutions can be obtained as the columns of the matrix $F(t)T$, where

$$F(t) = \begin{pmatrix} 2e^{2it} & 2e^{-2it} \\ (i-1)e^{2it} & -(i+1)e^{-2it} \end{pmatrix}$$

and

$$T = \begin{pmatrix} \dfrac{1}{2} & \dfrac{1}{2i} \\ \dfrac{1}{2} & \dfrac{-1}{2i} \end{pmatrix}.$$

19 If λ is an eigenvalue of A, then show that the **eigenspace** of all eigenvectors of A corresponding to λ is a real vector space.

20 Show by example that the eigenvectors of an n-by-n matrix A need not span all of $\boldsymbol{R}^n$.

3 NONHOMOGENEOUS LINEAR SYSTEMS

Having learned how to solve the homogeneous linear system

$$\frac{d\boldsymbol{x}}{dt} = A\boldsymbol{x}(t),$$

we can now complete the solution of the general linear system

$$\boxed{\frac{d\boldsymbol{x}}{dt} = A\boldsymbol{x}(t) + \boldsymbol{b}(t)} \tag{1}$$

if we can find a single *particular solution* $\boldsymbol{x} = \boldsymbol{x}_p(t)$ of (1). This section, then, is devoted to the problem of finding a particular solution for (1).

As you recall, the easiest technique for finding a particular solution to an n-th order linear differential equation with constant coefficients is the method of undetermined coefficients, when that method can be used. Since the coefficients in (1) are constants, we should be able to use this technique to find a particular solution to (1) if each coordinate of $\boldsymbol{b}(t)$ is a solution to some linear differential equation. We should in fact be able to use Table 8.7.1 to furnish the expected general form of $\boldsymbol{x}_p(t)$. Once we have this general form, we then can determine the constants A, B, C, etc., occurring in it by substituting $\boldsymbol{x}_p(t)$ into the given system. The following example shows how this scheme is carried out in practice.

3.1 EXAMPLE. Solve the system

(2)
$$\begin{cases} \dfrac{dx}{dt} = 2x - 9y + te^t \\ \dfrac{dy}{dt} = -x + 2y - te^t + e^t \end{cases}$$

Solution. In Example 2.9, we found that the associated homogeneous system has general solution

$$\boldsymbol{x}_0(t) = c_1\begin{pmatrix}-3\\1\end{pmatrix}e^{5t} + c_2\begin{pmatrix}3\\1\end{pmatrix}e^{-t}.$$

Here

$$\boldsymbol{b}(t) = \begin{pmatrix}b_1(t)\\b_2(t)\end{pmatrix} = \begin{pmatrix}te^t\\-te^t+e^t\end{pmatrix}.$$

Consulting Table 8.7.1, we find that we should seek a particular solution of the form

$$\boldsymbol{x}_p(t) = \begin{pmatrix}(At+B)e^t\\(Ct+D)e^t\end{pmatrix},$$

since $b_1(t)$ and $b_2(t)$ both have the form $p(t)e^t$ where $p(t)$ is a polynomial of degree one. Differentiation of $\boldsymbol{x}_p(t)$ with respect to t gives us

$$\frac{d\boldsymbol{x}_p}{dt} = \begin{pmatrix}Ae^t + Ate^t + Be^t\\Ce^t + Cte^t + De^t\end{pmatrix}.$$

Substituting into (2), we get

$$\begin{pmatrix}(A+B)e^t + Ate^t\\(C+D)e^t + Cte^t\end{pmatrix} = \begin{pmatrix}(2A-9C+1)te^t + (2B-9D)e^t\\(-A+2C-1)te^t + (-B+2D+1)e^t\end{pmatrix}.$$

If we equate coefficients of e^t and te^t, then we have

$$\begin{cases} A+B = 2B-9D \\ A = 2A-9C+1 \\ C+D = -B+2D+1 \\ C = -A+2C-1 \end{cases} \rightarrow \begin{cases} A-B+9D = 0 \\ -A+9C = 1 \\ B+C-D = 1 \\ A-C = -1. \end{cases}$$

Upon adding the second and fourth equations, we find $C=0$. Hence $A=-1$, and the remaining two equations become

$$\begin{aligned} -B+9D&=1 \\ B-\ D&=1 \end{aligned}$$

Addition of these gives us $D=\frac{1}{4}$. Hence $B=\frac{5}{4}$. So our particular solution is

$$\boldsymbol{x}_{\mathrm{p}}(t)=\begin{pmatrix} -te^t+\frac{5}{4}e^t \\ \frac{1}{4}e^t \end{pmatrix}.$$

The general solution to (2) is therefore given by

$$\boldsymbol{x}(t)=c_1\begin{pmatrix} -3 \\ 1 \end{pmatrix}e^{5t}+c_2\begin{pmatrix} 3 \\ 1 \end{pmatrix}e^{-t}+\begin{pmatrix} -t+\frac{5}{4} \\ \frac{1}{4} \end{pmatrix}e^t.$$

You will recall that we have to *modify* the form of the particular solution suggested by Table 8.7.1 in case that suggested form is a solution of the associated homogeneous problem. We expect then that a similar modification will be needed here, but that it will be along the more complicated lines seen in Example 2.12 rather than simple multiplication of the suggested $\boldsymbol{x}_{\mathrm{p}}(t)$ by powers of t. This is exactly the case, and the next example provides an illustration.

3.2 **EXAMPLE.** Solve the system

$$\left\{ \begin{aligned} \frac{dx}{dt}&=2x-9y+3e^{-t} \\ \frac{dy}{dt}&=-x+2y-2e^{-t} \end{aligned} \right. \tag{3}$$

Solution. Table 8.7.1 would lead us to hunt for a particular solution of the form

$$\begin{pmatrix} Ae^{-t} \\ Be^{-t} \end{pmatrix}.$$

But since this form includes solutions of the associated homogeneous problem, we can't be sure that such an expression will provide a solution of (3). If we look for a particular solution of the form

$$\begin{pmatrix} Ate^{-t} \\ Bte^{-t} \end{pmatrix},$$

as would be indicated in the case of a single n-th order equation, we will find (Exercise 19 below) that there is no particular solution of this form, either. So, following the idea of Example 2.12, we look for an $\boldsymbol{x}_{\mathrm{p}}(t)$ of the form

$$\boldsymbol{x}_{\mathrm{p}}(t)=\begin{pmatrix} A \\ C \end{pmatrix}te^{-t}+\begin{pmatrix} B \\ D \end{pmatrix}e^{-t}.$$

We have

$$\frac{d\boldsymbol{x}_{\mathrm{p}}}{dt}=\begin{pmatrix} A-B & -At \\ C-D & -Ct \end{pmatrix}e^{-t}.$$

Substituting in (3), we get

$$(3')\qquad \begin{pmatrix} A-B & -At \\ C-D & -Ct \end{pmatrix} e^{-t} = \begin{pmatrix} (2A-9C)t & 2B-9D+3 \\ (-A+2C)t & -B+2D-2 \end{pmatrix} e^{-t}.$$

Equating coefficients of e^{-t} and te^{-t}, we obtain

$$\begin{cases} A-B=2B-9D+3 \\ -A=2A-9C \\ -C=-A+2C \\ C-D=-B+2D-2 \end{cases} \rightarrow \begin{cases} A-3B+9D=3 \\ A-3C=0 \\ A-3C=0 \\ -B-C+3D=2. \end{cases}$$

Solving this system by Gaussian elimination, we get

$$\begin{pmatrix} 1 & -3 & 0 & 9 & 3 \\ 1 & 0 & -3 & 0 & 0 \\ 1 & 0 & -3 & 0 & 0 \\ 0 & -1 & -1 & 3 & 2 \end{pmatrix} \xrightarrow[\text{Add } -R_1 \text{ to } R_3]{\text{Add } -R_1 \text{ to } R_2} \begin{pmatrix} 1 & -3 & 0 & 9 & 3 \\ 0 & 3 & -3 & -9 & -3 \\ 0 & 3 & -3 & -9 & -3 \\ 0 & -1 & -1 & 3 & 2 \end{pmatrix} \xrightarrow[\frac{1}{3}R_3]{\frac{1}{3}R_2}$$

$$\begin{pmatrix} 1 & -3 & 0 & 9 & 3 \\ 0 & 1 & -1 & -3 & -1 \\ 0 & 1 & -1 & -3 & -1 \\ 0 & -1 & -1 & 3 & 2 \end{pmatrix} \xrightarrow[\substack{\text{Add } -R_2 \text{ to } R_3 \\ \text{Add } R_2 \text{ to } R_4}]{\text{Add } 3R_2 \text{ to } R_1} \begin{pmatrix} 1 & 0 & -3 & 0 & 0 \\ 0 & 1 & -1 & -3 & -1 \\ 0 & 0 & 0 & 0 & 0 \\ 0 & 0 & -2 & 0 & 1 \end{pmatrix} \xrightarrow[\substack{\text{Interchange } R_3 \\ \text{and } R_4}]{-\frac{1}{2}R_4}$$

$$\begin{pmatrix} 1 & 0 & -3 & 0 & 0 \\ 0 & 1 & -1 & -3 & -1 \\ 0 & 0 & 1 & 0 & -\frac{1}{2} \\ 0 & 0 & 0 & 0 & 0 \end{pmatrix} \xrightarrow[\text{Add } 3R_3 \text{ to } R_1]{\text{Add } R_3 \text{ to } R_2} \begin{pmatrix} 1 & 0 & 0 & 0 & -\frac{3}{2} \\ 0 & 1 & 0 & -3 & -\frac{3}{2} \\ 0 & 0 & 1 & 0 & -\frac{1}{2} \\ 0 & 0 & 0 & 0 & 0 \end{pmatrix}$$

We have then $A=-\frac{3}{2}$, $C=-\frac{1}{2}$, and $B=-\frac{3}{2}+3D$. Here D is arbitrary. If we take $D=1$, we get $B=\frac{3}{2}$. So one particular solution (among many) is

$$\begin{aligned} \boldsymbol{x}_p(t) &= \begin{pmatrix} -\frac{3}{2}te^{-t}+\frac{3}{2}e^{-t} \\ -\frac{1}{2}te^{-t}+e^{-t} \end{pmatrix} \\ &= -\frac{1}{2}\begin{pmatrix} 3 \\ 1 \end{pmatrix} te^{-t} + \frac{1}{2}\begin{pmatrix} 3 \\ 2 \end{pmatrix} e^{-t}. \end{aligned}$$

The general solution is then

$$\boldsymbol{x}(t) = c_1\begin{pmatrix} -3 \\ 1 \end{pmatrix} e^{5t} + c_2\begin{pmatrix} 3 \\ 1 \end{pmatrix} e^{-t} + \frac{1}{2}\begin{pmatrix} -3t+3 \\ -t+2 \end{pmatrix} e^{-t}.$$

Notice that in (3′), equating coefficients of te^{-t} gives us

$$-\begin{pmatrix} A \\ C \end{pmatrix} = \begin{pmatrix} 2 & -9 \\ -1 & 2 \end{pmatrix}\begin{pmatrix} A \\ C \end{pmatrix}.$$

Thus $\begin{pmatrix} A \\ C \end{pmatrix}$ is an eigenvector belonging to the eigenvalue -1. We could then have taken $\boldsymbol{x}_p(t)$ of the slightly simpler form

$$\boldsymbol{x}_p(t) = A\begin{pmatrix} 3 \\ 1 \end{pmatrix} te^{-t} + \begin{pmatrix} B \\ D \end{pmatrix} e^{-t}.$$

and thereby saved a bit of work.

The method of undetermined coefficients is, as in the previous chapter, of rather limited applicability. The components of $\boldsymbol{b}(t)$ in (1) must be of one of the forms given in Table 8.7.1. We are thus led to consider extending the method of variation of parameters of Section 9.3 to first order linear systems (1). As before, the method can actually be used on systems more general than (1), namely first order linear systems of the form $\frac{d\boldsymbol{x}}{dt} = A(t)\boldsymbol{x}(t) + \boldsymbol{b}(t)$, where the matrix $A(t)$ of coefficients has variable entries $a_{ij}(t)$ that are continuous. Since we have no methods for solving the associated homogeneous problem $\frac{d\boldsymbol{x}}{dt} = A(t)\boldsymbol{x}(t)$, however, we won't be in a position to exploit this wider applicability. So we will develop the method just for constant coefficient systems (1). We need the following analogue of Theorem 9.3.4, its proof was asked for in Exercise 17 of Exercises 10.2.

3.3 LEMMA. Suppose that $\{\boldsymbol{x}_1(t), \ldots, \boldsymbol{x}_n(t)\}$ is a basis for the solution space of

(4) $$\frac{d\boldsymbol{x}}{dt} = A\boldsymbol{x}(t)$$

on an interval I. Then the Wronskian determinant

(5) $$\det W(t) = \det(\boldsymbol{x}_1(t), \ldots, \boldsymbol{x}_n(t))$$

is never 0 on the interval I.

The matrix $W(t)$ in (5), which is formed by putting the basis vectors $\boldsymbol{x}_1(t), \ldots, \boldsymbol{x}_n(t)$ into its columns, is called a **fundamental matrix** for the system (4) (or (1)). We need the following fact about this matrix, which follows immediately from Lemma 3.3.

3.4 LEMMA. A fundamental matrix $W(t)$ for (1) is invertible on any interval I on which $\{\boldsymbol{x}_1(t), \ldots, \boldsymbol{x}_n(t)\}$ is a basis for the solution space of (4).

Proof. Since $\det W(t) \neq 0$ on I, $W(t)$ is invertible on I by Theorem 6.7.3. QED

It is now easy to give and justify the method of variation of parameters. The following result is exactly analogous to Theorem 9.3.5.

3.5 **THEOREM.** The system (1) has a particular solution of the form

(6) $$\boldsymbol{x}_p(t) = W(t)\boldsymbol{c}(t)$$

where

$$\boldsymbol{c}(t) = \begin{pmatrix} c_1(t) \\ \cdot \\ \cdot \\ \cdot \\ c_n(t) \end{pmatrix}$$

and $W(t)$ is any fundamental matrix for (1) on I. Moreover, $\boldsymbol{c}(t)$ is given by

(7) $$\boldsymbol{c}(t) = \int W^{-1}(t)\boldsymbol{b}(t)\,dt.$$

Proof. We let $\boldsymbol{x}_p(t) = W(t)\boldsymbol{c}(t)$ and try to determine $\boldsymbol{c}(t)$ so that $\boldsymbol{x}_p(t)$ will be a solution of (6). Note that we have

$$\boldsymbol{x}_p(t) = W(t)\boldsymbol{c}(t) = c_1(t)\boldsymbol{x}_1(t) + \ldots + c_n(t)\boldsymbol{x}_n(t).$$

Hence,

$$\begin{aligned}\frac{d\boldsymbol{x}_p(t)}{dt} &= c_1(t)\dot{\boldsymbol{x}}_1(t) + \dot{c}_1(t)\boldsymbol{x}_1(t) + \ldots + c_n(t)\dot{\boldsymbol{x}}_n(t) + \dot{c}_n(t)\boldsymbol{x}_n(t)\\ &= c_1(t)\dot{\boldsymbol{x}}_1(t) + \ldots + c_n(t)\dot{\boldsymbol{x}}_n(t) + \dot{c}_1(t)\boldsymbol{x}_1(t) + \ldots + \dot{c}_n(t)\boldsymbol{x}_n(t)\\ &= c_1(t)A\boldsymbol{x}_1(t) + \ldots + c_n(t)A\boldsymbol{x}_n(t) + W(t)\dot{\boldsymbol{c}}(t),\end{aligned}$$

since each $\boldsymbol{x}_i(t)$ is a solution of (4),

(8) $$= AW(t)\boldsymbol{c}(t) + W(t)\dot{\boldsymbol{c}}(t).$$

We want $\boldsymbol{x}_p(t)$ to be a solution of (1), i.e.,

(9) $$\frac{d\boldsymbol{x}_p}{dt} = A\boldsymbol{x}_p(t) + \boldsymbol{b}(t) = AW(t)\boldsymbol{c}(t) + \boldsymbol{b}(t).$$

Equating the expressions for $\dfrac{d\boldsymbol{x}_p}{dt}$ in (8) and (9), we want

(10) $$W(t)\dot{\boldsymbol{c}}(t) = \boldsymbol{b}(t).$$

Now by Lemma 3.4, $W(t)$ is invertible. So we can solve (10) for $\dot{\boldsymbol{c}}(t)$, and then integrate to get (7):

(11) $$\dot{\boldsymbol{c}}(t) = W^{-1}(t)\boldsymbol{b}(t)$$

$$\boldsymbol{c}(t) = \int W^{-1}(t)\boldsymbol{b}(t)\,dt.$$

We have thus shown that if (6) is a particular solution of (1), then $\boldsymbol{c}(t)$ *has* to be given by (7). It remains only to verify that (6) really *is* a solution of (1) if $\boldsymbol{c}(t)$ is given by (7). But if (7) holds, then we can differentiate to get (11), and multiply by $W(t)$ to get (10). Then differentiating (6), we arrive at (8). So

$$\begin{aligned}\frac{d\boldsymbol{x}_p}{dt} &= AW(t)\boldsymbol{c}(t) + W(t)\boldsymbol{c}(t)\\ &= A\boldsymbol{x}_p(t) + \boldsymbol{b}(t) \quad \text{by} \quad (10).\end{aligned}$$

Hence $\boldsymbol{x}_p(t)$ actually *is* a solution of (1) if $\boldsymbol{c}(t)$ is given by (7). QED

While (7) is a simple formula for $\boldsymbol{c}(t)$, actually *evaluating* it is not a simple matter in general. We first have to invert the matrix $W(t)$ using the techniques of Section 6.4, and then we have to work out the vector integral in (7). The following example shows how the work goes in a fairly simple problem.

3.6 EXAMPLE. Solve the system

(12)
$$\begin{cases}\dfrac{dx}{dt} = 2x + 4y + \csc 2t\\[2mm] \dfrac{dy}{dt} = -2x - 2y + \sec 2t\end{cases}$$

on the interval $I = \left[\frac{\pi}{12}, \frac{\pi}{6}\right]$.

Solution. In Example 2.10 we found that the solution of the associated homogeneous system has a basis made up of

$$\boldsymbol{x}_1(t) = \begin{pmatrix} 2\cos 2t \\ -\cos 2t - \sin 2t\end{pmatrix} \quad \text{and} \quad \boldsymbol{x}_2(t) = \begin{pmatrix} 2\sin 2t \\ \cos 2t - \sin 2t\end{pmatrix}.$$

So here our fundamental matrix is

$$W(t) = \begin{pmatrix} 2\cos 2t & 2\sin 2t \\ -\cos 2t - \sin 2t & \cos 2t - \sin 2t\end{pmatrix}.$$

We must invert $W(t)$. Using Theorem 6.4.9, we have first

$$\begin{aligned}\det W(t) &= 2\cos^2 2t - 2\sin 2t\cos 2t + 2\sin 2t\cos 2t + 2\sin^2 2t\\ &= 2.\end{aligned}$$

Then

$$W^{-1}(t) = \tfrac{1}{2}\begin{pmatrix}\cos 2t - \sin 2t & -2\sin 2t \\ \cos 2t + \sin 2t & 2\cos 2t\end{pmatrix}.$$

Since

$$\boldsymbol{b}(t) = \begin{pmatrix}\csc 2t \\ \sec 2t\end{pmatrix}, \text{ we have}$$

$$W^{-1}(t)\boldsymbol{b}(t) = \tfrac{1}{2}\begin{pmatrix}\cot 2t - 1 - 2\tan 2t \\ \cot 2t + 1 + 2\end{pmatrix}.$$

Then from (7),

$$\mathbf{c}(t) = \tfrac{1}{2}\int \begin{pmatrix} \cot 2t - 2\tan 2t - 1 \\ \cot 2t + 3 \end{pmatrix} dt$$

$$= \tfrac{1}{2}\begin{pmatrix} -t + \tfrac{1}{2}\ln|\sin 2t| + \ln|\cos 2t| \\ 3t + \tfrac{1}{2}\ln|\sin 2t| \end{pmatrix}.$$

So the general solution of (12) is

$$\boldsymbol{x}(t) = c_1\begin{pmatrix} 2\cos 2t \\ -\cos 2t - \sin 2t \end{pmatrix} + c_2\begin{pmatrix} 2\sin 2t \\ \cos 2t - \sin 2t \end{pmatrix}$$

$$+ \tfrac{1}{2}(\boldsymbol{x}_1(t), \boldsymbol{x}_2(t))\begin{pmatrix} -t + \tfrac{1}{2}\ln|\sin 2t| + \ln|\cos 2t| \\ 3t + \tfrac{1}{2}\ln|\sin 2t| \end{pmatrix}.$$

As remarked in Section 9.3, the main drawback in using the method of variation of parameters is that it may be difficult or even impossible to carry out the integration in (7). However, it is clear from Examples 3.2 and 3.6 that, when the integration *can* be performed, the method of variation of parameters may well be *less work* than the method of undetermined coefficients. An additional advantage, of course, is the greater applicability of the former method. It *always* produces a particular solution for (1), even if we have to be content with (6) and (7) unsimplified as a formula for $\boldsymbol{x}_p(t)$.

EXERCISES 10.3

In Exercises 1 to 10, use the method of undetermined coefficients to solve the given first order system. (In Exercises 1 to 8, use the results of pertinent exercises of Exercises 10.2 to lessen your work.)

1 $\begin{cases} \dfrac{dx}{dt} = 3x + 8y - 2e^t \\ \dfrac{dy}{dt} = x + y + 3e^t \end{cases}$

2 $\begin{cases} \dfrac{dx}{dt} = 3x + 8y - 5t \\ \dfrac{dy}{dt} = x + y + 3 \end{cases}$

3 $\begin{cases} \dfrac{dx}{dt} = 3x + 8y - \cos t \\ \dfrac{dy}{dt} = x + y + \sin t \end{cases}$

4 $\begin{cases} \dfrac{dx}{dt} = 3x + 8y + 4\cos t \\ \dfrac{dy}{dt} = x + y - 3\sin t \end{cases}$

5 $\begin{cases} \dfrac{dx}{dt} = 3x + 8y - e^{-t} \\ \dfrac{dy}{dt} = x + y + 2e^{-t} \end{cases}$

6 $\begin{cases} \dfrac{dx}{dt} = 3x + 8y + 2e^{5t} \\ \dfrac{dy}{dt} = x + y - 3e^{5t} \end{cases}$

7 $\begin{cases} \dfrac{dx}{dt} = x - y + 2e^{3t} \\ \dfrac{dy}{dt} = x + 2y + z - 5e^{3t} \\ \dfrac{dz}{dt} = -2x + y - z + 2e^{3t} \end{cases}$

8 $\begin{cases} \dfrac{dx}{dt} = y + z - 2e^t \\ \dfrac{dy}{dt} = x + z + 3e^t \\ \dfrac{dz}{dt} = x + y - 5e^t \end{cases}$

9 $\begin{cases} \dfrac{dx}{dt} = 3x - 2y - e^{-t}\sin t \\ \dfrac{dy}{dt} = 4x - y + 2e^{-t}\cos t \end{cases}$

10 $\begin{cases} \dfrac{dx}{dt} = 2x + 4y + 2e^{-t}\sin t \\ \dfrac{dy}{dt} = -2x - 2y + e^{-t}\cos t \end{cases}$

In Exercises 11 to 18, use variation of parameters to solve the given system.

11 $\begin{cases} \dfrac{dx}{dt} = x - y + \cos^2 t \\ \dfrac{dy}{dt} = 2x - y - \cos t \sin t \end{cases}$

12 $\begin{cases} \dfrac{dx}{dt} = x - y + \sin^2 t \\ \dfrac{dy}{dt} = 2x - y + \sin t \cos t \end{cases}$

13 $\begin{cases} \dfrac{dx}{dt} = 2x - 5y + \csc t \\ \dfrac{dx}{dt} = x - 2y + \sec t \end{cases}$ on $\left[\dfrac{\pi}{6}, \dfrac{\pi}{3}\right]$.

14 $\begin{cases} \dfrac{dx}{dt} = 2x - 5y \\ \dfrac{dy}{dt} = x - 2y + \cot t \end{cases}$ on $\left[\dfrac{\pi}{6}, \dfrac{\pi}{3}\right]$

15 $\begin{cases} \dfrac{dx}{dt} = 4x - 2y + \dfrac{1}{t^3} \\ \dfrac{dy}{dt} = 8x - 4y - \dfrac{1}{t^2} \end{cases}$ on $[1, 10]$

16 $\begin{cases} \dfrac{dx}{dt} = 4x - 2y + \dfrac{1}{t^3} \\ \dfrac{dy}{dt} = 8x - 4y + \dfrac{1}{t^2} \end{cases}$ on $[1, 10]$

17 $\begin{cases} \dfrac{dx}{dt} = 4x - 2y + 2z + 2te^{6t} + 1 \\ \dfrac{dy}{dt} = -2x + y - z - te^{6t} + 2 \\ \dfrac{dz}{dt} = 2x - y + z + 3te^{6t} - 1 \end{cases}$

(*Hint:* See Example 2.11.)

18 $\begin{cases} \dfrac{dx}{dt} = 4x - 2y + 2z + t^2e^{6t} - e^{6t} \\ \dfrac{dy}{dt} = -2x + y - z - 3t^2e^{6t} + 2e^{6t} \\ \dfrac{dz}{dt} = 2x - y + z - t^2e^{6t} + e^{6t} \end{cases}$

19 Show that (3) has no solution of the form $\begin{pmatrix} Ate^{-t} \\ Bte^{-t} \end{pmatrix}$.

20 Does (3) have a solution of the form $\begin{pmatrix} Ae^{-t} \\ Be^{-t} \end{pmatrix}$?

21 Given that

$$\begin{cases} \dfrac{dx}{dt} = y + t^3 \\ \dfrac{dy}{dt} = \dfrac{2}{t^2}x + \dfrac{1}{t^2} \end{cases}$$

has fundamental matrix

$$\begin{pmatrix} t^2 & t^{-1} \\ 2t & -t^{-2} \end{pmatrix},$$

find the general solution using variation of parameters.

22 Repeat Exercise 21 for

$$\begin{cases} \dfrac{dx}{dt} = y + t^2 \\ \dfrac{dy}{dt} = \dfrac{2}{t^2}x - \dfrac{2}{t^3}. \end{cases}$$

23 A first order **Cauchy-Euler system** is a system of the form

$$t\frac{d\mathbf{x}}{dt} = A\mathbf{x} + \boldsymbol{b}.$$

Show that the change of variable $u=\ln t$, $t=e^u$, $t>0$, reduces this to an ordinary first order linear system. $\left[\textit{Hint: } \text{Show that if } \mathbf{X}(u)=\mathbf{x}(t)=\mathbf{x}(e^u), \text{ then } \frac{d\mathbf{X}}{dt}=t\frac{d\mathbf{x}}{dt}.\right]$

24 Use Exercise 23 to solve

$$t\frac{d\mathbf{x}}{dt}=\begin{pmatrix}3x-2y\\2x-2y\end{pmatrix}.$$

25 Solve

$$t\frac{d\mathbf{x}}{dt}=\begin{pmatrix}3x-2y+3t\\2x-2y-2t\end{pmatrix}.$$

4 APPLICATIONS

Systems of differential equations arise in a wide variety of applied settings. In this section we briefly discuss three important applications for which useful models involving linear systems exist. The first example was introduced in Exercises 27 to 29 of Exercises 9.2, where *ad hoc* methods were used to study the motion. It represents a more involved version of the mechanical systems studied in Section 8.8.

4.1 **EXAMPLE.** Analyze the motion of the two-mass vibrating system in Figure 4.1. The system is activated by pulling the mass m_2 to the right and releasing it at time $t=0$.

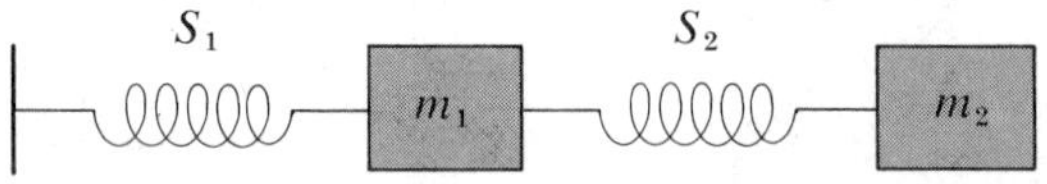

FIGURE 4.1

Solution. Let $x(t)$ be the distance by which m_1 is to the right of its equilibrium position at time t, and $y(t)$ the distance by which m_2 is to the right of its equilibrium position at time t. Let k_1 be the spring constant of S_1 and k_2 the spring constant of S_2. The force on m_1 due to S_1 at time t is, by Hooke's Law, $-k_1x(t)$. The force on m_1 due to S_2 is $k_2(y(t)-x(t))$, since $y(t)-x(t)$ gives the net stretching of S_2 from a position of equilibrium. Similarly, the force on m_2 at time t is $-k_2(y(t)-x(t))=-k_2y(t)+k_2x(t)$. Using Newton's law of motion in the form $F_1=\frac{d^2x}{dt^2}$ and $F_2=\frac{d^2y}{dt^2}$, we thus have the following system of differential equations giving the total forces on m_1 and m_2 at time t:

$$\begin{cases} F_1=-k_1x(t)+k_2(y(t)-x(t))=m_1\dfrac{d^2x}{dt^2} \\ F_2= \qquad\qquad k_2x(t)-k_2y(t)=m_2\dfrac{d^2y}{dt^2} \end{cases}$$

Collecting terms we have

$$\begin{cases} m_1\dfrac{d^2x}{dt^2}+(k_1+k_2)x(t)-k_2y(t)=0 \\ m_2\dfrac{d^2y}{dt^2} \qquad -k_2x(t)+k_2y(t)=0. \end{cases}$$

We proceed to solve this second order system by the elimination technique of Section 1. First we divide the equations through by m_1 and m_2 respectively to obtain

(*) $$\begin{cases} \left(D^2+\dfrac{k_1+k_2}{m_1}\right)x-\dfrac{k_2}{m_1}\,y=0 \\ -\dfrac{k_2}{m_2}\,x+\left(D^2+\dfrac{k_2}{m_2}\right)y=0. \end{cases}$$

Next we operate on the first equation by $D^2+\dfrac{k_2}{m_2}$ and multiply the second equation through by k_2/m_1. This gives

$$\begin{cases} \left(D^4+\left(\dfrac{k_1+k_2}{m_1}+\dfrac{k_2}{m_2}\right)D^2+\dfrac{k_1k_2+k_2^2}{m_1m_2}\right)x-\dfrac{k_2}{m_1}\left(D^2+\dfrac{k_2}{m_2}\right)y=0 \\ \qquad\qquad -\dfrac{k_2^2}{m_1m_2}\,x+\dfrac{k_2}{m_1}\left(D^2+\dfrac{k_2}{m_2}\right)y=0. \end{cases}$$

Addition of these equations gives us

(1) $$\left(D^4+\frac{m_2(k_1+k_2)+m_1k_2}{m_1m_2}\,D^2+\frac{k_1k_2}{m_1m_2}\right)x=0.$$

The auxiliary polynomial is

$$r^4+\frac{m_2(k_1+k_2)+m_1k_2}{m_1m_2}\,r^2+\frac{k_1k_2}{m_1m_2}.$$

Solving for r^2 using the quadratic formula, we get the intermediate result

$$\begin{aligned} b^2-4ac &= \frac{1}{m_1^2m_2^2}\,(m_2^2(k_1+k_2)^2+2m_1m_2k_2(k_1+k_2)+m_1^2k_2^2)-4\,\frac{k_1k_2}{m_1m_2} \\ &= \frac{1}{m_1^2m_2^2}\,(m_2^2(k_1^2+2k_1k_2+k_2^2)+2m_1m_2k_1k_2 \\ &\quad +2m_1m_2k_2^2+m_1^2k_2^2-4m_1m_2k_1k_2) \\ &= \frac{1}{m_1^2m_2^2}\,(m_2^2k_1^2-2m_1m_2k_1k_2+m_1^2k_2^2+m_2^2k_2^2 \\ &\quad +2m_1m_2k_2^2+2m_2^2k_1k_2) \\ &= \frac{1}{m_1^2m_2^2}\,[(m_2k_1-m_1k_2)^2+m_2k_2(m_2k_2+2m_1k_2+2m_2k_1)]. \end{aligned}$$

This is a positive quantity which is certainly less than b^2. So its square root is less than

$$b=\frac{m_2(k_1+k_2)+m_1k_2}{m_1m_2},$$

which is also positive. Thus $r^2=-\tfrac{1}{2}b\pm\tfrac{1}{2}\sqrt{b^2-4ac}<0$. If we then put

$$-\ell_1=-\tfrac{1}{2}b+\tfrac{1}{2}\sqrt{b^2-4ac} \text{ and } -\ell_2=-\tfrac{1}{2}b-\tfrac{1}{2}\sqrt{b^2-4ac},$$

then the roots of the auxiliary polynomial are the four pure imaginary numbers $\sqrt{\ell_1}\, i$, $-\sqrt{\ell_1}\, i$, $\sqrt{\ell_2}\, i$, and $-\sqrt{\ell_2}\, i$. So the general solution of (1) is

$$x(t) = c_1 \cos\sqrt{\ell_1}\, t + c_2 \sin\sqrt{\ell_1}\, t + c_3 \cos\sqrt{\ell_2}\, t + c_4 \sin\sqrt{\ell_2}\, t$$

We then have, from the first equation in (*), that

$$\begin{aligned} y(t) &= \frac{m_1}{k_2}\left(-c_1\ell_1 \cos\sqrt{\ell_1}\, t - c_2\ell_1 \sin\sqrt{\ell_1}\, t - c_3\ell_2 \cos\sqrt{\ell_2}\, t - c_4\ell_2 \sin\sqrt{\ell_2}\, t + \frac{k_1+k_2}{m_1}\, x(t)\right) \\ &= d_1 \cos\sqrt{\ell_1}\, t + d_2 \sin\sqrt{\ell_1}\, t + d_3 \cos\sqrt{\ell_2}\, t + d_4 \sin\sqrt{\ell_2}\, t \end{aligned}$$

If we introduce phase angles ϕ_i as in Section 8.8, then we have

(2) $$\begin{cases} x(t) = a_1 \cos(\sqrt{\ell_1}\, t - \phi_1) + a_2 \cos(\sqrt{\ell_2}\, t - \phi_2). \\ y(t) = a_3 \cos(\sqrt{\ell_1}\, t - \phi_3) + a_4 \cos(\sqrt{\ell_2}\, t - \phi_4) \end{cases}$$

The graph of the motion is thus obtained by adding in each case two periodic cosine curves of generally different periods $1/\sqrt{\ell_1}$ and $1/\sqrt{\ell_2}$. Since the phase angles are in general also different, the resulting graph may be somewhat complicated. We do see from (2), however, that the system *oscillates in a bounded manner.* Two specific systems of this type are treated in Exercises 1 and 2 below.

In view of the close correspondence between mechanical systems and electrical circuits seen in Section 8.8, we expect that systems of differential equations would also be useful in modeling electrical systems of a more complex nature than those treated in Section 8.8. Consider, for example, the network shown in Figure 4.2.

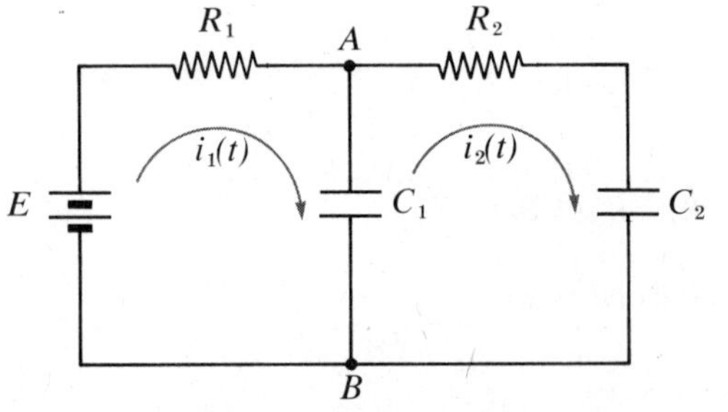

FIGURE 4.2

This network consists of two loops, ER_1AC_1BE and $AR_2C_2BC_1A$. If we want to analyze the current flowing in each loop at time t, then in addition to Kirchhoff's voltage law (p. 522 or p. 569) we need another law formulated by Kirchhoff, his *current law.* This states that at any *branch point* (like A or B in Figure 4.2) the total current flowing *into* the point is the same as the total current flowing *out of* the point.

In Figure 4.2 we have arbitrarily assigned a clockwise direction of current flow in each loop. By applying Kirchhoff's current law at A or B, we see that the current flowing from A to B must be $i_1(t) - i_2(t)$. We can now apply Kirchhoff's voltage law to complete the analysis of the network.

4.2 EXAMPLE. Write the system of linear differential equations for the currents $i_1(t)$ and $i_2(t)$ in each loop of Figure 4.2.

Solution. The voltage drop across R_1 is (cf. p. 569) $R_1 i_1(t)$, and the drop across R_2 is $R_2 i_2(t)$. The voltage drop across the capacitor C_2 is $\frac{1}{C_2} q_2(t)$, where $q_2(t)$ is the charge on the capacitor at time t. In Loop 1, the voltage drop across C_1 at time t is $\frac{1}{C_1}(q_1(t)-q_2(t))$ where $q_1(t)$ is the charge in the first loop at time t. [The charge on C_1 at time t is $q_1(t)-q_2(t)$, since the current flowing from A to B, $i_1(t)-i_2(t)$, must be the derivative of this charge.] In Loop 2, the voltage drop across C_1 is similarly $\frac{1}{C_1}(q_2(t)-q_1(t))$. We have then, from Kirchhoff's voltage law applied to each loop,

$$\begin{cases} E(t)=R_1 i_1(t)+\dfrac{1}{C_1}(q_1(t)-q_2(t)) \\ 0=R_2 i_2(t)+\dfrac{1}{C_2}q_2(t)+\dfrac{1}{C_1}(q_2(t)-q_1(t)). \end{cases}$$

If we differentiate with respect to t, we get

$$\begin{cases} R_1\dfrac{di_1}{dt}+\dfrac{1}{C_1}i_1(t)-\dfrac{1}{C_1}i_2(t)=\dot{E}(t) \\ R_2\dfrac{di_2}{dt}-\dfrac{1}{C_1}i_1(t)+\left(\dfrac{1}{C_1}+\dfrac{1}{C_2}\right)i_2(t)=0. \end{cases}$$

This first order nonhomogeneous system in standard form is

$$\text{(3)}\qquad \boxed{\begin{cases} \dfrac{di_1}{dt}=-\dfrac{1}{C_1R_1}i_1(t)+\dfrac{1}{C_1R_1}i_2(t)+\dfrac{\dot{E}(t)}{R_1} \\ \dfrac{di_2}{dt}=\dfrac{1}{C_1R_2}i_1(t)-\left(\dfrac{1}{R_2C_1}+\dfrac{1}{R_2C_2}\right)i_2(t). \end{cases}}$$

If $E(t)$ is constant, then (3) becomes a homogeneous first order system.

4.3 **EXAMPLE.** If in Figure 4.2 $E(t)$ is produced by a 12 volt battery, R_1 and R_2 are resistors of 5 ohms, C_1 is a capacitor of 10^{-4} farads, and C_2 is a capacitor of $\frac{2}{3}\times 10^{-4}$ farads, then find $i_1(t)$ and $i_2(t)$.

Solution. The system (2) becomes

$$\text{(4)}\qquad \begin{cases} \dfrac{di_1}{dt}=-2000i_1+2000i_2 \\ \dfrac{di_2}{dt}=\ \ 2000i_1-(3000+2000)i_2 \end{cases}$$

since for a 12 volt battery $E(t)=12$ for all t. In matrix form (4) is

$$\begin{pmatrix} \dfrac{di_1}{dt} \\ \dfrac{di_2}{dt} \end{pmatrix}=\begin{pmatrix} -2000 & 2000 \\ 2000 & -5000 \end{pmatrix}\begin{pmatrix} i_1 \\ i_2 \end{pmatrix}.$$

Using the method of Section 10.2, we find the eigenvalues by solving the characteristic equation. We have

$$\det\begin{pmatrix}-2000-\lambda & 2000\\ 2000 & -5000-\lambda\end{pmatrix}=\lambda^2+7000\lambda+10(1000)^2-4(1000)^2=0$$
$$\lambda^2+7000\lambda+6(1000)^2=0$$
$$(\lambda+1000)(\lambda+6000)=0.$$

The roots of this equation are $\lambda=-1000$ and $\lambda=-6000$. Eigenvectors belonging to $\lambda=-1000$ satisfy

$$\begin{pmatrix}-2000 & 2000\\ 2000 & -5000\end{pmatrix}\begin{pmatrix}x\\ y\end{pmatrix}=\begin{pmatrix}-1000x\\ -1000y\end{pmatrix},$$

which reduces to $-x+2y=0$. So a basic eigenvector is $\begin{pmatrix}2\\ 1\end{pmatrix}$. Eigenvectors belonging to $\lambda=-6000$ satisfy

$$\begin{pmatrix}-2000 & 2000\\ 2000 & -5000\end{pmatrix}\begin{pmatrix}x\\ y\end{pmatrix}=\begin{pmatrix}-6000x\\ -6000y\end{pmatrix},$$

which reduces to $2x+y=0$. So a basic eigenvector is $\begin{pmatrix}1\\ -2\end{pmatrix}$. The general solution to (4) is then

(5) $$\boldsymbol{i}(t)=k_1\begin{pmatrix}2\\ 1\end{pmatrix}e^{-1000t}+k_2\begin{pmatrix}1\\ -2\end{pmatrix}e^{-6000t}.$$

If we are given initial conditions in a problem like Example 4.3, then we can determine values of k_1 and k_2 affording the unique solution of (4).

4.4 EXAMPLE. Determine $i_1(t)$ and $i_2(t)$ in Example 4.3 if $i_1(0)=0$ and $i_2(0)=2$.

Solution. We simply substitute the initial values into (5) and solve for k_1 and k_2. We have $0=2k_1+k_2$, $0=4k_1+2k_2$ and $2=k_1-2k_2$. Adding these gives $k_1=2/5$. The first equation then tells us that $k_2=-4/5$. So the unique solution in this case is

$$\boldsymbol{i}(t)=\frac{1}{5}\begin{pmatrix}4\\ 2\end{pmatrix}e^{-1000t}+\frac{1}{5}\begin{pmatrix}-4\\ 8\end{pmatrix}e^{-6000t}.$$

We conclude this section with a consideration of the population of two competing species, one of which feeds upon the other. The model we give is due to the American biologist A. J. Lotka (1880–1949) and the Italian mathematician Vito Volterra (1860–1940), and is now considered a classical model for a *predator-prey* situation. Let $x(t)$ be the population of the prey species and $y(t)$ the population of the predator species at time t. Since these species are related by a feeding process, it is clear that the rate of growth of each at time t depends on the two respective populations at time t. That is, we must model the population change by a first order system of the form

(6) $$\begin{cases}\dfrac{dx}{dt}=f(x, y)\\[2mm] \dfrac{dy}{dt}=g(x, y)\end{cases}$$

where we add the *extinction* constraints $f(0, y) = 0$ and $g(x, 0) = 0$. (Once a species becomes extinct, it remains extinct!) We know that if $y(t)$ is small, then $x(t)$ should grow at a flourishing rate, while if $y(t)$ is large, then $x(t)$ will grow slowly or even decline. How can we model this mathematically? Lotka proposed the following nonlinear system as a realistic model in 1925.

4.5 **DEFINITION.** The **Lotka–Volterra predator-prey model** is given by the nonlinear system

(7)
$$\begin{cases} \dfrac{dx}{dt} = ax - bxy \\ \dfrac{dy}{dt} = cxy - dy \end{cases}$$

where a, b, c, and d are positive constants.

How reasonable is this model? Well, observe that we can rewrite (7) as

(8)
$$\begin{cases} \dfrac{dx}{dt} = x(a - by) \\ \dfrac{dy}{dt} = y(cx - d) \end{cases}$$

In this form the model seems quite reasonable. Notice that if $y = 0$, then x grows exponentially in (8); and if $x = 0$, then y declines exponentially. These extremes seem to reflect well the vital relationship existing between a food source and a feeder, as exemplified by what happened in Australia when rabbits were introduced into an environment where no species existed to prey upon them. Also notice that if x is large, then y grows rapidly with an abundant food source present. But even then, the growth is *damped* in proportion to the size of y, i.e., in response to overcrowding, or growth beyond even a large food supply. Concurrently, the growth rate of x is slowed markedly from a by the rapidly growing damping term $-by$. Finally, if y is large, then $x(t)$ will decline. This in turn will slow down the rate of growth of y.

This is the first time we have come across a nonlinear system (aside from the Cauchy-Euler systems in Exercises 10.3, which readily reduced to linear systems anyhow). And its reasonableness seems to imply that we are *stuck* with it. Since we have no solution techniques for nonlinear systems like (7), what can we do? One answer was suggested in Section 8.0. We can try to *approximate* (7) *by a linear system* in the spirit of multivariable differential calculus (cf. Chapters 4 and 6). This is based on (8). Notice that when $a - by = 0$ and $cx - d = 0$, we have *stable* populations (*zero population growth* in x and y, a *closed ecological system*). The values $y_0 = a/b$ and $x_0 = d/c$ are termed *stable* values. What we do is linearly approximate (7) near the point $\boldsymbol{x}_0 = \left(\dfrac{d}{c}, \dfrac{a}{b}\right)$.

4.6 **EXAMPLE.** (a) Show that the change of variables

(9)
$$\begin{cases} \bar{x} = x - \dfrac{d}{c} \\ \bar{y} = y - \dfrac{a}{b} \end{cases}$$

transforms (7) to a system that is approximately linear near $\boldsymbol{x}_0$.

(b) Analyze that approximating linear system.

Proof. (a) From (9) we have

$$x = \bar{x} + \frac{d}{c}, \; y = \bar{y} + \frac{a}{b}.$$

Hence $\frac{dx}{dt} = \frac{d\bar{x}}{dt}$ and $\frac{dy}{dt} = \frac{d\bar{y}}{dt}$. So (7) transforms to

$$\begin{cases} \frac{d\bar{x}}{dt} = a\left(\bar{x} + \frac{d}{c}\right) - b\left(\bar{x} + \frac{d}{c}\right)\left(\bar{y} + \frac{a}{b}\right) \\ \frac{d\bar{y}}{dt} = c\left(\bar{x} + \frac{d}{c}\right)\left(\bar{y} + \frac{a}{b}\right) - d\left(\bar{y} + \frac{a}{b}\right). \end{cases}$$

Collecting terms, we get

$$\text{(10)} \qquad \begin{cases} \frac{d\bar{x}}{dt} = -\frac{bd}{c}\,\bar{y} - b\bar{x}\bar{y} \\ \frac{d\bar{y}}{dt} = \frac{ac}{b}\,\bar{x} + c\bar{x}\bar{y}. \end{cases}$$

Now, (10) is not linear. But if (x, y) is near $\mathbf{x}_0 = \left(\frac{d}{c}, \frac{a}{b}\right)$, then the terms $\bar{x} = x - \frac{d}{c}$ and $\bar{y} = y - \frac{a}{b}$ are small. Hence their product is *very* small. So for (x, y) near $\mathbf{x}_0$, (10) is closely approximated by the linear system

$$\text{(11)} \qquad \begin{cases} \frac{d\bar{x}}{dt} = -\frac{bd}{c}\,\bar{y} \\ \frac{d\bar{y}}{dt} = \frac{ac}{b}\,\bar{x} \end{cases}$$

(b) The system (11) is easy to solve. The characteristic equation is

$$\det\begin{pmatrix} -\lambda & -bd/c \\ ac/b & -\lambda \end{pmatrix} = \lambda^2 + ad = 0.$$

Since a and d are positive, we get $\lambda = \pm\sqrt{ad}\, i$, a pure imaginary number. Thus the reasoning of Example 4.1 shows that

$$\begin{pmatrix} x(t) \\ y(t) \end{pmatrix} = \begin{pmatrix} a_1 \cos(\sqrt{ad}\, t - \phi_1) \\ a_2 \cos(\sqrt{ad}\, t - \phi_2) \end{pmatrix}.$$

Hence $(x(t), y(t))$ *oscillates* around the point (x_0, y_0).

The result of Example 4.6(b) seems quite satisfactory when interpreted in the real world. Oscillation around the stable point is a reasonable expectation once the populations are close to the stable values of d/c and a/b. The linear approximation (11) to the Volterra model (7) seems then not only to be helpful mathematically but also to be effective in terms of generating good results for the phenomenon modeled by (7). For a more detailed study of (7), see W. Boyce and

R. DiPrima, *Elementary Differential Equations and Boundary Value Problems*, 3rd Edition, Wiley, New York, 1977, pp. 428–432.

EXERCISES 10.4

1 Suppose that in Example 4.1 $m_1 = m_2 = 1$, $k_1 = 3$, and $k_2 = 2$. Suppose further that $\boldsymbol{x}(0) = (x(0), y(0)) = (-1, 2)$ and $\dot{\boldsymbol{x}}(0) = (0, 0)$. Find an explicit formula for $\boldsymbol{x} = \boldsymbol{x}(t)$.

2 Repeat Exercise 1 when $m_1 = m_2 = 1$, $k_1 = 3$, $k_2 = 2$, $\boldsymbol{x}(0) = (1, 1)$, and $\dot{\boldsymbol{x}}(0) = (0, 0)$.

3 Write a system of differential equations for the mechanical system shown in Figure 4.3. Here the spring S_i has spring constant $k_i > 0$.

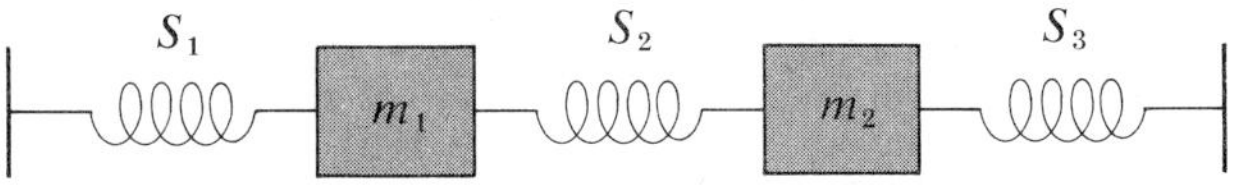

FIGURE 4.3

4 In Exercise 3, solve for the displacements $x(t)$ of m_1 and $y(t)$ of m_2 if $m_1 = m_2 = k_1 = k_2 = k_3 = 1$.

5 In Example 4.1, suppose that we add a dashpot D as shown in Figure 4.4. This exerts a damping force on the motion proportional to the velocities of m_1 and m_2. If the constant of proportionality is c, then write a system of differential equations for the motion.

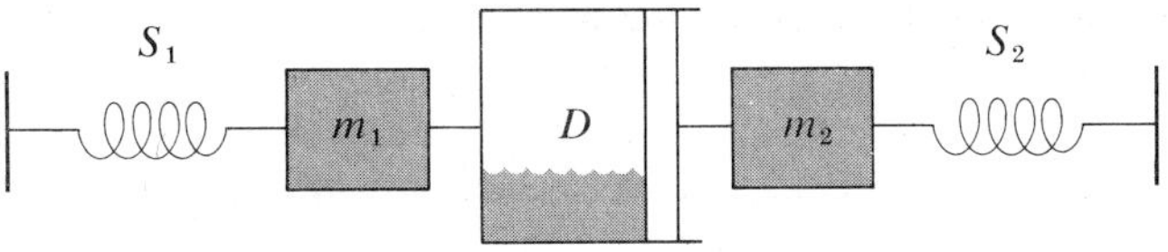

FIGURE 4.4

6 In Exercise 5, solve for the displacements $x(t)$ and $y(t)$ if $m_1 = m_2 = k_1 = k_2 = c = 1$.

7 In Example 4.1, suppose that an external force $\boldsymbol{F}(t) = (f_1(t), f_2(t))$ acts on the masses. Write the resulting nonhomogeneous system of second order linear equations for the motion.

8 Repeat Exercise 7 for the system in Figure 4.4.

9 Suppose that a projectile is fired from a gun that has an angle of elevation θ from the horizontal. Suppose that its initial speed is v_0. Assume that the only forces acting on the projectile are those due to gravity and air resistance. Suppose also that air resistance is proportional to the velocity. Write a system of second order linear differential equations for the motion of the projectile.

10 Solve the system in Exercise 9. Discuss the nature of the motion predicted by your solution.

11 (a) Consider the electrical network in Figure 4.5. Write the first order linear system of differential equations for this network. Assume $i_1(0) = i_2(0) = 0$.
(b) Solve for i_1 and i_2 in (a) if $L_1 = 0.02$, $L_2 = 0.04$, $R_1 = 10$, $R_2 = 20$, and $E(t) = 30$.

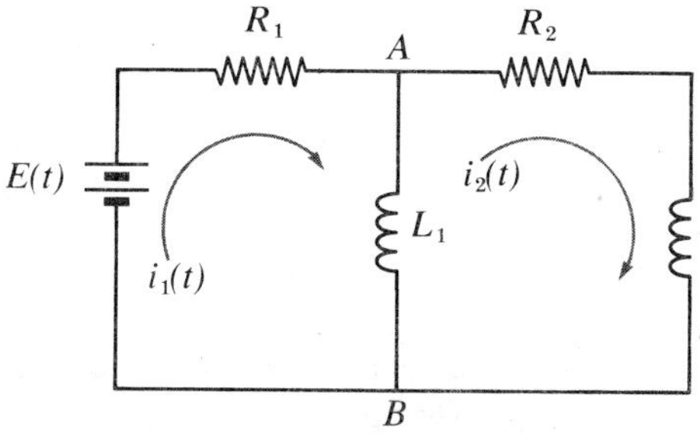

FIGURE 4.5

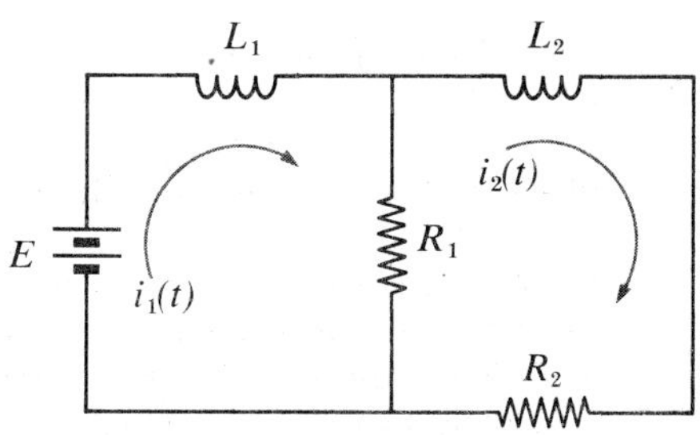

FIGURE 4.6

12 Repeat Exercise 11 for the network shown in Figure 4.6. Here $E = 100$, $R_1 = 20$, $R_2 = 40$, $L_1 = 0.01$, $L_2 = 0.02$, and $i_1(0) = i_2(0) = 0$.

13 Write a system of three first order linear equations for the network in Figure 4.7.

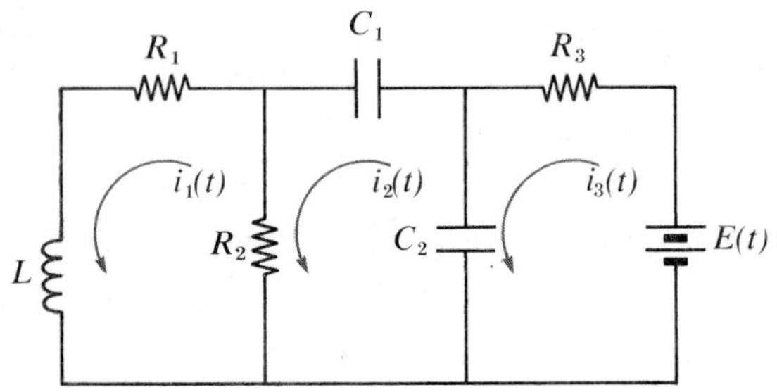

FIGURE 4.7

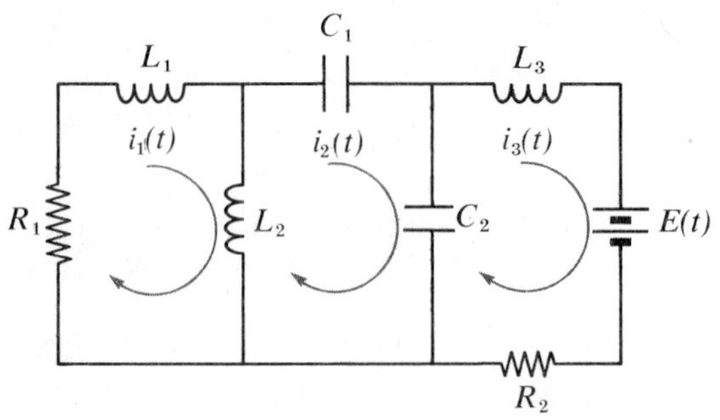

FIGURE 4.8

14 Write a system of three linear equations for the network shown in Figure 4.8.

15 Suppose we model the predator-prey relationship of Example 4.6 by the simpler *linear* model

$$\begin{cases} \dfrac{dx}{dt} = ax - by \\ \dfrac{dy}{dt} = cx - dy \end{cases}$$

where a, b, c, and d are positive constants such that $ad - bc < 0$.

(a) What happens if x or y is zero?

(b) What happens if x or y is large?

(c) By finding the eigenvalues of the system, discuss the various long-term possibilities for $x(t)$ and $y(t)$.

16 A more refined version of the Lotka–Volterra model, in which each species can impede its own growth by overpopulation, is given by

$$\begin{cases} \dfrac{dx}{dt} = x(a - bx - cy) \\ \dfrac{dy}{dt} = y(d - ex - fy) \end{cases}$$

where a, b, c, d, e, and f are positive.

(a) What happens if x and y are both small?

(b) What happens if x and/or y becomes large?

(c) What point $\mathbf{x}_0$ would be a stable point near which one could approximate the system by a linear system?

17 Suppose that three species x, y, and z compete in such a way that both y and z feed on

x. For the linear model

$$\begin{pmatrix} \dfrac{dx}{dt} \\ \dfrac{dy}{dt} \\ \dfrac{dz}{dt} \end{pmatrix} = \begin{pmatrix} 5 & -1 & -1 \\ 1 & 2 & 0 \\ 1 & 0 & 2 \end{pmatrix} \begin{pmatrix} x \\ y \\ z \end{pmatrix}$$

what is the long-term growth pattern? Is this reasonable?

18 Suppose that in Exercise 17 we alter the model slightly by lowering the growth rate of x:

$$\frac{dx}{dt} = 4x - y - z.$$

Then what is the new growth pattern? Is this reasonable?

19 Two fifty gallon tanks A and B (see Figure 4.9) contain brine. Each minute, Tank A receives one gallon of brine whose salt content is one pound, and Tank B receives one gallon of pure water. Also, three gallons per minute of solution are pumped from Tank B into Tank A, and two gallons per minute of solution are pumped from Tank A to Tank B. Each tank is kept thoroughly mixed, and two gallons per minute are drained off from Tank A. Give a system of differential equations for the amounts $x(t)$ and $y(t)$ of salt in each tank after t minutes.

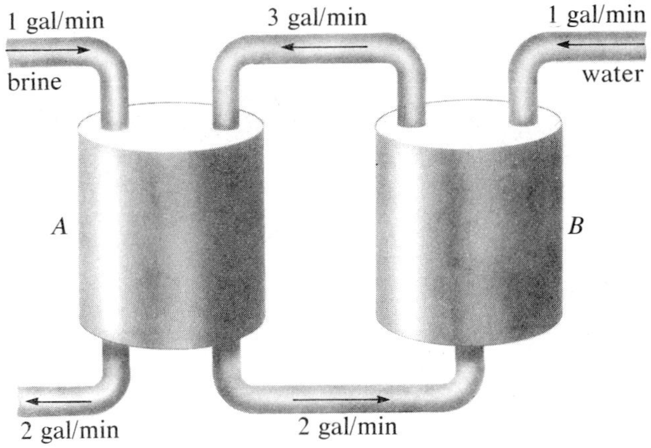

FIGURE 4.9

20 Solve the system in Exercise 19. What are the long-term amounts of salt in the two tanks?

5 THE LAPLACE TRANSFORM FOR SYSTEMS

While Theorem 1.6 assures us that any initial value problem for a first order linear system has a unique solution, the only technique we have at our disposal for actually *finding* that unique solution is rather cumbersome. In the case of constant coefficients we have to first find the general solution to $\dfrac{d\boldsymbol{x}}{dt} = A\boldsymbol{x}(t)$ by the method of Section 2, and then try to determine the appropriate values of the arbitrary constants needed to satisfy

$$\text{(1)} \qquad \boxed{\frac{d\boldsymbol{x}}{dt} = A\boldsymbol{x}(t), \qquad \boldsymbol{x}(t_0) = \boldsymbol{x}_0.}$$

Recalling our experience in Section 9.4 with the Laplace transform for an n-th order linear differential equation, it is not surprising that in the present context there is an alternative to the cumbersome procedure just described. This lies in extending the notion of the Laplace transform to $\mathbf{E}^n[0, +\infty)$, the set of all vector functions $\mathbf{f}:[0, +\infty) \to \mathbf{R}^n$ each of whose real valued coordinate functions $f_i:[0, +\infty) \to \mathbf{R}$ is an integrable function of exponential order (Definition 9.4.1). Since $\mathbf{E}^n[0, +\infty)$ just consists of all n-tuples of functions in $E[0, +\infty)$, $\mathbf{E}^n[0, +\infty)$ can be shown to be a real vector space just as $\mathbf{R}^n$ is a real vector space.

5.1 **PROPOSITION.** $\mathbf{E}^n[0, +\infty)$ is a real vector space.

Proof. By Proposition 9.4.3, $E[0, +\infty)$ is a real vector space. Since all operations in $\mathbf{E}^n[0, +\infty)$ are performed coordinate-wise, $\mathbf{E}^n[0, +\infty)$ will satisfy all the necessary requirements. For example, if $\mathbf{f}, \mathbf{g} \in \mathbf{E}^n[0, +\infty)$, then $\mathbf{f}+\mathbf{g} \in \mathbf{E}^n[0, +\infty)$ since

$$\mathbf{f}+\mathbf{g} = (f_1+g_1, \ldots, f_n+g_n),$$

and each $f_i+g_i \in E[0, +\infty)$. Similarly, for any $a \in \mathbf{R}$, we have $a\mathbf{f} \in \mathbf{E}^n[0, +\infty)$. All required identities hold in $\mathbf{E}^n[0, +\infty)$ since they hold separately for each coordinate in $E[0, +\infty)$ by Proposition 9.4.3. QED

Now we can define the *Laplace transform* $\mathcal{L}$ on $\mathbf{E}^n[0, +\infty)$ by analogy with Definition 9.4.4. Given

$$\mathbf{f} = (f_1, f_2, \ldots, f_n) \in \mathbf{E}^n[0, +\infty),$$

$\mathcal{L}(\mathbf{f})$ will be a function $\mathbf{F} = (F_1, F_2, \ldots, F_n)$ in the real vector space $\mathcal{F}(\mathbf{R}, \mathbf{R}^n)$ of all functions $\mathbf{F}:\mathbf{R} \to \mathbf{R}^n$ with common domain $(\alpha, +\infty)$.

5.2 **DEFINITION.** The **Laplace transform** $\mathcal{L}: \mathbf{E}^n[0, +\infty) \to \mathcal{F}(\mathbf{R}, \mathbf{R}^n)$ is given by $\mathcal{L}(\mathbf{f}) = \mathbf{F}$, where

(2)
$$\begin{aligned}\mathbf{F}(s) &= \int_0^{+\infty} e^{-st}\mathbf{f}(t)\,dt \\ &= \left(\int_0^{+\infty} e^{-st}f_1(t)\,dt, \ldots, \int_0^{+\infty} e^{-st}f_n(t)\,dt\right),\end{aligned}$$

for $s \in (\alpha, +\infty)$. Here α is as in Definition 9.4.1.

To calculate the Laplace transforms of vector functions in $\mathbf{E}^n[0, +\infty)$, we can use Table 9.4.1.

5.3 **EXAMPLE.** If $\mathbf{f}(t) = (t^5e^{4t}, e^{3t}\cos t)$, then find $\mathcal{L}(\mathbf{f})$.

Solution. The Laplace transforms of the coordinate functions $f_1(t) = t^5e^{4t}$ and $f_2(t) = e^{3t}\cos t$ are found in entries 6 and 11 of Table 9.4.1: $\mathcal{L}(f_1)(s) = \dfrac{5!}{(s-4)^6}$ and $\mathcal{L}(f_2)(s) = \dfrac{s-3}{(s-3)^2+1^2}$. So

$$\mathcal{L}(\mathbf{f})(s) = \left(\frac{5!}{(s-4)^6}, \frac{s-3}{(s-3)^2+1}\right).$$

Our definition of $\mathcal{L}$ guarantees that it is a linear transformation.

5.4 COROLLARY. $\mathscr{L}: \boldsymbol{E}^n[0, +\infty) \to \mathscr{F}(\boldsymbol{R}, \boldsymbol{R}^n)$ is a linear transformation.

Proof. Each coordinate function $\mathscr{L}_i$ of $\mathscr{L}$ is linear by Theorem 9.4.6. So the reasoning of Theorem 6.1.8 shows that $\mathscr{L}$ is a linear transformation. QED

$\mathscr{L}$ treats derivatives as we expect also.

5.5 COROLLARY. If $\dot{\boldsymbol{x}}(t) = \dfrac{d\boldsymbol{x}}{dt}$ is in $\boldsymbol{E}^n[0, +\infty)$, then

(3) $$\mathscr{L}(\dot{\boldsymbol{x}}(t))(s) = s\mathscr{L}(\boldsymbol{x}(t))(s) - \boldsymbol{x}(0).$$

Proof. By Theorem 9.4.7, we have for each coordinate function $x_i(t)$ of $\boldsymbol{x}(t)$ that

$$\mathscr{L}(\dot{x}_i(t))(s) = s\mathscr{L}(x_i(t))(s) - x_i(0).$$

Putting these coordinate equations together for $i = 1, 2, \ldots, n$, we have (3). QED

We can now use $\mathscr{L}$ and Table 9.4.1 to efficiently solve initial value problems for constant coefficient linear systems given by

(1) $$\frac{d\boldsymbol{x}}{dt} = A\boldsymbol{x}(t) + \boldsymbol{b}(t), \qquad \boldsymbol{x}(t_0) = \boldsymbol{x}_0.$$

The procedure is just the natural one of putting the solution together coordinate-by-coordinate, using the method of Section 9.4 in each coordinate. The following example shows how this is carried out in practice.

5.6 EXAMPLE. Solve the initial value problem

$$\begin{cases} \dfrac{dx}{dt} = 2x - y + e^t \\ \dfrac{dy}{dt} = 3x - 2y - e^t, \end{cases} \qquad \boldsymbol{x}(0) = (1, 1)$$

using the Laplace transform.

Solution. We apply $\mathscr{L}$ by applying the scalar Laplace transform $\mathscr{L}$ to each equation. Using Corollary 5.5 and Table 9.4.1, we get

$$\begin{cases} s\mathscr{L}(x(t)) - x(0) = 2\mathscr{L}(x(t)) - \mathscr{L}(y(t)) + \dfrac{1}{s-1} \\ s\mathscr{L}(y(t)) - y(0) = 3\mathscr{L}(x(t)) - 2\mathscr{L}(y(t)) - \dfrac{1}{s-1}, \end{cases}$$

(4) $$\begin{cases} (s-2)\mathscr{L}(x(t)) + \mathscr{L}(y(t)) = 1 + \dfrac{1}{s-1} = \dfrac{s}{s-1} \\ -3\mathscr{L}(x(t)) + (s+2)\mathscr{L}(y(t)) = 1 - \dfrac{1}{s-1} = \dfrac{s-2}{s-1}. \end{cases}$$

We solve the algebraic system (4) for $\mathscr{L}(x(t))$ and $\mathscr{L}(y(t))$. Multiplying the first equation by $-(s+2)$ and then adding the result to the second equation, we have

$$\begin{aligned}
(-s^2+4-3)\mathscr{L}(x(t)) &= -(s+2)\left(\frac{s}{s-1}\right)+\frac{s-2}{s-1},\\
-(s^2-1)\mathscr{L}(x(t)) &= \frac{-s^2-2s+s-2}{s-1}\\
&= -\frac{s^2+s+2}{s-1},\\
\mathscr{L}(x(t)) &= \frac{s^2+s+2}{(s-1)^2(s+1)}.
\end{aligned}$$

From the first equation in (4),

$$\begin{aligned}
\mathscr{L}(y(t)) &= \frac{s}{s-1}-\frac{(s-2)(s^2+s+2)}{(s-1)^2(s+1)}\\
&= \frac{s(s^2-1)-(s-2)(s^2+s+2)}{(s-1)^2(s+1)}.
\end{aligned}$$

To find $x(t)$ and $y(t)$, we need to use Table 9.4.1. So we first decompose $\mathscr{L}(x(t))$ and $\mathscr{L}(y(t))$ into partial fractions as in Section 9.4. We have

$$\begin{aligned}
\mathscr{L}(x(t)) &= \frac{A}{s-1}+\frac{B}{(s-1)^2}+\frac{C}{(s+1)},\\
s^2+s+2 &= A(s-1)(s+1)+B(s+1)+C(s-1)^2.
\end{aligned}$$

From $s=1$ we find $B=2$, and from $s=-1$ we find $C=\frac{1}{2}$. Then $s=0$ gives us $A=\frac{1}{2}$. Next,

$$\begin{aligned}
\mathscr{L}(y(t)) &= \frac{D}{s-1}+\frac{E}{(s-1)^2}+\frac{F}{s+1},\\
s(s^2-1)-(s-2)(s^2+s+2) &= D(s-1)(s+1)+E(s+1)+F(s-1)^2.
\end{aligned}$$

From $s=1$ we find $E=2$, and from $s=-1$ we find $F=\frac{3}{2}$. Then $s=0$ gives us $D=-\frac{1}{2}$. So we have

$$\begin{aligned}
\mathscr{L}(x(t)) &= \frac{\frac{1}{2}}{s-1}+\frac{2}{(s-1)^2}+\frac{\frac{1}{2}}{s+1},\\
\mathscr{L}(y(t)) &= \frac{-\frac{1}{2}}{s-1}+\frac{2}{(s-1)^2}+\frac{\frac{3}{2}}{s+1}.
\end{aligned}$$

From Table 9.4.1 (entries 4 and 5) we get

$$\begin{aligned}
x(t) &= \tfrac{1}{2}e^t+2te^t+\tfrac{1}{2}e^{-t}\\
y(t) &= -\tfrac{1}{2}e^t+2te^t+\tfrac{3}{2}e^{-t}.
\end{aligned}$$

So the given initial value problem has solution

$$\mathbf{x}(t) = \begin{pmatrix}\frac{1}{2}\\ -\frac{1}{2}\end{pmatrix}e^t+\begin{pmatrix}2\\2\end{pmatrix}te^t+\begin{pmatrix}\frac{1}{2}\\ \frac{3}{2}\end{pmatrix}e^{-t}.$$

As you can confirm by trying other methods, the Laplace transform is probably the easiest way to solve Example 5.6.

EXERCISES 10.5

In Exercises 1 to 10, use the Laplace transform to solve the given initial value problem.

1 $\begin{cases} \dfrac{dx}{dt} = 3x - 2y \\ \dfrac{dy}{dt} = 2x - 2y, \end{cases}$ $\boldsymbol{x}(0) = (1, 0)$

2 $\begin{cases} \dfrac{dx}{dt} = 2x + 4y \\ \dfrac{dy}{dt} = -2x - 2y, \end{cases}$ $\boldsymbol{x}(0) = (1, 3)$

3 $\begin{cases} \dfrac{dx}{dt} = x - 5y \\ \dfrac{dy}{dt} = x - 3y, \end{cases}$ $\boldsymbol{x}(0) = (1, 1)$

4 $\begin{cases} \dfrac{dx}{dt} = x - 5y \\ \dfrac{dy}{dt} = x - 3y, \end{cases}$ $\boldsymbol{x}(0) = (2, 1)$

5 $\begin{cases} \dfrac{dx}{dt} = x + y + 2z \\ \dfrac{dy}{dt} = 2y + 2z \\ \dfrac{dz}{dt} = -x + y + 3z, \end{cases}$ $\boldsymbol{x}(0) = (2, 0, 1)$

6 $\begin{cases} \dfrac{dx}{dt} = 3x - z \\ \dfrac{dy}{dt} = -2x + 2y + z \\ \dfrac{dz}{dt} = 8x - 3z, \end{cases}$ $\boldsymbol{x}(0) = (-1, 2, 8)$

7 $\begin{cases} \dfrac{dx}{dt} = 6x - 3y + 8e^t \\ \dfrac{dy}{dt} = 2x + y + 4e^t, \end{cases}$ $\boldsymbol{x}(0) = (-1, 0)$

8 $\begin{cases} 2\dfrac{dx}{dt} + \dfrac{dy}{dt} = x + y + e^{-t} \\ \dfrac{dx}{dt} + \dfrac{dy}{dt} = -2x - y + e^t, \end{cases}$ $\boldsymbol{x}(0) = (2, 1)$

9 $\begin{cases} \dfrac{d^2x}{dt^2} - 3\dfrac{dx}{dt} + \dfrac{dy}{dt} = -2x + y \\ \dfrac{dx}{dt} + \dfrac{dy}{dt} = 2x - y, \end{cases}$

$\boldsymbol{x}(0) = (0, -1)$, $\dot{\boldsymbol{x}}(0) = 0$
(*Hint:* Use Theorem 9.4.7.)

10 $\begin{cases} \dfrac{d^2x}{dt^2} - 3\dfrac{dx}{dt} + \dfrac{dy}{dt} = -2x + y \\ \dfrac{dx}{dt} + \dfrac{dy}{dt} = 2x - y, \end{cases}$

$\boldsymbol{x}(0) = (0, 0)$, $\dot{\boldsymbol{x}}(0) = 1$

11 Suppose that $W(t)$ is a fundamental matrix (p. 653) for $\dfrac{d\boldsymbol{x}}{dt} = A\boldsymbol{x}(t)$ that satisfies $W(0) = I_n$. Show that $\dot{W}(t) = AW(t)$.

12 Apply $\mathscr{L}$ to the equation $\dot{W}(t) = A\,W(t)$ in Exercise 11 and show that this leads to

$$(A - sI_n)\mathscr{L}(W(t)) = -I_n.$$

13 In Exercise 12, show that for s sufficiently large we have $W(t) = \mathscr{L}^{-1}(-(A - sI_n)^{-1})$.

REVIEW EXERCISES 10.6

In Exercises 1 to 12, solve the given system.

1 $\begin{cases} \dfrac{dx}{dt} = 2x - y - 2e^{-3t} \\ \dfrac{dy}{dt} = 2y - 3e^{-3t} \end{cases}$

2 $\begin{cases} \dfrac{d^2x}{dt^2} + \dfrac{dy}{dt} = -2y + 2e^{-2t} \\ \dfrac{dx}{dt} - \dfrac{dy}{dt} = 2y \end{cases}$

3 $\begin{cases} \dfrac{dx}{dt} = 3x - 4y + e^t \\ \dfrac{dy}{dt} = x - y - e^t \end{cases}$

where $x(0) = 1$ and $y(0) = -1$.

4 $\begin{cases} \dfrac{dx}{dt} = -3x - y \\ \dfrac{dy}{dt} = 2x - y \end{cases}$

5 $\begin{cases} \dfrac{dx}{dt} = 9x - 3y \\ \dfrac{dy}{dt} = -3x + 12y - 3z \\ \dfrac{dz}{dt} = -3y + 9z \end{cases}$

6 $\begin{cases} \dfrac{dx}{dt} = 2x + y + z \\ \dfrac{dy}{dt} = x + 2y + z \\ \dfrac{dz}{dt} = -2x - 2y - z \end{cases}$

7 $\begin{cases} \dfrac{dx}{dt} = x + y - 3e^t \\ \dfrac{dy}{dt} = 4x + y + 5e^t \end{cases}$

8 $\begin{cases} \dfrac{dx}{dt} = x + y - \cos t \\ \dfrac{dy}{dt} = 4x + y - \sin t \end{cases}$

9 $\begin{cases} \dfrac{dx}{dt} = x + y - e^{-t} \\ \dfrac{dy}{dt} = 4x + y + 2e^{-t} \end{cases}$

10 $\begin{cases} \dfrac{dx}{dt} = x + y + 2e^{-t} \\ \dfrac{dy}{dt} = 4x + y - 3e^{-t} \end{cases}$

11 $\begin{cases} \dfrac{dx}{dt} = y + \cos^2 t \\ \dfrac{dy}{dt} = -x - \cos t \sin t \end{cases}$

12 $\begin{cases} \dfrac{dx}{dt} = -3x - y + \csc t \\ \dfrac{dy}{dt} = 2x - y + \sec t \end{cases}$ on $\left[\dfrac{\pi}{6}, \dfrac{\pi}{3}\right]$

13 Suppose in Example 4.1 that $m_1 = 1$, $m_2 = 1$, $k_1 = 2$, and $k_2 = 3$. Find a formula for $\mathbf{x} = \mathbf{x}(t)$.

14 Write the system of differential equations for the system in Figure 6.1 if $R_1 = 2$, $R_2 = 1$, $L = 4$, and $E = 6$.

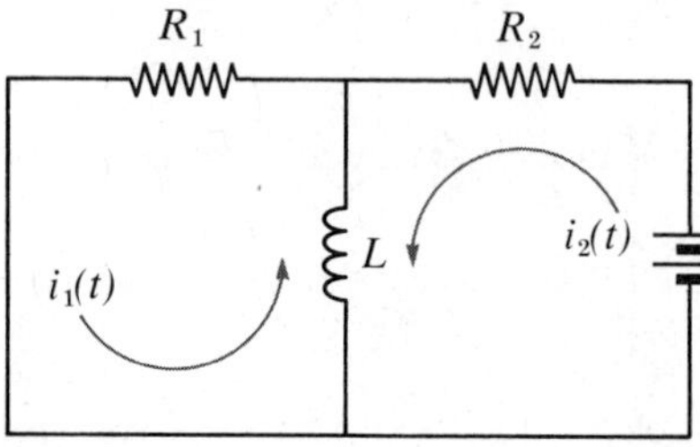

FIGURE 6.1

15 Suppose that a predator-prey relationship is modeled linearly by

$$\begin{cases} \dfrac{dx}{dt} = 3x - 4y \\ \dfrac{dy}{dt} = x - y. \end{cases}$$

Then what is the long-term trend for $x(t)$ and $y(t)$?

16 Use the Laplace transform to solve

$$\begin{cases} \dfrac{dx}{dt} = 2x - 2y \\ \dfrac{dy}{dt} = 4x - 2y, \end{cases}$$

if $x(0) = y(0) = 1$.

17 Repeat Exercise 16 for

$$\begin{cases} \dfrac{dx}{dt} = 2x - y + e^t \\ \dfrac{dy}{dt} = 3x - 2y - e^t, \end{cases}$$

where $x(0) = y(0) = 0$.

11 INFINITE SERIES

0 INTRODUCTION

Material in this chapter is prerequisite to Sections 3.1, 4.8, 4.9, 5.11, and 8.9, and some of its ideas occur elsewhere in a less essential way. You may have already studied infinite series in elementary calculus. If so, then this chapter can be used for reference and review. If this is your first experience with infinite series, though, then you may wish to refer to texts such as *The Calculus with Analytic Geometry* by Louis Leithold (Harper & Row, New York), *Calculus with Analytic Geometry* by Richard H. Crowell and William E. Slesnick (Norton, New York), or *Advanced Calculus* by R. Creighton Buck (McGraw–Hill, New York) for some of the omitted proofs.

We begin with the notion of an infinite sequence and a very basic sufficient condition for convergence of an infinite sequence. This condition involves the *Axiom of Completeness*, the property of the real numbers that distinguishes them from the rational numbers and underlies most of the ideas of calculus. We emphasize the very important idea of *approximating the infinite* (in this case infinite nonrepeating decimals, i.e., *irrational* numbers) *by the finite* (in this case, finite decimals, which are *rational* numbers). The second section introduces the notions of *infinite series* and *convergence* in terms of convergence of a naturally associated infinite sequence. The roots of this go all the way back to Greek philosophers and mathematicians. We then present various *tests for convergence* of series of positive terms, series of alternating positive and negative terms, and finally series of arbitrary terms.

The rest of the chapter (Sections 6 to 10) is devoted to applying the ideas of convergence of infinite series to the problem of *approximating arbitrary functions* by those functions that have the easiest calculus—*polynomials in x*. Section 6 introduces the *Taylor polynomial* approximation $p_n(x)$ to an arbitrary function f near a fixed point x_0. Then, in order to justify the process of approximating the derivative and integral of f near x_0 by the derivative and integral of $p_n(x)$, we consider *power series* in Section 7 and Taylor series expansions of functions in Section 8. In Section 9 we are ready to justify using polynomial approximations $p_n(x)$ in place of $f(x)$ when differentiating and integrating. The final section presents an interesting and classical extension of the *Binomial Theorem* of elementary algebra, which goes back to Newton.

1 INFINITE SEQUENCES

We have all dealt with finite sequences. A shopping list, a telephone number, a ZIP code all are simple examples. Much of this text has been concerned with vectors in $\boldsymbol{R}^n$, that is, with finite sequences

$$\boldsymbol{x} = (x_1, x_2, \ldots, x_n).$$

In the calculus of sequences, one removes the restriction that the sequence be finite. We are thus going to consider *infinite* sequences.

To see what should be meant by such a thing, we once again return to the known idea for a clue. What mathematically is a finite sequence? Concretely, it seems to be a list with a first entry, second entry, third entry, etc. We can capture the sense of the order of occurrence of the entries if we view a sequence

$$\boldsymbol{a} = (a_1, a_2, \ldots, a_n)$$

as a *function* $f:\{1, 2, \ldots, n\} \rightarrow \boldsymbol{R}$ where

$$f(i) = a_i \quad \text{for} \quad i = 1, 2, \ldots, n.$$

Thus a finite sequence is a real-valued function whose domain is the set of the first n positive integers. The *entries* of the sequence are

$$f(1) = a_1, f(2) = a_2, \ldots, f(n) = a_n.$$

So the first entry is the value of f at 1, the second entry is the value of f at 2, etc. With this point of view, we can regard the sequence

$$(-2, \pi, 3, \tfrac{1}{2}, -1)$$

as a function

$$f:\{1, 2, 3, 4, 5\} \rightarrow \boldsymbol{R}$$

whose range is the set $\{-2, -1, \frac{1}{2}, 3, \pi\}$, since $f(1) = -2$, $f(2) = \pi$, $f(3) = 3$, $f(4) = \frac{1}{2}$, and $f(5) = -1$. [**NOTE:** *Sequences* are ordered; *sets* are not.] This point of view now suggests a simple way to extend the notion of a finite sequence to infinite sequences.

1.1 **DEFINITION.** An **infinite sequence** $\boldsymbol{a} = (a_1, a_2, a_3, \ldots, a_n, \ldots)$ is a function $f:\boldsymbol{N} \rightarrow \boldsymbol{R}$, where $\boldsymbol{N}$ is the set $\{1, 2, 3, \ldots, n, \ldots\}$ of positive integers. The numbers $a_1 = f(1)$, $a_2 = f(2)$, $a_3 = f(3)$, etc., are called the **entries** of the sequence.

Hence the set of entries of a given sequence is just the range of the associated function $f:\boldsymbol{N} \rightarrow \boldsymbol{R}$. In writing a sequence, we usually list its first few entries, and then its n-th entry $a_n = f(n)$. This serves to specify the formula for the function f. Occasionally, you will be expected to come up with a formula for f from knowledge of the first few terms, as you have doubtless done on I.Q. tests.

1.2 **EXAMPLE.** Show that the sets of entries of the sequences

$$\boldsymbol{a} = \left(1, \tfrac{1}{2}, \tfrac{1}{4}, \tfrac{1}{8}, \ldots, \frac{1}{2^{n-1}}, \ldots\right)$$

and

$$\boldsymbol{b} = (1, \tfrac{1}{2}, 1, \tfrac{1}{4}, 1, \tfrac{1}{8}, \ldots, g(k), \ldots),$$

where $g(k) = \dfrac{1}{2^{k/2}}$ for k even and $g(k) = 1$ for k odd, are the same. Are the sequences the same?

Solution. The range of the first sequence consists of all rational numbers of the form $a_n = \dfrac{1}{2^{n-1}}$ for n a positive integer. Clearly the range of the second sequence is contained in this set, since only $1 = \dfrac{1}{2^0}$ and $\dfrac{1}{2^m}$ for $m \geq 1$ can occur as entries. Conversely, all entries of the first sequence occur in the second: Any $a_n = \dfrac{1}{2^{n-1}}$ is the $2(n-1)$-st term of $\boldsymbol{b}$, since

$$b_{2(n-1)} = \frac{1}{2^{2(n-1)/2}} = \frac{1}{2^{n-1}}.$$

So the respective sets of entries are identical. However, the sequences are *not* the same, since their respective defining functions f and g differ. For example, $f(3) = a_3 = \frac{1}{4}$, but $g(3) = b_3 = 1$. Since we have defined "infinite sequence" to *mean* "defining function," two infinite sequences are equal *only* in case their defining functions are identical.

As you may well be expecting, the kinds of sequences we can handle in calculus are the well-behaved ones, those whose entries a_n approach a limit a as $n \to \infty$. This should mean that a_n is arbitrarily close to a for large enough n. The formal definition is the following.

1.3 **DEFINITION.** A sequence $\boldsymbol{a} = (a_1, a_2, \ldots, a_n, \ldots)$ is said to **converge** (or **be convergent**) to a if $\lim_{n\to\infty} a_n = a$ in the following sense: For every $\varepsilon > 0$, there is some natural number N such that $|a_n - a| < \varepsilon$ for all $n > N$. A sequence that does not converge is called **divergent** (or is said to **diverge**).

In Example 1.2 above,

$$\boldsymbol{a} = (1, \tfrac{1}{2}, \tfrac{1}{4}, \ldots, 1/2^{n-1}, \ldots),$$

$$\boldsymbol{b} = (1, \tfrac{1}{2}, 1, \tfrac{1}{4}, \ldots, g(n), \ldots),$$

we see that $\boldsymbol{a}$ converges to $0 = \lim_{n\to\infty} \dfrac{1}{2^{n-1}}$. However, $\boldsymbol{b}$ diverges, since $\lim_{n\to\infty} g(n)$ fails to exist.

1.4 **EXAMPLE.** If

$$s_n = \frac{2n^2 - 5n}{3n^2 - 2},$$

then determine whether $\mathbf{s} = (s_1, s_2, \ldots, s_n, \ldots)$ is convergent or divergent. If it converges, then find its limit.

Solution. Here

$$\lim_{n\to\infty} s_n = \lim_{n\to\infty} \frac{2n^2-5n}{3n^2-2} = \lim_{n\to\infty} \frac{2-\dfrac{5}{n}}{3-\dfrac{2}{n^2}} = \frac{2}{3}$$

(dividing numerator and denominator by n^2, the highest power of n that occurs in the fraction s_n). Thus $\mathbf{s}$ converges to $\frac{2}{3}$.

You will notice that we calculated $\lim_{n\to\infty} s_n$ in Example 1.4 just as we would have calculated

$$\lim_{x\to+\infty} \frac{2x^2-5x}{3x^2-2}.$$

This is an illustration of the following useful general theorem.

1.5 **THEOREM.** Suppose that $\boldsymbol{a} = (a_1, a_2, \ldots, a_n, \ldots)$ is a sequence with $a_n = f(n)$. Extend f to a function $\bar{f}: \boldsymbol{R} \to \boldsymbol{R}$ by using the formula for $f: \boldsymbol{N} \to \boldsymbol{R}$ with the real variable x in place of the integer variable n. If $\bar{f}$ is continuous and $\lim_{x\to+\infty} \bar{f}(x) = a$, then $\boldsymbol{a}$ converges to a. If $\lim_{x\to+\infty} \bar{f}(x) = \pm\infty$, then $\boldsymbol{a}$ diverges.

1.6 **EXAMPLE.** Decide whether the sequences $\boldsymbol{a}$, $\boldsymbol{b}$, and $\boldsymbol{c}$ are convergent, if

$$a_n = n \sin\frac{1}{n}, \quad b_n = ne^{-n}, \quad \text{and} \quad c_n = e^n/n^2.$$

Solution. (a) Let $\bar{f}(x) = x \sin\dfrac{1}{x}$. Then

$$\lim_{x\to+\infty} \bar{f}(x) = \lim_{x\to+\infty} \frac{\sin\dfrac{1}{x}}{\dfrac{1}{x}} = \lim_{t\to0+} \frac{\sin t}{t}, \quad \text{where} \quad t = \frac{1}{x}.$$

An important fact from elementary calculus is that $\lim_{t\to0} \dfrac{\sin t}{t} = 1$. Thus, $\lim_{x\to+\infty} \bar{f}(x) = 1$. So by Theorem 1.5, the given sequence $\boldsymbol{a}$ converges to 1.

(b) Put $\bar{g}(x) = xe^{-x} = x/e^x$. Then L'Hôpital's Rule allows us to compute

$$\lim_{x\to+\infty} \bar{g}(x) = \lim_{x\to+\infty} \frac{x}{e^x} = \lim_{x\to+\infty} \frac{(x)'}{(e^x)'} = \lim_{x\to+\infty} \frac{1}{e^x} = 0.$$

So the given sequence $\boldsymbol{b}$ converges to 0.

(c) Finally, we put $\bar{h}(x) = e^x/x^2$. We can use L'Hôpital's Rule twice to get

$$\lim_{x\to+\infty} \bar{h}(x) = \lim_{x\to+\infty} \frac{e^x}{2x} = \lim_{x\to+\infty} \frac{e^x}{2} = +\infty,$$

so the given sequence $\boldsymbol{c}$ diverges to $+\infty$.

The following result is sometimes useful, especially in theoretical work. The proof is virtually the same as the proof of Theorem 3.2.2.

1.7 **THEOREM.** If $\boldsymbol{a}$ and $\boldsymbol{b}$ are convergent sequences with limits a and b respectively, and c is a constant, then

(a) $c\boldsymbol{a} \equiv (ca_1, ca_2, \ldots, ca_n, \ldots)$ converges to ca.

(b) $\boldsymbol{a}+\boldsymbol{b} \equiv (a_1+b_1, a_2+b_2, \ldots, a_n+b_n, \ldots)$ converges to $a+b$ and $\boldsymbol{a}-\boldsymbol{b}$ converges to $a-b$.

(c) $\boldsymbol{ab} \equiv (a_1b_1, a_2b_2, \ldots, a_nb_n, \ldots)$ converges to ab.

(d) If $b \neq 0$ and $b_i \neq 0$ for all i, then

$$\frac{\boldsymbol{a}}{\boldsymbol{b}} \equiv \left(\frac{a_1}{b_1}, \frac{a_2}{b_2}, \ldots, \frac{a_n}{b_n}, \ldots\right) \quad \text{converges to } \frac{a}{b}.$$

An important criterion for convergence exists in the case of *monotonic sequences* and is related to the basic nature of the real number system.

1.8 **DEFINITION.** A sequence $\boldsymbol{a} = (a_1, a_2, \ldots, a_n, \ldots)$ is called **monotonic** if one of the following properties is true:

(i) $a_n \leq a_{n+1}$ for all n ($\boldsymbol{a}$ is **increasing**)

or

(ii) $a_n \geq a_{n+1}$ for all n ($\boldsymbol{a}$ is **decreasing**).

If $a_n < a_{n+1}$ (respectively, $a_n > a_{n+1}$) for all n, then we say that the monotonic sequence $\boldsymbol{a}$ is *strictly increasing* (respectively, *strictly decreasing*).

1.9 **EXAMPLE.** Determine whether $\boldsymbol{a}$ and $\boldsymbol{b}$ are monotonic if

$$a_n = \frac{n^2+1}{2n+3} \quad \text{and} \quad b_n = \frac{(-1)^n}{n}.$$

Solution. (a) The first few entries of $\boldsymbol{a}$ are $\frac{2}{5}, \frac{5}{7}, \frac{10}{9}, \frac{17}{11}$, etc. So we suspect that the series *is* monotonic increasing. To confirm this, we consider the ratio $\dfrac{a_{n+1}}{a_n}$. If this is always ≥ 1, then since a_n is positive, we can multiply through by a_n to get $a_{n+1} \geq a_n$, and therefore conclude that $\boldsymbol{a}$ is increasing. We have

$$\begin{aligned}\frac{a_{n+1}}{a_n} &= \frac{(n+1)^2+1}{2(n+1)+3}\Big/\frac{n^2+1}{2n+3} \\ &= \frac{n^2+2n+2}{n^2+1} \cdot \frac{2n+3}{2n+5} \\ &= \frac{2n^3+7n^2+10n+6}{2n^3+5n^2+2n+5} > 1 \quad \text{for all } n \geq 1.\end{aligned}$$

Since $\dfrac{a_{n+1}}{a_n} > 1$ for all n, we have $a_{n+1} > a_n$, so the sequence is monotonic (strictly increasing).

(b) The first few entries of $\boldsymbol{b}$ are $-1, \frac{1}{2}, -\frac{1}{3}, \frac{1}{4}, -\frac{1}{5}, \frac{1}{6}$, etc. It is clear that this sequence is *not* monotonic: each even-numbered entry is positive and each odd-numbered entry is negative.

To give the criterion we mentioned for convergence of monotonic sequences, we need the notion of *bounded sets.*

1.10 DEFINITION. Let $S \subseteq \boldsymbol{R}$ be a set of real numbers. A number ℓ is called a **lower bound** of S if $\ell \leq s$ for every $s \in S$. A number u is called an **upper bound** of S if $u \geq s$ for all $s \in S$. S is **bounded** if it has both an upper bound and a lower bound. A **greatest lower bound** (glb) of a set S is a number g such that
(1) g is a lower bound of S
(2) $g \geq \ell$ for every lower bound ℓ of S.
A **least upper bound** (lub) of a set S is a number b such that
(1) b is an upper bound of S
(2) $b \leq u$ for every upper bound u of S.
If S is the set of entries of a sequence $\boldsymbol{a}$, we say that **$\boldsymbol{a}$ is bounded** if S is bounded, b is a **least upper bound of $\boldsymbol{a}$** if b is a least upper bound of S, and so forth.

1.11 EXAMPLE. (a) Let $\boldsymbol{a} = (1, \frac{1}{2}, 1, \frac{1}{4}, 1, \frac{1}{8}, \ldots)$. Is $\boldsymbol{a}$ bounded? Does it have a least upper bound? A greatest lower bound?

(b) Answer the same questions for

$$\boldsymbol{b} = \left(\frac{2}{5}, \frac{5}{7}, \frac{10}{9}, \ldots, \frac{n^2+1}{2n+3}, \ldots\right).$$

Solution. (a) Here *any* number $\ell \leq 0$ will do for a lower bound. The *greatest* lower bound is 0, since any larger number fails to be a lower bound: $\dfrac{1}{2^k}$ becomes arbitrarily close to zero as k becomes larger and larger. Also, any number $u \geq 1$ will do as an upper bound. The least upper bound is then 1. The sequence *is* bounded, since its set of entries has both an upper bound and a lower bound.

(b) We saw in Example 1.9(a) that this sequence is strictly increasing. Since

$$\lim_{n\to\infty} \frac{n^2+1}{2n+3} = \lim_{n\to\infty} \frac{1+\dfrac{1}{n^2}}{\dfrac{2}{n}+\dfrac{3}{n^2}} = +\infty,$$

we see that it cannot have any upper bounds. Since it is strictly increasing, any number $\ell \leq \frac{2}{5} = b_1$ will do for a lower bound. The greatest lower bound is then $\frac{2}{5}$. Since the set of entries is bounded only below (and not both above and below), it is *not* a bounded sequence.

As you can see from Example 1.11, the least upper bound b of a set or sequence S is the *most efficient* upper bound u that can be used. It represents the irreducible minimum point you must reach on the x-axis to put the set S to your left. Clearly, symmetric remarks show that the greatest lower bound g of S represents the most efficient lower bound: the point farthest to the right that has the entire set S to its right. See Figure 1.1.

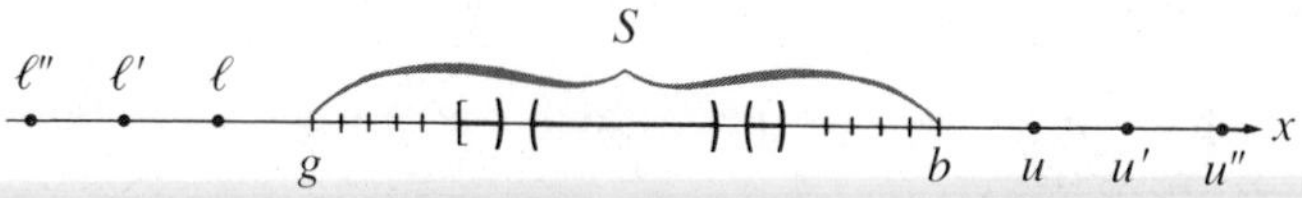

FIGURE 1.1

The least upper bound b and the greatest lower bound g of a set S are respectively an upper bound and a lower bound for S, but just "by the skin of their teeth," as the saying goes!

We do calculus using *real* variables x, not *rational* variables, even though, algebraically speaking, the real and rational numbers *seem* to have identical properties. If you have ever wondered why calculus is done on the real number system exclusively, then you will want to read the following discussion.

The real number system $\boldsymbol{R}$ is a *continuum*, a number line with no breaks in it such as occur in the rational number system $\boldsymbol{Q}$ at the places where $\sqrt{2}$, π, $\sqrt{3}$, e, and the other irrational numbers "should" be. Just as in arithmetic the natural number system $\boldsymbol{N}$ had to be *expanded* so that we could always be assured of getting answers when we subtracted, or divided by nonzero quantities, so the rational number system has to be expanded to $\boldsymbol{R}$ if we want to be sure of getting greatest lower bounds and least upper bounds of bounded sets. For example, if we let $L=\{r\in \boldsymbol{Q} \mid r^2<2\}$, then L is a perfectly nice set of rational numbers that is bounded (4, 2, $\frac{3}{2}$, etc. are upper bounds and -10, -2, $-\frac{3}{2}$, etc. are lower bounds). But it has no least upper bound *in the set of rational numbers*: the obvious candidate for the least upper bound, $\sqrt{2}$, is not in $\boldsymbol{Q}$! This turns out to account for the fact that the sequence $(r_1, r_2, r_3, \ldots)$ of rational approximations to $\sqrt{2}$ that comes from the square root algorithm does not have a limit in $\boldsymbol{Q}$. The limit of $(1, 1.4, 1.41, 1.414, 1.4142, \ldots)$ "should" be $\sqrt{2}$, but $\sqrt{2}\notin \boldsymbol{Q}$. So we don't do calculus in $\boldsymbol{Q}$ because *we can't take limits satisfactorily if we restrict ourselves to rational numbers only*. That this sort of problem can't arise in $\boldsymbol{R}$ is guaranteed by the fundamental *axiom of continuity* (or *completeness*) for $\boldsymbol{R}$.

1.12 AXIOM OF COMPLETENESS. Every nonempty set S of real numbers that has an upper bound has a least upper bound.

This is a fundamental *assumption* about the real numbers. It is not a theorem to prove, nor is it a definition. It is rather a description of the inherent property of $\boldsymbol{R}$ that distinguishes $\boldsymbol{R}$ from $\boldsymbol{Q}$. The following consequence of the axiom is left as an exercise for you (Exercise 15) below.

1.13 COROLLARY. Every nonempty set T of real numbers that has a lower bound has a greatest lower bound.

Our set $L=\{r\in \boldsymbol{Q} \mid r^2<2\}$ is a nonempty set of real numbers that is bounded above, and it does have a least upper bound, namely $\sqrt{2}$. The real number $\sqrt{2}$ is certainly bigger than any rational r whose square is less than 2, and it is fairly clear that you can't do better than $\sqrt{2}$. (For any smaller number s, the proposition that $s<r_j$ for some rational number r_j obtained from the square root algorithm can be proved rigorously, but we will leave that for later courses.) See Figure 1.2.

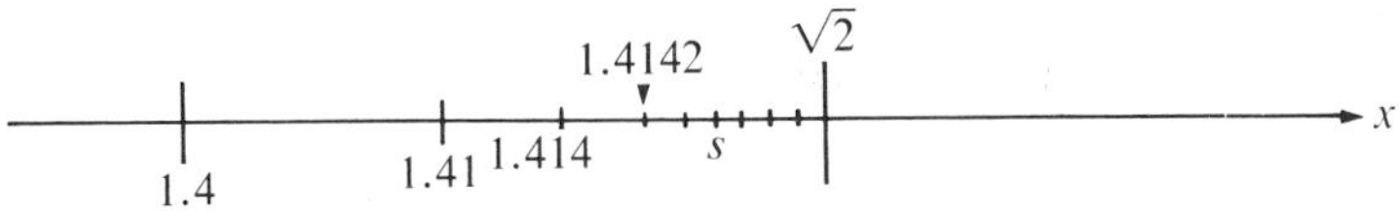

FIGURE 1.2

While the set $\boldsymbol{R}$ has the completeness property 1.12, it pays a price for this in terms of representing its elements as decimals. You may recall that every rational number $r=\dfrac{m}{n}$ (where m and $n\neq 0$ are integers) has a decimal representation that is either *finite* (e.g., $\frac{3}{20}=0.15$) or *infinite repeating* (e.g., $\frac{3}{11}=0.272727\ldots$). In either case it is easy to perform arithmetic with these decimals according to the familiar rules of elementary school. But irrational real numbers (like $\sqrt{2}$, e, π, and $\sqrt{3}$) have *infinite nonrepeating* decimal representations. How do we do arithmetic with, for example, $e=2.718281828459\ldots$? In practice, we *approximate* e by a rational number by "rounding off" to a finite decimal. For example, $e \approx 2.7183$ is a 4-decimal-place approximation that is adequate for many calculations. Similarly, $\pi \approx 3.1416$ and $\sqrt{2} \approx 1.4142$. As you know, tables of functions like ln, exp, and sine are simply tabulations of rational approximations to usually irrational quantities, all to four or five decimal places. This sort of thing allows us to have the best of both worlds: *we have completeness* (hence limits of convergent sequences in $\boldsymbol{R}$ are also in $\boldsymbol{R}$), and *we have a finitely calculable arithmetic* of real numbers using *rational approximations.* The latter idea of approximating infinite quantities (in this case, nonterminating nonrepeating decimals) by finite quantities (in this case finite decimals) is really the basic theme of this chapter. Keep it in mind as you read on.

We are now ready to give the convergence criterion promised on p. 677.

1.14 **THEOREM.** Any bounded monotonic sequence $\boldsymbol{a}$ of real numbers converges to a limit $a \in \boldsymbol{R}$. In fact, a bounded increasing sequence converges to its least upper bound. A bounded decreasing sequence converges to its greatest lower bound.

Looking back to the sequence $\boldsymbol{a} = \left(1, \frac{1}{2}, \frac{1}{4}, \ldots \frac{1}{2^{n-1}}, \ldots\right)$ in Example 1.2, we see that it is monotonic decreasing, since for all n we have $\frac{1}{2^n} < \frac{1}{2^{n-1}}$. It is clearly bounded below by 0 (or any negative number), so it has a greatest lower bound. Hence $\boldsymbol{a}$ converges, as we saw by other means following Definition 1.3.

The following example suggests the power of Theorem 1.14.

1.15 **EXAMPLE.** If $a_n = \frac{n!}{n^n}$, then decide whether $\boldsymbol{a} = (a_1, a_2, a_3, \ldots, a_n, \ldots)$ converges.

Solution. Since computation of $\lim_{n\to\infty} a_n$ appears to be a bit involved, we try to determine whether the sequence is monotonic by studying the ratio of successive terms. We have $a_{n+1} = \frac{(n+1)!}{(n+1)^{n+1}}$, so

$$\frac{a_{n+1}}{a_n} = \frac{(n+1)!}{(n+1)^{n+1}} \Big/ \frac{n!}{n^n} = \frac{(n+1)!}{(n+1)^{n+1}} \cdot \frac{n^n}{n!}$$

$$= \frac{(n+1)!}{n!} \frac{n^n}{(n+1)^n(n+1)} = (n+1)\left(\frac{n}{n+1}\right)^n \frac{1}{n+1}$$

$$= \left(\frac{n}{n+1}\right)^n < 1 \quad \text{for all} \quad n \geq 1.$$

So $a_{n+1} \leq a_n$. So the sequence is monotonic *decreasing.* It is clearly bounded: above by the first term, below by any negative number. Hence it converges by Theorem 1.14 to its greatest lower bound.

Notice that Theorem 1.14 permits us to conclude that $\boldsymbol{a}$ converges *without* finding $\lim_{n\to\infty} a_n$. (It is a bit more work to show that $\lim_{n\to\infty} a_n = 0$. See Exercise 22 on the opposite page.)

EXERCISES 11.1

In Exercises 1 to 4, give as simple a formula for $f(n) = a_n$ as you can.

1 $(2, 2, \frac{8}{6}, \frac{16}{24}, \frac{32}{120}, \frac{64}{720}, \ldots)$

2 $(-1, 1, -1, 1, -1, 1, \ldots)$

3 $(1, \frac{1}{4}, \frac{1}{9}, \frac{1}{16}, \frac{1}{25}, \frac{1}{36}, \ldots)$

4 $\left(1, \frac{\sqrt{2}}{4}, \frac{\sqrt{3}}{9}, \frac{1}{8}, \frac{\sqrt{5}}{25}, \frac{\sqrt{6}}{36}, \ldots\right)$

In Exercises 5 to 10, determine whether the given sequence converges or diverges by computing $\lim_{n\to\infty} a_n$.

5 $a_n = \dfrac{2n^2+5n+2}{3n^2-7}$

6 $a_n = \dfrac{(n^2+1)3^n}{n2^{(n+100)}}$

7 $a_n = \dfrac{\ln n}{n}$

8 $a_n = \dfrac{\sin n\pi/2}{n}$

9 $a_n = \dfrac{n}{n+1} \sin \frac{1}{2}n\pi$

10 $a_n = \sqrt{n} - \sqrt{n-1}$

In Exercises 11 to 14, show that each sequence is monotonic and determine whether it converges.

11 $a_n = \dfrac{2^n}{n!}$

12 $a_n = \dfrac{n!}{3n}$

13 $a_n = \dfrac{n!}{3 \cdot 5 \cdot 7 \ldots (2n+1)}$

14 $a_n = \dfrac{1 \cdot 4 \cdot 7 \cdot \ldots \cdot (3n-2)}{n!}$

15 Prove Corollary 1.13.

16 Show that if a set $S \subseteq \boldsymbol{R}$ has a least upper bound ℓ, then ℓ is unique.

17 Prove that if $|r| < 1$, then $\boldsymbol{r} = (r, r^2, r^3, \ldots, r^n, \ldots)$ converges to 0. **This result is needed in the next section.**

18 Prove that if $|r| > 1$, then the sequence $\boldsymbol{r} = (r, r^2, r^3, \ldots, r^n, \ldots)$ diverges by showing that $\lim_{n\to\infty} |r|^n = +\infty$. What happens if $|r| = 1$?

19 Give an example to show that if the hypothesis "monotonic" is removed from Theorem 1.14, then the sequence need no longer be convergent.

20 Give an example of a sequence that is not monotonic but diverges to $+\infty$.

21 Give an example of a sequence that is not monotonic but converges to 0.

22 Show that the sequence in Example 1.15 actually converges to 0.

2 INFINITE SERIES

The ancient gadfly Zeno of Elea (c. 495–430 B.C.) formulated a number of tantalizing paradoxes, which have proved troublesome to philosophers from Socrates (who in Plato's *Dialogues* gives a wholly unsatisfactory "refutation") down to modern times. Consideration of one of his paradoxes leads us to the ideas of this section.

According to Zeno, you cannot ever leave the room where you are reading this book. For, to reach the door, which is at distance d from you, say, you will first have to cover the distance $\frac{1}{2}d$, which will take some time, say T. From that point, you will need to cover half the *remaining* distance, $\frac{1}{4}d$, which will take some more time (say $\frac{1}{2}T$). Then you *still* need to cover half the remaining distance, $\frac{1}{8}d$,

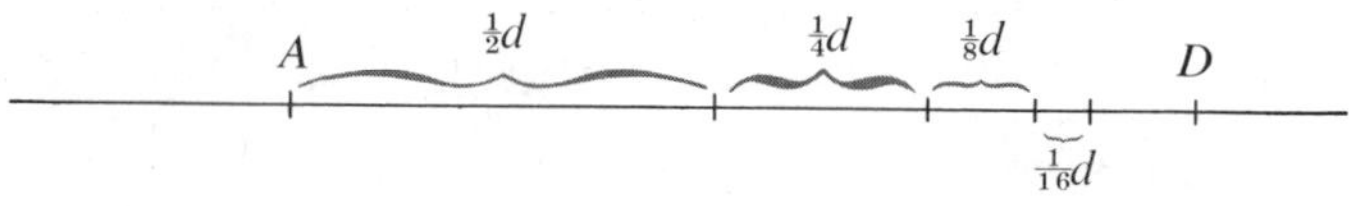

FIGURE 2.1

again expending time, say $\frac{1}{4}T$. (The assumption here is that you traverse these discrete pieces all at the same rate of speed, which seems reasonable.) Refer to Figure 2.1, where we label your starting point A and the door D. It is clear, says Zeno, that you will *never* stop adding to the time used up in reaching the points at distances $\frac{1}{2}d$, $\frac{1}{4}d$, $\frac{1}{8}d$, $\frac{1}{16}d$, etc., from D, because you can always divide the distance ℓ from where you are to the door in half, and you will need some amount $\frac{1}{2^{n-1}}T$ of time to cover the distance $\frac{1}{2}\ell$. Since you have only a finite life span, you will never survive to reach the door! Mathematically, Zeno's contention then is that

$$T+\tfrac{1}{2}T+\tfrac{1}{4}T+\ldots+\frac{1}{2^{n-1}}T$$

becomes arbitrarily large (at least becomes larger than any human life span) as n increases without bound.

How can we answer Zeno's contention? The only hope seems to lie in an investigation of $T+\frac{1}{2}T+\frac{1}{4}T+\ldots+\frac{1}{2^{n-1}}T$. This is the sum of a finite *geometric sequence.*

2.1 **DEFINITION.** A **geometric sequence** is a sequence $(a, ar, ar^2, ar^3, \ldots, ar^{n-1}, \ldots)$ where r is a fixed real number called the **common ratio** of the sequence.

In our example of Zeno, the sequence $\left(T, \frac{1}{2}T, \frac{1}{4}T, \ldots, \frac{T}{2^{n-1}}, \ldots\right)$ is geometric, with $r=\frac{1}{2}$. A classical formula exists for summing a finite geometric sequence.

2.2 **THEOREM.** $a+ar+ar^2+\ldots+ar^{n-1}=\dfrac{a(1-r^n)}{1-r}$ if $r\neq 1$.

Let's apply Theorem 2.2 to our paradox. We see that $T+\frac{1}{2}T+\frac{1}{4}T+\ldots+\frac{1}{2^{n-1}}T=\dfrac{T(1-(\frac{1}{2})^n)}{1-\frac{1}{2}}=2T\left(1-\dfrac{1}{2^n}\right)$.

What happens as Zeno keeps dividing your distance from the door into halves? The index n grows without bound, i.e., $n\to\infty$. The time required to reach the door then approaches

$$\lim_{n\to\infty} 2T\left(1-\frac{1}{2^n}\right)=2T,$$

which is *finite*! Hence you *can* get to the door even if your lifespan is finite! Zeno tricked us into believing that if we added an infinite number of positive terms, we should get ∞ as an answer, but this doesn't necessarily happen. In order to avoid any more pitfalls like Zeno's, we had better investigate what *does* happen when you add infinitely many terms together.

2.3 **DEFINITION.** An **infinite series** $\sum_{n=1}^{\infty} a_n$ is a sequence $(s_1, s_2, \ldots, s_n, \ldots)$ where $s_1 = a_1$, $s_2 = a_2 + a_1$, $s_3 = a_3 + s_2 = a_1 + a_2 + a_3, \ldots, s_n = a_n + s_{n-1} = a_1 + \ldots + a_n$, etc. The numbers a_n are called the **terms** of the series. The numbers s_n are called the **partial sums** of the series. The series is said to **converge** (or be **convergent**) to the sum s if the sequence of partial sums converges in the sense of Definition 1.3, i.e., if $\lim_{n\to\infty} s_n = s$. Otherwise, the series is said to **diverge** (or be **divergent**).

Our example above says that

$$\sum_{n=1}^{\infty} \frac{1}{2^{n-1}} T = T + \tfrac{1}{2}T + \tfrac{1}{4}T + \ldots$$

converges to $2T$. This can be generalized as follows.

2.4 **THEOREM.** If $a \neq 0$, and $|r| < 1$, then the geometric series

$$\sum_{n=1}^{\infty} ar^{n-1} = a + ar + ar^2 + \ldots + ar^{n-1} + \ldots$$

converges to

$$s = \frac{a}{1-r}.$$

For all other r, the series diverges.

We now have some examples both of convergent series and of divergent series: any geometric series converges if its common ratio is strictly between -1 and $+1$, and diverges otherwise. Notice that if $-1 < r < 1$, then the terms ar^{n-1} of the geometric series tend to zero as $n \to \infty$, by Exercise 17 of the preceding section. The next theorem says that for any given series $\sum_{n=1}^{\infty} a_n$, the series will diverge unless $a_n \to 0$ as $n \to \infty$.

2.5 **THEOREM.** If $\lim_{n\to\infty} a_n \neq 0$, then $\sum_{n=1}^{\infty} a_n$ diverges.

Once we know that the n-th term of a series must approach 0 if there is to be *any chance* for it to converge, it is natural to ask whether a series whose n-th term *does* approach 0 will in fact converge. The answer is a resounding NO!

2.6 **EXAMPLE.** Show that $\sum_{n=1}^{\infty} \dfrac{1}{n}$, the **harmonic series**, is divergent.

Proof. Our proof is a preview of the Integral Test of the next section. Let s_n be the n-th partial sum of the harmonic series, so

$$s_n = 1 + \tfrac{1}{2} + \tfrac{1}{3} + \ldots + \frac{1}{n}.$$

Now $1/n$ represents the area of a rectangle R_n whose base is the interval $[n, n+1]$

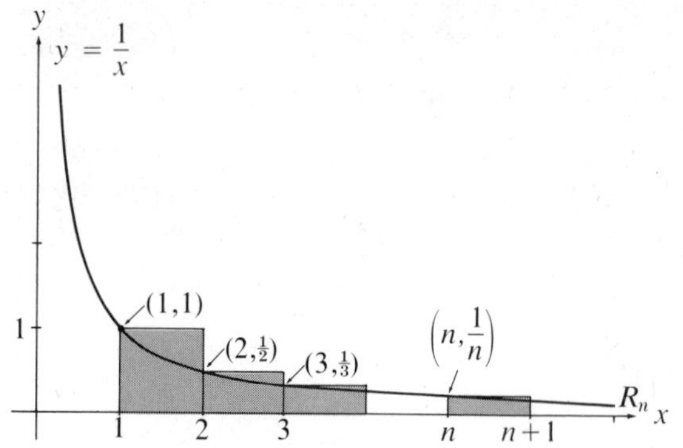

FIGURE 2.2

and whose height is $1/n$. (See Figure 2.2.) So S_n gives the area of the union of all such rectangles $R_1, R_2, \ldots, R_n$. Note that the sum of the areas of these rectangles is a Riemann upper sum for the function $f(x) = 1/x$ between $x = 1$ and $x = n+1$. This Riemann sum is bigger than

$$\int_1^{n+1} \frac{1}{x}\,dx = \ln x\Big]_1^{n+1} = \ln(n+1),$$

which is the area under $y = 1/x$ from $x = 1$ to $x = n+1$. Thus $s_n \geq \ln(n+1)$. Hence

$$\lim_{n\to\infty} s_n \geq \lim_{n\to\infty} \ln(n+1) = +\infty,$$

so the sequence of partial sums $(s_1, s_2, \ldots, s_n, \ldots)$ diverges to $+\infty$. Thus $\sum_{n=1}^{\infty} 1/n$ is divergent.

QED

The preceding example is classical and means that the theory of convergence of infinite series is not trivial, as it would have been if convergence *were* equivalent to the n-th term of the series approaching 0 as $n \to \infty$. So, *in testing a given series* $\sum_{n=1}^{\infty} a_n$ *for convergence, first see if* $\lim_{n\to\infty} a_n = 0$. *If not, then Theorem 2.5 guarantees that the series diverges. If* $\lim_{n\to\infty} a_n = 0$, *then try to use some other tool (like Theorem 2.4) to decide whether the series converges or diverges.* We will consider some applications of this rule as soon as we have a few more tools to work with. The next result states that convergence is a question about the *ultimate* behavior of the series—the first few terms have no effect on convergence.

2.7 **THEOREM.** The series $\sum_{n=1}^{\infty} a_n$ converges if and only if the series $\sum_{n=k}^{\infty} a_n = a_k + a_{k+1} + a_{k+2} + \ldots$ converges for any $k > 1$.

2.8 **EXAMPLE.** If $a_n = 1000$ for $n \leq 100{,}000$ and $a_n = \dfrac{1}{2^n}$ for $n > 100{,}000$, then does $\sum_{n=1}^{\infty} a_n$ converge?

Solution. Yes, it does converge, since $\sum_{n=k}^{\infty} a_n$ converges for $k = 100{,}001$ as a geometric series with common ratio $\frac{1}{2}$.

The next result says that convergent series possess a property whose importance you will recognize if you have studied earlier chapters. To simplify notation, we will, in dealing with convergent series, write $\sum_{n=1}^{\infty} a_n$ both for the series itself and also for its sum s.

2.9 THEOREM (LINEARITY). If $\sum_{n=1}^{\infty} a_n$ and $\sum_{n=1}^{\infty} b_n$ are convergent series and $c \in \boldsymbol{R}$, then

$$\sum_{n=1}^{\infty} (a_n + b_n) = \sum_{n=1}^{\infty} a_n + \sum_{n=1}^{\infty} b_n$$

and

$$\sum_{n=1}^{\infty} ca_n = c \sum_{n=1}^{\infty} a_n.$$

2.10 COROLLARY. If $\sum_{n=1}^{\infty} a_n$ diverges, then so does any series $\sum_{n=1}^{\infty} ca_n$ for $c \neq 0$.

2.11 EXAMPLE. Decide whether or not the following series converge. (a) $\sum_{n=1}^{\infty} \frac{3}{2n}$, (b) $\sum_{n=1}^{\infty} \left(\frac{1}{2^n} - \frac{1}{4^n}\right)$.

Solution. (a) First, $\lim_{n\to\infty} \frac{3}{2n} = 0$, so Theorem 2.5 does not apply. However, we can use Corollary 2.10:

$$\sum_{n=1}^{\infty} \frac{3}{2n} = \sum_{n=1}^{\infty} \frac{3}{2}\left(\frac{1}{n}\right).$$

Thus, the given series must diverge, since its terms are multiples of the terms of the divergent harmonic series of Example 2.6.

(b) Here

$$\sum_{n=1}^{\infty} \left(\frac{1}{2^n} - \frac{1}{4^n}\right) = \sum_{n=1}^{\infty} \left[\frac{1}{2^n} + \left(-\frac{1}{4^n}\right)\right] = \sum_{n=1}^{\infty} \frac{1}{2^n} + \sum_{n=1}^{\infty} \left(-\frac{1}{4^n}\right)$$

since

$$\sum_{n=1}^{\infty} \frac{1}{2^n} \quad \text{and} \quad \sum_{n=1}^{\infty} -\frac{1}{4^n}$$

are convergent geometric series ($r = \frac{1}{2}$ and $r = \frac{1}{4}$ respectively). Thus the given series converges by Theorem 2.9.

EXERCISES 11.2

In Exercises 1 to 10, determine whether or not the given series is convergent. If it is convergent, then find its sum.

1 $\sum_{n=1}^{\infty} \frac{1}{3n}$

2 $\sum_{n=1}^{\infty} \left(\frac{3}{2}\right)^n$

3 $\sum_{n=1}^{\infty} 5\left(\frac{3}{4}\right)^{n-1}$

4 $1 - \frac{1}{2} + \frac{1}{4} - \frac{1}{8} + - \ldots$

5 $\frac{1}{2}\sin\theta - \frac{1}{4}\sin^2\theta + \frac{1}{8}\sin^3\theta - \frac{1}{16}\sin^4\theta + - \ldots$

6 $\sum_{n=1}^{\infty} \left(\frac{4}{3^n} + \frac{(-2)^n}{3^n}\right)$

7 $\sum_{n=1}^{\infty} \frac{3^{n-2}}{4^n}$

8 $\sum_{n=1}^{\infty}\left(\frac{1}{3n}-\frac{1}{4n}\right)$

9 $\sum_{n=1}^{\infty}\frac{n^2}{10n-5}$

10 $-1-2-\ldots-1000+\sum_{n=1}^{\infty}\frac{3}{4n}$

11 Prove the following part of *Cauchy's criterion*, which extends Theorem 2.5. If $\sum_{n=1}^{\infty} a_n$ converges, then for any $\varepsilon>0$, there is an N such that $|s_m-s_n|<\varepsilon$ for all m, $n>N$. Then show that the choice $m=n+1$ yields Theorem 2.5 as a special case.

12 Use the Cauchy criterion of the preceding exercise to give another proof that the harmonic series $\sum_{n=1}^{\infty} 1/n$ diverges. [*Hint:* Note that

$$\frac{1}{n+k}\geq\frac{1}{2n}$$

for $0\leq k\leq n$. Use this to show that $|s_{2n}-s_n|\geq\frac{1}{2}$, i.e., that the criterion in Exercise 11 fails for the particular case $m=2n$.]

13 Prove that if $\sum_{n=1}^{\infty} a_n$ converges and $\sum_{n=1}^{\infty} b_n$ diverges, then $\sum_{n=1}^{\infty}(a_n+b_n)$ diverges. [*Hint:* Use Theorem 2.9.]

14 Show that if $\sum_{n=1}^{\infty} a_n$ and $\sum_{n=1}^{\infty} b_n$ both diverge, then we can't conclude that $\sum_{n=1}^{\infty}(a_n+b_n)$ diverges.

15 A **telescoping series** is one of the form $\sum_{n=1}^{\infty}(a_n-a_{n+1})$. Show that such a series converges if and only if the sequence $\boldsymbol{a}=(a_1, a_2, \ldots, a_n, \ldots)$ converges.

16 Show that $\sum_{n=1}^{\infty}\frac{1}{n(n+1)}$ converges. $\left[\textit{Hint: } \text{Compute } \frac{1}{n}-\frac{1}{n+1}.\right]$

17 A rubber ball is dropped from a height of six feet. Each time it hits the floor, it rebounds 5/8 of the distance it fell. What is the total distance traveled by the ball?

18 **Cantor's middle third set** [named for the nineteenth century German mathematician Georg Cantor (1835–1918), the founder of modern set theory] is constructed as follows. The open interval $(\frac{1}{3},\frac{2}{3})$ is erased from the closed interval $[0, 1]$. Next the two open intervals $(\frac{1}{9},\frac{2}{9})$ and $(\frac{7}{9},\frac{8}{9})$ are erased from the remnant. One continues in this way, at each stage erasing the open middle third of each remaining subinterval of $[0, 1]$. (a) Find the total length of all the subintervals erased. (b) Give an example of a point that is never erased.

19 Zeno claimed that the tortoise would always outrace the swiftest hare if only the tortoise could have a head start. Suppose, he said, that the tortoise starts at P, at some positive distance from the starting point of the hare. Then by the time the hare has arrived at P, the tortoise has proceeded on to Q, at some positive distance from P. Again, by the time the hare has reached Q, the tortoise has lumbered on to a new point R, still a positive distance from Q. "Clearly" we can continue this analysis, and thus prove that the hare can never catch the tortoise! Can you explain this paradox?

20 Show that $1-1+1-1+-\ldots+(-1)^{n-1}+\ldots$ diverges.

21 Refer to Exercise 20. Explain the fallacy in the following "proof" that the series there converges, to sum $\frac{1}{2}$. Let $s=1-1+1-1+-\ldots$. Then

$$s-1=-1+1-1+-\ldots=-s.$$

Hence, adding $s+1$ to both sides, we obtain $2s=1$, i.e., $s=\frac{1}{2}$.

3 SERIES OF POSITIVE TERMS

While Theorems 2.4 and 2.5, Corollary 2.10, and Example 2.6 provide some tools for trying to determine whether a given series $\sum_{n=1}^{\infty} a_n$ converges, there are many series for which these tools are inadequate. Consider, for example, the series

$$\sum_{n=1}^{\infty} \frac{1}{3n+2}.$$

It seems that for large values of n the terms of this series are more or less the same as the corresponding terms of

$$\sum_{n=1}^{\infty} \frac{1}{3n} = \sum_{n=1}^{\infty} \left(\frac{1}{3}\right)\left(\frac{1}{n}\right),$$

which we know to be divergent by Corollary 2.10 and Example 2.6. Thus it seems reasonable to guess that $\sum_{n=1}^{\infty} \frac{1}{(3n+2)}$ diverges. Unfortunately, we don't have any tool that permits us to verify this guess (which *is* a correct guess, by the way). The aim of this section is to provide some tools to handle convergence questions for series of positive terms $\sum_{n=1}^{\infty} a_n$ for $a_n > 0$. (Clearly, all the results we state have duals for series $\sum_{n=1}^{\infty} b_n$ for $b_n < 0$, but we won't bother to give these explicitly.) First we give two important *comparison tests.*

3.1 **THEOREM (TERM SIZE COMPARISON TEST).** Let $\sum_{n=1}^{\infty} a_n$ and $\sum_{n=1}^{\infty} b_n$ be series of positive terms.

(a) If $\sum_{n=1}^{\infty} b_n$ converges and if there is some $k \geq 1$ such that $a_n \leq b_n$ for all $n \geq k$, then $\sum_{n=1}^{\infty} a_n$ also converges.

(b) If $\sum_{n=1}^{\infty} b_n$ diverges and if there is some $k \geq 1$ such that $a_n \geq b_n$ for all $n \geq k$, then $\sum_{n=1}^{\infty} a_n$ also diverges.

3.2 **EXAMPLE.** Decide whether or not

$$\sum_{n=1}^{\infty} \frac{1}{3n+2}$$

converges.

Solution. We guessed above that it diverges. This would follow immediately from Theorem 3.1 if $\frac{1}{3n+2} > \frac{1}{3n}$ for *all* $n \geq$ some k, since $\sum_{n=1}^{\infty} \frac{1}{3n}$ is known to diverge. But, alas, $\frac{1}{3n+2} < \frac{1}{3n}$ for all n. Let's not go looking for more tools too quickly, however. The problem is that

$$\frac{1}{3n+2} < \frac{1}{3n}$$

and we need $\frac{1}{3n+2}$ *bigger* than some b_n, where $\sum_{n=1}^{\infty} b_n$ diverges. What about $b_n = \frac{1}{4n}$? Well,

$$\frac{1}{3 \cdot 1 + 2} = \frac{1}{5} < \frac{1}{4} = \frac{1}{4 \cdot 1} \quad \text{and} \quad \frac{1}{3 \cdot 2 + 2} = \frac{1}{8} = \frac{1}{4 \cdot 2},$$

but

$$\frac{1}{3\cdot 3+2}=\frac{1}{11}>\frac{1}{12}=\frac{1}{4\cdot 3}, \quad \text{etc.}$$

For all $n\geq 2$, then, we have

$$\frac{1}{3n+2}\geq\frac{1}{3n+n}=\frac{1}{4n}.$$

Since

$$\sum_{n=1}^{\infty}\frac{1}{4n}=\sum_{n=1}^{\infty}\frac{1}{4}\left(\frac{1}{n}\right)$$

diverges, we conclude from Theorem 3.1(b) that $\sum_{n=1}^{\infty} 1/(3n+2)$ diverges.

3.3 **EXAMPLE.** Decide whether or not

$$\sum_{n=1}^{\infty}\frac{3}{4^n+1}$$

is convergent.

Solution. We have

$$\frac{3}{4^n+1}<\frac{3}{4^n} \quad \text{for all } n,$$

and

$$\sum_{n=1}^{\infty}\frac{3}{4^n}=\sum_{n=1}^{\infty}3\left(\tfrac{1}{4}\right)^n$$

converges as a geometric series with common ratio $r=\frac{1}{4}$. So by Theorem 3.1(a), the given series converges.

Notice that, in order to successfully apply Theorem 3.1, we have to first guess correctly whether the series converges or diverges, and then show that its general term is properly related to the general term of a known series. We now give a second comparison test, which in many cases is simpler to use than Theorem 3.1.

3.4 **THEOREM (TERM RATIO COMPARISON TEST).** Suppose that $\sum_{n=1}^{\infty} a_n$ and $\sum_{n=1}^{\infty} b_n$ are series of positive terms. If

$$\lim_{n\to\infty}\frac{a_n}{b_n}=\ell>0,$$

then either both series converge or both diverge.

3.5 **EXAMPLE.** Determine whether or not

$$\sum_{n=1}^{\infty}\frac{3n+1}{5n^2-2}$$

converges.

Solution. Since the numerator of each term has a degree that is one less than that of the denominator, we can reason that

$$a_n = \frac{3n+1}{5n^2-2} \approx \frac{3}{5n}$$

for large n, so that the given series is probably divergent. We can show this by letting $b_n = 3/5n$ or even $1/n$, and computing $\lim_{n\to\infty} a_n/b_n$. We have, using $b_n = 1/n$,

$$\frac{a_n}{b_n} = \frac{3n+1}{5n^2-2} \Big/ \frac{1}{n} = \frac{3n^2+n}{5n^2-2} = \frac{3+\dfrac{1}{n}}{5-\dfrac{2}{n^2}}.$$

Hence,

$$\lim_{n\to\infty} \frac{a_n}{b_n} = \frac{3}{5}.$$

Since $\sum_{n=1}^{\infty} 1/n$ diverges, so does

$$\sum_{n=1}^{\infty} \frac{3n+1}{5n^2-2}.$$

Observe that Theorem 3.4 could also be used in Examples 3.2 and 3.3.

We come next to the promised generalization of the method used in Example 2.6 to show that the harmonic series diverges. [If you did not study improper integrals in elementary calculus, then before proceeding you will need to study Section 5.11 (pp. 314–318) and to work some of Exercises 6 to 10 of Exercises 5.11.] This test was put on a firm basis by the French mathematician Augustin-Louis Cauchy (1789–1857) but goes back at least as far as the Scottish mathematician Colin Maclaurin (1698–1746).

3.6 **THEOREM (INTEGRAL TEST).** Suppose that $f:[k, +\infty) \to \boldsymbol{R}$ is a continuous, decreasing function such that $f(x) > 0$ for $x \in [k, +\infty)$. Suppose that $\sum_{n=1}^{\infty} a_n$ is a series for which $a_n = f(n)$ for all $n \geq k$. Then $\sum_{n=1}^{\infty} a_n$ is convergent if and only if $\int_k^{+\infty} f(x)\,dx$ is convergent.

This result allows us to give a far-reaching extension of Example 2.6.

3.7 **EXAMPLE.** The **hyperharmonic series** (or ***p*-series**) $\sum_{n=1}^{\infty} \frac{1}{n^p}$ converges if and only if $p > 1$.

Proof. Let $f(x) = 1/x^p = x^{-p}$. This is positive and continuous on $[1, +\infty)$ and also decreasing, since

$$f'(x) = -px^{-p-1} < 0 \quad \text{for} \quad x \in [1, +\infty).$$

Also $f(n) = 1/n^p$. So we can use the integral test: the given series converges if and only if the improper integral $\int_1^{+\infty} f(x)\,dx$ converges. If $p = 1$, we know that the series diverges by Example 2.6, so we can assume $p \neq 1$. We have

$$\begin{aligned}\int_1^{+\infty} x^{-p}\,dx &= \lim_{b\to+\infty}\int_1^b x^{-p}\,dx \\ &= \lim_{b\to+\infty}\left[\frac{x^{-p+1}}{-p+1}\right]_1^b \\ &= \lim_{b\to+\infty}\frac{1}{1-p}[b^{-p+1}-1] \\ &= \lim_{b\to+\infty}\frac{1}{1-p}\left[\frac{1}{b^{p-1}}-1\right].\end{aligned}$$

This limit exists if and only if $p-1>0$, that is, if and only if $p>1$. Thus $\sum_{n=1}^{\infty} 1/n^p$ converges if and only if $p>1$. QED

3.8 EXAMPLE. Determine whether or not

$$\sum_{n=1}^{\infty}\frac{1}{n^{3/2}-5}$$

converges.

Solution. We know that $\sum_{n=1}^{\infty} 1/n^{3/2}$ converges by Example 3.7, since $p=\frac{3}{2}>1$. We can then use the Term Ratio Comparison Test (Theorem 3.4). Letting $a_n = 1/n^{3/2}$ and $b_n = 1/(n^{3/2}-5)$, we have

$$\frac{a_n}{b_n}=\frac{n^{3/2}-5}{n^{3/2}},$$

so

$$\lim_{n\to\infty}\frac{a_n}{b_n}=\lim_{n\to\infty}\frac{n^{3/2}-5}{n^{3/2}}=\lim_{n\to\infty}\left(1-\frac{5}{n^{3/2}}\right)=1.$$

Hence, since $\sum_{n=1}^{\infty} 1/n^{3/2}$ converges, so does $\sum_{n=1}^{\infty} 1/(n^{3/2}-5)$.

We now have a number of tests for convergence of an infinite series for which all the terms have the same sign. Theorem 2.4 completely classifies geometric series, and Example 3.7 covers hyperharmonic series. Theorem 2.5 disposes of any series whose n-th term doesn't approach zero. The integral test gives a criterion for convergence for series whose defining function is readily integrable. And the two comparison tests (Theorems 3.1 and 3.4) allow us to dispose of many series that closely resemble standard series. With a number of tools now available, some thought about which tool best fits a given problem is usually helpful.

3.9 EXAMPLE. Decide whether or not $\sum_{n=1}^{\infty} ne^{-n^2}$ converges.

Solution. This doesn't look very much like any of the standard known series, so no obvious comparison suggests itself. We might try taking the limit of its n-th term as $n\to\infty$. We saw in Example 1.6 that this limit is 0, so we can't conclude anything so far. Having eliminated virtually every other tool, let's try the Integral

Test. Let $f(x) = xe^{-x^2}$. Then f is positive and continuous on $[1, +\infty)$ and $f'(x) = e^{-x^2} - 2x^2e^{-x^2} = e^{-x^2}(1-2x^2) < 0$ for $x \in [1, +\infty)$, so f is decreasing. We have

$$\begin{aligned}\int_1^{+\infty} xe^{-x^2}\,dx &= \lim_{b\to+\infty} \int_1^b xe^{-x^2}\,dx \\ &= \lim_{b\to+\infty} \left[-\tfrac{1}{2}\int_1^b -2xe^{-x^2}\,dx\right] \\ &= -\tfrac{1}{2}\lim_{b\to+\infty} e^{-x^2}\Big]_1^b \\ &= -\tfrac{1}{2}\lim_{b\to+\infty} [e^{-b^2} - e^{-1}] = \frac{1}{2e}.\end{aligned}$$

Since $\int_1^{+\infty} xe^{-x^2}\,dx$ converges, so does $\sum_{n=1}^{+\infty} ne^{-n^2}$.

EXERCISES 11.3

In Exercises 1 to 10, determine whether or not the given series converges.

1 $\sum_{n=1}^{\infty} \frac{1}{n^2+n}$

2 $\sum_{n=2}^{\infty} \frac{1}{n^2-n}$

3 $\sum_{n=1}^{\infty} \frac{1}{\sqrt{2n+1}}$

4 $\sum_{n=1}^{\infty} \frac{\ln n}{2n}$

5 $\sum_{n=2}^{\infty} \frac{\mathrm{Tan}^{-1}\, n}{3n^2+3}$

6 $\sum_{n=1}^{\infty} \frac{3e^{1/n}}{n^2}$

7 $\sum_{n=1}^{\infty} \frac{3n^2-4}{n^3+2n^2+1}$

8 $\sum_{n=1}^{\infty} \frac{n-1}{n^3+4n^2+1}$

9 $\sum_{n=1}^{\infty} \frac{1}{\sqrt{n^3+2n^2-1}}$

10 $\sum_{n=1}^{\infty} \frac{1}{\sqrt{4n^2-3}}$

11 (a) Show that $\frac{1}{n!} \le \frac{1}{n^2-n}$ for $n > 2$.

(b) Show that $\sum_{n=1}^{\infty} \frac{1}{n!}$ converges.

This exercise is used in Section 8 (Example 8.10).

12 Decide whether $\sum_{n=1}^{\infty} n^2/n!$ converges. (See Exercise 11(b).)

13 (a) Show that $\ln n < n$ for any positive integer n.

(b) Decide whether $\sum_{n=1}^{\infty} \frac{\ln n}{n^3}$ converges.

14 Decide whether $\sum_{n=1}^{\infty} e^{-n^2}$ converges. (See Example 3.9.)

15 Show that $\sum_{n=1}^{\infty} \frac{1}{n^3} \sin(1/n^2)$ converges.

16 Prove that if $\sum_{n=1}^{\infty} a_n$ and $\sum_{n=1}^{\infty} b_n$ are convergent series, then so is $\sum_{n=1}^{\infty} a_nb_n$. Hence, in particular, if $\sum_{n=1}^{\infty} a_n$ converges, then so does $\sum_{n=1}^{\infty} a_n^2$.

17 If $\sum_{n=1}^{\infty} a_n^2$ converges, then must $\sum_{n=1}^{\infty} a_n$ converge? If so, prove it. If not, give a counterexample.

18 If in Theorem 3.4, $\lim_{n\to\infty} \frac{a_n}{b_n} = 0$, then show by example that it no longer follows that either both series converge or both diverge.

4 ALTERNATING SERIES

In the last section we gave a number of tools to use in testing series of positive (or negative) terms for convergence. Now we want to investigate series containing *both* positive and negative terms. In this section we treat series whose terms are alternately positive and negative, and in the next section consider arbitrary series of positive and negative terms. That alternating series are easier to handle is suggested by the fact that the main result of this section goes all the way back to Leibniz in the late 17th century, whereas those of the next section are usually credited to Cauchy, and appear to go back no further than the mid-18th century at the earliest.

4.1 DEFINITION. An **alternating series** $\sum_{n=1}^{\infty} a_n$ is a series for which $a_k a_{k+1} < 0$ for all $k = 1, 2, 3, \ldots$.

Our condition is just that the signs of every pair of successive terms are $+-$ or $-+$. Thus, an alternating series has one of the two forms

(1) $$\sum_{n=1}^{\infty} (-1)^n b_n = -b_1 + b_2 - b_3 + b_4 - + \ldots,$$

or

(2) $$\sum_{n=1}^{\infty} (-1)^{n+1} b_n = b_1 - b_2 + b_3 - b_4 + - \ldots,$$

where $b_i > 0$ for all i. It is clearly sufficient to consider a series of type (1), since (2) can be obtained from (1) simply by multiplying by -1.

It turns out that alternating series whose n-th terms approach 0 as $n \to \infty$ often converge. The precise result is the following.

4.2 THEOREM (ALTERNATING SERIES TEST). If $0 \le b_{n+1} \le b_n$ for all $n \ge k$ (where $k \ge 1$ is fixed) and $\lim_{n\to\infty} b_n = 0$, then $\sum_{n=1}^{\infty} (-1)^n b_n$ converges.

4.3 EXAMPLE. Determine whether or not the **alternating harmonic series**

$$\sum_{n=1}^{\infty} (-1)^n \frac{1}{n} \quad \text{and} \quad \sum_{n=1}^{\infty} (-1)^{n+1} \frac{1}{n}$$

converge.

Solution. We try to apply the test of Theorem 4.2. In each series $b_n = \frac{1}{n}$, so $b_{n+1} \le b_n$ holds since

$$\frac{1}{n+1} < \frac{1}{n} \quad \text{for all } n.$$

Also,

$$\lim_{n\to\infty} b_n = \lim_{n\to\infty} \frac{1}{n} = 0.$$

So without further ado, we can conclude that both of the alternating harmonic series converge.

In the introduction and also in Section 1 (p. 679) we talked about the importance of finitely approximating infinite quantities. We now have enough tools to start considering this process. You probably realize that although we have a number of tests that will show whether a given series converges, we have almost nothing that will tell us *to what* the series converges. Indeed, if you look back over the tests we've developed, you will notice that only in the case of geometric series (Theorem 2.4) do we have a formula for the sum of a convergent series. What about the computational question of finding the sum s of a convergent series? In practice, one usually *approximates* s by s_n for as large a value of n as necessary for the degree of accuracy desired. To determine *how* large an n is required, then, we need to be able to somehow calculate the *error* made in using this approximation.

4.4 DEFINITION. The **error in approximating** $s = \sum_{n=1}^{\infty} a_n$ **by** $\mathbf{s_n}$ is $e_n = |s - s_n|$.

In the case of an alternating series that satisfies the hypotheses of Theorem 4.2, there is a particularly simple bound on the size of e_n.

4.5 THEOREM. Suppose that $0 \le b_{n+1} \le b_n$ for all $n \ge$ some k and $\lim_{n\to\infty} b_n = 0$. Then the error e_n made in approximating

$$\sum_{n=1}^{\infty} (-1)^n b_n \quad \text{or} \quad \sum_{n=1}^{\infty} (-1)^{n+1} b_n$$

by s_n is at most b_{n+1}, the absolute value of the first term omitted from s_n.

4.6 DEFINITION. An approximation s_n to a quantity s is **correct to k decimal places** if $e_n = |s - s_n| < \frac{1}{2}(10^{-k})$.

This means that the true value of s lies between $s_n - \frac{1}{2}(10^{-k})$ and $s_n + \frac{1}{2}(10^{-k})$. As an example, 1.41 is an approximation to $\sqrt{2}$ that is correct to 2 decimal places, since $|\sqrt{2} - 1.41| < 0.0043 < 0.005$. Asserting that 1.41 is correct to two decimal places is the same as asserting

$$1.41 - 0.005 < \sqrt{2} < 1.41 + 0.005$$

$$1.405 < \sqrt{2} < 1.415.$$

4.7 EXAMPLE. Approximate the sum of the series

$$\sum_{n=1}^{\infty} \frac{(-1)^n}{n^4}$$

correct to three decimal places.

Solution. First we have to check that this series satisfies the hypotheses of Theorem 4.5. Here $b_n = \frac{1}{n^4}$, so $\lim_{n\to\infty} b_n = 0$. Also,

$$b_{n+1} = \frac{1}{(n+1)^4} < \frac{1}{n^4} = b_n.$$

So the series converges by Theorem 4.2, and $e_n = |s - s_n| < b_{n+1}$. So we need to take n large enough so that

$$b_{n+1} = \frac{1}{(n+1)^4} < 0.0005.$$

This is a good job for a hand-held calculator if you have one. In any case, you can try $n = 2, 3, 4$, etc. until you find the first n such that $\frac{1}{(n+1)^4} < 0.005$. We find that for $n = 6$ we get

$$e_n < b_{n+1} = \frac{1}{7^4} = \frac{1}{2401} < 0.00042.$$

So we get the sum correct to three decimal places if we approximate it by s_6:

$$\begin{aligned} s_6 &= -1 + \tfrac{1}{16} - \tfrac{1}{81} + \tfrac{1}{256} - \tfrac{1}{625} + \tfrac{1}{1296} \\ &\approx -1.0000 + 0.0625 - 0.0123 + 0.0039 - 0.0016 + 0.0008 \\ &\approx -0.9467. \end{aligned}$$

Since s_6 is correct to three decimal places, we express our answer as $s \approx -0.947$.

Warning. Be careful to take into account the *accuracy* of your calculations. Since in the preceding example our approximation to s is accurate only to *three* decimal places, a reasonable procedure is to calculate s_6 using four-decimal-place approximations, and then round off to three places to get the approximation to s. If you have a hand-held calculator, be sure you aren't lured into expressing your approximation for s as a 6, 8, 10, or 12-place decimal. Since the approximation is accurate *only* to three decimal places, any places after that are not significant and can only lead to a mistaken impression of the accuracy of calculations you might make with the approximated s.

EXERCISES 11.4

In Exercises 1 to 10, determine whether or not the given alternating series converges.

1 $\sum_{n=1}^{\infty} (-1)^n \frac{n^2-1}{3n^2+1}$

2 $\sum_{n=1}^{\infty} (-1)^{n+1} \frac{n^3-10}{2n^3+5n+1}$

3 $\sum_{n=1}^{\infty} \frac{(-1)^n}{\sqrt{n}}$

4 $\sum_{n=1}^{\infty} (-1)^n n e^{-n}$

5 $\sum_{n=1}^{\infty} \cos n\pi$

6 $\sum_{n=1}^{\infty} \sin \frac{1}{2}(2n-1)\pi$

7 $\sum_{n=1}^{\infty} \frac{\cos n\pi}{n}$

8 $\sum_{n=1}^{\infty} \frac{\sin \frac{1}{2}(2n-1)\pi}{n}$

9 $\sum_{n=1}^{\infty} \frac{(-1)^n n!}{3 \cdot 5 \cdots (2n+1)}$

10 $\sum_{n=1}^{\infty} \frac{(-1)^n \ln n}{n}$

In Exercises 11 and 12, approximate the sum of the given series correct to three decimal places.

11 $\sum_{n=1}^{\infty} \frac{(-1)^n}{n!}$

12 $\sum_{n=1}^{\infty} \frac{(-1)^n}{n2^n}$

Exercise 13 is designed to obtain an upper bound for e_n in case $\sum_{n=1}^{\infty} a_n$ is a series of positive terms.

13 Suppose that $f:[k, +\infty) \to \boldsymbol{R}$ is a continuous, decreasing, positive valued function. If $\sum_{n=1}^{\infty} a_n$ is a series for which $a_n = f(n)$ for $n \geq k$, then show that

$$e_k \leq \int_k^{+\infty} f(x)\,dx.$$

(**Hint:** Use the technique in the proof of Example 2.6 on the interval $[k, n]$.)

14 What is an upper bound on e_{20} for $\sum_{n=1}^{\infty} \frac{(-1)^n}{n^3}$? Compare the usefulness of Exercise 13 with that of Theorem 4.5.

15 Using the estimate in Exercise 13, how large an n do you need in order to be sure that s_n approximates $\sum_{n=1}^{\infty} \frac{n}{2^{n^2}}$ correct to three decimal places?

16 If you have access to a hand-held calculator, approximate $\sum_{n=1}^{\infty} \frac{n}{2^{n^2}}$ correct to three decimal places, using Exercises 13 and 15.

5 RATIO TEST

This section takes its name from a result fully established by Cauchy in 1837, but which seems to have been used by Jean d'Alembert (1717–1783) much earlier. This result is of far-reaching importance in studying power series representations for functions, which will be the major topic in the rest of the chapter. Before we can give this result, however, we need some new ideas.

We have seen in Example 4.3 that the alternating harmonic series $\sum_{n=1}^{\infty} (-1)^n/n$ and $\sum_{n=1}^{\infty} (-1)^{n+1}/n$ converge even though the series of absolute values fail to converge. The convergence is rather slow, however. To compute the sum correct to three decimal places, we must, by Theorem 4.5, take n so large that $1/n < 0.0005$. That is, we must take $n > 1/0.0005 = 2000$. We need to add more than 2000 terms!. Compare this with Example 4.7, where to compute the sum of the hyperharmonic series

$$\sum_{n=1}^{\infty} \frac{(-1)^n}{n^4}$$

accurate to three decimal places, we need to add only *six* terms! You might suspect that the reason for such a marked difference in the rate of convergence is related to the fact that the associated series of absolute values

$$\sum_{n=1}^{\infty} \frac{1}{n^4}$$

converges in the case of the alternating hyperharmonic series, whereas $\sum_{n=1}^{\infty} 1/n$ diverges. Thus it seems reasonable to distinguish the two types of convergence demonstrated by these series.

5.1 **DEFINITION.** An infinite series $\sum_{n=1}^{\infty} a_n$ is called **absolutely convergent** if $\sum_{n=1}^{\infty} |a_n|$ converges. If $\sum_{n=1}^{\infty} a_n$ converges, but $\sum_{n=1}^{\infty} |a_n|$ diverges, then $\sum_{n=1}^{\infty} a_n$ is said to be **conditionally convergent**.

Thus the alternating hyperharmonic series

$$\sum_{n=1}^{\infty} \frac{(-1)^n}{n^4}$$

is absolutely convergent, but the alternating harmonic series

$$\sum_{n=1}^{\infty} \frac{(-1)^n}{n}$$

is conditionally convergent.

Our first result about absolute convergence is that an absolutely convergent series is in particular convergent, as the terminology itself seems to suggest.

5.2 **THEOREM.** If $\sum_{n=1}^{\infty} a_n$ is absolutely convergent, then it is convergent.

There is a further result about absolutely convergent series which states that the absolute value of the sum is less than or equal to the sum of the absolute values, and which may be thought of as an extension of the triangle inequality for absolute value to infinite sums.

5.3 **THEOREM.** If $\sum_{n=1}^{\infty} a_n$ converges, then $|\sum_{n=1}^{\infty} a_n| \leq \sum_{n=1}^{\infty} |a_n|$ (where we regard any real number s as less than $+\infty$).

Thus absolute convergence is a stronger property than convergence. The extra strength in this concept will allow us to carry out a great deal of the work of the rest of this chapter. The next result is the main tool in determining absolute convergence.

5.4 **THEOREM (RATIO TEST).** Let $\sum_{n=1}^{\infty} a_n$ be an infinite series such that $a_n \neq 0$ for all sufficiently large n and

$$\lim_{n\to\infty} \left| \frac{a_{n+1}}{a_n} \right| = \ell.$$

(i) If $\ell < 1$, then the series is absolutely convergent.
(ii) If $\ell > 1$ or $\ell = +\infty$, then the series diverges.
(iii) If $\ell = 1$, then no conclusion follows from this information alone.

While we omit the proofs of parts (i) and (ii), part (iii) is easy to see. Consider the hyperharmonic series $\sum_{n=1}^{\infty} \frac{1}{n^p}$. We have

$$\frac{|a_{n+1}|}{|a_n|} = \frac{n^p}{(n+1)^p} = \left(\frac{n}{n+1}\right)^p.$$

So

$$\lim_{n\to\infty}\frac{|a_{n+1}|}{|a_n|}=\lim_{n\to\infty}\left(\frac{n}{n+1}\right)^p=\lim_{n\to\infty}\left(\frac{1}{1+\dfrac{1}{n}}\right)^p=1.$$

We know from Example 3.7 that sometimes (for $p\leq 1$) this series diverges and sometimes (for $p>1$) it converges. So no conclusion can be drawn simply from the fact that

$$\lim_{n\to\infty}\frac{|a_{n+1}|}{|a_n|}=1.$$

5.5 **EXAMPLE.** Determine whether $\sum_{n=1}^{\infty}(-1)^n\dfrac{2^n}{n!}$ is absolutely convergent, conditionally convergent, or divergent.

Solution. Here we have

$$\frac{|a_{n+1}|}{|a_n|}=\frac{2^{n+1}}{(n+1)!}\bigg/\frac{2^n}{n!}=\frac{2}{n+1},$$

$$\lim_{n\to\infty}\left|\frac{a_{n+1}}{a_n}\right|=\lim_{n\to\infty}\frac{2}{n+1}=0<1.$$

Thus, by Theorem 5.4(i), the given series is absolutely convergent.

5.6 **EXAMPLE.** Determine whether

$$\sum_{n=1}^{\infty}(-1)^n\frac{3^n}{5\cdot 2^{(3/2)n}}$$

is absolutely convergent, conditionally convergent, or divergent.

Solution. Here

$$\left|\frac{a_{n+1}}{a_n}\right|=\frac{3^{n+1}}{5\cdot 2^{(3/2)(n+1)}}\bigg/\frac{3^n}{5\cdot 2^{(3/2)n}}=\frac{3}{2^{3/2}}=\frac{3}{2\sqrt{2}},$$

$$\lim_{n\to\infty}\left|\frac{a_{n+1}}{a_n}\right|=\frac{3}{2\sqrt{2}}>1.$$

Thus the given series diverges.

The series in the following example can be thought of as an analogue to geometric series and will be of use in Section 9.

5.7 **EXAMPLE.** Show that $\sum_{n=1}^{\infty}nr^{n-1}$ converges absolutely if $|r|<1$ and diverges if $|r|\geq 1$.

Proof. We have

$$\left|\frac{a_{n+1}}{a_n}\right|=\frac{n+1}{n}|r|=\left(1+\frac{1}{n}\right)|r|,$$

$$\lim_{n\to\infty}\left|\frac{a_{n+1}}{a_n}\right|=\lim_{n\to\infty}\left(1+\frac{1}{n}\right)|r|=|r|.$$

So by Theorem 5.4, the given series diverges if $|r|>1$, and converges absolutely if $|r|<1$. What about $|r|=1$? In this case

$$\lim_{n\to\infty} |a_n| = \lim_{n\to\infty} n = \infty,$$

so by Theorem 2.5 the given series diverges. QED

At this point we have quite a number of tests to use in determining whether or not a given series converges. You may not be that sure about which method to try first. Here are some suggestions for what to do when you are given a series $\sum_{n=1}^{\infty} a_n$.

(1) *Check whether or not* $\lim_{n\leftarrow\infty} a_n = 0$. If not, then you can conclude that the series diverges by Theorem 2.5. If $\lim_{n\to\infty} a_n = 0$, then go on.

(2) *Try the ratio test.* If

$$\lim_{n\to\infty} \left|\frac{a_{n+1}}{a_n}\right| < 1,$$

then the series is absolutely convergent. If

$$\lim_{n\to\infty} \left|\frac{a_{n+1}}{a_n}\right| > 1$$

or this limit is $+\infty$, then the series diverges. If

$$\lim_{n\to\infty} \left|\frac{a_{n+1}}{a_n}\right| = 1,$$

then go on.

(3) *If the series is an alternating series,* see whether the Alternating Series test can be applied to it. If it can't, go on.

(4) *If it is not alternating, then try one of the comparison tests* (Theorems 3.1 and 3.4).

(5) *If you can't come up with a useful series to compare the given one to, then see if the integral test can be applied.*

EXERCISES 11.5

In Exercises 1 to 10, classify the given series as absolutely convergent, conditionally convergent, or divergent.

1 $\sum_{n=1}^{\infty} \frac{(-1)^n n}{2^n}$

2 $\sum_{n=1}^{\infty} \frac{(-1)^n n^2}{3^n}$

3 $\sum_{n=1}^{\infty} \frac{(-1)^n 4^n}{3^{(4/3)n}}$

4 $\sum_{n=1}^{\infty} \frac{(-1)^n\, 1000^n}{n!}$

5 $\sum_{n=2}^{\infty} \frac{(-1)^n n}{n^2-1}$

6 $\sum_{n=1}^{\infty} (-1)^n \frac{1\cdot 3\cdot 5 \cdots (2n-1)}{n!}$

7 $\sum_{n=1}^{\infty} \frac{(-1)^n}{e^n - n}$

8 $\sum_{n=1}^{\infty} \frac{(-1)^n}{n\sqrt{n+1}}$

9 $\sum_{n=1}^{\infty} (-1)^n \frac{2^{n^2}}{n!}$

10 $\sum_{n=1}^{\infty} \frac{(-1)^n \ln n}{n^3}$

6 TAYLOR POLYNOMIALS

We are now ready to begin the task of developing polynomial approximations $p_n(x)$ of degree n near a point x_0 for arbitrarily given functions $f: \boldsymbol{R} \to \boldsymbol{R}$. In doing this, we shall assume that f can be differentiated at least n times near x_0. Furthermore, to obtain some estimates of how closely f can be approximated by $p_n(x)$ near x_0, we will sometimes need to know that $f^{(n+1)}$ is defined near x_0 also. Our polynomial approximations will both allow us to compute rational approximations to $f(x)$ at points x near x_0 and also later allow us to approximately differentiate and integrate f near x_0 by differentiating and integrating $p_n(x)$.

The methods of this section go back to early 18th century mathematicians, prominent among whom were James Stirling (1692–1770) and Brooke Taylor (1685–1731) of England, Maclaurin of Scotland, and later, Joseph Lagrange (1736–1813) and Cauchy of France.

The main idea is a simple extension of the differential (or tangent) approximation discussed in Sections 4.2 and 4.3. If $f: \boldsymbol{R} \to \boldsymbol{R}$ is differentiable at x_0, then we know that for x near x_0 we can approximate the graph of f by its tangent line at x_0, whose equation is

(1) $$y = p_1(x) = f(x_0) + f'(x_0)(x - x_0).$$

See Figure 6.1. This can be thought of as an approximation of f by means of a *first degree polynomial* $p_1(x)$. This is a good approximation near x_0. Note that

(2) $$p_1(x_0) = f(x_0) \quad \text{and} \quad p_1'(x_0) = f'(x_0)$$

so p_1 agrees with f at x_0, and p_1' agrees with f' at x_0 also. Moreover, $p_1(x)$ approaches $f(x)$ as $x \to x_0$ at a rate *faster* than the rate at which x approaches x_0, as shown on p. 161.

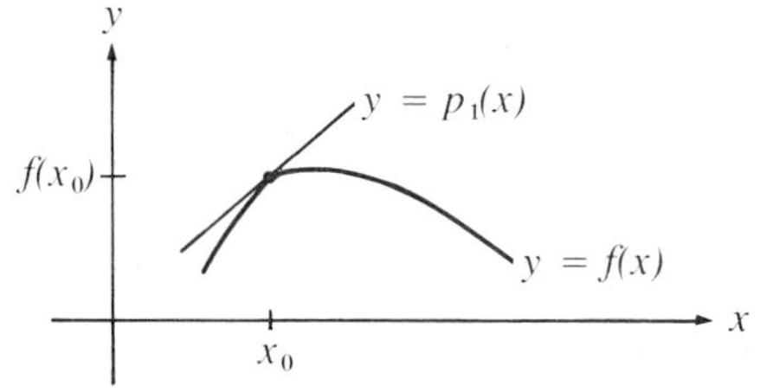

FIGURE 6.1

We now want to see whether for $n > 1$, we can find a polynomial $p_n(x)$ of degree n which approximates f in a corresponding way. To begin with, we want the analogue of Equation (2) to hold for the approximating polynomial $p_n(x)$. Thus we require

(3) $$p_n(x_0) = f(x_0),\ p_n'(x_0) = f'(x_0),\ p_n''(x_0) = f''(x_0), \ldots, p_n^{(n)}(x_0) = f^{(n)}(x_0).$$

Note that this requires that f be at least n-times differentiable at x_0. This requirement also says that $p_n(x)$ has the same value as f at x_0, the same slope as f

at x_0, the same concavity as f at x_0, the same rate of change of the concavity as f at x_0, and so forth. Hence near $(x_0, f(x_0))$ the graphs of f and $p_n(x)$ should be nearly identical.

It turns out that conditions (3) are enough to determine a formula for $p_n(x)$. Suppose that

(4) $$p_n(x) = a_0 + a_1(x-x_0) + a_2(x-x_0)^2 + \dots + a_n(x-x_0)^n.$$

[We write $p_n(x)$ as a polynomial in powers of $x - x_0$ to correspond to the form of (1). Every polynomial $q(x)$ in x can be so written if we let $u = x - x_0$. Then $x = u + x_0$, so that the powers of x are sums of powers of u times constants. So $q(x)$ is a polynomial in $u = x - x_0$.] From (4), our requirement that $p_n(x_0) = f(x_0)$ implies that

$$a_0 = f(x_0).$$

If we differentiate (4), we get

(5) $$p_n'(x) = a_1 + 2a_2(x-x_0) + 3a_3(x-x_0)^2 + 4a_4(x-x_0)^3 + \dots + na_n(x-x_0)^{n-1}.$$

Thus, our requirement that $p_n'(x_0) = f'(x_0)$ gives

$$a_1 = f'(x_0).$$

Differentiation of (5) produces

(6) $$p_n''(x) = 2a_2 + 3 \cdot 2a_3(x-x_0) + 4 \cdot 3a_4(x-x_0)^2 + \dots + n(n-1)a_n(x-x_0)^{n-2}.$$

Thus, our requirement that $p_n''(x_0) = f''(x_0)$ gives

$$a_2 = \tfrac{1}{2}f''(x_0).$$

If we continue in this way, differentiation of (6) gives

$$a_3 = \frac{1}{3!}f'''(x_0),$$

and finally we get

$$a_n = \frac{1}{n!}f^{(n)}(x_0).$$

We thus have from (3) that

(7) $$p_n(x) = f(x_0) + f'(x_0)(x-x_0) + \tfrac{1}{2}f''(x_0)(x-x_0)^2 + \frac{1}{3!}f'''(x_0)(x-x_0)^3 + \dots + \frac{1}{n!}f^{(n)}(x_0)(x-x_0)^n.$$

6.1 DEFINITION. Let f be at least n-times differentiable at x_0. Then the ***n*-th degree Taylor polynomial of f near x_0** is the polynomial $p_n(x)$ defined by (7).

6.2 EXAMPLE. Find the 5th degree Taylor polynomial for the sine function near $x_0 = 0$. (When $x_0 = 0$, the Taylor polynomial is sometimes called the *Maclaurin polynomial.*)

Solution. We use (7). Here

$$\begin{aligned} f(x) &= \sin x, & f(0) &= 0; \\ f'(x) &= \cos x, & f'(0) &= 1; \\ f''(x) &= -\sin x, & f''(0) &= 0; \\ f'''(x) &= -\cos x, & f'''(0) &= -1; \\ f^{(4)}(x) &= \sin x, & f^{(4)}(0) &= 0; \\ f^{(5)}(x) &= \cos x, & f^{(5)}(0) &= 1. \end{aligned}$$

From (7),

$$\begin{aligned} p_5(x) &= f(0)+f'(0)x+\tfrac{1}{2}f''(0)x^2+\frac{1}{3!}f'''(0)x^3+\frac{1}{4!}f^{(4)}(0)x^4+\frac{1}{5!}f^{(5)}(0)x^5 \\ &= 0+x+0-\frac{1}{3!}x^3+0+\frac{1}{5!}x^5 \\ &= x-\tfrac{1}{6}x^3+\tfrac{1}{120}x^5. \end{aligned}$$

(For an illustration with $x_0 \neq 0$, see Example 6.8.)

The n-th degree Taylor polynomial $p_n(x)$ appears to be an appropriate extension of the polynomial $p_1(x)$, so we would expect that as $x \to x_0$, $p_n(x)$ approaches $f(x)$ faster than $(x-x_0)^n \to 0$. That is, we expect that

$$\lim_{x \to x_0} \frac{p_n(x)-f(x)}{(x-x_0)^n} = 0.$$

This turns out to be true if $f^{(n+1)}(x)$ is continuous and bounded near x_0. In order to establish this fact, we need a formula for the numerator.

6.3 THEOREM (CAUCHY REMAINDER THEOREM). Suppose that $f: \boldsymbol{R} \to \boldsymbol{R}$ is at least $(n+1)$-times differentiable and that $f^{(n+1)}$ is continuous on an open interval I containing x_0. Let $r_n(x) = f(x) - p_n(x)$, the **remainder** when $f(x)$ is approximated by $p_n(x)$. Then for $x \in I$,

(8) $$r_n(x) = \frac{1}{n!}\int_{x_0}^{x} (x-t)^n f^{(n+1)}(t)\, dt.$$

6.4 COROLLARY (REMAINDER ESTIMATE). If f and I are as in Theorem 6.3 and $|f^{(n+1)}(x)| \leq B$, a fixed real number, for all x in the interval I, then

(9) $$|r_n(x)| \leq \frac{B}{(n+1)!}|x-x_0|^{n+1}$$

for $x \in I$.

We can now show that $p_n(x)$ approaches $f(x)$ faster than $(x-x_0)^n$ approaches 0 when the hypotheses of Corollary 6.4 hold.

6.5 **THEOREM.** Suppose that f and I are as in Corollary 6.4. Suppose also that $f^{(n+1)}(x)$ is bounded by some real number B on the interval I. Then

$$\lim_{x \to x_0} \frac{r_n(x)}{(x-x_0)^n} = 0.$$

Proof. From (9), we have

$$0 \le \left| \frac{r_n(x)}{(x-x_0)^n} \right| \le \frac{B}{(n+1)!} |x-x_0|.$$

Since the right-hand term approaches 0 as $x \to x_0$, so also does the middle term. Thus so does $\frac{r_n(x)}{(x-x_0)^n}$. QED

We have now shown that $p_n(x)$ is a very good approximation to f near x_0. It and its first n derivatives coincide with f and the first n derivatives of f at x_0, and $p_n(x)$ approaches $f(x)$ very rapidly as x approaches x_0. In order to use $p_n(x)$ to approximate $f(x)$, however, we still need a formula for the remainder (or error) $r_n(x)$ that can be more easily applied than (8). This is supplied by our next result.

6.6 **THEOREM (LAGRANGE REMAINDER THEOREM).** If f is as in Theorem 6.3, then

(10) $$r_n(x) = \frac{f^{(n+1)}(c)}{(n+1)!}(x-x_0)^{n+1}$$

for some c between x_0 and x.

Notice that the Lagrange Remainder Theorem gives the same estimate as (9) for $r_n(x)$ if $|f^{(n+1)}(x)| \le B$ on the interval I. The following examples indicate how Taylor polynomials and our expressions and estimates for the remainder $r_n(x)$ can give us rational approximations to the values of transcendental functions.

6.7 **EXAMPLE.** Use a Taylor polynomial for $f(x) = e^x$ to approximate e correct to five decimal places.

Solution. We have $f^{(i)}(x) = e^x$ for every positive integer i, so $f^{(i)}(0) = 1$. If we use a Taylor polynomial with $x_0 = 0$, then

$$p_n(x) = f(0) + f'(0)x + \tfrac{1}{2}f''(0)x^2 + \frac{1}{3!}f'''(0)x^3 + \ldots + \frac{1}{n!}f^{(n)}(0)x^n$$

$$= 1 + x + \tfrac{1}{2}x^2 + \frac{1}{3!}x^3 + \ldots + \frac{1}{n!}x^n.$$

We need to know how large to take n so that $p_n(1)$ will differ from $f(1) = e^1 = e$ by less than 0.000005, i.e., so that $p_n(1)$ will be an approximation for e accurate to five decimal places. So we want $|r_n(1)| < 0.000005$. From (10),

$$r_n(1) = \frac{f^{(n+1)}(c)}{(n+1)!}x^{n+1}$$

where c is between 0 and 1. Thus $1 < e^c < e$ and

$$|r_n(1)| \leq \frac{e}{(n+1)!} < \frac{3}{(n+1)!} \quad \text{since } e < 3.$$

Calculating $\frac{3}{(n+1)!}$ for $n = 1, 2, 3, \ldots$ (a hand-held calculator was used), we find

$$r_8(1) < \frac{3}{9!} = \frac{1}{3 \cdot 8!} \approx 0.000008$$

and

$$r_9(1) < \frac{3}{10!} = \frac{1}{30 \cdot 8!} \approx 0.0000008 < 0.000005.$$

So we can use $p_9(1)$ to get the desired approximation. We have

$$\begin{aligned} p_9(1) &= 1 + 1 + \tfrac{1}{2} + \frac{1}{3!} + \frac{1}{4!} + \frac{1}{5!} + \frac{1}{6!} + \frac{1}{7!} + \frac{1}{8!} + \frac{1}{9!} \\ &\approx 2.500000 + 0.166667 + 0.041667 + 0.008333 \\ &\quad + 0.001389 + 0.000198 + 0.000025 + 0.000003 \\ &\approx 2.718282 \end{aligned}$$

which rounds off to

$$p_9(1) \approx 2.71828.$$

Taylor polynomials for e^x can be used to generate tables of values for e^x in the same way we have used the Taylor polynomial to approximate $e = e^1$. In fact, this is how such tables of values were first obtained. This was done before the advent of calculating machines, so you can readily visualize how formidable a job it was!

Suppose that we wanted to generate a 5-place table of the sine and cosine functions. As you probably recall, the magnitudes of all values assumed by $\sin x$ and $\cos x$ occur between $x = 0$ and $x = \frac{1}{2}\pi$. Also recall that $\cos(\frac{1}{2}\pi - x) = \sin x$ and $\sin(\frac{1}{2}\pi - x) = \cos x$, so we only need to tabulate the functions for $x \in [0, \frac{1}{4}\pi]$. Consider the sine function. (Similar remarks apply to the cosine.) How large should n be so that $p_n(x)$ differs from $\sin x$ by less than 0.000005, i.e., so that $|r_n(x)| < 0.000005$? From Example 6.2 we have

$$p_3(x) = x - \frac{x^3}{6} = p_4(x).$$

Thus $|r_3(x)| = |r_4(x)| = \left| \frac{f^{(5)}(c)}{5!} x^5 \right|$ where c is between 0 and x. Hence,

$$|r_3(x)| = \left| \frac{\cos c}{5!} x^5 \right| < x^5/5!$$

Thus $|r_3(x)| < 0.000005$ when $x^5 < 120(0.000005) = 0.0006$, i.e., when $x < 0.2268$.

So for $x<\frac{\pi}{15}\approx 0.2094$ radians we can use $p_3(x)$ to generate 5-place tables. A similar analysis shows that

$$p_5(x)=x-\frac{x^3}{3!}+\frac{x^5}{5!}$$

gives 5-place accuracy for $x<0.59\approx 33.8^\circ$ and

$$p_7(x)=x-\frac{x^3}{3!}+\frac{x^5}{5!}-\frac{x^7}{7!}$$

gives 5-place accuracy for the remaining x up to 45°. So *low degree Taylor polynomials* (of degree 3, 5, and 7) are sufficient to generate a table of the sine function accurate to five decimal places.

Our final example shows that there are times when a choice of $x_0\neq 0$ is appropriate.

6.8 **EXAMPLE.** Obtain an approximation of sin 32° accurate to five decimal places.

Solution. Since 32° is close to $30^\circ=\pi/6$, it seems useful to use $x_0=\pi/6$. This should enable us to use a polynomial of lower degree than we would need for $x_0=0$. We have

$$f(x)=\sin x, \qquad f\left(\frac{\pi}{6}\right)=\frac{1}{2},$$

$$f'(x)=\cos x, \qquad f'\left(\frac{\pi}{6}\right)=\frac{1}{2}\sqrt{3},$$

$$f''(x)=-\sin x, \qquad f''\left(\frac{\pi}{6}\right)=-\frac{1}{2},$$

$$f'''(x)=-\cos x, \qquad f'''\left(\frac{\pi}{6}\right)=-\frac{1}{2}\sqrt{3},$$

$$f^{(4)}(x)=\sin x, \qquad f^{(4)}\left(\frac{\pi}{6}\right)=\frac{1}{2}.$$

Also

$$r_3(x)=\frac{f^{(4)}(c)}{4!}\left(x-\frac{\pi}{6}\right)^4.$$

Thus for $x=32^\circ=\frac{\pi}{6}+\frac{\pi}{90}$, we see that

$$|r_3(x)|\leq\frac{|\sin c|}{24}\left(\frac{\pi}{90}\right)^4.$$

Since c is between $x_0=\frac{\pi}{6}$ and x, we have $|\sin c|<0.6$. Therefore,

$$|r_3(x)|<\frac{1}{40}\left(\frac{\pi}{90}\right)^4\approx\frac{0.000001}{40}<0.000005.$$

We will then have an approximation of sin 32° accurate to five decimal places if we use

$$p_3(x) = f\left(\frac{\pi}{6}\right) + f'\left(\frac{\pi}{6}\right)\left(x - \frac{\pi}{6}\right) + \tfrac{1}{2}f''\left(\frac{\pi}{6}\right)\left(x - \frac{\pi}{6}\right)^2 + \frac{1}{3!}f'''\left(\frac{\pi}{6}\right)\left(x - \frac{\pi}{6}\right)^3.$$

This gives us (with the help of a hand-held calculator)

$$\begin{aligned}\sin 32° &= \sin\left(\frac{\pi}{6} + \frac{\pi}{90}\right) \approx p_3\left(\frac{\pi}{6} + \frac{\pi}{90}\right)\\ &\approx \tfrac{1}{2} + \tfrac{1}{2}\sqrt{3}\left(\frac{\pi}{90}\right) - \frac{1}{4}\left(\frac{\pi}{90}\right)^2 - \frac{\sqrt{3}}{12}\left(\frac{\pi}{90}\right)^3\\ &\approx 0.500000 + 0.030230 - 0.000305 - 0.000006\\ &\approx 0.529919 \approx 0.52992.\end{aligned}$$

EXERCISES 11.6

In Exercises 1 to 10, obtain the indicated Taylor polynomial of $f(x)$ near the given x_0. Give a formula for $r_n(x)$.

1 $f(x) = \sin x$; $p_9(x)$; $x_0 = 0$.

2 $f(x) = \cos x$; $p_8(x)$; $x_0 = 0$.

3 $f(x) = \ln(1+x)$; $p_6(x)$; $x_0 = 0$.

4 $f(x) = \sin x + \cos x$; $p_5(x)$; $x_0 = 0$.

5 $f(x) = x^2 \ln x$; $p_4(x)$; $x_0 = 5$.

6 $f(x) = \sin x$; $p_5(x)$; $x_0 = \pi/3$.

7 $f(x) = \sqrt{x}$; $p_4(x)$; $x_0 = 4$.

8 $f(x) = \sqrt[3]{x}$; $p_4(x)$; $x_0 = 8$.

9 $f(x) = 4x^3 + 5x^2 - 2x + 1$; $p_3(x)$; $x_0 = 2$.

10 $f(x) = \mathrm{Tan}^{-1}\, x$; $p_3(x)$; $x_0 = 0$.

In Exercises 11 to 16, approximate the given expressions correct to four decimal places. Use results of Exercises 1 to 10 as needed.

11 sin 12°

12 ln 1.02

13 $e^{1/2}$

14 cos 58°

15 $\sqrt{4.2000}$

16 $\sqrt[3]{7.9000}$

17 Show that $p_4(x)$ with $x_0 = 0$ satisfies $|e^x - p_4(x)| < 0.001$ on the interval $[-\frac{1}{2}, \frac{1}{2}]$.

18 Show that $p_{10}(x)$ with $x_0 = 0$ satisfies $|e^x - p_{10}(x)| < 0.001$ on the interval $[-2, 2]$.

19 Show that we need $n = 12$ if we want a Taylor polynomial $p_n(x)$ near $x_0 = 0$ such that $|e^{-x^2} - p_n(x)| < 0.001$ on the interval $[-1, 1]$.

20 What degree Taylor polynomial $p_n(x)$ near $x = 0$ is needed to approximate cos 40° correct to five decimal places?

7 POWER SERIES

We have seen how to approximate any sufficiently differentiable function f near a point x_0 by a polynomial, namely the Taylor polynomial $p_n(x)$. In the preceding section the *closeness* of the approximation was improved when we increased the degree of $p_n(x)$. For instance, in the discussion of approximating $\sin x$ correct to five decimal places, it turned out that for x near $\frac{1}{4}\pi$, we needed to use a seventh degree Taylor polynomial instead of a fifth degree one. It seems then that as n

increases, the Taylor polynomial

$$p_n(x) = x - \frac{x^3}{3!} + \frac{x^5}{5!} - \frac{x^7}{7!} + - \ldots + \frac{(-1)^{n-1}x^{2n-1}}{(2n-1)!}$$

approximates $\sin x$ with greater and greater accuracy. In view of this, it seems that the infinite series

$$\sum_{n=1}^{\infty} (-1)^{n-1} \frac{x^{2n-1}}{(2n-1)!} = x - \frac{x^3}{3!} + \frac{x^5}{5!} - \frac{x^7}{7!} + - \ldots$$

converges to $\sin x$. This turns out to be what happens, but before we can pursue this idea we need to study the convergence of such power series.

7.1 **DEFINITION.** A **power series in $x - x_0$** is a series

(1) $$a_0 + a_1(x - x_0) + a_2(x - x_0)^2 + \ldots + a_n(x - x_0)^n + \ldots$$

that we represent by the notation $\sum_{n=0}^{\infty} a_n(x - x_0)^n$. Here x_0 is fixed and x is a real variable.

In particular, if $x_0 = 0$, then we have a *power series* in x, written $\sum_{n=0}^{\infty} a_n x^n$. For each value assumed by x, we get an ordinary infinite series whose convergence or divergence can be investigated by the methods of Sections 2 through 5. That sort of procedure is not very efficient, of course, since we cannot test for convergence separately at each of the infinitely many different values x can assume. What we need is a criterion that will give us all x for which *a given power series does converge.* There is a criterion which, while not quite as good as that, is very efficient. Before giving it, we need to develop some preliminary results.

We remark that we can, without losing any generality, restrict attention to power series in x (i.e., $x_0 = 0$). For any other x_0 in (1), we can let $y = x - x_0$, and transform (1) to

(2) $$a_0 + a_1 y + a_2 y^2 + \ldots + a_n y^n + \ldots,$$

a power series in y.

7.2 **LEMMA.** Suppose that a power series $\sum_{n=0}^{\infty} a_n x^n$ converges for some $x = c > 0$. Then it is absolutely convergent for every real number x such that $|x| < c$. That is, it is absolutely convergent for x in the entire interval $(-c, c)$.

Looked at slightly differently, this result says that if a power series *diverges* for some $x = d$, then it must be divergent for all x of larger absolute value. Put in this form, it yields our next result.

7.3 **LEMMA.** If the power series $\sum_{n=0}^{\infty} a_n x^n$ is divergent for some $x = d$, then it is divergent for every x such that $|x| > |d|$.

7.4 **THEOREM.** For a power series $\sum_{n=0}^{\infty} a_n(x - x_0)^n$, exactly one of the following alternatives holds.

(a) The series converges only for $x = x_0$.

(b) The series is absolutely convergent for every $x \in \boldsymbol{R}$.

(c) There is a real number $r > 0$ such that the series is absolutely convergent for all x such that $|x - x_0| < r$ and is divergent for all x such that $|x - x_0| > r$.

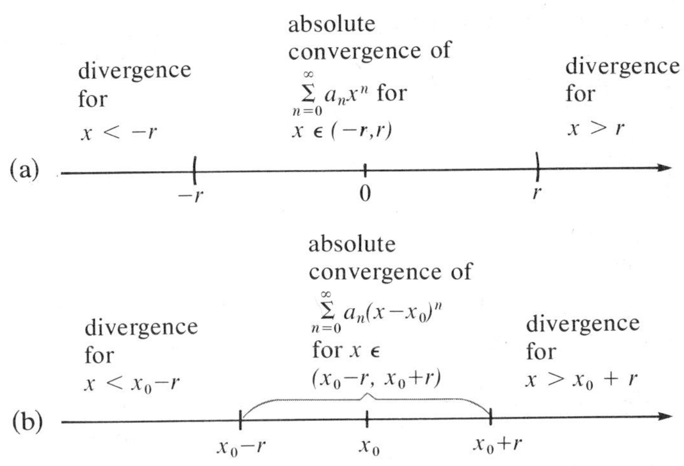

FIGURE 7.1

Figure 7.1 illustrates the situation in Theorem 7.4 for the cases $x_0 = 0$ and $x_0 \neq 0$.

7.5 **DEFINITION.** The number r in Theorem 7.4(c) is called the **radius of convergence** of the series $\sum_{n=0}^{\infty} a_n(x-x_0)^n$. (In case alternative (a) holds in Theorem 7.4, we say that $r=0$. In case alternative (b) holds, we say that $r=\infty$.) The **interval of convergence** is the set of all real numbers x such that $\sum_{n=0}^{\infty} a_n x^n$ converges.

Note that Theorem 7.4 tells us that the interval of convergence of a given power series will be either
(a) $\{x_0\}$, or
(b) $(-\infty, +\infty) = \boldsymbol{R}$, or
(c) (x_0-r, x_0+r), $[x_0-r, x_0+r)$, $(x_0-r, x_0+r]$, or $[x_0-r, x_0+r]$.
The theorem does not tell us *which* of the four possibilities in (c) holds. That must be investigated in each individual example. However, a slight refinement of Theorem 7.4 will give us the radius of convergence r in many common examples.

7.6 **THEOREM.** Given a power series $\sum_{n=0}^{\infty} a_n(x-x_0)^n$, suppose that

$$\lim_{n\to\infty} \left|\frac{a_n}{a_{n-1}}\right| = \ell.$$

Then
(a) If $\ell = +\infty$, then $r=0$.
(b) If $\ell = 0$, then $r=\infty$.
(c) If $\ell \in (0, +\infty)$, then $r = \dfrac{1}{\ell}$.

We now illustrate how the precise interval of convergence of a given power series can be found using Theorem 7.6 and our accumulated knowledge of convergence of series of constants.

7.7 **EXAMPLE.** Find all real numbers x such that

$$\sum_{n=1}^{\infty} \frac{(x-2)^n}{n}$$

converges. Is the series absolutely convergent for all x for which it is convergent?

Solution. To find r we compute the limit of the ratio

$$\left|\frac{a_{n+1}}{a_n}\right| = \left|\frac{\frac{1}{n+1}}{\frac{1}{n}}\right| = \frac{n}{n+1}.$$

We get

$$\lim_{n\to\infty}\left|\frac{a_{n+1}}{a_n}\right| = \lim_{n\to\infty}\frac{n}{n+1} = 1.$$

So $\ell = 1$. By Theorem 7.6, then, $r = 1/1 = 1$. Hence we have absolute convergence for x such that $|x-2| < 1$, i.e., $-1 < x-2 < 1$, i.e., $1 < x < 3$. And we have divergence for $x > 3$ or $x < 1$. The only points not accounted for are $x = 1$ and $x = 3$. We have to examine the given series for each of these values. For $x = 1$ we get

$$\sum_{n=1}^{\infty}\frac{(-1)^n}{n},$$

an alternating harmonic series that converges but does not converge absolutely. (See p. 696, following Definition 5.1.) For $x = 3$ we get $\sum_{n=1}^{\infty}\frac{1}{n}$, which is the divergent harmonic series. Thus, the complete interval of convergence is $[1, 3)$ with conditional convergence at $x = 1$ and absolute convergence elsewhere.

We can now return to the series we considered in the first paragraph of this section.

7.8 EXAMPLE. Find all real numbers x such that the series

$$x - \frac{x^3}{3!} + \frac{x^5}{5!} - \frac{x^7}{7!} + - \dots + \frac{(-1)^{n-1}x^{2n-1}}{(2n-1)!} + \dots$$

converges. Is the convergence always absolute?

Solution. Here

$$a_{n+1} = \frac{(-1)^n}{(2n+1)!} \quad \text{and} \quad a_n = \frac{(-1)^{n-1}}{(2n-1)!},$$

so

$$\lim_{n\to\infty}\left|\frac{a_{n+1}}{a_n}\right| = \lim_{n\to\infty}\frac{1/(2n+1)!}{1/(2n-1)!} = \lim_{n\to\infty}\frac{1}{(2n+1)2n} = 0.$$

Hence, by Theorem 7.6(b), $r = \infty$ so the given series converges for *all* values of x. The convergence is absolute by Theorem 7.4(b).

As you know, the Taylor polynomial $p_n(x)$ for $f(x) = \sin x$ near $x_0 = 0$ is the n-th partial sum of the series in Example 7.8. You may recall that in the opening paragraph we mentioned that the series in Example 7.8 actually converges to $\sin x$ for every real x. We will pursue this idea and explore its implications in the rest of this chapter.

7.9 EXAMPLE. Find all real numbers x such that the series

$$1+\frac{x}{3^2+1}+\frac{x^2}{3^4+2}+\frac{x^3}{3^6+3}+\dots+\frac{x^n}{3^{2n}+n}+\dots$$

converges. Is the convergence always absolute?

Solution. Here we have

$$a_{n+1}=\frac{1}{3^{2n+2}+n+1} \quad \text{and} \quad a_n=\frac{1}{3^{2n}+n}.$$

So

$$\lim_{n\to\infty}\left|\frac{a_{n+1}}{a_n}\right|=\lim_{n\to\infty}\frac{3^{2n}+n}{3^{2n+2}+n+1}=\lim_{n\to\infty}\frac{1+\dfrac{n}{3^{2n}}}{3^2+\dfrac{n+1}{3^{2n}}}$$

$$=\frac{1}{3^2}=\frac{1}{9},$$

since

$$\lim_{n\to\infty}\frac{n}{3^{2n}}=\lim_{x\to\infty}\frac{x}{3^{2x}}=\lim_{x\to\infty}\frac{1}{3^{2x}(2\ln 3)}=0.$$

using L'Hôpital's Rule. Thus $r=9$ by Theorem 7.6(c). So we have absolute convergence for $|x|<9$, i.e., for $-9<x<9$, and divergence for $x<-9$ and $x>9$. Let us consider $x=9$ and $x=-9$ now. When $x=9$, we have

$$1+\frac{9}{3^2+1}+\frac{9^2}{3^4+2}+\dots+\frac{9^n}{3^{2n}+n}+\dots$$

i.e.,

$$1+\frac{3^2}{3^2+1}+\frac{3^4}{3^4+2}+\dots+\frac{3^{2n}}{3^{2n}+n}+\dots$$

Note that

$$\lim_{n\to\infty}\frac{3^{2n}}{3^{2n}+n}=\lim_{n\to\infty}\frac{1}{1+\dfrac{n}{3^{2n}}}=1\neq 0,$$

so the series diverges by Theorem 2.5. Similarly, at $x=-9$ we get

$$1-\frac{9}{3^2+1}+\frac{9^2}{3^4+2}+\dots+\frac{(-1)^n 9^n}{3^{2n}+n}+\dots$$

which also diverges since its n-th term fails to approach 0 as $n\to\infty$. Thus the complete interval of convergence is $(-9,+9)$ and the series is absolutely convergent whenever it converges.

EXERCISES 11.7

In Exercises 1 to 18, determine the interval of convergence of the given series and determine where the convergence is absolute.

1 $\sum_{n=0}^{\infty} \frac{(x-x_0)^{n+1}}{(n+1)!}$

(This exercise is used in the next section.)

2 $\sum_{n=1}^{\infty} \frac{(-1)^{n-1}x^{2n}}{(2n)!}$

3 $\sum_{n=1}^{\infty} \frac{n!\,x^{2n}}{2n-1}$

4 $\sum_{n=1}^{\infty} \frac{n!\,x^{2n}}{n+1}$

5 $\sum_{n=1}^{\infty} \frac{(-1)^{n-1}2^n x^n}{n3^n}$

6 $\sum_{n=1}^{\infty} \frac{(-1)^n 3^n x^n}{n5^n}$

7 $\sum_{n=1}^{\infty} n(x-1)^n$

8 $\sum_{n=1}^{\infty} \frac{n^2}{n+1}(x-2)^n$

9 $\sum_{n=1}^{\infty} \frac{x^n}{n^2+1}$

10 $\sum_{n=1}^{\infty} \frac{(-1)^n(x-1)^n}{n+1}$

11 $\sum_{n=1}^{\infty} \frac{nx^n}{3n^2+1}$

12 $\sum_{n=1}^{\infty} \frac{(n^2+1)x^n}{n^3+1}$

13 $\sum_{n=1}^{\infty} \frac{n!\,(x-1)^n}{n^n}$

14 $\sum_{n=1}^{\infty} \frac{(\ln n)(x-2)^n}{n}$

15 $\sum_{n=1}^{\infty} \frac{(-1)^{n-1}1\cdot 3\cdot 5\cdots(2n-1)}{2^n n!}x^{2n-1}$

16 $\sum_{n=0}^{\infty} \frac{x^{3n}}{n+5}$

17 $1+3x+x^2+3x^3+x^4+3x^5+\ldots$ (*Hint:* Consider the odd and even terms separately.)

18 $1+2x+x^2+2x^3+x^4+2x^5+\ldots$ (See Exercise 17.)

19 In Theorem 7.4, show that if $\sum_{n=0}^{\infty} a_n(x-x_0)^n$ converges absolutely at either end of its interval of convergence, then it converges absolutely on the entire closed interval $[x_0-r, x_0+r]$.

20 In Theorem 7.4, show that if $\sum_{n=0}^{\infty} a_n(x-x_0)^n$ converges at only one endpoint of its interval of convergence, then it is conditionally convergent at that endpoint.

8 TAYLOR SERIES

In Section 6 you learned a technique for approximating any sufficiently differentiable function $f:\mathbf{R}\to\mathbf{R}$ by a polynomial $p_n(x)$ near a point $x=x_0$. In Section 7 you learned how to determine for which values of x a given power series $\sum_{n=0}^{\infty} a_n(x-x_0)^n$ converges. We are now ready to put these two ideas together in the notion of a *power series representation for a function f near a point* x_0.

This is approached as follows. Suppose we are given a function $f:\mathbf{R}\to\mathbf{R}$ that is infinitely differentiable at the point x_0. Then we can form the power series

(1)
$$\sum_{n=0}^{\infty} \frac{f^{(n)}(x_0)}{n!}(x-x_0)^n = f(x_0)+f'(x_0)(x-x_0)+\frac{1}{2!}f''(x_0)(x-x_0)^2 + \frac{1}{3!}f'''(x_0)(x-x_0)^3+\ldots+\frac{1}{n!}f^{(n)}(x_0)(x-x_0)^n+\ldots$$

whose n-th partial sum is the n-th degree Taylor polynomial approximation $p_n(x)$.

8.1 DEFINITION. If $f: \boldsymbol{R} \to \boldsymbol{R}$ has derivatives of every order at $x = x_0$, then the power series (1) is called the **Taylor series of f near x_0**. If $x_0 = 0$, the Taylor series is also known as the **Maclaurin series** of f.

We know from the preceding section that a power series like (1) generally has an interval of convergence, which may be the entire real line. We have previously used the n-th partial sum of (1), i.e., the Taylor polynomial $p_n(x)$, to approximate f near x_0. Hence it is natural to suppose that whenever (1) converges for some x, it will converge to $f(x)$. Unfortunately, this very reasonable supposition is *not* accurate, as the following classical example shows.

8.2 EXAMPLE. If

$$f(x) = \begin{cases} e^{-1/x^2} & \text{for } x \neq 0 \\ 0 & \text{when } x = 0, \end{cases}$$

then find the Maclaurin series of f. Show that its interval of convergence is all of $\boldsymbol{R}$ but that it converges to $f(x)$ only for the one value $x = 0$.

Solution. By definition $f(0) = 0$. We compute $f'(0)$ from the definition of the derivative.

$$f'(0) = \lim_{x \to 0} \frac{e^{-1/x^2} - 0}{x - 0} = \lim_{x \to 0} \frac{1}{xe^{1/x^2}}.$$

Let $t = \dfrac{1}{x}$, so $x = \dfrac{1}{t}$; then, as $x \to 0$, $t \to \infty$. Then

$$f'(0) = \lim_{t \to \infty} \frac{1}{\dfrac{1}{t} e^{t^2}} = \lim_{t \to \infty} \frac{t}{e^{t^2}} = \lim_{t \to \infty} \frac{1}{2te^{t^2}} = 0$$

using L'Hôpital's rule. Similarly,

$$f''(0) = \lim_{x \to 0} \frac{f'(x) - f'(0)}{x - 0}.$$

For $x \neq 0$, we have

$$f'(x) = \frac{d}{dx}(e^{-1/x^2}) = -\frac{2}{x^3} e^{-1/x^2}.$$

So

$$f''(0) = \lim_{x \to 0} \frac{-\dfrac{2}{x^3} e^{-1/x^2}}{x} = \lim_{x \to 0} \frac{-2}{x^4 e^{1/x^2}}.$$

Again we let $t = 1/x$. Then

$$f''(0) = \lim_{t \to \infty} \frac{-2t^4}{e^{t^2}} = 0,$$

again using L'Hôpital's rule. If we continue in this way we will find that $f^{(n)}(0) = 0$

for all n. Hence, the Maclaurin series of f near 0 is

$$0+0x+0\frac{x^2}{2!}+0\frac{x^3}{3!}+\ldots+0\frac{x^n}{n!}+\ldots.$$

This series obviously converges to 0 for all x, so its interval of convergence is $(-\infty, +\infty)=\boldsymbol{R}$. But $f(x)=e^{-1/x^2}\neq 0$ for $x\neq 0$. Hence, only for $x=0$ does the Maclaurin series converge to $f(x)$. QED

This example does not mean that all our hopes of finding a power series that converges to a given infinitely differentiable function near x_0 are now dashed. It *does* mean that the simple procedure of computing the Taylor series of f near x_0 and determining its interval of convergence is not good enough by itself. We must develop a criterion to determine whether

$$\sum_{n=0}^{\infty} \frac{f^{(n)}(x_0)}{n!}(x-x_0)^n$$

converges to $f(x)$ for all those x at which it converges. The following result provides us with one such criterion.

8.3 **THEOREM.** Let f be infinitely differentiable on an interval $I=(x_0-r, x_0+r)$. Suppose that its Taylor series is convergent on I. Then the Taylor series converges to $f(x)$ on I if and only if $\lim_{n\to\infty} r_n(x)=0$ for $x\in I$.

[Here $r_n(x)=f(x)-p_n(x)$ as in Theorems 6.3 and 6.6.]

We can apply Corollary 6.4 to obtain a workable test for deciding whether the Taylor series of a given function f converges to $f(x)$ near x_0.

8.4 **COROLLARY.** If f and I are as in Theorem 8.3 and $|f^{(n+1)}(x)|\leq B$ on the interval I, then the Taylor series converges to $f(x)$ for all $x\in I$.

Proof. From Corollary 6.4, we have in this case that

$$|r_n(x)|\leq\frac{B}{(n+1)!}|x-x_0|^{n+1}.$$

Then

$$\lim_{n\to\infty}|r_n(x)|\leq B\lim_{n\to\infty}\frac{|x-x_0|^{n+1}}{(n+1)!} \tag{2}$$

But $\dfrac{(x-x_0)^{n+1}}{(n+1)!}$ is the n-th term of the power series

$$\sum_{n=0}^{\infty}\frac{(x-x_0)^{n+1}}{(n+1)!}=(x-x_0)+\frac{(x-x_0)^2}{2!}+\ldots \tag{3}$$

By Exercise 1 of Exercises 11.7, (3) converges for all x. Hence by Theorem 2.5, its n-th term must approach 0 as $n \to \infty$. That is,

$$\lim_{n\to\infty} \frac{(x-x_0)^{n+1}}{(n+1)!} = 0.$$

It therefore follows from (2) that

$$0 \le \lim_{n\to\infty} |r_n(x)| \le B \lim_{n\to\infty} \frac{|x-x_0|^{n+1}}{(n+1)!} = 0.$$

Thus $\lim_{n\to 0} |r_n(x)| = 0$. Hence, $\lim_{n\to\infty} r_n(x) = 0$. Then, by Theorem 8.3, the Taylor series converges to $f(x)$ on I. QED

8.5 EXAMPLE. Show that the Taylor series for $f(x) = \cos x$ near $\pi/3$ converges to $\cos x$ for all $x \in \boldsymbol{R}$.

Solution. We have

$$\begin{aligned}
f(x) &= \cos x, & f(\pi/3) &= \tfrac{1}{2},\\
f'(x) &= -\sin x, & f'(\pi/3) &= -\sqrt{3}/2,\\
f''(x) &= -\cos x, & f''(\pi/3) &= -\tfrac{1}{2},\\
f'''(x) &= \sin x, & f'''(\pi/3) &= \sqrt{3}/2,\\
f^{(4)}(x) &= \cos x, & f^{(4)}(\pi/3) &= \tfrac{1}{2}, \quad \text{etc.}
\end{aligned}$$

So the Taylor series is

$$\frac{1}{2} - \frac{\sqrt{3}}{2}\left(x - \frac{\pi}{3}\right) + \frac{1}{2!}\left(-\frac{1}{2}\right)\left(x - \frac{\pi}{3}\right)^2 + \frac{1}{3!}\frac{\sqrt{3}}{2}\left(x - \frac{\pi}{3}\right)^3 + \frac{1}{4!}\frac{1}{2}\left(x - \frac{\pi}{3}\right)^4 + \dots.$$

Here $|f^{(n)}(x)| \le 1$ for all x since $f^{(n)}(x) = \pm\sin x$ or $\pm\cos x$. Hence, with $B = 1$ in Corollary 8.4, we conclude that the Taylor series converges to $\cos x$ for all $x \in \boldsymbol{R}$. QED

8.6 EXAMPLE. Show that the Maclaurin series for $f(x) = e^x$ converges to $f(x)$ for all $x \in \boldsymbol{R}$.

Solution. Here $f^{(i)}(x) = e^x$ for all i, so $f^{(i)}(0) = 1$ for all i. Hence, the Maclaurin series is

$$\sum_{n=0}^{\infty} \frac{x^n}{n!} = 1 + x + \frac{x^2}{2!} + \frac{x^3}{3!} + \dots + \frac{x^n}{n!} + \dots.$$

We have

$$\begin{aligned}
r_n(x) &= \frac{f^{(n+1)}(c)}{(n+1)!}(x-x_0)^{n+1}\\
&= \frac{e^c}{(n+1)!}(x-x_0)^{n+1}
\end{aligned}$$

where c is between 0 and x. If $x>0$, then $e^c<e^x$. If $x<0$, then $e^c<e^0=1$. Let $B=\max(1, e^x)$. Then B is independent of n, and

$$|r_n(x)|\le B\frac{|x-x_0|^{n+1}}{(n+1)!}.$$

Hence,

$$\lim_{n\to\infty}|r_n(x)|\le B\lim_{n\to\infty}\frac{|x-x_0|^{n+1}}{(n+1)!}=0.$$

Therefore, by Theorem 8.3, the Maclaurin series for e^x converges to e^x for all real numbers x.

QED

EXERCISES 11.8

In Exercises 1 to 12, show that the Taylor series for the given function f near the given point x_0 converges to $f(x)$ for x in the interval I.

1 $f(x)=\sin x$; $x_0=0$; $I=(-\infty, +\infty)$.

2 $f(x)=\cosh x=\frac{1}{2}(e^x+e^{-x})$; $x_0=0$; $I=(-\infty, +\infty)$.

3 $f(x)=\sin x$; $x_0=\frac{1}{4}\pi$; $I=(-\infty, +\infty)$.

4 $f(x)=\cos x$; $x_0=\frac{1}{4}\pi$; $I=(-\infty, +\infty)$.

5 $f(x)=\ln(1+x)$; $x_0=0$; $I=[0, 1]$.

6 $f(x)=\ln x$; $x_0=5$; $I=[4, 6]$.

7 $f(x)=\sqrt{x}$; $x_0=4$; $I=(1, 7)$.

8 $f(x)=\sin^2 x$; $x_0=0$; $I=(-\infty, +\infty)$. (*Hint:* $\cos 2x=1-2\sin^2 x$.)

9 $f(x)=\cos^2 x$; $x_0=0$; $I=(-\infty, +\infty)$. (See Exercise 8.)

10 $f(x)=e^{-x^2}$; $x_0=0$; $I=(-\infty, +\infty)$.

11 $f(x)=\dfrac{1}{1-x^2}$; $x_0=0$; $I=(-1, 1)$.

12 $f(x)=\dfrac{1}{1+x}$; $x_0=0$; $I=(-1, 1)$.

9 CALCULUS OF POWER SERIES

Let $f:\boldsymbol{R}\to\boldsymbol{R}$ be infinitely differentiable on an interval I containing x_0. Suppose that its Taylor series

(1) $$f(x_0)+f'(x_0)(x-x_0)+\frac{f''(x_0)}{2!}(x-x_0)^2+\ldots$$

converges to $f(x)$ on I. We are now ready to justify using $p_n'(x)$ to approximate $f'(x)$ and $\int_a^b p_n(x)\,dx$ to approximate $\int_a^b f(x)\,dx$ if x, a, and b belong to the interior of I. Here $p_n(x)$ is as usual the n-th degree Taylor polynomial of f near x_0, i.e., the n-th partial sum of the Taylor series (1).

9.1 **THEOREM.** Suppose that the power series $\sum_{n=0}^{\infty} a_n(x-x_0)^n$ converges to $f(x)$ on $I=(x_0-r, x_0+r)$. Suppose that $[a, b]\subseteq I$. Then

$$\int_a^b f(x)\,dx = \lim_{n\to\infty} \int_a^b s_n(x)\,dx,$$

where $s_n(x)=\sum_{k=0}^{n} a_k(x-x_0)^k$. In other words,

$$\int_a^b \lim_{n\to\infty} s_n(x)\,dx = \lim_{n\to\infty} \int_a^b s_n(x)\,dx,$$

i.e., the integral of the limit is the limit of the integral.

Since

$$s_n(x)=a_0+a_1(x-x_0)+a_2(x-x_0)^2+\ldots+a_n(x-x_0)^n$$

is a polynomial, the n-th partial sum of the series $\sum_{k=0}^{\infty} a_k(x-x_0)^k$, we see that *we can approximate the integral of $f(x)$ by the integral of a polynomial approximation $s_n(x)$ to $f(x)$ on $[a, b]$*. This is of enormous importance in practical work, and the next example suggests just how valuable it can be.

9.2 **EXAMPLE.** Compute $\int_0^{1/3} e^{-t^2}\,dt$ correct to five decimal places.

Solution. The function $f(x)=e^{-t^2}$ does not have an antiderivative expressible in terms of elementary functions. But this function is used to define the normal probability distribution in statistics, so it is necessary to work with it.

We know that, replacing x by $-t^2$ in the Taylor series for e^x,

$$e^{-t^2}=\sum_{n=0}^{\infty} \frac{(-1)^n t^{2n}}{n!} \quad \text{for all } t\in \mathbf{R}$$

from Example 8.6. This series converges on any finite interval, and in particular on $(-1, 1)\supseteq[0, 1/3]$. Thus, according to Theorem 9.1,

$$\begin{aligned}\int_0^{1/3} e^{-t^2}\,dt &= \lim_{n\to\infty} \int_0^{1/3} \sum_{k=0}^{n} \frac{(-1)^k t^{2k}}{k!}\,dt \\ &= \lim_{n\to\infty} \sum_{k=0}^{n} \left[\frac{(-1)^k t^{2k+1}}{(2k+1)k!}\right]_0^{1/3} \\ &= \sum_{k=0}^{\infty} \frac{(-1)^k (\frac{1}{3})^{2k+1}}{(2k+1)\cdot k!} \\ &= \frac{1}{3^1\cdot 1\cdot 0!}-\frac{1}{3^3\cdot 3\cdot 1}+\frac{1}{3^5\cdot 5\cdot 2!}-\frac{1}{3^7\cdot 7\cdot 3!}+\frac{1}{3^9\cdot 9\cdot 4!}-+\ldots\end{aligned}$$

This is an alternating series, so the error made in approximating it by a partial sum s_n is (by Theorem 4.5) at most $|a_{n+1}|$. Here we find

$$|a_3|=\frac{1}{3^7\cdot 7\cdot 3!}\approx 0.000011$$

$$|a_4|=\frac{1}{3^9\cdot 9\cdot 4!}\approx 0.0000002<0.000005.$$

So we will have a value that is accurate to five decimal places if we sum the first four terms. Then

$$\begin{aligned}\int_0^{1/3} e^{-t^2}\,dt &\approx 0.333333-0.012346+0.000412-0.000011\\ &\approx 0.321388\\ &\approx 0.32139.\end{aligned}$$

(The decimal arithmetic was done on a hand-held calculator.)

Notice that what we have *really* done here is to approximate $\int_0^{1/3} e^{-t^2}\,dt$ by $\int_0^{1/3} p_3(t)\,dt$, where $p_3(t)$ is the third degree Taylor polynomial of e^{-t^2} near 0. Thus, as promised, we are now able to approximate a hitherto intractable integral $\int_a^b f(x)\,dx$ by integrating a *polynomial approximation* to f over $[a, b]$. The techniques of elementary calculus for approximating definite integrals, such as Simpson's rule, involve *many* more calculations to achieve an approximation of such a high degree of accuracy.

Our next result is also of far-reaching importance.

9.3 **THEOREM.** Suppose that the power series $\sum_{n=0}^{\infty} a_n(x-x_0)^n$ converges to $f(x)$ on $I=(x_0-r, x_0+r)$. Then f is differentiable on I, and moreover

$$f'(x)=\sum_{n=1}^{\infty} na_n(x-x_0)^{n-1} \quad \text{for} \quad x\in I.$$

Whereas Theorem 9.1 had obvious practical implications for evaluating integrals, you may be thinking that Theorem 9.3 might be rather unimportant. After all, you can usually evaluate derivatives more easily than integrals, so it may not really save much work to use Theorem 9.3. This observation is sound, as far as it goes. But it turns out that Theorem 9.3 in its own way may be an even *bigger* labor saver than Theorem 9.1, for it enables us to obtain the following very useful result.

9.4 **THEOREM.** If $\sum_{n=0}^{\infty} a_n(x-x_0)^n$ converges to a function f for all x in $I=(x_0-r, x_0+r)$, then

$$a_n=\frac{f^{(n)}(x_0)}{n!} \quad \text{for all } n.$$

That is, $\sum_{n=0}^{\infty} a_n(x-x_0)^n$ is the Taylor series for f near x_0.

Theorem 9.4 says that once you find, *by any method*, a power series that converges to $f(x)$ for $x\in(x_0-r, x_0+r)$ then you have found the Taylor series, since any power series that converges to $f(x)$ near x_0 *must* coincide with the Taylor series.

9.5 **EXAMPLE.** Find the Taylor series for $\ln(1+x)$ near $x=0$. [This series was first obtained by Nicholas Mercator (1620–1687) in 1668.]

Solution. We take advantage of the fact that

$$\int_0^x \frac{1}{1+t}\,dt=\ln(1+x) \quad \text{for} \quad x\in(-1, 1)$$

and

(2) $$\frac{1}{1+t} = 1 - t + t^2 - t^3 + - \ldots + (-1)^n t^n + - \ldots$$

for $t \in (-1, 1)$ since this power series is geometric with common ratio $-t$. Then by Theorem 9.1

$$\begin{aligned}\ln(1+x) &= \int_0^x \frac{1}{1+t}\,dt = \int_0^x (1 - t + t^2 - t^3 + - \ldots + (-1)^n t^n + - \ldots)\,dt \\ &= t - \frac{t^2}{2} + \frac{t^3}{3} - \frac{t^4}{4} + - \ldots + \frac{(-1)^n t^{n+1}}{n+1} + - \ldots \Big]_0^x \\ &= x - \frac{x^2}{2} + \frac{x^3}{3} - \frac{x^4}{4} + - \cdots + \frac{(-1)^n x^{n+1}}{n+1} + - \cdots\end{aligned}$$

for $x \in (-1, 1)$. The latter power series must be the Taylor series for $\ln(1+x)$ near $x = 0$ by Theorem 9.4.

9.6 EXAMPLE. Find the Taylor series for $\ln \frac{1+x}{1-x}$ near $x = 0$.

Solution. From the preceding example,

(3) $$\ln(1+x) = x - \frac{x^2}{2} + \frac{x^3}{3} - \frac{x^4}{4} + - \ldots + \frac{(-1)^n x^{n+1}}{n+1} + \ldots,$$

for $x \in (-1, 1)$. Replacement of x by $-x$ gives

(4) $$\ln(1-x) = -x - \frac{x^2}{2} - \frac{x^3}{3} - \frac{x^4}{4} - \ldots - \frac{x^{n+1}}{n+1} - \ldots,$$

for $x \in (-1, 1)$. We can subtract the two series by Theorem 2.9, to get

$$\begin{aligned}\ln \frac{1+x}{1-x} &= \ln(1+x) - \ln(1-x) \\ &= 2x + \frac{2x^3}{3} + \ldots + \frac{2x^{2n-1}}{2n-1} + \ldots \\ &= 2\left(x + \frac{x^3}{3} + \ldots + \frac{x^{2n-1}}{2n-1} + \ldots\right)\end{aligned}$$

for $x \in (-1, 1)$. Again, this series must be the Taylor series for $\ln \frac{1+x}{1-x}$ by Theorem 9.4.

The series in Example 9.6 can be used to compute a rational approximation for $\ln y$ for any positive real number y. One lets

$$x = \frac{y-1}{y+1}.$$

Then $-1<x<1$ and

$$xy+x=y-1,$$
$$x+1=y-xy=y(1-x).$$

Therefore,

$$y=\frac{1+x}{1-x} \quad \text{for} \quad x\in(-1, 1).$$

So $\ln y$ is given by the series for

$$f(x)=\ln\frac{1+x}{1-x}.$$

It is possible to obtain an upper bound for the error when the n-th partial sum of this series is used to approximate $\ln\frac{1+x}{1-x}$. We can then compute $\ln y$ to any degree of accuracy desired. This is explored in Exercises 18 to 20 below.

EXERCISES 11.9

In Exercises 1 to 6, approximate the given integral correct to four decimal places.

1 $\int_0^{1/2} e^{-t^2}\,dt$

2 $\int_0^{1/2} \frac{dt}{1+t^3}$
[*Hint:* See Equation (2) in Example 9.5.]

3 $\int_0^1 \sin t^2\,dt$

4 $\int_0^1 \ln(1+\sin x)\,dx$

[*Hint:* Use Example 9.5, with $\sin x$ in place of x.]

5 $\int_0^{0.1} e^{x^3}\,dx$

6 $\int_{1/2}^1 \frac{\sin x}{x}\,dx$

7 (a) Show that $1-x^2+x^4-x^6+-\ldots+(-1)^n x^{2n}+\ldots=\frac{1}{1+x^2}$, for $x\in(-1, 1)$.
(b) Obtain the Taylor series for $\mathrm{Tan}^{-1}\,x$ where $x\in(-1, 1)$. [This was discovered by James Gregory (1638–1675).]
(c) If $f(x)=\mathrm{Tan}^{-1}\,x$, then find $f^{(5)}(0)$ and $f^{(6)}(0)$.

8 (a) Find the first three terms of the Taylor series for $\tan x$ near 0.
(b) Find the first three terms of the Taylor series for $\sec^2 x$ near 0.
(c) Find the first three terms of the Taylor series for $\ln|\cos x|$ near 0.

9 (a) Show that $\frac{1}{1-x}=1+x+x^2+\ldots+x^n+\ldots$, for $x\in(-1, 1)$.
(b) Obtain the Taylor series for $\frac{1}{1-2x+x^2}$, for $x\in(-1, 1)$.

10 Using Exercise 7(a), find a power series for $\frac{x}{(1+x^2)^2}$ if $x\in(-1, 1)$.

11 (a) Find the interval of convergence for $\sum_{n=1}^{\infty}(-1)^{n-1}\frac{x^n}{n}$.

(b) Find the interval of convergence of $\sum_{n=1}^{\infty} (-1)^{n-1}x^{n-1}$. (Thus, the series obtained by differentiating a given power series term-by-term need not have the same interval of convergence as the original series.)

12 Show that term-by-term differentiation of the Taylor series for $\cosh x = \frac{1}{2}(e^x + e^{-x})$ near 0 gives the Taylor series for $\sinh x = \frac{1}{2}(e^x - e^{-x})$ near 0. What does term-by-term integration give?

13 Find the Taylor series for $\dfrac{e^x - 1}{x}$ near $x = 0$. Differentiate term-by-term to show that $\displaystyle\sum_{n=1}^{\infty} \frac{n}{(n+1)!}$ converges to sum 1.

14 Find the Taylor series for x^2e^x near $x = 0$. Integrate term-by-term to show that

$$2 + \left(\frac{1}{3} + \frac{1}{4} + \frac{1}{5 \cdot 2!} + \frac{1}{6 \cdot 3!} + \ldots + \frac{1}{(n+3)n!} + \ldots\right)$$

converges to e.

15 The **error function** erf(x) is defined by

$$\operatorname{erf}(x) = \frac{2}{\sqrt{\pi}} \int_0^x e^{-t^2}\, dt.$$

Find the Taylor series for erf(x) near 0. (The error function is widely applied in scientific and statistical work.)

16 (a) Show that

$$\int \ln(1-x)\, dx = x + (1-x)\ln(1-x) + C.$$

(b) Integrate the Taylor series for $\ln(1-t)$ to show that

$$\sum_{n=2}^{\infty} \frac{x^n}{(n-1)n} = x + (1-x)\ln(1-x).$$

17 (a) Show that

$$\int x \operatorname{Tan}^{-1} x\, dx = \tfrac{1}{2}(x^2+1)\operatorname{Tan}^{-1} x - \tfrac{1}{2}x + C.$$

(b) Integrate the Taylor series for $t \operatorname{Tan}^{-1} t$ to show that

$$\sum_{n=1}^{\infty} (-1)^{n+1} \frac{x^{2n+1}}{(2n-1)(2n+1)} = \tfrac{1}{2}(x^2+1)\operatorname{Tan}^{-1} x - \tfrac{1}{2}x.$$

18 Let $t \in (-1, 1)$. Show that

$$\frac{1}{1-t} = 1 + t + t^2 + \ldots + t^n + \frac{t^{n+1}}{1-t}$$

and

$$\frac{1}{1+t} = 1 - t + t^2 - + \ldots + (-1)^n t^n + \frac{(-1)^{n+1}t^{n+1}}{1+t}.$$

19 Integrate the formulas in Exercise 18 and add to show that

$$\ln\frac{1+x}{1-x}=2x+\tfrac{2}{3}x^3+\ldots+\frac{[1+(-1)^n]x^{n+1}}{n+1}+r_n(x),$$

where

$$r_n(x)=\int_0^x t^{n+1}\left(\frac{1}{1-t}+\frac{(-1)^{n+1}}{1+t}\right)dt,$$

for $x\neq 1$ in the interval $(-1, 1)$. (See Exercise 20.)

20 If we want to compute $\ln 2$ correct to four decimal places, then we let $x=\dfrac{2-1}{2+1}=\dfrac{1}{3}$.
(a) Show that

$$|r_n(x)|\leq\int_0^{1/3}\tfrac{5}{2}t^{n+1}\,dt=\frac{5}{2(n+2)3^{n+2}}$$

using the formula from Exercise 19.
(b) Show that $n=6$ gives $|r_n(x)|<0.00005$.
(c) Compute $\ln 2$ correct to four decimal places.

10 BINOMIAL SERIES[1]

The main result of this section is an extension of the important Binomial Theorem of elementary algebra for the expansion of $(1+x)^p$, where p is a positive integer. You probably recall that this theorem states that

(1) $$(1+x)^p=\sum_{n=0}^{p}\binom{p}{n}x^n,$$

where

(2) $$\binom{p}{n}=\frac{p(p-1)(p-2)\ldots(p-n+1)}{n!}\quad\text{for}\quad n=1,2,\ldots,p,\quad\text{and}\quad\binom{p}{0}=1.$$

is called the *n-th binomial coefficient.* The quantity (2) is of great importance in probability and statistics, and in combinatorial mathematics generally. In the early eighteenth century, mathematicians [led by Leonhard Euler (1707–1783) of Switzerland, who was a master of formal manipulation] began investigating what would happen if one attempted to compute $(1+x)^p$ using (1) and (2) even if p was not a positive integer. This produces

(3) $$1+px+\frac{p(p-1)}{2!}x^2+\frac{p(p-1)(p-2)}{3!}x^3+\ldots+\frac{p(p-1)(p-2)\ldots(p-n+1)}{n!}x^n+\ldots.$$

As you can readily verify, (3) is the Maclaurin series for the function $f(x)=(1+x)^p$. The series (3) led to an anomaly in the work of Euler that became famous.

1. This section is optional.

Euler *assumed* that (3) converged to $(1+x)^p$ for all x and any p, and proceeded to do voluminous calculations on this basis. Sometimes he obtained paradoxes that apparently did not trouble him, but which *did* trouble others. For example, if we take $x=-2$ and $p=-1$, then we have

$$(1+x)^p=(1-2)^{-1}=(-1)^{-1}=\frac{1}{-1}=-1.$$

And yet (3) becomes

$$1+2+4+8+\ldots+2^n+\ldots,$$

which obviously diverges to $+\infty$. Euler shrugged this off with a remark about the wonders of the infinite: where else but in the realm of the infinite could we get -1 by adding together a bunch of positive integers! Having begun our study of infinite series with a paradox of Zeno, it is perhaps fitting that we conclude our study of series by straightening out this situation.

10.1 **THEOREM.** The bionomial series (3) is absolutely convergent for $|x|<1$, and divergent for $|x|>1$.

Proof. We use the Ratio Test (Theorem 5.4). Here

$$a_{n+1}=\frac{p(p-1)\ldots(p-n+1)(p-n)}{(n+1)!}x^{n+1}$$

and

$$a_n=\frac{p(p-1)\ldots(p-n+1)}{n!}x^n.$$

Then

$$\left|\frac{a_{n+1}}{a_n}\right|=\frac{|p-n|}{n+1}|x|,$$

so

$$\lim_{n\to\infty}\left|\frac{a_{n+1}}{a_n}\right|=\lim_{n\to\infty}\frac{\left|\frac{p}{n}-1\right|}{1+\frac{1}{n}}|x|=|x|.$$

Thus the series is absolutely convergent for $|x|<1$, and divergent for $|x|>1$, by the Ratio Test 5.4. QED

Thus, referring to our example above, *it makes no sense to try computing the binomial series for* $x=-2$, or any other number of absolute value greater than 1. What about $x=1$ and $x=-1$? It can be shown, but requires more work than we can spare here, that when $x=1$ the series converges to 2^p provided that $p>-1$; that when $x=-1$, the series converges to 0 provided that $p>0$; and that (3) converges to $(1+x)^p$ for any p, if $|x|<1$.

10.2 **THEOREM.** The binomial series

(3) $$1+\sum_{n=1}^{\infty}\frac{p(p-1)\dots(p-n+1)}{n!}x^n$$

converges to $(1+x)^p$ if $|x|<1$.

We can use Theorem 10.2 to generate a number of useful Taylor series for functions that involve fractional powers.

10.3 **EXAMPLE.** Find the Taylor series for $f(x)=(1-x)^{-1/2}$ if $x\in(-1, 1)$.

Solution. For $x\in(-1, 1)$ we know that the binomial series for $(1+(-x))^{-1/2}$ will converge to $(1-x)^{-1/2}$. So by Theorem 9.7, this will be the Taylor series for f near 0. We can then use Theorem 10.2 with $-x$ in place of x. We have from (3), with $p=-\frac{1}{2}$,

$$(1-x)^{-1/2}=1-\tfrac{1}{2}(-x)+\frac{(-\frac{1}{2})(-\frac{1}{2}-1)}{2!}(-x)^2+\frac{(-\frac{1}{2})(-\frac{1}{2}-1)(-\frac{1}{2}-2)}{3!}(-x)^3$$
$$+\dots+\frac{(-\frac{1}{2})(-\frac{3}{2})(-\frac{5}{2})\cdots(-\frac{1}{2}-n+1)}{n!}(-x)^n+\dots$$

Consider the general term.

$$\frac{(-\frac{1}{2})(-\frac{3}{2})(-\frac{5}{2})\cdots\left(\frac{-1-2n+2}{2}\right)}{n!}(-x)^n=\frac{(-\frac{1}{2})(-\frac{3}{2})(-\frac{5}{2})\cdots\left(\frac{-2n+1}{2}\right)}{n!}(-1)^nx^n$$
$$=\frac{(-1)^n\left(\frac{1}{2}\cdot\frac{3}{2}\cdot\frac{5}{2}\cdots\frac{2n-1}{2}\right)}{n!}(-1)^nx^n$$
$$=(-1)^{2n}\frac{1\cdot3\cdot5\cdots(2n-1)}{2^nn!}x^n$$
$$=\frac{1\cdot3\cdot5\cdots(2n-1)}{2^nn!}x^n.$$

Thus, for $x\in(-1, 1)$, we have

$$(1-x)^{-1/2}=1+\sum_{n=1}^{\infty}\frac{1\cdot3\cdot5\cdots(2n-1)}{2^nn!}x^n.$$

10.4 **EXAMPLE.** Find a power series for $\mathrm{Sin}^{-1}\,x$.

Solution. We recall that

$$\mathrm{Sin}^{-1}\,x=\int_0^x\frac{dt}{\sqrt{1-t^2}}=\int_0^x(1-t^2)^{-1/2}\,dt.$$

We can thus get the Taylor series for $\mathrm{Sin}^{-1}\,x$ by using Theorem 9.1. That is, we can give the Taylor series for $(1-t^2)^{-1/2}$, with $t\in(-1, 1)$, by replacing x in the preceding example by t^2. This gives

$$(1-t^2)^{-1/2}=1+\sum_{n=1}^{\infty}\frac{1\cdot3\cdot5\cdots(2n-1)}{2^nn!}t^{2n}$$

Then if $x \in (-1, 1)$ we can use Theorem 9.1 to integrate this series. We get

$$\int_0^x (1-t^2)^{-1/2}\,dt = \mathrm{Sin}^{-1}\,x = \int_0^x \left(1+\sum_{n=1}^{\infty} \frac{1\cdot 3\cdot 5\cdots(2n-1)}{2^n n!}\,t^{2n}\right) dt$$
$$= x + \sum_{n=1}^{\infty} \frac{1\cdot 3\cdot 5\cdots(2n-1)}{2^n n!}\,\frac{t^{2n+1}}{2n+1}\Bigg]_0^x$$
$$= x + \sum_{n=1}^{\infty} \frac{1\cdot 3\cdot 5\cdots(2n-1)}{2^n n!\,(2n+1)}\,x^{2n+1}$$

for $x \in (-1, 1)$.

Computational applications of the binomial series use the approach of the previous section. There is nothing new in that regard. What the binomial series gives us is a new and often simple way to generate the Taylor series of functions. We close with an indication of how tables of n-th roots could be generated using binomial series.

10.5 **EXAMPLE.** Compute $\sqrt[3]{5/4}$ correct to four decimal places.

Solution. Since $\frac{5}{4} = 1 + \frac{1}{4}$, we use the binomial series $(1+x)^{1/3}$ with $x = \frac{1}{4}$. We then have

$$(1+x)^{1/3} = 1 + \tfrac{1}{3}x + \frac{\frac{1}{3}(\frac{1}{3}-1)}{2!}x^2 + \frac{\frac{1}{3}(\frac{1}{3}-1)(\frac{1}{3}-2)}{3!}x^3 + \dots$$
$$+ \frac{\frac{1}{3}(\frac{1}{3}-1)(\frac{1}{3}-2)\cdots(\frac{1}{3}-n+1)}{n!}x^n + \dots$$
$$= 1 + (\tfrac{1}{3})x + \frac{(\frac{1}{3})(-\frac{2}{3})}{2!}x^2 + \frac{(\frac{1}{3})(-\frac{2}{3})(-\frac{5}{3})}{3!}x^3 + \dots$$
$$+ \frac{(\frac{1}{3})(-\frac{2}{3})(-\frac{5}{3})\cdots\left(\frac{4-3n}{3}\right)}{n!}x^n + \dots$$

Consider the n-th term. It is

$$\frac{(\frac{1}{3})(-\frac{2}{3})(-\frac{5}{3})\cdots[(4-3n)/3]}{n!}x^n = \frac{(-1)^{n-1}2\cdot 5\cdots(3n-4)}{n!\,3^n}x^n.$$

For $x = \frac{1}{4}$, we have

$$(1+\tfrac{1}{4})^{1/3} = 1 + \tfrac{1}{3}x + \sum_{n=2}^{\infty} (-1)^{n-1}\frac{2\cdot 5\cdots(3n-4)}{n!\,3^n 4^n}.$$

The error made in approximating $(1+\frac{1}{4})^{1/3}$ by using the n-th partial sum of this alternating series is, by Theorem 4.5, at most the absolute value of the first term omitted. By trial and error we find that the fifth term is the first one that is less than 0.00005 in absolute value:

$$(-1)^4 \cdot \frac{2\cdot 5\cdot 8\cdot 11}{5!\,3^5 4^5} = \frac{2\cdot 5\cdot 8\cdot 11}{5\cdot 24\cdot 3^5\cdot 4\cdot 4^4} = \frac{5\cdot 8\cdot 2\cdot 11}{5\cdot 8\cdot 3\cdot 3^5\cdot 2\cdot 2\cdot 4^4}$$
$$= \frac{11}{2\cdot 3^6\cdot 4^4} = \frac{11}{373248} < 0.00003.$$

So we get $\sqrt[3]{5/4}$ correct to 4 decimal places by using the terms of the infinite series from $n=0$ through $n=4$. We get

$$\begin{aligned}\sqrt[3]{5/4}=(1+\tfrac{1}{4})^{1/3}&\approx 1+\frac{1}{12}-\frac{1}{144}+\frac{5}{3^4 4^3}-\frac{5}{2\cdot 3^5 4^3}\\ &\approx 1.00000+0.08333-0.00694+0.00096-0.00016\\ &\approx 1.07719\\ &\approx 1.0772.\end{aligned}$$

EXERCISES 11.10

In Exercises 1 to 6, find a power series for the given function. Give the largest open interval on which you know that the power series converges to the given function.

1 $\sqrt{1+x}$

2 $\dfrac{1}{\sqrt[3]{1+x}}$

3 $\sqrt[3]{1-x}$

4 $\sqrt{2+x}=\sqrt{2}\sqrt{1+\frac{1}{2}x}$

5 $\sqrt[3]{1+2x^2}$

6 $\dfrac{1}{\sqrt{1-\frac{1}{4}x^2}}$

In Exercises 7 and 8, compute the first four terms of the Taylor series for the functions.

7 $\sqrt{1+xe^x}$

8 $\sqrt{1-k^2\sin^2 t}$

In Exercises 9 to 12, compute the number correct to four decimal places.

9 $\sqrt{26}$

10 $\sqrt[3]{6}$. [*Hint:* $\sqrt[3]{6}=\sqrt[3]{8}\sqrt[3]{\frac{3}{4}}=2\sqrt[3]{1-\frac{1}{4}}$.]

11 $\sqrt[4]{630}$. [*Hint:* Proceed as in Exercise 10.]

12 $\sqrt[3]{1001}$

In Exercises 13 to 16, compute the value of the given definite integral correct to four decimal places. (Use the results of earlier exercises and examples as needed.)

13 $\displaystyle\int_0^{1/3}\frac{dx}{\sqrt[3]{1+x^2}}$

14 $\displaystyle\int_0^{1/2}\frac{dx}{\sqrt{1-x^3}}$

15 $\displaystyle\int_0^{1/4}\sqrt{1+x^3}\,dx$

16 $\displaystyle\int_0^{1/2}\frac{dx}{1+x^3}$

The following two exercises involve *elliptic integrals*, $\int_0^{\pi/2}\sqrt{1-k^2\sin^2 t}\,dt$, which arise in applied work.

17 Find, correct to four decimal places, the length of the first quadrant arc of the ellipse $\mathbf{x}(t)=(2\cos t, \sqrt{5}\sin t)$. Express your answer as a decimal multiple of $\sqrt{5}\pi$. (See Exercise 8.)

18 Find, correct to four decimal places, the length of the first quadrant arc of the ellipse $\mathbf{x}(t)=(2\sqrt{2}\cos t, 3\sin t)$. (Express your answer as a decimal multiple of π.)

REVIEW EXERCISES 11.11

1 Give a formula for a_n if the sequence $\mathbf{a}$ is
(a) $(-1, \frac{4}{9}, -\frac{5}{27}, \frac{6}{81}, -\frac{7}{243}, \ldots)$
(b) $(3, \frac{6}{5}, \frac{12}{25}, \frac{24}{125}, \frac{48}{625}, \ldots)$

2 Determine whether $\boldsymbol{a}$ converges by computing $\lim_{n\to\infty} a_n$ if

(a) $a_n = \dfrac{-n^3+5n^2+5n+4}{3n^3+n^2+1}$ (b) $a_n = \dfrac{n5^{n-1}}{4^{n+10}}$

(c) $a_n = (0, 1, \frac{1}{2}, \frac{1}{2}, \frac{1}{4}, \frac{3}{4}, \frac{1}{8}, \frac{7}{8}, \ldots)$ (d) $a_n = \dfrac{1}{n^2}\cos n\pi$

3 Show that each sequence $\boldsymbol{a}$ is monotonic and determine whether it converges if

(a) $a_n = \dfrac{(n+1)!}{4n}$ (b) $a_n = \dfrac{n!}{2 \cdot 4 \cdot 6 \cdot \cdots \cdot 2n}$

4 Determine whether the given series converges and, if it does, find its sum.

(a) $-\frac{1}{3}+\frac{1}{9}-\frac{1}{27}+\frac{1}{81}-+\ldots$ (b) $\displaystyle\sum_{n=1}^{\infty}\left(\frac{3}{2^n}-\frac{(-1)^n 5}{2^n}\right)$

(c) $\displaystyle\sum_{n=1}^{\infty}\frac{5^{n-1}}{4^{n+3}}$ (d) $\displaystyle\sum_{n=1}^{\infty}\left(\frac{2}{5n}-\frac{1}{n}\right)$

(e) $\displaystyle\sum_{n=1}^{\infty}\frac{3n^2-20-400}{1000n^2+1000n+595}$

In Exercises 5 to 17, determine whether the given series converges. If it converges, is the convergence absolute?

5 $\displaystyle\sum_{n=1}^{\infty}\frac{1}{n^3+n}$

6 $\displaystyle\sum_{n=1}^{\infty}\frac{1}{\sqrt{2n-1}}$

7 $\displaystyle\sum_{n=1}^{\infty}\frac{2n-5}{n^3-2n+5}$

8 $\displaystyle\sum_{n=1}^{\infty}\frac{1}{\sqrt[3]{2n^2+3n+1}}$

9 $\displaystyle\sum_{n=1}^{\infty}\frac{(-1)^{n+1}}{\sqrt[3]{n}}$

10 $\displaystyle\sum_{n=1}^{\infty}(-1)^{n+1}n^2e^{-n}$

11 $\displaystyle\sum_{n=1}^{\infty}\frac{(-1)^{n+1}n!}{2 \cdot 4 \cdots (2n)}$

12 $\displaystyle\sum_{n=1}^{\infty}\frac{(-1)^{n+1}}{n!+1}$

13 $\displaystyle\sum_{n=1}^{\infty}\frac{(-5)^{n+1}}{4^{4n/3}}$

14 $\displaystyle\sum_{n=1}^{\infty}\frac{(-1)^{n+1}100^{n+100}}{2n!}$

15 $\displaystyle\sum_{n=1}^{\infty}(-1)^{n+1}\frac{2 \cdot 4 \cdot 6 \cdots (2n)}{n!}$

16 $\displaystyle\sum_{n=1}^{\infty}\frac{(-1)^{n+1}}{(n+1)\sqrt{n+2}}$

17 $\displaystyle\sum_{n=1}^{\infty}\frac{(-1)^{n+1}\ln(2n+1)}{3n+5}$

18 Obtain the Taylor polynomial of $f(x)$ near the given x_0 and give a formula for $r_n(x)$.
(a) $f(x)=\ln x$, $x_0=5$. Find $p_6(x)$.
(b) $f(x)=e^x \sin x$, $x_0=0$. Find $p_5(x)$.

19 Approximate, using the Taylor polynomial, correct to four decimal places.
(a) $\ln 5.02$ (b) $e^{1/3}$

20 Determine the interval of convergence of the given series and find where the convergence is absolute.

(a) $\displaystyle\sum_{n=0}^{\infty}\frac{(2x)^n}{n!}$ (b) $\displaystyle\sum_{n=2}^{\infty}\frac{x^n}{n^2-1}$

(c) $\displaystyle\sum_{n=1}^{\infty}\frac{(-1)^{n-1}(x-3)^n}{n+2}$ (d) $\displaystyle\sum_{n=1}^{\infty}\frac{(-2)^n n! x^n}{1 \cdot 3 \cdot 5 \cdots (2n-1)}$

21 Show that the Taylor series for f near x_0 converges to $f(x)$ on the interval I.
(a) $f(x)=\cos x$, $x_0=0$, $I=(-\infty, +\infty)$.
(b) $f(x)=e^{-\frac{1}{2}x^2}$, $x_0=0$, $I=(-\infty, +\infty)$.

22 Approximate, correct to four decimal places:
(a) $\displaystyle\int_0^{1/3} \frac{dt}{1+t^3}$ (b) $\displaystyle\int_0^{1/5} e^{x^3}\, dx$

23 Find a power series for the given $f(x)$ and give the largest open interval on which you know that the power series converges to $f(x)$.

(a) $\sqrt{1-x}$ (b) $\dfrac{1}{\sqrt[3]{1-x^2}}$

24 Compute the first four terms of the Taylor series for the function $\sqrt{1-x^2e^{2x}}$.

25 Compute correct to four decimal places:
(a) $\sqrt{24}$ (b) $\sqrt[3]{7}$

TABLE OF INTEGRALS

1 $\int u\,dv = uv - \int v\,du$ (integration by parts)

2 $\int \frac{dx}{x} = \ln|x| + C$

3 $\int a^x\,dx = \frac{1}{\ln a}a^x + C$

4 $\int \sin x\,dx = -\cos x + C$

5 $\int \cos x\,dx = \sin x + C$

6 $\int \tan x\,dx = \ln|\sec x| + C = -\ln|\cos x| + C$

7 $\int \cot x\,dx = \ln|\sin x| + C$

8 $\int \sec x\,dx = \ln|\sec x + \tan x| + C$

9 $\int \csc x\,dx = \ln|\csc x - \cot x| + C$

10 $\int \sec^2 x\,dx = \tan x + C$

11 $\int \csc^2 x\,dx = -\cot x + C$

12 $\int \sin^2 x\,dx = \frac{1}{2}x - \frac{\sin 2x}{4} + C$

13 $\int \cos^2 x\,dx = \frac{1}{2}x + \frac{\sin 2x}{4} + C$

14 $\int \tan^2 x\,dx = \tan x - x + C$

15 $\int \cot^2 x\,dx = -\cot x - x + C$

16 $\int \ln ax\,dx = x\ln ax - x + C$

17 $\int xe^{ax}\,dx = \frac{1}{a^2}e^{ax}(ax-1) + C$

18 $\int x^n e^{ax}\,dx = \frac{1}{a}x^n e^{ax} - \frac{n}{a}\int x^{n-1}e^{ax}\,dx$

19 $\int \frac{dx}{x^2+a^2} = \frac{1}{a}\mathrm{Tan}^{-1}\frac{x}{a} + C \qquad (a \neq 0)$

20 $\int \frac{dx}{\sqrt{a^2-x^2}} = \mathrm{Sin}^{-1}\frac{x}{a} + C$

21 $\int \frac{dx}{x^2-a^2} = \frac{1}{2a}\ln\left|\frac{x-a}{x+a}\right| + C = -\frac{1}{a}\coth^{-1}\frac{x}{a} + C$

22 $\int \frac{dx}{a^2-x^2} = \frac{1}{2a}\ln\left|\frac{x+a}{x-a}\right| + C = \frac{1}{a}\tanh^{-1}\frac{x}{a} + C$

23 $\displaystyle\int \sqrt{x^2+a^2}\,dx = \frac{x}{2}\sqrt{x^2+a^2} + \frac{a^2}{2}\ln|x+\sqrt{x^2+a^2}| + C$

24 $\displaystyle\int \sqrt{x^2-a^2}\,dx = \frac{x}{2}\sqrt{x^2-a^2} - \frac{a^2}{2}\ln|x+\sqrt{x^2-a^2}| + C$

25 $\displaystyle\int \sqrt{a^2-x^2}\,dx = \frac{x}{2}\sqrt{a^2-x^2} + \frac{a^2}{2}\,\mathrm{Sin}^{-1}\frac{x}{a} + C$

26 $\displaystyle\int (a^2-x^2)^{3/2}\,dx = \tfrac{1}{4}x(a^2-x^2)^{3/2} + \tfrac{3}{8}a^2x\sqrt{a^2-x^2} + \tfrac{3}{8}a^4\,\mathrm{Sin}^{-1}\frac{x}{a} + C$

27 $\displaystyle\int \frac{dx}{\sqrt{x^2+a^2}} = \ln|x+\sqrt{x^2+a^2}| + C = \sinh^{-1}\frac{x}{a} + C$

28 $\displaystyle\int \frac{dx}{\sqrt{x^2-a^2}} = \ln|x+\sqrt{x^2-a^2}| + C = \cosh^{-1}\frac{x}{a} + C$

29 $\displaystyle\int e^{ax}\sin bx\,dx = \frac{e^{ax}}{a^2+b^2}[a\sin bx - b\cos bx] + C$

30 $\displaystyle\int e^{ax}\cos bx\,dx = \frac{e^{ax}}{a^2+b^2}[a\cos bx + b\sin bx] + C$

31 $\displaystyle\int \sec^3 x\,dx = \tfrac{1}{2}\sec x\tan x + \tfrac{1}{2}\ln|\sec x + \tan x| + C$

REFERENCES

In some places we have mentioned, but not pursued, topics usually treated in greater depth in later courses. A number of proofs have been omitted because of their length or sophistication. Such gaps can be filled in, should you wish to do so, by consulting the following advanced texts for the theorems listed below.

Buck, R. C.: *Advanced Calculus*, 3rd edition. McGraw-Hill Book Co., New York, 1978.

Curtis, C. W.: *Linear Algebra: An Introductory Approach*, 3rd edition. Allyn & Bacon, Inc., Boston, 1974.

Dettman, J. W.: *Introduction to Linear Algebra and Differential Equations.* McGraw-Hill Book Co., New York, 1974.

Edwards, C. H., Jr.: *Advanced Calculus of Several Variables.* Academic Press, New York, 1973.

Flanders, H., Korfhage, R., and Price, J.: *A Second Course in Calculus.* Academic Press, New York, 1974.

Goffman, C.: *Calculus of Several Variables.* Harper & Row Publishers, Inc., New York, 1965.

Williamson, R., Crowell, R., and Trotter, H.: *Calculus of Vector Functions*, 3rd edition. Prentice-Hall, Inc., Englewood Cliffs, NJ, 1972.

Chapter 1

Expansion of determinant by minors: Curtis, Chapter 5.

Chapter 3

Theorem 2.10: Buck, p. 91.

Chapter 4

Theorem 3.10, general case: Goffman, p. 39.
Theorem 6.2: Goffman, pp. 89–95.
Theorem 6.3: Goffman, pp. 95–97.
Theorem 8.6: Williamson, Crowell & Trotter, pp. 598–600.
Theorem 8.9: Flanders, Korfhage & Price, pp. 335–336.
Second degree Taylor polynomials of $f: \mathbf{R}^n \to \mathbf{R}$: Goffman, pp. 55–57.
Theorem 9.6: Goffman, p. 61.
Theorem 9.10, general case: Goffman, p. 61; Flanders, Korfhage & Price, pp. 252, 349.
Theorem 10.1, general case: Edwards, p. 92.
Example 10.3, second derivative test for Lagrange multipliers: Edwards, p. 154.
Theorem 10.5: Edwards, pp. 102–113.

Chapter 5

Theorem 1.5: Buck, pp. 169–172.

Exercise 20, Exercises 5.1, Intermediate Value Theorem for Continuous Functions $f:\boldsymbol{R}^n \to \boldsymbol{R}$: Buck, pp. 93–94.

Theorem 2.4, $\dfrac{dV}{dt} = A(t)$: Edwards, pp. 236–237.

Theorem 11.6, general case: Edwards, Section IV.6.

Chapter 6

Extension of Jacobian Chain Rule 6.1: Buck, pp. 349–350.
Theorem 7.2: Edwards, pp. 188–192.
Theorem 7.3: Curtis, p. 151.
Theorem 8.3: Edwards, pp. 188–192.
Theorem 8.5: Curtis, p. 147.
Theorem 8.10, general case: Edwards, p. 246.
Theorem 9.1: Edwards, pp. 245–255.

Chapter 7

Differential forms (notation $dx \wedge dy$) following Example 6.5: Buck, pp. 446–477.
Theorem 7.1, general case: Edwards, pp. 366–376.

Chapter 8

Theorem 9.1: Dettman, pp. 346–348.

Chapter 9

Theorem 1.4: Dettman, pp. 245, 250, 374–379.

Chapter 10

Theorem 1.6: Dettman, pp. 374–379.
Theorem 2.8(b): Curtis, pp. 274–275.
Exercises 4.19 and 4.20 and Figure 4.9 after R. E. Williamson and H. F. Trotter, *Multivariable Mathematics*, 2nd edition, Prentice-Hall, Inc., Englewood Cliffs, NJ, 1979. Adapted by permission of Prentice-Hall, Inc., Englewood Cliffs, New Jersey.

Chapter 11

Buck, Chapters 5 and 6.

ANSWERS TO ODD NUMBERED PROBLEMS

CHAPTER 1

EXERCISES 1.1, p. 15

3. (a) 7; $(-\frac{1}{2}, 1, 4)$. (b) 4; $(1, \frac{5}{2}, 0)$. **5.** (a) $(2, -2)$. (b) $(-3, 4)$. (c) $(3, -4)$, (d) $(-7, 10)$. (e) $(11, -14)$. (f) $\frac{\sqrt{5}}{2}$. (g) $\frac{\sqrt{61}}{2}$. (h) $\left(-\frac{1}{\sqrt{5}}, \frac{2}{\sqrt{5}}\right)$; $\left(\frac{5}{\sqrt{61}}, -\frac{6}{\sqrt{61}}\right)$. **7.** (a) $(0, -1, 2, 4)$. (b) $(2, -3, -2, 2)$. (c) $(-2, 3, 2, -2)$. (d) $(-5, 7, 6, -3)$. (e) $\frac{1}{3}(5, -6, -8, -1)$. (f) $\sqrt{14}$. (g) $\sqrt{7}$. (h) $\frac{1}{\sqrt{14}}(1, -2, 0, 3)$; $\frac{1}{\sqrt{7}}(-1, 1, 2, 1)$. **9.** (a) $(-3, 0, -2)$; $\sqrt{13}$. (b) $(1, -3, 2)$; $\sqrt{14}$. **13** (a) $(x-1)^2+(y+2)^2+(z+2)^2=4$. (b) $(x-2)^2+y^2+(z-1)^2=9$. (c) $(x-1)^2+(y+1)^2+(z-4)^2=\frac{1}{4}$. **21.** (a) $\boldsymbol{F}(\boldsymbol{x}) = F_1(\boldsymbol{x})\boldsymbol{i}+F_2(\boldsymbol{x})\boldsymbol{j}+F_3(\boldsymbol{x})\boldsymbol{k}$.

EXERCISES 1.2, p. 26

1. (a) $\frac{1}{4}\pi$ (45°). (b) $\text{Cos}^{-1}\frac{4}{5} \approx 0.6435$ rad($\approx$36.87°). **3.** (a) $\frac{5\pi}{6}$ (150°). (b) $\frac{\pi}{3}$ (60°). **5.** $\boldsymbol{x} = c\boldsymbol{y}$ for some $c \in \boldsymbol{R}$ (i.e., $\boldsymbol{x}$ parallel to $\boldsymbol{y}$) is necessary and sufficient for equality. **7.** (a) $\frac{1}{\sqrt{2}}(1, -1, 0)$; $\alpha = \frac{1}{4}\pi$, $\beta = \frac{3\pi}{4}$, $\gamma = \frac{1}{2}\pi$. (b) $\frac{1}{2}(\sqrt{3}, 0, 1)$; $\alpha = \frac{\pi}{6}$, $\beta = \frac{1}{2}\pi$, $\gamma = \frac{\pi}{3}$. **11.** $\frac{1}{\sqrt{27}}(5, 1, 1)$; infinitely many. **13.** $-\frac{4}{3}$; no. **15.** $(-\frac{1}{4}+\sqrt{2})(1, 1, \sqrt{2})$. **17.** 15.

EXERCISES 1.3, p. 34

1. (a) $\boldsymbol{x} = (2, 1)+t(3, -4)$. (b) $\boldsymbol{x} = t(0, 1)$. (c) $\boldsymbol{x} = (-7, -2)+t(11, 6)$. **3.** (a) $\boldsymbol{x} = (1, 2, 3)+t(2, 4, 6)$. (b) $\boldsymbol{x} = (0, 1, -2)+t(3, 0, 0)$. (c) $\boldsymbol{x} = t(0, 1, 0)$. (d) $\boldsymbol{x} = (1, -1, 4)+t(6, 0, -5)$. **5.** $\boldsymbol{x} = (1, 2, 3)+t(2, 5, -2)$; $\frac{x-1}{2} = \frac{y-2}{5} = \frac{z-3}{-2}$; $x = 1+2t$, $y = 2+5t$, $z = 3-2t$. **7.** $\boldsymbol{x} = (-1, 5, 4)+t(1, -1, -2)$; $\frac{x+1}{1} = \frac{y-5}{-1} = \frac{z-4}{-2}$; $x = -1+t$, $y = 5-t$, $z = 4-2t$. **9.** $x = 1+3t$, $y = 1-t$, $z = 2+2t$. **11.** No, they are skew lines. **13.** Yes. **15.** Yes, you should slouch down about 0.2 feet.

EXERCISES 1.4, p. 39

1. $3x-4y+z = 16$. **3.** $z = -5$; $y = 3$; $x = -1$. **5.** $3x-2y = 10$. **7.** (a) $x = 1$. (b) $y = -2$. **9.** (a) and (b) coincide, and are parallel to (d), perpendicular to (c).

11. $\boldsymbol{x}=(-2,0,3)+t(-4,1,3)$; $x=-2-4t$, $y=t$, $z=3+3t$; $\dfrac{x+2}{-4}=\dfrac{y}{1}=\dfrac{z-3}{3}$.
13. xy-plane: $x=5+5t$, $y=3t$, $z=0$; yz-plane: $x=0$, $y=-3+4t$, $z=10t$; xz-plane: $x=5+2t$, $y=0$, $z=-3t$. **15.** Yes; $(3,-2,-1)$. **19.** $\dfrac{3}{\sqrt{38}}$.

EXERCISES 1.5, p. 47

1. $2x+y-z=-2$. **3.** $(x-3,\ y,\ z-1)\cdot(1,\ 1,\ 1)=0$. **5.** $3\sqrt{38}$. **9.** $\frac{1}{2}\sqrt{257}$. **19.** $\dfrac{1}{\sqrt{2}}$.

EXERCISES 1.6, p. 59

1. (a) $\{(0,0,0)\}$. (b) $(-1,1,0)$. (c) $\{(-1,1,0)\}$. **3.** $\varnothing$. **5.** (a) $\{t(-1,1,1)\mid t\in \boldsymbol{R}\}$. (b) $(2,-5,0)$ (c) $\{(2,-5,0)+t(-1,1,1)\mid t\in\boldsymbol{R}\}$. **7.** (a) $\{t(-7,-5,1)\mid t\in\boldsymbol{R}\}$. (b) $(-10,-6,0)$. (c) $\{(-10,-6,0)+t(-7,-5,1)\mid t\in\boldsymbol{R}\}$. **9.** (a) $\{t(2,1,1,0)\mid t\in\boldsymbol{R}\}$. (b) $(0,0,0,1)$. (c) $\{(0,0,0,1)+t(2,1,1,0)\mid t\in\boldsymbol{R}\}$. **11.** (a) $\{t(1,2,0,1,0)+s(-1,1,3,0,1)\mid s,t\in\boldsymbol{R}$. (b) $(1,-1,2,0,0)$. (c) $\{(1,-1,2,0,0)+t(1,2,0,1,0)+s(-1,1,3,0,1)\mid s,t\in\boldsymbol{R}\}$. **13.** Pork 4, Beef 3, Cereal 9. **15.** $i_1=\frac{3}{2}$, $i_2=1$, $i_3=\frac{5}{2}$. **17.** (a) $a_{i1}x_1+a_{i2}x_2+a_{i3}x_3$. (b) $b_i+a_{i1}x_1+a_{i2}x_2+a_{i3}x_3$.

(c) $$\begin{cases}(a_{11}-1)x_1+a_{12}x_2+a_{13}x_3=-b_1\\ a_{21}x_1+(a_{22}-1)x_2+a_{23}x_3=-b_2.\\ a_{31}x_1+a_{32}x_2+(a_{33}-1)x_3=-b_2\end{cases}$$

EXERCISES 1.7, p. 65

1. Linearly independent; yes. **3.** Linearly independent; no. **5.** Linearly independent; yes. **7.** Linearly dependent; $-2(1,-1,1,-1)+(2,3,-4,1)+(0,-5,6,-3)=(0,0,0,0)$; no. **13.** (b) $(0,0,1)=\frac{1}{2}(1,0,1)+\frac{1}{2}(0,1,1)-\frac{1}{2}(1,1,0)$, (c) No. (d) No.

15. (a) $$\begin{cases}a+b-c=0\\ 2a+3b-2d=0\\ 9b-6c+2e-2f=0\\ 6b-3d+e-f=0\end{cases}$$

(b) $\boldsymbol{v}_1=(1,1,-1,0,0,0)$, $\boldsymbol{v}_2=(2,3,0,-2,0,0)$, $\boldsymbol{v}_3=(0,9,-6,0,2,-2)$, $\boldsymbol{v}_4=(0,6,0,-3,1,-1)$. (c) $\boldsymbol{v}_4=-3\boldsymbol{v}_1+\frac{3}{2}\boldsymbol{v}_2+\frac{1}{2}\boldsymbol{v}_3$.

17. (a) $$\begin{cases}2a+b-2c=0\\ a+b-2d=0\\ a+2b-2e=0\\ a+2b-f=0\\ 2e-f=0\end{cases}$$

(b) $\boldsymbol{v}_1=(2,1,-2,0,0,0)$, $\boldsymbol{v}_2=(1,1,0,-2,0,0)$, $\boldsymbol{v}_3=(1,2,0,0,-2,0)$, $\boldsymbol{v}_4=(1,2,0,0,0,-1)$, $\boldsymbol{v}_5=(0,0,0,0,2,-1)$. (c) $\boldsymbol{v}_3-\boldsymbol{v}_4+\boldsymbol{v}_5=\boldsymbol{0}$.

EXERCISES 1.8, p. 75

1. Yes; 9. **3.** No; 0. **5.** 6. **7.** 4. **9.** 5 **11.** 6. **19.** Left-handed. **21.** $x=3$, $y=2$, $z=4$.

REVIEW EXERCISES 1.9, p. 77

1. (a) $(-13,-1,-1)$. (b) $\frac{1}{15}(46,7,-2)$. (c) 3. (d) $(\frac{2}{3},-\frac{1}{3},\frac{2}{3})$; $\dfrac{1}{\sqrt{11}}(-3,-1,1)$.
5. $\left(\dfrac{1}{2},\dfrac{1}{2},\dfrac{\sqrt{2}}{2}\right)$; $\alpha=\dfrac{\pi}{3}$, $\beta=\dfrac{\pi}{3}$, $\gamma=\dfrac{\pi}{4}$. **7.** $(-\frac{2}{7},\frac{3}{7},-\frac{6}{7})$. **9.** $x=2-2t$, $y=-2t$, $z=-1+\frac{3}{2}t$. **11.** No, they are skew. **13.** $3y-4z+17=0$. **15.** $5x-2y+z+12=0$.

17. 0. **19.** $4x+3y+3z=11$. **21.** $\boldsymbol{x}=(-1,-1,2)+s(1,4,-3)+t(3,-2,3)$; $x=-1+s+3t$, $y=-1+4s-2t$, $z=2-3s+4t$; $3x-6y-7z+11=0$. **23.** 11. **25.** $\{(-10,-6,0)+t(-7,-5,1) \mid t\in \boldsymbol{R}\}$. **27.** (a) Linearly dependent; $-(1,3,4)-2(4,0,1)+3(3,1,2)=(0,0,0)$. (b) Linearly independent. **29.** 14. **31.** Yes, of volume 25.

CHAPTER 2

EXERCISES 2.1, p. 90

1. (b) The unit circle $x^2+y^2=1$. (c) $\dot{\boldsymbol{x}}(0)=(0,1)$. **3.** $\boldsymbol{x}=(1,5,7)+t(1,4,12)$; $\sqrt{161}$. **5.** $\boldsymbol{x}=(2,1,1)+t(4,1,3)$; $\sqrt{26}$. **7.** $\boldsymbol{x}=\left(\frac{\pi}{8},1,0\right)+t(1,0,-4)$. **9.** $(1,0,0)$; not continuous. **11.** (a) $(-4\sin t-4t\cos t-3t^2,\ 2+4\cos t-4t\sin t,\ 2t\cos t-t^2\sin t-2\cos t)$. (b) $-2\sin t+2t\sin t+t^2\cos t-8t$. **21.** $\left(-1,-1,\frac{15\pi^2}{8}\right)$.

EXERCISES 2.2, p. 99

1. $\boldsymbol{v}(t)=(-\frac{1}{2}\sin\frac{1}{2}t, \frac{1}{2}\cos\frac{1}{2}t, 1)$; $\boldsymbol{a}(t)=(-\frac{1}{4}\cos\frac{1}{2}t, -\frac{1}{4}\sin\frac{1}{2}t, 0)$; speed is $\frac{\sqrt{5}}{2}$. **3.** $\boldsymbol{v}(t)=(-t\sin t+\cos t, t\cos t+\sin t, 1)$; $\boldsymbol{a}(t)=(-t\cos t-2\sin t, -t\sin t+2\cos t, 0)$; speed $=\sqrt{2+t^2}$. **5.** $\boldsymbol{x}(t)=t^2\boldsymbol{c}+t\boldsymbol{d}+\boldsymbol{e}$, for $\boldsymbol{c},\boldsymbol{d},\boldsymbol{e}\in\boldsymbol{R}^3$. **7.** $\frac{8}{27}[10\sqrt{10}-1]$. **9.** $\boldsymbol{x}(t(u))=\left(\frac{u}{\sqrt{17}}, \sin\frac{4u}{\sqrt{17}}, \cos\frac{4u}{\sqrt{17}}\right)$. **11.** (a) $\boldsymbol{v}(\theta)=\left(\frac{df}{d\theta}\cos\theta-f(\theta)\sin\theta, \frac{df}{d\theta}\sin\theta+f(\theta)\cos\theta\right)$ (b) $|\boldsymbol{v}(\theta)|=\sqrt{\left(\frac{dr}{d\theta}\right)^2+r^2}$. **13.** (a) $\boldsymbol{x}=(v_0t\cos\alpha, -\frac{1}{2}gt^2+v_0t\sin\alpha)$; the parabola $y=-\frac{1}{2}\frac{g}{v_0^2\cos^2\alpha}x^2+x\tan\alpha$. (b) $\frac{v_0^2\sin 2\alpha}{g}$. **15.** $\frac{8\sqrt{2}}{3}\pi^3$.

EXERCISES 2.3, p. 110

1. $K=\frac{1}{2}$; $\boldsymbol{T}=\frac{1}{\sqrt{2}}(-\sin t, \cos t, 1)$; $\boldsymbol{N}=(-\cos t, -\sin t, 0)$. **3.** $K=\frac{\sqrt{19}}{7\sqrt{14}}$; $\boldsymbol{T}=\frac{1}{\sqrt{14}}(1,2,3)$. **5.** $\boldsymbol{T}=\frac{1}{5}(-3\cos t, -3\sin t, 4)$; $\boldsymbol{N}=(-\sin t, -\cos t, 0)$; $K=\frac{3}{25}$, **7.** $\boldsymbol{a_T}=\boldsymbol{0}$; $\boldsymbol{a_N}=3\boldsymbol{N}=(-3\cos t, -3\sin t, 0)=\boldsymbol{a}$. **9.** $\boldsymbol{v}(t)=(2t,1)$; $\boldsymbol{a}(t)=(2,0)$; $\boldsymbol{a}(0)=2\boldsymbol{N}$; $\boldsymbol{a}(1)=\frac{4}{\sqrt{5}}\boldsymbol{T}+\frac{2}{\sqrt{5}}\boldsymbol{N}$ where $\boldsymbol{T}(1)=\frac{1}{\sqrt{5}}(2,1)$ and $\boldsymbol{N}(1)=\left(\frac{1}{\sqrt{5}}, -\frac{2}{\sqrt{5}}\right)$. **15.** 1.

REVIEW EXERCISES 2.4, p. 111

1. $\boldsymbol{x}=(0,1,\frac{1}{4}\pi)+t(-2,0,1)$; speed is $\sqrt{5}$. **3.** $\left(\frac{1}{2},\frac{1}{2},\frac{\pi^2}{32}\right)$. **5.** $\boldsymbol{x}(t)=\frac{1}{m}(\cos t, \sin t)+\frac{1}{m}\boldsymbol{c}_1t+\frac{1}{m}\boldsymbol{c}_2$. **9.** $K=\frac{4}{5}$; $\boldsymbol{T}=\frac{1}{\sqrt{5}}(2\cos 2t, 1, -2\sin 2t)$; $\boldsymbol{N}=(-\sin 2t, 0, -\cos 2t)$. **11.** (a) $\boldsymbol{v}(t)=(t\cos t+\sin t, -t\sin t+\cos t, 2t)$; $\boldsymbol{v}(\pi)=(-\pi,-1,2\pi)$; $\frac{ds}{dt}=\sqrt{1+5t^2}=\sqrt{1+5\pi^2}$ when $t=\pi$; $\boldsymbol{a}(t)=(-t\sin t+2\cos t, -t\cos t-2\sin t, 2)=(-2,\pi,2)$ at $t=\pi$. (b) $\pi\sqrt{1+20\pi^2}+\frac{1}{2\sqrt{5}}\ln(2\pi\sqrt{5}+\sqrt{1+20\pi^2})$. (c) $\boldsymbol{a_T}=\frac{(-5\pi^2,-5\pi,10\pi^2)}{1+5\pi^2}$; $\boldsymbol{a_N}=\frac{(-2-5\pi^2, 6\pi+5\pi^3, 2)}{1+5\pi^2}$.

CHAPTER 3

EXERCISES 3.1, p. 120

1. Closed. **3.** Open. **5.** Open. **7.** Closed. **9.** Closed. **11.** Open. **13.** Open. **15.** Closed. **17.** Closed. **19.** Closed. **21.** Open and closed. **23.** Closed. **27.** (b) $\left\{(x, y) \,\middle|\, \frac{x^2}{4}+\frac{y^2}{9}<1\right\} \cup \left\{(x, y) \,\middle|\, \frac{x^2}{4}+\frac{y^2}{9}>1\right\}$. **29.** Converges to $(0, 0)$. **31.** Converges to $(\frac{1}{2}\pi, 3)$. **33.** Converges to $(1, 0)$. **35.** Diverges. **37.** Converges to $(-1, 0, 0)$.

EXERCISES 3.2, p. 126

1. ln 2. **3.** No limit. **5.** Continuous on $\boldsymbol{R}^2$ except at $(0, 0)$. **7.** Continuous on $\boldsymbol{R}^2$. **9.** Continuous on $\boldsymbol{R}^2$ except at $(0, 0)$. **11.** Continuous on $\boldsymbol{R}^2-\{(0, 0)\}$. **17.** Yes.

EXERCISES 3.3, p. 137

1. Parabolic cylinder. **3.** Circular cylinder. **5.** Hyperbolic cylinder. **7.** Ellipsoid of revolution. **9.** $y^2+z^2=x^4$. **11.** $x^2+z^2=\frac{1}{16}y^4$. **13.** $\frac{x^2}{4}-\frac{y^2}{9}-\frac{z^2}{9}=1$. **15.** $-\frac{x^2}{9}+\frac{y^2}{16}-\frac{z^2}{9}=1$. **17.** $x^2+y^2+z^2=4$. **19.** $y^2+z^2=9x^2$. **21.** No. **23.** Yes, of the hyperbola $4x^2-3y^2=9$ in the xy-plane about the y-axis. **25.** Hyperbolas. **27.** Ellipses. **31.** Spheres. **33.** Ellipsoids of revolution.

EXERCISES 3.4, p. 145

1. Elliptical cylinder. **3.** Parabolic cylinder. **5.** Circular cone. **7.** Elliptical paraboloid. **9.** Hyperbolic paraboloid. **11.** Hyperbolic paraboloid. **13.** Ellipsoid of revolution. **15.** Hyperboloid of one sheet. **17.** Hyperboloid of two sheets.

REVIEW EXERCISES 3.5, p. 146

1. (a) Closed. (c) Neither. (e) Open. (g) Neither. **3.** $\boldsymbol{R}^2$ except for the line $y=-x$. **5.** $\boldsymbol{R}^2-\{(0, b) \mid b\neq 0\}$. **7.** Hyperbolic cylinder. **9.** Hyperboloid of revolution of one sheet. **11.** (a) $4x^2-9y^2+4z^2+36=0$. (b) $9x^2+9y^2-4z^2=36$. **13.** Ellipsoids. **15.** Parabolic cylinder. **17.** Right circular cone. **19.** Hyperbolic paraboloid. **21.** Hyperboloid of one sheet.

CHAPTER 4

EXERCISES 4.1, p. 153

1. $\frac{\partial f}{\partial x}=\frac{x}{\sqrt{x^2+y^2}}=\frac{1}{\sqrt{5}}$ at $(1, 2)$; $\frac{\partial f}{\partial y}=\frac{y}{\sqrt{x^2+y^2}}=\frac{2}{\sqrt{5}}$ at $(1, 2)$. **3.** $f_x=-6e^y \sin x-5ye^{xy} \ln x^2y^2-\frac{10e^{xy}}{x}=-6e-5e^{\pi/2} \ln \frac{\pi^2}{4}-\frac{20}{\pi}e^{\pi/2}$ at $(\frac{1}{2}\pi, 1)$; $f_y=6e^y \cos x-5xe^{xy} \ln x^2y^2-\frac{10e^{xy}}{y}=\frac{-5\pi}{2}e^{\pi/2} \ln \frac{\pi^2}{4}-10e^{\pi/2}$ at $(\frac{1}{2}\pi, 1)$. **5.** $f_x=\frac{x}{\sqrt{x^2+y^2+z^2}}=\frac{1}{\sqrt{3}}$ at $(1, 1, 1)$; $f_y=\frac{y}{\sqrt{x^2+y^2+z^2}}=\frac{1}{\sqrt{3}}$ at $(1, 1, 1)$; $f_z=\frac{z}{\sqrt{x^2+y^2+z^2}}=\frac{1}{\sqrt{3}}$ at $(1, 1, 1)$. **7.** $6x+10y-z=8$. **9.** $3x-2y+3z=0$. **11.** $\frac{1}{\sqrt{137}}(6, 10, -1)$. **13.** $\frac{1}{\sqrt{22}}(-3, 2, -3)$. **15.** $\frac{1}{\sqrt{6}}$. **17.** $\frac{\partial P}{\partial T}=\frac{P}{T}$; $\frac{\partial P}{\partial V}=-\frac{P}{V}$. P increases by about $\frac{P}{T}$ for each unit increase in T, decreases by about $\frac{P}{V}$ for each unit increase in V. **19.** $\frac{\partial d}{\partial p}=-10, \frac{\partial d}{\partial x}=5, \frac{\partial d}{\partial y}=1$. **21.** $\frac{\partial d_1}{\partial y}$ and $\frac{\partial d_2}{\partial x}$ should be positive: as the price of competition rises, demand for the competition should drop, increasing demand for the products whose price didn't rise.

EXERCISES 4.2, p. 160

1. Nonlinear. **3.** Linear. **5.** Linear. **7.** Linear. **9.** Nonlinear. **11.** Nonlinear. **13.** Linear. **15.** Linear. **17.** Exercises 4, 8, 11, 16.

EXERCISES 4.3, p. 171

1. $(4, 3)$. **3.** $(0, 2e)$. **5.** $\left(\frac{1}{\sqrt{2}}, \frac{1}{\sqrt{2}}\right)$ **7.** $e^2(\sqrt{3}, \sqrt{3}, \frac{1}{2})$. **9.** $\mathbf{f}'(2, 1)=(4, 2)$. **11.** 4.99. **13.** 32.00. **15.** 17 feet; 0.03 feet. **17.** 2.96. **19.** 0.64π (≈ 2.01) cubic inches. **25.** If $f(\mathbf{x})=\mathbf{m}\cdot\mathbf{x}$, then $\mathbf{f}'(\mathbf{x})=\mathbf{m}$. **27.** At all points except the origin. **29.** Such a limit is not unique, hence is ambiguous.

EXERCISES 4.4, p. 178

1. $3\sqrt{2}$. **3.** $-\frac{3}{5}$. **5.** $\frac{2}{3}$. **7.** 0. **9.** Maximum rate of increase is $2\sqrt{5}$ in direction $\left(\frac{1}{\sqrt{5}}, \frac{2}{\sqrt{5}}\right)$; maximum rate of decrease is $-2\sqrt{5}$ in direction $\left(-\frac{1}{\sqrt{5}}, -\frac{2}{\sqrt{5}}\right)$. **11.** Maximum rate of increase is $2\sqrt{6}$ in direction $\left(\frac{1}{\sqrt{6}}, \frac{2}{\sqrt{6}}, \frac{1}{\sqrt{6}}\right)$; maximum rate of decrease is $-2\sqrt{6}$ in direction $\left(-\frac{1}{\sqrt{6}}, -\frac{2}{\sqrt{6}}, -\frac{1}{\sqrt{6}}\right)$. **13.** $(-\frac{1}{3}, -\frac{2}{3}, -\frac{2}{3})$. **15.** Heat flows outward across the surface and at right angles to it. **17.** $\left(\frac{1}{\sqrt{2}}, \frac{1}{\sqrt{2}}\right)$.

EXERCISES 4.5, p. 187

1. -1. **3.** $\frac{6}{5}$. **5.** $\frac{2\pi}{4+\pi^2}$. **7.** $(1, \frac{1}{2})$. **9.** $\frac{\partial w}{\partial s}=2s$; $\frac{\partial w}{\partial t}=0$. **11.** $\frac{\partial w}{\partial s}= 2xyz^2t-\frac{x^2z^2t}{s^2}+\frac{2x^2yz}{t}$; $\frac{\partial w}{\partial t}=2xyz^2s+\frac{x^2z^2}{s}-\frac{2x^2yzs}{t^2}$. **13.** $\frac{\partial w}{\partial s}=\frac{st^2}{x^2+y^2+z^2}$; $\frac{\partial w}{\partial t}=\frac{s^2t}{1+s^2t^2}$. **15.** $(4xst-4syu, 2xs^2-2y, 2x-2ys^2)$; $\mathbf{0}$. **17.** $(y\cos\theta+x\sin\theta, -y^2+x^2)$; $(0, -1)$. **19.** $(tyze^s+xze^t+txy, yze^s+sxze^t+sxy)$; $e^2(3, 3)$. **21.** $\frac{1}{x^2+y^2+z^2}(x+ytu+ze^{tu}, x+ysu+zsue^{tu}, x+yst+zste^{tu})$; $(\frac{2}{5}, \frac{3}{10}, \frac{1}{2})$. **23.** $4x+y+2z=0$. **25.** $2x-3y-2z=5$. **27.** $3x-2y+2z=1$. **31.** $(1, 0, 0)$.

EXERCISES 4.6, p. 193

1. $\frac{dy}{dx}=-\frac{14xy^6-5y^5}{42x^2y^5-25xy^4+9y^2}$; valid where denominator is not zero; $y(1.05)\approx 0.98$.

3. $\frac{dy}{dx}=\frac{x\cos y-2x^2e^{xy}-x^3ye^{xy}+y^2-15x^5}{x^2\sin y+x^4e^{xy}-2xy\ln x}$; valid where denominator is not zero; $y(0.95)\approx 0.80$. **5.** When $z\neq 0$, $\frac{\partial z}{\partial x}=-\frac{x}{z}$ and $\frac{\partial z}{\partial y}=-\frac{y}{z}$; 2.00.

7. $\frac{\partial z}{\partial x}=\frac{2x\cos xyz-x^2yz\sin xyz+y^3z\cos xyz}{x^3y\sin xyz-xy^3\cos xyz+2x}$;

$\frac{\partial z}{\partial y}=\frac{-x^3z\sin xyz+2y\sin xyz+xy^2z\cos xyz}{x^3y\sin xyz-xy^3\cos xyz+2z}$; valid where denominator is not zero. **9.** $\frac{\partial z}{\partial x}=\frac{2x-z}{ye^z+x}$, $\frac{\partial z}{\partial y}=\frac{2y-e^z}{ye^z+x}$; valid where denominator is not zero.
11. $\frac{\partial z}{\partial x}=\frac{-3x^2e^{y+z}+y\cos(x-z)}{x^3e^{y+z}+y\cos(x-z)}$; $\frac{\partial z}{\partial y}=\frac{-x^3e^{y+z}+\sin(x-z)}{x^3e^{y+z}+y\cos(x-z)}$; valid where denominator is not zero. **13.** $3x+y+3z=1$. **15.** $4x+9y-2\sqrt{23}z=36$. **21.** $(10, 14, -3)$.

EXERCISES 4.7, p. 201

1. $\begin{pmatrix} 6 & -5 \\ -5 & -4 \end{pmatrix}$. **3.** $\begin{pmatrix} -2\sin(x+y)-x\cos(x+y) & -\sin(x+y)-x\cos(x+y) \\ -\sin(x+y)-x\cos(x+y) & -x\cos(x+y) \end{pmatrix}$. **5.** $f_x=2x\sin xy+x^2y\cos xy$, $f_y=x^3\cos xy$; $f_{xx}=2\sin xy+4xy\cos xy-x^2y^2\sin xy$, $f_{yy}=-x^4\sin xy$, $f_{yx}=f_{xy}=3x^2\cos xy-x^3y\sin xy$; $f_{xxx}=6y\cos xy-6xy^2\sin xy-x^2y^3\cos xy$, $f_{yxx}=f_{xyx}=f_{xxy}=6x\cos xy-6x^2y\sin xy-x^3y^2\cos xy$, $f_{xyy}=f_{yxy}=f_{yyx}=-4x^3\sin xy-x^4y\cos xy$, $f_{yyy}=-x^5\cos xy$. **7.** $f_x=\frac{2x}{x^2+y^2}$, $f_y=\frac{2y}{x^2+y^2}$; $f_{xx}=\frac{2y^2-2x^2}{(x^2+y^2)^2}$, $f_{xy}=f_{yx}=-\frac{4xy}{(x^2+y^2)^2}$, $f_{yy}=\frac{2x^2-2y^2}{(x^2+y^2)^2}$; $f_{xxx}=\frac{4x^3-12xy^2}{(x^2+y^2)^3}$, $f_{xxy}=f_{xyx}=f_{yxx}=\frac{12x^2y-4y^3}{(x^2+y^2)^3}$, $f_{yyx}=f_{yxy}=f_{xyy}=\frac{12xy^2-4x^3}{(x^2+y^2)^3}$, $f_{yyy}=\frac{4y^3-12x^2y}{(x^2+y^2)^3}$.

9. $\begin{pmatrix} 2 & 0 & 0 \\ 0 & -2 & 0 \\ 0 & 0 & 6 \end{pmatrix}$. **11.** $\begin{pmatrix} 2 & -8 & -5 \\ -8 & 4 & -4 \\ -5 & -4 & 10 \end{pmatrix}$.

EXERCISES 4.8, p. 210

1. $\begin{pmatrix} x-\ y \\ -x+2y \end{pmatrix}$. **3.** $\begin{pmatrix} x-x_0-4(y-y_0) \\ -4(x-x_0)+5(y-y_0) \end{pmatrix}$. **5.** $\begin{pmatrix} 12 \\ -8 \end{pmatrix}$. **7.** $\begin{pmatrix} x-\ y+3z \\ -x-2y+\ z \\ 3x+\ y+2z \end{pmatrix}$.
9. $\begin{pmatrix} x-x_0-(y-y_0)+2(z-z_0) \\ -(x-x_0)-3(z-z_0) \\ 2(x-x_0)-3(y-y_0)+z-z_0 \end{pmatrix}$. **11.** $\begin{pmatrix} -4 \\ 10 \\ 5 \end{pmatrix}$. **13.** xy. **15.** No second degree Taylor polynomial. **17.** $2+(x-1)+3(y-1)+(x-1)(y-1)+(y-1)^2$. **19.** $1+\frac{1}{2}(x^2+y^2)$. **21.** y^2. **23.** x. **25.** $1-\frac{1}{2}\left(x-\frac{\pi}{6}\right)^2-\frac{1}{2}\left(y-\frac{\pi}{6}\right)^2-\frac{1}{2}\left(z-\frac{\pi}{6}\right)^2-\left(x-\frac{\pi}{6}\right)\times\left(y-\frac{\pi}{6}\right)-\left(x-\frac{\pi}{6}\right)\left(z-\frac{\pi}{6}\right)-\left(y-\frac{\pi}{6}\right)\left(z-\frac{\pi}{6}\right)$. **27.** No second degree Taylor polynomial. **29.** 4.99004 (=4.99 to two decimal places). **31.** 17.00; 0.0312 (=0.03 to two decimal places). **33.** 0.6456π (≈ 2.0282).

EXERCISES 4.9, p. 218

1. Absolute minimum at (0, 0). **3.** Saddle point at (1, 2). **5.** Saddle point at (0, 3). Local minima at (2, 3) and (−2, 3). **7.** Local minimum at (0, 0, 0). **9.** Local minimum at $(-\frac{1}{2}, -1, \frac{3}{2})$. **11.** 2 by 2 by 1; volume 4. **13.** 2 by 2 by 1. **17.** (a) $P=2p(7-p+q)+q(16+2p-6q)-6(7-p+q)-4(8+p-3q)$; (b) $p=\$8.75$, $q=\$4.75$. **19.** $z=\sum_{i=1}^{n}(mx_i+b-y_i)^2$. **21.** $y=\frac{62}{35}x+\frac{18}{35}$. **23.** (0, 0) is a stable equilibrium point. (Note: p is not differentiable at (0, 0); an extreme value can occur, as in elementary calculus, at a point where p is not differentiable.) **27.** If $D=0$ and $f_{yy}(\mathbf{x}_0)>0$, then $f(\mathbf{x}_0)$ is a local minimum of $z=f(x_0, y)$.

EXERCISES 4.10, p. 228

1. Maximum is $\sqrt{5}$; minimum is $-\sqrt{5}$. **3.** Maximum is $\frac{1}{2}$; minimum is $-\frac{1}{2}$. **5.** Maximum is $\sqrt{3}$; minimum is $-\sqrt{3}$. **7.** 2 by 2 by 1; volume 4. **9.** 2 by 2 by 1. **11.** Maximum is 6; minimum is 2. **13.** $\dfrac{1}{\sqrt{2}}$ at $(1, \frac{1}{2}, \frac{3}{2})$. **15.** $\left(\dfrac{-2}{\sqrt{3}}, \dfrac{2}{\sqrt{3}}, \dfrac{-2}{\sqrt{3}}\right)$. **17.** $\dfrac{16}{\sqrt{3}}$. **19.** Maximum is 3; minimum is 1. **21.** A square of side 1. **23.** $x = 225$; $y = \frac{225}{6} = 37.5$ units of capital.

REVIEW EXERCISES 4.11, p. 229

1. $\left(\dfrac{2}{5}, \dfrac{\sqrt{5}}{5}, \dfrac{4}{5}\right)$; $2x + \sqrt{5}y + 4z = 25$. **3.** (a) Nonlinear. (c) Nonlinear (affine). (e) Nonlinear. (g) Nonlinear (affine). **5.** $(\frac{9}{2}\pi, 6, \frac{1}{2}\pi)$; all entries of ∇f are continuous at $(1, \frac{1}{2}\pi, 3)$. **7.** $\frac{160}{3}$; $\dfrac{1}{\sqrt{61}}(6, 3, 4)$; $8\sqrt{61}$. **9.** $\frac{1}{3}(5 - 2\sqrt{7}, 1 + 2\sqrt{7}, 2 + \sqrt{7})$. **11.** $-8x + y + 20z = 16$. **13.** $\dfrac{dy}{dx} = \dfrac{12x^3 - 2xy^3}{3x^2y^2 + 6y^2}$ if $y \neq 0$; 0.98. **15.** $\begin{pmatrix} 2y^3 + 8y^2 - 2 & 6xy^2 + 16xy + 1 \\ 6xy^2 + 16xy + 1 & 6x^2y + 8x^2 \end{pmatrix}$. **17.** (a) Saddle points at (1, 1) and (1, −1); local maximum at (0, 0); local minimum at (2, 0). (b) Absolute minimum at (0, 0). **19.** Maximum $2\sqrt{37}$; minimum $-2\sqrt{37}$.

CHAPTER 5

EXERCISES 5.1, p. 238

1. $1 - e$. **3.** 1. **5.** $1 - \dfrac{1}{\sqrt{2}}$. **7.** −3. **9.** $\frac{1}{2}$. **11.** $\dfrac{\pi}{12}$.

EXERCISES 5.2, p. 245

1. $\dfrac{-27}{2}$. **3.** $\ln\frac{4}{3}$. **5.** 154. **7.** 2. **9.** 2. **11.** $\ln\frac{27}{16} = 3\ln 3 - 4\ln 2$. **13.** $\frac{44}{3}$. **15.** $\frac{20}{3}$. **17.** $\frac{1}{2}e^2 - e + \frac{1}{2}$. **21.** 2.140. **23.** $\frac{7}{6}$.

EXERCISES 5.3, p. 254

1. $\int_0^2 x\,dx \int_0^{2x} y\,dy = 8$. **3.** $\int_0^2 dx \int_0^{4-x^2} (4 - x^2 - y)\,dy = \frac{128}{15}$. **5.** $\int_0^1 dx \int_{1-x}^{1+x} (3x + 2y)\,dy = 4$. **7.** $\int_{1/2}^1 dx \int_0^x e^{-x-y}\,dy = \frac{1}{2}e^{-2} - \frac{3}{2}e^{-1} + e^{-1/2}$. **9.** $\int_{-1}^1 dx \int_{-1}^1 (1 - x^2 - y^2)\,dy = \frac{4}{3}$. **11.** $\int_0^3 dx \int_{2x^2/3}^{2\sqrt{3x}} dy = 6$. **13.** $\int_0^1 dx \int_{x^4}^{x^2} dy = \frac{2}{15}$. **15.** $\int_0^1 dx \int_0^{e^x} dy = e - 1$. **17.** $\int_{\frac{-1-\sqrt{5}}{2}}^{\frac{-1+\sqrt{5}}{2}} dy \int_y^{1-y^2} dx = \dfrac{5\sqrt{5}}{6}$. **19.** $\int_0^2 dx \int_0^1 (5 - 2x - y)\,dy = 5$. **21.** $\int_0^1 dx \int_0^{2x} (x^2 + y^2)\,dy = \frac{7}{6}$. **23.** $\int_0^2 dx \int_{x^2}^{2x} xy\,dy = \frac{8}{3}$. **25.** $\int_0^3 dx \int_0^{\sqrt{9-x^2}} \sqrt{9 - x^2}\,dy = 18$. **27.** $\int_0^3 dy \int_0^y \sqrt{9 - y^2}\,dx = 9$. **29.** $\int_0^4 dy \int_{\sqrt{y}}^2 f(x, y)\,dx$. **31.** $\int_1^e dx \int_0^{\ln x} f(x, y)\,dy$. **33.** $\int_0^1 dy \int_{-\sqrt{y}}^{\sqrt{y}} f(x, y)\,dx + \int_1^4 dy \int_{y-2}^{\sqrt{y}} f(x, y)\,dx$.

EXERCISES 5.4, p. 264

3. $r=3$. **5.** $r^2=4\sec 2\theta$. **11.** Parabola $\frac{1}{4}x^2=y+1$.

13. Hyperbola $225\left(x-\frac{8}{15}\right)^2-15y^2=4$. **17.** $r=\dfrac{4}{2-\cos\theta}$. **19.** $r=\dfrac{12}{1-3\cos\theta}$.

21. $\frac{8}{3}[(\pi^2+1)^{3/2}-1]$. **23.** $\dfrac{3\pi}{2}=\displaystyle\int_0^{3\pi}\sin^2\frac{\theta}{3}\,d\theta$.

25. $1+\dfrac{3\pi}{8}$. **27.** $\displaystyle\int_0^{\pi/3}2\sin^2 3\theta\,d\theta=\frac{\pi}{3}$.

29. $\displaystyle\int_{\pi/6}^{\pi/2}9\sin^2\theta\,d\theta-\int_{\pi/6}^{5\pi/6}(4-4\sin\theta+\sin^2\theta)\,d\theta=3\sqrt{3}$.

EXERCISES 5.5, p. 270

1. $\displaystyle\int_0^{2\pi}d\theta\int_0^{\sqrt{5}}r\,dr=5\pi$. **3.** $\displaystyle\int_0^{\pi/3}d\theta\int_0^{2\sin 3\theta}r\,dr=\frac{\pi}{3}$.

5. $\displaystyle 2\int_0^{\pi}d\theta\int_0^{1+\cos\theta}r\,dr=\frac{3\pi}{2}$. **7.** $\displaystyle 2\int_0^{\pi/6}d\theta\int_0^{4\sin\theta}r\,dr+2\int_{\pi/6}^{\pi/2}d\theta\int_0^{2}r\,dr=\frac{8\pi}{3}-2\sqrt{3}$.

9. $\displaystyle\int_0^{2\pi}d\theta\int_0^{1}r^2\,dr=\frac{2\pi}{3}$.

11. $\displaystyle\int_0^{2\pi}d\theta\int_1^{2}r^3\,dr=\frac{15\pi}{2}$. **13.** $\displaystyle 4\int_0^{2\pi}d\theta\int_0^{1}(r-r^3)\,dr=2\pi$.

15. $\displaystyle\frac{4}{3}\pi a^3-2\int_0^{2\pi}d\theta\int_b^{a}\sqrt{a^2-r^2}\,r\,dr=\frac{4}{3}\pi a^3-\frac{4}{3}\pi(a^2-b^2)^{3/2}$.

17. $\displaystyle\int_0^{\pi/4}d\theta\int_1^{2\sec\theta}dr=2\ln(\sqrt{2}+1)-\frac{1}{4}\pi$.

19. $\displaystyle\int_0^{\pi/2}d\theta\int_0^{2\cos\theta}r^2\,dr=\frac{16}{9}$. **21.** $\displaystyle\int_0^{2\pi}\cos^2\theta\sin\theta\,d\theta\int_0^{1}r^4\,dr=0$.

23. $\displaystyle\int_0^{2\pi}(\sin\theta\cos\theta-1)\,d\theta\int_1^{3}r^3\,dr=-40\pi$.

EXERCISES 5.6, p. 277

1. $\displaystyle\int_{-1}^{1}x\,dx\int_1^{2}dy\int_0^{3}dz=0$. **3.** $\displaystyle\int_0^{1}dx\int_0^{1-x}dy\int_0^{1-x^2}(1-z)\,dz=\frac{7}{30}$.

5. $\displaystyle 24\int_0^{1}dx\int_{x^2}^{1}y\,dy\int_0^{1-y^2}z\,dz=\frac{256}{195}$.

7. $\displaystyle\int_0^{1}dx\int_0^{1-x}dy\int_0^{1-x-y}(xy-z)\,dz=-\frac{1}{30}$.

9. $\displaystyle\int_0^{1}dx\int_{x^2}^{x}dy\int_0^{x}(x+y)\,dz=\frac{11}{120}$. **11.** $\displaystyle\int_1^{2}x^2\,dx\int_0^{x^2}y\,dy\int_0^{1/x}z\,dz=\frac{31}{20}$.

13. $\displaystyle 4\int_0^{1}z^2\,dz\int_0^{\sqrt{1-z}}dx\int_0^{\sqrt{1-z}}dy=\frac{1}{3}$.

15. $\displaystyle\int_0^{a}dx\int_0^{b(1-x/a)}dy\int_0^{c(1-x/a-y/b)}dz=\frac{abc}{6}$.

17. $\displaystyle\int_0^{2}dx\int_0^{4-x^2}dy\int_0^{x}dz=4$. **19.** $\displaystyle 4\int_0^{2}dx\int_0^{\sqrt{2-(x^2/2)}}dy\int_{x^2+3y^2}^{8-x^2-y^2}dz=8\pi\sqrt{2}$.

EXERCISES 5.7, p. 285

1. (a) $[\sqrt{5},\operatorname{Tan}^{-1}2,3]$. (c) $[\sqrt{13},\operatorname{Tan}^{-1}\frac{3}{2},-1]$. (e) $[\sqrt{13},\operatorname{Tan}^{-1}(-\frac{3}{2}),-1]$. **3.** (a) $(-1,\sqrt{3},1)$. (c) $(-1,0,2)$. (e) $(1,-\sqrt{3},-1)$. **5.** (a) $r=4$. (c) $r^2+9z^2=9$. **7.** (a) $x^2+y^2=9$. (c) $y=2$. (e) $4x^2+4y^2=z^2$, where $z\geq 0$ if we require $r\geq 0$.

9. $\int_0^{2\pi} d\theta \int_0^a r\,dr \int_0^{h(a-r)/a} dz = \frac{1}{3}\pi a^2 h.$

11. $\int_0^{2\pi} d\theta \int_0^2 r\,dr \int_0^{4-r^2} dz = 8\pi.$ **13.** $\int_0^{2\pi} d\theta \int_0^3 r\,dr \int_{3-r}^{\sqrt{9-r^2}} dz = 9\pi.$

15. $\int_0^{2\pi} d\theta \int_0^1 r\,dr \int_0^{2r^2} dz = \pi.$

17. $\int_0^{\pi} d\theta \int_0^2 r^2\,dr \int_0^{1-1/2r} z\,dz = \dfrac{2\pi}{15}.$ **19.** $\int_0^{\pi/2} \sin\theta \cos\theta\,d\theta \int_0^a r^3\,dr \int_0^4 z\,dz = a^4.$

21. $\int_0^{\pi/2} d\theta \int_0^1 r\,dr \int_{-r}^{r} z^4\,dz = \dfrac{\pi}{35}.$

EXERCISES 5.8, p. 295

1. (a) $\left\{\sqrt{14}, \operatorname{Cos}^{-1}\dfrac{3}{\sqrt{14}}, \operatorname{Tan}^{-1} 2\right\}$. (c) $\left\{2\sqrt{2}, \dfrac{3\pi}{4}, \dfrac{1}{2}\pi\right\}$.
(e) $\left\{\sqrt{14}, \operatorname{Cos}^{-1}\left(\dfrac{-1}{\sqrt{14}}\right), \pi - \operatorname{Tan}^{-1}(-\frac{3}{2})\right\}$. **3.** (a) $\left(\dfrac{3\sqrt{2}}{4}, \dfrac{3\sqrt{2}}{4}, \dfrac{3\sqrt{3}}{2}\right)$.
(c) $\left(\dfrac{\sqrt{3}}{2}, \dfrac{1}{\sqrt{2}}, -\sqrt{2}\right)$. (e) $(0, 0, -2)$. **5.** (a) $\rho = 3$. (b) $\rho = 6\cos\phi$.
(c) $\rho = 4\csc\phi$. **7.** (a) $x^2+y^2+z^2 = 25$. (c) $x^2+y^2+(z-4)^2 = 16$.

9. $\frac{1}{3}\pi a^2 h + \int_0^{2\pi} d\theta \int_{\operatorname{Tan}^{-1} a/h}^{\pi/2} \sin\phi\,d\phi \int_0^{a\csc\phi} \rho^2\,d\rho = \pi a^2 h.$

11. $\int_0^{2\pi} d\theta \int_0^{\pi/6} \sin\phi\,d\phi \int_0^1 \rho^2\,d\rho = \dfrac{2\pi}{3}\left(1 - \dfrac{\sqrt{3}}{2}\right).$

13. $\int_0^{2\pi} d\theta \int_0^{\pi} \sin\phi\,d\phi \int_1^2 \rho^2\,d\rho = \dfrac{28\pi}{3}.$

15. $\int_0^{2\pi} d\theta \int_0^{\pi} \sin\phi\,d\phi \int_0^{1-\cos\phi} \rho^2\,d\rho = \dfrac{8\pi}{3}.$

17. $\int_0^{2\pi} d\theta \int_0^{\pi} \sin\phi\,d\phi \int_0^1 \rho^3\,d\rho = \pi.$

19. $\int_0^{2\pi} d\theta \int_0^{\pi} \sin\phi\,d\phi \int_1^2 d\rho = 4\pi.$

21. $\int_0^{\pi/2} d\theta \int_0^{\pi/2} \sin^3\phi\,d\phi \int_0^1 \rho^4\,d\rho = \dfrac{\pi}{15}.$

25. $\frac{1}{2}\pi\rho_0$, where ρ_0 is the radius of the earth.

27. $\frac{1}{2}\int_0^1 t\sqrt{16+t^2}\,dt = \frac{1}{6}(17\sqrt{17} - 64)$ **29.** $\frac{1}{2}\pi^2 a^2$.

EXERCISES 5.9, p. 306

1. $M = \frac{1}{2}acd$; $\bar{x} = \frac{1}{3}(a+b)$; $\bar{y} = \frac{1}{3}c$. **3.** $M = \int_1^2 x\,dx \int_1^3 y\,dy = 6$,
$\bar{x} = \frac{1}{6}\int_1^2 x^2\,dx \int_1^3 y\,dy = \frac{14}{9}$, $\bar{y} = \frac{1}{6}\int_1^2 x\,dx \int_1^3 y^2\,dy = \frac{13}{6}$.
5. $M = k\int_0^1 dx \int_{x^2}^{\sqrt{x}} y\,dy = \dfrac{3k}{20}$; $\bar{x} = \dfrac{20}{3}\int_0^1 x\,dx \int_{x^2}^{\sqrt{x}} y\,dy = \dfrac{5}{9}$;
$\bar{y} = \dfrac{20}{3}\int_0^1 dx \int_{x^2}^{\sqrt{x}} y^2\,dy = \dfrac{4}{7}$. **7.** $M = \frac{3}{4}\pi d$;
$\bar{x} = \dfrac{4}{3\pi}\int_0^{\pi/2} \cos\theta\,d\theta \int_1^2 r^2\,dr = \dfrac{28}{9\pi}$; $\bar{y} = \dfrac{4}{3\pi}\int_0^{\pi/2} \sin\theta \int_1^2 r^2\,dr = \dfrac{28}{9\pi}$.
9. $72\pi^2$. **11.** $\frac{1}{3}\pi ac(b+4a)$. **13.** $M = \frac{1}{2}$; $\bar{\mathbf{x}} = (\frac{1}{2}, \frac{1}{2}, \frac{2}{3})$.

15. $M=\frac{16}{3}\pi$; $\bar{\mathbf{x}}=(0,0,\frac{3}{4})$. **17.** $M=4\pi d$; $\bar{x}=\bar{y}=0$,
$\bar{z}=\dfrac{1}{4\pi d}\int_0^{2\pi} d\theta \int_0^{\mathrm{Tan}^{-1}2/3} \sin\phi\cos\phi\, d\phi \int_0^{3\sec\phi} \rho^3\, d\rho=\frac{9}{4}$.
19. $R=\sqrt{\frac{2}{3}}$; $I_x=2\delta$.
21. $I_z=k\int_0^{2\pi} d\theta \int_0^{\pi} \sin^3\phi\, d\phi \int_0^{3} \rho^5\, d\rho=324\pi k$; $R=2$.
23. $K=648\pi k$; $R=2$.
25. $K=\frac{1}{2}\omega^2 d\int_0^{2\pi} d\theta \int_0^{6} r^3\, dr \int_0^{h} dz=324\pi\omega^2 dh$, where $\omega=33\frac{1}{3}$ r.p.m., $h=$ thickness. **27.** $\left(\dfrac{2a}{\pi},\dfrac{2a}{\pi}\right)$. **29.** (a) $4\pi a^2$; (b) $4\pi^2 ab$.

EXERCISES 5.10, p. 313

9. $\dfrac{1}{y}[\sin 2\pi y-\sin \pi y]$.
11. $(x+1)^{-2}[(x+1)2^{x+1}\ln 2-2^{x+1}+1]$.
13. $\dfrac{1}{x}(\sin x-3\sin x^3)$
15. $\dfrac{1}{y}(2\sin y^2-3\sin y^2)$. **17.** $\dfrac{1}{x}(3e^{x^3}-2e^{x^2})$.

EXERCISES 5.11, p. 323

1. Convergent to $\frac{1}{2}\ln 2$. **3.** Convergent to $3(1+\sqrt[3]{2})$. **5.** Divergent. **7.** Convergent to 1. **9.** Divergent. **11.** Convergent to 1. **13.** Convergent to 2π. **15.** Divergent. **17.** Divergent. **19.** Divergent. **21.** Convergent to 0.

REVIEW EXERCISES 5.12, p. 326

1. $\frac{1}{2}(e^2-1)$. **3.** $\int_{-1}^{1} dx \int_{x^2}^{1} (x+y^2)\, dy=\frac{4}{7}$.
5. $\int_{\pi/6}^{\pi/2} d\theta \int_0^{2} r^4\, dr=\dfrac{32\pi}{15}$. **7.** $\int_0^{1} x\, dx \int_{x^2}^{1} y\, dy \int_0^{1-y^2} dz=\frac{1}{15}$.
9. $\int_0^{\pi/2} \sin\theta\cos\theta\, d\theta \int_0^{a} r^3\, dr \int_0^{3} z\, dz=\frac{9}{16}a^4$.
11. $\int_0^{2\pi} d\theta \int_0^{\pi} \sin\phi\, d\phi \int_0^{2} \rho^3\, d\rho=16\pi$. **13.** $\int_0^{4} dy \int_0^{y} \sqrt{16-y^2}\, dx=\frac{64}{3}$.
15. $4\int_0^{2\pi} d\theta \int_0^{1} (r-r^3)\, dr=2\pi$.
17. $\frac{8}{3}\sqrt{5}\pi+\int_0^{2\pi} d\theta \int_0^{\mathrm{Cos}^{-1}\sqrt{5}/3} \sin\phi\, d\phi \int_0^{3} \rho^2\, d\rho=\frac{8}{3}\sqrt{5}\pi+18\pi\left(1-\dfrac{\sqrt{5}}{3}\right)$.
19. $\int_1^{e} dx \int_0^{\ln x} f(x,y)\, dy$. **21.** $\int_{-1}^{0} dx \int_0^{e^x} dy=1-\dfrac{1}{e}$.
23. $M=\int_0^{3} x\, dx \int_0^{4-4x/3} y\, dy=6$; $\bar{x}=\dfrac{1}{6}\int_0^{3} x^2\, dx \int_0^{4-4x/3} y\, dy=\frac{6}{5}$;
$\bar{y}=\dfrac{1}{6}\int_0^{3} x\, dx \int_0^{4-4x/3} y^2\, dy=\frac{8}{5}$.
25. (a) $M=\int_0^{2\pi} d\theta \int_0^{\pi/2} \sin\phi\, d\phi \int_0^{4} \rho^3\, d\rho=128\pi$; $\bar{x}=\bar{y}=0$;
$\bar{z}=\dfrac{1}{128\pi}\int_0^{\pi/2} d\theta \int_0^{2\pi} \sin\phi\cos\phi\, d\phi \int_0^{4} \rho^4\, d\rho=\dfrac{8}{5}$. (b) $I_z=\dfrac{\pi 4^7}{9}$;

$R=\frac{8\sqrt{2}}{3}$. (c) $K=\frac{1}{2}\pi 4^7$. **27.** (a) $\frac{3}{y}\sin y^3-\frac{\sin y}{y}$.

(b) $\frac{4}{x}\sin x^4-\frac{\sin xe^{x^3}}{x}-3x^2\sin xe^{x^3}$. **29.** $\frac{1}{4}\pi$.

31. $\frac{1}{4}\pi$ linear units, about π linear miles of water.

CHAPTER 6

EXERCISES 6.1, p. 333

1. $(3, -\frac{1}{4}\pi, \frac{1}{2})$. **3.** $(8, 2e^4, 8, 0)$. **5.** No limit. **7.** No limit. **13.** Linear. **15.** Not linear. **17.** Not linear. **19.** Linear.

EXERCISES 6.2, p. 340

1. $\begin{pmatrix} 2 & 0 & 0 \\ 0 & 3 & 0 \\ 0 & 0 & -1 \end{pmatrix}$ **3.** $\begin{pmatrix} -2 & 1 & -5 \\ \frac{1}{2} & -\frac{1}{4} & 1 \\ 3 & -1 & -1 \end{pmatrix}$

5. $\begin{pmatrix} 3 & -1 & 0 \\ 2 & 0 & 3 \end{pmatrix}$ **7.** $\begin{pmatrix} -1 & 1 \\ 0 & 2 \\ 3 & 1 \end{pmatrix}$ **9.** $\begin{pmatrix} 2 & -3 & 1 & 1 \\ 1 & 1 & -2 & -5 \\ \frac{1}{2} & 0 & -\frac{3}{2} & \frac{1}{3} \end{pmatrix}$

11. $\begin{pmatrix} 1 & -1 & 2 \\ -1 & 1 & 1 \\ 2 & 0 & 1 \end{pmatrix}$ **13.** $\begin{pmatrix} 1 & -1 & -1 \\ 3 & 1 & 1 \end{pmatrix}$ **15.** $\begin{pmatrix} 1 & -2 \\ -\frac{1}{2} & \frac{1}{4} \end{pmatrix}$ **17.** $\begin{pmatrix} 1 & 0 \\ -1 & -3 \\ 0 & 0 \end{pmatrix}$

19. $\begin{pmatrix} 2 & -\frac{1}{2} & 0 \\ -1 & 0 & \frac{1}{3} \\ 1 & 1 & -\frac{1}{2} \\ 3 & 1 & 6 \end{pmatrix}$ **21.** $\begin{pmatrix} -18 \\ \frac{15}{4} \\ 1 \end{pmatrix}$

23. $\begin{pmatrix} \frac{7}{2} \\ -\frac{1}{2} \\ 5 \end{pmatrix}$ **25.** $\begin{pmatrix} -3 \\ -2 \\ 5 \end{pmatrix}$ **27.** $\begin{pmatrix} \frac{1}{2} \\ -\frac{3}{2} \\ 0 \end{pmatrix}$

29. $(-2, -9, -19/6)^t$. **31.** Grower #2 should switch.

33. $\frac{dx}{dt}=\begin{pmatrix} 0 & 1 & -1 \\ 1 & 1 & 0 \\ 1 & 0 & 2 \end{pmatrix}\begin{pmatrix} x_1 \\ x_2 \\ x_3 \end{pmatrix}$ **35.** 16,000 of Radio A, 19,500 of Radio B.

EXERCISES 6.3, p. 348

1. (a) $\begin{pmatrix} 2 & 7 & -5 \\ 11 & -2 & -1 \end{pmatrix}$ (b), (c). Not defined. **3.** (a) (b) Not defined. (c) $\begin{pmatrix} 2 & 1 & -1 \\ 1 & 6 & 8 \end{pmatrix}$ **5.** (a) $\begin{pmatrix} 4 & -1 & -2 \\ 3 & 4 & 0 \\ -4 & 5 & 15 \end{pmatrix}$

(b) $\begin{pmatrix} 5 & -3 & 2 \\ -7 & 3 & -1 \\ 3 & -2 & -9 \end{pmatrix}$ (c) $\begin{pmatrix} 3 & -2 & 3 \\ -7 & 5 & 0 \\ 5 & -7 & -9 \end{pmatrix}$

7. $\begin{pmatrix} 8 & -5 & 8 \\ 5 & -7 & 0 \end{pmatrix}$ **9.** $\begin{pmatrix} -7 & 9 \\ 0 & -1 \end{pmatrix}$

11. $\begin{pmatrix} 2 & 1 & -3 & -1 \\ 6 & -4 & 12 & 4 \\ -2 & 3 & -9 & -3 \end{pmatrix}$ **13.** $\begin{pmatrix} 55 \\ -18 \end{pmatrix}$ **15.** $\begin{pmatrix} 22 \\ -9 \end{pmatrix}$

17. $(AB)C = A(BC) = \begin{pmatrix} 11 & 8 \\ 6 & -5 \end{pmatrix}$ **21.** (a) $\begin{pmatrix} c & 0 & 0 \\ 0 & c & 0 \\ 0 & 0 & c \end{pmatrix}$

EXERCISES 6.4, p. 358

1. $\begin{pmatrix} 2 & 3 \\ 1 & 2 \end{pmatrix}$ **3.** $\frac{1}{4}\begin{pmatrix} 1 & -1 \\ 1 & 3 \end{pmatrix}$ **5.** No inverse. **7.** $\begin{pmatrix} -40 & 16 & 9 \\ 13 & -5 & -3 \\ 5 & -2 & -1 \end{pmatrix}$

9. $\frac{1}{5}\begin{pmatrix} -1 & 2 & -1 \\ 4 & -3 & -1 \\ 6 & -2 & 1 \end{pmatrix}$ **11.** No inverse. **13.** $\frac{1}{5}\begin{pmatrix} 4 & 3 & 2 & 1 \\ 3 & 6 & 4 & 2 \\ 2 & 4 & 6 & 3 \\ 1 & 2 & 3 & 4 \end{pmatrix}$

17. See Exercise 3. **19.** See Exercise 9. **21.** It is invertible if and only if $d_1 \neq 0$ and $d_2 \neq 0$ and $d_3 \neq 0$.

23. $A^{-1} = \begin{pmatrix} 1 & 0 & 0 \\ 0 & \cos\theta & \sin\theta \\ 0 & -\sin\theta & \cos\theta \end{pmatrix}$ **24.** *Hint:* $\mathbf{x} = A^{-1}\mathbf{b}$ is the unique solution if A is invertible.

27. $\begin{pmatrix} 1 \\ 3 \\ 6 \end{pmatrix}$

EXERCISES 6.5, p. 364

1. $J_{\mathbf{F}}(1, 2) = \begin{pmatrix} 2 & 4 \\ 2 & -4 \end{pmatrix}$. **3.** $J_{\mathbf{F}}(1, 2) = \begin{pmatrix} 1 & 1 \\ 1 & -1 \\ 2 & 1 \end{pmatrix}$

5. $J_{\mathbf{F}}(1, 0, 1) = \begin{pmatrix} 1 & 0 & 1 \\ 0 & 3 & 0 \end{pmatrix}$.

7. $J_{\mathbf{F}}(1, 1, 0) = \begin{pmatrix} 1 & 0 & 0 \\ \frac{1}{\sqrt{5}} & \frac{4}{\sqrt{5}} & 0 \\ \pi & \pi & 0 \end{pmatrix}$.

9. $J_{\mathbf{F}}(1, -1, \frac{1}{2}) = \begin{pmatrix} -2 & 1 & -1 \\ 2 & 0 & -1 \\ -\frac{\pi}{2} & \frac{\pi}{2} & -\pi \end{pmatrix}$ **11.** $\mathbf{L}' = M_{\mathbf{L}}$.

19. (0.64, 0.20, −3.04).

EXERCISES 6.6, p. 371

1. $J_{\mathbf{F}\circ\mathbf{G}}(1, 1) = \begin{pmatrix} 4 & -4 \\ 4 & 0 \end{pmatrix}$. **3.** $\begin{pmatrix} -1 \\ -1 \end{pmatrix}$. **5.** $\begin{pmatrix} 4 & 0 \\ 4 & -4 \\ -2 & -6 \end{pmatrix}$.

7. $\begin{pmatrix} -5 & -2 & 0 \\ 1 & 1 & 0 \end{pmatrix}$ **9.** $\begin{pmatrix} -3 & 0 \\ 4 & 2 \end{pmatrix}$.

11. $\begin{pmatrix} 2 & 0 & -2 \\ 4 & 0 & -2 \\ 2 & 0 & 0 \end{pmatrix}$. **13.** $p \approx 10.20$.

15. $\dfrac{\partial w}{\partial x} = 2x(u+v) = 2xz, \quad \dfrac{\partial z}{\partial y} = 0.$

17. $\dfrac{\partial p}{\partial x}=2x+2y,\quad \dfrac{\partial q}{\partial y}=2xu-2yv,\quad \dfrac{\partial r}{\partial x}=2x-4y.$

19. $\dfrac{\partial u}{\partial r}=e^{x+y^2}+6r^2ye^{x+y^2}=u+6r^2yu,$

$\dfrac{\partial v}{\partial s}=4xse^{x^2+y}+e^{x^2+y}=(4xs+1)v,\quad \dfrac{\partial u}{\partial t}=2tu+4tyu.$

21. $\dfrac{\partial w}{\partial s}=\dfrac{\partial w}{\partial x}\dfrac{\partial x}{\partial s}+\dfrac{\partial w}{\partial y}\dfrac{\partial y}{\partial s}+\dfrac{\partial w}{\partial z}\dfrac{\partial z}{\partial s},\quad \dfrac{\partial w}{\partial t}=\dfrac{\partial w}{\partial x}\dfrac{\partial x}{\partial t}+\dfrac{\partial w}{\partial y}\dfrac{\partial y}{\partial t}+\dfrac{\partial w}{\partial z}\dfrac{\partial z}{\partial t}.$

23. $\dfrac{\partial g}{\partial r}=\cos\theta\dfrac{\partial f}{\partial x}+\sin\theta\dfrac{\partial f}{\partial y},\quad \dfrac{\partial g}{\partial \theta}=-r\sin\theta\dfrac{\partial f}{\partial x}+r\cos\theta\dfrac{\partial g}{\partial y},\quad \dfrac{\partial g}{\partial z}=\dfrac{\partial f}{\partial z}.$

EXERCISES 6.7, p. 380

1. $\dfrac{\partial u}{\partial x}=1,\quad \dfrac{\partial u}{\partial y}=\dfrac{3}{4},\quad \dfrac{\partial v}{\partial x}=0,\quad \dfrac{\partial v}{\partial y}=-\dfrac{5}{4}.$

3. $\dfrac{\partial u}{\partial x}=-8,\quad \dfrac{\partial u}{\partial y}=-14,\quad \dfrac{\partial u}{\partial z}=-25,\dfrac{\partial v}{\partial x}=-5,\quad \dfrac{\partial v}{\partial y}=-8,\quad \dfrac{\partial v}{\partial z}=-16.$

5. $\dfrac{\partial u}{\partial x}=\dfrac{13}{32},\quad \dfrac{\partial u}{\partial y}=\dfrac{5}{32},\quad \dfrac{\partial v}{\partial x}=\dfrac{7}{16},\quad \dfrac{\partial v}{\partial y}=-\dfrac{1}{16}.$

7. $\dfrac{\partial u}{\partial x}=-\dfrac{1}{2},\quad \dfrac{\partial u}{\partial y}=\dfrac{1}{2},\quad \dfrac{\partial u}{\partial z}=-\dfrac{1}{2}.$

9. $\dfrac{\partial u}{\partial y}=4,\quad \dfrac{\partial v}{\partial x}=3,\quad \dfrac{\partial v}{\partial y}=-5,\quad \dfrac{\partial w}{\partial y}=-1.$

11. $u_x=2,\quad u_y=1,\quad v_x=-2,\quad v_y=0,\quad w_x=-2.$

13. $u_x=-\frac{11}{8}$, $u_y=\frac{7}{8}$, tangent plane is $z=\frac{3}{2}-\frac{11}{8}(x+1)+\frac{7}{8}(y-1)$, $u(-0.80, 0.80)\approx 1.05$.

15. $u_x=\frac{13}{32}$, $u_y=\frac{5}{32}$, tangent plane is $z=2+\frac{13}{32}(x-2)+\frac{5}{32}(y+1)$, $u(1.98, -1.01)\approx 1.99$.

17. $x=2+t,\ y=1+\frac{1}{2}t,\ z=-2-\frac{3}{2}t.$

19. $x=2+t,\ y=-1+\frac{1}{2}t,\ z=1-\frac{3}{2}t.$

21. $u_x=\dfrac{3xu+yv}{u^2-v^2},\quad u_y=\dfrac{yu+xv}{u^2-v^2},\quad v_x=-\dfrac{yu+3xv}{u^2-v^2},\quad v_y=-\dfrac{xu+yv}{u^2-v^2}.$

23. $u_x=-\dfrac{3x}{u},\quad u_{xx}=-\dfrac{3}{u}-\dfrac{9x^2}{u^3}.$

25. $v_x=\dfrac{2x}{v},\quad \dfrac{\partial^2 v}{\partial x^2}=\dfrac{2}{v}-\dfrac{4x^2}{v^3}.$ **27.** $\dfrac{-3xy}{u^3}.$

EXERCISES 6.8, p. 390

1. $J_T=\begin{pmatrix}3 & -2\\ 1 & 1\end{pmatrix},\ \dfrac{\partial(x, y)}{\partial(u, v)}=5,\quad \dfrac{\partial(u, v)}{\partial(x, y)}=\dfrac{1}{5}.$

3. $\dfrac{\partial(x, y)}{\partial(u, v)}=-8uv,\quad J_T=\begin{pmatrix}2u & 2v\\ 2u & -2v\end{pmatrix},\quad \dfrac{\partial(u, v)}{\partial(x, y)}=-\dfrac{1}{8uv}.$

5. $\boldsymbol{J_T}=\begin{pmatrix}2 & -3\\ -1 & 2\end{pmatrix}=\boldsymbol{M_T}$, $\det \boldsymbol{J_T}=0$, $\boldsymbol{T}^{-1}$ does not exist.

7. $J_T=\begin{pmatrix}-1 & 2 & 1\\ 2 & -1 & -2\\ 1 & 1 & -1\end{pmatrix}$, $\det J_T=0$, $\boldsymbol{T}^{-1}$ does not exist.

9. $J_T=\begin{pmatrix}\cos\theta & -r\sin\theta & 0\\ \sin\theta & r\cos\theta & 0\\ 0 & 0 & 1\end{pmatrix}$, $\det J_T=r$, $\det J_{T^{-1}}=\dfrac{1}{r}$ if $r\neq 0$.

11. $J_T = \begin{pmatrix} \sin\phi\cos\theta & \rho\cos\phi\cos\theta & -\rho\sin\phi\sin\theta \\ \sin\phi\sin\theta & \rho\cos\phi\sin\theta & \rho\sin\phi\cos\theta \\ \cos\phi & -\rho\sin\phi & 0 \end{pmatrix}$,

$\det J_T = \rho^2 \sin\phi$, $\det J_{T^{-1}} = \dfrac{1}{\rho^2 \sin\phi}$.

13. 2. **15.** 4. **17.** *Hint:* See Exercise 29, Exercises 6.7.

EXERCISES 6.9, p. 398

1. $\frac{4}{3}$. **3.** $\frac{42}{5}$. **5.** 2. **7.** $\dfrac{62\pi}{5}$. **9.** 0. **11.** $\frac{4}{9}$. **13.** 4π. **19.** $\dfrac{64\pi}{5}$.

REVIEW EXERCISES 6.10, p. 400

1. (a) $\lim_{x \to x_0} \boldsymbol{F}(\boldsymbol{x}) = (\frac{1}{4}\pi, \sqrt{2}, 1)$; no. (b) No limit; not continuous. **3.** (a) Not linear. (b) Not linear. **5.** (a) $(\frac{5}{2}, -1)$. (b) $(-5, \frac{7}{2}, -8)$. **7.** (a) $(\frac{2}{26})$, (b) $(-1, -2, -12)$. **9.** Not invertible.

11. $A^{-1} = \frac{1}{4}\begin{pmatrix} 3 & 2 & 1 \\ 2 & 4 & 2 \\ 1 & 2 & 3 \end{pmatrix}$. **13.** Not invertible.

15. $\begin{pmatrix} 4\pi^2 e^2 & -2\pi e^2 & -2\pi^2 e^2 \\ 2\pi & 2\pi & 2\pi \end{pmatrix}$. **17.** $\dfrac{\partial p}{\partial x} = 20$, $\dfrac{\partial q}{\partial y} = 184$, $\dfrac{\partial r}{\partial x} = 140$.

19. $\dfrac{\partial u}{\partial x}(x_0, y_0) = 3$, $\dfrac{\partial u}{\partial y}(x_0, y_0) = 4$, $z - 1 = 3(x-1) + 4(y-2)$, $u(0.98, 2.02) \approx 1.02$.

21. 5. **23.** $\frac{1}{8}\int_{-1}^{1} du \int_0^2 (v-u)^2\, dv = \frac{5}{6}$.

25. $\frac{1}{2}\int_{-1}^{1} u\, du \int_0^2 dv \int_1^3 w\, dw = 0$.

27. $108\int_0^1 \rho^4\, d\rho \int_0^{2\pi} d\theta \int_0^{\pi} \sin\phi \cos^2\phi\, d\phi = \dfrac{144\pi}{5}$.

CHAPTER 7

EXERCISES 7.1, p. 411

1. $\frac{1}{4}$ **3.** -2 **5.** $\frac{26}{15}$ **7.** 2π **9.** 1 **11.** 2 **13.** $\frac{31}{2}$ **15.** -2π **17.** $1 - \dfrac{1}{\sqrt{5}}$ **19.** $(\bar{x}, \bar{y}, \bar{z})$, where $\bar{x}M = \int_\gamma x f(\boldsymbol{x})\, ds$, $\bar{y}M = \int_\gamma y f(\boldsymbol{x})\, ds$, $\bar{z}M = \int_\gamma z f(\boldsymbol{x})\, ds$. **21.** $2\sqrt{2}\pi(1 + \frac{4}{3}\pi^2)$. **23.** M. **25.** (a) 3, (b) -1. **27.** 6π.

EXERCISES 7.2, p. 424

1. 12. **3.** $\frac{-243}{2}\pi$. **5.** 0. **7.** $12 - \pi$. **9.** $\frac{1}{2}\pi$. **11.** 2π. **13.** 0. **15.** 8π. **17.** 24π. **19.** 7π.

23. $\bar{x} = \dfrac{1}{4A(D)} \oint_{\partial D} x^2\, dy - 2xy\, dx$.

29. $A(E) = \int_{\partial E} x\, dy = \int_a^b x(u(t), v(t)) \cdot \left(\dfrac{\partial y}{\partial u} \dot{u} + \dfrac{\partial y}{\partial v} \dot{v} \right) dt$.

EXERCISES 7.3, p. 436

1. $1 + y, -1$. **3.** $y^2 z^2 + z^2 \cos y, 4\pi^2 - 4$. **5.** $y\boldsymbol{k}, 4\boldsymbol{k}$. **7.** $\boldsymbol{0}, \boldsymbol{0}$. **19.** $\nabla^2 f = \text{Trace } f'' = \text{Trace } J_{f'}$ **33.** $(0, 0, 2)$.

EXERCISES 7.4, p. 448

1. $\frac{5}{2}$,. **3.** $\frac{1}{4}\pi$. **5.** 0. **7.** -8. **9.** $-e^{-\pi/2}+e^{9\pi/2}$.
11. Yes, $=\nabla(x^3+xy+e^y)$. **13.** Yes, $=\nabla(\sqrt{x^2+y^2})$.
15. Yes, $=\nabla(x^2yz+xz^2-2xy^2+x-2z)$.
17. Yes, $=\nabla(x^3+y^3-3xyz)$. **23.** $-mgz$.
25. $T=\frac{1}{2}m\,|\boldsymbol{v}|^2$.

EXERCISES 7.5, p. 456

1. $\boldsymbol{X}(u, v)=(u, v, \frac{1}{2}(u^2+v^2))$, $(u, v)\in[1, 2]\times[2, 5]$.
3. $\boldsymbol{X}(r, \theta)=(r\cos\theta, r\sin\theta, 4-r)$, $(r, \theta)\in[0, 4]\times[0, 2\pi]$.
5. $\boldsymbol{X}(r, \theta)=(r\cos\theta, r\sin\theta, 3r\cos\theta-2r\sin\theta)$, $(r, \theta)\in[0, 2]\times[0, 2\pi]$.
7. $\boldsymbol{X}(\phi, \theta)=(\sin\phi\cos\theta, \sin\phi\sin\theta, \cos\phi)$, $(\phi, \theta)\in[0, \frac{1}{4}\pi]\times[0, 2\pi]$.
9. $\boldsymbol{X}(u, v)=((b+a\cos v)\cos u, (b+a\cos v)\sin u, a\sin v)$, $(u, v)\in[0, 2\pi]\times[0, 2\pi]$.
11. $\frac{\pi}{6}(5\sqrt{5}-1)$. **13.** $\frac{\pi}{6}(17\sqrt{17}-5\sqrt{5})$.
15. $\frac{1}{2}\pi\sqrt{2}+\frac{1}{2}\pi\ln(1+\sqrt{2})$. **17.** $\sqrt{14}\int_0^{2\pi}d\theta\int_0^2 r\,dr=4\sqrt{14}\pi$.
19. $4\pi^2ab$. **21.** $2\pi ah$. **23.** $\frac{1}{2}\sqrt{a^2b^2+b^2c^2+a^2c^2}$.
27. $\boldsymbol{X}(u, v)=(u\cos v, u\sin v, f(u))$, $(u, v)\in[a, b]\times[0, 2\pi]$.

EXERCISES 7.6, p. 467

1. $\frac{1}{2}\pi$. **3.** 12π. **5.** $2\pi\left[\frac{2}{3}-\frac{5}{6\sqrt{2}}\right]$. **7.** πha^3. **9.** -4π. **11.** $3e-18$. **13.** $\frac{81\pi}{2}$. **15.** 2π.
17. 3π. **19.** $4\pi\,\delta a$. **21.** 4π. **23.** 4π. **25.** No (Consider Example 6.5 with u and v interchanged.)

EXERCISES 7.7, p. 477

1. 0. **3.** 8π. **5.** 0. **7.** $-\pi$. **9.** $-\pi$. **11.** $-\pi$. **13.** 0. **15.** -4π. **19.** $(a+b+c)\pi r^2$.

EXERCISES 7.8, p. 488

1. $32\pi/3$. **3.** 0. **5.** $\frac{11}{24}$. **7.** $7\pi/4$. **9.** 216. **11.** $7\pi/4$. **13.** $2\pi=2\int_0^{\pi}d\theta\int_0^{2\cos\theta}r\,dr\int_{r^2}^{2r\cos\theta}(r\cos\theta+1)\,dz$.

REVIEW EXERCISES 7.9, p. 491

1. $-\frac{3}{4}$. **3.** -24π. **5.** 0. **7.** 0. **9.** -4. **13.** Yes, $\boldsymbol{F}(x, y)=\nabla\left(\frac{1}{\sqrt{x^2+y^2}}\right)$. **15.** No.
17. (a) $\boldsymbol{X}(r, \theta)=(r\cos\theta, r\sin\theta, r\cos\theta+r\sin\theta)$, $(r, \theta)\in[0, 1]\times[0, 2\pi]$.
(b) $\sqrt{3}\pi=\int_0^1\sqrt{3}r\,dr\int_0^{2\pi}d\theta$.
19. 0. **21.** $\frac{1}{3}=\int_0^1 dy\int_0^{1-y}(x, -2y, 1)\cdot(1, 1, 1)\,dx$.
23. $\frac{3\pi}{2}=3\int_0^1 dz\int_0^1 r^3\,dr\int_0^{2\pi}(\frac{1}{2}+\cos^2\theta)\,d\theta$. **25.** 0.

CHAPTER 8

EXERCISES 8.1, p. 500

1. $y=\frac{1}{2}\ln|\frac{1}{2}x^4+c|$. **3.** $y^2-2y=x^3+2x^2+2x+c$.
5. $y=1+k(x^2+1)$. **7.** $y=\frac{1}{5}(ce^{5x}+1)$.
9. $y^3+e^y=\sin x+e^2+8$.
11. $-\frac{1}{x}+\frac{2}{x^2}+\frac{1}{y}-\frac{1}{y^3}=\frac{11}{8}$. **13.** $\ln|x+2|=y-\ln|1+y|+2\ln 2-1$.
17. $y=\tan(x+c)-(x+5)$. **19.** (a) $y=ke^{x^2}$. (b) $y=e^{(x^2-49)/2}$. **21.** $y=cx^2$; $y=0$.
23. $y^2-2x^2=c$; $y^2-2x^2=-8$.

EXERCISES 8.2, p. 510

1. $xy=k$. **3.** $y=k(2x-c)$. **5.** $y=ke^{-x/c}$.

7. 5.3902%. **9.** $v=\sqrt{\frac{mg}{k}}\frac{e^{ct}-1}{e^{ct}+1}$, $s=\sqrt{\frac{m}{k}}\ln(e^{ct/2}+e^{-ct/2})$, $\lim_{t\to\infty} v(t)=\sqrt{\frac{mg}{k}}$.

11. (a) $k=\frac{-gR^2}{M}$. (b) $v^2=v_0^2-2gR+\frac{2gR^2}{x}$. (c) $v_0>\sqrt{2gR}$.

13. 1910: 87,150,000, error = 5.27%
1920: 99,938,000, error = 5.54%
1930: 114,600,000, error = 6.67%
1940: 131,417,000, error = 0.21%
1960: 172,810,000, error = 3.6%
1970: 198,167,000, error = 2.85%

15. $k=0.0378526$; 1970: 199,400,000, error about 2 25%; 1980: 215,900,000; 2000: 238,800,000; $L=263{,}400{,}000$.

17. About 9958 years ago. **19.** About 29%. **21.** About 9:42 P.M.

23. $y=ab(e^{k(a-b)t}-1)/(ae^{k(a-b)t}-b)$.

25. (a) $y=y_0e^{kt}/(1-y_0+y_0e^{kt})$. (b) Everyone ultimately becomes infected.

EXERCISES 8.3, p. 521

1. $y=ce^{\cos x}$. **3.** $y=c\cos x$. **5.** $y=\frac{1}{2}e^{3x}+ce^x$.

7. $y=-x^2-\frac{1}{3}+cx^3$. **9.** $y=-\frac{1}{3x^2}\cos 3x+cx^{-2}$.

11. $y=1-x^2$. **13.** $y=-\frac{19}{3}x^2+\frac{1}{3}x^5$.

15. $y=1+(x+1)\ln|x+1|$. **17.** $y=\frac{1}{cx^3-x^3\ln|x|}$.

19. $\frac{1}{y^2}=x+\frac{1}{2}-\frac{1}{4}c^{2x}$. **21.** (a) $L\frac{di}{dt}+Ri=E(t)$. (b) $i(t)=\frac{E}{R}+ce^{-Rt/L}$.

23. $\frac{ds}{dt}=75-\frac{3}{10}s(t)$, $s(t)=750e^{-0.3t}+250\to 250$ as $t\to\infty$.

25. $y(t)=20(1-e^{-t/5000})$; $2\frac{1}{2}$ minutes.

29. $p(t)=\left[p(0)-\frac{b-c}{\ell}-\frac{k^2\ell c}{k^2\ell^2+a^2}\right]e^{-k\ell t}+\frac{b-c}{\ell}+\frac{kc}{k^2\ell^2+a^2}(k\ell\cos at+a\sin at)$.

Let $\theta_0=\mathrm{Tan}^{-1}\frac{k\ell}{a}$. Then $p(t)\approx\frac{b-c}{\ell}+\frac{kc}{(k^2\ell^2+a^2)^{\frac{1}{2}}}\sin(at+\theta_0)$ for large t; $p(t)$ oscillates about $(b-c)/\ell$: min. at $t=\frac{2n\pi-(\frac{1}{2}\pi+\theta_0)}{a}$, max. at $t=\frac{2n\pi+(\frac{1}{2}\pi-\theta_0)}{a}$.

EXERCISES 8.4, p. 530

1. $y^2=x^2\ln x^2+cx^2$. **3.** $y=x^2-\sqrt{x^2+y^2}$.

5. $y=2x(1-8\ln x^2)^{-1/2}$. **7.** $y=-x\ln\left|1-\ln\frac{|x|}{2}\right|$.

9. $xy+\cos y=\pi-1$. **11.** $x^3+2x^2y+y^2=c$.

13. $x^2=cx^4(2y^2+4xy)$. **15.** $xy+y\cos y-\sin y=c$.

17. $x^2y+\sin y=\pi$. **19.** $xy^3+2x^2y^2=c$.

25. $x^2+4xy+4y^2+6x-2y=c$. **29.** The single solution is tangent at each point to some member of the one-parameter family of solutions: $y=cx+c^2$; $y=-\frac{1}{4}x^2$.

EXERCISES 8.5, p. 540

3. $z_1+z_2=4+i$; $z_1z_2=5-i$, $z_1\div z_2=\frac{1}{13}-\frac{5}{13}i$.

5. $\pm i, \pm\sqrt{2}i$. **9.** The vector $2-3i$ is multiplied by $\sqrt{2}$ and rotated 45° counterclockwise. **11.** $-8-8i\sqrt{3}$.

13. $3\left(\cos\frac{\pi}{9}+i\sin\frac{\pi}{9}\right)$, $3\left(\cos\frac{7\pi}{9}+i\sin\frac{7\pi}{9}\right)$, $3\left(\cos\frac{13\pi}{9}+i\sin\frac{13\pi}{9}\right)$.

15. $\frac{\sqrt{3}}{2}+\frac{1}{2}i, i, -\frac{\sqrt{3}}{2}+\frac{1}{2}i, -\frac{\sqrt{3}}{2}-\frac{1}{2}i, -i, \frac{\sqrt{3}}{2}-\frac{1}{2}i.$
19. $e^i=\cos 1+i\sin 1,\quad e^{2\pi i}=1.$
23. $(1+2i)e^{x(1+2i)}, (-3+4i)e^{x(1+2i)}, (-11-2i)e^{x(1+2i)}.$
25. $e^{x/\sqrt{2}}\left(\cos\frac{x}{\sqrt{2}}+i\sin\frac{x}{\sqrt{2}}\right).$
27. $\int e^{ax}\cos bx\,dx=\frac{e^{ax}}{a^2+b^2}(a\cos bx+b\sin bx)+c,$
$\int e^{ax}\sin bx\,dx=\frac{e^{ax}}{a^2+b^2}(a\sin bx-b\cos bx)+c.$
29. $m=n$ gives π.

EXERCISES 8.6, p. 550
5. $y=c_1e^{-2x}+c_2e^{-3x}$. **7.** $y=c_1e^{-3x}+c_2xe^{-3x}$.
9. $x=e^{-t}(c_1\cos 2t+c_2\sin 2t)$.
11. $y=c_1\cos 3x+c_2\sin 3x$. **13.** $y=\cos 3x-\sin 3x$.
15. $y=-\frac{2}{3}\cos 3x+\sin 3x$. **17.** $y=e^{-3x}+4xe^{-3x}$.
19. No solution. **21.** $y=\cos 3x-\sin 3x$.
25. $y=c,\quad x=y-c\ln|y+c|-k$.
27. $y=c,\quad y=-2x+c,\quad y=\frac{1}{c_1}(c_2e^{c_1x}+2)$. **29.** $y=\frac{1}{2}x^2+\frac{3}{2}$.

EXERCISES 8.7, p. 559
1. $y_p(x)=-x^2-6x-19$. **3.** $\frac{1}{17}(3\cos x-5\sin x)$.
5. $-\frac{1}{3}e^{-x}$. **7.** $e^x(\frac{1}{6}x^2-\frac{1}{9}x)$.
9. $c_1e^{2x}+c_2e^{3x}+\frac{3}{4}e^x+\frac{1}{2}xe^x$.
11. $y=c_1e^{-x}+c_2xe^{-x}+\frac{3}{2}x^2e^{-x}$. **13.** $c_1e^{3x}+c_2e^{-x}-\frac{1}{2}e^x+2\sin x-\cos x$.
15. $\frac{3}{2}e^{3x}+2e^{-x}-\frac{1}{2}e^x-\cos x+2\sin x$. **17.** $e^x+\frac{1}{2}x^2e^x$.
19. $c_1e^x+c_2e^{2x}+x^2+3x+\frac{7}{2}+2e^{3x}-x^2e^x-3xe^x$.
21. $-\frac{1}{2}\sqrt{\pi}e^x\operatorname{erf}(x)+c_1e^x+c_2xe^x$.
23. $c_1\cos 3x+c_2\sin 3x+\frac{1}{5}\cos 2x$.
25. $y=c_1x+c_2x^2+3\ln x+\frac{9}{2}y$.

EXERCISES 8.8, p. 573
3. (a) $\frac{F_0}{2m\omega_0}t\sin\omega_0 t$. **5.** $x=8+\frac{1}{2}\cos 2t$.
7. $x=(\frac{1}{2}+2t)e^{-4t}$. Critically damped.
9. $x(t)=\frac{10}{3}\sin\frac{1}{2}t\sin\frac{3}{2}t=\frac{5}{3}(\cos t-\cos 2t)$.
11. $\omega=\frac{1}{\sqrt{CL}}$. Current is also periodic.
15. $q=10^{-6}(2e^{-500t}-e^{-1000t})$, and $i=\frac{dq}{dt}$.
17. $i=\frac{dq}{dt}$ where $q=e^{-400t}(2.4\times10^{-3}\cos 300t+3.2\times10^{-3}\sin 300t)+2.4\times10^{-3}$.
19. $q=\frac{6e^{-10t}}{61}(6\sin 50t+5\cos 50t)-\frac{5}{61}(5\sin 60t+6\cos 60t)$.
$i(t)=\frac{dq}{dt},\ i_p(t)=\frac{-1500}{61}\cos 60t+\frac{1800}{61}\sin 60t.$

EXERCISES 8.9, p. 586
1. $y=a_0\left(1+\frac{(2x)^2}{2!}+\dots+\frac{(2x)^{2n}}{(2n)!}+\dots\right)+\frac{a_1}{2}\left(2x+\frac{(2x)^3}{3!}+\dots+\frac{(2x)^{2n+1}}{(2n+1)!}+\dots\right)$
$=a_0\cosh 2x+a_1\sinh 2x.$

3. $y = a_0\left(1 - \frac{(2x)^2}{2!} + \frac{(2x)^4}{4!} - \frac{(2x)^6}{6!} + - \ldots\right) + \frac{a_1}{2}\left(2x - \frac{(2x)^3}{3!} + \frac{(2x)^5}{5!} - + \ldots\right)$

5. $y = a_0\left(1 - x^2 + \frac{x^4}{3} - \frac{x^6}{3 \cdot 5} + - \ldots\right) + a_1\left(x - \frac{1}{2}x^3 + \frac{x^5}{2 \cdot 4} - \frac{x^7}{2 \cdot 4 \cdot 6} + - \ldots\right)$

7. $y = a_0(1 - x^2 + \frac{1}{4}x^4 - + \ldots) + a_1(x - \frac{1}{2}x^3 + \frac{3}{40}x^5 - + \ldots)$

9. $y = a_0(1 - x^2 + x^3 - \frac{13}{12}x^4 + - \ldots) + a_1(x - \frac{1}{2}x^2 + \frac{1}{6}x^3 - \frac{1}{8}x^4 + - \ldots)$

11. $y = a_0\left[1 - \frac{x^3}{6} + \frac{3x^5}{40} + \ldots\right] + a_1\left(x - \frac{x^3}{6} - \frac{x^4}{12} + \frac{3x^5}{40} + \ldots\right)$

13. $y = a_0\left[1 + \frac{(x-1)^2}{2} + \frac{(x-1)^3}{6} + \frac{(x-1)^4}{24} + \frac{(x-1)^5}{30} + \ldots\right]$

$+ a_1\left[x - 1 + \frac{(x-1)^3}{6} + \frac{(x-1)^4}{12} + \frac{(x-1)^5}{120} + \ldots\right]$

15. $y = a_0\Bigg[1 + \sum_{m=1}^{\infty} (-1)^m$

$\times \frac{\lambda(\lambda-2)(\lambda-4)\cdots(\lambda-2m+2)(\lambda+1)(\lambda+3)\cdots(\lambda+2m-1)}{(2m)!} x^{2m}\Bigg]$

$+ a_1\left[x + \sum_{m=1}^{\infty} (-1)^m \frac{(\lambda-1)(\lambda-3)\cdots(\lambda-2m+1)(\lambda+2)(\lambda+4)\cdots(\lambda+2m)}{(2m+1)!} x^{2m+1}\right]$

17. $y = x - \frac{x^3}{2} + \frac{x^5}{2 \cdot 4} - \frac{x^7}{2 \cdot 4 \cdot 6} + \ldots$

19. $y = 2(1 - \frac{1}{6}x^3 + \frac{3}{40}x^5 + \ldots) + 3(x - \frac{1}{6}x^3 - \frac{1}{12}x^4 + \ldots)$

21. 1.0000, 1.5000, 2.1900, 3.1460, 4.4744, 6.3242. *Exact solution* $y = \frac{1}{4}x - \frac{3}{16} + \frac{19}{16}e^{4x}$ gives 1.0000, 1.6090, 2.5053, 3.8301, 5.7942, 8.7120.

23. 1.0000, 1.0951, 1.1810, 1.2580, 1.3264, 1.3866.

25. 1.0000, 1.5950, 2.4636, 3.7371, 5.6099, 8.3697. See Exercise 21 for exact solution values.

27. 1.5475, 2.3249.

31. 1.000, 1.000, 0.940, 0.818, 0.634, 0.385.

REVIEW EXERCISES 8.10, p. 588

1. $(2cy^2 + 1)e^{2x} = y^2$. **3.** $y = k \csc x$. **5.** $\frac{1}{y^2} = -\frac{x^4}{3} + \frac{7}{12x^2}$

7. $y^2 = x^2 + cx^3$. **9.** $y = c_1e^{3x} + c_2xe^{3x}$.

11. $y = -e^x + 2e^{2x}$. **13.** $y = c_1e^{((-3+\sqrt{13})x)/2} + c_2e^{((-3-\sqrt{13})x)/2} - \frac{6}{61}\cos 2x - \frac{5}{61}\sin 2x$.

15. $y = c_1e^{-6x} + c_2e^x + \frac{3}{8}e^{2x}$.

17. $y = c_0\left(1 + \sum_{m=1}^{\infty} \frac{(-1)^m 2^m !x^{2m}}{1 \cdot 3 \cdot 5 \cdots (2m-1)}\right)$

$+ c_1 \sum_{m=1}^{\infty} (-1)^{m+1} \frac{1 \cdot 3 \cdot 5 \cdots (2m-1)}{4^{m-1}(m-1)!} x^{2m-1}.$

19. 1.000, 1.0000, 1.0100, 1.0304, 1.0623, 1.1074.

21. Approximately 955,793,280,000 dollars.

23. After 10 days.

25. $x(t) = \frac{2}{87}[3(9\cos 2t + 8\sin 2t) - e^{-2t}(27\cos 3t + 34\sin 3t)]$; $\sqrt{13}$.

27. $i_p(t) = \frac{13{,}200\pi}{I}[(40{,}000 - 10\omega_0^2)\cos 120\pi t + 3000\omega_0 \sin 120\pi t]$, where $I^2 = 100(40{,}000 - \omega_0^2)^2 + 9{,}000{,}000\omega_0^2$ and $\omega_0 = 120\pi$.

29. (a) $\frac{\sqrt{3}}{2} + \frac{1}{2}i$; $3i$. (b) $\frac{e^{x/4}}{512}(\sin \frac{1}{4}x - i\cos \frac{1}{4}x)$.

CHAPTER 9

EXERCISES 9.2, p. 605

1. $y=c_1+c_2e^{2x}+c_3e^{3x}$. **3.** $y=c_1+c_2x+c_3e^{2x}+c_4xe^{2x}$.
5. $y=c_1e^{-x}+c_2xe^{-x}+c_3x^2e^{-x}$. **7.** $y=c_1e^{x}+c_2e^{-x}+c_3\cos 2x+c_4\sin 2x$.
9. $y=c_1e^{x}+c_2e^{-x}+c_3e^{3x}$. **11.** $y=c_1e^{2x}+c_2xe^{2x}+c_3x^2e^{2x}+c_4e^{-x}$.
13. $y=c_1+c_2x+c_3\cos x+c_4x\cos x+c_5\sin x+c_6x\sin x$.
15. $y=e^{x/\sqrt{2}}\left(c_1\cos\dfrac{x}{\sqrt{2}}+c_2\sin\dfrac{x}{\sqrt{2}}\right)+e^{-x/\sqrt{2}}\left(c_3\cos\dfrac{x}{\sqrt{2}}+c_4\sin\dfrac{x}{\sqrt{2}}\right)$.
17. $y=3-e^{-x}$. **19.** $y=-\frac{1}{2}\cos x-\frac{1}{2}\sin x+\frac{1}{2}e^{x}$.
21. $y=e^{x}-2e^{2x}+e^{3x}$. **23.** $y=\cos x-\frac{1}{2}x\cos x+\frac{3}{2}\sin x+\frac{1}{2}x\sin x$.

EXERCISES 9.3, p. 614

1. $y=c_1+c_2\cos x+c_3\sin x-2x\cos x$.
3. $y=c_1+c_2x+c_3\sin x+c_4\cos x+x^4-12x^2$.
5. $y=c_1+c_2e^{x}+c_3e^{-x}-\frac{1}{2}x^2-x$. **7.** $y=c_1+c_2e^{2x}+c_3e^{-2x}-\frac{1}{8}x^2-\frac{3}{5}\sin x+\frac{1}{8}xe^{-2x}$.
9. $y=c_1+c_2x+c_3\sin x+c_4\cos x+\frac{1}{4}x^4-3x^2+x\sin x-2x\cos x$.
11. $y=c_1\cos x+c_2\sin x-(\cos x)\ln|\sec x+\tan x|$.
13. $y=c_1e^{x}+c_2e^{2x}+e^{x}\ln(e^{x}+1)+e^{2x}\ln(1+e^{-x})$.
15. $y=c_1\cos x+c_2\sin x+\frac{1}{3}\cos^2 x+\frac{2}{3}\sin^2 x$.
17. $y=c_1+c_2\cos x+c_3\sin x-\ln|\cos x|-(\sin x)\ln|\sec x+\tan x|$.
19. $y=(\frac{1}{2}\pi+1)\cos x+\sin x-x\cos x+(\sin x)\ln|\sin x|$.
21. $y=2+\frac{1}{2}\sin x-\cos x+\frac{1}{2}xe^{x}-e^{x}$.
25. (a) $y=c_1e^{x}+c_2xe^{x}+c_3e^{-x}+3e^{2x}$. (b) $y=c_1e^{-x}+c_2e^{2x}+c_3xe^{2x}+x^2e^{2x}$.

EXERCISES 9.4, p. 624

1. $\dfrac{1}{t^2}$ for $t>0$. **3.** $\dfrac{1}{t-a}$, $t>a$. **5.** $\dfrac{t}{t^2+b^2}$, $t>0$.
7. $\dfrac{t-a}{(t-a)^2+b^2}$, $t>a$. **9.** $\dfrac{t}{t^2-a^2}$, $t>a$. **11.** $\dfrac{x^n}{n!}$.
13. $\dfrac{1}{a}\sin ax$. **15.** $\dfrac{1}{2a^3}\sin ax-ax\cos ax$. **17.** $-\frac{1}{2}e^{x}+xe^{x}+\frac{1}{2}\cos x-\frac{1}{2}\sin x$.
19. $\frac{1}{2}e^{-3x}\sin 2x$. **21.** $y=3e^{-2x}$. **23.** $y=3e^{x}-x-1$. **25.** $y=2e^{4x}+e^{-2x}$.
27. $y=\cos 2x+\frac{7}{8}\sin 2x-\frac{3}{4}x\cos 2x$. **29.** $y=x^2-x+\frac{1}{2}-\frac{1}{2}e^{-2x}$.
31. $y=-e^{-2x}+2e^{-x}-2xe^{-x}-\cos x+2\sin x$. **33.** $y=xe^{x}-x^2e^{x}+\frac{2}{3}x^3e^{x}$.

REVIEW EXERCISES 9.5, p. 625

5. $y=c_1+c_2e^{3x}+c_3xe^{3x}$. **7.** $y=c_1+c_2x+c_3e^{3x}+c_4xe^{3x}$.
9. $y=c_1+c_2\cos x+c_3\sin x-x\sin x$.
11. $y=c_1e^{x}+c_2e^{-x/2}\cos\dfrac{\sqrt{3}}{2}x+c_3e^{-x/2}\sin\dfrac{\sqrt{3}}{2}x-x-1$.
13. $y=c_1\cos x+c_2\sin x+\cos^2 x-\frac{1}{3}\cos^4 x+\frac{1}{3}\sin^4 x$.
15. $y=c_1+c_2\cos x+c_3\sin x+\frac{1}{2}\ln|\sec x+\tan x|$.
17. $y=\frac{1}{3}e^{2x}+\frac{2}{3}e^{-x}$. **19.** $y=\frac{5}{3}\sin x-\frac{1}{3}\sin 2x$.
21. $y=-\frac{1}{5}e^{-2x}+\frac{1}{4}e^{-3x}-\frac{1}{20}e^{-x}\cos 2x+\frac{3}{20}e^{-x}\sin 2x$. **23.** $\dfrac{2}{t^3}+\dfrac{1}{t}$.

CHAPTER 10

EXERCISES 10.1, p. 634

1. $\mathbf{x}(t)=\begin{pmatrix}-1\\2\end{pmatrix}+c_1\begin{pmatrix}e^{t}\\0\end{pmatrix}+c_2\begin{pmatrix}0\\e^{-t}\end{pmatrix}$. **3.** $\mathbf{x}(t)=\begin{pmatrix}1\\4\end{pmatrix}3e^{-2t}+c_1\begin{pmatrix}-\frac{1}{2}\\t\end{pmatrix}e^{-t}+c_2\begin{pmatrix}0\\e^{-t}\end{pmatrix}$.

5. $\boldsymbol{x}(t)=\frac{1}{13}\begin{pmatrix}4e^{-t}\sin t-7e^{-t}\cos t\\-2e^{-t}\sin t-16e^{-t}\cos t\end{pmatrix}+c_1\begin{pmatrix}\cos 2t\\\cos 2t+\sin 2t\end{pmatrix}e^t$

$+c_2e^t\begin{pmatrix}\sin 2t\\-\cos 2t+\sin 2t\end{pmatrix}.$

7. $\boldsymbol{x}(t)=\begin{pmatrix}1+3t^2+3t\\\frac{3}{2}t^2\end{pmatrix}e^t+c_1\begin{pmatrix}2\\1\end{pmatrix}e^t+c_2\begin{pmatrix}1+2t\\t\end{pmatrix}e^t.$

9. $\boldsymbol{x}(t)=c_1\begin{pmatrix}7\\-2\\13\end{pmatrix}e^{-t}+c_2e^{2t}\begin{pmatrix}1\\1\\1\end{pmatrix}+c_3e^t\begin{pmatrix}1\\0\\1\end{pmatrix}.$

11. $\boldsymbol{x}(t)=c_1e^t\begin{pmatrix}7\\2\\3\end{pmatrix}+c_2\begin{pmatrix}\cos 2t\\\cos 2t\\\cos 2t-\sin 2t\end{pmatrix}+c_3\begin{pmatrix}\sin 2t\\\sin 2t\\\cos 2t+\sin 2t\end{pmatrix}.$

13. $c_1e^t\begin{pmatrix}\frac{5}{2}\cos\sqrt{3}t+\frac{\sqrt{3}}{2}\sin\sqrt{3}t\\\cos\sqrt{3}t\end{pmatrix}+c_2e^t\begin{pmatrix}\frac{5}{2}\sin\sqrt{3}t-\frac{\sqrt{3}}{2}\cos\sqrt{3}t\\\sin\sqrt{3}t\end{pmatrix}$

15. $\boldsymbol{x}(t)=c_1e^{-t}\begin{pmatrix}-3\\1\end{pmatrix}+c_2e^{-t}\begin{pmatrix}1-3t\\t\end{pmatrix}+c_3e^{-2t}\begin{pmatrix}-5\\1\end{pmatrix}+\begin{pmatrix}-14t+33\\6t-13\end{pmatrix}.$

19. $\boldsymbol{x}(t)=e^{-2t}\begin{pmatrix}-6\\2\end{pmatrix}+e^{-3t}\begin{pmatrix}-3\\2\end{pmatrix}.$

EXERCISES 10.2, p. 648

1. $c_1e^{5t}\begin{pmatrix}4\\1\end{pmatrix}+c_2e^{-t}\begin{pmatrix}-2\\1\end{pmatrix}.$ **3.** $c_1e^t\begin{pmatrix}-1\\0\\1\end{pmatrix}+c_2e^{-t}\begin{pmatrix}1\\2\\-7\end{pmatrix}+c_3e^{2t}\begin{pmatrix}1\\-1\\-1\end{pmatrix}.$

5. $c_1e^{2t}\begin{pmatrix}1\\1\\1\end{pmatrix}+e^{-t}\left[c_3\begin{pmatrix}-1\\0\\1\end{pmatrix}+c_2\begin{pmatrix}-1\\1\\0\end{pmatrix}\right].$

7. $c_1\begin{pmatrix}-1\\1\end{pmatrix}e^{2t}+c_2\left[\begin{pmatrix}-1\\1\end{pmatrix}te^{2t}+\begin{pmatrix}0\\1\end{pmatrix}e^{2t}\right].$

9. $c_1\begin{pmatrix}-3/2\\2\\1\end{pmatrix}e^{-t}+c_2\begin{pmatrix}0\\-1\\1\end{pmatrix}e^{2t}+c_3\left[\begin{pmatrix}0\\-1\\1\end{pmatrix}te^{2t}+\begin{pmatrix}-1\\-2\\1\end{pmatrix}e^{2t}\right].$

11. $\frac{1}{2}e^{-t}\begin{pmatrix}1\\1\end{pmatrix}+\frac{1}{2}e^{3t}\begin{pmatrix}1\\5\end{pmatrix}.$ **13.** $e^t\begin{pmatrix}0\\-2\\1\end{pmatrix}+2\begin{pmatrix}1\\1\\0\end{pmatrix}e^{2t}.$

EXERCISES 10.3, p. 656

1. $\boldsymbol{x}(t)=c_1\begin{pmatrix}4\\1\end{pmatrix}e^{5t}+c_2\begin{pmatrix}2\\-1\end{pmatrix}e^{-t}+\begin{pmatrix}-3\\1\end{pmatrix}e^t.$

3. $\boldsymbol{x}(t)=c_1\begin{pmatrix}4\\1\end{pmatrix}e^{5t}+c_2\begin{pmatrix}2\\-1\end{pmatrix}e^{-t}+\frac{1}{26}\begin{pmatrix}15\cos t-29\sin t\\-6\cos t+9\sin t\end{pmatrix}.$

5. $\boldsymbol{x}(t)=c_1\begin{pmatrix}4\\1\end{pmatrix}e^{5t}+c_2\begin{pmatrix}-2\\1\end{pmatrix}e^{-t}+\begin{pmatrix}-3te^{-t}-\frac{7}{2}e^{-t}\\\frac{3}{2}te^{-t}+\frac{3}{2}e^{-t}\end{pmatrix}.$

7. $\boldsymbol{x}(t)=c_1\begin{pmatrix}-1\\0\\1\end{pmatrix}e^t+c_2\begin{pmatrix}1\\2\\-7\end{pmatrix}e^{-t}+c_3\begin{pmatrix}1\\-1\\-1\end{pmatrix}e^{2t}+\begin{pmatrix}3\\-4\\-2\end{pmatrix}e^{3t}.$

9. $\boldsymbol{x}(t)=c_1\begin{pmatrix}\cos 2t\\\cos 2t+\sin 2t\end{pmatrix}e^t+c_2\begin{pmatrix}\sin 2t\\-\cos 2t+\sin 2t\end{pmatrix}e^t$

$+\frac{1}{13}\begin{pmatrix}4\\-2\end{pmatrix}e^{-t}\sin t+\frac{1}{13}\begin{pmatrix}-7\\-16\end{pmatrix}e^{-t}\cos t.$

11. $\boldsymbol{x}(t)=c_1\begin{pmatrix}\cos t\\ \cos t+\sin t\end{pmatrix}+c_2\begin{pmatrix}\sin t\\ \sin t-\cos t\end{pmatrix}+(\boldsymbol{x}_1(t),\boldsymbol{x}_2(t))\begin{pmatrix}\frac{1}{3}\sin^3 t+\frac{2}{3}\cos^3 t+\sin t\\ \frac{2}{3}\sin^3 t-\frac{1}{3}\cos^3 t+\sin t\end{pmatrix}.$

13. $\boldsymbol{x}(t)=c_1\begin{pmatrix}2\cos t-\sin t\\ \cos t\end{pmatrix}+c_2\begin{pmatrix}\cos t+2\sin t\\ \sin t\end{pmatrix}+(\boldsymbol{x}_1(t),\boldsymbol{x}_2(t))\begin{pmatrix}\ln\sec^2 t\\ \ln|\tan t|-2t\end{pmatrix}.$

15. $\boldsymbol{x}(t)=c_1\begin{pmatrix}1\\2\end{pmatrix}+c_2\begin{pmatrix}t\\2t-\frac{1}{2}\end{pmatrix}-2\begin{pmatrix}1\\2\end{pmatrix}\ln t+\begin{pmatrix}-\dfrac{1}{2t^2}+\dfrac{2}{t}-2\\ \dfrac{5}{t}-4\end{pmatrix}.$

17. $\boldsymbol{x}(t)=c_1\begin{pmatrix}2\\-1\\1\end{pmatrix}e^{6t}+c_2\begin{pmatrix}\frac{1}{2}\\1\\0\end{pmatrix}+c_3\begin{pmatrix}-\frac{1}{2}\\0\\1\end{pmatrix}+\frac{1}{6}(\boldsymbol{x}_1(t),\boldsymbol{x}_2(t),\boldsymbol{x}_3(t))\begin{pmatrix}\frac{1}{6}e^{-6t}+4t^2\\ \frac{1}{18}e^{6t}(6t-1)+11t\\ \frac{5}{18}e^{6t}(6t-1)-5t\end{pmatrix}.$

21. $\boldsymbol{x}(t)=c_1\begin{pmatrix}t^2\\2t\end{pmatrix}+c_2\begin{pmatrix}t^{-1}\\-t^{-2}\end{pmatrix}+\frac{1}{10}\begin{pmatrix}3t^4-5\\2t^3\end{pmatrix}.$

25. $\boldsymbol{x}(t)=c_1\begin{pmatrix}1\\2\end{pmatrix}t^{-1}+c_2\begin{pmatrix}2\\1\end{pmatrix}t^2-\begin{pmatrix}13/2\\5\end{pmatrix}t.$

EXERCISES 10.4, p. 665

1. $\boldsymbol{x}(t)=\frac{3}{5}\begin{pmatrix}1\\2\end{pmatrix}\cos t+\frac{4}{5}\begin{pmatrix}-2\\1\end{pmatrix}\cos\sqrt{6}t.$

3. $\begin{cases}\dfrac{d^2x}{dt^2}=-\dfrac{k_1+k_2}{m_1}x+\dfrac{k_2}{m_1}y\\ \dfrac{d^2y}{dt^2}=\dfrac{k_2}{m_2}x-\dfrac{k_2+k_3}{m_2}y.\end{cases}$

5. $\begin{cases}\ddot{x}=-\dfrac{k_1}{m_1}x+\dfrac{c}{m_1}(\dot{y}-\dot{x})\\ \ddot{y}=-\dfrac{k_2}{m_2}y-\dfrac{c}{m_2}(\dot{y}-\dot{x}).\end{cases}$

7. $\begin{cases}m_1\ddot{x}=-k_1x(t)+k_2(y(t)-x(t))+f_1(t)\\ m_2\ddot{y}=k_2x(t)-k_2y(t)+f_2(t).\end{cases}$

9. $\begin{cases}\ddot{x}+\dfrac{k}{m}\dot{x}=0\\ \ddot{y}+\dfrac{k}{m}\dot{y}+g=0,\end{cases}$ $\quad\dot{\boldsymbol{x}}(0)=(v_0\cos\theta,v_0\sin\theta),$ $\quad\boldsymbol{x}(0)=\boldsymbol{0}.$

11. (a) $\begin{cases}L_1\left(\dfrac{di_1}{dt}-\dfrac{di_2}{dt}\right)+R_1i_1=E(t)\\ L_1\left(\dfrac{di_2}{dt}-\dfrac{di_1}{dt}\right)+L_2\dfrac{di_2}{dt}+R_2i_2=0.\end{cases}$

(b) $\begin{pmatrix}i_1\\i_2\end{pmatrix}=\begin{pmatrix}-1\\1\end{pmatrix}e^{-250t}-\begin{pmatrix}2\\1\end{pmatrix}e^{-1000t}+\begin{pmatrix}3\\0\end{pmatrix}.$

13. $\begin{cases}L_1\dfrac{di_1}{dt}+(R_1+R_2)i_1-R_2i_2=0\\ -R_2\dfrac{di_1}{dt}+R_2\dfrac{di_2}{dt}+\left(\dfrac{1}{C_1}+\dfrac{1}{C_2}\right)i_2-\dfrac{1}{C_2}i_3=0\\ R_3\dfrac{di_3}{dt}-\dfrac{1}{C_2}i_2+\dfrac{1}{C_2}i_3=\dot{E}(t).\end{cases}$

15. (a) If $x=0$, then y decreases exponentially. If $y=0$, then x grows exponentially. (b) If x is large, x grows almost exponentially and y also will grow. If y is large, then x will decrease and so will y, almost exponentially. (c) If $(a-d)^2+4ad-4bc<0$, then both species oscillate. If $(a-d)^2+4ad-4bc>0$, then both eigenvalues have the sign of $a-d$. If $a-d<0$, both populations become extinct, tending toward extinction exponentially. If $a-d>0$, both populations grow exponentially.

17. $\boldsymbol{x}=c_1\begin{pmatrix}0\\-1\\1\end{pmatrix}e^{2t}+c_2\begin{pmatrix}1\\1\\1\end{pmatrix}e^{3t}+c_3\begin{pmatrix}2\\1\\1\end{pmatrix}e^{4t}.$

All three species grow exponentially. (Not very reasonable.)

19. $\begin{pmatrix}\dfrac{dx}{dt}\\ \dfrac{dy}{dt}\end{pmatrix}=\begin{pmatrix}-\frac{2}{25} & \frac{3}{50}\\ \frac{1}{25} & -\frac{3}{50}\end{pmatrix}\begin{pmatrix}x\\y\end{pmatrix}+\begin{pmatrix}1\\0\end{pmatrix}.$

EXERCISES 10.5, p. 671

1. $\boldsymbol{x}(t)=\begin{pmatrix}-\frac{1}{3}\\-\frac{2}{3}\end{pmatrix}e^{-t}+\begin{pmatrix}\frac{4}{3}\\\frac{2}{3}\end{pmatrix}e^{2t}.$

3. $\boldsymbol{x}(t)=\begin{pmatrix}1\\1\end{pmatrix}e^{-t}\cos t-\begin{pmatrix}3\\1\end{pmatrix}e^{-t}\sin t.$

5. $\boldsymbol{x}(t)=\begin{pmatrix}0\\-2\\1\end{pmatrix}e^{t}+2\begin{pmatrix}1\\1\\0\end{pmatrix}e^{2t}.$

7. $\boldsymbol{x}(t)=\begin{pmatrix}-2\\-\frac{2}{3}\end{pmatrix}e^{t}+\begin{pmatrix}1\\\frac{2}{3}\end{pmatrix}e^{4t}.$

9. $\boldsymbol{x}(t)=\begin{pmatrix}2\\1\end{pmatrix}e^{t}+\begin{pmatrix}-1\\0\end{pmatrix}e^{2t}+\begin{pmatrix}-1\\-2\end{pmatrix}.$

REVIEW EXERCISES 10.6, p. 672

1. $\begin{pmatrix}x\\y\end{pmatrix}=c_1\begin{pmatrix}1\\0\end{pmatrix}e^{2t}+c_2\begin{pmatrix}t+1\\-1\end{pmatrix}e^{2t}+\begin{pmatrix}-\frac{13}{25}\\\frac{3}{5}\end{pmatrix}e^{-3t}.$

3. $\begin{pmatrix}x\\y\end{pmatrix}=e^{t}\begin{pmatrix}1\\-1\end{pmatrix}+te^{t}\begin{pmatrix}7\\2\end{pmatrix}+t^2e^{t}\begin{pmatrix}3\\\frac{3}{2}\end{pmatrix}.$

5. $c_1e^{6t}\begin{pmatrix}1\\1\\1\end{pmatrix}+c_2e^{9t}\begin{pmatrix}1\\0\\-1\end{pmatrix}+c_3e^{15t}\begin{pmatrix}1\\-2\\1\end{pmatrix}.$

7. $c_1\begin{pmatrix}1\\2\end{pmatrix}e^{3t}+c_2\begin{pmatrix}1\\-2\end{pmatrix}e^{-t}-\frac{1}{8}e^{-2t}(\boldsymbol{x}_1(t),\boldsymbol{x}_2(t))\begin{pmatrix}-1\\11e^{4t}\end{pmatrix}$

9. $c_1\begin{pmatrix}1\\2\end{pmatrix}e^{3t}+c_2\begin{pmatrix}1\\-2\end{pmatrix}e^{-t}+te^{-t}\begin{pmatrix}-1\\2\end{pmatrix}$

11. $\boldsymbol{x}(t)=c_1\begin{pmatrix}\cos t\\-\sin t\end{pmatrix}+c_2\begin{pmatrix}\sin t\\\cos t\end{pmatrix}+(\boldsymbol{x}_1(t),\boldsymbol{x}_2(t))\begin{pmatrix}\sin t\\0\end{pmatrix}.$

13. $x(t)=c_1\cos\sqrt{\ell_1}t+c_2\sin\sqrt{\ell_1}t+c_3\cos\sqrt{\ell_2}t+c_4\sin\sqrt{\ell_2}t,$
$y(t)=\frac{1}{3}[(5-\ell_1)c_1\cos\sqrt{\ell_1}\,t+(5-\ell_1)c_2\sin\sqrt{\ell_1}\,t+(5-\ell_2)c_3\cos\sqrt{\ell_2}\,t$
$+(5-\ell_2)c_4\sin\sqrt{\ell_2}\,t]$
where $\ell_1=4-\sqrt{10}$, $\ell_2=4+\sqrt{10}$. **15.** $x=(2c_1+c_2+2c_2t)e^{t}$,
$y=(c_1+c_2t)e^{t}$. (Thus x will be roughly twice as numerous as y.)

17. $\begin{pmatrix}x\\y\end{pmatrix}=\frac{1}{2}e^{-t}\begin{pmatrix}1\\3\end{pmatrix}-\frac{1}{2}e^{t}\begin{pmatrix}1\\3\end{pmatrix}+2te^{t}\begin{pmatrix}1\\1\end{pmatrix}.$

CHAPTER 11

EXERCISES 11.1, p. 681

1. $2^n/n!$ **3.** $\dfrac{1}{n^2}$. **5.** Converges to $\frac{2}{3}$. **7.** Converges to 0. **9.** Diverges.

11. Decreasing, convergent. **13.** Decreasing, convergent. **21.** $(-1)^n \frac{1}{n}$.

EXERCISES 11.2, p. 685

1. Divergent. **3.** Converges to 20. **5.** Converges to $\frac{\sin\theta}{2+\sin\theta}$, if $\theta \neq \frac{\pi}{2}, \frac{3\pi}{2}$. **7.** Converges to $\frac{1}{3}$. **9.** Diverges. **17.** 26 feet. **21.** We can't let s be the nonexistent sum and then treat it as a number.

EXERCISES 11.3, p. 691

1. Converges. **3.** Diverges. **5.** Converges. **7.** Diverges. **9.** Converges. **13.** (b) Converges **17.** No. Consider the harmonic series.

EXERCISES 11.4, p. 694

1. Diverges. **3.** Converges. **5.** Diverges. **7.** Converges. **9.** Converges. **11.** -0.632. **15.** $n \geq 4$.

EXERCISES 11.5, p. 698

1. Absolutely convergent. **3.** Divergent. **5.** Conditionally convergent. **7.** Absolutely convergent. **9.** Divergent. **13.** Divergent.

EXERCISES 11.6, p. 705

1. $x - \frac{x^3}{3!} + \frac{x^5}{5!} - \frac{x^7}{7!} + \frac{x^9}{9!}$, $r_9(x) = \frac{\sin c}{10!} x^{10}$, c between 0 and x.

3. $x - \frac{x^2}{2} + \frac{x^3}{3} - \frac{x^4}{4} + \frac{x^5}{5} - \frac{x^6}{6}$, $r_6(x) = \frac{x^7}{7(1+c)^7}$, c between 0 and x.

5. $25\ln 5 + (5+10\ln 5)(x-5) + (\frac{3}{2}+\ln 5)(x-5)^2 + \frac{1}{15}(x-5)^3 - \frac{1}{300}(x-5)^4$, $r_4(x) = \frac{1}{30c^3}(x-5)^5$, c between 5 and x. **7.** $2 + \frac{1}{4}(x-4) - \frac{1}{64}(x-4)^2 + \frac{3}{512}(x-4)^3 - \frac{5}{16{,}384}(x-4)^4$, $r_4(x) = \frac{7}{256}c^{-9/2}(x-4)^5$, c between 4 and x. **9.** $49 + 66(x-2) + 29(x-2)^2 + 4(x-2)^3 = f(x)$, i.e., $r_3(x) = 0$. **11.** 0.2079. **13.** 1.6487. **15.** 2.0494.

EXERCISES 11.7, p. 710

1. Absolutely convergent on $(-\infty, +\infty)$.
3. Absolutely convergent only at $x_0 = 0$.
5. Convergent on $(-\frac{3}{2}, \frac{3}{2}]$. Absolutely convergent on $(-\frac{3}{2}, \frac{3}{2})$.
7. Convergent (and absolutely) on $(0, 2)$.
9. Convergent (and absolutely) on $[-1, 1]$.
11. Convergent on $[-1, 1)$. Absolutely convergent on $(-1, 1)$.
13. Convergent (and absolutely) on $(1-e, 1+e)$.
15. Convergent (and absolutely) on $[-1, 1]$.
17. Convergent (and absolutely) on $(-1, 1)$.

EXERCISES 11.9, p. 718

1. 0.4613. **3.** 0.3103. **5.** 0.1000.
7. (b) $x - \frac{x^3}{3} + \frac{x^5}{5} - \frac{x^7}{7} + - \ldots + (-1)^n \frac{x^{2n+1}}{2n+1} + \ldots$ (c) $f^{(6)}(0) = 0$, $f^{(5)}(0) = 24$.
9. (b) $1 + 2x + 3x^2 + \ldots + nx^{n-1} + \ldots$, $x \in (-1, 1)$.
11. (a) $(-1, 1]$, (b) $(-1, 1)$. **13.** $1 + \frac{1}{2}x + \frac{1}{3!}x^2 + \frac{1}{4!}x^3 + \ldots + \frac{x^{n-1}}{n!} + \ldots$

EXERCISES 11.10, p. 724

1. $1 + \sum_{n=1}^{\infty} (-1)^n \frac{(-1)\cdot 1 \cdot 3 \cdots (2n-3)}{2^n \cdot n!} x^n$, $x \in (-1, 1)$.

3. $1-\frac{1}{3}x+\sum_{n=2}^{\infty}(-1)^{2n-1}\frac{2\cdot 5\cdots(3n-4)}{n!\cdot 3^n}x^n,\ x\in(-1,1)$.

5. $1+\frac{2}{3}x^2+\sum_{n=2}^{\infty}(-1)^{2n-1}\frac{2\cdot 5\cdots(3n-4)}{n!\cdot 3^n}(-2x^2)^n,\ x\in(-1,1)$.

7. $1+\frac{1}{2}x+\frac{3}{8}x^2+\frac{1}{16}x^3$. **9.** 5.0990. **11.** 5.0100. **13.** 0.3361. **15.** 0.2505. **17.** $(0.4740)\sqrt{5\pi}\approx 3.3297$.

REVIEW EXERCISES 11.11, p. 724

1. (a) $a_n=\frac{(-1)^n(n+2)}{3^n}$, (b) $a_n=\frac{3\cdot 2^{n-1}}{5^{n-1}}$.

3. (a) Increasing, divergent. (b) Decreasing, convergent. **5.** Absolutely convergent. **7.** Absolutely convergent. **9.** Conditionally convergent. **11.** Absolutely convergent. **13.** Absolutely convergent. **15.** Divergent. **17.** Conditionally convergent. **19.** (a) 1.6134, (b) 1.3956.

23. (a) $1+\sum_{n=1}^{\infty}(-1)\frac{(-1)\cdot 1\cdot 3\cdots(2n-3)}{2^n n!}x^n,\ x\in(-1,1)$.

(b) $1+\sum_{n=1}^{\infty}\frac{1\cdot 4\cdots(3n-2)}{3^n n!}x^{2n},\ x\in(-1,1)$.

25. (a) 4.8990, (b) 1.9129.

INDEX OF SYMBOLS

INDEX

An italicized page number indicates that the term occurs in a figure or in an exercise (or both).

D

Z